Examples and Applications

P9-CIS-759

Chapter 5 Examples

Chapter 5 Applications

Chapter 6 Examples

Chapter 7 Examples

Chapter 7 Applications

Chapter 8 Examples

Make Statistics Work for You

If you want to help yourself through this course and learn how to use statistics effectively for business and financial economics, look for the *Study Guide* (0-669-24599-2) at your college bookstore. This unique guide offers additional practice and a deeper understanding of the material presented in this text.

You'll get:

- an intuitive introduction to each chapter of *Statistics for Business and Financial Economics*
- a detailed review of every text chapter
- helpful sample problems with worked-out solutions
- sample tests covering a variety of problem types (with solutions at the end of the chapter)

Using the *Study Guide* really can help you improve your grades in this course. You can get a copy from your college bookstore. Understanding statistics can make a real difference!

Statistics for Business and Financial Economics

Cheng F. Lee
Rutgers University

D. C. Heath and Company
Lexington, Massachusetts   Toronto

Address editorial correspondence to:

D. C. Heath and Company
125 Spring Street
Lexington, MA 02173

Acquisitions Editor: George Lobell
Developmental Editor: Patricia Wakeley
Production Editor: Andrea Cava
Designer: Kenneth Hollman
Art Editor: Diane B. Grossman
Production Coordinator: Richard Tonachel
Permissions Editor: Margaret Roll

I would like to dedicate this book to one of my most respected professors, Emanuel Parzen, who taught me a lot about statistics, both theory and application.

Published simultaneously in Canada.

Printed in the United States of America.

International Standard Book Number: 0-669-24598-4

Library of Congress Catalog Number: 92-70357

10   9   8   7   6   5   4   3   2   1

Preface

When I first began writing *Statistics for Business and Financial Economics,* my goal was to develop a text that would give my students at the University of Illinois and at Rutgers University the basic statistical tools they need not only for a general business school education but also for the statistics that a finance major needs. Over time, that original purpose has evolved into a broad statistical approach that integrates concepts, methods, and applications. The scope has widened to include all students of business and economics, especially upper-level undergraduates and MBA students, who want a clear and comprehensive introduction to statistics. This book is written for them.

A distinguishing feature of the text is the creative ways in which it weaves useful and interesting concepts from general business (accounting, marketing, management, and quality control), economics, and finance into the text. It actively shows how various statistical methods can be applied in business and financial economics.

More specifically, the text incorporates the following pedagogical features:

Usefulness of statistical methods. This text features an unusually large number of real-life examples that show students how statistical methods can help them.

Non-calculus approach. Extensive use of examples and applications (more than 300) in the text and problem sets at the end of the chapters (more than 1,500) shows students how statistical methodology can be effectively implemented and applied. All the examples, applications, and problems can be worked out using only high-school algebra and geometry. Calculus, which offers an alternative and intellectually satisfying perspective, is presented only in footnotes and appendixes.

Emphasis on data analysis. Most statistics texts, in their justifiable need to demonstrate to students how to use the various statistical tests, focus all too often on the mechanical aspects of problem-solving. Lost is the simple but important notion that statistics is the study of data. Data analysis is an important theme of this text. In particular, one set of financial economic data for GM and Ford is used continuously throughout the text for various types of statistical analysis.

Use of computers. After students understand the step-by-step processes, the text shows how computers can make statistical analysis more efficient and less time-consuming. Examples utilizing MINITAB, Lotus 1-2-3, and SAS are shown, and a supplementary manual based entirely on MINITAB is available.

Straightforward language. Not least, the text employs clear and simple language to guide the reader to a knowledge of the basic statistical methods used in business decision-making and financial economics.

Additionally, this text explores in slightly greater depth many of the standard statistical topics: There is more coverage of regression analysis than in other texts (see Chapters 13–16 and part of Chapters 18, 19, 20, and 21). Quality control is explicitly integrated with point and interval estimation (Chapter 10). Stock market indexes and the index of leading economic indicators are both treated as an expanded portion of regular index numbers (Chapter 19).

Many chapters have appendixes that develop useful financial applications of the standard topics found in the chapter body. Some appendixes may be used as case studies. See especially:

Financial Statements and Financial Ratio Analysis (Appendixes 2C, 3A, and 4C may be used together as a single case study)
Applications of the Binomial Distribution to Evaluate Call Options (Appendix 6B)
Cumulative Normal Distribution Function and the Option Pricing Model (Appendix 7B)
Control Chart Approach for Cash Management (Appendix 10A)

Organization of the Text

The text has 21 chapters divided into five parts. Part I, Introduction and Descriptive Statistics, consists of four chapters. Following the introductory chapter, Chapter 2 addresses Data Collection and Presentation. Chapter 3 delves into Frequency Distributions and Data Analyses. It is followed by Numerical Summary Measures in Chapter 4.

Probability and Important Distributions, Part II, includes five chapters, the first of which, Chapter 5, is entitled Probability Concepts and Their Analysis. Discrete Random Variables and Probability Distributions are discussed in Chapter 6, after which Chapter 7 covers The Normal and Lognormal Distributions. Sampling and Sampling Distributions are covered in Chapter 8. Chapter 9 closes Part II of the text by discussing Other Continuous Distributions and Moments for Distributions.

Part III, Statistical Inferences Based on Samples, is comprised of three chapters. Chapter 10 covers Estimation and Statistical Quality Control. Chapter 11 explores Hypothesis Testing, and Chapter 12 provides an Analysis of Variance and Chi-Square Tests.

Chapters 13 through 16 make up Part IV, which is entitled Regression and Correlation: Relating Two or More Variables. The first of these chapters is Simple Linear Regression and the Correlation Coefficient. From a discussion of Simple Linear Regression and Correlation: Analyses and Applications in Chapter 14, the book moves on to address Multiple Linear Regression in Chapter 15. Finally, Chapter 16 closes Part IV with a look at Other Topics in Applied Regression Analysis.

The last part of the text, Part V, considers Selected Topics in Statistical Analysis for Business and Economics. Nonparametric Statistics is the subject of Chapter 17, which is followed by an exploration of Time-Series: Analysis, Model, and Forecasting in Chapter 18. Chapters 19 and 20 discuss Index Numbers and Stock Market Indexes, and Sampling Surveys: Methods and Applications, respectively. Statistical Decision Theory: Methods and Applications is the topic of the final chapter, Chapter 21.

There are four appendixes. Appendix A provides fourteen statistical tables. Appendix B gives a full description of the data sets available on computer disk. Appendix C briefly describes the use of MINITAB, especially the microcomputer version, and Appendix D introduces the microcomputer version of SAS. Finally, to make sure they are on the right track in working the problems, students can consult the section at the end of the book that gives short Answers to Selected Odd-Numbered Questions and Problems. (Full solutions are given in the Instructor's Guide.)

About This First Edition

One legitimate concern with a new statistics text is that the first edition will contain errors (too many errors!) that must await correction only in the second edition. We have taken action to confront this problem by carrying out a thorough and detailed accuracy check of the entire text: every problem in the text has been reworked by "outsiders" to the project. So confident are we that this is an error-free book that the publisher is willing to pay $10 for the first report (in writing) of each substantive error.

Alternative Ways to Use the Text

Based upon my own teaching experience, I would like to suggest 3 alternative ways to use this textbook.

Alternative One: The goal of this alternative is to demonstrate to students the basic applications of statistics in general business, economics, and finance. This goal can be achieved by skipping all appendixes, technical footnotes, optional sections, and other sections at the instructor's discretion. Using this alternative, the student needs only basic algebra, geometry, and business and economic common sense to understand how statistics can be used in general business, economics, and finance applications.

Alternative Two: The goal of this alternative is not only to illustrate basic overall business, economic, and finance applications but to show how to use statistics in financial analysis and decision making. This goal can be achieved by omitting all the technical appendixes, technical footnotes, and most optional sections but covering all or most of the following topics:

Chapter	Topic
2	Appendixes 2B and 2C on stock and market rates of return and on financial statements and financial ratio analysis
3	Appendix 3A, financial ratio analysis
4	Appendix 4C, financial ratios for three auto firms. As mentioned earlier, Appendixes 2C, 3A, and 4C can be treated as a single case study
6	Appendix 6B, applications of the binomial distribution to evaluate call options
7	Appendix 7B, cumulative normal distribution function and the option pricing model

Chapter	Topic
9	Section 9.8, analyzing the first four moments of rates of return of the 30 DJI firms
10	Appendix 10A, a control chart approach for cash management
19	Section 19.5, stock market indexes
21	Sections 21.7 and 21.8 on mean and variance trade-off analysis and the mean and variance method for capital budgeting decisions; Appendixes 21B, 21C, and 21D on the graphical derivation of the capital market line, present value and net present value, and derivation of the standard deviation for NPV

Alternative Three: The objective of the third approach is to show students how calculus can be used in statistical analysis. To achieve this goal, the instructor can try to cover all optional sections and as many of the technical footnotes and appendixes as possible. To do this, of course, the instructor may have to skip many application examples, such as the finance applications discussed in Alternative Two.

Supplementary Materials

Study Guide, by Ahyee Lee, Monmouth College, and Ronald L. Moy, St. John's University. This fine workbook encourages learning by doing. Each chapter begins with a section describing the basic import of each chapter in intuitive terms. Then the student goes on to a formal review of the chapter and several worked-out problems that show in detail how the solution is obtained. A variety of multiple-choice, true-false, and open-ended questions and problems follows, and finally a brief sample test. All answers are included at the end of each chapter.

MINITAB Manual, by John C. Lee, University of Illinois. This manual, keyed to the text chapter by chapter, is designed to help students use MINITAB throughout the course. Each chapter includes a variety of specific applications and ends with both a statistical summary and a summary of MINITAB commands.

Data Sets. A wide variety of macroeconomic, financial, and accounting data is available on computer disks to facilitate student practice. A complete listing of these data sets is given at the end of this book. The disks themselves are free of charge.

Instructor's Guide. The three main parts of the Instructor's Guide are the Overview and Objectives by Cheng F. Lee, the complete Solutions to the text problems by Ahyee Lee and Ronald L. Moy, and the Test Bank, with more than 1,000 multiple-choice and true-false problems, by Alice C. Lee, University of Pennsylvania. Most instructors will find the Instructor's Guide indispensable.

Computerized Testing Program. With the Test Bank on disk for either IBM or Macintosh computers, instructors can select, rearrange, edit, or add problems as they wish.

Acknowledgments

I am very grateful to my colleagues across the country who have contributed to the development of this book. In particular I would like to thank Kent Becker, Temple

University, and Edward L. Bubnys, Suffolk University, who not only reviewed parts of the manuscript but also class-tested several chapters; John Burr, Mobil Oil Company; H. H. Liao, my research assistant at Rutgers; D. Y. Huang and C. C. Young, both of National Taiwan University; and Kimberly Catucci, my editorial assistant.

I am also indebted to many other people who reviewed all or part of the manuscript:

Richard T. Baillie
Michigan State University

Abdul Basti
Northern Illinois University

Philip Bobko
Rutgers University

Warren Boe
University of Iowa

Y. C. Chang
University of Notre Dame

Shaw K. Chen
The University of Rhode Island

Whewon Cho
Tennessee Technological University

Daniel S. Christiansen
Portland State University

James S. Ford
University of Southern California

Mel H. Friedman
Kean College

R. A. Holmes
Simon Fraser University

James Freeland Horrell
University of Oklahoma

Der Ann Hsu
University of Wisconsin–Milwaukee

Dongcheol Kim
Rutgers University

Bharat Kolluri
University of Hartford

Supriya Lahiri
University of Lowell

Leonard Lardaro
The University of Rhode Island

Ahyee Lee
Monmouth College

Keh Shin Lii
University of California

Chao-nan Liu
Trenton State College

Tom Mathew
The Troy State University in
    Montgomery

Richard McGowan
Boston College

Ronald L. Moy
St. John's University

Hassan Pourbabaee
University of Central Oklahoma

Jean D. Powers
The Ohio State University

Peter E. Rossi
The University of Chicago

William E. Stein
Texas A&M University

William Wei
Temple University

Jeffrey M. Wooldridge
Massachusetts Institute of Technology

Gili Yen
National Central University, Taiwan

Not least, I would like to thank and salute my family—my wife, Schwinne, for her good humor and patience, and my two children, John and Alice, whose contributions are described elsewhere in this preface.

C.F.L.

About the Author

Cheng F. Lee

Cheng F. Lee is a Distinguished Professor and Chair of the Finance Department in the School of Business at Rutgers University. He has also served on the faculty of the University of Georgia (1972–1976) and the University of Illinois (1976–1988). While at the University of Illinois, he was the IBE distinguished professor (1982–1988). He has maintained academic and consulting ties in Taiwan, Hong Kong, and China for the past decade. Currently, he is a consultant to the American Insurance Group (AIG), the World Bank, and the United Nations.

Dr. Lee has two master's degrees, one in statistics from West Virginia University and one in economics from National Taiwan University. His Ph.D. in economics and finance is from the State University of New York at Buffalo. He founded the *RQFA* in 1990 and serves as its editor. He was also a co-editor of *The Financial Review* (1985–1991) and of the *Quarterly Review of Economics and Business* (1987–1989).

In the past ten years, Dr. Lee has written four textbooks in addition to *Statistics for Business and Financial Economics.* Those texts range in subject matter from corporate finance and urban econometrics to security analysis and the theory of financial analysis and planning.

During the past 20 years, Dr. Lee has published more than 100 articles in more than 20 different finance, economics, statistics, and management journals including the *Journal of Finance, Management Science, JASA, Journal of Business and Economic Statistics, JFQA, Journal of Business,* and the *Journal of Econometrics.* He uses statistics methodology and theory extensively in his research.

Brief Contents

Contents

Chapter 11 Hypothesis Testing   424

Chapter 12 Analysis of Variance and Chi-Square Tests   473

Part IV *Regression and Correlation: Relating Two or More Variables*   527

Chapter 13 Simple Linear Regression and the Correlation Coefficient   528

Chapter 14 Simple Linear Regression and Correlation: Analyses and Applications   569

Chapter 15 Multiple Linear Regression   **622**

Chapter 16 Other Topics in Applied Regression Analysis   **670**

Part V *Selected Topics in Statistical Analysis for Business and Economics*   717

Chapter 17 Nonparametric Statistics   718

Chapter 18 Time-Series: Analysis, Model, and Forecasting   757

Appendix A Statistical Tables   A1

Appendix B Description of Data Sets   A32

Appendix C Introduction to MINITAB: Microcomputer Version   A34

Appendix D Introduction to SAS: Microcomputer Version   A38

Answers to Selected Odd-Numbered Questions and Problems   A43

Index   A69

Statistics for Business and Financial Economics

PART I

Introduction and Descriptive Statistics

Part I of this book describes how statistical data can be effectively presented. The effective presentation of data is often very important, whether the presentation itself is a final goal or is to be used as background for further analysis and inference.

Chapter 1 discusses the role of statistics and introduces the basic concepts of descriptive, inferential, deductive, and inductive statistics. Chapter 2 covers data collection and the presentation of data in tables and/or graphs (charts). Chapter 3 discusses how data sets can be organized in a frequency distribution. Finally, in Chapter 4, important statistical measures of various data characteristics are developed and presented; these measures are then used in statistical analyses.

The examples in Part I deal with macroeconomic data, financial ratios, and the rates of return on shares of stock. Other, related topics are also discussed.

1

CHAPTER 1

Introduction

CHAPTER OUTLINE

1.1 The Role of Statistics in Business and Economics
1.2 Descriptive Versus Inferential Statistics
1.3 Deductive Versus Inductive Analysis in Statistics

Key Terms

statistics
data
descriptive statistics
inferential statistics
population
sample

hypothesis testing
deduction
deductive statistical analysis
induction
inductive statistical analysis

1.1 THE ROLE OF STATISTICS IN BUSINESS AND ECONOMICS

Statistics is a body of knowledge that is useful for collecting, organizing, presenting, analyzing, and interpreting **data** (collections of any number of related observations) and numerical facts. Applied statistical analysis helps business managers and economic planners formulate management policy and make business decisions more effectively. And statistics is an important tool for students of business and economics. Indeed, business and economic statistics has become one of the most important courses in business education, because a background in applied statistics is a key ingredient in understanding accounting, economics, finance, marketing, production, organizational behavior, and other business courses.

We may not realize it, but we deal with and interpret statistics every day. For example, the Dow Jones Industrial Average (DJIA) is the best-known and most widely watched indicator of the direction in which stock market values are heading. When people say, "The market was up 12 points today," they are probably referring to the DJIA. This single statistic summarizes stock prices of 30 large companies. Rather than listing the prices at which all of the approximately 2,000 stocks traded on the New York Stock Exchange are currently selling, analysts and reporters often cite this one number as a measure of overall market performance.

Let's take another example. Before elections, the media sometimes present surveys of voter preference in which a sample of voters instead of the whole population of voters is asked about candidate preferences. The media usually give the results of the poll and then state the possible margin of error. A margin of error of 3 percent means that the actual extent of a candidate's popular support may differ from the poll results by as much as 3 percentage points in either direction ("plus or minus"). Anyone who conducts a survey must understand statistics in order to make such decisions as how many people to contact, how to word the survey, and how to calculate the potential margin of error.

In business and industry, managers frequently use statistics to help them make better decisions. A shoe manufacturer, for instance, needs to produce a forecast of future sales in order to decide whether to expand production. Sales forecasts provide statistical guidance in most business decision making.

On a broader scale, the government publishes a variety of data on the health of the economy. Some of the most popular measures are the gross national product (GNP), the index of leading economic indicators, the unemployment rate, the money supply, and the consumer price index (CPI). All these measures are statistics that are used to summarize the general state of the economy. And, of course, business, government, and academic economists use statistical methods to try to *predict* these macroeconomic and other variables.

The following additional examples are presented to show that the use of statistics is widespread not only in business and economics but in everyday life as well.

EXAMPLE 1.1   *TV Show Ratings*

Television executives and advertisers use the ratings provided by A. C. Nielson to determine which television shows are the most popular. The Nielson organization regularly surveys a sample of television viewers in the United States about their viewing habits. Their responses are then used to draw conclusions about the viewing habits of the entire U.S. population.

EXAMPLE 1.2   *ABC-GPA*

In order to assign letter grades at the end of the semester, a teacher may calculate each student's grade point average to determine how well that student has performed in the class. In so doing, the teacher is calculating the mean or average of a series of grade points. The teacher might also want to know how widely dispersed the scores are across students in that class. In Chapter 4 we will discuss measures that describe the dispersion, or spread, of a group of data.

EXAMPLE 1.3   *One, Two, Three, "Fore!"*

To improve their golf scores, golfers often compute the average distance they can hit a ball with each golf club. These golfers then use the mean of a series of measurements to select the best club and thus fine-tune their game.

EXAMPLE 1.4   *Health Benefits of Oat Bran*

To determine the health benefits of eating oat bran, a doctor who has access to a large data base to which many physicians have contributed compares the average cholesterol level of people who eat oat bran with that of similar people who don't eat oat bran. The doctor is using statistics to evaluate the health benefits of different diets.

EXAMPLE 1.5   *Fertilizer Choice and Plant Growth Rate*

Refusing to accept on blind faith the advertising claims of either supplier, a farmer compares the average growth of plants fed with fertilizer A with the average growth of plants fed with fertilizer B to determine which fertilizer is more effective.

1.2   DESCRIPTIVE VERSUS INFERENTIAL STATISTICS

Having gotten a feel for the use of statistics by looking at several illustrations, we can now refine our definition of the term. **Statistics** is the collection, presentation, and summary of numerical information in such a way that the data can be easily interpreted.

There are two basic types of statistics: descriptive and inferential. **Descriptive statistics** deals with the presentation and organization of data. Measures of central tendency, such as the mean and median, and measures of dispersion, such as the standard deviation and range, are descriptive statistics. These types of statistics summarize numerical information. For example, a teacher who calculates the mean, median, range, and standard deviation of a set of exam scores is using descriptive statistics. Descriptive statistics is the subject of the first part of this book.

The following are examples of the use (or misuse) of descriptive statistics.

EXAMPLE 1.6   *Baseball Players' Batting Averages*

Descriptive statistics can be used to provide a point of reference. The batting averages of baseball players are commonly reported in the newspapers, but to people unfamiliar with baseball these numbers may be misleading. For example, Wade Boggs of the Boston Red Sox hit .366 in 1988; that is, he got a hit in almost 37 percent of his official at-bats. Because he was unsuccessful over 63 percent of the time, however, a person with little knowledge of baseball might conclude that Boggs is an inferior hitter. Comparing Boggs's average to the mean batting average of all players in the same year, which was .285, reveals that Boggs is among the best hitters.

EXAMPLE 1.7   *Monthly Unemployment Rates*

Graphical statistical analysis can be used to summarize small amounts of information. Figure 1.1 displays the U.S. unemployment rates for each month from November 1990 to October 1991. It shows, for instance, that the unemployment rates for November 1990, September 1991, and October 1991 were 5.7 percent, 6.7 percent, and 6.8 percent, respectively.[1]

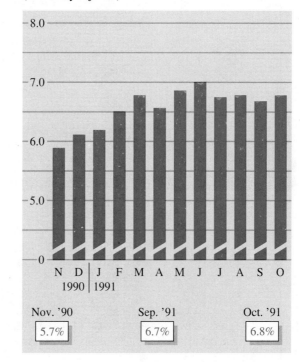

Figure 1.1

Monthly Unemployment Rate for the United States (November 1990–October 1991)
Source: U.S. Department of Labor. Reprinted by permission of The Associated Press.

[1]This graph is called a bar chart. Bar charts will be discussed in detail in Chapter 2.

EXAMPLE 1.8 *Comparison of Male and Female Earnings*

Descriptive statistics can also be used to compare different groups of data. For example, the mean earnings of full-time working men of different age groups who have had a four-year college education can be compared with the mean earnings of full-time working women of the same age groups with the same educational background to see whether any differences exist between their earnings. Drawing on 1990 Bureau of the Census data, the *Home News* of central New Jersey used the graph reproduced in Figure 1.2 to show that mean earnings for full-time working men are higher than those for full-time working women. This figure also shows that the pay gap between full-time working men and women is wider in older age groups. A college-educated woman between the ages of 18 and 24 earns an average of 92 cents for every dollar earned by a man of the same age and educational background. The gap widens steadily as we look at older age groups. Between ages 55 and 64, the average female worker in 1991 was making only 54 cents for every dollar earned by a man of like age and education.

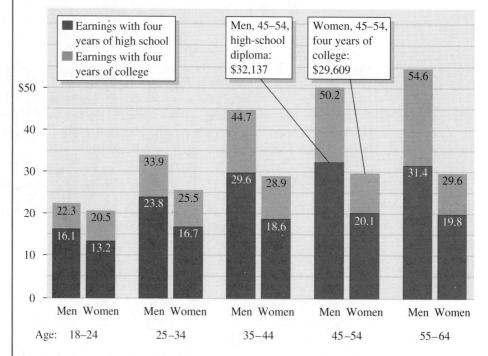

Figure 1.2

Mean Earnings for Full-Time Working Men and Women, by Age and Education
Source: Home News, November 14, 1991. Reprinted by permission of The Associated Press.

EXAMPLE 1.9   *Returns on Stocks and Bonds*

A financial analyst computes financial returns on stocks, corporate bonds, and government bonds to compare their performance during, say, the past 20 years. Because the analyst is collecting and summarizing data, we say that he or she is using descriptive statistics.

EXAMPLE 1.10   *Pitfalls of Comparing the Earnings of Males and Females*

We must always be careful when interpreting descriptive statistics. For example, it is sometimes noted that, on average, women earn 70 cents for each dollar that men earn. That is, the mean earnings for women are compared to the mean earnings for men to suggest that women experience wage discrimination. These descriptive statistics, however, may not tell the whole story. Differences in earnings may result from different occupational choices (which are perhaps influenced by social perceptions of the role women play), from different educational levels and choices (such as which subject to major in) or from career interruptions (which women may experience when choosing to leave their jobs to raise a family). Attributing wage differences entirely to discrimination, then, is an example of the misuse of statistics.[2]

Inferential statistics deals with the use of sample data to infer, or reach, general conclusions about a much larger population. In statistics, we define a **population** as the entire group of individuals we are interested in studying. A **sample** is any subset of such a population. In the election example presented earlier, the pollsters took a sample because it would have been too expensive and time-consuming to contact every voter. This is an example of the use of inferential statistics, because conclusions about a population were made on the basis of sample information. There might be differences between the characteristics of the actual population and the information gained from a sample, so errors can result. Inferential statistics—and in particular **hypothesis testing,** in which sample information is used to test a hypothesis about a population—is the subject of another major part of this book. Here we will merely look at several examples of inferential statistics.

EXAMPLE 1.11   *Sampling Survey of Residents' Voting Decision*

To obtain information on how residents will vote, the *Jericho Clarion* takes a sample and asks the people selected as part of the sample for whom they will vote. This newspaper is using inferential statistics because it is inferring, from a sample, information about a larger population. Again, the newspaper samples the population, rather than contacting all its members, because taking a sample is a lot cheaper and less time-consuming.

[2]Whether or not career interruptions should be a factor is a matter of debate among behavioral scientists.

EXAMPLE **1.12**   *Unemployment Rate*

The federal government releases information on the unemployment rate every month, which has been discussed in Example 1.7. To arrive at this figure, it samples households across the United States to determine the employment status of the members of those households. Extrapolating from the sample results to the general population is an example of applying inferential statistics.

EXAMPLE **1.13**   *Quality Control Via Sampling Data*

Suppose a production manager of Ford Motor Company compares two samples of a piston produced by different methods to find out whether the two methods result in different fractions of defective units. This production manager takes a sample of 100 pistons produced by one method and checks to see how many are defective and then compares this number to the number of defectives generated by the second production method. One hypothesis is that the number of defectives from the two methods are equal; an alternative hypothesis is that they are not. Inferential statistics can be used here to determine whether the proportions are sufficiently different for the first of these hypotheses to be rejected. This cost-conscious manager has taken a sample to gain information on a much larger quantity.

EXAMPLE **1.14**   *A Record Drop in Stock Prices*

Statistics is used to summarize the performance of the stock market on a given day. The Dow Jones Industrial Average, an average of the stocks of 30 major firms traded on the New York Stock Exchange, is used as a barometer of the performance of the overall stock market. Other indexes, such as Standard & Poor's 500 Com-

Table 1.1 Ten Largest Dow Jones Drops (10/6/87–11/15/91)

508.00 pts. to 1,738.74 (22.6%)	Oct. 19, 1987
190.58 pts. to 2,569.26 (6.91%)	Oct. 13, 1989
156.83 pts. to 1,793.93 (8.4%)	Oct. 26, 1987
140.58 pts. to 1,911.31 (6.9%)	Jan. 8, 1988
120.31 pts. to 2,943.20 (3.93%)	Nov. 15, 1991
108.35 pts. to 2,246.74 (4.6%)	Oct. 16, 1987
101.46 pts. to 2,005.64 (4.8%)	April 14, 1988
95.46 pts. to 2,005.64 (3.8%)	Oct. 14, 1987
93.31 pts. to 2,716.34 (3.32%)	Aug. 6, 1990
91.55 pts. to 2,548.63 (3.5%)	Oct. 6, 1987

Source: *Home News,* November 16, 1991. Reprinted by permission of The Associated Press.

posite Index (S&P 500), the Value Line Index, and the American Stock Exchange Index, are also calculated to generate summary measures of stock market performance. Each of these measures is derived through inferential statistics, because a sample is used to provide representative—though incomplete—information about the stock market at large.[3] For example, the Dow Jones Industrial Average dropped 120.31 points on November 15, 1991. It was the fifth largest point drop in history, as indicated in Table 1.1. Out of 1,815 stocks listed on the New York Stock Exchange, 1,476 declined in share price that day.

1.3 DEDUCTIVE VERSUS INDUCTIVE ANALYSIS IN STATISTICS

We also encounter another dichotomy in statistical analysis. **Deduction** is the use of general information to draw conclusions about specific cases. For example, probability tells us that if a student is chosen by lottery from a calculus class composed of 60 mathematics majors and 40 business administration majors, then the odds against picking a mathematics major are 4 to 6. Thus we can deduce that about 40 percent of such single-member samples of the students in this calculus class will be business administration majors. As another example of deduction, consider a firm that learns that 1 percent of its auto parts are defective and concludes that in any random sample, 1 percent of its parts are therefore going to be defective. The use of probability to determine the chance of obtaining a particular kind of sample result is known as **deductive statistical analysis.**

In Chapters 5, 6, and 7, we will learn how to apply deductive techniques when we know everything about the population in advance and are concerned with studying the characteristics of the possible samples that may arise from that known population.

Induction involves drawing general conclusions from specific information. In statistics, this means that on the strength of a specific sample, we infer something about a general population. The sample is all that is known; we must determine the uncertain characteristics of the population from the incomplete information available. This kind of statistical analysis is called **inductive statistical analysis.** For example, if 56 percent of a sample prefer a particular candidate for a political office, then we can estimate that 56 percent of the population prefers this candidate. Of course, our estimate is subject to error, and statistics enables us to calculate the possible error of an estimate. In this example, if the error is 3 percentage points, it can be inferred that the actual percentage of voters preferring the candidate is 56 percent plus or minus 3 percent; that is, it is between 53 percent and 59 percent.

Deductive statistical analysis shows how samples are generated from a population, and inductive statistical analysis shows how samples can be used to infer the

---

[3]Such is not the case, however, with the NYSE Composite Index. This index is a weighted average (mean) of *all* the firms on the NYSE and is thus the value of a *population* characteristic of a population.

characteristics of a population. Inductive and deductive statistical analyses are fully complementary. We must study how samples are generated before we can learn to generalize from a sample.

Summary

This chapter introduced the concept of statistics by presenting examples from everyday life. We saw that statistics can serve as a fundamental tool for decision making not only in business and economics but also in teaching, sports, medicine, quality control, and politics. Finally, we noted that statistics can be classified as either descriptive or inferential, and we drew the distinction between deductive and inductive analysis in statistics.

In the next chapter, we explain the process of collecting data and discuss how to present these data so that they can be interpreted easily and effectively.

Questions and Problems

1. Briefly explain why learning statistical inference is important for students of business and economics. Give two examples.

2. Define the term *statistics.* What are the two basic types of statistics? Describe them and give examples.

3. Explain the difference between deductive and inductive statistics. Illustrate your answer with examples taken from everyday life.

4. You are assigned by your general manager to examine each of last month's sales transactions, find their average, find the difference between the highest and lowest sales figures, and construct a chart showing the differences between charge account and cash customers. Is this a problem in descriptive or inferential statistics?

5. Suppose you are dealing with problems in probability and statistical inference. Which is usually the larger value in the problem, the population size or the sample size? Why?

6. State which type of statistical problem (deductive or inductive) makes each of the following assumptions.
   a. You know what the population characteristics are.
   b. You know what the sample characteristics are.

7. When a cosmetics manufacturer tests the market to determine how many women will buy eye liner that has been tested for safety without subjecting animals to injury, is it involved in a descriptive statistics problem or an inferential statistics problem? Explain your answer.

8. As controller of the Hamby Corporation, you are directed by the chairman of the board to investigate the problem of overspending by employees who have expense accounts. You ask the accounting department to provide you with records of the number of dollars spent by each of 25 top employees during the past month. The following record is provided:

$292	$494	$600	$807	$535
435	870	725	299	602
322	397	390	420	469
712	520	575	670	723
560	298	472	905	305

The question the board of directors wants answered is "How many of our 25 top executives spent more than $600 last month?" What will be your answer? Are you dealing with descriptive statistics or inferential statistics?

9. A teacher has just given an algebra exam. What are some of the statistics she could compute?

10. Suppose the teacher in question 9 is teaching four algebra classes. She would like to predict the average course grade of all her students from only the midterm scores. Should she use inferential or descriptive statistics?

11. Suppose the teacher in question 10 would like to predict the average grade of all her students by using only the midterm scores from one class. Should she use inferential or descriptive statistics?

12. A bullet manufacturer would like to keep the number of duds (bullets that won't fire) to a maximum of 2 per box of 100. Should inferential or descriptive statistics be used to decide this issue? Why?

13. Explain why election pollsters use inferential statistics rather than descriptive statistics to predict the outcome of an election.

14. Explain whether each of the following was arrived at via descriptive or inferential statistics.
    a. Ted Williams's lifetime batting average.
    b. The number of people watching the Super Bowl, based on A. C. Nielson's ratings.
    c. The number of people who favor teacher-led prayer in school, based on a survey of church goers.
    d. The average rate of return for IBM stock over the last 10 years.

15. Suppose you are interested in purchasing AT&T stock. You know that AT&T stock has had an average rate of return of 8% over the last 5 years. Explain how you could use descriptive statistics to help you decide whether to purchase AT&T.

16. Using any newspaper of your choice, find examples of statistics from the following sections:
    a. Sports section
    b. Business section
    c. Entertainment section

17. Are the statistics you found in answering question 16 inferential or descriptive statistics?

18. A popular commercial claims that "Four out of five dentists prefer sugarless gum." Is this conclusion drawn from a sample or a population?

19. The most commonly reported indicator of stock market performance is the Dow Jones Industrial Average. Explain whether the firms whose share price are included in the DJIA represent a sample or a population.

20. Use the information given in the accompanying figure to answer the following questions.
    a. Which month has the greatest amount of help-wanted advertising?
    b. How did help-wanted advertising fluctuate during the period of October 1988 through September 1991?

Help-Wanted Advertising

In percent, seasonally adjusted (1967 = 100)

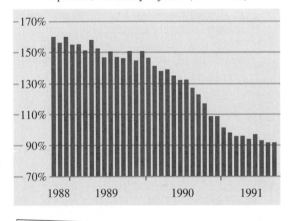

HELP-WANTED advertising remained unchanged in September from a month earlier at 91% of the 1967 average, the Conference Board reports.

Source: Wall Street Journal, November 7, 1991, p. A1.

21. Suppose a Gallup Poll is to be conducted to predict the outcome of the 1992 presidential election. Should the pollsters survey a sample or a population?

22. If managers at Weight Watchers are interested in the average number of pounds that people on Weight Watchers diets lose, should they use a sample or the population to find out?

23. Suppose that in question 22, Weight Watchers uses a sample. Does this represent the use of inferential or descriptive statistics? How would your answer change if Weight Watchers used the population?

24. The following list gives the seven highest-paid baseball players for 1992 and their salaries.

Dwight Gooden	$5,166,666
George Brett	4,700,000
Roger Clemens	4,300,000
Will Clark	4,250,000
Andy Van Slyke	4,250,000
Darryl Strawberry	4,050,000
Fred McGriff	4,000,000

   a. Does this list represent a sample or population of baseball players' salaries?
   b. If you were the agent for a top baseball star, how could you use the foregoing information?

25. Suppose Greg Norman has the following scores in his last 8 rounds of golf: 71, 68, 64, 73, 69, 62, 75, 69.
   a. If Greg computed his average score over these 8 rounds, would he be computing a descriptive or an inferential statistic?
   b. If Greg used these scores to predict his overall scoring average for the 1993 season, would he be using descriptive or inferential statistics?

26. Suppose a real estate broker in Albany, New York, is interested in the average price of a home in a development comprising 100 homes.
   a. If she uses 12 homes to predict the average price of all 100 homes, is she using inferential or descriptive statistics?
   b. If she uses all 100 homes, is she using inferential or descriptive statistics?

27. J. D. Power is a consulting firm that assesses consumer satisfaction for the auto industry. Do you think this company uses a sample or the population to conduct its survey?

CHAPTER 2

Data Collection and Presentation

CHAPTER OUTLINE

2.1 Introduction

2.2 Data Collection

2.3 Data Presentation: Tables

2.4 Data Presentation: Charts and Graphs

2.5 Applications

Appendix 2A Using Lotus 1-2-3 to Draw Graphs

Appendix 2B Stock Rates of Return and Market Rates of Return

Appendix 2C Financial Statements and Financial Ratio Analysis

Key Terms

primary data
secondary data
census
sample
random error
systematic error
time series graph
line chart
component-parts line chart
component-parts line graph
bar charts

pie charts
balance sheet
income statement
assets
liabilities
net worth
liquidity ratios
leverage ratios
activity ratios
profitability ratios
market value ratios

2.1 INTRODUCTION

The collection, organization, and presentation of data are basic background material for learning descriptive and inferential statistics and their applications. In this chapter we first discuss sources of data and methods of collecting them. Then we explore in detail the presentation of data in tables and graphs. Finally, we use both accounting and financial data to show how the statistical techniques discussed in this chapter can be used to analyze the financial condition of a firm and to analyze the recent deterioration of the financial health of the U.S. banking industry. In addition, we use a pie chart to examine how Congress voted on the Gulf Resolution in 1991.

2.2   DATA COLLECTION

After identifying a research problem and selecting the appropriate statistical methodology, researchers must collect the data that they will then go on to analyze. There are two sources of data: primary and secondary sources. **Primary data** are data collected specifically for the study in question. Primary data may be collected by methods such as personal investigation or mail questionnaires. In contrast, **secondary data** were not originally collected for the specific purpose of the study at hand, but rather for some other purpose. Examples of secondary sources used in finance and accounting include the *Wall Street Journal, Barron's, Value Line Investment Survey, Financial Times,* and company annual reports. Secondary sources used in marketing include sales reports and other publications. Although the data provided in these publications can be used in statistical analysis, they were not specifically collected for that use in any particular study.

EXAMPLE 2.1   *Primary and Secondary Sources of Data*

Let us consider the following cases and then characterize each data source as primary or secondary.

1. (Finance) To determine whether airline deregulation has increased the return and risk of stocks issued by firms in the industry, a researcher collects stock data from the *Wall Street Journal* and the Compustat data base. (The Compustat data base contains accounting and financial information for many firms.)
2. (Production) To determine whether ball bearings meet measurement specifications, a production engineer examines a sample of 100 bearings.
3. (Marketing) Before introducing a hamburger made with a new recipe, a firm gives 25 customers the new hamburger and asks them on a questionnaire to rate the hamburger in various categories.
4. (Political science) A candidate for political office has staff members call 1,000 voters to determine what candidate they prefer in an upcoming election.
5. (Marketing) A marketing firm looks up, in *Consumer Reports,* the demand for different types of cars in the United States.
6. (Economics) An economist collects data on unemployment from a Department of Labor report.
7. (Accounting) An accountant uses sampling techniques to audit a firm's accounts receivable or its inventory account.
8. (Economics) The staff from the Department of Labor uses a survey to estimate the current unemployment rate in the United States.

The cases numbered 1, 5, and 6 illustrate the use of secondary sources; these researchers relied on existing data sets. The remainder involve primary sources, because the data involved were generated specifically for that study.

The main advantage of primary data is that the investigator directly controls how the data are collected; therefore, he or she can ensure that the information is relevant to the problem at hand. For example, the investigator can design the questionnaires and surveys to elicit the most relevant information. The disadvantage of this method is that developing appropriate surveys or questionnaires requires considerable time, money, and experience. In addition, mail questionnaires are usually plagued by a low response rate. What response rate is acceptable varies with context and with other factors. A response rate of 50 percent is often considered acceptable, but it is rarely achieved with mail questionnaires.

Fortunately, there are many good secondary sources of information in business, economics, and finance. Financial information such as stock prices and accounting data is easy to locate but tedious to organize. As an alternative, data bases such as Compustat and CRSP (Center for Research on Securities Prices) tapes can be used. Economic data can be found in many government publications, such as the *Federal Reserve Bulletin,* the *Economic Report of the President,* and the *Statistical Abstract of the United States.* In addition, macroeconomic variables are found in data bases such as that of Data Resources. Of course, not all secondary sources are unimpeachable. Possible problems include outdated data, the restrictive definitions used, and unreliability of the source.

A sample or a census may be taken from either primary or secondary data. A **census** contains information on *all* members of the population; a **sample** contains observations from a *subset* of it. A census of primary data results, for example, from the polling of all voters in a city to determine their preference for mayor. If a subset of voters in the city is asked about their preference for mayor, a sample of primary data results. These are both examples of using primary data, because the data are collected for purposes of the study that is under way.

If a researcher records the prices of all the securities traded on the New York Stock Exchange for one day as they are listed in the *Wall Street Journal,* he or she is taking a census from a secondary source. However, if he or she takes a subset from the population—say, every fifth price—he or she is developing a sample of secondary data. Note that taking stock prices from the newspaper is an example of using secondary data because the data were not collected specifically for the study.

Given that the purpose of taking a sample is to gain information on a population, why do we not take a census every time we need information? The first reason is the high cost of taking a census. It would be extremely expensive for a pollster who wanted information on the outcome of a presidential election to contact all the registered voters in the country. Of course, the costs of obtaining the names of voters, hiring people to conduct the survey, performing computer analysis, and carrying out research must also be incurred when taking a sample, but because the sample is usually much smaller than the population, these costs are substantially reduced.

For example, to determine Illinois voters' preferences in the 1988 presidential election, the *Chicago Tribune* sampled 766 Illinois residents who said they would vote in the election. Obviously, sampling was cheaper than contacting all Illinois adults. The poll was accurate to within five percentage points, which is an acceptable margin of error. In Chapters 8 and 20, we will return to the topic of calculating the error in sampling.

The second advantage of sampling is accuracy. Because fewer people are contacted in a sample, the interviewers can allot more time to each respondent. In addition, the need for fewer workers to conduct the study may make it possible to select and train a more highly qualified staff of researchers. This, in turn, may result in a study of higher quality.

Another problem in taking a large census is the time involved. For example, suppose it would take at least 2 months for the *Tribune* to contact all the adults in Illinois. If the election were only 1 month away, the poll would not be of any use. In cases where the population is very large and will take a long time to reach, a sample is the more timely method of obtaining information.

This is not to suggest that a sample is always better than a census. A census is appropriate when the population is fairly small. For example, a census would be feasible if you wanted information on how the members of a small high school class intended to vote for Student Council president, because the cost and time of contacting every member of the class would be relatively low. In contrast, a sample is more cost- and time-effective when the population is a city, state, nation, or other large entity.

There are two types of errors that can arise when we are dealing with primary or secondary data. The first is **random error,** which is the difference between the value derived by taking a random sample and the value that would have been obtained by taking a census. This error arises from the random chance of obtaining the specific units that are included in the sample. Happily, random error can be reduced by increasing the sample size, and it can be reduced to zero by taking a census. Random error can also be estimated. Using statistics, the *Chicago Tribune* was able to determine that this poll was subject to random error of plus or minus 5 percent. This issue will be discussed in Part III, on sampling and statistical inference.

Systematic error results when there are problems in measurement. Unlike random error, which can occur only in sampling, systematic error can occur in both samples and census. For example, suppose that a basketball coach measures the heights of his players with an imprecise ruler. The resulting error is "systematic": the ruler distorts all measurements equally. As another example, when a researcher uses an incorrect computer program that calculates an arithmetic mean by dividing by the number of observations plus 5, a systematic error results because the divisor should have been the number of observations.

Let's use the measurement of basketball players' heights to compare random and systematic errors. Suppose the basketball coach selects a sample of five players, measures their heights with a "good" ruler, and finds (by dividing properly) that the mean of the sample is 6 ft 1 in. If the actual average height of all the players is 6 ft 2 in., the mean random error is −1 in. A random error will result. Now suppose the coach uses a ruler that is 2 in. too short. When measuring all the players (a census), he comes up with a population mean of 6 ft even. In this case, a systematic error of -2 in. results.

2.3 DATA PRESENTATION: TABLES

All data tables have four elements: a caption, column labels, row labels, and cells. The caption describes the information that is contained in the table. The column labels identify the information in the columns, such as the gross national product, the inflation rate, or the Dow Jones Industrial Average. Examples of row labels include years, dates, and states. A cell is defined by the intersection of a specific row and a specific column.

EXAMPLE 2.2   *Annual GNP, CPI, T-Bill Rate, and Prime Rate*

To illustrate, Table 2.1 gives macroeconomic information from 1950 to 1990. The caption is "GNP, CPI, T-Bill Rate, and Prime Rate (1950–1990)." The row labels are the years 1950 to 1990. The column labels are GNP (gross national product), CPI (consumer price index), 3-Month T-Bill Rate, and Prime Rate. The gross national product is the total market value of final goods and services produced in the United States for each 1-year period and is the most general measure of the overall state of the economy. Changes in the consumer price index, the most commonly used indicator of the economy's price level, are a measure of inflation or

Table 2.1 GNP, CPI, T-Bill Rate, and Prime Rate (1950–1990)

Year	GNP[a]	CPI[b]	3-Month T-Bill Rate	Prime Rate
50	1203.7	24.1	1.218%	2.07%
51	1328.2	26.0	1.552	2.56
52	1380.0	26.5	1.766	3.00
53	1435.3	26.7	1.931	3.17
54	1416.2	26.9	.953	3.05
55	1494.9	26.8	1.753	3.16
56	1525.6	27.2	2.658	3.77
57	1551.1	28.1	3.267	4.20
58	1539.2	28.9	1.839	3.83
59	1629.1	29.1	3.405	4.48
60	1665.3	29.6	2.928	4.82
61	1708.7	29.9	2.378	4.50
62	1799.4	30.2	2.778	4.50
63	1873.3	30.6	3.157	4.50
64	1973.3	31.0	3.549	4.50
65	2087.6	31.5	3.954	4.54
66	2208.3	32.4	4.881	5.63

(*continued*)

Table 2.1 (Continued)

Year	GNP[a]	CPI[b]	3-Month T-Bill Rate	Prime Rate
67	2271.4	33.4	4.321	5.61
68	2365.6	34.8	5.339	6.30
69	2423.3	36.7	6.677	7.96
70	2416.2	38.8	6.458	7.91
71	2484.8	40.5	4.348	5.72
72	2608.5	41.8	4.071	5.25
73	2744.1	44.4	7.041	8.03
74	2729.3	49.3	7.886	10.81
75	2695.0	53.8	5.838	7.86
76	2826.7	56.9	4.989	6.84
77	2958.6	60.6	5.265	6.83
78	3115.2	65.2	7.221	9.06
79	3192.4	72.6	10.041	12.67
80	3187.1	82.4	11.506	15.27
81	3248.8	90.9	14.029	18.87
82	3166.0	96.5	10.686	14.86
83	3279.1	99.6	8.63	10.79
84	3501.4	103.9	9.58	12.04
85	3618.7	107.6	7.48	9.93
86	3717.9	109.6	5.98	8.33
87	3845.3	113.6	5.82	8.22
88	4016.9	118.3	6.69	9.32
89	4117.7	124.9	8.12	10.87
90	4155.8	132.1	7.51	10.01

Source: *Economic Report of the President,* January 1991.
[a]Millions of 1982 dollars.
[b]CPI base: 1982–1984 = 100.

deflation. (For a more detailed description of the CPI, see Chapter 19). The three-month T-bill interest rate is the interest rate that the U.S. Treasury pays on 91-day debt instruments, and the prime rate is the interest rate that banks charge on loans to their best customers, usually large firms. This table, then, presents macroeconomic information for any year indicated. For example, the CPI for 1980 was 82.4, and the prime rate in 1990 was 10.01 percent.

2.4 DATA PRESENTATION: CHARTS AND GRAPHS

It is sometimes said that a picture is worth a thousand words, and nowhere is this statement more true than in the analysis of data. Tables are usually filled with highly specific data that take time to digest. Graphs and charts, though they are often less detailed than tables, have the advantage of presenting data in a more accessible and memorable form. In most graphs and charts, the independent variable is plotted on the horizontal axis (the X axis) and the dependent variable on the vertical axis (the Y axis). Frequently, "time" is plotted along the X axis. Such a graph is known as a **time series graph** because on it, changes in a dependent variable (such as GNP, inflation rate, or stock prices) can be traced over time.

Line charts are constructed by graphing data points and drawing lines to connect the points. Figure 2.1 shows how the rate of return on the S&P 500 and the 3-month T-bill rate have varied over time.[1] The independent variable is the year (ranging from 1970 to 1990), so this is a time series graph. The dependent variables are often in percentages.

Figure 2.2 is a graph of the components of the gross national product (GNP)— private consumption expenditures, government purchases of goods and services, gross private domestic investment, and net exports—over time. This is also a time series graph because the independent variable is time. It is a **component-parts line chart.** These series have been "deflated" by expressing dollar amounts in constant 1982 dollars. (Chapter 19 discusses the deflated series in further detail.)

Figure 2.2 is also called a **component-parts line graph** because the four parts of the GNP are graphed. The sum of the four components equals the GNP. Using this

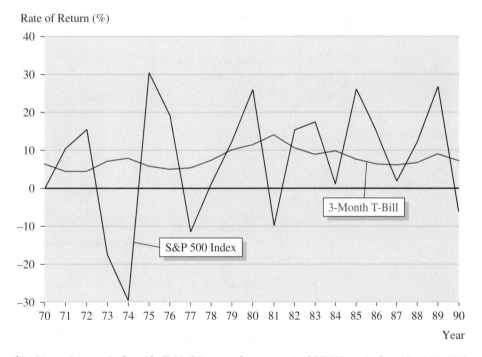

Figure 2.1

Rates of Return on S&P 500 and 3-Month T-Bills

[1]T-bill rate data can be found in Table 2.1; rates of return on the S&P 500 can be found in Table 2B.2 in Appendix 2B of this chapter. Most of the figures in this book are drawn with the Lotus 1-2-3 PC program. The procedure for using the Lotus program to draw these graphs can be found in Appendix 2A of this chapter.

type of graph makes it possible to show the sources of increases or declines in the GNP. (The data used to generate Figure 2.2 are found in Table 2.2.)

Bar charts can be used to summarize small amounts of information. Figure 2.3 shows the average annual returns for Tri-Continental Corporation for investment periods of seven different durations ending on September 30, 1991. This figure shows that Tri-Continental has provided investors double-digit returns during a 50-year period. It also shows that the investment performance of this company was better than that of the Dow Jones Industrial Average (DJIA) and the S&P 500.[2] As this example illustrates, using a bar graph is most appropriate when we are comparing only a few items.

Pie charts are used to show the proportions of component parts that make up a total. Figure 2.4 shows how the U.S. soft drink market was broken down in 1985. The two industry leaders, Coca-Cola and PepsiCo, enjoyed 40% and 28% of the market share, respectively. The next four largest firms (Seven-Up, Dr Pepper, Royal Crown, and Cadbury Schweppes) accounted for 21.8% of the market, and the remaining 10.2% of the market was divided among still smaller companies.

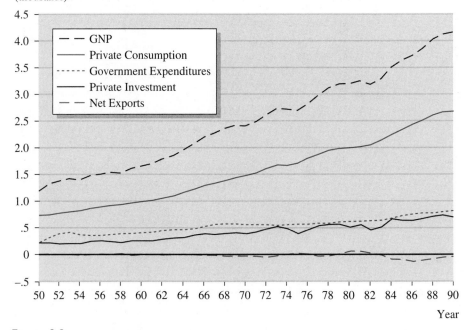

Dollars (Thousand Billions)
(thousands)

Figure 2.2

Components of GNP (Billions of 1982 Dollars)

[2]Both the DJIA and the S&P 500 will be discussed in Chapter 19 of this book.

Table 2.2 Annual Macroeconomic Data 1950–1990 (in 1982 dollars)

Year	GNP[a]	CPI[b]	3-Month T-Bill Rate	Prime Rate	Private Consumption[a]	Private Investment	Net Exports[a]	Government Expenditures[a]
50	1203.7	24.1	1.218%	2.07%	733.2	234.9	4.7	230.8
51	1328.2	26.0	1.552	2.56	748.7	235.2	14.6	329.7
52	1380.0	26.5	1.766	3.00	771.4	211.8	6.9	389.9
53	1435.3	26.7	1.931	3.17	802.5	216.6	−2.7	419.0
54	1416.2	26.9	0.953	3.05	822.7	212.6	2.5	378.4
55	1494.9	26.8	1.753	3.16	873.8	259.8	0	361.3
56	1525.6	27.2	2.658	3.77	899.8	257.8	4.3	363.7
57	1551.1	28.1	3.267	4.20	919.7	243.4	7.0	381.1
58	1539.2	28.9	1.839	3.83	932.9	221.4	10.3	395.3
59	1629.1	29.1	3.405	4.48	979.4	270.3	−18.2	397.7
60	1665.3	29.6	2.928	4.82	1005.1	260.5	−4.0	403.7
61	1708.7	29.9	2.378	4.50	1025.2	259.1	−2.7	427.1
62	1799.4	30.2	2.778	4.50	1069.0	288.6	−7.5	449.4
63	1873.3	30.6	3.157	4.50	1108.4	307.1	−1.9	459.8
64	1973.3	31.0	3.549	4.50	1170.6	325.9	5.9	470.8
65	2087.6	31.5	3.954	4.54	1236.4	367.0	−2.7	487.0
66	2208.3	32.4	4.881	5.63	1298.9	390.5	−13.7	532.6
67	2271.4	33.4	4.321	5.61	1137.7	374.4	−16.9	576.2
68	2365.6	34.8	5.339	6.30	1405.9	391.8	−29.7	597.6
69	2423.3	36.7	6.677	7.96	1456.7	410.3	−34.9	591.2
70	2416.2	38.8	6.458	7.91	1492.0	381.5	−30.0	572.6
71	2484.8	40.5	4.348	5.72	1538.8	419.3	−39.8	566.5
72	2608.5	41.8	4.071	5.25	1621.9	465.4	−49.4	570.7
73	2744.1	44.4	7.041	8.03	1689.6	520.8	−31.5	565.3
74	2729.3	49.3	7.886	10.81	1674.0	481.3	.8	573.2
75	2695.0	53.8	5.838	7.86	1711.9	383.3	18.9	580.9
76	2826.7	56.9	4.989	6.84	1803.9	453.5	−11.0	580.3
77	2958.6	60.6	5.265	6.83	1883.8	521.3	−35.5	589.1
78	3115.2	65.2	7.221	9.06	1961.0	576.9	−26.8	604.1
79	3192.4	72.6	10.04	12.67	2004.4	575.2	3.6	609.1
80	3187.1	82.4	11.51	15.27	2000.4	509.3	57.0	620.5
81	3248.8	90.9	14.03	18.87	2024.2	545.5	49.4	629.7
82	3166.0	96.5	10.69	14.86	2050.7	447.3	26.3	641.7
83	3279.1	99.6	8.63	10.79	2146.0	504.0	−19.9	649.0
84	3501.4	103.9	9.58	12.04	2249.3	658.4	−84.0	677.7
85	3618.7	107.6	7.48	9.93	2354.8	637.0	−104.3	731.2
86	3717.9	109.6	5.98	8.33	2446.4	639.6	−129.7	761.6
87	3845.3	113.6	5.82	8.22	2515.8	669.0	−118.5	779.1
88	4016.9	118.3	6.69	9.32	2606.5	705.7	−75.9	780.5
89	4117.7	124.9	8.12	10.87	2656.8	716.9	−54.1	798.1
90	4155.8	132.1	7.51	10.01	2682.2	690.3	−37.5	820.8

Source: *Economic Report of the President,* January 1991.

[a]Millions of 1982 dollars.
[b]CPI base: 1982–1984 = 100.

Rate of Return (%)

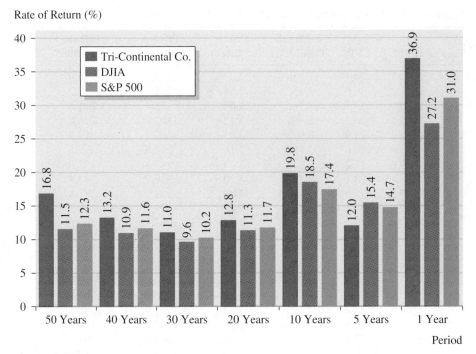

Figure 2.3

Average Annual Returns for Tri-Continental Corporation for Investment Periods of Seven Different Durations Ending on September 30, 1991)
Source: *Wall Street Journal,* November 18, 1991, p. C5.

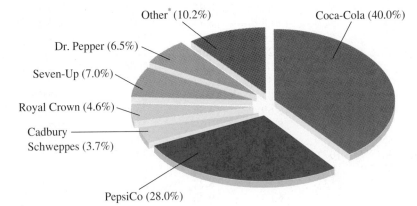

Figure 2.4

U.S. Soft Drink Market Breakdown (1985)
Data: *Beverage Digest,* Montgomery Securities.
Source: *Business Week.*

<table>
<tr><td>**2.5**</td><td></td></tr>
</table>

APPLICATIONS

In the last several sections, we have drawn primarily on macroeconomic data to show how tables and graphs can be used to examine various economic variables. In this section, we will use the same tabular and graphical tools to analyze financial and accounting data that are important in financial analysis and planning. We also will see how Congress voted on the Gulf Resolution.

■ **APPLICATION 2.1** Analysis of Stock Rates of Return and Market Rates of Return

Stock prices and stock indexes are two familiar measures of stock market performance. In addition to these indicators, percentage rates of return can be calculated to determine how well a particular stock—or the stock market overall—is doing.

Figure 2.5 is a line graph of yearly rates of return for GM, Ford, and the S&P 500, which, as we have noted, is a market index. The yearly rates of return have been similar for the three. However, this indicator has fluctuated less for GM stock than for Ford, and the overall market (as gauged by the S&P 500) has varied least of all. Ford's huge capital gain of 132 percent in 1982 accounts for the rate of return of the stock more than doubling from 1981 to 1982.[3]

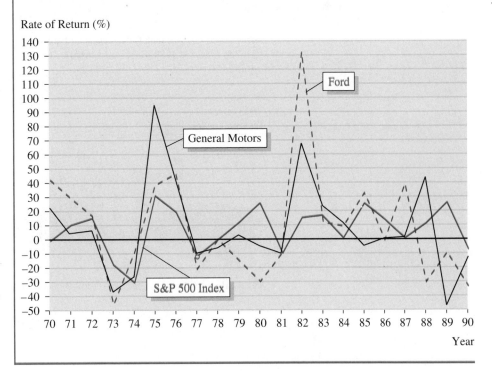

Figure 2.5

Rates of Return for S&P 500, Ford, and GM

[3]Rates of return for GM and Ford and market rates of return are analyzed in more detail in Appendix 2B.

■ | **APPLICATION 2.2** Financial Health of the Banking Industry: 1981 Versus 1989

On January 6, 1991, the business section of the *Home News* (a central New Jersey newspaper) printed an Associated Press article that used two pie charts prepared by Veribanc Inc., a financial rating service. The pie charts, presented in Figure 2.6, compare the financial condition of U.S. commercial banks in 1981 to their condition in 1989. These two pie charts show that the percentage of non-problem banks (those whose equity is 5% or more of their assets) has fallen from 98 percent to 77.8 percent, revealing that the probability that depositors are dealing with a problem-plagued bank has increased about 11 times.[4] In view of this deterioration, the article offers the following five tips to anyone shopping for a new financial institution.

1. Determine whether deposits are protected by federal deposit insurance, which covers deposits of up to $100,000.
2. Research any state deposit insurance funds.
3. Investigate the institution's history.
4. Check new reports for the health of specific banks and other industry trends.
5. Ask the bank for its yearly financial statement. Or contact federal bank regulators for the institution's quarterly statement of financial condition and its income statement.

Figure 2.6

How Healthy
Is Your Bank?
Source: *Home News,*
January 6, 1991.
Reprinted by
permission of The
Associated Press.

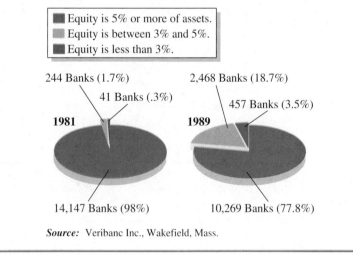

Federal Reserve Board data for commercial banks, December 1981 and 1989. Banks are classified according to equity.

■ Equity is 5% or more of assets.
■ Equity is between 3% and 5%.
■ Equity is less than 3%.

244 Banks (1.7%)   2,468 Banks (18.7%)
41 Banks (.3%)   457 Banks (3.5%)
1981   **1989**
14,147 Banks (98%)   10,269 Banks (77.8%)

Source: Veribanc Inc., Wakefield, Mass.

[4]Eleven times can be calculated as $\dfrac{1.0 - .778}{1.0 - .98} = .111$

APPLICATION 2.3 How Congress Voted on the Gulf Resolution

Following a heated debate, Congress voted to grant President Bush the power to go to battle against Iraq if the Iraqis did not withdraw from Kuwait by January 15, 1991.

As indicated in the pie chart in Figure 2.7, the Senate vote of 52 to 47 and the House vote of 250 to 183 authorized President Bush to use military force against Iraq. Among those voting *yes* were 43 Republican senators, 9 Democratic senators, 164 Republicans in the house of representatives, and 86 Democrats in the house of representatives; among those voting *no* were 2 Republican senators and 45 Democratic senators. In the House, 3 Republicans, 179 Democrats, and 1 independent voted no. In terms of percentages, about 52.53 percent of the Senate and 57.74 percent of the House voted to support the Gulf Resolution. One senator and two representatives were not present.

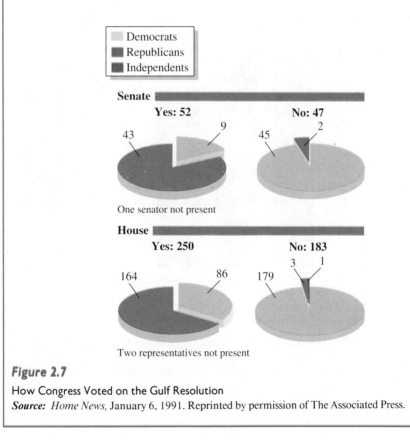

Figure 2.7

How Congress Voted on the Gulf Resolution

Source: *Home News,* January 6, 1991. Reprinted by permission of The Associated Press.

■ | **APPLICATION 2.4** Bar Charts Reveal How Several Economic
                         Indicators Are Related

In *Time* magazine, November 1991, eight bar charts showed how economic conditions fluctuated during 1989–1991.

To stimulate the economy, policy decision makers at the Fed lowered interest rates. Mortgage rates dropped from 10.32 percent in 1989 to 8.76 percent in 1991, while auto loan rates dropped from 12.27 percent to 11.78 percent. Despite these lower interest rates, both housing starts and auto sales experienced surprising declines. Figures 2.8(a–h) give a clear picture of these relationships.

Because interest rates such as that on 1-year CDs declined (specifically, from 8.67 percent to 6.24 percent) over the 3-year period, the proportion of savers' income derived from interest also went down (from 15 percent in 1989 and 1990 to 14 percent in 1991).

Mortgage Rate (%)

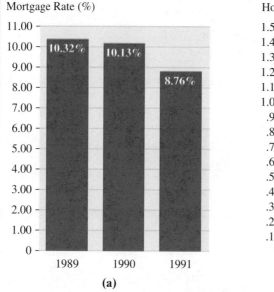

(a)

Housing Starts (millions)

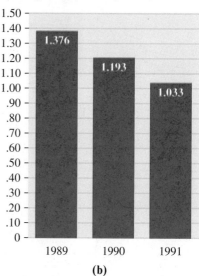

(b)

Figure 2.8

Eight Macroeconomic Indicators: (a) Mortgage Rates, (b) Housing Starts, (c) Consumer Finance Rate for Auto Loans, (d) Domestic Auto Sales, (e) 1-Year CD Interest Yield, (f) Portion of Income Derived from Interest, (g) Time and Savings-Account Deposits, and (h) NASDAQ Composite Index
Source: Adapted from "Statistics for Business and Economics," *TIME* Magazine, November 25, 1991. Copyright 1991 The Time, Inc. Magazine Company. Reprinted by permission.

Figure 2.8 (Continued)

Consumer Finance Rate (%)

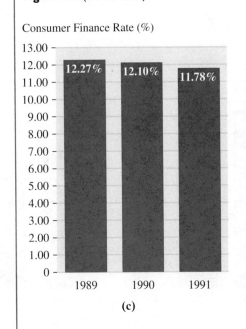

(c)

Auto Sales (millions)

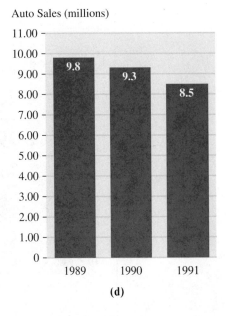

(d)

Interest Yield (%)

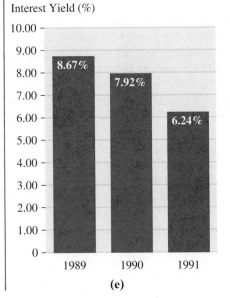

(e)

Portion of Income (%)

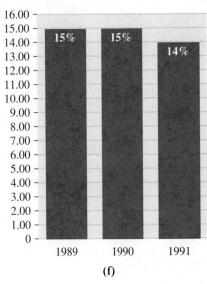

(f)

Figure 2.8 (Continued)

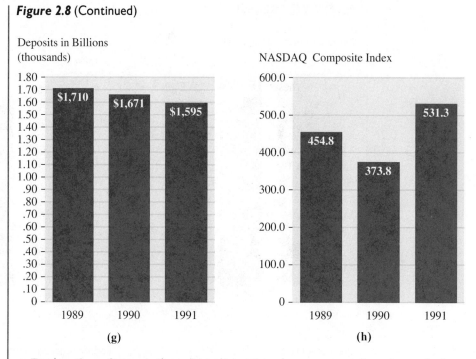

Deposits in Billions
(thousands)

NASDAQ Composite Index

(g)   (h)

During these 3 years, time deposits and savings account deposits sank from
$1,710 billion to $1,595 billion. Many savers withdrew their savings to invest them
in the stock market. Consequently, the stock market indexes went up dramatically.
For example, the NASDAQ Composite Index (an index compiled from over-the-
counter stock) fluctuated from 454.8 in 1989 to 373.8 in 1990 to 531.3 in 1991.

As we can see in Figure 2.8, bar charts can very effectively and clearly show
changes in economic conditions. Common sense is all that the viewer needs to
interpret the charts.

We discussed some of the data analysis related to Figure 2.8(a–h) in this chapter;
in later chapters we will discuss it further, using more sophisticated statistical methods. These bar charts give us a great deal of information. And other statistical analysis related to this set of information will improve our understanding of macroeconomic analysis.

Summary

Good data are essential in business and economic decision making. Hence it is important to be familiar with the sources of business and economic data and to know how these data can be collected.

Data for a census or a sample can be gleaned from both primary and secondary sources. However, we must guard against random error when using a sample and against systematic error in all our data collection.

Because we want to use sample data to make inferences about the population from which they are drawn, it is important for us to be able to present the data effectively. Tables and charts are two simple methods for presenting data. Line charts, bar charts, and pie charts are three basic and important graphical methods for describing data. In the next chapter, we will discuss other tabular and graphical methods for describing data in a more sophisticated and detailed manner.

Appendix 2A Using Lotus 1-2-3 to Draw Graphs

This appendix explains how to use Lotus 1-2-3 to draw graphs. Three steps are involved: entering Lotus 1-2-3, preparing the data, and drawing the graph(s).

Stage 1: Entering Lotus 1-2-3

1. C:\> cd\123 (hit return)

2. C:\123> lotus (hit return)

```
-------------------------------------------------------------
1-2-3  PrintGraph  Translate  Install  View  Exit
Enter 1-2-3 -- Lotus Worksheet/Graphics/Database program
=============================================================
                      1-2-3 Access System
                    Copyright © 1986, 1987
                  Lotus Development Corporation
                      All Rights Reserved
                         Release 2.01

The Access System lets you choose 1-2-3, PrintGraph, the Translate utility,
the Install program, and A View of 1-2-3 from the menu at the top of this
screen.  If you're using a diskette system, the Access System may prompt
you to change disks. Follow the instructions below to start a program.

○ Use [RIGHT] or [LEFT] to move the menu pointer (the highlight bar at
   the top of the screen) to the program you want to use.

○ Press [RETURN] to start the program.

You can also start a program by typing the first letter of the menu
choice. Press [HELP] for more information.

----------------------- Press [NUM LOCK]-----------------------
```

Figure 2A.1
Source: Lotus 1-2-3, Lotus Development, reprinted by permission.

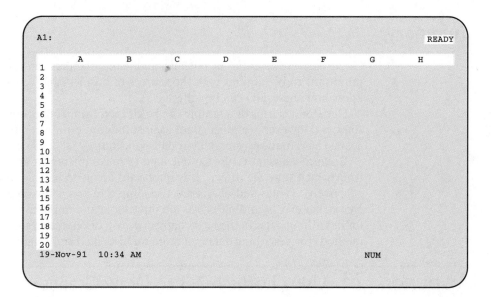

Figure 2A.2

Figure 2A.3

3. Then the screen will look like Figure 2A.1.

4. Move cursor to 1-2-3 on the bar at the top of the screen, and hit return. The screen as shown in Figure 2A.2 will appear. When we reach the screen shown in Figure 2A.2, it means we have entered the Lotus 1-2-3 worksheet and are ready to enter our data.

```
A1:                                                                  READY

           D     E     F     G         H         I     J    K     L
 1
 2              AVERAGE ANNUAL RETURN (%)
 3              Tri-Continental   DJIA           S&P 500
 4              Co. Market Value
 5
 6    50 Years                  16.8      11.5        12.3
 7    40 Years                  13.2      10.9        11.6
 8    30 Years                  11.0       9.6        10.2
 9    20 Years                  12.8      11.3        11.7
10    10 Years                  19.8      18.5        17.4
11     5 Years                  12.0      15.4        14.7
12     1 Year                   36.9      27.2        31.0
13
14
15
16
17
18
19
20
19-Nov-91   11:14 AM                                       NUM
```

Figure 2A.4

```
K11: [W18] Other                                                    READY

           K         L       M      N      O      P      Q
 1    SOFT DRINK: U.S. MARKET SPLIT UP
 2
 3    COMPANY         MARKET SHARES (%)
 4
 5    Coca-Cola            40
 6    PepsiCo              28
 7    Cadbury Schweppes    3.7
 8    Royal Crown          4.6
 9    Seven-Up             7
10    Dr. Pepper           6.5
11    Other                10.2
12
13
14
15
16
17
18
19
20
19-Nov-91   11:39 AM                                       NUM
```

Figure 2A.5

Stage 2: Preparing the Data

Now that we have entered the worksheet, we can fill it in with data. One observation can be entered into each cell. Figure 2A.3 shows the data for Figure 2.1. The screen shows that the data range of rate of return of 3-month T-bills is B5 to B25 (B5. .B25), and that it is C5 to C25 (C5. .C25) for the S&P 500. On the same worksheet, the screens for Figure 2.3 and Figure 2.4 are shown in Figures 2A.4 and 2A.5.

Stage 3: Drawing the Graphs

With Lotus 1-2-3, we can draw graphs directly from the data on the worksheet. A simple six-step procedure should be followed.

1. Type "/".

After finishing the data preparation, we can start to draw graphs by typing "/", which brings us into Lotus Manual. After we have typed "/", the screen will look like Figure 2A.6.

2. Select Graph

Move the cursor to "Graph" on the menu line at the top of the screen, and hit return. The manual part of the upper worksheet is shown in Figure 2A.7.

3. Select Type

Move the cursor to "Type" on the menu line and hit return. The manual will change to the screen shown in Figure 2A.8. We can choose any of five different types of graphs: line chart, bar chart, XY chart, stacked bar chart, or pie chart. In the case of Figure 2.1, we choose a line chart. To create Figures 2.3 and 2.4, choose a bar chart and a pie chart, respectively.

```
A1:                                                                 MENU
Worksheet  Range   Copy   Move   File   Print   Graph   Data  System  Quit
Global,    Insert, Delete, Column, Erase, Titles, Window, Status, Page
        A        B       C       D       E      F       G       H       I       J
 1
 2           RATE OF RETURN               AVERAGE ANNUAL RETURN (%)
 3    Year    TB-3M  S&P 500              Tri-Continental  DJIA    S&P 500
 4                                        Co. Market Value
 5    1970    6.46   0.0010
 6    1971    4.35   0.1080   50 Years                16.8    11.5    12.3
 7    1972    4.07   0.1557   40 Years                13.2    10.9    11.6
 8    1973    7.04  -0.1737   30 Years                11.0     9.6    10.2
 9    1974    7.89  -0.2964   20 Years                12.8    11.3    11.7
10    1975    5.84   0.3149   10 Years                19.8    18.5    17.4
11    1976    4.99   0.1918    5 Years                12.0    15.4    14.7
12    1977    5.27  -0.1153    1 Year                 36.9    27.2    31.0
13    1978    7.22   0.0105
14    1979   10.04   0.1228
15    1980   11.51   0.2586
16    1981   14.03  -0.0994
17    1982   10.69   0.1549
18    1983    8.63   0.1706
19    1984    9.58   0.0115
20    1985    7.48   0.2633
21    1986    5.98   0.1462
22    1987    5.82   0.0203
23    1988    6.69   0.1240
24    1989    8.12   0.2725
25    1990    7.51  -0.0656
```

Figure 2A.6

```
A1:
Type  X  A  B  C  D  E  F  Reset  View  Save  Options  Name  Quit
Set graph type
```

	A	B	C	D	E	F	G	H
1								
2		RATE OF RETURN					AVERAGE ANNUAL RETURN (%)	
3	Year	TB-3M	S&P 500				Tri-Continental	DJIA
4							Co. Market Value	
5	1970	6.46	0.0010					
6	1971	4.35	0.1080	50 Years			16.8	11.5
7	1972	4.07	0.1557	40 Years			13.2	10.9
8	1973	7.04	-0.1737	30 Years			11.0	9.6
9	1974	7.89	-0.2964	20 Years			12.8	11.3
10	1975	5.84	0.3149	10 Years			19.8	18.5
11	1976	4.99	0.1918	5 Years			12.0	15.4
12	1977	5.27	-0.1153	1 Year			36.9	27.2
13	1978	7.22	0.0105					
14	1979	10.04	0.1228					
15	1980	11.51	0.2586					
16	1981	14.03	-0.0994					
17	1982	10.69	0.1549					
18	1983	8.63	0.1706					
19	1984	9.58	0.0115					
20	1985	7.48	0.2633					
21	1986	5.98	0.1462					
22	1987	5.82	0.0203					
23	1988	6.69	0.1240					
24	1989	8.12	0.2725					
25	1990	7.51	-0.0656					

Figure 2A.7

```
A1:
Line  Bar  XY  Stacked-Bar  Pie
Line graph
```

	A	B	C	D	E	F	G	H
1								
2		RATE OF RETURN					AVERAGE ANNUAL RETURN (%)	
3	Year	TB-3M	S&P 500				Tri-Continental	DJIA
4							Co. Market Value	
5	1970	6.46	0.0010					
6	1971	4.35	0.1080	50 Years			16.8	11.5
7	1972	4.07	0.1557	40 Years			13.2	10.9
8	1973	7.04	-0.1737	30 Years			11.0	9.6
9	1974	7.89	-0.2964	20 Years			12.8	11.3
10	1975	5.84	0.3149	10 Years			19.8	18.5
11	1976	4.99	0.1918	5 Years			12.0	15.4
12	1977	5.27	-0.1153	1 Year			36.9	27.2
13	1978	7.22	0.0105					
14	1979	10.04	0.1228					
15	1980	11.51	0.2586					
16	1981	14.03	-0.0994					
17	1982	10.69	0.1549					
18	1983	8.63	0.1706					
19	1984	9.58	0.0115					
20	1985	7.48	0.2633					
21	1986	5.98	0.1462					
22	1987	5.82	0.0203					
23	1988	6.69	0.1240					
24	1989	8.12	0.2725					
25	1990	7.51	-0.0656					

Figure 2A.8

4. Enter Data Range

After selecting the type of graph, press return to go back to the graph manual. Select X to specify the data range for the X axis. For Figure 2.1, the X range is A5 to A25 (A5. .A25). Select A and B individually to specify the data range for observations of rates of return on 3-month T-bills and the S&P 500. These ranges are B5 to B25 (B5. .B25) and C5 to C25 (C5. .C25), respectively.

5. View the Graph

After finishing steps 1 through 4, you can view the graph by selecting "view" in the graph manual.

6. Supply Titles

There are other commands you can use to specify the titles of the graphs and legends for each variable (see the Lotus 1-2-3 *Manual*).

Appendix 2B Stock Rates of Return and Market Rates of Return

Table 2B.1 presents data on earnings per share (EPS), dividends per share (DPS), and price per share (PPS) for GM, Ford, and the S&P 500 during the period 1969 to 1990. Table 2B.2 shows rates of return for GM, Ford, and the S&P 500, calculated from the data in Table 2B.1.

The formula for calculating the rate of return, R_{jt}, on the jth individual stock in period t is

$$R_{jt} = \frac{P_{jt} - P_{jt-1} + d_{jt}}{P_{jt-1}} \tag{2B.1}$$

where P_{jt} represents price per share for the jth stock in period t, and d_{jt} represents dividends per share for the jth stock. The market rate of return, R_{mt}, in period t is

$$\frac{SP_t - SP_{t-1}}{SP_{t-1}} \tag{2B.2}$$

where SP_t represents the S&P 500 in period t.

The rate of return on an individual stock can be rewritten as

$$R_{jt} = \frac{(P_{jt} - P_{jt+1})}{P_{jt-1}} + \frac{d_{jt}}{P_{jt-1}} = \text{Capital gain yield} + \text{dividend yield} \tag{2B.3}$$

Table 2B.1 EPS, DPS, and PPS for GM, Ford, and the S&P 500

Year	General Motors			Ford			S&P 500 Index
	EPS	*DPS*	*PPS*	*EPS*	*DPS*	*PPS*	
69	5.95	4.30	69.13	5.03	2.40	41.13	92.06
70	2.09	3.40	80.50	4.77	2.40	56.25	92.15
71	6.72	3.40	80.50	6.18	2.50	70.25	102.10
72	7.51	4.45	81.13	8.52	2.68	79.63	118.00
73	8.34	5.25	46.13	9.13	3.20	40.50	97.50
74	3.27	3.40	30.75	3.86	3.20	33.38	68.60
75	4.32	2.40	57.63	3.46	2.60	44.00	90.20
76	10.08	5.55	78.50	10.45	2.80	61.50	107.50
77	11.62	6.80	62.88	14.16	3.04	45.75	95.10
78	12.24	6.00	53.75	13.35	3.50	42.13	96.10
79	10.04	5.30	50.00	9.75	3.90	32.00	107.90
80	−2.65	2.95	45.00	−12.83	2.60	20.00	135.80
81	1.07	2.40	38.50	−8.81	1.20	16.75	122.30
82	3.09	2.40	62.38	−5.46	0.00	38.88	141.24
83	11.84	2.80	74.38	10.29	0.50	43.38	165.34
84	14.22	4.75	78.38	15.79	2.00	45.63	167.24
85	12.28	5.00	70.38	13.63	2.40	58.00	211.28
86	8.22	5.00	66.00	12.32	2.22	56.25	242.17
87	10.06	5.00	61.38	18.10	3.15	75.38	247.08
88	13.64	5.00	83.50	10.96	2.30	50.50	277.72
89	6.33	3.00	42.25	8.22	3.00	43.63	353.40
90	−4.09	3.00	34.38	1.86	3.00	26.63	330.22

Source: EPS, DPS, and PPS for General Motors and Ford are from Industrial Compustat, 1991 Version. S&P 500 data are from ISL Daily Stock Price Publication.

The first term of the rewritten equation is the capital gain yield (in percent); the second is the dividend yield (also in percent). As an example, let us calculate the rate of return for GM in 1970. From Table 2B.1, we know that PPS_{69} = \$69.13, PPS_{70} = \$80.50, and DPS_{70} = \$3.40. Thus the rate of return for GM in 1970 equals

$$R_{GM70} = \frac{80.50 - 69.13 + 3.4}{69.13} = .2137.$$

As another example, from Table 2B.1 we know that the S&P 500 was 247.08 and 277.72 in 1987 and 1988, respectively. Thus the market rate of return in 1988 equaled

$$R_{M88} = \frac{277.72 - 247.08}{247.08} = .124008$$

Figure 2.5 in the text compares the rates of return for GM, Ford, and the S&P 500 over time, as discussed in Section 2.5 of this chapter.

Table 2B.2 Rates of Return for GM and Ford Stock and the S&P 500

Year	GM	Ford	S&P 500
70	.2137	.4260	.0010
71	.0422	.2933	.1080
72	.0631	.1717	.1557
73	−.3667	−.4512	−.1737
74	−.2597	−.0968	−.2964
75	.9522	.3960	.3149
76	.4584	.4614	.1918
77	−.1124	−.2067	−.1153
78	−.0498	−.0026	.0105
79	.0288	−.1479	.1228
80	−.0410	−.2938	.2586
81	−.0911	−.1025	−.0994
82	.6826	1.3212	.1549
83	.2373	.1286	.1706
84	.1176	.0980	.0115
85	−.0383	.3237	.2633
86	.0088	.0081	.1462
87	.0058	.3961	.0203
88	.4418	−.2995	.1240
89	−.4581	−.0766	.2725
90	−.1153	−.3209	−.0656

Appendix 2C Financial Statements and Financial Ratio Analysis

Review of Balance Sheets and Income Statements

Accounting concepts are used to understand a firm's financial condition. We will discuss two basic sources of accounting information: the **balance sheet,** which reveals the assets, liabilities, and owners' (stockholders') equity of a firm *at a point* in time, and the **income statement,** which shows the firm's profit or loss *over* a given *period* of time. **Assets,** which are things that the firm owns, can be classified as current, fixed, or "other" assets. Current assets consist of cash and of property that can be turned into cash quickly. Examples of current assets include cash, stocks, bonds, inventory, and accounts receivable (cash that customers owe the firm). Fixed assets are not easily convertible into cash; they include land, machinery, and buildings. Fixed assets are generally valued at their historical value (purchase price) minus depreciation, not at their current market value. Other assets include intangibles such as goodwill, trademarks, patents, copyrights, and leases.

Liabilities, which are debts of the firm, are divided into current and long-term debts. Current debts come due within 1 year, whereas long-term debts are due in more than 1 year. Current debts include accounts payable (unpaid bills), notes payable, debts on agreements, accrued expenses, expenses incurred but not yet paid, and taxes payable. Examples of long-term liabilities include mortgages, payable corporate bonds, and capitalized leases.

Stockholders' equity makes up the second part of the liabilities section of a balance sheet. It consists of funds invested by shareholders plus retained earnings. The **net worth** of the firm is calculated by subtracting total liabilities from total assets.

Whereas a balance sheet looks at the firm at a point in time, the income statement, as we noted earlier, evaluates the firm over a period of time. The end product of the income statement is the profit or loss, which is calculated by taking the sales for a period and subtracting the cost of goods sold and such expenses as research and development, interest, and selling, general, and administrative expenses.

Table 2C.1 presents GM's balance sheet for 1990 and 1989. Total assets are divided into current and fixed assets. Again, the current assets are those assets that can be converted into cash in a year or less; fixed assets such as land cannot be quickly turned into cash. The liabilities section is separated into liabilities and stockholders' equity. GM's liabilities include long-term debt, current liabilities such as accounts payable, and deferred taxes. Stockholders' equity consists of common stock and paid-in surplus, preferred stock, retained earnings, and other adjustments. Note that *total assets equal the sum of total liabilities and equity.*

Table 2C.1 General Motors Corporation Balance Sheet ($ Million)

	December 31, 1990		December 31, 1989	
Assets				
Current Assets				
Cash and marketable securities	4,606.5	4.48%	7,070.7	7.31%
Receivables	19,689.1	19.16%	21,796.2	22.52%
Inventories	9,331.3	9.08%	7,991.7	8.26%
Other current assets	6,316.8	6.15%	4,447.7	4.60%
Total current assets	39,943.7	38.86%	41,306.3	42.69%
Fixed Assets				
Property, plant, and equipment	9,752.2	9.49%	9,000.1	9.30%
Other assets	53,082.9	51.65%	46,458.7	48.01%
Total fixed assets	62,835.1	61.14%	55,458.8	57.31%
Total assets	102,778.8	100.00%	96,765.1	100.00%
Liabilities and Stockholders' Equity				
Current Liabilities				
Accounts and loans payable	11,306.6	11.00%	9,960.9	10.29%
Income tax payable	1,148.5	1.12%	706.4	0.73%
Accrued liabilities	15,851.5	15.42%	13,409.2	13.86%
Stocks subject to repurchase	822.0	0.80%	0.0	0.00%
Total current liabilities	29,128.6	28.34%	24,076.5	24.88%

(continued)

Table 2C.1 (Continued)

	December 31, 1990		December 31, 1989	
Long-Term Liabilities				
Long-term debt	4,614.5	4.49%	4,254.7	4.40%
Other liabilities	23,027.6	22.41%	15,584.3	16.11%
Payable to GMAC	12,918.0	12.57%	14,460.5	14.94%
Capitalized leases	309.3	0.30%	311.0	0.32%
Deferred credits	1,449.1	1.41%	1,445.6	1.49%
Common stock subject to repurchase	1,284.3	1.25%	1,650.0	1.71%
Total long-term liabilities	43,602.8	42.42%	37,706.1	38.97%
Stockholders' Equity				
Preferred stock	235.4	0.23%	236.4	0.24%
Common stock and paid-in surplus	3,231.0	3.14%	3,631.9	3.75%
Retained earnings	27,148.6	26.41%	31,230.7	32.27%
Other adjustments	(567.6)	(0.55%)	(116.5)	(0.12%)
Total stockholder's equity	30,047.4	29.24%	34,982.5	36.15%
Total liabilities and equity	102,778.8	100.00%	96,765.1	100.00%

Table 2C.2 displays GM's income statement for both 1990 and 1989, showing the firm's profit after expenses are subtracted from revenues. To determine gross profits, the costs of goods sold are subtracted from net sales. Operating income is then calculated by deducting selling and administrative expenses, debt amortization, and depreciation. Earnings before interest and taxes (EBIT) are next obtained by adding other income to operating income, and interest is subtracted to get earnings before taxes (EBT). Finally, net income is obtained by subtracting the provision for income taxes and adding earnings in unconsolidated subsidiaries and associates. Both earnings per share and dividends per share also are reported.

Financial Ratio Analysis

To help them analyze balance sheets and income statements, financial managers construct various financial ratios. There are five basic types of these ratios: leverage ratios, activity ratios, liquidity ratios, profitability ratios, and market value ratios. Let's use General Motors (GM) and Ford data to calculate a number of these ratios and discuss their significance.

Table 2C.3 shows how financial ratios are derived from the 1989 balance sheet and income statement. The information for the current ratio comes from the assets side of the balance sheet. The ratio for GM is 1.716, which means that 1 dollar in current liabilities is matched by 1.716 dollars in current assets. To calculate the inventory turnover ratio, we use net sales from the income statement and inventories from the current assets in the balance sheet. The resulting ratio (14.081) reveals how often the average value of goods in inventory was sold in 1989.

Table 2C.2 Income Statement for General Motors Corporation[a]

	1990[b]		1989[b]	
Net sales and revenues	110,797.3	100.00%	112,533.2	100.00%
Cost of sales	96,366.3	86.98%	94,049.2	83.57%
Gross margin on sales	14,431.0	13.02%	18,484.0	16.43%
Operating Expense				
Selling, general, and administrative expenses	8,667.1	7.82%	8,104.7	7.20%
Depreciation	5,486.8	4.95%	5,087.6	4.52%
Amortization of specified tools	3,314.0	2.99%	—	0%
Amortization of intangible assets	373.1	.34%	505.0	.45%
Operating income	(3,410.0)	(3.08%)	4,786.7	4.25%
Other Revenue and Expense				
Interest received	0	0%	0	0%
Interest expense	2,049.0	1.85%	2,228.4	1.98%
Net other income	1,814.4	1.64%	2,331.2	2.07%
Income before tax (EBT) and earnings in unconsolidated subsidiaries and associates	(3,644.6)	(3.29%)	4,889.5	4.34%
Foreign income taxes	0	0%	0	0%
Income tax	(889.7)	(.80%)	1,733.2	1.54%
Earnings in unconsolidated subsidiaries and associates	769.2	.69%	1,068.5	.95%
Net income	(1,985.7)	(1.79%)	4,224.8	3.75%
Earnings per share on common stock	(4.09)		6.33	
Dividends per share	3.00		2.50	

[a]In millions of dollars.
[b]Year's end December 31.

The total debt to total assets ratio is derived from the balance sheet; it indicates that about 63.8 percent of GM's assets are financed by debt. Data to calculate the net profit margin come from the income statement. GM's profit margin is .038; that is, about 4 cents out of every dollar of sales is profit (net income). ROI (return on investment) is more accurately described as return on total assets; it is calculated from information in both the income statement and the balance sheet. The resulting figure for GM is 4.4 percent.

The price/earnings (P/E) ratio is calculated by taking the price per share divided by the earnings per share (EPS). Although the price per share does not appear in the balance sheet or the income statement, it can be found in stock reports in newspapers. The EPS is then found by dividing net income by the number of common shares. (The ratio cannot be calculated if the firm experienced losses.) The P/E ratio for GM is 7.367.

Table 2C.3 Financial and Market Ratio Calculations for GM, 1989 (Dollar amounts for 1–5 are in $ millions)

1. Current ratio $= \dfrac{\text{current assets}}{\text{current liabilities}} = \dfrac{\$41,306.3}{\$24,076.5} = 1.716$ (liquidation)

2. Inventory turnover $= \dfrac{\text{Net sales}}{\text{inventories}} = \dfrac{\$112,533.2}{\$7,991.7} = 14.081$ (activity ratio)

3. Total debt to total assets ratio $= \dfrac{\text{total debt}}{\text{total assets}} = \dfrac{\$61,782.6}{\$96,765.1} = .638$ (leverage ratio)

(total debt = total current liabilities + total long-term liabilities)

4. Net profit margin $= \dfrac{\text{net income}}{\text{net sales}} = \dfrac{\$4,224.8}{\$112,533.2} = .038$ (profitability ratio)

5. ROI $= \dfrac{\text{net income}}{\text{total assets}} = \dfrac{\$4,224.8}{\$96,765.1} = .044$ (profitability ratio)

6. Price/earnings ratio $= \dfrac{\text{price per share}}{\text{earnings per share}} = \dfrac{\$46.63}{\$6.33} = 7.367$ (market value ratio)

7. Payout ratio $= \dfrac{\text{dividends per share}}{\text{earnings per share}} = \dfrac{\$2.50}{\$6.33} = .395$ (market value ratio)

Finally, the payout ratio is calculated by dividing the price of the stock by the dividends per share (DPS). DPS is the value of dividends paid out divided by the number of shares of common stock. This ratio reveals that GM paid out about 39.5 percent of its earnings in dividends.

As mentioned above, there are five basic types of financial ratios. **Liquidity ratios** measure how quickly or effectively the firm can obtain cash. If the bulk of a firm's assets are fixed (such as land), the firm may not be able to obtain enough cash to finance its operations. The *current ratio,* defined as current assets divided by current liabilities, is used to gauge the firm's ability to meet current obligations. If the firm's current assets do not significantly exceed current liabilities, the firm may not be able to pay current bills, because although current assets are expected to generate cash within 1 year, current liabilities are expected to use cash within that same 1-year period. In Figure 2C.1, this ratio is graphed for both Ford and General Motors for the years 1966 to 1990. As the figure reveals, GM had a higher current ratio for almost the entire period. This implies that GM was in the better position to pay current obligations. Remember that the national recession of 1981–1982 had an adverse effect on the auto industry. It is one of the reasons for the drop in the current ratio from 1980 to 1982.

Leverage ratios measure how much of the firm's operation is financed by debt. Although some debt is expected, too much debt can be a sign of trouble. One indicator of how much debt the firm has incurred is the ratio of total debt to total assets, which measures the percentage of total assets financed by debt. Figure 2C.2 shows that Ford had a greater share of its assets financed by debt than did GM over the 1966–1990 period. This fact is not necessarily a reason for concern unless Ford's

. . . .

Appendix 2C Financial Statements and Financial Ratio Analysis   **41**

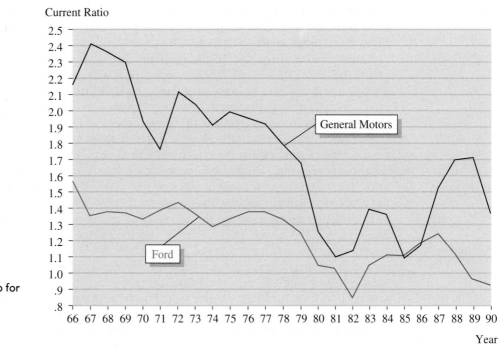

Current Ratio

Figure 2C.1

Current Ratio for
Ford and GM

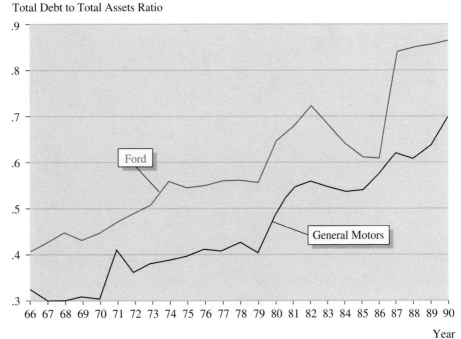

Total Debt to Total Assets Ratio

Figure 2C.2

Total Debt to
Total Assets Ratio
for Ford and GM

leverage ratio was too high in absolute terms. The general trend shows that both firms increased their debt during the period, particularly from 1979 to 1982, and that a sharp increase occurred from 1987 to 1990.

Activity ratios measure how efficiently the firm is using its assets. Figure 2C.3 graphs the *inventory turnover* ratio for each firm; it is found by dividing net sales by average inventory. This ratio measures how quickly a firm is turning over its inventories. A high ratio usually implies efficiency because the firm is selling inventories quickly. This ratio varies greatly with the line of business, however. A supermarket must have a high turnover ratio because it is dealing with perishable goods; in contrast, a jewelry store selling diamonds has a much lower turnover ratio. The seasonality of the product must also be considered. Auto dealers have high inventories in the fall, when the new autos arrive, and lower inventories in other seasons. On the other hand, Christmas tree dealers have rather low inventories in August!

Profitability ratios measure the profitability of the firm's operations. One of these ratios is the *return on total assets,* defined as net income divided by total assets. This ratio, often abbreviated ROA, measures how much the company has earned on its total investment of financial resources. Looked at in another way, it measures how well the firm used funds, regardless of how the firm's assets are divided into fixed and current assets. As Figure 2C.4 suggests, GM had a higher ROA than Ford until 1984–85 and from 1988 to 1990. The ROA for both firms was negative in the early 1980s; both companies incurred losses.

The *net profit margin,* defined as net income divided by net sales, is another measure of profitability. This ratio gauges the percentage of sales revenue that consists

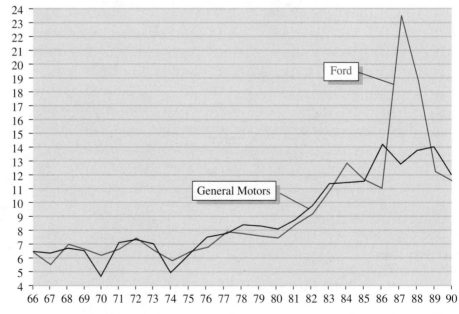

Inventory Turnover Ratio

Figure 2C.3

Inventory Turnover for Ford and GM

Return on Total Assets

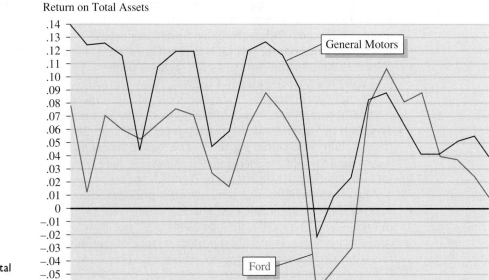

Figure 2C.4

Return on Total
Assets for
Ford and GM

Net Profit Margin

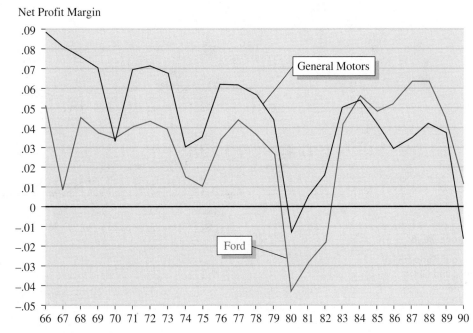

Figure 2C.5

Net Profit Margin
for Ford and GM

of profit. This ratio varies for different industries; a successful supermarket might have a ratio of 20 percent, whereas most manufacturing firms tend to have ratios around 8 percent. Although many Americans believe that corporations make a profit of 25 cents or more on each dollar of sales, the average net profit ratio for the *Fortune* 500 industrial firms in 1981 was 4.6 percent. The net profit margins for GM and Ford are presented in Figure 2C.5.

An indirect profitability indicator is the *payout ratio,* which measures the proportion of current earnings paid out in dividends. This ratio, which is expressed as dividends per share divided by earnings per share, can fluctuate widely because of the variability in earnings per share. The reason, for example, why the payout ratio was so high for GM in 1981 is that earnings per share were low. The payout ratios for GM and Ford are illustrated in Figure 2C.6.

Market value ratios measure how stock price per share is related to either earnings per share or book value per share. The *price/earnings ratio,* or P/E, is shown in Figure 2C.7. This ratio, defined as the price per share of a stock divided by the earnings per share, is usually reported in stock quotations in newspapers such as the *Wall Street Journal* every day. However, you should be careful in looking at P/E ratios, because a high ratio can be the result of low earnings. This seems to have been the case for GM in 1981. Moreover, firms calculate earnings per share differently, making comparisons of P/E ratios between firms difficult or even misleading.

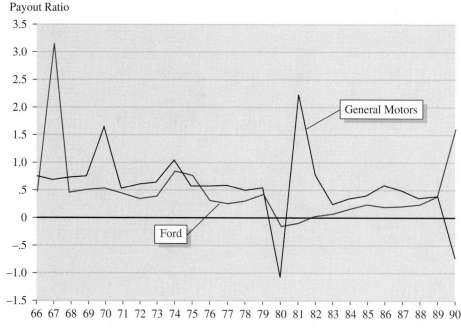

Figure 2C.6

Payout Ratio for Ford and GM

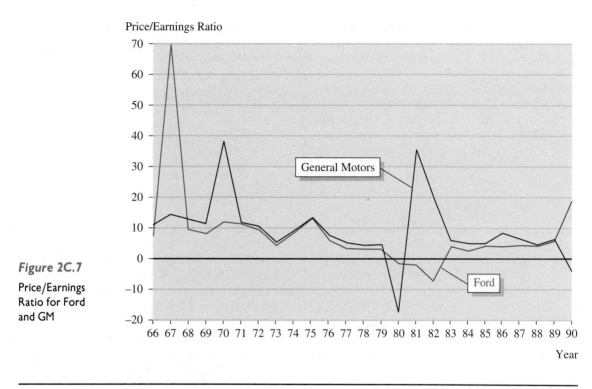

Figure 2C.7

Price/Earnings
Ratio for Ford
and GM

Questions and Problems

1. What is a primary source of data? Give two examples of primary sources of data.

2. What is a secondary source of data? Give two examples of secondary sources of data.

3. What is a sample? What is a census? What advantages does using a sample have over using a census? Are there any advantages to using a census?

4. What two types of error might we encounter when dealing with primary and secondary sources of data?

5. Explain how the following can be used to present data.
   a. Line chart
   b. Component-parts line chart
   c. Bar chart

6. Frederick Hallock is approaching retirement with a portfolio consisting of cash and money market fund investments worth $135,000, bonds worth $165,000, stocks worth $185,000 and real estate worth $1,200,000. Present these data in a bar chart.

7. LaPoint Glass Company has the following earnings before interest and taxes (EBIT) and profits (EBIT and profits are in millions of dollars).

Year	1988	1989	1990	1991
EBIT	3.3	3.3	4.1	5.5
Profits	1.6	1.8	2.1	2.8

Present these data in a bar chart by hand and by using Lotus 1-2-3.

8. Of 350 MBA students, the following numbers chose to concentrate their study in these fields: 35 in finance, 63 in accounting, 70 in marketing, 35 in operations management, 52 in management information systems, 56 in economics, and 43 in organizational behavior. Present these data in a pie chart.

9. Use the data in Table 2.1 to draw line charts for the following:
   a. GNP
   b. CPI
   c. GNP and CPI
   d. 3-month T-bill rate and prime rate

10. Study Figure 2.2, and comment on the relationship between GNP and private consumption.

11. Using the data in Figure 2.3, analyze the average rates of return for
    a. the DJIA
    b. the S&P 500
    c. Tri-Continental Corporation

12. Using the graph in Figure 2C.1, answer the following questions:
    a. Which company has the higher current ratio?
    b. Which company's current ratio appears to be more stable over time?

13. Using the graph in Figure 2C.3, carefully explain the relationship between Ford's inventory turnover and GM's.

14. You are given the following information about a certain company's current assets over the past 4 years.

Current Assets	Years			
	1988	1989	1990	1991
Cash and marketable securities	4,215	5,341	6,325	5,842
Receivables	6,327	6,527	7,725	6,750
Inventories	9,254	9,104	10,104	11,100
Other current assets	2,153	3,277	4,331	3,956

Use a component-lines graph to plot this firm's current assets.

15. Explain under what conditions it is best to use a pie chart to present data.

16. Using the data given in question 14, present the components of total current assets for 1990 in two pie charts, one drawn by hand and one by using Lotus 1-2-3.

17. A statistics teacher has given the following numbers of the traditional grades to her class of 105 students.

Number of Students	Grade
10	A
30	B
50	C
10	D
5	E

a. Use a bar graph to show the distribution of grades.
b. Use a pie chart to show the distribution of grades.
c. Which of these graphs do you think is best for presenting the distribution of grades? Why?

18. Using the data in Table 2C.1, show the distribution of current assets for 1990 in a pie chart and a bar chart. Which of these graphs do you think is best for presenting the data?

19. The following table gives the sales figures for five products manufactured by Trends Clothing Company, your employer.

Item	Sales
Sweaters	$ 5 million
Shirts	12 million
Pants	9 million
Blazers	16 million
Overcoats	7 million

The president of the company asks you for a report showing how sales are distributed among the five goods. What type of chart would you use?

20. Explain the benefits of graphs over tables in presenting data.

21. In the course of researching the benefits of diversification, you collect the information given in the table on page 47 (top), which presents rates of return for different portfolios.
    a. Use a line chart to plot the 20-year return for all five portfolios.
    b. What information do these plots provide?

Year	Stocks[a]	Bonds[b]	60% Stocks, 40% Bonds	$\frac{1}{3}$ Stocks, $\frac{1}{3}$ Bonds, $\frac{1}{3}$ Cash	BB&K Index[c]
1970	4.01%	12.10%	7.52%	7.98%	4.7%
1971	14.31	13.23	14.14	10.83	13.7
1972	18.98	5.68	13.54	9.38	15.1
1973	−14.66	−1.11	−9.11	−3.03	−2.2
1974	−26.47	4.35	−14.88	−5.44	−6.6
1975	37.20	9.19	25.65	17.04	19.6
1976	23.84	16.75	21.18	15.19	11.5
1977	−7.18	−.67	−4.57	−0.94	6.1
1978	6.56	−1.16	3.65	4.40	13.0
1979	18.44	−1.22	10.28	9.14	11.5
1980	32.42	−3.95	17.45	13.17	17.9
1981	−4.91	1.85	−1.99	4.06	6.4
1982	21.41	40.35	28.98	23.97	14.4
1983	22.51	.68	13.43	10.52	15.4
1984	6.27	15.43	10.11	10.75	10.4
1985	32.16	30.97	31.85	23.38	25.4
1986	18.47	24.44	21.11	16.61	23.3
1987	5.23	−2.69	3.59	3.92	8.6
1988	16.81	9.67	13.97	11.01	13.2
1989	31.49	18.11	26.24	19.22	14.3
Compound annual return	11.55	9.00	10.89	9.78	11.54

Source: Bailard, Biehl & Kaiser, Ibbotson Associates, Inc. This figure was printed in the *Wall Street Journal* on January 25, 1990, p. C1.

[a]Standard & Poor's 500 index.
[b]Long-Term Treasury Bonds.
[c]20% U.S. stocks, 20% bonds, 20% cash, 20% real estate, 20% foreign stocks.

22. Use the data given in question 21 to construct a bar graph.

23. You are given the following exchange rate information for the number of dollars it takes to buy 1 British pound and the number of dollars it takes to buy 100 Japanese yen.

Month	$/BP	$/100 yen
Jan 88	1.7505	.7722
Feb 88	1.7718	.7782
Mar 88	1.8780	.8042
Apr 88	1.8825	.8015
May 88	1.8410	.7995
Jun 88	1.7042	.7475
Jul 88	1.7160	.7533

Month	$/BP	$/100 yen
Aug 88	1.6808	.7307
Sep 88	1.6930	.7477
Oct 88	1.7670	.7951
Nov 88	1.8505	.8227
Dec 88	1.8075	.8013

a. Draw a line chart showing the exchange rates between British pounds (BP) and U.S. dollars during this period.

b. Draw a line chart showing the exchange rates between Japanese yen and U.S. dollars.

c. Use Lotus 1-2-3 to draw a line chart containing the exchange rates in *a* and a line chart representing the exchange rates in *b*.

24. You are given the following financial ratios for Johnson & Johnson and for the pharmaceutical industry.

Year	Current Ratio		Inventory Turnover	
	Industry	J&J	Industry	J&J
79	2.30	2.73	2.17	2.71
80	2.29	2.55	2.22	2.70
81	2.18	2.50	2.34	2.78
82	2.12	2.50	2.30	2.38
83	2.12	2.66	2.34	2.28
84	2.09	2.41	2.40	2.37
85	2.19	2.47	2.27	2.45
86	1.91	1.40	2.38	2.33
87	1.86	1.86	2.24	2.27
88	1.93	1.88	2.27	2.32

Year	ROA		Price/Earnings	
	Industry	J&J	Industry	J&J
79	.11	.12	12.04	13.76
80	.11	.12	15.03	15.35
81	.10	.12	28.02	14.79
82	.11	.12	15.19	17.79
83	.11	.11	14.56	15.90
84	.10	.11	14.88	13.14
85	.10	.12	17.98	15.66
86	.10	.06	23.46	35.47
87	.09	.13	35.09	15.50
88	.12	.14	15.98	14.88

a. Draw a line chart showing the current ratio over time for the industry and for J&J, and compare the two.

b. Use a bar graph to present the data for the industry and J&J's current ratio.

25. Repeat question 24 for inventory turnover.

26. Repeat question 24 for return on total assets (ROA).

27. Repeat question 24 for the price/earnings ratio.

28. An August 27, 1991, *Wall Street Journal* article reported that increasing numbers of small software firms are being absorbed by that industry's biggest companies. According to WSJ, the result of this dominance by a few giants is that the industry has become much tougher for software entrepreneurs to break into. The newspaper printed the chart given in the accompanying figure to depict the breakdown of market share among software companies. Refer to this chart to answer the following questions.

a. List the companies in descending order of market share.

b. What is the combined market share for Lotus Development and WordPerfect?

c. What is the combined market share for Microsoft, Lotus Development, and Novell?

From Entrepreneurs to Corporate Giants: Market Share Among the Top 100 Software Companies, Based on Total 1990 Revenue of $5.7 Billion

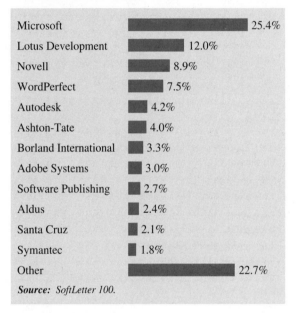

Source: SoftLetter 100.

29. The results of the 1991 city council election in Monroe Township, New Jersey, were

Nalitt	4,656
Riggs	4,567
Anderson	4,140
Miller-Paul	4,142

Use a pie chart to present the results of the election.

30. Redo question 29 using a bar chart. Which method is better for presenting these election results?

To answer questions 31–37, refer to the table, which gives the rankings for team defense and offense for NFC teams for the first 9 weeks of the 1991 season.

Rankings of Team Defense and Offense for NFC Teams in the 1991 Season (rankings based on averages a game)

NFC Team Defense				
	Yds	*Rush*	*Pass*	*Avg.*
Philadelphia	1955	715	1240	217.2
New Orleans	2035	562	1473	226.1
Washington	2325	830	1495	258.3
San Francisco	2460	851	1609	273.3
New York	2551	959	1592	283.4
Tampa Bay	2652	1065	1587	294.7
Chicago	2665	950	1715	296.1
Green Bay	2684	775	1909	298.2
Atlanta	2728	1202	1526	303.1
Dallas	2736	863	1873	304.0
Minnesota	3097	1147	1950	309.7
Detroit	2799	932	1867	311.0
Phoenix	3277	1381	1896	327.7
Los Angeles	2986	959	2027	331.8

NFC Team Offense				
	Yds	*Rush*	*Pass*	*Avg.*
San Francisco	3392	1178	2214	376.9
Washington	3019	1337	1682	335.4
Dallas	2969	970	1999	329.9
New York	2842	1254	1588	315.8
Minnesota	3095	1328	1767	309.5
Atlanta	2768	998	1770	307.6
Detroit	2705	1070	1635	300.6
New Orleans	2665	918	1747	296.1
Chicago	2662	1006	1656	295.8
Los Angeles	2515	748	1767	279.4
Phoenix	2636	897	1739	263.6
Philadelphia	2319	688	1631	257.7
Green Bay	2250	650	1600	250.0
Tampa Bay	2142	779	1363	238.0

Source: *USA Today,* November 7, 1991, p. 11C.

31. Use a pie chart to show how San Francisco's total team offense is divided between rush and pass.

32. Use a pie chart to show how Phoenix's total team defense is divided between rush and pass.

33. Use a bar chart to show the total pass offense for the 14 NFC teams.

34. Repeat question 33 for rush offense.

35. Repeat question 33 for pass defense.

36. Repeat question 33 for rush defense.

37. Use the graphs from questions 33–36 to answer the following questions.
   a. Which team has the best pass offense?
   b. Which team has the best pass defense?
   c. Which team has the best rush offense?
   d. Which team has the best rush defense?

38. The following table is a table of salaries for the same top NHL forwards and defensemen.
   a. Use a bar chart to show the players' salaries.
   b. Do you think the bar chart is a better vehicle than a table for comparing players' salaries?

Salary Comparisons for Top NHL Forwards and Defensemen

Posi-tion	Name	Team	Gross Salary ($ Millions)
C	Wayne Gretzky	Los Angeles Kings	$3
C	Mario Lemieux	Pittsburgh Penguins	$2.338
RW	Brett Hull	St. Louis Blues	$1.5
C	Pat LaFontaine	Buffalo Sabres	$1.4
C	Steve Yzerman	Detroit Red Wings	$1.4
LW	Kevin Stevens	Pittsburgh Penguins	$1.4
LW	Luc Robitaille	Los Angeles Kings	$1.3
C	John Cullen	Hartford Whalers	$1.2
D	Ray Bourque	Boston Bruins	$1.2
D	Scott Stevens	New Jersey Devils	$1.155

Source: *USA Today,* October 7, 1991, p. 8C.

39. The accompanying pie chart presents data on why teen-agers drink. Use information shown in the pie chart to answer the following questions:
  a. For what reason do the highest numbers of teen-agers drink?
  b. What percentage of teen-agers drink because they are bored or because they are upset?

Drink When Upset (41%)   Drink to Get High (25%)

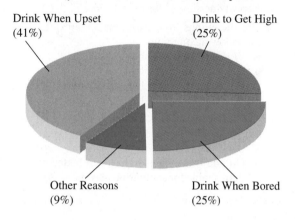

Other Reasons (9%)   Drink When Bored (25%)

Source: National Council on Alcoholism and Drug Dependence. Surgeon General survey. *USA TODAY,* November 5, 1991. Copyright 1991, USA TODAY. Reprinted with permission.

To answer questions 40–42, use the following results of the election to the General Assembly from one New Jersey district in 1991.

Batten	17,026
Lookabaugh	17,703
LoBiondo	27,452
Gibson	24,735

40. Use a bar graph to show the distribution of votes.

41. Use a pie chart to show the distribution of votes.

42. Which type of graph presents these data more effectively?

43. The following bar graph shows net purchases of bond mutual funds. Does this graph tell us anything?

Net Bond Fund Purchases* (in billions)

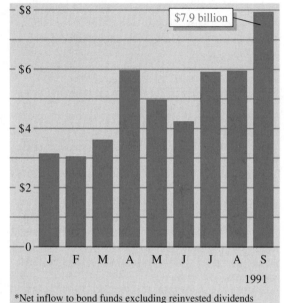

*Net inflow to bond funds excluding reinvested dividends

Source: Investment Company Institute. *USA TODAY,* November 6, 1991. Copyright 1991, USA TODAY. Reprinted with permission.

44. On November 9, 1991, the *Home News* of central New Jersey used the bar chart given in the accom-

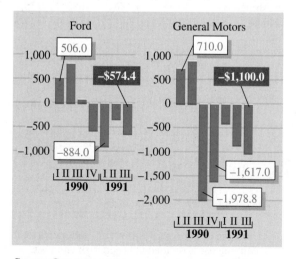

Source: Company reports, news reports. Reprinted by permission of *Knight-Ridder Tribune News.*

panying figure to show quarterly net income or losses for both Ford and GM.

a. Comment on the possible implications of this bar chart.

b. If you were a stock broker, would you recommend that your client buy either Ford's or GM's stock now?

45. The two pie charts given here present household income for new first-time homeowners and all other homeowners, by income group, in 1989.

a. Describe these two pie charts.

b. Recent first-time homeowners are most likely to be in which income group?

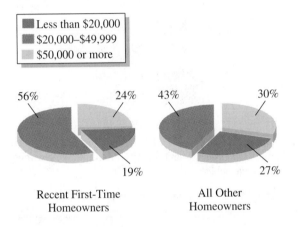

56% 24% 43% 30%

19% 27%

Recent First-Time   All Other
Homeowners   Homeowners

Source: U.S. Census Bureau. AP. *Home News,* October 31, 1991.

46. The three pie charts in the next column (top) present the racial composition of new first-time homeowners, other recent movers, and all owners in 1989.

a. Describe these three pie charts.

b. Do members of minority groups show any gains among new first-time homeowners?

47. On March 20, 1991, the *Home News* (a central New Jersey newspaper) used the bar charts given here (next column, bottom) to show the amount of money pledged to, and the amount received by, the United States from allied countries as financial support for the Gulf War. Use the information in this chart to draw implications and do related analysis. (Hint: Use the pie chart.)

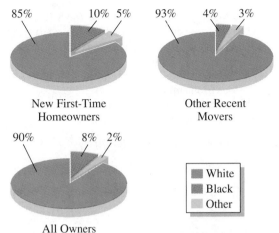

85% 10% 5%   93% 4% 3%

New First-Time   Other Recent
Homeowners   Movers

90% 8% 2%

■ White
■ Black
■ Other

All Owners

Source: U.S. Census Bureau. *Home News,* October 31, 1991. AP/Ed De Gasero, reprinted by permission of The Associated Press.

Figures in billions of dollars

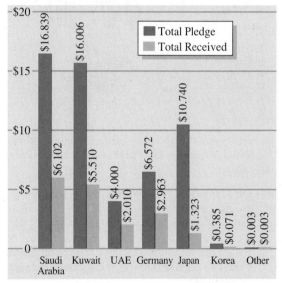

Source: Senate Appropriations Committee. *Home News,* March 20, 1991. Reprinted by permission of The Associated Press.

48. On March 14, 1991, the *Home News* (a central New Jersey newspaper) used the bar chart given here to show what problems New Jerseyans considered "very serious."
   a. What do New Jerseyans consider the most serious problem?
   b. Is traffic congestion regarded as more serious than crime? Explain your answer.

Main Cause of Air Pollution in New Jersey

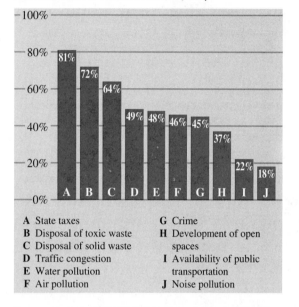

A State taxes
B Disposal of toxic waste
C Disposal of solid waste
D Traffic congestion
E Water pollution
F Air pollution
G Crime
H Development of open spaces
I Availability of public transportation
J Noise pollution

Source: Project: CLEAN AIR/Eagleton Institute of Politics. *Home News,* March 14, 1991. Reprinted with permission of the publisher.

49. On March 14, 1991, the *Home News* used two pie charts (next column, top) to show (1) the main causes of air pollution in New Jersey and (2) people's attitudes toward using increased taxes to reduce air pollution. Discuss the implications of these two pie charts and of the bar chart given in question 48.

50. The *Home News* used this bar chart (next column, bottom) on page D1 of its November 20, 1991, issue to depict the increasing popularity of turkey not just at holiday meals but throughout the year.
   a. How much turkey was consumed per person in 1960 and 1990, respectively?
   b. How much has per-person consumption of turkey increased from 1970 to 1990?

Attitudes Toward Using Increased Taxes to Reduce Air Pollution

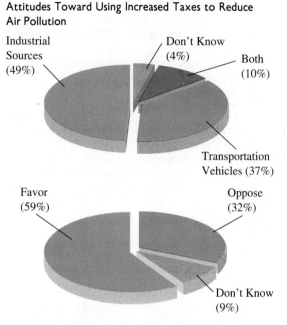

Source: Project: CLEAN AIR/Eagleton Institute of Politics. From *Home News,* March 14, 1991. Reprinted with permission of the publisher.

Eating More Turkey per Person (in pounds)

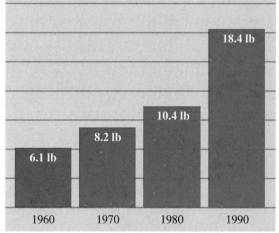

51. The following line chart was printed in the *Home News* on page A1 of its November 22, 1991, issue to show the increase in the number of college students over age 35 during the period 1972 to 1989.

  a. Did more men or more women over 35 years old attend college during this 18-year period?

  b. What was the percentage increase for older female college students from 1979 to 1989?

  c. What was the percentage increase for college students over age 35 from 1972 to 1989?

Percentage of College Enrollees Aged 35 and Older

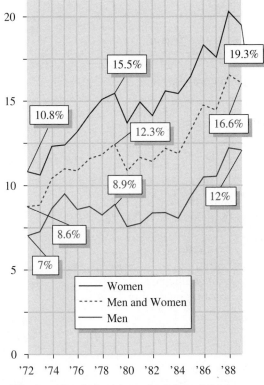

Note: Several methods used on estimated population base by the U.S. Census Bureau to calculate percentage.

Source: U.S. Census Bureau. *Home News,* November 22, 1991. Reprinted by permission of The Associated Press.

CHAPTER 3

Frequency Distributions and Data Analyses

CHAPTER OUTLINE

3.1 Introduction
3.2 Tally Table for Constructing a Frequency Table
3.3 Three Other Frequency Tables
3.4 Graphical Presentation of Frequency Distributions
3.5 Further Economic and Business Applications
Appendix 3A Financial Ratio Analysis

Key Terms

grouped data
raw (nongrouped) data
frequency
frequency table
frequency distribution
cumulative frequencies
relative frequency
cumulative relative frequency

histograms
stem-and-leaf displays
frequency polygon
cumulative frequency polygon
pie chart
Lorenz curve
Gini coefficient
absolute inequality

3.1 INTRODUCTION

Using the tabular and graphical methods discussed in Chapter 2, we will now develop two general ways to describe data more fully. We discuss first the tally table approach to depicting data frequency distributions and then three other kinds of frequency tables. Next we explore alternative graphical methods for describing frequency distributions. Finally, we study further applications for frequency distributions in business and economics.

3.2 TALLY TABLE FOR CONSTRUCTING A FREQUENCY TABLE

Before conducting any statistical analysis, we must organize our data sets. One way to organize data is by using a tally table as a worksheet for setting up a frequency table. To set up a tally table for a set of data, we split the data into equal-sized classes in such a way that each observation fits into one and only one class of numbers (that is, the classes are mutually exclusive). Sometimes data are reported in a frequency table with class intervals given, but with actual values of observations in the classes unknown; data presented in this manner are called **grouped data.** The analyst assigns each data point to a class and enters a tally mark made by that class. Let's see how this works.

EXAMPLE 3.1   *Tallying Scores from a Statistics Exam*

Suppose a statistics professor wants to summarize how 20 students performed on an exam. Their scores are as follows: 78, 56, 91, 59, 78, 84, 65, 97, 84, 71, 84, 44, 69, 90, 73, 77, 80, 90, 68, and 75. Data in this form are called **nongrouped data** or **raw data.** We can use a tally table like Table 3.1 to list the number of occurrences, of **frequency,** of each score. A corresponding diagram is shown in Figure 3.1.

This table presents nongrouped data, and no pattern emerges from them. As an alternative, the data can be grouped into classes by letter grade. If the professor uses a straight grading scale, the classes might be 90–99, 80–89, 70–79, 60–69, and below 60. After establishing the classes, the professor counts scores in each class and

Table 3.1 Student Exam Scores

Score	Tallies	Frequency
44	/	1
56	/	1
59	/	1
65	/	1
68	/	1
69	/	1
71	/	1
73	/	1
75	/	1
77	/	1
78	//	2
80	/	1
84	///	3
90	//	2
91	/	1
97	/	1
Total		20

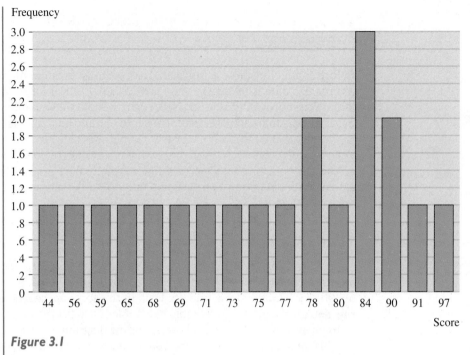

Figure 3.1

Bar Graph for Nongrouped Student Exam Scores Given in Table 3.1

Table 3.2 Tally Table for Statistics Exam Scores

Class	Tally	Frequency
Below 60	///	3
60–69	///	3
70–79	///////	6
80–89	////	4
90–99	////	4
		20

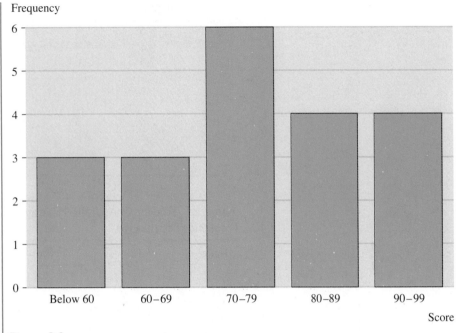

Figure 3.2

Bar Graph for Grouped Student Exam Scores Given in Table 3.2

Table 3.3 Frequency Table for Statistics Exam Scores

Class	Grade	Frequency
Below 60	F	3
60–69	D	3
70–79	C	6
80–89	B	4
90–99	A	4
		20

records these numbers to obtain a tally sheet, as shown in Table 3.2 and Figure 3.2.

Note that each observation is included in one and only one class. The tallies are counted and a **frequency table** is constructed as shown in Table 3.3, where letter grades are assigned to each class.

EXAMPLE **3.2**   *A Frequency Distribution of Grade Point Averages*

Suppose that there are 30 students in a classroom and that they have the grade point averages listed in Table 3.4. A tally table is constructed, in which classes are (arbitrarily) defined at every half-point and each tally marked next to a particular class

Table 3.4 Student GPAs: Raw Data

1.2	3.9	1.9
3.8	2.4	2.7
2.3	2.3	2.6
.7	3.1	3.7
3.6	2.9	4.0
2.2	2.7	1.2
1.9	.8	1.8
2.1	.3	2.4
3.1	3.2	3.2
.8	3.1	3.6

Table 3.5 Student GPAs: Tally Table and Frequency Distribution

Range	Tallies	Frequency
Below 1.5	//////	6
1.5–1.9	///	3
2.0–2.4	//////	6
2.5–2.9	////	4
3.0–3.4	/////	5
3.5–4.0	//////	6
Total		30

accounts for one data entry. The entries are then counted to obtain a **frequency distribution,** as shown in Table 3.5. A frequency distribution simply shows how many observations fall into each class. We will discuss this concept in further detail in the next chapter.

Generally, a data set should be divided into 5 to 15 classes. Having too few or too many classes gives too little information. Imagine a frequency distribution with only two classes: 0.0–2.0 and 2.1–4.0. With such broadly defined classes, it is difficult to distinguish among GPAs. Similarly, if the class interval were only one-tenth of a point, the large number of classes, each with only one or a few tallies, would make summarizing the data almost impossible.

In the GPA example, it was relatively easy to construct the classes because GPA cutoffs were used. However, in most examples there are no natural dividing lines between classes. The following guidelines can be used to construct classes:

1. Construct from 5 to 15 classes. This step is the most difficult, because using too many classes defeats the purpose of grouping the data into classes, whereas having too few classes limits the amount of information obtained from the data. As a general rule, when the range and number of observations are large, more classes can be defined. Fewer classes should be constructed when the number of observations is only around 20 or 30.

2. Make sure each observation falls into only one class. This can often be accomplished by defining class boundaries in terms of several decimal places. If the percentage return on stocks is carried to one decimal place, for example, then defining the classes by using two decimal places will ensure that each observation falls into only one class.

3. Try to construct classes with equal class intervals. This may not be possible, however, if there are outlying observations in the data set.

EXAMPLE 3.3   *A Frequency Distribution of 3-Month Treasury Bill Rates*

Table 3.6 presents another example, and here the data presented are the interest rates on three-month Treasury bills (T-bills) from 1950 to 1990. (T-bills are debt

Table 3.6 T-Bill Interest Rates, 1950–1987

Class (%)	Tallies	Frequency
0–1.99	///////	7
2.00–3.99	/////////	9
4.00–5.99	//////////	10
6.00–7.99	////////	8
8.00–9.99	///	3
10.00–11.99	///	3
12.00 and above	/	1
Total		40

instruments sold by the U.S. government to finance its budgetary needs.) The annual data for interest rates (average daily rates for a year) are taken from Table 2.1 in Chapter 2.

As we have noted, a frequency distribution gives the total number of occurrences in each class. In the next chapter we will talk about using a frequency distribution to present data.

By setting up a tally table and a frequency table, we can scrutinize data for errors. For example, if the data value 123 appears in a column for the rate in the T-bill example, a mistake has clearly been made—one that could be due to a missing decimal point. Probably the data point could be 12.3 percent instead, which makes more sense because it is in the range of the other data points. Data should also be checked for accuracy. Otherwise, invalid conclusions could be reached.

3.3   THREE OTHER FREQUENCY TABLES

In this section, using the frequency table discussed in the Section 3.2, we move ahead to cumulative frequency tables, relative frequency tables, and relative cumulative frequency tables.

EXAMPLE 3.4   *A Frequency Distribution for Statistics Exam Scores*

Suppose that for the data listed in Table 3.3, the professor wants to know how many students receive a C or below, the proportion of students who receive a B, and the proportion of students who receive a D or an F. To obtain this information, she calculates cumulative, relative, and cumulative relative frequencies.

By constructing **cumulative frequencies,** the professor determines the number of students who scored in a particular class *or* in one of the classes before it (Table 3.7). Obviously, the cumulative frequency for the first class is the frequency itself (3): there are no classes before it. The cumulative frequency for the second class is calculated by taking the frequency in the first class and adding it to the frequency in the second class (3) to arrive at a cumulative frequency of 6. This means that 6 students were in the first two classes. Then 6 is added to the frequency of the third class (6) to derive a cumulative frequency of 12. Thus 12 students scored a C or a worse grade. The remaining cumulative frequencies are calculated in a similar

Table 3.7 Cumulative Frequency Table for Grade Distribution

Class	Grade	Frequency	Cumulative Frequency
Below 60	F	3	3
60–69	D	3	6
70–79	C	6	12
80–89	B	4	16
90–99	A	4	20

manner. Note that the cumulative observation in the last class equals the total number of sample observations, because all frequencies have occurred in that class or in previous classes.

Another important concept is the **relative frequency,** which measures the proportion of observations in a particular class. It is calculated by dividing the frequency in that class by the total number of observations. For the data summarized in Table 3.7, the relative frequency for both the first and second classes is .15, and the relative frequencies for the remaining three classes are .30, .20, and .20, respectively, as shown in Table 3.8. The sum of the relative frequencies always equals 1.

This table indicates that 15 percent of the class received an F, 15 percent a D, 30 percent a C, and so on. The professor can calculate the cumulative relative frequency for any class by adding the appropriate relative frequencies. **Cumulative rel-**

Table 3.8 Relative Frequency Table for Grade Distribution

Class	Grade	Relative Frequency	Cumulative Relative Frequency
Below 60	F	.15	.15
60–69	D	.15	.30
70–79	C	.30	.60
80–89	B	.20	.80
90–99	A	.20	1.00

ative frequency measures the percentage of observations in a particular class and all previous classes. Thus, if she wants to determine what percentage of the students scored below a B, our conscientious professor can add the relative frequencies associated with grades C, D, and F to arrive at 60 percent.

EXAMPLE 3.5   *Frequency Distributions of Current Ratios for GM and Ford*

The current ratios for Ford and GM from 1969 to 1990 are shown in Table 3.9. A frequency distribution for the current ratios of Ford and General Motors is shown in Table 3.10. This ratio is a measure of liquidity, which (as we noted in Chapter 2) indicates how quickly a firm can obtain cash for operations. The first column defines the classes. Note that the use here of class boundaries ensures that each observation will fall into only one class.

The next column shows the frequency—that is, the number of times that an observation appears in each class. In Table 3.10 we see that Ford experienced four current ratios between .801 and 1.000, six between 1.001 and 1.200, and so on. The third column presents the cumulative frequency. Because there are 22 observations in the population, the cumulative frequency for the last class is 22.

Table 3.9 Current Ratio for Ford and GM

Year	Ford	GM
69	1.37	2.30
70	1.33	1.93
71	1.39	1.76
72	1.44	2.12
73	1.37	2.04
74	1.28	1.91
75	1.33	1.99
76	1.37	1.95
77	1.38	1.92
78	1.33	1.79
79	1.25	1.68
80	1.04	1.26

(continued)

Table 3.9 (Continued)

Year	Ford	GM
81	1.02	1.09
82	.84	1.13
83	1.05	1.40
84	1.11	1.36
85	1.10	1.09
86	1.18	1.17
87	1.24	1.56
88	1.00	1.00
89	.97	1.72
90	.93	1.37

Table 3.10 Cumulative Distributions of Current Ratios for Ford and General Motors

Class	Frequency	Cumulative Frequency	Relative Frequency	Cumulative Relative Frequency
		Ford		
.81–1.00	4	4	.18	.18
1.01–1.20	6	10	.27	.45
1.21–1.40	11	21	.50	.95
1.41–1.60	1	22	.05	1.00
Total	22		1.00	
		General Motors		
.81–1.00	1	1	.05	.05
1.01–1.20	4	5	.18	.23
1.21–1.40	4	9	.18	.41
1.41–1.60	1	10	.05	.46
1.61–1.80	4	14	.18	.64
1.81–2.00	5	19	.22	.86
2.01–2.20	2	21	.09	.95
2.21–2.40	1	22	.05	1.00
Total	22		1.00	

The fourth column presents the relative frequency, which measures the percentage of observations in each class. Relative frequencies can be thought of as probabilities. For example, the probability that an observation is in the first class is .18.

The last column indicates the cumulative relative frequency, which measures the percentage of observations in a particular class and all previous classes. The cumulative relative frequency for GM's fifth class is calculated by adding the relative frequencies of the first five classes to arrive at .64. That is, 64 percent of the observations occur in the first five classes. The cumulative relative frequency of the last class always equals 1, because the last class includes all the observations.

3.4 GRAPHICAL PRESENTATION OF FREQUENCY DISTRIBUTIONS

We have spoken before of the special effectiveness of using graphs to present data. In this section, we discuss four different graphical approaches to presenting frequency distributions.

Histograms

Frequency distributions can be represented on a variety of graphs. The **histogram,** which is one of the most commonly used types, is similar to the bar charts discussed in Chapter 2 except that

1. Neighboring bars touch each other.
2. The area inside any bar (its height times its width) is proportional to the number of observations in the corresponding class.

To illustrate these two points, suppose the age distribution of personnel at a small business is as shown in Table 3.11.

To construct a histogram, we need to enter a scale on the horizontal axis. Because the data are discrete, there is a gap between the class intervals, say between 20–29 and 30–39. In such a case, we will use the midpoint between the end of one class and the beginning of the next as our dividing point. Between the 20–29 interval and 30–39 interval, the dividing point will be $(29 + 30)/2 = 29.5$. We find the dividing point between the remaining classes similarly.

Table 3.11 Age Distribution of Personnel

Class	Frequency
20–29	3
30–39	6
40–49	7
50–59	4
60–69	1
70–79	1

Frequency

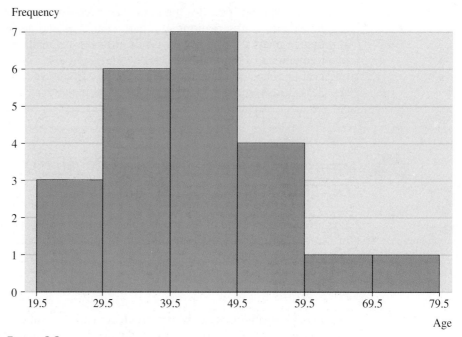

Figure 3.3

Histogram of Age Distribution Given in Table 3.11

To satisfy the second condition, we note that all five classes have an interval width of 10 years. Figure 3.3 is the histogram that reflects these data.

Drawn from the data of Tables 3.10, Figures 3.4a and 3.4b are histograms of Ford's and GM's current ratios for the years 1969–1990. The X axis indicates the classes, the Y axis the frequencies. As the histograms show, Ford's current ratios have tended to fall in the .8–1.4 range, whereas those of GM show no clear pattern. (In Chapter 4, we will cover measures of skewness, which give us more insight into the shape of a distribution.)

Most standard statistical software packages will construct a histogram from these data. Using MINITAB, we can specify the class width and the starting class midpoint, or we can let MINITAB select these values. The output will contain the frequency distribution as well as a graphical representation in the form of a histogram (without the bars). MINITAB will provide each class frequency next to the corresponding class midpoint (not class limits). Figure 3.5a contains the necessary MINITAB commands and the resulting output for the current ratio of GM where the class width (CW) and the midpoint of the first class were not specified. Figure 3.5b specified CW as .2000 and the first midpoint as .905. We can use the output as it appears or use this information to construct Figure 3.4b, which is a graphical representation of GM's current ratios as given in Table 3.10.

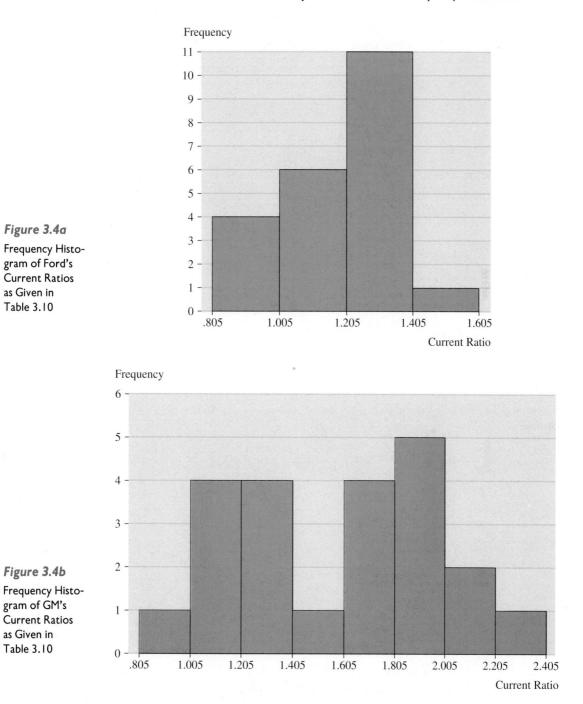

Figure 3.4a

Frequency Histogram of Ford's Current Ratios as Given in Table 3.10

Figure 3.4b

Frequency Histogram of GM's Current Ratios as Given in Table 3.10

```
MTB > SET INTO C1
DATA> 2.3 1.93 1.76 2.12 2.04 1.91 1.99 1.95 1.92 1.79 1.68 1.26 1.09 1.13 1.41
DATA> 1.36 1.09 1.17 1.56 1.00 1.72 1.37
DATA> END
MTB > NAME C1'GMCA/CL'
MTB > PRINT C1

GMCA/CL
   2.30    1.93    1.76    2.12    2.04    1.91    1.99    1.95    1.92    1.79    1.68
   1.26    1.09    1.13    1.41    1.36    1.09    1.17    1.56    1.00    1.72    1.37

MTB > HISTOGRAM C1


Histogram of GMCA/CL    N = 22

Midpoint   Count
      1.0      3    ***
      1.2      3    ***
      1.4      3    ***
      1.6      2    **
      1.8      3    ***
      2.0      6    ******
      2.2      1    *
      2.4      1    *

MTB > PAPER
```

Figure 3.5a

Histogram Using MINITAB, Where the Class Width and the Midpoint of the First Class Are Not
Specified

```
MTB > HISTOGRAM C1, FIRST MIDPOINT AT 0.905, CLASS WIDTH IS 0.2000


Histogram of GMCA/CL    N = 22

Midpoint    Count
    0.905      1    *
    1.105      4    ****
    1.305      3    ***
    1.505      2    **
    1.705      4    ****
    1.905      5    *****
    2.105      2    **
    2.305      1    *

MTB > PAPER
```

Figure 3.5b

Histogram Using MINITAB Using Specified Classes, Where the Class Width Is 0.2000 and the First
Midpoint Is 0.905

Histograms can also be used to chart the companies' relative and cumulative frequencies, as shown in Figures 3.6 and 3.7. Note the similarity between the frequency and relative frequency histograms (Figures 3.4 and 3.6) and between the

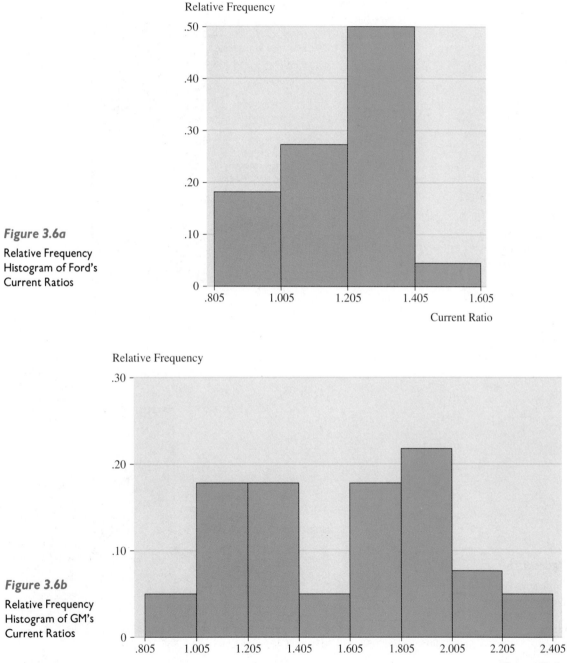

Figure 3.6a

Relative Frequency Histogram of Ford's Current Ratios

Figure 3.6b

Relative Frequency Histogram of GM's Current Ratios

cumulative frequency and the relative cumulative frequency graphs (Figures 3.7 and 3.8); the only difference between them is the variable on the Y axis. Note also that geometrically, the relative frequency of each class in a frequency histogram

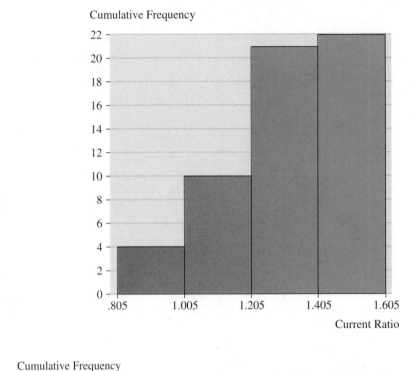

Figure 3.7a

Cumulative Frequency Histogram of Ford's Current Ratios

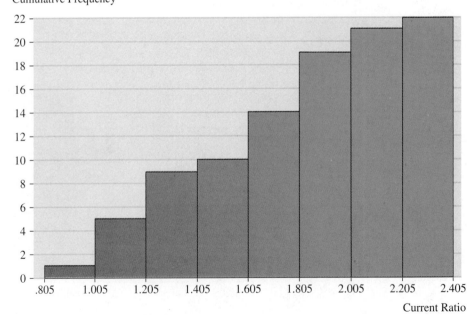

Figure 3.7b

Cumulative Frequency Histogram of GM's Current Ratios

equals its area divided by the total area of all the classes. For example, the area for the first class for GM's current ratio (Figure 3.4b) is equal to the base of the bar times its height (.2 × 1 = .2), and the sum of all the areas is 4.4. The relative frequency for the first class is thus .2/4.4 = .0455.

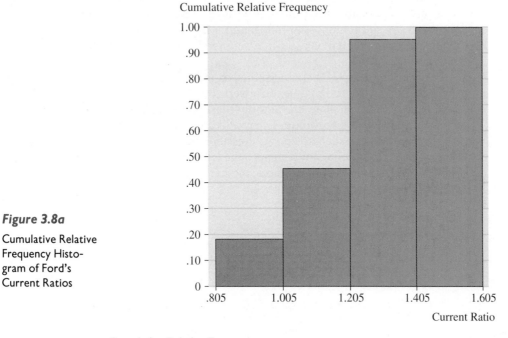

Figure 3.8a

Cumulative Relative Frequency Histogram of Ford's Current Ratios

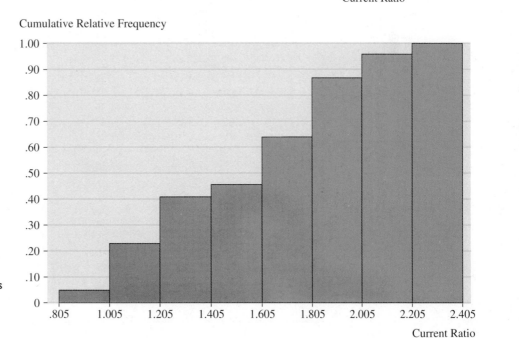

Figure 3.8b

Cumulative Relative Frequency Histogram of GM's Current Ratios

Stem-and-Leaf Display

An alternative to histograms for the presentation of either nongrouped or grouped data is the stem-and-leaf display. **Stem-and-leaf displays** were originally developed by John Tukey of Princeton University. They are extremely useful in summarizing data sets of reasonable size (under 100 values as a general rule), and unlike histograms, they result in no loss of information. By this we mean that it is possible to reconstruct the original data set in a stem-and-leaf display, which we cannot do when using a histogram.

For example, suppose a financial analyst is interested in the amount of money spent by food product companies on advertising. He or she samples 40 of these food product firms and calculates the amount that each spent last year on advertising as a percentage of its total revenue. The results are listed in Table 3.12.

Let's use this set of data to construct a stem-and-leaf display. In Figure 3.9 each observation is represented by a stem to the left of the vertical line and a leaf to the right of the vertical line. For example, the stems and leaves for the first three observations in Table 3.12 can be defined as

Stem	Leaf
12	.5
8	.8
11	.5

Table 3.12 Percentage of Total Revenue Spent on Advertising

Company	Percentage	Company	Percentage
1	12.5	21	6.4
2	8.8	22	7.8
3	11.5	23	8.5
4	9.1	24	9.5
5	9.4	25	11.3
6	10.1	26	8.9
7	5.3	27	6.6
8	10.3	28	7.5
9	10.2	29	8.3
10	7.4	30	13.8
11	8.2	31	12.9
12	7.8	32	11.8
13	6.5	33	10.4
14	9.8	34	7.6
15	9.2	35	8.6
16	12.8	36	9.4
17	13.9	37	7.3
18	13.7	38	9.5
19	9.6	39	8.3
20	6.8	40	7.1

Stems	Leaves	Frequency
5	3	1
6	4 5 6 8	4
7	1 3 4 5 6 8 8	7
8	2 3 3 5 6 8 9	7
9	1 2 4 4 5 5 6 8	8
10	1 2 3 4	4
11	3 5 8	3
12	5 8 9	3
13	7 8 9	3
Total		40

Figure 3.9

Stem-and-Leaf Display for Advertising Expenditure

In other words, stems are the integer portions of the observations, whereas leaves represent the decimal portions.

The procedure used to construct a stem-and-leaf display is as follows:

1. Decide how the stems and leaves will be defined.
2. List the stems in a column in ascending order.
3. Proceed through the data set, placing the leaf for each observation in the appropriate stem row. (You may want to place the leaves of each stem in increasing order.)

The percentage of revenues spent on advertising by 40 production firms listed in Table 3.12 are represented by a stem-and-leaf diagram in Figure 3.9. From this diagram we observe that the minimum percentage of advertising spending is 5.3 percent of total revenue, the maximum percentage of advertising spending is 13.9 percent, and the largest group of firms spends between 9.1 and 9.8 percent of total revenue on advertising. Also, the 7 leaves in stem row 7 indicate that 7 firms' advertising spending is at least 7 percent but less than 8 percent. The 3 leaves in stem row 13 tell us at a glance that 3 firms spend more than 13 percent of total revenue on advertising. A MINITAB version of the stem-and-leaf diagram generated by these data is shown in Figure 13.10. A stem-and-leaf diagram is presented in the last portion of Figure 3.10. In the first column of this diagram, (8) represents the total observation in the middle group with a stem of 9; 1, 5, 12, and 19 represent the cumulative frequencies from the first group up to the fourth group; 3, 6, 9, and 13 represent the cumulative frequencies from the ninth group up to the sixth group.

Frequency Polygon

A **frequency polygon** is obtained by linking the midpoints indicated on the X axis of the class intervals from a frequency histogram. A **cumulative frequency polygon** is derived by connecting the midpoints indicated on the X axis of the class intervals from a cumulative frequency histogram. Figures 3.11 and 3.12 show the frequency polygon and the cumulative frequency polygon, respectively, for Ford's current ratio.

```
MTB > SET INTO C1
DATA> 12.5 8.8 11.5 9.1 9.4 10.1 5.3 10.3 10.2 7.4 8.2 7.8 6.5 9.8 9.2 12.8
DATA> 13.9 13.7 9.6 6.8 6.4 7.8 8.5 9.5 11.3 8.9 6.6 7.5 8.3 13.8 12.9 11.8
DATA> 10.4 7.6 8.6 9.4 7.3 9.5 8.3 7.1
DATA> END
MTB > PRINT C1

GMCA/CL
   12.5    8.8   11.5    9.1    9.4   10.1    5.3   10.3   10.2    7.4    8.2
    7.8    6.5    9.8    9.2   12.8   13.9   13.7    9.6    6.8    6.4    7.8
    8.5    9.5   11.3    8.9    6.6    7.5    8.3   13.8   12.9   11.8   10.4
    7.6    8.6    9.4    7.3    9.5    8.3    7.1

MTB > STEM AND LEAF USING C1

Stem-and-leaf of GMCA/CL   N = 40
Leaf Unit = 0.10

     1    5 3
     5    6 4568
    12    7 1345688
    19    8 2335689
    (8)   9 12445568
    13   10 1234
     9   11 358
     6   12 589
     3   13 789

MTB > PAPER
```

Figure 3.10

Stem-and-Leaf Diagram for Advertising Expenditure Using MINITAB

Figure 3.11

Frequency Polygon
of Ford's Current
Ratios

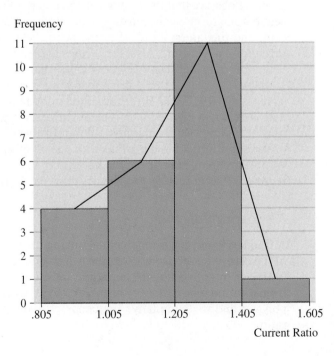

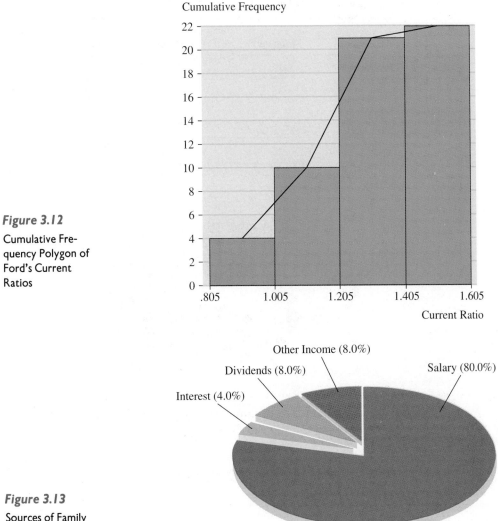

Figure 3.12

Cumulative Frequency Polygon of Ford's Current Ratios

Figure 3.13

Sources of Family Income

Pie Chart

Histograms are perhaps the graphical forms most commonly used in statistics, but other pictorial forms, such as the **pie chart,** are often used to present financial and marketing data. For example, Figure 3.13 depicts a family's sources of income. This pie chart indicates that 80 percent of this family's income comes from salary.

For data already in frequency form, a pie chart is constructed by converting the relative frequencies of each class into their respective arcs of a circle. For example, a pie chart can be used to represent the student grade distribution data originally

Table 3.13 Grade Distribution for 20 Students

Class	Frequency	Relative Frequency	Arc (Degrees)
Below 60	3	.15	54
60–69	3	.15	54
70–79	6	.30	108
80–89	4	.20	72
90–99	4	.20	72
Total	20	1.00	360

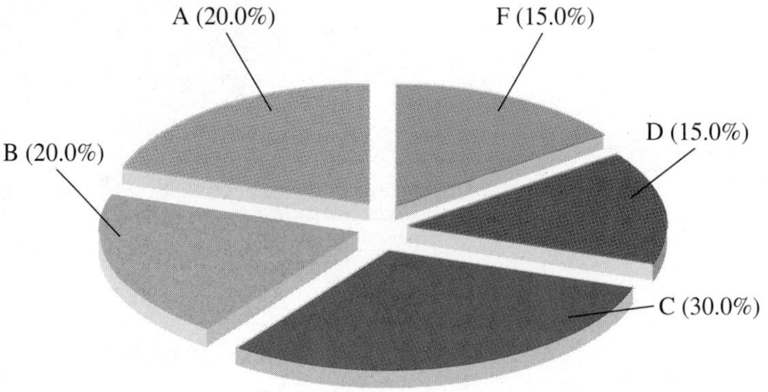

Figure 3.14

Grade Distribution Pie Chart

presented in Table 3.3. In Table 3.13, the arcs (in degrees) for the five slices shown in Figure 3.14 were obtained by multiplying each relative frequency by 360°.

3.5 FURTHER ECONOMIC AND BUSINESS APPLICATIONS

Lorenz Curve

The **Lorenz curve,** which represents a society's distribution of income, is a cumulative frequency curve used in economics (Figure 3.15a). The cumulative percentage of families (ranked by income) is measured on the X axis, and the cumulative percentage of family income received is measured on the Y axis. For example, suppose there are 100 families and each earns \$100—that is, the distribution of income is perfectly equal. The resulting Lorenz curve will be a 45° line (OP), because the cumulative percentage of families (20 percent, for example) and the cumulative share of family income received are always equal.

Now suppose that one family receives 100 percent of total family income—that is, the income distribution is absolutely *un*equal. The resulting Lorenz curve (ONP) coincides with the X axis until point N, where there is a discontinuous jump to

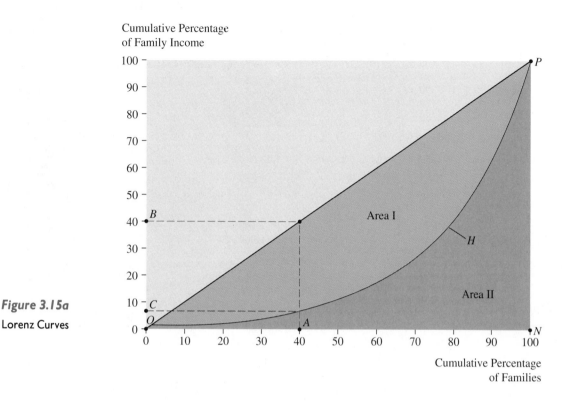

Cumulative Percentage
of Family Income

Figure 3.15a

Lorenz Curves

Cumulative Percentage
of Families

point P. This is because, with the exception of that single family (represented by point N), each family receives 0 percent of total family income. Therefore, these families' cumulative share of total family income is also 0 percent.

The shape the Lorenz curve is most likely to assume is curve H, which lies between absolute inequality and equality. This curve indicates that the lowest-income families, who comprise 40 percent of families (point A), receive a disproportionately small share (about 7 percent) of total family income (point C). If every family had the same income, the share going to the lowest 40 percent would be represented by point B (40 percent).

Note that with a more equitable distribution of income, the Lorenz curve is less bowed, or flatter. Curve S in Figure 3.15b is the Lorenz curve after a progressive income tax is imposed. Because S is flatter than H (which is reproduced from Figure 3.15a), we can conclude that the distribution of income (after taxes) is more nearly equal than before, as would be expected.

One way to measure the inequality of income from the Lorenz curve is to use the Gini coefficient.

$$\text{Gini coefficient for curve } H = \frac{\text{area I}}{\text{area (I + II)}}$$

The **Gini coefficient** can range from 0 (perfect equality) to 1 (**absolute inequality,** wherein one family receives all the income).

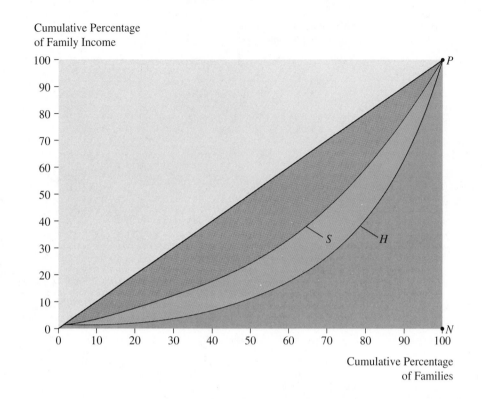

Cumulative Percentage
of Family Income

Figure 3.15b

Lorenz Curves

Cumulative Percentage
of Families

Examining Figure 3.15b reveals that the Gini coefficient will be smaller for curve *S* than it is for curve *H*. In other words, the progressive income tax makes the distribution of income more nearly equal.

Stock and Market Rate of Return

Table 3.14 presents the frequency tables for the rate of return for Ford, GM, and the stock market overall. (The data are drawn from Table 2B.2 in Appendix 2B.) Because the two firms have similar frequency distributions, we can conclude that the performances of Ford and GM stocks have been similar. However, Ford's highest class is .600–1.400, whereas GM's is .600–1.000, so Ford's rate of return has had a greater range.

The stock market overall has the same lowest class, but its highest class was only .201–.400. Thus the overall market has fluctuated less than the return of the two auto firms. And although Ford and GM have a higher top class, the market suffered through fewer negative returns. Moreover, Ford had eight and GM nine years of losses while the market had only five. In other words, the auto firms offered the potential of higher returns but also threatened the investor with a greater risk of loss.

Table 3.14 Annual Rate of Return on Stock, 1970–1990

Class	Frequency (Years)	Cumulative Frequency	Relative Frequency	Cumulative Relative Frequency
		Ford		
−.200 and below	4	4	.190	.190
−.199–.000	4	8	.190	.380
.001–.200	6	14	.286	.666
.201–.400	4	18	.190	.856
.401–.600	2	20	.095	.951
.600–1.400	1	21	.048	.999
Total	21		.999	
		General Motors		
−.200 and below	3	3	.143	.143
−.199–.000	6	9	.286	.429
.001–.200	6	15	.286	.715
.201–.400	2	17	.095	.81
.401–.600	2	19	.095	.905
.600–1.000	2	21	.095	1.000
Total	21		1.000	
		Market		
−.200 and below	1	1	.048	.048
−.199–.000	4	5	.190	.238
.001–.200	12	17	.571	.809
.201–.400	4	21	.190	.999
Total	21		.999	

Interest Rates

Histograms can be used to summarize movements in such interest rates as the prime rate and the Treasury bill rate. The prime rate is the interest rate that banks charge to their best customers; Treasury bills are short-term debt instruments issued by the U.S. government. Let us examine how these rates have moved over the period 1970 to 1990, as shown in Table 3.15.

As can be seen in Table 3.16 and Figure 3.16, the prime rate is skewed to the right, with 58 percent of the observations appearing in the ranges made up of the lowest interest rates (4–6 percent, 6–8 percent, and 8–10 percent). If you were to predict a future value for the prime rate, your best guess would be in the 4–10 per-

Table 3.15 3-Month T-Bill Rate and
Prime Rate (1970–1990)

Year	3-Month T-Bill Rate	Prime Rate
70	6.458	7.91
71	4.348	5.72
72	4.071	5.25
73	7.041	8.03
74	7.886	10.81
75	5.838	7.86
76	4.989	6.84
77	5.265	6.83
78	7.221	9.06
79	10.041	12.67
80	11.506	15.27
81	14.029	18.87
82	10.686	14.86
83	8.63	10.79
84	9.58	12.04
85	7.48	9.93
86	5.98	8.33
87	5.82	8.22
88	6.69	9.32
89	8.12	10.87
90	7.51	10.01

Figure 3.16

Frequency Histogram of Prime Lending Rates Given in Table 3.15

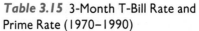

cent range. This wide range would probably not be of much use. Better methods for prediction, such as multiple regression and time series analysis, will be discussed later (Chapters 15 and 18).

The frequency table for the Treasury bill rate is shown in Table 3.16. This distribution, like that of the prime rate, is skewed to the right. Sixty-six percent of the observations appear in the first two classes, 4–6 percent and 6–8 percent. This distribution is depicted in the histogram shown in Figure 3.17.

Table 3.16 Frequency Distributions of Interest Rates

Class	T-Bill		Prime Rate	
	Frequency	*Relative Frequency*	*Frequency*	*Relative Frequency*
4–5.99%	7	.33	2	.10
6–7.99%	7	.33	4	.19
8–9.99%	3	.14	6	.29
10–11.99%	3	.14	4	.19
12–13.99%	0	0	2	.10
14–15.99%	1	.05	2	.10
16–17.99%	0	0	0	0
18–19.99%	0	0	1	.05
Total	21	.99	21	1.02

Figure 3.17

Frequency Histogram of T-Bill Rates Given in Table 3.15

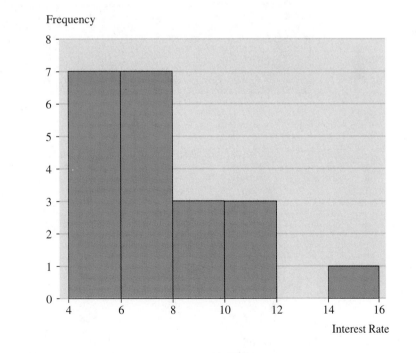

If you were to make a prediction of the Treasury bill rate, it would probably be in the 4–8 percent range. Again, better methods for predicting observations will be discussed later.

Quality Control

Figure 3.18 depicts the quality control data on electronic parts given in Table 3.17. This control chart shows the percentage of defects for each sample lot. Figure 3.18 indicates that both lots 5 and 7 have exceeded the allowed maximum defect level of 3 percent. Therefore, the product quality in these two lots should be improved.

Percentage Defective

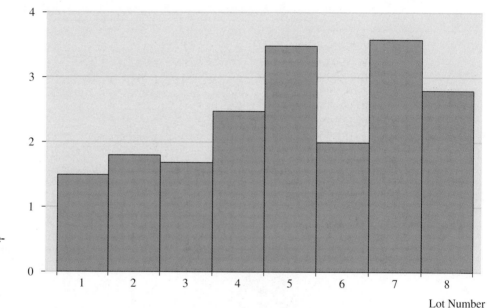

Figure 3.18

Frequency Bar Graph of the Percentage of Defects for Each Sample Lot

Lot Number

Table 3.17 Quality Control Report on Electronic Parts

Sample Lot	Sample	Defects	Percentage
1	1000	15	1.5
2	1000	20	2.0
3	1000	17	1.7
4	1000	25	2.5
5	1000	35	3.5
6	1000	20	2.0
7	1000	36	3.6
8	1000	28	2.8
Total	8000	196	2.45 (mean)

Summary

In this chapter, we extended the discussion of Chapter 2 by showing how data can be grouped to make analysis easier. After the data are grouped, frequency tables, histograms, stem-and-leaf displays, and other graphical techniques are used to present them in an effective and memorable way.

Our ultimate goal is to use a sample to make inferences about a population. Unfortunately, neither the tabular nor the graphical approach lends itself to measuring the *reliability* of an inference in data analysis. To do this we must develop numerical measures for describing data sets. Therefore, in the next chapter we show how data can be described by the use of descriptive statistics such as the mean, standard deviation, and other summary statistical measures.

Appendix 3A Financial Ratio Analysis

The financial ratios that lenders and investors need to analyze the health of a firm are constructed from information in the firm's income statement and balance sheet. There are five basic types of ratios: leverage ratios, activity ratios, liquidity ratios, profitability ratios, and market value ratios. We have already covered the current ratio, which is a liquidity ratio, in Example 3.5, so we will start with leverage ratios.

Leverage ratios measure how much of the firm's operation is financed by debt. Dividing total debts by total assets yields the proportion of total assets financed by debt. Table 3A.1 presents a frequency distribution of Ford's and GM's leverage

Table 3A.1 Frequency Distribution of Leverage Ratio for Ford and General Motors

Class	Frequency	Cumulative Frequency	Relative Frequency	Cumulative Relative Frequency
		Ford		
.301–.350	0	0	0	0
.351–.400	0	0	0	0
.401–.450	5	5	.2	.2
.451–.500	2	7	.08	.28
.501–.550	3	10	.12	.40
.551–.600	4	14	.16	.56
.601–.650	4	18	.16	.72
.651–.700	2	20	.08	.80
.701–.750	1	21	.04	.84
.751–.800	0	21	0	.84
.801–.850	1	22	.04	.88
.851–.900	3	25	.12	1.00
Total	25		1.00	

Table 3A.I (Continued)

Class	Frequency	Cumulative Frequency	Relative Frequency	Cumulative Relative Frequency
General Motors				
.301–.350	5	5	.2	.2
.351–.400	4	9	.16	.36
.401–.450	5	14	.2	.56
.451–.500	1	15	.04	.60
.501–.550	4	19	.16	.76
.551–.600	2	21	.08	.84
.601–.650	3	24	.12	.96
.651–.700	0	24	0	0
.701–.750	1	25	.04	1.00
Total	25		1.00	

ratio. This ratio is widely dispersed for Ford, with most value falling in the .401–.450 interval. The cumulative relative frequencies increase greatly until the .651–.700 class. For GM, 60 percent of the ratios occur between .301 and .50. Sixteen percent of the observations occur in the .501–.550 class. The lowest class in which there are observations for Ford is .401–.450 and the highest is .851–.900. The lowest class for GM is .301–.350, the highest .701–.750. Thus we can conclude that the range of values is greater for Ford than for GM.

Activity ratios measure how efficiently the firm employed its assets. The inventory turnover ratio, which is calculated by dividing net sales by average inventory, measures how quickly the firm is selling inventory. It is illustrated in Table 3A.2.

Table 3A.2 Frequency Distribution of Inventory Turnover Ratio for Ford and General Motors

Class	Frequency	Inventory Cumulative Frequency	Turnover Relative Frequency	Cumulative Relative Frequency
Ford				
4.01–5.00	0	0	.00	.00
5.01–6.00	2	2	.08	.08
6.01–7.00	8	10	.32	.40
7.01–8.00	5	15	.20	.6
8.01–9.00	2	17	.08	.68
9.01–10.00	0	17	.00	.68
10.01–11.00	2	19	.08	.76
11.01–12.00	2	21	.08	.84
12.01–13.00	2	23	.08	.92
13.01–14.00	2	25	.08	1.00
Total	25		1.00	

Table 3A.2 (Continued)

Class	Frequency	Inventory Cumulative Frequency	Turnover Relative Frequency	Cumulative Relative Frequency
		General Motors		
4.01–5.00	2	2	.08	.08
5.01–6.00	0	2	.00	.08
6.01–7.00	6	8	.24	.32
7.01–8.00	5	13	.20	.52
8.01–9.00	3	16	.12	.64
9.01–10.00	2	18	.08	.72
10.01–11.00	0	18	.00	.72
11.01–12.00	4	22	.16	.88
12.01–13.00	1	23	.04	.92
13.01–14.00	1	24	.04	.96
14.01–15.00	1	25	.04	1.00
Total	25		1.00	

Most of the observations appear between 6.01 and 9.00 for Ford and GM. Thus a good prediction for this variable would be in this interval. The lowest class for Ford is 5.01–6.00 and the highest is 12.01–13.00. The lowest for GM is 4.01–5.00 and the highest is 13.01–14.00.

One *profitability ratio* is the return on total assets, defined as net income divided by total assets. This ratio measures how much the company has earned on its investment of all the financial resources committed to the firm. It reveals how well the firm used its funds, regardless of how the firm split up the assets into fixed and current assets. Table 3A.3 shows that for Ford, 80 percent of the ratios were between .001 and .090, and a full 40 percent occurred in the .061–.090 class. For GM, 36

Table 3A.3 Frequency Distribution of Return on Total Assets for Ford and General Motors

Class	Frequency	Cumulative Frequency	Relative Frequency	Cumulative Relative Frequency
		Ford		
−.030 and less	3	3	.12	.12
−.029–.000	0	3	.00	.12
.001–.030	7	10	.28	.40
.031–.060	3	13	.12	.52
.061–.090	10	23	.40	.92
.091–.120	2	25	.08	1.00
Total	25		1.00	

Table 3A.3 (Continued)

Class	Frequency	Cumulative Frequency	Relative Frequency	Cumulative Relative Frequency
		General Motors		
.000 and less	1	1	.04	.04
.001–.030	2	3	.08	.12
.031–.060	8	11	.32	.44
.061–.090	4	15	.16	.60
.091–.120	9	24	.36	.96
.121–.140	1	25	.04	1.00
Total	25		1.00	

percent of the ratios were in the .091–.120 class. Ford experienced some negative values and thus had the lowest class, whereas GM enjoyed the highest class, .120–.140.

Table 3A.4 Frequency Distribution of Net Profit Margin for Ford and General Motors

Class	Frequency	Cumulative Frequency	Relative Frequency	Cumulative Relative Frequency
		Ford		
−.040 and less	1	1	.04	.04
−.039–.020	1	2	.04	.08
−.19–.00	1	3	.04	.12
.001–.020	3	6	.12	.24
.021–.040	8	14	.32	.56
.041–.060	11	25	.44	1.00
Total	25		1.00	
		General Motors		
.000 and less	2	2	.08	.08
.001–.020	2	4	.08	.16
.021–.040	6	10	.24	.40
.041–.060	6	16	.24	.64
.061–.080	7	23	.28	.92
.081–.100	2	25	.08	1.00
Total	25		1.00	

The net profit margin, defined as net income divided by sales, is another profit-ability measure. This ratio measures the percentage of sales that make their way into profits. It is shown in Table 3A.4. The best forecast for this ratio for Ford would be in the .001 to .060 range because 88 percent of previous ratios were in this range. The estimate for GM might be in the .021 to .080 range; 76 percent of the observations appeared there. Again, Ford suffered through some negative values and had the lowest class, whereas GM enjoyed ratios in the range of .081 to .100. Managers want high net profit margin ratios because they signal that the firm is operating profitably.

One of the *market value ratios* (Table 3A.5) is the payout ratio, which measures the percentage of earnings paid out in dividends. Most of these ratios are in the .01 to .60 range for Ford, and 76 percent were in the .31 to .90 range for GM. GM's ratios were more dispersed than Ford's, because GM had the lowest and highest classes.

Another market value ratio is the price/earnings ratio, defined as price per share divided by earnings per share. The majority of P/E ratios for Ford and GM have been in the zero to 15 range. The frequency distribution is shown in Table 3A.6. Ford experienced more widely dispersed P/E ratios with a high class of 35.01–70.00 whereas GM's highest class was 35.01–40.00. Both had the same lowest class.

Table 3A.5 Frequency Distribution of Payout Ratio for Ford and General Motors

Class	Frequency	Cumulative Frequency	Relative Frequency	Cumulative Relative Frequency
Ford				
−.31–.00	2	2	.08	.08
.01–.30	10	12	.40	.48
.31–.60	9	21	.36	.84
.61–.90	2	23	.08	.92
.91–3.20	2	25	.08	1.00
Total	25		1.00	
General Motors				
Less than .01	2	2	.08	.08
.01–.30	1	3	.04	.12
.31–.60	12	15	.48	.60
.61–.90	7	22	.28	.88
Greater than .90	3	25	.12	1.00
Total	25		1.00	

Table 3A.6 Frequency Distribution of P/E Ratio for Ford and General Motors

Class	Frequency	Cumulative Frequency	Relative Frequency	Cumulative Relative Frequency
Ford				
Less than .01	3	3	.12	.12
.01–5.00	10	13	.40	.52
5.01–10.00	7	20	.28	.80
10.01–15.00	3	23	.12	.92
15.01–20.00	1	24	.04	.96
20.01–25.00	0	24	.00	.96
25.01–30.00	0	24	.00	.96
30.01–35.00	0	24	.00	.96
35.01–70.00	1	25	.04	1.00
Total	25		1.00	
General Motors				
Less than .01	2	2	.08	.08
.01–5.00	2	4	.08	.16
5.01–10.00	11	15	.44	.60
10.01–15.00	7	22	.28	.88
15.01–20.00	0	22	.00	.88
20.01–25.00	1	23	.04	.92
25.01–30.00	0	23	.00	.92
30.01–35.00	1	24	.04	.96
35.01–40.00	1	25	.04	1.00
Total	25		1.00	

Questions and Problems

1. Explain the difference between grouped and non-grouped data.

2. Explain the difference between frequency and relative frequency.

3. Explain the difference between frequency and cumulative frequency.

4. Carefully explain how the concept of cumulative frequency can be used to form the Lorenz curve.

5. Suppose you are interested in constructing a frequency distribution for the heights of 80 students in a class. Describe how you would do this.

6. What is a frequency polygon? Why is a frequency polygon useful in data presentation?

7. Use the current ratio data for Ford given in Table 3.9 in the text to construct relative frequency and cumulative relative frequency tables.

8. Use the percentage of total revenue spent on advertising listed in Table 3.12 of the text to draw a frequency polygon and a cumulative frequency polygon.

9. On November 17, 1991, the *Home News* of central
New Jersey used the bar chart given here to show
that foreign investors are taxed at a lower rate than
U.S. citizens.
   a. Construct a table to show frequency, relative
   frequency and cumulative frequency.

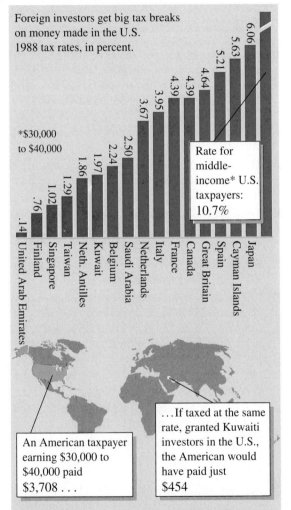

Foreign investors get big tax breaks
on money made in the U.S.
1988 tax rates, in percent.

*$30,000
to $40,000

Rate for middle-income* U.S. taxpayers:
10.7%

.14 United Arab Emirates
.76 Finland
1.02 Singapore
1.29 Taiwan
1.86 Neth. Antilles
1.97 Kuwait
2.24 Belgium
2.50 Saudi Arabia
3.67 Netherlands
3.95 Italy
4.39 France
4.39 Canada
4.64 Great Britain
5.21 Spain
5.63 Cayman Islands
6.06 Japan

An American taxpayer earning $30,000 to $40,000 paid $3,708 . . .

. . . If taxed at the same rate, granted Kuwaiti investors in the U.S., the American would have paid just $454

Source: Philadelphia Inquirer, Internal Revenue Service.

Source: *Home News,* November 17, 1991. Reprinted by permission of Knight-Ridder Tribune News.

b. Draw a frequency polygon and a cumulative
frequency polygon.

10. Use the EPS and DPS data given in Table 2B.1 in
Chapter 2 to construct frequency distributions.

11. Use the data from question 10 to construct a relative frequency graph and a cumulative relative frequency graph for both EPS and DPS.

12. On November 17, 1991, the *Home News* of central
New Jersey used the bar chart in the accompanying figure to show the 1980–1991 passenger traffic
trends for Newark International Airport.
   a. Use these data to draw a line chart and interpret
   your results.
   b. Use these data to draw a stem-and-leaf diagram
   and interpret your results.

Passengers (in millions)

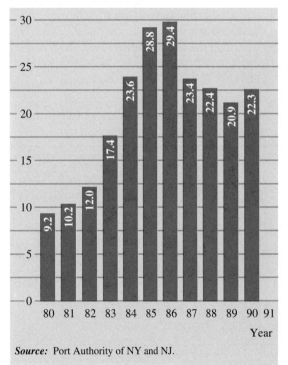

Year	Passengers
80	9.2
81	10.2
82	12.0
83	17.4
84	23.6
85	28.8
86	29.4
87	23.4
88	22.4
89	20.9
90 91	22.3

Source: Port Authority of NY and NJ.

Source: *Home News,* November 17, 1991. Reprinted by permission of the publisher.

13. An advertising executive is interested in the age distribution of the subscribers to *Person* magazine. The age distribution is as follows:

Age	Number of Subscribers
18–25	10,000
26–35	25,000
36–45	28,000
46–55	19,000
56–65	10,000
Over 65	7,000

a. Use a frequency distribution graph to present these data.

b. Use a relative frequency distribution to present these data.

14. Use the data from question 13 to produce a cumulative frequency graph and a cumulative relative frequency graph.

15. Construct stem-and-leaf displays for the 3-month T-bill rate and the prime rate, using the data listed in Table 3.15.

Use the goaltenders' salaries for the 1991 NHL season given in the following table to answer questions 16–20.

Name	Team	Gross Salary
Patrick Roy	Montreal Canadiens	$1.056M[a]
Ed Belfour	Chicago Blackhawks	$925,000
Ron Hextall	Philadelphia Flyers	$735,000
Mike Richter	New York Rangers	$700,000
Kelly Hrudey	Los Angeles Kings	$550,000
Mike Liut	Washington Capitals	$525,000
Mike Vernon	Calgary Flames	$500,000
Grant Fuhr	Toronto Maple Leafs	$424,000
John Venbiesbrouck	New York Rangers	$375,000
Ken Wregget	Philadelphia Flyers	$375,000
Tom Barrasso	Pittsburgh Penguins	$375,000

[a]Roy's salary is $500,000 Canadian, and $700,000 Canadian deferred. The salary listed is U.S. equivalent.

16. Group the data given in the table into the following groups: $351,000–400,000; 401,000–450,000; 451,000–500,000; 501,000–550,000; 551,000–600,000; 601,000–650,000; 651,000–700,000; over 701,000.

17. Use your results from question 16 to construct a cumulative frequency table.

18. Use your results from question 16 to construct a relative frequency table and a cumulative relative frequency table.

19. Use a bar graph to plot the frequency distribution.

20. Use a bar graph to plot the cumulative relative frequency.

21. Briefly explain why the Lorenz curve shown in Figure 3.15b has the shape it does.

22. The students in an especially demanding history class earned the following grades on the midterm exam: 86, 75, 92, 98, 71, 55, 63, 82, 94, 90, 80, 62, 62, 65, and 68. Use MINITAB to draw a stem-and-leaf graph of these grades.

23. Use the data given in question 22 to construct a tally table for the grades.

24. Construct a cumulative frequency table for the tally table you constructed in question 23.

25. Use the data in question 24 to graph the cumulative frequency on a bar chart by using Lotus 1-2-3.

26. Suppose the Gini coefficient of the United States were equal to 0. What would that tell us about income in this country?

27. Suppose the Gini coefficient for the United States were equal to 1. What would that tell us about income in this country?

Use the following information to answer questions 28–34. Suppose Weight Watchers has collected the following weight loss data, in pounds, for 30 of its clients.

15, 20, 10, 6, 8, 18, 32, 17, 19, 7, 9, 12, 14, 9, 25, 18, 21, 3, 2, 18, 12, 15, 14, 28, 34, 30, 18, 12, 11, 8

28. Construct a tally table for weight loss. Use 5-pound intervals beginning with 1–5 pounds, 6–10 pounds, etc.

29. Construct a cumulative frequency table for weight loss.

30. Construct a frequency histogram for weight loss using MINITAB.

31. Construct a frequency polygon for weight loss.

32. Construct a table for the relative frequencies and the cumulative relative frequencies.

33. Graph the relative frequency.

34. Graph the cumulative relative frequency.

35. The following graph shows the Lorenz curves for two countries, Modestia and Richardonia. Which country has the most nearly equal distribution of income?

Cumulative Percentage
of Family Income

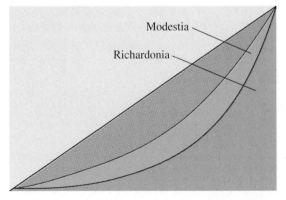

Cumulative Percentage
of Families

Use the following information to answer questions 36–41. Suppose a class of high school seniors had the following distribution of SAT scores in English.

SAT Score	Number of Students
401–450	8
451–500	10
501–550	15
551–600	6
601–650	4
651–700	1

36. Construct a cumulative frequency table.

37. Use a histogram to graph the cumulative frequencies.

38. Construct a frequency polygon.

39. Compute the relative frequencies and the cumulative relative frequencies.

40. Construct a relative frequency histogram.

41. Construct a cumulative relative frequency histogram.

Use the following prices of Swiss stocks to answer questions 42 through 49.

Switzerland (in Swiss Francs)

	Close	Prev. Close
Alusuisse	976	982
Brown Boveri	3960	4080
Ciba-Geigy br	3190	3240
Ciba-Geigy reg	3080	3110
Ciba-G ptc ctf	3020	3040
CS Holding	1920	1915
Hof LaRoch br	8280	8300
Roce div rt	5360	5330
Nestle bearer	8420	8450
Nestle reg	8310	8310
Nestle ptc ctf	1570	1585
Sandoz	2390	2410
Sulzer	465	470
Swiss Bnk Cp	301	299
Swiss Reinsur	2520	2530
Swissair	667	680
UBS	3230	3230

(*continued*)

Switzerland (in Swiss Francs) (Continued)

	Close	Prev. Close
Winterthur	3390	3420
Zurich Ins	4080	4090

Source: Wall Street Journal, November 1, 1991.

42. Construct a tally table for the closing stock prices "Close" column.

43. Compute the change in prices by subtracting the previous closing price from the current closing price.

44. Use your answer to question 43 to construct a tally table.

45. Use your answer to question 44 to compute the cumulative frequencies.

46. Use your answer to question 44 to compute the relative and cumulative frequencies.

47. Use your answer to question 46 to graph the relative frequency.

48. Use your answer to question 46 to graph the cumulative frequency.

49. Create a frequency polygon using data from question 44.

CHAPTER 4

Numerical Summary Measures

CHAPTER OUTLINE

4.1 Introduction

4.2 Measures of Central Tendency

4.3 Measures of Dispersion

4.4 Measures of Relative Position

4.5 Measures of Shape

4.6 Calculating Certain Summary Measures from Grouped Data (Optional)

4.7 Applications

Appendix 4A Short-Cut Formulas for Calculating Variance and Standard Deviation

Appendix 4B Short-Cut Formulas for Calculating Group Variance and Standard Deviation

Appendix 4C Financial Ratio Analysis for Three Auto Firms

Key Terms

measure of central tendency
arithmetic mean
geometric mean
median
mode
dispersion
variance
standard deviation
mean absolute deviation
range
coefficient of variation
percentile

quartiles
interquartiles range
box and whisker plots
z score
Tchebysheff's theorem
skewness
coefficient of skewness
Pearson coefficient
zero skewness coefficient
positive skewness coefficient
negative skewness coefficient
kurtosis

4.1 INTRODUCTION

In this chapter, we extend the graphical descriptive method in data analysis by examining measures of central tendency, dispersion, position, and shape. All these numerical summary measures are important because they enable us to describe a set of data with only a small number of summary statistics. One use of these summary statistics is to compare individual observations from a data set. For example, a student in a statistics class could use one measure of central tendency, the class average, or mean, to determine how well her performance stacks up to the rest of the class. Measures of central tendency can also be used to compare two different sets of data. For example, a statistics teacher interested in comparing the performances of two different statistics classes could take the average, or mean, for each class and compare the two.

We first address four measures of *central tendency,* discussing how they are computed from a data set and how they help us locate the center of a distribution (see Figure 4.1a). Similarly, we examine measures of *dispersion,* which describe the dispersion, or spread, of a set of observations and therefore of its distribution (see Figure 4.1b). The coefficient of variation (a measure of relative dispersion) is also investigated. Next we explore measures of a distribution's *position.* Numerical descriptive measures have also been devised to measure *shape:* the skewness of a distribution (the tendency of a relative frequency distribution to stretch out in one direction or another) and its kurtosis (peakedness). Here we discuss only the numerical measurement of skewness. The numerical measurement of kurtosis will be discussed in Chapter 9. Finally, we present applications of numerical descriptive statistics in business and economics.

Figure 4.1

Numerical Summary Measures: (a) Central Tendency and (b) Dispersion

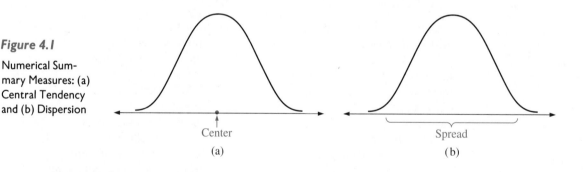

Center

(a)

Spread

(b)

4.2 MEASURES OF CENTRAL TENDENCY

The purpose of a **measure of central tendency** is to determine the "center" of a distribution of data values or possibly the "most typical" data value. Measures of central tendency include the arithmetic mean, geometric mean, median, and mode.

Using a quality control example, we will illustrate each of these measures with the following data, which represent the number of defective parts in each of four samples.[1]

$$5, 8, 14, 3$$

The Arithmetic Mean

Most of you have calculated your grade point average or average test score in a course by adding all your grade points or scores and dividing by the number of courses or tests. You might not have realized it, but you were calculating the **arithmetic mean.**

The arithemetic mean of a set of raw data is denoted by $x_1, x_2, \ldots, x_N$ (N represents the total number of observations in a population) or $x_1, x_2, \ldots, x_n$ (n represents the sample size). We find it by adding together all the observations and dividing by the number of observations. A sample mean is denoted by $\bar{x}$, a population mean by μ. For the quality control data set, $n = 4$, so

$$\bar{x} = (5 + 8 + 14 + 3)/4 = 30/4 = 7.5$$

Thus when the observations are $x_1, x_2, \ldots, x_n$, the sample mean is

$$\bar{x} = \frac{\sum_{i=1}^{n} x_i}{n} \tag{4.1}$$

The population mean is

$$\mu = \sum_{i=1}^{N} x_i/N \tag{4.2}$$

where N is the total number of observations of the population, and the observations are $x_1, x_2, \ldots, x_N$. The summation notation (Σ) used in Equations 4.1 and 4.2 simply means that the first observation is added to the second, and so on, until all the observations have been added.

EXAMPLE 4.1 *Six from Nine to Five*

Say we want to find the average annual salary of all secretaries. We believe we can do this on the basis of our knowledge of the annual salaries of 6 particular secretaries, who each year earn $10,400, $34,000, $14,000, $18,500, $27,000, and $25,800, respectively. This is a sample of $n = 6$, where $x_1 = 10,400$, $x_2 = 34,000$, $x_3 = 14,000$, $x_4 = 18,500$, $x_5 = 27,000$, and $x_6 = 25,800$. We find the sample mean by adding all the observations and dividing by 6:

---

[1]Quality control was addressed briefly in Table 3.17 of Chapter 3. This issue will be discussed further in Chapters 10 and 20.

$$\bar{x} = (x_1 + x_2 + x_3 + x_4 + x_5 + x_6)/6 = 129{,}700/6$$
$$= \$21{,}616.67$$

Our result is a *sample* mean because we are interested in finding the mean annual income of all secretaries on the basis of the annual income of a smaller sample consisting of only 6 secretaries.

EXAMPLE 4.2   *Arithmetic Average of Stock Rates of Return*

As an example of computing the mean of a *population,* suppose an individual owns 5 stocks that last year returned 15 percent, 10 percent, -4 percent, 7 percent, and -10 percent. We find the mean of this population by adding all the returns and dividing by $N = 5$. Thus the population mean is $\mu = (15 + 10 + -4 + 7 + -10)/5 = 18/5 = 3.6$ percent.

The Geometric Mean

The **geometric mean** of a set of observations is another measure of central tendency. It can be calculated by multiplying all the observations and taking the product to the $1/N$ or the $1/n$ power, depending on whether the observations come from a finite population or a sample. The sample mean ($\bar{x}_g$) and the population geometric mean (μ_g) can be expressed as follows:

$$\bar{x}_g = (x_1 \cdot x_2 \cdot \cdots \cdot x_n)^{1/n} \tag{4.3}$$
$$\mu_g = (x_1 \cdot x_2 \cdot \cdots \cdot x_N)^{1/N} \tag{4.4}$$

Using the quality control data set, we find that

$$\bar{x}_g = (5 \cdot 8 \cdot 14 \cdot 3)^{1/4} = (1680)^{1/4} = 6.40$$

Note that the geometric mean, 6.4, is smaller than the arithemetic mean, 7.5.

All the observations in Equations 4.3 and 4.4 must be positive. It should be noted that the geometric mean is less sensitive to extreme values than is the arithmetic mean. The geometric mean is frequently used in finance to calculate the rate of return on a stock or bond. The reason why the geometric mean is popular in calculating average rates of return is that this method explicitly incorporates the concept of compound interest (interest received on interest).[2] To avoid negative and zero returns, holding period returns (HPR) are used. An HPR is calculated by taking the rate of return and adding 1. Adding 1 avoids negative numbers and makes it possible to calculate an average return. Now let's use the data given in Example 4.2 to calculate the geometric average of stock rates of return.

---

[2]The advantage of using the geometric average rather than the arithmetic average is discussed by Lee *et al.* (1990), *Security Analysis and Portfolio Management,* Scott, Foresman, Little, Brown (Chapter 3).

EXAMPLE **4.3** *Geometric Average of Stock Rates of Return*

Here we must calculate a geometric mean of the following rates of return: 15 percent, 10 percent, −4 percent, 7 percent, and −10 percent. To obtain the HPR, we add 1 to each of the returns, which yields 1.15, 1.10, .96, 1.07, and .90. To obtain the geometric mean of the HPR, we multiply the individual HPRs and take the product to the $1/N$ power:

$$\mu_g = [(1.15)(1.10)(.96)(1.07)(.90)]^{1/5} = (1.169)^{1/5} = 1.032$$

To obtain the geometric mean for the conventional rate of return, we subtract 1 from the geometric-mean HPR, arriving at .032, or 3.2 percent. Note that the geometric mean is 3.2 percent, whereas the arithmetic mean (calculated in Example 4.2) is 3.6 percent. In general, the geometric mean is smaller than the arithmetic mean and less sensitive to extreme observations.

The Median

The **median** (Md) is the middle observation of a set of ordered observations if the number of observations is odd; it is the average of the middle pair if the number of observations is even. In other words, if there are N observations, where N is an odd number, the median is the $[(N + 1)/2]$th observation. If N is even, the median is the average of the $(N/2)$th and the $[(N + 2)/2]$th observations. Sometimes the median is a preferred measure of central tendency, particularly when the data include extreme observations that could affect the geometric or arithmetic mean.

Consider again our quality control data. We find the median Md by first constructing an order array:

$$3, 5, 8, 14$$

Because N is an even number (4), Md = $(5 + 8)/2 = 6.5$. The median (6.5) is smaller than the mean (7.5), as indicated in Figure 4.2. This difference is essentially caused by the extreme value 14. The relationship between mean and median will be discussed in Section 4.5.

Figure 4.2

Sample Quality Control Data with Mean and Median Shown

EXAMPLE **4.4** *Median of Stock Rates of Return*

Arrange the rate-of-return data in Example 4.2 in numerical order: −10 percent, −4 percent, 7 percent, 10 percent, and 15 percent. There is an odd number of observations, so the median is the third observation—$[(5 + 1)/2] = 3$—which in this case is 7 percent.

EXAMPLE 4.5 *What Does "Average" Mean? Arithmetic Mean Versus Median for Sample Family Income*

The median can be particularly useful when there are a few extreme observations. Consider the following incomes for 6 sample families: $10,000, $13,400, $15,000, $17,000, $19,000, and $120,000. Although the arithmetic mean of the series is $32,400, the median is ($15,000 + $17,000)/2 = $16,000. The substantial difference between the two means is due mainly to the extreme observation of $120,000. The median is the better measure of central tendency in this example. Note that the median would not change if the fifth and sixth observations were larger. For example, the last number could be $5,000,000 and the median would remain unchanged. Thus using the median is preferred when outlying data could lead to a distorted picture of the mean of a distribution.

Calculations of average income are especially vulnerable to such distortions. Consider the effect of that single $120,000 income if, say, federal assistance for day care were being made available in communities where the average income was under $20,000—and the "average income" of our community of 6 families were being interpreted as the mean.

EXAMPLE 4.6 *This Teacher Is Really Mean*

Students may complain that one or two very high scores raise the class average on an exam and thus lower their letter grade. See Table 4.1, where the "rank on exam" in the third column is obtained by ranking all scores in order from lowest to highest. If the mean is taken as the average score that translates into a grade of C, 5 of these 7 students have scored "below average." Students who see this as unfair are in effect arguing against using the mean exam score as a measure of central tendency. Are they right?

Well, Albert and Sue did score exceptionally high on the exam, and they do in fact raise the mean score dramatically. Should the teacher base the class grades on

Table 4.1 Student Exam Scores

Student	Score	Rank on Exam
Kim	60	2
Mary	62	4
Tom	55	1
Ann	61	3
Juan	70	5
Albert	99	6
Sue	100	7
Total	507	

Mean = $\bar{x}$ = 72.43

Median = 62

the mean of 72.43 or use some other measure of central tendency? The median (here it is 62), which lies in the middle and is not altered by the extreme values that affect the mean, may be a better measure of central tendency in this case. (Juan and Mary, whose grades have just risen from D to B and C, respectively, will certainly think so.)

The Mode[3]

The **mode** of a set of observations is the value that occurs the most times. In cases of a tie, it may assume more than one value. The mode is most useful when we are dealing with data that are in categories where the mean and median are not useful. For example, suppose that a computer sales representative sells the brands and numbers of computers shown in Table 4.2. Here it makes no sense to take the mean or median of the data, because the categories are mutually exclusive. Instead, the sales rep is interested in the most popular and the least popular products. Thus he or she wants to know which is the *modal* class (it is the Compaq computer) because that class contains the highest number of computers solds.

The main disadvantage of the mode is that it does not take the nonmodal observations into consideration. Thus, in the computer example, the mode does not reflect the facts that the IBM M30 has almost as many sales as the Compaq model and that the IBM M70 has almost the same amount of sales as the IBM M30. As another example, suppose a sample of the incomes of workers is taken and the arithmetic mean is $25,746. Suppose further that the observation $38,500 appears the most times and therefore is the mode. Obviously, the mode is not a good measure of central tendency here, because it is so far away from the mean. This problem occurs often, and researchers must be aware of the limitations of this and other statistical measures.

Table 4.2 Sales of Personal Computers

Type	Number Sold
IBM PS2/M30	487
IBM PS2/M50	201
IBM PS2/M70	432
Compaq	506

EXAMPLE 4.7   *The Model Wears 4, But the Modal Is 7*

Suppose a clerk in a shoe store sells eight pairs of shoes in the following sizes: 5, 7, 7, 7, 4, 5, 10, and 11. The modal shoe size is 7, because it appears the greatest number of times. If this result is obtained regularly, it is certainly something the purchasing manager wants to know. Although the mean and median are more widely

[3]Relationships among mean, median, and mode will be discussed in Section 4.5.

used as measures of central tendency in business and economics, the mode gives useful information on the most numerous value in a set of observations.

The numerical example discussed in Example 4.7 is a unimodal distribution. The following is a bimodal distribution: 5, 5, 7, 7, 7, 8, 9, 10, 10, 10 (7 and 10 are the modes). It should be noted that if each different number has the same frequency, there is no mode. An example of a case of no mode is 1, 1, 2, 2, 4, 4.

4.3   MEASURES OF DISPERSION

The mean, median, and mode all give us information about the central tendency of a set of observations, but these measures shed no light on the **dispersion,** or spread, of the data. For example, suppose a professor gives a test to two classes and the mean for each class is 75. However, suppose that all the students in the first class scored in the 70s, with a high of 79 and a low of 70. In the second class, the lowest score was 42 and the highest 97. It is obvious that the scores in the second class are more widely dispersed, or spread, around the mean than the scores in the first. In this section, we discuss measures of dispersion: the variance, standard deviation, mean absolute deviation, range, and coefficient of variation. We will use our now-familiar quality control data (3, 5, 8, 14) to illustrate these different measures.

The Variance and the Standard Deviation

Suppose we have a set of observations from a population $x_1, x_2, \ldots, x_N$. We are interested in finding a dispersion measure, so it would seem natural to calculate the deviations from the mean $(x_1 - \mu), (x_2 - \mu), \ldots, (x_N - \mu)$. But the negative deviations from the mean cancel out the positive deviations,[4] so the sum of these deviations will always be zero, which sheds no light on the extent of dispersion. To avoid this problem, we square and sum the deviations (distances) to give an indication of the total dispersion:

$$(x_1 - \mu)^2 + (x_2 - \mu)^2 + \cdots + (x_N - \mu)^2$$

If we take this sum and divide by the number of observations, N, we arrive at the **variance,** which represents the average squared deviation (distance) from the mean. The *population variance* is denoted by σ^2, as indicated in Equation 4.5, and the *sample variance* by s^2, as indicated in Equation 4.7. The **standard deviation** is the square root of the variance; it is denoted by σ for the population (Equation 4.6) and by s for the sample (Equation 4.8).

---

[4] Because $(x_1 - \mu) + (x_2 - \mu) + \cdots + (x_N - \mu) = \sum_{i=1}^{N} x_i - N\mu = N\mu - N\mu = 0.$

Population Variance		Population Standard Deviation	

$$\sigma^2 = \frac{\sum_{i=1}^{N} (x_i - \mu)^2}{N}. \qquad (4.5)$$

$$\sigma = \sqrt{\frac{\sum_{i=1}^{N} (x_i - \mu)^2}{N}} \qquad (4.6)$$

Sample Variance		Sample Standard Deviation	

$$s^2 = \frac{\sum_{i=1}^{n} (x_i - \bar{x})^2}{n - 1} \qquad (4.7)$$

$$s = \sqrt{\frac{\sum_{i=1}^{n} (x_i - \bar{x})^2}{n - 1}} \qquad (4.8)$$

A short-cut formula that can be used to compute the variance and standard deviation for samples and populations is given in Appendix 4A.

Using the quality control data, we calculate the sample variance and standard deviation in accordance with Equations 4.7 and 4.8. Recall that $\bar{x} = 7.5$.

	x	$(x_i - \bar{x})$	$(x_i - \bar{x})^2$
	3	−4.5	20.25
	5	−2.5	6.25
	8	.5	.25
	14	6.5	42.25
Total	30	$\Sigma(x_i - \bar{x}) = 0$	$\Sigma(x_i - \bar{x})^2 = 69$

Substituting $\Sigma(x - \bar{x})^2 = 69$ into Equations 4.7 and 4.8, we obtain

$$s^2 = 69/(4 - 1) = 23$$
$$s = 4.80$$

Note that we use the divisor $(n - 1)$ instead of n to calculate the sample variance. This is because using the divisor $(n - 1)$ yields a more precise estimate of σ^2 than dividing the sum of squared distances by n.[5]

For purposes of comparison, let's calculate the variance and the standard deviation of another set of quality control data (3, 4, 5, 6).

$$\bar{x} = (3 + 4 + 5 + 6)/4 = 4.5$$

$$s^2 = \frac{(3 - 4.5)^2 + (4 - 4.5)^2 + (5 - 4.5)^2 + (6 - 4.5)^2}{4 - 1} = 1.67$$

$$s = 1.29$$

We see, then, that the variance of the second set of quality control data is smaller than the variance of the first. The smaller variance of the second set of data is graphically represented in Figure 4.3.

---

[5]"More precise" means that s^2 with a divisor of $(n - 1)$ instead of n has the mean σ^2. See Section 8.4 for the proof.

Figure 4.3

Two Sets of Quality Control Data: (a) First Set of Data and (b) Second Set of Data

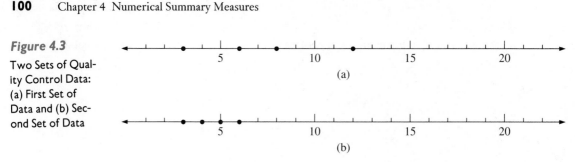

EXAMPLE 4.8   *Variability of Profit Margin*

Suppose we want to calculate the variance and standard deviation for the net profit margins indicated, for a certain firm over a 5-year period, in Table 4.3. Because this is a sample, $n = 5$.

Table 4.3 Worksheet for Data on Net Profit Ratios

Net Profit Margin (x)	$(x - \bar{x})$	$(x - \bar{x})^2$	x^2
5.6	1.778	3.16	31.36
2.7	-1.122	1.26	7.29
7.3	3.478	12.10	53.29
3.5	$-.322$	.103	12.25
.01	-3.812	14.53	.00
19.11	0	31.15	104.19

Mean $= \bar{x} = 19.11/5 = 3.822$
Variance $= s^2 = 31.153/4 = 7.79$
Standard deviation $= s = \sqrt{7.79} = 2.79$

The next computations show how to calculate the variance and standard deviation by using the short-cut formulas of Equations 4.7a and 4.8a, as indicated in Appendix 4A.

$$\text{Variance} = s^2 = \frac{\sum_{i=1}^{n} x_i^2 - n\bar{x}^2}{n - 1}$$

$$= \frac{104.19 - 5(3.822)^2}{4} = 7.79$$

$$\text{Standard deviation} = s = \sqrt{7.79} = 2.79$$

Note that the answers are the same as those derived with the standard formulas as indicated in Table 4.3.

EXAMPLE **4.9**   *Variability of Sales*

Suppose a sample of sales is taken from 4 firms, and that the figures are $2.3 million, $1.1 million, $.7 million and $6.8 million (see Table 4.4). Substituting related information into Equations 4.1, 4.7, and 4.8, we obtain

$$\text{Mean} = \bar{x} = 10.9/4 = 2.725$$
$$\text{Variance} = s^2 = 23.53/(4 - 1) = 7.8$$
$$\text{Standard deviation} = s = \sqrt{7.8} = 2.8$$

Table 4.4 Worksheet for Sales Data

Sales (x)	$x - \bar{x}$	$(x - \bar{x})^2$	x^2
2.3	$-.425$	.181	5.29
1.1	-1.625	2.64	1.21
.7	-2.025	4.100	.49
6.8	4.075	16.60	46.24
10.9	0	23.53	53.23

Calculating the variance and standard deviation via the short-cut formula of Equations 4.7a and 4.8a, we get

$$s^2 = \frac{53.23 - 4(2.725)^2}{3} = 7.8$$
$$s = \sqrt{7.8} = 2.8$$

Again, the results are identical to those obtained with the standard formula.

The Mean Absolute Deviation

Rather than squaring the deviations from the mean, we can arrive at another useful measure by calculating the absolute deviations from the mean or median and then dividing by the number of observations to obtain the average absolute deviation from the mean or median. This measure, called the **mean absolute deviation** (MAD), is defined as follows:

$$\text{MAD} = \frac{\sum_{i=1}^{n} |x_i - \bar{x}|}{n} \quad \text{or} \quad \frac{\sum_{i=1}^{n} |x_i - \text{Md}_s|}{n} \tag{4.9}$$

where n is the sample size, and $\bar{x}$ is the sample mean, and Md_s is the sample median.

If population data instead of sample data are used, then

$$
\text{MAD} = \frac{\displaystyle\sum_{i=1}^{N} |x_i - \mu|}{N} \quad \text{or} \quad \frac{\displaystyle\sum_{i=1}^{N} |x_i - \text{MAD}_p|}{N} \tag{4.10}
$$

where N is the total observations of population, μ is the population mean, and MAD_p is the population median. Let us calculate $|x_i - \bar{x}|$ and $|x_i - \text{Md}_s|$ for our quality control data. Recall that $\bar{x} = 7.5$ and $\text{Md}_s = 6.5$.

	x_i	$|x_i - \bar{x}|$	$|x_i - \text{Md}_s|$
	3	4.5	3.5
	5	2.5	1.5
	8	.5	1.5
	14	6.5	7.5
Total	30	14	14

Substituting $\Sigma|x_i - \bar{x}| = 14$ and $\Sigma|x_i - \text{Md}_s| = 14$ into Equation 4.9, we obtain

$$\text{MAD} = 14/4 = 3.5 \quad \text{if sample mean is used}$$
$$\text{MAD} = 14/4 = 3.5 \quad \text{if sample median is used}$$

One advantage of this measure is that it is not influenced so much as the variance by extreme observations. A second advantage is that the MAD is easier to interpret than the standard deviation. It is much easier to form a mental picture of the average deviation from the mean than to visualize the square root of the squared deviation from the mean! The MAD is not used much in statistical analysis, however, because complications can arise from its use in making inferences about a population on the basis of sample observations alone.

EXAMPLE 4.10   *Variability of Inflation Forecast*

Assume that a population of inflation forecasts for next year consists of the following values: 7 percent, 5 percent, 4 percent, 2 percent, and 1 percent. The worksheet for calculating MAD by using Equation 4.10 is given in Table 4.5.

Table 4.5 Inflation Forecasts

Forecast (x, %)	$x - \mu = x - 3.8$	$|x - \mu|$	$|x - \text{median}|$
7	3.2	3.2	3
5	1.2	1.2	1
4	.2	.2	0
2	−1.8	1.8	2
1	−2.8	2.8	3
		9.2	9

$\mu = 19/5 = 3.8$, $\text{Md}_p = 4$
$\text{MAD} = 9.2/5 = 1.84$
$\text{MAD} = 9/5 = 1.80$

The MAD we find by using the mean is 1.84, and the MAD we find by using the median is 1.80. In this case, the results are not identical because the distribution is not symmetric. If the distribution were highly skewed (like that of the student exam scores given in Table 4.1), the MAD found in terms of the mean would be very different from that found in terms of the median. Should we use the mean or the median, then, in calculating the mean absolute deviation? That depends on which measure of central tendency we believe is best for describing our distribution.

The Range

The range is one of the easiest measures of dispersion to calculate and interpret. The **range** is simply the difference between the highest and lowest values:

$$R = x_{max} - x_{min} \qquad (4.11)$$

where R = range, x_{max} = the largest value of all observations, and x_{min} = the smallest value of all observations. The range of our quality control data is

$$R = 14 - 3 = 11$$

The disadvantage of using this measure is that it takes into consideration only these two values. Thus it is easily thrown off by extreme values. In contrast, the variance, standard deviation, and MAD use all the observations. Despite this problem, the range has some value; for example, the typical range of temperatures in New England during the winter tells us a lot about that area's climate. The *Wall Street Journal* and other newspapers use the range when they report the 52-week high and low for each stock price per share. For example, on January 9, 1991, the 52-week high and low for IBM and Digital Equipment Corporation were $123\frac{1}{8}$–95 and $95\frac{1}{8}$–$45\frac{1}{2}$, respectively (see Figure 4.4). Comparing these two ranges reveals that the price range of Digital Equipment ($49\frac{5}{8}$) was much greater than the price range of IBM ($28\frac{1}{8}$).

Price Range for Digital Equipment

Figure 4.4

Ranges of Stock Prices per Share for Digital Equipment and IBM (in dollars)

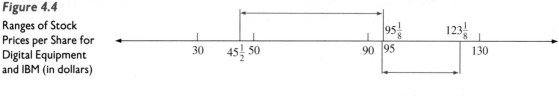

Price Range for IBM

The Coefficient of Variation

The **coefficient of variation** (CV) is the ratio of the standard deviation to the mean. The coefficient of variation for sample data can be defined as

$$CV_x = s/\bar{x} \tag{4.12}$$

For our quality control data,

$$CV_x = 4.8/7.5 = .64$$

The coefficient of variation is particularly useful when we must compare the variabilities of data sets that are measured in different units. For example, suppose a researcher wants to see how Japanese and American wage incomes compare in variability. Because workers are paid in yen in Japan and in dollars in the United States, it would be difficult to make this comparison using, say, the standard deviation. Being a *relative* expression—that is, a ratio—the coefficient of variation neatly avoids this problem.

The coefficient of variation is also useful when we are comparing data of the same type from different time periods. Suppose, for example, that the mean sales for a firm from 1980 to 1985 were $.5 million with a standard deviation of $50,000. The mean sales for the same firm from 1986 to 1991 were $3.2 million with a standard deviation of $100,000. If we compared the standard deviations and concluded that sales in the 1980s were less variable, we would obtain a distorted picture. The value of sales during the earlier period was much lower than during the later period, so the resulting standard deviation is almost certain to be smaller. Therefore, we use the coefficient of variation, which we calculate as .1 for the earlier period and .03 for the later. Sales were actually less variable from 1986 to 1991.

EXAMPLE **4.11**   *Using Coefficient of Variation to Analyze the Volatility of Stocks*

Whenever we compare two different stocks (A and B), it is useful to know two particular statistics: (1) the average, or mean, of the stocks' rates of return and (2) the standard deviation of the returns, which is an indicator of risk. A high rate of return and low risk are desirable, so having a high mean and a low standard deviation is the most desirable combination. Let's say the average rates of return ($\bar{R}$) and standard deviations (σ) for these two stocks are

	$\bar{R}$	σ
Stock A	10%	1%
Stock B	12%	1.5%

Stock B has a higher mean return ($\bar{R}$), but it also has a larger standard deviation (σ), which means it is more risky. Because A is less risky but B has a higher expected return, the choice between the two is not obvious. By using the coefficient of variation, however, we can find the amount of risk (standard deviation) per unit of expected return, which is an appropriate measure of relative variability that combines risk and rate of return. Substituting the values of $\bar{R}$ and σ into Equation 4.12 yields

$$CV_A = .01/.10 = .10$$
$$CV_B = .015/.12 = .125$$

Thus stock B has a greater risk per unit of expected return than stock A.

4.4 MEASURES OF RELATIVE POSITION

In some situations, we may want to describe the relative position of a particular measurement in a set of data. In this section, we discuss three measures of relative standing: percentiles, quartiles, and Z scores.

To illustrate these measures, suppose the personnel managers of Johnson & Johnson have administered an aptitude test to 40 job applicants. Their scores are presented in Table 4.6. The mean of the data is $\bar{x} = 58.45$, and the standard deviation is $s = 22.99$. The sample size is $n = 40$.

Table 4.6 Ordered Array of Aptitude Test Scores for 40 Job Applicants ($\bar{x} = 58.45$, $s = 22.99$)

i	x	i	x	i	x	i	x
1.	20	11.	42	21.	56	31.	78
2.	21	12.	43	22.	58	32.	80
3.	23	13.	43	23.	59	33.	81
4.	25	14.	46	24.	61	34.	85
5.	30	15.	48	25.	62	35.	90
6.	35	16.	50	26.	65	36.	92
7.	36	17.	51	27.	68	37.	96
8.	39	18.	52	28.	70	38.	98
9.	40	19.	54	29.	71	39.	99
10.	41	20.	55	30.	75	40.	100

Percentiles, Quartiles, and Interquartile Range

One useful way of describing the relative standing of a value in a set of data is through the use of percentiles. **Percentiles** give valuable information about the rank of an observation. Most of you are familiar with percentiles from taking standardized college admissions tests such as the SAT or ACT. These tests assign each student not only a raw score but also a percentile to indicate his or her relative performance. For example, a student scoring in the 85th percentile scored higher than 85 percent of the students who took the test and lower than $(100 - 85) = 15$ percent of those who took it.

Let $x_1, x_2, \ldots,$ be a set of measurements arranged in ascending (or descending) order. The Pth percentile is a number x such that P percent of the measurement fall below the Pth percentile and $(100 - P)$ percent fall above it.

Quartiles are merely particular percentiles that divide the data into quarters. The 25th percentile is known as the first quartile (Q_1), the 50th percentile is the second (Q_2), and the 75th percentile is the third (Q_3).

To approximate the quartiles from a population containing N observations, the following positioning point formulas are used:

$$Q_1 = \text{value corresponding to the } \frac{N + 1}{4} \text{ ordered observation}$$

$$Q_2 = \text{median, the value corresponding to the } \frac{2(N + 1)}{4} = \frac{N + 1}{2}$$

ordered observation

$$Q_3 = \text{value corresponding to the } \frac{3(N + 1)}{4} \text{ ordered observation}$$

The formulas given for Q_1 and Q_3 sometimes are defined as the $(N + 3)/4^{th}$ and $(3N + 1)/4^{th}$ observations, respectively. If Q_1, Q_2, or Q_3 in not an integer, then the interpolation method can be used to estimate the value of the corresponding observation.

Interquartiles range (IQR), a measure commonly used in conjunction with quartiles, can be defined as

$$\text{IQR} = Q_3 - Q_1 \qquad\qquad (4.13)$$

The interquartile range has an easy and sometimes convenient interpretation. For large data sets, it is the range that contains the middle half of all the observations.

Now we use the Johnson & Johnson applicant data to determine the first quartile (Q_1), second quartile (Q_2), third quartile (Q_3), and interquartiles range (IQR). First we find the locations $Q_1 = 41(.25) = 10.25$, $Q_2 = 41(.5) = 20.5$, and $Q_3 = 41(.75) = 30.75$. On the basis of these locations and the information in Table 4.6, we find that the scores for Q_1, Q_2, and Q_3 are 41.25, 55.5, and 77.25. Then, according to Equation 4.13, IQR $= 77.25 - 41.25 = 36$.

EXAMPLE 4.12   *Finding One Applicant's Percentile*

If James Fleetdeer received an aptitude test score of 92, what is the percentile value?

Because Table 4.6 is arranged in ascending order, Mr. Fleetdeer's 5th-highest score is the 36th-smallest value (out of a total of 40). Hence the percentile is

$$P = \frac{36}{40} \cdot 100 = 90$$

Box and Whisker Plots: Graphical Descriptions Based on Quartiles

A **box and whisker plot** is a graphical representation of a set of sample data that illustrates the lowest data value (L), the first quartile (Q_1), the median (Q_2, Md), the third quartile (Q_3), the interquartile range (IQR), and the highest data value (H).

In the last section, the following values were determined for the aptitude test scores in Table 4.6: $L = 20$, $Q_1 = 41 + .25(42 - 41) = 41.25$, $Q_2 = $ Md $= 55.5$, $Q_3 = 75 + .75(78 - 75) = 77.25$, $IQR = 36$, and $H = 100$.

A box and whisker plot of these values is shown in Figure 4.5. The ends of the box are located at the first and third quartiles, and a vertical bar is inserted at the median. Consequently, the length of the box is the interquartile range. The dotted lines are the whiskers; they connect the highest and lowest data values to the end of the box. This means that approximately 25 percent of the data values will lie in each

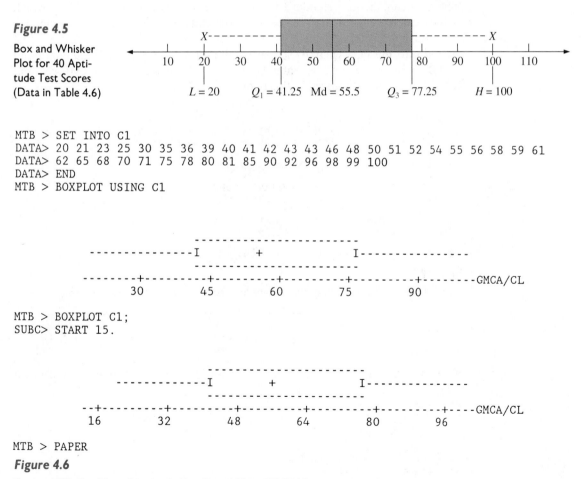

Figure 4.5

Box and Whisker Plot for 40 Aptitude Test Scores (Data in Table 4.6)

```
MTB > SET INTO C1
DATA> 20 21 23 25 30 35 36 39 40 41 42 43 43 46 48 50 51 52 54 55 56 58 59 61
DATA> 62 65 68 70 71 75 78 80 81 85 90 92 96 98 99 100
DATA> END
MTB > BOXPLOT USING C1



               --------------------------
 ---------------I        +            I----------------
               --------------------------
 --------+----------+---------+---------+---------+--------+-------GMCA/CL
        30         45        60        75        90

MTB > BOXPLOT C1;
SUBC> START 15.



                 ------------------------
 ------------I        +            I--------------
                 ------------------------
 --+---------+---------+---------+---------+---------+----GMCA/CL
   16        32        48        64        80        96

MTB > PAPER
```

Figure 4.6

Box and Whisker Plot of Aptitude Test Scores Using MINITAB

whisker and in each portion of the box. If the data are symmetric, the median bar should be located at the center of the box. Consequently, the location of the bar informs us about any skewness of the data; if the bar is located in the left (or right) half of the box, the data are skewed right (or left), as defined in the next section.

In Figure 4.5, the distribution of the data is skewed to the right because the median bar is located in the left. A box and whisker plot using MINITAB is shown in Figure 4.6. In this figure, a rectangle (the box) is drawn with the ends (the hinges) drawn at the first and third quartiles (Q_1 and Q_3). The median of the data is shown in the box by the symbol $+$. There are two boxes in Figure 4.6. The only difference is that the second specifies the starting value at 15.

EXAMPLE 4.13   *Using MINITAB to Compute Some Important Statistics of 40 Aptitude Test Scores*

The MINITAB/PC input and printout are presented in Figure 4.7. This printout presents mean, median, standard deviation, L (MIN), Q_1, Q_3, and H (MAX), which we have calculated and analyzed before. Note that the MINITAB/PC can calculate this information very effectively. In Figure 4.7, 40 aptitude test scores are first entered into the PC. Then 10 statistics will automatically print if the command "MTB > describe C1" is entered. Two of those statistics, TRMEAN and SEMEAN, are not discussed in this book.

```
MTB > SET INTO C1
DATA> 20 21 23 25 30 35 36 39 40 41 42 43 43 46 48 50 51 52 54 55 56 58 59 61
DATA> 62 65 68 70 71 75 78 80 81 85 90 92 96 98 99 100
DATA> END
MTB > DESCRIBE C1

                    N      MEAN    MEDIAN    TRMEAN    STDEV    SEMEAN
GMCA/CL            40     58.45     55.50     58.28    22.99      3.63

                  MIN       MAX        Q1        Q3
GMCA/CL         20.00    100.00     41.25     77.25

MTB > PAPER
```

Figure 4.7

The MINITAB/PC Input and Printout of Some Important Statistics of 40 Aptitude Test Scores

Z Scores

A sample **Z Score,** which is based on the mean $\bar{x}$ and standard deviation s of a data set, is defined as

$$Z = \frac{x - \bar{x}}{s} \tag{4.14}$$

Like a percentile, a Z score expresses the relative position of any particular data value in terms of the number of standard deviations above or below the mean. Recall from Example 4.12 that Mr. Fleetdeer had a score of 92 on the test. For this score, $\bar{x} = 58.45$ and $s = 22.99$, as indicated in Table 4.6. His score of 92 is in the 90th percentile. The corresponding Z score is

$$Z = \frac{92 - 58.45}{22.99} = 1.46$$

This means that Mr. Fleetdeer's score of 92 is 1.46 standard deviations to the *right* of (above) the mean. Thus if Z is positive, it indicates how many standard deviations x is *above* the mean.

A negative value implies that x is to the left of (below) the mean. Look at Table 4.6 again. What is the Z score for the person who got a score of 30 on Johnson & Johnson's aptitude examination?

$$Z = \frac{30 - 58.45}{22.99} = -1.24$$

This individual's score is 1.24 standard deviations *below* the mean.

As a rule of thumb, for mound-shaped data sets, approximately 68 percent of the observations have a Z score between -1 and 1, and approximately 95 percent of the observations have a Z score between -2 and 2.[6]

Z scores of the aptitude test scores indicated in Table 4.6 are calculated and listed in Table 4.7. From Table 4.7, we find that 67.5 percent (27/40) of these observations have a Z score between -1 and 1. All of the observations have Z scores between -2 and 2.

Table 4.7 Z Scores of Aptitude Test Scores for 40 Applicants

i	x_i	Z_i	i	x_i	Z_i
1	20	-1.6728	21	56	$-.1066$
2	21	-1.6293	22	58	$-.0196$
3	23	-1.5422	23	59	.0239
4	25	-1.4552	24	61	.1109
5	30	-1.2377	25	62	.1544
6	35	-1.0202	26	65	.2850
7	36	$-.9767$	27	68	.4155
8	39	$-.8462$	28	70	.5025
9	40	$-.8027$	29	71	.5460
10	41	$-.7592$	30	75	.7200
11	42	$-.7157$	31	78	.8505
12	43	$-.6721$	32	80	.9375
13	43	$-.6721$	33	81	.9810
14	46	$-.5416$	34	85	1.1551
					(continued)

[6] Z scores and their application will be explored further in Chapter 7. This rule of thumb is derived from **Tchebysheff's theorem,** defined as follows: In any data set the proportion of items within $\pm k$ standardized deviations of the mean is at least $1 - (1/k)^2$, where k is any number greater than 1.0.

Table 4.7 (Continued)

i	x_i	Z_i	i	x_i	Z_i
15	48	$-.4546$	35	90	1.3726
16	50	$-.3676$	36	92	1.4596
17	51	$-.3241$	37	96	1.6336
18	52	$-.2806$	38	98	1.7206
19	54	$-.1936$	39	99	1.7641
20	55	$-.1501$	40	100	1.8076

4.5   MEASURES OF SHAPE

A basic question in many applications is whether data exhibit a symmetric pattern. Skewness and kurtosis are two important characteristics that determine the shape of a distribution.

Skewness

In addition to measures of central tendency and dispersion, there are measures that give information on the skewness of the distribution. The **skewness** indicates whether the distribution is skewed to the left or right in relation to the mean or is symmetric about the mean. The population skewness for *raw* data is given by

$$\mu_3 = \frac{\sum_{i=1}^{N} (x_i - \mu)^3}{N} \tag{4.15}$$

We can scale the result by dividing μ_3 by σ^3. This gives us the **coefficient of skewness** (CS):

$$CS = \frac{\mu_3}{\sigma^3} \tag{4.16}$$

The estimate of μ_3 for a sample can be defined as

$$\text{Skewness} = \sum_{i=1}^{n} (x_i - \bar{x})^3 / n \tag{4.15a}$$

The sample coefficient of skewness (SCS) can be defined as

$$SCS = \frac{\sum_{i=1}^{n} (x_i - \bar{x})^3 / n}{s^3} \tag{4.16a}$$

An alternative measure of skewness is given by the **Pearson coefficient,** which is defined as:

Pearson coefficient = 3(mean − median)/(standard deviation)   (4.16b)

Returning to our quality control data, we can calculate the skewness as follows (recall that $\bar{x} = 7.5$ and $s = 4.8$).

x	$(x - \bar{x})^3$
3	−91.13
5	−15.63
8	.13
14	274.63
Total 30	168

Substituting $\Sigma(x_i - \bar{x})^3 = 168$, $n = 4$, and $s = 4.8$ into Equations 4.15a and 4.16a, we obtain the skewness and the sample coefficient of skewness as

$$\text{Skewness} = \frac{168}{4} = 42$$

$$\text{SCS} = \frac{42}{(4.8)^3} = .38$$

This implies that the quality control data are skewed to the right.

A **zero skewness coefficient** means that the distribution is symmetric with mean = median (see Figure 4.8), which is also equal to the mode if the distribution is unimodal.

A **positive skewness coefficient** means that the distribution is skewed to the right, or positively skewed, and that the mode (most observations) and median lie below the mean (see Figure 4.9). A **negative skewness coefficient** means that the distribution is skewed to the left, or negatively skewed, and that the mode and median lie above the mean (see Figure 4.10).

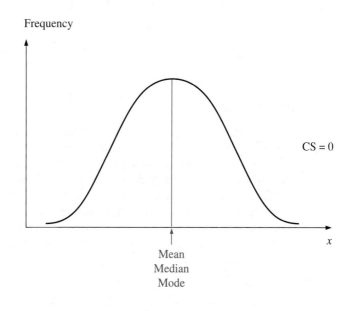

Figure 4.8

Symmetric
Distribution

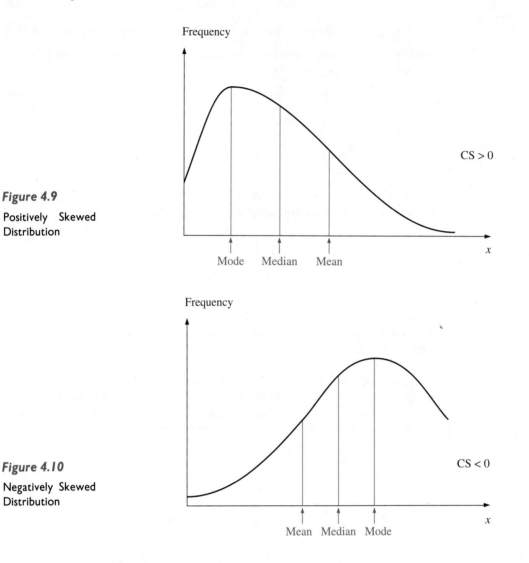

Figure 4.9

Positively Skewed
Distribution

Figure 4.10

Negatively Skewed
Distribution

Kurtosis

Skewness reflects the tendency of a distribution to be stretched out in a particular
direction. Another measure of shape, referred to as **kurtosis,** measures the peakedness of a distribution. In principle, the kurtosis value is small if the frequency of
observations close to the mean is high and the frequency of observations far from
the mean is low. Because kurtosis is not so frequently used as other numerical summary measures, it is not pursued in further detail here. However, the numerical
calculation of kurtosis will be discussed in Chapter 9. Both skewness and kurtosis
measures are useful in analyzing stock rates of return (see Chapter 9).

4.6   CALCULATING CERTAIN SUMMARY MEASURES FROM GROUPED DATA (Optional)

In this section, we discuss how to calculate mean (arithmetic average), median, mode, variance, standard deviation, and percentiles for grouped data.

The Mean

Sometimes data are in grouped form, so calculating the mean from raw data is impossible. Recall from Chapter 3 that raw data are sometimes grouped into classes. For example, a teacher might group exam scores into A's (90–100), B's (80–89), and so on. In cases such as this, we can estimate the mean from the grouped data by multiplying the midpoint of each class by the number of observations and dividing by the total number of observations (N):

$$\mu = \sum_{i=1}^{k} f_i m_i / N \tag{4.17}$$

where f_i = the frequency or number of observations in the ith group, m_i = the midpoint of the ith group, and k = the number of groups. Note that

$$\sum_{i=1}^{k} f_i = N$$

Although this is the formula for estimating the population mean from grouped data, we estimate the sample mean in the same manner by substituting n for N.

EXAMPLE 4.14   *Finding the Mean of Market Rates of Return in Terms of Grouped Data*

Table 4.8 presents a frequency distribution for the rate of return on the S&P Composite Stock Index from 1970 to 1990; it has seven classes or groups. Suppose we

Table 4.8 Frequency Distribution of Annual Market Rates of Return

Class	Midpoint (m_i)	Class Frequency (f_i)	$m_i f_i$
$-.30$ to $-.20$	$-.25$	1	$-.25$
$-.19$ to $-.10$	$-.145$	3	$-.435$
$-.09$ to $.00$	$-.045$	2	$-.09$
01 to $.10$	$.055$	3	$.165$
$.11$ to $.20$	$.155$	8	1.24
$.21$ to $.30$	$.255$	3	$.765$
$.31$ to $.40$	$.355$	1	$.355$

$$\sum_{i=1}^{n} m_i f_i = 1.750$$

do *not* have access to the raw data that underlie this frequency distribution (which, however, are shown in Table 4.9). Will our calculated group mean be reasonably

Table 4.9 Annual Market Rates of Return in Terms of S&P 500 (1970–1990)

Year	Rate of Return
70	.0010
71	.1080
72	.1557
73	−.1737
74	−.2964
75	.3149
76	.1918
77	−.1153
78	.0105
79	.1228
80	.2586
81	−.0994
82	.1549
83	.1706
84	.0115
85	.2633
86	.1462
87	.0203
88	.1240
89	.2725
90	−.0656
Sum	1.5760

close to the true mean? In this example,

$$\sum_{i=1}^{7} f_i = 21.$$

Following Equation 4.17, we calculate the group mean as

$$\sum_{i=1}^{7} f_i m_i / 21 = 1.750/21 = .0833.$$

The actual mean of the raw data (see Table 4.9) is $1.5760/21 = .0750$. The outcome of our test suggests that the mean of the grouped data is a fairly accurate measure of the true mean of the series.

The Median

Although a median can also be calculated for grouped data, it may be impossible to determine the exact value of the median if individual data values are not available. However, we can approximate the median value by first assuming that data fall equally throughout the median class. For example, suppose we have 5 students who scored between 90 and 100 on the exam. By assuming that the observations are equally spaced, we can hazard an educated guess of the 5 students' scores. Because the width of this class is 10, and because there are 5 students in the class, an assumption of equal spacing means each pair of "adjacent" scores should be separated by 2 points. Making this assumption for all the classes enables us to find what is *approximately* the median or middle score.

To obtain the median for grouped data, find the class in which the median observation appears. Then apply the following formula to estimate the median:

$$m = L + \frac{(N/2 - F)}{f}(U - L) \qquad (4.18)$$

where L and U are the lower and upper boundaries, respectively, of the class that contains the median; f is the frequency in this class; and F is the cumulative frequency of the observations in the classes prior to this class.

EXAMPLE 4.15   *Finding the Median of Stock Rates of Return in Terms of Grouped Data*

Referring to the grouped data in Example 4.14, we know that the median is the 10.5th observation. Thus we know that the median is in the .11 to .20 class. F is equal to 9, because 9 observations occurred before the class; f is equal to 8, because there are 8 observations in the class; and the lower (L) and upper (U) boundaries of the class are .105 and .205, respectively. Hence the median estimate is

$$m = .110 + \frac{(21/2 - 9)}{8}(.200 - .110) = .1269$$

The median for the raw data is .1240, which is similar to that calculated from grouped data.

The Mode

For nongrouped data, the mode of a set of observations is the value that occurs the most times; for grouped data, the modal class is the one with the highest frequency. Like nongrouped data, grouped data can have more than one class as modal classes.

In Example 4.14 on market rates of return, the modal class, .11 to .20, contains 8 observations.

Variance and Standard Deviation

Note that both of the variance formulas yield the same answer.

We can calculate the standard deviation and variance for *grouped* data by using the following formulas:

<table>
<tr><td align="center">Population Variance</td><td align="center">Population Standard Deviation</td></tr>
</table>

$$\sigma^2 = \frac{\sum_{i=1}^{k} f_i(m_i - \mu)^2}{N} \qquad (4.19)$$

$$\sigma = \sqrt{\frac{\sum_{i=1}^{k} f_i(m_i - \mu)^2}{N}} \qquad (4.20)$$

<table>
<tr><td align="center">Sample Variance</td><td align="center">Sample Standard Deviation</td></tr>
</table>

$$s^2 = \frac{\sum_{i=1}^{k} f_i(m_i - \bar{x})^2}{n - 1} \qquad (4.21)$$

$$s = \sqrt{\frac{\sum_{i=1}^{k} f_i(m_i - \bar{x})^2}{n - 1}} \qquad (4.22)$$

where f_i = frequency or number of obervations in the ith group

m_i = midpoint of the ith group

k = number of groups

The short-cut formulas found in Appendix 4B can be used to arrive at the same answer.

EXAMPLE **4.16**   *Analyzing a GNP Forecast*

Suppose we want to calculate the mean, variance, and standard deviation for a sample of forecasts for next year's GNP growth rate. We record our data in Table 4.10.

Using the short-cut formulas of Equation 4.21a, we can calculate the sample standard deviation as follows:

Table 4.10 Forecasts of GNP Growth Rate

Forecast Class	Class Midpoint (m_i)	Frequency (f_i)	$f_i m_i$	m_i^2	$f_i m_i^2$
−2–0%	−1	4	−4	1	4
0–2%	1	12	12	1	12
2–4%	3	8	24	9	72
4–6%	5	6	30	25	150
		30	62		238

$$\bar{x} = \frac{-4 + 12 + 24 + 30}{30} = 62/30 = 2.07$$

$$s^2 = \frac{238 - [30 \times (2.07)^2]}{30 - 1} = \frac{109.45}{29} = 3.77$$

$$s = \sqrt{3.77} = 1.94$$

Percentiles

The calculation of a particular percentile boundary (B) for grouped data for percentiles is similar to that of the median for grouped data:

$$B = L + \frac{(pN - F)}{f}(U - L) \tag{4.23}$$

Here N is the number of observations, p is the percentile desired, and the product pN gives the corresponding observation. F is the number of observations up to the lower limit of the class that contains the observation, and f is the frequency of the class. L and U are the lower and upper boundaries, respectively.

EXAMPLE 4.17   *Playing with Percentiles for the Prime Rate*

Suppose we have the grouped data given in Table 4.11 for the prime interest rate for the past 31 years (1960–1990).

Table 4.11   Prime Rate, 1960–1990

Class (%)	Frequency	Cumulative Frequency
3.1 to 6	13	13
6.1 to 9	9	22
9.1 to 12	6	28
12.1 to 15	2	30
15.1 to 18	1	31

Source: *Economic Report of the President*, 1991.

To determine the 60th percentile boundary, we reason as follows: $p = .60$, and $.60 \times 31 = 18.6$. The 19th observation is in the 6.1 to 9 class. Thus $L = 6.10, f = 9, F = 13$, and $U = 9.00$. Our estimate of the 60th percentile boundary is therefore

$$6.10 + \frac{(.60)(31) - 13}{9}(9.00 - 6.10) = 7.90$$

By this estimate, 60 percent of the observations are below 7.90 percent.

EXAMPLE 4.18   *Examining the Skewness of ACT Scores*

Suppose the grouped data in Table 4.12 represent the ACT scores for a high school class. The population skewness for *grouped* data is

$$\mu_3 = \frac{\sum_{i=1}^{k} f_i(x_i - \mu)^3}{N} \qquad (4.24)$$

Table 4.12 ACT Scores

ACT	x	f	xf	$x - \mu$	$(x - \mu)^2$	$f(x - \mu)^2$	$(x - \mu)^3$	$f(x - \mu)^3$
14–18	16	8	128	−8.5	72.25	578.0	−614.13	−4913.00
19–23	21	34	714	−3.5	12.25	416.5	−42.88	1457.75
24–28	26	20	520	1.5	2.25	45.0	3.38	67.50
29–33	31	10	310	6.5	42.25	422.5	274.63	2746.25
34–38	36	8	288	11.5	132.25	1058.0	1520.88	12,167.00
		80	1960			2520		8610

Using the information listed in Table 4.12, we can calculate the summary statistics as follows:

$$\mu = \frac{1960}{80} = 24.5$$

$$\sigma^2 = \frac{2520}{80} = 31.5$$

$$\sigma = 5.61$$

$$\text{Skewness} = \mu_3 = \sum \frac{f(x - \mu)^3}{N} = \frac{8610}{80} = 107.63$$

$$\text{CS} = \frac{\mu_3}{\sigma^3} = \frac{107.63}{176.56} = .610$$

The skewness μ_3 is positive and equal to 107.63. The coefficient of skewness (.610) is positive, indicating that the distribution is skewed to the right. We can use the same formula (Equation 4.24) for a sample if we replace μ by $\bar{x}$ and N by n.

4.7 APPLICATIONS

In this section, we will demonstrate how measures of central tendency, dispersion, position, and skewness can be used to analyze sample market survey data, rates of return on a stock, and economic data.[7] First, a sample of survey data is used to show how statistical analysis can be applied in making an inventory decision. Second, these same concepts are used to examine the market rates of return for Ford and GM stock and for the stock market overall. The T-bill and prime interest rates are explored in the third application, and the fourth application involves the macro-economic variables GNP, personal consumption, and disposable income. Finally, in Appendix 4C, we return to the seven accounting ratios for the auto industry presented in Chapter 3 and analyze them in terms of mean, median, MAD, variance, standard deviation, and coefficient of variation, as well as percentiles and skewness.

APPLICATION 4.1   Statistical Analysis of a Survey

Suppose Jack Miller, a manager at A&P, wants to determine the average monthly purchase, per dwelling unit, of six-packs of soda in the central New Jersey area. This average monthly purchase information will help A&P establish an inventory policy.

Miller has hired you to conduct a survey and perform statistical analyses in accordance with simple random survey procedures. Let us see how you proceed. First you conduct a survey and assemble the results in Table 4.13. Then, using the random sample data, in Table 4.13, you calculate the related summary statistics and list them in Table 4.14. You have computed all the descriptive statistics discussed in this chapter except the geometric mean. (Because some of the values of x are equal to zero, it would make no sense to compute the geometric average, wherein the data are multiplied together.)

Your measures of central tendency—the mean, median, and mode—indicate where the center of the data is. In addition, because the mean is greater than the median, you know the data are positively skewed. The fact that the coefficient of skewness is positive confirms this.

The variance, standard deviation, and mean absolute deviation provides information on how the data are spread out around their average value.

Finally, you show percentiles for the data. The nth percentile reveals that n percent of the data will be below that value. For example, the 50th percentile has a value of 4, so 50 percent of the data will have a value of 4 or less. Likewise, because the 90th percentile has a value of 8, 90 percent of the data will have a value of 8 or less. The interquartile range is just the difference between the third and first quartiles.

Using the information you have provided, the manager can make better decisions about how many six-packs of soda to keep in inventory.

---

[7]Financial ratio analysis for three auto firms is carried out in Appendix 4C.

Table 4.13 Monthly Purchase of Six-Packs of Soda per Dwelling Unit in the Central New Jersey Area as of January 1991

i	x_i	i	x_i
1	8	21	9
2	4	22	8
3	4	23	1
4	9	24	4
5	3	25	6
6	3	26	5
7	1	27	4
8	2	28	2
9	0	29	1
10	4	30	0
11	2	31	8
12	3	32	7
13	5	33	5
14	7	34	6
15	10	35	4
16	6	36	3
17	5	37	2
18	7	38	1
19	3	39	0
20	2	40	8

Table 4.14 Summary Measures of Monthly Purchase of Six-Packs of Soda

Arithmetic mean	4.30
Median	4.00
Mode	4.00
Range	10.00
Standard deviation	2.77
Variance	7.65
Mean absolute deviation from mean	2.30
Mean absolute deviation from median	2.25
Coefficient of variation	.64
Coefficient of skewness	1.32
Percentiles	
10th	1.00
25th	2.00
50th	4.00
75th	6.00
90th	8.00
Interquartiles ranges	4.00

■ *APPLICATION 4.2* Stock Rates of Return for GM, Ford, and the Market

Central tendency and dispersion statistics can also be used to analyze the rates of return for GM and Ford stock, as well as for the general auto market.

The rates of return listed in Table 4.15 are calculated on a yearly basis. With these statistics, we can determine whether the stock rates of return for the two firms fluctuated more than the market. And we can determine whether the stocks have generally outperformed or underperformed the market over this period.

Much useful information can be obtained by merely perusing Table 4.15. The greatest gain for Ford occurred in 1982, when the price of its stock increased by 132 percent. Ford's worst year occurred in 1973, when its stock lost 45 percent of its value. The range of the three stock rates of return—the difference between the highest and lowest values—was 1.7724 (1.3212 + .4512) for Ford, 1.4103 (.9522 + .4581) for GM, and .6113 (.3149 + .2964) for the overall market. Note that the three tended to move together; when the market went up, the stocks also tended to rise, and vice versa. This does not always hold, though. In 1980 the market went up by over 25 percent and the two auto stocks went down. The relationship between the market rate of return and the rate of return for individual firms will be analyzed in Chapter 14 when simple regression analysis is discussed.

Examining the ranges alone makes it appear that the overall market was less volatile then the two stocks. However, recall that because the range takes into consideration only the highest and lowest observations, it is strongly influenced by outlying observations. Therefore, we must examine more sophisticated measures of dispersion, such as the standard deviation, MAD, and CV, to obtain a sense of the volatility of the observations. Ford had the highest standard deviation (.3923), followed by GM (.3325) and the market (.1605). The same rankings hold for the MAD. GM has the highest CV, followed by Ford and the market.

For the two auto firms, the mean return is greater than the median because a few extreme observations on the high side of the distribution are pushing up the mean. For example, Ford had a return of 132 percent in 1982, and GM enjoyed a return of 95 percent in 1975 and 68 percent in 1982. These observations affect the mean but do not influence the median. Thus the median is probably a more accurate indicator of central tendency. The opposite is true of the market, where the median is larger than the mean.

Investors would have preferred owning Ford stock to owning GM stock because of Ford's mean and median returns of 9.64 percent and .81 percent, respectively. When we compare GM to the market, we find that GM has a higher mean but a lower median. We also note that both GM's and Ford's means are higher than their medians, indicating that their returns are positively skewed, whereas the market's mean is below its median, indicating negative skewness. The results imply that stocks for both GM and Ford have more upside potential than that of the overall market. See Section 9.7 in Chapter 9 for further discussion on this implication.

Table 4.15 Rates of Return for GM, Ford, and S&P 500

Year	GM	Ford	S&P 500
1970	0.2137	0.4260	0.0010
1971	0.0422	0.2933	0.1080
1972	0.0631	0.1717	0.1557
1973	-0.3667	-0.4512	-0.1737
1974	-0.2597	-0.0968	-0.2964
1975	0.9522	0.3960	0.3149
1976	0.4584	0.4614	0.1918
1977	-0.1124	-0.2067	-0.1153
1978	-0.0498	-0.0026	0.0105
1979	0.0288	-0.1479	0.1228
1980	-0.0410	-0.2938	0.2586
1981	-0.0911	-0.1025	-0.0994
1982	0.6826	1.3212	0.1549
1983	0.2373	0.1286	0.1706
1984	0.1176	0.0980	0.0115
1985	-0.0383	0.3237	0.2633
1986	0.0088	0.0081	0.1462
1987	0.0058	0.3961	0.0203
1988	0.4418	-0.2995	0.1240
1989	-0.4581	-0.0767	0.2725
1990	-0.1154	-0.3209	-0.0656

	GM	Ford	S&P 500
MEAN	0.081903	0.096445	0.075049
MEDIAN	0.0088	0.0081	0.1228
STD	0.332507	0.392281	0.160488
CS.	0.864367	1.240275	-0.51004
CV	4.059450	4.067375	2.138428
MAD	0.240980	0.290616	0.131683

PERCENTILES

	GM	Ford	S&P 500
10th	-0.35599	-0.31878	-0.16789
25th	-0.10704	-0.19196	-0.04894
50th	0.007283	0.002714	0.115382
75th	0.189651	0.316101	0.166906
90th	0.456781	0.422978	0.262857

■ *APPLICATION 4.3* 3-Month Treasury Bill Rate and Prime Rate

Table 4.16 shows two key interest rates, the 3-month T-bill rate and the prime rate, for the period 1950–1990. As we have noted, a T-bill is a short-term debt instrument issued by the United States government. T-bills are backed by the full faith and credit of the U.S. government, which makes these investments the safest in the world and the closest thing to a risk-free asset. The prime rate is the rate that banks charge their best customers, such as large corporations. This rate may differ slightly from bank to bank.

As Table 4.16 shows, the T-bill rate fluctuated between 4 percent and 7 percent from 1966 to 1978. Then inflation pushed the rate to double digits from 1979 to 1982, and declining inflation caused the rate to drop after 1983. The trend for the prime rate is similar, although the prime is several percentage points higher.

The T-bill rate is lower largely because the T-bill is close to a risk-free asset, whereas the prime rate includes a risk premium. This relationship is illustrated statistically by the means and medians of the two rates. The mean for the T-bill rate from 1950 to 1990, for example, is 5.35 percent; it is 7.21 percent for the prime rate. For the same period, the median for the T-bill rate is 4.989 percent; it is 6.30 percent for the prime rate. The fact that the man is higher than the median indicates that a few extreme observations are pushing up the mean and that both distributions are positively skewed. The coefficient of skewness for the T-bill rate is .69, which is lower than that for the prime rate, .975.

Table 4.16 T-Bill Rate and Prime Rate (%), 1950–1990

Year	3-Month T-Bill Rate	Prime Rate
50	1.218	2.07
51	1.552	2.56
52	1.766	3.00
53	1.931	3.17
54	.953	3.05
55	1.753	3.16
56	2.658	3.77
57	3.267	4.20
58	1.839	3.83
59	3.405	4.48
60	2.928	4.82
61	2.378	4.50
62	2.778	4.50
63	3.157	4.50
64	3.549	4.50
65	3.954	4.54
66	4.881	5.63
67	4.321	5.61
68	5.339	6.30

(continued)

Table 4.16 (Continued)

Year	3-Month T-Bill Rate	Prime Rate
69	6.677	7.96
70	6.458	7.91
71	4.348	5.72
72	4.071	5.25
73	7.041	8.03
74	7.886	10.81
75	5.838	7.86
76	4.989	6.84
77	5.265	6.83
78	7.221	9.06
79	10.041	12.67
80	11.506	15.27
81	14.029	18.87
82	10.686	14.86
83	8.63	10.79
84	9.58	12.04
85	7.48	9.93
86	5.98	8.33
87	5.82	8.22
88	6.69	9.32
89	8.12	10.87
90	7.51	10.01
Mean	5.3535	7.2107
Median	4.989	6.30
Std Dev	3.0783	3.8383
CV	.5750	.5323
CS	.6869	.9753

The range for the T-bill over the period 1970 to 1990 is $14.03 - 1.22 = 12.81$; the range for the prime is $18.87 - 2.07 = 16.80$. The standard deviation for the prime (3.8383) is also higher than the T-bill standard deviation (3.0783). Finally, the coefficient of variation is lower for the prime rate. From these three descriptive statistics, we can conclude that the prime rate has varied more than the T-bill rate. One reason why the prime rate has fluctuated more than the 3-month T-bill rate during this 20-year period may be the fact that 3-month Treasury bills, in general, have shorter maturities than loans made at the prime rate. Because this time period was marked by high and extremely volatile interest rates (especially in the late seventies and early eighties), the prime rate may have been adjusted to reflect the added risk associated with longer-term loans. In addition, because T-bills are marketable securities, the rate they return is determined by the market. Loans made at the prime rate, in contrast, are not marketable, so the prime rate is adjusted (by bankers) only periodically. This may make it more volatile.

■ | *APPLICATION 4.4* GNP, Personal Consumption, and Disposable Income

Here we will examine accounting data on national income. The GNP is one of the most popular economic indicators because it measures the market value of all final goods and services produced in the United States within a given time period. GNP is a key indicator of the health of the economy: the occurrence of consecutive quarters of decline in real (inflation-adjusted) GNP is sometimes used to define a recession. The GNP is calculated by adding personal consumption expenditures, gross private domestic investment, government purchases, and net exports. Disposable income is the amount of after-tax income that individuals have available to spend.

Data on GNP, personal consumption, and disposable income in constant 1982 dollars for the period 1950–1990 are shown in Table 4.17. Clearly, there is a tendency for the indicators to increase steadily. Personal consumption declined in only two years (1973–1974 and 1979–1980). Disposable income declined in only 1973–1974, and GNP declined in seven years (1953–1954, 1957–1958, 1969–1970, 1973–1974, 1974–1975, 1979–1980, and 1981–1982). The mean for GNP is 2500.1 billion dollars, the median 2423.3. Because the mean is greater than the median, the distribution is positively skewed. The mean is also greater than the median for consumption. However, there is a large difference between the two measures of central tendency for GNP, indicating that there is a great deal of skewness and that, therefore, the distribution is not symmetric. In the case of consumption, too, there is a great deal of difference between the mean and median, indicating that this distribution is also not symmetric. For disposable income, the mean is once again larger than the median, which results in a skewness coefficient that is positive. As in the two previous cases, we conclude that the distribution is not symmetric.

Table 4.17 GNP, Personal Consumption, and Disposable Income, 1950–1990

Year	GNP	Personal Consumption	Disposable Income
50	1203.7	733.2	791.8
51	1328.2	748.7	819.0
52	1380.0	771.4	844.3
53	1435.3	802.5	880.0
54	1416.2	822.7	894.0
55	1494.9	873.8	944.5
56	1525.6	899.8	989.4
57	1551.1	919.7	1012.1
58	1539.2	932.9	1028.8
59	1629.1	979.4	1067.2
60	1665.3	1005.1	1091.1
61	1708.7	1025.2	1123.2
62	1799.4	1069.0	1170.2
63	1873.3	1108.4	1207.3
64	1973.3	1170.6	1291.0
65	2087.6	1236.4	1365.7

(*continued*)

Table 4.17 (Continued)

Year	GNP	Personal Consumption	Disposable Income
66	2208.3	1298.9	1431.3
67	2271.4	1337.7	1493.2
68	2365.6	1405.9	1551.3
69	2423.3	1456.7	1599.8
70	2416.2	1492.0	1668.1
71	2484.8	1538.8	1728.4
72	2608.5	1621.9	1797.4
73	2744.1	1689.6	1916.3
74	2729.3	1674.0	1896.6
75	2695.0	1711.9	1931.7
76	2826.7	1803.9	2001.0
77	2958.6	1883.8	2066.6
78	3115.2	1961.0	2167.4
79	3192.4	2004.4	2212.6
80	3187.1	2000.4	2214.3
81	3248.8	2024.2	2248.6
82	3166.0	2050.7	2261.5
83	3279.1	2146.0	2331.9
84	3501.4	2249.3	2469.8
85	3618.7	2354.8	2542.8
86	3717.9	2446.4	2635.3
87	3845.3	2515.4	2670.7
88	4016.9	2606.5	2800.5
89	4117.7	2656.8	2869.0
90	4155.8	2682.2	2893.3
Median	2423.3	1492.0	1668.1
Mean	2500.1	1554.0	1705.3
Std Dev	874.7	602.4	651
CS	.2625	.3182	.2287
CV	.3499	.3876	.3818

The standard deviation for GNP is higher than the standard deviation for consumption. However, the CV is lower for GNP. The CV is probably a better measure of variability than the standard deviation, because the level of GNP is always higher than consumption. The standard deviation for disposable income is lower than the standard deviation for GNP and greater than the standard deviation for consumption. However, the CV of disposable income is lower than the CV for consumption and greater than the CV for GNP. Thus we must use the CV when comparing the variability of data that are different in range of values. Because GNP is greater than consumption and disposable income, it usually has a greater variance. But using the CV makes it possible to compare the dispersion because the dispersion is standardized.

Summary

In this chapter, we showed how a series of data can be described by using only a few summary statistics.

1. Measures of central tendency such as the mean, median, and mode provide information on the center of the distribution.
2. Measures of dispersion such as the variance, standard deviation, mean absolute deviation, and coefficient of variation provide information on how spread out the data are.
3. Measures such as percentiles, quartiles, interquartiles, and Z scores provide information on the relative position of a data set.
4. Shape measures such as skewness are used to measure the degree of a distribution's asymmetry.

In the next chapter, we introduce the concepts of probability that are required for making statistical inference. However, we will continue to use descriptive statistics such as the mean and variance to describe the central tendency and the dispersion of a distribution.

Appendix 4A Short-Cut Formulas for Calculating Variance and Standard Deviation

Population Variance

$$\sigma^2 = \frac{\sum_{i=1}^{N} x_i^2}{N} - \mu^2 \qquad (4.5a)$$

Population Standard Deviation

$$\sigma = \sqrt{\frac{\sum_{i=1}^{N} x_i^2}{N} - \mu^2} \qquad (4.6a)$$

Sample Variance

$$s^2 = \frac{\sum_{i=1}^{n} x_i^2 - n\bar{x}^2}{n - 1} \qquad (4.7a)$$

Sample Standard Deviation

$$s = \sqrt{\frac{\sum_{i=1}^{n} x_i^2 - n\bar{x}^2}{n - 1}} \qquad (4.8a)$$

where N = number of observations in the population
  n = number of observations in the sample
  $\bar{x}$ = sample mean of x
  μ = population mean of x

Appendix 4B Short-Cut Formulas for Calculating Group Variance and Standard Deviation

Population Variance

$$\sigma^2 = \frac{\sum_{i=1}^{k} f_i m_i^2}{N} - \mu^2 \qquad (4.19a)$$

Population Standard Deviation

$$\sigma = \sqrt{\frac{\sum_{i=1}^{k} f_i m_i^2}{N} - \mu^2} \qquad (4.20a)$$

Sample Variance

$$s^2 = \frac{\sum_{i=1}^{k} f_i m_i^2}{n-1} - \frac{n\bar{x}^2}{n-1} \qquad (4.21a)$$

Sample Standard Deviation

$$s = \sqrt{\frac{\sum_{i=1}^{k} f_i m_i^2}{n-1} - \frac{n\bar{x}^2}{n-1}} \qquad (4.22a)$$

where f_i = frequency, or number of observations in the ith group
m_i = midpoint of the ith group
k = number of groups

Appendix 4C Financial Ratio Analysis for Three Auto Firms

Summary statistics for seven accounting ratios for the three U.S. auto manufacturers over the time period 1966–1990 are presented in Table 4C.1. The mean and median enable us to compare the central tendencies of the various ratios. Recall that the mean is calculated by adding all the observations in the sample and divid-

Table 4C.1 Statistics for Selected Financial Ratios of Three Auto Companies

	CA/CL	Inventory Turnover	TD/TA	NI/SAL	NI/TA	P/E Ratio	Payout Ratio
				Chrysler			
MAD	.207	2.131	.122	.034	.055	16.622	.565
Mean	1.321	7.943	.672	.006	.010	13.908	.139
Median	1.372	7.113	.619	.013	.019	4.771	.108
STD	.255	2.556	.140	.054	.085	35.274	1.204
Variance	.065	6.533	.020	.003	.007	1244.284	1.450
Skewness	−.005	10.064	.001	−.000287	−.00075	167486.163	−1.86
CV	.193	.322	.209	8.878	8.279	2.536	8.664
CS	−.283	.603	.428	−1.821	−1.230	3.816	−.107
Percentiles							
10th	.956	4.924	.527	−.068	−.121	−2.387	−.881
25th	1.087	5.910	.552	−.010	−.018	.283	.000
50th	1.372	7.113	.619	.013	.019	4.771	.108
75th	1.494	10.273	.843	.033	.048	10.468	.339
90th	1.570	11.695	.855	.053	.082	17.991	.664

Table 4C.1 (Continued)

	CA/CL	Inventory Turnover	TD/TA	NI/SAL	NI/TA	P/E Ratio	Payout Ratio
			Ford				
MAD	.155	3.090	.114	.021	.034	6.899	.366
Mean	1.232	9.135	.599	.030	.044	8.283	.453
Median	1.281	7.385	.561	.039	.059	4.700	.314
STD	.183	4.276	.142	.027	.043	13.810	.657
Variance	.033	18.284	.020	.001	.002	190.716	.432
Skewness	−.002	146.629	.002	−.000023	−.000076	8971.861	.790
CV	.149	.468	.237	.899	.970	1.667	1.450
CS	−.359	1.876	.579	−1.159	−.960	3.407	2.783
Percentiles							
10th	.951	5.790	.428	−.023	−.038	−1.730	−.068
25th	1.061	6.418	.476	.012	.018	3.244	.176
50th	1.281	7.385	.561	.039	.059	4.700	.314
75th	1.373	10.987	.681	.047	.076	9.274	.439
90th	1.387	12.442	.846	.054	.084	12.255	.790
			General Motors				
MAD	.337	2.471	.104	.021	.038	6.666	.340
Mean	1.726	8.880	.460	.045	.076	9.728	.558
Median	1.757	7.983	.412	.044	.082	7.788	.556
STD	.407	2.915	.119	.027	.043	10.718	.615
Variance	.166	8.499	.014	.001	.002	114.878	.378
Skewness	−.005	12.396	.001	−.000012	−.000027	838.685	−.029
CV	.236	.328	.260	.609	.570	1.102	1.103
CS	−.080	.500	.312	−.609	−.342	.681	−.125
Percentiles							
10th	1.114	5.582	.305	−.004	.016	.216	−.248
25th	1.363	6.546	.366	.031	.042	5.436	.398
50th	1.757	7.983	.412	.044	.082	7.788	.556
75th	2.004	11.463	.549	.068	.118	12.270	.716
90th	2.231	13.312	.613	.074	.124	17.337	.909

Source: *Moody's Industrial Manual,* 1991.

ing by the number of observations in the sample (here, 25). That is, $\bar{x} = \Sigma_{i=1}^{25} x_i/25$. For the current ratio, for example, the ratios for all the 25 years are added and divided by 25. The means for the other ratios are calculated in a similar manner.

The second measure of central tendency presented is the median, the middle observation of an ordered set of observations. In this example there are 25 observations, so the median is the mean of the 10th and 11th observations. As noted previously, the median is not affected by extreme observations, because it takes into consideration only the middle observations. In contrast, the mean uses all the observations. The mean and the median are usually close if the data contain no outliers, but if there *are* extreme observations, the mean and median may differ substantially.

The measures of variability presented are MAD, variance, and standard deviation (STD). The variance measures the average squared deviation from the mean, and the standard deviation is the square root of the variance. The MAD is the average absolute deviation from the mean. The difference between the variance and the MAD is that we square the deviations from the mean when calculating the variance and use the absolute value of the deviations when calculating the MAD.

The boundaries for the 10th, 25th, 50th, 75th, and 90th percentiles are presented to suggest the rank of the data. Recall that the percentiles indicate the percentages of observations below a certain score. For example, the boundary for the 25th percentile for GM's current ratio (CA/CL) is 1.363. This means that 25 percent of GM's current ratios were below 1.363.

The skewness coefficient enables us to measure the shape of the distribution. A positive coefficient means that the distribution is skewed to the right, and a negative value indicates left skewness.

These statistics can be used to compare the performance of one firm to that of other firms. A high mean and a high median for the current ratio, net profit margin, and return on total assets indicate that the firm is performing well. A high current ratio (current assets divided by current liabilities) means that the firm has enough liquidity to meet current obligations, such as accounts payable and wages payable. The net profit margin measures the percentage of each dollar that goes to profits. High profit margins are a sign of efficiency. The return on total assets is another measure of efficiency; specifically, it measures how effectively the total assets are being utilized.

A high inventory turnover ratio is good up to a point, but very high ratios indicate that the firm may be running out of certain items in stock and losing sales to competition. Thus inventory levels must be reasonable for the firm to maintain profitability. Generally, a high P/E ratio is also good because it means that investors believe that the firm has good growth opportunities; however, a firm may have a very high P/E ratio (approaching infinity) because of low earnings.

The total debt/total assets ratio (TD/TA in Table 4C.1) is a measure of the relative amount of debt the firm has assumed. A high ratio indicates that the firm may be taking on too much debt and could be a credit risk.

The debt that a firm takes on is related to financial activity. In periods of recession, the firm is unlikely to invest in new plant and equipment, because if it did so, its existing plants and equipment would be underutilized. The firm is more likely to incur debt and invest in new projects when the economy is expanding and its plant is operating at full capacity.

Current Ratio

To determine which of the four firms was in the best liquidity position during the 25-year period, let us examine the mean and median of each firm's current ratio. GM had the highest current ratios, with a mean of 1.726 and a median of 1.757. Figure 2C.1 in Appendix 2C shows the current ratio of GM over time. During this period, GM's current ratios were close to 2.000 in the early years, but the ratio declined in the 1980s, reaching a low of 1.088 in 1985. GM's current ratio increased after 1986. Ford and Chrysler had similar measures of central tendency, in the 1.200–1.300 range. Like GM, the other auto firms experienced low current ratios in the first half of the 1980s, reflecting the deep recession of 1980 and 1981–1982.

To examine the variability of the current ratio, let us look at the variance, standard deviation, coefficient of variation, and MAD. Ideally, firms like to have a high current ratio with low variability so that managers can plan from year to year. Ford had the lowest variance, .033, followed by Chrysler and GM. The MAD follows the same pattern. However, it would be a mistake to conclude that Ford's current ratio is less volatile merely on the basis of the variance and the MAD, because Ford may have a lower mean and therefore a lower variance.

To eliminate this scaling problem, we use the coefficient of variation, which is calculated by dividing the standard deviation by the mean. By this standard, Ford had the lowest variability, .149, again followed by Chrysler and GM. On the basis of the central tendency and dispersion statistics, we can conclude that GM has experienced higher current ratios but also greater variability. The higher ratios are beneficial, but managers do not welcome the greater dispersion because it makes planning difficult.

The skewness coefficient can give useful information on the shape of the distribution. The coefficient of skewness (CS) for Chrysler, Ford, and GM are $-.283$, $-.359$, and $-.080$, respectively.

The percentiles give the rank of the data. Again, recall that the values listed for each percentile show the percentage of observations *below* that value. For example, the 10th percentile for Chrysler's current ratio is .956, indicating that 10 percent of the current ratios for this 20-year period lie below .956. Likewise, the 10th percentile for Ford's current ratio is .951. By comparing the percentiles of Chrysler and Ford's current ratios, we can get a feeling for how the current ratios of these two companies have fluctuated over time. With the highest values for each of the percentiles, GM enjoyed superior ratios. Chrysler and Ford had similar current ratios up to the 25th percentile; then Chrysler had the edge. GM had higher values than both Ford and Chrysler for all percentiles.

Inventory Turnover Ratio

The second ratio examined is the inventory turnover ratio, which is calculated by dividing the cost of goods sold by the average value of inventory. A high inventory turnover ratio is a sign of efficiency. All the firms have similar ratios, though Ford has the highest mean (9.135) and Chrysler the lowest (7.943). The medians are similar to the means for each of the firms, so the distributions exhibit very little skewness. From these measures of central tendency, we can conclude that although Ford has been slightly more efficient than its competitors, the differences among auto firms with respect to inventory turnover have been small.

Chrysler had the lowest variance, standard deviation, MAD, and coefficient of variation for this ratio, indicating that its inventory turnover has been less volatile than that of the other firms. A stable inventory turnover ratio makes it easier for managers to plan inventory levels. Chrysler performed the worst in terms of level. However, it has the smallest volatility of inventory turnover. All 3 companies have a positive coefficient of skewness.

Total Debt/Total Asset Ratio

Let's look more closely at the total debt/total assets ratio (TD/TA). Chrysler relied heavily on debt over the 25-year time period (the mean and median ratios are above .60). All firms increased debt substantially in the 1980s. Chrysler had major financial difficulties in the late 1970s and early 1980s and relied heavily on debt. Ford was in better shape with a mean ratio of .599, and GM, with a mean ratio of .460, enjoyed the lowest debt ratios in the industry. Because GM relied less on debt to finance operations than did the other firms, it has a lower credit risk.

GM's debt ratio has fluctuated less than its competitors', with a variance of .014. However, it has the highest coefficient of variation at .260. This implies that GM's debt ratio has greater variability, even after we adjust for the mean of each company's debt ratio. GM is lower than all percentiles for both Chrysler and Ford. GM's superior position can be seen in the percentiles, because the firm has the lowest values in all the percentiles presented. All 3 companies have a positive coefficient of skewness.

Net Profit Margin

The net profit margin is one of the most important ratios for managers and shareholders, because it reveals what percentage of each sales dollar goes to profits. This ratio, calculated by dividing net income by sales (NI/SAL in Table 4C.1), is one measure of a firm's overall profitability. By this measure of profitability, GM has been operating much more profitably than the other firms. It has a mean profit ratio of .045 and a median measure of .044. This means that about 4.5 percent of GM's income winds up as profits. As can be seen in Figure 2C.5 in Appendix 2C, GM suffered two negative net profit margins, in 1980 and 1990. The highest ratio was .089 in 1966, and the low was −.013 in 1980. A negative profit margin indicates that the firm suffered losses.

Ford comes in second place with a mean net profit margin of .030 and a median of .039. By examining the data, you can see that the values for Ford have tended to be lower and that the firm suffered losses from 1980 to 1982. Chrysler's mean and median for the 25-year period (.006 and .013, respectively) reflect the fact that Chrysler lost money in 8 of 25 of those years, particularly during the 1980s.

GM also had the lowest variability as measured by the variance, MAD, and coefficient of variation. The coefficient of variation for Chrysler is 8.878.

All of the firms have a negative skewness coefficient. GM's is negative at $-.609$. The rest of the firms have distributions skewed even more to the left.

Return on Total Assets

The return on total assets is important to managers because it indicates how effectively the firm is using assets. This ratio is calculated by dividing net income by total assets (NI/TA). The higher the ratio, the greater the income generated from each dollar of total assets. As in the case of the net profit ratio, GM outpaced its competitors with a mean of .076 and a median of .082, followed by Ford's mean and median of .044 and .059, respectively. Chrysler's mean and median were .010 and .019. GM and Ford have the lowest variance at .002. GM's CV is .570. All 3 companies have a negative coefficient of skewness.

Price/Earnings (P/E) Ratios

Let us turn now to the price/earnings ratio. For Chrysler there is a great discrepancy between the mean and the median P/E ratio. The mean for Chrysler is 13.908, the median 4.771. The reason for this difference is that some extreme observations on the high side for this firm boosted its mean. Chrysler's P/E ratio in 1984 was 87.5. Although these high ratios pushed up the mean, they had no effect on the median. Here, as in all cases where there are outlying observations, the median is the better measure of central tendency. Because Ford and GM did not have so many outliers, their mean and median P/Es are similar.

The wide dispersion in Chrysler's P/E ratios is reflected in a large variance, CV, and MAD. Chrysler's dispersion ratios are higher than those of the other firms; that is, this firm's P/E ratio has shown the greatest fluctuation. The skewness coefficient for the three firms is positive, indicating that most of the observations occurred on the left side of the distribution and that the median is smaller than the mean.

Payout Ratio

The last ratio presented is the payout ratio, which reveals what percentage of the firms' earnings are paid out in dividends. This ratio is calculated by dividing dividends per share by earnings per share. For obvious reasons, investors pay a great deal of attention to this ratio. GM's investors enjoyed the highest payout ratio, with a mean of .558 and a median of .556. In other words GM paid out, on average, about 56 percent of earnings in dividends. Ford had the second-highest payout ratio, with a mean of .453 and a median of .314. Chrysler's mean and median were .139 and .108, respectively. Both Chrysler and GM have a negative coefficient of skewness; Ford, however, has a positive coefficient of skewness.

From the foregoing discussion, we can conclude that of the three auto companies, GM was in the best financial condition for the period 1966–1990. GM had the highest mean current ratio, net profit margin, and return on total assets. In addition, it relied less on debt than the other firms. Ford was second in terms of net profit ratio and return on total assets. Chrysler, on the other hand, had major problems during this period.

Questions and Problems

1. The midterm scores from an honors seminar in accounting are 25, 84, 82, 83, 90, 91, 99, 100, and 100. Find the mean, median, and mode. Is one measure preferable to another? Why or why not?

2. What is your mean speed if you drive 35 miles per hour for 2 hours and 55 miles per hour for 3 hours?

3. What are descriptive statistics? Why are they important? Give some examples of descriptive statistics.

4. The following sample annual starting salaries were offered to 12 college seniors in 1992.

$21,400	$15,600	$16,500	$24,200
22,300	20,000	17,000	21,750
18,750	19,250	14,900	15,750

   a. Calculate the mean and median for these observations.
   b. Calculate the variance and standard deviation for these observations.
   c. Use MINITAB to construct a box and whisker plot and explain the result.

5. A $250 suit is on sale for $190, and a $90 pair of shoes is on sale for $65. Find the average percent decrease in price for the 2 items.

6. The following are the average daily reported share volumes traded on the NYSE, in thousands, for the years listed.

1971	15,381	1981	46,882
1972	16,487	1982	64,859
1973	16,084	1983	85,336
1974	13,904	1984	91,229
1975	18,551	1985	109,132
1976	21,186	1986	141,489
1977	20,928	1987	188,796
1978	28,591	1988	161,509
1979	32,233	1989	165,568
1980	44,867	1990	156,777

   a. Calculate the mean and median share volume for these observations.
   b. Calculate the variance and standard deviation for these observations.

7. The following data are annual rates of return on the DJIA and the S&P 500.

	DJIA	**S&P 500**
1960	−9.34	−2.97
1961	18.71	23.13
1962	−10.91	−11.81
1963	17.12	18.89
1964	14.57	12.97
1965	10.88	9.06
1966	−18.94	−13.09
1967	15.20	20.09
1968	5.24	7.66
1969	−15.19	−11.36
1970	4.82	.10
1971	6.11	10.80
1972	14.58	15.57
1973	−16.58	−17.37
1974	−27.57	−29.64
1975	38.34	31.49
1976	17.86	19.18
1977	17.27	−11.53
1978	−3.15	1.05
1979	4.19	12.28
1980	5.57	25.86
1981	4.66	−9.94
1982	−5.21	15.49
1983	34.60	17.06
1984	−1.00	1.15
1985	12.71	26.33
1986	34.97	14.62
1987	26.95	2.03
1988	−9.45	12.40
1989	21.74	27.25
1990	6.78	−6.56

a. Calculate the arithmetic mean and standard deviation of the DJIA and the S&P 500 for the years 1960–1979, 1970–1989, and 1981–1990.

b. Calculate the geometric mean for these same years.

8. A sample of 20 workers in a small company earned the following weekly wages:

$175, 175, 182, 175, 175, 200, 250, 225, 250, 200, 195, 200, 200, 190, 325, 300, 310, 325, 400, 225

a. Calculate the mean and standard deviation.

b. Calculate the mode.

c. Calculate the median.

9. You are given the following information about stock A and stock B.

State of World Next Year	Chance of Occurrence	Returns Next Year A	B
Recession	.30	10%	9.8%
Normal growth	.40	11%	11.2%
Inflation	.30	12%	13.0%

a. Calculate the mean, standard deviation, and coefficient of variation for each stock.

b. If you could purchase only one stock, which would you choose? Why?

10. Consider the following annual data on profit rates for Cherry Computers, Lemon Motors, and Orange Electronics.

Year	Cherry Computers	Lemon Motors	Orange Electronics
1983	14.2	−6.2	37.5
1984	12.3	13.3	−10.6
1985	−16.2	−8.4	40.3
1986	15.4	27.3	5.4
1987	17.2	28.2	6.2
1988	10.3	14.5	10.2
1989	−6.3	−2.4	13.8
1990	−7.8	−3.1	11.5
1991	3.4	15.6	−6.2
1992	12.2	18.2	27.5

a. Calculate the mean and standard deviation of each company's profits.

b. Compare the performance of these 3 companies. Which company do you believe was the best performer over these 10 years?

11. The following table gives the price of Charleston Corporation's stock under different economic conditions.

Economic Condition	Chance of Occurrence	Price per Share
Depression	.25	$65
Recession	.25	$80
Normal growth	.3	$95
Inflation	.2	$100

a. Sketch a relative frequency diagram for Charleston's stock.

b. Calculate the mean and standard deviation of the stock's price.

12. The final scores from an honors seminar in Marketing were

65   55   70   80   90   100   50   75

Find the mean, median, and mode. Is one measure preferable to another. Why or why not?

13. a. Briefly compare the arithmetic mean with the geometric mean. Cite some cases where the geometric mean would be preferred.

b. Use data given in Table 4.9 to calculate the arithmetic mean and the geometric mean of market rates of return during 1970–1990.

c. Analyze the results which you obtained in part (b).

14. Compare the use of the mean to the use of the median as a measure of central tendency. If you were taking a tough calculus class where 3 brilliant students out of 20 nevertheless received perfect scores of 100 on the midterm, would you prefer that the professor use the mean or the median to determine the average grade? Or would it make no difference? (*Hint:* If you said it doesn't matter, you must be very good at calculus; you're one of the 3 who got a perfect score!)

15. In major league baseball, rookies earn a minimum salary of $100,000, whereas superstar players earn as much as $5 million per year. Do you think the mean or the median of major league salaries would be higher?

16. Suppose you are a market researcher and have been asked to assess the popularity of 4 brands of coffee. Should you construct your test on the basis of the mean, median, or mode?

17. Why is the standard deviation sometimes preferred to the variance as a measure of dispersion, even though they measure the same thing?

18. In finance, we generally use a measure of dispersion such as the variance to measure the risk of a stock's returns. Explain why the variance may not, however, be the best measure of risk of a stock's returns.

19. Carefully explain the difference between a population and a sample. Why is the formula we use to calculate the population standard deviation different from the one we use to calculate the sample standard deviation?

20. The members of the offensive line of the Denver Broncos weigh 275, 281, 285, 265, and 292 pounds, respectively.

    a. Calculate the mean and median weight of the offensive line.
    b. Calculate the variance and standard deviation for these observations.

21. A quality control manager finds the following number of defective light bulbs in 10 cases of light bulbs.

Case	Number Defective	Case	Number Defective
1	5	6	6
2	3	7	3
3	7	8	4
4	1	9	5
5	0	10	2

    a. Draw a frequency diagram for the class intervals

       0–2   3–5   6–8   9 and over

    b. Draw a relative frequency diagram for these data.

22. Calculate the mean, variance, and skewness for the observations given in question 21, and do some analysis.

23. You are given the following information about two stocks.

Economic Condition	Chance of Occurrence	Return on A	Return on B
Recession	.25	7%	0%
Normal growth	.50	8%	10%
Inflation	.25	9%	20%

    a. Calculate the mean and standard deviation for each stock.
    b. Compare the mean and standard deviation of each stock. If you had to purchase only one stock, which would you choose?

24. What is the coefficient of variation? What does it measure? Use your results from question 23 to calculate the coefficient of variation. Explain how the coefficient of variation can be used to decide which of these two stocks to purchase.

25. Suppose you are an efficiency expert who is concerned with the absentee rate for workers in a factory. You collect the following information:

Days Absent per Month	Number of Employees
0	10
1	17
2	25
3	28
4	30
5	27

    a. Calculate the mean and standard deviation for days absent.
    b. Calculate the median and mode of the distribution.
    c. Is the distribution symmetric?

26. When a distribution is skewed to the right, which measure of central tendency—the mean, median, or mode—has the highest value? Which has the lowest value?

27. Calculate the mean, variance, and skewness coefficient for the data given in Table 4.9. Is the distribution symmetric?

28. On November 17, 1991, the *Home News* used the information in this figure to show that the U.S. Congress taxes foreigners at lower rates than it taxes American citizens.

    a. Calculate the mean and standard deviation of the tax rates for the 16 foreign countries.
    b. Calculate the Z scores for the 16 foreign countries and then draw related conclusions.

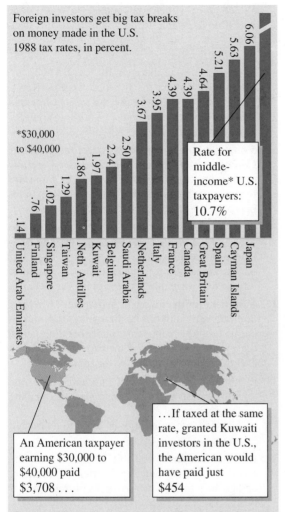

Foreign investors get big tax breaks on money made in the U.S. 1988 tax rates, in percent.

*$30,000 to $40,000

Rate for middle-income* U.S. taxpayers: 10.7%

6.06
5.63
5.21
4.64
4.39
4.39
3.95
3.67
2.50
2.24
1.97
1.86
1.29
1.02
.76
.14

Japan
Cayman Islands
Spain
Great Britain
Canada
France
Italy
Netherlands
Saudi Arabia
Belgium
Kuwait
Neth. Antilles
Taiwan
Singapore
Finland
United Arab Emirates

...If taxed at the same rate, granted Kuwaiti investors in the U.S., the American would have paid just $454

An American taxpayer earning $30,000 to $40,000 paid $3,708 . . .

Source: Philadelphia Inquirer, Internal Revenue Service.

Congress Taxes Foreigners at Lower Rates Than U.S. Citizens
Source: Reprinted by permission of Knight-Ridder Tribune News.

29. Compare the following measures of dispersion: variance, standard deviation, mean absolute deviation, and range.
    a. What are the benefits and disadvantages of each measure?
    b. Which measure is the easiest to compute? Which is the most difficult?

30. Calculate the range and mean absolute deviation for the data given in question 21.

31. Explain whether each of the following distributions is symmetric, skewed to the left, or skewed to the right.

a.
Frequency

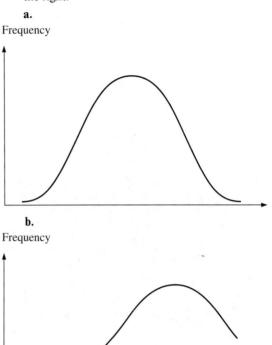

b.
Frequency

X

c.
Frequency

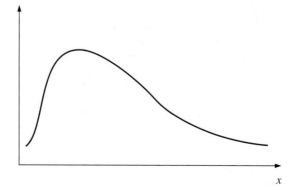

x

32. a. Use the data from the following figure, which are reprinted from the *Home News* of November 17, 1991, to calculate the mean, standard deviation, and *Z* score of Newark International Airport's passenger traffic trends for the period 1980–1991.

   b. What do the *Z* scores you obtained in part (a) suggest?

Passengers (in millions)

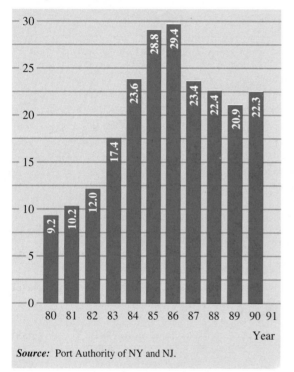

Source: Port Authority of NY and NJ.

33. Your long-lost great uncle has recently died, leaving you $5000, but stipulating that you must invest it in the stock of either XYZ Company or ABC Company. To compare their rates of return you calculate that over the last 10 years, they had the following means and standard deviations:

	XYZ	ABC
Mean	8%	10%
Standard deviation	2%	3%

Which stock would you choose? Have you done everything possible to help you make this decision? That is, are there any other statistics that would be helpful?

34. Suppose you were the agent for Ralph "Boomer" Smith, the punter for the Los Angeles Rams. Explain how you could use the mean yards and standard deviation for Boomer's punts to argue for a pay increase for Boomer.

35. You are a quality control specialist for Brite Lite Company, a light bulb manufacturer. Carefully explain why the standard deviation is as important to you as the mean number of defective light bulbs per case.

36. Use the data given in question 23 of Chapter 2 to compute the mean, standard deviation, and coefficient of variation for
   a. The exchange rate between dollars and pounds.
   b. The exchange rate between dollars and yen.

37. Comment on the following statement: "Investors don't care about the variability of a stock's returns, because they have the same chance of falling below the median as above the median. Therefore, on average, their returns will be the same."

38. a. Use the 3-month T-bill rate and prime rate information given in Table 4.16 to calculate the *Z* score.
   b. Use this information on *Z* score to do related analysis.

39. Use the data given in Table 3.9 to compute the mean, standard deviation, coefficient of variation, and coefficient of skewness for the current ratio of GM.

40. Repeat question 39 using the current ratio data for Ford.

41. You would like to compare the risk and return of two mutual funds. You have the following information:

	Fund A	Fund B
Expected return	10%	7%
Standard deviation	3%	2.5%

Which fund do you think is more desirable? Explain.

42. You are given the following information about two stocks.

State of Economy	Chance of Occurrence	Return on A	Return on B
Poor	.35	5%	0%
Good	.20	6%	10%
Excellent	.45	9%	20%

a. Calculate the mean and standard deviation for each stock.

b. Compare the mean, standard deviation, and coefficient of variation of each stock. Is the coefficient of variation or the standard deviation a better measure of risk here?

c. If you could buy only one stock, which would you choose?

43. Calculate the skewness coefficient for the data given in question 42.

The following information is for questions 44–52. The following table gives the current ratio and inventory turnover for Chrysler, Ford, GM, and the auto industry from 1969 to 1990.

Current Ratio

Year	Chrys	Ford	GM	Indus
69	1.32	1.37	2.30	1.66
70	1.40	1.33	1.93	1.55
71	1.46	1.39	1.76	1.54
72	1.49	1.44	2.12	1.68
73	1.55	1.37	2.04	1.65
74	1.36	1.28	1.91	1.52
75	1.27	1.33	1.99	1.53
76	1.37	1.37	1.95	1.56
77	1.34	1.38	1.92	1.55
78	1.43	1.33	1.79	1.52
79	.97	1.25	1.68	1.30
80	.94	1.04	1.26	1.08
81	1.08	1.02	1.09	1.06
82	1.12	.84	1.13	1.03
83	.80	1.05	1.40	1.08
84	.97	1.11	1.36	1.15
85	1.12	1.10	1.09	1.10
86	1.05	1.18	1.17	1.13
87	1.74	1.24	1.53	1.50
88	1.76	1.29	1.71	1.59
89	1.59	.97	1.72	1.43
90	1.50	.94	1.37	1.27

Inventory Turnover

Year	Chrys	Ford	GM	Indus
69	5.76	6.46	6.46	6.27
70	5.03	6.03	4.56	5.21
71	5.68	6.47	7.08	6.41
72	7.11	7.26	7.25	7.21
73	6.53	6.41	6.92	6.62
74	4.47	5.55	4.93	4.98
75	5.61	6.31	6.28	6.07
76	6.60	6.62	7.46	6.89
77	6.37	7.70	7.66	7.24
78	6.88	7.58	8.34	7.60
79	6.41	7.39	8.21	7.34
80	4.82	7.23	7.98	6.68
81	6.76	8.24	8.68	7.89
82	8.87	8.99	9.71	9.19
83	10.17	10.81	11.26	10.75
84	12.04	12.73	11.40	12.06
85	11.41	11.47	11.65	11.51
86	13.29	10.83	14.21	12.78
87	11.46	23.51	12.82	15.93
88	11.93	18.70	13.81	14.81
89	10.57	12.16	14.08	12.27
90	8.39	11.50	11.87	10.59

44. Draw a frequency and cumulative frequency histogram for Chrysler's current ratio. Is this frequency histogram symmetric or skewed?

45. Compute the mean, variance, and coefficient of variation for Chrysler's current ratio.

46. Compute the mean, variance, and coefficient of variation for Chrysler's current ratio for the period 1969 to 1978 and the period 1979 to 1990. Compare these descriptive statistics to those you computed in question 45. Have any changes occurred over the two different time periods? If so, can you propose an explanation?

47. Draw a frequency and cumulative frequency histogram for Chrysler's inventory turnover. Is this frequency histogram symmetric or skewed?

48. Compute the mean, variance, and coefficient of variation for Chrysler's inventory turnover.

49. Compute the mean, variance, and coefficient of variation for Chrysler's inventory turnover for the period 1969 to 1978 and the period 1979 to 1990. Compare these descriptive statistics to those you computed in question 48. Have any changes occurred over the two different time periods? If so, can you propose an explanation?

50. Answer the following questions by referring to the MINITAB output of Ford's current ratio.
   a. Is the frequency histogram of Ford's current ratio symmetric or skewed?
   b. Compare and analyze the means, variances, and coefficients of variance of Ford's current ratio calculated from different periods.

51. Using a calculator and the MINITAB program, answer questions 44–49 again, using the data for GM.

52. Answer questions 44–49 again, using the data for the auto industry.

The following information from *Best's Aggregates and Averages* can be used for questions 53–55. You are given the following information on the property-casualty insurance industry. NPW (net premiums written) is a measure of the dollar value of premiums written in property-casualty insurance (such as auto insurance, home insurance, and so on). PHS (policyholders' surplus) is a measure of the net worth, or equity, of an insurer.

Insurance Industry

Year	NPW	PHS
1967	23583	14802
1968	25766	16192
1969	28956	13964
1970	32578	15499

MINITAB Output of Ford's Current Ratio (Question 50)

```
MTB > SET INTO C1
DATA> 1.37 1.33 1.39 1.44 1.37 1.28 1.33 1.37 1.38 1.33 1.25 1.04 1.02 0.84
DATA> 1.05 1.11 1.10 1.18 1.24 1.29 0.97 0.94
DATA> END
MTB > HISTOGRAM C1


Histogram of GMCA/CL   N = 22

Midpoint   Count
   0.85       1   *
   0.90       0
   0.95       2   **
   1.00       1   *
   1.05       2   **
   1.10       2   **
   1.15       0
   1.20       1   *
   1.25       2   **
   1.30       2   **
   1.35       6   ******
   1.40       2   **
   1.45       1   *

MTB > SET INTO C1

DATA> 1.37 1.33 1.39 1.44 1.37 1.28 1.33 1.37 1.38 1.33
DATA> END
MTB > MEAN C1
   MEAN    =      1.3590
MTB > STDEV C1
   ST.DEV. =      0.043576
MTB > SET INTO C2
DATA> 1.25 1.04 1.02 0.84 1.05 1.11 1.10 1.18 1.24 1.29 0.97 0.94
DATA> END
MTB > MEAN C1
   MEAN    =      1.3590
MTB > MEAN C2
   MEAN    =      1.0858


MTB > STDEV C2
   ST.DEV. =      0.13648
MTB > PAPER
```

Year	NPW	PHS
1971	35860	19065
1972	38930	23812
1973	42075	21389
1974	44704	16270
1975	49605	19712
1976	60439	24631
1977	72406	29300
1978	81699	35379
1979	90169	42395
1980	95702	52174
1981	99373	53805
1982	104038	60395
1983	109247	65606
1984	118591	63809
1985	144860	75511
1986	176993	94288

53. Draw a relative and cumulative relative frequency histogram for NPW and PHS.

54. Compute the mean, standard deviation, and coefficient of variation for NPW and PHS.

55. If you were interested in comparing the dispersion of NPW to the dispersion of PHS, should you use the variance, the standard deviation, or the coefficient of variation? Which variable—NPW or PHS—has the greater dispersion around its mean?

56. Suppose the variance of a population is 0. What can you say about the members of that population?

57. Suppose you have three populations containing two members each. Suppose the means and the variances of the three populations are the same. Are the numerical values of the members of the first population necessarily identical to the numerical values of the members of the second or third population?

58. Reconsider question 57, but this time assume that each population has three members.

59. In a class of 10 students, we find that the students spent the following amounts of money on textbooks for the semester.

$225 $178 $272 $310 $190 $145 $150
$220 $285 $112

a. Find the median dollar value spent on books.
b. Find the mean and standard deviation for the dollar value spent on books.

60. Suppose you are in a statistics class of 12 students. Your score on the test is a 75 out of 100. The scores for the entire class (including your score) were

100   75   100   50   45   100   60   65   72   70
66   74

Would you prefer that the teacher use the median or the mean as the average score when deciding how to draw the curve that she or he will use in determining grades?

61. The *Home News* used the chart reproduced here in its September 29, 1991, issue to show the economic growth record for 12 different periods since World War II.

−4.9%		FDR/Truman
	Truman	6.1%
	Eisenhower I	2.6%
	Eisenhower II	2.3%
	Kennedy	4.1%
	Johnson	5.3%
	Nixon I	2.6%
	Nixon/Ford	2.1%
	Carter	3.2%
	Reagan I	2.5%
	Reagan II	3.7%
	Bush	1.5%

Annual Average GNP Growth by Presidential Term Since World War II
Source: Knight-Ridder Tribune News/Marty Westman and Judy Treible as found in the *Home News,* September 29, 1991. Reprinted by permission of Knight-Ridder Tribune News.

a. Calculate the mean and the median.
b. Calculate the standard deviation and mean absolute deviation.
c. Calculate the Z scores and do related analysis.

62. The following table, reprinted from the November 20, 1991, *Wall Street Journal,* shows the percentage change of stock prices for 10 Dow Jones sectors over 2 different periods.

a. Calculate the arithmetic mean and geometric mean of the two sets of data.

b. Calculate the standard deviation and coefficient of variation of these two sets of data.

c. Calculate the *Z* scores and do related analysis.

Of the Market's Slide: Performance of the DJ Sectors

	% Change 11/13/91 to 11/19/91	% Change 12/31/90 to 11/19/91
Conglomerates	−2.73%	19.50%
Utilities	−3.50	3.73
Energy	−3.65	2.63
Consumer Non-Cyclical	−3.97	24.35
DJ Equity Index	−4.52	16.53
Technology	−5.04	15.32
Consumer Cyclicals	−5.16	23.11
Industrial	−5.36	13.94
Basic Materials	−5.63	15.64
Financial	−5.73	31.12

Source: Wall Street Journal, November 20, 1991. Reprinted by permission of the Wall Street Journal, © 1991 Dow Jones & Company, Inc. All Rights Reserved Worldwide.

PART II

Probability and Important Distributions

Chapter 5 introduces and explains such basic probability concepts as conditional, marginal, and joint probabilities. Chapter 6 discusses the analysis of discrete random variables and their distributions. Chapter 7 deals with the normal and lognormal distributions and their applications. In Chapter 8, we introduce sampling and sampling distributions. Other important continuous distributions are discussed in Chapter 9.

The examples and applications presented in Part II involve determining commercial lending rates, finding the expected value of stock price by means of the decision tree approach, determining option value via binomial and normal distributions and stock rates of return distribution, and studying auditing sampling. Other business decision applications also are explored.

CHAPTER 5

Probability Concepts and Their Analysis

CHAPTER OUTLINE

5.1 Introduction

5.2 Random Experiment, Outcomes, Sample Space, Event, and Probability

5.3 Alternative Events and Their Probabilities

5.4 Conditional Probability and Its Implications

5.5 Joint Probability and Marginal Probability

5.6 Independent, Dependent, and Mutually Exclusive Events

5.7 Bayes' Theorem

5.8 Business Applications

Appendix 5A Permutations and Combinations

Key Terms

random experiment	intersection
basic outcomes	addition rule
sample points	partition
subset	complement
event	combinatorial mathematics
occur	conditional probability
simple event	multiplication rule of probability
basic event	joint probability
sample space	marginal probability
Venn diagram	unconditional probability
mutually exclusive events	simple probability
probability	independent
a priori probability	Bayes' Theorem
subjective probability	prior probability
simple event	posterior (revised) probability
composite event	number of permutations
compound event	number of combinations
union	outcome trees

5.1 INTRODUCTION

In Part I of this book we discussed the use of descriptive statistics, which is concerned mainly with organizing and describing a set of sample measurements via graphical and numerical descriptive methods. We now begin to consider the problem of making inferences about a population from sample data. Probability and the theory that surrounds it are discussed in this chapter. These topics provide an essential foundation for the methods of making inferences about a population on the basis of a sample. A well-known example is the election poll, in which pollsters select at random a small number of voters to question in order to predict the winner of an election. Probability is also used in daily decision making. For example, investment decisions are based on the investor's assessment of the probable future returns of various investment opportunities, and such assessments are often based on some sample information.

In this chapter, we first discuss how basic concepts such as random experiment, outcomes, sample space, and event can be used to analyze probability. Then we investigate alternative events and their probabilities. In probability theory, conditional, joint, and marginal probabilities are the most important concepts in analyzing statistical business and economic problems. Therefore, they are explored in detail in this chapter. We also discuss independent and dependent events and Bayes' theorem. Finally, four business applications of probability are demonstrated.

5.2 RANDOM EXPERIMENT, OUTCOMES, SAMPLE SPACE, EVENT, AND PROBABILITY

A **random experiment** is a process that has at least two possible outcomes and is characterized by uncertainty as to which will occur.

Each of the following examples involves a random experiment.

1. A die is rolled.
2. A voter is asked which of four candidates he or she prefers.
3. A person is asked whether President Bush should order U.S. troops to liberate Kuwait.
4. The daily change in the price of silver per ounce is observed.

When a die is rolled, the set of basic outcomes comprises 1 through 6; these basic outcomes represent the various possibilities that can occur. In other words, the possible outcomes of a random experiment are called the **basic outcomes.** The set of all basic outcomes is called the **sample space.** Thus basic outcomes are equivalent to sample points in a sample space.

Suppose you are interested in getting an even number in rolling a die; in this case the event is rolling a 2, 4, or 6, which is a **subset** of $\{1, 2, 3, 4, 5, 6\}$. In other words, an **event** is a set of basic outcomes from the sample space, and it is said to **occur** if the random experiment gives rise to one of its constituent basic outcomes. Each basic outcome within each event ({2} {4} {6}, for example) can also be called a **simple event.** Hence an event is a collection of one or more simple events. Finally, a **basic event** is a subset of the sample space. The concepts of random experiment, outcomes, sample space, and event, then, are fundamental to an understanding of probability.

Properties of Random Experiments

The starting point of probability is the random experiment. Random experiments have three properties:

1. They can be repeated physically or conceptually.
2. The set consisting of all of possible outcomes—that is, the sample space—can be specified in advance.
3. Various repetitions do not always yield the same outcome.

Simple examples of conducting a random experiment include rolling dice, tossing a coin, and drawing a card from a deck of 52 playing cards.

Because of uncertainty in the business environment, business decision making is a tricky and an important skill. If the executive knew the exact outcomes of the courses of action available, he or she would have no difficulty making optimal decisions. However, the executive generally does not know the exact outcome of a decision. Thus business executives spend much time evaluating the probabilities of various alternative outcomes. For example, an executive may need to determine the probability of extensive employee turnover if the firm moves to another area. Or a business decision maker may want to evaluate the impact of changes in economic indicators such as interest rate, inflation, and gross national product (GNP) on a company's future earnings.

Sample Space of an Experiment and the Venn Diagram

For convenience, we can represent each outcome of a random experiment by a set of symbols. The symbol S is used to represent the *sample space* of the experiment. As we have noted, the sample space is the set of all basic outcomes (simple events) of the random experiment. In the foregoing die-rolling example, the sample space is $S = \{1, 2, 3, 4, 5, 6\}$. When a person takes a driver's license test, the sample space contains only two elements: $S = \{P, F\}$, where P indicates a pass and F a failure. In a stock price forecast, the sample space could contain three elements: $S = \{U, D, N\}$, where U, D, and N represent movement up, movement down, and no change in the price of a stock. In sum, the different basic outcomes of an experiment are often referred to as **sample points** (simple events), and the set of all possible outcomes is called the **sample space.** Thus the sample points (simple events) form the sample space.

A **Venn diagram** can be used to describe graphically various basic outcomes (simple events) in a sample space. The rectangle represents the sample space, and the points are basic outcomes. Events are usually represented by circles or rectangles. Figure 5.1 shows a Venn diagram. The elements labeled represent the six basic outcomes of rolling a die. In Figure 5.2, the circle indicates the event of all even numbers that can result from rolling a single die. Let event $A = \{2, 4, 6\}$. Again, the sample space is the possible outcomes of rolling a die. Figure 5.3 shows events $A = \{1, 3\}$ and $B = \{4, 5\}$. When two events have no basic outcome in common, they are said to be **mutually exclusive events.** When events have some elements in common, the **intersection** of the events is the event that consists of the common elements. Say we have one event $A = \{2, 3, 4, 6\}$ and another event $B = \{2, 3, 5\}$. The intersection of these events is shown in Figure 5.4. The common elements are 2 and 3.

Probabilities of Outcomes

The **probability** of an event is a real number on a scale from 0 to 1 that measures the likelihood of the event's occurring. If an outcome (or event) has a probability of 0, then its occurrence is impossible; if an outcome (or event) has a probability of 1.0, then its occurrence is certain. Getting *either* a head *or* a tail in a coin toss is an

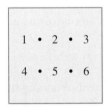

Figure 5.1

Venn Diagram Showing Six Different Sample Points

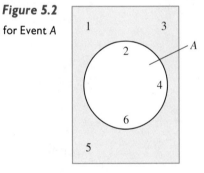

Figure 5.2

Venn Diagram for Event A

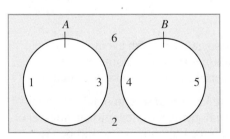

Figure 5.3

Venn Diagram for Mutually Exclusive Events

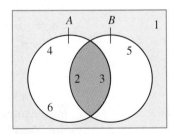

Figure 5.4

Venn Diagram for the Intersection of Events A and B

example of an event that has a probability of 1.0. Because there are only two possibilities, either one event or the other is certain to occur. An event with a zero probability is an impossible event, such as getting both a head and a tail when tossing a coin once.

When we roll a fair die, we are just as likely to obtain any face of the die as any other. Because there are six faces to a die, we generally say the "outcome" of the toss can be one of six numbers: 1, 2, 3, 4, 5, 6.

The probability of an outcome can be calculated by the classical approach, the relative frequency approach, or the subjective approach. The first two approaches are discussed in this section, the third approach in the next.

Classical probability is often called **a priori probability,** because if we keep using orderly examples, such as fair coins and unbiased dice, we can state the answer in advance *(a priori)* without tossing a coin or rolling a die. In other words, we can make statements based on logical reasoning before any experiments take place. Classical probability defines the probability that an event will occur as

$$\text{Probability of an event} = \frac{\text{number of outcomes contained in the event}}{\text{total number of possible outcomes}} \qquad (5.1)$$

Note that this approach is applicable only when all basic outcomes in the sample space are equally probable.

For example, the probability of getting a tail upon tossing a fair coin is

$$P(\text{tail}) = \frac{1}{1 + 1} = \frac{1}{2}$$

And for the die-rolling example, the probability of obtaining the face 4 is

$$P(4) = \tfrac{1}{6}$$

The relative frequency approach to calculating probability requires the random experiment to take place as defined in Equation 5.2.

$$P(o = e_i) = \frac{n_i}{N} \quad \text{or} \quad P(e_i) = \frac{n_i}{N} \qquad (5.2)$$

where o = outcome
e_i = outcome associated with ith event
n_i = number of times the ith outcome occurs
N = total number of times the trial is repeated

From Equation 5.2, we know that we can obtain the relative frequency by dividing the total number of trials being repeated into the number of ith outcomes occurring. Another explanation for Equation 5.2 would be $P(e_i) = f_i/N$, where f_i equals the number of favorable outcomes for event e_i and N equals the total outcome in sample space, S.

The credit cards issued by Citicorp in 1984 are listed here (the data is from *Fortune* magazine, February 4, 1985, page 21):

Credit Card	Number of Cards Issued (in millions)
Visa and MasterCard	6.0
Diners Club	2.2
Carte Blanche	.3
Choice	1.0

Visa and MasterCard credit cards are issued by thousands of banks, including Citicorp. The other three credit cards listed above are issued by Citicorp only. If one Citicorp credit card customer is selected randomly, the probability that the customer selected uses one of Citicorp's own credit cards is

$$\frac{2.2 + .3 + 1.0}{2.2 + .3 + 1.0 + 6.0} = .368.$$

EXAMPLE 5.1   *Toss a Fair Coin*

Suppose a random experiment is to be carried out and we are interested in the chance of occurrence of a particular event. The concept of probability can help us, because it provides a numerical measure for the likelihood of an event or a set of events occurring.

We conduct 50 experiments of tossing a fair coin in different sample sizes ($N = 10, 20, \ldots, 500$). The results of these experiments are presented in Table 5.1. In Table 5.1, the first column represents that the ith ($i = 1, 2, \ldots, 50$) experiment has been done; the second column, N, shows the number of times a coin has been tossed for experiment; the third column lists the number of times heads appeared; and the fourth column gives the proportion of heads that appeared.[1] Figure 5.5 is the figure generated by MINITAB in terms of the data given in the fourth column.

Table 5.1 Frequency and Proportion in Tossing a Fair Coin

```
MTB > PRINT C51-C53

    ROW      N       f           P

      1     10       7      0.700000
      2     20      12      0.600000
      3     30      17      0.566667
      4     40      19      0.475000
      5     50      22      0.440000
      6     60      34      0.566667
      7     70      30      0.428571
      8     80      44      0.550000
      9     90      52      0.577778
     10    100      37      0.370000
     11    110      57      0.518182
     12    120      64      0.533333
```

(*continued*)

[1]This set of data was generated by the Bernoulli process, which will be discussed in the next chapter.

Table 5.1 (Continued)

13	130	59	0.453846
14	140	63	0.450000
15	150	78	0.520000
16	160	78	0.487500
17	170	95	0.558824
18	180	88	0.488889
19	190	90	0.473684
20	200	100	0.500000
21	210	117	0.557143
22	220	109	0.495455
23	230	115	0.500000
24	240	119	0.495833
25	250	130	0.520000
26	260	128	0.492308
27	270	137	0.507407
28	280	126	0.450000
29	290	140	0.482759
30	300	158	0.526667
31	310	161	0.519355
32	320	157	0.490625
33	330	160	0.484848
34	340	165	0.485294
35	350	168	0.480000
36	360	188	0.522222
37	370	201	0.543243
38	380	199	0.523684
39	390	190	0.487179
40	400	198	0.495000
41	410	220	0.536585
42	420	192	0.457143
43	430	207	0.481395
44	440	210	0.477273
45	450	226	0.502222
46	460	217	0.471739
47	470	239	0.508511
48	480	243	0.506250
49	490	233	0.475510
50	500	243	0.486000

```
                MTB > PAPER
```

Figure 5.5

MINITAB Output
for the Proportion
of Heads in *N* Toss-
es of a Fair Coin

```
MTB > GPLOT C53 C51;

SUBC> LINE C53 C51;
SUBC> YINCREMENT = 0.1;
SUBC> YSTART = 0.2;
SUBC> XLABEL 'SAMPLE SIZE';
SUBC> YLABEL 'PROPORTION'.
* Increment increased to cover range
```

Figure 5.5 (Continued)

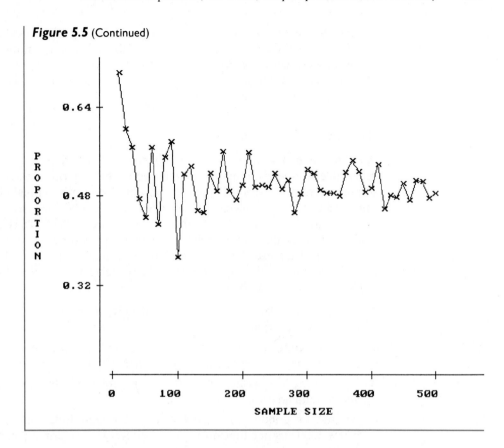

In this example, we know that if the coin is fair, anytime we toss the coin, the probability of our getting heads is $\frac{1}{2}$. One important property of probability is that *the sum of the probabilities of all outcomes must be equal to 1*. The sum of the probabilities must equal 1 because the possible outcomes are collectively exhaustive and mutually exclusive. From our previous example of the toss of a coin, the outcomes are mutually exclusive and collectively exhaustive because we have included all the possible basic outcomes (simple events) that can occur. Mathematically, we can express this property for our roll of the die as

$$P(H) + P(T) = 1 \qquad (5.3)$$

where $P(H)$ and $P(T)$ represent the probability of getting heads and that of getting tails, respectively.

Note that the results of Table 5.1 are obtained by tossing a coin again and again. Tossing it only 4 (or even 20) times would not be enough to average out the chance fluctuations shown in Table 5.1. When N is large enough, however, the relative frequency of tossing a coin for heads moves toward $\frac{1}{2}$, as indicated in Figure 5.5.

So far, we have used both the classical and the relative frequency approaches to define probability. We now summarize the definition of probability in terms of relative frequency. Let n_i be the number of occurrences of event i in N repeated trials.

Then under the relative frequency concept of probability, the probability that event i will occur is the relative frequency (the ratio n_i/N) as the number of trials N becomes infinitely large. Alternatively, the probability can be interpreted as the proportion of times the ith event (n_i/N) occurs in the long run (N becomes infinitely large) when conditions are stable.

Subjective Probability

An alternative view about probability, which does not depend on the concept of repeatable random experiments, defines probability in terms of a subjective, or personalistic, concept. According to this concept of **subjective probability,** the probability of an event is the degree of belief, or degree of confidence, an individual places in the occurrence of an event on the basis of whatever evidence is available. This evidence may be data on the relative frequency of past occurrences, or it may be just an educated guess. The individual may assign an event the probability of 1, 0, or any other number between those two. Here are a few examples of situations that require a subjective probability:

1. An individual consumer assigns a probability to the event of purchasing a TV during the next quarter.

2. A quality control manager asserts the probability that a future incoming shipment will have 1.5 percent or fewer defective items.

3. An auditing firm wishes to determine the probability that an audited voucher will contain an error.

4. An investor ponders the probability that the Dow Jones closing index will be below 3000 at some time during a 3-month period beginning on November 10, 1992.

5.3   ALTERNATIVE EVENTS AND THEIR PROBABILITIES

As we have stated, an event is the result of a random experiment consisting of one or more basic outcomes. If an event consists of only one basic outcome, it is a **simple event;** if it consists of more than one basic outcome, it is a **composite event.** In the die-rolling experiment discussed in Example 5.1, the sample space is $S = \{1, 2, 3, 4, 5, 6\}$.

Suppose we are interested in the event E, where the outcome is 1 or 6. We can clearly describe the event E as $E = \{1, 6\}$. An event E is a subset of the sample space S. This is a composite event because it includes the simple events $\{1\}$ and $\{6\}$. The subset definition enables us to define an event in general.

In the tossing of a fair die, suppose that event A represents the faces 1, 2, 3, 4, and 5 and event B the faces of 4, 5, and 6. Graphically, the relationship between

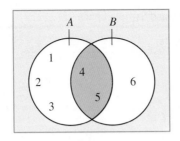

Figure 5.6

Venn Diagram for the Intersection of Events A and B

basic outcomes and events can be represented as shown in Figure 5.6. The intersection of these two events is the faces 4 and 5, because these faces are common to both events.

Probabilities of Union and Intersection of Events

An event can often be viewed as a composite of two or more other events. Such an event, called a **compound event,** can be classified as union or as intersection. The **union** of two events A and B is the event that occurs when either A or B or both occur on a single performance of the experiment. For example, if event B is getting an even number (2, 4, or 6) on a die toss, and event A is getting a number 1 or 2, then the union of events A and B, which we represent as $A \cup B$, is 1, 2, 4, and 6. The union $A \cup B$ is indicated in Figure 5.7.

The **intersection** of two events A and B is the event that occurs when both A and B occur on a single performance of the experiment. That is, the common members make up the intersection of two events. Because 2 is the only common sample point in our two events, the intersection of events A and B, which we represent as $A \cap B$, is 2. This intersection is indicated by the shaded area in Figure 5.8.

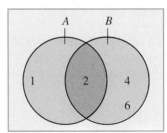

Figure 5.7

Venn Diagram for $A \cup B$

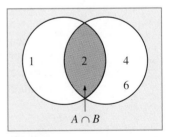

Figure 5.8

Venn Diagram for $A \cap B$

EXAMPLE 5.2 *Pick a Card, Any Card*

To illustrate the union and intersection of events, we shall use a standard deck of cards. We know that there are 52 cards in a deck (13 spades, 13 hearts, 13 diamonds, and 13 clubs) and 4 cards for each number (see Table 5.2). Using this information,

Table 5.2 Playing Card Sample Space

Spades	Hearts	Diamonds	Clubs
A	A	A	A
2	2	2	2
3	3	3	3
4	4	4	4
5	5	5	5
6	6	6	6
7	7	7	7
8	8	8	8
9	9	9	9
10	10	10	10
J	J	J	J
Q	Q	Q	Q
K	K	K	K

let's calculate the probability of the union and intersection of events in the sample space of a deck of playing cards.

Probability of Union

To assess the probability of union, first imagine we randomly select one card from the deck. Let event $A = \{$club$\}$ and event $B = \{$heart or diamond$\}$. Let $A \cup B$ denote the union, so $A \cup B = \{$club, heart, diamond$\}$.

The union of A and B means the event "A or B" occurs. We can now compute the mathematical probability of A or B:

$$P(A) = \frac{13}{52} = \frac{1}{4} \quad \text{and} \quad P(B) = \frac{13 + 13}{52} = \frac{1}{2}$$

The probability of getting a club, a heart, or a diamond is obtained by adding the number of club, heart, and diamond cards and dividing by the total number of cards, 52. As a result, the probability of drawing a card that is a member of the union of these two events is

$$P(A \cup B) = P(A) + P(B) = \tfrac{1}{4} + \tfrac{1}{2} = \tfrac{3}{4}$$

Thus we have a $\tfrac{3}{4} = 75$ percent chance of randomly drawing a single card that is a club *or* a heart *or* a diamond.

If A and B are mutually exclusive, the probability formula for a union of A and B is

$$P(A \cup B) = P(A) + P(B) \tag{5.4}$$

The rule for obtaining the probability of the union of A and B as indicated in Equation 5.4 is the **addition rule** for two events that are mutually exclusive. This addition

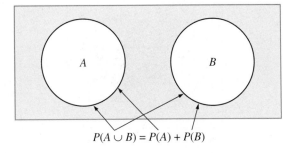

$$P(A \cup B) = P(A) + P(B)$$

Figure 5.9

Venn Diagram for the Probability of Two Mutually Exclusive Events

rule is illustrated by the Venn diagram in Figure 5.9, where we note that the area of two circles taken together (denoting $A \cup B$) is the sum of the areas of the two circles.

As another example, if A = all clubs and B = all diamonds, then

$$P(A \cup B) = \tfrac{13}{52} + \tfrac{13}{52} = \tfrac{26}{52} = \tfrac{1}{2}$$

A new pharmaceutical product is about to be introduced commercially, and both Upjohn and Merck want to be the first to put the product on the market. An industrial analyst believes that the probability is .40 that Upjohn will be first and .25 that Merck will be first. If the analyst's beliefs are correct, what is the probability that either Upjohn or Merck will be first, assuming that a tie does not occur?

Let A be the event that Upjohn is first. Let B be the event that Merck is first. Then, from Equation 5.4 we have $P(A \cup B) = .40 + .25 = .65$. Consequently, the probability that either Upjohn or Merck will be first is .75.

If A and B are not mutually exclusive, then the simple probability of union defined in Equation 5.4 must be modified to take the intersection into account and thereby avoid double counting:

$$P(A \cup B) = P(A) + P(B) - P(A \cap B) \tag{5.5}$$

The rule for obtaining the probability of the union of A and B as indicated in Equation 5.5 is the addition rule for two events that are *not* mutually exclusive. This addition rule is illustrated by Figure 5.10. In Figure 5.10a, the event $A \cup B$ is the sum of the areas of circles A and B. The event $A \cap B$ is the shaded area in the middle, as indicated in Figure 5.10b. When we add the areas of circles A and B, we count the shaded area twice, so we must subtract it to make sure it is counted only once.

If, instead, A = all diamonds and B = all diamonds or all hearts, then

$$P(A \cup B) = \tfrac{1}{4} + \tfrac{1}{2} - \tfrac{1}{4} = \tfrac{1}{2}$$

Midlantic Bank in New Jersey gives summer jobs to two Rutgers University business school students, Mary Smith and Alice Wang. The bank personnel manager hopes that at least one of these students will decide to work for the bank upon

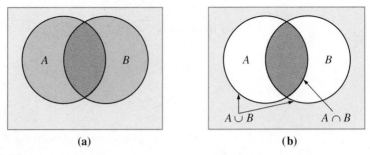

Figure 5.10

Venn Diagram for the Probability of Two Events That Are Not Mutually Exclusive: (a) $A \cup B$ and (b) $A \cap B$

graduation. Assume that the probability that Mary will decide to work for the bank is .4, the probability for Alice is .3, and the probability that both will decide to work for the bank is .2. Then the probability that the personnel manager's hopes will be fulfilled is

$$P(A \cup B) = .4 + .3 - .2 = .5.$$

EXAMPLE 5.3 *Probability Analysis of Family Size*

Table 5.3 contains data on the size of families in a certain town in the United States in 1992. If we randomly choose a family from this town, what is the probability that this family includes 3 or more children?

Using Equation 5.4, we can calculate the answer as

$$
\begin{aligned}
P(3, 4, 5, 6 \text{ or more}) &= P(3) + P(4) + P(5) + P(6) + P\,(7 \text{ or more}) \\
&= .26 + .14 + .10 + .05 + .01 \\
&= .56
\end{aligned}
$$

Table 5.3 Family Size Data

Number of children	0	1	2	3	4	5	6	7 or more
Proportion of families having this many children	.04	.11	.29	.26	.14	.10	.05	.01

Probability of Intersection

If $A = \{\text{diamond}\}$ and $B = \{\text{diamond or heart}\}$, then $A \cap B = \{\text{diamond}\} = $ set of points that are in both A and B. Using Table 5.2, we obtain

$$
\begin{aligned}
P(A) &= \tfrac{13}{52} = \tfrac{1}{4} \\
P(B) &= (13 + 13)/52 = \tfrac{1}{2} \\
P(A \cap B) &= \tfrac{13}{52} = \tfrac{1}{4}
\end{aligned}
$$

Thus the probability of drawing a diamond *and* drawing a diamond or a heart is the probability of drawing a diamond, which is $\frac{1}{4}$, or 25 percent.

From Equation 5.5 we can define the probability of an intersection as

$$P(A \cap B) = P(A) + P(B) - P(A \cup B) \tag{5.6}$$

If, instead A = all diamonds and B = all diamonds or all hearts, then

$$P(A \cap B) = \tfrac{1}{4} + \tfrac{1}{2} - \tfrac{1}{2} = \tfrac{1}{4}$$

Partitions, Complements, and Probability of Complements

Now suppose we randomly choose a card from the deck. Let A = (red suit) and B = (black suit). A card cannot be a member of both a red suit *and* a black suit. Therefore, we say A and B are *mutually exclusive events;* they have no basic outcomes in common. In addition, if mutually exclusive events A and B cover the whole sample space S, we call the collection of events A and B a **partition** of S. Another alternative is to partition the card deck sample space as follows:

$$A = \{club\} \qquad B = \{diamond\} \qquad C = \{heart\}, \qquad D = \{spade\}$$

Those four events—A, B, C, and D—are mutually exclusive and collectively exhaustive; the collection of these events is called a partition of sample space S, which can be explicitly defined as

$$S = \{A, B, C, D\} \tag{5.7}$$

Equation 5.7 can itself be partitioned again into G_1 and G_2 as

$$S = \{G_1, G_2\}, \tag{5.8}$$

where $G_1 = \{A, B\}$ and $G_2 = \{C, D\}$. G_1 consists of exactly those cards that are not in G_2. We therefore call G_2 the **complement** of G_1, denoted by $\overline{G}_1$ (which is read "not G_1"). In other words, $\overline{G}_1$ represents a set of cards that are not in G_1: $\overline{G}_1 = \{C, D\}$.

Figure 5.11 depicts three different types of partitions. Figure 5.11a depicts two mutually exclusive sets, G_1 and G_2. Figure 5.11b shows mutually exclusive events, A, B, C, and D. $\overline{G}_1$ is the complement of G_1 in Figure 5.11c. $\overline{G}_1$ and G_1 are mutually exclusive for a simple partition.

Figure 5.11

Venn Diagrams of Mutually Exclusive Events, Showing Partitions and Complements

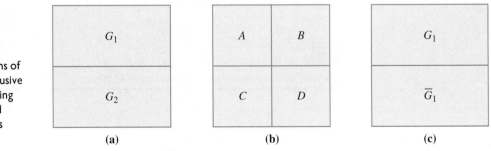

(a)   (b)   (c)

Probability of Complement

Because an event and its complement, $\{E, \overline{E}\}$, constitute a simple partition, these events are mutually exclusive. By Equation 5.4,

$$P(E \cup \overline{E}) = P(E) + P(\overline{E}) \tag{5.9}$$

E and $\overline{E}$ constitute all of the sample space, so

$$P(E \cup \overline{E}) = 1 \tag{5.10}$$

Substituting Equation 5.10 into Equation 5.9, we obtain

$$1 = P(E) + P(\overline{E})$$
$$P(\overline{E}) = 1 - P(E) \tag{5.11}$$

Recalling Equation 5.8 about playing cards, where G_1 represents a club or a diamond, we find that the probability of $\overline{G}_1$ (neither a club nor a diamond) is

$$P(\overline{G}_1) = 1 - P(G_1)$$
$$= 1 - 26/52$$
$$= 1/2$$

In 1987, 112,440,000 workers were employed in the United States. Table 5.4 shows the relative frequencies of these employed workers, classified by different types of occupations.

Table 5.4 Employed Workers in 1987

Occupation	Relative Frequency	
Male Worker	.552	
Managerial/professional		.137
Technical/sales/administrative		.110
Service		.053
Precision production, craft, and repair		.110
Operators/fabricators		.115
Farming, forestry, and fishing		.027
Female Worker	.448	
Managerial/professional		.109
Technical/sales/administrative		.202
Service		.081
Precision production, craft, and repair		.010
Operators/fabricators		.040
Farming, forestry, and fishing		.006

Source: *Statistical Abstract of the United States: 1989*, p. 388.

If we need to select a worker randomly from the population and determine his or her occupation, the probability that the worker will not be in a technical/sales/administrative occupation can be calculated as follows:

P(non- technical/sales/administrative occupation)
$= 1 - P$(technical, sales, or administrative occupation)
$= 1 - P$(male, technical, sales, or administrative occupation)
$\quad - P$(female, technical, sales, or administrative occupation)
$= 1 - .110 - .202 = .688.$

Using Combinatorial Mathematics to Determine the Number of Simple Events

The purpose of introducing combinatorial mathematics here is to show how the number of simple events can be determined and the probability computed. **Combinatorial mathematics** is the mathematics that develops counting principles and techniques in terms of permutations and combinations, which are discussed in Appendix 5A. For example, a simple rule for finding the number of different samples of r auto part items selected from n auto part items in doing quality control sampling can be derived from the combination formula discussed in this section. According to the combination formula developed in Appendix 5A, the total number of possible combinations of samples is

$$\binom{n}{r} = \frac{n!}{n!(n-r)!} \tag{5.12}$$

where n is the number of possible objects (items), r is the number of objects to be selected, and the factorial symbol (!) means that, say,

$$n! = n(n-1)(n-2) \cdots 3 \cdot 2 \cdot 1$$
$$r! = r(r-1)(r-2) \cdots 3 \cdot 2 \cdot 1$$

For example $6! = 6 \cdot 5 \cdot 4 \cdot 3 \cdot 2 \cdot 1$. (The quantity of $0!$ is defined as equal to 1.)

EXAMPLE 5.4 *Possible Combinations in Selecting Gifts*

The United Jersey Bank in New Jersey is giving out gifts to depositors. If eligible, depositors may choose any 2 out of 6 gifts. How many possible combinations of gifts can different depositors select? This question can be answered either manually or by combinatorial mathematics.

Manual method

Let g_1, g_2, g_3, g_4, g_5, and g_6 represent first gift, second gift, third gift, fourth gift, fifth gift, and sixth gift. The number of possible combinations of 2 gifts chosen from among these 6 gifts is 15:

g_1, g_2	g_2, g_3	g_3, g_5
g_1, g_3	g_2, g_4	g_3, g_6
g_1, g_4	g_2, g_5	g_4, g_5
g_1, g_5	g_2, g_6	g_4, g_6
g_1, g_6	g_3, g_4	g_5, g_6

Combinatorial mathematics method

Or, if we have less paper, we can use Equation 5.5 and find the number of possible combinations as follows:

$$\binom{6}{2} = \frac{6!}{(2!)(6-2)!} = \frac{6 \cdot 5 \cdot 4 \cdot 3 \cdot 2 \cdot 1}{(2 \cdot 1)(4 \cdot 3 \cdot 2 \cdot 1)} = 15$$

This result agrees with the result we obtained manually.

If both n and r are large, combinatorial mathematics is the far better method for counting the number of outcomes in the sample space. Trust me.

EXAMPLE 5.5 *Just Take the Toaster*

If the two gifts are randomly selected, what is the probability of gift 1 being selected? Well, there are 5 combinations that include gift 1, and there are 15 possible combinations. The probability of gift 1 being selected, then, is $P = 5/15 = 1/3$.

5.4 CONDITIONAL PROBABILITY AND ITS IMPLICATIONS

Basic Concept of Conditional Probability

Conditional probability is the probability that an event will occur, given that (on the condition that) some other event *has* occurred. The concept of conditional probability is relatively simple. In the example involving playing cards that was discussed in Section 5.3, we have 13 spades, 13 hearts, 13 diamonds, and 13 clubs. Suppose we put 13 spades on the table and then select a card randomly from that group.[2] What is the probability that the card's face value will be 2, $P(S2)$, given that it is a spade? Here we have changed the condition under which the experiment is performed, because now we are now considering only a subset of the population: just the spades. To obtain a new probability for each element of this subpopulation, we simply find the total probability of the subpopulation (spades) and then divide the probability of each event in the subpopulation by the total probability. We know that the total probability of the subpopulation is 13/52 [(1/52)(13)]. The new probabilities we assign are

[2]This is equivalent to randomly drawing a card from the deck and finding that it is a spade.

$$P(\text{S2} \mid \text{spades}) = \frac{P(\text{S2})}{P(\text{spades})} = \frac{1/52}{13/52} = \frac{1}{13}$$

$$P(\text{S3} \mid \text{spades}) = \frac{1/52}{13/52} = \frac{1}{13}$$

.

.

.

$$P(\text{SA} \mid \text{spades}) = \frac{1/52}{13/52} = \frac{1}{13}$$

where S2, S3, and SA represent the 2, 3, and ace of spades. The notation | means "given." For example, $P(\text{S2} \mid \text{spades})$ means the probability of drawing a 2 of spades, given that the card is a spade.

If we let A = the event of picking a spade from the deck, and B = the card being a 2, the conditional probability of this drawing is written

$$P(B \mid A) = \frac{P(B \cap A)}{P(A)} = \frac{1/52}{13/52} = \frac{1}{13}$$

where $P(B \cap A)$ is the probability that the card is a 2 of spades, and $P(A)$ is the probability that the card is a spade. Now we can give the formula for conditional probability as

$$P(B \mid A) = \frac{P(B \cap A)}{P(A)} \qquad (5.13)$$

Assume that J, Q, and K are greater than 10, as defined in Table 5.2. If we let A represent the event that the card we draw is a spade and let B represent the event that the card is a jack, queen, or king, then

$$P(B \cap A) = 1/52 + 1/52 + 1/52 = 3/52$$

$P(A)$ is the probability that the card we pick up is a spade, or $13/52 = 1/4$. Then

$$P(B \mid A) = \frac{P(B \cap A)}{P(A)} = \frac{3/52}{1/4} = \frac{3}{13} = 23.08 \text{ percent}$$

The conditional probability $P(B \mid A) = 3/13$ can be shown on a Venn diagram as indicated in Figure 5.12, where $A \cap B$ takes 23.08 percent of the total area of A, which means that $P(B \mid A) = 23.08$ percent.

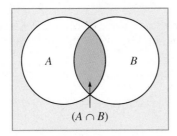

$(A \cap B)$

Figure 5.12 Venn Diagram of Conditional Probability

Multiplication Rule of Probability

An immediate consequence of the definition of conditional probability is the **multiplication rule of probability,** which expresses the probability of an intersection in terms of the probability of an individual event and the conditional probability. It can be derived as follows. From Equation 5.13 we know that

$$P(A \mid B) = \frac{P(A \cap B)}{P(B)} \tag{5.13a}$$

From Equations 5.13 and 5.13a we obtain

$$P(B \cap A) = P(B \mid A)P(A) \tag{5.14}$$

$$P(A \cap B) = P(A \mid B)P(B) \tag{5.15}$$

Clearly, $B \cap A = A \cap B$. Thus, from Equations 5.14 and 5.15 we obtain

$$P(A \cap B) = P(B \mid A)P(A) = P(A \mid B)P(B) \tag{5.16}$$

For example, suppose 30 percent of all students receive a grade of C (event A). Of all students who receive C, 60 percent are male (event B). This is a conditional probability because we are limiting ourselves to male students. In symbols, $P(B \mid A)$ = .60. What is the probability of a randomly selected student who is a male having a grade of C?

The event of a male student with a grade of C is the intersection of A and B. We also know that

$$P(A) = .3$$
$$P(B \mid A) = .6$$

Then, from our rule of Equation 5.16, we find that

$$P(A \cap B) = P(A)P(B \mid A) = (.3)(.6) = .18$$

EXAMPLE 5.6 *Joint Probability on Wall Street*

Let A be the event that the stock market will be bullish next year, and let B be the event that the stock price of Meridian Company will increase by 10 percent next year. An investment analyst would like to estimate the probability that the stock price of Meridian Company will increase *and* that the stock market will be bullish next year. Let

$P(A)$ = probability that the stock market will be bullish next year
   = 60 percent
$P(B \mid A)$ = probability that the stock price of Meridian Company will
      increase by 10 percent, given that the stock market will be bullish
   = 30 percent

Using Equation 5.15 we obtain

$$P(A \cap B) = (.6)(.3) = .18$$

This implies that there is about an 18 percent chance that the stock price of Meridian Company will increase and that the stock market will be bullish.

The probability $P(A \cap B)$ of Equation 5.15 is called the joint probability, which is discussed in Section 5.5 in further detail. Equation 5.15 can also be used to derive Bayes' theorem, which is discussed in Section 5.7.

5.5 *JOINT PROBABILITY AND MARGINAL PROBABILITY*

In this section, we will examine joint and marginal probabilities and their relationships to the conditional probability we discussed in Section 5.4.

Joint Probability

In many applications, we are interested in **joint probability,** the probability of two or more events occurring simultaneously. To illustrate joint probabilities, consider the data in Table 5.5. These figures represent the results of a market survey in which 500 persons were asked which of two competitive soft drinks they preferred, soft drink 1 from company I or soft drink 2 from company II. To simplify the discus-

Table 5.5 500 Persons Classified by Sex and Product Preference

Sex	Product Preference		Total
	S_1	S_2	
Male	100	160	260
Female	200	40	240
Total	300	200	500

sion, as shown in Table 5.5, we use M, F, S_1, and S_2, to represent male, female, prefers soft drink 1, and prefers soft drink 2. Hence the joint outcome that an individual is both male and prefers soft drink 1 is denoted as "M and S_1," and the joint probability that a randomly selected individual is male and prefers soft drink 1 is $P(M \text{ and } S_1) = P(M \cap S_1)$. This probability is

$$P(M \cap S_1) = \frac{100}{500} = 0.20$$

Other joint probabilities can be calculated similarly. Table 5.6 is a joint probability table obtained by dividing all entries in Table 5.5 by the total number of individuals.

Table 5.6 Joint Probability Table for 500 Persons Classified by Sex and Product Preference

Sex	Product Preference		Marginal Probability
	S_1	S_2	
Male	.20	.32	.52
Female	.40	.08	.48
Marginal probability	.60	.40	1.00

For two events, a joint probability is the probability of the intersection of two events—in other words, the probability that both events will occur at the same time. As we saw in Equation 5.16, the joint probability can be defined as

$$P(A \cap B) = P(B|A)P(A) = P(A|B)P(B) \qquad (5.16)$$

$P(A \cap B)$ can be also represented as $P(A$ and $B)$. It is used to denote the probability that both events A and B will occur. This equation implies that a joint probability is the product of a marginal probability [either $P(A)$ or $P(B)$] and a conditional probability [either $P(B|A)$ or $P(A|B)$]. In Table 5.6, marginal probabilities are presented in the last row and the last column. The concept of marginal probability is discussed later in this section.

In the case where we drew a spade from among 13 spades, $P(A) = 13/52$ and $P(B|A) = 1/13$, so the joint probability is

$$P(B \cap A) = (1/13)(13/52) = 1/52$$

Marginal Probabilities

In addition to joint probability, we can also obtain from Table 5.6 probabilities for the two classifications "sex" and "product preference." These probabilities, which are shown in the margins of the joint probability table, are referred to as **marginal probabilities** or **unconditional probabilities.** For example, the marginal probability that a randomly chosen individual is female is $P(F) = .48$, and the marginal probability that a person prefers soft drink 1 is $P(S_1) = .60$. In these cases, the marginal probabilities for each classification are obtained by summing the appropriate joint probabilities. Because marginal probability is a probability of a simple event, it is often called the **simple probability.**

Armed with this information on joint probability and marginal probability, we can calculate conditional probability as indicated in Equation 5.13. The probability that the individual is female *and* prefers soft drink 1, for example, can be calculated, in terms of data listed in Table 5.6, as $P(S_1 \cap F) = .40$. We also know that $P(F) = .48$. Substituting this information into Equation (5.13), we obtain

$$P(S_1|F) = \frac{P(S_1 \cap F)}{P(F)} = \frac{.40}{.48} = \frac{40}{48} = \frac{5}{6}$$

For comparison, we now calculate

$$P(F|S_1) = \frac{P(F \cap S_1)}{P(S_1)} = \frac{.40}{.60} = \frac{2}{3}$$

Note that $P(F|S_1) \neq P(S_1|F)$. Note that $P(S_1 \cap F)$ can be calculated by dividing 500 into 200 (see the data presented in Table 5.6).

To further illustrate this point, let's consider the following example.

EXAMPLE 5.7   *Classifying Students by Two Criteria*

Suppose we consider the problem of randomly selecting 1 student as a representative from a class of 80 students. In this class there are 5 black male students, 25 black female students, 35 white male students, and 15 white female students, as indicated in Table 5.7.

Table 5.7   50 Persons Classified by Sex and Race

Sex	Black	White	Total
Male	5	35	40
Female	25	15	40
Total	30	50	80

Let

$$B = \text{event that the student is black}$$
$$W = \text{event that the student is white}$$
$$M = \text{event that the student is male}$$
$$F = \text{event that the student is female}$$

Events B and W classify the students by race. Events M and F classify them by sex.

Because each event represents only one of the different classifications (sex or race), the probabilities of these events are called marginal probabilities. Marginal probabilities for race and sex can be calculated as

$$P(B) = \frac{30}{80}, \quad P(W) = \frac{50}{80}, \quad \text{and } P(B \cup W) = 1$$

$$P(M) = \frac{40}{80}, \quad P(F) = \frac{40}{80}, \quad \text{and } P(M \cup F) = 1$$

We can calculate the conditional and joint probabilities by using Table 5.7. The conditional probabilities are

$$P(B|M) = \frac{\frac{5}{80}}{\frac{40}{80}} = \frac{5}{40}$$

$$P(W \mid M) = \frac{\frac{35}{80}}{\frac{40}{80}} = \frac{35}{40}$$

$$P(F \mid W) = \frac{\frac{15}{80}}{\frac{50}{80}} = \frac{15}{50}$$

$$P(F \mid B) = \frac{\frac{25}{80}}{\frac{30}{80}} = \frac{25}{30}$$

The joint probabilities are

$$P(B \cap M) = \frac{5}{40} \frac{40}{80} = \frac{5}{80}$$

$$P(B \cap F) = \frac{25}{30} \frac{30}{80} = \frac{25}{80}$$

$$P(W \cap M) = \frac{35}{40} \frac{40}{80} = \frac{35}{80}$$

$$P(W \cap F) = \frac{15}{50} \frac{50}{80} = \frac{15}{80}$$

Suppose we do not know the exact numbers indicated in the table, but we know the joint probabilities and conditional probabilities. We can obtain the marginal probabilities from the joint probabilities by simply summing the joint probabilities. For example, if we want to know the probability of selecting a black student, we can sum all the probabilities we know to be associated with black students. The probability of selecting a black student is the probability of selecting a black male student plus the probability of selecting a black female student. That is,

$$P(B) = P(B \cap M) + P(B \cap F) = \frac{5}{80} + \frac{25}{80} = \frac{30}{80}$$

We can calculate other marginal probabilities in the same way:

$$P(W) = P(W \cap M) + P(W \cap F) = \frac{35}{80} + \frac{15}{80} = \frac{5}{8}$$

$$P(M) = P(B \cap M) + P(W \cap M) = \frac{5}{80} + \frac{35}{80} = \frac{1}{2}$$

$$P(F) = P(B \cap F) + P(W \cap F) = \frac{25}{80} + \frac{15}{80} = \frac{1}{2}$$

Hence the probabilities for individual events—$P(B)$, $P(W)$, $P(M)$, and $P(F)$—are known as marginal probabilities. In this example, say A represents sex and B represents color. The probabilities of individual events can then be represented as $P(A_i)$ and $P(B_j)$, where $i = B, W$ and $j = M, F$.

EXAMPLE **5.8**   *Marginal Probabilities on Wall Street*

Let A represent the state of economic conditions, and let B represent movement upward or downward of the stock price for Linden, Inc. The probabilities for Linden's stock price movement are presented in Table 5.8.

Table 5.8 Probabilities for Stock Prices and Economic Conditions

Economic Condition	Stock Price		Total
	Increase	*Decrease*	**Total**
Good	.28	.06	.34
Normal	.16	.15	.31
Poor	.05	.30	.35
Totals	.49	.51	1.00

Let us use I, D, G, N, and P to represent the events of stock price increase, stock price decrease, good economic conditions, normal economic conditions, and poor economic conditions. Then, from Table 5.8, we can calculate the marginal probabilities for the stock price movement of Linden, Inc.:

$$P(I) = P(G \cap I) + P(N \cap I) + P(P \cap I)$$
$$= .28 + .16 + .05$$
$$= .49$$
$$P(D) = P(G \cap D) + P(N \cap D) + P(P \cap D)$$
$$= .06 + .15 + .30$$
$$= .51$$

This outcome implies there is a 49 percent chance that the stock price will increase and a 51 percent chance that the stock price will decrease.

5.6 INDEPENDENT, DEPENDENT, AND MUTUALLY EXCLUSIVE EVENTS

Two events are referred to as **independent events** when the probability of one event is not affected by the occurrence of the other. For example, suppose a fair coin is flipped twice. The probability of getting a head on the second toss is not affected by having gotten a head or a tail on the first trial; thus the two trials are independent. However, many events are not independent. For example, the probability that a child in a less developed country will receive an advanced education is affected by his or her family's economic status. If the family is well off, the child probably will go on to higher education. Otherwise, the child may have to give up the opportunity for education to help support the family. Therefore, the event of the child's higher education *depends* on the event of his or her family's financial condition.

Suppose a fair coin is tossed once with the probability of 1/2 of obtaining a tail (event A). Let event B be the event of tossing the coin a second time and getting a tail. What is the probability of event B, given event A (one tail)? Or, in symbols, $P(B|A)$?

We observe that the occurrence of the second tail is not influenced by (is independent of) the occurrence of the first tail. In such a case, we say event B is statistically independent of event A. Here $P(B|A) = P(B) = .5$, because the occurrence of A has no influence on B.

For independent events, then,

$$P(A|B) = P(A) \tag{5.17}$$
$$P(B|A) = P(B)$$
$$P(A \cap B) = P(B \cap A) = P(B)P(A) \tag{5.18}$$

Equation 5.18 is a special case of Equation 5.16. From Equations 5.17 and 5.18 we know that the joint probability of two independent events is equal to the product of the marginal probabilities of these two events.

Let A and B be independent such that we know $P(A \cap B) = P(A)P(B)$. Under what circumstances could A and B also be mutually exclusive? If they were, $P(A \cap B)$ would be equal to 0, which implies either $P(A) = 0$ or $P(B) = 0$. Thus independent events with positive marginal probabilities can never be mutually exclusive. For example, the GMAT score of student A is independent of the score of student B. The events {A scores ≥ 700} and {B scores ≤ 650} are assumed independent, but they are not mutually exclusive, and two mutually exclusive events cannot be independent.

In sum, a pair of events are mutually exclusive if they cannot jointly occur—that is, if the probability of their intersection is zero.

5.7 BAYES' THEOREM

On the basis of Equation 5.13, we can incorporate additional information into probability analysis. From Equations 5.13 and 5.14, we define the conditional probability $P(B|A)$ as

$$P(B|A) = \frac{P(A \cap B)}{P(A)} = \frac{P(A|B)P(B)}{P(A)} \tag{5.19}$$

Equation 5.19 represents **Bayes' theorem,** which can be used to incorporate some extra information into the analysis.[3] The most interesting interpretation of Bayes' theorem is in terms of subjective probability, which was discussed in Section 5.2.

If we are interested in the event B and form a subjective view of the probability that B will occur, then $P(B)$ is called the **prior probability** in the sense that it is assigned *prior to* the observation of any empirical information. If we later acquire the information that the event A has occurred, this may cause us to modify to original judgment about the probability of event B. Because A is known to have occurred, the relevant probability for event B is now the conditional probability of B given A, and it is called the **posterior probability,** or the **revised probability,** because it is assigned *after* the observation of empirical evidence or additional information.

Bayes' theorem provides a method for incorporating new information into our probability beliefs. Formally, we use Bayes' theorem to update a prior probability to a posterior probability when additional information about event A becomes available. We do this by multiplying the prior probability by the adjustment factor $P(A|B)/P(A)$.

For example, financial analysts have observed stock prices declining when interest rates increase. They have also observed stock prices moving randomly.

If we collect historical data, we can obtain estimates of the probability of the event "a stock price increase and a decline in interest rates." The probability of a fall in stock prices is what we are most interested in. This probability is called the *prior probability,* because it is based on historical data and our own subjective judgment. If we see the interest rate rise (this is *additional information*), using Bayes' theorem gives us a better estimate of the probability that stock prices will fall. This forecasting method is described in Figure 5.13.

To use Equation 5.19 in the analysis, let

I_D = event that interest rates decline
I_U = event that interest rates increase
S_D = event that stock prices fall
S_U = event that stock prices increase

Then the conditional probability $P(S_D|I_U)$ can be defined as

$$P(S_D|I_U) = \frac{P(S_D \cap I_U)}{P(I_U)} = \frac{P(S_D)\,P(I_U|S_D)}{P(I_U)} \tag{5.19a}$$

[3]This theorem is attributed to an English clergyman, the Reverend Thomas Bayes (1702–1761).

Table 5.9 summarizes the historical data derived from 2,000 observations of changes in stock price.

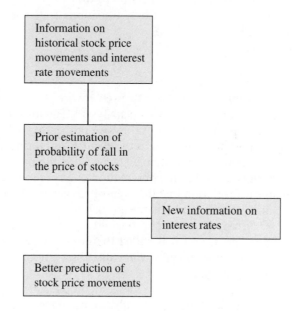

Figure 5.13

Flow Chart of Stock Price Forecasting

Table 5.9 Frequency Distribution of Changes in Stock Price

| Stock Price | Interest Rate | | Unit: Frequency |
	Decline	*Increase*	
Decline	100	850	950
Increase	900	150	1050
	1000	1000	2000

From Table 5.9, we can easily estimate $P(S_D | I_U)$ as

$$P(S_D | I_U) = \frac{P(S_D)P(I_U | S_D)}{P(I_U)} = \frac{(950/2{,}000)(850/950)}{1{,}000/2{,}000}$$

$$= \frac{850}{1{,}000} = .85 \text{ (revised probability)}$$

Using the new information that the interest rate will rise helps us predict a fall in stock prices more accurately. In other words, we are better able to predict the decline in the stock price.

Comparing $P(S_D | I_U)$ with $P(S_D)$ reveals the importance of Bayes' theorem, defined in Equation 5.19 or Equation 5.19a. Using Bayes' theorem, enables us to make better decisions by incorporating additional information into our probability estimates.

Here we have discussed using Bayes' theorem for just one basic event. The use of Bayes' theorem for two or more than two events, and the application of this technique in decision making, will be discussed in Chapter 21.

5.8 BUSINESS APPLICATIONS

APPLICATION 5.1 Determination of the Commercial Lending Rate

In this example we show a process for estimating the lending rate a financial institution would extend to a firm (or the lending rate that a borrower would feel is reasonable) on the basis of economic, industry, and firm-specific factors.

In standard banking practice, the lending rate depends in part on the interest rate on government Treasury bills.[4] Therefore, in order to determine the commercial lending rate, we need a forecast of the Treasury bill rate (R_f). This rate will be estimated for three types of economic conditions: boom, normal, and recession.

The second component of the lending rate, (R_p), is the risk premium.[5] It is possible to calculate R_p for each firm by examining the change in *earnings before interest and taxes* (EBIT) under the three types of economic conditions. The EBIT is used as an indicator of the ability of the borrower to repay borrowed funds. Table 5.10 lists all probability information we need to determine the commercial lending rate for Briarworth Company. Column (4) gives the marginal probability, and the probabilities listed in columns (1), (2), and (3) are the joint probabilities.

The probabilities shown in Table 5.10 can also be presented in terms of a tree diagram. Figure 5.14 shows that there are a total of 9 possible joint probabilities under the 3 different economic conditions and the 3 possible EBIT forecasts.

Table 5.10 Probabilities of the Lending Rate Determination for Briarworth Company

Economic Condition	Level of EBIT			Totals (4)
	High (1)	*Middle (2)*	*Low (3)*	
Boom	.15	.075	.025	.25
Normal	.20	.15	.15	.50
Poor	.025	.05	.175	.25
Totals	.375	.275	.350	1.00

[4]The Treasury bill rate was discussed in Chapter 3.

[5]The risk premium is the portion of the interest rate that is above the Treasury bill rate. This additional amount of interest is paid to compensate the lender for the risk it runs in making the loan.

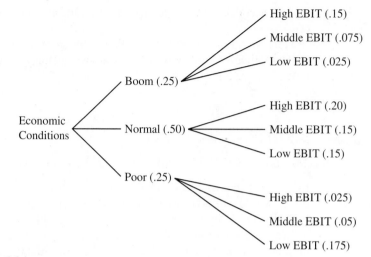

Figure 5.14

Tree Diagram of Events for Lending Rate Forecasting

We can use Table 5.10 and Equation 5.13 to calculate the conditional probabilities of EBIT level given the economic condition. For example, the probability that Briarworth Company has middle EBIT, given that the economic condition is boom, is

$$P(\text{middle EBIT})|\text{boom}) = \frac{P(\text{boom} \cap \text{middle EBIT})}{P(\text{boom})}$$

$$= \frac{.075}{.25} = .30$$

The conditional probabilities for EBIT level given economic condition are displayed in Table 5.11.

According to Section 5.6, a pair of events are independent if and only if their joint probability is the product of their marginal probabilities. In our example, for the events "normal" (normal economic condition) and "middle EBIT," we have, from Table 5.11,

Table 5.11 Conditional Probabilities of EBIT Levels, Given Economic Condition

Economic Condition	EBIT Level		
	High	*Middle*	*Low*
Boom	.60	.30	.10
Normal	.40	.30	.30
Poor	.10	.20	.70

$$P(\text{normal} \cap \text{middle EBIT}) = .15$$

and

$$P(\text{middle EBIT})P(\text{normal}) = (.275)(.50) = .1375$$

The product of the marginal probabilities is .1375, which differs from the joint probability .15. Hence the two events are not statistically independent.

In order to calculate the potential lending rate for Briarworth Company, the loan officer assumes that the predicted Treasury bill rates for boom, normal, and poor economic conditions are 12 percent, 10 percent, and 8 percent, respectively. In addition, the loan officer assumes that the risk premiums for the three different EBIT levels are 3 percent, 5 percent, and 8 percent. Using all this information, the loan officer constructs a worksheet such as Table 5.12.

Table 5.12 shows that during a boom, the Treasury bill rate is assumed to be 12 percent but the risk premium can take on different values. There is a 40 percent chance that it will be 3.0 percent, a 30 percent chance it will be 5.0 percent, and a 30 percent chance it will be 8.0 percent. According to the joint probability concepts we discussed in Section 5.5, the products of the conditional probability associated with R_p and the marginal probability associated with R_f are the joint probabilities of occurrence for the lending rates computed from these parameters. Therefore, during a boom, there is a 10 percent chance that a firm will be faced with a 15 percent lending rate, a 7.5 percent chance of a 17 percent rate, and a 7.5 percent chance of a 20 percent rate. This process also applies for the other conditions: normal economic ($R_f = 10$ percent) conditions and recession ($R_f = 8$ percent).

Table 5.12 Worksheet for Alternative Lending Rate Estimates for Briarworth Company

Economic Condition	(A) R_f	(B) Marginal Probability	(C) R_p	(D) Conditional Probability	(B × D) Joint Probability	(A + C) Lending Rate
Boom	12%	.25	3.0%	.40	.10	15%
			5.0	.30	.075	17
			8.0	.30	.075	20
Normal	10%	.50	3.0%	.40	.200	13%
			5.0	.30	.150	15
			8.0	.30	.150	18
Poor	8%	.25	3.0%	.40	.10	11%
			5.0	.30	.75	13
			8.0	.30	.75	16

From Table 5.12, the loan officer for Briarworth Company can estimate the potential lending rates and their probabilities as follows:

Potential Lending Rate (x_i), %	Probability (P_i), %
20	7.5
18	15.0
17	7.5
16	7.5
15	25.0
13	27.5
11	10.0
	100.0%

To calculate the estimated average lending rate, we generalized the simple arithmetic average indicated in Equation 4.2 in Chapter 4 as[6]

$$\overline{x} = \sum_{i=1}^{N} P_i x_i \tag{5.20}$$

where $\sum_{i=1}^{N} P_i = 1$. If P_1 and $P_2 = \cdots = P_N = \dfrac{1}{N}$, then Equation 5.20 reduces to Equation 4.1.

Substituting the information x_i and P_i into Equation 5.20 yields the estimated average lending rate.

$$\begin{aligned}
\overline{x} &= (.20)(.075) + (.18)(.15) + (.17)(.075) + (.16)(.075) \\
&\quad + (.15)(.25) + (.13)(.275) + (.11)(.10) \\
&= 15.1\%
\end{aligned}$$

The loan officer can use this estimated average lending rate as a guideline in determining the lending rate. The variance associated with this lending rate and other related analyses will be explored in Example 6.9 and Application 10.1.

■ **APPLICATION 5.2**   Analysis of a Personnel Data File

The personnel office of the J. C. Francis Company has files for 21,600 employees. These employees are broken down by age and sex in Table 5.13.

If one file is selected at random from the personnel office, what is the probability that it represents

1. An employee who is 40 years old or younger?

2. A female employee who is 40 years old or younger?

---

[6]We treat x as a measure of central tendency, as discussed in the last chapter. Alternatively, it can be treated as the expected value of a discrete random variable (lending rate), which will be discussed in Section 6.3.

3. Either a male employee or any employee over 40?

4. A male employee over 40?

5. A female employee or any employee 30 years old or older?

Table 5.13 Age and Sex Classification for J. C. Francis Company

	Sex		
	Female (F)	*Male (M)*	**Total**
Under 30 (*A*)	3,000	2,500	5,500
30–40 (*B*)	4,550	3,800	8,350
Over 40 (*C*)	2,850	4,900	7,750
Total	10,400	11,200	21,600

We shall denote the various events involved by A = under 30, B = 30–40, C = over 40, M = male, and F = female.

1. $P(40 \text{ or under}) = P(A \cup B)$

$$= \frac{5,500}{21,600} + \frac{8,350}{21,600} = .6412$$

2. $P(\text{female 40 or under}) = P(A \cap F) + P(B \cap F)$

$$= \frac{3,000 + 4,550}{21,600} = .3495$$

3. $P(\text{male or over 40}) = P(M \cup C)$

$$= P(M) + P(C) - P(M \cap C)$$

$$= \frac{11,200 + 7,750 - 4,900}{21,600} = .6505$$

4. $P(\text{male and over 40}) = P(M \cap C) = \dfrac{4,900}{21,600} = .2269$

5. $P(\text{female or 30 or older}) = P[F \cup (B \cup C)]$

$$= P(F) + P(B \cup C) - P[F \cap (B \cup C)]$$

$$= \frac{10,400 + 8,350 + 7,750 - (4,550 + 2,850)}{21,600} = .8843$$

■ *APPLICATION 5.3* Soda Purchase Survey

Mr. Mac Francis, manager of a Pathmark Supermarket in central New Jersey, would like to determine

1. The percentage of Kyle City families that did not purchase any soda during July of 1991.

2. The percentage of Kyle City families that purchased either diet or regular soda (or both) during July of 1991.

3. The percentage of Kyle City families that purchased only diet soda (no regular soda) during July of 1991.

4. The percentage of Kyle City families that purchased only regular soda (no diet soda) during July of 1991.

5. Whether diet soda purchases were related to regular soda purchases during the observed month, July of 1991.

Mr. Francis has asked you to conduct a study to answer these questions. He has provided you with sufficient Kyle City families for you to draw a random sample of 200. You conduct the survey and present Mr. Francis with Table 5.14, which summarizes the data you have accumulated.

Table 5.14 Summary Table of Diet Soda and Regular Soda Purchases for Kyle City Families During July of 1991

Purchases of Six-Packs of Regular Soda (RSODA)	Purchases of Six-Packs of Diet Soda (DSODA)		Total
	0	*1 or More*	**Total**
0	53	46	99
1 or more	62	39	101
Total	115	85	200

The data presented in Table 5.12 make it possible to answer all of Mr. Francis's questions.

1. $P[(DSODA = 0) \cap (RSODA = 0)] = \dfrac{53}{200} = .265$

   Hence it is inferred that 26.5 percent of Kyle City families did not purchase soda during July of 1991. Note that (DSODA = 0) is an event. It does not imply that P(DSODA) = 0.

2. $P[(DSODA > 0) \cup (RSODA > 0)] = P(DSODA > 0) + P(RSODA > 0)$
   $- P[(DSODA > 0) \cap (RSODA > 0)] = \dfrac{85 + 101 - 39}{200} = .735$

   Consequently, it is inferred that 73.5 percent of Kyle City families purchased either diet or regular soda (or both) during July of 1991.

3. $P[(DSODA > 0) \cap (RSODA = 0)] = \dfrac{46}{200} = .23$

   Hence it is inferred that 23 percent of Kyle City families purchased only diet soda (no regular soda) during July of 1991.

4. $P[(DSODA = 0) \cap (RSODA > 0)] = \dfrac{62}{200} = .31$

Consequently, it is inferred that 31 percent of Kyle City families purchased only regular soda (no diet soda) during July of 1991.

5. Last, we come to the joint probability of (DSODA $\geq$ 1) and (RSODA $\geq$ 1).

$$P[(\text{DSODA} \geq 1) \cap (\text{RSODA} \geq 1)] = \frac{39}{200} = .195$$

$$P(\text{DSODA} \geq 1) \times P(\text{RSODA} \geq 1) = \frac{85}{200} \times \frac{101}{200} = .2146$$

The fact that .195 $\neq$ .2146 implies that the joint probability is not equal to the product of two marginal probabilities. Hence, in accordance with the definition of dependence given in Section 5.6, it can be concluded that the purchase of diet soda was statistically dependent on the purchase of regular soda (and vice versa) during July of 1991.

■ | **APPLICATION 5.4** Ages and Years of Teaching Experience

Table 5.15 presents the age and number of years of teaching experience of 15 marketing professors. Figure 5.15 is the MINITAB printout of a Venn diagram of set *A* of marketing professors between 33 and 43 years of age, inclusive, and set *B* of marketing professors with more than 5 years of teaching experience.

Table 5.15 Age and Years of Teaching Experience of 15 Marketing Professors

Person	Age	Years of Teaching Experience
1	38	5
2	33	4
3	40	6
4	43	7
5	45	10
6	38	6
7	36	7
8	29	3
9	35	5
10	28	3
11	30	2
12	42	5
13	41	6
14	37	1
15	42	7

```
MTB > READ C1 C2
DATA> 38 5
DATA> 33 4
DATA> 40 6
DATA> 43 7
DATA> 45 10
DATA> 38 6
DATA> 36 7
DATA> 29 3
DATA> 35 5
DATA> 28 3
DATA> 30 2
DATA> 42 5
DATA> 41 6
DATA> 37 1
DATA> 42 7
DATA> END
       15 ROWS READ
MTB > NAME C1'AGE'
MTB > NAME C2'EXPERIEN'
MTB > PRINT C1-C2

 ROW   AGE  EXPERIEN

   1    38        5
   2    33        4
   3    40        6
   4    43        7
   5    45       10
   6    38        6
   7    36        7
   8    29        3
   9    35        5
  10    28        3
  11    30        2
  12    42        5
  13    41        6
  14    37        1
  15    42        7


MTB > PLOT C2 C1


            -                                             *
            -
        9.0+
            -
  EXPERIEN-
            -                          *              *  *
            -
        6.0+                               *    *  *
            -
            -                      *       *       *
            -                   *
            -
        3.0+      *  *
            -
            -         *
            -                        *
            -
            ----+---------+---------+---------+---------+---------+--AGE
              28.0      31.5      35.0      38.5      42.0      45.5

MTB > PAPER
```

Figure 5.15

MINITAB Printout of Venn Diagram for Age and Years of Teaching Experience

From the Venn diagram, we can calculate the following probabilities:

1. $P(A) = \dfrac{11}{15} = .73$

2. $P(B) = \dfrac{7}{15} = .47$

3. $P(A \cap B) = \dfrac{6}{15} = .40$

4. $P(A \cup B) = \dfrac{12}{15}$
$$= P(A) + P(B) - P(A \cap B)$$
$$= .73 + .47 - .4$$
$$= .80$$

5. $P(A \mid B) = \dfrac{P(A \cap B)}{P(B)} = \dfrac{\frac{6}{15}}{\frac{7}{15}} = \dfrac{6}{7} = .857$

Summary

In this chapter, we developed some of the basic tools of probability. The concept of probability enables us to assess the probabilities of various sample outcomes, given a specific population structure. In addition to discussing the basic concepts of probability, we explored more advanced topics such as conditional probability, joint probability, and marginal probability. We also showed how it is possible to use additional information to update probabilities by applying Bayes' theorem.

In Chapter 6 we extend the topics discussed in this chapter by introducing the concepts of discrete random variables and probability distributions. And in Chapters 7 and 8, we extend these concepts to the case of continuous random variables.

Appendix 5A Permutations and Combinations

In some probability problems, we encounter a finite set with distinct elements (objects):

$$S = \{e_1, e_2, \ldots, e_n\} \tag{5A.1}$$

If we want to know how many different ways there are of ordering these elements, then using permutation and combination techniques is the most effective way to proceed. For example, we know that 10 percent of Wakeley Company's accounts receivable contain errors. If 6 are selected at random, with replacement, then we can use permutation and combination techniques to calculate the probability that exactly 2 of those selected contained errors. (The solution of this problem appears in Example 5A.2.)

Permutations

The number of distinct arrangements that can be made from n elements of S, using r of them at a time, is denoted by $_nP_r$ and is called the **number of permutations** of n things taken r at a time ($r \leq n$). The number of permutations of a set of objects represents the number of ways the objects can be ordered. To obtain the result of $_nP_r$, we can apply the basic counting rule to the coin-tossing case. If a coin was tossed 4 times, then there are 4 steps (tosses) and each toss has 2 possible outcomes (heads and tails). The total number of outcomes in the experiment (N) is $N = 2 \cdot 2 \cdot 2 \cdot 2 = 16$.

To generalize this type of calculation, suppose we denote the number of outcomes in the first step of the experiment as n_1, the number of outcomes in the second step as n_2, and so on, where n_k denotes the number of outcomes in the last (the kth) step. The basic counting rule states that the total number of outcomes (N) equals the product of the numbers of outcomes in all steps.

$$N = n_1 \cdot n_2 \cdot n_3 \cdots \cdots n_k \tag{5A.2}$$

Suppose we have some number r of objects that are to be placed in order, and suppose each object can be used only once. How many different sequences are possible? This problem is similar to that defined in Equation 5A.2. It can readily be shown that $n_1 = r, n_2 = r - 1, \ldots, n_r = 1$. Hence the number of possible orderings of r objects is

$$r! = (r)(r - 1) \cdots (2)(1) \tag{5A.3}$$

Equation 5A.3 represents a factorial product.

Suppose now that we have n objects from which r are to be selected. The number of ways in which it is possible to select the r objects can be determined from the following product:

$$n \cdot (n - 1) \cdot (n - 2) \cdots \cdots (n - r + 1)$$

where

> n = number of choice for the first object
> $n - 1$ = number of choice for the second object
> $n - 2$ = number of choice for the third object
> $n - r + 1$ = number of choice for the $(n - r + 1)$th object

Thus the permutations $_nP_r$ can be defined as

$$_nP_r = n(n - 1) \cdots (n - r + 1)$$
$$= \frac{n!}{(n - r)!} \tag{5A.4}$$

For example, say we want to know in how many arrangements we can assign 4 students to 3 seats. We can put any of the 4 students in the first seat; there are 4 possibilities here. Then we can put any of the remaining 3 in the second seat. Finally, we must choose between the remaining 2 for the third seat. Thus $_4P_3 = (4)(3)(2) = 24$. This example illustrates that the "order," or arrangement, is important for a permutation.

EXAMPLE 5A.1 *Permutations of the Letters A, B, and C*

We are given the three letters A, B, and C. To determine the number of possible arrangements, note that we have 3 ways to select the first letter. Once the first letter has been selected, there are 2 ways to select the second letter from those that remain. There is only 1 way to select the third letter. Of course, no letter can be selected more than once in any arrangement. Using Equation 5A.2, we find that the total number of ways to make the selection (that is, to arrange the letters in order) is

$$3! = 3 \cdot 2 \cdot 1 = 6$$

Figure 5A.1 is a tree diagram showing 6 possible arrangements of the 3 letters. Each arrangement is a branch of the tree.

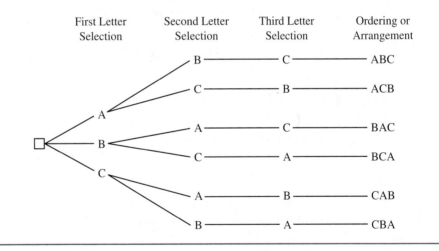

Figure 5A.1

Tree Diagram: Permutations of the Letters A, B, and C

Combinations

The number of permutations of a set of objects represents the number of ways the object can be ordered. Suppose we are interested in the number of different ways in which r objects can be selected from n, but we are not concerned with the order. Then the number of possible selections is called the **number of combinations** and is denoted by $\binom{n}{r}$. It can be shown that $\binom{n}{r}$ and $_nP_r$ are related by formula

$$r! \binom{n}{r} = {_nP_r} = \frac{n!}{(n-r)!}$$

Therefore

$$\binom{n}{r} = \frac{n!}{r!(n - r)!}$$

(5A.5)

For example, if $n = 5$ and $r = 3$, then

$$\binom{n}{r} = \frac{5!}{3!(5 - 3)!} = \frac{5 \cdot 4 \cdot 3 \cdot 2 \cdot 1}{(3 \cdot 2 \cdot 1)(2 \cdot 1)} = 10$$

Permutations and combinations can be used to simplify probability expressions and facilitate their evaluation.

EXAMPLE 5A.2   *Errors of Accounts Receivable*

Suppose 10 percent of Wakeley Company's accounts receivable are known to contain errors. If 6 accounts receivable are selected at random, with replacement, what is the probability that

1. None of those selected contains an error?
2. Exactly 2 of those selected contain errors?
3. At most 2 of those selected contain errors?
4. At least 2 of those selected contain errors?

Solutions

1. $P(\text{no errors}) = (9/10)^6 \approx .531$

2. We consider first the probability that the first 2 accounts receivable chosen contain errors and the remaining 4 do not. This is given by

$$\frac{1}{10} \cdot \frac{1}{10} \cdot \frac{9}{10} \cdot \frac{9}{10} \cdot \frac{9}{10} \cdot \frac{9}{10} = \left(\frac{1}{10}\right)^2 \left(\frac{9}{10}\right)^4$$

But there are $\binom{6}{2} = \binom{6}{4}$ combinations of the accounts with errors and 4 without. Each of these arrangements occurs with the probability

$$\left(\frac{1}{10}\right)^2 \cdot \left(\frac{9}{10}\right)^4$$

Thus, by using Equation 5A.5, we obtain

$$P(\text{exactly two errors}) = \binom{6}{2}(1/10)^2(9/10)^4$$

$$= \frac{6!}{2!4!} \cdot (.1)^2(.9)^4$$

$$= (15)(.0066)$$

$$= .098$$

3. By repeatedly employing the binomial formula as discussed in part 2, we obtain

$$P(\text{at most 2}) = P(\text{exactly 2}) + P(\text{exactly 1}) + P(0)$$

$$= \binom{6}{2}(1/10)(9/10)^4 + \binom{6}{1}(1/10)(1/9)^5 + \binom{6}{0}(9/10)^6$$

$$= .098 + .354 + .531$$

$$= .984$$

4. $P(\text{at least 2}) = P(\text{exactly 2}) + P(\text{more than 2})$

$$= .098 + [1 - P(\text{at most 2})]$$

$$= .098 + 1.000 - .984 = .114$$

Of course, this problem can also be solved by direct computation similar to the method used in part 3.

EXAMPLE 5A.3   *The Birthday Problem*

To compute probabilities, we often need to understand the concept of permutations and combinations. The "birthday problem" is a popular example of probability based on permutations. In the birthday problem, we are interested in the probability that at least 2 people in a given room have the same birthday. As we increase the number of people in the room, the number of possible combinations of people increases. With only 2 people in a room, there is only 1 possibility for a match. With 3 people (A, B, and C) in a room, there are 3 possible matches: A with B, A with C, and B with C. With 4 people in a room there are 6 possible matches, and so on.

What is the probability that in a class of $m = 50$ students, at least 2 students have the same birthday? To solve this, we assume that there are not twins among the m people in the class and that each of the 365 possible birthdays is equally likely. We therefore assume that anyone born on February 29 (leap year) will consider her or his birthday to be March 1 to make the problem manageable.

On the basis of these assumptions, we can see that there are 365 possible birthdays for each of the m people. Therefore, the sample space contains 365^m outcomes, all of which are equally probable. Now we proceed as though we were interested in the probability that *no* 2 people have the same birthday. We divide the number of permutations by the total number of outcomes. Letting B represent the event of m

students having different birthdays is precisely the same as asking in how many ways m birthdays can be selected from 365 possible birthdays and arranged in order. This is just the number of permutations, $_{365}P_m$. Then

$$P(\overline{B}) = \frac{_{365}P_m}{365^m}$$

is the probability that of our m people, no 2 have the same birthday. This is the complement of event B, at least 2 people having the same birthday.

Using Equations 5.11 and 5A.4, we find that the probability P that out of m students at least 2 people have the same birthday is

$$P(B) = 1 - \frac{_{365}P_m}{365^m} = 1 - \frac{365!}{(365 - m)!365^m} \tag{5A.6}$$

The following table shows the probability (P) for different values of m. We can see that with only 50 students in a class, there is a 97 percent probability that 2 or more students will have the same birthday.

m	P
10	.117
20	.411
30	.706
40	.891
50	.970
100	.9999997

Outcome Trees and Probabilities

The probability of an outcome is often much more difficult to calculate than that of the outcome of rolling a die once. For example, consider a biased coin that has a 1/3 probability of coming up heads and a 2/3 probability of coming up tails. If we flip this biased coin 4 times and list the possible outcomes toss by toss, we obtain the results shown in Figure 5A.2, which make up an **outcome tree.**

Let's consider the possible outcomes listed in the fifth column of Figure 5A.2. There are 16 distinct possible outcomes, which can be represented as

$$\{e_1, e_2, e_3, \ldots, e_{16}\}$$

where $e_1 = $ (HHHH), $e_2 = $ (HHHT), ..., $e_{15} = $ (TTTH) and $e_{16} = $ (TTTT). Using the relative frequency concept of probability as indicated in Equation 5.1 in the text, how do we find the probability of, for example, $e_1 = $ (HHHH)? If the probability of an individual outcome is independent, we can find $P(e_1)$ by multiplying together the probabilities of all outcomes. An event is said to be independent if its outcome does not depend on past outcomes in this case. For example, from our coin-tossing experiment, the probability of tossing a head is always 1/3, regardless

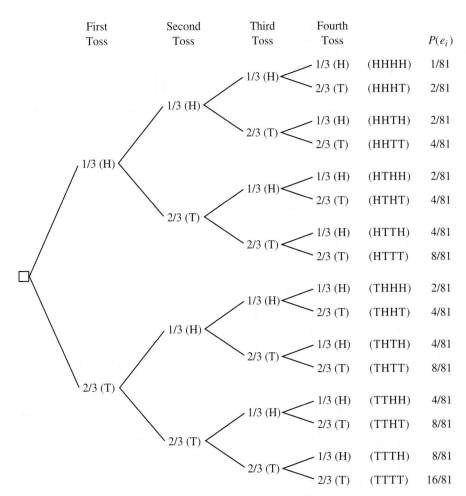

First Toss	Second Toss	Third Toss	Fourth Toss		$P(e_i)$
			1/3 (H)	(HHHH)	1/81
		1/3 (H)	2/3 (T)	(HHHT)	2/81
	1/3 (H)		1/3 (H)	(HHTH)	2/81
		2/3 (T)	2/3 (T)	(HHTT)	4/81
1/3 (H)			1/3 (H)	(HTHH)	2/81
		1/3 (H)	2/3 (T)	(HTHT)	4/81
	2/3 (T)		1/3 (H)	(HTTH)	4/81
		2/3 (T)	2/3 (T)	(HTTT)	8/81
			1/3 (H)	(THHH)	2/81
		1/3 (H)	2/3 (T)	(THHT)	4/81
	1/3 (H)		1/3 (H)	(THTH)	4/81
		2/3 (T)	2/3 (T)	(THTT)	8/81
2/3 (T)			1/3 (H)	(TTHH)	4/81
		1/3 (H)	2/3 (T)	(TTHT)	8/81
	2/3 (T)		1/3 (H)	(TTTH)	8/81
		2/3 (T)	2/3 (T)	(TTTT)	16/81

Figure 5A.2 An Outcome Tree for Four Tosses of a Biased Coin

of the previous toss. So the probability of tossing two heads in a row is the probability of tossing a head on the first toss multiplied by the probability of tossing a head on the second toss. Thus the probability of getting four heads is

$$\frac{1}{3} \cdot \frac{1}{3} \cdot \frac{1}{3} \cdot \frac{1}{3} = \frac{1}{81}$$

and the probability of getting two heads first and two tails later is

$$\frac{1}{3} \cdot \frac{1}{3} \cdot \frac{2}{3} \cdot \frac{2}{3} = \frac{4}{81}$$

From Figure 5A.2, we can calculate the probability of the event consisting of three heads and one tail as

$$\frac{2}{81} + \frac{2}{81} + \frac{2}{81} + \frac{2}{81} = \frac{8}{81}$$

Alternatively, this probability can be calculated as follows:

$$\binom{n}{r}(p)^r(1-p)^{n-r} = \binom{4}{3}\left(\frac{1}{3}\right)^3\left(\frac{2}{3}\right) = \frac{4!}{3!1!}\left(\frac{2}{81}\right) = \frac{8}{81} \qquad (5A.7)$$

Equation 5A.7 represents a binomial combination formula for calculating the probability. This formula was discussed in Example 5A.2. The concept will be used in developing binomial distribution in the next chapter.

Questions and Problems

1. Two cards are drawn from an ordinary deck of shuffled cards.
   a. What is the probability that they are both queens if the first card is replaced?
   b. What is the probability that they are both queens if the first card is not replaced?

2. Find the probability of a 5 turning up at least once in 2 tosses of a fair die.

3. Find the probability of rolling a 1 on the first roll of a die, a 2 on the second, and a 3 on the third.

4. A bag consists of 10 balls, 3 white and 7 red.
   a. What is the probability of drawing a white ball?
   b. What is the probability of drawing a white ball on the first draw and a red ball on the second draw when the first ball is replaced?
   c. How would your answer to part (b) change if there were no replacement?

5. You are given the sample space $S = \{a, b, c, d, e\}$ and the events $A = \{a, c, e\}$ and $B = \{b, d, e\}$.
   a. List the events $A \cup B, A \cap B, \overline{A}, \overline{B}, \overline{A} \cap B$, and $(\overline{A \cup B})$.
   b. Draw a Venn diagram for $A \cap B$.

6. Suppose you are flipping a fair coin.
   a. Find the probability of flipping 4 heads in a row.
   b. Find the probability of flipping H T H T H.
   c. Find the probability of flipping 5 heads in 6 flips.

7. What is the probability that at least 2 students in a class of 20 students will have the same birthday? Assume that there are no twins among the students and that all of the 365 birthdays are equally likely.

8. What is the probability of drawing 3 spades in a row from a standard deck of cards without replacement?

9. Roll a pair of dice 10 times and then calculate the mean and standard deviation for these rolls.
   a. What is the largest possible mean?
   b. What is the smallest possible mean?
   c. What is the smallest possible standard deviation?

10. Suppose a bag contains 12 balls distributed as follows: 5 red dotted balls, 2 red striped balls, 1 gray dotted ball, and 4 gray striped balls.
    a. Suppose you draw a red ball from the bag. What is the probability that it is striped?
    b. Suppose you draw a grey ball. What is the probability that it is dotted?
    c. Suppose you draw a dotted ball. What is the probability that it is red?

11. Determine the probability of betting on a winning number in a game of roulette. The numbers on the wheel are 0, 00, and 1 through 36. Each number is as likely as any other to become a winning number.

12. The probability that a car dealer will make a sale when he meets a prospective customer is 20 percent. If he meets 3 customers at random, what is the probability that all 3 customers will purchase a car?

13. Find the following joint probability: the probability that a sale will result in a sales commission, given that 75 percent of the sales representatives receive a commission on their sales and that 80 percent of the company's sales are made by sales reps.

14. Roll a die 25 times and construct a table showing the relative frequency of each of the 6 possible numbers.

15. Complete the probability for scores less than 600.

SAT Scores for Fiesta University

SAT	Number of Students
750–800	40
700–750	60
650–700	100
600–650	250
550–600	375
500–550	575
450–500	400
<450	100

16. In which of the following sets are the two events independent? In which are they mutually exclusive? In which are they neither?
    a. The Detroit Pistons win the NBA championship and the Oakland A's win the World Series.
    b. The Boston Red Sox win the pennant and the Boston Red Sox sell more than two million tickets in the same season.
    c. Both the New York Mets and the New York Yankees win the pennant in 1995.
    d. Both the New York Mets and the New York Yankees win the World Series in 1995.

17. A bag contains 3 balls: a black one, a white one, and a red one. A magician takes the balls out one by one. Draw an outcome tree. What is the probability of drawing the balls in the order of red-white-black?

18. Suppose that $P(E_1) = 0.3$ and $P(E_2) = 0.4$. Obtain $P(E_1 \cup E_2)$, $P(E_1|E_2)$, and $P(E_2|E_1)$, given that $P(E_1 \cap E_2) = .1$.

19. A baseball player has a lifetime batting average of .3. During a game, he has 5 at bats. A student of statistics argues that his chance of going 5 for 5 is $(.3)^5$. Do you agree? What assumption does the student make to come up with this answer?

20. The sales department wants to send two sales representatives on a business trip. There are 5 women and 5 men in the department. If the sales manager randomly selects two people, what is the probability that 1 woman and 1 man will be picked?

21. A survey of 200 students yields the following data:

	Own TV	Do Not Own TV
Own computer	50	30
Do not own computer	80	40

   a. What is the probability of drawing at random a student who owns both a computer and a TV?
   b. What is the probability of drawing at random a student who owns only a computer?
   c. In part (b), suppose we draw a student who owns a computer. What is the probability that this student also owns a TV?
   d. What is the marginal probability of owning a computer?

22. In question 21, if we draw two students at random without replacement, what is the probability of our getting a student who owns both and a student who owns neither?

23. A cereal company runs a certain advertisement in 3 media: newspaper, radio, and TV. Of the customers surveyed, 40 saw the advertisement on TV, 40 heard it on the radio, 30 read it in the newspaper, 20 saw it both in the newspaper and on TV, 20 know it both from TV and radio, 20 know it both from newspaper and radio, and 10 know it from all 3 media. How many customers were surveyed? (*Hint:* Use a Venn diagram.)

24. A city company has 300 employees. Among these employees, 2 out of every 3 take public transportation to work, 1 out of every 2 owns a car, and 1 out of every 3 owns a car but takes public transportation to work. How many employees do not own a car and take public transportation to work?

25. Draw 2 cards from a deck of cards without replacement. What is the probability of getting a diamond on the first draw and a club on the second? What is the probability of drawing 2 cards and getting a diamond and a club regardless of order?

26. What is the probability of getting the same outcome in 2 rolls of a die? What is the probability that the sum of 2 outcomes is 7?

27. Of the light bulbs delivered on May 25, 400 are produced in the morning shift, 300 in the evening shift, and 300 in the night shift. Say we pick a light bulb at random.
    a. What is the probability that we have a light bulb produced in the night shift?
    b. What is the probability that we have a light bulb produced in either the morning shift or the night shift?

28. In question 27, assume that $\frac{1}{10}$ of the light bulbs produced in the morning shift, $\frac{1}{10}$ of the light bulbs produced in the evening shift, and $\frac{1}{5}$ of the light bulbs produced in the night shift are defective. Say we pick a light bulb at random.
    a. What is the probability that the light bulb is defective?
    b. What is the chance that the light bulb is defective and was produced in the night shift?
    c. Suppose we get a defective light bulb. What is the chance that this light bulb was produced in the night shift?

29. A sports magazine wants to learn something about its subscribers. The subscribers are classified as teen-agers or older people and as being in school or holding a job. The magazine sends out a questionnaire to its readers and obtains the following results:

    40 percent are older than 20.
    60 percent are teen-agers.
    40 percent of the teen-agers who subscribe are in school.
    40 percent of the subscribers hold a job.

    What is the possibility that a subscriber is older than 20 and holds a job?

30. The business majors at Metropolitan University can be broken down as follows:

Major	Male	Female	Total
Accounting	200	400	600
Finance	400	250	650
Marketing	200	250	450
Management	200	100	300
Total	1000	1000	2000

a. We have randomly selected 4 students to attend a regional conference. What is the probability that we have a representative from each major?
b. We have randomly selected 4 students to attend a regional conference. What is the probability that we have a female student from each department?
c. The dean has randomly selected a student from each department to attend a regional conference. What is the probability that all 4 students selected are females?

31. A survey at Metropolitan College shows that among 750 economics majors, every student has taken at least 1 course in either economics or statistics. We also know that

    450 students have taken statistics.
    450 have taken microeconomics.
    450 have taken macroeconomics.
    250 have taken both micro- and macroeconomics.
    200 have taken both microeconomics and statistics.
    250 have taken both macroeconomics and statistics.

    a. How many students have taken all 3 courses?
    b. What is the probability that a student who we know has taken a course in microeconomics has also taken statistics?

32. A local factory has two shifts: day shift and night shift. The day shift produces $\frac{2}{3}$ of the total product. Of the day shift product, 1 percent are defective. Of the night shift product, 2 percent are defective. If we randomly select 1 product, what is the chance that it was produced during the day shift? If the selected product is defective, what is the probability that it was produced during the night shift?

33. The following picture helps you obtain the probability that events *A, B,* and *C,* happen jointly. Write down the formula for obtaining $P(A \cap B \cap C)$.

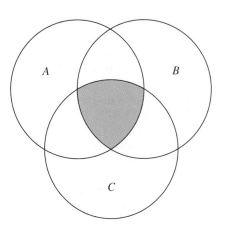

34. A hospital found that the probability of a power failure in a certain time period is .00001. To guarantee the functioning of the hospital, the hospital installed a back-up system that has a probability of .005 of breaking down. The two power systems operate independently. What is the probability that the hospital will completely stop functioning?

35. An insurance agent talks to 3 customers each day. Her probability of making a sale in the first meeting with a customer is .2. When she gets the second meeting with the same customer, the probability of making a sale increases to .8. In the past 2 days, this agent has talked to 3 customers twice. What is the probability that she made no sales?

36. Three different manuals were used to teach students how to type. Each manual was used by $\frac{1}{3}$ of the students. The results show that 30 percent of the students using manual A, 20 percent of the students using manual B, and 10% of the students using manual C can pass a typing test. We have found a student who passed the test. What is the probability that this student used manual A?

37. Fifty percent of the economists in the country are conservatives. The other 50 percent are liberals. Thirty percent of the conservative economists and 20 percent of the liberal economists believe that we will have a recession. We have found an economist who thinks we will see a recession in the next year. What is the probability that he or she is a conservative economist?

38. A training program is effective for 80 percent of the students whose mathematics background is strong, but it is effective for only 60 percent whose math background is not good. Assume that only 60 percent of a group of students are well trained in mathematics. What is the chance that the training program will be effective for this group of students?

39. A training program is effective for 80 percent of the students who are strongly motivated, but it is effective on only 60 percent of the students who are not strongly motivated. Assume that only 60 percent of the students are strongly motivated. We have selected a student who has benefited from the program. What is the probability that this student was strongly motivated?

40. Three percent of the products produced by the new assembly line are defective. Five percent of the product produced by the old machine are defective. The new machine produced 70 percent of the total product. The old machine produced 30 percent of the total product. We randomly draw a product and discover that it is defective. What is the probability that this defective item was produced by the old machine?

41. Mr. Doe wants to send 2 employees in his company on a business trip. He has 5 employees in the company. In how many different ways can he organize the trip?

42. When playing the Megabucks Lottery, a player is supposed to pick 6 numbers out of 48. If the lottery committee randomly picks the same 6 numbers, then the player hits the jackpot. What is the chance that a player will hit the jackpot?

43. An advertising agency wanted to find out what kinds of readers subscribed to a sports magazine. The agency sent out questionnaires with the magazine and received the following result:

	Blue-collar Job	White-collar Job
Teen-agers	20	30
The middle aged	30	30
Old people	30	10

a. If we know that a reader is a blue-collar worker, what is the probability that this reader is also an old person?

b. If we know that a reader is a teen-ager, what is the probability that this reader is also a white-collar worker?

44. Define the following terms: event, random experiment, subset, sample space, sample points.

45. Why is the concept of probability important to understanding statistics?

46. Explain what we mean when we say two events are independent.

47. Compare a simple event to a composite event. Give an example of each.

48. What do we mean by the union of two events? What do we mean by the intersection of two events? Use a Venn diagram to illustrate this point.

49. Explain what we mean by mutually exclusive events.

50. Briefly define conditional probability. Give some examples of conditional probability.

51. What is a joint probability? What is a marginal probability?

52. What is a prior probability? What is a posterior probability? Briefly explain how Bayes' theorem can be used to link the two.

53. What is the probability of obtaining a head in one toss of a fair coin? What is the probability of rolling a 5 in one roll of a fair die? What is the probability of tossing a head and rolling a 5?

54. You are dealt 4 cards from a standard 52-card deck. What is the probability that you will be dealt all 4 aces?

55. Again, you are dealt 4 cards. The first card is a spade, the second a heart, the third a diamond, and the fourth a club. What is the probability that you are dealt all 4 aces?

56. Consider the roll of a 6-sided die, its faces numbered 1, 2, 3, 4, 5 and 6. Draw a Venn diagram showing the 6 possible outcomes. Now draw circles showing the following rolls:
    a. An odd number
    b. 2 or an odd number
    c. 3 or an even number

57. Again, consider the roll of a 6-sided die. Given the following events A and B, find the intersection and the union for A and B if
    a. $A = \{1, 3, 5\}$ and $B = \{2, 4, 6\}$
    b. $A = \{1, 3\}$ and $B = \{1, 3, 5\}$
    c. $A = \{1, 2, 3\}$ and $B = \{2, 4, 5\}$
    d. $A = \{1, 2, 3, 4\}$ and $B = \{3, 4, 5, 6\}$

58. You have drawn 3 diamonds, 1 spade, and 1 heart from a deck of cards. If you discard the spade and the heart, what is the probability of your drawing 2 cards from the remaining 47 cards to obtain a flush (5 cards of the same suit)?

59. In poker, a royal flush consists of A, K, Q, J, and 10 of the same suit. What is the probability of drawing 5 cards and obtaining a royal flush? What is the probability of being dealt a royal flush in spades?

60. Suppose there are 6 unrelated people in a room. What is the probability that any 2 of them have the same birthday?

61. Suppose you toss a coin 3 times. What is the probability of tossing 3 heads in a row? What is the probability of tossing 3 tails in a row? What is the probability of tossing either 3 heads or 3 tails in a row?

62. An advertising executive decides that a television commercial should be shown on 2 television stations. If 3 television stations serve the area the company wants to reach, how many possible combinations does the executive have to choose from? If a fourth television station becomes available, how many combinations are there now?

63. An automobile manufacturer produces cars in 4 different colors and offers 3 different options packages. How many different combinations of color and options package can the auto manufacturer offer?

64. A basketball player makes 75 percent of his shots from the foul line. What is the probability of this player making 10 shots in a row from the foul line? Are there any assumptions we need to make to compute this answer?

65. Your investment advisor has a portfolio of 75 stocks: 40 high-growth stocks and 35 high-dividend stocks. Of the 40 high-growth stocks, 25 have increased in value over the last year, whereas 10 of the high-dividend stocks have increased in value.
    a. If a stock is selected at random, what is the probability that the stock will be a high-dividend stock that has increased in value?
    b. What is the probability of selecting a stock that has not increased in value?
    c. If the stock selected has increased in value, what is the probability that it is a high-growth stock?

66. The Whiter Smile Company is about to begin selling a new toothpaste. Company planners know that the probability of the new product being profitable is 10 percent. They also know from previous market research that when their test panel likes the product, it has an 80 percent chance of being profitable. Historically, panels like 10 percent of the new products. Using the Bayesian approach, find the probability that the panel liked the product if the toothpaste is profitable.

67. Suppose you flip a fair coin once and roll a 6-sided die once. What is the probability of tossing a tail and rolling an odd?

68. Suppose you flip a coin twice and roll a die twice. What is the probability that you will toss 1 head and 1 tail and will roll two 6's?

69. A top amateur bowler has a 70 percent chance of rolling a strike. What is the probability that this bowler will bowl a perfect game (12 strikes in a row)? Are there any assumptions we need to make to answer this question?

70. The Tastee Coffee Company is about to begin selling a new gourmet coffee. Company managers know that the probability that the test panel will like the product is 20 percent. They also know from previous market research that when the product is profitable, there is a 60 percent chance that their test panel liked the product and that when it is unprofitable, there is a 95 percent chance that the panel did not like it. Using the Bayesian approach, find the probability that the gourmet coffee is profitable if the test panel liked the product.

71. A real estate developer offers homes in 5 different colors and 3 different models. How many different combinations of color and model can the real estate developer offer?

72. Rah Rah College has a limited number of dorm rooms available to students, so every year students participate in a lottery to determine whether they will have school housing next year. Suppose that every year 25 percent of the students do not receive school housing. In his sophomore year, Bob Smith is one of the "losers" in the lottery and does not receive school housing. He consoles himself by noting that because 25 percent of the students do not receive housing each year "everyone should lose once in the 4 years of college." He therefore figures that he is assured of getting housing in his junior and senior years. Is Bob's assumption accurate?

73. A stock broker owns 5 suits and 12 ties. Assuming that all the suits and ties match, determine how many different outfits (combinations) the stock broker can wear.

74. An advertising agency suggests that a bicycle manufacturer advertise in 4 of the 7 bicycling magazines. How many different combinations of 4 magazines can be selected?

75. A Senate committee consists of 6 Democrats and 5 Republicans. In how many ways can a subcommittee consisting of 4 Democrats and 4 Republicans be formed?

76. You believe you have come up with a fool-proof way to win at roulette. Because the odds are nearly 50-50 that red will come up and nearly 50-50 that black will come up in roulette, you believe that whenever black comes up, red will come up next, and vice versa. Do you think this is a winning strategy?

77. At the beginning of each week, a company decides how much to spend on newspaper ads for that week. It spends either $250 or $500 each week on newspaper ads. Assuming there is an equal probability of spending either amount, find the probability that in a month (4 weeks), the total advertising expenditure is greater than $1250.

78. A car salesman meets 12 customers each week. His probability of making a sale in the first meeting with a customer is .3. When he gets the second meeting with the same customer, the probability of making a sale increases to .9. Over the last 2 weeks, he talked to 4 customers twice. What is the probability that he made no sales?

79. A life insurance company knows with certainty that all people will die someday. Does it make sense for the insurance company to use probability theory to set its life insurance rates?

80. Consider the sample space $S = \{A, B, C, D, E, F, G\}$ and the following events:

$I = \{A, C, E, G\}$
$II = \{B, D, E,\}$
$III = \{A, B, C, D\}$
$IV = \{E, F, G\}$
$V = \{B, F\}$
$VI = \{B, D, E, F\}$

Are the following sets of events mutually exclusive, collectively exhaustive, both, or neither?

a. I and II e. I and IV
b. III and IV f. II and III
c. I and III g. I and V
d. II and IV h. I and VI

81. State the complement of each of the following events:
a. Drawing a spade from a full deck of cards.
b. Inflation of less than 5 percent per year.
c. GNP growth of more than 4 percent per year.

82. The figure below is a plot of salary and experience of employees of the Endicott Company. Answer the following questions by using a Venn diagram.
a. The probability of experience between 3–5 years.
b. The probability of more than 4 years' experience and a salary of between $25,000 and $37,000.

83. Use the data in question 82 to construct histograms of salary and experience using the MINITAB program. Explain the results.

84. Use the data in question 82 to calculate the mean and standard deviation of salary and experience using the MINITAB program.

Figure for Question 82

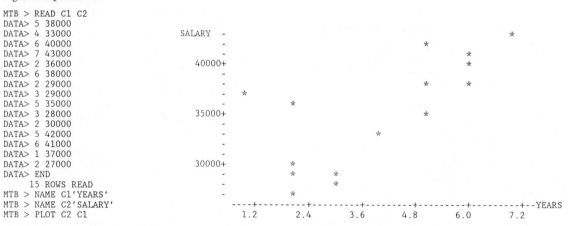

```
MTB > READ C1 C2
DATA> 5 38000
DATA> 4 33000              SALARY  -                                                      *
DATA> 6 40000                      -                                      *
DATA> 7 43000                      -                                               *
DATA> 2 36000              40000+                                                  *
DATA> 6 38000                      -
DATA> 2 29000                      -                                      *       *
DATA> 3 29000                      -      *
DATA> 5 35000                      -              *
DATA> 3 28000              35000+                                         *
DATA> 2 30000                      -
DATA> 5 42000                      -                      *
DATA> 6 41000                      -
DATA> 1 37000                      -
DATA> 2 27000              30000+          *
DATA> END                          -         *        *
      15 ROWS READ                 -                *
MTB > NAME C1'YEARS'               -           *
MTB > NAME C2'SALARY'              ----+---------+---------+---------+---------+---------+--YEARS
MTB > PLOT C2 C1                       1.2       2.4       3.6       4.8       6.0       7.2
```

CHAPTER 6

Discrete Random Variables and Probability Distributions

Key Terms

random variable
discrete random variable
continuous random variable
probability function
probability distribution
probability mass function
cumulative distribution function
step function
expected value
Bernoulli process
binomial distribution
binomial probability function
lot acceptance sampling

hypergeometric distribution
hypergeometric random variable
hypergeometric formula
Poisson distribution
joint probability function
joint probability distribution
marginal probability function
conditional probability function
conditional probability distribution
covariance
coefficient of correlation
option
random walk

6.1 INTRODUCTION

In Chapters 2, 3, and 4 we explored descriptive statistical measures, and we examined probability concepts and techniques in Chapter 5. Here we will build on this foundation as we establish the definitions of discrete and continuous random variables and discuss important discrete probability distributions in terms of specific numerical outcomes.

The binomial distribution, hypergeometric distribution, Poisson distribution, and joint probability functions are discussed in detail in this chapter. We also explore the Poisson approximation to the binomial distribution and examine joint probability functions and distributions. Finally, we investigate expected value and variance of the sum of both uncorrelated and correlated random variables.

In Appendix 6A, the mean and variance for the binomial distribution are derived. And in Appendix 6B, we explain how the binomial distribution can be used in developing the binomial option pricing model.

6.2 DISCRETE AND CONTINUOUS RANDOM VARIABLES

A random experiment generally results in numerical values that can be attached to the possible outcomes. In experiments such as throwing a die or measuring a firm's net earnings, the outcomes are naturally in numerical form. The possible outcomes of tossing a fair die are 1, 2, 3, 4, 5, and 6, and the corresponding probabilities are $\frac{1}{6}$ for each outcome, as we saw in Chapter 5. The result of a random experiment can be conveniently described by a random variable. A **random variable** is a variable that assigns a numerical value to each possible outcome of a random experiment. We can think of a random variable as a value or magnitude that changes from occurrence to occurrence in no predictable sequence. A breast-cancer screening clinic, for example, has no way of knowing exactly how many women will be screened on any one day. So tomorrow's number of patients is a random variable. For another example, say a company manufactures TV sets that are sometimes defective. Buyers return the defective sets for repair. A variable used to describe the number of TV sets that will be returned before the warranty runs out is a random variable.

Random variables are either *discrete* or *continuous*. A **discrete random variable** is one that can take on a countable number of values; usually it is an integer. The number of claims on an automobile policy in a particular year is a discrete random variable. Another discrete random variable is the number of defective parts produced in a particular run. Here the discrete random variable can take on the values 0, 1, 2, . . . , n. If we let X stand for a discrete random variable, then we can use x to represent one of its possible values. In other words, X is a quantity and x a value. For example, before the results of rolling a fair die are observed, the random variable can be used to denote the outcome. This random variable can assume the specific values $x = 1$, $x = 2$, ..., $x = 6$, and each value has a probability of $\frac{1}{6}$. Other discrete random variables include

1. The number of bids received in a stock offering: $x = 0, 1, 2, \ldots$.
2. The number of customers waiting to be served in a bank at a particular time: $x = 0, 1, 2, \ldots$.
3. The number of sales made by a salesperson in a given month: $x = 0, 1, 2, \ldots$.
4. The number of people in a sample of 800 who favor a particular presidential candidate: $x = 0, 1, 2, \ldots, 800$.

In contrast, a **continuous random variable** can take on an uncountable number of values within an interval. The amount of rainfall in a given area is a continuous random variable. This number can take on an infinite number of values—8.01 inches of rain is different from 8.012 inches. Measurement may stop at some number of decimal points, but the variable is theoretically continuous. Although it is impossible to attach a probability to the amount of rain equaling exactly 8.012000 . . . inches, it is possible to give the probability that the amount of rain will be within an interval. Continuous random variables can also represent the amount of time it takes to fill a food order at a restaurant or the length of a bolt used in the production of an automobile. Other continuous random variables appear in the following examples.

1. Let X be the arrival time at an airport between 8:00 and 9:00 A.M.: $8:00 \leq x \leq 9:00$.
2. For a new residential division, the length of time X from completion until a specified number of houses are sold: $a \leq x \leq b$, for $b > a$.
3. Let Y be the amount of orange juice loaded into a 24-ounce bottle in a bottling operation: $0 \leq y \leq 24$.
4. The depth at which a successful natural gas drilling venture first strikes natural gas.
5. The weight of a bag of rice bought in a supermarket.

6.3 PROBABILITY DISTRIBUTIONS FOR DISCRETE RANDOM VARIABLES

Probability Distribution

To analyze a random variable we must generally know the probability that the variable will take on certain values. The **probability function,** or the **probability distribution,** of a discrete random variable is a systematic listing of all possible values a discrete random variable can take on, along with their respective probabilities. The probability that the random variable X will assume the value x is symbolized by

$P(X = x)$ or simply $P(x)$. Note that X is a quantity and x a value. Because a discrete probability function takes nonzero values only at discrete points x, it is sometimes called a **probability mass function.**

EXAMPLE 6.1 *Probability Distribution for the Outcome of Tossing a Fair Coin*

Suppose a fair coin is tossed. Let the random variable X represent the outcome, where 1 denotes heads and 0 denotes tails. The probability that heads appears is $P(X = 1) = .5$, and the probability that tails appears is $P(X = 0) = .5$. Figure 6.1 shows a probability distribution where the possible outcomes are charted on the horizontal axis, and probabilities on the vertical axis. The spikes in the figure place the probability of heads and that of tails at .5. Note that the probabilities for both outcomes (heads and tails) are between 0 and 1 inclusive and that the sum of both probabilities is 1.

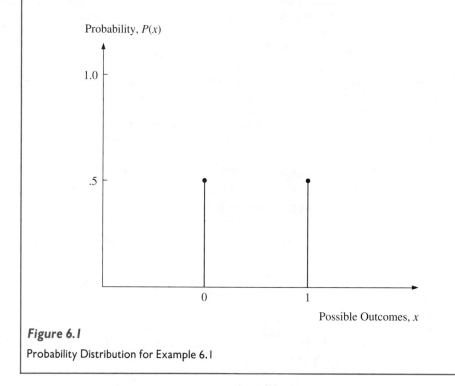

Figure 6.1

Probability Distribution for Example 6.1

EXAMPLE 6.2 *Probability Distribution for Section Assignment in a Marketing Course*

Suppose that 5 sections of a marketing course are offered and each section has a different number of openings (see Table 6.1). If students are assigned randomly to the sections, then a probability distribution can be drawn for section assignment. The probability that a student is assigned to section 1, $P(X = 1)$, is equal to $\frac{23}{179} = .128$; the probability that a student is assigned to section 2, $P(X = 2)$, is equal to $\frac{45}{179} = .251$. The rest of the probabilities are calculated in the same manner, as indicated in the third column of Table 6.1. Figure 6.2 shows this probability distribution.

Table 6.1 Probability Distribution of Marketing Course Openings

Section, x	Openings	Probability, $P(x)$
1	23	23/179 = .128
2	45	45/179 = .251
3	21	21/179 = .117
4	56	56/179 = .313
5	34	34/179 = .190
Total	179	1.00

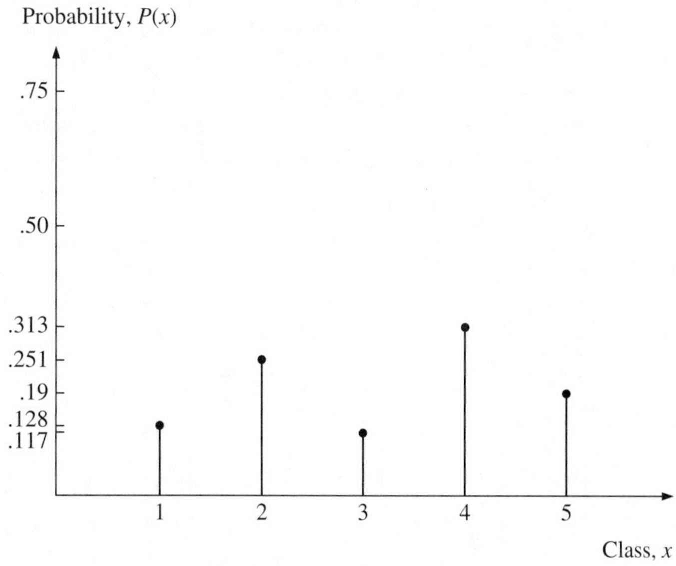

Probability, $P(x)$

Figure 6.2

Probability Distribution for Marketing Course Openings

EXAMPLE 6.3 *Probability Distribution for the Outcome of Rolling a Fair Die*

The probability distribution for the roll of a fair die is shown in Figure 6.3. Here all the spikes are equal to $\frac{1}{6}$ because $P(X = 1) = P(X = 2) = \cdots = P(X = 6) = \frac{1}{6}$.

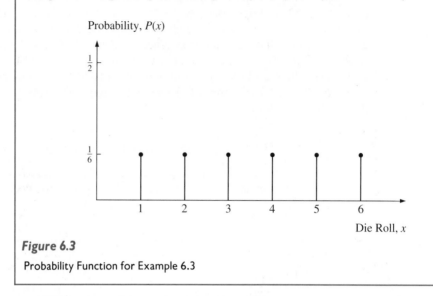

Figure 6.3

Probability Function for Example 6.3

Examples 6.1–6.3 show that the probability of a random variable X taking on the specific value x can be denoted as $P(X = x)$. The probability distribution of a random variable is a representation of the probabilities for *all* possible outcomes. $P(X = x)$ is the probability function of random variable X denoting the probability that X takes on the value of x. This expression can be rewritten as $P(X = x) = P(x)$ where the function is evaluated at all possible values of X.

For all probability functions of discrete random variables,

1. $P(x_i) \geq 0$ for all i

2. $\sum_{i=1}^{n} P(x_i) = 1$

where x_i is the ith observation of random variable X. Property 1 states that the probabilities cannot be negative. Property 2 implies that the individual probabilities add up to 1.

Probability Function and Cumulative Distribution Function

For some problems, we need to find the probability that X will assume a value less than or equal to a given number. A function representing such probabilities is called a **cumulative distribution function** (cdf) and is usually denoted by $F(x)$. If x_1, $x_2, \cdots, x_m$ are the m values of X given in increasing order (that is, if $x_1 < x_2 < \cdots < x_m$, then the cumulative distribution function of x_k, $1 \leq k \leq m$, is given by

$$F(x_k) = P(X \leq x_k) \tag{6.1}$$

In Equation 6.1, $P(X \leq x_k)$ gives us the probability that X will be less than or equal to x_k. The relationship between the probability function $P(x)$ and the cumulative distribution function $F(x_k)$ can be expressed as follows:

$$F(X_k) = P(x_1) + P(x_2) + \cdots + P(x_k) = \sum_{i=1}^{k} P(x_i) \tag{6.2}$$

Because the values outside the range of X (values smaller than x_1 or larger than x_m) occur only with probability equal to zero, we may equally well write

$$F(x_k) = \sum_{i=-\infty}^{k} P(x_i) \quad \text{for } k \leq m \tag{6.2a}$$

The following examples show how to calculate the cumulative distribution function.

EXAMPLE 6.4 *Cumulative Distribution Function for Rolling a Fair Die*

Reviewing Example 6.3, we find that the value of a random variable X and its probability of occurring upon the rolling of a fair die are listed in the first and second columns of Table 6.2, respectively. Because $x_1 < x_2 < \cdots < x_6$, the cumulative distribution function can be calculated in accordance with Equation 6.1 as follows:

$$F(1) = P(X = 1) = \tfrac{1}{6}$$
$$F(2) = P(X = 1) + P(X = 2) = \tfrac{1}{6} + \tfrac{1}{6} = \tfrac{1}{3}$$
$$F(3) = P(X = 1) + P(X = 2) + P(X = 3) = \tfrac{1}{6} + \tfrac{1}{6} + \tfrac{1}{6} = \tfrac{1}{2}$$

and so on, as listed in the last column of Table 6.2. The cumulative distribution function is shown in Figure 6.4.

Table 6.2 Cumulative Distribution Function for the Outcome of Tossing a Fair Die

x	Probability Function, $p(x)$	Cumulative Distribution Function, $F(x)$
1	1/6	1/6
2	1/6	1/3
3	1/6	1/2
4	1/6	2/3
5	1/6	5/6
6	1/6	1
Total	1	

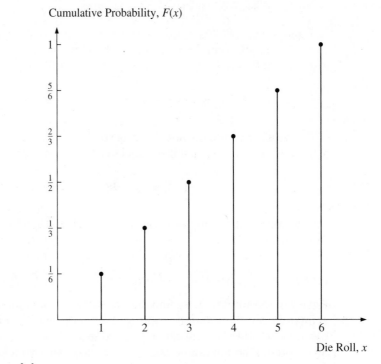

Figure 6.4

Cumulative Probability Distribution for Example 6.4

This graph is a **step function:** the values change in discrete "steps" at the indicated integral values of the random variable X. Thus $F(x)$ takes the value 0 to the left of the point $x = 1$, steps up to $F(x) = \frac{1}{6}$ at $x = 1$, and so on. The dot at the left of each horizontal line segment indicates the probability for that integral value of x. At these points, the values of the cumulative distribution function are read from the upper line segments.

6.4 EXPECTED VALUE AND VARIANCE FOR DISCRETE RANDOM VARIABLES

Probability distributions tell us a great deal about the probability characteristics of a random variable. Graphical depictions reveal at a glance the central tendency and dispersion of a discrete distribution, but numerical measures of central tendency and dispersion for a probability distribution are also useful. The mean (central location) of a random variable is called the **expected value** and is denoted by $E(X)$. The expected value of a random variable, which we also denote as μ, is calculated by summing the products of the values of the random variable and their corresponding probabilities.

$$\mu = E(X) = \sum_{i=1}^{N} x_i P(x_i) \tag{6.3}$$

John Kraft, a marketing executive for Computerland, Inc., must decide whether to use a new label on one of the company's personal computer products. The firm will gain $900,000 if Mr. Kraft adopts the new label and it turns out to be superior to the old label. The firm will lose $600,000 if Mr. Kraft adopts the new label and it proves to be inferior to the old one. In addition, Mr. Kraft feels that there is .60 probability that the new label is superior to the old one and .40 probability that it is not. The expected value of the firm's gain for adopting the new label is

$$E(X) = (\$900,000)(.6) + (-\$600,000)(.4)$$
$$= \$300,000$$

Therefore, Mr. Kraft should consider adopting the new label.

EXAMPLE 6.5   *Expected Value for Earnings per Share*

Suppose a stock analyst derives the following probability distribution for the earnings per share (EPS) of a firm.

EPS ($)	P(x)	EPS ($)	P(x)
1.50	.05	2.25	.15
1.75	.30	2.50	.10
2.00	.35	2.75	.05

To calculate the expected value (the mean of the random variable), we multiply each EPS by its probability and then add the products.

$$E(X) = 1.50(.05) + 1.75(.30) + 2.00(.35) + 2.25(.15)$$
$$+ 2.50(.10) + 2.75(.05) = 2.025$$

The expected value for the earnings per share is 2.025.

EXAMPLE 6.6   *Expected Value of Ages of Students*

Suppose the distribution of the ages of students in a class is

Age	P(x)	Age	P(x)
20	.06	24	.10
21	.10	25	.03
22	.28	26	.04
23	.39		

The expected age is

$$E(X) = 20(.06) + 21(.10) + 22(.28) + 23(.39)$$
$$+ 24(.10) + 25(.03) + 26(.04) = 22.62$$

In addition to calculating expected value for a probability distribution, we can compute the variance and standard deviation as measures of variability. The variance of a distribution is computed similarly to the variance for raw data, which we discussed in Chapter 4. The variance is the summation of the square of the deviations from the mean, multiplied by the corresponding probability:

$$\sigma^2 = \sum_{i=1}^{N} (x_i - \mu)^2 P(x_i) \qquad (6.4)$$

where σ^2 is the variance of X, μ is the mean of X, and $P(x_i)$ is the probability function of x_i. If $P(x_1) = P(x_2) = \cdots = P(x_N) = 1/N$, then Equation 6.4 reduces to

$$\sigma^2 = \frac{\sum_{i=1}^{N} (x_i - \mu)^2}{N} \qquad (6.4a)$$

The standard deviation is the square root of the variance. An alternative—and possibly easier—way to calculate the variance is to sum the product of the square of values of the random variables multiplied by the corresponding probabilities, and then subtract the expected value squared.[1]

$$\sigma^2 = \sum_{i=1}^{N} x_1^2 P(x_i) - \mu^2 \qquad (6.5)$$

[1]From Equation 6.4, the variance of X is

$$\sigma^2 = \sum_{i=1}^{N} (x_i^2 - 2\mu x_i + \mu^2) P(x_i)$$

$$= \sum_{i=1}^{N} x_i^2 P(x_i) - 2\mu \sum_{i=1}^{N} x_i P(x_i) + \mu^2 \sum_{i=1}^{N} P(x_i)$$

Because $\sum_{i=1}^{N} x_i P(x_i) = \mu$ and $\sum_{i=1}^{N} P(x_i) = 1$,

$$\sigma^2 = \sum_{i=1}^{N} x_i^2 P(x_i) - 2\mu^2 + \mu^2 = \sum_{i=1}^{N} x_i^2 P(x_i) - \mu^2$$

$$= \sum_{i=1}^{N} x_i^2 P(x_i) - \mu^2$$

EXAMPLE 6.7   *Expected Value and Variance: Defective Tires*

Suppose the following table gives the number of defective tires that roll off a production line in a day. Calculate the mean and variance.

Defects	Probability
0	.05
1	.15
2	.20
3	.25
4	.25
5	.10

The expected value is equal to $(0)(.05) + (1)(.15) + (2)(.20) + (3)(.25) + (4)(.25) + (5)(.10) = 2.8$. Thus the mean number of defective tires in a production run is 2.8 tires in a day.

The variance is

$$(0 - 2.8)^2(.05) + (1 - 2.8)^2(.15) + (2 - 2.8)^2(.20)$$
$$+ (3 - 2.8)^2(.25) + (4 - 2.8)^2(.25) + (5 - 2.8)^2(.10) = 1.86$$

The alternative formula yields the same answer for the variance:

$$[(0^2)(.05) + (1^2)(.15) + (2^2)(.20) + (3^2)(.25)$$
$$+ (4^2)(.25) + (5^2)(.10)] - (2.8)^2 = 1.86$$

EXAMPLE 6.8   *Expected Value and Variance: Commercial Lending Rate*

Returning to the example of commercial lending interest rates in Section 5.8, we can tabulate the possible lending rates x and the corresponding probabilities, $P(x)$ as follows:

x	$P(x)$	x	$P(x)$
15%	.100	18%	.150
17	.075	11	.100
20	.075	13	.075
13	.200	16	.075
15	.150		

From formulas for the expected value and the variance for discrete random variables, the mean of X is

$$E(X) = \sum_{i=1}^{N} x_i P(x_i) = \mu$$

$$= (.100)(.15) + (.075)(.17) + (.075)(.20) + (.200)(.13)$$
$$+ (.150)(.15) + (.150)(.18) + (.100)(.11) + (.075)(.13) \qquad (6.6)$$
$$+ (.075)(.16)$$
$$= 15.1\%$$

The standard deviation of X can be calculated from

$$\sigma = \left[\sum_{i=1}^{N} (x_i - \mu)^2 P(x_i) \right]^{1/2}$$

$$= \{(.100)(15 - 15.1)^2 + (.075)(17 - 15.1)^2 + (.075)(20 - 15.1)^2$$
$$+ (.200)(13 - 15.1)^2 + (.150)(15 - 15.1)^2 + (.150)(18 - 15.1)^2 \quad (6.7)$$
$$+ (.100)(11 - 15.1)^2 (.075)(13 - 15.1)^2 + (.075)(16 - 15.1)^2\}^{1/2}$$

$$= 2.51\%$$

A bank manager may use this information to make lending decisions, which is discussed in Application 7.4 in Chapter 7.

6.5 THE BERNOULLI PROCESS AND THE BINOMIAL PROBABILITY DISTRIBUTION

In this section, we examine first the Bernoulli process and then the binomial probability distribution and its applications.

The Bernoulli Process

The binomial distribution is based on the concept of a **Bernoulli process,** which has three important characteristics. First, a Bernoulli process is a repetitive random process consisting of a series of independent trials. This means that the outcome of one trial does not affect the probability of the outcome of another. Second, only two outcomes are possible in each trial: success or failure. The probability of success is equal to p, and the probability of failure is $(1 - p)$. Third, the probabilities of success and failure are the same in each trial. For example, suppose the owner of an oil firm believes that the probability of striking oil is .10. Success is defined as striking oil and failure as not striking oil. If the probability of striking oil is .10 on every trial, and all the trials are independent of each other, then this is a Bernoulli process. Note that the events of striking oil and not striking oil are mutually exclusive.

A simple example of a Bernoulli process is the tossing of a fair coin. The outcomes can be classified into the events success (heads, for example) and failure (tails). The outcomes are mutually exclusive, and the probability of success is constant at .5. The MINITAB output of the Bernoulli process for the first four experiments of Table 5.1 are presented in Figure 6.5. Columns C1, C2, C3 and C4 present the number and sequence of heads and tails occurring for random experiments with $N = 10, 20, 30,$ and 40.

```
MTB > RANDOM 10 C1;
SUBC> BERNOULI 0.5.
MTB > RANDOM 20 C2;
SUBC> BERNOULI 0.5.
MTB > RANDOM 30 C3;
SUBC> BERNOULI 0.5.
MTB > RANDOM 40 C4;
SUBC> BERNOULI 0.5.
MTB > PRINT C1-C4
```

ROW	C1	C2	C3	C4
1	1	0	1	0
2	1	0	1	1
3	1	1	0	0
4	1	1	0	1
5	0	0	1	1
6	1	0	1	0
7	1	1	0	0
8	1	0	0	0
9	1	1	1	1
10	1	1	1	1
11		1	1	1
12		0	1	0
13		0	0	1
14		0	1	1
15		1	0	1
16		0	1	0
17		0	0	0
18		0	0	0
19		1	0	1
20		1	1	0
21			1	0
22			1	0
23			1	1
24			1	0
25			1	1
26			1	0
27			0	0
28			1	1
29			1	0
30			1	0
31				1
32				0
33				0
34				1
35				0
36				0
37				1
38				1
39				0
40				0

Figure 6.5

MINITAB Output of Bernoulli Process for 4 Experiments ($N = 10$, $N = 20$, $N = 30$ and $N = 40$)

Binomial Distribution

If n trials of a Bernoulli process are observed, then the total number of successes in the n trials is a random variable, and the associated probability distribution is known as a **binomial distribution.** The number of successes, the number of trials, and the probability of success on a trial are the three pieces of information we need to generate a binomial distribution.

To develop the binomial distribution, assume that each of the n trials of an experiment will generate one of two outcomes, a success, S, or a failure, F. Suppose the trials generate x successes and $(n - x)$ failures. The probability of success on a particular trial is p, and the probability of failure is $(1 - p)$. Thus the probability of obtaining a specific sequence of outcomes is

$$p^x(1 - p)^{n-x} \tag{6.8}$$

Equation 6.8 presents the joint probability of x successes and $(n - x)$ failures occurring simultaneously. Because the n trials are independent of each other, the probability of any particular sequency of outcomes is, by the multiplication rule of probabilities (Section 5.6), equal to the product of the probabilities for the individual outcomes.

Probability Function

There are several ways in which x successes can be arranged among $(n - x)$ failures. Therefore, the probability of x successes in n trials can be defined as

$$
\begin{aligned}
P(x \text{ successes in } n \text{ trials}) &= \binom{n}{x} p^x(1 - p)^{n-x} \\
&= \frac{n!}{x!(n - x)!} p^x(1 - p)^{n-x}, \, x = 0, 1, \ldots, n, \tag{6.9}
\end{aligned}
$$

where

$$\binom{n}{x} = n \text{ combinations taken } x \text{ at a time}$$

$$n! = n(n - 1)(n - 2)(n - 3) \cdot \cdot \cdot (1)$$

The symbol $n!$ is read "n factorial." When $n = 0$, then $n! = 0! = 1$. Equation 6.9 is the **binomial probability function,** which gives the probability of x successes in n trials. Using this formula, we can evaluate a binomial probability.

EXAMPLE 6.9 *Probability Distribution for GM Stock*

Suppose that the price of a share of stock in General Motors Corporation in each year period in the future will either go up (U) or come down (D) in 1 day with the probabilities .40 and .60, respectively. Calculate the probability of each possible outcome of the stock price 4 days later.[2]

[2]Assume that the price movement of GM stock today is completely independent of its movement in the past. See Example 6B.1 in Appendix 6B for further discussion.

Using the outcome tree approach discussed in Appendix 5A, we find the possible outcomes e and probabilities $p(e)$ indicated in Table 6.3.

Table 6.3 Probability Distribution of GM Stock 4 Days Later

Outcome, e		Probability, $p(e)$
e_1	(UUUU)	$(.4)(.4)(.4)(.4) = .0256$
e_2	(UUUD)	$(.4)(.4)(.4)(.6) = .0384$
e_3	(UUDU)	$(.4)(.4)(.6)(.4) = .0384$
e_4	(UUDD)	$(.4)(.4)(.6)(.6) = .0576$
e_5	(UDUU)	$(.4)(.6)(.4)(.4) = .0384$
e_6	(UDUD)	$(.4)(.6)(.4)(.6) = .0576$
e_7	(UDDU)	$(.4)(.6)(.6)(.4) = .0576$
e_8	(UDDD)	$(.4)(.6)(.6)(.6) = .0864$
e_9	(DUUU)	$(.6)(.4)(.4)(.4) = .0384$
e_{10}	(DUUD)	$(.6)(.4)(.4)(.6) = .0576$
e_{11}	(DUDU)	$(.6)(.4)(.6)(.4) = .0576$
e_{12}	(DUDD)	$(.6)(.4)(.6)(.6) = .0864$
e_{13}	(DDUU)	$(.6)(.6)(.4)(.4) = .0576$
e_{14}	(DDUD)	$(.6)(.6)(.4)(.6) = .0864$
e_{15}	(DDDU)	$(.6)(.6)(.6)(.4) = .0864$
e_{16}	(DDDD)	$(.6)(.6)(.6)(.6) = .1296$

The probability of GM stock going up three times and coming down once is the sum of the probabilities associated with e_2, e_3, e_5, and e_9: $.0384 + .0384 + .0384 + .0384 = .1536$.

Alternatively, this probability can be calculated in terms of the binomial combination formula (Equation 6.9).

$$\binom{4}{3}(.4)^3(.6) = \frac{4!}{(4-1)!1!}(.0384) = .1536$$

Hence the binomial combination formula can be used to replace the diagram for calculating such a probability.

EXAMPLE 6.10 *Probability Function of Insurance Sales*

Assume that an insurance sales agent believes that the probability of her making a sale is .20. She makes 5 contacts and, eager to leave nothing to chance, calculates a binomial distribution.

$$P(0 \text{ success}) = \frac{5!}{0!5!}.2^0.8^5 = .3277$$

$$P(1 \text{ success}) = \frac{5!}{1!4!}.2^1.8^4 = .4096$$

$$P(2 \text{ successes}) = \frac{5!}{2!3!} .2^2.8^3 = .2048$$

$$P(3 \text{ successes}) = \frac{5!}{3!2!} .2^3.8^2 = .0512$$

$$P(4 \text{ successes}) = \frac{5!}{4!1!} .2^4.8^1 = .0064$$

$$P(5 \text{ successes}) = \frac{5!}{5!0!} .2^5.8^0 = .0003$$

Alternatively these numbers can be calculated by the MINITAB program as shown here.

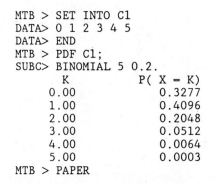

```
MTB > SET INTO C1
DATA> 0 1 2 3 4 5
DATA> END
MTB > PDF C1;
SUBC> BINOMIAL 5 0.2.
       K            P( X = K)
      0.00            0.3277
      1.00            0.4096
      2.00            0.2048
      3.00            0.0512
      4.00            0.0064
      5.00            0.0003
MTB > PAPER
```

Figure 6.6 gives the probability distribution for this sales agent's successes. Because the events of the sales agent's number of successes are mutually exclusive, the probability that she has 3 or more successes is equal to $P(3 \text{ successes}) + P(4 \text{ successes}) + P(5 \text{ successes}) = .0512 + .0064 + .0003 = .0579$.

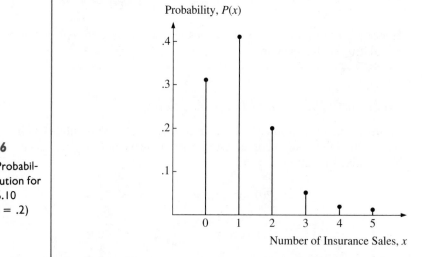

Figure 6.6

Binomial Probability Distribution for Example 6.10 ($n = 5$, $p = .2$)

EXAMPLE 6.11   *Cumulative Probability Distribution for Insurance Sales*

Suppose the sales agent we met in Example 6.10 wants to determine the probability of making between 1 and 4 sales.

$P(1 \text{ success}) + P(2 \text{ successes}) + P(3 \text{ successes}) + P(4 \text{ successes}) = .672$

Unless the number of trials n is very small, it is easier to determine binomial probabilities by using Table A1 in Appendix A of this book. All three variables listed in Equation 6.9 (n, p, and x) appear in the binomial distribution table extracted from the National Bureau of Standard Tables. Using probabilities from this table, we can calculate both individual probabilities and cumulative probabilities.

The individual probabilities drawn for Example 6.11 from the binomial table are listed in Table 6.4. These probabilities are identical to those we found with Equation 6.9.

Table 6.4 Part of Binomial Table
($n = 5$, $p = .2$)

x	$P(x)$
5	.0003
4	.0064
3	.0512
2	.2048
1	.4096
0	.3277

The cumulative binomial function can be defined as

$$B(n, p) = \sum_{x=0}^{n} \binom{n}{x} p^x (1 - p)^{n-x} \qquad (6.10)$$

Using Table 6.4, we can calculate the cumulative probabilities for the sales agent having two or more successes.

$$P(X \geq 2 \mid n = 5, p = .2) = P(x = 2) + P(x = 3) + P(x = 4) + P(x = 5)$$
$$= \sum_{x=2}^{5} \binom{n}{x} .2^x .8^{5-x}$$
$$= .2048 + .0512 + .0064 + .0003 = .2627$$

In a nationwide poll of 2,052 adults by the American Association of Retired Persons (*USA Today,* August 8, 1985), approximately 40 percent of those surveyed described the current version of the federal income tax system as fair. Suppose we randomly sample 20 of the 2,052 adults surveyed and record x as the number who think the federal income tax system is fair. To a reasonable degree of approximation, x is a binomial random variable. The probability that x is less than or equal to 10 can be defined as[3]

$$P(X \leq 10 \mid n = 20, p = 0.4)$$

$$= \sum_{x=1}^{10} \binom{n}{x} (0.4)^x (0.6)^{20-x}$$

$$= 0 + .005 + .0031 + .0123 + .0350 + .0746 + .1244$$
$$+ .1659 + .1797 + .1597 + .1171 = .8725$$

Another situation that requires the use of a binomial random variable is **lot acceptance sampling,** where we must decide, on the basis of *sample* information about the quality of the lot, whether to accept a lot (batch) of goods delivered from a manufacturer. It is possible to calculate the probability of accepting a shipment with any given proportion of defectives in accordance with Equation 6.9.

EXAMPLE 6.12   *Cumulative Probability Distribution: A Shipment of Calculator Chips*

A shipment of 800 calculator chips arrives at Century Electronics. The contract specifies that Century will accept this lot if a sample of size 20 drawn from the shipment has no more than 1 defective chip. What is the probability of accepting the lot by applying this criterion if, in fact, 5 percent of the whole lot (40 chips) turns out to be defective? What if 10 percent of the lot is defective?

This is a binomial situation where there are $n = 20$ trials, and $p =$ the probability of success (chip is defective) $= .05$. The shipment is accepted if the number of defectives is either 0 or 1, so the probability of the shipment being accepted is

$$P(\text{shipment accepted}) = P(X \leq 1)$$
$$= P(0) + P(1)$$

Using Table A1 in Appendix A ($n = 20, p = .05$), we obtain $P(0) = .3585$ and $P(1) = .3774$. Hence the probability that Century Electronics accepts delivery is

$$P(\text{shipment accepted}) = .3585 + .3774 = .7359$$

Similarly, if 10 percent of the items in the shipment are defective (that is, if $p = .10$), then

$$P(\text{shipment accepted}) = .1216 + .2702 = .3918$$

This implies that the higher the proportion of defectives in the shipment, the less likely is acceptance of the delivery. And that's as it should be.

---

[3]Refer to Table A1 in Appendix A at the end of the book.

Mean and Variance

The *expected value* (mean) of the binomial distribution is simply the number of trials times the probability of a success.

$$\mu_x = np \tag{6.11}$$

The variance of the binomial distribution is equal to

$$\sigma_x^2 = np(1 - p) \tag{6.12}$$

Thus the standard deviation of the binomial distribution is $\sqrt{np(1 - p)}$. The derivation of Equations 6.11 and 6.12 can be found in Appendix 6A.

EXAMPLE 6.13   *Probability Distribution of Insurance Sales*

In the insurance sales case we discussed in Examples 6.10 and 6.11, the expected number of sales can be calculated in terms of Equation 6.11 as $np = 5(.20) = 1$. The variance of the distribution can be calculated in terms of Equation 6.12 as $np(1 - p) = 5(.2)(.8) = .8$. Thus the expected number of sales by the sales agent is equal to 1, and the standard deviation is $\sqrt{.8} = .894$.

6.6   THE HYPERGEOMETRIC DISTRIBUTION

In the last section, we described the binomial distribution as the appropriate probability distribution for a situation in which the assumptions of a Bernoulli process are met. A major application of the binomial distribution is in the computation of probability for cases where the trials are independent. If the experiment consists of randomly drawing n elements (samples), with replacement, from a set of N elements, then the trials are independent.

In most practical situations, however, sampling is carried out *without replacement,* and the number sampled is not extremely small relative to the total number in the population. For example, when a researcher selects a sample of families in a city to estimate the average income of all families in the city, the sampling units are ordinarily not replaced prior to the selection of subsequent ones. That is, the families are not replaced in the original population and thus are not given an opportunity to appear more than once in the sample. Similarly, when a sample of accounts receivable is drawn from a firm's accounting records for a sample audit, sampling units are ordinarily not replaced before the selection of subsequent units. Sampling without replacement also takes place in quality control sampling and other sampling. Furthermore, if the number sampled is extremely small relative to the total number of items, then the trial is almost independent even if the sampling is without replacement (as in Example 6.12).[4] Under such circumstances, and in

---

[4]In that case, the probability on the first trial, p, is 50/800 = .0625. On the second trial, p is either 50/799 = .06258 (if the first chip was not defective) or 49/799 = .06133 (if the first chip was defective).

sampling with replacement, the binomial distribution can be used in the analysis.

The **hypergeometric distribution** is the appropriate model for sampling without replacement. To solve the following hypergeometric problems, let's divide our population (such as a group of people) into two categories: adults and children. For a population of size N, h members are S (successes) and $(N - h)$ members are F (failures). Let sample size $= n$ trials, obtained without replacement. Let $X =$ number of successes out of n trials (a hypergeometric random variable).

Suppose there are $h = 60$ adults and $N - h = 40$ children. Thus there are $N = 100$ persons. Numbers from 1 to 100 are assigned to these individuals and printed on identical discs, which are placed in a box. If 10 chips are randomly drawn from the box, then the hypergeometric problem involves calculating the probability of there being x adults and $(n - x)$ children in a sample of size 10. If $n = 10$ people are selected, what is the probability that exactly 4 adults will be included in the sample?

Because there are $h = 60$ adults, there are $\binom{h}{x}$ possible ways of selecting $x = 4$ adults. Of the $n = 10$ people, $n - x = 10 - 4 = 6$ are children. Hence there are $\binom{N-h}{n-x}$ possible ways of selecting $n - x = 6$ children. Thus the total number of ways of selecting a group of 10 persons that includes exactly 4 adults and 6 children is $\binom{h}{x}\binom{N-h}{n-x}$. There are $\binom{N}{n} = \binom{100}{10}$ possible ways of selecting 10 persons from 100 persons. Thus the probability of selecting a group of 10 persons that includes $x = 4$ adults is

$$\frac{\binom{60}{4}\binom{40}{10 - 4}}{\binom{100}{10}}$$

The Hypergeometric Formula

From the example we just outlined, we can state the general hypergeometric probability function.

$$P(x) = P[(x \text{ successes and } (n - x) \text{ failures})] = \frac{\binom{h}{x}\binom{N - h}{n - x}}{\binom{N}{n}} \tag{6.13}$$

The **hypergeometric formula** gives the probability of x successes when a random sample of n is drawn without replacement from a population of N within which h units have the characteristic denoting success. The number of successes achieved under these circumstances is the **hypergeometric random variable.**

EXAMPLE **6.14**   *Sampling Probability Function of Party Membership*

Consider a group of 10 students in which 4 are Democrats and 6 are Republicans. A sample of size 6 has been selected. What is the probability that there will be only 1 Democrat in this sample?

Using the hypergeometric probability function shown in Equation 6.13, we have

$$P(x = 1 \text{ and } n - x = 5) = \frac{\binom{4}{1}\binom{10-4}{6-1}}{\binom{10}{6}} = \frac{\dfrac{4!}{1!(4-1)!}\dfrac{6!}{5!(6-5)!}}{\dfrac{10!}{6!4!}}$$

$$= \frac{\dfrac{24}{(1)(6)}\dfrac{720}{(120)(1)}}{\dfrac{3,628,800}{(720)(24)}} = \frac{(4)(6)}{210} = .1143$$

Similarly, we can calculate other probabilities. All possible probabilities are:

$$\begin{array}{ll} P(x = 0) = .0048 & P(x = 3) = .3809 \\ P(x = 1) = .1143 & P(x = 4) = .0714 \\ P(x = 2) = .4286 & \end{array}$$

The hypergeometric probability distribution is shown in Figure 6.7.

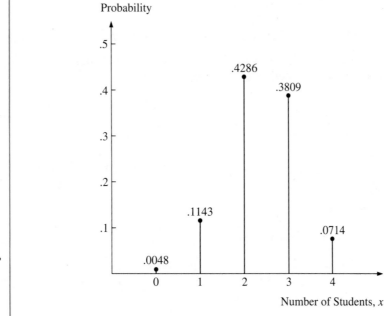

Figure 6.7

Probability Distribution for Example 6.14 (Hypergeometric Distribution for $N = 10$, $x = 4$, $n = 6$)

Mean and Variance

The mean of the hypergeometric probability distribution for Example 4.16 can be calculated by using Equation 6.3.

$$\mu = \sum_{i=1}^{n} x_i P(x_i) = 0(.0048) + 1(.1143) + 2(.4286) + 3(.3809) + 4(.0714)$$
$$= 2.40$$

On average, we expect 2.40 students to be Democrats. Alternatively, it can be shown that the mean of this distribution is

$$\mu = n(h/N) \tag{6.14}$$

The ratio h/N is the proportion of successes on the first trial. The product $n(h/N)$ is similar to the mean of the binomial distribution, np. It can be shown that the *variance of the hypergeometric distribution* is equal to

$$\sigma^2 = \left(\frac{N-n}{N-1}\right)\left[n\left(\frac{h}{N}\right)\left(1 - \frac{h}{N}\right)\right] \tag{6.15}$$

In other words, the variance of the hypergeometric distribution is the variance of the binomial distribution with an adjustment factor, $(\frac{N-n}{N-1})$. If the sample size is small relative to the total number of objects N, then $(\frac{N-n}{N-1})$ is very close to 1. Consequently, the binomial distribution can be used to replace the hypergeometric distribution.[5]

EXAMPLE 6.15 *Mean and Variance of a Hypergeometric Probability Function*

Using the data of Example 6.14, we can calculate the mean and variance of a hypergeometric function as follows:

$$\mu = 6\left(\frac{4}{10}\right) = 2.4$$

$$\sigma^2 = \left(\frac{10-6}{10-1}\right)\left[(6)\left(\frac{4}{10}\right)\left(1 - \frac{4}{10}\right)\right]$$
$$= .64$$

[5]The approximation is valid only when N is large. Usually we require $N/n \geq 20$.

6.7 THE POISSON DISTRIBUTION AND ITS APPROXIMATION TO THE BINOMIAL DISTRIBUTION

In the previous two sections, we have discussed two major types of discrete probability distributions, one for binomial random variables and the other for hypergeometric random variables. Both of these random variables were defined in terms of the number of success, and these successes were *obtained within a fixed number of trials* of some random experiment. In this section, we will discuss a distribution called the *Poisson distribution.* This distribution can be used to deal with a single type of outcome or "event," such as number of telephone calls that come through a switchboard and number of accidents. It is also possible to use a Poisson distribution to investigate the probability of, say, a certain number of defective parts in a plant in a 1-year period, a certain number of sales in a given week, and a certain number of customers entering a bank in a day.

The Poisson Distribution

The **Poisson distribution,** which is named after the French mathematician Simeon Poisson, is useful for determining the probability that a particular event will occur a certain number of times over a specified period of time or within the space of a particular interval. For example, the number of customer arrivals per hour at a bank or other servicing facility is a random variable with Poisson distribution. Here are some other random variables that may exhibit a Poisson distribution:

1. The number of days in a given year in which a 50-point change occurs in the Dow Jones Industrial Average.
2. The number of defects detected each day by a quality control inspector in a light bulb plant.
3. The number of breakdowns per month that a supercomputer experiences.
4. The number of car accidents that occur per month (or week or day) in the city of Princeton, New Jersey.

The formula for the Poisson probability distribution is

$$P(X = x) = e^{-\lambda}\lambda^x/x! \quad \text{for } x = 0, 1, 2, 3, \ldots \text{ and } \lambda > 0 \qquad (6.16)$$

where X represents the discrete Poisson random variable; x represents the number of rare events in a unit of time, space, or volume; λ is the mean value of x; e is the base of natural logarithms and is approximately equal to 2.71828; and ! is the factorial symbol.

It can be shown that the value of both the mean and the variance of a Poisson random variable X is λ. That is,

$$E(X) = \lambda \qquad (6.17a)$$
$$\text{Var}(X) = \lambda \qquad (6.17b)$$

We will explore this distribution further in Chapter 9 when we discuss the exponential distribution.

In studying a retailer's supply account at a large U.S. Air Force base, the Poisson probability distribution was used to describe the number of customers (x) in a 7-day lead time period (*Management Science*, April 1983). Here "lead time" is used to describe the time needed to replenish a stock item.

Items were divided into two categories for individual analysis. The first category was items costing $5 or less and the second category was items costing more than $5. The mean number of customers during lead time for the first category was estimated to be .09. For the second category, the mean was estimated to be .15.

From Equation 6.17, the mean and variance for the number x of customers who demand items that cost over $5 during lead time is

$$E(x) = \text{Var}(x) = \lambda = .15.$$

From Equation 6.16, the probability that no customers will demand an item that costs $5 or less during the lead time is

$$P(x = 0) = e^{-0.09}(.09)^0/0! = .9139$$

EXAMPLE 6.16 *Customer Arrivals in a Bank*

Suppose the average number of customers entering a bank in a 30-minute period is 5. The bank wants to determine the probability that 4 customers enter the bank in a 30-minute period. Substituting $\lambda = 5$ and $x = 10$ into Equation 6.16, we obtain

$$P(X = 10) = (e^{-5})(5^4)/4!$$

Table A2 in Appendix A of this book is a Poisson probability table that can be used to calculate probabilities. From this table, we find that $(e^{-5})(5^4)/4! = .1755$. As another example, say we know that the probability that 3 customers enter the bank is .1404. Using the Poisson probability table, we can calculate the other individual probabilities for $X = 0$, 1, and 2. Table 6.5 gives the probability function for $X = 0, 1, \ldots, 4$.

Our calculations can tell us such things as the probability that 0, 1, 2, 3, or 4 customers arrive within a 20-minute period. From Table 6.5, we know that the

Table 6.5 Probability Function for Example 6.16 ($\lambda = 5$)

x	$P(x)$
4	.1755
3	.1404
2	.0842
1	.0337
0	.0067

probability that 4 or fewer individuals enter the bank is .1755 + .1404 + .0842 + .0337 + .0067 = .4405.

We could continue by calculating the probabilities for more than 4 customers and eventually produce a Poisson probability distribution for this bank. Table 6.6 shows such a distribution. To produce this table, we used Equation 6.16. The probability of more than 4 customer arrivals can also be calculated from Table A2.

Table 6.6 Poisson Probability Distribution of Customer Arrivals per 3-Minute Period

x = Number of Customer Arrivals	$P(x)$ = Probability of Exactly That Number
0	.0067
1	.0337
2	.0842
3	.1404
4	.1755
5	.1755
6	.1462
7	.1044
8	.0653
9	.0363
	.9682
10 or more	.0318 (1 − .9682)
	1.0000

Alternatively, MINITAB can be used to calculate part of Table 6.6 as:

```
MTB > SET INTO C1
DATA> 0 1 2 3 4 5 6 7
DATA> END
MTB > PDF C1;
SUBC> POISSON 5.
        K              P( X = K)
       0.00              0.0067
       1.00              0.0337
       2.00              0.0842
       3.00              0.1404
       4.00              0.1755
       5.00              0.1755
       6.00              0.1462
       7.00              0.1044
MTB > PAPER
```

Figure 6.8 uses MINITAB to illustrate graphically the Poisson probability distribution of the number of customer arrivals.

```
MTB > SET INTO C1
DATA> 0 1 2 3 4 5 6 7 8 9
DATA> END
MTB > SET INTO C2
DATA> 0.0067 0.0337 0.0842 0.1404 0.1775 0.1775 0.1462 0.1044 0.0653 0.0363
DATA> END
MTB > GPLOT C2 C1;

SUBC> YLABEL 'PROBABILITY';
SUBC> XLABEL 'X'.
```

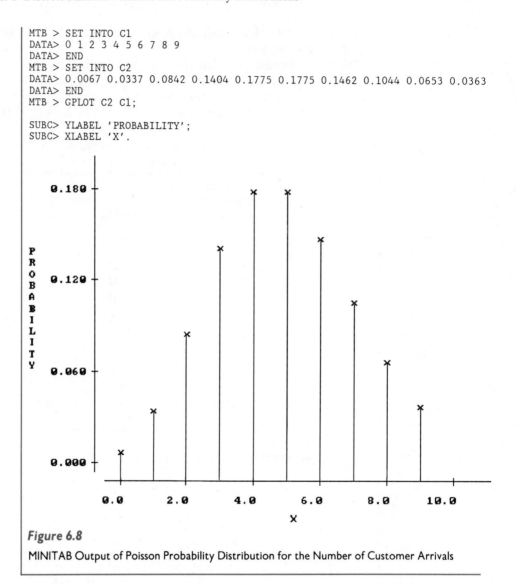

Figure 6.8

MINITAB Output of Poisson Probability Distribution for the Number of Customer Arrivals

EXAMPLE 6.17 *Defective Spark Plug*

In one day's work on a spark plug assembly line, the average number of defective parts is 2. The manager is concerned that more than 4 defectives could occur and wants to estimate the probability of that happening. Using the Poisson distribution table (Table A2 in Appendix A), she determines that the probability of 0 through 4 defective spark plugs is $P(X = 0) + P(X = 1) + P(X = 2) + P(X = 3) + P(X = 4) = .1353 + .2707 + .2707 + .1804 + .0902 = .9473$. Then the probability of having more than 4 defective spark plugs is $1 - .9473 = .0527$.

The Poisson Approximation to the Binomial Distribution

The Poisson distribution can sometimes be used to *approximate* the binomial distribution and avoid tedious calculations.

If the number of trials in a binomial, n, is large, then a Poisson random variable with $\lambda = np$ will provide a reasonable approximation. This is a good approximation, provided that n is large ($n > 20$) and p is small ($p < .05$).

$$P(X = x) = \frac{e^{-np}(np)^x}{x!} \tag{6.18}$$

EXAMPLE 6.18   *Comparison of the Poisson and Binomial Probability Approaches*

Suppose 20 parts are selected from a production process and tested for defects. The manager of the firm wants to determine the probability that 3 defectives are encountered. Previous experience indicates that the probability of a part being defective is .05. The mean is $np = (20)(.05) = 1$. Setting $\lambda = 1$, we can use the Poisson distribution formula of Equation 6.18 to calculate the probability.

$$P(X = 3) = 1^3 e^{-1}/3! = .0613$$

If we use the binomial distribution formula of Equation 6.9, then the probability is

$$P(x) = \frac{20!}{3!(20 - 3)!} (.05)^3 (.95)^{17} = .0596$$

The difference between .0613 and .05916 is slight (only about .2 percent).

6.8   JOINTLY DISTRIBUTED DISCRETE RANDOM VARIABLES

In Sections 5.4 and 5.5 we discussed conditional, joint, and marginal probabilities in terms of events. We now consider these probabilities for two or more related discrete random variables. For a single random variable, the probabilities for all possible outcomes can be summarized by using a probability function; for two or more possible related discrete random variables, the probability function must define the probabilities that the random variables of interest simultaneously take specific values.

Joint Probability Function

Suppose we want to know the probability of a worker being a member of a labor union *and* over age 50. We now concern ourselves with the distribution of random variables, age (X) and membership in a labor union (Y). In notation, the probability that X takes on a value x and that Y takes on a value y is given by

$$P(x, y) = P(X = x, Y = y) \tag{6.19}$$

Equation 6.19 represents the **joint probability function** of X and Y. Joint probabilities are usually presented in tabular form so that the probabilities can be identified easily. **Joint probability distributions** of discrete random variables are probability distributions of two or more discrete random variables. The next example illustrates the use of Equation 6.19.

EXAMPLE 6.19   *Joint Probability Distribution for 100 Students Classified by Sex and by Number of Accounting Courses Taken*

Table 6.7 shows the probability function for two random variables, X (the total number of accounting courses a student takes) and Y (the sex of the student, where 1 denotes a male student and 0 a female). The values in the cells of Table 6.7 are joint probabilities of the outcomes denoted by the column and row headings for X and Y. Also displayed in the margins of the table are separate univariate probability distributions of X and Y.

Table 6.7 Joint Probability Distribution for 100 Students Classified by Sex and Number of Accounting Courses Taken

			X		
Y	*2*	*3*	*4*	*5*	**Total**
0	.14	.17	.08	.12	.51
1	.16	.20	.12	.01	.49
Total	.30	.37	.20	.13	1

The joint probability that a student is female and takes three courses, $P(3,0) = P(X = 3, Y = 0)$, is equal to .17. The probability that a student is male and takes four courses, $P(4,1) = P(X = 4, Y = 1)$, is equal to .12. The probabilities inside the box are all joint probabilities, which are, again, probabilities of the intersection of two events.

The probability distribution of a single discrete random variable is graphed by displaying the value of the random variable along the horizontal axis and the corresponding probability along the vertical axis. In the case of a bivariate distribution, two axes are required for the values of random variables and a third for the probability. A graph of the joint probability of Table 6.7 is shown in Figure 6.9.

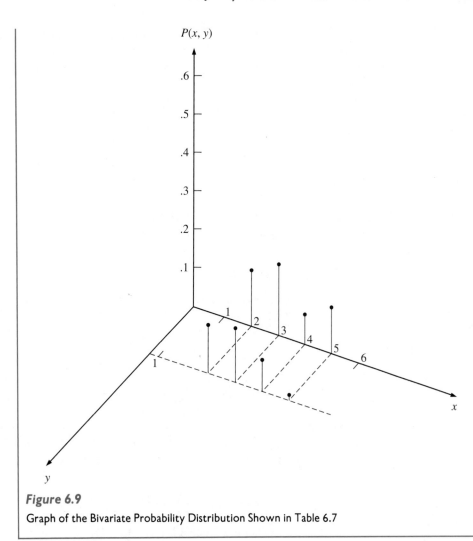

Figure 6.9

Graph of the Bivariate Probability Distribution Shown in Table 6.7

Marginal Probability Function

The *marginal probability* can be obtained by summing all the joint probabilities over all possible values. In other words, the probabilities in the margins of the table are the marginal probabilities. These probabilities form marginal probability functions. For example (see Table 6.7 in Example 6.19), the probability that a randomly selected student is female, $P(Y = 0)$, is found by adding the respective probabilities that a female student takes two courses (.14), three courses (.17), four courses (.08), and five courses (.12), for a total of .51. The probability that a randomly selected student is male, $P(Y = 1)$, is therefore .49 $(1 - .51)$. Similarly, the probability that a randomly selected student takes two courses, $P(X = 2)$, is equal to the probability that a female takes two courses (.14), plus the probability that a male takes two courses (.16), for a total of .30. Note that the sum of the marginal probabilities is 1. From these results, we can define **marginal probability functions** for X and Y as follows:

$$P(x_i) = \sum_{j=1}^{m} P(x_i, y_j), \quad i = 1, \ldots, n \tag{6.20}$$

$$P(y_j) = \sum_{i=1}^{n} P(x_i, y_j), \quad j = 1, \ldots, m \tag{6.21}$$

where

x_i = the ith observation of the X variable
y_j = the jth observation of the Y variable

Conditional Probability Function

Conditional probability functions can be calculated from joint probabilities. The conditional probability function of X, given $Y = y$, is

$$P(x|y) = P(x, y)/P(y) \tag{6.22}$$

The conditional probability is found by taking the intersection of the probability of $X = x$ and $Y = y$ and dividing by the probability of Y. For example, the probability from Table 6.7 that a student takes four courses, given that the student is female, is $P(4|0) = P(4,0)/P(0) = .08/.51 = .16$.

Similarly, the probability that a student is male, given that the student takes three courses, is $P(1|3) = P(3,1)/P(3) = .20/.37 = .54$. The **conditional probability distribution** for X given $Y = 1$ is shown in Table 6.8.

Table 6.8 Conditional Probability Distribution for Numbers of Accounting Courses, Given That the Student Is Male ($Y = 1$)

x	$P(X = x \mid Y = 1)$
2	.16/.49 = .3265
3	.20/.49 = .4082
4	.12/.49 = .2449
5	.01/.49 = .0204
	1.0000

Independence

Returning to the terminology of events explained in Chapter 5, we saw in Section 5.6 that if two events are statistically independent, then $P(B|A) = P(B)$ and $P(B$

$\cap\ A) = P(B)P(A)$. In random variable notation, the analogous statement is that if X and Y are independent random variables, then

$$P(X = x \mid Y = y) = P(X = x) \quad \text{for all } X$$
$$P(Y = y \mid X = x) = P(Y = y) \quad \text{and all } Y \tag{6.23}$$

Equation 6.23 implies that the conditional probability function of X given Y or of Y given X is the same as the marginal probability of X or Y. We will illustrate this definition of independence by returning to Table 6.7.

Suppose we consider the outcome pair (3, 1)—that is, $X = 3$ and $Y = 1$. In this case,

$$P(X = 3 \mid Y = 1) = \frac{.20}{.49} = .4082$$

and

$$P(X = 3) = .37$$

Because $P(X = 3 \mid Y = 1)$ is not equal to $P(X = 3)$, X and Y are not independent.

EXAMPLE 6.20   *Store Satisfaction*[6]

Table 6.9 shows the probability function for two random variables: X, which measures a consumer's satisfaction with food stores in a particular town, and Y, the number of years the consumer has resided in that town. Suppose X can take on the value 1, 2, 3, and 4, which reflect a satisfaction level ranging from low to high, and that Y takes on the value 1 if the consumer has lived in the town fewer than 6 years and 2 otherwise. The values in the cells of Table 6.9 are joint probabilities of the respective joint events denoted by the column and row headings for x and y. Also displayed in the margins of the table are separate univariate probability distributions of x and y.

Table 6.9 Joint Probability Distribution for Consumer Satisfaction (x) and Number of Years of Residence in a Particular Town (y)

			x		Total
y	*1*	*2*	*3*	*4*	
1	.04	.14	.23	.07	.48
2	.07	.17	.23	.05	.52
Total	.11	.31	.46	.12	1

[6]This example is based on the material discussed in J. H. Miller, "Store Satisfaction and Aspiration Theory," *Journal of Retailing,* 52 (Fall 1976), 65–84.

The joint probability that a consumer has satisfaction level 3 and has lived in town fewer than 6 years, $P(3,1)$, $= P(X = 3, Y = 1)$, is .23. The probability that a consumer has satisfaction level 4 and has lived in the town more than 6 years, $P(4,2)$, is .05. The probabilities inside the box are all joint probabilities, which are, again, the intersections of two events.

The marginal probability is obtained by summing the joint probabilities over all possible values, as discussed in Example 6.19. For example (see Table 6.9), the probability that a consumer has lived in town fewer than 6 years, $P(Y = 1)$, is found by adding the probabilities that a consumer has satisfaction level 1, 2, 3, and 4, a total of .48. The marginal probability that a consumer has lived in town 6 or more years, $P(Y = 2)$, is therefore .52. Similarly, the probability that a randomly selected consumer has satisfaction level 1, $P(X = 1)$, is equal to the probability that a consumer has lived in town fewer than 6 years, .04, plus the probability that a consumer has lived in the town more than 6 years, .07, for a total of .11. Note that the sum of the marginal probabilities is equal to 1.

Conditional probability functions can be calculated from joint probabilities as discussed in Example 6.19. The conditional probability is found by taking the intersection of the probability of $X = x$ and $Y = y$ and dividing by the probability of $Y = y$. For example, the probability (from Table 6.7) that a consumer has satisfaction level 4, given that the consumer has lived in town fewer than 6 years is $P(X = 4 | Y = 1) = P(X = 4, Y = 1)/P(Y = 1) = .07/.48 = .1458$.

Similarly, the probability that a consumer has lived in town 6 or more years, given that the consumer has satisfaction level 3, is $P(Y = 2 | X = 3) = P(Y = 2, X = 3)/P(X = 3) = .23/.46 = .5$. The conditional probability distribution for X given $Y = 2$ is shown in Table 6.10.

Table 6.10 Conditional Probability Distribution for Satisfaction Level for a Consumer Who Has Lived in Town 6 or More Years

x	$P(X = x | Y = 2)$
1	.07/.52 = .1346
2	.17/.52 = .3269
3	.23/.52 = .4423
4	.05/.52 = .0962
	1.0000

6.9 EXPECTED VALUE AND VARIANCE OF THE SUM OF RANDOM VARIABLES

Covariance and Coefficient of Correlation for the Sum of Two Random Variables

The concept of expected value and variance of discrete random variables discussed in Section 6.4, can be extended to measure the degree of relationship between two discrete random variables X and Y. Here we will discuss two alternative means of determining the possibility of a linear association between two random variables X and Y. These two measures are covariance and coefficient of correlation.

The **covariance** is a statistical measure of the linear association between two random variables X and Y. Its sign reflects the direction of the linear association. The covariance is positive if the variables tend to move in the same direction. If the variables tend to move in opposite directions, the covariance is negative. Specifically, the covariance between X and Y can be defined as

$$\text{Cov}(X, Y) = \sigma_{X,Y} = E[(X - \mu_X)(Y - \mu_Y)] \tag{6.24}$$

where μ_X and μ_Y are the means of X and Y, respectively. For discrete variables, Equation 6.24 can be defined as

$$\sigma_{X,Y} = \sum_{j=1}^{m} \sum_{i=1}^{n} (X_i - \mu_X)(Y_j - \mu_Y)P(X_i,Y_j) \tag{6.25}$$

Equation 6.25 can be written as a short-cut formula as follows:[7]

$$\text{Cov}(X,Y) = E(XY) - \mu_X\mu_Y$$

$$= \sum_{j=1}^{m} \sum_{i=1}^{n} (X_iY_j)P(X_i,Y_j) - \left[\sum_{i=1}^{n} X_iP(X_i)\right]\left[\sum_{j=1}^{m} Y_jP(Y_j)\right] \tag{6.26}$$

To illustrate, we evaluate the covariance between number of years of residence in the town and satisfaction level, as discussed in Example 6.20. Using the probabilities in Table 6.9, we calculate μ_X, μ_Y, and $E(XY)$ as

$$\mu_X = \sum_{i=1}^{4} X_iP(X_i) = 1(.11) + 2(.31) + 3(.46) + 4(.12) = 2.59$$

$$\mu_Y = \sum_{j=1}^{2} Y_jP(Y_j) = 1(.48) + 2(.52) = 1.52$$

$$E(XY) = \sum_{j=1}^{m} \sum_{i=1}^{n} (X_iY_i)P(X_iY_j)$$

[7]$\text{Cov}(X,Y) = E[(X - \mu_X)(Y - \mu_Y)]$
$= E(XY - Y\mu_X - X\mu_Y + \mu_X\mu_Y)$
$= E(XY) - \mu_XE(Y) - \mu_YE(X) + \mu_X\mu_Y$
$= E(XY) - \mu_X\mu_Y - \mu_Y\mu_X + \mu_X\mu_Y = E(XY) - \mu_X\mu_Y$

$$= (1)(1)(.04) + (1)(2)(.14) + (1)(3)(.23) + (1)(4)(.07) + (2)(1)(.07)$$
$$+ (2)(2)(.17) + (2)(3)(.23) + (2)(4)(.05)$$
$$= 3.89$$

Substituting this information into Equation 6.26, we obtain the covariance.

$$\text{Cov}(X,Y) = 3.89 - (2.59)(1.52) = -0.05$$

The negative value of covariance indicates some tendency toward a negative relationship between number of years of residence in the town and level of satisfaction.

In addition to the direction of the relationship between variables, we may want to measure its strength. We can easily do so by *scaling* the covariance to obtain the coefficient of correlation.

The **coefficient of correlation** ρ between X and Y is equal to the covariance divided by the product of the variables' standard deviations. That is,

$$\rho = \frac{\sigma_{X,Y}}{\sigma_X \sigma_Y} \tag{6.27}$$

where ρ = coefficient of correlation, σ_X = standard deviation of X, and σ_Y = standard deviation of Y.

It can be shown that ρ is always less than or equal to 1.0 and greater than or equal to -1.0.

$$-1 \leq \rho \leq 1$$

Again, let us use data given in Table 6.9 to show how to calculate the correlation coefficient between X and Y. We use Equation 6.5 to calculate the variances of X and Y.

$$\sigma_X^2 = \sum_{i=1}^{4} X_i^2 P(X_i) - (\mu_X)^2$$
$$= (1)^2(.11) + (2)^2(.31) + (3)^2(.46) + (4)^2(.12) - (2.59)^2$$
$$= .7019$$

$$\sigma_Y^2 = \sum_{j=1}^{2} Y_j^2 P(Y_j) - (\mu_Y)^2$$
$$= (1)^2(.48) + (2)^2(.52) - (1.52)^2$$
$$= .2496$$

Then we substitute $\sigma_{X,Y} = -.05$, $\sigma_X = \sqrt{.7019} = .8378$ and $\sigma_Y = \sqrt{.2496} = .5$ into Equation 6.27. We obtain

$$\rho = \frac{-.05}{(.8378)(.5)} = -.1194$$

This means the relationship between X and Y is negative, as indicated by the covariance.

As might be expected, the notions of covariance (and coefficient of correlation) and statistical independence are not unrelated. However, the precise relationship between these notions is beyond the scope of this book. Covariance and coefficient of correlation will be discussed in detail in Chapters 13 and 14.

Expected Value and Variance of the Summation of Random Variables X and Y

If X and Y are a pair of random variables with means μ_X and μ_Y and variances σ_X^2 and σ_Y^2, and the covariance between X and Y is $\text{Cov}(X,Y) = \sigma_{X,Y}$, then

1. The expected value of their sum (difference) is the sum (difference) of their expected values.

$$E(X + Y) = \mu_X + \mu_Y$$
$$E(X - Y) = \mu_X - \mu_Y$$
(6.28)

2. The variance of the sum of X and Y, $\text{Var}(X + Y)$, or the difference of X and Y, $\text{Var}(X - Y)$, is the sum of their variances plus (minus) two times the covariance between X and Y.

$$\text{Var}(X + Y) = \sigma_X^2 + \sigma_Y^2 + 2\sigma_{X,Y}$$
$$\text{Var}(X - Y) = \sigma_X^2 + \sigma_Y^2 - 2\sigma_{X,Y}$$
(6.29)

EXAMPLE 6.21 *Rates of Return for Two Stocks*

Rates of return for stocks A and B are listed in Table 6.11. Let X = rates of return for stock A, and let Y = rates of return for stock B. The worksheet for calculating μ_X, μ_Y, σ_X, σ_Y, $\rho_{X,Y}$, $E(X + Y)$, and $\text{Var}(X + Y)$ is presented in Table 6.12.

Table 6.11 Rates of Return for Stocks A and B

Time Period	Stock A	Stock B
1	.10	−.10
2	−.05	.05
3	.15	.00
4	.05	−.10
5	.00	.10

Table 6.12 Worksheet to Calculate Summary Statistics

Time Period	X_i	Y_i	$(X_i - \mu_X)$	$(Y_i - \mu_Y)$	$(X_i - \mu_X)^2$	$(Y_i - \mu_Y)^2$	$(X_i - \mu_X)(Y_i - \mu_Y)$
1	.10	−.10	.05	−.09	.0025	.0081	−.0045
2	−.05	.05	−.10	.06	.010	.0036	−.006
3	.15	.00	.10	.01	.010	.0001	.001
4	.05	−.10	.00	−.09	.00	.0081	.00
5	.00	.10	−.05	.11	.0025	.0121	−.0055
Total	.25	−.05	0	0	.0250	.032	−.015

$$\mu_X = \frac{.25}{5} = .05$$

$$\mu_Y = \frac{-.05}{5} = -.01$$

Substituting information into related formulas for variance, covariance, and correlation coefficient yields

$$\sigma_X^2 = .025/5 = .0050$$
$$\sigma_Y^2 = .032/5 = .0064$$
$$\sigma_{X,Y} = -.015/5 = -.003$$
$$\rho_{X,Y} = \frac{-.003}{\sqrt{(.0050)(.0064)}} = -.53$$
$$E(X + Y) = \mu_X + \mu_Y = .05 - .01 = .04$$
$$Var(X + Y) = \sigma_X^2 + \sigma_Y^2 + 2\sigma_{XY}$$
$$= .0050 + .0064 - .0060$$
$$= .0054$$

The MINITAB output of these empirical results is presented in Figure 6.10. Note that MINITAB uses a sample formula instead of a population formula to calculate variance and covariance. They are

$$s_X^2 = .025/(5 - 1) = .00625$$
$$s_Y^2 = .032/(5 - 1) = .008$$
$$s_{X,Y} = -.015/(5 - 1) = -.00375$$
$$r_{X,Y} = \frac{.00375}{\sqrt{(.00625)(.008)}} = -.5303$$
$$s_{X+Y}^2 = s_X^2 + S_Y^2 + 2s_{X,Y}^2$$
$$= .00625 + .008 - 2(.00375)$$
$$= .0067$$

where s_X^2 and s_Y^2 are the sample variances for X and Y, respectively; $s_{X,Y}$ and $r_{X,Y}$ are the sample covariance and sample correlation coefficient, respectively; and s_{X+Y}^2 is the sample variance for $(X + Y)$.

These statistics suggest several things.

1. The average rate of return for stock A is higher than that for Stock B, and the variance of rates of return for stock A is smaller than that for stock B.

2. The rates of return for stock A are negatively correlated with those of stock B.

3. $E(X + Y) = .04$ represents the average rate of return for a portfolio wherein the same percentage of the money is invested in stock A and in stock B.

4. $Var(X + Y) = .0054$ represents the variance of a portfolio wherein the same percentage of the money is invested in stock A and stock B. Note that $Var(X + Y)$ is .0067 instead of .0054 when the sample formula is used instead of the population formula.

```
MTB > READ C1-C2
DATA> 0.10 -0.10
DATA> -0.05 0.05
DATA> 0.15 0.00
DATA> 0.05 -0.10
DATA> 0.00 0.10
DATA> END
      5 ROWS READ
MTB > LET C3=C1+C2
MTB > PRINT C1-C3

  ROW     C1      C2      C3

   1     0.10   -0.10    0.00
   2    -0.05    0.05    0.00
   3     0.15    0.00    0.15
   4     0.05   -0.10   -0.05
   5     0.00    0.10    0.10

MTB > MEAN C1
   MEAN    =      0.050000
MTB > MEAN C2
   MEAN    =     -0.010000
MTB > MEAN C3
   MEAN    =      0.040000
MTB > STDEV OF C1 PUT INTO K1
   ST.DEV. =      0.079057
MTB > STDEV OF C2 PUT INTO K2
   ST.DEV. =      0.089443
MTB > STDEV OF C3 PUT INTO K3
   ST.DEV. =      0.082158
MTB > LET K4=K1**2
MTB > LET K5=k2**2
MTB > LET K6=K3**2
MTB > PRINT K4-K6
K4        0.00625000
K5        0.00800000
K6        0.00675000

MTB > COVARIANCE C1 C2

                    C1             C2
C1        0.00625000
C2       -0.00375000    0.00800000

MTB > CORRELATION C1 C2

Correlation of C1 and C2 = -0.530

MTB > PAPER
```

Figure 6.10

MINITAB Output for Example 6.21

Expected Value and Variance of Sums of Random Variables

For N random variables, $X_1, X_2, \ldots, X_n$, Equation 6.28 can be generalized as

$$E(X_1 + X_2 + \cdots + X_n) = E(X_1) + E(X_2) + \cdots + E(X_n) \quad (6.30)$$

Thus the expected value of a sum of N random variables is equal to the sum of the expected values of these random variables.

A somewhat analogous relationship exists for variances of *uncorrelated* random variables.[8] If $X_1, X_2, \ldots, X_n$ are n uncorrelated variables, then

$$\text{Var}(X_1 + X_2 + \cdots + X_n) = \text{Var}(X_1) + \text{Var}(X_2) + \cdots + \text{Var}(X_n) \quad (6.31)$$

Otherwise covariances are needed, as in Equation 6.29.

EXAMPLE 6.22 *Rates of Return, Variance, and Covariance for IBM and Merck*

Using monthly rates of return for both IBM and Merck during the period January 1988 to June 1990, we calculate the average rate of return, variance, covariance, and correlation coefficient by using MINITAB. The MINITAB outputs are presented in Figure 6.11.

Figure 6.11

MINITAB Output for Example 6.22

```
MTB > PRINT C2

IBM
 -0.027050    0.055394   -0.092530    0.053426    0.001984    0.121258   -0.012750
 -0.104570    0.024644    0.062838   -0.024660    0.019021    0.071794   -0.061430
 -0.109910    0.044673   -0.027760    0.009383    0.027932    0.029000   -0.076770
 -0.082370   -0.014110   -0.047650    0.047808    0.065500    0.009896    0.027090
  0.112018   -0.030600

MTB > PRINT C3

MERK
 -0.053670    0.103383   -0.056120    0.002388    0.017474    0.055862   -0.008540
 -0.015680    0.050031   -0.008650   -0.010940    0.019897    0.129870   -0.021070
  0.015838    0.051833    0.018485   -0.037930    0.136448   -0.032560    0.025161
  0.033167    0.012841   -0.015530   -0.066320   -0.056890    0.018647    0.051687
  0.131740    0.046441

MTB > PRINT C4

C4
 -0.080720    0.158777   -0.148650    0.055814    0.019458    0.177120   -0.021290
 -0.120250    0.074675    0.054188   -0.035600    0.038918    0.201664   -0.082500
 -0.094072    0.096506   -0.009275   -0.028547    0.164380   -0.003560   -0.051609
 -0.049203   -0.001269   -0.063180   -0.018512    0.008610    0.028543    0.078777
  0.243758    0.015841
```

[8]Independent variables imply uncorrelated variables, so Equation 6.31 also holds for independent random variables. Applications of Equations 6.30 and 6.31 will be discussed in Appendix 6A and Section 21.8 as well as in Appendix 21C.

```
MTB > MEAN C2
    MEAN    =    0.0023833
MTB > MEAN C3
    MEAN    =    0.017910
MTB > MEAN C4
    MEAN    =    0.020293
MTB > STDEV C2
    ST.DEV. =    0.060998
MTB > STDEV C3
    ST.DEV. =    0.055071
MTB > STDEV C4
    ST.DEV. =    0.096900
MTB > COVARIANCE C2 C3

                    IBM          MERK
IBM        0.00372071
MERK       0.00131804   0.00303279

MTB > CORRELATION C2 C3

Correlation of IBM and MERK = 0.392

MTB > PAPER
```

If we let X and Y represent rates of return for IBM and Merck, respectively, then we find from Figure 6.11 that $\overline{X} = .0024$, $\overline{Y} = .0179$, $\overline{X} + \overline{Y} = .0203$, $S_X^2 = .0037$, $S_Y^2 = .0030$, $S_{XY} = .0013$, and $S_{X+Y}^2 = .0094$. Means of these monthly rates of return can be used to measure the profitability of the investments; variances of these monthly rates of return can be used as a measure of the risk, or uncertainty, involved in the different investments.

Summary

In this chapter, we discussed basic concepts and properties of probability distributions for discrete random variables. Important discrete distributions such as the binomial, hypergeometric, and Poisson distributions are discussed in detail. Applications of these distributions in business decisions are also examined.

Using the probability distribution for a random variable, we can calculate the probabilities of specific sample observations. If the probabilities are difficult to calculate, then the means and standard deviations can be used as numerical descriptive measures that enable us to visualize the probability distributions and thereby to make some approximate probability statements about sample observations.

In this chapter we also discussed joint probability of two random variables. The covariance and coefficient of correlation were presented as means of measuring the degree of relationship between two random variables X and Y.

Appendix 6A   The Mean and Variance of the Binomial Distribution

Let X_i represent the Bernoulli random variable; then the random variables X_1, X_2, ..., X_n are independent Bernoulli variables. From Equations 6.3 and 6.4, we know that the mean of a Bernoulli variable is

$$E(X_i) = \sum_{i=1}^{2} x_i P(x_i) = 0(1 - p) + 1(p) = p \qquad (6A.1)$$

and that the variance is

$$\begin{aligned} \text{Var}(X_i) &= \sum_{i=1}^{2} (x_i - \mu_X)^2 P(x_i) \\ &= (0 - p)^2(1 - p) + (1 - p)^2 p = p(1 - p) \end{aligned} \qquad (6A.2)$$

To find the mean and variance of the binomial distribution, we use the fact that the binomial random variable can be expressed as the sum of independent Bernoulli random variables (X_i):

$$X = X_1 + X_2 + \cdots + X_n \qquad (6A.3)$$

From Equations 6.30 and 6A.1, we can derive Equation 6.11 as

$$\begin{aligned} E(X) = \mu_X &= E(X_1 + X_2 + \cdots + X_n) \\ &= E(X_1) + E(X_2) + \cdots + E(X_n) = np \quad (6A.4) \end{aligned}$$

Because the X_i variables are statistically independent of one another, the variance of their sum is equal to the sum of their variances. Therefore, following Equation 6.31, we can derive Equation 6.12 as

$$\sigma_X^2 = \text{Var}\left(\sum_{i=1}^{n} X_i\right) = \text{Var}(X_1) + \text{Var}(X_2) + \cdots + \text{Var}(X_n) = np(1 - p)$$

Appendix 6B   Applications of the Binomial Distribution to Evaluate Call Options

In this appendix, we show how the binomial distribution is combined with some basic finance concepts to generate a model for determining the price of stock options.

What Is an Option?

In the most basic sense, an **option** is a contract conveying the right to buy or sell a designated security at a stipulated price. The contract normally expires at a predetermined date. The most important aspect of an option contract is that the purchaser is under no obligation to buy; it is, indeed, an "option." This attribute of an option contract distinguishes it from other financial contracts. For instance, whereas the holder of an option may let his or her claim expire unused if he or she so desires, other financial contracts (such as futures and forward contracts) obligate their parties to fulfill certain conditions.

A *call option* gives its owner the right to buy the underlying security, a *put option* the right to sell. The price at which the stock can be bought (for a call option) or sold (for a put option) is known as the exercise price.

The Simple Binomial Option Pricing Model

Before discussing the binomial option pricing model, we must recognize its two major underlying assumptions. First, the binomial approach assumes that trading takes place in discrete time—that is, on a period-by-period basis. Second, it is assumed that the stock price (the price of the underlying asset) can take on only two possible values each period; it can go up or go down.

Say we have a stock whose current price per share S can advance or decline during the next period by a factor of either u (up) or d (down). This price either will increase by the proportion $u - 1 \geq 0$ or will decrease by the proportion $1 - d$, $0 < d < 1$. Therefore, the value S in the next period will be either uS or dS. Next, suppose that a call option exists on this stock with a current price per share of C and an exercise price per share of X and that the option has one period left to maturity. This option's value at expiration is determined by the price of its underlying stock and the exercise price X. The value is either

$$C_u = \text{Max}(0, uS - X) \tag{6B.1}$$

or

$$C_d = \text{Max}(0, dS - X) \tag{6B.2}$$

Why is the call worth $\text{Max}(0, uS - X)$ if the stock price is uS? The option holder is not obliged to purchase the stock at the exercise price of X, so she or he will exercise the option only when it is beneficial to do so. This means the option can never have a negative value. When is it beneficial for the option holder to exercise the option? When the price per share of the stock is greater than the price per share at which he or she can purchase the stock by using the option, which is the exercise price, X. Thus if the stock price uS exceeds the exercise price X, the investor can exercise the option and buy the stock. Then he or she can immediately sell it for uS, making a profit of $uS - X$ (ignoring commission). Likewise, if the stock price declines to dS, the call is worth $\text{Max}(0, dS - X)$.

Also for the moment, we will assume that the risk-free interest rate for both borrowing and lending is equal to r percent over the one time period and that the exercise price of the option is equal to X.

To intuitively grasp the underlying concept of option pricing, we must set up a *risk-free portfolio*—a combination of assets that produces the same return in every state of the world over our chosen investment horizon. The investment horizon is assumed to be one period (the duration of this period can be any length of time, such as an hour, a day, a week, etc.). To do this, we buy h shares of the stock and sell the call option at its current price of C.[9] Moreover, we choose the value of h such that our portfolio will yield the same payoff whether the stock goes up or down.

$$h(uS) - C_u = h(dS) - C_d \qquad (6B.3)$$

By solving for h, we can obtain the number of shares of stock we should buy for each call option we sell.

$$h = \frac{C_u - C_d}{(u - d)S} \qquad (6B.4)$$

Here h is called the *hedge ratio*. Because our portfolio yields the same return under either of the two possible states for the stock, it is without risk and therefore should yield the risk-free rate of return, r percent, which is equal to the risk-free borrowing and lending rate. The condition must be true; otherwise, it would be possible to earn a risk-free profit without using any money. Therefore, the ending portfolio value must be equal to $(1 + r)$ times the beginning portfolio value, $hS - C$.

$$(1 + r)(hS - C) = h(uS) - C_u = h(dS) - C_d \qquad (6B.5)$$

Note that S and C represent the beginning values of the stock price and the option price, respectively.

Setting $R = 1 + r$, rearranging to solve for C, and using the value of h from Equation 6B.4, we get

$$C = \left[\left(\frac{R - d}{u - d} \right) C_u + \left(\frac{u - R}{u - d} \right) C_d \right] \bigg/ R \qquad (6B.6)$$

where $d < r < u$. To simplify this equation, we set

$$p = \frac{R - d}{u - d} \quad \text{so} \quad 1 - p = \left\{ \frac{u - R}{u - d} \right\} \qquad (6B.7)$$

Thus we get the option's value with one period to expiration:

$$C = [pC_u + (1 - p)C_d]/R \qquad (6B.8)$$

This is the binomial call option valuation formula in its most basic form. In other words, this is the binomial option valuation formula with one period to expiration of the option.

---

[9]To sell the call option means to write the call option. If a person writes a call option on stock A, then he or she is obliged to sell at exercise price X during the contract period.

Table 6B.1 Possible Option Values at Maturity

	Today	
Stock (S)	*Option (C)*	**Next Period (Maturity)**

$uS = \$110$ $C_u = \text{Max}(0, uS - X)$
$= \text{Max}(0, 110 - 100)$
$= \text{Max}(0, 10)$
$= \$10$

$\$100$ C

$dS = \$90$ $C_d = \text{Max}(0, dS - X)$
$= \text{Max}(0, 90 - 100)$
$= \text{Max}(0, -10)$
$= \$0$

To illustrate the model's qualities, let's plug in the following values, while assuming the option has one period to expiration. Let

$$X = \$100$$
$$S = \$100$$
$$u = (1.10), \text{ so } uS = \$110$$
$$d = (.90), \text{ so } dS = \$90$$
$$R = 1 + r + 1 + .07 = 1.07$$

First we need to determine the two possible option values at maturity, as indicated in Table 6B.1.

Next we calculate the value of p as indicated in Equation 6B.7.

$$p = \frac{1.07 - .90}{1.10 - .90} = .85 \quad \text{so} \quad 1 - p = \frac{1.10 - 1.07}{1.10 - .90} = .15$$

Solving the binomial valuation equation as indicated in Equation 6B.8, we get

$$C = [.85(10) + .15(0)]/1.07$$
$$= \$7.94$$

The correct value for this particular call option today, under the specified conditions, is $7.94. If the call option does not sell for $7.94, it will be possible to earn arbitrage profits. That is, it will be possible for the investor to earn a risk-free profit while using none of his or her own money. Clearly, this type of opportunity cannot continue to exist indefinitely.

The Generalized Binomial Option Pricing Model[10]

Suppose we are interested in the case where there is more than one period until the option expires. We can extend the one-period binomial model to consideration of

[10]This section is essentially based on Cheng F. Lee, Joseph E. Finnerty, and Donald H. Wort (1990) *Security Analysis and Portfolio Management* (Glenview, Ill.: Scott, Foresman), Chapter 15. Copyright © 1990 by Cheng F. Lee, Joseph E. Finnerty, and Donald H. Wort. Reprinted by permission of Harper Collins Publishers.

two or more periods. Because we are assuming that the stock follows a binomial process, from one period to the next it can only go up by a factor of u or go down by a factor of d. After one period the stock's price is either uS or dS. Between the first and second periods, the stock's price can once again go up by u or down by d, so the possible prices for the stock two periods from now are uuS, udS, and ddS. This process is demonstrated in tree diagram form (Figure 6B.1) in Example 6B.1 later in this appendix.

Note that the option's price at expiration, two periods from now, is a function of the same relationship that determined its expiration price in the one-period model. More specifically, the call option's maturity value is always

$$C_T = \text{Max}[0, S_T - X] \tag{6B.9}$$

where T designates the maturity date of the option.

To derive the option's price with two periods to go ($T = 2$), it is helpful as an intermediate step to derive the value of C_u and C_d with one period to expiration when the stock price is uS and dS, respectively.

$$C_u = [pC_{uu} + (1 - p)C_{ud}]/R \tag{6B.10}$$

$$C_d = [pC_{du} + (1 - p)C_{dd}]/R \tag{6B.11}$$

Equation 6B.10 tells us that if the value of the option after one period is C_u, the option will be worth either C_{uu} (if the stock price goes up) or C_{ud} (if stock price goes down) after one more period (at its expiration date). Similarly, Equation 6B.11 shows that if the value of the option is C_d after one period, the option will be worth either C_{du} or C_{dd} at the end of the second period. Replacing C_u and C_d in Equation 6B.8 with their expressions in Equations 6B.10 and 6B.11, respectively, we can simplify the resulting equation to yield the two-period equivalent of the one-period binomial pricing formula, which is

$$C = [p^2C_{uu} + 2p(1 - p)C_{ud} + (1 - p)^2C_{dd}]/R^2 \tag{6B.12}$$

In Equation 6B.12, we used the fact that $C_{ud} = C_{du}$ because the price will be the same in either case.

We know the values of the parameters S and X. If we assume that R, u, and d will remain constant over time, the possible maturity values for the option can be determined exactly. Thus deriving the option's fair value with two periods to maturity is a relatively simple process of working backwards from the possible maturity values.

Using this same procedure of going from a one-period model to a two-period model, we can extend the binomial appoach to its more generalized form, with n periods to maturity:

$$C = \frac{1}{R^n} \sum_{k=0}^{n} \frac{n!}{k!(n-k)!} p^k(1-p)^{n-k} \text{Max}[0, u^k d^{n-k} S - X] \quad (6B.13)$$

To actually get this form of the binomial model, we could extend the two-period model to three periods, then from three periods to four periods, and so on. Equation 6B.13 would be the result of these efforts. To show how Equation 6B.13 can be used to assess a call option's value, we modify the example as follows: $S = \$100$, $X = \$100$, $R = 1.07$, $n = 3$, $u = 1.1$, and $d = .90$.

First we calculate the value of p from Equation 6B.7 as .85, so $1 - p$ is .15. Next we calculate the four possible ending values for the call option after three periods in terms of $\text{Max}[0, u^k d^{n-k} S - X]$.

$$C_1 = \text{Max}[0, (1.1)^3(.90)^0(100) - 100] = 33.10$$
$$C_2 = \text{Max}[0, (1.1)^2(.90)(100) - 100] = 8.90$$
$$C_3 = \text{Max}[0, (1.1)(.90)^2(100) - 100] = 0$$
$$C_4 = \text{Max}[0, (1.1)^0(.90)^3(100) - 100] = 0$$

Now we insert these numbers (C_1, C_2, C_3, and C_4) into the model and sum the terms.

$$\begin{aligned} C &= \frac{1}{(1.07)^3} \left[\frac{3!}{0!3!} (.85)^0(.15)^3 \times 0 + \frac{3!}{1!2!} (.85)^1(.15)^2 \times 0 \right. \\ &\quad \left. + \frac{3!}{2!1!} (.85)^2(.15)^1 \times 8.90 + \frac{3!}{3!0!} (.85)^3(.15)^0 \times 33.10 \right] \\ &= \frac{1}{1.225} \left[0 + 0 + \frac{3 \times 2 \times 1}{2 \times 1 \times 1} (.7225)(.15)(8.90) \right. \\ &\quad \left. + \frac{3 \times 2 \times 1}{3 \times 2 \times 1 \times 1} \times (.61413)(1)(33.10) \right] \\ &= \frac{1}{1.225} [(.32513 \times 8.90) + (.61413 \times 33.10)] \\ &= \$18.96 \end{aligned}$$

As this example suggests, working out a multiple-period problem by hand with this formula can become laborious as the number of periods increases. Fortunately, programming this model into a computer is not too difficult.

Now let's derive a binomial option pricing model in terms of the cumulative binomial density function. As a first step, we can rewrite Equation 6B.13 as

$$C = S \left[\sum_{k=m}^{n} \frac{n!}{k!(n-k)!} p^k(1-p)^{n-k} \frac{u^k d^{n-k}}{R^n} \right]$$
$$- \frac{X}{R^n} \left[\sum_{k=m}^{n} \frac{n!}{k!(n-k)!} p^k(1-p)^{n-k} \right] \quad (6B.14)$$

This formula is identical to Equation 6B.13 except that we have removed the Max operator. In order to remove the Max operator, we need to make $u^k d^{n-k} S - X$ positive, which we can do by changing the counter in the summation from $k = 0$ to $k = m$. What is m? It is the minimum number of upward stock movements necessary for the option to terminate "in the money" (that is, $u^k d^{n-k} S - X > 0$). How can we interpret Equation 6B.14? Consider the second term in brackets; it is just a cumulative binomial distribution with parameters of n and p.[11] Likewise, via a small algebraic manipulation we can show that the first term in the brackets is also a cumulative binomial distribution. This can be done by defining $p' \equiv (u/R)p$ and $1 - p' \equiv (d/R)(1 - p)$.[12] Thus

$$p^k (1 - p)^{n-k} \frac{u^k d^{n-k}}{R^n} = p'^k (1 - p')^{n-k}$$

Therefore the first term in brackets is also a cumulative binomial distribution with parameters of n and p'. Using Equation 6.10 in the text, we can write the binomial call option model as

$$C = SB_1(n, p', m) - \frac{X}{R^n} B_2(n, p, m) \qquad (6B.15)$$

where

$$B_1(n, p', m) = \sum_{k=m}^{n} {_nC_k} p'^k (1 - p')^{n-k}$$

$$B_2(n, p, m) = \sum_{k=m}^{n} {_nC_k} p^k (1 - p)^{n-k}$$

and m is the minimum amount of time the stock has to go up for the investor to finish *in the money* (that is, for the stock price to become larger than the exercise price).

In this appendix, we showed that by employing the definition of a call option and by making some simplifying assumptions, we can use the binomial distribution to find the value of a call option. In the next chapter, we will show how the binomial distribution is related to the normal distribution and how this relationship can be used to derive one of the most famous valuation equations in finance, the Black–Scholes option pricing model.

---

[11]Note that this is not exactly a cumulative binomial distribution as defined by a statistician. Strictly speaking,

$$1 - [\,] = \sum_{k=0}^{m-1} \frac{n!}{k!(n - k)!} p^k (1 - p)^{m-k}$$

is a cumulative binomial distribution.

[12]Because $u < R < d$,

$$(u/R)P = \frac{1 - d/R}{1 - d/u} < 1$$

EXAMPLE 6B.1   *A Decision Tree Approach to Analyzing Future Stock Price*

By making some simplifying assumptions about how a stock's price can change from one period to the next, it is possible to forecast the future price of the stock by means of a decision tree. To illustrate this point, let's consider the following example.

Suppose the price of Company A's stock is currently $100. Now let's assume that from one period to the next, the stock can go up by 17.5 percent or go down by 15 percent. In addition, let us assume that there is a 50 percent chance that the stock will go up and a 50 percent chance that the stock will go down. It is also assumed that the price movement of a stock (or of the stock market) today is completely independent of its movement in the past; in other words, the price will rise or fall today by a random amount. A sequence of these random increases and decreases is known as a **random walk.**

Given this information, we can lay out the paths that the stock's price may take. Figure 6B.1 shows the possible stock prices for company A for four periods.

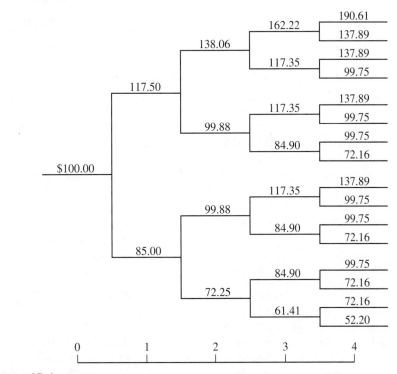

Figure 6B.1

Price Path of Underlying Stock

Source: R. J. Rendelman, Jr., and B. J. Bartter (1979), "Two-State Option Pricing," *Journal of Finance* 34 (December), 1096.

Note that in period 1 there are two possible outcomes: the stock can go up in value by 17.5 percent to $117.50 or down by 15 percent to $85.00. In period 2 there are four possible outcomes. If the stock went up in the first period, it can go up again to $138.06 or down in the second period to $99.88. Likewise, if the stock went down in the first period, it can go down again to $72.25 or up in the second period to $99.88. Using the same argument, we can trace the path of the stock's price for all four periods.

If we are interested in forecasting the stock's price at the end of period 4, we can find the average price of the stock for the 16 possible outcomes that can occur in period 4.

$$\overline{P} = \frac{\sum_{i=1}^{16} P_i}{16} = \frac{190.61 + 137.89 + \cdots + 52.20}{16}$$

$$= \$105.09$$

We can also find the standard deviation for the stock's return.

$$\sigma_P = \left[\frac{(190.61 - 105.09)^2 + \cdots + (52.20 - 105.09)^2}{16} \right]^{1/2}$$

$$= \$34.39$$

$\overline{P}$ and σ_P can be used to predict the future price of stock A.

Questions and Problems

1. A team of students participates in a project. The results show that all students are able to finish the project in 7 days. The distribution for the finishing time is given in the following table.

Finishing time, hours	1	2	3	4	5	6	7
Students	21	43	23	48	31	29	35

Define x as the finishing time.
   a. Obtain $P(X = 1)$, $P(X = 2)$, ..., $P(X = 7)$.
   b. Draw the probability distribution.
   c. Calculate the cumulative function $F_x(x)$.
   d. Draw the cumulative function.

2. An investment banker estimates the following probability distribution for the earnings per share (EPS) of a firm.

x (EPS)	2.25	2.50	2.75	3.00	3.25	3.50	3.75
$P(x)$	.05	.10	.20	.35	.15	.10	.05

Calculate the expected value of the EPS.

3. The following table gives the number of unpainted machines in a container. Calculate the mean and standard deviation.

Unpainted	0	1	2	3	4	5	6	7
Probability	.05	.09	.15	.30	.25	.10	.05	.01

4. The following table gives the probability distribution for two random variables: x, which measures the total number of times a person will be ill during a year, and y, the sex of this person, where 1 represents male and 0 female.

x	1	2	3	4	Total
y					
0	.10	.11	.11	.17	.49
1	.13	.16	.07	.15	.51
	.23	.27	.18	.32	1.00

a. Calculate the expected value and standard deviation of the number of times a person will be ill during a year, given that the sex of the person is male. *Hint:* The conditional expectation and conditional standard deviation, which we have not addressed, can be defined as follows:

$$F(X_k) = P(X \le X_k \mid Y = 0)$$
$$F(X_k) = P(X \le X_k \mid Y = 1)$$
$$\mu_0 = \Sigma X_k P(X = X_k \mid Y = 0)$$
$$\mu_1 = \Sigma X_k P(X = X_k \mid Y = 1)$$
$$\sigma_0 = \left[\sum_{i=1}^{n} (X_i - \mu_0)^2 P(X_i \mid Y = 0) \right]^{1/2}$$
$$\sigma_1 = \left[\sum_{i=1}^{n} (X_i - \mu_1)^2 P(X_i \mid Y = 1) \right]^{1/2}$$

b. Calculate the mean and standard deviation of the number of times a person will be ill during a year, given that the sex of the person is female.

5. The rate of defective items in a production process is 15 percent. Assume a random sample of 10 items is drawn from the process. Find the probability that 2 of them are defective. Calculate the expected value and variance.

6. Find the mean and standard deviation of the number of successes in binomial distributions characterized as follows:
a. $n = 20, p = .5$ **c.** $n = 30, p = .7$
b. $n = 100, p = .09$ **d.** $n = 50, p = .4$

7. A fair die is rolled 10 times. Find the probability of getting exactly 4 aces, of getting 5 aces, of getting 6 aces, and of getting 4 aces or more.

8. A fair coin is tossed eight times.
a. Use MINITAB to construct a probability function table.
b. What is the probability that you will have exactly four heads?

9. Consider a group of 12 employees of whom 5 are in management and 7 do clerical work. Select at random a sample of size 4. What is the probability that there will be 1 manager in this sample?

10. A survey was conducted. Of 20 questionnaires that were sent, 12 were completed and returned. We know that 8 of the 20 questionnaires were sent to students and 12 to nonstudents. Only 2 of the returned questionnaires were from students.
a. What is the response rate for each group?
b. Assume we have a response in hand. What is the probability that it comes from a student?

11. The number of people arriving at a bank teller's window is Poisson-distributed with a mean rate of .75 persons per minute. What is the probability that 2 or fewer people will arrive in the next 6 minutes?

12. The Wicker company has one repair specialist who services 200 machines in the shop and repairs machines that break down. The average breakdown rate is $\lambda = .5$ machine per day (or 1 breakdown every 2 days). This technician can fix 2 machines a day.
a. Wicker is interested in determining the probability that there will be more than 2 breakdowns in a day. Assuming a Poisson distribution, answer the company's question.
b. Use MINITAB to construct a probability function, including breakdown frequency from 0 to 10 cases per day.

13. Two teams enter the NBA championship playoff. Team A is considered to have a 60 percent to 40 percent edge over team B. Whichever team wins 4 of 7 games will win the championship. What is the probability that team A will win the championship?

14. A baseball player usually has 4 at-bats each game. Suppose the baseball player is a lifetime 0.25 hitter. Find the probability that this player will have
a. 2 hits out of 4 at-bats
b. No hits out of 4 at-bats
c. At least 1 hit out of 4 at-bats

15. A certain insurance salesman sees an average of 5 customers in a week. Each time he speaks to a customer, he has a 30 percent chance of making a deal. What is the probability that he makes 5 deals after speaking with 5 customers in a week?

16. A student takes an exam that consists of 10 multiple-choice questions. Each question has 5 possible answers. Suppose the student knows nothing about the subject and just guesses the answer on each question. What is the probability that this student will answer 4 out of the 10 questions correctly?

17. A hospital has 3 doctors working on the night shift. These doctors can handle only 3 emergency cases in a time period of 30 minutes. On average, $\frac{1}{2}$ an emergency case arises in each 30-minute period. What is the probability that 4 emergency cases will arise in a 30-minute period?

18. An average of 3 small businesses go bankrupt each month. What is the probability that 5 small businesses will go bankrupt in a certain month?

19. During each hour, 0.1 percent of the total production of paperclips is defective. For a random sample of 500 pieces of the product, what is the chance of finding more than 1 defective item?

20. The local bank manager has found that 1 out of every 400 bank loans ends up in default. Last year the bank made 400 loans. What is the probability that 2 bank loans will end up in default?

21. Every week a truckload of springs is delivered to the warehouse you supervise. Every time the springs arrive, you have to measure the strength of 400 springs. You accept the shipment only when there are fewer than 20 bad springs. One day a truckload of springs arrives that contains 10 percent bad springs. What is the probability that you will accept the shipment? (Just set up the question. Do not try to solve it.)

22. Returning to question 21, say (1), your company's policy is to accept the shipment only when fewer than 2 springs (out of 400 springs examined) are bad, and (2) the proportion of the bad springs in the truck is only 0.0001. Under these conditions, what is the probability that you will accept the shipment?

23. Despite your discomfort with statistics, you find yourself employed by a dog food manufacturer to do statistical research for quality control purposes. Your job is to weigh the dog food to determine whether the cans contain the 16 ounces of dog food that the label will claim they contain. You pick 25 cans from each hour's production and weigh them. If there are more than 2 cans that contain less than 16 ounces, you are to discard the production from that hour. If in a certain hour, 5 percent of the cans of dog food produced actually contain less than 16 ounces, what is the probability that the whole hour's production will be discarded?

24. A medical report shows that 5 percent of stock brokers suffer stress and need medical attention. There are 10 brokers working for your brokerage house. What is the probability that 3 of them will need medical attention as a result of stress?

25. There are 38 numbers in the game of roulette. They are 00, 0, 1, 2, . . . , 36. Each number has an equal chance of being selected. In the game, the winning number is found by a spin of the wheel. Say a gambler bets $1 on the number 35 three times.
   a. What is the probability that the gambler will win the second bet?
   b. What is the probability that the gambler will win two of the three bets?

26. In the game of roulette, a gambler who wins the bet receives $36 for every dollar she or he bet. A gambler who does not win receives nothing. If the gambler bets $1, what is the expected value of the game?

27. A company found that on average, on a given day, .5 percent of its employees call in sick. Assume a Poisson distribution. What is the probability that fewer than 2 of 300 employees will call in sick?

28. Billings Company is considering leasing a computer for the next 3 years. Two computers are available. The net present value of leasing each computer in the next 3 years, under different business conditions, is summarized in the following table.

	Business Is	
	Good	**Bad**
Plan A: Big computer	200,000	20,000
Plan B: Small computer	150,000	100,000

A consulting company estimates that the chances of having good and of having bad business are 20 percent and 80 percent, respectively. Compute the expected net present value of leasing a big computer. Compute the expected value of leasing a small computer. What are the variances of these two plans?

29. The makers of two kinds of cola are having a contest in the local shopping mall. Assume that 60 percent of the people in this region prefer brand A and 40 percent prefer brand B. Ten local residents were randomly selected to test the colas. What is the probability that 5 of these 10 testers will prefer brand A?

30. In a certain statistics course, the misguided professor is very lenient. He fails about 1 percent of the students in the class. Assume that the probability of failing the course follows a Poisson distribution. In a certain year, the professor teaches 400 students. What is the probability that no one fails the course? What is the average number of failing students?

31. A factory examines its work injury history and discovers that the chance of there being an accident on a given work day follows a Poisson distribution. The average number of injuries per work day is .01. What is the probability that there will be 3 work injuries in a given month (30 days)?

32. The state highway bureau found that during the rush hour, in the treacherous section of a highway, an average of 3 accidents occur. The probability of there being an accident follows a Poisson distribution. What is the probability that there is no accident in a given day?

33. In question 32, what is the probability that there are no traffic accidents in all 5 work days of a week?

34. Of 7 prominent financial analysts who are attending a meeting, 3 are pessimistic about the future of the stock market, and 4 are optimistic. A newspaper reporter interviews 2 of the 7 analysts. What is the probability that one of these interviewees takes an optimistic view and the other a pessimistic view?

35. After assembly, a finished TV is left turned on for one full day (24 hours) to determine whether the product is reliable. On average, 2 TVs break down each day. Yesterday 500 TVs were produced. What is the probability that less than 1 TV broke down?

36. A soft drink company argues that its new cola is the favorite soft drink of the next generation. Ten

teenagers were picked to test-drink the cola one by one. Assume that 5 of them liked the new cola and the rest did not.
   a. What is the probability that the first test-drinker liked the new cola?
   b. What is the probability that the second test-drinker liked the new cola?
   c. What is the probability that after 5 test-drinks, the new product received 3 yes votes and 2 no votes?

37. Suppose school records reveal that historically, 10 percent of the students in Milton High School have dropped out of school. What is the probability that more than 2 students in a class of 30 will drop out?

38. An insurance company found that 1 of 5,000 fifty-year-old, nonsmoking males will suffer a heart attack in a given year. The company has 50,000 fifty-year-old, nonsmoking male policyholders. What is the probability that fewer than 3 such policyholders will suffer a heart attack this year?

39. Suppose that of 40 salepersons in a company, 10 are females and the rest males. Five of them are randomly chosen to attend a seminar. What is the probability that 3 females and 2 males are chosen?

40. Consider a single toss of a fair coin, and define X as the number of heads that come up on that toss. Then X can be 0 or 1, with a probability of 50 percent. The expected value of X is $\frac{1}{2}$. Can we "expect" to get $\frac{1}{2}$ a head when we toss the coin? If not, how should we interpret the concept of the expected value?

41. What is a random variable? What is a discrete random variable? What is a continuous random variable? Give some examples of discrete random variables.

42. Tell whether each of the following is a discrete or a continuous random variable.
   a. The number of beers sold at a bar during a particular week.
   b. The length of time it takes a person to drive 50 miles.
   c. The interest rate on 3-month Treasury bills.
   d. The number of products returned to a store on a particular day.

43. An analyst calculates the probability of McGregor stock going up in value for any month as .6 and the

probability of the same stock going down in any month as .4. Calculate the probability that the stock will go up in value in exactly 7 months during a year. (Assume independence.)

44. Using the information from question 43, compute the probability that the stock will go up in at least 7 months during the year.

45. What is a Bernoulli trial? Give some examples of a Bernoulli trial related to the binomial distribution?

46. What is the Poisson distribution? Give some examples of situations wherein it would be appropriate to use the Poisson distribution. Compare the Poisson approximation to the binomial distribution.

47. Suppose Y represents the number of times a homemaker stops by the local convenience store in a week. The probability distribution of Y follows. Find the expected value and variance of Y.

y	Probability
0	.15
1	.25
2	.25
3	.20
4	.15

48. The managers of a grocery store are interested in knowing how many people will shop in their store in a given hour. Suppose they collect data and find that the average number of people who enter the store in any 15-minute period is 12. Find the probability that 8 people will enter the store in any 15-minute period. What is the probability that no more than 8 people will enter the store in any 15-minute period?

49. Suppose you are tossing a fair coin 25 times. What is the probability that you will toss exactly 5 heads? What is the probability that you will toss 5 or fewer heads?

50. Calculate the mean and variance for the distribution given in question 49.

51. You are rolling a six-sided fair die 8 times. What is the probability that you will roll exactly 2 sixes? What is the probability that you will roll 2 or fewer sixes?

52. Calculate the mean and variance for the distribution given in question 51.

53. Doctors at the Centers for Disease Control estimate that 30 percent of the population will catch the Tibetan flu. What is the probability that in a sample of 10 people, exactly 3 will catch the flu? What is the probability that 3 or fewer people in this sample will catch the flu? (Assume that the conditions of a Bernoulli process apply.)

54. A golfer enters a long-driving contest in which he wins if he drives the golf ball 300 yards or more and loses if he drives it less than 300 yards. Assume that every time he hits a golf ball, he has a 40 percent chance of driving it over 300 yards. If the golfer gets to hit 4 balls and needs only one 300-yard drive to win, what is the probability that he will win?

55. A phone marketing company knows that the number of people who answer the phone between 10:00 and 10:15 A.M. has a Poisson distribution. The average number is 8. What is the probability that the phone company will reach exactly 10 people when it calls during this period? What is the probability that it will reach exactly 3 people?

56. A market survey shows that 75 percent of all households own a VCR. Suppose 100 households are surveyed.
   a. What is the probability that none of the households surveyed owns a VCR?
   b. What is the probability that exactly 75 of the households surveyed own a VCR?
   c. Suppose X is the number of households that own a VCR. Compute the mean and variance for X.

57. You are given the following information about a stock:

$S = \$100$ price of stock
$X = \$101$ exercise price for a call option on the stock
$r = .005$ interest rate per month
$n = 5$ number of months until the option expires
$u = 1.10$ amount of increase if stock goes up
$d = .95$ amount of decrease if stock goes down

Calculate the value of the call option if the stock goes up in 3 out of the 5 months.

58. Answer question 57 when $u = 1.20$. How does a change in the amount of increase if the stock goes up affect the value of the call option?

59. Answer question 57 when $d = .85$. How does a change in the amount of decrease if the stock goes down affect the value of the call option?

60. Answer question 57 when $X = \$95$. How does a change in the exercise price affect the value of the call option?

61. Answer question 57 when $S = \$110$. How does a change in the value of the stock affect the value of the call option?

62. Answer question 57 again, finding the value of the option if the stock goes up in 4 out of the 5 months.

63. Suppose a box is filled with 25 white balls and 32 red balls. Find the probability of drawing 6 red balls and 4 white balls in 10 draws without replacement.

64. Redo question 63 *with* replacement.

65. You are drawing 5 cards from a standard deck of cards with replacement. You win if you draw at least 3 red cards in 5 draws. What is the probability of your winning?

66. Two tennis players are playing in a final-set tie breaker. Player A is considered to have a 70 percent to 30 percent edge over player B. The player who wins 7 of 13 points will win the championship. What is the probability that player B will win the championship?

67. A car sales representative sees an average of 4 customers in a day. Each time she talks to a customer, she has a 25 percent chance of making a deal. What is the chance that she will make 4 deals after talking to 4 customers in a day?

68. Again consider the car sales rep in question 67. What is the probability that she will make at least 2 sales after speaking to 4 customers?

69. Again consider the car sales rep in question 67. What is the probability that she will make exactly 2 sales after speaking to four customers?

70. The following test is given to people who claim to have extrasensory perception (ESP). Five cards with different shapes on them are hidden from the person. A card is randomly drawn, and the person is then supposed to guess (or use ESP to deter-

mine) the shape on the card. Suppose that this test is administered 10 times with replacement. What is the probability that the subject will get 5 correct? Do you think getting more than 5 out of 10 correct supports the subject's claim to be endowed with ESP?

71. Again consider the test in question 70. What is the probability that a person taking the test will get 8 out of 20 correct? Do you believe that a person who gets 8 out of 20 correct has ESP?

72. Say we toss 2 six-sided dice and let the random variable be the total number of dots observed.
  a. Calculate both the probability and the cumulative probability distributions.
  b. Draw the graphs associated with the distributions you obtained in part (a).

73. **a.** Use the monthly rates of return for both IBM and Merck listed in Figure 6.12 to calculate the correlation coefficient between the monthly rates of return for these two companies.
  b. Using the results you obtained in part (a) and some other statistics listed in Figure 6.12, discuss how these two securities are related and how this information can be used to make investment decisions.

74. The following table exhibits the monthly rate of return of S&P 500 and American Express. Use the MINITAB program to
  a. Calculate the mean and standard deviation of both returns.
  b. Calculate the correlation coefficient between these two sets of monthly returns.
  c. Explain the results you get from the two questions above.

		S&P 500	AMEX
1989	9	$-.65$	-2.69
	10	-2.52	12.73
	11	16.54	-2.97
	12	21.42	-1.11
1990	1	-6.88	-15.25
	2	8.54	-2.10
	3	24.26	-11.59
	4	-2.69	6.23
	5	9.20	6.93
	6	$-.89$	5.91

CHAPTER 7

The Normal and Lognormal Distributions

CHAPTER OUTLINE

7.1 Introduction

7.2 Probability Distributions for Continuous Random Variables

7.3 The Normal and Standard Normal Distributions

7.4 The Lognormal Distribution and Its Relationship to the Normal Distribution (Optional)

7.5 The Normal Distribution as an Approximation to the Binomial and Poisson Distributions

7.6 Business Applications

Appendix 7A Mean and Variance for Continuous Random Variables

Appendix 7B Cumulative Normal Distribution Function and the Option Pricing Model

Key Terms

continuous random variable

probability mass function

cumulative probability

cumulative distribution function

probability density function

normal distribution

normal probability density function

cumulative uniform density function

uniform distribution

coefficient of variation

standard normal distribution

Z score

lognormal distribution

option pricing model

7.1 INTRODUCTION

In Chapter 6 we discussed discrete random variables and their distributions. Particularly, we focused on the means and variances of binomial, hypergeometric, and Poisson distributions. Although the distributions derived from these discrete random variables are useful, they are limited. And therefore, statisticians have derived several important continuous distributions to substitute for and/or complement the discrete distributions. The normal distribution is the first important continuous distribution discussed in this chapter. Examples of continuous random variables include the number of miles a car travels on 1 gallon of gas and the exact weight of a box of cereal.

The lognormal distribution, a transformation of the normal distribution, is the second important continuous distribution we will examine. It is useful in many business and economic analyses. Because the lognormal distribution is valid only for nonnegative values of the random variable, it is more appropriate than the normal distribution for describing the distribution of a stock's price.

In this chapter we also discuss how the normal distribution can be used to approximate both binomial and Poisson distributions when the sample size is large.

7.2 PROBABILITY DISTRIBUTIONS FOR CONTINUOUS RANDOM VARIABLES

Continuous Random Variables

Unlike the values of discrete random variables, which are limited to a finite or countable number of distinct (integer) values, values of continuous variables are *not* limited to being integers; theoretically, they are infinitely divisible. A **continuous random variable** may take on any value within an interval, as we noted in Section 6.2. Measures of height, weight, time, distance, and temperature fit naturally into this category. In general, specific probabilities cannot be assigned to individual values of continuous random variables. The probability that any one specific value will occur for a continuous random variable is zero. For example, the probability that today's temperature is exactly 83.231 degrees is zero, because temperature is regarded as a continuous variable.

One may argue that in the real world all data are discrete. For example, if a scale permits determination of weight only to the nearest thousandth of a pound, then any resulting data will be discrete in units of thousandths of pounds. Despite the limitations of measuring instruments, however, it is useful in many instances to use continuous mathematical models that treat certain discrete variables as continuous. If we use a continuous mathematical model of heights of individuals, where the underlying data are discrete, we may conceive of this model not as a convenient approximation but rather as a model of reality that is more accurate than the discrete data from which the model was derived. In sum, even though measurement limitations make continuous data discrete, we are going to treat data as continuous because the model that results when we do so is more accurate.

Probability Distribution Functions for Discrete and Continuous Random Variables

We shall now consider experiments for which the theoretical set of possible outcomes forms a continuous interval on the real number line. Note that such observations are often rounded off so that the set of observations may seem to come from a finite set of real numbers. For such an experiment, we should consider conceptual sample spaces that are intervals of finite or infinite length. In this section, we contrast probability distribution functions for continuous random variables with those for discrete random variables, which we discussed in Section 6.3.

Approximation of a Histogram by a Continuous Curve

In Chapter 6 we showed that the probability distribution of a discrete random variable can be represented by a histogram. It can also be shown that a histogram can be approximated by a continuous curve. Now we use a fair coin-tossing example to demonstrate the meaning of a graph of the probability distribution of a continuous random variable. This example demonstrates the relationship between the probability distribution of a discrete variable and the probability distribution of a continuous variable.

EXAMPLE 7.1   *Using a Continuous Curve to Approximate the Histogram of a Fair Coin-Tossing Experiment*

The binomial distribution discussed in Chapter 6 is an example of a probability distribution of a discrete random variable. To get better insight into the meaning of a graph of the probability distribution of a discrete versus a continuous random variable, we first graph a binomial distribution as a histogram.

If we toss a fair coin four times, the probabilities of our getting 0, 1, 2, 3, and 4 heads, respectively, can be calculated by using the binomial formula, Equation 6.9.

$$P(x_1 = 4 \text{ tails}) = \binom{4}{0}\left(\frac{1}{2}\right)^0\left(\frac{1}{2}\right)^4 = \frac{1}{16}$$

$$P(x_2 = 1 \text{ head and 3 tails}) = \binom{4}{1}\left(\frac{1}{2}\right)\left(\frac{1}{2}\right)^3 = \frac{4}{16}$$

$$P(x_3 = 2 \text{ heads and 2 tails}) = \binom{4}{2}\left(\frac{1}{2}\right)^2\left(\frac{1}{2}\right)^2 = \frac{6}{16}$$

$$P(x_4 = 3 \text{ heads and 1 tail}) = \binom{4}{3}\left(\frac{1}{2}\right)^3\left(\frac{1}{2}\right) = \frac{4}{16}$$

$$P(x_5 = 4 \text{ heads}) = \binom{4}{4}\left(\frac{1}{2}\right)^4\left(\frac{1}{2}\right)^0 = \frac{1}{16}$$

This is a binomial distribution with $p = \frac{1}{2}$ and $n = 4$. The graph of this histogram is shown in Figure 7.1. Using this histogram, we interpret 0, 1, 2, 3, and 4 heads not as discrete values, but as midpoints of five classes whose respective limits are $-\frac{1}{2}$ to $\frac{1}{2}$, $\frac{1}{2}$ to $1\frac{1}{2}$, $1\frac{1}{2}$ to $2\frac{1}{2}$, $2\frac{1}{2}$ to $3\frac{1}{2}$, $3\frac{1}{2}$ to $4\frac{1}{2}$. The probabilities or relative frequencies associated with these classes are represented in the graph by the areas of rectangles or bars. Thus, because the rectangle for the class interval $1\frac{1}{2}$ to $2\frac{1}{2}$ has 1.5 times the area of that for the interval $2\frac{1}{2}$ to $3\frac{1}{2}$, it represents 1.5 times the probability. We can draw a continuous curve over the histogram and make the total area of this curve equal to the total area of the sum of five rectangles, which is 1.

. . . .

7.2 Probability Distributions for Continuous Random Variables   **249**

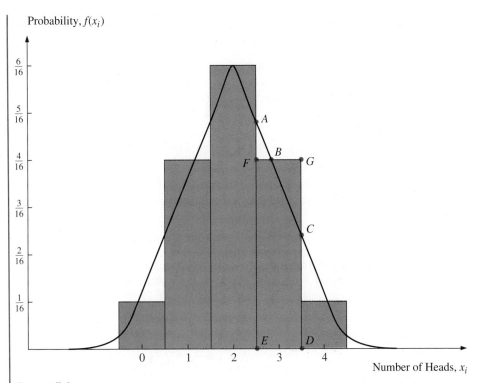

Figure 7.1

Approximation of a Binomial Distribution Histogram by a Continuous Curve ($n = 4$, $p = \frac{1}{2}$)

The curve would pass through the rectangle at B for 3 heads, as shown in Figure 7.1. For example, area $ABCDE$ represents the probability of the class with 3 heads in terms of a continuous-variable curve. This is due to the fact that $\triangle ABF$ is approximately equal to $\triangle BCG$. The area under the curve bounded by the class limits for any given class represents the probability of occurrence of that class.

If n increased (say, to 6 or 200), the width of the rectangles would decrease, and the corresponding shape of the histogram would approach that of a continuous curve more closely. Just as the total area of the rectangles in a histogram, representing a discrete random variable distribution, is equal to 1, so is the total area under the continuous curve representing a continuous random variable distribution.

We will use this example first to review the probability distributions for discrete variables. Then we will develop probability distributions for continuous variables by contrasting them with the probability functions of discrete variables discussed here.

Cumulative Probability and Cumulative Distribution Function for Discrete Random Variables

Let the value of the **probability mass function** (PMF) of a discrete random variable X at x be denoted as $P(x)$. In accordance with Equation 6.2, the **cumulative probability** for X, which is the probability that X will assume a value less than or equal to a given number x_k, can be defined as

$$F(x_k) = P(X \le x_k) = P(x_1) + P(x_2) + \cdots + P(x_k) = \sum_{i=1}^{k} P(x_i) \quad (6.2)$$

where $x_1 < x_2 < \cdots < x_k$.

Now let us see how cumulative probabilities are calculated.

EXAMPLE 7.2   *Cumulative Probability for Fair Coin-Tossing Experiments*

In the coin-tossing case discussed in Example 7.1, $P(x_1) = \frac{1}{16}$, $P(x_2) = \frac{4}{16}$, $P(x_3) = \frac{6}{16}$, $P(x_4) = \frac{4}{16}$, and $P(x_5) = \frac{1}{16}$. We calculate the cumulative probabilities $F(x_2)$, $F(x_4)$, and $F(x_5)$ by using Equation 6.2.

$$F(x_2) = \frac{1}{16} + \frac{4}{16} = \frac{5}{16}$$
$$F(x_4) = \frac{1}{16} + \frac{4}{16} + \frac{6}{16} + \frac{4}{16} = \frac{15}{16}$$
$$F(x_5) = \frac{1}{16} + \frac{4}{16} + \frac{6}{16} + \frac{4}{16} + \frac{4}{16} + \frac{1}{16} = 1$$

If the values outside the range of X (that is, the values smaller than x_1 or larger than x_k) occur with probability equal to zero, then in accordance with Equation 6.2a, the **cumulative distribution function** (CDF) of X can be written as

$$F(x_k) = \sum_{i=-\infty}^{k} P(x_i) \quad (6.2a)$$

The probability that X lies between a and b is

$$P(a \le X \le b) = F(b) - F(a) \quad (7.1)$$
$$= \sum_{i=1}^{b} P(x_i) - \sum_{i=1}^{a} P(x_i)$$

where $F(a)$ and $F(b)$ are cumulative probabilities at $X = b$ and $X = a$, respectively.

EXAMPLE 7.3   *Cumulative Distribution Function for the Tossing of a Fair Coin*

For the fair coin-tossing case discussed in Example 7.1, the CDFs calculated in accordance with Equation 6.2 are presented in Table 7.1. Using the probabilities of Table 7.1, we calculate the probability that lies between x_4 and x_2 as

$$P(x_2 \le X \le x_4) = F(x_4) - F(x_2) = \frac{15}{16} - \frac{5}{16} = \frac{10}{16}$$

Table 7.1 Cumulative Distribution Function for Coin Tossing

Possible Values of X	$F(x)$
0	1/16
1	5/16
2	11/16
3	15/16
4	1

Probability Distributions for Continuous Random Variables

The **probability density function** (PDF) for a continuous random variable is a curve, $f(x)$, that shows the probability of a range of values as the area under the curve. For example, the probability of the birth weights of infants being between 2 and 12 pounds can be written $P(2 < X < 12)$. Graphically, it is represented by the shaded area of Figure 7.2.

For continuous random variables, the probability that X has a value between a and b is written $P(a \leq X \leq b)$. This probability is equal to $P(a < X < b)$, because the probability at a point is considered to be zero; that is, $P(X = a) = 0$ and $P(X = b) = 0$. Thus we can write $P(a \leq X \leq b) = P(a < X \leq b) = P(a \leq X < b) = P(a < X < b)$.

Analogously to Equation 7.1, we define the probability that X lies between a and b for a continuous random variable as

$$P(a < X < b) = F(b) - F(a) \tag{7.2}$$

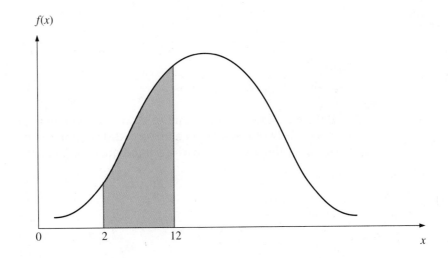

Figure 7.2

Probability Density Function

To show how Equation 7.2 can be used to calculate the probability of a continuous variable, we first discuss the simplest continuous cumulative density function, the so-called **cumulative uniform density function.** The density and cumulative density functions for this random variable can be defined as

$$f(X) = \begin{cases} \dfrac{1}{d-c} & \text{if } c \leq X \leq d \\ 0 & \text{elsewhere} \end{cases} \tag{7.3a}$$

$$F(X \leq x) = \begin{cases} 0 & x < c \\ \dfrac{x-c}{d-c} & c \leq x \leq d \\ 1 & x > d \end{cases} \tag{7.3b}$$

where $X = x$ is a continuous random variable that represents a point in the interval $c \leq x \leq d$, as described in Figure 7.3. Any one value of a uniform random variable is as likely to occur as any other, so the distribution is evenly spread over the entire region of possible values. In Chapter 9, we will discuss the **uniform distribution** further.

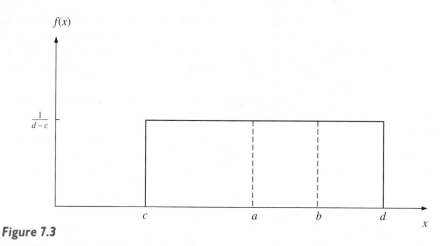

Figure 7.3

The Uniform Probability Distribution

If there exist two points, a and b, between c and d in Figure 7.3, the probability of x being between a and b for a uniform distribution can be calculated as follows: First, substituting $x = a$ and $x = b$ into Equation 7.3b, we get

$$F(a) = \frac{a-c}{d-c}$$

$$F(b) = \frac{b-c}{d-c}$$

Then, substituting both $F(b)$ and $F(a)$ into Equation 7.2, we obtain the probability of x being between a and b.

$$P(a < X < b) = \frac{b - a}{d - c}$$

For other types of continuous random variables, the calculation of Equation 7.2 generally requires knowledge of calculus.[1] For example, if $c = 10$, $d = 20$, $a = 15$, and $b = 19$, then $P(a < X < b) = (19 - 15)/(20 - 10) = .30$.

7.3 THE NORMAL AND STANDARD NORMAL DISTRIBUTIONS

The Normal Distribution

The **normal distribution** is the most widely used continuous density distribution in statistics. Many random variables have been found to be normally distributed, including measurements of weight, height, age, time, snowfall, yields, dimension, and other measures of interest to managers in both the public and private sectors.[2] When attempting to make an assertion about a population by using sample information, a major assumption we often make is that the population has a normal distribution.

From Figure 7.4, it is obvious that the normal distribution is centered on its mean. Because the distribution is symmetric, the mean and median also occur at the same point. In addition, the bell-shaped normal curve has a single peak; it is unimodal. The **normal probability density function** (PDF) for a normal variable X gives the height of an observation such as cd in Figure 7.4.[3]

[1] The probability that X lies between a and b for a continuous variable can be defined as

$$P(a < X < b) = \int_a^b f(x)\, dx = F(b) - F(a) \qquad (A)$$

where $f(x)$ represents the PDF of a continuous random variable X being valued at x.

From integral calculus, we know that the integration ($\int$) for the continuous case is the counterpart of the summation (Σ) in the discrete case. From Figure 7.2 and Equation A we know that the probability for a continuous PDF is represented by the area bounded by the curve whose value at x is $f(x)$, by the x-axis, and by the lines $x = a$ and $x = b$. Discussion of areas under the continuous PDF and of the mean and variance of a continuous variable can be found in Appendix 7A.

[2] Karl F. Gauss (1777–1855) discovered that the measurement of errors often follows a normal distribution.

[3] The PDF of a normal random variable can be defined as

$$f(x) = \frac{1}{\sqrt{2\pi}\sigma} e^{-(x-\mu)^2/2\sigma^2}, \quad -\infty < x < \infty$$

where $\pi = 3.14159$, $e = 2.71828$, and $\mu\,(-\infty < \mu < \infty)$ and $\sigma^2\,(0 < \sigma^2 < \infty)$ are the mean and variance of the normal random variable X. To graph the normal curve, we must know the numerical values of μ and σ^2.

There is no closed-form expression for $P(a < X < b) = \int_a^b f(x)\, dx$ for the normal probability distribution. However, the value of the definite integral can be obtained by numerical approximation procedures. The areas in Table A3 in Appendix A were obtained by using such a procedure.

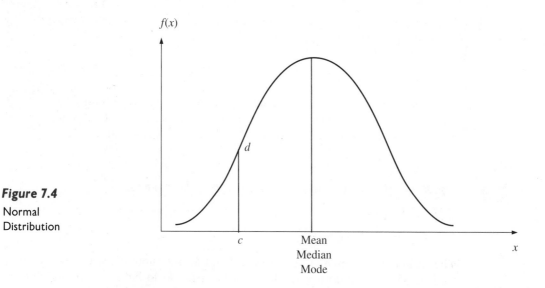

Figure 7.4

Normal
Distribution

Normally distributed populations with different shapes may nevertheless have many characteristics in common. The factor that determines shape is standard deviation: the larger the standard deviation, the wider the curve. Figure 7.5 shows MINITAB results for three normal distributions with mean 0 and three different standard deviations. Note that the mean μ and standard deviation σ completely characterize the normal PDF.

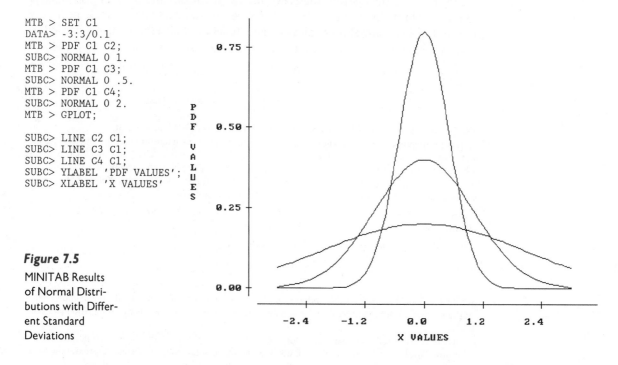

Figure 7.5

MINITAB Results
of Normal Distri-
butions with Differ-
ent Standard
Deviations

Areas Under the Normal Curve

Measuring the Area Under a Normal Curve

No matter what the values of μ and σ for a normal probability distribution, the total area under the normal curve is 1.00, so we may think of areas under the curve as representing probability. Table 7.2 shows how the area under the curve within a certain interval can be determined mathematically.[4]

Table 7.2 Using the Mean and Standard Deviation to Determine the Area of a Normal Distribution

---

Approximately 68.26 percent of the area (probability) under the normal curve lies between $\mu - \sigma$ and $\mu + \sigma$.

Approximately 95.45 percent of the area (probability) under the normal curve lies between $\mu - 2\sigma$ and $\mu + 2\sigma$.

Approximately 99.73 percent of the area (probability) under the normal curve lies between $\mu - 3\sigma$ and $\mu + 3\sigma$.

EXAMPLE 7.4   *The Normal Distribution: An Application to EPS*

Firms frequently use debt to fund various projects. The overall level at which a company employs debt throughout its operations directly affects the expected level and range of its earnings per share (EPS). This is because of the increased risk to which debt exposes the firm within the capital markets.[5] A firm that employs debt is said to be leveraged. The more debt it uses, the higher the firm is leveraged. A firm that does not use any debt financing is said to be without leverage.

If we calculate the means and standard deviations for a hypothetical firm with and without leverage, we can estimate the interval of the possible EPS in the future. The means and standard deviations of the firm are listed in Table 7.3. Figure 7.6 shows the distributions in terms of the parameter values given in Table 7.3.

When we say that EPS is normally distributed, we mean that as the number of observations becomes very large, graphing them yields a normal curve. We can predict EPS intervals for the firm from the information given in Table 7.3 by using the empirical rule described in Table 7.2. Over a large number of observations, x percent will lie within a specified interval that we can determine via the mean and standard deviation. Table 7.4 shows the results of predictions about the firm's EPS. In

---

[4]It is virtually impossible to capture all observations under the curve, because theoretically, the tails continue indefinitely in both directions, never touching the horizontal axis. Integrating the probability density function over the range from $-\infty$ to $+\infty$ would yield the total area of a normal distribution, which is equal to 1.

[5]Increased risk results because interest and principal payments on debt represent legal obligations. Because stock represents ownership in a company, dividends are *not* a legal obligation of the firm.

Table 7.3 Mean and Standard Deviation of EPS

With Leverage	Without Leverage
$\mu_{EPS} = \$1.98$	$\mu_{EPS} = \$1.80$
$\sigma_{EPS} = \$0.32$	$\sigma_{EPS} = \$0.20$

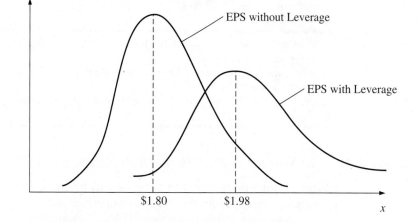

Figure 7.6

EPS Distributions

Table 7.4 Interval Statements About the EPS for a Hypothetical Firm

Chance	With Leverage	Without Leverage
68%	$\$1.66 \leq X \leq \$2.30$	$\$1.60 \leq X \leq \$2.00$
95.5%	$\$1.34 \leq X \leq \$2.62$	$\$1.40 \leq X \leq \$2.20$
99.7%	$\$1.02 \leq X \leq \$2.94$	$\$1.20 \leq X \leq \$2.40$

Table 7.4, X is the EPS for the firm. For example, the last interval statement implies that without leverage, approximately 99.7 percent of EPS should lie between $1.20 and $2.40.

The Standard Normal Distribution and the Z Statistic

There is an infinitely large number of normal curves—one for each pair of values for μ and σ. It is neither possible nor necessary to have different tables for every possible normal curve. The **standard normal distribution** is a transformation of the normal distribution. In the standard normal curve, $\mu = 0$ and $\sigma = 1$. This standard normal curve can be displayed in terms of Z scores (presented in Section 4.5) as indicated in Figure 7.7.

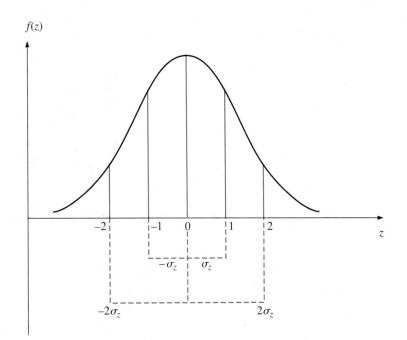

Figure 7.7

Normal Probability Distribution with $\mu = 0$ and $\sigma = 1$

The standard normal curve helps simplify the calculation of probabilities for normally distributed populations. Because not all normally distributed random variables have $\mu = 0$ and $\sigma = 1$, we need to transform the variable so that $\mu = 0$ and $\sigma = 1$. We do this by using the Z score, which is calculated as follows:

$$Z = \frac{X - \mu}{\sigma} \tag{7.4}$$

The **Z score** represents the distance, or deviation, between a given value of the continuous random variable X and its mean μ in standard units. With this information in hand, we can construct the standard normal area table as presented in Table A3 in Appendix A to calculate the area under the curve associated with the value of Z. It is important to note that for any positive value of Z, we are looking at only half the curve. We must therefore add .5 to that value to find the total area under the curve at or below that point.

How to Use the Normal Area Table

Assume that the IQs of undergraduate students at your school are normally distributed with $\mu = 120$ and $\sigma = 15$. What proportion of these undergraduates have an IQ between 120 and 142.5? In this case, we have to find the area of the shaded portion in Figure 7.8.

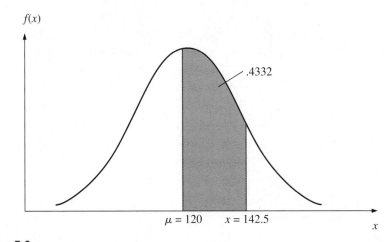

Figure 7.8

Normal Distribution for Student IQs with Interval Between 120 and 142.5 in Shaded Area

To use the normal area table, we need to calculate the Z value for $x = 142.5$.

$$Z = \frac{x - \mu}{\sigma} = \frac{142.5 - 120}{15} = 1.5$$

This implies that the value 142.5 lies 1.5 standard deviations above the mean. Using this information and the normal area table (Table A3 of Appendix A), we find that the corresponding portion in the table is .4332, as indicated in Figure 7.8.

Note that the portions in the normal area table show the area under the upper tail of the normal curve. Because the total area under the normal curve is 1.0, each half is .5. Hence the shaded area that we seek in Figure 7.9 is .5 − .0668 = .4332.

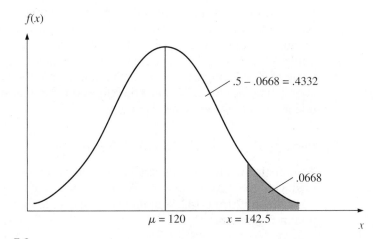

Figure 7.9

Normal Distribution for Student IQs with Intervals Above 142.5 in Shaded Area

The shaded area is 43.32 percent of the total area. From the probability concepts discussed in Chapter 5, we can write the event "IQ between 120 and 142.5" as E, and we conclude that $P(E) = .4332$. In other words, 43.32 percent of the undergraduate students at your school have an IQ between 120 and 142.5.

Alternatively, the MINITAB program can be used to calculate the percentage of undergraduate students at your school who have IQ scores between 120 and 142.5, as indicated here.

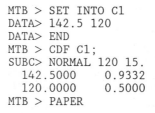

```
MTB > SET INTO C1
DATA> 142.5 120
DATA> END
MTB > CDF C1;
SUBC> NORMAL 120 15.
    142.5000      0.9332
    120.0000      0.5000
MTB > PAPER
```

From the example, we know that 93.32 percent of the students at your school have IQ scores of 142.5 or below and 50 percent of the students have IQ scores of 120 or below. By subtracting 50 percent from 93.32 percent, we obtain 34.32 percent.

The marketing manager of a chain of supermarkets needed to know the weekly sales of extra large eggs. He asked one of his staff to do a survey over a 25-week period. The survey revealed that the weekly sales of extra large eggs was normally distributed, with a mean of 743 cartons and a standard deviation of 254 cartons (*Journal of Marketing Research,* August 1984).

From this information, we can calculate the probability that a supermarket will sell between 550 and 850 cartons of extra large eggs in a randomly selected week as:

$$P\{550 < X < 850\} = P\left\{ \frac{550 - 743}{254} \le X \le \frac{850 - 743}{254} \right\}$$
$$= P\{z \le .42\} + P\{z \le -.76\}$$
$$= P\{z \le .42\} + P\{z \ge .76\}$$
$$= .1628 + .2764 = .4392 \qquad \text{(From Table A3)}$$

EXAMPLE 7.5 *Determining Daily Donut Demand (in Dozens)*

A Dunkin' Donuts shop located in New Brunswick, New Jersey, sells dozens of fresh donuts. Any donuts remaining unsold at the end of the day are either discarded or sold elsewhere at a loss. The demand for the Dunkin' Donuts at this shop has followed a normal distribution with $\mu = 50$ dozen and $\sigma = 5$ dozen. How many dozen donuts should this Dunkin' Donuts shop make each day so that it can meet the demand 95 percent of the time?

Let the random normal variable X represent the demand for fresh Dunkin' Donuts (measured in dozens). To meet the demand 95 percent of the time, the Dunkin' Donuts shop must determine an amount—say, A dozen—such that

$$P(X \leq A) = .95$$

Similarly to the student IQ case, we can express this probability statement in terms of Z.

$$P\left(\frac{X - \mu}{\sigma} \leq \frac{A - \mu}{\sigma}\right) = P\left(Z \leq \frac{A - \mu}{\sigma}\right) = .95$$

Because we know that $\mu = 50$ and $\sigma = 5$, we can rewrite the probability statement as

$$P\left(Z \leq \frac{A - 50}{5}\right) = .95$$

In accordance with Table A3 in Appendix A, a Z curve having an area to the left equals .95, as shown in Figure 7.10. From Figure 7.10, we know that

$$P(0 \leq Z \leq 1.64) = .4495 \doteq .45$$

which means that

$$P(Z \leq 1.64) = .5 + .45 = .95$$

Thus

$$\frac{A - 50}{5} = 1.64$$

$$A = 58.2$$

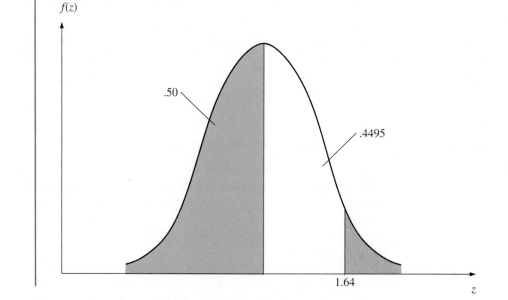

Figure 7.10

$P(Z \leq 1.64)$

and

$$A = 50 + 8.20 = 58.20 \text{ dozen}$$

To be conservative, we round this value up to 59 dozen and assume that any occasional surplus will be welcome at a nearby shelter for the homeless. By stocking 59 dozen donuts each day, the Dunkin' Donuts shop will meet the demand for donuts 95 percent of the time.

7.4 THE LOGNORMAL DISTRIBUTION AND ITS RELATIONSHIP TO THE NORMAL DISTRIBUTION (Optional)[6]

The Lognormal Distribution

Before we discuss the **lognormal distribution,** we must briefly review and expand on three topics covered in Chapter 4: mean, variance, and skewness. For a continuous random variable we can generally calculate the mean, variance, and skewness. Values of these parameters affect the shape of a distribution. Lognormally distributed random variables are related to the normally distributed continuous variables, but normally distributed random variables have zero skewness, whereas lognormally distributed continuous random variables have positive skewness.

If a continuous random variable Y is normally distributed, then the continuous variable X defined in Equation 7.5 is lognormally distributed.

$$X = e^Y \tag{7.5}$$

By performing a logarithmic transformation on this variable X, we obtain a normally distributed variable Y:

$$\ln(X) = \ln(e^Y) = Y$$

where ln denotes the natural logarithm, and e is a constant approximately equal to 2.71828. Lognormal continuous random variables have the following properties:

Mean: $E[\ln(X)] = E(Y) = \mu$
Variance: $\text{Var}[\ln(X)] = \text{Var}(Y) = \sigma^2$

Our discussion in the next section of the mean and variance for a lognormal variable X is based on these relationships.

---

[6]This section can be omitted without affecting the continuity of the text. Further discussion on the lognormal distribution can be found in J. Aitchison and J. A. C. Brown (1957), *The Lognormal Distribution with Special Reference to Its Uses in Economics* (London: Cambridge University Press).

Mean and Variance of Lognormal Distribution

Because of the relationship between X and Y indicated in Equation 7.5, the mean and variance of variable X can be defined as follows:[7]

$$\mu_X = e^{\mu + 1/2\sigma^2} \tag{7.6}$$

$$\sigma_X^2 = e^{2\mu + \sigma^2}(e^{\sigma^2} - 1), \tag{7.7}$$

where $\mu = E[\ln X]$, $\sigma^2 = \text{Var}[\ln X]$, and $e = 2.71828$.

Equations 7.6 and 7.7 indicate that both mean and variance of a lognormal variable are functions of the mean and variance of a normal variable. The normal and the corresponding lognormal frequency curves are illustrated in Figure 7.11. Note that the mean of the lognormal is larger than the mode of the lognormal, because it is a positively skewed distribution. In addition, the shape of a lognormal is

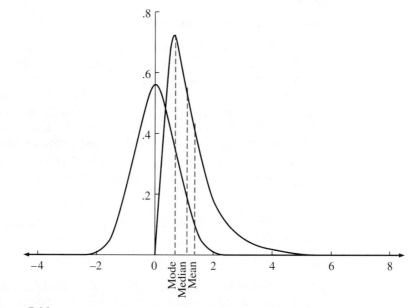

Figure 7.11

Frequency Curves of the Normal and Lognormal Distributions
Source: Charles R. Nelson, *Applied Time Series Analysis,* (Oakland, Calif.: Holden-Day, 1973), p. 164.

---

[7]The density function of a lognormal distribution can be defined as

$$f(x) = \frac{1}{x\sigma\sqrt{2\pi}} \exp\left[-\frac{1}{2\sigma^2}(\ln x - \mu)^2 \right], \quad x > 0$$

This is similar to the density function of a normal distribution as defined in footnote 3 of this chapter. Mean, variance, and skewness of the lognormal distribution will be discussed and derived in Section 9.7.

affected by the values of both mean μ and variance σ^2, as indicated in Figures 7.12 and 7.13. Furthermore, the lognormal distribution differs from the normal distribution in that its mean, median, and mode are not identical.

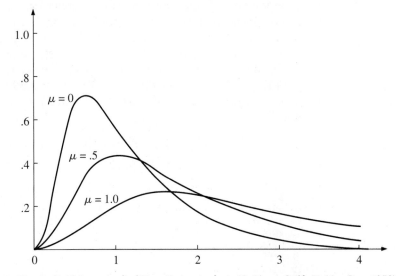

Figure 7.12

Frequency Curves of the Lognormal Distribution for Three Values of μ from the Parent Normal

Source: Charles R. Nelson, *Applied Time Series Analysis,* (Oakland, Calif.: Holden-Day, 1973), p. 164.

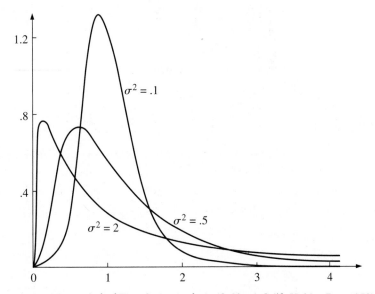

Figure 7.13

Frequency Curves of the Lognormal Distribution for Three Values of σ^2

Source: Charles R. Nelson, *Applied Time Series Analysis,* (Oakland, Calif.: Holden-Day, 1973), p. 165.

In surveys of household income and in the examination of consumer behavior, the lognormal distribution is useful in the income distribution analysis.[8] In addition, the lognormal distribution is more suitable than the normal distribution for cost–volume–profit analysis, which is discussed in Application 7.2. Furthermore, in the option pricing model discussed in Appendix 7B, it is assumed that stock price per share is lognormally distributed (see Equation 7B.2).

To show how Equations 7.6 and 7.7 can be used to calculate the mean and standard deviation of a lognormal distribution, we let X represent the stock price per share of GM as presented in Table 2B.1 in Chapter 2. Then, we can calculate $E(\log X) = E(Y) = 4.0749$ Var$[\ln(X)] =$ Var$(Y) = \sigma^2 = .08995$ by MINITAB as shown here. Substituting $E(Y) = \mu = 4.0749$ and Var$(Y) = \sigma^2 = .08995$ into Equations 7.6 and 7.7, we obtain

$$\mu_X = e^{4.11988} = 61.5516$$
$$\sigma_X^2 = (61.5516)^2(e^{.08995} - 1) = 356.5814.$$

```
MTB > SET INTO C1
DATA> 69.13 80.50 80.50 81.13 46.13 30.75 57.63 78.50 62.88 53.75 50.00
DATA> 45.00 38.50 62.38 74.38 78.38 70.38 66.00 61.38 83.50 42.25 34.38
DATA> END
MTB > MEAN C1
    MEAN    =       61.247
MTB > STDEV C1
    ST.DEV. =       16.599
MTB > LET C2=LOGE(C1)
MTB > PRINT C2

C2
    4.23599    4.38826    4.38826    4.39605    3.83146    3.42589    4.05404
    4.36310    4.14123    3.98434    3.91202    3.80666    3.65066    4.13324
    4.30919    4.36157    4.25391    4.18965    4.11708    4.42485    3.74360
    3.53748

MTB > MEAN C2
    MEAN    =       4.0749
MTB > STDEV C2
    ST.DEV. =      0.29992
MTB > PAPER
```

[8]This is because household income is generally lognormally distributed. See Aitchison and Brown (1957), Chapters 11 and 12.

7.5 THE NORMAL DISTRIBUTION AS AN APPROXIMATION TO THE BINOMIAL AND POISSON DISTRIBUTIONS

In Chapter 6 we discussed binomial and Poisson distributions. Recall that when we were interested in deriving the cumulative probability from the binomial distribution, the computations could be quite burdensome. For example, if we toss a coin 100 times to test the probability that the number of heads will be 50 or less, we need to compute $P(X = 0)$, $P(X = 1)$, $P(X = 2)$, . . . , $P(X = 50)$. An analogous situation arises for the Poisson distribution. Fortunately, it is possible to reduce the job of computation greatly by approximating the binomial or Poisson distribution with a normal distribution.

Normal Approximation to the Binomial Distribution

As the sample size gets large, we can use a normal distribution to approximate the binomial distribution. For an experiment that does n independent trials each having probability of success p, the distribution of the number of successes, X, is binomial and has the following mean and variance.

$$\text{Mean:} \qquad E(X) = \mu_X = np \qquad (7.8)$$

$$\text{Variance:} \qquad \text{Var}(X) = np(1 - p) \qquad (7.9)$$

From Equation 7.4, we substitute for the mean μ and standard deviation σ and get the following expression for the Z statistic.

$$Z = \frac{x - np}{\sqrt{np(1 - p)}} \qquad (7.10)$$

We say that the distribution of the random variable Z is approximately standard normal. This approximation works well when $np > 5$ and $n(1 - p) > 5$. Because x stands for the number of successes of the binomial trials, we can now determine the probability that the actual number of successes will lie within a certain interval.

If the range we wish to examine is between a and b, inclusive, then we may obtain the following probability:

$$P(a \leq x \leq b) = P\left(\frac{a - np}{\sqrt{np(1 - p)}} \leq Z \leq \frac{b - np}{\sqrt{np(1 - p)}}\right) \qquad (7.11)$$

EXAMPLE 7.6   *Using a Normal Distribution to Approximate a Binomial Distribution*

Suppose a very bumpy conveyor belt in a brewery transports beer bottles from the point where they are capped to the point where they are boxed for shipping. Furthermore, let us suppose there is a 16 percent chance that each beer bottle will fall off the conveyor belt. In 1 hour, exactly 1,000 beer bottles travel from one end of the belt to the other. Because n is large, we can make probability statements about whether the actual number of bottles, X, that fall off the conveyor will be within a certain range. Using Equations 7.8 and 7.9, we get the following results for a binomially distributed random variable with the mean and variance shown.

$$E(X) = np = 160$$

$$\text{Var}(X) = np(1 - p) = 134.4$$

Suppose we wish to know the probability that the actual number of beer bottles that fall off the conveyor belt will be between 142 and 185. Using Equation 7.11 yields

$$P(142 \leq x \leq 185) = P\left(\frac{142 - 160}{\sqrt{134.4}} \leq Z \leq \frac{185 - 160}{\sqrt{134.4}}\right)$$
$$= P(-1.553 \leq Z \leq 2.156)$$

Now we can use the values of the Z statistic and the standard normal distribution table to compute the probability. We calculate the area beneath the curve between these two numbers. Then we let the symbol F_Z represent the value of cumulative probability as taken from the standard normal distribution table. Then

$$
\begin{aligned}
P(-1.55 \leq Z \leq 2.16) &= F_Z(2.16) - F_Z(-1.55) \\
&= F_Z(2.16) - [1 - F_Z(1.55)] \\
&\qquad \text{Because } F_Z(-w) = 1 - F_Z(w) \\
&= .9846 - [1 - .9394] \\
&= .9240
\end{aligned}
$$

Thus, there is a 92.4 percent chance that the number of beer bottles that fall off the conveyor belt during the period will be between 142 and 185.

This example illustrates how the normal distribution can be used to approximate the binomial distribution. Further applications of the normal distribution are given in Appendix 7B.

Normal Approximation to the Poisson Distribution

Recall from Section 6.6 that the Poisson random variable measures the probability of X occurrences of some event in the time interval between 0 and t. Therefore, this distribution measures successes when they occur within specified units of time. Recall the Poisson probability function:

$$P(X = x) = e^{-\lambda}\lambda^x/x! \qquad \text{for } x = 0, 1, 2, 3, \ldots \text{ and } \lambda > 0 \qquad (6.16)$$

where λ is the average number of successes in the unit of time, and e is the base of the natural logarithms (2.71828). The mean and variance of this distribution are

$$\text{Mean:} \qquad E(x) = \mu_x = \lambda \qquad\qquad (7.12)$$

$$\text{Variance:} \qquad \text{Var}(x) = \sigma^2 = \lambda \qquad\qquad (7.13)$$

Note that both the variance and the mean are equal to λ.

The Poisson probabilities can be approximated by the normal distribution when the sample size is large—say, greater than 30. To calculate Poisson probabilities in this manner, we can develop the Z statistic by substituting for the mean and variance in Equation 7.4 as follows:

$$Z = \frac{X - \lambda}{\sqrt{\lambda}} \qquad (7.14)$$

Now we can examine the probability that the number of successes is within a certain range.

$$P(a \leq X \leq b) = P\left(\frac{a - \lambda}{\sqrt{\lambda}} \leq Z \leq \frac{b - \lambda}{\sqrt{\lambda}}\right) \qquad (7.15)$$

EXAMPLE 7.7 *Using a Normal Distribution to Approximate a Poisson Distribution*

In Example 6.16 we examined the average number of customers entering a bank in a 10-minute period. Now let's assume that in a 20-minute period we have an average of 50 customers instead of 5; X is still a Poisson random variable. This change is made so that normal approximation will hold. Equations 7.12 and 7.13 reveal that the mean and variance of this Poisson random variable are both 50.

Suppose we wish to find the probability that the number of people entering the bank in a 20-minute period will be between 42 and 57, inclusive. We can calculate this probability by using the standard normal distribution and its related Z statistic.

$$P(42 \leq X \leq 57) = P\left(\frac{42 - 50}{\sqrt{50}} \leq Z \leq \frac{57 - 50}{\sqrt{50}}\right) = P(-1.13 \leq Z \leq .99)$$

Now the F_Z's are computed in the same way as in Example 7.6. Then

$$\begin{aligned}
P(-1.13 \leq Z \leq .99) &= F_Z(.99) - F_Z(-1.13) \\
&= .8389 - (.1292) \\
&= .7097
\end{aligned}$$

There is a 70.97 percent chance that the number of customers who arrive in a 20-minute period will fall between 42 and 57, inclusive.

7.6   *BUSINESS APPLICATIONS*

■ | *APPLICATION 7.1* Analyzing Earnings per Share and Rates of Return

In financial analysis, historical data are often used to forecast future values. Table 7.5 presents data on earnings per share and rates of return on stock in General Motors and Ford over a 21-year period (1970–1990). Mean, standard deviation, and skewness estimates for EPS and rates of return are presented in Table 7.6.

Table 7.5 EPS and Rates of Return for GM and Ford

Year	EPS		Rate of Return	
	GM	*Ford*	*GM*	*Ford*
1970	2.09	4.77	.2137	.4261
1971	6.72	6.18	.0422	.2933
1972	7.51	8.52	.0630	.1715
1973	8.34	9.13	−.3667	−.4512
1974	3.27	3.86	−.2596	−.0969
1975	4.32	3.46	.9520	.3963
1976	10.08	10.45	.4586	.4614
1977	11.62	14.61	−.1124	−.2067
1978	12.24	13.35	−.0497	−.0027
1979	10.04	9.75	.0288	−.1478
1980	−2.65	−12.83	−.0410	−.2938
1981	1.07	−8.81	−.0911	−.1025
1982	3.09	−5.46	.6825	1.3209
1983	11.84	10.29	.2373	.1029
1984	14.22	15.79	.1176	.1239
1985	12.28	13.63	−.0383	.3238
1986	8.22	12.32	.0088	.0081
1987	10.06	18.10	.0058	.3961
1988	13.64	10.96	.4418	−.2995
1989	6.33	8.22	−.4581	−.0767
1990	−4.09	1.86	−.1154	−.3209

Table 7.6 Mean, Standard Deviation, and Skewness Estimates for EPS and Rates of Return

	EPS		Rate of Return	
	GM	*Ford*	*GM*	*Ford*
Mean	7.154	7.033	.082	.096
Std. Dev.	5.195	7.973	.333	.392
Skewness	−.624	.323	.864	1.240

By analyzing past data, and by assuming that the data on GM and Ford are distributed normally, we can make interval statements about EPS and return for GM and Ford, as indicated in Table 7.7.

First consider the return on GM and Ford stock. The average return on Ford (see Table 7.6) is higher than that on GM. However, the standard deviation measure presented in Table 7.6 makes it clear that GM's stock is less volatile than Ford's. An investor seeking a higher return might choose Ford but, in doing so, would incur a greater risk of losing money.

Table 7.7 Probability Distribution for EPS and Return

	EPS	
Chance	*GM*	*Ford*
68.3%	$1.959 \leq x \leq 12.349$	$-.940 \leq x \leq 15.006$
95.4%	$-3.236 \leq x \leq 17.544$	$-8.913 \leq x \leq 22.979$
99.7%	$-8.431 \leq x \leq 22.739$	$-16.886 \leq x \leq 30.952$
	Rate of Return	
68.3%	$-.251 \leq x \leq .415$	$-.296 \leq x \leq .488$
95.4%	$-.584 \leq x \leq .748$	$-.688 \leq x \leq .88$
99.7%	$-.917 \leq x \leq 1.081$	$-1.08 \leq x \leq 1.272$

Now we analyze the mean and standard deviation of EPS presented in Table 7.6. EPS is an absolute measure, so the EPS data for GM and Ford are not directly comparable; one share of GM stock does not cost the same as one share of Ford stock. Therefore, EPS are earnings on different amounts of investment per share. The problem can be resolved by comparing the **coefficient of variation** (CV) of EPS for GM and Ford. Recall that the coefficient of variation, which we discussed in Chapter 4, divides the standard deviation by the mean and gives an indication of relative volatility. Substituting the related data of Table 7.6 into Equation 4.11, we obtain the coefficients of variation of EPS for both GM and Ford.

$$\text{CV(GM)} = 5.195/7.154 = .726 \qquad \text{CV(Ford)} = 7.973/7.033 = 1.134$$

Comparing these two coefficient of variation, we see that Ford's EPS is much more volatile than GM's. Therefore, the risk-averse investor might prefer GM stock to Ford.

■ | **APPLICATION 7.2** Cost–Volume–Profit Analysis Under Uncertainty: The Normal Versus the Lognormal Approach

Cost–volume–profit (CVP) analysis is one of the most important concepts in accounting, economics, finance, marketing, and production management. The total profit w of a firm can be defined as

$$w = TR - TC = Q(P - V) - F \qquad (7.16)$$

where

$$
\begin{aligned}
TR &= \text{total revenue} \\
TC &= \text{total cost} \\
Q &= \text{unit sales} \\
P &= \text{price per unit} \\
V &= \text{variable cost per unit} \\
(P - V) &= \text{contribution margin per unit} \\
F &= \text{fixed cost}
\end{aligned}
$$

This model can be analyzed in terms of the certainty approach or the uncertainty approach. Under certainty analysis, we assume that future Q, P, and V are known for sure and that, accordingly, the specified future total profit will occur with 100 percent certainty.

Uncertainty analysis is more complicated and (not surprisingly) less certain. Hilliard and Leitch (1975) have suggested two alternative assumptions.[9]

1. Q is not known for certain, and it is normally distributed with mean μ_q and variance σ_q^2. Thus the random variable w is normally distributed with mean μ_w and variance σ_w^2 as follows:

$$\mu_w = \mu_q(P - V) - F$$
$$\sigma_w^2 = \sigma_q^2(P - V)^2 \tag{7.17}$$

Following Hilliard and Leitch (1975), we suppose that μ_q = 5,000 units, σ_q = 400 units, price = \$3,000/unit, variable cost = \$1,750/unit, and fixed costs = \$5,800,000. Thus μ_w = \$450,000 and σ_w = \$500,000. We can calculate the probability of a profit greater than \$200,000 (A) by using Table A3 in Appendix A.

$$
\begin{aligned}
P(w > \$200,000) &= 1 - P(w \le \$200,000] \\
&= 1 - F_w[(A - \mu_w)/\sigma_w] \\
&= 1 - F_w\left(\frac{200,000 - 450,000}{500,000}\right) \\
&= 1 - F_w(-.5) = 1 - .3085 = .6915
\end{aligned}
$$

This implies there is a 69.2 percent chance that the firm's profit will exceed \$200,000. Similarly, we can calculate the probability of meeting the break-even point by setting w_0 equal to 0. Note that w_0 = 0 implies w = 0.

2. Q is not known for certain, and it is lognormally distributed. Following Equations 7.6 and 7.7, we can define the mean and variance of Q as

$$\mu_q = E(Q) = e^{\mu + 1/2\sigma^2}$$
$$\sigma_q^2 = \text{Var}(Q)e^{2\mu + \sigma^2}(e^{\sigma^2} - 1)$$

where $\mu = E(\ln Q)$ and $\sigma^2 = \text{Var}(\ln Q)$.

Logical assumptions for a CVP model would require that sales be nonnegative. In addition, it would be somewhat surprising if the distribution of sales (Q) were perfectly symmetric about its mean. Thus assuming the lognormal distribution might be more suitable than assuming the normal.

The relationships between the normal and lognormal distributions under conditions of large and small coefficients of variation are presented in Figures 7.14 and 7.15, respectively. In Figure 7.14, both the normal and the lognormal distribution have identical means, identical variances, and a coefficient of variation of .50. This relatively large coefficient of variation emphasizes the difference between the nor-

---

[9]J. E. Hilliard and R. A. Leitch (1975). "Cost–Volume–Profit Analysis Under Certainty: A Lognormal Approach," *The Accounting Review,* January 1975, 69–80.

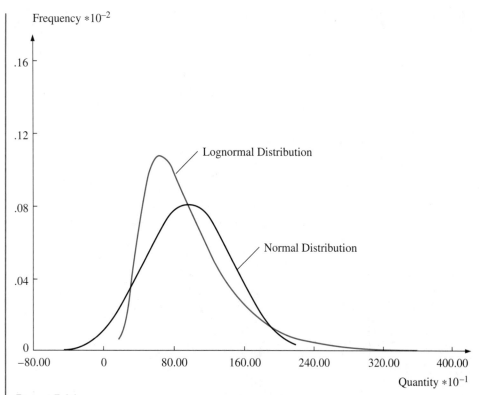

Figure 7.14

Comparison of Normal and Lognormal Distributions: Coefficient of Variation = .50
Source: Hilliard and Leitch, "CVP Under Uncertainty," *The Accounting Review,* January 1975, 71.

mal and lognormal distributions. The figure indicates that the major differences between these two distributions are the skewness of the lognormal and the possibility of the occurrence of negative values for the normal distribution. In Figure 7.15, the distributions are nearly coincident, as we expect for small coefficients of variation (CV = .08). The important observation, however, is that the lognormal assumption is an intuitive choice for the CVP model inputs, regardless of the values of the coefficient of variation.

Under the lognormal distribution assumption, Hilliard and Leitch (1975) have derived the probability of a given level, say A, as

$$\begin{aligned} P[w > A] &= 1 - P[w \le A] \\ &= 1 - F_z\{[\ln(A + F) - \mu_n]/\sigma_n\} \end{aligned} \tag{7.18}$$

where

$$\mu_n = \ln[\mu_q^2/(\sigma_q^2 + \mu_q^2)^{1/2}] + \ln(P - V)$$
$$\sigma_n^2 = \ln[(\sigma_q/\mu_q)^2 + 1]$$

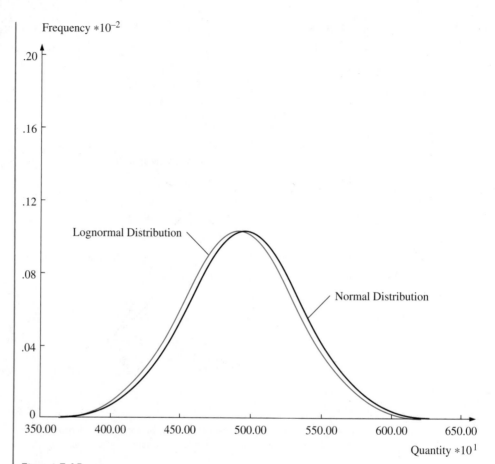

Figure 7.15

Comparison of Normal and Lognormal Distributions: Coefficient of Variation = .08
Source: Hilliard and Leitch, "CVP Under Uncertainty," *The Accounting Review,* January 1975, p. 72.

Substituting the information used in case 1 into these two formulas, we obtain

$$\mu_w = 2 \ln(5{,}000) - \frac{1}{2} \ln[(400)^2 + (5{,}000)^2] + \ln(3{,}000 - 1{,}750)$$

$$= 17.034 - 8.521 + 7.131 = 15.644$$

$$\sigma_w^2 = \ln\left[\left(\frac{400}{5{,}000}\right)^2 + 1\right] = .0064$$

$$\sigma_w = .08$$

Substituting A, F, μ_n, and σ_n into Equation 7.18 yields

$$P[w > 200{,}000] = 1 - F_w\{[\ln(200{,}000 + 5{,}800{,}000) - 15.644]/.08\}$$

$$= 1 - F_w[-.4625]$$

$$= 1 - .3219$$

$$= .6781$$

The probability we get when we make the lognormal assumption is 67.81 percent, which is lower than that in terms of the normal assumption, 69.15 percent.

■ | ***Application 7.3*** Investment Decision Making Under Uncertainty

Professor Hillier (1963) suggested several easy and effective ways for a business firm to evaluate risky investment projects.[10] In one of his approaches, Hillier assumes that the **net cash inflow** from an investment to the firm in the tth future year after the investment is made is normally distributed.[11] He has also shown that the net present value (NPV) of a proposed investment is normally distributed with mean μ_{NPV} and variance σ_{NPV}^2.[12] Using the assumption that NPV is normally distributed, Hillier has provided an example of how management can evaluate the risk of an investment.

> Suppose that, on the basis of the forecasts regarding prospective cash flows from a proposed investment of $10,000, it is determined that $\mu_p = \$1,000$ and $\sigma_p = \$2,000$. Ordinarily, the current procedure would be to approve the investment since $\mu_p > 0$. However, the additional information available ($\sigma_p = \$2,000$) regarding the considerable risk of the investment, the executive can analyze the situation further. Using widely available tables for the normal distribution, he could note that the probability that NPV < 0, so that the investment won't pay, is 0.31. Furthermore, the probability is 0.16, 0.023, and 0.0013, respectively, that the investment will lose the present-worth equivalent of at least $1,000, $3,000, and $5,000, respectively. Considering the financial status of the firm, the executive can use this and similar information to make his decision. Suppose, instead, that the executive is attempting to choose between this investment and a second investment with $\mu_p = \$500$ and $\sigma_p = \$500$. By conducting a similar analysis for the second investment, the executive can decide whether the greater expected earnings of the first investment justifies the greater risk. A useful technique for making this comparison is to superimpose the drawing of the probability distribution of NPV for the second investment upon the corresponding drawing for the first investment. This same approach generalizes to the comparison of more than two investments.

Let's see how Hillier obtained his probabilities:

$$P(NPV < 0) = .31 \qquad P(NPV < -\$1,000)$$
$$P(NPV < -\$3,000) \qquad P(NPV < -\$5,000)$$

by using the information $\mu_{NPV} = \$1,000$ and $\sigma_{NPV} = \$2,000$.

$$P(NPV < 0) = P\left(Z \leq \frac{0 - 1,000}{2,000}\right) = P(Z \leq -.5) = P(Z \geq .5)$$
$$= .5 - .1915 \doteq .31$$

---

[10]F. S. Hillier (1963), "The Derivation of Probabilities Information for the Evaluation of Risky Investments," *Management Science,* April 1963, 443–457.

[11]Via Equation 7.16, net cash inflow can be defined as net profit + depreciation.

[12]How to calculate the NPV is explained in Appendix 21B. In Hillier's example discussed below, he used P to represent NPV.

$$P(\text{NPV} < -\$1,000) = P\left(Z \leq \frac{-1,000 - 1,000}{2,000}\right) = P(Z \leq -1.0) = P(Z \geq 1.0)$$

$$= .5 - .3413 \doteq .16$$

$$P(\text{NPV} < -\$3,000) = P\left(Z \leq \frac{-3,000 - 1,000}{2,000}\right) = P(Z \leq -2.0) = P(Z \geq 2.0)$$

$$= .5 - .4772 \doteq .023$$

$$P(\text{NPV} < -\$5,000) = P\left(Z \leq \frac{-5,000 - 1,000}{2,000}\right) = P(Z \leq -3.0) = P(Z \geq 3.0)$$

$$= .5 - .4987 \doteq .0013$$

The above probability can be calculated directly by using the MINITAB program as indicated here.

```
MTB > SET INTO C1
DATA> 0 -1000 -3000 -5000
DATA> END
MTB > CDF C1;
SUBC> NORMAL 1000 2000.
    0.0000      0.3085
   -1.0E+03     0.1587
   -3.0E+03     0.0228
   -5.0E+03     0.0013
MTB > PAPER
```

APPLICATION 7.4 Determination of Commercial Lending Rates[13]

The loan officers of a bank and the financial analysts of a firm seeking to borrow money consider the firm's total risks when analyzing the lending rate to the firm or—what is the same thing—the firm's cost of borrowing.

The lending rate is based partly on the risk-free rate. (For example, the federal government bond rate is free from default risk.) First we have to forecast the risk-free rate (R_f) for three economic conditions: boom, normal, and recession.

The second component of the lending rate is the risk premium (R_p). Risk premium is the bank's reward for taking risk. It can be calculated individually for each firm by examining the change in EBIT (earnings before interest and tax) under the three types of economic conditions. The EBIT is used as an indicator of the ability of the prospective borrower to repay borrowed funds.

Table 7.8 contains the information on R_f, EBIT, and R_p required for the analysis. It also shows the probability that each economic conditions will prevail (column B) and the probability of various levels of EBIT for the firm (column D).

A total of nine lending rates under the three different economic conditions are given in column F of Table 7.9. The probabilities of their occurrence under the different conditions are shown in column E.

---

[13]This application is similar to Applications 5.1 and Example 6.8. Note that the conditional probability used here is different from that used in Application 5.1 and Example 6.8.

Table 7.8 Worksheet for Interest Rate Calculation

Economic Condition	(A) R_f	(B) Marginal Probability	(C) EBIT	(D) Conditional Probability	(E) R_p
Boom	12.0%	.25	$2.5m	.60	3%
			1.5	.30	5
			.5	.10	8
Normal	10.0	.50	$2.5m	.40	3%
			1.5	.30	5
			.5	.30	8
Poor (recession)	8.0	.25	$2.5m	.10	3%
			1.5	.20	5
			.5	.70	8

Table 7.9 Worksheet for Interest Rate Calculation

Economic Condition	(A) R_f	(B) Marginal Probability	(C) R_p	(D) Conditional Probability	(E) Joint Probability (B × D)	(F) Lending Rate (A + C)
Boom	12%	.25	3.0%	.60	.150	15%
			5.0	.30	.075	17
			8.0	.10	.025	20
Normal	10%	.50	3.0%	.40	.200	13%
			5.0	.30	.150	15
			8.0	.30	.150	18
Poor (recession)	8%	.25	3.0%	.10	.025	11%
			5.0	.20	.050	13
			8.0	.70	.175	16
					1.000	

Let us see how the numbers for columns E and F of Table 7.9 are calculated. During a period of normal economic conditions, the risk-free rate is at 10 percent, as indicated in column A, but the risk premium can take on different values. There is a 40 percent chance it will be 3.0 percent, a 30 percent chance it will be 5.0 percent, and a 10% chance it will be 8.0 percent, as indicated in column D. We must multiply the probability for the risk-free rate by the conditional probability for the risk premium to get the probability of their occurring jointly (column E). The chance that the firm will receive a 13 percent lending rate during normal economic conditions is 20 percent; for a 15 percent or 18 percent rate the probability is 15 percent (see column F). This process also applies for the other two conditions (boom and recession).

For the problem set up in Table 7.9, the weighted-average lending rate is

$$\overline{R} = (.150)(.150) + (.075)(.17) + (.025)(.20) + (.200)(.13) +$$
$$(.150)(.15) + (.150)(.180) + (.025)(.11) + (.050)(.13) +$$
$$(.175)(.16)$$
$$= .1531 = 15.31\%$$

with a standard deviation of

$$\sigma = [(.15)(.15 - .153)^2 + (.075)(.17 - .153)^2 +$$
$$(.025)(.20 - .153)^2 + (.20)(.13 - .153)^2 +$$
$$(.15)(.15 - .153)^2 + (.15)(.18 - .153)^2 +$$
$$(.025)(.11 - .153)^2 + (.050)(.13 - .153)^2 +$$
$$(.175)(.16 - .153)^2]^{1/2}$$
$$= [.000001 + .000021 + .000055 + .000107 +$$
$$.000001 + .000109 + .000046 + .000027 + .000008]^{1/2}$$
$$= .0194$$

We assume that the distribution of the lending rate is normal. Given the mean and standard deviation for such a distribution from Table 7.2, we see that 68.3 percent of the observations of a normal distribution are within one standard deviation of the mean, 95.4 percent are within two standard deviations, and 99.7 percent are within three.

On the basis of the mean and standard deviation of the estimated lending rate, we can depict the expected lending rate and its standard deviation as shown in Figure 7.16.

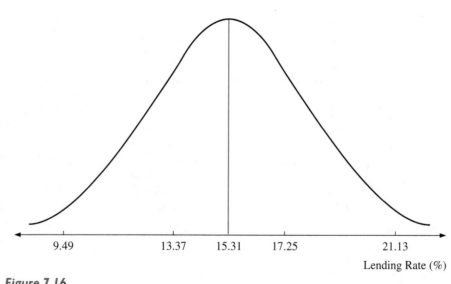

9.49   13.37   15.31   17.25   21.13

Lending Rate (%)

Figure 7.16

Probability Distribution for Estimated Lending Rate

The percentages in Figure 7.16, along with the mean and standard deviations, are an illustration of the normal distribution. The average lending rate is normally distributed with a mean of 15.31 percent and a standard deviation of 1.94 percent. This implies that almost all rates (99.7 percent) will lie in the range of 9.49 percent to 21.13 percent. We also know that 68.3 percent of the rates will lie in the range of 13.37 percent to 17.25 percent.

The MINITAB output of the empirical calculation procedure for this problem is presented in Figure 7.17.

```
MTB > SET INTO C1
DATA> .15  .17  .20  .13  .15  .18  .11  .13  .16
DATA> END
MTB > SET INTO C2
DATA> .150 .075 .025 .200 .150 .150 .025 .050 .175
DATA> END
MTB > LET C3=C1*C2
MTB > SUM C3 INTO K1
    SUM     =       0.15300
MTB > LET C4=C1-K1
MTB > LET C5=C4**2
MTB > LET C6=C5*C2
MTB > SUM C6 INTO K2
    SUM     =  0.00037600
MTB > LET K3=K2**.5
MTB > PRINT C1-C6 K1-K3
K1        0.153000
K2        0.000376000
K3        0.0193907

ROW    C1      C2      C3        C4          C5           C6

  1   0.15   0.150   0.02250  -0.0030000   0.0000090    0.0000014
  2   0.17   0.075   0.01275   0.0170000   0.0002890    0.0000217
  3   0.20   0.025   0.00500   0.0470000   0.0022090    0.0000552
  4   0.13   0.200   0.02600  -0.0230000   0.0005290    0.0001058
  5   0.15   0.150   0.02250  -0.0030000   0.0000090    0.0000014
  6   0.18   0.150   0.02700   0.0270000   0.0007290    0.0001093
  7   0.11   0.025   0.00275  -0.0430000   0.0018490    0.0000462
  8   0.13   0.050   0.00650  -0.0230000   0.0005290    0.0000265
  9   0.16   0.175   0.02800   0.0070000   0.0000490    0.0000086

MTB > PAPER
```

Figure 7.17 MINITAB Output for Application 7.4

Summary

In this chapter we introduced two of the most important continuous distributions, the normal and lognormal distributions. These distributions can be used to describe a wide variety of random variables in business, economics, and finance. In fact, even when our distribution is not normally distributed, it may be possible to transform our random variables (e.g., the logarithmic transformation discussed in Section 7.4) so they are approximately normally distributed. This means that many of the analyses we will perform throughout the rest of the book will be based on the normal distribution. We illustrated this point by showing how the binomial and Poisson distributions could be approximated by using the normal distribution.

We also showed how the normal distribution could be applied to a variety of business problems, including EPS forecasting, CVP analysis, determining of commercial lending rates, and option pricing (see Appendix 7B).

Appendix 7A Mean and Variance for Continuous Random Variables

In this appendix, we will discuss areas under a continuous PDF and explore the variance for continuous random variables.

Areas Under Continuous Probability Density Function

In accordance with Equation 6.2a, the **cumulative distribution function** (CDF) for a continuous variable X is given by

$$F(x_0) = \int_{-\infty}^{x_0} f(x)\, dx \tag{7A.1}$$

where x_0 is any value that the random variable X can take. Equation 7A.1 implies that the area under curve $f(x)$ is to the left of x_0.

Using Equation 7A.1, we can calculate the probability that X lies between a and b for a continuous random variable.

$$P(a \le x \le b) = \int_{a}^{b} f(x)\, d(x)$$

$$= \int_{-\infty}^{b} f(x)\, dx - \int_{-\infty}^{a} f(x)\, dx = F(b) - F(a) \tag{7A.2}$$

For any $a \le b$, $F(a) \le F(b)$. Equation 7A.2 is similar to Equation 7.1 for a discrete variable case.

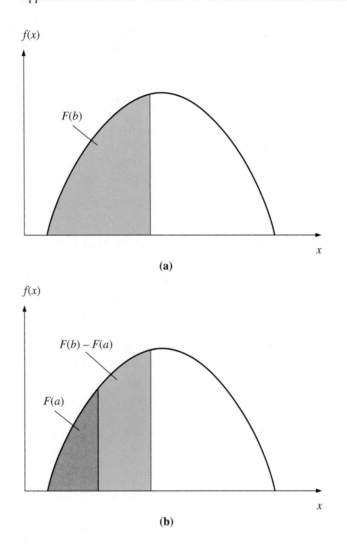

Figure 7A.1

A diagrammatic representation of cumulative probability is given in Figure 7A.1. The total area under the curve $f(x)$ is 1. In integral calculus notation,

$$\int_{-\infty}^{\infty} f(x)\, dx = 1 \tag{7A.3}$$

Mean of Discrete and Continuous Random Variables

From Chapter 6, we know that the expected value of a discrete random variable can be defined as

$$\mu = E(X) = \sum_{i=1}^{N} x_i P(x_i) \tag{6.3}$$

The expected value of a continuous variable can be defined in a similar fashion. If X is a continuous random variable with probability density $f(x)$, its expected value is

$$E(X) = \int_{-\infty}^{\infty} xf(x)\, dx \qquad\qquad (7A.4)$$

We carry out the integration from $-\infty$ to $+\infty$ to make sure that all possible values of x are covered.

Let's look at an example of how Equation 7A.4 can be used to calculate the mean of a continuous variable. Suppose we let

$$
\begin{aligned}
f(x) &= 2(1 - x), \quad 0 < x < 1 \\
&= 0 \qquad\qquad \text{otherwise}
\end{aligned}
$$

Substituting $f(x) = 2(1 - x)$ into Equation 7A.4, we obtain

$$\int_{-\infty}^{\infty} 2(1 - x)\, dx = \int_{0}^{1} 2(1 - x)\, dx$$

$$= 2x - x^2 \Big|_{0}^{1} = 1$$

Therefore, $f(x)$ is a PDF is $x \geq 0$ and $x \leq 1$. This PDF is shown in Figure 7A.2.

The expected value of X in terms of $f(x) = 2(1 - x)$ can be calculated as

$$E(x) = \int_{-\infty}^{\infty} xf(x)\, dx = \int_{0}^{1} (x)2(1 - x)\, dx$$

$$= x^2 - \frac{2}{3}x^3 \Big|_{0}^{1} = 1 - \frac{2}{3} = \frac{1}{3}$$

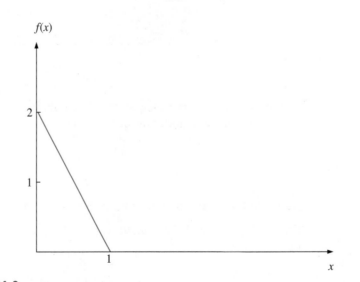

Figure 7A.2

Probability Density Function of $2(1 - x)$

Variance for Discrete and Continuous Random Variables

For discrete variables, we calculate the variance by averaging the squares of all possible individual deviations about the mean. The variance is a measure of how spread out the observations are, and it indicates the general shape of a distribution. When all members of the population are obtainable and are used, we can define the variance of a discrete random variable as

$$\sigma_x^2 = \sum_{i=1}^{N} (x_i - \mu_x)^2 P(x) \tag{6.4}$$

For a continuous random variable, the variance is defined as

$$\sigma_x^2 = \int_{-\infty}^{\infty} (x - \mu_x)^2 f(x)\, dx \tag{7A.5}$$

Equation 7A.5 can be rewritten as[14]

$$\sigma_x^2 = E(X^2) - \mu_x^2 \tag{7A.6}$$

where

$$E(X^2) = \int_{-\infty}^{\infty} x^2 f(x)\, dx \tag{7A.7}$$

Now let's see how we can use Equation 7A.6 to calculate the variance for a continuous random variable. For $f(x) = 2(1 - x)$, $E(X^2)$ can be calculated in accordance with Equation 7A.6 as

$$E(X^2) = \int_{-\infty}^{\infty} x^2 f(x)\, dx = \int_{0}^{1} (x^2) 2(1 - x)\, dx$$

$$= \frac{2}{3}x^3 - \frac{2}{4}x^4 \Big|_{0}^{1} = \frac{2}{3} - \frac{2}{4} = \frac{1}{6}$$

Substituting $E(X^2) = \frac{1}{6}$ and $E(X) = \frac{1}{3}$ into Equation 7A.6, we obtain

$$\sigma_x^2 = E(X^2) - [E(X)]^2 = \tfrac{1}{6} - \tfrac{1}{9} = \tfrac{1}{18}$$

[14]From Equation 7A.5 we obtain

$$\sigma_x^2 = \int_{-\infty}^{\infty} (x^2 - 2x\mu_x + \mu_x^2) f(x)\, dx$$

$$= \int_{-\infty}^{\infty} (x^2) f(x)\, dx - 2\mu_x \int_{-\infty}^{\infty} x f(x)\, dx + \mu_x^2 \int_{-\infty}^{\infty} f(x)\, dx$$

$$= \int_{-\infty}^{\infty} (x^2) f(x)\, dx - 2\mu_x^2 + \mu_x^2$$

$$= \int_{-\infty}^{\infty} (x^2) f(x)\, dx - \mu_x^2 = E(X^2) - \mu_x^2$$

Appendix 7B Cumulative Normal Distribution Function and the Option Pricing Model

The cumulative normal density function tells us the probability that a random variable Z will be less than some value x. Note in Figure 7B.1 that $P(Z < x)$ is simply the area under the normal curve from $-\infty$ up to point x.

One of the many applications of the cumulative normal distribution function is in valuing stock options. Recall from the Appendix 6B that a call option gives the option holder the right to purchase, at a specified price known as the exercise price, a specified number of shares of stock during a given time period. A call option is a function of the following five variables:

1. Current price of the firm's common stock (S)
2. Exercise price of the option (X)
3. Term to maturity in years (T)
4. Variance of the stock's price (σ)
5. Risk-free rate of interest (r)

From Appendix 6B, the binomial **option pricing model** can be written as[15]

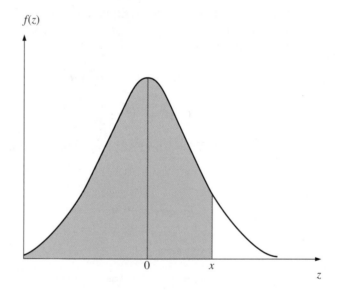

[15]See R. J. Rendleman, Jr. and B. J. Barter (1979), "Two-State Option Pricing," *Journal of Finance* 34 (December 1979), 1093–1110.

Figure 7B.1

$P(Z < x)$

$$C = S \left[\sum_{k=m}^{T} \frac{T!}{k!(T-k)!} p'^{k}(1-p')^{T-k} \right]$$
$$- \frac{X}{(1+r)^{T}} \left[\sum_{k=m}^{T} \frac{T!}{k!(T-k)!} p^{k}(1-p)^{T-K} \right]$$

(7B.1)

where

T = term to maturity in years

m = minimum number of upward movements in stock price that is necessary for the option to terminate "in the money"

$$p = \frac{R-d}{u-d} \quad \text{and} \quad 1-p = \frac{u-R}{u-d}$$

where

$R = 1 + r = 1 + $ risk-free rate of return

$u = 1 + $ percentage of price increase

$d = 1 + $ percentage of price decrease

$$p' = \left(\frac{u}{R}\right) p$$

Following Section 7.5, we know that a binomial distribution can be approximated by the normal distribution. Therefore, it can be shown that Equation 7B.1 approaches Equation 7B.2 if T becomes infinite (the time interval approaches zero).

$$C = SN(d_1) - Xe^{-rT}N(d_2)$$

(7B.2)

where

C = price of the call option

S = current price of the stock

X = exercise price of the option

$e = 2.71828\ldots$

r = short-term interest rate (T-bill rate) = R_f

T = time to expiration of the option, in years

$N(d_i) = F_z(d_i)$ = value of the cumulative standard normal distribution ($i = 1, 2$)

σ^2 = variance of the stock rate of return

$d_1 = [\ln(S/X) + (r + \frac{1}{2}\sigma^2 T]/\sigma\sqrt{T}$

$d_2 = d_1 - \sigma\sqrt{T}$

If future stock price is constant over time, then $\sigma^2 = 0$. It can be shown that both $N(d_1)$ and $N(d_2)$ are equal to 1 and that Equation 7B.2 becomes

$$C = S - Xe^{-rT}$$

(7B.3)

Alternatively, Equations 7B.2 and 7B.3 can be understood in terms of the following steps:

Step 1: The future price of the stock is constant over time.

Because a call option gives the option holder the right to purchase the stock at the exercise price E, the value of the option, C, is just the current price of the stock less the present value of the stock's purchase price. The concept of present value is discussed in Appendix 21C in detail. Mathematically, the value of the call option is

$$C = S - \frac{X}{(1 + r)^T} \tag{7B.4}$$

Note that Equation 7B.4 assumes discrete compounding of interest, whereas Equation 7B.3 assumes continuous compounding of interest. To adjust Equation 7B.4 for continuous compounding, we substitute e^{-rT} for $1/(1 + r)^T$ to get

$$C = S - Xe^{-rT} \tag{7B.3}$$

Step 2: Assume the price of the stock fluctuates over time (S_t).

In this case, we need to adjust Equation 7B.3 for the uncertainty associated with that fluctuation. We do this by using the cumulative normal distribution function. In deriving Equation 7B.2, we assume that S_t follows a lognormal distribution, as discussed in Section 7.4.[16]

The adjustment factors $N(d_1)$ and $N(d_2)$ in the Black–Scholes option valuation model are simply adjustments made to Equation 7B.3 to account for the uncertainty associated with the fluctuation of the price of the stock.

Equation 7B.2 is a continuous option pricing model. Compare this to the binomial option pricing model given in Appendix 6B, which is a discrete option pricing model. The adjustment factors $N(d_1)$ and $N(d_2)$ are cumulative normal density functions. The adjustment factors B_1 and B_2 are cumulative binomial probabilities.

We can use Equation 7B.2 to determine the theoretical value, as of November 29, 1991, of one of IBM's options with maturity on April 1992. In this case we have $X = \$90$, $S = \$92.5$, $\sigma = 0.2194$, $r = 0.0435$, and $T = \frac{5}{12} = .42$ (in years).[17] Armed with this information we can calculate the estimated d_1 and d_2.

$$d_1 = \frac{\{\ln(92.5/90) + [(.0435) + \frac{1}{2}(.2194)^2](.42)\}}{(.2194)(.42)^{1/2}}$$

$$= .392$$

$$d_2 = d_1 - (0.2194)(0.42)^{1/2}$$

$$= .25$$

---

[16]See Cheng F. Lee et al. (1990), *Security Analysis and Portfolio Management,* (Glenview, Ill.: Scott, Foresman/Little, Brown), 754–760.

[17]Values of $X = 90$, $S = 92.5$, and $r = .0435$ were obtained from Section C of the *Wall Street Journal* on December 2, 1991. And $\sigma = .2194$ is estimated in terms of monthly rates of return during the period January 1989 to November 1991.

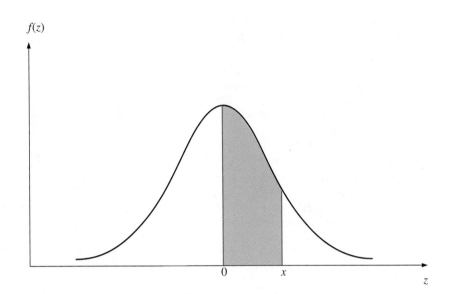

Figure 7B.2

$P(0 < Z < x)$

In Equation 7B.2, $N(d_1)$ and $N(d_2)$ are the probabilities that a random variable with a standard normal distribution takes on a value less than d_1 and a value less than d_2, respectively. The values for $N(d_1)$ and $N(d_2)$ can be found by using the tables in the back of the book for the standard normal distribution, which provide the probability that a variable Z is between 0 and x (see Figure 7B.2).

To find the cumulative normal density function, we need to add the probability that Z is less than zero to the value given in the standard normal distribution table. Because the standard normal distribution is symmetric around zero, we know that the probability that Z is less than zero is .5, so

$$P(Z < x) = P(Z < 0) + P(0 < Z < x) = .5 + \text{value from table}$$

We can now compute the values of $N(d_1)$ and $N(d_2)$.

$$N(d_1) = P(Z < d_1) = P(Z < 0) + P(0 < Z < d_1)$$
$$= P(Z < .392) = .5 + .1517$$
$$= .6517$$
$$N(d_2) = P(Z < d_2) = P(Z < 0) + P(0 < Z < d_2)$$
$$= P(Z < .25) = .5 + .0987$$
$$= .5987$$

Then the theoretical value of the option is

$$P_o = (92.5)(.6517) - [(90)(.5987)]/e^{(.0435)(0.42)]}$$
$$= 60.282 - 53.883/1.0184$$
$$= \$7.373$$

and the actual price of the option on November 29, 1991, was \$7.75.

Questions and Problems

1. A study indicates that an assembly line task should take an average of 3.20 minutes to complete, with a standard deviation of 0.75 minute. What is the probability that the task will take between 1.80 and 3.80 minutes to complete? Graph the area being determined by assuming that the completion time is normally distributed.

2. Indicate whether each of the following random variables is continuous or discrete.
   a. The time it takes a mechanic to service a car.
   b. The number of new housing starts in New Jersey this year.
   c. The age of an applicant for an MBA program.
   d. The sex of a new company chief executive officer.

3. The random variable Z is normally distributed with a mean of $\mu_Z = 0$ and a standard deviation of $\sigma_Z = 1$. Find the following probabilities.
   a. $P(Z > 1.65)$
   b. $P(Z > -2.38)$
   c. $P(Z > 2.95)$
   d. $P(Z < -1.37)$
   e. $P(1.05 < Z < 2.82)$
   f. $P(-2.43 < Z < 1.72)$

4. The random variable Z is normally distributed with $\mu_Z = 0$ and $\sigma_Z = 1$. Find the following values of b.
   a. $P(Z < b) = .9280$
   b. $P(Z > b) = .9949$
   c. $P(Z > b) = .0074$
   d. $P(Z < b) = .0130$
   e. $P(-b < Z < b) = .5390$
   f. $P(0 < Z < b) = .1860$

5. A random variable X is normally distributed with $\mu_X = 5$ and $\sigma_X = 2.7$. Use the MINITAB program to find the following probabilities.
   a. $P(X < 7.40)$
   b. $P(X > -1.50)$
   c. $P(X > 8.32)$
   d. $P(X < .95)$
   e. $P(2.35 < X < 7.05)$
   f. $P(-2.80 < X < 0)$

6. The random variable Y is normally distributed with $\mu_Y = 28.00$ and $\sigma_Y = 10$. Find the following values of b.
   a. $P(Y < b) = .8964$
   b. $P(Y > b) = .8099$
   c. $P(Y > b) = .0100$
   d. $P(Y < b) = .3405$
   e. $P(b < Y < 38) = .0230$
   f. $P(b < Y < 25) = .1498$

7. Suppose that X represents the number of cars arriving at a toll booth in 1 minute. Further, suppose that X can assume the values 1, 2, 3, 4, and 5 and has the following distribution.

r	1	2	3	4	5
$P(X = r)$	.10	.20	.30	.25	.15

Calculate the expected value of this random variable and explain your result.

8. The following table gives the amount of time X, in seconds, by which an automated manufacturing process misses the designed completion time when performing a certain task. Negative values indicate early completion, positive values late completion.

r	-1	0	1	2
$P(X = r)$	.1	.2	.3	.4

   a. Find the mean and the variance of X.
   b. On average, how do the completion times for this particular task compare with the designed completion times?

9. Find the probability density function of $Y = e^X$ when x is normally distributed with parameters μ and σ^2. The random variable Y is said to have a lognormal distribution (because log Y has a normal distribution) with parameters μ and σ^2.

10. Suppose that 35 percent of the employees of the Harrison Company belong to unions. To determine union members' attitudes toward management, the company's personnel manager takes a random sample of 100 employees. The selection of a union member in this random sample is a "success." Calculate the probability that the number X of successes will be between 20 and 40, inclusive.

11. What is the probability that the number X of successes in the personnel manager's sample of 100 employees in question 10 will be 48 or more?

12. Suppose that a batch of $n = 80$ items is taken from a manufacturing process that produces a fraction $p = .16$ of defectives. What is the probability that this batch will contain exactly 20 defectives? (Finding a defective is considered a success.)

13. The IQ scores of human beings are scaled to follow a normal distribution with mean 100 and standard deviation 16. If those with IQ scores higher than 154 are regarded as geniuses, how many geniuses are there among 20,000 children?

14. A college professor teaches corporate finance every semester. The tests for the course are standardized so that the test scores exhibit a normal distribution with a mean of 75 and a standard deviation of 12. The professor gives 15 percent A, 25 percent B, 30 percent C, 20 percent D, and 10 percent F.
    a. What letter grade will a student who scores 79 points on the test receive?
    b. What letter grade will a student who scores 58 points receive?
    c. How many points does a student need to score to get an A?
    d. How many points does a student need to score to pass the course?

15. The manager in the local bank discovers that people come in to cash their paychecks on Friday. The amount of money withdrawn on Friday follows a normal distribution with 5 million as the mean and 1 million as the standard deviation. The bank manager wants to make sure that the amount of money in the bank can cover 99.9 percent of the Friday withdrawals. What is the minimum amount of money he or she should plan to have on hand?

16. A local bakery found that it was throwing out too many cookies every night, so the manager conducted a study on the sales of cookies and found that on an ordinary day, the sales of cookies follows a normal distribution with 30 pounds as the mean and 12 pounds as the standard deviation. The manager then decided to prepare only 35 pounds of cookies each day. What is the chance that the bakery will run out of cookies on a certain day?

17. A soft drink producer has just installed a new assembly line. The assembly line is adjusted to dispense an average of 12.05 ounces of soda into the 12-ounce soda can with a standard deviation of .02. What is the probability that a certain can will contain less than 12 ounces of soda?

18. In question 17, what is the probability that 2 cans out of a six-pack will contain less than 12 ounces?

19. In question 17, the average amount of soda dispensed into the cans is adjustable. If we want to make sure that 99.9 percent of the soda cans contain more than 12 ounces, to what should we adjust the average?

20. A battery producer invents a new product. The life of the new battery is found to follow a normal distribution with a mean of 72 months and a standard deviation of 12 months. The producer guarantees that the new battery will last longer than 60 months or the full price will be refunded. Last year the producer sold 1 million batteries, how many refunds will be claimed?

21. A car manufacturer designs a fuel-efficient car for 1993. The company argues that the car can attain an average of 45 miles per gallon. The miles per gallon of the car follows a normal distribution with a standard deviation of 5. What is the probability that a certain car will reach 40 miles per gallon?

22. You work for a furniture factory that procures springs from an outside supplier. Every month a truckload of springs comes in. From each shipment, you randomly inspect 400 springs. If there are 10 or more bad springs, then you send the shipment back. One day a shipment arrives that actually contains 3 percent bad springs. What is the probability of your accepting this shipment?

23. A consumer rights organization wants to find out whether a local dairy farm actually puts 16 ounces of milk into the container that is labeled 16 ounces. Assume the milk put into the container by the local dairy farm follows a normal distribution with a mean of 16.05 and a standard deviation of .03.
    a. What is the probability that a certain container contains more than 16 ounces of milk?
    b. The consumer rights organization bought 400 bottles of milk. What is the probability that among them, it found fewer than 12 bottles that do not contain enough milk?

24. A name-brand TV dinner boasts that its pot roast has no more than 120 calories per serving. Suppose 95 percent of the servings of this product actually contain fewer than 120 calories. Find the probability that out of a random sample of 500 packs of pot roast, fewer than 10 packs (servings) actually contain more than 120 calories.

25. The Food and Drug Administration randomly tests 1,000 of a certain brand of cigarettes to see whether the nicotine content reaches a dangerous level. If 20 or more cigarettes contain more nicotine that a prespecified level, the production of the cigarettes is suspended. Assume that in this month, as a result of either machine failure or worker discontent, 3 percent of the cigarettes contain more than the prespecified level. What is the probability that cigarette production will be suspended?

26. The light bulbs produced by Edison Lighting Corporation last an average of 300 hours. The life of the light bulbs is believed to follow a normal distribution with a standard deviation of 10. A customer buys 2 dozen light bulbs during a sale. What is the probability that the fifth light bulb used will last longer than 315 hours?

27. The number of phone calls that reach 1-800 numbers in a certain time period follows a Poisson distribution. Assume that there are about 15,000 potential callers. Each caller has a probability of 0.001 of making such a phone call. What is the probability that we have less than 20 callers who make phone calls during this time period?

28. A camcorder is sold with a 1-year warranty. The probability that a camcorder is brought back for service under the warranty is 2 percent. It costs the manufacturer $20 on average to repair a camcorder brought back under warranty. Last year 5,000 camcorders were sold. What is the probability that fewer than 80 of them will be brought back to be repaired under warranty. How much should the company expect to spend living up to the warranty?

29. What are the advantages of using the lognormal distribution over using the normal distribution to describe stock prices?

30. Suppose that X is distributed as normal with a mean of 5 and a standard deviation of 2. Compute the standard normal values of X, given the following values of X.
   a. 3   b. 2   c. 9
   d. 11   e. 5   f. 10

31. Use the standard normal values you computed in question 30, and find the probability that Z is less than those values.

32. Calculate the area under the normal curve between
   a. $z = 0$ and $z = 2.0$
   b. $z = -3.5$ and $z = -1$
   c. $z = 1.2$ and $z = 3$
   d. $z = -1.3$ and $z = 1.3$
   e. $z = -1$ and $z = 1$
   f. $z = 3$ and $z = 4$

33. Find the value for z_0 for the following probabilities.
   a. $P(z > z_0) = .10$
   b. $P(z > z_0) = .75$
   c. $P(-z_0 < z < z_0) = .95$
   d. $P(z < z_0) = .95$
   e. $P(-z_0 < z < z_0) = .90$
   f. $P(-z_0 < z < z_0) = 1.00$

34. Suppose that X is normally distributed with a mean of 5 and a standard deviation of 2. Find the following probabilities.
   a. X is between 5 and 9.
   b. X is between 0 and 8.
   c. X is greater than 6.
   d. X is less than 10.
   e. X is between -1 and 3.

35. Briefly explain why it is useful for us to be able to approximate the binomial and Poisson distributions by using a normal distribution. Explain how we make this approximation.

36. Use the normal approximation to the binomial distribution with $n = 100$ and $p = .3$.
   a. What is the probability that a value from the binomial distribution will have a value greater than 35?
   b. What is the probability that a value from the binomial distribution will have a value less than 20?
   c. What is the probability that a value from the binomial distribution will have a value between 15 and 45, inclusive?

37. Use the normal approximation to the binomial distribution with $n = 500$ and $p = .7$.
    a. What is the probability that a value from the binomial distribution will have a value greater than 325?
    b. What is the probability that a value from the binomial distribution will have a value less than 325?
    c. What is the probability that a value from the binomial distribution will have a value between 325 and 375, inclusive?

38. Use the normal approximation to the Poisson distribution with $\lambda = 75$.
    a. What is the probability that a value from the Poisson distribution will be greater than 50?
    b. What is the probability that a value from the Poisson distribution will be between 50 and 80, inclusive?
    c. What is the probability that a value from the Poisson distribution will be less than 60?

39. The time a customer waits for service at a bank is distributed normally with a mean of 4 minutes and a standard deviation of 1 minute. Compute the probability that a customer must wait for
    a. More than 10 minutes
    b. Less than 5 minutes
    c. Between 2 and 6 minutes
    d. Between 3 and 9 minutes

40. The time it takes to get a car's oil changed at Speedy Lube is distributed normally with a mean of 12 minutes and a standard deviation of 2 minutes. Compute the probability that a customer will have her or his oil changed
    a. In less than 9 minutes
    b. In between 9 and 15 minutes

41. A quality control manager has determined that the number of defective light bulbs in a case of 1,000 follows a normal distribution with a mean of 10 and a standard deviation of 3. Compute the probability that the number of defective light bulbs in a case is
    a. Greater than 10
    b. Less than 9

42. A survey of recent masters of business administration (MBAs) reveals that their starting salaries follow a normal distribution with mean $48,000 and standard deviation $9,000. Find the probability that a randomly selected MBA degree holder will begin his or her career earning
    a. More than $50,000   b. Less than $35,000

43. Use the Black–Scholes option pricing formula to compute the value of a call option, given the following information.

    $S = \$55$ — price of stock
    $X = \$50$ — exercise price
    $r = .065$ — risk-free interest rate
    $t = .5$ — time until the option expires, in years
    $\sigma = .25$ — standard deviation of the stock's return

44. Answer question 43 again for an option with an exercise price of $55. How does the exercise value of the call option affect the option's price?

45. Answer question 43 again for an option with $t = .3$ years. How does the time until the option expires affect the value of the call option?

46. Answer question 43 again for an option whose price is $60. How does the price of the stock affect the value of the call option?

47. Answer question 43 again when $r = .10$. How does a change in the risk-free rate of interest affect the value of the call option?

48. Answer question 43 again when $\sigma^2 = .50$. How does a change in the variance of the stock's return affect the value of the call option?

49. Draw a standard normal probability function and show the area under the curve for
    a. Plus or minus one standard deviation from the mean
    b. Plus or minus two standard deviations from the mean
    c. Plus or minus three standard deviations from the mean

50. A company has a mean earnings per share (EPS) of $3.25 with a standard deviation of $1.21. Assume that the earnings are normally distributed. Compute the probability that EPS will be
    a. Between $1.50 and $6.00
    b. Above $5.00

51. An investment analyst is following the stock of High Flyer Company. She believes that in any month, the stock has a 65 percent chance of going

up and a 35 percent chance of going down. Using the binomial distribution, compute the probability that the stock goes up in 18 or more months during a 36-month period. Now use the normal approximation to the binomial distribution to recompute your answer. Compare the two results. Which method was easier to use?

52. A gas station finds that the mean number of people buying gas in any 30-minute period is 32. Use the Poisson distribution to compute the probability that between 25 and 35 people, inclusive, will buy gas in any 30-minute period. Use the normal approximation to the Poisson distribution to recalculate your result. Which method is easier to use?

53. Calculate e^y for the following values of y.
   a. $y = 1$ f. $y = -1$
   b. $y = .5$ g. $y = .05$
   c. $y = -.5$ h. $y = .32$
   d. $y = -2.5$ i. $y = 6.1$
   e. $y = 3.1$ j. $y = -5.4$

54. Suppose $x = e^y$. Compute the value of y, given the following values of x.
   a. $x = 2$ e. $x = .5$
   b. $x = 3$ f. $x = .002$
   c. $x = 1.5$ g. $x = 10$
   d. $x = .3$ h. $x = 1$

55. Briefly explain what a cumulative distribution function is. Give some examples of occasions when the cumulative distribution function is useful.

56. A quality control manager has found that the mean number of ounces of cereal in a 16-ounce box is 16 ounces with a standard deviation of 2 ounces. Calculate the probability that a randomly selected box of cereal will contain
   a. More than 16 ounces of cereal
   b. Less than 15 ounces of cereal
   c. Between 14 and 18 ounces of cereal
   d. Between 15 and 17 ounces of cereal

57. An investment analyst calculates that the mean price of gold is $392 per ounce with a standard deviation of $12. Assume the price of gold follows a normal distribution. Compute the probability that the price of gold will be
   a. Greater than $400 an ounce
   b. Less than $350 an ounce

58. You know that a certain stock's dividend yield has a mean of 6 percent and a standard deviation of 2 percent. Assume the dividends follow a normal distribution. Compute the probability that the dividend yield will be
   a. Less than 2%
   b. Greater than 10%
   c. Between 4% and 8%

59. Use the Black–Scholes option pricing formula to compute the value of a call option, given the following information.

   $S = \$105$ price of stock
   $X = \$110$ exercise price
   $r = .055$ risk-free interest rate
   $t = .9$ time until the option expires, in years
   $\sigma = .45$ standard deviation of the stock's return

60. The number of claims filed each week with Security Insurance Company has a mean of 700 and a standard deviation of 250. Calculate the probability that the number of claims this week will be
   a. Greater than 1,000
   b. Less than 500
   c. Between 300 and 800
   d. Between 1,000 and 1,250

61. From past history, we know that 60 percent of people audited by the IRS owe money to the government. If we take a random sample of 500 people who are being audited, what is the probability that between 280 and 320, inclusive, owe the IRS money?

62. The probability is .1 that a customer entering a food store will buy a can of coffee. If 1,000 customers enter the store, what is the minimum number of cans of coffee the store must have on hand to prevent the probability of running out of coffee from being higher than 5 percent?

63. Determine the following probabilities. Assume that X follows a normal distribution.
   a. $P(80 \leq X \leq 95 \mid \mu = 92, \sigma = 10)$
   b. $P(X \geq 150 \mid \mu = 99, \sigma = 25)$

64. A public library has observed that the fine for overdue books is approximately normally distributed with a mean of $2.72 and a standard deviation of $.37.

a. What is the probability that a fine will be greater than $3?

b. What is the probability that a fine will be less than $2?

65. The breaking strength for paper bags used in a grocery store is approximately normally distributed with a mean of 15 pounds and a standard deviation of 2 pounds.

a. What proportion of these bags have a breaking strength less than 10 pounds?

b. What proportion of the bags have a breaking strength greater than 17 pounds?

66. A newspaper publisher has mean sales of 28,200 copies per day with a standard deviation of 3,100. If the publisher distributes 32,000 copies of the paper to the newsstands, what is the probability that at least 6,000 or more copies will go unsold?

67. Value Line ranks 1,700 stocks according to their timeliness and riskiness. In other words, Value Line classifies these 1,700 stocks into five ranks (groups) on the basis of their return potential and the degree of riskiness as follows:

Rank 1	Top 100	Rank 4	Next 300
Rank 2	Next 300	Rank 5	Bottom 100
Rank 3	Middle 900		

These five groups are classified by assuming that they are normally distributed.

a. Find what percentage of the 1,700 stocks is classified in each group.

b. Calculate the mean and standard deviation of this ranking.

68. The following MINITAB output displays the cumulative distribution function curves of three normal distributions. Their mean and variance, respectively, are (0, .5), (0, 1), (0, 2). Please compare the three cumulative distribution curves indicated in the figure.

69. The following MINITAB output exhibits the cumulative distribution function curves of three lognormal distributions. Their mean and variance, respectively, are (1, .5), (1, 1), (1, 2). Compare the three cumulative distribution curves indicated in the figure.

```
MTB > SET C1

DATA> 0:10/0.1
DATA> END
MTB > CDF C1 C2;
SUBC> LOGNORMAL 1 0.5.
MTB > CDF C1 C3;
SUBC> LOGNORMAL 1 1.
MTB > CDF C1 C4;
SUBC> LOGNORMAL 1 2.
MTB > GPLOT;

SUBC> LINE C2 C1;
SUBC> LINE C3 C1;
SUBC> LINE C4 C1.
```

```
MTB > SET C1
DATA> -5:5/0.1
DATA> END
MTB > CDF C1 C2;
SUBC> NORMAL 0 0.5.
MTB > CDF C1 C3;
SUBC> NORMAL 0 1.
MTB > CDF C1 C4;
SUBC> NORMAL 0 2.
MTB > GPLOT;

SUBC> LINE C2 C1;
SUBC> LINE C3 C1;
SUBC> LINE C4 C1.
```

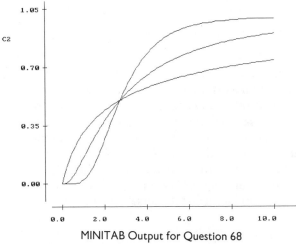

MINITAB Output for Question 68

MINITAB Output for Question 69

CHAPTER 8

Sampling and Sampling Distributions

Key Terms

census	systematic error
population	sampling costs
sample	cost–benefit analysis of sampling
simple random sampling	sampling distribution
random sample	central limit theorem
sampling errors	finite population multiplier
random errors	sample proportion
nonsampling errors	confidence interval

8.1 INTRODUCTION

In this chapter we take an in-depth look at the operational end of statistical analysis. Statistical analysis primarily involves selecting parts of populations (known as samples) and analyzing them in order to make inferences about the populations. Inferences made about a population by using sample data are widespread in business, economics, and finance. For example, the A. C. Nielsen Company infers the number of people who watch each television show on the basis of a sample of TV viewers. The use of political polls to project election winners is another example of statistical inference. And when you fill out a warranty card on an appliance you have bought, you are often asked to provide information about yourself that the warrantor compiles (and probably sells to someone who will later try to convince you to buy a magazine subscription). These data are also sample data.

First, sampling from a population is discussed. Second, we explore the issue of sampling costs versus sampling errors. Next, sampling distributions for sample means and sample proportions are illustrated. Then one of the most important principles in statistics, the central limit theorem, and confidence intervals are discussed in detail. Finally, an accounting application illustrates how sampling and sampling distributions can be used in auditing. The sampling distribution concept also is used to do patient waiting-time analysis.

8.2   *SAMPLING FROM A POPULATION*

In previous chapters we have discussed many different topics in statistics. Among these are distributions, probabilities, measurements of dispersion and symmetry, and data collection and analysis. The topic most closely related to sampling is data collection, organization, and presentation, which we discussed in Chapter 2. Either a census or a sampling survey approach can be used in data collection. A **census** is a survey that attempts to include every element in the universe, or population, in which we are interested. Sampling is used to count or measure only a subset of the population; these collected data are called sample data. In this book we will return again and again to problems whose solutions depend on making inferences about a population from a sample.

The management, analysis, and interpretation of data are the foundation of statistics. In order to make full use of the information that data can yield, the statistician must start with clear objectives and follow well-defined pathways to a desired result. Along these pathways, there are points where the analyst must make decisions on the basis of an evaluation of costs and benefits. Key considerations include how much information is appropriate, how specific this information should be, and whether statistical inferences drawn from the data are analytically sound.

Now let us formally define the terms *population* and *sample*. A **population** consists of all members, objects, or observations that fall into a certain category. A **sample** is a subset of the members, objects, or observations in a given population.

In analyzing the characteristics of a population, a researcher can analyze the entire population or draw conclusions about the entire population on the basis of a random sample selected from the population. Using sampling to determine the true characteristics of a population offers several advantages:

1. The cost is less.
2. The data are more manageable.
3. It is less time-consuming.
4. Sample observation can be more accurate.
5. It makes analysis possible even when not all population elements are accessible.

As in other areas of statistics, the goal of the researcher is to choose methods that will lead to informative and useful results. These issues are discussed in the following section.

Sampling enables a statistician (or researcher) to make inferences about a given population from a more manageable segment of the population. It follows that the sample must be chosen in a manner that will ensure that it represents the original population. Two kinds of errors can arise in a sampling experiment: sampling errors and nonsampling errors. Before we can examine samples, we must thoroughly understand these two types of errors.

Sampling Error and Nonsampling Error

Sampling errors are errors that result from the chance selection of sampling units. They occur only when a sample, rather than the entire population, is observed. They are **random errors** (or chance errors), as discussed in Chapter 2. For example, if the sample from a given population had a mean of .6 and the true population mean was .5, there is an average sampling error of .1. Sampling error can be reduced by taking more observations, and it can be eliminated by taking all observations. Sampling error can usually be analyzed by first identifying the source of the error and then making the needed inferences. The relationship between sampling cost and sampling error is discussed in the next section of this chapter.

Nonsampling errors are errors that result from inaccurate measurement of the data or improper selection of sample observations. For example, if you measure flour with a cup that holds 15 ounces rather than 16, the bread you make will contain less flour than the recipe intended. This kind of error is not related to the number of observations but rather is due to the inaccurate measurement of data. If a given section of the population has an unduly low or an unduly high chance of being selected for a sample, then sampling data can result in systematic, rather than random, sampling error. Other examples of nonsampling error include faulty questions and choosing observations that do not pertain to the population being examined. Nonsampling error is **systematic error** (or bias), as discussed in Chapter 2.

Unlike sampling error, nonsampling error cannot be reduced by increasing the sample size. (This issue will be discussed further in Chapter 20.) Although it is possible to minimize this type of error by carefully specifying the criteria by which observations are selected or measured, nonsampling error persists to a certain extent in almost all cases. The more complicated the data set, the greater the chance that nonsampling error will creep in.

Selection of a Random Sample

For a sample to be drawn from a population representatively, each member, object, or observation must have an independent and equal chance of being selected for the sample. Suppose a sample of n elements must be selected from a population of N elements. A **simple random sampling** procedure is one in which every possible combination $[\binom{N}{n}]$ of n elements in the population has an equal probability of being selected. The *n* elements obtained from simple random sampling constitute a simple random sample or **random sample.** Random selection is the key to this process; it significantly reduces nonsampling errors due to improper selection of sample observations.

There are two useful methods for carrying out simple random sampling: drawing chips from a box and using random-number tables.

Drawing from a Box

If we want to draw, with replacement, a simple random sample of 5 students from a business statistics class made up of 80 students, we assign the numbers 1 to 80 to the students and place these numbers on physically similar balls, slips of paper, or poker chips. We then put all the balls (or whatever) in a box, shake the box to mix them thoroughly, and proceed to draw the sample. The first ball is drawn, and we record the number written on it. We then replace the ball and shake the box again, draw the second ball, and record the result. We repeat the process until we have drawn 5 distinct numbers. The students corresponding to these 5 numbers constitute the required simple random sample.

Using a Random-Number Table

If the population size is large, the method just described becomes unwieldy and time-consuming. Furthermore, it may introduce biases if the balls are not thoroughly mixed. Using such random-number tables as Table 8A in Appendix A to draw random samples is much easier. A table of random digits is simply a table of digits generated by a random process. The application of random number tables to draw random samples will be thoroughly discussed in Chapter 20.

MINITAB, SAS, and other computer programs can be used to generate random numbers. Both Table 8.1A and Table 8.1B are generated from MINITAB. Table 8.1A contains the instructions for generating 200 random numbers between 1 and 1,000 in terms of a uniform distribution. Table 8.1B contains the instructions for generating 200 numbers between 0 and 1 in terms of a uniform distribution.

Whichever sampling method is used, the analyst must be sure the population under consideration is appropriate for the analysis; taking this precaution largely eliminates another kind of nonsampling error.

Table 8.1A Generating 200 Random Numbers Between 1 and 1,000 in Terms of a Uniform Distribution by Using MINITAB

```
MTB > RANDOM 200 VALUES INTO C1;
SUBC> UNIFORM A=1 AND B=1000.
MTB > PRINT C1

YEARS
   666.413    260.611    484.343    478.919    799.820    952.465     53.150
   525.752    660.005    458.566    253.788    630.495    765.841    701.103
   600.888     62.057    640.185    978.076    257.514     97.286     48.718
   970.805    347.574    365.780    643.547    399.360    844.980    603.558
   395.703    387.898    411.238    331.706    380.261    455.612    883.460
   418.486    238.788    753.496    157.008    521.058     73.233     39.111
   768.885     82.685    221.648    609.036     81.501     73.604     85.556
```

(continued)

Table 8.1A (Continued)

580.526	513.775	161.908	134.553	711.120	853.987	730.316
256.503	969.812	223.512	842.021	233.604	105.542	81.738
103.258	795.234	379.314	337.311	81.823	113.887	125.833
620.133	469.647	639.880	939.982	490.772	283.553	355.151
313.904	152.959	14.851	733.331	633.423	132.819	494.436
741.440	647.953	950.181	766.646	801.809	202.081	161.075
30.364	674.505	273.087	45.917	620.635	532.393	491.077
321.647	121.893	127.636	845.526	671.752	927.947	984.352
43.035	260.373	454.601	757.076	604.445	506.591	262.865
766.140	738.527	283.895	397.890	661.210	609.200	102.044
643.484	391.500	861.487	668.927	574.875	806.318	765.764
691.457	394.102	187.693	360.577	992.092	11.478	311.753
883.068	369.540	114.536	206.978	773.257	629.175	600.827
54.332	673.439	139.826	371.194	321.261	73.662	92.702
474.702	272.712	998.025	753.171	116.369	436.122	445.279
590.883	809.433	156.144	412.939	544.371	989.397	673.572
156.528	460.992	556.976	567.047	826.934	345.709	132.617
469.114	573.249	603.121	341.086	553.688	155.963	390.431
727.868	949.545	687.154	855.231	885.907	724.399	515.864
422.984	800.953	95.078	771.772	443.483	366.428	724.558
535.737	909.128	629.996	703.497	900.147	506.364	234.462
212.768	497.953	182.105	661.158			

```
MTB > PAPER
```

Table 8.1B Generating 200 Random Numbers Between 0 and 1 in Terms of a Uniform Distribution by Using MINITAB

```
MTB > RANDOM 200 VALUES INTO C1;
SUBC> UNIFORM A=0 B=1.
MTB > PRINT C1
```

YEARS

0.602372	0.296482	0.060301	0.537632	0.204035	0.504407	0.050903
0.362914	0.364285	0.535634	0.954188	0.273440	0.179967	0.495818
0.977305	0.163073	0.384071	0.008853	0.106679	0.334913	0.864182
0.022692	0.836480	0.560050	0.006228	0.778511	0.313884	0.235528
0.440951	0.118852	0.856530	0.066306	0.288204	0.025460	0.182530
0.816253	0.031621	0.952637	0.079605	0.950681	0.835167	0.395854
0.481744	0.217954	0.244311	0.538913	0.364118	0.514802	0.350210
0.776239	0.029838	0.729790	0.223751	0.968821	0.102572	0.821494
0.686751	0.843821	0.477633	0.704177	0.022177	0.772107	0.513424
0.177967	0.245926	0.740749	0.593644	0.205494	0.686753	0.844089
0.511161	0.895128	0.891063	0.382829	0.853598	0.699782	0.472726
0.090801	0.350124	0.765465	0.683146	0.393255	0.156839	0.604845
0.605562	0.695282	0.910267	0.783406	0.925696	0.712003	0.000378
0.047296	0.912035	0.004375	0.546858	0.357206	0.650773	0.346606
0.325806	0.725807	0.725933	0.741579	0.697356	0.169503	0.187839
0.479921	0.990101	0.762608	0.325965	0.745656	0.206976	0.871959
0.994903	0.362886	0.360754	0.094187	0.773332	0.666444	0.305468
0.183521	0.940082	0.510200	0.775055	0.881870	0.233726	0.215735
0.966837	0.854602	0.825220	0.152471	0.058926	0.365703	0.712838
0.104761	0.095124	0.890499	0.312421	0.052601	0.575122	0.890211
0.276435	0.554313	0.289184	0.148037	0.504647	0.080899	0.112419
0.052315	0.539359	0.419864	0.482956	0.369544	0.192998	0.124768
0.596060	0.507511	0.438840	0.855012	0.876539	0.567421	0.927567
0.945847	0.230865	0.858152	0.268946	0.618312	0.289006	0.125730
0.716268	0.533467	0.683419	0.427319	0.414845	0.855632	0.954055
0.256899	0.112431	0.053835	0.729349	0.168586	0.073309	0.163632
0.453987	0.748383	0.547929	0.491138	0.392225	0.028182	0.522768
0.346025	0.253074	0.634238	0.279696	0.962054	0.256737	0.092180
0.522500	0.312542	0.067711	0.463844			

```
MTB > PAPER
```

8.3 SAMPLING COST VERSUS SAMPLING ERROR

This section deals primarily with the costs associated with selecting a sample and with how those **sampling costs** can affect sampling errors. This type of analysis is often referred to as **cost–benefit analysis** of sampling. Cost–benefit analysis in this context involves comparing the benefits of sampling with its disadvantages (costs). The underlying need for the information is the gauge by which incurred costs and allowable error are measured. This issue will be explored further in Chapter 20.

The aim of drawing a random sample from a population is to measure indirectly population attributes such as mean and variance without having to include all possible data. We have all heard the saying "time is money." This is the heart of this issue. It takes people (and usually machines as well) to work through detailed analyses, and neither of these resources is free. A researcher must pay employees to collect the data and enter them into a computer; it also costs a lot to buy, use, and run the computer. The computer costs are numerous: hardware, software, electricity, paper, maintenance, operators, storage, and so on. If the computer is rented, these costs are included in the rental fee. Either way, the more data collected and analyzed, the higher the costs of the study. It is obvious that the statistician faces a crucial question: How much data are actually necessary?

The Gallup Organization and National Opinion Research Center used a sampling survey approach in their poll to obtain Americans' views on their work ethic in a timely manner. The results of the poll were published in the *Wall Street Journal* (February 13, 1992, page B1).

The three questions asked and the results are presented here.

Would you welcome or not welcome less emphasis on working hard?

Would not	67%
Would	30%

Are you satisfied or dissatisfied with Americans' willingness to work hard to better themselves?

Dissatisfied	45%
Satisfied	52%

Would you strongly agree, agree, disagree, or strongly disagree with the following: "I am willing to work harder than I have to in order to help this organization succeed."

Strongly disagree	1%
Disagree	9%
Agree	52%
Strongly agree	38%

Note that the percentages for questions 1 and 2 add up to only 97 percent because the category of "No Response" was omitted during the poll.

Sample Size and Accuracy

If an entire population is used as a sample for analysis, then such numerical characteristics as the mean and variance of the sample are identical to those of the population. However, suppose the population is very large—say, 10 million units. To collect all the observations and analyze them would be a ponderous task. Fortunately, if only some of the members are chosen at random and analyzed, the population mean and variance can be estimated with some precision from the sample. Even though it is possible to estimate the population parameters by analyzing a random sample of the observations, the results are only estimates. In general, the fewer the observations used in the sample, the larger the sample error. Significantly, sample error is not necessarily a linear function; that is, there is not necessarily an equal trade-off between additional data and greater accuracy. A relatively small sample of the entire population may yield estimates close to the true population values. However, it generally takes increasing amounts of data to make sample estimates closer to the true population value. Consequently, to have the sample estimates equal the population parameters is very expensive. The following two applications may shed some light on the problem of whether large or small samples should be used in the real world.

■ | *APPLICATION 8.1* A Case for a Large Sample

Suppose a pharmaceutical firm wishes to test a new shampoo formulated to help control dandruff for an acceptable amount of a certain active ingredient. If there is not enough of the active ingredient, the shampoo is not effective, yet if too much of the active ingredient is present, the shampoo may cause harmful side effects, including hair loss. Although there is a great need for accuracy, it is not economically feasible to test an entire batch of the shampoo. A sample can be used to test the content of the shampoo and to conduct related analyses and make inferences. In this case it is particularly important to work with a large sample in order to reduce sampling errors, because hair loss among users would be an intolerable outcome.

■ | *APPLICATION 8.2* A Case for a Small Sample

Suppose a company manufactures a crude grade of cement mix to be used as a foundation for sidewalks. The company wishes to check that a certain amount of small stones is included in each 50-pound bag. All components of the cement mix are equally valuable, so the only reason for conducting this test is to ensure the most durable mixture possible. A few stones more or less in a bag of cement mix will negligibly affect the performance of the cement mix. Therefore, the producer will want only a small sample of the total number of cement mix bags to be examined to ensure that the stone content of each is within a certain range. Here the underlying need for accuracy is small, so the company can save money by examining its product infrequently. When extremely high accuracy is not required, cost considerations and time constraints usually hold down sample size.

Time Constraints

If there is a deadline to be met, that in itself may limit the number of observations that are analyzed in a given study. For example, if we want to know the monthly inflation rate of the United States of America for an economic policy decision, we can use only a small number of sample data to calculate the monthly inflation rate in time. (How to use a price index to calculate the inflation rate will be discussed in Chapter 19.)

In general, sampling cost and sampling error are traded off according to the needs of the situation. The greater the accuracy required, the lower the allowable sampling error and the higher the cost of analysis. The issue of trade-offs between sampling cost and sampling error will be analyzed in more detail in Chapter 20.

In addition to the examples discussed so far, we turn to the real-world example of a telephone sampling survey used to find out about the different opinions among Americans and Japanese regarding their trade relationship.

Infoplan/Yankelovich International polled 500 Japanese adults via telephone on January 28 and 29 of 1992; 1,000 American adults were surveyed via phone by Yankelovich Clancy Shulman on January 30. The results of the TIME/CNN sponsored poll that posed questions about how Japanese and Americans feel about each other were published in the February 10, 1992, issue of *TIME*. Sampling errors are plus or minus 4.5 percent for the survey of Japanese and 3 percent for the survey of Americans. Responses of "not sure" were omitted.

Here are the results of 1 of the 5 questions in the *TIME* article.

Which is the main reason for the large trade imbalance between the United States and Japan?
1. Sixty six percent of the Americans and 33 percent of the Japanese surveyed responded that Japan unfairly keeps American products out of the country.
2. "American products are not as good as Japanese products," according to 22 percent of the American respondents and 44 percent of the Japanese.

8.4 SAMPLING DISTRIBUTION OF THE SAMPLE MEAN

In previous chapters we examined population distributions. Now we will examine **sampling distributions** of the sample mean. The sampling distribution is derived from a set of values taken at random from the population. In short, the population distribution represents the distribution of the members of a population, whereas the sample distribution represents the distribution of a sample statistic for certain randomly chosen members of a population.

All Possible Random Samples and Their Mean

The following example shows how to calculate all possible random samples and their mean.

EXAMPLE 8.1 *Sampling Distribution: Three Cases*

Consider the data in Table 8.2, which consist of the numbers of years of work experience for 6 secretaries in Francis Engineering, Inc.
The mean of this population is

$$\mu_X = (1 + 2 + 3 + 4 + 5 + 6)/6 = 3.5$$

A sample mean as indicated in Equation 8.1 can be used to estimate this population mean.

$$\overline{X} = \left(\frac{1}{n}\right) \sum_{i=1}^{n} X_i \tag{8.1}$$

where $X_1, X_2, \ldots, X_n$ denote the sample observations.

Table 8.2 **Work Experience for 6 Secretaries in Francis Engineering, Inc.**

Secretary	Mary	Gerry	Alice	Debbie	Elizabeth	Kimberly
Years of Experience	1	2	3	4	5	6

This example will show how the distribution of the sample mean can be affected by the sample size n as in the cases here.

Case 1: $n = 2$

Table 8.3 shows the possible values for a sample consisting of 2 observations from the above-mentioned population. Table 8.3 indicates that there are 15 possible samples. Because all are equally likely to be selected, the probability that any specific sample will be selected is 1/15. Using this information, we can summarize the probability distribution associated with $\overline{X}$ indicated in Table 8.3 as shown in Table 8.4. Now look at Figure 8.1. Part (a) is the population distribution of work experience for 6 secretaries, which is a uniform distribution. Part (b) shows the sampling distribution of the mean for a sample size of 2; the information on which it is based is taken from Table 8.4. Note the difference between these probability distributions. That in part (b) looks more like the bell shape of the normal distribution. The numbers in the population range from 1 to 6, and the sample means have a more narrow range—from 1.5 to 5.

Table 8.3 Possible Samples and Sample Means ($n = 2$)

Sample	Sample Mean	Sample	Sample Mean
1, 2	1.5	2, 6	4
1, 3	2	3, 4	3.5
1, 4	2.5	3, 5	4
1, 5	3	3, 6	4.5
1, 6	3.5	4, 5	4.5
2, 3	2.5	4, 6	5
2, 4	3	5, 6	5.5
2, 5	3.5		

Table 8.4 Probability Function of X for $n = 2$

$\overline{X}_i$	$P(\overline{X}_i)$
1.5	1/15
2	1/15
2.5	2/15
3	2/15
3.5	3/15
4	2/15
4.5	2/15
5	1/15
5.5	1/15

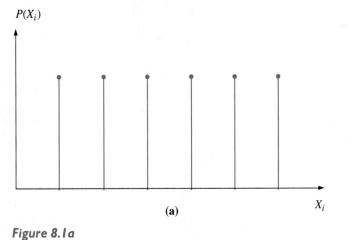

(a)

Figure 8.1a
Population Distribution

Figure 8.1b

Sampling Distribution of Mean Work Experience ($n = 2$)

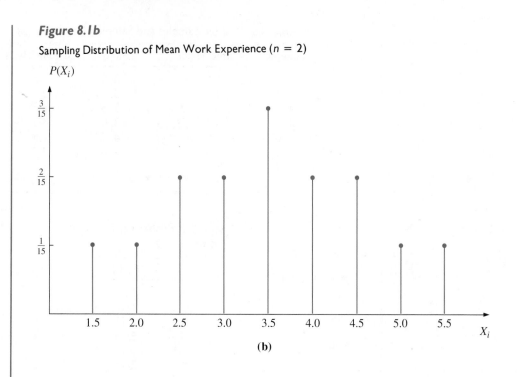

(b)

Case 2: $n = 3$

If the sample size is increased from 2 to 3, then the sample means and probabilities are as shown in Tables 8.5 and 8.6. Comparing Table 8.6 with Table 8.4 reveals that the range of possible values of the sample mean has been reduced from (5.5 − 1.5) to (5 − 2)—that is, from 4 to 3. Note that Figure 8.2 looks more like a bell-shaped normal distribution than Figure 8.1b.

Table 8.5 Possible Samples and Sample Means ($n = 3$)

Sample	Sample Mean	Sample	Sample Mean
1, 2, 3	2	2, 3, 4	3
1, 2, 4	2.33	2, 3, 5	3.33
1, 2, 5	2.67	2, 3, 6	3.67
1, 2, 6	3	2, 4, 5	3.67
1, 3, 4	2.67	2, 4, 6	4
1, 3, 5	3	2, 5, 6	4.33
1, 3, 6	3.33	3, 4, 5	4
1, 4, 5	3.33	3, 4, 6	4.33
1, 4, 6	3.67	3, 5, 6	4.67
1, 5, 6	4	4, 5, 6	5

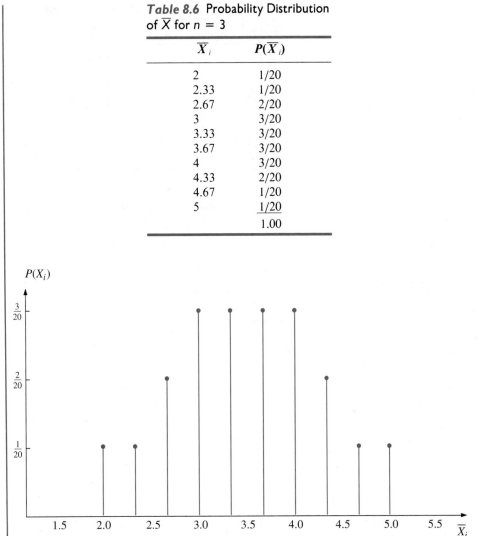

Table 8.6 Probability Distribution of $\overline{X}$ for $n = 3$

$\overline{X}_i$	$P(\overline{X}_i)$
2	1/20
2.33	1/20
2.67	2/20
3	3/20
3.33	3/20
3.67	3/20
4	3/20
4.33	2/20
4.67	1/20
5	1/20
	1.00

Figure 8.2

Sample Distribution of Mean Work Experience for $n = 3$

Case 3: $n = 4$

If the sample size is increased to 4, then the sample means and probabilities are as shown in Tables 8.7 and 8.8. Comparing Table 8.8 with Tables 8.4 and 8.6 reveals that the range of the possible values of the sample mean has been further reduced, from $(5 - 2)$ to $(4.5 - 2.5)$—that is, from 3 to 2. The sampling distribution shown in Figure 8.3 looks almost like a normal distribution.

Table 8.7 Possible Samples and Sample Means ($n = 4$)

Sample	Sample Mean	Sample	Sample Mean
1, 2, 3, 4	2.5	1, 3, 5, 6	3.75
1, 2, 3, 5	2.75	1, 4, 5, 6	4
1, 2, 3, 6	3	2, 3, 4, 5	3.5
1, 2, 4, 5	3	2, 3, 4, 6	3.75
1, 2, 4, 6	3.25	2, 3, 5, 6	4
1, 2, 5, 6	3.5	2, 4, 5, 6	4.25
1, 3, 4, 5	3.25	3, 4, 5, 6	4.5
1, 3, 4, 6	3.5		

$X_1 =$ 1, 2, 3, 4
$X_2 =$ 1, 2, 3, 5
$X_n =$ 3, 4, 5, 6

Table 8.8 Probability Distribution of $\overline{X}$ for $n = 4$

$\overline{X}_i$	$P(\overline{X}_i)$
2.5	1/15
2.75	1/15
3	2/15
3.25	2/15
3.5	3/15
3.75	2/15
4	2/15
4.25	1/15
4.5	1/15
	1.00

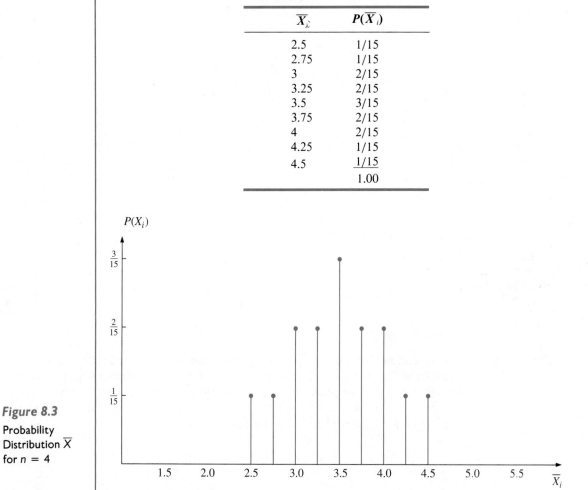

Figure 8.3

Probability Distribution $\overline{X}$ for $n = 4$

In Example 8.1 we saw how the sampling distribution can be identified, how the sample size can affect the variation of a sample mean distribution, and how the sample distribution approaches a bell-shaped normal distribution when sample size increases. It remains to consider how the sample size affects the number of possible sample means and sample variances. Let N be the size of a population with mean μ_X and standard deviation σ_X. A random sample of n observations is drawn from this population, so there are $\binom{N}{n}$ sample means $\overline{X}_i$, and $\binom{N}{n}$ sample variances S_i^2, where $i = 1, 2, \ldots, \binom{N}{n}$. Let's use the information related to Example 8.1 to illustrate this concept.

EXAMPLE 8.2 *Sizes of Sample Means and Their Distributions*

In Example 8.1, $N = 6$, and random samples of size 2, 3, and 4 were used to show how all possible sample means can be calculated when sampling without replacement. The possible numbers of sample means and sample variances for these three alternative samples are

$$\binom{6}{2} = \frac{6!}{2!(6-2)!} = \frac{(6)(5)(4)(3)(2)(1)}{(2)(1)(4)(3)(2)(1)} = 15$$

$$\binom{6}{3} = \frac{6!}{3!(6-3)} = \frac{(6)(5)(4)(3)(2)(1)}{(3)(2)(1)(3)(2)(1)} = 20$$

$$\binom{6}{4} = \frac{6!}{4!(6-4)!} = \frac{(6)(5)(4)(3)(2)(1)}{(2)(1)(4)(3)(2)(1)} = 15$$

If sampling with replacement, the number of samples will be $(6)^2 = 36$, $(6)^3 = 216$, and $(6)^4 = 1296$.

In the next section, we will discuss the concepts of mean and variance analytically for the sample mean distribution in accordance with the results we got in Examples 8.1 and 8.2.

Mean and Variance for a Sample Mean

Example 8.1 shows how a random sample of n observations is drawn from a population with mean μ_X and variance σ_X^2, where the sample members are denoted X_1, $X_2, \ldots, X_n$. The sample mean is obtained from a random sample drawn from the population, so the expected value of the sample mean $\overline{X}$ of Equation 8.1 is the population mean μ_X.[1]

$$\mu_{\overline{X}} = E(\overline{X}) = \mu_X \tag{8.2}$$

[1]This is because

$$E(\overline{X}) = E\left(\frac{1}{n}\sum_{i=1}^{n} X_i\right)$$

$$= \frac{1}{n}[E(X_1) + E(X_2) + \cdots + E(X_n)]$$

$$= \frac{1}{n}(n\mu_X) = \mu_X$$

The variance of the sample mean is equal to the variance of the summation of the individual observations of X divided by the number of observations in the sample. This can be written and simplified as follows:[2]

$$\text{Var}(\overline{X}) = \text{Var}\left[\left(\frac{1}{n}\right)\sum_{i=1}^{n} X_i\right] = \frac{1}{n^2}\text{Var}\left(\sum_{i=1}^{n} X_i\right) = \frac{\sigma_X^2}{n} \tag{8.3}$$

The variance of the sampling distribution of $\overline{X}$ decreases as the sample size n increases. In other words, the more observations in the sample, the more concentrated is the sampling distribution of the sample mean around the population mean, as we saw in Example 8.1. Using Equation 8.3, we find the standard deviation of the sample mean as follows:

$$\sigma_{\overline{X}} = \sigma_X/\sqrt{n} \tag{8.4}$$

Equation 8.2 is applicable to both an infinite sample or a finite sample, with and without replacement. Equation 8.4, however, is applicable only to either an infinite sample or a finite sample with replacement.

Sample Without Replacement from a Finite Sample

In the case of a sample drawn without replacement, it is important to consider the size of the sample relative to the population size N. If the sample size is less than 5 percent of the population ($n \leq .05N$), then Equation 8.4 may be used as it appears here. If the population is large, and if the sample size is larger than 5 percent of the total population ($n > .05N$), then a correction factor must be incorporated into Equation 8.4. When samples are drawn from populations without replacement, each observation can be chosen only once. Therefore, as the available choices for new sample members becomes large, the chance that a given sample member will be chosen is still random, but there is a larger probability of its being chosen than in sampling *with* replacement because fewer members remain in the population. This has been shown to bias sample variance and standard deviation. The bias can be corrected as follows:

---

[2]$X_1, X_2, \ldots, X_n$ are independent of each other, so we can use Equation 6.31 in Chapter 6 to obtain

$$\text{Var}\left(\sum_{i=1}^{n} X_i\right) = \text{Var}(X_1) + \text{Var}(X_2) + \cdots + \text{Var}(X_n)$$
$$= n\sigma_X^2$$

Therefore,

$$\frac{1}{n^2}\text{Var}\left(\sum_{i=1}^{n} X_i\right) = \frac{1}{n^2}(n\sigma_x^2) = \frac{\sigma_X^2}{n}$$

Because σ_X^2 generally is not known, it can be estimated by s_X^2, the sample variance.

$$s_X^2 = \frac{\sum_{i=1}^{n}(X_1 - \overline{X})^2}{n - 1}$$

$$\text{Var}(\overline{X}) = \frac{\sigma_X^2}{n} \cdot \frac{N-n}{N-1} \qquad (8.5)$$

$$\sigma_{\overline{X}} = \frac{\sigma_X}{\sqrt{n}} \cdot \sqrt{\frac{N-n}{N-1}} \qquad (8.6)$$

for all samples where $n > .05N$. Here $(N-n)/(N-1)$ is called the **finite population multiplier**.[3] Equations 8.5 and 8.6 are the variance and standard deviation in cases of finite population.

EXAMPLE 8.3 *Sample Mean Distribution with Samples of Different Size*

Suppose a class has 6 students with the following grade points: 1.5, 2, 3, 3.5, 4, 5. The population mean and standard deviation of this set of data are

$$\mu = (1.5 + 2 + 3 + 3.5 + 4 + 5)/6 = 3.167$$
$$\sigma = [(1.5 - 3.167)^2 + \cdots + (5 - 3.167)^2/6]^{1/2} = (8.334/6)^{1/2} = 1.179$$

If samples of 2 were drawn from this population, 15 combinations would be possible. Fifteen samples of 2 students each and the calculated $\overline{X}$ for each sample are listed in Table 8.9. These sample means are not all 3.167, but they are close to

Table 8.9 All Possible Sample Means and Associated Probabilities

Number of Sample	Combinations of Grade Points	Mean ($\overline{X}_i$)
1	1.5, 2	1.75
2	1.5, 3	2.25
3	1.5, 3.5	2.50
4	1.5, 4	2.75
5	1.5, 5	3.25
6	2, 3	2.50
7	2, 3.5	2.75
8	2, 4	3.00
9	2, 5	3.50
10	3, 3.5	3.25
11	3, 4	3.50
12	3, 5	4.00
13	3.5, 4	3.75
14	3.5, 5	4.25
15	4, 5	4.50
		$E(\overline{X}_i) = 3.167$

[3]We encountered this issue in Chapter 6, where we found that the hypergeometric distribution considered the population size N but the binomial distribution did not. Equation 6.15 can be redefined as

$$\begin{bmatrix} \text{Variance of hypergeometric} \\ \text{random variable} \end{bmatrix} = \begin{bmatrix} \text{Variance of corresponding} \\ \text{binomial random variable} \end{bmatrix} \cdot \begin{bmatrix} \dfrac{N-n}{N-1} \end{bmatrix}$$

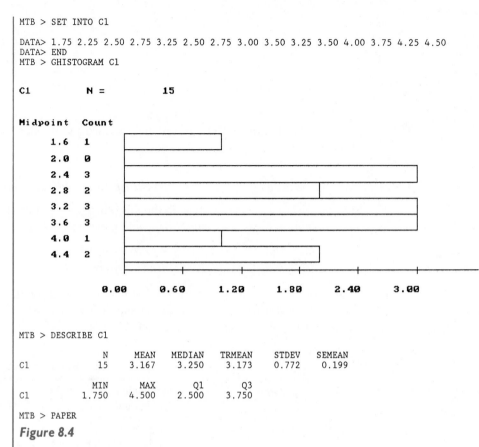

```
MTB > SET INTO C1

DATA> 1.75 2.25 2.50 2.75 3.25 2.50 2.75 3.00 3.50 3.25 3.50 4.00 3.75 4.25 4.50
DATA> END
MTB > GHISTOGRAM C1


C1          N =          15


Midpoint  Count
    1.6   1
    2.0   0
    2.4   3
    2.8   2
    3.2   3
    3.6   3
    4.0   1
    4.4   2

         0.00      0.60      1.20      1.80      2.40      3.00


MTB > DESCRIBE C1
               N      MEAN    MEDIAN    TRMEAN     STDEV    SEMEAN
C1            15     3.167     3.250     3.173     0.772     0.199

              MIN       MAX        Q1        Q3
C1          1.750     4.500     2.500     3.750

MTB > PAPER
```

Figure 8.4

Sample Distribution on Sample Mean ($N = 6$, $n = 2$)

3.167. The mean and standard deviation and other related information for 15 sample means generated by MINITAB are presented in Figure 8.4. Here we learn that (1) the average mean of these 15 values is 3.167 (this is equal to $\mu = 3.167$) and (2) the standard deviation of these 15 values is .772 or .745, which depends on whether $n - 1$ or n is used as the denominator for calculating the standard deviation. Substituting $N = 6$, $n = 2$, and $\sigma_X = 1.179$ into Equation 8.6, we obtain

$$\sigma_{\overline{X}} = \frac{1.179}{\sqrt{2}} \sqrt{\frac{6 - 2}{6 - 1}} = .7456$$

This result also proves that Equation 8.6 holds approximately true. We have proved the $E(\overline{X}) = \mu_X$, indicated in Equation 8.2, holds true.

Now we offer two other examples to show how Equations 8.2, 8.4, and 8.6 can be applied.

Suppose an extremely large population has mean $\mu_X = 90.0$ and standard deviation $\sigma_X = 15.0$. We already know that the expected value of the sample mean is equal to the population mean. Therefore, the sampling distribution of the sample means for a sample size of $n = 25$ has the following parameters:

$$E(\overline{X}) = \mu_X = 90.0$$
$$\sigma_{\overline{x}} = \sigma/\sqrt{n} = 15.0/\sqrt{25} = 15.0/5.0 = 3.0$$

Suppose this time that the population is $N = 50$ firms in an industry. Further, let's assume that the population represents earnings per share (EPS) observations for all firms in a given industry with mean \$10 and standard deviation \$2. A financial analyst takes a random sample of 20 of these firms. Because the sample size $n > .05N$, our estimate of the standard deviation of the sample mean must take the correction factor $(N - n/N - 1)$ into account. We use Equations 8.2 and 8.6 to calculate the sample mean and sample standard deviation.

$$E(\overline{X}) = \mu_X = \$10$$
$$\sigma_{\overline{x}} = (\sigma_X/\sqrt{n})\sqrt{(N - n)/(N - 1)}$$
$$= (2/\sqrt{20})\sqrt{(50 - 20)/(50 - 1)}$$
$$= \$0.35$$

If the population is either normally distributed or large and $n \geq 30$, then the random variable Z is distributed standard normally and is defined as follows:

$$Z = \frac{(\overline{X} - \mu)}{\sigma_X/\sqrt{n}} \tag{8.7}$$

Researchers then can use Equation 8.7 and the standard normal distribution table (Table A3 in Appendix A at the end of the book) to do statistical analysis in terms of $\overline{X}$ and $\sigma_X/\sqrt{n}$.

■ | **APPLICATION 8.3** Probability and Sampling Distributions
                         of Radial Tires' Lives

Suppose there is a population of radial tires whose lives are normally distributed and have a mean of 26,000 miles with a standard deviation of 3,000 miles. A random sample of 36 of these tires was taken and found to have a mean life of 25,000 miles. If the population parameters are correct, what is the probability of finding a sample mean less than or equal to 25,000? Following Section 7.4 of Chapter 7 on the use of the normal area table, and using Equation 8.7, we find that the probability is

$$P(\overline{X} \leq 25,000) = P[(\overline{X} - \mu_X)/\sigma_{\overline{x}} \leq P(25,000 - \mu_X)/\sigma_{\overline{x}}]$$

But, from Equation 8.4, we know that the standard deviation of the sample mean is

$$\sigma_{\bar{X}} = \sigma_X/\sqrt{n} = 3{,}000/\sqrt{36} = 500$$

and

$$P(\bar{X} \le 25000) = P[Z \le (25000 - 26000)/500]$$
$$= P[Z \le -2]$$

Because the distribution of Z is a standard normal distribution, we use Table A3 to calculate the probability.

$$\begin{aligned} P(\bar{X} \le 25{,}000) &= F_Z(-2.0) \\ &= 1 - F_Z(2.0) \\ &= 1 - .9772 \\ &= .0228 \end{aligned}$$

Thus the probability that the sample mean for the life of the radial tires is less than or equal to 25,000 miles is approximately 2.3 percent. The normal probability curves for the Z and $\bar{X}$ statistics for the population distribution are shown in Figure 8.5.

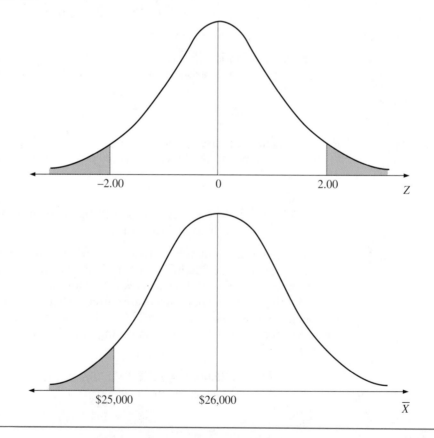

Figure 8.5

The Normal Probability Curve for Z and $\bar{X}$ statistics

To further investigate the relationship between sample size and the sampling distribution of $\overline{X}$, we use the information of Application 8.3. The expected value and standard deviation of the sampling distribution of $\overline{X}$ for 5 different sample sizes can be calculated as follows:

Sample Size	$E(\overline{X}) = \mu_{\overline{x}}$	$\sigma_{\overline{x}} = \sigma/\sqrt{n}$
$n = 1$	26,000	$\dfrac{3,000}{1} = 3,000$
$n = 2$	26,000	$\dfrac{3,000}{\sqrt{2}} = 2121.3407$
$n = 8$	26,000	$\dfrac{3,000}{\sqrt{8}} = 1060.6703$
$n = 16$	26,000	$\dfrac{3,000}{\sqrt{16}} = 750$
$n = 32$	26,000	$\dfrac{3,000}{\sqrt{32}} = 530.3258$

On the basis of this information, 5 different normal distributions with mean 26,000 and 5 different standard deviations are displayed in Figure 8.6. Sample size does not affect the expected value μ of the sample mean, but the standard deviation $\sigma_{\overline{x}}$ of the sample mean becomes smaller when the sample size increases.

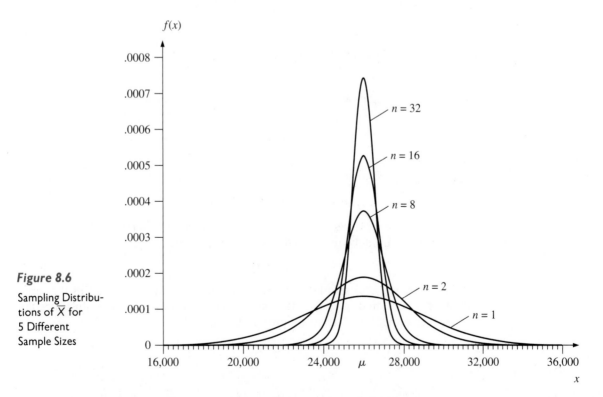

Figure 8.6

Sampling Distributions of $\overline{X}$ for 5 Different Sample Sizes

8.5 *SAMPLING DISTRIBUTION OF THE SAMPLE PROPORTION*

Sometimes in statistical analysis, it is important to estimate the proportion of a certain characteristic in a population. For example, it may be of interest to estimate the proportion of people in New York City who are unemployed or the proportion of students at Rutgers University who favor changing the grading system. The **sample proportion,** $\hat{P}$, is simply the number of sample members X with the specified characteristic divided by the sample size n.

$$\hat{P} = \frac{X}{n} \tag{8.8}$$

The mean and variance of a sample proportion can be derived from the binomial distribution discussed in Chapter 6. Recall that the mean and standard deviation of a binomially distributed random variable X are

$$\mu_X = np \tag{6.12}$$

$$\sigma_X = \sqrt{np(1 - p)} \tag{6.13}$$

where p is the probability of success.

From Equations 6.12 and 6.13, the mean and standard deviation of a sample proportion $\hat{P}$ can be calculated as

$$\mu_{\hat{p}} = \frac{np}{n} = p \tag{8.9}$$

$$\sigma_{\hat{p}} = \frac{\sqrt{np(1 - p)}}{n} = \sqrt{\frac{p(1 - p)}{n}} \tag{8.10}$$

For the same reasons as stated in Section 8.4, we need the finite population correction if $n > .05N$. The corrected variance for large samples (relative to population size) is

$$\sigma_{\hat{p}}^2 = [p(1 - p)/n](N - n)/(N - 1) \tag{8.11}$$

and

$$\sigma_{\hat{p}} = \sqrt{[p(1 - p)/n} \cdot \sqrt{[(N - n)/(N - 1)]} \tag{8.12}$$

Finally, if the sample size is large—say, greater than 30—then the following Z statistic is approximately distributed as standard normal.

$$Z = (\hat{p} - p)/\sigma_{\hat{p}} \tag{8.13}$$

Mean, variance, and standard deviation calculations are performed in the same manner as in Equation 8.7 of section 8.4. The following example illustrates the inferential use of the Z statistic shown in Equation 8.13.

EXAMPLE 8.4   *Calculating the Probability of Defective Chips*

A shipment of 1,000 calculator chips arrives at Kraft Electronics. Suppose the company takes a random sample of 50 chips. The company claims that the proportion of defective chips in this shipment is about 25 percent. Assume the claim is correct. What is the probability that this shipment will contain between 23 percent and 27 percent defective chips?

We begin with the information available and calculate the associated probability with Equation 8.16. The population proportion is $P = .25$, and the related probability can be defined as

$$P(.23 < \hat{P} < .27) = P\left(\frac{.23 - p}{\sigma_{\hat{p}}} < \frac{\hat{P} - p}{\sigma_{\hat{p}}} < \frac{.27 - p}{\sigma_{\hat{p}}}\right)$$

and

$$\sigma_{\hat{p}} = \sqrt{p(1 - p)/n} = \sqrt{(.25)(.75)/50} = .061$$

Therefore,

$$P(.23 < \hat{P} < .27) = P\left(\frac{.23 - .25}{.061} < \frac{\hat{P} - p}{\sigma_{\hat{p}}} < \frac{.27 - .25}{.061}\right)$$

Using the cumulative distribution function $F_Z(Z)$ of the standard normal random variable and the standard normal distribution table (Table A3 in Appendix A), we obtain

$$
\begin{aligned}
P(.23 < \hat{P} < .27) &= P\{-.33 < Z < .33\} \\
&= F_Z(.33) - F_Z(-.33) \\
&= F_Z(.33) - [1 - F_Z(.33)] \\
&= .6293 - [1 - .6293] \\
&= .2586
\end{aligned}
$$

There is a 25.86 percent chance that between 23 percent and 27 percent of the chips in this shipment will be defective.

8.6   THE CENTRAL LIMIT THEOREM

As we found in Section 8.4, the sample means of $\binom{N}{n}$ possible samples have the following properties.

1. If the population is normally distributed, the distribution of the sample mean is normal.

2. If the population is large but not normally distributed—for example, if the distribution is uniform or U-shaped—the distribution of sample mean approaches a normal distribution provided that the sample is large, as indicated in Figure 8.7.[4]

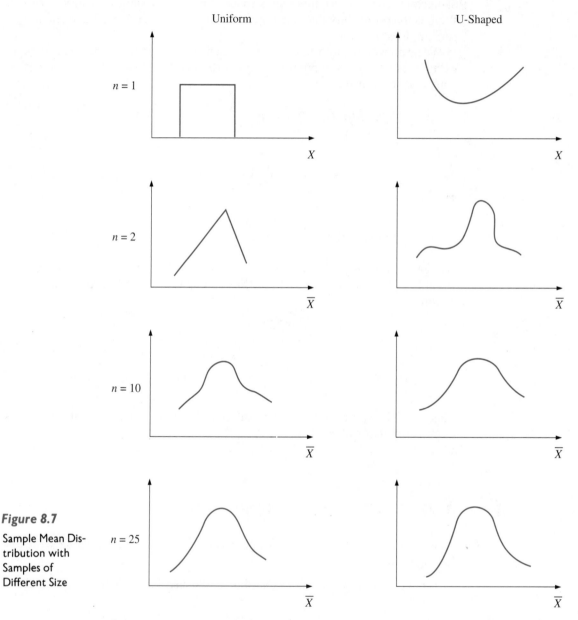

Figure 8.7

Sample Mean Distribution with Samples of Different Size

[4]Random samples from a uniform distribution for sample size $n = 2, 5, 10, 25$, and 50 are presented in Appendix 8A.

Following these results is one of the most important theorems in statistics, the **central limit theorem:**

> As the sample size (n) from a given population gets "large enough," the sampling distribution of the mean, $\overline{X}$, can be approximated by a normal distribution with mean μ and standard deviation $\sigma/\sqrt{n}$, regardless of the distribution of the individual values in the population.

Alternatively, the central limit theorem can be stated in the following way. Let $X_1, X_2, \ldots, X_N$ be independent and identically distributed random variables with mean μ and standard deviation σ. Let $\overline{X}$ represent the sample mean with sample size n. Then, as n becomes large, the distribution of the following Z statistic as indicated in Equation 8.7 approaches the standard normal distribution.

$$Z = \frac{\overline{X} - \mu}{\sigma/\sqrt{n}} \qquad (8.7)$$

Many useful calculations can be made via the central limit theorem. It is worthwhile to know that the central limit theorem can be employed to justify using the normal distribution as an approximation for both binomial and Poisson distributions, as discussed in Chapter 6.

Why is the central limit theorem so important in statistics? It enables us to analyze the means of many different random variables even when we don't know the actual population distributions of these variables. For instance, in Application 8.3 we computed the probability that the mean tire life was less than or equal to 25,000 miles. Even though we assumed that tire life was normally distributed, we could have conducted this analysis without making that assumption simply by using the central limit theorem.

Other possible uses of the central limit theorem include quality control analysis (such as examining the mean number of defective parts in a car, which will be discussed in Chapter 10), investment analysis (such as examining the mean rates of return for stocks, which was discussed in Chapters 3 and 4), and educational analysis (such as examining mean IQ scores).

EXAMPLE 8.5   *Illustrating the Central Limit Theorem*

The distribution of annual earnings of all marketing assistant professors in the United States with five years of experience is skewed negatively, as shown in part (a) of Figure 8.8. This distribution has a mean of $55,000 and a standard deviation of $4,000. Say we draw a random sample of 50 assistant professors of marketing. What is the probability that their annual earnings will average more than $56,500? Part (b) of Figure 8.8 shows the sampling distribution of the mean that will result. It also indicates the area representing "earnings over $56,500."

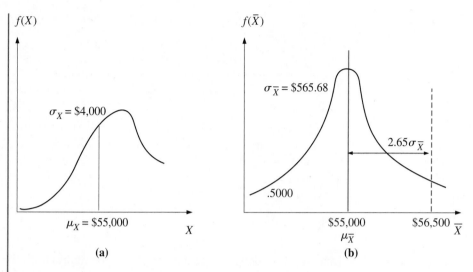

Figure 8.8

(a) Population and (b) Sampling Distributions for Marketing Assistant Professors' Annual Earnings

First we calculate the standard deviation of the mean from the population standard deviation in accordance with Equation 8.4.

$$\sigma_{\overline{X}} = \frac{\sigma}{\sqrt{n}}$$

$$= \frac{\$4,000}{\sqrt{50}}$$

$$= \$565.68$$

From Equation 8.2, we know that

$$E(\overline{X}) = \mu_{\overline{X}} = \mu_X = \$55,000$$

Because the sample mean $\overline{X}$ is normally distributed, we can use Equation 8.7 to calculate

$$Z = \frac{\overline{X} - \mu}{\sigma/\sqrt{n}}$$

$$= \frac{\$56,000 - \$55,000}{\$565.68}$$

$$= 2.65$$

Finally, we use the Z statistics given in Table A3 in Appendix A to obtain the desired probability.

$$P(\overline{X} > \$56,500) = P(Z > 2.65)$$
$$= .5000 - .4960 = .0040$$

We have determined that there is .4 percent chance of average annual earnings being more than $56,500 in a group of 50 assistant marketing professors.

8.7 OTHER BUSINESS APPLICATIONS

■ **APPLICATION 8.4** Audit Sampling

It is possible in accounting to make inferences about an entire large, finite population by drawing samples of size *n* and thus using only a small portion of the data. The information in Table 8.10 on a sample of 30 accounts was taken from the pop-

Table 8.10 Sample of Trade Accounts Receivable Balances

Account Number	Customer Name	Book Amount[a] X_i	Rank by Dollar Size
101	Beekmans, F.M.	$ 195.81	10
102	Morsby, A.F.	152.65	2
103	Sack, I.E.	225.74	25
104	Hoschke, K.R.	190.73	8
105	Hosken, A.J.	207.66	18
106	Manitzky, A.A.	207.57	17
107	Worner, C.J.	210.21	19
108	Walsh, A.	147.75	1
109	Ryland, K.L.	217.73	22
110	Nolde, J.P.	206.47	15
111	Rehn, L.M.	222.12	24
112	Argent, A.	204.26	14
113	Mollison, A.M.	247.35	30
114	Conolly, E.W.J.	230.24	27
115	England, A.G.	198.12	11
116	Brown, C.	220.03	23
117	Luther, E.	216.36	21
118	Sarikas, A.D.	241.62	29
119	Martinez, B.P.	169.53	6
120	Beech, D.F.	228.98	26
121	Bedford, B.A.	159.57	4
122	Apps, A.J.	194.75	9
123	Hamlyn-Harris, T.H.	181.01	7
124	Mangan, M.R.	157.60	3
125	Topel, Z.H.	198.15	12
126	Westaway, W.R.	203.73	13
127	A-Izzedin, T.B.	206.47	16
128	Alrey, R.C.	239.12	28
129	Biment, W.	165.76	5
130	Dimick, M.C.	215.95	20
		$6,063.04	

Source: Andrew D. Bailey, Jr., *Statistical Auditing: Review, Concepts and Problems*, pp. 138–42. Copyright © 1981 by Harcourt Brace Jovanovich, Inc., reprinted by permission of the publisher.

[a]These amounts were generated by using a mean of $200.00, a standard deviation of $30.00, and an assumed normal distribution. Rounding is to the nearest cent.

ulation of 3,000 trade accounts receivable for a given company.[5] Using Equation 8.1, we can calculate the mean of the sample in Table 8.8 as follows:

$$\overline{X} = \left(\frac{1}{n}\right)\sum_{i=1}^{n} x_i$$

$$= \left(\frac{1}{30}\right)[195.81 + 152.65 + \cdots + 215.95] \qquad (8.1)$$

$$= \$202.10$$

Using Equation 8.3b, we can calculate the variance of the sample as follows:

$$s_X^2 = \sum_{i=1}^{n} \frac{(x_i - \overline{x})^2}{n-1}$$

$$= (1/29)[(195.81 - 202.10)^2 + (152.65 - 202.10)^2 + \cdots$$
$$+ (215.95 - 202.10)^2]$$

$$= 719.164$$

Armed with this information and with information on the population standard deviation σ_X, we can make inferences about the population mean.[6] This is achieved by using the same structure as in Table 7.2.

The fact that the population mean is unknown is the motivation for this analysis. By taking a sample of 30 observations from a population of 3,000, the auditor can calculate the mean and standard deviation of this sample. From this type of information, the auditor can use the central limit theorem to determine the possible ranges that should include the true population mean if the population standard deviation is known. Suppose the population standard deviation is $25.560. We can use Equation 8.4 to estimate the standard deviation of the sample mean.[7]

$$S_{\overline{X}} = \sigma_X/\sqrt{n} = 25.56/\sqrt{30} = 4.667$$

The guidelines listed in Table 7.2 give us a rule for determining **confidence intervals.** We can use this rule to make the three confidence-interval statements listed in Table 8.11. For example, the first confidence-interval statement can be expressed as follows: "We can be about 68 percent confident that the population mean μ will fall between $197.43 to $206.77."

---

[5]Andrew D. Bailey, Jr., *Statistical Auditing: Review, Concepts and Problems* (New York: Harcourt, Brace Jovanovich, 1981), pp. 138–42.

[6]If population standard deviation is not available, we can substitute s_X for σ_X, but in this case the Z statistics defined in Equation 8.7 can no longer be used. A different statistic can be used, however. See Section 9.3 for the discussion and application.

[7]If the population standard deviation is not known, then we can use the information on sample mean and sample variance to do a similar analysis. This kind of analysis will be done in Section 9.3.

Table 8.11 Confidence Intervals of Accounts Receivable Population Mean (μ)

Confidence Level	Confidence Interval
68.0%	$197.43 < \mu < $206.77
95.5%	$192.77 < \mu < $211.43
99.7%	$188.10 < \mu < $216.10

APPLICATION 8.5 Patient Waiting Time[8]

Sloan and Lorant (1977) studied the relationship between the length of time patients wait in a physician's office and certain demand and cost factors. They obtained data on typical patient waiting times for 4,500 physicians and reported a mean waiting time of 24.7 minutes and a standard deviation of 19.3 minutes.

Suppose a pediatrician does not have this set of data and has one of the nurses in the office monitor the waiting times for 64 randomly selected patients during the year. Applying the central limit theorem, we know that the sample mean, $\overline{X}$, is approximately normally distributed and that the mean $\mu_{\overline{X}}$ and standard deviation $\sigma_{\overline{X}}$ are

$$\mu_{\overline{X}} = \mu = 24.7 \text{ minutes}$$

$$\sigma_{\overline{X}} = \frac{\sigma}{\sqrt{n}} = \frac{19.3}{\sqrt{64}} = 2.4 \text{ minutes}$$

The chance of the sample mean falling between 18 minutes and 26 minutes can be calculated as follows. Because $\overline{X}$ is normally distributed, we can use Equation 8.7 to calculate

$$Z_1 = \frac{\overline{X} - \mu}{\sigma/\sqrt{n}} = \frac{18 - 24.7}{2.4} = -2.79$$

$$Z_2 = \frac{26 - 24.7}{2.4} = .54$$

By using Table A3 in Appendix A, we can calculate the probability that the sample mean $\overline{X}$ falls between 18 minutes and 26 minutes as

$$P(18 \leq \overline{X} \leq 26) = P(-2.79 \leq Z \leq .54)$$
$$= .4974 + .2054$$
$$= .7028$$

There is about a 70.28 percent chance that the sample mean will fall between 18 and 26 minutes. The pediatrician can use this information to determine how efficiently the office is operating.

---

[8]F. A. Sloan and J. H. Lorant (1977), "The Role of Patient Waiting Time: Evidence from Physicians' Practices," *Journal of Business,* October, 486–507.

Summary

In this chapter, we began our treatment of inferential statistics by discussing the concept of sampling and sampling distributions. Inferential statistics deals with drawing inferences about population parameters by looking at a sample of the population. We considered the costs and benefits of sampling and how to draw a random sample. In addition, we discussed the distribution of the sample mean and one of the most important theorems in statistics, the central limit theorem.

In Chapter 9 we will examine other important continuous distributions. In Chapter 10 we continue our discussion of inferential statistics by introducing the concepts of point estimation and confidence intervals.

Appendix 8A Sampling Distribution from a Uniform Population Distribution

To show how sample size can affect the shape and standard deviation of a sample distribution, consider samples of size $n = 2, 5, 10, 25$ and 50 taken from the uniform distribution shown in Figure 8A.1.

To generate different random samples with different sample sizes, we use the MINITAB random variable generator with uniform distribution. Portions of this output are shown in Figure 8.1b in the text discussion. First we generate 40 random samples with a sample size of 2. Similarly, we generate 40 random samples for $n = 5, n = 10, n = 25$, and $n = 50$.

Forty sample means for sample sizes equal to 2, 5, 10, 25, and 50 are presented in Table 8A.1. Histograms based on the five sets of data given in Table 8A.1 are presented in Figures 8A.2, 8A.3, 8A.4, 8A.5, and 8A.6, respectively. The means associated with Figures 8A.2 to 8A.6 are .4458, .4857, .4776, .48688 and .49650,

Figure 8A.1

Uniform Distribution from 0 to 1

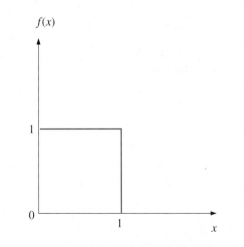

respectively; the standard deviations associated with Figures 8A.2 to 8A.6 are .1927, .1300, .0890, .06235, and .04414. By comparing these five figures, we can draw two important conclusions. First, when sample size increases from 2 to 50, the shape of the histogram becomes more similar to the bell-shaped normal distribution. Second, as the sample size increases, the standard deviation of the sample mean falls drastically. In sum, this data simulation reinforces the central limit theorem discussed in Section 8.6.

Table 8A.1 Sample Means for 5 Different Sample Sizes

	n = 2		*n = 5*		*n = 10*
MTB >	PRINT K1-K40	MTB >	PRINT K1-K40	MTB >	PRINT K1-K40
K1	0.766039	K1	0.547407	K1	0.508664
K2	0.307122	K2	0.631557	K2	0.572561
K3	0.328305	K3	0.263628	K3	0.449149
K4	0.275898	K4	0.503923	K4	0.444761
K5	0.407282	K5	0.372527	K5	0.514358
K6	0.783204	K6	0.676701	K6	0.467277
K7	0.568790	K7	0.345473	K7	0.449323
K8	0.339255	K8	0.579194	K8	0.319622
K9	0.363014	K9	0.697247	K9	0.404897
K10	0.594667	K10	0.337961	K10	0.425953
K11	0.676628	K11	0.565731	K11	0.739907
K12	0.313933	K12	0.412871	K12	0.364374
K13	0.209292	K13	0.364198	K13	0.407843
K14	0.185163	K14	0.409698	K14	0.497028
K15	0.167203	K15	0.403446	K15	0.421893
K16	0.546706	K16	0.444514	K16	0.558162
K17	0.284282	K17	0.589411	K17	0.279482
K18	0.413619	K18	0.489910	K18	0.464799
K19	0.298615	K19	0.372339	K19	0.559232
K20	0.355693	K20	0.751555	K20	0.572471
K21	0.709642	K21	0.594944	K21	0.642801
K22	0.661339	K22	0.522054	K22	0.526308
K23	0.418546	K23	0.513973	K23	0.439639
K24	0.275210	K24	0.311244	K24	0.419171
K25	0.661684	K25	0.525532	K25	0.495180
K26	0.805315	K26	0.449068	K26	0.400065
K27	0.542858	K27	0.462357	K27	0.509161
K28	0.149722	K28	0.491276	K28	0.497628
K29	0.411248	K29	0.731628	K29	0.511615
K30	0.250607	K30	0.243674	K30	0.442089
K31	0.236916	K31	0.514128	K31	0.521398
K32	0.316803	K32	0.579376	K32	0.474808
K33	0.546881	K33	0.660470	K33	0.361446
K34	0.508401	K34	0.604544	K34	0.537571
K35	0.260588	K35	0.384347	K35	0.341068
K36	0.687794	K36	0.321559	K36	0.419929
K37	0.783227	K37	0.294595	K37	0.464406
K38	0.423624	K38	0.408459	K38	0.503744
K39	0.618296	K39	0.508032	K39	0.600979
K40	0.379861	K40	0.545842	K40	0.571873

(continued)

Table 8A.1 (Continued)

	n = 25		n = 50
MTB > PRINT K1-K40		MTB > PRINT K1-K40	
K1	0.477483	K1	0.467134
K2	0.491987	K2	0.486903
K3	0.440873	K3	0.525870
K4	0.529234	K4	0.546720
K5	0.395570	K5	0.491633
K6	0.483190	K6	0.561301
K7	0.496404	K7	0.470143
K8	0.469850	K8	0.580831
K9	0.489090	K9	0.487812
K10	0.553480	K10	0.495207
K11	0.381792	K11	0.527002
K12	0.592549	K12	0.446332
K13	0.456389	K13	0.623809
K14	0.475342	K14	0.425805
K15	0.442597	K15	0.563418
K16	0.403973	K16	0.443062
K17	0.411229	K17	0.478683
K18	0.463406	K18	0.484854
K19	0.611943	K19	0.516529
K20	0.624390	K20	0.507905
K21	0.518310	K21	0.480531
K22	0.453991	K22	0.432330
K23	0.483019	K23	0.539987
K24	0.449016	K24	0.527544
K25	0.503539	K25	0.497368
K26	0.498799	K26	0.524032
K27	0.534755	K27	0.482675
K28	0.547716	K28	0.494472
K29	0.489730	K29	0.516521
K30	0.401362	K30	0.444785
K31	0.431294	K31	0.500245
K32	0.501472	K32	0.434281
K33	0.641570	K33	0.471526
K34	0.436615	K34	0.492110
K35	0.534256	K35	0.473473
K36	0.554134	K36	0.474246
K37	0.433804	K37	0.464606
K38	0.485380	K38	0.545458
K39	0.465627	K39	0.514041
K40	0.420188	K40	0.418817

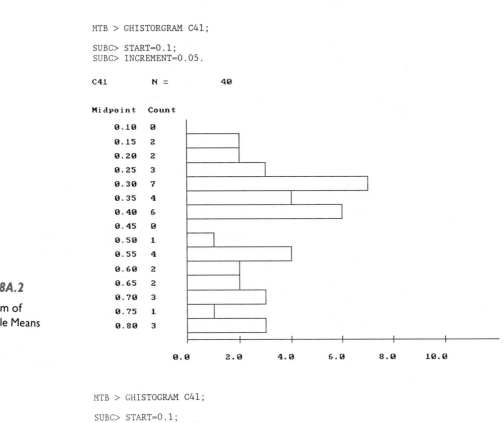

Figure 8A.2

Histogram of
40 Sample Means
(*n* = 2)

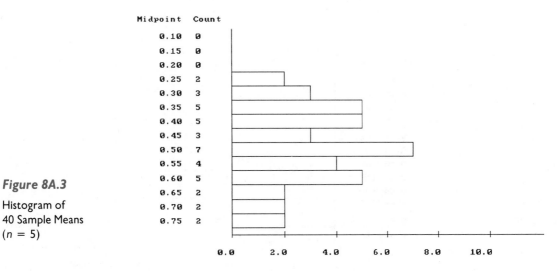

Figure 8A.3

Histogram of
40 Sample Means
(*n* = 5)

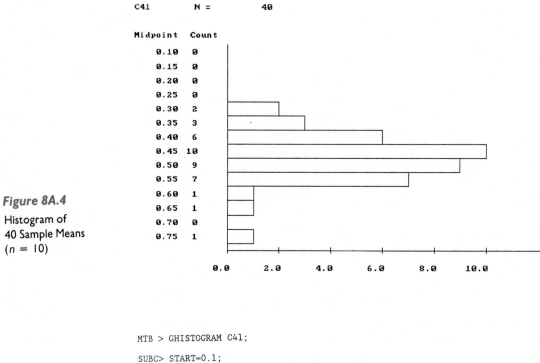

```
MTB > GHISTOGRAM C41;

SUBC> START=0.1;
SUBC> INCREMENT=0.05.


C41        N =         40

Midpoint  Count
    0.10   0
    0.15   0
    0.20   0
    0.25   0
    0.30   2
    0.35   3
    0.40   6
    0.45  10
    0.50   9
    0.55   7
    0.60   1
    0.65   1
    0.70   0
    0.75   1
```

Figure 8A.4

Histogram of
40 Sample Means
($n = 10$)

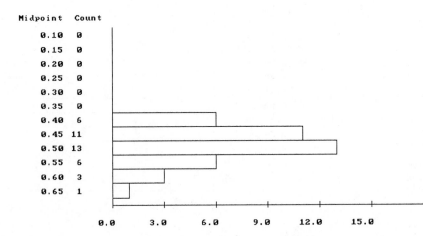

```
MTB > GHISTOGRAM C41;

SUBC> START=0.1;
SUBC> INCREMENT=0.05.


C41        N =         40

Midpoint  Count
    0.10   0
    0.15   0
    0.20   0
    0.25   0
    0.30   0
    0.35   0
    0.40   6
    0.45  11
    0.50  13
    0.55   6
    0.60   3
    0.65   1
```

Figure 8A.5

Histogram of
40 Sample Means
($n = 25$)

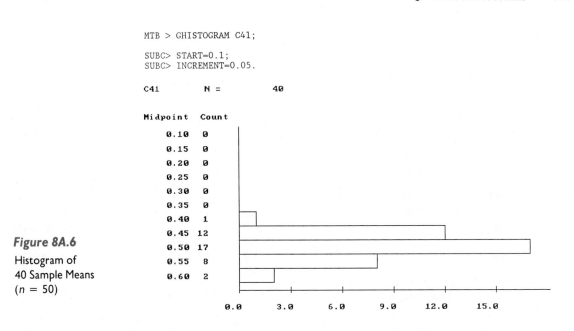

```
MTB > GHISTOGRAM C41;

SUBC> START=0.1;
SUBC> INCREMENT=0.05.


C41          N =          40


Midpoint  Count

  0.10    0
  0.15    0
  0.20    0
  0.25    0
  0.30    0
  0.35    0
  0.40    1
  0.45    12
  0.50    17
  0.55    8
  0.60    2
```

Figure 8A.6

Histogram of
40 Sample Means
($n = 50$)

Questions and Problems

1. A grocery store sells an average of 478 loaves of bread each week. Sales (X) are normally distributed with a standard deviation of 17.

    a. If a random sample of size $n = 1$ (week) is drawn, what is the probability that the $\overline{X}$ value will exceed 495?

    b. If a random sample of size $n = 4$ (weeks) is drawn, what is the probability that the $\overline{X}$ value will exceed 495?

    c. Why does your response in part (a) differ from that in part (b)?

2. A random variable s measures the daily balances in customers' savings accounts. It is normally distributed, with a mean of $\mu_s = \$108$ and a standard deviation of $\$15$.

    a. If a random sample of size $n = 4$ is drawn, what is the probability that the s_4 value exceeds $\$116$?

    b. If a random sample of size $n = 16$ is drawn, what is the probability that the s_{16} value exceeds $\$116$?

    c. What happened to the standard deviation of $\overline{s}$ when the sample size increased from $n = 4$ to $n = 16$?

    d. What happened to the probability of observing $\overline{s} \geq \$116$ as the sample size increased from $n = 4$ to $n = 16$?

3. On average, a book distributor fills orders for 1,000 books per day. If daily orders are normally distributed and the standard deviation is 100, what is the probability that a 5-day average will be between 900 and 1,100 books?

4. A company makes a pastry called a chocco. During the manufacturing process, the individual choccos are placed in a baking oven. The time it takes to bake them is normally distributed around a mean of 64 minutes with a standard deviation of 5 minutes. Thus distribution of the population of baking times is normal in shape.

    a. When choccos are baked, what is the probability that the mean baking time of 4 choccos will be 64 minutes and 45 seconds or longer?

    b. What proportion of the individual choccos bake in 57 minutes or less?

5. In a large group of corporate executives, 20 percent have no college education, 10 percent have exactly 2 years of college, 20 percent have exactly

4 years, and 50 percent have 6 years. A sample size of 2 (with replacement) is to be taken from this population. Find the sampling distribution of the mean number of years in college of the executives in the sample.

6. Out of 10 pay telephones located in a municipal building, 2 phones are to be picked at random, with replacement, for a study of phone use. The actual usage of the phones on a particular day is shown in the accompanying table.

Number of Calls	Number of Phones with This Number of Calls
10	2
12	5
16	3

a. Find the sampling distribution of the average number of calls per phone in the sample of 2 phones.

b. Find the variance of this distribution.

7. To demonstrate the central limit theorem, draw 100 samples of size 5 from a random-number table and calculate the sample mean for each of the 100 samples. Construct a frequency distribution of sample means. Do the same for 100 samples of size 10 and compare the two frequency distributions. Does the central limit theorem appear to be working?

8. The accompanying probability density function is a uniform distribution showing that a certain delicate new medical device will fail between 0 and 10 years after it is implanted in the human body. The

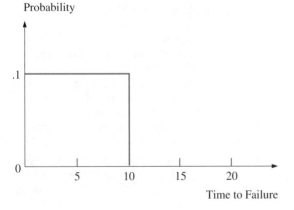

Time to Failure

mean time to failure is $\mu = 5$ years, and the standard deviation is $\sigma = 2.88$ years.

a. Verify that the total area beneath this density function is 1.0.

b. Find the probability that an individual device will fail 8 years or more after implantation.

c. Find the probability that in a sample of $n = 36$ of these devices, the sample mean time of failure $\bar{x}$ will be 8 years or less.

9. The daily catch of a small tuna-fishing fleet averages 130 tons. The fleet's logbook shows that the weight of the catch varies from day to day, and this variation is measured by the standard deviation of the daily catch, $\sigma = 42$ tons. What is the probability that during a sample of $n = 36$ fishing days, the total weight of the catch will be 4,320 tons or more?

10. A type of cathode ray tube has a mean life of 10,000 hours and a variance of 3,600. If we take samples of 25 tubes each, and for each sample we find the mean life, between what limits (symmetric with respect to the mean) will 50 percent of the sample means be expected to lie?

11. The population of times measured by 3-minute egg timers is normally distributed with $\mu = 3$ minutes and $\sigma = .2$ minutes. We test samples of 25 timers. Find the time that would be exceeded by 95 percent of the sample means.

12. A light bulb manufacturer claims that 90 percent of the bulbs it produces meet tough new standards imposed by the Consumer Protection Agency. You just received a shipment containing 400 bulbs from this manufacturer. What is the probability that 375 or more of the bulbs in your shipment meet the new standards? (*Hint:* Use the continuity approximation.)

13. At the beginning of every decade, the U.S. government conducts a census. Why does it take a census? What are the advantages of a census over a sample?

14. Briefly explain the relationship between inferential statistics and sampling.

15. Suppose a town consists of 2,000 people, 1,100 of whom are registered voters. You are interested in how the people in this town will vote on a bond issue. What group constitutes the population? Give an example of a sample from this population.

16. What is sampling error? Give an example. Is there any way to eliminate sampling error?

17. What is nonsampling error? Give an example. Is there any way to eliminate nonsampling error?

18. State whether each of the following represents sampling or nonsampling error.
    a. The sample weight of newborn babies is taken with a scale that is inaccurate by 1 pound.
    b. Sample data suggest that the price of a new home in New Jersey is $150,000 when the actual new home price is $180,000.
    c. A movie theater owner asks the first 100 people leaving the theater whether they liked the movie. By chance, however, the first 100 people to leave are all women. (This in itself may say something about the movie!)

19. What is a representative sample? Why is getting a representative sample important?

20. Briefly explain the relationship between sampling cost and sampling error. Give some examples of sampling costs.

21. The mean life of light bulbs produced by the Brite Lite Bulb Company is 950 hours with a standard deviation of 225 hours. Assume that the population is normally distributed. Suppose you take a random sample of 12 light bulbs.
    a. What is the mean of the sample mean life?
    b. What is the standard deviation of the sample mean?

22. Suppose the mean amount of money spent by students on textbooks each semester is $175 with a standard deviation of $25. Assume that the population is normally distributed. Suppose you take a random sample of 25 students.
    a. What is the mean of the sample mean amount spent on textbooks?
    b. What is the standard deviation of the sample mean?

23. The Better Health Cereal Company produces Healthy Oats cereal. The true mean weight of a box of cereal is 24 ounces with a standard deviation of 1 ounce. Assume the population is normally distributed. Suppose you purchase 8 boxes of cereal.
    a. What is the mean of the sample mean weight?
    b. What is the standard deviation of the sample mean?

24. Explain the relationship between a probability distribution and a sampling distribution.

25. Suppose the average time a customer waits at the check-out line in a grocery store is 12 minutes with a standard deviation of 3 minutes. If you take a random sample of 5 customers, what is the probability that the average check-out time will be at least 10 minutes? What is the mean of the sample check-out time? What is the standard deviation of the sample mean?

26. Historically, 65 percent of the basketball players from Slam Dunk University graduate in 4 years. If a random sample of 50 former players is taken, what proportion of the samples is likely to have at least 25 basketball players graduating?

27. A coffee machine is set so that it dispenses a normally distributed amount of coffee with a mean of 6 ounces and a standard deviation of .4 ounces. Samples of 12 cups of coffee are taken. What is the probability that the sample means will be more than 6.2 ounces?

28. Suppose that historically, 61 percent of the companies on the NYSE have prices that go up each year. If random samples of 100 stocks are taken, what proportion of samples is likely to have between 55 percent and 65 percent of stock prices going up.

29. Suppose the mean life for a company's batteries is 12 hours with a standard deviation of 3 hours. If you take a sample of 20 batteries, what is the standard deviation of the sampling distribution of the mean?

30. The mean useful life of Better Traction tires is 40,000 miles with a standard deviation of 4,000 miles. If you purchase 4 of these tires for your car, what is the probability that the mean useful life of the 4 tires is less than 35,000 miles?

31. The mean interest rate of 500 money market mutual funds is 7.98 percent with a standard deviation of 1.01 percent. Suppose you draw a sample of 25 mutual funds.
    a. What is the mean of the sample mean rate?
    b. What is the variance of the sample mean?
    c. What is the probability that this sample will have a mean rate above 8.2 percent?

32. Of 500 students in a high school, 72 percent have indicated that they are interested in attending college. What is the probability of selecting a random sample of 50 students wherein the sample proportion indicating interest in college is greater than 80 percent?

33. The professor in a statistics course takes a random sample of 100 students from campus to determine the number in favor of multiple-choice tests. Suppose that 50 percent of the entire college population are actually in favor of the multiple-choice test. What is the probability that more than 50% of the students sampled will favor the multiple-choice test?

34. Suppose 40 percent of the students in Genius High School scored above 650 on the math portion of the SAT. What is the probability that more than 50 percent of a random sample of 150 students will score less than 650?

35. The Sorry Charlie Tuna Company produces canned tuna fish. The true mean weight of a can of tuna is 6 ounces with a standard deviation of 1 ounce. Assume the population is normally distributed, and suppose you purchase 9 cans of tuna.
    a. What is the mean of the sample mean weight?
    b. What is the variance of the sample mean?

36. Suppose the time a customer waits in line at a bank averages 8 minutes with a standard deviation of 2 minutes. In a random sample of 5 customers, what is the probability that the average time in line will be at least 10 minutes? What is the mean of the sample waiting time? What is the standard deviation of the sample mean?

37. In Freeport High School, 40 percent of the seniors who are eligible to vote indicated that they plan to vote in the upcoming election. What is the probability of selecting a random sample of 400 students with a sample proportion of voting greater than 35 percent?

38. A credit card company accepts 70 percent of all applicants for credit cards. A random sample of 100 applications is taken.
    a. What is the probability that the sample proportion of acceptance is between .60 and .80?
    b. What is the probability that the sample proportion is greater than .75?
    c. What is the probability that the sample proportion is less than .65?

39. From past history, a bookstore manager knows that 25 percent of all customers entering the store make a purchase. Suppose 200 people enter the store.
    a. What is the mean of the sample proportion of customers making a purchase?
    b. What is the variance of the sample proportion?
    c. What is standard deviation of the sample proportion?
    d. What is the probability that the sample proportion is between .25 and .30?

40. Suppose 60 percent of the members in a lifeguards' union favor certification tests for lifeguards. If a random sample of 100 lifeguards is taken, what is probability that the sample proportion in favor of certification tests is greater than 70 percent?

41. A bank knows that its demand deposits are normally distributed with a mean of $1,122 and a standard deviation of $393. A random sample of 100 deposits is taken.
    a. What is the probability that the sample mean will be greater than $1,000?
    b. Compute the mean of the sample mean demand deposits.
    c. Compute the variance of the sample mean.

42. A company claims that its accounts receivable follow a normal distribution with a mean of $500 and a standard deviation of $75. An auditor will certify the bank's claim only if the mean of a random sample of 50 accounts lies within $25 of the mean. Assume that the bank has accurately reported its mean accounts receivable. What is the probability that the auditor will certify the bank's claim?

43. Consider the members of a group with ages 23, 19, 25, 32, and 27. If a random sample of 2 is to be taken without replacement, what is the sampling distribution for their mean age? What is the mean and variance for the distribution?

44. Answer question 31 again, assuming that the sample is taken with replacement.

45. Consider a population of 6 numbers, 1, 2, 3, 4, 5, and 6. What is the mean of this population? Suppose you roll a pair of dice. Construct a table showing the different possible combinations of the 2 numbers you will obtain. Construct a probability function for this sample. Find the mean of the sample.

46. Answer question 45 again, assuming that the sample is taken from a population of 4 numbers, 1, 2, 3, and 4.

47. Answer question 45 again, assuming that the sample is taken from a population of 3 numbers, 1, 2, and 3.

48. Compute the values for $\binom{N}{n}$ if
    a. $N = 5, n = 2$
    b. $N = 6, n = 3$
    c. $N = 6, n = 2$
    d. $N = 4, n = 3$

49. Why are we interested in the sample mean and its distribution?

50. Consider the members of a weight-loss group who weigh 225, 231, 195, 184, and 131 pounds. If a simple random sample is to be taken without replacement, what is the sampling distribution for their mean weight? What are the mean and variance for the distribution?

51. Suppose there are 2,000 members in a construction workers' union and 40 percent of the members favor ratifying the union contract. If a random sample of 100 construction workers is taken, what is the probability that the sample proportion in favor of ratifying the contract is greater than 50 percent?

52. From past history, a service manager at Honest Abe's Auto Dealership knows that 35 percent of all customers entering the dealership will have service work done that is under warranty. Suppose 200 people enter the dealership for service work on their cars.
    a. What is the mean of the sample proportion of customers having work done that is covered by the warranty?
    b. What is the variance of the sample proportion?
    c. What is standard error of the sample proportion?
    d. What is the probability that the sample proportion is between .25 and .40?

53. A quality control engineer knows from past experience that the mean weight for ball bearings is 7.4 ounces with a standard deviation of 1.2 ounces. Suppose the engineer draws a random sample of 20 ball bearings. What is the probability that the mean of the sample will be greater than 8.0 ounces?

54. Suppose you take an ordinary deck of 52 cards and randomly select 5 cards without replacement. How many different combinations of cards can you have?

55. Suppose you draw 3 balls without replacement from a bag of balls numbered 1 to 10. How many different possible combinations can you have?

56. Suppose the ages of members of a senior citizens' bridge club are 63, 71, 82, 60, 84, 75, 77, 65, and 70.
    a. Compute the population mean and standard deviation for the age of the bridge club members.
    b. If you were to select a sample of 4 members from the bridge club, how many possible samples could you select?

57. Use the information given in question 43 to randomly select 5 samples of 4 people and determine the mean and standard deviation for each sample.

58. In each of the following cases, find the mean and standard deviation of the sampling distribution for the sample mean, for a sample of size n from a population with mean μ and standard deviation σ.
    a. $n = 5, \mu = 10, \sigma = 2$
    b. $n = 10, \mu = 10, \sigma = 3$
    c. $n = 10, \mu = 5, \sigma = 3$
    d. $n = 20, \mu = 5, \sigma = 2$

59. Suppose the cost of sampling is 50 cents per observation. If the population has zero variance, how large a sample should be collected to estimate the mean of the population?

60. Suppose you would like to randomly select 4 of the following 6 companies for a study: IBM, Apple Computer, AT&T, MCI, Ford, and Chrysler. What is the probability that Apple Computer will be in the sample? What is the probability that at least one company from the auto industry, one from the computer industry, and one from the telecommunications industry will be included in the sample?

61. Suppose a population is normally distributed. What is the probability that the sample mean will be less than the population mean?

62. Suppose an obstetrician knows from past experience that the mean weight of a newborn baby is 7.5 pounds with a standard deviation of 2 pounds. The doctor randomly chooses 5 newborn babies.

What is the expected value of the sample mean weight? What is the expected value of the sample mean variance?

63. Review the information given in question 62. What is the probability that a sample of 50 babies will have a mean weight greater than 8 pounds?

64. A cigarette manufacturer came up with a new brand of cigarettes called Long Life. The nicotine content of the cigarettes follows a normal distribution with a mean of 20 and a standard deviation of 5. A consumer bought a pack of Long Life that contains 25 cigarettes. Consider these 25 cigarettes as a random sample.
    a. What is the probability that a cigarette contains over 23 units of nicotine?
    b. What is the probability that the average nicotine content for the whole pack of cigarettes is higher than 23?

65. Table 8.4 shows the probability distribution of X for $n = 2$. Show that the average of the random variable X is 3.5. What is the standard deviation of X?

66. Assume the tips received by 5 waitresses in a given weeknight are $25, $27, $28, $29, and $30. We draw 2 numbers randomly and take the average. Write the probability distribution of the sample mean. What are the expected value and standard deviation of the sample mean?

67. In a big university, 70 percent of the faculty members like to give plus and minus grades (such as B plus and C minus). The other 30 percent of the faculty members do not like the plus and minus system. The school newspaper randomly surveyed 200 faculty members for their opinions. What is the probability that more than half of the faculty members interviewed will be in favor of plus and minus grades? What is the expected number of faculty interviewed who answer the question positively?

68. Assume that the amount of milk in a 16-ounce bottle follows a normal distribution with a mean of 16 and a standard deviation of 1. A consumer protection agency bought 30 bottles of milk and weighed them. What is the probability that the average weight of these 30 bottles of milk falls between 15.9 and 16.1 ounces?

69. If, in question 68, 90 percent of the bottles contain more than 16 ounces of milk, what is the proba-

bility that fewer than 3 of the 30 bottles that the agency bought contain more than 16 ounces of milk?

70. The newly produced 1992 Honda boasts 45 miles per gallon on the highway. Assume that the distribution of the miles per gallon is a normal distribution with a mean of 40 and a standard deviation of 5. The Environmental Protection Agency randomly draws 100 1992 Hondas to test-drive.
    a. What is the probability that a certain car can achieve 45 miles per gallon?
    b. What is the probability that the average of 100 cars exceeds 45 miles per gallon?

71. In question 70, what is the probability that of the 100 cars test-driven, more than 35 cars get more than 45 miles per gallon? How many of the 100 cars tested would you expect to get more than 45 miles per gallon?

72. The National Treasury Bank wants to approve, at random, 2 of 5 loan applications that have been submitted. The loan amounts are $5,000, $8,000, $9,000, $10,000, and $12,000. Obtain the sampling distribution of average loans. What is the expected amount of loans?

73. Recently the State Education Department of New Jersey wanted to determine the competence in math of the state's fourth-grade students. Assume that 20 percent of the students are actually incompetent in mathematics. A test was given to 120 fourth-grade students in New Jersey. What is the probability that at least 20 percent of the students who took the test failed it?

74. Assume that 80 percent of the employees are union members, whereas 20 percent are not. In the last year, 100 of 500 employees were randomly selected to receive a working bonus. If the company does not discriminate against the union members, what is the probability that 30 or more bonus recipients are union members?

75. Suppose the sampling distribution of a sample mean that was developed from a sample of size 40 has a mean of 20 and a standard deviation of 10. Assuming that the population exhibits a normal distribution, find the mean and standard deviation of the population distribution.

76. A natural food company is marketing a new yogurt that it advertises as having only half the fat

of regular yogurt. The average amount of fat in a cup of regular yogurt is 1 unit. The Food and Drug Administration has asked us to investigate the product to see whether the company has engaged in false advertising. The test results are as follows:

Amount of Yogurt Tested	400 Cups
Average amount of fat contained per cup	.52 units
Number of cups containing more than half the fat of regular yogurt	12 cups
Standard deviation of the amount of fat	.2 units

What is the probability of our observing .52 units of average fat, as shown in the report if the population average fat is .5, as stated in the advertisement?

77. On the basis of your answer to question 76 do you believe the advertisement is accurate?

78. The company in question 76 further claims that only 2 percent of the cups contain more than half the fat of regular yogurt. What is the probability of our seeing more than 12 cups out of 400 (which is what we saw in the report) that contain more than half the fat of regular yogurt?

79. In a game, a player rolls two dice and counts how many points he gets between them. Write out the sampling distribution.

80. What are the expected value and standard deviation of the random variable generated in question 79?

81. In a local factory, 20 percent of the assembly line workers make $5 per hour and 80 percent earn $8 per hour. The union computes the mean hourly wage by randomly drawing 5 workers. Write the sampling distribution for the 5 workers' average wage.

82. Write out the sampling distribution for rolling a die and flipping a coin.

83. Suppose you play a game in which you flip three coins. If the flip is a head, you receive 1 point; if the flip is a tail, you receive 2 points. Write out the sampling distribution. What are the expected value and standard deviation of this random variable?

84. Suppose you draw 2 cards from a standard 52-card deck with replacement. Write out the sampling distribution for the suit drawn.

85. The MINITAB output in the figure (see below and page 332) is 20 random samples drawn from a uniform distribution between 0 and 1. Calculate the sample means and sample standard deviations by using the MINITAB program.

86. Use MINITAB to draw histograms for both the sample means and the sample standard deviations, which have been calculated in question 85. Explain the results.

87. Use the results you got in question 85 to plot sample means against sample standard deviation. What is the probability of the range, the sample means between .45 and .55, and the sample standard deviation between .2 and .35?

MINITAB for Question 85

```
MTB > RANDOM 10  C1;
SUBC> UNIFORM A=0 B=1.
MTB > RANDOM 10 C2;
SUBC> UNIFORM A=0 B=1.
MTB > RANDOM 10 C3;
SUBC> UNIFORM A=0 B=1.
MTB > RANDOM 10 C4;
SUBC> UNIFORM A=0 B=1.
MTB > RANDOM 10 C5;
SUBC> UNIFORM A=0 B=1.
MTB > RANDOM 10 C6;
SUBC> UNIFORM A=0 B=1.
MTB > RANDOM 10 C7;
SUBC> UNIFORM A=0 B=1.
MTB > RANDOM 10 C8;
SUBC> UNIFORM A=0 B=1.
MTB > RANDOM 10 C9;
SUBC> UNIFORM A=0 B=1.
MTB > RANDOM 10 C10;
SUBC> UNIFORM A=0 B=1.
MTB > RANDOM 10 C11;
SUBC> UNIFORM A=0 B=1.
MTB > RANDOM 10 C12;
SUBC> UNIFORM A=0 B=1.
MTB > RANDOM 10 C13;
SUBC> UNIFORM A=0 B=1.
MTB > RANDOM 10 C14;
SUBC> UNIFORM A=0 B=1.
MTB > RANDOM 10 C15;
SUBC> UNIFORM A=0 B=1.
MTB > RANDOM 10 C16;
SUBC> UNIFORM A=0 B=1.
MTB > RANDOM 10 C17;
SUBC> UNIFORM A=0 B=1.

MTB > RANDOM 10 C18;
SUBC> UNIFORM A=0 B=1.
MTB > RANDOM 10 C19;
SUBC> UNIFORM A=0 B=1.
MTB > RANDOM 10 C20;
SUBC> UNIFORM A=0 B=1.
MTB > PAPER
```

(continued)

MINITAB for Question 85 (Continued)

```
MTB > PRINT C1-C20
```

ROW	C1	C2	C3	C4	C5	C6
1	0.452103	0.090459	0.917629	0.840114	0.886693	0.085789
2	0.512886	0.307387	0.703627	0.014282	0.836597	0.723605
3	0.110689	0.423400	0.953333	0.785233	0.574623	0.450559
4	0.836084	0.924940	0.166688	0.154176	0.827895	0.319901
5	0.510518	0.617455	0.835978	0.271938	0.486853	0.987574
6	0.814796	0.181896	0.497286	0.992212	0.856678	0.446778
7	0.849544	0.737038	0.160768	0.026492	0.084723	0.847206
8	0.192978	0.129801	0.095957	0.311544	0.590433	0.900752
9	0.122310	0.225083	0.994614	0.943033	0.804101	0.593980
10	0.288724	0.135341	0.326721	0.879094	0.512686	0.247515

ROW	C7	C8	C9	C10	C11	C12
1	0.939411	0.451744	0.465591	0.029541	0.042475	0.517031
2	0.426426	0.468048	0.198877	0.692686	0.309432	0.628847
3	0.303265	0.505990	0.859649	0.585765	0.679014	0.605836
4	0.908147	0.248761	0.456120	0.220587	0.876751	0.729525
5	0.518403	0.095114	0.014995	0.573374	0.593840	0.190625
6	0.800419	0.889203	0.874372	0.671701	0.230036	0.828161
7	0.052330	0.150371	0.296517	0.962609	0.754529	0.520113
8	0.541236	0.796337	0.064682	0.326162	0.316114	0.014101
9	0.654557	0.542174	0.085250	0.770243	0.514209	0.762658
10	0.819614	0.771725	0.656236	0.280340	0.276136	0.332268

ROW	C13	C14	C15	C16	C17	C18
1	0.533555	0.598764	0.433456	0.695772	0.127340	0.239718
2	0.694326	0.845469	0.181957	0.971466	0.917523	0.964697
3	0.790775	0.683627	0.744593	0.433288	0.690350	0.587096
4	0.846856	0.453381	0.074162	0.161036	0.293717	0.387018
5	0.856987	0.672611	0.270205	0.129440	0.714609	0.377255
6	0.123392	0.076380	0.775624	0.179977	0.326133	0.156850
7	0.423996	0.547456	0.952948	0.497115	0.766577	0.606230
8	0.999550	0.431995	0.118542	0.139355	0.822129	0.778788
9	0.943718	0.999388	0.817773	0.419336	0.766144	0.348529
10	0.964790	0.923468	0.221566	0.417019	0.767993	0.566177

ROW	C19	C20
1	0.772133	0.680139
2	0.516598	0.017388
3	0.574711	0.173483
4	0.838847	0.685389
5	0.855897	0.673635
6	0.987180	0.204366
7	0.397559	0.545713
8	0.694890	0.214155
9	0.861228	0.769382
10	0.653441	0.172694

CHAPTER 9

Other Continuous Distributions and Moments for Distributions

Key Terms

uniform distribution	exponential distribution
Student's *t* distribution	moments
degree of freedom	coefficient of variation
chi-square distribution	coefficient of skewness
F distribution	coefficient of kurtosis
F variable	

9.1 INTRODUCTION

Two very useful continuous distributions, the normal and lognormal distributions, were discussed in Chapter 7. Because many random variables have distributions that are not normal, in this chapter we explore five other important continuous distributions and their applications. These five distributions are the uniform distribution, Student's t distribution, the chi-square distribution, the F distribution, and the exponential distribution. All are directly or indirectly used in analyzing business and economic data. The relationship between moments and distributions is also discussed in this chapter. Finally, we explore business applications of statistical distributions in terms of the first four moments for stock rates of return.

9.2 THE UNIFORM DISTRIBUTION

The simplest continuous probability distribution is called the **uniform distribution.** This probability distribution provides a model for continuous random variables that are evenly (or randomly) distributed over a certain interval. To picture this distribution, assume that the random variable X can take on any value in the range from, for example, 5 to 15, as indicated in Figure 9.1. In a uniform distribution, the probability that the variable will assume a value within a given interval is proportional to the length of the interval. For example, the probability that X will assume a value in the range from 6 to 8 is the same as the probability that it will assume a value in the range from 9 to 11, because these two intervals are equal in length.

The uniform distribution has the following probability density function:

$$f(X) = \begin{cases} \dfrac{1}{b-a} & \text{if } a \le X \le b \\ 0 & \text{elsewhere} \end{cases} \tag{9.1}$$

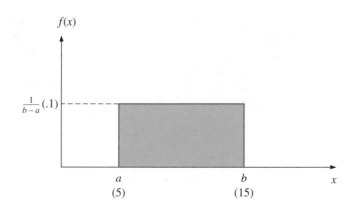

Figure 9.1

The Uniform Probability Distribution

If the foregoing condition holds, then X is uniformly distributed, and the shape under the density function forms a rectangle, as shown in Figure 9.1. The rectangle's area is equal to 1, which means that X is sure to take on some value between $a = 5$ and $b = 15$. Mathematically, we can express this as $P(5 \leq X \leq 15) = 1$.

Figure 9.1 shows a density function for a set of values between a and b. Each density is a horizontal line segment with constant height $1/(b - a)$ over the interval from a to b. Outside the interval, $f(X) = 0$. This means that for a uniformly distributed random variable X, values below a and values above b are impossible. Substituting $b = 15$ and $a = 5$ into Equation 9.1, we obtain $1/(b - a) = 1/(15 - 5) = .1$, as indicated in Figure 9.1.

From Chapters 5 and 7, we know that the probability that X will fall below a point is provided by the area under the density curve and to the left of that point. In other words, the cumulative probability distribution function, $P(X \leq x) = (x - a)/(b - a)$, is represented by this area. The cumulative function for values of X between a and b is the area of the rectangle, which, again, is found by multiplying the height, $1/(b - a)$, times the base, $x - a$. To the left of a, the cumulative probabilities must be zero, whereas the probability that X lies "below points beyond b" must be 1.

The cumulative probabilities for a uniform distribution are

$$P(X \leq x) = \begin{cases} 0 & \text{if } x < a \\ \dfrac{x - a}{b - a} & \text{if } a \leq x \leq b \\ 1 & \text{if } x > b \end{cases} \tag{9.2}$$

Figure 9.2 shows the cumulative distribution function in terms of data indicated in Figure 9.1. It presents the cumulative probabilities for $x = 5$, $x = 10$, $x = 15$, and $x = 20$ at points A, B, C, and D, respectively. Cumulative probabilities for these three points can be calculated as follows.

At point A: $\quad P(X \leq 5) = \dfrac{5 - 5}{15 - 5} = 0$

At point B: $\quad P(X \leq 10) = \dfrac{10 - 5}{15 - 5} = \dfrac{1}{2}$

At point C: $\quad P(X \leq 15) = \dfrac{15 - 5}{15 - 5} = 1$

At point D: $\quad P(X \leq 20) = P(X \leq 15) + P(15 \leq X \leq 20) = 1 + 0 = 1$

The mean and standard deviation of a uniform distribution (see Appendix 9A) can be shown as

$$\mu_X = E(X) = \frac{a + b}{2}$$
$$\sigma_X = \frac{b - a}{\sqrt{12}} \tag{9.3}$$

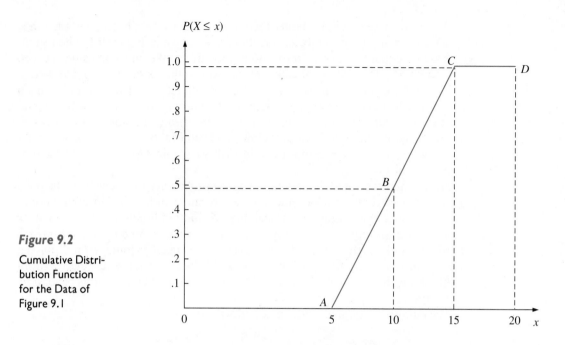

Figure 9.2

Cumulative Distribution Function for the Data of Figure 9.1

EXAMPLE 9.1 *An Application of the Uniform Distribution in Quality Control*

A quality control inspector for Gonsalves Company, which manufactures aluminum water pipes, believes that the product has varying lengths. Suppose the pipes turned out by one of the production lines of Gonsalves Company can be modeled by a uniform probability distribution over the interval 29.50 to 30.05 feet. The mean and standard deviation of X, the length of the aluminum water pipe, can be calculated as follows. Substituting $b = 30.05$ feet and $a = 29.50$ feet in Equation 9.3, we obtain

$$\mu_X = \frac{30.05 + 29.50}{2} = 29.775 \text{ feet}$$

and

$$\sigma_X = \frac{30.05 - 29.50}{\sqrt{12}} = .1588 \text{ feet}$$

This information can be used to create a control chart to determine whether the quality of the water pipes is acceptable. The control chart and its use in statistical quality control will be discussed in Chapter 10.

Computer simulation is an application of statistics that frequently relies on the uniform distribution. In fact, the uniform distribution is the underlying mechanism for this often-complex procedure. Thus, although not so many "real world" populations resemble this distribution as resemble the normal, the uniform distribution is important in applied statistics. For example, managers may use the uniform distribution in a simulation model to help them decide whether the company should undertake production of a new product.[1] Basic concepts of investment decision making can be found in Section 21.8.

9.3   *STUDENT'S t DISTRIBUTION*

Student's *t* distribution was first derived by W. S. Gosset in 1908. Because Gosset wrote under the pseudonym "A Student," this distribution became known as Student's *t* distribution.

If the sampled population is normally distributed with mean μ and variance σ_X^2, the sample size n is equal to or larger than 30, and σ_X^2 is known, then from the last chapter we know that the Z score for sample mean $\overline{X}$ defined as

$$Z = \frac{\overline{X} - \mu}{\sigma_X / \sqrt{n}} \qquad (8.7)$$

which we met as Equation 8.7, has a normal distribution with mean 0 and variance 1. Under most circumstances, however, the population variance is not known. In order for us to conduct various types of statistical analysis, we need to know what happens to Equation 8.7 when we replace the population standard deviation σ_X by the sample standard deviation s_X. We then have the following equation for the t statistic:

$$t = \frac{\overline{X} - \mu}{s_X / \sqrt{n}} \qquad (9.4)$$

Thus the Z of Equation 8.7 has only one source of variation: each sample has a different $\overline{X}$. Equation 9.4, however, has two sources of variation: both the sample mean $\overline{X}$ and the sample standard deviation s_X change from sample to sample. Thus the term on the right-hand side of Equation 9.4 follows a sampling distribution different from the normal distribution, which is the distribution followed by the term on the right-hand side of Equation 8.7. Equation 9.4 is used only when the population from which the n sample items are drawn is normally distributed and the sample size (n) is smaller than 30.

The t distribution forms a family of distributions that are dependent on a parameter known as the **degrees of freedom.** For the t variable in Equation 9.4, the degrees of freedom (ν) are ($n - 1$), where n is the sample size. In general, the degrees of freedom for a t statistic are the degrees of freedom associated with the sum of squares used to obtain an estimate of the variance. The variance estimate depends not only on the size of sample but also on how many parameters must be estimated

[1]See C. F. Lee (1985), *Financial Analysis and Planning: Theory and Application,* (Reading, MA: Addison-Wesley), pp. 358–363.

with the sample. The more data we have, the more confidence we can have in our results; the more parameters we have to estimate, the less confidence we have. Statisticians keep track of these two factors by calculating the degrees of freedom as follows:

$$\begin{array}{ccc} \text{Degrees of} \\ \text{freedom} \end{array} = \begin{array}{c} \text{number of} \\ \text{observations} \end{array} - \begin{array}{c} \text{number of parameters that must} \\ \text{be estimated beforehand} \end{array}$$

Here we calculate s_X by using n observations and estimating one parameter (the mean). Thus there are $(n - 1)$ degrees of freedom.

The t distribution is a symmetric distribution with mean 0. Its graph is similar to that of the standard normal distribution, as Figure 9.3 shows. However, the tail areas are greater for the t distribution, and the standard normal distribution is higher in the middle. The larger the number of degrees of freedom, the more closely the t distribution resembles the standard normal distribution. As the number of degrees of freedom increases without limit, the t distribution approaches the standard normal distribution. In fact, the standard normal distribution *is* a t distribution with an infinite number of degrees of freedom.

To determine whether the normal distribution or the Student's t distribution is more suitable for describing stocks' rates of return, Blattberg and Gonedes (1975, *Journal of Business,* pp. 244–280) used both daily and weekly stock rates of return for Dow Jones 30 companies to estimate the degrees of freedom for these two kinds of rates of return. They found, for example, that the degrees of freedom for Allied Chemical is 5.04 when daily data is used and 89.98 when weekly data is used. This indicates that the student's t distribution is more suitable for daily data for Allied Chemical, whereas the normal distribution better describes weekly data for Allied Chemical.

In addition, they found that the average degrees of freedom for daily rates of return for these 30 companies is 4.79. The average degrees of freedom in terms of weekly rate of return for these 30 companies is 11.22. They concluded that Student's t distribution is more suitable for describing daily stock rate of return distri-

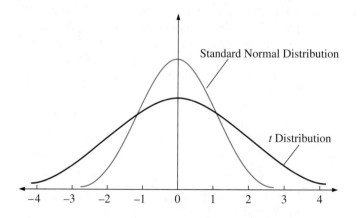

Figure 9.3

The t Distribution and the Standard Normal Distribution

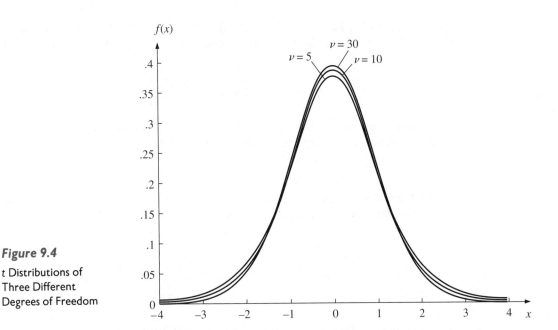

Figure 9.4

t Distributions of
Three Different
Degrees of Freedom

bution, and normal distribution is more suitable for weekly rate of return distribution. Hence, *t* distribution is an important distribution for describing daily stock rate of return.

The *t* table, as presented in Table A4 at the end of the book, gives the value, t_α, such that the probability of the *t* value larger than t_α is equal to α. The percentage cutoff point t_α is defined as that point at which

$$P(t > t_\alpha) = \alpha \tag{9.5}$$

Because the distribution is symmetric around 0, only positive *t* values (upper-tail areas) are tabulated. The lower α cutoff point is $-t_\alpha$, because

$$P(t < -t_\alpha) = P(t > t_\alpha) = \alpha$$

In general, we denote a cutoff point for *t* by $t_{\alpha,\nu}$, where α is the probability level and ν is the degrees of freedom. The number of degrees of freedom determines the shape of the *t* distribution. Figure 9.4 shows *t* distributions of varying degrees of freedom.

EXAMPLE 9.2 *Using the t Distribution to Analyze Audit Sampling Information*

Let's borrow information presented in Section 8.7 to see how the *t* distribution can be used to do audit sampling analysis.

The sample mean and the sample variance for 30 trade accounts receivable balances are

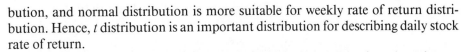

$$\overline{X} = \$202.10 \quad \text{and} \quad s_X^2 = \$719.164$$

From Table A4, we know that the t statistic with $30 - 1 = 29$ degrees of freedom and $\alpha = .05$ is 1.6991. Substituting related information into Equation 9.4, we obtain

$$1.699 = \frac{\$202.10 - \mu}{\sqrt{719.164/30}} = \frac{\$202.10 - \mu}{4.896}$$

This implies that there is a 5 percent chance that the average population account receivable value will be larger than $\$202.10 + \$(1.699)(4.896) = \$210.42$.

Other applications of the t distribution appear in Chapters 10 and 11, and we will encounter more when we discuss regression analysis in Chapters 13–16.

9.4 THE CHI-SQUARE DISTRIBUTION AND THE DISTRIBUTION OF SAMPLE VARIANCE

In this section we first show how a chi-square distribution can be derived from a standard normal distribution and then derive the distribution of a sample variance.

The Chi-Square Distribution

The **chi-square distribution** (χ^2), is a continuous distribution ordinarily derived as the sampling distribution of a sum of squares of independent standard normal variables. For instance, let $X_1, X_2, \ldots, X_n$ denote a random sample of size n from a normal distribution with mean μ and variance σ_X^2. Because these variables are not standardized, we can standardize them as

$$Z_i = \frac{X_i - \mu}{\sigma_X}$$

where Z_i is normally distributed with mean 0 and variance 1.

Now, if we define a new variable Y such that

$$Y = Z_1^2 + Z_2^2 + \cdots + Z_n^2 = \sum_{i=1}^{n} \left(\frac{X_i - \mu}{\sigma_X} \right)^2 \tag{9.6}$$

it can be shown that this new variable is distributed as χ^2 with n degrees of freedom.[2]

---

[2]First, it can be proved that $(X_i - \mu)^2/\sigma^2$ is a χ^2 distribution with 1 degree of freedom. Then, by using the additive property of χ^2 distribution, we can prove that

$$\sum_{i=1}^{n} \left(\frac{X_i - \mu}{\sigma} \right)^2 \text{ is also a } \chi^2 \text{ distribution with } n \text{ degrees of freedom.}$$

Equation 9.6 can be rewritten as[3]

$$\sum_{i=1}^{n} \frac{(X_i - \mu)^2}{\sigma_X^2} = \frac{n(\overline{X} - \mu)^2}{\sigma_X^2} + \frac{(n - 1)s_X^2}{\sigma_X^2} \tag{9.7}$$

where

$$s_X^2 = \frac{\sum_{i=1}^{n} (X_i - \overline{X})^2}{n - 1} \; ; \quad \sum_{i=1}^{n} \frac{(X_i - \mu)^2}{\sigma^2}$$

has a χ^2 distribution with n degrees of freedom, as discussed in Equation 9.6. In addition, from the last chapter, we know that $\overline{X}$ is normally distributed with mean μ and variance σ^2/n, so $\sqrt{n}(\overline{X} - \mu)/\sigma$ is normally distributed with mean 0 and variance 1. It can be shown that $n(\overline{X} - \mu)^2/\sigma_X^2$ has a χ^2 distribution with 1 degree of freedom. From this information it can be proved that

$$\frac{(n - 1)s_X^2}{\sigma_X^2}$$

defined in Equation 9.7, has a χ^2 distribution with $(n - 1)$ degrees of freedom.[4]

$$\frac{(n - 1)s_X^2}{\sigma_X^2}$$

can be redefined as expressed in Equation 9.8.

$$\frac{(n - 1)s_X^2}{\sigma_X^2} = \sum_{i=1}^{n} \left(\frac{X_i - \overline{X}}{\sigma_X} \right)^2 \tag{9.8}$$

where s_X^2 and σ_X^2 are sample variance and population variance, respectively. The left-hand side of Equation 9.8 implies that the ratio of sample variance to population variance, multiplied by $(n - 1)$, has a χ^2 distribution with $(n - 1)$ degrees of

---

[3]Since

$$\sum_{i=1}^{n} (X_i - \mu)^2 = \sum_{i=1}^{n} (X_i - \overline{X} + \overline{X} - \mu)^2$$

$$= \sum_{i=1}^{n} (X_i - \overline{X})^2 + 2(\overline{X} - \mu) \sum_{i=1}^{n} (X_i - \mu) + \sum_{i=1}^{n} (\overline{X} - \mu)^2 \tag{A}$$

$$= \sum_{i=1}^{n} (X_i - \overline{X})^2 + n(\overline{X} - \mu)^2$$

because

$$2(\overline{X} - \mu) \sum_{i=1}^{n} (X_i - \overline{X}) = 0.$$

By dividing Equation A by σ_X^2, we obtain Equation 9.7.

[4]In addition to the condition described here, it is also necessary to assume that $\overline{X}$ is independent of s_X^2.

freedom. The χ^2 distribution defined in Equation 9.8 can be used to describe the distribution of s^2, which will be discussed later in this section.

The χ^2 distribution is a skewed distribution, and only nonnegative values of the variable χ^2 are possible. It depends on a single parameter, the degrees of freedom $\nu = n - 1$. The χ^2 distributions for degrees of freedom 5, 10, and 30 are graphed in Figure 9.5. The figure shows that the skewness decreases as the degrees of freedom increase. In fact, as the degrees of freedom increase to infinity, the χ^2 distribution approaches a normal distribution.[5]

Critical values of the χ^2 distributions are given in Table A5 in Appendix A. They are defined by

$$P(\chi^2 \geq \chi^2_{\alpha,\nu}) = \alpha \tag{9.9}$$

where $\chi^2_{\alpha,\nu}$ is that value for the χ^2 distribution with ν degrees of freedom such that the area to the right (the probability of a larger value) is equal to α. For example, the upper 5 percent point for χ^2 with 10 degrees of freedom, $\chi^2_{.05,10}$, is 18.307 (see Figure 9.6).

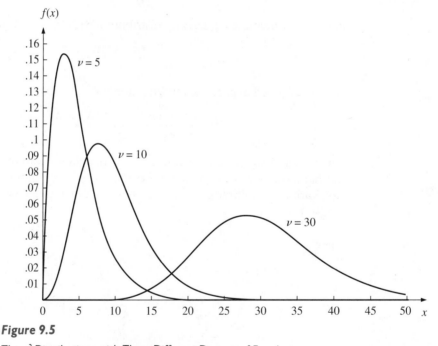

Figure 9.5

The χ^2 Distributions with Three Different Degrees of Freedom

[5]W. L. Johnson and S. Katz, in *Continuous Univariate Distribution I* (Boston: Houghton Mifflin, 1970), pp. 170–181, show that a normalized χ^2 distribution approaches a standard normal distribution when the number of degrees of freedom approaches infinity. The normalized statistic is defined as $(\chi^2_\nu - \nu)/\sqrt{2\nu}$.

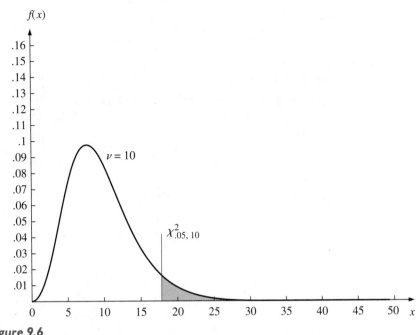

Figure 9.6

The χ^2 Distribution with 10 Degrees of Freedom

The mean and variance of this distribution are equal to the number of degrees of freedom and twice the number of degrees of freedom. That is,

$$E(\chi_\nu^2) = \nu \qquad \mathrm{Var}(\chi_\nu^2) = 2\nu \qquad (9.10)$$

where ν is the degrees of freedom of a χ^2 distribution.

The Distribution of Sample Variance

The properties of the χ^2 distribution can be used to find the mean and variance of the sampling distribution of the sample variance (s_X^2).

The Mean of s_X^2

From the definition of the mean for a χ^2 distribution, we obtain

$$E\left[\frac{(n-1)s_X^2}{\sigma_X^2}\right] = n - 1$$

Because $E(a \cdot X) = a \cdot E(X)$, we have

$$\frac{(n-1)}{\sigma_X^2} E(s_X^2) = n - 1$$

Thus[6]

$$E(s_X^2) = \sigma_X^2 \tag{9.11}$$

The Variance of s_X^2

Equation 9.11 implies that the mean of the sample variance is equal to the population variance. On the basis of the definition of the variance for a χ^2 distribution, we have

$$\text{Var}\left[\frac{(n-1)S_X^2}{\sigma_X^2}\right] = 2(n-1)$$

Because $\text{Var}(aX) = a^2 \cdot \text{Var}(X)$, we have

$$\frac{(n-1)^2}{\sigma_X^4}\text{Var}(s_X^2) = 2(n-1)$$

so

$$\text{Var}(s_X^2) = \frac{2\sigma_X^4}{n-1} \tag{9.12}$$

This is the variance of the sample variance. In sum, if X is normally distributed, then the mean and variance of S_X^2 are σ_X^2 and $2\sigma_X^4/(n-1)$; respectively. We will explore applications of the χ^2 distribution and the distribution of sample variance in Chapters 10 and 11 when we discuss confidence intervals and hypothesis testing for population variances.

Drawing on the concepts of the χ^2 distribution and the normal distribution, we can interpret the t distribution by rewriting Equation 9.4 as

$$t = \frac{(\overline{X} - \mu)/(\sigma_X/\sqrt{n})}{s_X/\sigma_X} \tag{9.4'}$$

In Equation 9.4', $(\overline{X} - \mu)/(\sigma_X/\sqrt{n})$ is normally distributed with mean 0 and variance 1; it is a standard normal distribution. S_X/σ_X is a square root of a χ^2-distributed variable with $(n-1)$ degrees of freedom divided by $\nu = n - 1$. Hence a t distribution with ν degrees of freedom is the ratio between a standard normal variable and a transformed χ^2 variable.

$$t_\nu = \frac{Z}{\sqrt{\chi_\nu^2/\nu}} \tag{9.13}$$

---

[6]This result suggests why $\sum_{i=1}^{n}(X_i - \overline{X})^2/n - 1$ instead of $\sum_{i=1}^{n}(X_i - \overline{X})^2/n$ is an unbiased estimator for the population variance, σ_X^2. Unbiased estimators will be discussed in Chapter 10.

9.5   *THE F DISTRIBUTION*

Some problems revolve around the value of a single population variance, but often it is a comparison of the variances of two populations that is of interest. This will be discussed in Chapters 13, 14, and 15. In addition, we may want to know whether the means of three or more populations are equal. This will be discussed in Chapter 12. The **F distribution** is used to make inferences about these kinds of issues.

Assume two populations, each having a normal distribution. We draw two independent random samples with sample sizes n_X and n_Y and population variances σ_X^2 and σ_Y^2. From each sample, we can compute sample variances S_X^2 and S_Y^2. Then the random variable of Equation 9.14 follows a distribution known as the *F* distribution.

$$F = \frac{S_X^2/\sigma_X^2}{S_Y^2/\sigma_Y^2} \tag{9.14}$$

Equation 9.14 can be rewritten as

$$F = \frac{\chi_{\nu_1}^2(X)/(n_X - 1)}{\chi_{\nu_2}^2(Y)/(n_Y - 1)} \tag{9.14'}$$

where $\chi_{\nu_1}^2(X) = (n_X - 1)S_X^2/\sigma_X^2$ and $\chi_{\nu_2}^2(Y) = (n_Y - 1)S_Y^2/\sigma_Y^2$; $\nu_1 = n_X - 1$; $\nu_2 = n_Y - 1$

In other words, a random variable formed by the ratio of two independent chi-square variables, each divided by its degrees of freedom, is called an **F variable.**

The *F* distribution has an asymmetric probability density function defined only for nonnegative values. It should be observed that the *F* distribution is completely determined by two parameters, ν_1 and ν_2, which are degrees of freedom. These density functions with different sets of degrees of freedom are illustrated in Figure 9.7.

The cutoff points $F_{\nu_1, \nu_2, \alpha}$, for α equal to .05, .025, .01, and .005 are provided in Table A6 at the end of this book. For example, in the case of 10 numerator degrees of freedom and 6 denominator degrees of freedom,

$$F_{10,6,.05} = 4.06 \qquad F_{10,6,.025} = 5.46$$
$$F_{10,6,.01} = 7.87 \qquad F_{10,6,.005} = 10.25$$

MINITAB output for $F_{10,6}$ is presented in Figure 9.8. Hence

$$P(F_{10,6} > 4.06) = .05 \qquad P(F_{10,6} > 5.46) = .025$$
$$P(F_{10,6} > 7.87) = .01 \qquad P(F_{10,6} > 10.25) = .005$$

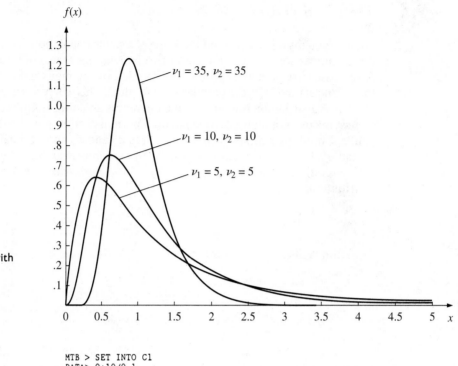

Figure 9.7

F Distributions with
Three Different
Sets of Degrees
of Freedom

```
MTB > SET INTO C1
DATA> 0:10/0.1
DATA> END
MTB > PDF C1 C2;
SUBC> F U=10 V=6.
MTB > GPLOT;

SUBC> LINE C2 C1;
SUBC> TITLE 'PDF OF F DISTRIBUTION';
SUBC> YLABEL 'PROBABILITY DENSITY';
SUBC> XLABEL 'F VALUE'.
```

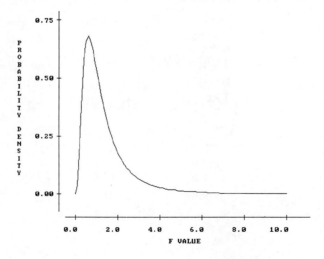

Figure 9.8

MINITAB output
for $F_{10,6}$

These probabilities also can be calculated by using MINITAB as shown here.

```
MTB > SET C1
DATA> 4.06 5.46 7.87 10.25
DATA> END
MTB > CDF C1;
SUBC> F 10 6.
    4.0600    0.9500
    5.4600    0.9750
    7.8700    0.9900
   10.2500    0.9950
MTB > PAPER
```

By subtracting 1 from .95 we obtain .05; by subtracting 1 from .975 we obtain .025; by subtracting 1 from .99 we obtain .01; finally, by subtracting 1 from .9950 we obtain .005. In practice, we usually place the larger sample variance in the numerator. The four significance levels listed here are the cutoff points that are often used to test the hypothesis of equality of population variances, which will be discussed in Chapters 11 and 12. When the population variances are equal, Equation 9.14 becomes

$$F = \frac{S_X^2}{S_Y^2} \tag{9.15}$$

The right-hand side of Equation 9.15 is the ratio of two sample variances.

Applications of the F distribution will be discussed in Chapters 11 and 12 and in the chapters related to regression analysis.

9.6 THE EXPONENTIAL DISTRIBUTION

The **exponential distribution** is related to the Poisson distribution, which, as we noted in Chapter 6, is often applied to occurrences of an event over time. The Poisson distribution is the distribution of the number of occurrences of an event in a given time interval of length t. The single parameter of the Poisson distribution is λ, the intensity of the process. Think of the number as the average occurrence of the event being counted. For example, say the average arrival rate of customers at the Brownell Bank is 5 per 100 seconds. Suppose that instead of the number of occurrences in a given time period, we are interested in the amount of time until the first customer arrives at the bank. This is a problem to be solved by the exponential distribution instead of the Poisson distribution. As another example, if the number of traffic accidents in an interval of time follows the Poisson distribution, the length of time from one accident to another follows the exponential distribution. The exponential distribution can also be applied to (1) the length of time that must pass before the first incoming telephone call and (2) the length of time someone must wait for a cab in a given location, such as Penn Station in New York City.

Denoting the mean rate at which events occur over time by λ and denoting the time until the first event occurs by t, we can use the Poisson probability density function to derive the exponential probability density function (PDF).[7] It is

$$
\begin{aligned}
f(t) &= \lambda e^{-\lambda t}, & t \geq 0 \\
&= 0, & t < 0
\end{aligned}
\tag{9.16}
$$

where $\lambda > 0$ is the only parameter.

From Equation 9B.3 we know that the cumulative probability function is given by

$$
\begin{aligned}
F(t) = P(T \leq t) &= 1 - e^{-\lambda t}, & t \geq 0 \\
&= 0, & t < 0
\end{aligned}
\tag{9.17}
$$

where T is a random variable representing time and t is a specific value.

Figure 9.9 represents four exponential functions for which λ equals 3, 2, 1, and $\frac{1}{2}$. From Appendix 9B we know that

$$
E(T) = \frac{1}{\lambda}
\tag{9.18a}
$$

$$
\mathrm{Var}(T) = \frac{1}{\lambda^2}
\tag{9.18b}
$$

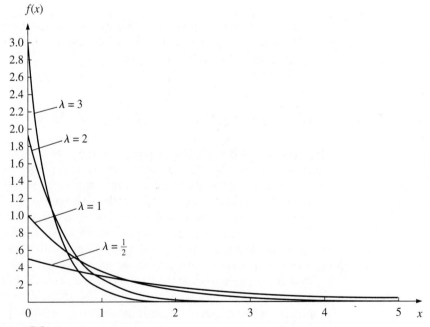

Figure 9.9

Four Exponential Density Functions Specified by Four Alternative Values of λ

[7]See Appendix 9B for the derivation.

EXAMPLE 9.3   *"No More Than 8 Items in This Line, Please!"*

Under fairly plausible assumptions about the behavior of clerks at supermarket check-out counters, it is possible to show that the time t (in minutes) a customer spends at a check-out counter is a random variable with the exponential distribution described by Equation 9.16.

Suppose a supermarket check-out counter has a mean number of customers per minute $= \frac{1}{3}$; that is, $\lambda = \frac{1}{3}$. Our task is to find the probability that the length of time between a pair of customer arrivals is less than 6 minutes.

Substituting $\lambda = \frac{1}{3}$ and $t = 6$ into Equation 9.17, we obtain $F(T \le 6) = 1 - e^{-6/3}$. And referring to Table A7 of Appendix A (or to a hand calculator), we find $P(T \le 6) = 1 - .1353 = .8647$. Thus the probability that the service time available between two customer arrivals at the check-out counter will be less than 6 minutes is approximately .86. Alternatively, the probability .8647 can be obtained by MINITAB as shown here.

```
MTB > CDF 6;
SUBC> EXPONENTIAL 3.
      6.0000      0.8647
MTB > PAPER
```

9.7   MOMENTS AND DISTRIBUTIONS

The properties of a distribution can be described in many ways, but the most popular approach is by means of a set of measurements called moments. **Moments** describe the central tendency, degree of dispersion, asymmetry, peakedness, and many other aspects of a distribution. This section discusses only the first four moments of a distribution; they are the most important statistical characteristics.

The first k moments can be defined either as

$$\mu'_k = E(X^k) \tag{9.19}$$

or

$$\mu_k = E[(X - \mu)^k] \tag{9.20}$$

Equation 9.19 defines the k moments about the origin, and Equation 9.20 defines the moments about the population mean μ. (The relationship between μ'_k and μ_k is discussed in Appendix 9C.) The *population mean* is the first moment about the origin. We obtain the first moment of a distribution about the origin by letting $k = 1$ in Equation 9.19. It is defined as follows:

$$\mu'_1 = E(X) = \mu$$

This is the population mean of X. Following equation 4.1, we can define μ for a discrete variable as

$$\mu = \sum_{i=1}^{N} X_i/N \qquad (4.2)$$

where N is the total number of observations in the population. The sample mean $\overline{X}$ associated with μ can be defined as

$$\overline{X} = \frac{\sum_{i=1}^{n} X_i}{n} \qquad (4.1)$$

where n is the sample size.

The Second Moment and the Coefficient of Variation

The second moment about the mean, the *variance,* is a measure of the dispersion of the random variable around the mean. The larger the variance, the more dispersed the distribution. Letting $k = 2$ in Equation 9.20, we obtain

$$\mu_2 = \sigma_X^2 = E[X - E(X)]^2$$

This is the population variance of X. Following Equation 4.5, we can define the population variance for a discrete variable as

$$\sigma^2 = \sum_{i=1}^{N} (X_i - \mu)^2/N \qquad (4.5)$$

The sample variance (s_X^2) associated with X can be defined as

$$s_X^2 = \sum_{i=1}^{n} (X_i - \overline{X})^2/(n - 1) \qquad (4.7)$$

Following Equation 4.12, we can define the sample **coefficient of variation** (CV) as

$$\text{CV} = \frac{s_X}{\overline{X}} \qquad (4.12)$$

The Third Moment and the Coefficient of Skewness

The third moment about the mean—*skewness,* which characterizes the asymmetry of the distribution—is given by

$$\mu_3 = E[X - E(X)]^3$$

Following Equation 4.15, we can define the population skewness for a discrete variable as

$$\mu_3 = \sum_{i=1}^{N} (X_i - \mu)^3/N \tag{4.15}$$

Following Equation 4.16, we can define the **coefficient of skewness** (CS), which is a relative measure of asymmetry, as

$$CS = \frac{\mu_3}{\sigma^3} \tag{4.16}$$

Following Equation 4.16a, we can define the sample coefficient of skewness (SCS) as

$$SCS = \frac{\sum_{i=1}^{n} (X_i - \overline{X})^3/(n-1)}{s^3} \tag{4.16a}$$

where

$$s^2 = \sum_{i=1}^{n} (X_i - \overline{X})^2/(n-1)$$

Figures 9.10a, 9.10b, and 9.10c present graphs of distributions with differing degrees of symmetry. Figure 9.10a shows a symmetrical distribution—that is, a distribution with zero skewness. Note that the symmetrical distribution's measures of central tendency (the mean, median, and mode) all coincide. We can also see that the half of the distribution above the mode is a mirror image of the half of the distribution below the mode.

Figure 9.10b presents a distribution that is said to be positively skewed because the distribution tapers off more slowly to the right of the mode than to the left. It is clear that the mean, median, and mode do not coincide. Here, the mode is smaller than the median and the mean.

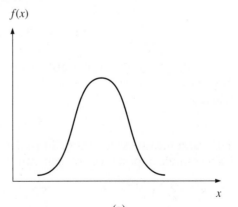

Figure 9.10
(a) Zero Skewness

(a)

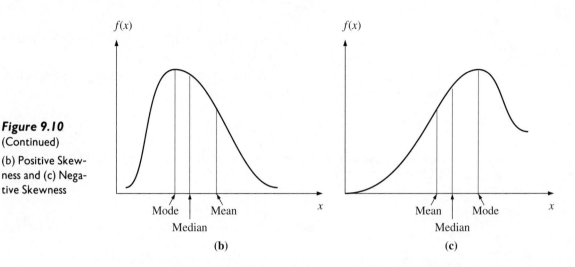

Figure 9.10
(Continued)

(b) Positive Skewness and (c) Negative Skewness

Figure 9.10c presents a distribution that is said to be negatively skewed because the distribution tapers off more slowly to the left of the mode than to the right. Once again the mean, median, and mode do not coincide. Here the median and mean lie to the left of the mode.

Kurtosis and the Coefficient of Kurtosis

The fourth moment about the mean—*kurtosis,* which characterizes the degree of peakedness—is defined by

$$\mu_4 = E[X - E(X)]^4$$

For discrete variables, the population kurtosis can be defined as

$$\mu_4 = \sum_{i=1}^{N} (X_i - \mu)^4/N \tag{9.21}$$

and can be estimated in terms of sample data as follows:

$$\text{Sample kurtosis} = \sum_{i=1}^{n} (X_i - \overline{X})^4/n$$

The relative peakedness of a distribution is expressed by the ratio of the fourth moment to the square of the second moment. It is called **coefficient of kurtosis** (CK).

$$\text{CK} = \mu_4/\mu_2^2 \tag{9.22}$$

This ratio measures the degree of peakedness relative to the level of dispersion. Using sample information, we can estimate the coefficient of kurtosis by

$$\text{SCK} = \frac{\displaystyle\sum_{i=1}^{n} (X_i - \overline{X})^4/(n-1)}{s^4} \tag{9.23}$$

Of two distributions having the same dispersion, the one with the larger kurtosis ratio has more observations concentrated near the mean and also at the tails of the distribution (at the expense of the intermediate area).

Skewness and Kurtosis for Normal and Lognormal Distributions

The bell-shaped normal curve is characterized by the *mesokurtic* shape: a value of 3 for the coefficient of kurtosis as defined in Equation 9.23. Distributions with values of the kurtosis ratio greater than 3 are *leptokurtic*. These distributions are more peaked than the standard mesokurtic (normal curve) shape. Distributions with values of the coefficient of kurtosis less than 3 are *platykurtic*—flatter in shape than the standard normal distribution. Each of these types of coefficients of kurtosis is illustrated in Figure 9.11. Sometimes the sample *coefficient of kurtosis* (SCK) can be redefined as

$$\text{SCK} = \frac{\sum_{i=1}^{n} (X_i - \overline{X})^4}{[\Sigma\, (X_i - \overline{X})^2]^2} - 3 \tag{9.23a}$$

The value for the redefined CK for a normal distribution is 0 instead of 3.

If X is lognormally distributed, then from Section 7.6, the mean and variance of X can be defined as

$$\mu_1 = \mu_X = e^{\mu + 1/2\sigma^2} \tag{7.6}$$

$$\mu_2 = \sigma_X^2 = e^{2\mu + \sigma^2}(e^{\sigma^2} - 1) \tag{7.7}$$

where $\mu = E(\log X)$ and $\sigma^2 = \text{Var}(\log X)$.

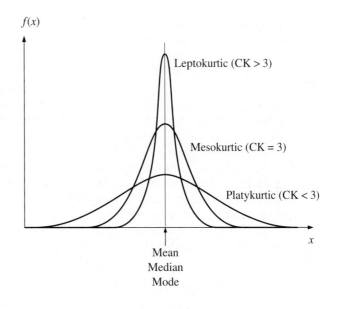

Figure 9.11

Three Types of Kurtosis

From Equations 7.6 and 7.7, the coefficient of variation (η) for X can be defined as

$$\eta = (e^{\sigma^2} - 1)^{1/2} \tag{9.24}$$

The third and fourth moments about the mean for lognormal distributions are

$$\mu_3 \text{ (skewness of } X) = (\mu_X)^3(\eta^6 + 3\eta^4) \tag{9.25}$$

$$\mu_4 \text{ (kurtosis of } X) = (\mu_X)^4(\eta^{12} + 6\eta^{10} + 15\eta^8 + 16\eta^6 + 3\eta^4) \tag{9.26}$$

where $\eta^2 = e^{\sigma^2} - 1$. (See Appendix 9D for the derivation of Equations 9.25 and 9.26.)

Substituting μ_1, μ_3, and μ_4 into Equations 4.16 and 9.22, we obtain the following equations for the coefficient of skewness (CS) and the coefficient of kurtosis (CK).

$$CS = \eta^3 + 3\eta \tag{9.27}$$

$$CK = \eta^8 + 6\eta^6 + 15\eta^4 + 16\eta^6 \tag{9.28}$$

where $\eta^2 = e^{\sigma^2} - 1$.

From Equations 9.26, 9.27, 9.28, and 9.29, we know that the coefficient of variation is the key variable in determining the magnitude of both skewness and kurtosis for a lognormal distribution.

In the next section, we will see how Equations 4.1, 4.7, 4.12, 4.16a, and 9.23a are applied with data on stock rates of return.

9.8 ANALYZING THE FIRST FOUR MOMENTS OF RATES OF RETURN OF THE 30 DJI FIRMS

In Table 9.1 we have listed the first four moments of the monthly returns of the 30 companies included in the Dow Jones Industrial (DJI) Average. These moments describe the central tendency, variability, asymmetry, and peakedness of the monthly return distributions between January 1988 and June 1990, inclusive. The Mean column gives us a measure of central tendency. The average mean of these 30 companies is .012246. The highest monthly return mean was from Boeing, followed by Philip Morris Inc., Procter & Gamble, and Woolworth. The lowest performances were for Goodyear and Allied Signal, which had returns of $-.2056$ and $-.00757$, respectively.

The measure of variability is given by the standard deviation. The average standard deviation was .069119. The two companies that showed the highest variability were Eastman Kodak and Navistar. The lowest variability was achieved by Exxon Corp., followed by Westinghouse.

In fact, we usually observe that higher rates of return are associated with higher levels of risk. Note that those companies that generated high rates of return tend to have high variability. The principle is simple: the higher the return you seek, the

Table 9.1 Statistical Estimates for the Dow Jones Industrial Average

	Monthly Statistical Estimates				
Name of Firm	*Mean*	*Standard Deviation*	*Skewness*	*Kurtosis*	*Coefficient of Variation*
Allied Signal Corp.	−.00757	.075899	−.45923	1.310551	−10.0170
Alcoa	.013308	.076758	.074372	−.07495	5.767458
American Express	.012326	.065994	.000480	.474129	5.353967
American Telephone and Telegraph Co.	.013974	.065712	−.22873	−.19194	4.702470
Bethlehem Steel Corp.	.004008	.103332	.314286	.141733	25.77891
Boeing	.045615	.071555	.515446	.480939	1.568660
Chevron Corporation	.020783	.054615	.240182	−.49224	2.627897
Coca-Cola Co.	.010645	.093494	−2.71903	11.12200	8.782336
Dupont	.010937	.056940	.262491	−.63075	5.205891
Eastman Kodak Co.	.006177	.171218	2.609095	12.77865	27.7169
Exxon Corp.	.008316	.037673	.435101	−.27083	4.529775
General Electric Co.	.016865	.059623	.362670	−.45102	3.535144
General Motors Corp.	.017107	.053222	.184619	−1.08948	3.111073
Goodyear Tire and Rubber Co.	−.02056	.063449	−.84169	.241885	−3.08579
IBM	.002381	.059974	−.12021	−.65913	25.18670
Int'l. Paper Co.	.008926	.074358	.268750	−.98769	8.330218
McDonald's Corp.	.017397	.051889	.146424	−.27720	2.982507
Merck & Co. Inc.	.017908	.054146	.644653	−.11844	3.023575
Minnesota Mining & Manufacturing Co.	.011003	.046323	.070076	−1.13956	4.210074
Navistar	.004948	.115958	2.014702	6.152188	23.43348
Philip Morris Inc.	.027382	.053230	.029738	−.49869	1.944003
Primerica	.016955	.076135	.617651	.745959	4.490340
Procter & Gamble Co.	.025563	.055382	.058830	−.00945	2.166438
Sears, Roebuck & Co.	.004315	.053842	−.10039	−.57631	12.47672
Texaco Inc.	.015298	.050928	.957945	.569231	3.328916
USX	.004941	.054235	−.18874	.022316	10.97667
Union Carbide	−.00030	.086639	.108428	−.75628	−282.533
United Technologies	.019132	.055516	.533348	−.51193	2.901662
Westinghouse Electric	.014104	.045681	.662122	−.22531	3.238720
Woolworth, F.W., Co.	.025522	.089838	.763562	−.07333	3.520031
Mean	.012246	.069119	.240563	.833498	−2.82486

more risk you have to take. There is a trade-off between risk and return, which will be discussed in Chapter 21 in some detail.

The skewness can be used to evaluate the stock's upside potential and downside risk. Positive skewness indicates the upside potential for a stock, because such a stock has a greater probability of very large payoffs. On the other hand, negative skewness is associated with downside risk; it indicates that the stock has a greater probability of very small payoffs.[8] Only seven companies in Table 9.1 exhibit the downside risk associated with negative skewness. The others, led by Eastman Kodak and Navistar, exhibit the upside potential associated with positive skewness.

The Kurtosis column shows that 11 companies here have a leptokurtic distribution (kurtosis ratio > 0);[9] these companies have more monthly returns concentrated near the mean. Only 3 companies have a distribution close to a mesokurtic distribution; Alcoa with SCK$' = -.07495$; Procter & Gamble, SCK$' = -.00945$; and Woolworth, SCK$' = -.07333$.

The last column, showing the coefficient of variation, enables us to compare monthly returns for the different companies. Remember that the coefficient of variation is a unitless figure that expresses the standard deviation as a percentage of the mean. High coefficients of variation show volatile monthly returns. We have 3 negative coefficients of variation because their mean is negative. The companies that show high volatility are Union Carbide, Eastman Kodak, Bethlehem Steel, IBM, and Navistar. The companies with the lowest volatility are Boeing, Philip Morris Inc., and Procter & Gamble.

Summary

In this chapter we discussed five continuous distributions. Four of these—Student's t distribution and the exponential, F, and χ^2 distributions—are closely related to the normal distribution discussed in Chapter 7. These five distributions, along with the normal and lognormal distributions, are the primary distributions we will use throughout the rest of the text for conducting statistical analyses such as determination of confidence intervals, hypothesis testing, and goodness-of-fit tests.

In Chapters 11, 12, 13, 14 and 15, we will begin to apply these distributions in alternative statistical analyses.

Appendix 9A Derivation of the Mean and Variance for a Uniform Distribution

On the basis of the definitions of $E(X)$ and $E(X^2)$ for a continuous variable given in Appendix 7A of Chapter 7, we can derive the mean and the variance of a uniform distribution as follows. First, substituting Equation 9.1 into Equation 7A.4, we get

$$
\begin{aligned}
E(X) &= \int_a^b xf(x)dx = \int_a^b \frac{x}{b-a}\,dx \\
&= \frac{1}{b-a} \cdot \frac{x^2}{2}\bigg|_a^b = \frac{b^2 - a^2}{2(b-a)} = \frac{a+b}{2}
\end{aligned}
\tag{9A.1}
$$

[8]This is so because a positively skewed distribution has more observations above the mode and a negatively skewed distribution more observations below.

[9]We use Equation 9.23a to calculate the coefficient of kurtosis.

Then, substituting Equation 9.1 into Equation 7A.7 yields

$$E(X^2) = \int_a^b x^2 f(x)dx = \frac{1}{b-a} \int_a^b x^2 dx = \frac{1}{b-a} \cdot \frac{x^3}{3}\Big|_a^b \quad (9A.2)$$

$$= \frac{b^3 - a^3}{3(b-a)} = \frac{(b-a)(b^2 + ab + a^2)}{3(b-a)} = \frac{b^2 + ab + a^2}{3}$$

Finally, substituting Equations 9A.1 and 9A.2 into the definition of variance given in Equation 7A.6, we obtain

$$\sigma_X^2 = E(X^2) - [E(X)]^2 = \frac{b^2 + ab + a^2}{3} - \left(\frac{a+b}{2}\right)^2 \quad (9A.3)$$

$$= \frac{4b^2 + 4ab + 4a^2 - 3b^2 - 6ab - 3a^2}{12} = \frac{(b-a)^2}{12}$$

This implies that $\sigma_X = (b-a)/\sqrt{12}$.

The following example shows how the formulas for both the mean and the variance of a continuous variable, as discussed in Appendix 7A, can be applied for a uniform distribution.

EXAMPLE 9A.1   *Calculating the Mean and Variance of a Uniform Distribution*

Let us look at an example of a continuous random variable in terms of the uniform distribution. Consider the density function of Equation 9A.4 as depicted in Figure 9A.1.

$$f(x) = \begin{cases} 1.55 - .06x & \text{if } 20 \leq x \leq 25 \\ 0 & \text{otherwise} \end{cases} \quad (9A.4)$$

For every value of x between 20 and 25 we get $f(x) > 0$, and for every x value outside of this range we have $f(x) = 0$. Therefore, for every x we have $f(x) \geq 0$. Furthermore, the area under the curve equals 1.

Figure 9A.1

The Density Function $f(x)$

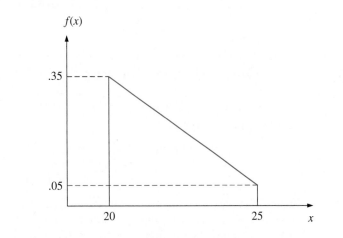

$$\int_{20}^{25} (1.55 - .06x)dx = 1.55x \Big|_{20}^{25} - \frac{.06x^2}{2}\Big|_{20}^{25} = 1$$

This confirms that $f(x)$ is a density function. Now let us calculate the expected value and variance of X.

$$E(X) = \int_{20}^{25} xf(x)dx = \int_{20}^{25} x(1.55 - .06x)dx$$

$$= \frac{1.55x^2}{2}\Big|_{20}^{25} - \frac{.06x^3}{3}\Big|_{20}^{25}$$

$$= 174.375 - 152.5 = 21.875$$

Next let us calculate $E(X^2)$.

$$E(X^2) = \int_{20}^{25} x^2(1.55 - .06x)dx$$

$$= \frac{1.55x^3}{3}\Big|_{20}^{25} - \frac{.06x^4}{4}\Big|_{20}^{25}$$

$$= 3939.583 - 3459.375 = 480.208$$

From this result we obtain

$$V(X) = E(X^2) - (EX)^2 = 480.208 - (21.875)^2 = 1.692$$

To find the probability, such as $P(22 \le x \le 24.5)$, we calculate

$$P(22 \le x \le 24.5) = \int_{22}^{24.5} f(x)dx = \int_{22}^{24.5} (1.55 - .06x)$$

$$= 1.55x \Big|_{22}^{24.5} - \frac{.06x^2}{2}\Big|_{22}^{24.5}$$

$$= 3.875 - 3.4875 = .3875$$

Appendix 9B Derivation of the Exponential Density Function

The cumulative distribution function (CDF) for the first event to occur in time interval t can be written as

$$
\begin{aligned}
P(T \le t) &= P(\text{wait until next arrival} \le t) \\
&= P(\text{at least one arrival in time } t) \qquad \text{(9B.1)}\\
&= 1 - P(\text{non-arrival in time } t)
\end{aligned}
$$

where T is the random variable of which t is a specific value. P(non-arrival in time t) can be obtained by letting $x = 0$ in the Poisson function as defined in Equation 6.16. We obtain P(non-arrival in time interval $[0,t]$) as

$$f(0) = P(T \geq t) = \frac{(\lambda t)^0 e^{-\lambda t}}{0!} = e^{-\lambda t} \text{ for } t \geq 0$$
$$= 0 \quad \text{for } t < 0 \tag{9B.2}$$

where λ denotes the mean rate at which events occur over time. Substituting Equation 9B.2 into Equation 9B.1, we obtain the CDF as

$$F(t) = P(T \leq t) = 1 - e^{-\lambda t} \tag{9B.3}$$

If we differentiate $F(t)$ with respect to t, we obtain the PDF as[10]

$$f(t) = \lambda e^{-\lambda t}, \quad t \geq 0$$
$$= 0, \qquad t < 0 \tag{9B.4}$$

The probability that the waiting time lies between a and b is

$$P(a) < T < b) = \int_a^b \lambda e^{-\lambda t} \tag{9B.5}$$

From the definition of $E(t)$ in Appendix 7A of Chapter 7, we obtain

$$E(T) = \int_{-\infty}^{\infty} t f(t) dt = \lambda \int_0^{\infty} t e^{-\lambda t} dt$$

The integral can be evaluated by parts. Let $U = t$ and $dv = e^{-\lambda t} dt$, so $dU = dt$ and $v = -e^{-\lambda t}/\lambda$. Then

$$E(T) = \lambda [(-te^{-\lambda t}/\lambda)_0^{\infty} + \frac{1}{\lambda} \int_0^{\infty} e^{-\lambda t} dt]$$

$$= \lambda [(-0 + 0) + \frac{1}{\lambda^2}(-0 + 1)] = \frac{1}{\lambda}$$

Similarly, we can prove that

$$\text{Var}(T) = \frac{1}{\lambda^2} \tag{9B.6}$$

This appendix shows how a mean value formula of a continuous variable, which was discussed in Appendix 7A, can be applied to an exponential distribution.

EXAMPLE **9B.1** *The Average Time Required to Find the Next Computer Program Error*

In finding and correcting errors in a computer program (debugging) and determining the program's reliability, Schick and others have noted the importance of the distribution of the time until the next program error is found. The cumulative expo-

[10]This is because

$$\frac{dF(t)}{dt} = \frac{d(1 - e^{-\lambda t})}{dt} = 0 - \left[\frac{d(-\lambda t)}{dt}\right] e^{-\lambda t} = \lambda e^{-\lambda t}$$

nential probability function of Equation 9B.2 is most useful in analyzing this problem.

By using the computer debugging data supplied by the U.S. Navy, Schick (1974, *Decision Sciences,* Vol. 5, 529–544) estimated the value of λ. After 26 of 31 program errors were found, Schick estimated λ to be .042. Accordingly, $1/\lambda = 23.8$ days. This means that the average time it would take to find 1 of the remaining errors (the 27th error) would be about 24 days. From this information, we can estimate, for example, that the probability of taking 50 or more days to find the next error is

$$P(T \geq 50) = e^{-(.042)(50)} = e^{-2.1} = .1125.$$

The second equality is obtained by using Table A7 in Appendix A.

EXAMPLE 9B.2   *The Probability of Truck Arrivals*

Rutgers Trucking Company had 15,600 trucks to unload at the receiving warehouse during the last calendar year. The warehouse was open from 8 A.M. to 8 P.M. each weekday. There was no noticeable pattern of truck arrivals each day. It is known that approximately 5 trucks arrived to unload cargo each hour. What is the probability that on September 20, 1991, the first truck arrived between 8:15 and 8:30 A.M.?

To use exponential distribution to solve this problem, we first use a time interval of 15 minutes (8:15–8:30) for which $\lambda = (5/60)(15) = 1.25$.

Substituting $\lambda = 1.25$, $a = 1$, and $b = 2$ into Equation 9B.5,[11] we obtain the probability that the first truck arrived between 8:15 and 8:30 A.M.

$$P(1 < T < 2) = \int_{1}^{2} e^{-1.25t}(1.25dt) = -e^{-1.25t}\bigg|_{1}^{2}$$
$$= -e^{-2.5} + e^{-1.25} = .2$$

---

[11]We regard 15 minutes as 1 time unit that can be expressed as a time interval between $a = 1$ and $b = 2$.

Appendix 9C The Relationship Between the Moment About the Origin and the Moment About the Mean

Let $k = 1$ in Equation 9.21. Then

$$\mu_1 = E(X - \mu_1') = E(X) - \mu_1' = 0$$

This implies that the first moment about the population mean is zero.

Alternatively, if we let $k = 2$ in Equation 9.21 and let $\mu_1 = \mu_1$, we obtain

$$\begin{aligned}
\mu_2 &= E(X) - \mu_1')^2 = E(X^2 - 2X\mu_1' + \mu_1') \\
&= E(X^2) - 2\mu_1'E(X) + \mu_1'^2 = \mu_2' - \mu_1'^2
\end{aligned} \tag{9C.1}$$

where μ_2' and μ_1' are second and first moments, respectively. Equation 9C.1 is identical to Equation 7A.6 in Appendix 7A. It is a short-cut formula to calculate variance.

Now, if we let $k = 3$ in Equation 9.21 and substitute $\mu_1 = \mu_1'$, we obtain

$$\begin{aligned}
\mu_3 &= E(X - \mu_1')^3 = E(X^3 - 3X^2\mu_1' + 3X\mu_1'^2 - \mu_1'^3) \\
&= E(X^3) - 3\mu_1'E(X^2) + 3\mu_1'^2E(X) - \mu_1'^3 \\
&= \mu_3' - 3\mu_1'\mu_2' + 2\mu_1'^3
\end{aligned} \tag{9C.2}$$

where μ_1' and μ_2' are defined in Equation 9C.1 and μ_3' is the third moment about the origin.

Finally, letting $k = 4$ in Equation 9.21 and substituting $\mu_1 = \mu_1$, we obtain

$$\begin{aligned}
\mu_4 &= E(X - \mu_1')^4 \\
&= E(X^4 - 4X^3\mu_1' + 6X^2\mu_1'^2 - 4E(X)\mu_1'^3 + \mu_1'^4) \\
&= E(X^4) - 4E(X^3)\mu_1' + 6E(X^2)\mu_1'^2 - 4E(X)\mu_1'^3 + \mu_1'^4 \\
&= \mu_4' - 4\mu_3'\mu_1' + 6\mu_2'\mu_1'^2 - 3\mu_1'^4
\end{aligned} \tag{9C.3}$$

where μ_1', μ_2', and μ_3' have been defined in Equation 9C.2 and μ_4' is the fourth moment about the origin.

In Appendix 9D, Equations 9C.1, 9C.2, and 9C.3 will be used to derive variance, skewness, and kurtosis of the lognormal distribution.

Appendix 9D Derivations of Mean, Variance, Skewness, and Kurtosis for the Lognormal Distribution

Following Aitchison and Brown (1963), we express the moments about the origin for the lognormal distribution as

$$\mu_k' = e^{k\mu + 1/2k^2\sigma^2}, \quad k = 1, 2, \cdots \tag{9D.1}$$

In accordance with definitions given in Appendix 9C, the mean, variance, skewness, and kurtosis of a lognormal distribution can be derived as follows.

Mean

Substituting $k = 1$ into Equation 9D.1 yields

$$\mu_1' = e^{\mu + 1/2\sigma^2}$$

This is Equation 7.6.

Variance

Substituting $\mu_2' = e^{2\mu + 2\sigma^2}$ and $\mu_1' = e^{\mu + 1/2\sigma^2}$ into Equation 9C.1 in Appendix 9C, we obtain

$$e^{2\mu + 2\sigma^2} - e^{2\mu + \sigma^2} = e^{2\mu + \sigma^2}(e^{\sigma^2} - 1)$$

This is Equation 7.7.

Skewness

Substituting μ_1', μ_2', and $\mu_3' = e^{3\mu + 9/2\sigma^2}$ into Equation 9C.2 gives

$$
\begin{aligned}
\mu_3 &= (\mu_1)^3[e^{3\sigma^2} - 3e^{\sigma^2} + 2] \\
&= (\mu_1)^3[(e^{3\sigma^2} - 3e^{2\sigma^2} + 3e^{\sigma^2} - 1) + 3(e^{2\sigma^2} - 2e^{\sigma^2} + 1)] \\
&= (\mu_1)^3(\eta^6 + 3\eta^4)
\end{aligned}
$$

where $\eta^2 = e^{\sigma^2} - 1$. This is Equation 9.25.

Kurtosis

Substituting μ_1', μ_2', μ_3', and $\mu_4' = e^{3\mu + 8\sigma^2}$ into Equation 9C.3, we get

$$\mu_4 = (\mu_1)^4[e^{6\sigma^2} - 4e^{3\sigma^2} + 6e^{\sigma^2} - 3]$$

By considerable mathematical rearrangement of terms, it can be shown that

$$\mu_4 = (\mu_x)^4[\eta^{12} + 6\eta^{10} + 15\eta^8 + 16\eta^6 + 3\eta^4]$$

where $\eta^2 = e^{\sigma^2} - 1$. This is Equation 9.26.

Questions and Problems

1. Briefly discuss the cumulative distribution function of the uniform distribution presented in Figure 9.2.

2. Briefly discuss the relationship between the Poisson distribution and the exponential distribution.

3. X is normally distributed, and the sample variance $s^2 = 20$ is calculated from 20 observations. Calculate $E(s^2)$ and $\text{Var}(s^2)$.

4. W is a normally distributed random variable with mean 0 and variance 1, and V is a χ^2-distributed random variable with degrees of freedom $(n - 1)$. How can both t and F distributions be defined in terms of the variables W and V?

5. Briefly discuss how F statistics can be used to test the difference between two sample variances.

6. Briefly discuss how mean, variance, skewness, kurtosis, and the coefficient of variation can be used to analyze stock rates of return.

7. Suppose a random variable X can take on only values in the range from 2 to 10 and that the probability that the variable will assume any value within any interval in this range is the same as the probability that X will assume another value in another interval of similar width in the range. What is the distribution of X? Draw the probability density function for X.

8. Use the information given in question 7 to find $P(3 \le X \le 7)$.

9. Use the information given in question 7 to find $P(X \le 8)$.

10. Use the information given in question 7 to find $P(X < 2 \text{ or } X > 10)$.

11. Draw the cumulative distribution function for the distribution given in question 7.

12. Suppose a random variable x is best described by a uniform distribution with $a = 8$ and $b = 20$.
   a. Find $f(x)$.
   b. Find $F(x)$.
   c. Find the mean and variance of x.

13. Suppose a random variable Y is best described by a uniform distribution with $a = 3$ and $b = 32$.
   a. Find $f(y)$.
   b. Find $F(y)$.
   c. Find the mean and variance of y.

14. A very observant art thief (who should probably be teaching statistics instead) notices that the frequency of security guards passing by a museum is uniformly distributed between 15 and 60 minutes. Therfore, if X denotes the time (in minutes) before the guard passes by, the probability density function of X is

$$f_X(x) = \begin{cases} 1/(60 - 15) & \text{for } 15 < x < 60 \\ 0 & \text{for all other values} \\ & \text{of } x \end{cases}$$

   a. Draw the probability density function.
   b. Find and draw the cumulative distribution function.

15. Use the information given in question 14.
   a. Find the probability that the guard passes by within 35 minutes of the thief's arrival.
   b. Find the probability that the guard does not pass by within 30 minutes.
   c. Find the probability that the guard passes by between 30 minutes and 45 minutes after the thief's arrival.

16. An art dealer at an auction believes that the bid on a certain painting will be a uniformly distributed random variable between $500 and $2,000.
   a. What is the probability density function for this random variable?
   b. Find the probability that the painting will sell for less than $675.
   c. Find the probability that the painting will sell for more than $1,000.

17. Suppose x has an exponential distribution with $\lambda = 5$. Find the following probabilities:
   a. $P(x > 4)$
   b. $P(x > .7)$
   c. $P(x > .50)$

18. Suppose x has an exponential distribution with $\lambda = 4$. Find the following probabilities:
   a. $P(x \le .3)$
   b. $P(x \le .5)$
   c. $P(x \le 1.6)$

19. Suppose x has an exponential distribution with $\lambda = \frac{1}{3}$. Find the following probabilities:
    a. $P(3 \le x \le 5)$
    b. $P(5 \le x \le 10)$
    c. $P(2 \le x \le 7)$

20. Suppose the random variable X is best approximated by an exponential distribution with $\lambda = 8$. Find the mean and the variance of X.

21. Suppose the random variable Y is best approximated by an exponential distribution with $\lambda = 3$. Find the mean and the variance of Y.

22. Briefly compare the normal distribution discussed in Chapter 7 with the t distribution discussed in this chapter.

23. Find t_α for the following:
    a. $\alpha = .05$ and $\nu = 10$
    b. $\alpha = .025$ and $\nu = 4$
    c. $\alpha = .10$ and $\nu = 7$

24. Find the value t_0 such that:
    a. $P(t \ge t_0) = .025$, where $\nu = 6$
    b. $P(t \ge t_0) = .05$, where $\nu = 12$
    c. $P(t \le t_0) = .10$, where $\nu = 9$

25. Find the value t_0 such that:
    a. $P(t \le t_0) = .10$, where $\nu = 25$
    b. $P(t \ge t_0) = .025$, where $\nu = 14$
    c. $P(t \le t_0) = .01$, where $\nu = 17$

26. Find the following probabilities for the t distributions.
    a. $P(t > 3.078)$ if $\nu = 1$
    b. $P(t < 1.943)$ if $\nu = 6$
    c. $P(t > 2.492)$ if $\nu = 24$

27. Find the following probabilities for the t distributions.
    a. $P(t > 1.734)$ if $\nu = 18$
    b. $P(t > 1.943)$ if $\nu = 6$
    c. $P(t < 1.645)$ if $\nu = \infty$

28. Find the following $\chi^2_{\alpha,\nu}$ values.
    a. $\alpha = .05$ and $\nu = 25$
    b. $\alpha = .025$ and $\nu = 5$
    c. $\alpha = .10$ and $\nu = 50$
    d. $\alpha = .01$ and $\nu = 60$

29. Find the following $\chi^2_{\alpha,\nu}$ values.
    a. $\alpha = .025$ and $\nu = 30$
    b. $\alpha = .01$ and $\nu = 70$
    c. $\alpha = .10$ and $\nu = 10$
    d. $\alpha = .01$ and $\nu = 20$

30. Find the following probabilities.
    a. $P(\chi^2 > 10.8564)$ when $\nu = 24$
    b. $P(\chi^2 < 10.8564)$ when $\nu = 24$
    c. $P(\chi^2 < 48.7576)$ when $\nu = 70$
    d. $P(\chi^2 > 59.1963)$ when $\nu = 90$

31. Find the following probabilities.
    a. $P(\chi^2 \le 3.84146)$ when $\nu = 1$
    b. $P(\chi^2 \ge 15.9871)$ when $\nu = 10$
    c. $P(\chi^2 < 140.169)$ when $\nu = 100$
    d. $P(\chi^2 > 1.61031)$ when $\nu = 5$

32. Find the following $F_{\nu_1,\nu_2,\alpha}$ values.
    a. $\nu_1 = 8$, $\nu_2 = 10$, and $\alpha = .01$
    b. $\nu_1 = 3$, $\nu_2 = 11$, and $\alpha = .005$
    c. $\nu_1 = 12$, $\nu_2 = 9$, and $\alpha = .05$
    d. $\nu_1 = 24$, $\nu_2 = 19$, and $\alpha = .025$

33. Find the following $F_{\nu_1,\nu_2,\alpha}$ values.
    a. $\nu_1 = 10$, $\nu_2 = 10$, and $\alpha = .05$
    b. $\nu_1 = 15$, $\nu_2 = 3$, and $\alpha = .01$
    c. $\nu_1 = 12$, $\nu_2 = 15$, and $\alpha = .025$
    d. $\nu_1 = 20$, $\nu_2 = 10$, and $\alpha = .005$

34. Find the probabilities, given ν_1 and ν_2 as shown.
    a. $\nu_1 = 1$ and $\nu_2 = 3$; $P(F > 17.44)$
    b. $\nu_1 = 3$ and $\nu_2 = 1$; $P(F > 864.2)$
    c. $\nu_1 = 3$ and $\nu_2 = 1$; $P(F < 215.7)$
    d. $\nu_1 = 30$ and $\nu_2 = 12$; $P(F < 4.33)$

35. Using the MINITAB program, find the probabilities, given ν_1 and ν_2 as shown.
    a. $\nu_1 = 120$ and $\nu_2 = 120$; $P(F > 1.35)$
    b. $\nu_1 = \infty$ and $\nu_2 = \infty$; $P(F > 1.00)$
    c. $\nu_1 = 6$ and $\nu_2 = 17$; $P(F < 3.28)$
    d. $\nu_1 = 3$ and $\nu_2 = 23$; $P(F > 4.76)$

36. Find the probability that an exponentially distributed random variable X with mean $1/\lambda = 8$ will take on the values:
    a. Between 2 and 7
    b. Less than 9
    c. Greater than 6
    d. Between 1 and 15

37. Suppose the lifetime of a television picture tube is distributed exponentially with a standard deviation of 1,400 hours. Find the probability that the tube will last:
    a. More than 3,000 hours
    b. Less than 1,000 hours
    c. Between 1,000 and 2,000 hours

38. Suppose the time you wait at a bank is exponentially distributed with mean $1/\lambda = 12$ minutes.

What is the probability that you will wait between 10 and 20 minutes?

39. Suppose the length of time people wait at a fast-food restaurant is distributed exponentially with a mean of 1/7 minutes. Use MINITAB to answer the following questions.

   a. What percentage of people will be served within 4 minutes?

   b. What percentage of people will be served between 3 and 8 minutes after they arrive?

   c. What percentage of people will wait more than 9 minutes?

40. Suppose the length of time a student waits to register for courses is distributed exponentially with a mean of 1/15 minute.

   a. What percentage of students will register within 10 minutes?

   b. What percentage of students will register after waiting between 10 and 20 minutes?

   c. What percentage of students will wait more than 20 minutes to register?

41. Suppose a random variable is distributed as a χ^2 distribution with n degrees of freedom. Consider the probability $P(\chi^2 \leq 9)$. Explain the relationship between the probability and the degrees of freedom.

42. Suppose a random variable is distributed as Student's t distribution with $(n-1)$ degrees of freedom. Consider the probability $P(t \geq .7)$. Explain the relationship between the probability and the degrees of freedom.

43. The incomes of families in a town is assumed to be uniformly distributed between $15,000 and $85,000. What is the probability that a randomly selected family will have an income above $40,000?

44. At an antiques auction, the winning bids were found to be uniformly distributed between $500 and $2,500. What is the probability that a winning bid was less than $1,000? What is the probability that a winning bid was between $750 and $1,500?

45. The manager of a department store notices that the amount of time a customer must wait before being helped is distributed uniformly between 1 minute and 4 minutes. Find the mean and variance of the time a customer must wait to be helped.

46. A quality control expert for the Healthy Time Cereal Company notices that in a 16-ounce package of cereal, the amount in the box is uniformly distributed between 15.3 ounces and 17.1 ounces. Find the mean and standard deviation for the weight of this cereal in a package of cereal.

47. The shelf life of hearing aid batteries is found to be approximated by an exponential distribution with a mean of 1/12 day. What fraction of the batteries would be expected to have a shelf life greater than 9 days?

48. A computer programmer has decided to use the exponential distribution to evaluate the reliability of a computer program. After 10 programming errors were found, the time (measured in days) to find the next error was determined to be exponentially distributed with a $\lambda = .25$.

   a. Graph this distribution.

   b. Find the mean time required to find the 11th error.

49. Use the information given in question 48 to find the probability that it will take more than 5 days to find the 11th error. Find the probability that it will take between 3 and 10 days to find the 11th error.

50. An advertising executive believes that the length of time a television viewer can recall a commercial is distributed exponentially with a mean of .25 days. Find how long it will take for 75 percent of the viewing audience to forget the commercial.

51. Use the information given in question 50 to find the proportion of viewers who will be able to recall the commercial after 7 days.

52. An investment advisor believes that the rate of return for Horizon Company's stock is uniformly distributed between 3 percent and 12 percent. Find the probability that the return will be greater than 5 percent. Find the probability that the return will be between 6 percent and 8 percent.

53. The mean life of a computer's hard disk is found to be exponentially distributed with a mean of 12,000 hours. Find the proportion of hard disks that will have a life greater than 20,000 hours.

54. Suppose the life of a car battery is assumed to be uniformly distributed between 3.9 years and 7.3 years. Find the mean and variance of the life of a car battery.

55. Use the information given in question 54 to find the probability that the life of the car battery will be greater than 5 years. Find the probability that the life of the battery will be between 4 years and 6 years.

56. The chief financial officer at Venture Corporation believes that an investment in a new project will have a cash flow in year 1 that is uniformly distributed between $1 million and $10 million. What is the probability that the cash flow in year 1 will be greater than $1.7 million?

57. A hospital collects data on the number of emergency room patients in during a certain period. It is estimated that in an hour, the average number of emergency room patients to arrive is 1.2. If the time between two consecutive arrivals of patients follows an exponential distribution, what is the probability that a patient will show up in the next hour?

58. The campus bus at Haverford College is scheduled to arrive at the business school at 8:00 A.M. Usually the bus arrives at the bus stop during the interval 7:56 to 8:03. Assume that the arrival time follows a uniform distribution.
    a. What is the probability that the bus arrives at the business school before 8:00?
    b. What is the average arrival time?
    c. What is the standard deviation of arrival time?

59. A gas station's owner found that about 2 cars come into the station every minute. If the arrival time follows an exponential distribution, what is the probability that the next car will arrive in 1.5 minutes?

60. A college professor gives a standardized test to her students every semester. She finds that the students' grades follow a uniform distribution with 100 points as the maximum and 65 points as the minimum.
    a. Find the mean score.
    b. Compute the standard deviation of the score.
    c. If the passing grade is 70, what percentage of students will fail the course?

61. Suppose the weight of a football team is uniformly distributed with a minimum weight of 175 pounds and a maximum weight of 285 pounds.
    a. Find the mean weight of the team.
    b. Compute the standard deviation of the weight.
    c. Find the percentage of players with a weight of less than 195 pounds.

62. Briefly explain how the mean, standard deviation, coefficient of variation, and skewness can be used to analyze the returns of IBM and Boeing in Table 9.1.

63. A bank manager finds that about 6 customers enter the bank every 5 minutes. If the customer arrival time follows an exponential distribution, what is the probability that the next customer will arrive in 2 minutes?

64. Suppose the life of a steel-belted radial tire is uniformly distributed between 30,000 and 45,000 miles.
    a. Find the mean tire life.
    b. Find the standard deviation of tire life.
    c. What percentage of these tires will have a life of more than 40,000 miles?

65. Briefly discuss the relationship among t, χ^2, and F distributions.

66. Given $v_1 = 5$ and $\alpha = .05$, find v_2 for the following F values.
    a. 5.05   b. 3.33   c. 2.53

67. In their study, Vardeman and Ray (*Technometrics,* May 1985, 145–150) found that the number of accidents per hour at an industrial plant is exponentially distributed with a mean $\lambda = .5$. Use the formula $f(t) = \lambda e^{-\lambda t}$ to determine each of the following.
    a. $f(1)$ b. $f(4)$ c. $E(t)$

PART III

Statistical Inferences Based on Samples

In the next three chapters, we will discuss statistical inference based on samples and the applications of such statistical inference. So far, our discussion has focused on descriptive statistics, sampling and sampling distributions, and the analytical techniques and distributions used to describe statistical data.

Inferential statistics, on the other hand, is used to make inferences about a population by looking at a subset of that population. In Chapter 10, we continue the discussion of the previous five chapters by looking at point estimation, confidence intervals, and statistical quality control. In Chapter 11, we apply these techniques to testing hypotheses about a population. In Chapter 12, we discuss the analysis of variance for sample data and the use of chi-square tests in analyzing sample data.

CHAPTER 10

Estimation and Statistical Quality Control

Key Terms

population parameters
parameter
statistic
estimate
estimator
point estimate
point estimator
point estimation
unbiasedness
bias
efficiency
consistency
sufficient statistic
mean-squared error
interval estimation
confidence interval

confidence level
probability content
risk probability
significant level
acceptance sampling
lot
convenience lots
single-sampling plans
double-sampling plans
control chart
upper control limit
lower control limit
$\overline{X}$-chart
$\overline{R}$-chart
S-chart
P-chart

10.1 INTRODUCTION

In the previous two chapters, we discussed the basic principles of sampling and sampling distributions—techniques that enable us to make inferences about a population by looking at a subset of that population. In this chapter, we continue our discussion of inferential statistics by examining point estimation, confidence intervals, and statistical quality control. Note that this chapter draws heavily on your understanding of the standard normal distribution discussed in Chapter 7, the fundamental concepts of sampling discussed in Chapter 8, and the *t* distribution and chi-square distribution discussed in Chapter 9.

We first examine point estimates for population parameters and then discuss desirable attributes of point estimators. Second, basic concepts and the necessity of using interval estimates are discussed in detail. Third, we explain how to compute confidence intervals for population means both when the population variance is known and when it is unknown. Fourth, confidence intervals for the population proportion and the population variance are explored. Finally, we present applications of the use of confidence intervals for quality control. An application of confidence intervals for a cash management model appears in Appendix 10A. Appendix 10B shows how MINITAB can be used to generate control charts.

10.2 POINT ESTIMATION

As we have said, statistical inference enables us to make judgments about a population on the basis of sample information. The mean, standard deviation, and proportions of a population are called **population parameters;** in other words, they serve to define the population. Estimating a population's parameters is essential to statistical analysis, and sometimes sampling is the best (fastest and most economical) way to approach the study.

Point Estimate, Estimator, and Estimation

A **parameter** is a characteristic of an entire population; a **statistic** is a summary measure that is computed to describe a characteristic for only a sample of the population. An **estimate** is a specific observed value of a statistic. The rule that specifies how a sample statistic can be obtained for estimating the population parameter is called an **estimator.** For example, if a professor wants information on central tendency in a list of test scores, she can calculate a sample mean. The number for the sample mean is called the estimate, and the sample mean is the estimator for the population mean. The **point estimate** is the single number that is obtained from the estimator.

The symbols we use to represent several important population parameters and their sample counterparts follow.

	Population Parameter	Sample Statistic
Mean	μ_X	$\overline{X}$
Standard deviation	σ_X	s_X
Variance	σ_X^2	s_X^2
Proportion	p	$\hat{p}$

EXAMPLE 10.1   *Sample Mean and Sample Variance: Point Estimate*

Suppose that a professor, whose course has an enrollment of 50 students, wants information on the performance of his class. He takes a sample of 10 scores:

$$95, 67, 89, 70, 56, 97, 68, 78, 50, 79$$

The estimator for the population mean is the sample mean, $\overline{X}$. The estimate for the population mean, on the basis of the 10 sample scores, is $\overline{X} = 74.9$.

The estimator for the population variance is the sample variance. The estimate of the population variance is

$$s_X^2 = \frac{(95^2 + 67^2 + \cdots + 79^2) - 10(74.9)^2}{10 - 1} = 247.65$$

The professor can use $\overline{X} = 74.9$ and $s_X^2 = 247.65$ to do his or her class performance analysis.

The relationship among the point estimate, point estimator, and point estimation can be summarized as follows. A point estimate is a single value that is calculated from only one sample. In Example 10.1, $\overline{X} = 74.9$ is an estimate for population mean μ_X, and $s_X^2 = 247.65$ is an estimate for population variance σ_X^2. Using the formula for combinations reveals that there are $\binom{50}{10} = 10,272,278,000$ possible sample estimates for Example 10.1.[1] The random variable that is defined by a formula, and from which we obtain all possible estimates, is called the point estimator. A **point estimate** is a single value that is used to estimate a population parameter. A **point estimator** is a sample statistic used to estimate a population parameter. **Point estimation** is a process that generates specific numbers, each of which is a point estimate.

[1] From the combination formula discussed in Appendix 5A of Chapter 5, we obtain

$$\binom{50}{10} = \frac{50!}{10!(50 - 10)!} = \frac{(50)(49) \cdots (41)}{(10)(9) \cdots (1)} = 10,272,278,000$$

EXAMPLE 10.2   *Population Mean: Point Estimate*

We can use a sampling approach to obtain the point estimate of a population mean μ_X. In Example 9.1, we demonstrated the sampling results of taking samples of 2, 3, or 4 elements out of a uniformly distributed population that represents the numbers of years of working experience of six secretaries (1, 2, 3, 4, 5, and 6) at Francis Engineering Inc. If samples of 3 elements are randomly taken from this population, then there are 20 possible samples, as listed in Table 10.1.

Table 10.1 Possible Samples and Their Sample Means (Sample Size = 3)

Possible Samples	Elements in Sample	Sample Mean ($\overline{X}_i$)	Possible Samples	Elements in Sample	Sample Mean ($\overline{X}_i$)
1	1, 2, 3	2	11	2, 3, 4	3
2	1, 2, 4	2.33	12	2, 3, 5	3.33
3	1, 2, 5	2.67	13	2, 3, 6	3.67
4	1, 2, 6	3	14	2, 4, 5	3.67
5	1, 3, 4	2.67	15	2, 4, 6	4
6	1, 3, 5	3	16	2, 5, 6	4.33
7	1, 3, 6	3.33	17	3, 4, 5	4
8	1, 4, 5	3.33	18	3, 4, 6	4.33
9	1, 4, 6	3.67	19	3, 5, 6	4.67
10	1, 5, 6	4	20	4, 5, 6	5
		Sum = 30			Sum = 40

The population mean and population standard deviation are

$$\mu_X = \frac{1 + 2 + 3 + 4 + 5 + 6}{6} = 3.5$$

$$\sigma_X = \left[\frac{(1 - 3.5)^2 + (2 - 3.5)^2 + \cdots + (6 - 3.5)^2}{6} \right]^{1/2} = 1.71$$

All possible sample means listed in Table 10.1 are point estimates of the population mean μ_X. MINITAB output is given in Figure 10.1, which indicates that both the mean and the median of this set of sample means are equal to 3.5.

Figure 10.1

MINITAB Output for Example 10.2

```
MTB > SET INTO C1
DATA> 2 2.33 2.67 3 2.67 3 3.33 3.33 3.67 4 3 3.33 3.67 3.67 4 4.33 4 4.33 4.67
DATA> 5
DATA> END
MTB > DESCRIBE C1
```

(*continued*)

	N	MEAN	MEDIAN	TRMEAN	STDEV	SEMEAN
C1	20	3.500	3.500	3.500	0.784	0.175

	MIN	MAX	Q1	Q3
C1	2.000	5.000	3.000	4.000

MTB > PAPER

Figure 10.1

(Continued)

Four Important Properties of Estimators

A number of different estimators are possible for the same population parameter, but some estimators are better than others. To understand how, we need to look at four important properties of estimators: unbiasedness, efficiency, consistency, and sufficiency.

Unbiasedness

An estimator exhibits **unbiasedness** when the mean of the sampling estimator $\hat{\theta}$ is equal to the population parameter θ. In other words, the expected value of the estimator is equal to the population parameter: $E(\hat{\theta}) = \theta$. Let's use data given in Table 10.1 as an example:

$$E(\overline{X}_i) = \Sigma P(X_i)X_i = \frac{\sum_{i=1}^{20} \overline{X}_i}{20} = \frac{2 + 2.33 + \cdots + 4.67}{20} = \frac{70}{20}$$
$$= 3.5 = \mu_X$$

Note that $P(X_i) = \frac{1}{20}$ because each sample of 3 is equally likely. Figure 10.2 shows the sampling distributions of two estimators, $\hat{\theta}_1$ and $\hat{\theta}_2$. $\hat{\theta}_1$ is an unbiased estimator and $\hat{\theta}_2$ a biased estimator. Figure 10.2 indicates that $E(\hat{\theta}_1) = \theta$ and $E(\hat{\theta}_2) > \theta$.

In general, unbiasedness is a desirable property for an estimator. The sample mean is an unbiased estimator of the population mean because the mean of the sampling distribution of $\overline{X}$, $E(X)$, is equal to the population mean μ_X. Similarly, the sample variance is an unbiased estimator of the population variance because the mean of the sample distribution of s_X^2, $E(s_X^2)$, is equal to population variance σ_X^2.[2] And the sample proportion is an unbiased estimator of the population proportion; $E(\hat{p}) = p$. However, because standard deviation is a *nonlinear* function of variance, the sample standard deviation is not an unbiased estimator of population standard deviation.

The **bias** of a point estimator is defined in Equation 10.1.

$$\text{Bias} = E(\hat{\theta}) - \theta \tag{10.1}$$

For example, in Figure 10.2 the bias of using $\hat{\theta}_2$ as an estimator of θ is equal to $E(\hat{\theta}_2) - \theta$.

---

[2]If we divide the sum of squared discrepancies from $\overline{X}$ by $(n - 1)$ rather than n. Equation 9.11 in Chapter 9 can be used to demonstrate this point.

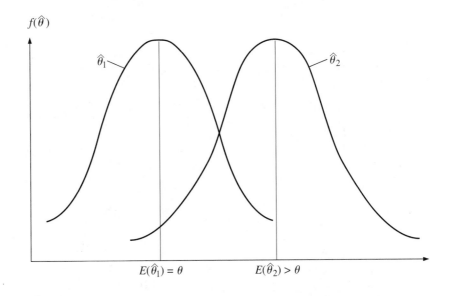

Figure 10.2

Probability Density Functions for $\hat{\theta}_1$ and $\hat{\theta}_2$

Unbiasedness, then, is an important attribute of estimators. But suppose we have a number of unbiased estimators to choose from. Here are three other criteria that could be used to select an estimator.

Efficiency

Efficiency is another standard that can be used to evaluate estimators. **Efficiency** refers to the size of the standard error of the statistics. The most efficient estimator is the one with the smallest variance. Thus if there are two estimators for θ with variances $\text{Var}(\hat{\theta}_1)$ and $\text{Var}(\hat{\theta}_2)$, then the first estimator $\hat{\theta}_1$ is said to be more efficient than the second estimator $\hat{\theta}_2$ if $\text{Var}(\hat{\theta}_1) < \text{Var}(\hat{\theta}_2)$ although $E(\hat{\theta}_1) = E(\hat{\theta}_2) = \theta$. Figure 10.3 shows the distributions of the two density functions.

The relative efficiency of one estimator compared with another is simply the ratio of their variances. Given two unbiased estimators $\hat{\theta}_1$ and $\hat{\theta}_2$ with variances $\text{Var}(\hat{\theta}_1)$ and $\text{Var}(\hat{\theta}_2)$, the relative efficiency of $\hat{\theta}_2$ with respect to $\hat{\theta}_1$ is

$$\text{Relative efficiency} = \frac{\text{Var}(\hat{\theta}_1)}{\text{Var}(\hat{\theta}_2)} \tag{10.2}$$

Why is the variance of the benchmark estimator ($\hat{\theta}_1$) placed in the numerator? Well, suppose two estimators are calculated for the population mean. The first is the sample mean $\hat{\theta}_1$, and the second is the sample median $\hat{\theta}_2$. It can be shown that the variance of the sample median of a normal distribution is $\text{Var}(\hat{\theta}_2) = \pi(\sigma^2/2n)$. The variance for the sample mean is σ^2/n. The relative efficiency of the sample median with respect to the sample mean is

$$\text{Efficiency} = \frac{\text{Var}(\hat{\theta}_1)}{\text{Var}(\hat{\theta}_2)} = \frac{\sigma^2/n}{\pi\sigma^2/2n} = \frac{2}{\pi} = 63.66\%$$

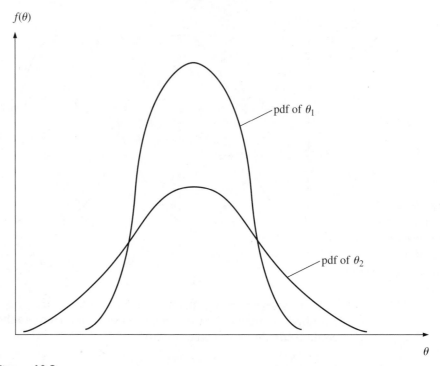

Figure 10.3

Probability Density Functions of Two Unbiased Estimators Θ_1 and Θ_2; $Var(\Theta_1) < Var(\Theta_2)$

The sample mean, rather than the sample median, is the preferred estimator of the population mean, because the amount of variability associated with the sample mean is about 64 percent of that associated with the sample median. Note that the sample mean is the best estimate of central tendency for symmetric distributions and that the sample median is generally used for skewed distributions.

Consistency

A third property of estimators, **consistency,** is related to their behavior as the sample size gets large. A statistic is a consistent estimator of a population parameter if, as the sample size increases, it becomes almost certain that the value of the statistic comes very close to the value of the population parameter. For example, suppose we are tossing a coin and are interested in rolling a head. The sample proportion X/n is an estimator for the population proportion, where X is the number of heads tossed and n is the number of trials. We know that the population proportion of heads tossed is equal to $1/2$, so we would expect the sample proportion to get closer to $1/2$ as the number of trials n increases. (This result was demonstrated by a computer simulation in Chapter 5.) We need information on the probability that the absolute difference between the estimator and the parameter will be less than some positive number ϵ. In other words, we need $P(|X/n - p| \leq \epsilon)$, and this probability should be close to 1 as n gets large. If it is, then X/n is said to be a consistent estimator of p.

It can be shown that an unbiased estimator $\hat{\theta}_n$ for θ is a consistent estimator if the variance approaches 0 as n increases. For example, we can show that the sample mean is a consistent estimator of the population. The sample mean is unbiased because $E(\overline{X}) = \mu$. The variance of $\overline{X}$ is σ_X^2/n. As n becomes large, the variance gets closer to 0; this estimator is consistent. Finally, it should be noted that the sample standard deviation is a consistent estimator of population standard deviation, although it is not an unbiased estimator of population standard deviation.

Following this approach, we can see that the sample proportion $X/n = \hat{p}$ is also consistent. From the last chapter we know that $E(\hat{p}) = P$, which establishes unbiasedness. Because the variance is equal to $\hat{p}(1 - \hat{p})/n$, the variance approaches 0 as n gets large. Thus X/n is a consistent estimator of p. $\hat{\theta}_n$ is a consistent estimator of θ if, for any positive number ϵ,

$$\lim_{n \to \infty} P(|\hat{\theta}_n - \theta| \leq \epsilon) = 1 \qquad (10.3)$$

where $n \to \infty$ means that sample size approaches infinity.

Sufficiency

The last property of a good estimator that we will consider is sufficiency, which was developed by Sir R. A. Fisher, a famous statistician, in 1922.[3] A **sufficient statistic** (such as $\overline{X}$) is an estimator that utilizes all the information a sample contains about the parameter to be estimated. For example, $\overline{X}$ is a sufficient estimator of the population mean μ_X. This means that no other estimator of μ_X from the same sample data, such as the sample median, can add any further information about the parameter μ that is being estimated.

It can be shown that the sample mean $\overline{X}$ and the sample proportion $\hat{p}$ are sufficient statistics (estimators) for μ_X and p.

Mean-Squared Error for Choosing Point Estimator

Frequently a trade-off must be made between bias and efficiency for a point estimator. Sometimes there is much to be gained by accepting some biases for the sake of increasing the efficiency of an estimator. A statistic called the **mean-squared error** (MSE), the expectation of the squared difference between the estimators and parameters as indicated in Equation 10.4, can be used to measure the trade-off between bias and efficiency for an estimator.

$$\text{MSE}(\hat{\theta}) = E(\hat{\theta} - \theta)^2 \qquad (10.4)$$

[3]R. A. Fisher (1922), "On the Mathematical Foundations of Theoretical Statistics," *Phil. Trans. Roy. Soc. London,* Series A, Vol. 222.

It can be shown[4] that

$$MSE(\hat{\theta}) = Var(\hat{\theta}) + [Bias(\hat{\theta})]^2 \qquad (10.5)$$

where $Var(\hat{\theta}) = E[\hat{\theta} - E(\hat{\theta})]^2$ and $Bias(\hat{\theta}) = E(\hat{\theta}) - \theta$.

Equations 10.4 and 10.5 imply that an estimator's variance is a measure of the dispersion of the sampling distribution around the estimator's expected value, $E(\hat{\theta})$, whereas the MSE is a measure of dispersion around the true population parameters, θ. If the estimator is unbiased, then $E(\hat{\theta}) = \theta$ and $MSE(\hat{\theta}) = Var(\theta)$. For example, in Figure 10.1 the expected variability of $\hat{\theta}_2$ around the true parameter, θ, is greater than that around $E(\hat{\theta}_2)$, which is the center of the sample distribution. Both nonsampling error and systematic sampling error bias an estimator.

10.3 INTERVAL ESTIMATION

In the last section, we discussed point estimation of a population parameter. We investigated methods for estimating population mean, variance, standard deviation, and proportion and methods for evaluating desirable features of estimators. Although these estimators give us much information about a population parameter, more information is usually desired. Many times, an interval estimate is needed. For example, a manager may want to know how likely it is that the mean number of defects is between 1 percent and 3 percent, or a professor may want to know how likely it is that between 10 percent and 20 percent of her class will get an A on the final exam. Sample statistics such as the mean and variance do not provide any information on the range of values the population parameters are likely to fall in.

We now wish to estimate a parameter μ_X by the interval

$$a < \mu_X < b$$

where a and b are obtained from sample observation. The estimation of a and b, values between which the parameter of interest will lie with a certain probability, is called **interval estimation.**

Suppose θ is a parameter to be estimated. A random sample is taken, and two random variables a and b are computed. The interval from a to b is called a **confidence interval;** its probability is $(1 - \alpha)$. In other words, if all of the population is repeatedly sampled and the intervals are calculated in the same fashion, then the probability is $(1 - \alpha)$ that the confidence interval will contain the population parameter.

$$P(a < \theta < b) = 1 - \alpha \qquad (10.6)$$

---

[4] $E(\hat{\theta} - \theta)^2 = E[\hat{\theta} - E(\hat{\theta}) + E(\hat{\theta}) - \theta]^2$
$= E[\hat{\theta} - E(\hat{\theta})]^2 + [E(\hat{\theta}) - \theta]^2 + 2E[\hat{\theta} - E(\hat{\theta})][E(\hat{\theta}) - \theta]$
$= E[\hat{\theta} - E(\hat{\theta})]^2 + [E(\hat{\theta}) - \theta]^2 \quad$ because $E[\hat{\theta} - E(\hat{\theta})] = 0$

For example, if $1 - \alpha = .95$, then the probability that a is less than Θ and b is greater than Θ is .05. Because $1 - \alpha = .95$, $\alpha = .05$. The term $(1 - \alpha)$ is called the **confidence level (probability content)**. The quantity α is often termed the **risk probability** or **significant level**. In the next four sections, we will use Equation 10.6 to estimate the confidence interval for population mean, population proportion, and population variance.

10.4 *INTERVAL ESTIMATES FOR μ WHEN σ^2 IS KNOWN*

In this section, we construct confidence intervals for the population mean. We assume that the random sample is taken from a normal distribution and that the population variance is known. The latter assumption is somewhat unrealistic, because the population variance is rarely known. However, these assumptions enable us to illustrate concepts that we will need later.

Suppose a random sample is taken with an unknown mean and known variance. The confidence interval uses the fact that the random variable Z, where

$$Z = \frac{\overline{X} - \mu}{\sigma/\sqrt{n}}$$

has a standard normal distribution. Suppose a $100(1 - \alpha)$ percent confidence interval is set up, so that $\alpha/2$ is the area of the right tail of the normal distribution, $\alpha/2$ is the area of the left tail, and $(1 - \alpha)$ is the area in the center, as shown in Figure 10.4. The cutoff points on the normal distribution are $z_{\alpha/2}$ and $-z_{\alpha/2}$. The confidence interval is derived as follows:

$$1 - \alpha = P(-z_{\alpha/2} < Z < z_{\alpha/2})$$
$$= P\left(-z_{\alpha/2} < \frac{\overline{X} - \mu}{\sigma/\sqrt{n}} < z_{\alpha/2}\right)$$

or

$$1 - \alpha = P\left\{\overline{X} - z_{\alpha/2}\left[\frac{\sigma}{\sqrt{n}}\right] < \mu < \overline{X} + z_{\alpha/2}\left[\frac{\sigma}{\sqrt{n}}\right]\right\} \qquad (10.7)$$

Equation 10.7 implies that confidence intervals have the following characteristics.

1. As the standard deviation increases, the length of the confidence interval increases. This result is understandable: the wider the deviation, the more uncertain the estimate of the mean.

2. The bigger the sample size, the smaller the confidence interval for a given variance. This is because more information decreases the interval, making a better interval possible.

3. The confidence interval is larger for smaller confidence levels (α). A 99 percent confidence interval has a smaller α than a 95 percent interval because a 99 percent interval has more certainty.

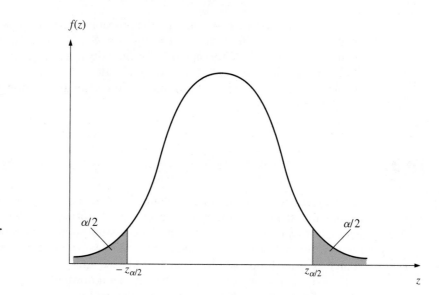

Figure 10.4

Risk Probability for
Sample Mean
Estimate

EXAMPLE 10.3 *Confidence Intervals in Terms of 20 Samples*

Let us now refer to the 20 samples in Table 10.1 and use the information $\sigma_X = 1.71$ and the 20 random samples given there. We calculate 20 different 95 percent confidence intervals in terms of Equation 10.7 by using MINITAB as presented in Figure 10.5. The 95 percent confidence interval (CI) results listed in Figure 10.5 reveal that all 20 samples resulted in a confidence interval containing $\mu_X = 3.5$.

Figure 10.5

MINITAB Output
for Example 10.3

```
MTB > READ C1-C20
DATA> 1 1 1 1 1 1 1 1 1 1 2 2 2 2 2 2 3 3 3 4
DATA> 2 2 2 2 3 3 3 4 4 5 3 3 3 4 4 5 4 4 5 5
DATA> 3 4 5 6 4 5 6 5 6 6 4 5 6 5 6 6 5 6 6 6
DATA> END
       3 ROWS READ
MTB > PRINT C1-C20

ROW   C1   C2   C3   C4   C5   C6   C7   C8   C9   C10   C11   C12   C13   C14

  1    1    1    1    1    1    1    1    1    1    1     2     2     2     2
  2    2    2    2    2    3    3    3    4    4    5     3     3     3     4
  3    3    4    5    6    4    5    6    5    6    6     4     5     6     5


ROW   C15    C16    C17    C18    C19    C20

  1    2      2      3      3      3      4
  2    4      5      4      4      5      5
  3    6      6      5      6      6      6

MTB > STORE
STOR> ZINTERVAL USING 95%, SIGMA=1.71, DATA IN CK1
STOR> LET K1=K1+1
STOR> END
MTB > LET K1=1
MTB > EXECUTE 'MINITAB' 20
MTB > ZINTERVAL USING 95%, SIGMA=1.71, DATA IN CK1
```

Figure 10.5
(Continued)

```
THE ASSUMED SIGMA =1.71

                N      MEAN    STDEV  SE MEAN   95.0 PERCENT C.I.
C1              3     2.000    1.000   0.987  (  0.062,   3.938)

MTB > LET K1=K1+1
MTB > END
MTB > ZINTERVAL USING 95%, SIGMA=1.71, DATA IN CK1

THE ASSUMED SIGMA =1.71

                N      MEAN    STDEV  SE MEAN   95.0 PERCENT C.I.
C2              3     2.333    1.528   0.987  (  0.396,   4.271)

MTB > LET K1=K1+1
MTB > END

MTB > ZINTERVAL USING 95%, SIGMA=1.71, DATA IN CK1

THE ASSUMED SIGMA =1.71

                N      MEAN    STDEV  SE MEAN   95.0 PERCENT C.I.
C3              3     2.667    2.082   0.987  (  0.729,   4.604)

MTB > LET K1=K1+1
MTB > END
MTB > ZINTERVAL USING 95%, SIGMA=1.71, DATA IN CK1

THE ASSUMED SIGMA =1.71

                N      MEAN    STDEV  SE MEAN   95.0 PERCENT C.I.
C4              3     3.000    2.646   0.987  (  1.062,   4.938)

MTB > LET K1=K1+1
MTB > END
MTB > ZINTERVAL USING 95%, SIGMA=1.71, DATA IN CK1

THE ASSUMED SIGMA =1.71

                N      MEAN    STDEV  SE MEAN   95.0 PERCENT C.I.
C5              3     2.667    1.528   0.987  (  0.729,   4.604)

MTB > LET K1=K1+1
MTB > END
MTB > ZINTERVAL USING 95%, SIGMA=1.71, DATA IN CK1

THE ASSUMED SIGMA =1.71

                N      MEAN    STDEV  SE MEAN   95.0 PERCENT C.I.
C6              3     3.000    2.000   0.987  (  1.062,   4.938)

MTB > LET K1=K1+1
MTB > END
MTB > ZINTERVAL USING 95%, SIGMA=1.71, DATA IN CK1

THE ASSUMED SIGMA =1.71

                N      MEAN    STDEV  SE MEAN   95.0 PERCENT C.I.
C7              3     3.333    2.517   0.987  (  1.396,   5.271)

MTB > LET K1=K1+1
MTB > END
MTB > ZINTERVAL USING 95%, SIGMA=1.71, DATA IN CK1

THE ASSUMED SIGMA =1.71

                N      MEAN    STDEV  SE MEAN   95.0 PERCENT C.I.
C8              3     3.333    2.082   0.987  (  1.396,   5.271)
```

Figure 10.5

(Continued)

```
MTB > LET K1=K1+1
MTB > END
MTB > ZINTERVAL USING 95%, SIGMA=1.71, DATA IN CK1

THE ASSUMED SIGMA =1.71

                  N      MEAN    STDEV   SE MEAN    95.0 PERCENT C.I.
C9                3      3.667   2.517   0.987    (  1.729,   5.604)

MTB > LET K1=K1+1
MTB > END
MTB > ZINTERVAL USING 95%, SIGMA=1.71, DATA IN CK1

THE ASSUMED SIGMA =1.71

                  N      MEAN    STDEV   SE MEAN    95.0 PERCENT C.I.
C10               3      4.000   2.646   0.987    (  2.062,   5.938)

MTB > LET K1=K1+1
MTB > END
MTB > ZINTERVAL USING 95%, SIGMA=1.71, DATA IN CK1

THE ASSUMED SIGMA =1.71

                  N      MEAN    STDEV   SE MEAN    95.0 PERCENT C.I.
C11               3      3.000   1.000   0.987    (  1.062,   4.938)

MTB > LET K1=K1+1
MTB > END

MTB > ZINTERVAL USING 95%, SIGMA=1.71, DATA IN CK1

THE ASSUMED SIGMA =1.71

                  N      MEAN    STDEV   SE MEAN    95.0 PERCENT C.I.
C12               3      3.333   1.528   0.987    (  1.396,   5.271)

MTB > LET K1=K1+1
MTB > END
MTB > ZINTERVAL USING 95%, SIGMA=1.71, DATA IN CK1

THE ASSUMED SIGMA =1.71

                  N      MEAN    STDEV   SE MEAN    95.0 PERCENT C.I.
C13               3      3.667   2.082   0.987    (  1.729,   5.604)

MTB > LET K1=K1+1
MTB > END
MTB > ZINTERVAL USING 95%, SIGMA=1.71, DATA IN CK1

THE ASSUMED SIGMA =1.71

                  N      MEAN    STDEV   SE MEAN    95.0 PERCENT C.I.
C14               3      3.667   1.528   0.987    (  1.729,   5.604)

MTB > LET K1=K1+1
MTB > END
MTB > ZINTERVAL USING 95%, SIGMA=1.71, DATA IN CK1

THE ASSUMED SIGMA =1.71

                  N      MEAN    STDEV   SE MEAN    95.0 PERCENT C.I.
C15               3      4.000   2.000   0.987    (  2.062,   5.938)

MTB > LET K1=K1+1
MTB > END
MTB > ZINTERVAL USING 95%, SIGMA=1.71, DATA IN CK1
```

Figure 10.5

(Continued)

```
THE ASSUMED SIGMA =1.71

            N     MEAN   STDEV  SE MEAN   95.0 PERCENT C.I.
C16         3    4.333   2.082   0.987   ( 2.396,   6.271)

MTB > LET K1=K1+1
MTB > END
MTB > ZINTERVAL USING 95%, SIGMA=1.71, DATA IN CK1

THE ASSUMED SIGMA =1.71

            N     MEAN   STDEV  SE MEAN   95.0 PERCENT C.I.
C17         3    4.000   1.000   0.987   ( 2.062,   5.938)

MTB > LET K1=K1+1
MTB > END
MTB > ZINTERVAL USING 95%, SIGMA=1.71, DATA IN CK1

THE ASSUMED SIGMA =1.71

            N     MEAN   STDEV  SE MEAN   95.0 PERCENT C.I.
C18         3    4.333   1.528   0.987   ( 2.396,   6.271)

MTB > LET K1=K1+1
MTB > END
MTB > ZINTERVAL USING 95%, SIGMA=1.71, DATA IN CK1

THE ASSUMED SIGMA =1.71

            N     MEAN   STDEV  SE MEAN   95.0 PERCENT C.I.
C19         3    4.667   1.528   0.987   ( 2.729,   6.604)

MTB > LET K1=K1+1
MTB > END
MTB > ZINTERVAL USING 95%, SIGMA=1.71, DATA IN CK1

THE ASSUMED SIGMA =1.71

            N     MEAN   STDEV  SE MEAN   95.0 PERCENT C.I.
C20         3    5.000   1.000   0.987   ( 3.062,   6.938)

MTB > LET K1=K1+1
MTB > END
```

EXAMPLE 10.4 *Sandbags We Can Have* Real *Confidence In: 95 Percent and 99 Percent Confidence Intervals*

Suppose a machine dispenses sand into bags. The population standard deviation is 9.0 pounds, and the weights are normally distributed. A random sample of 100 bags is taken, and the sample mean is 105 pounds. Let's calculate a 95 percent confidence interval by using Equation 10.7.

The estimate of sample standard deviation is $9/\sqrt{100}$. The confidence level is $(1 - \alpha) = .95$. Thus $\alpha = .05$ and $\alpha/2 = .025$. The Z value that corresponds to this area is 1.96. This is due to the fact that the area of the right tail is .025. The area of the left tail is .025. The 95 percent confidence interval is

$$105 - (1.96)\,\frac{(9)}{\sqrt{100}} < \mu < 105 + (1.96)\,\frac{(9)}{\sqrt{100}}$$

or $103.24 < \mu < 106.76$

Figure 10.6a and b

The Effects of Sample Size and Probability Content on Confidence Intervals in Cases of Sample Mean of 105

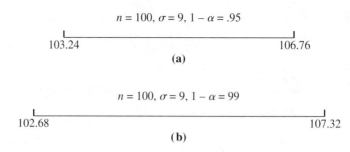

$n = 100, \sigma = 9, 1 - \alpha = .95$

103.24 ——————— 106.76

(a)

$n = 100, \sigma = 9, 1 - \alpha = 99$

102.68 ——————— 107.32

(b)

The 95 percent confidence interval for the mean weight of the bags ranges from 103.24 to 106.76, as presented in Figure 10.6a. In other words, we are certain that 95 percent of such intervals contain the population mean.

Suppose a 99 percent confidence interval is needed for the case discussed in Example 10.3. The confidence level is $1 - \alpha = .99$. Thus $\alpha = .01$ and $\alpha/2 = .005$. The z value is 2.575. The confidence interval is

$$105 - (2.575)\frac{(9)}{\sqrt{100}} < \mu < 105 + (2.575)\frac{(9)}{\sqrt{100}}$$

or

$$102.68 < \mu < 107.32$$

This 99 percent confidence interval is wider than the 95 percent confidence interval, as indicated in Figure 10.6b. Now we are certain that the interval generated by our statistics will include the true population mean 99 percent of the time.

EXAMPLE 10.5   *95 Percent Confidence Interval for the Sandbag Sample with a Smaller Sample Size*

Assume that rather than taking a sample of 100, we take a sample of 30. The 95 percent confidence interval becomes

$$105 - (1.96)\frac{(9)}{\sqrt{30}} < \mu < 105 + (1.96)\frac{(9)}{\sqrt{30}}$$

Figure 10.6c

The Effects of Sample Size and Probability Content on Confidence Interval in the Case of Sample Mean of 105

or

$$101.779 < \mu < 108.22$$

This interval is wider than the interval with a sample size of 100, as indicated in Figure 10.6c.

$n = 30, \sigma = 9, 1 - \alpha = 95$

101.779 ——————— 108.22

(c)

EXAMPLE 10.6 *95 Percent Confidence Interval for the Mean External Audit Fees for 32 Diverse Companies*

To study the effect of internal audit departments on external audit fees, W. A. Wallace recently conducted a survey of the audit departments of 32 diverse companies (*Harvard Business Review,* March–April 1984). She found that the mean annual external audit paid by the 32 companies was $779,030 and the standard deviation was $1,083,162.

Because this is a large sample case, we can replace the sample standard deviation for the population standard deviation. Substituting both the sample mean and the sample standard deviation and other information into Equation 10.7, we obtain the 95 percent confidence interval as

$$779,030 - (1.96)\frac{(1,083,162)}{\sqrt{32}} < \mu < 779,030 + (1.96)\frac{(1,083,162)}{\sqrt{32}}$$

or

$$403,733.23 < \mu < 1,154,326.77$$

A 95 percent confidence interval for the mean external audit fees paid by all companies during the year ranges from $403,733.23 to $1,154,326.77.

10.5 CONFIDENCE INTERVALS FOR μ WHEN σ^2 IS UNKNOWN

In the previous section, we constructed confidence intervals for known population variances. For a large sample size, the assumption of known population variance can be relaxed. In this section, we construct confidence intervals for small sample sizes ($n < 30$) and unknown population variance.

In some cases, it is not possible to obtain a large sample size. For example, we might be interested in constructing a confidence interval for sales in a particular industry that contains only 10 firms. Because of the size of the sample, the normal distribution cannot be used; the central limit theorem applies only to large sample sizes ($n \geq 30$).

Remember that the random variable

$$Z = \frac{\overline{X} - \mu}{\sigma/\sqrt{n}}$$

has a standard normal distribution. However, in our example the sample size is small and the population standard deviation is unknown, so the t statistic discussed in Chapter 9 must be used. Recall that the t statistic has a shape that is very similar to the normal distribution, which is bell-shaped and symmetrically distributed.

However, the t distribution has fatter tails than the Z (the standardized normal distribution) distribution. This is because using the sample standard deviation rather than the population standard deviation introduces uncertainty. The similarities and differences between the two distributions are shown in Figure 10.7.

The probability in the tail and the shape of the distribution depend on the number of degrees of freedom, $n - 1$, where n is the number of observations in the sample. Table A4 in Appendix A shows the relationship between the number of degrees of freedom and the t value of the distribution. For example, if the number of degrees of freedom equals 6 and the area in both tails combined is .100, then the t value is 1.943. If the degrees of freedom equal 12, then the t value is 1.782. In addition, the t value will be greater than the Z value for the same area under the curve. As the number of degrees of freedom approaches infinity, the t distribution approaches the Z distribution. This is due to the fact that as n becomes larger, the sample standard deviation s approaches the population standard deviation σ. We use a cutoff of $n = 30$ to distinguish between large and small samples because there is little difference between the two distributions at that sample size.

```
MTB > SET C1
DATA> -4:4/0.1
DATA> END
MTB > PDF C1 C2;
SUBC> NORMAL 0 1.
MTB > PDF C1 C3;
SUBC> T 3.
MTB > GPLOT;

SUBC> LINE 0 1 C2 C1;
SUBC> LINE 3 1 C3 C1;
SUBC> YLABEL 'PDF VALUES';
SUBC> XLABEL 'z or t Values'.
```

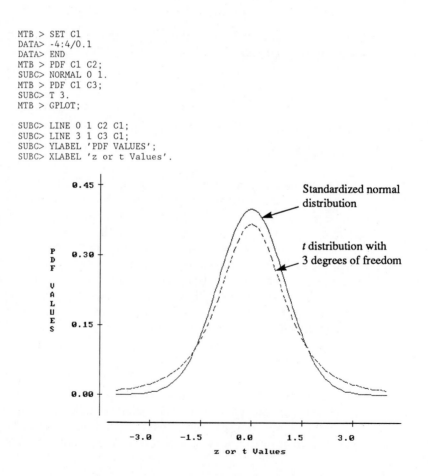

Figure 10.7

Standardized Normal Distribution Versus t Distribution

The random variable t with $\nu = (n-1)$ degrees of freedom can be defined as

$$t_\nu = \frac{\overline{X} - \mu}{s/\sqrt{n}} \tag{10.8}$$

which follows the t distribution with $(n-1)$ degrees of freedom. Note that this is similar to the Z statistic, but the sample standard deviation is used instead of the population standard deviation.

To use the t distribution for confidence intervals, we need to take areas on both sides of the distribution. Thus if we want a 95 percent confidence interval with a sample size of 10, we divide 5 percent by 2 (.05/2 = .025) and look up .025 with $(10-1)$ degrees of freedom in the t tables to arrive at a t value. The positive value will correspond to the right-side tail and the negative value to the left-side tail. This is shown in Figure 10.8.

Using the t distribution, we find that the confidence interval is

$$1 - \alpha = P\left[-t_{n-1,\alpha/2} < \frac{\overline{X} - \mu}{s/\sqrt{n}} < t_{n-1,\alpha/2} \right] \tag{10.9}$$

$$= P\left[\overline{X} - t_{n-1,\alpha/2} \frac{s}{\sqrt{n}} < \mu < \overline{X} + t_{n-1,\alpha/2} \frac{s}{\sqrt{n}} \right]$$

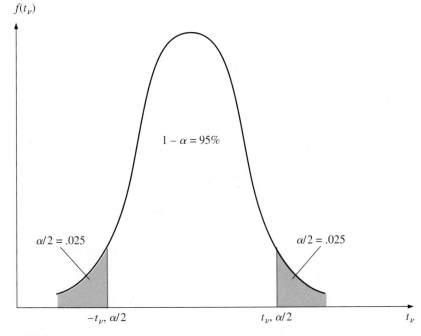

$f(t_\nu)$

$1 - \alpha = 95\%$

$\alpha/2 = .025$ $\alpha/2 = .025$

$-t_\nu, \alpha/2$ $t_\nu, \alpha/2$ t_ν

Figure 10.8

A t Distribution with $n = 10$ and $1 - \alpha = 95$ Percent

EXAMPLE 10.7 *95 Percent Confidence Interval for the Average Weight of Football Players*

A random sample yields the following weights of 8 football players, in pounds.

$$250 \ 210 \ 185 \ 242 \ 190 \ 200 \ 220 \ 205$$

The sample mean is $\overline{X} = 212.75$, and the sample standard deviation is $s = 23.34$.

Suppose we want a 95 percent confidence interval. The number of degrees of freedom is $8 - 1 = 7$, and the corresponding α is .05. The t value that is desired is $t_{n-1,\alpha/2} = t_{7,.05/2} = 2.365$.

Substituting all this information into Equation 10.7, we get

$$212.75 - (2.365)\frac{(23.34)}{\sqrt{8}} < \mu < 212.75 + (2.365)\frac{(23.34)}{\sqrt{8}}$$

or

$$193.23 < \mu < 232.27$$

Thus we can say with 95 percent certainty that the true mean weight of the football players is between 193.2 and 232.3 pounds. The MINITAB solution for this example is shown in Figure 10.9.

```
MTB > SET INTO C1
DATA> 250 210 185 242 190 200 220 205
DATA> END
MTB > TINTERVAL WITH 95% CONFIDENCE USING C1

          N      MEAN    STDEV   SE MEAN   95.0 PERCENT C.I.
C1        8     212.75   23.34    8.25   ( 193.23,   232.27)

MTB > PAPER
```

Figure 10.9

MINITAB Solution to Example 10.7

EXAMPLE 10.8 *90 Percent Confidence Interval for the Average Weight of Football Players*

Suppose a 90 percent confidence interval is constructed for the information given in Example 10.6. The t value is $t_{7,.10/2}$. The 90 percent confidence interval is

$$212.75 - (1.895)\frac{(23.34)}{\sqrt{8}} < \mu < 212.27 + (1.895)\frac{(23.34)}{\sqrt{8}}$$

or

$$197.11 < \mu < 228.39$$

Here, because we have chosen a 90 percent confidence interval, the confidence interval has gotten narrower.

EXAMPLE 10.9 *Estimate for Waiting Time at a Bank*

As part of an effort to improve customer service, a bank pledges not to keep customers waiting in line an unreasonable time. To determine the time interval of waiting in line, the bank collects the following data for 9 customers.

Customer	Waiting Time, minutes
A	4
B	3
C	6
D	2
E	7
F	1
G	3
H	4
I	2

The mean is $\overline{X} = 3.56$, and the standard deviation is $s = 1.94$. The t value is $t_{8,.05/2} = 2.306$. The bank constructs a 95 percent confidence interval for the mean waiting time per customer. It is

$$3.56 - (2.306)\frac{(1.94)}{\sqrt{9}} < \mu < 3.56 + (2.306)\frac{(1.94)}{\sqrt{9}}$$

or

$$2.069 < \mu < 5.051$$

The bank concludes that the true mean number of minutes a customer must wait is between 2.069 and 5.051 minutes with 95 percent probability.

EXAMPLE 10.10 *95 Percent Confidence Interval for the True Mean Incremental Profit of "Successful" Trade Promotion*

Each year, thousands of manufacturers' sales promotions are conducted by North American packaged goods companies. A sample of Canadian packaged goods companies provided information on examples of past sales promotion, including trade promotion. By interviewing the company managers, K. G. Hardy (*Journal of Marketing*, July 1986, Vol. 50, No. 7) identified 21 "successful" sample trade promotions with the mean incremental profit $53,000 and the standard deviation $95,000.

If the population from which the sample is selected has an approximate normal distribution, then the 95 percent confidence interval for the true mean incremental profit of "successful" trade promotion can be calculated in terms of Equation 10.9 as

$$53,000 - (2.086)\frac{(95,000)}{\sqrt{21}} < \mu < 53,000 + (2.086)\frac{(95,000)}{\sqrt{21}}$$

or

$$9,755.99 < \mu < 96,244.01$$

Hardy concluded that the true mean incremental profit of "successful" trade promotions is between $9,755.99 and $96,244.01 with 95 percent probability.

10.6 CONFIDENCE INTERVALS FOR THE POPULATION PROPORTION

Suppose a quality control expert needs to determine the proportion of defective parts for a company—that is, the proportion of a particular item that is returned by the company's customers. Or suppose a political analyst would like to report the proportion of voters who support a particular candidate for a U.S. Senate race. In this section we will derive confidence intervals for population proportions. The concepts are similar to those used in the section on large-sample mean confidence intervals, because the standard normal distribution is used in both.

In Chapter 8 we found that for large sample sizes, the random variable

$$Z = \frac{\hat{p} - p}{\sqrt{p(1 - p)/n}} \qquad (10.10a)$$

$$Z = \frac{\hat{p} - p}{\sqrt{\hat{p}(1 - \hat{p})/n}} \qquad (10.10b)$$

has a normal distribution, where $\hat{p}$ and p are the sample proportion and the population proportion, respectively. Equation 10.10a is defined in terms of population standard deviation, and Equation 10.10b is defined in terms of sample standard deviation. We will use Equation 10.10b to develop a confidence interval for the population proportion.

$$1 - \alpha = P(-z_{\alpha/2} < Z < z_{\alpha/2})$$
$$= P\left[-z_{\alpha/2} < \frac{\hat{p} - p}{\sqrt{\hat{p}(1 - \hat{p})/n}} < z_{\alpha/2}\right]$$

where $z_{\alpha/2}$ is the number such that $P(Z > z_{\alpha/2}) = \alpha/2$. We now move all the terms except the population proportion to the right and left sides of P, which gives a $(1 - \alpha)$ confidence interval.

$$1 - \alpha = P \left\{ \hat{p} - z_{\alpha/2} \sqrt{\frac{\hat{p}(1 - \hat{p})}{n}} < p < \hat{p} + z_{\alpha/2} \sqrt{\frac{\hat{p}(1 - \hat{p})}{n}} \right\} \quad (10.11)$$

In order for us to use the Z statistic, the sample size must be large. For most purposes, a sample size that is greater than 30 will do. By looking at the confidence interval, we can see that the larger the α value, the smaller the Z's (and the confidence interval) will be. In addition, as the sample size increases, the confidence interval gets narrower.

EXAMPLE 10.11 *95 Percent Confidence Interval for Voting Proportion*

Suppose that a random sample of 100 voters is taken, and 55 percent of the sample supports the incumbent candidate. Construct a 95 percent confidence interval for this proportion. The sample size is large, so we can use Equation 10.11 to obtain the interval.

The Z value for a 95 percent confidence interval is $z_{.05/2} = 1.96$.

$$.55 - 1.96 \sqrt{\frac{.55(1 - .55)}{100}} < p < .55 + 1.96 \sqrt{\frac{.55(1 - .55)}{100}}$$

or

$$.452 < p < .648$$

The 95 percent confidence interval for the true proportion of voters supporting the incumbent goes from 45.2 percent to 64.8 percent.

Now suppose we want a 90 percent confidence interval.

$$.55 - 1.645 \sqrt{\frac{.55(1 - .55)}{100}} < p < .55 + 1.645 \sqrt{\frac{.55(1 - .55)}{100}}$$

or

$$.468 < p < .632$$

As we have come to expect, the 95 percent confidence interval is wider than the 90 percent confidence interval.

EXAMPLE 10.12 *95 Percent Confidence Interval for Commodity Preference Proportion*

A marketing firm discovers that 65 percent of the 30 customers who participated in a blind taste test prefer brand A to brand B. The firm develops a 95 percent confidence interval in terms of Equation 10.10b for the number of people who prefer brand A.

$$.65 - 1.96 \sqrt{\frac{.65(1 - .65)}{30}} < p < .65 + 1.96 \sqrt{\frac{.65(1 - .65)}{30}}$$

or

$$.479 < p < .821$$

The firm can be 95 percent certain that the true proportion of those who prefer brand A lies between 47.9 percent and 82.1 percent. If the sample size were increased, the confidence interval would be narrower.

EXAMPLE 10.13 *95 Percent Confidence Interval for the Proportion of Working Adults Who Use Computer Equipment*

A recent study (*Journal of Advertising Research,* April/May 1984) to find the proportion of working adults using computer equipment (personal computers, microcomputers, computer terminals, or word processors) on the job employed the random sample approach to survey 616 working adults. The survey revealed that 184 of the adults now regularly use computer equipment on the job.

A 95 percent confidence interval for working adults' computer usage can be calculated in accordance with Equation 10.11. In this case, $n = 616$; $\hat{p} = 184/616 = .299$; $z_{.05/2} = 1.96$.

$$\sqrt{\frac{\hat{p}(1 - \hat{p})}{n}} = \sqrt{\frac{(.299)(.701)}{616}} = .018.$$

Substituting all of this information into Equation 10.11, we obtain

$$.299 - (1.96)(.018) < p < .299 + (1.96)(.018)$$

or

$$.264 < p < .334$$

In other words, the 95 percent confidence interval for the true proportion of all working adults who regularly use computer equipment on the job is between .264 and .334.

10.7 CONFIDENCE INTERVALS FOR THE VARIANCE

Despite an increased awareness of the importance of quality, and despite the subsequent introduction of robots and other precision tools into factories, some variance is inevitable in any manufacturing process. Manufacturers need to know whether the variance falls within an acceptable range. To determine this, they construct confidence intervals. We have seen that confidence intervals can be constructed by using the normal distribution (large-sample population means) and the

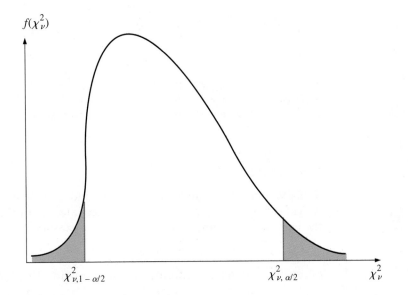

Figure 10.10

χ^2 Distribution

t distribution (small-sample population means). For variance we must use the chi-square distribution, because, as we noted in Chapter 9, the variance is χ^2-distributed.

Figure 10.10 shows the chi-square distribution and its confidence interval. The area of the middle part is $(1 - \alpha)$, the area of the right tail is $\alpha/2$, and the area of the left tail is $\alpha/2$. The number corresponding to the right tail is $\chi^2_{\nu,\alpha/2}$; the number for the left tail is $\chi^2_{\nu,1-\alpha/2}$, where ν is the degrees of freedom (the number in the sample less one, $n - 1$). For example, if a 90 percent confidence interval is desired with a sample size of 20, then the critical values are $\chi^2_{19,.05} = 30.1435$ and $\chi^2_{19,.95} = 10.1170$.

As discussed in Chapter 9, the random variable

$$\chi^2_\nu = \frac{(n - 1)s_X^2}{\sigma^2}$$

is a chi-square random variable with $\nu = (n - 1)$ degrees of freedom. The confidence interval for the population variance is derived as follows:

$$1 - \alpha = P(\chi^2_{\nu,1-\alpha/2} < \chi^2_\nu < \chi^2_{\nu,\alpha/2})$$

$$= P\left(\chi^2_{\nu,1-\alpha/2} < \frac{(n - 1)s_X^2}{\sigma^2} < \chi^2_{\nu,\alpha/2} \right)$$

$$= P\left(\frac{(n - 1)s_X^2}{\chi^2_{\nu,\alpha/2}} < \sigma^2 < \frac{(n - 1)s_X^2}{\chi^2_{\nu,1-\alpha/2}} \right) = 1 - \alpha \qquad (10.12)$$

Hence if s_X^2 is the sample variance estimate, it follows that a $100(1 - \alpha)$ percent confidence interval for a population variance is given by

$$\frac{(n-1)s_X^2}{\chi_{\nu,\alpha/2}^2} < \sigma^2 < \frac{(n-1)s_X^2}{\chi_{\nu,1-\alpha/2}^2} \qquad (10.13)$$

This formula gives us a $(1 - \alpha)$ confidence interval for the population variance. We find a confidence interval for the standard deviation simply by taking the square root of the upper and lower limits.

EXAMPLE 10.14 *Confidence Intervals for σ^2*

Suppose a random sample of 30 bags of sand is taken and the sample variance of weight is 5.5. Find a 95 percent confidence interval for the population variance of the bags. In this example, $\alpha = .05$ and

$$\chi_{\nu,\alpha/2}^2 = \chi_{29,.025}^2 = 45.7222 \quad \text{and} \quad \chi_{\nu,1-\alpha/2}^2 = \chi_{29,.975}^2 = 16.0471$$

Substituting all related information into Equation 10.13, we obtain

$$\frac{(29)(5.5)}{45.7222} < \sigma^2 < \frac{(29)(5.5)}{16.0471}$$

or

$$3.488 < \sigma^2 < 9.94$$

This implies that the 95 percent confidence interval for the sample variance of sand-bag weight is between 3.488 and 9.94.

10.8 AN OVERVIEW OF STATISTICAL QUALITY CONTROL[5]

Consumers are generally looking for a product that offers reasonable quality at a reasonable price. The quality of a good or service is often perceived by the consumer in terms of appearance, operation, and reliability. Examples of these three dimensions are listed in Table 10.2. Therefore, product or service quality should generally be managed and controlled in accordance with these criteria.

Stevenson, Grant and Leavenworth, Griffith, Evans and Lindsay, and others have shown that statistical methods are key ingredients for the management and

---

[5]This and the next section are essentially drawn from J. R. Evans and W. M. Lindsay (1989), *The Management and Control of Quality* (St. Paul, MN: West), Chapters 12, 13, and 15. Reprinted by permission by West Publishing Company. All rights reserved.

Table 10.2 Examples of Dimensions of Quality

Product/Service	Appearance	Operation	Reliability
Color TV	Cabinetry, position of controls, exterior workmanship	Clarity, sound, ease of adjustment, reception, realistic colors	Frequency of repair
Clothing	Seams matched, no loose threads or missing buttons, pattern matched, fit, style	Warm/cool, resistance to wrinkles, color-fastness	Durability
Restaurant meal	Color, arrangement, atmosphere, cleanliness, friendliness of servers	Taste and consistency of food	Indigestion?

Source: Stevenson (1990), Table. 16.1, p. 808.

control of product or service quality.[6] Three basic statistical quality control issues are

How much to inspect and how often
Acceptance sampling
Process control

The first two issues involve determination of the sample size and the sampling methods used for statistical quality control, which will be discussed in this section. Process control consists of (1) the construction and application of control charts in doing quality control and (2) related statistical analysis and testing of control charts. The construction and application of control charts will be discussed in the next section. Further statistical analysis and testing of control charts will be discussed in Chapter 11.

The Sample Size of an Inspection

The amount of inspection can range from conducting no inspection at all to scrutinizing each item many times. Low-cost, high-volume items such as paperclips, paper cups, and wooden rulers often require little inspection because the cost associated with defectives is low and the processes of production are usually very reliable. On the other hand, high-cost, low-volume items such as critical components

---

[6]W. J. Stevenson (1990), *Production/Operations Management,* 3rd ed. (Homewood, IL: Irwin); E. L. Grant and R. S. Leavenworth (1988), *Statistical Quality Control,* 6th ed. (New York: McGraw-Hill); G. K. Griffith (1989), *Statistical Process Control Methods for Long and Short Runs* (Milwaukee, WI: ASQC Quality Press); and J. R. Evans and W. M. Lindsay (1989), *The Management and Control of Quality,* (St. Paul, MN: West).

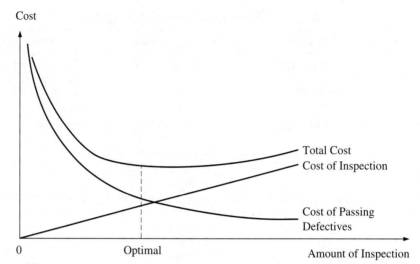

Figure 10.11

The Amount of Inspection Is Optimal When the Sum of the Cost of Inspection and the Cost of Passing Defectives Is Minimized
Source: W. J. Stevenson, *Production/Operations Management,* 3rd ed., 1990, Figure 16.7, p. 826, reprinted by permission of Richard D. Irwin.

of an occupied space vehicle are closely scrutinized because of the risk to human safety and high cost of mission failure. The majority of quality control applications lie somewhere between these two extremes, and here sampling comes into play.

The sample size of sampling surveys is determined by finding the proper trade-off between the costs and the benefits of inspection. The amount of inspection is optimal when the total cost of conducting the inspection *and* of passing defectives is minimized, as indicated in Figure 10.11.[7]

Acceptance Sampling and Its Alternative Plans

Statistical quality control generally uses only the sampling approach to examine the quality of a product. In **acceptance sampling,** the decision whether to accept an entire lot of a product or service is based only on a sample of the lot. By a **lot** we generally mean an amount of material that can be conveniently handled. It may consist of a certain number of items, a case, a day's production, a car load, or such similar quantity. These lots might be described as **convenience lots.** The following two sampling plans are customarily based on convenience lots.

[7]The formula for determining optimal sample size can be found in Section 20.3.

Single-Sampling Plans

In a **single-sampling plan,** one random sample with sample size n is drawn from each lot with N items, and every item in the sample is examined and classified as either good or defective. If any sample contains more than a specified number of defectives, c, then that lot is rejected.

Double-Sampling Plans

Double-sampling plans provide for taking a second sample when the results of a first sample are marginal, as is often the case when lots are of borderline quality. Such plans are commonly based on five statistics:

n_1 = size of the first sample
c_1 = acceptance number of defectives for the first sample (n_1)
n_2 = size of the second sample
c_2 = acceptance number of defectives for $n_1 + n_2$
k_1 = retest number for the first sample

For example, say $c_1 = 3$, $c_2 = 8$, $k_1 = 6$, $n_1 = 25$, and $n_2 = 40$. This sample plan dictates the lot size (the size of the initial sample), $n_1 = 25$ items, and it specifies the accept/reject criteria for the initial sample, $c_1 = 3$ and $k_1 = 6$. If 3 or fewer defectives are found, it tells us, accept the lot; if more than 6 defectives are found, reject the lot; and if 4, 5, or 6 defectives are found, take a second sample with sample size $n_2 = 40$.

The Advantages of Double-Sampling Plans

A double-sampling plan makes n_1 smaller than the sample size for a single-sampling plan that has essentially the same ability to discriminate between lots of high quality and lots of low quality. This means that good-quality lots of product or service are accepted most of the time on the basis of smaller samples than a comparable single-sampling plan requires. Also, bad lots are generally rejected on the basis of the first sample. A double-sampling plan also gives lots of marginal quality a second chance. This feature appeals to practical-minded production managers.

How well it discriminates between lots of high quality and lots of low quality is an important feature of a sampling plan. The ability of a sampling plan to discriminate can be analyzed and tested, as we will see in Appendix 11A.

Process Control

Process control is concerned with ensuring that *future* output is acceptable. Toward that end, periodic samples of process output are taken and evaluated. If the output is acceptable, the process is allowed to continue; if the output is not acceptable, the process is stopped and corrective action is instituted. The basic elements of control for quality, costs, labor power, accidents, and just about anything else are the same:

1. Define what is to be controlled.
2. Consider how measurement for control will be accomplished.
3. Define the level of quality that is to be the standard of comparison.
4. Distinguish between random and nonrandom variability, and determine what process is out of control.
5. Take corrective action and evaluate that action.

Among these elements of quality control, determining whether an output process is *in control* or *out of control* is the most important task. To do so, we need to analyze the statistical product distribution of the process. If the process variation is due to random variability (common causes of variation), then the process is in control. If the process variation is due to nonrandom variability (special causes of variation), then the process is out of control. The control charts discussed in the next section can help us differentiate between process variation attributable to common causes and variation due to special causes.

10.9 CONTROL CHARTS FOR QUALITY CONTROL

Control charts were first proposed by Walter Stewart at Bell Laboratories in the 1920s. More recently the control chart has become a principal tool in assisting businesses in Japan, the United States, and elsewhere in their quality and productivity efforts.

A **control chart** is a graphical tool for describing the state of control of a process. Figure 10.12 illustrates the general structure of a control chart. Time is measured on the horizontal axis, which usually corresponds to the average value of the quality characteristic being measured on the vertical axis. Two other horizontal lines (usually dashed) represent the **upper control limit** (UCL) and the **lower control limit** (LCL). These limits are chosen such that there is a high probability (generally greater than .99) that sample values will fall between them if the process is in control. Samples are chosen over time, plotted on the appropriate chart, and analyzed. Basic statistical concepts used to draw the control charts include the expected value, standard deviation, and confidence interval, which we discussed earlier in this chapter. The control charts we will examine in this section are the $\overline{X}$-chart, the $\overline{R}$-chart, and the S-chart.

$\overline{X}$-Chart

A statistical quality control chart for means ($\overline{X}$-chart) relies on the interval estimate concept discussed in Sections 10.4 and 10.5. The $\overline{X}$-**chart** is used to depict the variation in the centering process. Say copper rods that have a sample mean diameter

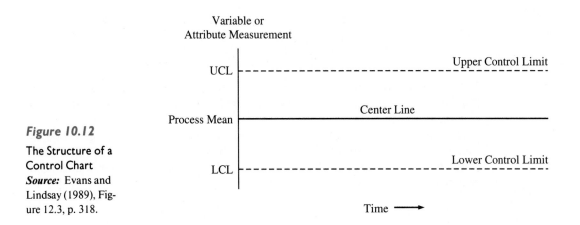

Figure 10.12

The Structure of a Control Chart
Source: Evans and Lindsay (1989), Figure 12.3, p. 318.

$\overline{X}$ of 3 centimeters and a given[8] standard deviation $\sigma_{\overline{X}}$ of .15 centimeters are being produced by a particular process. It is known that the diameter measurements are normally distributed. The quality control manager might like to determine what control limits will include 99.73 percent of the sample mean if the process is generating random output (around the mean) for sample size $n = 36$.

To solve this problem, the quality control department establishes an upper control limit (UCL) and a lower control limit (LCL) in accordance with Equation 10.7. For the upper control limit,

$$P(\overline{X} > \text{UCL}) = P\left(\frac{\overline{X} - \mu}{\sigma_{\overline{X}}} > \frac{\text{UCL} - \mu}{\sigma_{\overline{X}}}\right)$$

$$= P\left(z > \frac{\text{UCL} - 3}{.15/\sqrt{36}}\right) = .00135$$

For the lower control limit,

$$P(\overline{X} < \text{LCL}) = \left(\frac{\overline{X} - \mu}{\sigma_{\overline{X}}} < \frac{\text{LCL} - \mu}{\sigma_{\overline{X}}}\right)$$

$$= \left(z < \frac{\text{LCL} - \mu}{.15/\sqrt{36}}\right) = .00135$$

From Table A3 of Appendix A, we can solve for UCL and LCL as follows:[9]

$$\frac{\text{UCL} - 3}{.025} = 3$$

$$\frac{\text{LCL} - 3}{.025} = -3$$

[8]In quality control, *given standard deviation* means the quality standards of a product are given.

[9]In quality control work, control limits are three standard errors on either side of the mean of the sampling distribution. These limits are called $3 - \sigma$ limits.

From these two equations, we obtain

$$UCL = .075 + 3 = 3.075 \text{ centimeters}$$
$$LCL = 3 - .075 = 2.925 \text{ centimeters}$$

This example shows that we can calculate control limits for the $\overline{X}$-chart for given standards by using the interval estimate for the population mean μ when σ^2 is known, as was discussed in Section 10.4.

In quality control, a sequence of k samples with m_j observations each is taken over time on a measurable characteristic of the output of a production process. From this sample, the sample mean $\overline{X}_i$ ($i = 1, 2, \ldots, k$) and the overall mean can be defined as

$$\overline{X}_i = \sum_{j=1}^{n} X_{ij}/n \quad \text{and}$$

$$\overline{X} = \sum_{i=1}^{k} \overline{X}_i/k$$

Taking into account $\overline{X}$, the given standard deviation σ_X, and the logic illustrated in the foregoing example, we see that UCL and LCL can be defined as

$$UCL_{\overline{X}} = \overline{X} + A\sigma_X \tag{10.14a}$$

$$LCL_{\overline{X}} = \overline{X} - A\sigma_X \tag{10.14b}$$

where $A = 3/\sqrt{n}$, $\overline{X}$ = mean of sample means, and σ_X = given process standard deviation. Equations 10.14a and 10.14b can be used to construct the $\overline{X}$-chart when standards are given (σ_X is assumed to be known). The value of A can be found in Table A13 of Appendix A.

If the process standard deviation is not known, then it must be estimated from sample standard deviations as indicated in Equation 10.15.

$$\overline{s} = \sum_{i=1}^{k} s_i/k \tag{10.15}$$

where

$$s_i = \sqrt{\sum_{j=1}^{n} \frac{(X_{ij} - \overline{X}_j)}{n - 1}}$$

and X_{ij} is the jth observation in the ith sample. The sample standard deviation, of course, is a biased estimator of population standard deviation. If the population distribution is normal, it can be shown that

$$E(\overline{s}) = C_4\sigma_X \tag{10.16}$$

where C_4 is a number that can be calculated as a function of the sample size n. Then the standard deviation estimate for $\overline{X}$ can be defined as $\bar{s}/(C_4\sqrt{n})$. This information enables us to write Equations 10.14a and 10.14b as[10]

$$\text{UCL}_{\overline{X}} = \overline{X} + A_3\bar{s} \tag{10.17a}$$

$$\text{LCL}_{\overline{X}} = \overline{X} - A_3\bar{s} \tag{10.17b}$$

where $A_3 = 3\bar{s}/(C_4\sqrt{n})$, which can be found in Table A13.

In quality control, we often use sample range instead of sample standard deviation to estimate both upper and lower control limits for our $\overline{X}$ chart. Recall from Equation 4.11 in Chapter 4 that the range R of a sample is the difference between the maximum and minimum measurement in the sample. The range can be used to obtain an unbiased estimator for σ defined as follows[11]

$$\hat{\sigma} = \frac{\overline{RG}}{d_2}$$

where

$$\overline{RG} = \sum_{i=1}^{k} R_i/k, \qquad R_i$$

equals the range in ith sample, and d_2 is a constant that can be found in Table A13.

If sample range instead of sample standard deviation is used to replace the process standard deviation, then the control limits can be defined as

$$\text{UCL}_{\overline{X}} = \overline{X} + A_2\overline{RG} \tag{10.18a}$$

$$\text{LCL}_{\overline{X}} = \overline{X} - A_2\overline{RG} \tag{10.18b}$$

where $A_2 = 3/d_2\sqrt{n}$ can be found in Table A13 of Appendix A, and d_2 is a function of sample size n.

$\overline{R}$-Chart and S-Chart

Besides the variation of centering (mean), we are also interested in the variation of dispersion (standard deviation or range) in quality control. The $\overline{R}$-chart is used to depict the variation of the ranges of the samples. The S-chart is used to depict the variation of standard deviation. In other words, both the S-chart and $\overline{R}$-chart can be used to detect changes in process variation. The $\overline{R}$-chart is used more frequently than the S-chart because the range is much easier to calculate than the standard deviation.

---

[10]If the underlying sampling is the Poisson distribution as discussed in Section 6.7, then the $\bar{x}$ and $\bar{s}$ can be defined as $\bar{c}$ and $\sqrt{\bar{c}}$, respectively ($\bar{c}$ is defined as the mean number of defects per unit). In this situation the $\overline{X}$-chart defined in Equations 10.17a and 10.17b is called the C-chart (see Evans and Lindsay, 1989, pp. 366–368).

[11]See T. T. Ryan (1989), *Statistical Methods for Quality Improvement* (New York: Wiley) for a detailed discussion of this relationship.

It can be shown that $E(\bar{s}) = C_4\sigma_X$, as we have noted, and that the standard deviation of $\bar{s}$ is $\sigma_X\sqrt{1 - C_4^2}$. When no standards are given, we use $\bar{s}$ as an estimate of $C_4\sigma_{\bar{X}}$. Then the upper and lower limits of the S-chart can be defined as

$$\text{UCL}_s = B_4\bar{s} \tag{10.19a}$$

$$\text{LCL}_s = B_3\bar{s} \tag{10.19b}$$

where $B_4 = 1 + 3\sqrt{1 - C_4^2}/C_4$ and $B_3 = 1 - 3\sqrt{1 - C_4^2}/C_4$. Both can be found in Table A13.

If no standards are given, the upper and lower limits of the $\bar{R}$-chart can be defined as

$$\text{UCL}_{\bar{R}} = D_4\overline{\text{RG}} \tag{10.20a}$$

$$\text{LCL}_{\bar{R}} = D_3\overline{\text{RG}} \tag{10.20b}$$

where $\overline{\text{RG}}$ is the average of sample ranges and where $D_4 = 1 + 3d_3/d_2$ and $D_3 = 1 - 3d_3/d_2$ can be found in Table A13.

The $\bar{X}$-chart, S-chart, and $\bar{R}$-chart, then, all use the confidence interval concept to construct upper and lower limits. Now we will use quality control data on Consolidated Auto Supply Company to show how these control charts are constructed.

■   **APPLICATION 10.1** $\bar{X}$-Chart, $\bar{R}$-Chart, and S-Chart for Consolidated Auto Supply Company[12]

The quality control manager has measured the size of U-bolts by taking samples of 5 every hour over 3 shifts. The sample is presented in Table 10.3, which also shows the mean and range of each sample. How do we perform this statistical quality control analysis?

To construct our $\bar{X}$-chart and $\bar{R}$-chart, we first compute the average mean $\bar{x}$ and average range $\overline{\text{RG}}$ as follows:

$$\bar{X} = \frac{10.7 + 10.77 + \cdots + 10.66}{24} = 10.7171$$

$$\overline{\text{RG}} = \frac{.20 + .20 + \cdots + .10}{24} = .1792$$

Using the information on $\bar{X}$, $\overline{\text{RG}}$, and $n = 5$, we calculate control limits for our $\bar{X}$-chart and $\bar{R}$-chart in accordance with Equations 10.18a, 10.18b, 10.20a, and 10.20b.

$$\text{UCL}_{\bar{X}} = 10.7171 + .58(.1792) = 10.8210$$

$$\text{LCL}_{\bar{X}} = 10.7171 - .58(.1792) = 10.6132$$

$$\text{UCL}_{\bar{R}} = 2.11(.1792) = .3782$$

$$\text{LCL}_{\bar{R}} = 0(.1792) = 0$$

[12]This example is drawn from J. R. Evans and W. M. Lindsay (1989), *The Management and Control of Quality* (St. Paul, MN: West), pp. 317–323 and pp. 359–360.

Table 10.3 Sample Means and Ranges for Consolidated Auto Supply Company

Sample	Observations					Mean	Range
1	10.65	10.70	10.65	10.65	10.85	10.70	.20
2	10.75	10.85	10.75	10.85	10.65	10.77	.20
3	10.75	10.80	10.80	10.70	10.75	10.76	.10
4	10.60	10.70	10.70	10.75	10.65	10.68	.15
5	10.70	10.75	10.65	10.85	10.80	10.75	.20
6	10.60	10.75	10.75	10.85	10.70	10.73	.25
7	10.60	10.80	10.70	10.75	10.75	10.72	.20
8	10.75	10.80	10.65	10.75	10.70	10.73	.15
9	10.65	10.80	10.85	10.85	10.75	10.78	.20
10	10.60	10.70	10.60	10.80	10.65	10.67	.20
11	10.80	10.75	10.90	10.50	10.85	10.76	.40
12	10.85	10.75	10.85	10.65	10.70	10.76	.20
13	10.70	10.70	10.75	10.75	10.70	10.72	.05
14	10.65	10.70	10.85	10.75	10.60	10.71	.25
15	10.75	10.80	10.75	10.80	10.65	10.75	.15
16	10.90	10.80	10.80	10.75	10.85	10.82	.15
17	10.75	10.70	10.85	10.70	10.80	10.76	.15
18	10.75	10.70	10.60	10.70	10.60	10.67	.15
19	10.65	10.65	10.85	10.65	10.70	10.70	.20
20	10.60	10.60	10.65	10.55	10.65	10.61	.10
21	10.50	10.55	10.65	10.80	10.80	10.66	.30
22	10.80	10.65	10.75	10.65	10.65	10.70	.15
23	10.65	10.60	10.65	10.60	10.70	10.64	.10
24	10.65	10.70	10.70	10.60	10.65	10.66	.10

Source: Evans and Lindsay (1989), Table 12.2, p. 319.

The $\overline{X}$-chart and $\overline{R}$-chart for Consolidated Auto Supply Company are displayed in Figures 10.13 and 10.14, respectively. These control charts can be used to do statistical quality control analysis.

The location of points and patterns of points in a control chart makes it possible to determine, with only a small chance of error, whether a process is in a state of statistical control. In both Figures 10.13 and 10.14, the chance that a sample mean or range will fall outside the control limits is only .27 percent. Therefore, the first indication that a process may be out of control is a point lying outside the control limits. In the $\overline{R}$-chart, sample 11 is outside the UCL limit, indicating that the variability of the process has changed. In this case, it is found that the change in process variability is due to the fact that a substitute operator was used. In the $\overline{X}$-chart, sample 21 is outside the LCL, and samples 18 through 24 are all on one side. This indicates that the process mean has shifted. In this case, it is found that the shift in process mean occurred because nonconforming material was used.

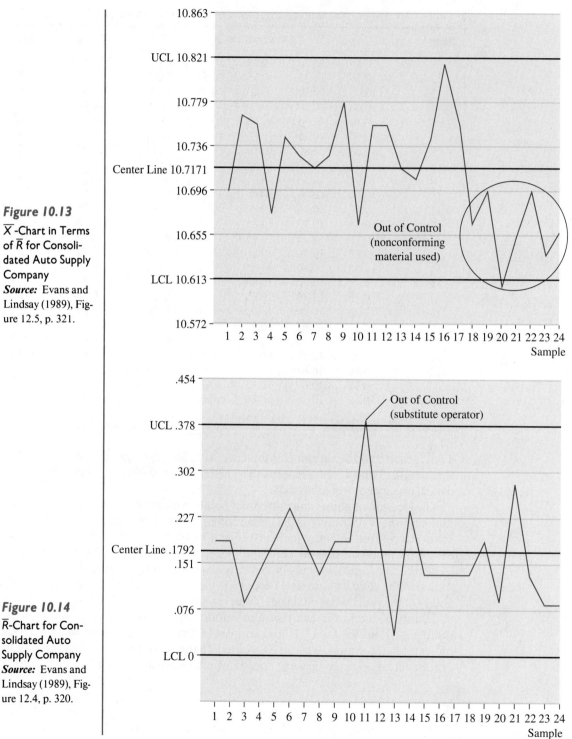

Figure 10.13
$\overline{X}$-Chart in Terms
of $\overline{R}$ for Consoli-
dated Auto Supply
Company
Source: Evans and
Lindsay (1989), Fig-
ure 12.5, p. 321.

Figure 10.14
$\overline{R}$-Chart for Con-
solidated Auto
Supply Company
Source: Evans and
Lindsay (1989), Fig-
ure 12.4, p. 320.

From Table 10.3, we find that the standard deviation of the observation is $\bar{s} = .07958$. Substituting $\bar{X} = 10.7171$, $\bar{s} = .07958$, $n = 5$, $A_3 = 1.427$, $B_3 = 0$, and $B_4 = 2.089$ (from Table A13 in Appendix A) into Equations 10.16a, 10.16b, 10.17a, and 10.17b, we obtain the following control limits for the S-chart and $\bar{X}$-chart.

$$\text{UCL}_s = 2.089(.07958) = .1662$$

$$\text{LCL}_s = 0(.07958) = 0$$

$$\text{UCL}_{\bar{X}} = 10.7171 + 1.427(.07958) = 10.8307$$

$$\text{LCL}_{\bar{X}} = 10.7171 - 1.427(.07958) = 10.6035$$

The $\bar{X}$-chart and S-chart are displayed in Figures 10.15 and 10.16, respectively. Both charts indicate that the product process is in a state of statistical control.

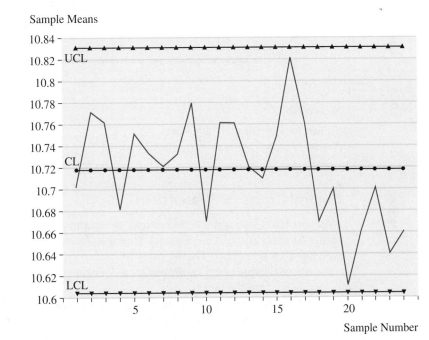

Figure 10.15

$\bar{X}$-Chart in Terms of $\bar{s}$ for Consolidated Auto Supply Company
Source: Evans and Lindsay (1989), Figure 12A.2, p. 377.

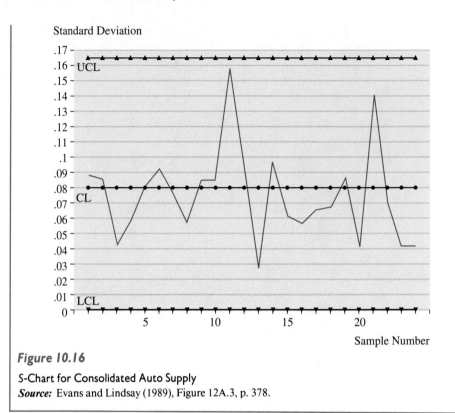

Figure 10.16

S-Chart for Consolidated Auto Supply

Source: Evans and Lindsay (1989), Figure 12A.3, p. 378.

Control Charts for Proportions

Control charts for proportions are used when the process characteristic is counted rather than measured. The **P-chart** is used to measure the percentage of defectives generated by a process. The theoretical basis for a *P*-chart is the binomial distribution (see Chapter 6). Conceptually, a *P*-chart is constructed and used in much the same way an $\overline{X}$-chart is.

Let $\hat{P}_i$ be the fraction of defectives in the *i*th sample with *n* observations, then the center line on a *P*-chart is the average fraction of defectives for *k* samples as defined as:

$$\overline{P} = \frac{\sum\limits_{i=1}^{k}}{k} \hat{P}_i$$

The standard deviation associated with $\overline{P}$ is

$$s_{\overline{P}} = \sqrt{\overline{P}(1 - \overline{P})/n}$$

and the upper and lower control limits are

$$\text{UCL}_{\bar{P}} = \bar{P} + 3s_{\hat{P}} \qquad (10.21a)$$

$$\text{LCL}_{\bar{P}} = \bar{P} - 3s_{\hat{P}} \qquad (10.21b)$$

◼ | **APPLICATION 10.2** *P*-Chart for Quality Control at the Newton Branch Post Office[13]

In the post office, operators use automated sorting machines that read the ZIP code on a letter and divert the letter to the proper carrier route. Over 1 month's time, 25 samples of 100 letters were chosen, and the numbers of errors were recorded. This information is summarized in Table 10.4. The fraction nonconforming is found by dividing the number of errors by 100. The average fraction nonconforming, $\bar{P}$, is determined to be

$$\bar{P} = \frac{.03 + .01 + \cdots + .01}{25} = .022$$

Table 10.4 Sorting Errors at the Newton Branch Post Office

Sample	1	2	3	4	5	6	7	8	9	10	11	12	13	14	15
Errors	3	1	0	0	2	5	3	6	1	4	0	2	1	3	4
Sample	16	17	18	19	20	21	22	23	24	25					
Errors	1	1	2	5	2	3	4	1	0	1					

Source: Evans and Lindsay (1989), Table 12.4, p. 333.

The standard deviation is

$$s_{\bar{P}} = \sqrt{\frac{.022(1 - .022)}{100}} = .01467$$

Thus the upper control limit, $\text{UCL}_{\bar{P}}$, is $.022 + 3(.01467) = .066$, and the lower control limit, $\text{LCL}_{\bar{P}}$, is $.022 - 3(.1467) = -.022$. Because this latter figure is negative, zero is used instead. The control chart for the Newton Branch Post Office is shown in Figure 10.17. The sorting process appears to be in control. If any values had been found above the upper control limit or if an upward trend were evident, it might indicate a need for operators with more experience or for more training of the operators.

[13]This example is drawn from Evans and Lindsay (1989), pp. 332–333.

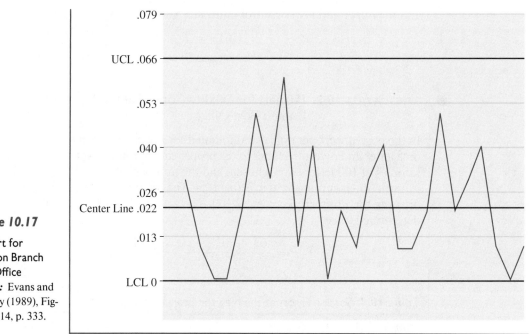

Figure 10.17

P-Chart for Newton Branch Post Office
Source: Evans and Lindsay (1989), Figure 12.14, p. 333.

10.10 FURTHER APPLICATIONS

In the last section we showed how interval estimates for the mean, proportion, and standard deviation can be used to construct quality control charts. The next two examples show how interval estimates can be used for other business applications.

■ | **APPLICATION 10.3** Using Interval Estimates to Evaluate Donors and Donations Models

Britto and Oliver (1986) developed models to forecast (1) the total numbers of donors, gifts, and donations by the end of each year and (2) the cumulative numbers of donors, gifts, and donations received up to and including the *t*th month for the Berkeley Engineering Fund.[14]

Forecasts are based on data from previous campaigns because identical mailings were used from 1982 to 1984. Monthly proportions of total giving have been stable from year to year, as shown in Tables 10.5A and 10.5B. For each mailing date, the forecasters determined the distribution for the number of gifts for each of the four subgroups (see Table 10.6), as well as estimates of the mean and standard deviation of gift size (see Table 10.7).

[14]M. Britto and R. M. Oliver (1986), "Forecasting Donors and Donations," *Journal of Forecasting* 5, 39–55.

Table 10.5A Fraction of Donations Arriving in or Before Month t

Month	Alumni	Parents	Faculty	Friends
1	.09	.04	.01	.05
2	.15	.10	.01	.12
3	.21	.19	.04	.20
4	.29	.30	.16	.28
5	.41	.37	.28	.37
6	.65	.58	.77	.56
7	.74	.75	.90	.71
8	.77	.77	.94	.72
9	.79	.87	.96	.73
10	.84	.88	.97	.75
11	.94	.93	.98	.95
12	1.00	1.00	1.00	1.00

Source: M. Britto and R. M. Oliver (1986), "Forecasting Donors and Donations," *Journal of Forecasting,* 5, 39–55. Reprinted by permission of John Wiley & Sons, Ltd.

Table 10.5B Fraction of Gifts Arriving in or Before Month t

Month	Alumni	Parents	Faculty	Friends
1	.08	.04	.01	.05
2	.13	.10	.01	.12
3	.19	.18	.04	.20
4	.26	.28	.16	.29
5	.38	.35	.27	.39
6	.61	.56	.75	.68
7	.70	.73	.89	.76
8	.72	.76	.93	.78
9	.76	.86	.95	.78
10	.80	.87	.95	.80
11	.90	.90	.96	.95
12	1.00	1.00	1.00	1.00

Source: M. Britto and R. M. Oliver (1986), "Forecasting Donors and Donations," *Journal of Forecasting,* 5, 39–55. Reprinted by permission of John Wiley & Sons, Ltd.

Table 10.6 Prior Expected Numbers of Donations and Gifts in 1984–1985

	Donors	Gifts
Alumni	2807	3265
Parents	248	277
Faculty	117	129
Friends	87	93

Source: M. Britto and R. M. Oliver (1986), "Forecasting Donors and Donations," *Journal of Forecasting,* 5, 39–55. Reprinted by permission of John Wiley & Sons, Ltd.

Table 10.7 1983–1984 Mean and Standard Deviation of Gift Size

	Mean	**Standard Deviation**
Alumni	215	1820
Parents	200	930
Faculty	225	445
Friends	505	1215

Source: M. Britto and R. M. Oliver (1986), "Forecasting Donors and Donations," *Journal of Forecasting,* 5, 39–55. Reprinted by permission of John Wiley & Sons, Ltd.

Parent data from 1982–1983 and 1983–1984 were used to test whether the Poisson distribution on which the model is based is acceptable. Using both Poisson tables and a normal approximation (discussed in Chapter 7), Britto and Oliver constructed 95 percent confidence intervals as shown in Figures 10.18 and 10.19. For both years, the actual donor counts for all months except September fell within 95 percent confidence limits. This leads us to believe that the Poisson distribution assumption is a good one.

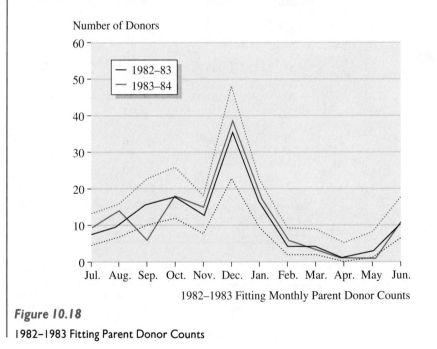

1982–1983 Fitting Monthly Parent Donor Counts

Figure 10.18

1982–1983 Fitting Parent Donor Counts

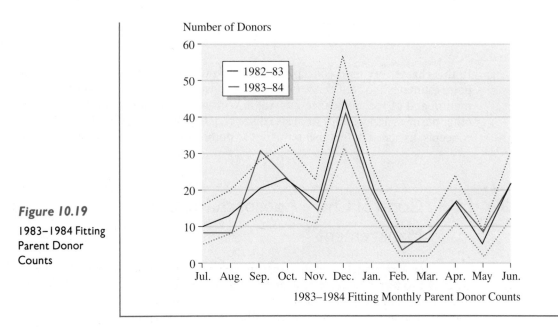

Figure 10.19

1983–1984 Fitting Parent Donor Counts

1983–1984 Fitting Monthly Parent Donor Counts

■ | *APPLICATION 10.4* Shoppers' Attitudes Toward Shoplifting and Shoplifting Prevention Devices

Guffey, Harris, and Laumer (1979) studied attitudes of shoppers toward shoplifting and devices for its prevention.[15] They sampled 403 shopping center patrons. Twenty-four percent of this sample expressed awareness of and uncomfortableness with the use of TV cameras as a device to prevent shoplifting.

Employing this information, we now find a 95 percent confidence interval of the population proportion that was used to describe the attitudes of shoppers about relying on TV devices to prevent shoplifting. Following Equation 10.11, we can define the 95 percent confidence interval for the true proportion p as

$$.24 - z_{.025} \sqrt{\frac{(.24)(.76)}{403}} < p < z_{.025} \sqrt{\frac{(.24)(.76)}{403}}$$

From Table A3 Appendix A, we know that $z_{.025} = 1.96$. Substituting $z_{.025} = 1.96$ into the previous equation, we obtain the 95 percent confidence interval for p as

$$.240 - (1.96)(.021) < p < .240 + (1.96)(.021)$$

or

$$.199 < p < .281$$

Hence, the interval for p ranges from 19.9 percent to 28.1 percent.

[15]H. L. Guffey, J. R. Harris, and J. F. Laumer (1979), "Shopper Attitudes Toward Shoplifting and Shoplifting Prevention Devices," *Journal of Retailing* 55, 75–99.

Summary

In this chapter, we used concepts discussed in Chapters 7, 8, and 9 to show how point estimates and confidence intervals are constructed. Applications of point estimation and of the confidence interval in constructing quality control charts and other business applications were also discussed. In Chapter 11, we will draw on the concepts discussed in this chapter to test hypotheses about sample point estimates.

Appendix 10A Control Chart Approach for Cash Management[16]

The Miller–Orr model for cash management starts with the assumption that there are only two forms of assets: cash and marketable securities. It also allows for cash balance movement in both positive and negative directions and for the optimal cash balance to be a range of values rather than a single point estimate. In other words, the Miller–Orr model uses the control chart approach we discussed in Section 10.9 to do cash management. This model is especially useful for firms that are unable to predict day-to-day cash inflows and outflows.

Figure 10A.1 shows the functioning of the Miller–Orr model. Note that the cash balance is allowed to meander undisturbed as long as it remains within the prede-

Figure 10A.1

The Workings of the Miller–Orr Model
Source: Cheng F. Lee and Joseph E. Finnerty, *Corporate Finance: Theory, Method, and Applications,* Figure 20.3, p. 595. Copyright © 1990 by Harcourt Brace Jovanovich, Inc., reprinted by permission of the publisher, Harcourt Brace Jovanovich, Inc.

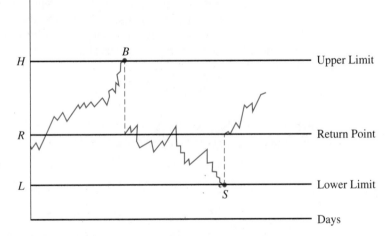

[16]This section on Miller and Orr's model for cash management is taken from Chang F. Lee and Joseph E. Finnerty (1990), *Corporate Finance: Theory, Method, and Applications* (New York: Harcourt) pp. 595–598.

termined boundary range shown by the upper limit H and the lower limit L. At point B, however, the cash balance reaches the maximum allowable level. At this point, the firm could purchase marketable securities in an amount equal to the dashed line, which would lower the cash balance to the "return point" from which it would again be allowed to fluctuate freely. At point S, the firm's cash balance reaches the minimum allowable level. At this point, the firm could sell marketable securities to investors or borrow to bring the cash level back up to the return point.

In how much of a range $(H - L)$ should the cash balance be allowed to fluctuate? According to the Miller–Orr model, the higher the day-to-day variability in cash flows and/or the higher the fixed-transactions cost associated with buying and selling securities, the farther apart the control limits should be set. On the other hand, if the opportunity cost of holding cash (the interest foregone by *not* purchasing marketable securities) is high, the limits should be set closer together. Management's objective is to minimize total costs associated with holding cash. Minimization procedures establish that the spread between the upper and lower cash limit S, the return point R, the upper limit H, and the average cash balance (ACB) are

$$S = 3 \left[\frac{3F\sigma^2}{4k} \right]^{1/3} \tag{10A.1}$$

$$R = \frac{S}{3} + L \tag{10A.2}$$

$$H = S + L \tag{10A.3}$$

$$\text{ACB} = \frac{4R - L}{3} \tag{10A.4}$$

where L = lower limit, F = fixed-transactions cost, k = opportunity cost on a daily basis, and σ^2 = variance of net daily cash flow.

The firm always returns to a point one-third of the spread between the lower and upper limits. With the return point set here, the firm is likely to bump against its lower limit more frequently than against its upper limit. Although this lower point does not minimize the number of transactions and their resulting cost (as the middle point would), it is an optimal point in that it minimizes the sum of transactions cost and foregone-interest cost, the latter of which the firm incurs whenever it holds excessive cash.

To use the Miller–Orr model, the financial manager takes three steps:

1. Set the lower limit.
2. Estimate the variance of cash.
3. Determine the relevant transactions cost and lost-interest cost.

Setting the lower limit is essentially a subjective task, but common sense and experience help. The lower limit is likely to be some minimum safety margin above 0. An important consideration in setting this limit is any bank requirements that must be satisfied.

To estimate the variance of cash flows, the manager can record the net cash inflows and outflows for each of the preceding 100 days and then compute the variance of those 100 observations. This approach requires regular updating, particularly if net cash flows have been unstable over time. One additional aspect to consider in this calculation is the impact of seasonal effects (see Chapter 18), which may also require adjusting the variance estimate.

To determine the relevant transactions cost, the financial manager need only observe what the firm currently pays to buy or sell a security. Interest foregone can be derived from current available market returns on short-term, high-grade securities. The financial manager may want to use a forecasted interest rate for the planning period if a significant change from current interest-rate levels is expected.

We now demonstrate the actual calculations for the Miller–Orr model. First, assume the following:

Minimum cash balance = $20,000
Variance of daily cash flows = $9,000,000 (hence the standard deviation σ = $3,000 per day)
Interest rate = .0329 percent per day
Transactions cost (average) of buying or selling one security = $20

Utilizing these data, we first compute the spread between the lower and upper limits in accordance with Equation 10A.1.

$$\text{Spread} = 3 \left(\frac{3 \times 20 \times 9,000,000}{4 \times .000329} \right)^{1/3}$$
$$= \$22,293$$

Next we compute the upper limit and return point in accordance with Equations 10A.3 and 10A.2.

$$\text{Upper limit} = \text{lower limit} + \text{spread}$$
$$= \$20,000 + \$22,293$$
$$= \$42,293$$

$$\text{Return point} = 20,000 + \left(\frac{22,293}{3} \right)$$
$$= \$27,431$$

Using Equation 10A.4, we find the average cash balance.

$$\text{Average cash balance} = \frac{4(\$27,431) - \$20,000}{3}$$
$$= \$29,908$$

Then, on the basis of our assumed input values and model calculations, we can establish the following rule:

If the cash balance rises to $42,293, invest $42,293 − $27,431 = $14,862 in marketable securities; if the cash balance falls to $20,000, sell $27,431 − $20,000 = $7,431 of marketable securities. Both will restore the cash balance to the return point.

Appendix 10B Using MINITAB to Generate Control Charts

This appendix shows how MINITAB may be used to generate an $\overline{X}$-chart, an $\overline{R}$-chart, and an S-chart based on the following data:

```
MTB  > SET INTO C1
DATA> 10.65  10.70  10.65  10.65  10.85
DATA> 10.75  10.85  10.75  10.85  10.65
DATA> 10.75  10.80  10.80  10.70  10.75
DATA> 10.60  10.70  10.70  10.75  10.65
DATA> 10.70  10.75  10.65  10.85  10.80
DATA> 10.60  10.75  10.75  10.85  10.70
DATA> 10.60  10.80  10.70  10.75  10.75
DATA> 10.75  10.80  10.65  10.75  10.70
DATA> 10.65  10.80  10.85  10.85  10.75
DATA> 10.60  10.70  10.60  10.80  10.65
DATA> 10.80  10.75  10.90  10.50  10.85
DATA> 10.85  10.75  10.85  10.65  10.75
DATA> 10.70  10.70  10.75  10.75  10.70
DATA> 10.65  10.70  10.85  10.75  10.60
DATA> 10.75  10.80  10.75  10.80  10.65
DATA> 10.90  10.80  10.80  10.75  10.85
DATA> 10.75  10.70  10.85  10.70  10.80
DATA> 10.75  10.70  10.60  10.70  10.60
DATA> 10.65  10.65  10.85  10.65  10.70
DATA> 10.60  10.60  10.65  10.55  10.65
DATA> 10.50  10.55  10.65  10.80  10.80
DATA> 10.80  10.65  10.75  10.65  10.65
DATA> 10.65  10.60  10.65  10.60  10.70
DATA> 10.65  10.70  10.70  10.60  10.65
DATA> END
MTB  > PAPER
```

Step 1: We input the data into MINITAB, storing it in Column 1 (C1) as presented in the data above.

Step 2: We can use different commands to ask MINITAB to generate an $\overline{X}$-chart, an $\overline{R}$-chart, or an S-chart.

The command to generate the $\overline{X}$-chart is "GXBARCHART C1 5"; the command to generate the $\overline{R}$-chart is "GRCHART C1 5"; and the command to generate the S-chart is "GSCHART C1 5".

In these three commands, "C1" indicates where the data is located and "5" indicates that there are 5 observations in each sample. The output for the $\overline{X}$-chart, the $\overline{R}$-chart, and the S-chart is presented in Figures 10B.1, 10B.2, and 10B.3.

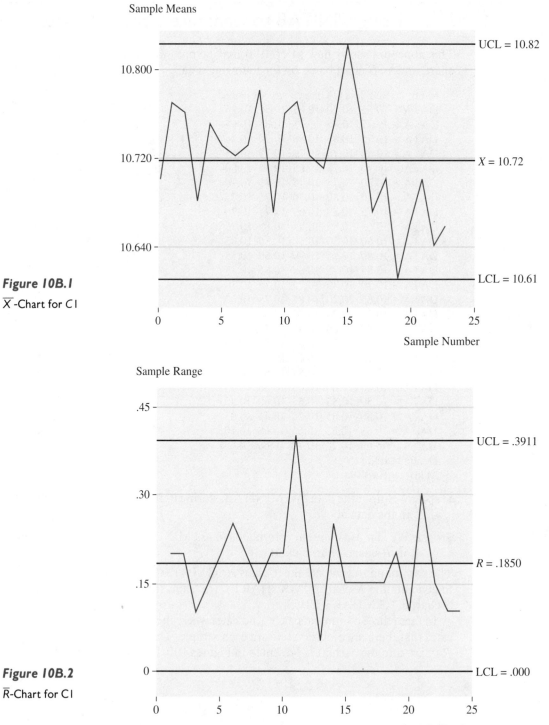

Figure 10B.1

$\overline{X}$-Chart for C1

Figure 10B.2

$\overline{R}$-Chart for C1

Sample Standard Deviation

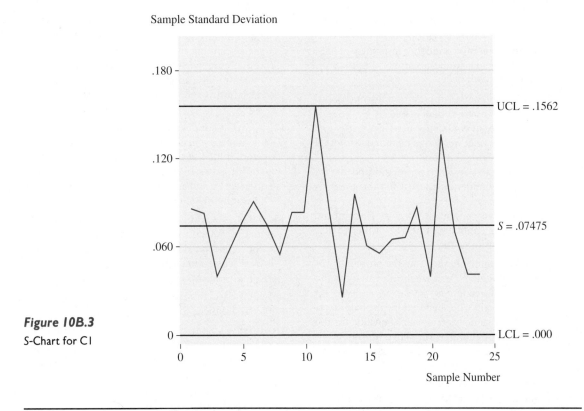

Figure 10B.3
S-Chart for C1

Questions and Problems

1. For the following results from samples drawn from normal populations, what are the best estimates for the mean, the variance, the standard deviation, and the standard deviation of the mean?

   a. $n = 9$, $\Sigma X_i = 36$, $\Sigma(X_i - \overline{X})^2 = 288$
   b. $n = 16$, $\Sigma X_i = 64$, $\Sigma(X_i - \overline{X})^2 = 180$
   c. $n = 25$, $\Sigma X_i = 500$, $\Sigma X_i^2 = 12,400$

2. For each of the following samples drawn from normal populations, find the best estimates for μ, σ^2, σ, and the standard deviation of $\overline{X}$.

   a. 4, 10, 2, 8, 4, 14, 12, 8, 10
   b. -4, 2, -6, 0, 6, 2, 4, 0, -4
   c. 6, 15, 13, 21, 10, 17, 12

3. Find the value of $z_{\alpha/2}$ for the following values of α.

   a. $\alpha = .01$ **b.** $\alpha = .03$ **c.** $\alpha = .002$

4. A stockbroker has taken a random sample of 4 stocks from a large population of low-priced stocks. Stock prices for this population are normally distributed. The sample prices of the 4 stocks are $5, $12, $17, and $10.

   a. Calculate a point estimate of the population mean.
   b. Calculate a point estimate of the population variance. What is your estimate for a population standard deviation?
   c. Calculate a point estimate of the proportion of stocks in this population that are priced at $10 or more.

5. A 90 percent confidence interval for the population mean time (in minutes) needed to finish a certain assembly process is $90 < \mu_x < 130$.

   a. Sketch this interval, indicating the margin for sampling error.

b. If the sample size was $n = 25$, what was the sample standard deviation?

c. To interpret this confidence interval, what did you have to assume about the population? Why?

6. Assuming you have samples from normal populations with known variance, find

  a. The degree of confidence used if $n = 16$, $\sigma = 8$, and the total width of a confidence interval for the mean is 3.29 units.

  b. The sample size when $\sigma^2 = 100$ and the 95 percent confidence interval for the mean is from 17.2 to 22.8.

  c. The known variance when $n = 100$ and the 98 percent confidence interval for the mean is 28.26 units in width.

7. a. Find the value of t such that the probability of a larger value is .005 when the value for the degrees of freedom is very large.

  b. Find the value of t such that the probability of a smaller value is .975 when the value for the degrees of freedom is very large (infinite).

  c. Are the t values essentially the same as corresponding Z values when the value of the degrees of freedom is very large?

8. A poll reported that 48 percent of probable voters seem determined to vote against the President. Assume that this sample was based on a random selection of 789 probable voters. Construct a 99 percent confidence interval for the probable voters who seem determined to vote against the President.

9. A survey of low-income families in New Jersey was designed to determine the average heating costs for a family of 4 during January and February. Heating costs are known to have a standard deviation of $25.75. The economists conducting the study wish to construct a 95 percent confidence interval with a margin for sampling error of no more than $3.95. Find the appropriate sample size.

10. A company has just installed a new automatic milling machine. The time it takes the machine to mill a particular part is recorded for a sample of 9 observations. The mean time is found to be $\overline{X} = 8.50$ seconds, and $S^2 = .0064$. Find a 90 percent confidence interval for the unknown mean time for milling this part.

11. A survey indicated that companies with fewer than 1,000 employees are expected to increase their spending by 20.4 percent. Form a 99 percent confidence interval for the unknown mean increase, assuming that the sample standard deviation is 6.8 percent and the sample size is 346.

12. A study conducted in 1984 reported that the median pay in the United States was $18,700. What difficulties do you see in using this type of study for asessing incomes? Would you be willing to use $18,700 as a point estimate of the central location of U.S. incomes?

13. What is a point estimate? What is a point estimator? What is point estimation? How are these concepts related to the concepts of sampling that we discussed in Chapter 9?

14. What is an unbiased estimator? What is an efficient estimator? What is a consistent estimator? Why are these concepts important?

15. Briefly explain why we sometimes construct confidence intervals for the population mean.

16. Explain what happens to the size of the confidence interval when

  a. The standard deviation increases.

  b. The standard deviation decreases.

  c. The probability content $(1 - \alpha)$ increases from 95 percent to 99 percent.

  d. The sample size increases from 100 to 1,000.

17. Labor economists at the Department of Labor say they have 95 percent confidence that factory workers' earnings will lie between $22,000 and $61,000. Explain what this means.

18. A real estate agent in Connecticut is interested in the mean home price in the state. A random sample of 50 homes shows a mean home price of $175,622 and a sample standard deviation of $37,221. Construct a 95 percent confidence interval for the mean home price.

19. Reconstruct the confidence interval for the mean home prices given in question 18, but this time construct a 99 percent confidence interval. What happens to the size of the confidence interval?

20. Again, use the information given in question 18. This time assume that 100 homes are randomly sampled instead of 50. Construct a 95 percent confidence interval for the mean home price. What happens to the size of the confidence interval?

21. Again, use the information given in question 18. This time, assume that the sample standard deviation is $28,000. Construct a 95 percent confidence interval for the mean home price. What happens to the size of the confidence interval?

22. An auditor randomly samples 75 accounts receivable of a company and finds a sample mean of $128 with a sample standard deviation of $27. Construct a 90 percent confidence interval for the mean accounts receivable.

23. A random sample of 300 residents of a town shows that 55 percent believe the mayor is doing a good job. Construct a 95 percent confidence interval for the proportion of all residents who believe the mayor is doing a good job.

24. A random sample of 200 students at Academic University finds the sample mean grade point average to be 3.10 with a standard deviation of .80. Construct a 99 percent confidence interval for the mean grade point average.

25. An insurance company is interested in the average claim on its auto insurance policies. It believes the claims are normally distributed. Using the last 37 claims, it finds the mean claim to be $1,270 with a standard deviation of $421. Construct a 95 percent confidence interval for the mean claim on all policies.

26. A random sample of the luggage of 30 passengers of Fly Me Airlines finds that the mean weight of the luggage is 47 pounds with a standard deviation of 8 pounds. Construct a 90 percent confidence interval for the mean weight of Fly Me Airlines luggage.

27. A bank manager finds from reviewing her records that the amount of money deposited on Saturday morning is normally distributed with a standard deviation of $150. A random sample of 7 customers reveals the following amounts deposited on Saturday morning.

$825 $972 $311 $1,212 $150 $1,800 $725

   a. Find a 95 percent confidence interval for the mean amount of deposits by using the MINITAB program.

   b. Find a 90 percent confidence interval for the mean amount of deposits by using MINITAB

again. Compare your answer to the confidence interval you computed in part (a). Which is larger?

28. Redo question 27, parts (a) and (b), this time assuming the population standard deviation is unknown. Use MINITAB.

29. A quality control engineer believes that the life of light bulbs for his company is normally distributed with a standard deviation of 100 hours. A random sample of 10 light bulbs gives the following information on the life of the light bulbs.

   1,000 hours; 1,200 hours; 600 hours; 400 hours; 900 hours; 500 hours; 1,520 hours; 1,800 hours; 300 hours; 525 hours

   a. Find a 90 percent confidence interval for the mean life of the light bulbs.

   b. Suppose the standard deviation is not known. Construct a 90 percent confidence interval for the mean life of the light bulbs.

30. Managers at the Smooth Ride Car Rental Company are interested in the mean number of miles that people drive per day. From past experience, they know that the standard deviation is 75 miles. A random sample of 6 car rentals shows that the people drove the following numbers of miles: 152, 222, 300, 84, 90, 122. Construct a 99 percent confidence interval for the mean number of miles driven.

31. A credit manager at the Bargain Basement Department Store is interested in the proportion of customers who pay their credit card balances in full each month. A random sample of 200 customers indicates that 95 paid their balance in full each month. Construct a 99 percent confidence interval for the proportion of customers who pay their balances in full each month.

32. Construct point estimates for the following situations.

   a. A labor union randomly samples 75 of its members and finds that 40 favor the new contract. Estimate the proportion of all workers who favor the new contract.

   b. An economics professor randomly samples 100 students in her class and finds that 70 do not know the meaning of *elasticity*. Estimate the proportion of all students in her class who cannot define this term.

33. An auditor randomly samples 50 accounts payable of a company and finds a sample mean of $1,100 with a sample standard deviation of $287. Construct a 90 percent confidence interval for the mean accounts payable.

34. A random sample of 250 residents of a town shows that 55 percent favor a bond issue to finance new school construction. Construct a 99 percent confidence interval for the proportion of all residents who favor the bond issue.

35. A random sample of 500 students at Average College finds the sample mean combined-SAT score to be 1,050 with a standard deviation of 120. Construct a 90 percent confidence interval for the mean SAT score.

36. Reviewing his records, a grocery store manager finds that the amount of money spent shopping on Friday evenings is normally distributed with a standard deviation of $22. A random sample of 5 customers reveals the following amounts spent shopping on Friday night: $125, $72, $15, $88, $96.
   a. Find a 95 percent confidence interval for the mean amount of money spent shopping.
   b. Find a 90 percent confidence interval for the mean amount of money spent shopping. Compare your answer to the confidence interval you computed in part (a). Which is larger?

37. A random sample of 75 observations from a population yielded the following summary statistics:

$$\Sigma x = 1,270 \qquad \Sigma x^2 = 21,520$$

Construct a 95 percent confidence interval for the population mean μ.

38. A random sample of 100 observations from a population yielded the following summary statistics:

$$\Sigma x = 375 \qquad \Sigma(x_i - \bar{x})^2 = 972$$

Construct a 99 percent confidence interval for the population mean μ.

39. Suppose a random sample of 40 professional golfers is taken and the mean scoring average of the sample is found to be 72.8 strokes per round with a standard deviation of 1.2 strokes per round. Construct a 99 percent confidence interval for the population's mean strokes per round.

40. Suppose a random sample of 10 professional golfers is taken and the mean scoring average of the sample is found to be 71.8 strokes per round with a standard deviation of 1.3 strokes per round. Construct a 90 percent confidence interval for the population mean strokes per round.

41. Reconsider the information given in question 40. Suppose now that the population standard deviation is known to be 1.3 strokes per round. Construct a 90 percent confidence interval for the population mean strokes per round. Compare your answer to your answer in question 40. Why are they different?

42. Suppose you construct a 95 percent confidence interval for the mean of an infinite population. Will the interval always be narrower when σ is known than when σ is unknown?

43. A random sample of 75 observations reveals that the sample mean is 20. You know that the population standard deviation is 5. Construct a 90% confidence interval for the population mean.

44. In a national survey, 200 cola drinkers were asked to compare Yum Yum Cola to Yuk Yuk Cola. Of the 200 people sampled, 120 preferred Yum Yum. Construct a 95 percent confidence interval for the actual proportion of consumers who prefer Yum Yum Cola.

45. The 80 members of a random sample of graduates of Mary's Typing School indicate that their mean salary is $22,500 with a sample standard deviation of $3,100. Construct a 99 percent confidence interval for the true mean salary.

46. Suppose a random sample of 30 college students reveals that the mean amount of money spent on textbooks each semester is $145 with a standard deviation of $25. Construct a 90 percent confidence interval for the mean amount of money that students spend on textbooks each semester.

47. The Better Health Cereal Company produces Healthy Oats cereal. A sample of 100 boxes of this cereal indicates that the mean weight of a box of cereal is 24 ounces with a standard deviation of 1 ounce. Construct a 99 percent confidence interval for the population's mean weight.

48. The Better Health Cereal Company produces Healthy Oats cereal. A sample of 15 boxes of this cereal indicates that the mean weight of a box of

cereal is 24 ounces with a standard deviation of 1 ounce. Construct a 99 percent confidence interval for the population mean weight. Compare your answer to your answer in question 47. Why are they different?

49. A sample of 100 former basketball players from Slam Dunk University shows that 55 of the players graduated in 4 years. Construct a 90 percent confidence interval for the proportion of basketball players graduating in 4 years from Slam Dunk U.

50. A sample of 20 cups of coffee from a coffee machine has a mean amount of coffee of 6 ounces. The standard deviation is known to be .5 ounces. Construct a 99 percent confidence interval for the mean amount of coffee per cup.

51. Reconsider question 50. This time, assume that the standard deviation is not known and that .5 ounces is the sample standard deviation. Again construct a 99 percent confidence interval for the mean amount of coffee per cup. Compare your answer to your answer in question 50.

52. Suppose a sample of 500 companies listed on the NYSE is found to contain 327 companies paying dividends that have increased over the last year. Construct a 95 percent confidence interval for the mean proportion of companies that paid dividends that increased over the last year.

53. A sample of 100 steel-belted radial tires yields a mean life of 35,000 miles with a sample standard deviation of 4,000 miles. Construct a 90 percent confidence interval for the mean life of steel-belted radial tires.

54. Suppose a bowler takes a random sample of 15 games she has bowled and finds the sample mean to be 172. She knows that the standard deviation of her score is 8. Construct a 99 percent confidence interval for her score.

55. Flip a coin 50 times and record the number of tails. Construct a 99 percent confidence interval for the proportion of tails in the tossing of a coin.

56. A random sample of 450 people who took Dollar Dave's CPA review course reveals that 310 of them passed the CPA exam on the first try. Construct a 90 percent confidence interval for the proportion of people who pass the CPA exam on the first try after taking Dollar Dave's course.

57. A random sample of 225 people who went to the Match Maker Dating Service finds that 100 of those people found their spouse through the service. Construct a 95 percent confidence interval for the proportion of people who find a spouse through this dating service.

58. A random sample of 200 observations from a population yielded the following summary statistics:

$$\Sigma x = 1,202 \qquad \Sigma x^2 = 121,020$$

Construct a 90 percent confidence interval for the population mean μ.

59. A random sample of 80 observations from a population yielded the following summary statistics:

$$\Sigma x = 475 \qquad \Sigma(x_i - \bar{x})^2 = 772$$

Construct a 95 percent confidence interval for the population mean μ.

60. A random sample of 100 bullets in a case of 1,000 includes 5 that are defective. Construct a 99 percent confidence interval for the proportion of defective bullets in a case.

61. Suppose a golfer on the University of Houston golf team plays 70 rounds of golf and breaks par 32 times. Construct a 90 percent confidence interval for the proportion of rounds in which this golfer will break par.

62. You roll a die 100 times and get the following results.

Number on Die	Number of Rolls
1	13
2	16
3	15
4	14
5	22
6	20

Construct a 90 percent confidence interval for the proportion of rolls that will be 1's.

63. Use the information given in question 62 to construct a 90 percent confidence interval for the proportion of rolls that will come up 6.

64. A surge in health insurance premiums imposes an additional burden on a business. A random sample of 10 employees indicates that the average cost increase per employee is about $2,345 with a standard deviation of $245. Assuming a normal distribution for the per-employee increase, construct a 90 percent confidence interval for the average increase.

65. The owner of a local bakery feels that too many bagels are thrown out every night, so he decides to estimate the demand for bagels. After a month's observation, he collected 30 days' sales and ascertained that the average sales were 120 and the standard deviation of daily sales was 10. Assume that the daily bagel sales follow a normal distribution. Construct a 90 percent confidence interval for the demand for bagels.

66. Suppose the owner in question 65 observed the sales for 60 days and found the average sales to be 115 with a standard deviation of 12. Obtain the 90 percent confidence interval for the demand for bagels. Compare your answer with the answer you got in question 65. Can you explain why the interval is smaller?

67. The manager in the local shoe factory wants to estimate the productivity of the midnight shift. He draws a random sample of 10 nights and records the productivity as follows:

124 124 145 132 123 124 122 141 133 122

a. Estimate the average productivity.
b. Assuming that the data follow a normal distribution, derive a 95 percent confidence interval.

68. A local dairy farm has just installed a new machine that pumps milk into 16-ounce bottles. The manager of the farm wants to make sure that the amount of milk put in the bottles is 16 ounces, so he randomly selects 12 bottles of milk each hour and weighs the milk. The results obtained in the last hour were

16.01   16.03   15.89   15.99   16.02   16.03
16.04   16.01   15.99   16.03   16.04   16.05

a. Obtain the average weight of the milk.
b. Obtain a 95 percent confidence interval for the average amount of milk in the bottles.

69. The personnel office found that in the last 5 years the average cost of recruiting management trainees has been $500. The cost varies but follows a normal distribution. The standard deviation is estimated to be 25. Assume that the cost of recruitment will remain the same next year and that the company will hire 50 new employees. How much money should the company allocate for recruitment? Construct a 90 percent confidence interval to estimate the recruitment expenses.

70. A survey wherein 90 employees were randomly drawn shows that the average number of sick days taken by employees each year is 5.4 days. The number of sick days follows a normal distribution with a standard deviation of 1.5. Obtain a 90 percent confidence interval for the average number of sick days.

71. A recent poll shows that 53 percent of the voters interviewed strongly support the incumbent and are willing to vote for her in the coming election. The poll was taken by asking 1,000 voters. Estimate the proportion of support for the incumbent with a 95 percent confidence interval.

72. A consumer rights organization tests a new car to estimate the car's average gasoline mileage. Because its budget is limited, the organization can test only 25 cars. The standard deviation of the cars tested is 2. What is the range of the 90 percent confidence interval?

73. A poll is conducted to predict whether new municipal bonds should be issued. Assume that 230 out of 500 interviewees voted for issuance of the new bonds. How precise is this prediction? Construct a 95 percent confidence interval for the proportion of yes votes.

74. In question 73, assume that 45 percent of the entire population of voters support issuance of the new bonds. Under this condition, if the pollsters want to stay within 2 percent error (plus and minus 1 percent), how many voters should they interview?

75. A multinational company wants to find out how society perceives it. The company sends a questionnaire to 2,000 people and learns that 893 have favorable opinions, others either have an unfavorable opinion or no opinion.
a. What percent of the people surveyed have favorable opinions of the company? Construct a 90 percent confidence interval.

b. What is the percentage of people who have favorable opinions of the company? Construct a 95 percent confidence interval.

76. When we construct a 90 percent confidence interval for, say, a mean, we build a range that has an upper bound and a lower bound, and we write the confidence interval as

$P(\text{lower bound} < \text{mean} < \text{upper bound})$
$$= 90\%$$

Comment on the following statement: Would you say the probability that the mean occurs between the upper and lower bounds is 90 percent?

77. A new machine was designed to cut a metal part at a length of .24 inches. Although the machine is well designed, for some uncontrolled reasons the machine cuts the metal with a standard deviation of .01 inch. For quality control purposes, the company wants to draw a sample from each hour's production and measure the average length of the sample metal parts. If the company wants to control the 99 percent confidence interval in a range of .01, how many parts should the company sample every hour?

78. The trains scheduled to arrive at the New Brunswick train station at 7:35 A.M. every weekday do not always arrive at 7:35. A commuter carefully recorded the arrival time for the last 200 working days and found that late arrivals follow a normal distribution with a mean delay of 0 minutes and a standard deviation of 1 minute.
a. Estimate the average arrival time for the train.
b. Estimate the average arrival time using a 90 percent confidence interval.
c. If you plan to arrive at the train station at 7:34 regularly for the next 200 working days, how many trains should you expect to miss?

79. A marketing consulting company wants to estimate the percentage of students holding credit cards by sending questionnaires to students. The sponsor of this research wants to establish a 95 percent confidence interval and a ± 1 percent error margin. To achieve this precision, how many questionnaires should the company send out if every student responds?

80. In a survey of 2,000 voters, 36 percent were found to support increasing taxes to build a new school system. Obtain the 95 percent confidence interval for the proportion supporting the tax increase.

81. The manager in the local supermarket wanted to know whether it is worth the trouble to keep the store open 24 hours a day. He randomly sampled and recorded 20 nights' sales and got

245	145	123	178	125	175	182	130
214	192	120	187	163	148	198	192
129	134	139	271				

Use MINITAB to answer the following questions.
a. Estimate the average sales per night.
b. Construct the 90 percent confidence level for the average sales.

82. A large mail-order company wants to find the effect of sending catalogs to potential customers. Of the 600 potential customers who have just received the new catalogs, 123 responded with an order within a month. Estimate the proportion of responses and establish a 90 percent confidence interval.

83. The personnel department wants to estimate the cost of hiring a new secretary. The following data are collected on 8 new secretaries.

| $2,100 | $2,135 | $2,545 | $2,433 |
| $2,344 | $2,564 | $2,457 | $2,556 |

Estimate the average cost of hiring. Construct a 90 percent confidence interval.

84. The dean of student activities wants to estimate the average spending on beer per week by a student. From a previous study, the standard deviation of spending was estimated to be $39. If the dean wants to control the 90 percent confidence interval within $\pm \$5$, how many students should he survey?

85. The dean of student activities wants to know students' reaction to the new student center. Of the 500 students queried, 350 report that they like the new building. Estimate the proportion of the students who like the building. Construct a 90 percent confidence interval.

86. In question 85, if the dean wants to narrow the 90 percent confidence interval to ±1 percent, how many students should he ask?

87. A soft drink producer installs a new assembly line to fill 12-ounce soda cans. After a week of operation, the plant manager randomly samples 120 cans of soda and weighs the soda. He finds that the soda cans contained an average of 12.05 ounces of soda. The standard deviation of the weight is .02 ounce. Construct a 95 percent confidence interval for the average amount of soda pumped into the cans.

88. In question 87, what is the 95 percent confidence interval for the variance of the soda pumped into the cans?

Use the following information to answer questions 89 to 91. In an airline company, a committee was formed to study the seriousness of late arrivals of freight. The following report was compiled about the arrival record.

Total number of freight shipments   625
Total number of late arrivals   159
Average late time   34 minutes
Standard deviation of late time   25 minutes

89. Construct a 90 percent confidence interval to estimate the average late time.

90. Construct a 90 percent confidence interval to estimate the percentage of late arrivals.

91. Construct a 90 percent confidence interval to estimate the standard deviation of late time.

92. A potential candidate in the third borough conducted a poll to decide whether he should challenge the incumbent. From a previous poll, he knows that the current incumbent has the support of 45 percent of the people. He wants to construct a 90 percent confidence interval with a ±3 percent error margin. How many voters should he survey?

Use the following information to answer questions 93 to 96. An automobile manufacturer wants to study the repair record of its own cars. The performance of 1,000 cars and their maintenance records were monitored after they were sold to consumers. In a span of 3 years, 3,560 repairs occurred among the 1,000 cars monitored. The standard deviation of the number of repairs for 1 car is 2.5. A total of $303,000 was spent to repair the cars. The standard deviation of repair costs for 1 car is $60. There are 205 cars that did not have any repairs in the 3 years.

93. Compute the average cost of 1 repair. Construct an 80 percent confidence interval.

94. Compute the average number of repairs for each car. Construct an 80 percent confidence interval.

95. Construct a 90 percent confidence interval for the standard deviation of costs.

96. Construct a 90 percent confidence interval for the proportion of trouble-free cars. What is the error margin?

97. Define the following:
   a. Convenience lot
   b. Single-sampling plan
   c. Double-sampling plan
   d. upper control limit
   e. lower control limit
   f. acceptance sampling

98. Discuss the similarities and differences among $\overline{X}$-charts, $\overline{R}$-charts, S-charts, and P-charts.

99. Thirty samples of 100 items each were inspected, and 68 were found to be defective. Compute control limits for a P-chart.

100. The following table gives the fraction defective for an automotive piston for 20 samples. Three hundred units are inspected each day. Construct a P-chart and interpret the results.

Sample	Fraction Defective	Sample	Fraction Defective
1	.11	11	.16
2	.16	12	.25
3	.12	13	.15
4	.10	14	.12
5	.09	15	.11
6	.12	16	.11
7	.12	17	.14
8	.15	18	.18
9	.09	19	.10
10	.13	20	.13

101. One hundred insurance forms are inspected daily over 25 working days, and the numbers of forms with errors are recorded below. Construct a P-chart.

Day	Number Defective	Day	Number Defective
1	4	14	4
2	3	15	1
3	3	16	3
4	2	17	4
5	0	18	0
6	3	19	1
7	0	20	1
8	1	21	0
9	6	22	2
10	3	23	6
11	2	24	2
12	0	25	1
13	0		

Hypothesis Testing

CHAPTER OUTLINE

Key Terms

hypotheses
hypothesis testing
null hypothesis
alternative hypothesis
mutually exclusive
exhaustive
Type I error
Type II error
one-tailed test
two-tailed test
acceptance region
rejection region
lower-tailed test
upper-tailed test
critical value
simple hypothesis
parameter

composite hypothesis
probability value (*p*-value)
observed level of significance
the power of a test
chi-square test
power function
power curve
operating-characteristic curve (OC curve)
acceptance sampling
operating characteristic
lot tolerance, percentage defective (LTPD)
consumer's risk
acceptable quality level (AQL)
producer's risk

11.1 INTRODUCTION

Business managers must always be ready to make decisions and take action on the basis of available information. During the process of decision making, managers form hypotheses that they can scientifically test by using that available information. They then make decisions in the light of the outcome. In this chapter, we use the concepts of point estimate and interval estimate discussed in Chapters 8, 9, and 10 to test hypotheses made about population parameters on the basis of sample data.

Hypotheses are assumptions about a population parameter. **Hypothesis testing** involves judging the correctness of the hypotheses. In fact, we often rely heavily on sample data in decision making. For example, the results of public opinion polls may actually dictate whether a presidential candidate decides to keep running or to drop out of the primary race. Similarly, a firm may use a market sampling survey to gauge consumer interest in a given product and thus determine whether to allocate funds for research and development of that product. And a plant manager may use a sample of canned food products produced by a food canning machine to determine whether the quality of this year's products is the same as that of the previous year's offering.

In this chapter, we first discuss the basic concepts of hypothesis testing and the errors it is subject to. Second, methods of constructing an hypothesis test and testing procedures are explored. Third, we examine in detail one-tailed tests and two-tailed tests for large samples. Small-sample hypothesis tests for means and chi-square tests of a normal distribution variance are discussed next. Then we investigate hypothesis testing for a population proportion and compare the variances of two normal populations. Finally, we present some business applications of hypothesis testing. The power of a test and the power function are discussed in Appendix 11A.

11.2 CONCEPTS AND ERRORS OF HYPOTHESIS TESTING

Concepts

The information obtained from the sample can be used to make inferential statements about the characteristics of the population from which the sample is drawn. One way to do this is to estimate unknown population parameters by calculating point estimates and confidence interval estimates.

Alternatively, we can use sample information to assess the validity of an hypothesis about the population. For example, the production manager in charge of a cereal box filling process hypothesizes that the average weight of a box of cereal is 30 ounces.

In statistics, hypotheses always come in pairs: the null hypothesis and the alternative hypothesis. The statistical hypothesis that is being tested is called the **null hypothesis.** Our cereal production manager can use a sample of 35 boxes and calculate their average weight and variance to ascertain the validity of the following null and alternative hypotheses.

1. The average weight of cereal per box is 30 ounces (the null hypothesis).

2. The average weight of cereal per box is not 30 ounces (the alternative hypothesis). This implies that it is less than 30 ounces, or it is more than 30 ounces.

Rejection of the null hypothesis that is tested implies acceptance of the other hypothesis. This other hypothesis is called the **alternative hypothesis.** These two hypotheses represent mutually exclusive and exhaustive theories about the value of a population parameter such as population mean μ, population variance σ^2, or population proportion P. When hypotheses are **mutually exclusive,** it is impossible for both to be true. When they are **exhaustive,** they cover all the possibilities; that is, either the null hypothesis or the alternative hypothesis must be true.

The null hypothesis is traditionally denoted as H_0, the alternative hypothesis as H_1. Each of these symbols is always followed by a colon and then by the statement about a population parameter.

The first problem we encounter in hypothesis testing is how to construct the test. To construct an hypothesis test, we first need to specify the null and alternative hypotheses. Because our goal in hypothesis testing is to find out whether we can reject the null hypothesis, we set up the null hypothesis so that it is consistent with the status quo. In addition, H_0 has to be a specific value so that the sampling distribution under H_0 can be determined for the test. By constructing our hypothesis test in this way, we ensure that the status quo is maintained unless we have sufficient information to prove otherwise (that is, unless we are able to reject H_0). For example, to minimize the risk of sending an innocent person to jail, our legal system is set up so that the accused is "innocent until proven guilty." We can restate this principle as the following null and alternative hypothesis.

$$H_0: \text{Not guilty} \tag{11.1}$$
$$H_1: \text{Guilty}$$

The hypothesis test is set up in this way so that the status quo (innocence) is upheld unless the test results show "beyond a reasonable doubt" that the null hypothesis should be rejected.

Another example of hypothesis testing is testing whether the average weight of a package of cookies is equal to the required weight. In this case the hypotheses are

$$H_0: \mu_x = w^* \tag{11.2}$$
$$H_1: \mu_x \neq w^*$$

where w^* is the required weight for each pack of cookies. The manufacturer does not want more or less than the required amount of cookies in each package. Therefore, it conducts a test to determine whether the statistics from the sample show any severe deviation from the required weight. If H_0 is rejected, then the manufacturer must impose tighter control on the packing process.

Type I and Type II Errors

It seems like a very simple idea that if we can't reject the null hypothesis, we accept it. But we must think twice before accepting H_0. When H_0 cannot be rejected, there are two possibilities: (1) H_0 is indeed true and (2) H_0 is wrong anyway. Maybe the sample size was not large enough, or for some other reason test results did not enable us to reject H_0. In any case, we cannot conclude from the fact that H_0 can't be rejected that H_0 is necessarily true. We can only say that, on the basis of the sample under study, we can't reject the null hypothesis. For example, suppose we are interested in testing the null hypothesis that there is no life on Mars. It is clear from this example that we will never be able to show that this statement is true. Why? Because even if astronauts are unable to find life on Mars, it doesn't mean that there are no living things on Mars—only that the astronauts were unable to find any living things. However, it *will* be possible to reject the null hypothesis if these astronauts *do* find life on Mars. In other words, we can never prove that the null hypothesis is true, but only that we are able or unable to reject it.

Table 11.1 illustrates the relationship between the actions we take concerning a null hypothesis and the truth or falsity of that hypothesis (which is called the state of nature). This table shows that the errors made in testing hypotheses are of two types. We make a **Type I error** when H_0 is true but we reject it. We make a **Type II error** when H_0 is false but we accept it.

Table 11.1 Actions and the States of Nature of the Null Hypothesis

	State of Nature	
Action	*H_0 Is True*	*H_0 Is False*
Do not reject H_0	Correct decision	Type II error
Reject H_0	Type I error	Correct decision

11.3 HYPOTHESIS TEST CONSTRUCTION AND TESTING PROCEDURE

Two Types of Hypothesis Tests

There are two types of hypothesis testing that we will be interested in: (1) testing whether or not the population mean is equal to a specific value (including zero) and (2) testing whether the population mean is greater than (or less than) a specific value. The first test is a **two-tailed test;** the other is a **one-tailed test.** These two concepts and the related testing procedures will be discussed in detail in Sections 11.4 and 11.5.

The first step in our hypothesis testing procedure is to divide the sample space into two mutually exclusive areas, the **acceptance region** and the **rejection** (or critical) **region.** We begin by assuming that we have a large sample ($n > 30$) so that we can use the central limit theorem. Later we will examine how our hypothesis testing procedure can be modified to account for small samples ($n < 30$). Where the acceptance and rejection regions lie depends on two things: whether the test is a one- or a two-tailed test and the significance level we assign to our test. The significance level, α, refers to the size of the Type I error that we are willing to accept. In other words, α represents the probability of Type I error.

$$\alpha = P(\text{reject } H_0 \,|\, H_0 \text{ is correct})$$
$$= P(\text{Type I error})$$

Similarly, the probability of Type II error can be defined as

$$\beta = P(\text{fail to reject } H_0 \,|\, H_0 \text{ is false})$$
$$= P(\text{Type II error})$$

How large a significance level we choose depends on the costs associated with making a Type I error. For example, the significance level used by a cookie company interested in the average weight of a package of cookies should differ from the significance level used by a pharmaceutical company interested in the average amount of an active ingredient in one of its medications. Clearly, the cost to the cookie company of having too many or too few cookies in a package is small compared to the cost to the pharmaceutical company of using too much or too little of an active ingredient in one of its products. (Too little may render the product ineffective; too much may cause the product to lead to harmful side effects or even death.) Similarly, there are costs associated with making a Type II error (failing to reject the null hypothesis even though it is false). The cost associated with making a Type II error is also smaller for the cookie company than for the pharmaceutical firm.

Figure 11.1 illustrates the sampling distribution of the mean, showing the acceptance and rejection regions for a null hypothesis. Here we display only Type I error. We will discuss both Type I and Type II errors and the trade-off between these two types of errors in the next section.

In Figure 11.1 C_L is the critical value for the lower-tailed test, and C_U is the critical value for the upper-tailed test. C_L and C_U are the critical values for the two-tailed test. The **critical value** is the cutoff point for hypothesis testing; its value depends on a level of probability, such as 5 percent, 1 percent or some other percentage.

Figure 11.1a presents the case of a **lower-tailed test.** We conduct a one-tailed test in the lower tail of the distribution when we are concerned only with when the population mean μ is smaller than some specified value μ_0. For example, an investor who is trying to evaluate a stockbroker's performance may be concerned only with below-par performance. In this case, a lower-tailed test is in order, and the investor rejects the null hypothesis of average or above-average performance on the part of the stockbroker if the broker's mean return $\overline{X}$ is less than the critical value C_L.

An **upper-tailed test** (Figure 11.1b) is in order when we are concerned only with when the population mean μ is larger than the specified value μ_0. For example, a

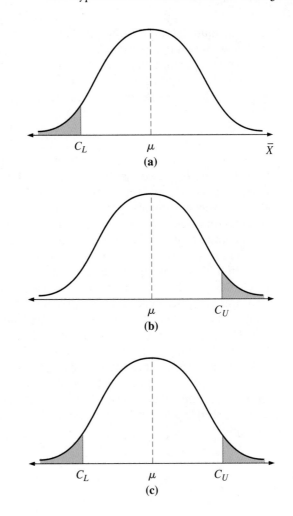

Figure 11.1

Different Types of Hypothesis Testing: (a) Lower-Tailed Test, (b) Upper-Tailed Test, and (c) Two-Tailed Test

pharmaceutical company might be interested in the average amount of an active ingredient in its sleeping pills. Because too much of the active ingredient may lead to harmful side effects, the company may choose to conduct an upper-tailed test. In Figure 11.1b, we can see that the company rejects the null hypothesis of an average or below-average amount of the active ingredient if the population mean of the sleeping pills tested, μ, is greater than the critical value C_U.

A two-tailed test is called for when we are interested in the population mean μ being either much larger or much smaller than the specified value μ_0. For example, a cookie manufacturer is interested in the average number of cookies per package. Too few cookies in a package may lead to complaints from consumers; too many will reduce the company's profits. In this case (Figure 11.1c), the company rejects the null hypothesis of the correct number of cookies in a package if the sample mean $\overline{X}$ falls below the lower critical value C_L or above the upper critical value C_U.

The Trade-off Between Type I and Type II Errors

One way to visualize the trade-off between Type I and Type II errors is to assume that there are only two distributions in which we are interested. One distribution corresponds to H_0, and the other is consistent with H_1. In this case, we are assuming that both the null and alternative hypotheses are simple. A **simple hypothesis** is one wherein we specify only a single value for the population parameter, θ. A **parameter** is a summary measure that is computed to describe a characteristic of an entire population. For example, we might be interested in testing H_0: $\mu = 5$ versus H_1: $\mu = 8$. In this example, both the null and the alternative hypotheses are simple.

On the other hand, we may choose to specify a range of values for the parameter θ. In this case, the hypothesis is called a **composite hypothesis.** For example, we could test a simple null hypothesis, H_0: $\mu = 5$ and a composite alternative hypothesis, H_1: $\mu > 5$. Here the alternative hypothesis is composite because H_1 is consistent with a range of values for μ. For both simple and composite hypotheses, we need to choose a significance level such as $\alpha = .10, .05,$ or $.01$.

In order to present the relationship between Type I and Type II errors in the simplest fashion, let's examine the case of testing a simple null hypothesis, H_0: $\mu = \mu_0$ and a simple alternative hypothesis, H_1: $\mu = \mu_1$. Figure 11.2 is a graph for this example.

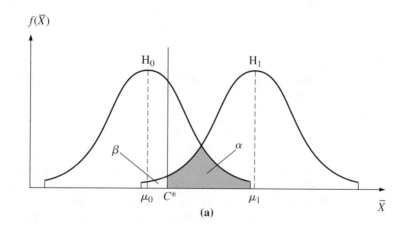

Figure 11.2a

α and β When Sample Size $= n$

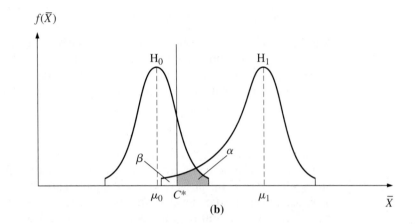

Figure 11.2b

α and β When Sample Size $= n'\,(n' > n)$

In Figure 11.2a, we can see that there are two distinct locations for the distributions $\overline{X}$. One corresponds to the null hypothesis, the other to the alternative hypothesis. Because the two distributions overlap, two possible errors can result from our hypothesis test. Type I error occurs when we reject H_0 when it is true. That is, even though the distribution of $\overline{X}$ is consistent with H_0, we reject H_0 because our sample mean is larger than the critical value C^*. Type I error is represented by α, the area under the H_0 distribution curve that lies to the right of the critical value C^*. Type II error occurs when we fail to reject H_0 when H_1 instead of H_0 is correct. In this case, even though the distribution of $\overline{X}$ is consistent with H_1, we accept H_0 because our sample mean is smaller than the critical value C^*. Type II error is represented by β, the area under the H_1 distribution curve that lies to the left of the critical value C^*.

As we can see, the areas α and β are related. If we choose to make α smaller (that is, reduce the chance of a Type I error), we must settle for a larger β (that is, increase the chance of a Type II error). This is the trade-off between α and β.

Does this trade-off imply that the only way for us to reduce the chance of making a Type II error is to settle for a larger chance of making a Type I error? The answer to this question is no. By increasing our sample size, it is possible to reduce our chance of making a Type II error without increasing our chance of making a Type I error. The sample standard deviation of both H_0 and H_1 distributions can be defined as

$$s_{\overline{x}} = s_x / \sqrt{n}$$

where s_x and n represent sample standard deviation and sample size, respectively. If sample size increases from n to n', then the standard deviations of both H_0 and H_1 distributions become smaller. Hence both α and β decrease, as indicated in Figure 11.2b.

The P-Value Approach to Hypothesis Testing

Another approach to hypothesis testing is through the use of a **probability value (p-value).** Under this approach, rather than testing an hypothesis at such preassigned levels of significance as $\alpha = .05$ or $.01$, investigators often determine the smallest level of significance at which a null hypothesis can be rejected. The p-value is this significance level. In other words, it is the probability of getting a value of the test statistic as extreme as or more extreme than that which is actually obtained, given that the tested null hypothesis is true. Using the p-value in hypothesis testing enables us to determine how significant or insignificant our test results are. Did we barely reject the null hypothesis or did we reject it overwhelmingly? The p-value is often referred to as the **observed level of significance.** If the p-value is smaller than or equal to significance level α, the null hypothesis is rejected; if the p-value is greater than α, the null hypothesis is not rejected. The advantage of the p-value approach is that it frees us from having to choose a value of α. The disadvantage is that we may obtain an inconclusive test. Applications of the p-value will be discussed further in the next two sections.

So far, our discussion of hypothesis testing has focused on determining the level of significance, α, of our test. In addition, we discussed the method of computing a critical value in terms of α and p-value and examined the relationship between the p-value and α. In all cases, our tests involved controlling the Type I error, α. We have also discussed the trade-off between Type I error and Type II error. It is important to investigate how well the hypothesis test controls Type II errors. The **power of a test,** which is defined as $1 - \beta$, can be used to measure how well Type II error has been controlled. This issue and related concepts will be discussed in Appendix 11A.

11.4 ONE-TAILED TESTS OF MEANS FOR LARGE SAMPLES

As we noted in Section 11.3, hypothesis tests can be conducted as one-tailed or two-tailed tests. In this section, we further examine one-tailed tests of means. We begin by examining the case where only one sample is drawn and where that sample is large. Using a large sample offers two important advantages. A large sample makes it possible to apply the central limit theorem. And it enables us, through our choice of significance level (α), to reduce our chance of making a Type II error.

One-Sample Tests of Means

In this section, we examine the one-tailed test of means where only one large random sample is taken. In this case, the null hypothesis is that the population mean is equal to some specified value μ_0. This hypothesis is denoted $H_0: \mu = \mu_0$. Suppose the alternative hypothesis of interest is that the population mean is smaller than this specified value; that is, $H_1: \mu < \mu_0$.

It is natural to test base tests of population mean μ on the sample mean $\overline{X}$. In particular, we would like to know whether the observed sample mean is greatly smaller than the specified value of μ_0. To do this, we require the format of a test with some preassigned significance level α. As described in the previous section, α is used to denote the Type I error.

By using the central limit theorem we saw in Chapter 8 that when the sample size is large, the sample mean $\overline{X}$ is approximately normally distributed. Therefore, the random variable Z, is defined in Equation 11.3, follows a standard normal distribution.

$$Z = \frac{\overline{X} - \mu_0}{\sigma_x/\sqrt{n}} \qquad (11.3)$$

This equation implies that the sampling distribution of the sample mean $\overline{X}$ is normally distributed with mean μ_0 and standard deviation $\sigma_x/\sqrt{n}$ when the null hypothesis is true. For large samples, the sample standard deviation s can be used in place of σ in Equation 11.3. The null hypothesis is to be rejected if the sample mean $\overline{X}$ is greatly smaller than the hypothesized value μ_0. Thus we will reject H_0 if we observe a large absolute value of the random variable Z, as indicated in Equation 11.3.[1]

If the Type I error α is fixed, then we can follow Chapter 10 in using z_α, for which $P(Z < z_\alpha) = \alpha$. If the null hypothesis is true, then the probability that the random variable as indicated in Equation 11.3 is smaller than z_α is α. In terms of sample mean $\overline{x}$, the decision rule is

$$\text{Reject } H_0 \text{ if } \frac{\overline{x} - \mu_0}{\sigma_x/\sqrt{n}} < -z_\alpha$$

This situation is illustrated in Figure 11.3. In this case, because z_α is in the lower tail, we have a lower-tailed hypothesis test.

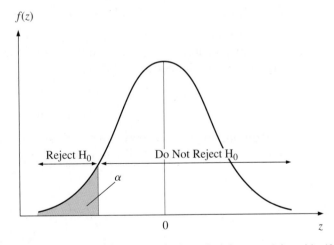

Figure 11.3

Lower-Tailed Hypothesis Test at the α Significance Level

[1] The observed value of Z is negative if the alternative hypothesis is $\mu < \mu_0$. It is positive if the alternative hypothesis is $\mu > \mu_0$.

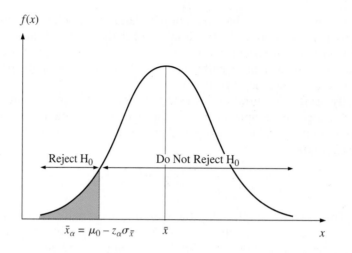

Figure 11.4

Lower-Tailed Hypothesis Test at the $\bar{x}_\alpha$ Significance Level

Alternatively, by letting

$$\frac{\bar{x} - \mu_0}{\sigma_x/\sqrt{n}} = -z_\alpha$$

we can obtain x_α as indicated in Equation 11.4.

$$\bar{x}_\alpha = \mu_0 - z_\alpha\sigma_{\bar{x}} = \mu_0 - z_\alpha\sigma_x/\sqrt{n} \qquad (11.4)$$

$\bar{x}_\alpha$ can be used as an acceptance limit for performing the null hypothesis test (see Figure 11.4). From the normal distribution in Table A3 of Appendix A at the end of the book, we find that $P(z \leq -1.645) = .05$. If $\alpha = .05$, then $\bar{x}_\alpha$ can be estimated as $\mu_0 - (1.645)\sigma_x/\sqrt{n}$. Similarly, when $\alpha = .01$, we find that $z = -1.96$ and $\bar{x}_\alpha = \mu_0 - (1.96)\sigma_x/\sqrt{n}$. If $\bar{x}$ is smaller than $\bar{x}_\alpha$, then we reject the null hypothesis.

EXAMPLE 11.1   *Testing the Average Weight of Cat Food per Bag*

Say we want to test whether the average weight of 60-ounce bags of cat food is less than, equal to, or smaller than 60 ounces at significance level $\alpha = .05$. The null and alternative hypotheses can be stated as

$$H_0: \mu = 60$$
$$H_1: \mu < 60$$

In addition, suppose we know that sample size $n = 100$, sample mean $\bar{x} = 59$, and standard deviation $s_x = 5$. We now will use three different approaches to do the test.

Figure 11.5

The Location of the Critical Value in a Lower-Tailed Test When the Test Statistic Is $-z_{.05} = -1.645$

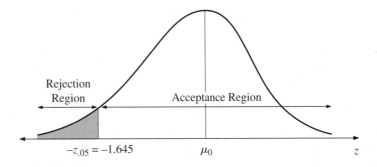

The z_α-value approach

Substituting related information into Equation 11.3, we obtain

$$z = \frac{59 - 60}{\dfrac{5}{10}} = -2$$

If $\alpha = .05$, the test statistic is $-z_{.05} = -1.645$, as indicted in Figure 11.5. A glance at Figure 11.5 reveals that -2 is in the rejection region, so we reject the null hypothesis.

The $\bar{x}_\alpha$-value approach

Substituting this information into Equation 11.5 in terms of $\alpha = .05$, we obtain $\bar{x}_{.05} = 60 - (1.645 \times 5)/10 = 60 - 8.225/10 = 59.1775$.

Figure 11.6 reveals that the observed sample mean of 59 ounces is in the rejection region, so the null hypothesis, H_0, is rejected.

Figure 11.6

The Location of the Critical Value in a Lower-Tailed Test When the Test Statistic Is $\bar{x}_{.05} = 59.1775$

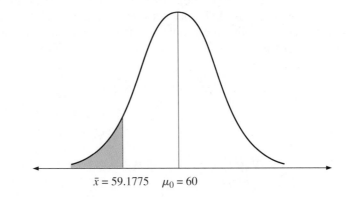

The p-value approach

Because this is only a one-tailed test, the p-value approach represents the probability in only one tail of the distribution. From the z-value approach, we know that

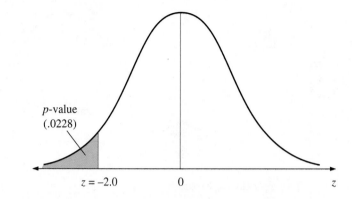

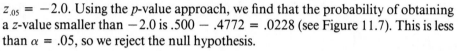

$z_{.05} = -2.0$. Using the p-value approach, we find that the probability of obtaining a z-value smaller than -2.0 is $.500 - .4772 = .0228$ (see Figure 11.7). This is less than $\alpha = .05$, so we reject the null hypothesis.

Thus we can choose among three approaches to doing one-tailed null hypothesis tests. Note that the z_α-value approach is equivalent to the $\bar{x}_\alpha$-value approach.

Two-Samples Tests of Means

Another important issue is how to test the difference between two population means, μ_1 and μ_2, of two normally distributed populations with variances σ_1^2 and σ_2^2. Because we will use large samples, the assumption of normality is not necessary.

We select two independent random samples from two different populations for n_1 and n_2 observations with observed sample means $\bar{x}_1$ and $\bar{x}_2$. Are we willing to attribute the difference between $\bar{x}_1$ and $\bar{x}_2$ to chance sampling errors, or should we conclude that the populations from which the two samples are drawn have different means? In this case, we have two options: the following one-sided tests with significance level α.

1. Upper-tailed null hypothesis

$$H_0: \mu_1 - \mu_2 = D$$
$$H_1: \mu_1 - \mu_2 > D$$

where D can be either zero or a positive number.

2. Lower-tailed null hypothesis

$$H_0: \mu_1 - \mu_2 = D$$
$$H_1: \mu_1 - \mu_2 < D$$

where D can be either zero or a positive number.

The z statistic in terms of the central limit theorem that is used to do the aforementioned one-tailed tests can be defined as follows:[2]

$$z = \frac{(\bar{x}_1 - \bar{x}_2) - D}{\sqrt{\dfrac{\sigma_1^2}{n_1} + \dfrac{\sigma_2^2}{n_2}}} \qquad (11.5)$$

If sample sizes n_1 and n_2 are large, tests of significance level α for the difference between μ_1 and μ_2 are obtained by replacing σ_1^2 and σ_2^2 by s_1^2 and s_2^2. Equation 11.5 can be rewritten as

$$z = \frac{(\bar{x}_1 - \bar{x}_2) - (\mu_1 - \mu_2)}{\sqrt{\dfrac{s_1^2}{n_1} + \dfrac{s_2^2}{n_2}}} \qquad (11.6)$$

The following example demonstrates how to test the difference between two population means.

EXAMPLE 11.2 *Comparing Unleaded Gasoline Prices at Texaco and Shell Stations*

David Smith conducts a market survey to compare the prices of unleaded gasoline at Texaco stations and Shell stations. A random sample of 32 Texaco stations and 38 Shell stations in central New Jersey is used. The cost of 1 gallon of unleaded gasoline is recorded, and the resulting data are summarized here.

Sample A (Texaco)

1.06	.97	.97	.96	1.02	1.09
1.08	1.04	1.11	1.12	1.19	1.07
1.14	1.17	1.22	.97	1.08	
1.05	1.21	.95	.99	1.18	
1.05	1.21	1.03	1.14	1.14	
1.13	1.00	1.16	.96	.98	

$$n_1 = 32 \qquad \bar{x}_1 = \$1.076 \qquad s_1 = \$.0085$$

Sample B (Shell)

1.08	.96	1.06	1.11	1.07
1.17	1.01	1.05	1.04	1.09
1.05	1.06	1.14	1.04	.94
1.01	.99	1.07	1.18	.94
1.08	1.13	1.16	1.00	.94

[2]Because $\bar{x}_1$ is independent of $\bar{x}_2$,

$$\mathrm{Var}(\bar{x}_1 - \bar{x}_2) = \mathrm{Var}(\bar{x}_1) + \mathrm{Var}(\bar{x}_2) = \frac{\sigma_1^2}{n_1} + \frac{\sigma_2^2}{n_2}$$

Sample B (Shell)

1.13	.91	1.13	.96	.95
1.00	1.09	1.15	1.13	
.98	1.04	1.03	1.17	

$$n_2 = 38 \qquad \bar{x}_2 = \$1.054 \qquad s_2 = \$.075$$

Is Texaco's average unleaded gasoline price per gallon ($\bar{x}_1$) more than Shell's average price per gallon ($\bar{x}_2$) at $\alpha = .05$? To perform the test, we can follow these steps:

Step 1: Define the hypotheses and evaluate the test statistic.

The question is whether the data support the claim that $\mu_1 > \mu_2$.[3]

$$H_0: \mu_1 \leq \mu_2 \text{ (Texaco is less expensive or equally expensive.)}$$

$$H_1: \mu_1 > \mu_2 \text{ (Texaco is more expensive.)}$$

From Equation 11.6, the test statistic can be calculated as

$$z = \frac{\bar{x}_1 - \bar{x}_2}{\sqrt{\dfrac{s_1^2}{n_1} + \dfrac{s_2^2}{n_2}}} = \frac{1.076 - 1.054}{\sqrt{\dfrac{(.085)^2}{32} + \dfrac{(.075)^2}{38}}}$$

$$= .022/.0188 = 1.1702.$$

Step 2: Define the rejection region and state a conclusion.

Figure 11.8 indicates that the null hypothesis is to be rejected if $z > 1.645$ under a significance level of .05. Because $z = 1.1702$ is smaller than 1.645, we accept H_0, and because $\bar{x}_1$ is not significantly larger than $\bar{x}_2$, we claim that $\mu_1 \leq \mu_2$. We con-

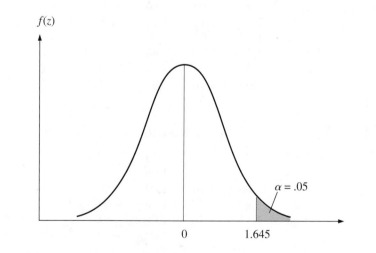

Figure 11.8

z Curve Showing Rejection Region for Example 11.2

---

[3]This kind of hypothesis is called composite null and alternative hypothesis. The decision rule is identical to that for a simple alternative hypothesis specified as $H_0: \mu_1 = \mu_2$ versus $H_1: \mu_1 > \mu_2$.

clude that the Texaco stations charge the same or less for gasoline (on the average) than the Shell stations. Alternatively, we input sample data into MINITAB and obtain the mean, standard deviation, t-statistic, and p-value as follows:

```
MTB > SET INTO C1
DATA> 1.06 1.05 0.97 1.21 0.97 0.95 0.96 0.99 1.02 1.18 1.09
DATA> 1.08 1.05 1.04 1.21 1.11 1.03 1.12 1.14 1.19 1.14 1.07
DATA> 1.14 1.13 1.17 1.00 1.22 1.16 0.97 0.96 1.08 0.98
DATA> END
MTB > SET INTO C2
DATA> 1.08 1.08 0.96 1.13 1.06 1.16 1.03 1.04 0.96 1.07 0.94
DATA> 1.17 1.13 1.01 0.91 1.05 1.13 1.11 1.18 1.13 1.09 0.94
DATA> 1.05 1.00 1.06 1.09 1.14 1.15 1.04 1.00 1.17 0.94 0.95
DATA> 1.01 0.98 0.99 1.04 1.07
DATA> END

MTB > TWOSAMPLE C1 C2;
SUBC> ALTERNATIVE=1.

TWOSAMPLE T FOR C1 VS C2
        N      MEAN     STDEV    SE MEAN
C1     32     1.0762    0.0846     0.015
C2     38     1.0537    0.0754     0.012

95 PCT CI FOR MU C1 - MU C2: (-0.016, 0.061)

TTEST MU C1 = MU C2 (VS GT): T= 1.17   P=0.12   DF=  62

MTB > PAPER
```

From this computer output, we find that the t-statistic is equal to 1.17 and the p-value equals .12. We conclude, therefore, that the average price per gallon of unleaded gasoline from Texaco stations is the same as that from Shell stations.

11.5 TWO-TAILED TESTS OF MEANS FOR LARGE SAMPLES

One-Sample Tests of Means

A cookie store sells individual cookies and cookies in packages. All the packages are sold for the same price, so the weights of the packages should be equal. If the weight is greater than the specified weight on the packing box, the store suffers a loss. If the weight is less, customers complain. Hence the store must periodically draw samples and test whether the average weight deviates from the required weight. The hypotheses tested are

$$H_0: \mu = D$$
$$H_1: \mu \neq D$$

(11.7)

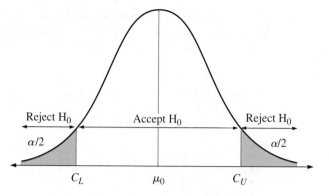

Figure 11.9

Two-Tailed
Hypothesis Testing

Figure 11.9 illustrates the hypothesis test. Note that the significance level is $\alpha/2$ instead of α.

The decision rule can be either of the following:

1. Reject H_0 if $\bar{x}$ is greater than the upper critical value C_U or less than the lower critical value C_L.

2. Reject H_0 if the p-value is less than $\alpha/2$, no matter which tail the sample mean falls in.

Using data from Example 11.1, we set up the two-tailed hypothesis test as

$$H_0: \mu = 60$$
$$H_1: \mu \neq 60$$

and calculate the Z statistic as

$$z = \frac{59 - 60}{5/10} = -2$$

From the standardized normal distribution table (Table A3 in Appendix A), we know that $z_{\alpha/2} = 1.96$ and $-z_{\alpha/2} = -1.96$. Our $z = -2$ is less than $-z_{\alpha/2}$, so our decision is to reject H_0.

Using the p-value approach, we could determine the probability of obtaining a Z-value smaller than -2.0. From Appendix A, that probability is $.5000 - .4772 = .0228$. Because we are performing a two-tailed test, we also need to find the probability of obtaining a value larger than 2.00. The normal distribution is symmetrical, so this value is also $.0228$. Thus the p-value for the two-tailed test is $.0456$ (see Figure 11.10). This result may be interpreted to mean that the probability of obtaining a more extreme result than the one observed is $.0456$. Because this value is smaller than $\alpha = .05$, the null hypothesis is rejected.

In addition, let's look at this real-world example based on C. S. Patterson's study of a sample of 47 large public electric utilities with revenues of $300 million or more according to Moody's Manual (*Financial Management,* Summer 1984). Patter-

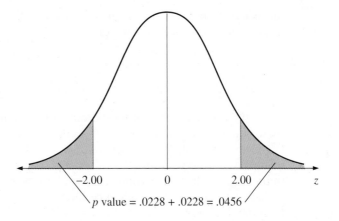

Figure 11.10

Determining the
p-Value for a
Two-Tailed Test

son's study focused on the financing practices and policies of these regulated utilities. Compilation of the actual debt ratios, or long-term debt divided by total capital, of the companies yielded the following results:

$$\bar{x} = .485 \qquad s = .029$$

Before giving their actual debt ratios, the companies cited .459 as the mean debt ratio at which they should operate to maximize shareholder wealth.

From this information, we can test whether the actual mean debt ratio of public utilities differed from the optimum value .459 at $\alpha = .01$. The two-tailed hypothesis test can be defined as

$$H_0\!: \mu = .459$$

$$H_1\!: \mu \neq .459$$

and the z statistic can be calculated as

$$z = \frac{\bar{x} - \mu_0}{s/\sqrt{n}} = \frac{.485 - .459}{.029/\sqrt{47}} = 6.146$$

From Table A3 in Appendix A, we know that $z_{.005} = 2.575$. Since $z > z_{.01}$, we reject the null hypothesis H_0.

Confidence Intervals and Hypothesis Testing

The hypothesis testing discussed in this chapter applies the same concepts as do the confidence intervals we discussed in the last chapter. We used confidence intervals to estimate parameters, whereas we used hypothesis testing to make decisions about specified values of population parameters.

In many situations, we can turn to confidence intervals to test a null hypothesis. This can be illustrated for the test of a hypothesis for a mean. In Example 11.1 (testing whether the average weight of packages of cat food was different from 60 ounces), we employed the formula

$$z = \frac{\bar{x} - \mu_0}{\dfrac{\sigma_x}{\sqrt{n}}} \qquad (11.3)$$

We could also solve the cat food problem by obtaining a confidence interval estimate of μ_0 in terms of sample mean $\bar{x}$. If the hypothesized value of $\bar{x} = 59$ did not fall in the interval, the null hypothesis would be rejected. That is, the value 59 would be considered unusual for the data observed. On the other hand, if it did fall in the interval, the null hypothesis would not be rejected because 59 would not be an unusual value. The confidence interval estimate in terms of data defined in Example 11.1 was

$$\bar{x} \pm z_\alpha \frac{s_x}{\sqrt{n}}$$
$$60 \pm (1.645)\tfrac{5}{10} = 60 \pm .8225$$

so that $59.1775 \le \mu \le 60.8225$. This interval does not include the hypothesized value of 59, so we would reject the null hypothesis. This, of course, is the same decision we reached by using the hypothesis testing technique.

Two-Samples Tests of Means

Two-samples tests involve testing the equality of two sample means. Two-tailed tests are similar to one-tailed tests, but the alternative hypothesis H_1 assumes that two population means are "unequal," and the significance level for each tail is now $\alpha/2$. The hypothesis test can be expressed as follows:

$$H_0: \mu_1 - \mu_2 = D \qquad (11.8)$$
$$H_1: \mu_1 - \mu_2 \ne D$$

where D can be either zero or a positive number.

In order to test, we can calculate either the p-value or the critical values on both tails. The decision rules are

1. Reject H_0 if the p-value is less than $\alpha/2$.
2. Reject H_0 if $(\bar{x}_1 - \bar{x}_2)$ is either greater than C_U or less than C_L, as shown in Figure 11.11.

C_L is calculated as follows.

$$C_L = -z_{\alpha/2} \sqrt{\frac{\sigma_1^2}{n_1} + \frac{\sigma_2^2}{n_2}} \qquad (11.9)$$

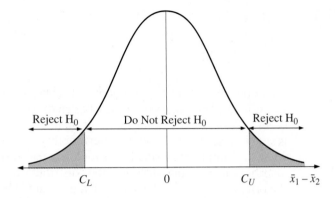

Figure 11.11

Rejection and Acceptance Regions for Two-Samples Case

If σ_1^2 and σ_2^2 are unknown, sample variances s_1^2 and s_2^2 can be used to approximate C_L, which can be defined as

$$C_L = -z_{\alpha/2} \sqrt{\frac{s_1^2}{n_1} + \frac{s_2^2}{n_2}} \tag{11.10}$$

Furthermore,

$$C_U = z_{\alpha/2} \sqrt{\frac{\sigma_1^2}{n_1} + \frac{\sigma_2^2}{n_2}}$$

can be approximated by

$$z_{\alpha/2} \sqrt{\frac{s_1^2}{n_1} + \frac{s_2^2}{n_2}} \tag{11.11}$$

With critical values C_L and C_U, we can perform the null hypothesis test as in the case of a one-sample test.

Using the unleaded gasoline prices in Example 11.2 as an example, we now show how Equations 11.10 and 11.11 can be used to do a two-tailed test at $\alpha = .05$. The question is whether data support the claim that $\mu_1 = \mu_2$:

$$H_0: \mu_1 - \mu_2 = 0$$
$$H_1: \mu_1 - \mu_2 \neq 0$$

From Table A3 in Appendix A, we find $z_{.025} = 1.96$. Substituting $z_{.025} = 1.96$, $s_1 = .085$, $s_2 = .075$, $n_1 = 32$ and $n_2 = 38$ into Equations 11.10 and 11.11, we obtain

$$C_L = -(1.96) \sqrt{\frac{(.085)^2}{32} + \frac{(.075)^2}{38}} = -(1.96)(.0188)$$

$$= -.0368$$
$$C_U = (1.96)(.0188) = .0368$$

Since $\overline{X}_1 - \overline{X}_2 = 1.076 - 1.054 = .022$ is smaller than C_U and larger than C_L, we cannot reject the null hypothesis $\mu_1 = \mu_2$.

11.6 SMALL-SAMPLE TESTS OF MEANS WITH UNKNOWN POPULATION STANDARD DEVIATIONS

So far our discussion of hypothesis testing has focused on cases where the sample size is large. This large sample size has enabled us to employ the central limit theorem and to use the normal distribution in our hypothesis tests. If our sample size is small ($n < 30$), however, we must modify our test. As we noted in Chapter 10, when the sample size is small, we should use the t distribution in place of the normal distribution. Note that the use of the t distribution with small samples requires that the original population be distributed normally. Using the Z test for small-sample hypothesis testing leads to inaccurate results. Table 11.2 shows how t_α approaches Z_α as the sample size increases. It gives some idea how "small" a sample should be for us to use the t test when population variances are unknown.

We will use both the one-tailed and the two-tailed tests to show how the t test can be employed for both one-sample and two-samples tests of means.

Table 11.2 Values of t_α Versus z_α

	t_α **Value**			
Sample Size	*.10*	*.05*	*.025*	*.01*
10	1.372	1.812	2.228	2.764
20	1.325	1.725	2.086	2.528
120	1.289	1.658	1.980	2.358
∞	1.282	1.645	1.960	2.326
z_α	1.282	1.645	1.960	2.326

One-Sample Tests of Means

If the population variance is unknown and the sample size is small, then we can use the t statistic defined in Equation 10.8 to test the null hypothesis associated with both one-tailed and two-tailed cases.

$$t_\nu = \frac{\overline{X} - \mu}{s/\sqrt{n}} \tag{10.8}$$

EXAMPLE **11.3**   *Average Mileage of a Moving Van*

United Van Lines Company is considering purchasing a large, new moving van. The sales agency agreed to lease the truck to United Van Lines for 4 weeks (24 working days) on a trial basis. The main concern of United Van Lines is the miles per gallon (mpg) of gasoline that the van obtains on a typical moving day. The mpg values for the 24 trial days were

8.5	9.5	8.7	8.9	9.1	10.1	12.0	11.5	10.5	9.6
8.7	11.6	10.9	9.8	8.8	8.6	9.4	10.8	12.3	11.1
10.2	9.7	9.8	8.1						

United Van Lines will purchase the van if it is convinced that the average value for miles per gallon is greater than 9.5.

To perform the hypothesis testing, we define the null and alternative hypothesis tests as

$$H_0: \mu \leq 9.5$$

$$H_1: \mu > 9.5$$

The significance level for this test is $\alpha = .05$.

The MINITAB output for Example 11.3 is presented in Figure 11.12. From this output, we calculate the test statistic as

$$t = \frac{\overline{X} - 9.5}{s/\sqrt{n}} = \frac{9.925 - 9.5}{.243}$$

$$= 1.75$$

From Table A4, we find that $t_{.05,23} = 1.714$. Because $1.75 > 1.714$, we reject H_0—and advise United Van Lines to buy the van.

Figure 11.12

MINITAB Output for Example 11.3

```
MTB > SET INTO C1
DATA> 8.5 9.5 8.7 8.9 9.1 10.1 12.0 11.5 10.5 9.6 8.7 11.6 10.9 9.8 8.8 8.6 9.4
DATA> 10.8 12.3 11.1 10.2 9.7 9.8 8.1
DATA> END
MTB > TINTERVAL WITH 95% CONFIDENCE USING C1

              N      MEAN    STDEV   SE MEAN    95.0 PERCENT C.I.
C1           24     9.925    1.189    0.243   (  9.423,  10.427)

MTB > TTEST 9.5 C1

TEST OF MU = 9.500 VS MU N.E. 9.500

              N      MEAN    STDEV   SE MEAN       T    P VALUE
C1           24     9.925    1.189    0.243     1.75     0.093

MTB > PAPER
```

Two-Samples Tests of Means

To test the difference between two means when the population variances are unknown and the samples are small, we use the t statistic of Equation 11.12, which is similar to Equation 11.8. Here two populations are normally distributed, and the two samples that are used to do the test are independent of each other. The hypotheses for a two-tailed case can be expressed as

$$H_0: \mu_1 - \mu_2 = D$$

$$H_1: \mu_1 - \mu_2 \neq D$$

where D can be either zero or a positive number. The statistic for testing the hypotheses can be defined as

$$t = \frac{(\bar{x}_1 - \bar{x}_2) - D}{s \sqrt{\dfrac{n_1 + n_2}{n_1 n_2}}} \tag{11.12}$$

This statistic has a t distribution with $(n_1 + n_2 - 2)$ degrees of freedom and where

$$s^2 = \frac{(n_1 - 1)s_1^2 + (n_2 - 1)s_2^2}{(n_1 + n_2 - 2)} \tag{11.13}$$

is the pooled variance. Note that the t statistic defined in Equation 11.12 also can be used in a one-tailed test. Two examples are used to show how Equation 11.12 can be used to do both two-tailed and one-tailed tests.

EXAMPLE 11.4   *Competitive Versus Coordinative Bargaining Strategies*

We now use a real-world example to show how the t statistics defined in Equations 11.12 and 11.13 can be used to test whether the competitive bargaining strategy differs in its results from the coordinative bargaining strategy. This example is adapted from S. W. Clopton's research (*Journal of Marketing Research,* February 1984) in which he compared so-called competitive and coordinative bargaining strategies in buyer-seller negotiations. Inflexibility in an effort to force concessions best defines competitive bargaining. A coordinative bargaining strategy, however, involves a great deal more cooperation and more of a problem-solving approach.

One of Clopton's negotiation experiments involved a sample of 16 organizational buyers. Clopton reported that in negotiations in which the maximum profit was fixed, the sample participants were perfectly divided in their choice of strategy; that is, 8 buyers employed a competitive bargaining strategy and the other 8 buyers used a coordinative approach.

The table lists the individual savings for the two groups of buyers. Using $\alpha = .05$, test to find if there is a difference in mean buyer savings for the two strategies.

	Competitive	Coordinative
Sample size	8	8
Mean savings	$1,706.25	$2,106.25
Standard deviation	$ 532.81	$ 359.99

$$H_0: (\mu_1 - \mu_2) = 0$$

$$H_1: (\mu_1 - \mu_2) \neq 0$$

Since sample size is only 8 for each, the t test statistic of Equation 11.12 should be used to do the test. Substituting related information into Equations 11.13 and 11.12, we obtain

$$s^2 = \frac{(8-1)^2(532.81)^2 + (8-1)(359.99)^2}{8+8-2}$$

$$= 206,739.648$$

$$t = \frac{1,706.25 - 2,106.25}{\sqrt{206,739.648\left(\dfrac{1}{8} + \dfrac{1}{8}\right)}}$$

$$= \frac{-400}{227.343} = -1.759$$

From Table A4 in Appendix A, we find $t_{14,.05} = 2.145$. Because 1.759 is smaller than 2.145, the null hypothesis cannot be rejected.

EXAMPLE 11.5   *The Effect of a Moderator on the Number of Ideas Generated*

Fern (1982) studied the impact of the presence of a moderator on the number of ideas generated by groups.[4] He first randomly sampled 4 groups that included a moderator. Then he independently and randomly sampled another 4 groups that lacked a moderator. The mean number of ideas generated and the sample standard deviation for the two sets of samples were

First set of samples: $\overline{X}_1 = 78.00$, $s_1 = 24.4$, $n_1 = 4$
Second set of samples: $\overline{X}_2 = 63.5$, $s_2 = 20.2$, $n_2 = 4$

Let μ_1 and μ_2 represent the respective population means for groups with and without a moderator. Then the test can be defined as

$$H_0: \mu_1 - \mu_2 = 0$$
$$H_1: \mu_1 - \mu_2 > 0$$

The significance level for this test is $\alpha = .05$. To perform the test, we substitute all related information into Equations 11.12 and 11.13 and obtain

$$s^2 = \frac{(3)(24.4)^2 + (3)(20.2)^2}{4+4-2} = 501.7$$

$$s = \sqrt{501.7} = 22.4$$

Then

$$t_6 = \frac{78.0 - 63.5}{22.4\sqrt{\dfrac{8}{16}}} = .915$$

[4]E. F. Fern (1982), "The Use of Focus Groups for Idea Generators: The Effect of Group Size, Acquaintanceship, and Moderator on Response Quantity and Quality," *Journal of Marketing Research* 19, pp. 1–13.

From Table A4 in Appendix A of this book, we find $t_{6,.05} = 1.943$. Because .915 is smaller than 1.943, the null hypothesis of equality of population means cannot be rejected.

11.7 HYPOTHESIS TESTING FOR A POPULATION PROPORTION

In Section 10.6 we discussed the confidence intervals for a population proportion. The Z statistic for the sample proportion ($\hat{P}$) and the confidence interval for the population proportion (P) are repeated here for convenience.

$$z = \frac{\hat{P} - P}{\sqrt{\hat{P}(1 - \hat{P})/n}} \tag{10.10}$$

$$1 - \alpha = P\left\{\hat{P} - z_{\alpha/2}\sqrt{\frac{\hat{P}(1 - \hat{P})}{n}} < P < \hat{P} + z_{\alpha/2}\sqrt{\frac{\hat{P}(1 - \hat{P})}{n}}\right\} \tag{10.11}$$

where $z_{\alpha/2}$ is the number such that $P(Z > z_{\alpha/2}) = \alpha/2$. The sample standard deviation used in Equation 10.11 can be defined as

$$s_{\hat{P}} = \sqrt{\frac{\hat{P}(1 - \hat{P})}{n}}$$

Note that $\hat{P}$ instead of P is used to estimate $s_{\hat{P}}$, because P is unknown and must be replaced by its estimate, $\hat{P}$.

The procedure discussed in Sections 11.4, 11.5, and 11.6 for testing population means for both one and two samples can be used to test the population proportion. Table 11.3 compares the null hypothesis for testing population means with that for testing population proportions.

Table 11.3 Null Hypothesis for Testing Population Means and Population Proportions

	Population Means		Population Proportions	
	One Sample	*Two Samples*	*One Sample*	*Two Samples*
1. Upper-tailed test	$H_0: \mu = D$	$\mu_1 - \mu_2 = D$	$P = C$	$P_1 - P_2 = C$
	$H_1: \mu > D$	$\mu_1 - \mu_2 > D$	$P > C$	$P_1 - P_2 > C$
2. Lower-tailed test	$H_0: \mu = D$	$\mu_1 - \mu_2 = D$	$P = C$	$P_1 - P_2 = C$
	$H_1: \mu < D$	$\mu_1 - \mu_2 < 0$	$P < C$	$P_1 - P_2 < C$
3. Two-tailed test	$H_0: \mu = D$	$\mu_1 - \mu_2 = D$	$P = C$	$P_1 - P_2 = C$
	$H_1: \mu \neq D$	$\mu_1 - \mu_2 \neq D$	$P \neq C$	$P_1 - P_2 \neq C$

In Table 11.3, both D and C can be zero or nonzero. If the sample size is large, the Z statistic should be used to do the null hypothesis test; if the sample size is small, the t statistic should be used. The Z statistic for testing one population proportion is

$$Z = \frac{\hat{P} - P_0}{\sqrt{P_0(1 - P_0)/n}} \qquad (11.14)$$

where P_0 is the value of P specified in H_0. Equation 11.14 is obtained by substituting P_0 for P in Equation 10.10. The Z statistic for testing the difference between two population proportions is defined as

$$Z = \frac{\hat{P}_1 - \hat{P}_2}{\sqrt{\frac{\overline{P}(1 - \overline{P})}{n_1} + \frac{\overline{P}(1 - \overline{P})}{n_2}}} \qquad (11.15)$$

where $\overline{P}$ is defined as

$$\frac{\hat{P}_1 n_1 + \hat{P}_2 n_2}{n_1 + n_2}$$

Now let's see how the Z statistic of Equations 11.14 and 11.15 for testing proportions can be applied.

EXAMPLE 11.6   *The Promotability of Company Employees*

Francis Company is evaluating the promotability of its employees—that is, determining the proportion of employees whose ability, training, and supervisory experience qualify them for promotion to the next level of management. The human resources director of Francis Company tells the president that 80 percent of the employees in the company are "promotable." However, a special committee appointed by the president finds that only 75 percent of the 200 employees who have been interviewed are qualified for promotion. Use this information to do a two-tailed null hypothesis test at $\alpha = 5$ percent.

$$H_0: P = .80$$

$$H_1: P \neq .80$$

From Table A3, we know that we should reject H_0 if $Z > Z_{.05} = 1.96$, or if $Z < -Z_{.05} = -1.96$.

Substituting $p = .75$, $p_0 = .80$, and $n = 200$ into Equation 11.14, we obtain

$$Z = \frac{.75 - .80}{\sqrt{\frac{(.8)(1 - .8)}{200}}} = \frac{-.05}{.0283}$$

$$= -1.7668$$

Because $-1.7668 > -1.96$, we cannot reject H_0. In other words, the percentage of "promotable" employees is 80 percent.

EXAMPLE 11.7 *Defects in Canned Food*

A food manufacturer has two canning plants. The company's management wants to know whether the mean defect rate of a canned food from the new plant is different than that of the same canned food from the old plant. The canned food is packed in a carton that holds 24 cans. There are 500 cartons in each lot. Table 11.4 gives the sample data obtained from each plant.

Table 11.4 Sample Data on Canned Food from Old and New Plants

Plant	Mean Defect Rate from Each Lot	Size of Sample
New	$\hat{P}_1 = .065$	$n_1 = 50$
Old	$\hat{P}_2 = .052$	$n_2 = 40$

The hypotheses to be tested in terms of Equation 11.15 are

$$H_0: P_1 - P_2 = 0$$
$$H_1: P_1 - P_2 \neq 0$$

First we calculate $\hat{P}_1 - \hat{P}_2$ and the standard derivation of $(\hat{P}_1 - \hat{P}_2)$ as follows:

$$\hat{P}_1 - \hat{P}_2 = .065 - .052 = .013$$

$$\overline{P} = \frac{\hat{P}_1 n_1 + \hat{P}_2 n_2}{n_1 + n_2} = \frac{(.065)(50) + (.052)(40)}{50 + 40} = .059$$

$$s = \sqrt{\frac{\overline{P}(1 - \overline{P})}{n_1} + \frac{\overline{P}(1 - \overline{P})}{n_2}}$$

$$= \sqrt{\frac{(.059)(.941)}{50} + \frac{(.059)(.941)}{40}} = .05$$

where s = standard deviation of $(\hat{P}_1 - \hat{P}_2)$. If we specify $\alpha = .05$ and $z_{.025} = 1.96$, then using the Z-value approach, we have

$$Z = \frac{.013}{.05} = .26$$

Z is smaller than 1.96, so we cannot reject H_0. In other words, the management confirms that the mean defect rate of the new plant is not statistically different from the mean defect rate of the old plant.

11.8 CHI-SQUARE TESTS OF THE VARIANCE OF A NORMAL DISTRIBUTION

In Chapter 10 we discussed confidence intervals for the variance. Now it is time to consider how to conduct hypothesis tests on the variance from a normal population. When we conducted tests on the population mean μ_x, we based our test on the sample mean $\overline{X}$. Thus it seems natural that when we conduct tests of the population variance σ^2, we base our tests on the sample variance s_x^2. From Chapters 9 and 10 we know that

$$\chi_{n-1}^2 = \frac{(n-1)s_x^2}{\sigma_x^2}$$

which follows a chi-square distribution with $(n-1)$ degrees of freedom. We are interested in testing whether the population variance is equal to some specific value, σ_x^{*2}; that is,

$$H_0: \sigma_x^2 = \sigma_x^{*2}$$

Thus, when the null hypothesis is true, the random variable defined in Equation 11.16 follows a chi-square distribution with $(n-1)$ degrees of freedom.

$$\chi_{n-1}^2 = \frac{(n-1)s_x^2}{\sigma_x^2} \tag{11.16}$$

For many applications, we are concerned that the variance of our population may be equal to, larger than, or smaller than some specified value, σ_x^{*2}. The hypothesis testing on σ_x^2 can be defined as follows:

1. Two-tailed test

$$H_0: \sigma_x^2 = \sigma_x^{*2}$$
$$H_1: \sigma_x^2 \neq \sigma_x^{*2}$$

Test substitute $\chi^2 = \dfrac{(n-1)s_x^2}{\sigma_0^2}$

Reject H_0 if $\chi^2 > \chi_{\alpha/2, n-1}^2$ or if $\chi^2 < \chi_{1-\alpha/2, n-1}^2$.

2. One-tailed test

$$H_0: \sigma_x^2 \leq \sigma_x^* \qquad\qquad H_0: \sigma_x^2 \geq \sigma_x^*$$
$$H_1: \sigma_x^2 > \sigma_x^* \qquad\qquad H_1: \sigma_x^2 < \sigma_x^*$$

Reject H_0 if $\chi^2 > \chi_{\alpha, n-1}^2$. Reject H_0 if $\chi^2 < \chi_{1-\alpha, n-1}^2$.

EXAMPLE 11.8 *Variability in Customer Waiting Time*

Suppose the manager of a bank is thinking of introducing a "single-line" policy that directs all customers to enter a single waiting line in the order of their arrival and "feeds" them to different tellers as the latter become available. Although such a policy does not change the average time customers must wait, the manager prefers it because it decreases waiting-time variability. The manager's critics, however, claim that this variability will be at least as great as for a policy of multiple independent lines (which in the past had a standard deviation of $\sigma_x^* = 6$ minutes per customer). All have agreed to use an hypothesis test at the 5 percent significance level to settle the issue. This test is to be based on the experience of a random sample of 20 customers on whom the new policy is "tried out." The two opposing hypotheses are

$$H_0: \sigma_x^2 \geq 36$$

$$H_1: \sigma_x^2 < 36$$

Here 36 is chosen as the H_0 value even though any number greater than 36 is in H_0. The bank's statistician selects the test statistic as

$$\chi_{n-1}^2 = \frac{(n-1)s_x^2}{\sigma_x^2}$$

For a desired significance level of $\alpha = .05$ and 19 degrees of freedom, Table A5 in Appendix A suggests a critical value of 10.117 (this being a lower-tailed test). Thus the decision rule must be as follows: Fail to reject H_0 if $\chi^2 \geq \sigma_{1-.05,19}^2 = 10.117$. After taking a sample of 20 customers, the statistician finds the sample single-line waiting times to have a standard deviation of $s_x = 4$ minutes per customer. Accordingly, the computed value of the test statistic is

$$\chi_{n-1}^2 = \frac{(n-1)s_x^2}{\sigma_x^2} = \frac{4^2(20-1)}{36} = 8.44$$

Because 8.44 is smaller than 10.117, the null hypothesis should be rejected at the 5 percent significance level, which means that the sample result is statistically significant. In other words, the observed divergence from the hypothesized value of $\sigma_x^* = 6$ minutes is not likely to be the result of chance factors operating during sampling.

11.9 COMPARING THE VARIANCES OF TWO NORMAL POPULATIONS

In the last section, we showed that a χ^2 distribution can be used to test whether the population variance of a normal distribution is equal to a specific value. In Chapter 9, we showed that the ratio of two independent χ^2 variables (each divided by its degrees of freedom) is an F random variable. The F random variable is defined as

$$F = \frac{s_x^2/\sigma_x^2}{s_y^2/\sigma_y^2} \tag{11.17}$$

The F random variable follows an F distribution. If $\sigma_x = \sigma_y$, then Equation (11.17) reduces to $F = s_x^2/s_y^2$. The F distribution has degrees of freedom $(n_x - 1)$ and $(n_y - 1)$.

If we want to test whether σ_x^2 is equal to σ_y^2, the hypotheses are

$$H_0: \sigma_x^2 = \sigma_y^2$$

$$H_1: \sigma_x^2 \neq \sigma_y^2$$

Using the data of Example 11.5, we define the F statistic as $F = (24.4)^2/(20.2)^2 = 1.46$. The degrees of freedom are $(n_x - 1) = 3$ and $(n_y - 1) = 3$ (from Table A6 in Appendix A), so we have $F_{3,3,.05} = 9.28$. Because the alternative hypothesis is two-sided, this is the appropriate critical value for testing at the 10 percent significance level. Clearly, 1.46 is much smaller than 9.28; the null hypothesis cannot be rejected. There is no evidence that variances are different in the two testing groups.

11.10 BUSINESS APPLICATIONS

■ APPLICATION 11.1 Rates of Return for GM Versus Those for Ford

In Chapter 7 we calculated descriptive statistics for the earnings per share (EPS) and rates of return (R) for GM and Ford. This information is presented in Table 11.5.

Using 21 years of EPS data, we can test whether GM's average EPS ($7.154) is significantly different from Ford's EPS ($7.033). Our sample consists of only 21 years of data, so we should use the t test instead of the Z test. The hypotheses to be tested are

$$H_0: \mu_1 - \mu_2 = 0$$

$$H_1: \mu_1 - \mu_2 \neq 0$$

where μ_1 and μ_2 are the average EPS for GM and Ford, respectively.

Table 11.5 Annual EPS and Returns for GM and Ford

	EPS		Return	
	GM	*Ford*	*GM*	*Ford*
Mean	7.154	7.033	.082	.096
Std. Dev.	5.195	7.973	.333	.392
Skewness	−.624	.323	.864	1.240

Because the sample is small, we use the pooled variance of Equation 11.13 and the test statistic of Equation 11.12. Substituting into Equations 11.12 and 11.13, we get[5]

$$t = \frac{7.154 - 7.033}{\sqrt{\dfrac{(21 - 1)(5.195)^2 + (21 - 1)(7.973)^2}{21 + 21 - 2} \left[\dfrac{21 + 21}{21 \times 21} \right]}}$$

$$= \frac{.121}{2.076} = .058$$

From Table A4 we can see that the critical t value for $\alpha = .05$ with 40 degrees of freedom is 2.021. Our test statistic has $n_1 + n_2 - 2 = 21 + 21 - 2 = 40$ degrees of freedom, so we compare .058 to 2.021 and learn that we are unable to reject the null hypothesis that the average EPS of Ford and GM are the same.

Alternatively, we input EPS data into MINITAB and obtain the means, the standard deviations, the t statistic, and the p-value as follows:

```
MTB > SET INTO C1
DATA> 2.09 6.72 7.51 8.34 3.27 4.32 10.08 11.62 12.24 10.04 -2.65
DATA> 1.07 3.09 11.84 14.22 12.28 8.22 10.06 13.64 6.33 -4.09
DATA> END
MTB > SET INTO C2
DATA> 4.77 6.18 8.52 9.13 3.86 3.46 10.45 14.16 13.35 9.75 -12.83 -8.81
DATA> -5.46 10.29 15.79 13.63 12.32 18.10 10.96 8.22 1.86
DATA> END
MTB > TWOSAMPLE C1 C2;
SUBC> POOLED

SUBC>

TWOSAMPLE T FOR C1 VS C2
        N      MEAN     STDEV    SE MEAN
C1   21       7.15      5.19       1.1
C2   21       7.03      7.97       1.7

95 PCT CI FOR MU C1 - MU C2: (-4.1, 4.3)

TTEST MU C1 = MU C2 (VS NE): T= 0.06   P=0.95   DF=  40

POOLED STDEV =        6.73

MTB > PAPER
```

From this computer output, we find that the t-statistic equals .06 and the p value equals .95. Again, we are unable to reject the null hypothesis that the average EPS of Ford and GM are the same.

---

[5]If the nonpooled variance is used, then the t value is

$$t = (7.154 - 7.033) \Big/ \sqrt{\frac{(5.195)^2}{21} + \frac{(7.973)^2}{21}} = \frac{.121}{2.077} = .058$$

■ | ***APPLICATION 11.2*** Analysis of the Bank Risk Premium[6]

The international banking crisis of 1974, involving the failure of the Franklin National Bank in New York, led the Federal Reserve System to guarantee the international as well as the domestic deposits of the bank. Giddy (1980) hypothesized that this "Franklin Message" would lead to a decrease in the risk premium attached to large American banks' deposits. (Risk premium here is taken to be measured by the excess of secondary-market certificate of deposit rates over Treasury bill yields.) For 48 months before the "Franklin Message," the mean risk premium was .899 and the variance was .247. For 48 months after the message, the mean and variance were .703 and .320. If μ_x and μ_y denote the means before and after the message, respectively, test the null hypothesis

$$H_0: \mu_x - \mu_y = 0$$

against the alternative hypothesis

$$H_1: \mu_x - \mu_y > 0$$

Assume that the data can be regarded as independent random samples from the two populations.

The decision rule is to reject H_0 in favor of H_1 if

$$\frac{\bar{x} - \bar{y}}{\sqrt{\dfrac{s_x^2}{n_x} + \dfrac{s_y^2}{n_y}}} > Z_\alpha$$

In this example, $\bar{x} = .899$, $s_x^2 = .247$, $n_x = 48$, $\bar{y} = .703$, $s_y^2 = .320$, and $n_y = 48$, so

$$\frac{\bar{x} - \bar{y}}{\sqrt{\dfrac{s_x^2}{n_x} + \dfrac{s_y^2}{n_y}}} = \frac{.899 - .703}{\sqrt{\dfrac{.247}{48} + \dfrac{.320}{48}}} = 1.80$$

From Table A3 in Appendix A, we find that the value of α corresponding to $z_\alpha = 1.80$ is .0359. Hence the null hypothesis can be rejected at all levels of significance greater than 3.59 percent. If the null hypothesis of equality of population means were true, the probability of observing a sample result as extreme as or more extreme than that found would be .0359. This is quite strong evidence against the null hypothesis of equality of these means, suggesting rather a decrease in the mean risk premium after the "Franklin Message."

---

[6]This example is drawn from a study by I. H. Giddy (1980), "Moral Hazard and Central Bank Rescues in an International Context," *The Financial Review* 15, 2, 50–60. Reprinted by permission of the publisher.

■ | **APPLICATION 11.3** Analysis of Rates of Return for Retail Firms[7]

In their study aimed at finding early warning signals of business failure, Sharma and Mahajan (1980) used a random sample of 23 failed retail firms that 3 years before showed a mean return on assets of .058 and a sample standard deviation .055. An independent random sample of 23 nonfailed retail firms showed a mean return of .146 and a standard deviation of .058 for the same period. If μ_x and μ_y denote the population means for failed and nonfailed firms, respectively, test the null hypothesis

$$H_0: \mu_x - \mu_y \geq 0$$

against the alternative hypothesis

$$H_1: \mu_x - \mu_y < 0$$

Assume that the two population distributions are normal and have the same variance.

The decision rule is to reject H_0 in favor of H_1 if

$$\frac{\bar{x} - \bar{y}}{s\sqrt{\dfrac{n_x + n_y}{n_x n_y}}} < -t_{\nu,\alpha}$$

For these data, we have $\bar{x} = .058$, $s_x = .055$, $n_x = 23$, $y = .146$, $s_y = .058$, and $n_y = 23$. Hence,

$$s^2 = \frac{(n_x - 1)s_x^2 + (n_y - 1)s_y^2}{n_x + n_y - 2}$$

$$= \frac{(22)(.055)^2 + (22)(.058)^2}{23 + 23 + -2} = .0031945$$

so that $s = \sqrt{.0031945} = .0565$. Then

$$\frac{\bar{x} - \bar{y}}{s\sqrt{\dfrac{n_x + n_y}{n_x n_y}}} = \frac{.058 - .146}{.0565\sqrt{\dfrac{23 + 23}{23 \times 23}}} = -5.282$$

For a 1 percent level test, we have by interpolation from Table A4, Student's t distribution with 44 ($23 + 23 - 2$) degrees of freedom, $t_{44,.01} = 2.414$. Because -5.282 is much less than -2.414, the null hypothesis is rejected at $\alpha = 1$ percent. The data cast considerable doubt on the hypothesis that the population mean return on assets is at least as large for failed than for nonfailed retail firms.

The test just discussed and illustrated is based on the assumption that the two population variances are equal. It is also possible to develop tests that are valid when this assumption does not hold.

---

[7]This example is taken from S. Sharma and V. Mahajan (1980), "Early Warning Indicators of Business Failure," *Journal of Marketing* 44, 80–89. Reprinted by permission of the American Marketing Association.

APPLICATION 11.4 Hypothesis Testing Approach to Interpret the Quality Control Chart

To use a quality control chart as discussed in Section 10.9 is to perform a statistical test of an hypothesis each time a sample is taken and plotted on the chart. In general, the null hypothesis H_0 is that the process is in control, and the alternative hypothesis H_1 is that the process is out of control. For example, in an $\overline{X}$-chart, to determine whether the process mean has shifted, we can test the hypothesis

$$H_0: \mu = \mu_0 \quad \text{versus} \quad H_1: \mu \neq \mu_0$$

where μ is the population mean and μ_0 is the specified value for μ. This null hypothesis can be converted into a confidence interval at a specified α value, as we noted in Section 11.5. This confidence interval defines the upper and lower limits of a control chart.

The $\overline{X}$-chart for Consolidated Auto Supply Company given in Application 10.1 illustrates how a hypothesis testing approach can be used to interpret the quality control chart. Under the assumption of $\alpha = .27$ percent, $UCL_X = 10.8210$, and $LCL_X = 10.6132$, it is found that 1 out of 24 sample means is smaller than 10.6132. Because $\frac{1}{24} = 4.17$ percent is larger than .27 percent, the null hypothesis should be rejected.

APPLICATION 11.5 Comparison of Organizational Values at Two Different Companies

Professor J. M. Liedtka used survey data from two firms, company A and company B.[8] Nine managers from each company were asked to rate the importance of each of a given list of organizational values on a scale of 1 to 7. Table 11.6 gives the ratings ascribed to these values by each company's managers. By using the t-statistic

Table 11.6 Organizational Values at Companies A and B[a]

Value	Company A Score	Standard Deviation	Company B Score	Standard Deviation	t-value
Industry leadership	6.4	.5	5.6	1.0	−2.3[c]
Reputation of the firm	6.1	.8	5.9	.6	−.7
Employee welfare	5.0	1.0	3.0	.9	−4.5[d]
Tolerance for diversity	5.0	.9	3.9	1.4	−2.1[b]
Service to the general public	3.6	1.9	3.7	.9	.2
Value to the community	3.6	1.8	3.8	1.1	.3
Stability of the organization	5.3	1.2	3.3	1.3	−3.3[d]

(*continued*)

[8] Jeanne M. Liedtka (1989), "Value Congruence: The Interplay of Individual and Organizational Value Systems," *Journal of Business Ethics* 8, 805–815. Reprinted by permission of Kluwer Academic Publishers.

Table 11.6 (Continued)

Value	Company A		Company B		t-value
	Score	Standard Deviation	Score	Standard Deviation	
Budget stability	4.3	1.5	4.6	1.3	.3
Organizational growth	5.2	1.6	4.9	1.1	−.5
Profit maximization	5.6	.7	6.7	.5	3.8[d]
Innovation	5.7	1.3	4.0	1.3	−2.9[c]
Honesty	5.9	1.1	4.7	1.8	−1.8[b]
Integrity	6.0	1.1	4.4	2.2	−1.9[b]
Product quality	6.0	.9	4.0	1.6	−3.3[d]
Customer service	5.0	1.3	4.0	1.9	−1.3
Average score	5.3	1.2	4.4	1.3	

Source: Adapted from Jeanne M. Liedtka (1989), "Value Congruence: The Interplay of Individual and Organizational Value Systems," *Journal of Business Ethics* 8. Reprinted by permission of Kluwer Academic Publishers.
[a]Score is based on a Likert Scale of 1 (of lesser importance) to 7 (of greater importance).
[b]Significant at alpha of .10
[c]Significant at alpha of .05
[d]Significant at alpha of .01

as indicated in Equations 11.12 and 11.13, Liedtka performed a test. The *t*-values are indicated in the last column of Table 11.6. From this table, we see that the managers of company A rated most important, followed by integrity and product quality. The managers of company B rated profit maximization most important, followed by organizational growth. From the *t*-values presented in Table 11.6, it is evident that the ratings of organizational values differed significantly at $\alpha = .10$ for most of the items, but not for budget stability, organizational growth, and customer service.

Summary

Using the concepts of statistical distributions and interval estimates, we showed how these concepts can be employed to test hypotheses about the parameters of a population. Hypothesis tests for one-tailed and two-tailed tests for both large and small samples were analyzed in detail. In addition to using the normal and *t* distributions for performing hypothesis testing, we discussed the use of the chi-square distribution to test null hypotheses about the sample variance from a normally distributed population.

The statistical concepts and methods discussed in the last 11 chapters will be used in the remaining 10 chapters to conduct further statistical analyses.

Appendix 11A The Power of a Test, the Power Function, and the Operating-Characteristic Curve

The main purpose of this appendix is to discuss the power of a test and the power function. The related concepts of the power curve and the operating-characteristic curve (OC curve) are also discussed.

The Power of a Test and the Power Function

The **power of a test** is the probability of rejecting H_0 when it is false. The probability is equal to $(1 - \beta)$, where β denotes the probability of Type II error. Other things being equal, the greater the power of the test, the better the test. The formula used to calculate $(1 - \beta)$ can be derived as follows:

$$
\begin{aligned}
\text{Power} &= 1 - \beta \\
&= P(\text{null hypothesis rejected when it is false}) \\
&= P\left(\frac{\overline{X} - \mu_0}{\sigma/\sqrt{n}} > z_\alpha\right) \\
&= P\left(\overline{X} > \mu_0 + \frac{z_\alpha \sigma}{\sqrt{n}}\right) \\
&= P\left(\frac{\overline{X} - \mu_1}{\sigma/\sqrt{n}} > \frac{\mu_0 - \mu_1}{\sigma/\sqrt{n}} + z_\alpha\right) \\
&= P\left(Z > \frac{\mu_0 - \mu_1}{\sigma/\sqrt{n}} + z_\alpha\right)
\end{aligned}
\tag{11A.1}
$$

where μ_1 is the true population mean when the null hypothesis is false.

The functional relationship defined in Equation 11A.1 is called the **power function.** This power function is derived from the assumption of an upper-tailed test. Similarly, the power function in terms of a lower-tailed test can be defined as Equation 11A.2, and the power in terms of a two-tailed test can be defined as Equation 11A.3.

$$
1 - \beta = P\left[Z < \frac{\mu_0 - \mu_1}{\sigma/\sqrt{n}} - z_\alpha\right]
\tag{11A.2}
$$

$$
1 - \beta = P\left[Z > \frac{\mu_0 - \mu_1}{\sigma/\sqrt{n}} + z_{\alpha/2}\right] + P\left[Z < \frac{\mu_0 - \mu_1}{\sigma/\sqrt{n}} - z_{\alpha/2}\right]
\tag{11A.3}
$$

EXAMPLE 11A.1 *Power Function and Type II Error*

From Example 11.1, investigating the average weight of a bag of cat food, we express the two hypotheses as

$$H_0: \mu = 60$$

$$H_1: \mu > 60$$

Using the data for Example 11.1 in the text, we have $Z_{.05} = 1.645$, $n = 100$, and $s_x = 5$. We calculate $(1 - \beta)$ for $\mu_1 = 60, 60.5, 61, 61.5,$ and 62 in accordance with Equation 11A.1.

1. $\mu_1 = 60$ $1 - \beta = P\left(Z > \dfrac{60 - 60}{5/10} + 1.645\right)$

$$= P(Z > 1.645)$$

$$= .05$$

2. $\mu_1 = 60.5$ $1 - \beta = P\left(Z > \dfrac{60 - 60.5}{5/10} + 1.645\right)$

$$= P(Z > .645)$$

$$= .25945$$

3. $\mu_1 = 61$ $1 - \beta = P\left(Z > \dfrac{60 - 61}{5/10} + 1.645\right)$

$$= P(Z > -.355)$$

$$= .6387$$

4. $\mu_1 = 61.5$ $1 - \beta = P\left(Z > \dfrac{60 - 61.5}{5/10} + 1.645\right)$

$$= P(Z > -1.355)$$

$$= .9123$$

5. $\mu_1 = 62$ $1 - \beta = P\left(Z > \dfrac{60 - 62}{5/10} + 1.645\right)$

$$= P(Z > -2.355)$$

$$= .99075$$

Using these results for $(1 - \beta)$, we can easily calculate the values of β. Values for both $(1 - \beta)$ and β are listed in Table 11A.1. In Figure 11A.1, we present the relationship between the power function $(1 - \beta)$ and the probability of Type II error (β). Curve I, which describes the relationship between $(1 - \beta)$ and μ_1, is called the **power curve**. It is an increasing function of the value of μ_1. Curve II, which describes the relationship between β and μ_1, is called the **operating-characteristic (OC) curve**.

Table 11A.1 The Relationship Among μ_1, β_1 and $(1 - \beta_1)$

μ_1	$1 - \beta$	β
60.0	.05	.95
60.5	.25945	.74055
61.0	.6387	.3613
61.5	.9123	.0877
62.0	.99075	.00925

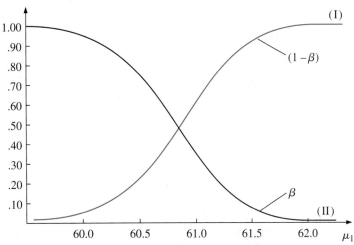

Figure 11A.1

The Power Function and the Probability of Type II Error

The OC curve is a decreasing function of μ_1. Overall, the relationship among μ_1, Type II error, and the power of the test can be summarized as follows: The larger the value of μ_1, the smaller the Type II error and the larger the power of the test. As we will see in the next section, the OC curve is frequently used in statistical quality control to analyze the risk involved in a sampling plan.

Operating-Characteristic Curve[9]

In Chapter 10 we constructed quality control charts by using sampling production process data. To collect sample data for quality control, we need sample plans that specify the lot size N, the sample size n, and the acceptance/rejection criterion. An important feature of a sampling plan is how well it discriminates between lots of high quality and lots of low quality. The **operating characteristic** can be used to describe the ability of a sample plan to differentiate high-quality lots from low-quality lots.

[9]The material in this section draws heavily on William J. Stevenson (1990), *Production/Operations Management,* 3rd ed. (Homewood, Ill.: Irwin), pp. 829–836.

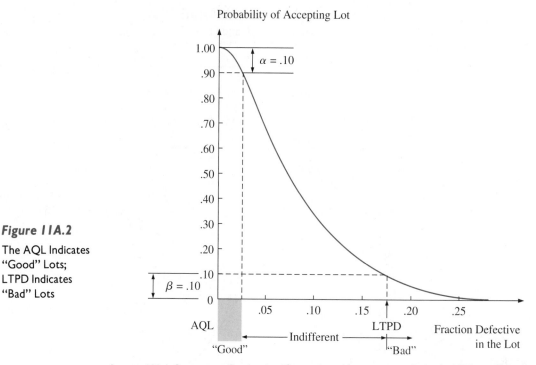

Probability of Accepting Lot

Figure 11A.2

The AQL Indicates "Good" Lots; LTPD Indicates "Bad" Lots

$\alpha = .10$

$\beta = .10$

AQL

"Good"

Indifferent

LTPD

"Bad"

Fraction Defective in the Lot

Source: W. J. Stevenson, *Production/Operations Management*, 3rd ed., 1990, p. 833, reprinted by permission of Richard D. Irwin.

Acceptance sampling is frequently the most desirable method for quality control. The inspector takes a statistically determined random sample and applies a decision rule to determine the acceptance or rejection of the lot on the basis of the observed number of nonconforming items. The Type II error and Type I error discussed in this chapter are bases for the decision rule for statistical quality control. The probability that a lot containing the **lot tolerance percentage defective** (LTPD) will be accepted is known as the **consumer's risk,** or beta (β), or a Type II error. The probability that a lot containing the **acceptable quality level** (AQL) will be rejected is known as the **producer's risk,** or alpha (α), or a Type I error. Sampling plans are frequently designed so that they have a producer's risk of 5 percent and a consumer's risk of 10 percent, although other combinations also are used. Figure 11A.2 shows an OC curve with the AQL, LTPD, producer's risk ($\alpha = 10$ percent) and consumer's risk ($\beta = 10$ percent).

EXAMPLE 11A.2 *Construction of an OC Curve with Sample Size n = 10 and Defective Items C = 1*

Suppose we want the OC curve for a situation in which a sample of $n = 10$ items is drawn from lots containing $N = 2{,}000$ items, and lots are accepted if no more than $C = 1$ defective is found.

Table IIA.2 β for Different Fractions Defective ($n = 10$, $C = 1$)

Fraction Defective (P)	Probability of Acceptance (β)
.05	.9139
.10	.7361
.15	.5443
.20	.3758
.25	.2440
.30	.1493
.35	.0860
.40	.0464
.45	.0233
.50	.0107
.55	.0045
.60	.0017

The sample size is $10/2,000 = .5$ percent, so it is small enough for us to use the binomial distribution.[10] In this case $n = 10$ and $C = 1$. Table 11A.2 presents the probability of acceptance (β) for 12 different fractions defective (P).

The β values indicated in Table 11A.2 are obtained from Table A1 in Appendix A. For example, when $n = 10$, $C = 1$, and $P = .20$, β is calculated as follows:

$$\binom{10}{0}(.20)^0(1 - .8)^{10} + \binom{10}{1}(.20)^1(1 - .20)^{10-1} = .1074 + .2684 = .3758$$

By plotting all 12 different β values on a graph and connecting them, we get the OC curve shown in Figure 11A.3.

In theoretical statistics, only the power curve is generally considered. But in practical statistics, in certain types of problems, the OC curve is much easier to interpret for practical purposes such as quality control. Hence the OC curve has been extensively used in quality control.

Up to this point, we have investigated the power function when the sample size is fixed. If sample size increases, then $\dfrac{(\mu_0 - \mu_1)}{(\sigma/\sqrt{n})}$ gets smaller if μ_1, σ, and α are held constant. From Equation 11A.1, it is obvious that the power $(1 - \beta)$ increases.

---

[10]Because the sampling is generally done without replacement, the hypergeometric distribution would be more appropriate if the ratio n/N exceeded 5 percent.

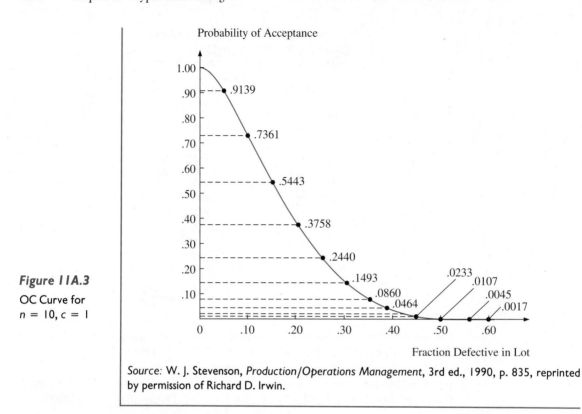

Probability of Acceptance

Figure 11A.3

OC Curve for
$n = 10, c = 1$

Fraction Defective in Lot

Source: W. J. Stevenson, *Production/Operations Management*, 3rd ed., 1990, p. 835, reprinted by permission of Richard D. Irwin.

Questions and Problems

1. For each of the following, test the indicated hypothesis.
   a. $n = 16, \bar{x} = 1{,}550, s^2 = 12, H_0: \mu = 1{,}500, H_1: \mu > 1{,}500, \alpha = .01$
   b. $n = 9, \bar{x} = 10.1, s^2 = .81, H_0: \mu = 12, H_1: \mu \neq 12, \alpha = .05$
   c. $n = 49, \bar{x} = 17, s = 1, H_0: \mu \geq 18, H_1: \mu < 18, \alpha = .05$

2. The estimated variance based on 4 measurements of a spring tension was .25 gram. The mean was 37 grams. Test the hypothesis that the true value is 35 grams. Use $\alpha = .10$ and $H_1: \mu > 35$.

3. A population has a variance σ^2 of 100. A sample of 25 from this population had a mean equal to 17. Can we reject $H_0: \mu = 21$ in favor of $H_1: \mu \neq 21$? Let $\alpha = .05$.

4. Suppose a sample of 15 rulers from a given supplier have an average length of 12.04 inches and the sample standard deviation is .015 inch. If α is .02, can we conclude that the average length of the rulers produced by this supplier is 12 inches, or should we accept $H_1: \mu \neq 12.00$?

5. The drained weights, in ounces, for a sample of 15 cans of fruit are given below. At a 5 percent level of significance, use MINITAB to test the hypothesis that on average a 12-ounce drained-weight standard is being maintained. Use $H_1: \mu \neq 12.0$ as the alternative hypothesis.

12.0	12.1	12.3	12.1	12.2
11.8	12.1	11.9	11.8	12.1
12.4	11.9	12.3	12.4	11.9

6. An advertisement for a brand-name camera stated that the cameras are inspected and that "60 percent are rejected for the slightest imperfections." To test this assertion, you observe the inspection of a random selection of 30 cameras and find that 15 are rejected. Construct a test, using $\alpha = .05$.

7. A 1984 study indicated that the average yearly housing cost for a family of 4 was $12,983. A random sample of 200 families in a U.S. city resulted in a mean of $14,039 with a standard deviation of $2,129. Is this city's sample mean significantly higher than the population mean? Use $\alpha = .05$.

8. The data entry operation in a large computer department claims that it gives its customers a turnaround time of 6.0 hours or less. To test this claim, one of the customers took a sample of 36 jobs and found that the sample mean turnaround time was $\bar{x} = 6.5$ hours with a sample standard deviation of $s = 1.5$ hours. Use $H_0: \mu = 6.0$, $H_1: \mu > 6.0$, and $\alpha = .10$ to test the data entry operation's claim.

9. The following data represent the time, in seconds, that it took the sand in a sample of timers to run out. At the 10 percent significance level, can we conclude that the mean for timers of this type is not equal to the nominal 3 minutes?
   a. Use $H_1: \mu \neq 180$ as the alternative hypothesis.
   b. Use MINITAB to test (1) $H_1: \mu \neq 180$ and (2) $H_1: \mu > 180$.

   | 190 | 199 | 198 | 176 | 180 | 174 |
   | 181 | 183 | 208 | 188 | 198 | 165 |

10. Independent random samples from normal populations with the same variance gave the results shown in the following table. Can we conclude that the difference between the means, $\mu_1 - \mu_2$, is less than 5? That is, test $H_0: \mu_1 - \mu_2 \geq 5$ with $\alpha = .05$.

Sample	n	Mean	Standard Deviation
1	15	22	9
2	9	25	7

11. What is hypothesis testing? Why are we interested in hypothesis testing? In hypothesis testing, is it possible to prove a hypothesis true?

12. What are the types of errors that can be made in hypothesis testing? Which type of error is generally regarded as more serious?

13. For each of the following pairs of hypotheses, explain what the null hypothesis should be.
    a. Not guilty versus guilty in a court case.
    b. Cage is safe versus cage is unsafe when testing the safety of lion cages.

c. New drug is safe to use versus new drug is unsafe when determining whether the FDA should allow a new arthritis medicine to be sold.
d. New treatment is safe versus new treatment is unsafe when determining whether the FDA should allow a new treatment for AIDS to be used.

14. Compare the concepts of interval estimation discussed in Chapter 10 with the concept of hypothesis testing discussed in this chapter. How are they related?

15. Compare a one-tailed test with a two-tailed test. Give some examples wherein a one-tailed test is preferable to a two-tailed test. Give some examples wherein a two-tailed test is preferable to a one-tailed test.

16. Briefly explain what is meant by the power of a test. Why is the power of the test important.?

17. What is a simple hypothesis? What is a composite hypothesis? Give some examples of a simple hypothesis. Give some examples of a composite hypothesis.

18. In 1981 the election for governor of the state of New Jersey in which Tom Kean defeated Jim Florio was so close that Florio demanded a recounting of the votes. If you were Florio and you were conducting an hypothesis test of who won the election, what would your null hypothesis be? How would your answer change if you were Kean?

19. In conducting an hypothesis test, how do we determine the rejection region?

20. Briefly explain why the central limit theorem is important in hypothesis testing.

21. Evaluate the following statement: "If we reject the null hypothesis that $\mu = \mu_0$ in a two-tailed test, we will also reject it in a one-tailed test (using the same α)."

22. Find the critical values for the following standard normal distributions:
    a. Two-tailed test for $\alpha = .05$
    b. One-tailed test for $\alpha = .05$
    c. Two-tailed test for $\alpha = .01$
    d. One-tailed test for $\alpha = .01$
    e. Two-tailed test for $\alpha = .10$
    f. One-tailed test for $\alpha = .10$

23. You are given the information $\bar{x} = 10$, $\sigma = 2$, and $n = 35$. Conduct the following hypothesis test at the .05 level of significance.

$$H_0: \mu = 0 \quad \text{versus} \quad H_1: \mu > 0$$

24. Use the information given in question 23 to test

$$H_0: \mu = 0 \quad \text{versus} \quad H_1: \mu \neq 0$$

at the .05 level of significance.

25. You are given the information $\bar{x} = 150$, $\sigma = 30$, and $n = 20$. Conduct the following hypothesis test at the .01 level of significance.

$$H_0: \mu = 100 \quad \text{versus} \quad H_1: \mu > 100$$

26. Use the information given in question 25 to test

$$H_0: \mu = 100 \quad \text{versus} \quad H_1: \mu \neq 100$$

at the .01 level of significance.

27. You are given the information $\bar{x} = 1{,}050$, $s_x = 250$, and $n = 20$. Conduct the following hypothesis test at the .10 level of significance.

$$H_0: \mu = 1100 \quad \text{versus} \quad H_1: \mu < 1100$$

28. Use the information given in question 27 to test

$$H_0: \mu = 1100 \quad \text{versus} \quad H_1: \mu \neq 1100$$

at the .10 level of significance.

29. A sample of 100 students in a high school have a sample mean score of 550 on the math portion of the SAT. Assuming that the sample standard deviation is 75, test, at the .05 level of significance, the hypothesis that the high school's mean SAT score is 500 against the alternative hypothesis that the school's mean SAT score does not equal 500.

30. Redo question 29, testing $H_0: \mu = 500$ against H_1: $\mu > 500$.

31. A sample of 20 students in a high school has a sample mean score of 520 on the English portion of the SAT. If the sample standard deviation is 65, test, at the .01 level of significance, the hypothesis that the school's mean SAT score is equal to 500 against the alternative hypothesis that the school's mean SAT score does not equal 500.

32. Redo question 31, substituting the alternative hypothesis $H_1: \mu > 500$.

33. Suppose a random sample of 25 people at a local weight-loss center is taken, and the mean weight loss is found to be 12 pounds. From past history, the standard deviation is known to be 3 pounds. Test the hypothesis that the mean weight loss for all the members of the weight-loss center is 10 pounds against the alternative that it is more than 10 pounds. Do the test at the .05 level of significance.

34. Redo question 33, but assume that the standard deviation is not known and that 3 pounds represents the sample standard deviation. Do the test at the 5 percent level of significance.

35. A quality control engineer is interested in testing the mean life of a new brand of light bulbs. A sample of 100 light bulbs is taken, and the sample mean life of these light bulbs is found to be 1,075 hours. Suppose the standard deviation is known and is 100 hours. Use a .05 level of significance to test the hypothesis that the mean life of the new bulbs is greater than 1,000 hours.

36. Suppose that the quality control engineer in question 35 does not know what the standard deviation is and therefore uses the sample standard deviation. Does your answer to question 35 change? Why or why not?

37. Suppose that the quality control engineer in question 35 does not know what the standard deviation is and that this time, he selects a random sample of only 25 light bulbs. Does your answer to question 35 change? Explain.

38. An auditor is interested in the mean value of a company's accounts receivable. He randomly samples 200 accounts receivable and finds that the mean accounts receivable is $231. From past experience he knows that the standard deviation is $25. Use a .01 level of significance to test whether the population mean accounts receivable is different from $200.

39. Use the information given in question 38 to test the hypothesis that the population mean accounts receivable is greater than $200 at the .05 level of significance.

40. An investment advisor is interested in determining whether a retirement community represents a potential clientele base. Of the 2,000 residents, he randomly samples 100 individuals and finds their

mean wealth to be $525,000 with a sample standard deviation of $52,000. Use a .10 level of significance to test the hypothesis that the mean wealth is greater than $500,000.

41. An automobile manufacturer claims that a new car gets an average of 35 miles per gallon. Assume that the distribution is known to be normal with a standard deviation of 3.2 miles per gallon. A random sample of 10 cars gives an average of 35.1 miles per gallon. Test, at the .01 level of significance, the alternative hypothesis that the population mean is at least 35 miles per gallon.

42. Use the information given in question 41, except this time assume that the standard deviation is not known and that 3.2 miles per gallon represents the sample standard deviation. Again test, at the .01 level of significance, the alternative hypothesis that the population mean is at least 35 miles per gallon.

43. An aspirin manufacturer claims that its aspirin stops headaches in less than 30 minutes. A random sample of 100 people who use the pain killer finds that the average time it takes to stop a headache is 28.6 minutes with a sample standard deviation of 4.2 minutes. Test, at the 5 percent level of significance, the manufacturer's claim that this product stops headaches in less than 30 minutes.

44. Bob's SAT Preparation Service claims that the course it offers enables students to score an average of 600 or better on the math portion of the SAT. Suppose a random sample of 25 people taking the course has a mean score of 650 with a sample standard deviation of 50. Would it be more appropriate to use a one-tailed or a two-tailed test? Test the company's claim at the 10 percent level of significance.

45. An advertising company claims that 80 percent of stores that use their advertisements show increased sales. A random sample of 100 stores that used the company's advertisements reveals that 80 showed increased sales. Test, at the 5 percent level of significance, whether at least 75 percent of stores using the advertisements had increased sales.

46. Flip a coin 40 times and count the number of heads. Test, at the 5 percent level of significance, whether the proportion of heads is .5.

47. A manufacturer claims that 95 percent of its parts are free of defects. A random sample of 100 parts finds that 92 are free of defects. Test the manufacturer's claim at the 1 percent level of significance.

48. An investment advisor claims that 70 percent of the stocks she recommends will increase in price. Suppose testing a random sample of 125 stocks she recommends reveals that 75 have increased in price. Test her claim at the 10 percent level of significance.

49. Ed's Bar Exam Review claims that 90 percent of the people who take its review course pass the bar exam on the first try. A random sample of 500 people who took the course reveals that 425 passed the bar exam on the first try. Test, at the 5 percent level of significance, the null hypothesis that at least 90 percent of those who take the review course pass the bar exam on the first try.

50. Use the information given in question 49 to test, at the 1 percent level of significance, the null hypothesis that less than 80 percent of those who take the course pass the bar exam on the first try.

51. In a taste test using 400 randomly selected people, 220 preferred a new brand of coffee to the leading brand. Test, at the 1 percent significance level, the alternative hypothesis that at least 52 percent prefer the new brand.

52. A popular commercial states that 4 out of 5 dentists who chew gum prefer sugarless gum. Suppose a random sample of 100 gum-chewing dentists is taken and 75 are found to prefer sugarless gum. Test, at the 10 percent level of significance, the null hypothesis that the commercial's claim is true.

53. A diet center claims that people subscribing to its program lose an average of 4 pounds in the first week of the diet. Suppose 25 people in the diet center's program are chosen at random and are found to have lost 4.3 pounds in the first week with a sample standard deviation of 1.1 pounds. Test, at the 5 percent level of significance, the hypothesis that the mean weight loss is 4 pounds.

54. Use the information given in question 53 and test, at the 5 percent level of significance, the alternative hypothesis that the mean weight loss is at least 4 pounds.

55. Suppose a farmer is interested in testing two fertilizers to see which is more effective. He uses the two fertilizers and gets the following results.

Fertilizer	Mean Growth	Standard Deviation	Size of Sample
A	7 inches	.5 inch	100
B	6 inches	.2 inch	125

Test, at the 10 percent level of significance, the hypothesis that the mean difference in growth between the two fertilizers is not significant.

56. A production manager is interested in the number of defects in batches derived from different production processes. He examines a random sample drawn from each process and records the following data.

Process	Mean Defects	Standard Deviation	Size of Sample
A	221	25	90
B	300	80	110

Test, at the 1 percent level of significance, the hypothesis that the mean difference in number of defects between the two production processes is not significant.

57. Suppose an attorney specializing in wage discrimination cases is interested in determining whether the earnings of men and women are significantly different. He collects the following data on earnings for a random sample of first-year accountants.

Sex	Mean Earnings	Standard Deviation	Size of Sample
Female	$39,217	$12,210	125
Male	$43,121	$17,020	100

Test, at the 5 percent level of significance, the hypothesis that the mean earnings of male and female first-year accountants do not differ significantly.

58. Suppose a political scientist is interested in whether wealth is a determining factor in the individual's propensity to vote. A random sample of 500 people who earned $100,000 or more showed that 390 voted, whereas a random sample of 400 people who earned less than $25,000 showed that 280 voted. Test, at the 10 percent level of significance, the null hypothesis that the two population voting rates are equal against the alternative hypothesis that the voting rate is higher for people earning $100,000 or more.

59. A mutual fund manager claims that the returns of stocks in her fund have a variance of no more than .50. A random sample of 25 stocks in her fund has a sample variance of .72. Assuming that the distribution is normal, test the fund manager's claim at the 5 percent level of significance.

60. Bob claims that the variance of the score for the people who took the SAT review course he offers is 100. Fred believes that Bob's students have a variance larger than 100. A random sample of 10 of Bob's students has a variance of 162. Test Fred's claim at the 10 percent level of significance.

61. A political science professor believes students majoring in political science are more likely to vote in elections than students majoring in other disciplines. He collects the following information from two random samples of students.

Major	Proportion Voting	Number of Students
Political science	.65	120
Other	.62	113

Test, at the 5 percent level of significance, this professor's hypothesis against a two-sided alternative that the population proportions are equal.

62. An education professor is interested in whether there is any difference between the proportion of students who have taken a review course that pass the bar exam and the proportion of those who have not taken a review course that pass the exam. She collects the following information from a random sample of students.

	Proportion Passing	Number of Students
Review course	.55	300
No review course	.49	400

Test the hypothesis that the population proportions are equal at the 10 percent level of significance in terms of a two-tailed test.

63. Use the information given in question 62, but this time test the hypothesis that the proportion of students passing the exam is greater for those who take the course than for those who do not.

64. The IRS is interested in knowing whether people who have an accountant prepare their tax returns have fewer errors than people who prepare their own returns. A random sample of 500 people who had their returns professionally prepared reveals that 125 had errors. A random sample of 450 returns of people who prepared their own returns reveals that 128 had errors. Test, at the 1 percent level of significance, the hypothesis that there is no difference between the number of errors for returns prepared by an accountant and the number of errors for returns prepared by the individual.

65. Use the information given in question 64 to test, at the 1 percent level of significance, the hypothesis that the number of errors is greater for individuals who prepare their own returns than for people who have their returns professionally prepared.

66. A muffler manufacturer claims that the variance of its product is no more than 200. A random sample of 25 mufflers has a sample variance of 391. Assuming a normal distribution, test the manufacturer's claim at the 10 percent level of significance.

67. An SAT review course claims that the variance of test scores of its graduates is less than 150. A random sample of 30 students who took the course is found to have a variance of 225. Assuming a normal distribution, test the review course's claim at the 10 percent level of significance.

68. From past experience, a teacher finds that the variance of midterm test scores is 76. A random sample of 21 midterms in her course has a sample variance of 110. Assuming that the population is distributed normally, test whether the sample variance is different from the population variance at a 5 percent level of significance.

69. Refer to question 41 to find the power of a 10 percent level test when the true population mean mileage is 36 miles per gallon.

70. Referring to question 43, find the power of a 5 percent level test when the true population mean time for headache relief is 35 minutes.

71. Assume that you're taking a part-time job in a zoo. You are called upon to inspect a new cage built to contain a ferocious lion. Do you set up the null hypothesis that the cage is safe or that the cage is dangerous?

Use the following information to answer questions 72 to 78. A college professor gives a test that has 10 true-false questions. Two students take the test. Student A, who does not know anything about the subject, answers the questions by tossing a coin. The college professor sets up the following hypothesis, where p represents the probability that a student gets an answer right.

H_0: The students do not know anything ($p = .5$).
H_1: The students know the subject ($p > .5$).

72. What is the consequence of a Type I error in this question?

73. What is the chance of student A getting exactly 6 correct answers when the null hypothesis is true?

74. If the professor decides to reject the null hypothesis (that means passing the student) when the students get 8 or more correct answers, what is the probability of a Type I error?

75. If the professor wants to raise the standard for passing the test to 9 or more correct answers, what is the probability of a Type I error?

76. Student B studies one night before the test, so he knows about 60 percent of the material. What is the probability that this student can pass the test when the standard for passing is 8 correct answers?

77. Plot the OC curve, assuming $p = .5, .6, .7, .8, .9, 1$.

78. Plot the power curve, assuming $p = .5, .6, .7, .8, .9, 1$.

79. A poll was done to predict the outcome of the upcoming election. Of the 900 potential voters who responded, 500 plan to vote for the incumbent. If a candidate needs 50 percent of the votes to win the election, can you reject the hypothesis that the incumbent will win? Do a 5 percent level of significance test.

80. On a given trading day, a financial economist randomly examines the stock prices of 500 companies and discovers that 205 went up and 295 went down. On this evidence, can he argue that more than 50 percent of all the stocks went down in price? Do a 10 percent level of significance test.

81. The head of the accounting department randomly examined some accounting entries and was upset with the high proportion of incorrect invoices. He instituted a new system to keep the proportion of bad invoices below 0.1 percent. A year later, 10,000 invoices were randomly examined and 6 were found to be incorrect. Can this manager reject the null hypothesis that the proportion of bad invoices is 0.1 percent? Do a 5 percent level of significance test.

82. You are working for a consumer rights organization. You are interested in knowing whether the milk contained in 16-ounce (1-pint) bottles really weighs 16 ounces. You do not want to accuse the packer of cheating its customers unless you obtain convincing evidence. You collect 60 bottles of milk. The average weight is 15.32, and the standard deviation is 1 ounce. Test at a 5 percent significance level.

83. You are working for a VCR manufacturer. There are three shifts in the plant: morning shift, evening shift, and midnight shift. The manager suspects that the midnight shift's productivity is lower than 70 units. He wants to shut down the midnight shift without causing any labor–management tension. That means he will take that action only when he has enough evidence. Your responsibility is to test whether the productivity of the midnight shift is really lower than 70 units. You obtain the production for 100 nights and compute the mean as 68 and the standard deviation as 15. Test at a 5 percent significance level. Propose your suggestion to the manager.

84. A college wants to increase its dormitory facilities to house 60 percent of the students enrolled. In order to make sure that more than 60 percent of the students want to live in the dormitory, the school randomly surveys 400 students and finds that 255 students intend to live in the dormitory. Can the school reject the null hypothesis of $p = 60$ percent? Test at a 5 percent significance level.

85. A cola company wants to change its formula for producing cola, but first it wants to make sure that more than 70 percent of its customers will like the new cola better than the old. Two thousand people taste-tested the cola, and 1,422 liked the new product better. Can the company reject the null hypothesis that only 70 percent of its customers will like the new cola more? Do a 5 percent test.

86. In order to control the job turnover ratio, the personnel department did a survey and found that out of the 500 employees who were hired in the last year, only 234 stayed. Does that provide enough evidence to support the hypothesis that the retention ratio is lower than 50 percent? Do a 5 percent test.

87. An insurance company wants to study the chances that a teen-aged driver who owns a sports car will have an auto accident. Two thousand teen-aged policyholders who own sports cars were sampled in the last year. Fifteen of them got into an accident and filed a claim for damages. Can the researcher reject the null hypothesis that less than 1 percent of the policyholders got into accidents last year? Do a 5 percent test.

88. A food company claims that its new product, low-fat yogurt, is 99 percent fat-free. The management wants to keep the proportion of bad (not 99 percent fat-free) products below 2 percent. Inspectors check 500 cups of yogurt every month. In September, 20 cups of yogurt were discovered to be bad. Can you reject the null hypothesis that less than 2 percent of the product is bad? Do a 5 percent test.

89. A questionnaire was sent to 500 of a dry cleaner's customers to solicit their opinions about service received. Twenty-three customers were found to be unhappy with the service. On this evidence, can you reject the null hypothesis that more than 10 percent of the customers are unhappy? Do a 5 percent test.

90. The dean of the school of business wants the proportion of A grades given out by his faculty members to be around 10 percent. He randomly surveys 2,000 students in 50 classes and finds that of the 2,000 grades given, 198 were A. Can he reject the null hypothesis that the proportion of A grades is about 10 percent? Do a 10 percent test.

91. The placement office in a college wants to know whether experience with personal computers is important in obtaining a job. The placement director randomly selects 600 job openings and finds that 313 jobs require computer experience. On this evidence, can he support the hypothesis that more than half of the jobs in the market today require computer experience? Do a 5 percent test.

92. The head accountant in a large corporation conducted a survey last year to study the proportion of incorrect invoices. Of the 2,000 invoices sampled, 25 were incorrect. To lower the proportion, he instituted a new system. A year later, he wants to know whether the new system worked. He collects 3,000 invoices and obtains 30 incorrect invoices. Can he argue that his new system has successfully lowered the proportion of incorrect invoices? Do a 5 percent test.

93. A new medicine was invented to treat hayfever, but the new drug was found to have unpleasant side effects. An experiment on 5,000 women and 4,000 men showed that 100 women and 60 men suffered side effects after they took the medicine. Does the evidence support the hypothesis that the drug causes side effects in more women than men? Test at the 5 percent level.

94. A company believed its new toothpaste to have an effect in controlling tooth decay among children. It randomly selected a group of 400 children and gave them the new toothpaste. Another 300 children were randomly selected also and given another brand of toothpaste. It was found that 30 children using the new toothpaste and 25 children using the other brand suffered tooth decay. Can the manufacturer legitimately argue that the new toothpaste is more effective in controlling tooth decay? Do a 5 percent test.

95. The PPP cola company wants to determine what age groups like its product. It surveyed 500 teen-agers and 600 middle-aged people and found that 300 teen-agers and 350 middle-aged people liked PPP cola. Can the company conclude that PPP cola is more popular among teen-agers than among middle-aged people? Do a 5 percent test.

96. Wood *et al.* (1979) studied the impact of comprehensive planning on the financial performance of banks. They used 4 random samples to perform their study. The sample size n, average annual percent return on owner's equity $\bar{x}$, and sample standard deviation s are presented in the table.

   a. Use the data in this table to construct a 90 percent confidence interval for the difference between the mean of the "comprehensive formal planners" group and that of the "no formal planning system" group.

   b. Perform a hypothesis test at $\alpha = 10$ percent.

Average Annual Percent Return on Net Income

Classification	n	$\bar{x}$, %	s
Comprehensive formal planners	26	11.928	3.865
Partial formal planners	6	9.972	7.470
No formal planning system	9	4.936	4.466
Control group	20	2.098	10.834

Source: D. R. Wood and R. L. LaForge (1979), "The Impact of Comprehensive Planning on Financial Performance," *Academy of Management Journal* 22, 516–526. Reprinted by permission of the publisher.

97. Professor Preston *et al.* (1978) studied the effectiveness of bank premiums (stoneware, calculators) given as an inducement to open bank accounts.[11] They randomly selected a sample of 200 accounts each for "premium offered" and "no premium offered." They found that 79 percent of the accounts opened when a premium was offered and 89 percent of accounts opened when a premium was not offered were retained over a 6-month period. Use these data to test whether $P_x = 79$ percent is statistically different from $P_y = 89$ percent. Do a 5 percent test.

98. Use the data in the table to answer the following questions by using MINITAB.

Current Ratios for GM and Ford

Year	Ford	GM
81	1.02	1.09
82	.84	1.13
83	1.05	1.40
84	1.11	1.36
85	1.10	1.09
86	1.18	1.17
87	1.24	1.56
88	1.00	1.00
89	.97	1.72
90	.93	1.37

[11] R. H. Preston, F. R. Dwyer, and W. Rudelius (1978), "The Effectiveness of Bank Premiums," *Journal of Marketing* 42, 3, 39–101.

a. Test whether the current ratio is equal to 1 for both GM and Ford respectively at $\alpha = .05$.

b. Construct a 95 percent confidence interval of current ratio for both GM and Ford.

c. Test whether there is a difference between current ratios of Ford and GM at $\alpha = .05$. (Assume that their variances are different.)

d. Test whether there is a difference between current ratios of Ford and GM at $\alpha = .05$. (Assume that their variances are equal.)

99. The result of a random sample of before and after weights of 11 participants in a weight-loss program are shown in the table to the right. Using MINITAB, test whether the average weight loss is at least 16 pounds at the 5 percent significance level.

Weights Before and After a Weight-Loss Program

Before	After
187	168
200	177
218	201
205	190
192	170
175	159
191	172
200	185
206	184
231	202
240	215

CHAPTER 12

Analysis of Variance and Chi-Square Tests

CHAPTER OUTLINE

Key Terms

analysis of variance (ANOVA)
factor
one-way ANOVA
two-way ANOVA
treatments
global (overall) mean
within-group variability
between-groups variability
sum of squares
between-treatments sum of squares
within-treatment sum of squares

between-treatments mean square
within-treatment mean square
treatment effect
F distribution
Scheffé's multiple comparison
blocking variable
goodness-of-fit tests
chi-square test
observed frequency
expected frequency
contingency table

12.1   INTRODUCTION

Both χ^2 and F distributions and their related testing statistics have been discussed in detail in the last three chapters. In this chapter, we will talk about how these two distributions can be used to do data analysis involving the means or the proportions of more than two populations. In other words, we will develop an understanding of (1) a technique known as **analysis of variance** (**ANOVA**), which enables us to test the significance of the differences among sample means in terms of an F distribution, and (2) tests of goodness of fit and independence in a χ^2 distribution. The ANOVA is used to test the equality of more than two population means. The goodness-of-fit test is used to test the equality of more than two population proportions or to assess the appropriateness of a distribution. The test of independence determines whether the differences among several sample proportions are significant or are instead likely to be due to chance alone.

First, we consider a one-way ANOVA model that has only one **factor** (characteristic) with several groups, such as different years of work experience or different types of tires. Then we explore both simple and simultaneous confidence intervals. Two-way analysis of variance with a single observation and more than one observation per cell is discussed in detail, as are tests of goodness of fit and independence. Finally, we consider applications of analysis of variance in business.

12.2   ONE-WAY ANALYSIS OF VARIANCE

In the analysis of variance the F statistic is used to test whether the means of two or more groups are significantly different. It operates by breaking down the variance of the two or more populations into components. These components are then used to construct the sample statistic—hence the term *analysis of variance.* The ANOVA can be used to analyze certain decisions, such as whether some products sell better when placed in certain sections of stores (as point-of-purchase, or impulse, sales, for example); whether advertising is more effective in selling some products than in selling others; whether some employees are more motivated by some incentives than by others; and whether technology is variously effective in different workplaces. Furthermore, an accountant can use this technique to test whether the mean value of one set of sample accounts receivable is significantly different from another set or other sets.

The groups of data used to do the analysis of variance can be defined in terms of a single basis of classification (location, design, region, company, or the like) or by a dual classification. An ANOVA based on group data that are defined by a single classification is called **one-way ANOVA.** An ANOVA based on group data that are defined by a dual classification is called **two-way ANOVA.** In principle, both one-way and two-way analyses of variance are used to find out whether the means of all the populations considered are equal to one another.

Defining One-Way ANOVA

Suppose we want to test whether number of years of work experience since graduation has an effect on beginning salary for economics majors. The three **treatments** (or groups) are

Treatment 1: Bachelor's degree with no work experience.
Treatment 2: Bachelor's degree with one year of work experience.
Treatment 3: Bachelor's degree with two years of work experience.

We also assume that each student in this sample graduated from Rutgers University and specialized in labor economics. In order to simplify the necessary computations, we have restricted this to a random sample of only 12 observations—3 samples (of 4 graduates) from each of the combinations. (A larger sample size would yield more convincing results.) Table 12.1 gives the 12 sample salaries, along with the respective means for the 3 treatments.

Let's consider individually the notations enclosed in parentheses in Table 12.1. There are 4 rows and 3 columns in Table 12.1. Salary observations in this table are represented by x_{ij}, where i stands for the number of rows (students) and j the number of columns (treatments). There are a total of $i \times j$ observations in the table; in this case, $4 \times 3 = 12$. For example, x_{32} denotes the salary of the third student who has 2 years of work experience. In this problem, different years of work experience are indicated in the columns of the table, and interest centers on the differences among salaries in the 3 columns. This is typical of *one-factor* (or *one-way*) analysis of variance, in which an attempt is made to assess the effect of only one factor (in this case, years of work experience) on the observations. Here we denote the values in the columns as x_{i1}, x_{i2}, and x_{i3} and the totals of these columns as $\Sigma_i x_{i1}$, $\Sigma_i x_{i2}$, and

Table 12.1 Salaries of 12 Graduates with Varying Work Experience (in thousands of dollars)

Student (i)	Years of Work Experience (j)		
	$1\ (x_{i1})$	$2\ (x_{i2})$	$3\ (x_{i3})$
1	16 (x_{11})	19 (x_{12})	24 (x_{13})
2	21 (x_{21})	20 (x_{22})	21 (x_{23})
3	18 (x_{31})	21 (x_{32})	22 (x_{33})
4	13 (x_{41})	20 (x_{42})	25 (x_{43})
Total $\left(\sum_i x_{ij}\right)$	68 $\left(\sum_i x_{i1}\right)$	80 $\left(\sum_i x_{i2}\right)$	92 $\left(\sum_i x_{i3}\right)$
Mean ($\bar{x}_j$)	17 ($\bar{x}_1$)	20 ($\bar{x}_2$)	23 ($\bar{x}_3$)

Overall mean $\bar{x} = \sum_{j=1}^{3} \bar{x}_j/3$

$$= (\$17 + \$20 + \$23)/3 = \$20$$

$\Sigma_i x_{i3}$. The subscript i under the summation signs indicates that the total of each column is obtained by summing the entries over the rows. We will refer to the means of the columns as $\bar{x}_1$, $\bar{x}_2$, and $\bar{x}_3$, or, in general, as $\bar{x}_j$. Finally, we denote the overall mean as $\bar{x}$, where $\bar{x}$ is the mean of all observations.

Specifying the Hypotheses

As stated earlier, we want to test whether the combination of a Bachelor's degree with 3 different levels of work experience affects beginning salaries. From Table 12.1 we calculated the following mean salaries of graduates from the 3 combinations: $\bar{x}_1 = \$17$, $\bar{x}_2 = \$20$, and $\bar{x}_3 = \$23$. Also included in the table is an overall average of the 12 graduates, $\bar{x} = \$20$, which is referred to as the *overall mean.* Hence we want to test whether these 3 sample means were drawn from populations that have identical means. In other words, we want to test the following null hypothesis:

$$H_0: \mu_1 = \mu_2 = \mu_3 \tag{12.1}$$

against the alternative hypothesis

H_1: At least two population means are not equal.

Thus we are testing whether the differences between the sample means are too large to be attributed solely to chance. If the test results indicate that the sample means are significantly different, then we can conclude that different years of work experience have an impact on beginning salaries. Note that we make inferences concerning the means of more than 2 populations here.

Generalizing the One-Way ANOVA

Table 12.1 can be generalized to resemble Table 12.2, where we see n observations and m populations. Here each of the m populations is a treatment. Table 12.2 illustrates how we initially set up a generalized matrix to perform the one-way analysis

Table 12.2 General Notation Corresponding to Table 12.1

Observation (i)	Population (j)				
	1	*2*	*3*	...	*m*
1	x_{11}	x_{12}	x_{13}	...	x_{1m}
2	x_{21}				
3					
4					
⋮					
n	x_{n1}	x_{n2}	x_{n3}	...	x_{nm}
Total	$\Sigma_i x_{i1}$	$\Sigma_i x_{i2}$	$\Sigma_i x_{i3}$		$\Sigma_i x_{im}$
Mean	$\bar{x}_1$	$\bar{x}_2$	$\bar{x}_3$		$\bar{x}_m$

of variance. Here the top row indicates that we will be testing the equality of m different means. Within each column there are n individual samples taken from each of the m treatments. In developing the one-way ANOVA model, our purpose is to specify the underlying relationships among the various treatments. Hence the first step is to calculate the sample means from the random observations taken from each of the m treatments. That is,

$$\overline{x}_j = \frac{\sum_{i=1}^{n_j} x_{ij}}{n_j}, \quad j = 1, \ldots, m \tag{12.2}$$

where

$\overline{x}_j$ = sample mean for the jth treatment

x_{ij} = ith sample observation for the jth treatment

n_j = number of sample observations in the jth treatment

As we have noted, the null hypothesis specifies that all of the j treatments have identical means. Thus the next step in our analysis is to obtain an estimate of a common mean, which we will call the *global* mean. The **global mean,** or **overall mean,** is the summation of the individual sample observations divided by the number of total observations. Stated formally,

$$\overline{x} = \frac{\sum_{j=1}^{m} \sum_{i=1}^{n_j} x_{ij}}{n} \tag{12.3}$$

where

$\overline{x}$ = global mean

$n = \sum_{j=1}^{m} n_j$

and variables x_{ij} and n_j have the same meaning as in Equation 12.2. Alternatively, we can restate the overall mean as

$$\overline{x} = \frac{\sum_{j=1}^{m} n_j \overline{x}_j}{n} \tag{12.3'}$$

To test the null hypothesis that the treatment means are equal, we need to assess two measures of variability. First, we are interested in the variability of the sample within each treatment. This classification of variability is called **within-group variability.** We are also interested in the variability between the m treatments, which is called **between-groups variability.** From our sample data, we can obtain measures of both.

Between-Treatments and Within-Treatment Sums of Squares

The aforementioned concepts are illustrated graphically in Figure 12.1. When sample data are combined, they appear to be observations from a single population with high dispersion, as shown in part (a). But when each treatment is viewed separately, these salary figures appear to belong to 3 separate populations with a smaller variance, as indicated in part (b). Under the null hypothesis, the treatment populations have identical frequencies, as shown in part (c).

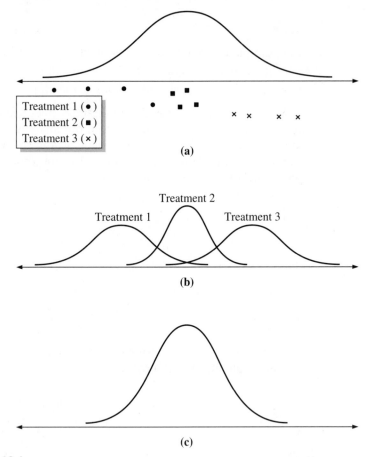

(a)

(b)

(c)

Figure 12.1

Distributions for ANOVA: (a) Sample Data Combined, (b) Each Treatment Viewed Separately, (c) The Assumption the Null Hypothesis Makes

The term *variation* refers to the sum of squared deviations, which is also called the **sum of squares.** We begin our analysis of variance by measuring the variation between the treatment means. The calculation is

$$\text{SST} = \sum_{j}^{m} n_j(\bar{x}_j - \bar{x})^2 \tag{12.4}$$

where

SST = between-treatments sum of squares (between-groups variability)
n_j = sample size of treatment j
$\bar{x}_j$ = sample mean of the jth treatment
$\bar{x}$ = overall mean

Table 12.3 illustrates calculation of the between-treatments variation for the data given in Table 12.1.

Table 12.3 Worksheet for Calculating Between-Treatments Sum of Squares

$$n_1(\bar{x}_1 - \bar{x})^2 = 4(17 - 20)^2 = 36$$
$$n_2(\bar{x}_2 - \bar{x})^2 = 4(20 - 20)^2 = 0$$
$$n_3(\bar{x}_3 - \bar{x})^2 = 4(23 - 20)^2 = 36$$

Substituting all squared-deviation values given in Table 12.3 into Equation 12.4, we obtain the **between-treatments sum of squares** as follows:

$$\sum_{j=1}^{3} n_j(\bar{x}_j - \bar{x})^2 = 4(9) + 4(0) + 4(9)$$

$$= 72$$

Again, this measure of variability may specify why treatment means are different.

On the other hand, the within-treatment variability specifies the treatment effect. That is, the **within-treatment sum of squares** indicates the unexplained variability that is due to the random sampling process. The calculation is

$$\text{SSW} = \sum_{j=1}^{m} \sum_{i=1}^{nj} (x_{ij} - \bar{x}_j)^2 \tag{12.5}$$

where

SSW = within-treatment sum of squares (within-group variability)
x_{ij} = value of the observation in the ith row and the jth column
$\bar{x}_j$ = mean of the jth treatment

Calculation of the within-treatment sum of squares is illustrated in Table 12.4.

Table 12.4 Worksheet for Calculating Within-Treatment Sum of Squares

Treatment 1	Treatment 2	Treatment 3
$(16 - 17)^2 = 1$	$(19 - 20)^2 = 1$	$(24 - 23)^2 = 1$
$(21 - 17)^2 = 16$	$(20 - 20)^2 = 0$	$(21 - 23)^2 = 4$
$(18 - 17)^2 = 1$	$(21 - 20)^2 = 1$	$(22 - 22)^2 = 1$
$(13 - 17)^2 = \underline{16}$	$(20 - 20)^2 = \underline{0}$	$(25 - 23)^2 = \underline{4}$
34	2	10

The between-treatments variation and within-treatment variation together represent the total variation of the ANOVA model. We calculate the total variation by summing the squared deviations of individual observations about the global mean. Formally, the total sum of squares can be written

$$\text{TSS} = \sum_{j=1}^{m} \sum_{i=1}^{nj} (x_{ij} - \bar{x})^2 \tag{12.6}$$

where

TSS = total sum of squares
x_{ij} = value of the observation in the ith row and the jth column
$\bar{x}$ = overall mean

We obtain the within-treatment sum of squares by substituting all squared-deviation values given in Table 12.4 into Equation 12.5.

$$\text{SSW} = \sum_{j=1}^{m} \sum_{i=1}^{nj} (x_{ij} - \bar{x})^2 = 34 + 2 + 10$$

$$= 46$$

To put it more simply, we find the total sum of squares by adding the between-treatments variation to the within-treatment variation.

$$\text{TSS} = \text{SST} + \text{SSW}$$
$$= 72 + 46$$
$$= 118$$

Even though the test of the null hypothesis for the one-way analysis of variance involves only between-treatments and within-treatment variation, it is useful to understand the relationship between total variation and its components.

Between-Treatments and Within-Treatment Mean Squares

The number of degrees of freedom associated with the between-treatments variation is $(m - 1)$. That is, because there are m treatments, or m sample means, there are m sums of squares used to measure the variation of these sample means around the overall mean. The overall mean is the only estimate of the population mean, so 1 degree of freedom is lost. Thus in our example, which consists of 3 treatments, there are $3 - 1 = 2$ degrees of freedom associated with the between-treatments mean square.

The number of degrees of freedom associated with the within-treatment variation is $(n - m)$. Because there are n ($n = \sum_{j=1}^{m} n_j$) observations, there are n sums of squares used to measure the within-treatment variation, with each deviation taken around its respective treatment mean. There are m treatment means, each an estimate of its respective population, so there is a loss of m degrees of freedom. Hence in our example, which contains 12 observations and 3 treatment means, there are $12 - 3 = 9$ degrees of freedom associated with the within-treatment mean square.

Again, the test of the null hypothesis is based on the assumption that all the m treatments have a common variance. If the null hypothesis is in fact true, then SSB and SSW can be used as a basis for an estimate of a common variance. To calculate these estimates, we can now divide each of the variability measures by its number of degrees of freedom. Hence the unbiased estimate of the **between-treatments mean square** can be obtained by dividing SSB by $(m - 1)$ degrees of freedom:

$$MST = SST/(m - 1)$$

where

$$MST = \text{between-treatments mean square (variance)}$$

In our example, the between-treatments mean square is MST $= 72/2 = 36$. Similarly, an unbiased estimate of the **within-treatment mean square** is found by dividing SSW by $(n - m)$ degrees of freedom.

$$MSW = SSW/(n - m)$$

where

$$MSW = \text{within-treatment mean square (variance)}$$

In our example, the within-treatment mean square is MSW $= 46/9 = 5.111$. We test the null hypothesis that the population treatment means are equal by comparing the between-treatments mean square with the within-treatment mean square. Read on.

The Test Statistic

Comparison of the between-treatments mean square with the within-treatment mean square is performed by computing a ratio:

$$F = (MST/MSW) \tag{12.7}$$

If the null hypothesis that the population treatment means are equal were true, the ratio given in Equation 12.7 would tend to equal 1. Alternatively, if the null hypothesis were not true, the ratio would be greater than 1 (MSB generally cannot be smaller than MSW), which implies that the treatment means do differ because the between-treatments variance exceeds the within-treatment variance. In the context of our example, this would imply that different amounts of work experience do have an impact on starting salaries for graduates. The ratio for our example can be calculated as $F = 36/5.111 = 7.04$.

From this calculation it appears that we can reject the null hypothesis that the population treatment means are equal. But first we need to determine how large the ratio must be in order for us to reject the null hypothesis. To do this, we must refer to the probability distribution of the F-distributed random variable discussed in Chapter 9 (the **F distribution**) and to the F table given as Table A6 in Appendix A at the end of the book. For our purposes, we will test the null hypothesis that the population treatment means are equal at the .05 level of significance. We refer to the F random variable as $F_{\nu_1, \nu_2, \alpha}$, where $\nu_1 = (m - 1)$ is the between-treatments degrees of freedom, $\nu_2 = (n - m)$ is the within-treatment degrees of freedom, and α is the level of significance. When the null hypothesis is true, the F variable in Equation 12.7 is distributed as F_{ν_1, ν_2}. From Table A6 we find that the critical value at the 5 percent level of significance is

$$F_{2,9,.05} = 4.26$$

Thus if the F ratio calculated for our example is greater than the critical value, then we can reject the null hypothesis that the population treatment means are equal. On the other hand, if the F ratio we calculate is less than the critical value, then we must accept the null hypothesis that the population treatment means are equal. Our sample F ratio, 7.04, is greater than the critical value, 4.26, so the null hypothesis is rejected at the .05 level of significance. We can conclude that the treatment means are significantly different. That is, work experience does affect starting salaries for graduates. A summary of this analysis of variance appears in Table 12.5, and Figure 12.2 presents MINITAB output related to Table 12.5.

Table 12.5 Summary of One-Way ANOVA Table

(1) Source of Variation	(2) Sum of Squares	(3) Degrees of Freedom	(4) Mean Square
Between-treatments	72 (SST)	2 ($m - 1$)	36 (MST)
Within-treatment	46 (SSW)	9 ($n - m$)	5.111 (MSW)

$F_{2,9} = \dfrac{36}{5.111} = 7.04$

$F_{2,9,.05} = 4.26$

```
MTB > READ C1-C3
DATA> 16 19 24
DATA> 21 20 21
DATA> 18 21 22
DATA> 13 20 25
DATA> END
      4 ROWS READ
MTB > AOVONEWAY C1-C3

ANALYSIS OF VARIANCE
SOURCE     DF        SS        MS        F         p
FACTOR      2     72.00     36.00     7.04     0.014
ERROR       9     46.00      5.11
TOTAL      11    118.00
                                     INDIVIDUAL 95 PCT CI'S FOR MEAN
                                     BASED ON POOLED STDEV
   LEVEL    N      MEAN     STDEV    ---------+---------+---------+-------
   C1       4    17.000     3.367   (-------*------)
   C2       4    20.000     0.816          (------*------)
   C3       4    23.000     1.826                 (-------*------)
                                     ---------+---------+---------+-------
POOLED STDEV =     2.261              17.5      21.0      24.5
MTB > PAPER
```

Figure 12.2

MINITAB Output
for Table 12.5

Population Model for One-Way ANOVA

The one-factor ANOVA model discussed in this section can also be described in a different type of specification. Let the random variable X_{ij} denote the ith observation from the jth population, and let μ_j denote the mean of this population. In addition, let μ denote the overall mean of m combined populations.

Then the population model for ANOVA states that any value X_{ij} is the sum of the grand mean μ, the **treatment effect** τ_j, and random error. In symbols, the one-factor ANOVA model is

$$X_{ij} = \mu + \tau_j + \epsilon_{ij}$$ (12.8)
$$= \mu + (\mu_j - \mu) + (\epsilon_{ij} - \mu_i)$$

where

X_{ij} = value of the dependent variable in the ith row and the jth column; this is the variable under investigation.

μ_j = mean of the jth column; this is the average value for the jth treatment. $\mu_j = \mu + \tau_j$

μ = grand mean; this is the mean of all the column means.

τ_j = treatment effect for the jth column, defined as $(\mu_j - \mu)$; this is the difference between a column mean and the grand mean. The value of τ_j indicates how much effect a particular treatment has on the grand mean. $\tau_j = \mu_j - \mu$

ϵ_{ij} = random error associated with X_{ij}, defined as the difference between X_{ij} and μ_j. This is the amount by which a particular value of the dependent variable differs from the mean of all values in that column. $\epsilon_{ij} = X_{ij} - \mu_j$

By using the model of Equation 12.8, we can redefine the null hypothesis defined in Equation 12.1 as Equation 12.1′. Then our null hypothesis is that every population mean μ_j is the same as the overall mean μ.

$$\text{H}_0: \tau_1 = \tau_2 = \tau_3 = 0 \tag{12.1′}$$

where $\tau_j = \mu_j - \mu \ (j = 1, 2, 3)$.

12.3   SIMPLE AND SIMULTANEOUS CONFIDENCE INTERVALS

In the salary study we have examined in this chapter, the analysis of variance was used to determine whether there was a difference in average salary among workers with different numbers of years of work experience. Once differences in the means of the groups are found, however, it is important to determine which particular groups are different. In other words, we are interested in establishing a confidence interval for the difference between two population means.

Simple Comparison

To compare the differences of the population means of group 1 and group 2, we can construct a confidence interval for $(\mu_1 - \mu_2)$ by using our estimates $(\overline{X}_1 - \overline{X}_2)$ as discussed in Chapters 10 and 11. Formally, the $(1 - \alpha)$ percent confidence interval for $\mu_1 - \mu_2$ is

$$(\overline{x}_1 - \overline{x}_2) \pm t_{\alpha/2,(n-m)} S_p \sqrt{\frac{1}{n_1} + \frac{1}{n_2}} \tag{12.9}$$

where

$\mu_1, \mu_2 =$ population means for treatments 1 and 2, respectively
$\overline{x}_1, \overline{x}_2 =$ sample means for treatments 1 and 2, respectively
$t_{\alpha/2,(n-m)} = t$ statistic at the $\alpha/2$ level of significance with $(n - m)$ degrees of freedom
$S_p^2 = \text{SSW}/(n - m)$, the within-treatment mean square, where SSW has been defined in Equation 12.5
$n_1, n_2 =$ number of observations for treatments 1 and 2, respectively.

Hence our pooled variance from the three treatments is calculated as follows:

$$S_p^2 = \frac{1}{(n - m)} \sum_{j=1}^{3} (S_j^2) = \frac{1}{(n - m)} (\text{SSW})$$

$$= \frac{1}{9} (34 + 2 + 10)$$

$$= 5.111$$

and the pooled standard deviation is $S_p = \sqrt{S_p^2} = \sqrt{5.111} = 2.261$. As we noted earlier for the within-treatment variation, the pooled standard deviation has $(n - m)$, or 9, degrees of freedom. From Table A4 in Appendix A, we have $t_{.025,9} = 2.262$. Therefore, according to Equation 12.9, a 95 percent confidence interval for the difference of the population means for treatments 1 and 2 can be determined as follows:

$$(17 - 20) \pm (2.262)(2.261) \cdot (\sqrt{\tfrac{1}{4} + \tfrac{1}{4}}) = -3 \pm 3.616 \text{ or } (-6.616, +.616)$$

Thus we conclude that the mean salary for treatment 2 is approximately between $6,616 higher or $616 less than that for treatment 1.

Accordingly, for our example, the 95 percent confidence intervals for the difference between two population treatment means are

$$(17 - 20) - 3.616 < \mu_1 - \mu_2 < (17 - 20) + 3.616, \text{ or } (-6.616, +.616)$$

$$(17 - 23) - 3.616 < \mu_1 - \mu_3 < (17 - 23) + 3.616, \text{ or } (-9.616, -2.384)$$

$$(20 - 23) - 3.616 < \mu_2 - \mu_3 < (20 - 23) + 3.616, \text{ or } (-6.616, +.616)$$

Note that each confidence interval has the same width. This is due to the fact that each interval contains the pooled variance and each treatment contains the same number of observations. Also note that not all 3 intervals overlap zero.

There is one problem with this approach. Although we may be 95 percent confident of the individual intervals listed, we are less confident that the whole *system* of intervals is true. The problem we face is to determine a simultaneous confidence level for the whole system given that the intervals are independent (all have the same S_p). We can achieve this goal by using Scheffé's multiple comparison.

Scheffé's Multiple Comparison

The problem we have just posed can be restated as determining how much wider the intervals must become in order for each interval simultaneously to yield a $(1 - \alpha)$ percent level of confidence. This can be done by employing **Scheffé's multiple comparison.** For the 95 percent confidence interval for the difference between the population means for treatment 1 and treatment 2, Scheffé's multiple comparison formula[1] is

$$(\bar{x}_1 - \bar{x}_2) \pm \sqrt{(m - 1)(F_{\alpha,m-1,n-m})(S_p^2)} \, \sqrt{\frac{1}{n_1} + \frac{1}{n_2}} \qquad (12.10)$$

where

$F_{\alpha,m-1,n-m}$ = critical value of F with $(m - 1)$ and $(n - m)$ degrees of freedom at the α level of significance

m = number of the means to be compared

[1] See H. Scheffé (1959), *The Analysis of Variance* (New York: Wiley). This method has been adjusted for the number of means to be compared. This is the simplest case of Scheffé's multiple comparison.

n_1, n_2 = number of observations for combination 1 and combination 2, respectively

S_p = sample standard deviation pooled from all samples

From Table A6, we have $F_{.05,2,9}$ = 4.26. Then, at a 95 percent level of confidence,

$$(17 - 20) \pm \sqrt{2(4.26)(2.261)^2} \sqrt{(\tfrac{1}{4}) + (\tfrac{1}{4})} = -3 \pm (2.919)(2.261)(.707)$$

$$= -3 \pm 4.666, \text{ or } (-7.666, +1.666)$$

For the entire system, we have

$$-7.666 < \mu_1 - \mu_2 < +1.666$$

$$-10.666 < \mu_1 - \mu_3 < -1.334$$

$$-7.666 < \mu_2 - \mu_3 < +1.666$$

As we expected, the increased width of each interval now makes us 95 percent confident that all the foregoing statements are simultaneously true. Again, not all 3 intervals overlap zero; only the second null hypothesis should be rejected.

12.4 TWO-WAY ANOVA WITH ONE OBSERVATION IN EACH CELL, RANDOMIZED BLOCKS

In this section, we extend one-way ANOVA to two-way ANOVA. We discuss first the case of two-way ANOVA with one observation per cell, then the case of two-way ANOVA with more than one observation per cell.

Basic Concept

This section offers a more in-depth interpretation of ANOVA technique. In the example we have been using, our primary interest focused on a single aspect of the one-way analysis of variance (years of work experience), but it is possible that another factor also affects the outcome. In the one-way analysis of variance, we concluded that number of years of work experience had a significant impact on starting salary. However, we may suspect that some of the variability of the model is due to the geographic location of the job. Hence we now want not only to look at **treatment effects** of number of years of work experience but also to isolate the impact of geographic location on the starting salaries of the graduates. By setting up a two-way ANOVA problem, we want to design a more accurate test to explain the differences in the mean population of the various treatments.

Our new model must be constructed in such a way as to test for the influence that a second factor may have on the starting salaries. Using the data from Table 12.6, we will have the 4 rows represent 4 geographic locations in the United States. Hence we will be able to acquire information about the various years of work experience as well as information about the geographic location of the job. This new

Table 12.6 Salaries of 12 Students with Varying Work Experience in 4 Different Geographic Locations (in thousands of dollars)

Region	Years of Work Experience			Row Sums	Row Means
	1	*2*	*3*		
1	16	19	24	59	19.67
2	21	20	21	62	20.67
3	18	21	22	61	20.33
4	13	20	25	58	19.33
Column Sums	68	80	92	240	
Column Means	17	20	23		
		Global Mean $\overline{x} = 20$			

factor in our analysis is called a **blocking variable.** To simplify our analysis, the blocks will contain only a single observation per cell. Thus, as in our one-way ANOVA problem, we will use only the 12 observations from Table 12.6. Each of the 4 rows will represent a geographic location.

Row 1: West   Row 3: Northeast
Row 2: Midwest   Row 4: South

Table 12.6 illustrates how to set up the two-way analysis of variance. The salary data of Table 12.6 are identical to those of Table 12.1. However, in Table 12.6 we interpret 4 students' salaries within each column, representing salaries from 4 different geographic regions. Therefore, different locations (regions) constitute an additional factor.

Specifying the Hypotheses

Our purpose is to test the following two hypotheses:

1. H_0: Population mean salaries among various years of work experience are equal.

2. H_0: Population mean salaries among various geographic locations are equal.

Again, the alternative hypotheses are that the mean population values are not equal.

Between and Residual Sum of Squares

The necessary calculations for the two-way analysis of variance are

SST = between-treatments sum of squares
SSB = between-blocks sum of squares
TSS = total sum of squares
SSE = error sum of squares

Here treatments and blocks represent different years of work experience and different locations, respectively.[2]

From the one-way analysis of variance, we have already calculated SSB (Table 12.3), SSW (Table 12.4), and TSS. The next step is to calculate the between-blocks (between-rows) sum of squares. In this case, the observation can be represented by x_{ijk}. The subscripts i, j, and k represent the kth salary observation in the ith row and the jth column. Then the between-rows sum of squares can be defined as

$$SSB = \sum_{i=1}^{I} JK(\bar{x}_{i..} - \bar{x})^2 \tag{12.11}$$

where

SSB = between-blocks sum of squares

$$\bar{x}_{i..} = \text{sample mean of the } i\text{th row} = \frac{\sum_{i=1}^{I} x_{i..}}{JK}$$

$\bar{x}$ = overall mean

Table 12.7 illustrates calculation of the between-blocks sum of squares for our example.

Table 12.7 Between-Blocks Sum of Squares

$(J)(\bar{x}_1 - \bar{x}) = (3)(19.667 - 20)^2 =$	.333
$(J)(\bar{x}_2 - \bar{x}) = (3)(20.667 - 20)^2 =$	1.335
$(J)(\bar{x}_3 - \bar{x}) = (3)(20.333 - 20)^2 =$	.333
$(J)(\bar{x}_4 - \bar{x}) = (3)(19.333 - 20)^2 =$	1.335
SSB $=$	3.336

The between-columns sum of squares can be defined as

$$SST = \sum_{k=1}^{K} IK(\bar{x}_{.j.} - \bar{x})^2 \tag{12.12}$$

where

SST = between-treatments sum of squares

$$\bar{x}_{.j.} = \text{sample mean of the } j\text{th column} = \frac{\sum_{j=1}^{J} x_{.j.}}{IK}$$

$\bar{x}$ = overall mean

From Table 12.3, we have SST = 72.

[2]Alternatively, we can use levels of factors (treatments) A and B to represent different years of work and different locations.

Finally, because TSS = SST + SSB + SSE, the residual sum of squares is calculated as follows:

$$SSE = TSS - SST - SSB \sum_{i=1}^{I} \sum_{j=1}^{J} (x_{ijk} - \bar{x}_{i..} - \bar{x}_{.j.} + \bar{x})^2 \qquad (12.13)$$

Hence SSE = $118 - 72 - 3.336 = 42.664$.

Before we can proceed with the test of our hypotheses, we must determine how many degrees of freedom are associated with the between-blocks variation and the residual variation.

The number of degrees of freedom associated with the between-blocks variation is $(I - 1)$. That is, because there are n blocks, or n sample factor means, there are n sums of squares used to measure the variation of these sample means around the global mean. The global mean is again the only estimate of the population mean, so 1 degree of freedom is lost. Thus for our example, which consists of 4 levels of the block, there are $4 - 1 = 3$ degrees of freedom associated with the between-blocks sum of squares.

Between Variance, Error Variance, and F Test

The number of degrees of freedom associated with the residual variation is $(J - 1)(I - 1)$. In this instance, the residual variation takes into account both the variation between the treatments and the variation between the blocks. Hence we must adjust the residual variation by the degrees of freedom associated with both the between-treatments degrees of freedom and the between-blocks degrees of freedom. Thus for our example, the number of degrees of freedom associated with the residual variation is $(2)(3) = 6$.

Now we can obtain unbiased estimates of the between-blocks variance and the residual variance. The between-blocks variance is calculated as follows:

$$MSB = \frac{SSB}{(I - 1)} = \frac{3.336}{3} = 1.112$$

Analogously, the residual variance is

$$MSE = \frac{SSE}{(J - 1)(I - 1)} = \frac{42.664}{6} = 7.111$$

To test our null hypothesis about the influence of various years of work experience, we must calculate the F ratio.

$$F(2,6) = \frac{MST}{MSE} = \frac{36}{7.111} = 5.063$$

The critical value associated with this test is 5.14 ($F_{2,6,.05}$), from Table A6 in Appendix A at the end of the book. Because the F ratio is less than the critical value, we cannot reject the null hypothesis that there is no difference between the population means of salaries associated with various years of work experience.

In testing the null hypothesis for the influence of geographic location on salaries, we find that the F ratio is

$$F(3,6) = \frac{MSB}{MSE} = \frac{1.112}{7.111} = .156$$

The critical value associated with this test is 4.76 ($F_{3,6,.05}$), from Table A5. Again we cannot reject the null hypothesis that the population means of salaries associated with geographic location are equal.

In conclusion, having accepted both hypotheses, we can state that there are no significant differences among various years of work experience or among various geographic locations in the effect they have on starting salaries for graduates. Note that the effect of work experience obtained from two-way ANOVA is different from that of one-way discussed in Section 12.2. Table 12.8 summarizes the data for the two-way analysis of variance. The MINITAB output of Table 12.8 is presented in Figure 12.3.

Because $F_{2,6} = 5.06 < 5.14$, we conclude that the null hypothesis cannot be rejected. In other words, different years of work experience do not affect starting salary. Similarly, $F_{3,6} = .156 < 4.76$, so we should conclude that no salary differences exist among different regions. This is a good illustration of the fact that MSB is generally smaller than MSE when the null hypothesis is true.

For the two-factor model we use three subscripts. As in the one-factor model, the latter j represents column treatments and runs from 1 to J. The letter i represents row treatments and runs from 1 to I. The letter k represents the number of the observations in a cell and runs from 1 to K.

Table 12.8 Two-Way ANOVA Summary

(1) Source of Variation	(2) Sum of Squares	(3) Degrees of Freedom	(4) Mean Square
Between-treatments	72 (SST)	2 ($J-1$)	36
Between-blocks	3.336 (SSB)	3 ($I-1$)	1.112
Residuals	42.664 (SSE)	6 [($J-1$)($I-1$)]	7.111

$$MST = \frac{SST}{J-1} = \frac{72}{2} = 36$$

$$MSB = \frac{SSB}{n-1} = \frac{3.336}{3} = 1.112$$

$$MSE = \frac{SSE}{(J-1)(n-1)} = \frac{42.664}{6} = 7.111$$

$$F_{2,6} = \frac{MST}{MSE} = \frac{36}{7.111} = 5.063$$

$$F_{2,6,.05} = 5.14$$

$$F_{3,6} = \frac{MSB}{MSE} = \frac{1.112}{7.111} = .156$$

$$F_{3,6,.05} = 4.76$$

```
MTB > READ C1-C3
DATA> 16 1 1
DATA> 21 1 2
DATA> 18 1 3
DATA> 13 1 4
DATA> 19 2 1
DATA> 20 2 2
DATA> 21 2 3
DATA> 20 2 4
DATA> 24 3 1
DATA> 21 3 2
DATA> 22 3 3
DATA> 25 3 4
DATA> END
     12 ROWS READ
MTB > TWOWAY USING DATA IN C1, LEVEL IN C2, BLOCK IN C3

ANALYSIS OF VARIANCE   C1

SOURCE        DF        SS         MS
C2            2       72.00      36.00
C3            3        3.33       1.11
ERROR         6       42.67       7.11
TOTAL        11      118.00

MTB > PAPER
```

Figure 12.3

MINITAB Output
for Table 12.8

Population Model for Two-Way ANOVA with One Observation in Each Cell

Following Equation 12.8 and assuming there is no interaction between treatment and block, we can construct a population model for two-way ANOVA without interaction. It is

$$X_{ijk} = \mu + \tau_j + \lambda_i + \epsilon_{ijk} \qquad (12.14)$$

where

X_{ijk} = kth population value in the jth column and the ith row
μ_j = population mean of the jth treatment
μ_i = population mean of the ith block
μ = grand mean of the population
λ_i = treatment effect of the ith row; $\lambda_i = \mu_i - \mu$
τ_j = treatment effect of the jth column; $\tau_j = \mu_j - \mu$.

So far we have discussed two-way ANOVA with only one observation in each cell. If there is more than one observation in each cell, then there exists an interaction effect in addition to treatment and block effects. And matters get more complicated.

12.5 TWO-WAY ANOVA WITH MORE THAN ONE OBSERVATION IN EACH CELL

Basic Concept and Hypothesis Testing

The data that we used to do the two-way ANOVA contained only one observation in each cell. Now we expand the data set of Table 12.6 by allowing two sample observations in each cell, as shown in Table 12.9. Here the total sums of squares

Table 12.9 Salaries of 24 Students with Varying Work Experience in 4 Different Geographic Locations (in thousands of dollars)

Region	Years of Work Experience		
	1	*2*	*3*
1	16 16.5	19 17	24 25
2	21 20.5	20 19	21 22.5
3	18 19	21 20.9	22 21
4	13 13.5	20 20.8	25 23

can be dissected into four components and can be defined as follows:

Total sums of squares (TSS) = between-treatments sum of squares (SST)

  + between-blocks sum of squares (SSB) + interaction sum of squares (SSI)

  + error sum of squares (SSE)

or, more briefly,

$$TSS = SST + SSB + SSI + SSE$$

In this case, we add an interaction sum of squares because there is more than one observation in each cell. On the basis of the data listed in Table 12.9, we calculate block means, treatment means, cell means, and overall mean as follows.

1. Block means

$$\bar{x}_{1..} = \frac{16 + 16.5 + \cdots + 25}{6} = 19.583$$

$$\bar{x}_{2..} = \frac{21 + 20.5 + \cdots + 22.5}{6} = 20.667$$

$$\bar{x}_{3..} = \frac{18 + 19 + \cdots + 21}{6} = 20.317$$

$$\bar{x}_{4..} = \frac{13 + 13.5 + \cdots + 23}{6} = 19.217$$

2. Treatment means

$$\bar{x}_{.1.} = \frac{16 + 21 + \cdots + 13.5}{8} = 17.188$$

$$\bar{x}_{.2.} = \frac{19 + 20 + \cdots + 20.8}{8} = 19.713$$

$$\bar{x}_{.3.} = \frac{24 + 21 + \cdots + 23}{8} = 22.938$$

3. Cell means

$$\bar{x}_{11.} = \frac{16 + 16.5}{2} = 16.25 \qquad \bar{x}_{21.} = \frac{19 + 17}{2} = 18$$

Similarly, we can obtain

$$\bar{x}_{21.} = 20.75 \qquad \bar{x}_{22.} = 19.50 \qquad \bar{x}_{13.} = 24.50 \qquad \bar{x}_{43.} = 24.00$$

$$\bar{x}_{31.} = 18.50 \qquad \bar{x}_{32.} = 20.95 \qquad \bar{x}_{23.} = 21.75$$

$$\bar{x}_{41.} = 13.25 \qquad \bar{x}_{42.} = 20.40 \qquad \bar{x}_{33.} = 21.50$$

4. Overall mean

We use the average of column means to calculate the overall mean.

$$\bar{x} = \frac{17.188 + 19.713 + 22.938}{3} = 19.946$$

Using all related data, we calculate TSS, SST, and SSB as follows:

$$\text{TSS} = \sum_i \sum_j \sum_k (x_{ijk} - \bar{x})^2 = (16 - 19.946)^2$$

$$+ (21 - 19.946)^2 + \cdots + (23 - 19.946)^2$$

$$= 226.380$$

$$\text{SST} = IK \sum_{j=1}^{J} (\bar{x}_{.j.} - \bar{x})^2 = (4)(2)[(17.188 - 19.946)^2$$

$$+ (19.713 - 19.946)^2 + (22.938 - 19.946)^2]$$

$$= 132.903$$

$$\text{SSB} = JK \sum_{i=1}^{I} (\bar{x}_{i..} - \bar{x})^2 = (3)(2)[(19.583 - 19.946)^2$$

$$+ (20.667 - 19.946)^2 + (20.317 - 19.946)^2$$

$$+ (19.217 - 19.946)^2]$$

$$= 7.921$$

Because there is more than one observation in each cell, the SSE given by Equation 12.13 can be dissected into interaction and error. The interaction term (SSI) is

Table 12.10 Two-Way Analysis of Variance with Interaction Summary

(1) Source of Variation	(2) Sum of Squares	(3) Degrees of Freedom	(4) Mean Square	(5) F Ratio
Between-treatments (SST)	132.903	2	66.452	101.91
Between-blocks (SSB)	7.921	3	2.640	4.05
Interaction (SSI)	77.730	6	12.955	19.87
Errors (SSE)	7.825	12	.652	
Total (TSS)	226.379			

identical to the SSE with only one observation in each cell, as defined in Equation 12.13. The error term, SSE, can be defined as

$$\text{SSE} = \sum_i \sum_j \sum_k (x_{ijk} - \bar{x}_{ij.})^2 \tag{12.15}$$

In terms of our data, the SSI and SSE are calculated as follows:

$$\text{SSI} = K \sum_i \sum_j (\bar{x}_{ij.} - \bar{x}_{.j.} - \bar{x}_{i..} + \bar{x})^2$$

$$= 2[(16.250 - 17.188 - 19.583 + 19.946)^2$$
$$+ (20.750 - 17.188 - 20.667 + 19.946)^2$$
$$+ \cdots + (21 - 22.938 - 20.317$$
$$+ 19.946)^2 + (24 - 22.938 - 19.217 + 19.946)^2]$$

$$= 77.730$$

$$\text{SSE} = \sum_i \sum_j \sum_k (x_{ijk} - \bar{x}_{ij.})^2$$

$$= (16 - 16.250)^2 + (16.500 - 16.250)^2$$
$$+ \cdots + (25 - 24)^2 + (23 - 24)^2$$

$$= 7.825$$

Using the foregoing data, we calculate the two-way ANOVA table with interaction. Our results are listed in Table 12.10.

From the F ratio shown in column (5) of Table 12.9, we can test whether the years-of-education effect, the regional effect, and the interaction effect are statistically significant. From Table A6 in Appendix A, we find that the critical values at $\alpha = .05$ are $F_{2,12,.05} = 3.89$, $F_{3,12,.05} = 3.49$, and $F_{6,12,.05} = 3.00$. Comparing these values with the F ratio listed in column (5) of Table 12.9 leads to the conclusion that number of years of education, geographic region, and their interaction all have significant impacts on the starting salary. The MINITAB output of Table 12.10 is presented in Figure 12.4.

```
MTB > READ C1-C3
DATA> 16 1 1
DATA> 16.5 1 1
DATA> 21 1 2
DATA> 20.5 1 2
DATA> 18 1 3
DATA> 19 1 3
DATA> 13 1 4
DATA> 13.5 1 4
DATA> 19 2 1
DATA> 17 2 1
DATA> 20 2 2
DATA> 19 2 2
DATA> 21 2 3
DATA> 20.9 2 3
DATA> 20 2 4
DATA> 20.8 2 4
DATA> 24 3 1
DATA> 25 3 1
DATA> 21 3 2
DATA> 22.5 3 2
DATA> 22 3 3
DATA> 21 3 3
DATA> 25 3 4
DATA> 23 3 4
DATA> END
      24 ROWS READ
MTB > TWOWAY USING DATA IN C1, A LEVEL IN C2, B LEVEL IN C3

ANALYSIS OF VARIANCE  C1

SOURCE        DF       SS        MS
C2             2   132.903    66.452
C3             3     7.921     2.640
INTERACTION    6    77.730    12.955
ERROR         12     7.825     0.652
TOTAL         23   226.380
```

Figure 12.4

MINITAB Output
for Table 12.10

Generalizing the Two-Way ANOVA

If there are several observations per cell, then the cell mean can be defined as

$$\overline{x}_{ij.} = \frac{\displaystyle\sum_{k=1}^{K} x_{ijk}}{K}$$

Here the column (group) mean $\overline{x}_{.j.}$ and the row (block) mean $\overline{x}_{i..}$ can be defined as

$$\overline{x}_{.j.} = \frac{\displaystyle\sum_{i=1}^{I}\sum_{k=1}^{K} x_{ijk}}{IK}$$

$$\overline{x}_{i..} = \frac{\displaystyle\sum_{j=1}^{J}\sum_{k=1}^{K} x_{ijk}}{JK}$$

Table 12.11 General Format of the Two-Way ANOVA Table with K Observations per Cell

(1) Source of Variation	(2) Sum of Squares	(3) Degrees of Freedom	(4) Mean Square	(5) F Ratio
Between-treatments	$SST = IK \sum\limits_{j=1}^{J} (\bar{x}_{.j.} - \bar{x})^2$	$J - 1$	$MST = \dfrac{SST}{J-1}$	$\dfrac{MST}{MSE}$
Between-blocks	$SSB = JK \sum\limits_{i=1}^{I} (\bar{x}_{i..} - \bar{x})^2$	$I - 1$	$MSB = \dfrac{SSB}{I-1}$	$\dfrac{MSB}{MSE}$
Interaction	$SSI = K \sum\limits_{i} \sum\limits_{j} (\bar{x}_{ij.} - \bar{x}_{.j.} - \bar{x}_{i..} + \bar{x})^2$	$(J-1)(I-1)$	$MSI = \dfrac{SSI}{(J-1)(I-1)}$	$\dfrac{MSI}{MSE}$
Error	$SSE = \sum\limits_{i} \sum\limits_{j} \sum\limits_{k} (x_{ijk} - \bar{x}_{ij.})^2$	$JI(K-1)$	$MSE = \dfrac{SSE}{IJ(K-1)}$	
Total	$SST = \sum\limits_{i} \sum\limits_{j} \sum\limits_{k} (\bar{x}_{ijk} - x)^2$	$IJK - 1$		

In addition, the overall mean ($\bar{x}$) can be defined as

$$\bar{x} = \sum_{i=1}^{I} \sum_{j=1}^{J} \sum_{k=1}^{K} x_{ijk}/IJK = \sum_{j=1}^{J} \bar{x}_{.j.}/J = \sum_{i=1}^{I} x_{i..}/I$$

Using the cell mean, treatment mean, block mean, overall mean, and other related concepts and notations discussed in this section, we can define the general format of the two-way ANOVA with K observations per cell as shown in Table 12.11. The population model for Table 12.11 is

$$x_{ijk} = \mu + \tau_j + \lambda_i + (\lambda\tau)_{ij} + \epsilon_{ijk} \tag{12.16}$$

where μ, τ_j, λ_i, and ϵ_{ijk} are as defined in Equation 12.14

$(\lambda\tau)_{ij}$ = interaction effect in the ith row and the jth column = $\mu_{ij} - \mu_i - \mu_j + \mu$
   μ_{ij} = population mean of the cell in the ith row and the jth column
   μ_i = population mean of the ith block
   μ_j = population mean of the jth treatment

12.6   CHI-SQUARE AS A TEST OF GOODNESS OF FIT

In this section, we will show how the chi-square statistic can be used to test the appropriateness of a distribution—its goodness of fit for a set of data. **Goodness-of-fit tests** are designed to study the frequency distribution to determine whether a set of data are generated from a certain probability distribution, such as the uniform, binomial, Poisson, or normal distribution.

Among the goodness-of-fit tests, the **chi-square test** is used to test the equality of more than two proportions if a probability distribution is assumed to be uniform. This is similar to using the F statistic to test the equality of more than two means in the analysis of variance.

If a marketing manager is interested in knowing whether 4 different brands of painkillers are recommended equally often by doctors (or enjoy the same market shares), the manager can set up the following hypotheses.

H_0: Same market shares

H_1: Different market shares

To test this hypothesis, the manager can send out questionnaires to 1,000 doctors asking what painkiller they usually recommend to their patients. The responses can be tallied to obtain the observed sample frequency distribution. The tallied responses are called the **observed frequency.** If the null hypothesis of equal market shares is true, we would expect to see that roughly 250 doctors recommended each brand. This frequency distribution is called the **expected frequency** because it is anticipated when the null hypothesis is true. In applying the goodness-of-fit test, we compare the expected frequency with the observed frequency to determine whether the observed frequency conforms to the expected frequency—and hence supports the null hypothesis. If the null hypothesis is true, the frequencies for four brands of painkillers will be equal. Therefore, we can regard this example as a test of uniform distribution.

To take another example, many statistical inferences drawn in studying stock rates of return are based on the assumption that the rates of return of a stock follow a normal distribution. It should be interesting to test whether the rates of return are really generated from a normally distributed population. Here the null hypothesis is that the data are from a normally distributed population, and the alternative hypothesis is that the data are not from a normally distributed population. Again we perform the goodness-of-fit test by comparing the anticipated frequency distribution when the null hypothesis is true with the frequency distribution that is actually observed.

The chi-square statistic for determining whether the data follow a specific probability distribution is

$$\chi^2_{k-1} = \sum_{i=1}^{k} \frac{(f_i^o - f_i^e)^2}{f_i^e} \tag{12.17}$$

where

f_i^o = observed frequency
f_i^e = expected frequency
k = number of groups
χ^2_{k-1} = chi-square statistic with $(k-1)$ degrees of freedom.

The following examples illustrate various applications of the goodness-of-fit test to deciding whether the population that generates the data follows a presumed distribution. This presumed distribution can be a uniform distribution, a binomial distribution or a Poisson distribution.

EXAMPLE 12.1   *Uniform Distribution: Market Shares of Different Types of Cars*

A marketing manager wants to test his belief that 4 different categories of cars share the auto market equally. These 4 categories of cars are brand A, brand B, brand C, and imported cars. He sends out 2,000 questionnaires to car owners throughout the nation and receives the following responses:

Brand	Number of Owners (observed frequency)
A	475
B	505
C	495
Imported	525
Total	2,000

Armed with these data, we can help him solve the problem. We first set up the hypotheses

H_0: Same market shares (uniform distribution)

H_1: Different market shares (nonuniform distribution)

The chi-square test statistic defined in Equation 12.17 can be used to perform the hypothesis test. When the null hypothesis is true, there should be 500 responses for each category of product. This implies that the expected frequency should be 500 for each category of product. Computation of the chi-square statistic in terms of Equation 12.17 is given in column (4) of Table 12.12.

Table 12.12 Computation of the Chi-Square Test Statistic for Example 12.1

(1) Brand	(2) Number of Owners	(3) f_i^e	(4) $(f_i^o - f_i^e)^2/f_i^e$
A	475	500	5/4
B	505	500	1/20
C	495	500	1/20
Imported	525	500	5/4
Sum	2000	2000	$\chi_3^2 = 52/20 = 2.6$

In this example, we divided the total sample into 4 groups. The frequencies of these 4 groups must add up to 2,000. This means that when any 3 groups' frequencies are known, the fourth group's frequency is also set. The number of degrees of freedom is therefore $(k - 1)$, so here it is $4 - 1 = 3$. From the χ^2-distribution table (Table A5 in Appendix A of this book), we obtain $\chi_{3,.05}^2 = 7.81$. Because

$$\chi^2 = \sum_{i=1}^{k} \frac{(f_i^o - f_i^e)^2}{f_i^e} = 2.6$$

which is smaller than 7.81, we fail to reject the null hypothesis at $\alpha = .05$. We conclude that we do not have enough evidence to argue that the frequency distribution of different car brands is not uniformly distributed. In other words, the differences in market share among these 4 different brands of automobiles are not statistically significant.

EXAMPLE **12.2**   *Binomial Distribution: Correct Picks in a Football Pool*

A football fan keeps track of the football betting record for the football betting pool in her company. In each bet, a player has to pick the winner for 10 games. In the last season 1,000 bets were placed. The numbers of correct picks are tallied in column (2) of Table 12.13; these figures are observed frequencies.

Table 12.13 Computation of the Chi-Square Test Statistic for Football Betting Pool Problem

(1) Number of Correct Picks	(2) Number of Bets, f_i^o	(3) Expected Binomial Probability	(4) Expected Frequency, f_i^e	(5) $(f_i^o - f_i^e)^2/f_i^e$
0	2	.001	1	1
1	8	.010	10	.4
2	39	.044	44	.57
3	123	.117	117	.31
4	207	.205	205	.02
5	250	.246	246	.07
6	203	.205	205	.02
7	115	.117	117	.03
8	40	.44	44	.36
9	13	.10	10	.9
10	0	.001	1	1
Sum	1,000	1.00	1,000	4.68

We would like to know whether the numbers of correct picks follow a binomial distribution with $P = .5$. Accordingly, we have

H_0: A binomial distribution with $P = .5$ is a good description of the number of correct picks.

H_1: A binomial distribution with $P = .5$ is not a good description of the number of correct picks.

To solve this problem, we must determine whether the discrepancies between the observed frequencies and those we would expect to observe if the binomial distribution were the proper model to use are actually due to chance. To calculate the expected frequencies, we find the probabilities of the numbers of correct picks in

Table A1 in Appendix A by looking for $n = 10$ and $P = .5$. The probabilities are listed in Column (3) of Table 12.13. Since the number of bets is 1,000, the expected frequencies f_i^e can be obtained by multiplying the probabilities listed in column (3) by 1,000; they are indicated in column (4). Again, comparing the observed and expected frequencies gives us the test statistic. Column (5) of Table 12.13 gives the results of computation of the test statistic in accordance with Equation 12.16. From column (5), we obtain

$$\sum_i^k (f_i^o - f_i^e)^2/f_i^e = 4.68$$

From Table A5 we find that $\chi^2_{10,.05} = 18.31$. Because $4.68 < 18.31$, we conclude that there is not enough evidence for us to reject, at $\alpha = .05$, the null hypothesis that the data are from a binomial distribution.

EXAMPLE 12.3 *Poisson Distribution: Number of Patient Arrivals*

Suppose a hospital has kept track of the number of patients arriving at the emergency room during a given hour for the last 480 hours (20 days). It was found that 960 patients came to the emergency room during that period. The observed distribution of the arrival of patients is given in row (1) of Table 12.14. We would like to know whether this distribution is a Poisson distribution. If the null hypothesis is true, the data are generated by the Poisson probability distribution—that is,

$$P(x) = \frac{e^{-\lambda}\lambda^x}{x!}$$

where λ is the expected number of patients arriving in a given hour. In the last 480 hours, there were 960 patients. The expected number of patients, λ, can be estimated as 2(960/480) per hour. Using the foregoing Poisson distribution formula, we compute the probability of $x = 0, 1, 2$ and $x \geq 3$. We use the Poisson probability table, Table A2 in Appendix A, to obtain the expected probabilities indicated in row (2) of Table 12.14. Multiplying the probabilities by 480, we obtain f_i^e as shown

Table 12.14 Computation of the Chi-Square Test Statistic for Patient Arrivals

	Number of Patients Arriving During 1 Hour				
	0	*1*	*2*	*3 or more*	*Sum*
(1) Number of Hours, f_i^o	60	140	125	155	480
(2) Probability	.135	.271	.271	.323	1.00
(3) Number of Hours, f_i^e	65	130	130	155	480
(4) $\dfrac{(f_i^o - f_i^e)^2}{f_i^e}$	.38	.77	.19	0	1.34

in row (3) of Table 12.14. Row (4) of this table gives the values of $(f_i^o - f_i^e)^2/f_i^e$. From row (4) we obtain

$$\chi_{4-1}^2 = .38 + .77 + .19 + 0 = 1.34 < \chi_{3,.05}^2 = 5.99$$

Therefore, we conclude that the null hypothesis that patient arrivals are generated by the Poisson probability distribution cannot be rejected.

12.7 CHI-SQUARE AS A TEST OF INDEPENDENCE

In this section, we show how to use the chi-square test introduced in Section 12.6 to test the independence of two variables (this was briefly discussed in Chapter 5). Suppose a sample is taken from a population each of whose members can be uniquely cross-classified according to a pair of attributes. To illustrate, Table 12.15 contains information on a sample of 300 students who are classified by major and by grade earned in a basic statistics course.

Table 12.15 300 Students Classified by Grades and Major

	Grade				
Major	*A*	*B*	*C*	*F*	**Sum**
Science	12	36	34	8	90
Humanities	10	24	46	10	90
Business	8	30	70	12	120
Sum	30	90	150	30	300

This type of table, which has one basis of classification across the columns (in this case, grade) and another across the rows (major), is known as a **contingency table.** Because Table 12.15 has 3 rows and 4 columns, it is called a three-by-four (often written 3 × 4) contingency table. In general notation, in an $r \times c$ contingency table (see Table 12.16), where r denotes the number of rows and c the number of columns, there are $r \times c$ cells.

We want to find out whether these data provide strong enough evidence to support the hypothesis that the majors and the grades are somehow related. To solve this problem, we need to compare the observed frequencies with the expected frequencies. When the expected frequencies are far away from what we observed, the test statistic yields a large value that leads to rejection of the null hypothesis. To compute the test statistic, we must find the expected frequencies.

First, we note that of the 300 students surveyed, 90 are science majors, 90 are humanities majors, and 120 are business majors. That means the distribution of students among the 3 majors is 30 percent, 30 percent, and 40 percent, respectively.

Table 12.16 Cross-Classification of n Observations in an $r \times c$ Contingency Table

| Attribute A | Attribute B | | | | |
	1	2	$\cdots$	c	Totals
1	0_{11}	0_{12}	$\cdots$	0_{1c}	RS_1
2	0_{21}	0_{22}	$\cdots$	0_{2c}	RS_2
$\vdots$	$\vdots$	$\vdots$	$\cdots$	$\vdots$	$\vdots$
r	0_{r1}	0_{r2}	$\cdots$	0_{rc}	RS_r
Totals	CS_1	CS_2		CS_r	n

If the students' majors are independent of their performance, the distribution of grades among the 3 majors should also be 30 percent, 30 percent, and 40 percent, respectively. In other words, because science majors make up 30 percent of the population, they would be expected to receive 30 percent of each grade. That means we expect science majors to account for 9 of the 30 A's, 27 of the 90 B's, 45 of the 150 C's, and 9 of the 30 F's. Similarly, we can obtain the expected frequencies for humanities and business majors. This process of obtaining expected frequencies is summarized in Table 12.17. From Tables 12.15 and 12.17, we can calculate the chi-square statistic of Equation 12.17 as indicated in Table 12.18. The chi-square statistic is $\chi^2 = 11.37$.

The degrees of freedom in this question can be obtained by the formula

$$(r - 1)(c - 1) = (3 - 1)(4 - 1) = 6$$

where r is the number of rows (majors) and c is the number of columns (grades). Note that once $(r - 1)(c - 1)$ cells are known, the remaining cells are determined (if marginals are known). From Table A5, we find $\chi^2_{6,.05} = 12.59$. Because $12.59 > 11.37$, we accept the null hypothesis that the majors and the grades received are independent of each other.

A MINITAB solution to this example is presented in Figure 12.5. The format of this table is similar to that of Table 12.18 with the observed value f_i^o and the expected value f_i^o in each cell.

Table 12.17 Expected Grade Frequency Distribution of 300 Students

	A	B	C	F	Sum
Science	9	27	45	9	90
Humanities	9	27	45	9	90
Business	12	36	60	12	120
Sum	30	90	150	30	300

Table 12.18 Computation of Chi-Square Test
Statistics for Student Performance in Different Majors

f_i^o	f_i^e	$(f_i^o - f_i^e)^2/f_i^e$
12	9	1
36	27	3.00
34	45	2.69
8	9	.11
10	9	.11
24	27	.33
46	45	.02
10	9	.11
8	12	1.33
30	36	1
70	60	1.67
12	12	0
Sum 300	300	11.37

```
MTB > READ INTO C1-C4
DATA> 12 36 34 8
DATA> 10 24 46 10
DATA> 8 30 70 12
DATA> END
       3 ROWS READ
MTB > CHISQUARE USING C1-C4

Expected counts are printed below observed counts

            C1       C2       C3       C4     Total
    1       12       36       34        8        90
          9.00    27.00    45.00     9.00

    2       10       24       46       10        90
          9.00    27.00    45.00     9.00

    3        8       30       70       12       120
         12.00    36.00    60.00    12.00

  Total     30       90      150       30       300
```

Figure 12.5

MINITAB Output
for Test for Indepen-
dence for Student
Grade Distribution

```
ChiSq = 1.000 +  3.000 +  2.689 +  0.111 +
        0.111 +  0.333 +  0.022 +  0.111 +
        1.333 +  1.000 +  1.667 +  0.000 = 11.378
df = 6

MTB > PAPER
```

12.8   BUSINESS APPLICATIONS

In this section we use 6 examples to show how ANOVA and the χ^2 test can be used to make business decisions.

■ APPLICATION 12.1 Comparing Cash Compensation for Different Groups of Corporate Executives

Business Week's Executive Compensation Scoreboard is *BW*'s annual report of the total cash compensation (salary and bonus) of the top corporate executives. The table lists the data from the 1986 report (*Business Week,* May 4, 1987, pp. 59–94). Assume that the data represent independent random samples of the 1986 total cash compensations for 8 corporate executives in each of 3 industries—banks, utilities, and office equipment/computers. Also assume that the experiment is a completely randomized design.

1986 Total Cash Compensation for Three Groups of Executives (thousands of dollars)

Banks and Bank Holding Companies	Utilities	Office Equipment and Computers
$ 755	$520	$438
712	295	828
845	553	622
985	950	453
1,300	930	562
1,143	428	348
733	510	405
1,189	864	938

To test whether there is a difference in the 1986 total cash compensation for the 3 groups of corporate executives, we use SAS to generate the following ANOVA table.

```
                                 SAS       22:22 Sunday, March 15, 1992    5

                        Analysis of Variance Procedure

Dependent Variable: COMPENS
                                    Sum of           Mean
Source                  DF          Squares          Square      F Value      Pr > F

Model                    2        685129.3333      342564.6667      6.45       0.0065

Error                   21       1115232.5000       53106.3095

Corrected Total         23       1800361.8333

              R-Square              C.V.           Root MSE            COMPENS Mean

              0.380551            31.95859         230.4481             721.083333
```

```
                                        SAS       22:22 Sunday, March 15, 1992   6

                        Analysis of Variance Procedure

Dependent Variable: COMPENS

Source               DF        Anova SS    Mean Square   F Value    Pr > F

GROUP                 2      685129.3333   342564.6667     6.45     0.0065
```

From Table A6 of Appendix A, we found $F_{.01,2,21} = 5.78$. This value is smaller than 6.451 indicated in the ANOVA table. Therefore, we cannot accept the hypothesis that the 1986 total cash compensation for the 3 groups of corporate executives is equal at $\alpha = .01$.

■ | *APPLICATION 12.2* Effects of Visual Display Scale on Estimates of Duration

Professor Bobko *et al.* investigated the effects of visual display scale on duration estimates.[3] They solicited 72 undergraduate volunteers (36 females, 36 males) from an introductory course in psychology. The experimental stimuli were 3 commercially available black-and-white television sets with 3 different screen sizes. The screens had diagonal measurements of .13 meters (small), .28 meters (medium), and .58 meters (large). Using ANOVA with interaction to analyze the empirical data, these researchers got the results listed in Table 12.19.

This value is larger than the critical value $F_{2,66,.05} = 3.11$ (obtained by interpolation from Table A5 in Appendix A), and the *p*-value for the factor is .005, so the null hypothesis that the display scale does not affect the duration estimates should be rejected. The effect of sex was marginally significant at $F_{1,66} = 3.73$, *p*-value = .06. The interaction of screen size and sex was not significant.

Table 12.19 Analysis of Variance for the Effects of Screen Size and Sex on Estimates of Duration

Source	Sum of Squares	df	Mean Square	F Ratio
Screen size	12.73	2	6.36	5.81[a]
Sex	4.08	1	4.08	3.73[b]
Interaction	.29	2	.15	.13
Residual	72.26	66	1.09	
Total	89.36	71		

[a]$p < .005$
[b]$p < .06$

---

[3]D. J. Bobko, P. Bobko, and M. A. Davis (1986), "Effects of Visual Display Scales on Duration Estimates," *Human Factor* 28, 153–158. Reprinted with permission. Copyright © 1986 by The Human Factors Society, Inc. All rights reserved. The duration is estimated by the length of time passing as a moving display is watched.

■ | *APPLICATION 12.3* Current Ratios for Failed and Nonfailed Firms

In our example from Section 12.2, involving starting salaries of economics graduates, each of the 3 treatments consisted of the same number of sample observations. Though it may be more convenient to work with samples of equal size, it is not always possible.

We apply the technique of one-way analysis of variance where the sample observations are not of equal size. In this application, we will test whether the population mean current ratio for two classifications of firms, failed and nonfailed, are significantly different.

Table 12.20 contains the sample current ratios for 6 nonfailed firms and 8 failed firms. The table also includes the sample mean for each treatment and a global

Table 12.20 Current Ratios for Nonfailed and Failed Firms

Nonfailed	Failed
2.0	1.8
1.8	1.9
2.3	1.7
3.1	1.5
1.9	1.2
2.5	1.8
	1.6
	1.4
$n = 6$	$n = 8$
$\bar{x}_1 = 2.267$	$\bar{x}_2 = 1.6125$

mean. Our purpose is to test the following null hypothesis against the following alternative.

H_0: The mean current ratio for nonfailed firms

= the mean current ratio for failed firms.

H_1: The mean current ratios for nonfailed and for failed firms are not equal.

The overall mean is calculated from the data as follows:

$$\bar{x} = \frac{(2.267)(6) + (1.6125)(8)}{14} = 1.892871$$

First, the within-group sum of squares is calculated; it is presented in Table 12.21. Accordingly, TSSW = 1.1736 + .3891 = 1.5627.

Next we pursue a measure of between-groups variability. In this example, the between-groups variability is determined as follows:

$$SSB_1 = (2.267 - 1.8928571)^2 = .1400$$
$$SSB_2 = (1.6125 - 1.8928571)^2 = .0786$$

Table 12.21 Within-Group Sum of Squares

Nonfailed	Failed
(2.0 − 2.267)	(1.8 − 1.6126)
(1.8 − 2.267)	(1.9 − 1.6125)
(2.3 − 2.267)	(1.7 − 1.6125)
(3.1 − 2.267)	(1.5 − 1.6125)
(1.9 − 2.267)	(1.2 − 1.6125)
(2.5 − 2.267)	(1.8 − 1.6125)
	(1.6 − 1.6125)
	(1.4 − 1.6155)
$SSW_1 = 1.1736$	$SSW_2 = .3891$

Therefore, TSSB = 6(.1400) + 8(.0786) = 1.4686983. Note that each squared discrepancy is weighted by the number of observations in each treatment.

Finally, we calculate the total sum of squares of the two treatments.

$$TSS = 1.5627 + 1.4686983 = 3.0313983$$

In order to test our hypothesis, we must calculate an unbiased estimate of the within-group and between-groups variances. Again, we find the estimate of the within-group variance by dividing the total sum-of-squares deviations of the within-groups variability (TSSW) by the appropriate degrees of freedom $(n - J)$.

$$MSW = \frac{1.5627}{12} = .130225$$

Similarly, we find the estimate of the between-groups variance by dividing the total sum-of-squares deviations of the between-groups variability (TSSB) by the appropriate degrees of freedom $(J - 1)$.

$$MSB = \frac{1.4686983}{1} = 1.4686983$$

Our calculated value of the F ratio is

$$F = \frac{MSB}{MSW} = \frac{1.4686983}{.130225} = 11.27816$$

From the F distribution in Table A6 with $(n - J)$ and $(J - 1)$ degrees of freedom and a .05 significance level, the critical value is 4.75. Because the calculated F ratio is greater than the critical value, we do not accept the null hypothesis that the mean current ratios of failed and nonfailed firms are equal.

■ | *APPLICATION 12.4* Distribution of Stock Rates of Return

In financial analysis, we are often interested in whether the rate of return of a certain stock follows a normal distribution. The example that follows demonstrates how we used the goodness-of-fit test to find out whether the rates of return of a mutual fund follow a normal distribution.

A stock analyst collected the annualized daily rates of return x_i of a mutual fund in the past 200 trading days and got a mean $\bar{x}$ of 15 percent and a standard deviation s_x of 5 percent. The rates of return are summarized in Table 12.22. Do the data support rejecting the hypothesis that the rates of return follow a normal distribution? A test at 5 percent level of significance follows.

To do the test, we first formulate the hypotheses.

H_0: The annualized daily average mutual fund rates of return are normally distributed with a mean of 15 percent and a standard deviation of 5 percent.

H_1: The annualized daily average mutual fund rates of return are not normally distributed with a mean of 15 percent and a standard deviation of 5 percent.

Table 12.22 Rate-of-Return Data for 200 Days

Rates of Return, x_i (%)	Observed Frequency, f_i^o
Under -5	20
-5 to under 0	33
0 to under 10	48
10 to under 20	41
20 to under 30	29
30 or above	29
Total	200

Second, we need to calculate the theoretical frequency (f_i^o) in accordance with the standard normal distribution table (Table A3 in Appendix A). The computation procedure is presented in Table 12.23.

Table 12.23 Computation of Theoretical Frequencies in Each Rates-of-Return Interval

Class Boundaries	x	$z = (x - 15)/5$	Area Under Standard Normal Curve Left of x	Area of Class Interval (P)	Expected Frequency If H_0 Is True, $f_i^o = 200P$	
Under -5	-5	-4	0	0	0	
-5 to under 0	0	-3	.0014	.0014	.28	31.74
0 to under 10	10	-1	.1587	.1573	31.46	
10 to under 20	20	1	.8413	.6826	136.52	
20 to under 30	30	3	.9986	.1573	31.46	31.74
30 and above	∞	∞	1.0000	.0014	.28	
Total				1.0000		

Finally, we calculate χ^2 in terms of Equation 12.17, as indicated in Table 12.24. The test statistic $\chi^2 = 281.92$. If a level of significance of $\alpha = .05$ is selected, the critical value of χ^2 with 2 degrees of freedom is 5.991. Because $281.92 > 5.991$, we conclude that the annualized daily mutual fund rates of return are not normally distributed with mean 15 percent and standard deviation 5 percent.

Table 12.24 Worksheet for Computing the Test Statistic χ^2

Class Boundaries	f_i^o	f_i^e	$(f_i^o - f_i^e)$	$(f_i^o - f_i^e)^2/f_i^e$
Under 10	101	31.74	69.26	47.49
10 to under 20	41	136.52	−95.52	222.54
20 and above	58	31.74	26.26	11.89
Total	200	200	0	281.92

APPLICATION 12.5 Market-Share Pattern of a New Cereal Product

G. A. Churchill proposed a goodness-of-fit technique to test the market-share pattern of a new cereal called Score produced by a breakfast food manufacturer.[4] The cereal was packaged in 3 standard sizes: small, large, and family size. The manufacturer's experience with other cereals suggested that for every small package, 3 of the large and 2 of the family size are also sold. The manufacturer wanted to know whether this same consumption pattern would hold with Score, because a change in consumption pattern could have significant implications for production and packaging. The manufacturer therefore decided to conduct a market test over a 1-week period. In this period, 1,200 boxes of the new cereal were sold. The distribution of sales by size is given in Table 12.25.

Table 12.25 Distribution of Boxes of New Cereal Sold

Small	Large	Family	Total
240	575	385	1,200

To test whether the relative frequencies of the various sizes of the new product are the same as those of the old product, we can test the hypotheses

$$H_0: P_1 = \frac{1}{1+3+2} = \frac{1}{6}, \quad P_2 = \frac{3}{1+3+2} = \frac{1}{2}, \quad P_3 = \frac{2}{1+3+2} = \frac{1}{3}$$

H_1: At least one of these P_i's is incorrect.

[4] G. A. Churchill, Jr. (1983), *Marketing Research: Methodological Foundations,* 3rd. ed., pp. 523–524. Copyright © 1983 by The Dryden Press, reprinted by permission of the publisher.

To perform this test, we first calculate the expected frequencies.

$$f_1^e = \frac{1{,}200}{6} = 200, \qquad f_2^e = \frac{1{,}200}{2} = 600, \qquad f_2^e = \frac{1{,}200}{3} = 400$$

Substituting both expected and observed frequencies into Equation 12.17, we obtain

$$\chi_{3-1}^2 = \frac{(240 - 200)^2}{200} + \frac{(575 - 600)^2}{600} + \frac{(385 - 400)^2}{400} = 9.60$$

From Table A5, we find that $\chi_{2,.05}^2 = 5.99$, which is smaller than 9.60. Hence we reject H_0 and accept H_1. In other words, the null hypothesis of sales in the ratio of 1:3:2 is rejected. This result suggests that the sale of the new cereal, Score, will follow a different pattern.

■ **APPLICATION 12.6** The Effect of Price Advertising on Alcoholic Beverage Sales

To study the effect of price advertising on alcoholic beverage sales, G. B. Wilcox examined the effects of price advertising on sales of beer in Lower Michigan. Wilcox used Michigan in his study because since 1975, except for the short period from March 1982 until May 1983, Michigan has banned retailers from advertising the price of beer products. The data he used covers 3 different periods and are presented in Table 12.26.

Table 12.26 Bimonthly Beer Sales for 3 Different Periods (units: thousands of 31-gallon barrels)

Period 1: Price Advertising Restricted (May/June 1981–Jan./Feb. 1982)	Period 2: No Restrictions (March/April 1982–May/June 1983)	Period 3: Price Advertising Restricted (July/August 1983–March/April 1984)
462	522	433
417	508	470
516	427	609
605	477	442
654	603	446
	692	
	584	
	496	

Source: G. B. Wilcox, "The Effect of Price Advertising on Alcoholic Beverage Sales," *Journal of Advertising Research,* Vol. 25, No. 5, October/November 1985, 33–37.

To examine whether there is sufficient evidence to indicate differences in the average total sales of beer in the 3 periods, we use SAS to generate the following ANOVA output.

```
                                      SAS        23:34 Sunday, March 15, 1992   2

                        Analysis of Variance Procedure

Dependent Variable: SALES
                                     Sum of            Mean
Source                   DF          Squares          Square      F Value     Pr > F

Model                     2       11357.82500       5678.91250      0.78      0.4760

Error                    15      109152.67500       7276.84500

Corrected Total          17      120510.50000

                  R-Square              C.V.        Root MSE            SALES Mean

                  0.094248           16.39944       85.30443           520.166667
                                      SAS        23:34 Sunday, March 15, 1992   3

                        Analysis of Variance Procedure

Dependent Variable: SALES

Source                   DF         Anova SS     Mean Square    F Value     Pr > F

PERIOD                    2       11357.82500     5678.91250      0.78      0.4760
```

From Table A6 in Appendix A, we found $F_{.05,2,15} = 3.68$. This number is larger than .78 indicated in the above ANOVA table; therefore, we cannot reject the hypothesis that price advertisements on alcoholic beverages did not affect sales of beer in the three different periods in the state of Michigan.

Summary

Using the basic concepts of mean, variance, and F statistics discussed in previous chapters, we explored a statistical method called analysis of variance for testing the difference between sample means. We also examined the use of the chi-square statistic in goodness-of-fit tests and in testing the assumption of independence. Several applications of analysis of variance in business decisions were discussed in some detail.

Questions and Problems

1. The following table shows the sales figures for 4 salespeople on 3 randomly selected days. Use analysis of variance to test the hypothesis that the mean daily sales figures (in thousands of dollars) are the same for all 4 salespeople. That is, test H_0: $\mu_1 = \mu_2 = \mu_3 = \mu_4$. Use $\alpha = .05$.

| | **Salesperson** | | | |
	1	**2**	**3**	**4**
Day	15	9	17	12
	17	12	20	13
	22	15	23	17
Mean	18	12	20	14
Variance	13	9	9	7

2. The yearly portfolio returns for 3 different investment firms over a 10-year period are listed in the accompanying table. Do these data show a statistically significant difference in the firms' performances? Assume that the population errors meet the conditions necessary for ANOVA. Let $\alpha = .01$.

	Firm 1	Firm 2	Firm 3
Mean return	11.5	12.0	10.0
Standard deviation	3.0	2.0	2.4
Number of investments	20	10	25

3. A consumer organization wants to compare the prices charged for a particular dishwasher in 3 types of stores in a suburban county: discount stores, department stores, and appliance stores. Random samples of 4 discount stores, 6 department stores, and 5 appliance stores were selected. The results were as shown.

Discount	Department	Appliance
12	15	15
14	18	18
16	18	16
15	14	16
	18	19
	15	

At the .05 level of significance, is there any evidence of a difference in the average price between the types of stores? Use the MINITAB commands presented in Figure 12.3 to answer this question.

4. Three packaging materials were tested for moisture retention by storing the same food product in each for a fixed period of time and then determining the moisture loss. Each material was used to wrap 10 food samples. The results are given in the accompanying table.
   a. Construct the ANOVA table.
   b. Can we reject the hypothesis that the materials are equally effective? Use $\alpha = .05$.

	Material 1	Material 2	Material 3
Number of packages	10	10	10
Mean loss	231	238	224
Sample variance	40	38	30

5. The quality control manager at a sugar refinery was worried that two packaging production lines might be filling the packages with different weights. Samples of size 16 were taken from each of the production lines, and the contents of these 10-pound packages were carefully weighed. The following sample results seem to indicate that the mean weights of the 10-pound packages from the two lines are the same, but there appears to be much more variation in the weights of the packages coming off the second line. Test the hypothesis that the two lines have the same variation in weights by testing $H_0: \sigma_1^2 = \sigma_2^2$ with $\alpha = .05$. Use $H_1: \sigma_1^2 < \sigma_2^2$. If the following data do not support rejection of the null hypothesis, what must we do to test $H_0: \mu_1 = \mu_2$?

$n_1 = 16$	$n_2 = 16$
$\overline{X}_1 = 10.15$	$\overline{X}_2 = 10.16$
$S_1 = .07$	$S_2 = .1$

6. The Environmental Protection Agency is studying coliform bacteria counts at the beaches of a large suburban county. Three types of beaches are to be considered (ocean, bay, and sound) in three geographic areas of the county (west, central, and east). Two beaches of each type are randomly selected in each region of the county. The coliform bacteria counts (in parts per thousand) at each beach on a particular day were as shown in the Table.

	Geographic Area		
Type of Beach	**West**	**Central**	**East**
Ocean	25	9	3
	20	6	6
Bay	32	18	9
	39	24	13
Sound	27	16	5
	30	21	7

   a. Use the .05 level of significance, and determine (1) whether there is an effect due to type of beach, (2) whether there is an effect due to geographic area, and (3) whether interaction between type of beach and geographic area has an effect.
   b. What conclusions about average bacteria count do the results support?

7. Explain the difference between the one-factor and the two-factor ANOVA models.

8. A researcher has an ANOVA problem with 5 columns (treatments). Would you recommend testing for differences between the columns by looking at pairs of columns? Explain why. How many pairs are there?

9. A researcher concludes that there is no difference in the column treatments in a one-factor model. Upon reanalysis of the same data via two-factor ANOVA, however, the researcher concludes that there is a difference in the column treatments. Did the researcher make a mistake? Which conclusion (if either of them) is true? Explain how these contradictory conclusions can (or cannot) be justified.

10. A secretarial training school is experimenting with 4 different manuals for a typing course. The school divided 20 students into 4 classes, and each class used a different manual. At the end of the training session, a test was given, and the scores shown in the table were reported. Do the data support the hypothesis that the 4 different manuals create different effects? Do a 5 percent test.

Manual

A	B	C	D
78	67	75	89
74	79	95	86
95	85	69	87
93	79	60	87
85	86	94	73

11. A research institution says that gasoline is gasoline. That is, there is no difference among different brands of gasoline in terms of mileage per gallon. An independent consumer rights organization did an experiment on 3 different brands of gasoline. It divided cars of the same make and the same condition into 3 groups, and the members of each group were filled with a different brand of gasoline. The test results follow. Do the data support the hypothesis of no difference among gasolines? Do a 5 percent test.

Gasoline

A	B	C
34	29	32
28	32	34
29	31	30
37	43	42
42	31	32
27	29	33
29	28	

12. A college professor taught an interdisciplinary course in the last 3 years to students of different majors. He believes that the students' majors do not have any impact on their performance in the class. He picked 24 students who represented 3 different majors and recorded their test scores. Do the data necessitate rejection of the professor's hypothesis? Do a 5 percent test.

Business	Humanities	Science
85	84	63
57	95	73
92	87	83
83	73	64
84	83	79
83	73	74
75	85	65
73	72	98

13. A doctor recruited 15 volunteers and put them on 3 different diets that were supposed to lower the subjects' cholesterol levels. The effects of the diet plans in terms of lowered cholesterol level are recorded in the table. Do the data cause us to reject the null hypothesis that the 3 different diet plans have the same effect? Do a 5 percent test.

Diet Plan

A	B	C
20	12	13
14	15	13
21	21	23
15	16	19
17	18	14

14. General Motors, the largest auto producer in this country, produces and sells its cars under 5 different brand names. Some of the cars sold under different names can be considered "sister cars" that

should turn in the same performance. An auto analyst wants to see whether the "sister cars" sold under different names do indeed have the same performance. He tested 20 cars from 3 different brands and recorded the mileages per gallon. Do the data require rejection of the null hypothesis that the mileages per gallon generated by three "sister cars" are the same? Do a 5 percent test.

Brand

A	B	C
32	31	34
29	28	25
32	30	31
25	34	37
35	39	32
33	36	
34	38	
31		

15. Market analysts want to investigate the popularity of 3 types of radio programs. Their market research yielded the following ratings over the last 5 months. Can the analysts conclude that the 3 types of programs have different ratings? Do a 5 percent test.

Time	Talk Show	Sports Show	Music
Early morning	20	30	17
Late morning	17	15	15
Afternoon	18	21	12
Evening	23	27	18
Midnight	25	22	11

16. In question 15, the market analyst hired a statistician to do the research. The statistician, realizing that these programs are broadcast at 5 different times of day, took the time factor into consideration. Do a test for the same hypothesis. Use 5 percent.

17. On a beach, there are 3 ice cream stands that are supposed to be occupying equally good locations. The management of the beach wants to know whether this assumption is true. The ice cream sales during the past few days, in hundreds of dollars, are recorded here. Do a 5 percent test to determine whether the ice cream sales are the same in the 3 locations.

Location

A	B	C
12	31	21
14	21	14
14	20	17
18	17	16
21	12	23

18. A stock analyst thinks 4 stock mutual funds generate about the same return. She collected the accompanying rate-of-return data on 4 different mutual funds during the last 5 years.

Fund

	A	B	C	D
1988	12	11	13	15
1989	12	17	19	11
1990	13	18	15	12
1991	18	20	25	11
1992	12	19	19	10

a. Do a one-way ANOVA to decide whether the funds give different performances. Use 5 percent.

b. Do a two-way ANOVA to decide whether the funds give different performances. Use 5 percent.

19. The personnel office recently produced a set of aptitude test questions designed to determine whether a potential employee can be a good salesperson. The test was tried on current employees before it was used for future employees. The test scores on employees in 3 different departments are listed here. Do the tests generate different scores for different departments? Do a 5 percent test.

Department

Accounting	Sales	Production
78	85	76
79	87	77
80	89	97
72	79	71
87	99	81
98	95	79

Use the following information to answer questions 20–23. An investor recorded the performance of stock mutual funds during the last 3 years. He classified the stock mutual funds into 3 categories, growth, income, and mixed. The rates of return during the last 3 years are presented here:

	Type		
Year	Growth	Income	Mix
	12	14	15
1990	17	12	16
	19	12	17
	17	13	15
1991	18	19	18
	21	14	15
	21	16	17
1992	22	13	15
	21	15	18

20. Use MINITAB to do a one-way ANOVA to determine whether the 3 different years have the same average rate of return. Use the 5 percent level of significance.

21. Use MINITAB to do a one-way ANOVA to determine whether the 3 different types of mutual funds have the same rate of return. Use the 5 percent level of significance.

22. Use MINITAB to do a two-way ANOVA to determine whether the 3 different types of mutual funds have the same rate of return. Use the 5 percent level of significance. (Hint: Follow the procedure presented in Figure 12.4.)

23. Use MINITAB to do a two-way ANOVA to determine whether the 3 different years have the same average rate of return. Use the 5 percent level of significance. (Hint: Follow the procedure presented in Figure 12.4.)

24. A researcher contacted 1,000 doctors and asked them what kind of pain reliever they would like to have with them if they were stranded on a desert island. The responses were

Brand A	Brand B	Brand C	Others
250	230	260	260

Do a test to determine whether the 4 categories of products receive the same number of recommendations from doctors. Use 5 percent.

25. Four instructors teach introductory-level economics courses during the same time period. The numbers of students taking their courses are

Instructors			
A	B	C	D
100	120	90	110

Can we reject that the 4 instructors are about equally popular among the students? Do a 5 percent test.

26. The personnel manager wants to know whether an equal number of employees call in sick on the 5 days of the regular work week. The sick days recorded during last year were distributed as follows:

M	T	W	Th	F
40	30	32	25	45

Can we conclude that the 5 different weekdays have different frequencies of sick calls? Do a 5 percent test.

27. An economics consulting company wants to study bank managers' opinions about what lending rate will prevail for the next 3 months. It sends questionnaires to 940 bank managers and gets the following responses.

Higher	Lower	Same	No Idea
210	220	210	300

a. Can we reject that all of the 4 opinions are held by about the same number of bank managers? Do a 5 percent test.

b. Can we reject that the 3 kinds of opinions (excluding "No Idea") are held by about the same number of bank managers? Do a 5 percent test.

28. A management consulting company is interested in how managers look at the prospects for the economy in this country. Questionnaires were sent to 2,000 managers in different areas of the country. The responses were

Future Prospects		
Optimistic	**Pessimistic**	**No Change**
Top-level managers 300	250	100
Middle managers 200	200	220

a. Does the evidence suggest that middle managers are equally split among the 3 different opinions? Do a 5 percent test.

b. Does the evidence suggest that middle management's opinion pattern is similar to that of top-level management? Do a 5 percent test.

29. Lotteries are getting more and more popular in this country. Many books claiming to teach people how to pick winning numbers have been published. According to an official in the state's lottery office, the game is fair in the sense that every number has the same chance of being drawn. Briefly explain how you can test this contention. (And remember that if it is true, any money spent on "systems" or "secrets" for winning the lottery is money wasted.)

30. Professor Maloy uses different textbooks to teach statistics to two different college classes. Book 1 is the standard textbook also used by other instructors. Book 2 is a more recently introduced text. Over the years, Maloy recorded the grade distribution of the students.

	A	B	C	D	F
Book 1	20	80	100	20	10
Book 2	5	22	30	4	1

Can the professor conclude that the grade distribution pattern of students using book 1 is different from the grade distribution pattern of students using book 2?

31. An insurance company reviewed its policyholders' records during last year and organized the data in the following table. Do the data dictate rejection of the null hypothesis that the accident pattern has a Poisson distribution? Do a 5 percent test.

Number of Accidents	Number of Policyholders
0	2000
1	150
2	10
3	1

32. The number of patients to arrive at an emergency room each day is recorded in the accompanying table. The average number of emergency room patients is approximately 10. Do the data support the hypothesis that the number of emergency room patients follows a Poisson distribution? Do a 5 percent test.

Number of patients	0	1	2
Number of days	400	14	1

33. An eight o'clock train that pulls into Penn Station in New York City every weekday has a 20 percent chance of being late. A supervisor from the Port Authority recorded the number of days that the train arrived late each week for the last 100 weeks. Does the evidence compel us to refute the null hypothesis that the data come from a binomial distribution? Do a 5 percent test.

Number of late arrivals in a week	0	1	2	3	4	5
Number of weeks	5	15	30	30	15	5

34. A nationwide testing service collected scores from a set of examination questions that were used during the last 3 years. The distribution is summarized in the accompanying table. Does the evidence support rejection of the hypothesis that the data come from a normal distribution? Do a 5 percent test.

Score	Frequency
≤ 300	10
301–400	25
401–500	40
501–600	45
601–700	44
701–800	35
801–900	20
More than 900	15

35. The daily rate of return for a stock (adjusted to an annual rate) is summarized in the following table. Can you show that these data do not come from a normal distribution? Do a 5 percent test.

Rate of Return	Frequency
Less than -3%	20
-3% to -2%	25
-2% to -1%	30
-1% to 0	50
0 to 1%	40
1% to 2%	25
More than 3%	5

36. The highway bureau records the following numbers of accidents during the past 365 days. Does this frequency distribution cause us to reject the null hypothesis that the accidents exhibit a Poisson distribution? Do a 5 percent test.

Number of Accidents in a Day	Number of Days
0	320
1	30
2	10
3	5

37. A college professor wants to use a normal distribution to analyze his students' grades. He randomly selects 200 previous grades and organizes them in the following table. Does the frequency distribution support the hypothesis that students' grades follow a normal distribution? Do a 5 percent test.

Grades	Frequency
≤ 40	2
41–50	15
51–60	35
61–70	70
71–80	40
81–90	28
>90	10

38. In a hospital, 100 patients were checked for their cholesterol level. Do the data collected support the hypothesis that the cholesterol levels follow a normal distribution? Do a 5 percent test.

Cholesterol Level	Frequency
≤ 160	4
161–180	21
181–200	25
201–220	30
221–240	10
241–260	8
>260	2

39. A college professor has taught business statistics for the last 5 years. He used standard tests every semester. The distribution of grades during the past 5 years was

A	B	C	D	F
15%	30%	40%	10%	5%

This professor has just finished grading students this semester and has found that the frequency distribution of the grades is

A	B	C	D	F
6	12	15	5	1

He feels that this semester's students have a different grade distribution pattern. Do you agree with him? Do a 5 percent test.

40. A travel agency was curious about whether the service a guest receives is related to the size of the hotel. The agency surveyed 300 customers and summarized their responses in the accompanying table. Determine whether the data support the hypothesis that the customer's opinion and the size of the hotel are related. Each customer gave only one opinion for one size hotel. Use 5 percent.

	Size of Hotel		
	Large	**Mid-size**	**Small**
Satisfied	80	40	30
So-so	60	30	10
Dissatisfied	20	20	10

Use the following information to answer questions 41–43. Four hundred and fifty economists of different ideologies were asked to forecast the prospects for the economy during the Bush administration. Here's what they said:

	Opinion		
Ideology	**Boom**	**So-So**	**Recession**
Conservative	80	60	60
Liberal	60	40	40
Radical	50	30	30

41. Do the data support the hypothesis that ideology and opinion are related? Do a 5 percent test.

42. Test, at the 5 percent level, the hypothesis that about equal numbers of economists hold each opinion.

43. Can you say that the opinion pattern of the liberal economists is the same as the opinion pattern of the radical economists? Do a 5 percent test.

44. A magazine is interested in knowing whether which newspaper is read and level of education of the reader are related.

	Newspaper		
Education	**Post**	**News**	**Tribune**
High school	300	200	100
College	200	300	100
Graduate school	100	200	300

Use MINITAB to determine whether newspaper read and education are related. Use the 5 percent level of significance. (Hint: Follow the procedures presented in Figure 12.5.)

45. The sales manager wants to know whether salespeople's performance is related to their zodiac sign. Three hundred salespeople were surveyed. Their performance is summarized in the following table.

	Performance		
Zodiac Sign	**Good**	**Mediocre**	**Bad**
Leo	80	30	20
Gemini	50	20	10
Virgo	40	40	10

Do the data support the belief that the performance and zodiac sign of a salesperson are related? Do a 5 percent test.

46. An advertising agency wants to know whether there is a relationship between TV shows and the age of the audience. The following data were compiled.

	Age of the Audience			
TV Show	**10 and Younger**	**Teen-ager**	**20–40**	**40 and Older**
Game show	100	120	200	400
Sitcom	20	120	400	200
News	2	40	48	50

Do the data support the hypothesis that the age of the audience is related to the type of show that is preferred? Do a 5 percent test.

Use the following information to answer questions 47–49. A developer asks visitors how they heard of the housing project they are looking at. Their responses are shown in the following table.

Distance from Construction Site	*Source of Information*			
	Referred by a Friend	**News-paper**	**Radio**	**TV**
Within 10 miles	40	200	120	150
10–30 miles	30	180	120	120
Farther than 30 miles	10	150	100	100

47. Is the distance from the construction sites independent of the way people hear of the housing project? Do a 5 percent test.

48. Can we conclude that the effects of publicizing the project in the 3 different media (newspaper, radio, and TV) are different? Do a 5 percent test.

49. If you live 10–30 miles away from the construction site, is your sources-of-information pattern the same as that of the people living within 10 miles? Do a test of 5 percent.

Use the following information to answer questions 50–55. A company operates 3 mutual funds. The managers of these mutual funds invest the money entrusted to them in different kinds of assets. The rates of return in the last 5 years are recorded in the following table.

	Real Estate Fund	Government Bond Fund	Stock Fund
1985	6%	7%	3%
1986	20	8	9
1987	6	12	8
1988	15	9	15
1989	3	10	20

The company also randomly sampled its customers and compiled the following table:

	Real Estate Fund	Government Fund	Stock Fund
Retirees	30	80	40
51–65	40	60	50
36–50	80	20	50
20–35	40	40	70

50. Do the 3 different kinds of mutual funds attract about the same numbers of investors? Do a 5 percent test.

51. Can we conclude that the 3 different kinds of mutual funds generate about the same average rate of return? Do a 5 percent test, using a one-way ANOVA.

52. Do a two-way ANOVA to determine whether the rates of return for the three kinds of mutual funds are about the same. Do a 5 percent test.

53. Determine whether the stock fund is equally popular among the 5 different age groups. Do a 5 percent test.

54. Determine whether the investment pattern of the age group 20–35 is the same as that of the age group 51–65. Do a 5 percent test.

55. Are fund preference and age group related? Do a 5 percent test.

56. A plant that runs 3 shifts would like to know whether the 3 shifts are equal in average productivity. The productivity breakdown is presented in the following table.

Day	Shift 1	Shift 2	Shift 3
Monday	30	40	20
Tuesday	40	50	30
Wednesday	40	40	30
Thursday	40	30	20
Friday	20	30	20

a. Are average productivities of the 3 different shifts the same? Do a 5 percent test, using a one-way ANOVA.

b. Are the average productivities of the 3 different shifts different? Do a 5 percent test. Consider the weekday factor in testing this hypothesis.

57. A nationwide real estate brokerage house wants to study the relationship between rent per square foot and size of the property. The data collected are summarized in the accompanying table. Using these data, can we reject the null hypothesis that the average rents per square foot are equal? Do a 5 percent test.

Size of the Property (in square feet)

Less than 1,000	1,000 to 2,000	2,000 or More
3	2	3
4	5	6
5	5	7
5	6	7
5	5	7
4	6	6

58. A consultant argues that location, the most important factor in the real estate business, was not considered in the test performed in question 57. He suggests redoing the test by controlling the location factor. Do a 5 percent test to see whether the conclusion changes when the data are presented as follows:

Size of the Property (in square feet)

Location	Less than 1,000	1,000 to 2,000	2,000 or more
Bad	3	2	3
	4	5	6
So-so	5	5	7
	5	6	7
Good	5	5	7
	4	6	6

59. The performances of 250 salespeople in a company are summarized in the following table.

Sales	Frequency
Less than 78	13
78–80	37
81–83	40
84–86	50
87–89	60
	20
	20
90 or more	10

Derive the expected frequencies, assuming that the data are from a normal distribution.

60. A chicken farm came up with 4 different ways of mixing chicken feeds. The feeds were tested on 20 chickens. The results, given in terms of the chickens' weight, are presented in the accompanying table. Do a 5 percent test of the hypothesis that the weights resulting from the different feeds are approximately the same.

Group			
A	**B**	**C**	**D**
4.5	4.2	4.1	4.6
4.4	4.3	4.6	4.2
4.5	4.3	4.2	4.9
4.3	4.2	4.3	4.4
4.9	4.9	4.8	4.7

Use the following information to answer questions 61–64. A survey was sent to 400 students to solicit their opinions about a new rule for using the student centers. Here are the results:

Year	Against	Indifferent	Agree
First-year	30	30	40
Sophomore	50	40	30
Junior	20	50	30
Senior	10	30	40
Total	110	150	140

61. Do you think the 3 different opinions have about the same number of responses? Do a 5 percent test.

62. Do you think the 3 different opinions receive about the same amount of support among first-year students? Do a 5 percent test.

63. Do you think opinion pattern and year in school are related? Do a 5 percent test.

64. Do you think there are equal amounts of support for the new rule from the 4 different classes? Do a 5 percent test.

65. An insurance company is interested in knowing the relationship between traffic accident claims and the type of cars that policyholders drive. The numbers of accidents per 1,000 automobiles during last year in 6 states are reported in the accompanying table. Determine whether the 3 kinds of cars have the same average accident rate. Use a 5 percent level of significance.

	Type of Automobile		
State	**Sports Car**	**Sedan**	**Wagon**
New Jersey	30	15	16
New York	20	15	17
Connecticut	15	12	11
Massachusetts	17	13	12
Vermont	18	21	15
New Hampshire	17	12	13

66. The accompanying table shows highway patrol data on the numbers of speeding tickets given in the last 3 months. Do the data show that all 3 months have about the same number of tickets? Do a 5 percent test.

	Sports Cars	**Sedans**	**Wagons**	**Trucks**
April	44	30	32	18
May	46	32	30	25
June	45	35	37	27

67. In a poll, people were asked their opinions about the death penalty. The breakdown of the responses is given in the accompanying table. Do the data show that educational level and opinion are independent of each other? Do a 5 percent test.

Educational Level	**Favor**	**Oppose**
Elementary school	400	200
High school	200	400
College	200	400

68. In a recent survey, people were asked whether they are happy with the current income tax structure. Do the results that follow support the hypothesis that how people feel about the tax structure depends on what tax bracket they are in? Do a 5 percent test.

	Satisfied	Dissatisfied	Very Dissatisfied
Low bracket	40	40	50
Middle bracket	50	30	30
High bracket	30	50	60

Use the following information to answer questions 69–71. A company puts vending machines in different locations. The numbers of sodas sold (in thousands) in the last 3 months are presented in the following table.

	Location		
	Beach	School Gymnasium	Gas Stations
April	3	4	4
	3	5	5
	2	5	6
May	6	4	5
	8	5	6
	6	4	7
June	10	4	8
	10	6	7
	12	6	8

69. Do the data support the hypothesis that April, May, and June have the same amount of sales? Do a 5 percent test, using one-way ANOVA.

70. Do a one-way ANOVA to see whether you can argue that the 3 different locations have different amounts of sales. Use 5 percent.

71. Do a two-way ANOVA to see whether the three different locations have different amounts of sales. Use 5 percent.

72. Hannah, a wine dealer, believes that the taste of wine depends on the year the wine was bottled. Do the data she collected from a recent wine-tasting contest support her belief? Do a 5 percent test.

	Year		
	1968	1973	1985
Excellent	40	45	25
Good	35	25	45
Fair	45	15	20
Yuck	10	15	20

73. The manager in a department store believes that whether a customer pays by cash, charge, or check depends on the amount of money spent. Do the following data support what the manager believes? Do a 5 percent test.

Amount Spent	Charge	Check	Cash
Less than $10	20	30	70
Between $10 and $100	40	40	40
Over $100	60	40	20

74. The demand for different types of automobiles should be related to their owners' needs. A manager in a local auto dealership randomly pulls samples from the dealership's customer files. Do the resulting data support the manager's belief? Do a 5 percent test.

	Auto Purchased		
	Sedan	Wagon	Sports Car
Single	30	5	20
Married, no children	40	15	20
Married, at least one child	30	40	20

75. An advertising agent believes that different types of programs attract audiences of different age groups. She collects the following data to study her claim.

	Program Type		
Age Group	Sitcom	Game Show	News
10–19	40	40	20
20–29	60	40	50
30–39	60	30	60
40 or older	40	20	40

Determine, at α = 5 percent, whether you can reject her claim.

76. A consumer rights organization wanted to check out different diet plans. It recruited 33 volunteers and sent them to 4 different programs. After the first 2 weeks, the weight losses, in pounds, were recorded and organized in the accompanying table. Do a 5 percent test to determine whether the 4 programs are equally effective.

A	B	C	D
8.0	9.9	8.9	7.6
8.8	9.1	8.2	7.7
8.7	9.8	8.1	7.5
8.6	9.8	8.0	7.8
8.0	9.9	8.6	7.6
8.8	9.6	8.6	7.3
8.5	9.2	8.6	7.1
	9.8	8.4	8.0
			7.5
			8.0

77. The dean of the business school wants to find out whether the instructors in 4 departments are grading students similarly. The following data are compiled. Do you think the grade distribution depends on the department? Do a 5 percent test.

	Finance	Management	Accounting	Marketing
A	35	45	35	25
B	50	60	55	35
C	15	30	25	10
F	30	45	35	20

78. A Consumer Protection Coalition decides to study the delay times, in minutes, for 4 different airlines: A, B, C, and D.

A	B	C	D
25	22	21	30
35	31	24	28
35	33	34	32
30	28	29	27
44	41	40	15
31	32	17	19

It is believed that the average delay times of the 4 airlines are about equal. Do a test at the 5 percent level to decide whether the data support rejecting this hypothesis.

79. In question 78, a statistician argues that the length of delay may depend on the airport from which the airplane departs. Accordingly, the data were regrouped to reflect departure sites X, Y, and Z. Here are the results:

	Airlines			
	A	B	C	D
X	25	22	21	30
	35	31	24	28
Y	35	33	34	32
	30	28	29	27
Z	44	41	40	15
	31	32	17	19

Redo the test to decide whether the airlines' delay times are about equal by considering the effect of departure location. Use 5 percent.

80. The delay times of 200 delayed flights were compiled in the following frequency distribution. Do the data follow a normal distribution? Do a 5 percent test.

Delay Time (in minutes)	Frequency
0–15	20
16–30	32
31–45	48
46–60	52
61–75	38
76–90	10

81. The numbers of missing pieces of luggage are compiled in the following table. Do the data follow a Poisson distribution? Do a 5 percent test.

Number of Missing Pieces of Luggage	Number of Flights
0	985
1	10
2	4
3	1
More than 3	0

82. A bank manager is interested in the amount of cash being withdrawn each Friday. He collects data on the last 90 Fridays and compiles them in the accompanying table. Determine whether the amount of cash withdrawn follows a normal distribution. Use a 5 percent significance level.

Cash Withdrawn	Frequency
Less than 250	5
250–300	21
301–350	25
351–400	20
401–450	15
More than 450	4

83. A financial analyst is interested in conducting an extensive study of credit card debt. He wants to know whether the income of cardholders is related to the size of the debt. He compiles the data in the accompanying table. Determine whether size of debt and income level are independent. Use a 5 percent level of significance.

	Size of Debt		
	$200	**$500**	**$1,000**
	to	**to**	**and**
Income	**$500**	**$1,000**	**above**
Less than $20,000	400	200	100
$20,000–$40,000	450	500	300
Higher than $40,000	100	200	500

84. There are many books to help people learn to use computer software packages. An instructor checked these books and found that they are all of similar quality. He picked four books and used them in his classes. If the students have the same average grades, he will use the cheapest book. On the basis of the test results that follow, do you think the 4 classes have about the same grades? Do a 5 percent test.

	Class		
W	**X**	**Y**	**Z**
43	77	72	72
45	72	73	74
67	75	71	75
68	69	65	65
73	67	68	66
72	66	69	68
55	65	73	74
62	63	72	81

85. It is believed that the quality of a certain product is related to the time of day the product is produced. The following table summarizes the results of tests on some random samples produced in a single day.

Time	Good	So-So	Bad
Morning shift	25	10	5
Afternoon shift	15	20	5
Evening shift	10	20	10

Use a 5 percent test to determine whether the quality of the product is independent of the time it was produced.

86. The placement office in a business school randomly sampled 10 graduates from 3 departments and recorded their starting salaries. Determine whether graduates of the 3 departments have about the same starting salaries. Use a 5 percent level of significance.

Management	Marketing	Accounting
$24,550	$25,200	$24,150
24,790	27,200	24,100
24,310	24,100	23,900
24,200	25,400	25,650
24,900	23,300	23,700
25,200	24,200	24,900
23,900	25,000	24,350

87. A magazine wants to know the relationship between people's voting behavior and their level of income. Questionnaires were sent to 200 voters, and the responses are summarized here. Do the data support the hypothesis that income and voting behavior are related? Use a 5 percent level of significance.

	Income		
	High	**Medium**	**Low**
Incumbent	35	22	10
Challenger	25	25	40
Did not vote	10	23	10

Use the following information to answer questions 88–91. A questionnaire was sent to 200 students on the campus, asking them to indicate their ethnic background and give their opinion about race relations on campus. The responses are summarized in the following table.

Ethnic Background	Opinion on Race Relations		
	Good	**So-So**	**Bad**
White	40	80	40
African–American	6	8	6
Asian–American	6	10	4

88. Are African–Americans' opinions equally split among the 3 categories? Do a 5 percent test.

89. The make-up of the student body is 75 percent white, 15 percent African–American, and 10 percent Asian–American. If the samples are randomly selected, the samples' ethnic distribution should be similar to the ethnic distribution on the campus as a whole. Do the data support that hypothesis? Do a 5 percent test.

90. Are ethnic background and opinion on this issue related? Do a 5 percent test.

91. Do the two minority groups have a similar opinion pattern? Do a 5 percent test. Assume we know that African–Americans' pattern is exactly 30 percent for good, 40 percent for so-so, and 30 percent for bad.

92. Two hundred and ten people were asked which TV news programs they usually watch. The answers are compiled in the following table. Can you say that the 3 networks have audiences of about the same size? Do a 5 percent test.

Network A	Network B	Network C
80	70	60

93. An election was held in a big city whose population is 50 percent white, 40 percent black, and 10 percent Hispanic. Among the elected, 40 council members are white, 30 are black, and 10 are Hispanic. Do we have enough evidence to say that the 3 ethnic groups are represented on the council in proportion to their representation in the population?

94. Do workbooks make a difference in students' performance? A statistics instructor uses her class as a sample. Do the results suggest that the grade patterns of those who own a workbook and of those who do not are different? Do a 5 percent test.

	Grade				
	A	**B**	**C**	**D**	**F**
Own a workbook	5	4	6	2	1
Don't own a workbook	2	6	4	3	2

95. The president of a local bank suspects that his employees care only about the big customers. He randomly sampled 325 loans made during the last year and asked the borrowers their opinion of the service they received. On the basis of the results, do you think loan size and service received are independent? Do a 5 percent test.

Service	Loan Size		
	Small	**Mid-Size**	**Large**
Satisfied	10	20	40
Acceptable	20	45	30
Dissatisfied	33	33	24

96. A gambler wants to know whether the dice used in a casino are fair. If the dice are fair, the probabilities of seeing 1, 2, ... , 6 are all $\frac{1}{6}$. The gambler recorded the outcomes of 600 rolls of the dice. Here are his results:

1	2	3	4	5	6
98	93	107	105	97	100

Do the data support the hypothesis that the dice are fair? Do a 5 percent test.

97. A magazine wants to study the relationship between people's education and the medium they are exposed to the most. Questionnaires were sent to 100 people of different educational backgrounds. The results are summarized here. Are educational background and medium used the most related? Do a 5 percent test.

	TV	Radio	Newspaper
Elementary school	20	15	15
High school	15	12	3
College	10	8	2

98. On November 18, 1980, the *Wall Street Journal* published a Gallup survey of the opinions of 782 chief executives of U.S. corporations. The 782 chief executives represent samples of 282 from large firms, 300 from medium-sized firms and 200 from small firms. Frank Allen, a staff reporter for the *Wall Street Journal,* used a questionnaire to ask "How many people in your company are capable of doing your job as chief executive?" The results are presented in the table.

A 6 × 3 Contingency Table for 782 Chief Executives' Responses

Suitable Successors	Large Firms	Medium Firms	Small Firms
1	6%	10%	22%
2	14	27	30
3	24	26	18
4 or 5	30	21	8
6 or more	22	11	4
Don't know	4	5	18

Source: Wall Street Journal, November 18, 1980. Reprinted by permission of the Wall Street Journal, © 1980 Dow Jones & Company, Inc. All Rights Reserved Worldwide.

Use the chi-square statistic to test whether "number of people capable of doing your job" is independent of "size of firm" at $\alpha = .05$.

PART IV

Regression and Correlation: Relating Two or More Variables

Part III of this book deals with statistical inference based on samples. This part continues the discussion of inferential statistics but focuses on the relationship between two or more variables, using regression and correlation analyses. Regression analysis is one of the analytical tools most frequently used in many areas of business and economics. Chapters 13 and 14 focus on simple regression and correlation analysis. Chapter 15 discusses regression analysis, and Chapter 16 explores the subject further.

Part IV includes applications and examples in accounting, economics, finance, marketing, and other areas of business.

CHAPTER 13

Simple Linear Regression and the Correlation Coefficient

CHAPTER OUTLINE

Key Terms

simple regression analysis
multiple regression analysis
dependent variable
response variable
independent variable
explanatory variable
linear model
regression model
intercept
slope
regression coefficient
scatter diagram
free-hand drawing method
method of least squares
standard deviation of error term

normal equations
sum of squared deviations
autocorrelated residuals
best linear unbiased estimator (BLUE)
standard error of residuals
coefficient of determination
total variation
explained variation
unexplained variation
sample standard deviation of error term
degrees of freedom
correlation analysis
bivariate normal distribution

13.1 INTRODUCTION

In Section 6.9, we used correlation to provide a measure of the strength of any linear relationship between a pair of random variables X and Y. The random variables are treated perfectly symmetrically; that is, "the correlation between X and Y" is equivalent to "the correlation between Y and X." In this chapter, we first discuss the linear relationship between a pair of variables without perfect symmetry. In other words, we assume that Y is a dependent variable and X an independent variable: Y depends on X. Then we discuss the bivariate normal relationship and concepts related to the correlation coefficient.

Regression analysis is perhaps the statistical technique used most frequently to analyze the relationship between two or more variables in business and economics. This technique deals with the way one variable tends to change as one or more other variables change. In this chapter and the next, we will consider a regression relationship in which Y depends on only one variable X. Examples of this relationship include how sales (Y) vary with advertising expenditures (X), how quantity demanded (Y) varies with prices (X), and the relationship between corporate profit (Y) and R&D spending (X). Because all these cases deal with the relationship between two variables only, we call this kind of relationship a **simple regression analysis.** In Chapter 15, we will extend regression analysis to cases where more than two variables come into play, such as the relationship among sales, price, advertising expenditures, and perhaps even growth of gross national product. A regression analysis that involves more than two variables is called a **multiple regression analysis.** In Chapter 16, other important techniques and issues related to simple and multiple regression are discussed in detail.

In this chapter, we first discuss the regression model and population parameters and then distinguish the sample regression model from the population regression model. The least-squares estimation of population parameters, standard assumptions for linear regression, the standard error of estimate, and the coefficient of determination are investigated. Finally, we explore the bivariate normal distribution and correlation analysis. The relationships among simple regression, slope, and correlation coefficient are also discussed.

13.2 POPULATION PARAMETERS AND THE REGRESSION MODELS

To study the relationship between two variables, we must distinguish between the dependent variable, denoted by Y, and the independent variable, denoted by X. Here the term **dependent variable,** means that the values of an estimated variable depend on the values of another variable. The dependent variable may also be known as the **response variable.** The **independent variable,** which is also known as the **explanatory variable,** is used to explain the dependent variable.[1] The value of an explanatory variable normally offers at least a partial explanation of the behavior of the dependent variable.

For example, in economic analysis when we investigate the relationship between income and consumption of goods and services, consumption is the dependent variable (Y) and income the independent variable (X). Consumption (consumer spending) depends on, and is determined by, level of income. Let's use regression analysis to consider the relationship between height and weight in a group of children. This set of common-sense data will be used in both Chapters 13 and 14 to demonstrate how simple regression analysis can be done intuitively. In Chapter 14, business and economic applications of simple regression are also discussed.

Data Description

Suppose we have a group of children who are classified according to their height, as shown in Table 13.1. The population consists of 30 pairs of observations: (55 in., 91 lb), (55 in., 92 lb), . . . , (60 in., 117 lb). Figure 13.1 is a graph of these observations. Note that these groups are formed according to fixed heights, such as 55 in. and 56 in., and that each group, or subpopulation, has 5 pairs of observations. There are 6 subpopulations corresponding to the fixed variable heights (X). We shall say that we have a collection, or family, of subpopulations. The average value of Y in each subpopulation is called the expected value of Y for a given height X. It is written $E(Y|X)$ and is given in the last column of Table 13.1. For example, the average value of Y for a height of 60 in. is

$$E(Y|X = 60) = \frac{94 + 99 + 101 + 104 + 117}{5} = 103 \text{ lb}$$

Using a similar approach, we can calculate the subpopulation means of all the other groups. They are graphed as the straight line ABC in Figure 13.1.

This e.g. is cooked, suppose 95 is 96 instead when x=55, then $E(Y|X=x)$ is NOT linear.

Table 13.1 Population Height and Weight Data for Children

x (inches)	y (pounds)					E(Y|X)
55	91	92	93	94	95	93
56	92	94	95	97	97	95
57	92	95	96	99	103	97
58	94	97	98	100	106	99
59	95	97	100	102	111	101
60	94	99	101	104	117	103

[1]For instance, the equation $y = x + 3$ is a linear model with x as the independent variable and y as the dependent variable. The variable x is considered independent because it is predetermined. For any given value of x we can find a corresponding value of y, so the value of y is dependent on the value of x. When x is equal to 4, y is equal to 7. Strictly speaking, the word *independent* implies that the values of this variable are preassigned and that the values of the dependent variable follow, at least in part, from this preassignment.

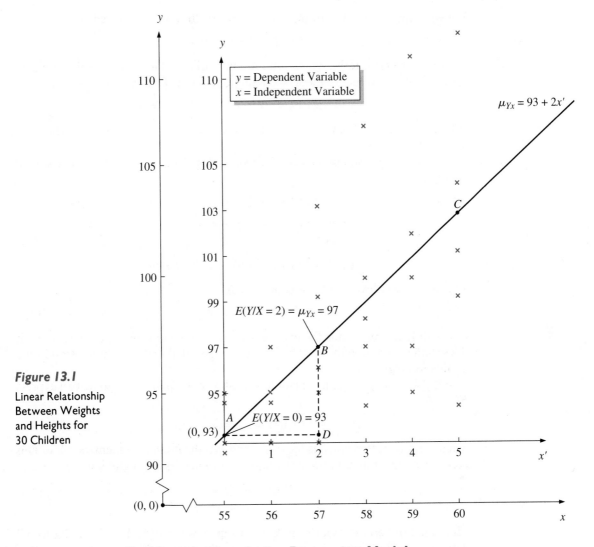

Figure 13.1

Linear Relationship
Between Weights
and Heights for
30 Children

Building the Population Regression Model

Let us now focus our attention on the subpopulation corresponding to $X = 57$ in.

$$E(Y|X = 57) = \frac{92 + 95 + 96 + 99 + 103}{5} = 97$$

The $y = 103$ lb in this subpopulation corresponding to $x = 57$ in. deviates from $E(Y|X)$ by

$$y - E(Y|X = 57) = 103 - 97 = 6 \text{ lb}$$

We will express such deviations as ϵ, so $y = 103$ lb can be expressed as

$$y = E(Y|X = 57) + \epsilon$$

where ϵ is the error term, which is a random variable. This is a general expression for individual Y values of the $X = 57$ subpopulation. That is, when $\epsilon = -5$,

$$y = E(Y|X = 57) + \epsilon = 97 - 5 = 92$$

When $\epsilon = -2$, $y = 95$; when $\epsilon = -1$, $y = 96$; when $\epsilon = 2$, $y = 99$; and when $\epsilon = 5$, $y = 102$.

The various y values in each subpopulation can be expressed in a similar manner.

$$y_1 = E(Y|X_1 = 55) + \epsilon_1 = 93 + \epsilon_1$$
$$y_2 = E(Y|X_2 = 56) + \epsilon_2 = 95 + \epsilon_2$$
$$\vdots$$
$$y_{30} = E(Y|X_{30} = 60) + \epsilon_{30} = 103 + \epsilon_{30}$$

In general, the ith value of Y is expressed as

$$y_i = E(Y_i|X = x_i) + \epsilon_i \tag{13.1}$$

where $E(Y|X_i)$ represents the expected value of those Y's for which X is equal to the specific value x_i, and ϵ_i is the error term associated with ith observation in the population regression.

$E(Y_i|X_i = x_i)$ gives us a straight line, as shown in Figure 13.1, so we can express $E(Y_i|X_i = x_i)$ as

$$E(Y_i|X_i = x_i) = \alpha + \beta x_i \tag{13.2}$$

where α is the y intercept, β is the slope, and x_i is the ith independent variable. This is called a linear function because the resulting curve is a straight line. Let

$$E(Y|X = x) = \mu_{Yx} \tag{13.3}$$
$$= \alpha + \beta x$$

This equation represents a linear relationship between $E(Y|X = X)$ and x for all data, whereas Equation 13.2 represents only the relationship for a specific pair of data. In addition, Equation 13.3 represents the conditional population mean as presented by line ABC in Figure 13.1.

Equation 13.3 represents a straight line with slope β and intercept α. Then the slope for line ABC can be interpreted as[2]

$$\beta = \frac{97 - 93}{57 - 55} = 2$$

The parameter β, the slope, measures the change in Y resulting from a change in X. It is calculated by dividing the change in Y by the change in X. One way of inter-

---

[2]From $\triangle ABD$, the slope of ABC can be defined as $\beta = BD/AD = (97 - 93)/(57 - 55) = 2$.

Table 13.2 Worksheet for Calculating μ_{Yx}

Heights, x_i (inches)	$x'_i = x_i - 55$	μ_{Yx}
55	0	93
56	1	95
57	2	97
58	3	99
59	4	101
60	5	103

preting a slope of 2 is to say that if the independent variable X is changed by 1 unit, the dependent variable changes by $+2$ units. To obtain the Y intercept, we shift the origin from $(0, 0)$ to A $(0, 93)$. In other words, we let the origin $x = 0$ for the height of 55 in.; then $\alpha = 93$. Hence the straight line used to describe ABC is $\mu_{Yx} = 93 + 2x'$. Substituting $x' = 0, 1, 2, 3, 4$, and 5 into μ_{Yx}, we obtain the results indicated in Table 13.2. By combining Equations 13.1 and 13.2, we can express an individual value of Y as

$$Y_i = \alpha + \beta x_i + \epsilon_i \tag{13.4}$$

where Y_i and x_i represent the ith value for Y and x, respectively.

Equations 13.1, 13.2, and 13.4 summarize all the data in the population and are called the **linear model** (or **regression model**). Equation 13.3 is called the *regression function;* it shows the relationship between the expected values of Y and the independent values X. The **Y intercept** α and the **slope** β are called **regression coefficients** (parameters).[3]

Using the population data listed in Table 13.2, we have our population regression line (Equation 13.2a) and our population regression model (Equation 13.2b) for describing the relationship between weights and heights:

$$\mu_{Yx} = 93 + 2x'_i \tag{13.2a}$$

$$Y_i = 93 + 2x'_i + \epsilon_i \tag{13.2b}$$

Population regression as indicated in Equation 13.2a represents conditional mean values. In Figure 13.1 the population mean value for a height of 57 in. is seen to be $\mu_{Y.57} = 97$ lb. In other words, the average weight for all children with a height of 57 in. is 97 lb. The value is calculated by substituting $x' = (57 - 55) = 2$ in. into the population regression line, as follows: $\mu_{Y.57} = 93 + 2(2) = 97$ lb.

In this section we have shown that in a simple regression analysis, two population regression parameters are to be calculated. Our assumption that α and β are known is, of course, an unrealistic one. Usually, α and β can only be estimated in terms of sample data. Read on.

---

[3]For an illustration of the meaning of the model, let x be the amount of advertising and Y be the amount of sales. Equation 13.3 tells us that, given a certain amount of advertising, the expected amount of sales is $\mu_{Yx} = \alpha + \beta x$.

Sample Versus Population Regression Model

If we have a large amount of information from a population to analyze, it may not be possible (or desirable) to obtain this specific information on each element in the population. Under these circumstances we generally use a *sample* to estimate the population parameters of the regression line in accordance with n pairs of observations $(x_1, y_1), (x_2, y_2), \ldots, (x_n, y_n)$. In the case of simple regression analysis, two population parameters, α and β, need to be estimated. Once we have estimates of α and β, we can derive an estimate of μ_{Yx} for any specified value of X. The sample regression line used to estimate α and β and to predict μ_{Yx} can be defined as

$$y = a + bx \tag{13.5}$$

where a and b are the intercept and slope to be estimated in terms of sample data.

Let us explore how this sample regression line is related to the population regression line described by Equation 13.3. The sample value of a is used to estimate α, and the sample value of b is used to estimate β. The values of a and b, together with a given value of X, yield an estimated value of Y that we can use to estimate the population value μ_{Yx} defined in Equation 13.3.

We can add the subscript i to these variables of Equation 13.5 to indicate specific values, just as we did for the population regression line. Thus if x_i is a specific value of X, the equation for estimating α and β is

$$y_i = a + bx_i + e_i \tag{13.6}$$

where a and b are intercept and slope in a sample regression with error term e_i.

Equation 13.6 yields a sample regression line that can be used to estimate parameters of the population regression line defined in Equation 13.4. As we will see in the next section, we take n pairs of sample observations to estimate α and β.

13.3 THE LEAST-SQUARES ESTIMATION OF α AND β

In this section we discuss scatter diagrams, the method of least squares, and how α and β are estimated.

Scatter Diagram

Using the hypothetical population we first met in Table 13.1, we select a random sample, for simplicity choosing one pair from each subpopulation. The sample is given in Table 13.3. Figure 13.2 is a graph of these observations that is called a **scatter diagram.** We can estimate the model without a diagram, but the scatter diagram gives us a preliminary idea of the shape of the regression function. For these six observations, we observe from the scatter diagram that the relationship is linear. In addition, the scatter diagram enables us to make rough estimates of α and β.

Table 13.3 Sample Height and Weight Data for Children

x_i (inches)	y_i (pounds)
55	92
56	95
57	99
58	97
59	102
60	104

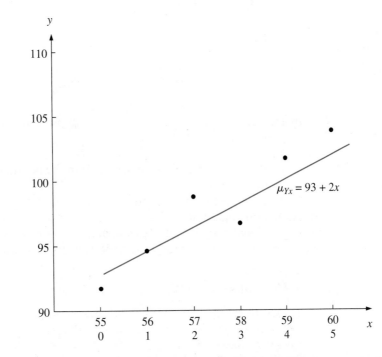

Figure 13.2

Scatter Diagram

We would like to use a line to show the relationship between x and y in Figure 13.2. A simple method of drawing a line to describe the relationship between x and y is the so-called **free-hand drawing method,** whereby we just draw a line in accordance with our best judgment about the relationship between x and y. However, the free-hand drawing method does not necessarily give systematic and objective estimates for α and β. Furthermore, the free-hand method provides no way of measuring sampling errors, which are always important in forming confidence intervals or doing tests of hypotheses on population parameters. From Equation 13.4,

$$\epsilon_i = Y_i - \alpha - \beta x_i$$

where ϵ_i represents the residual (error) term for the ith observation in population regression.

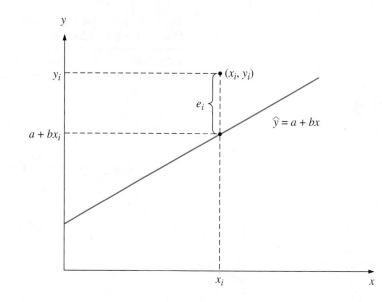

Figure 13.3

Measurement of ith Residual Term,

$e_i = y_i \; \leqq \; (a + bx_i)$

In a similar manner, we can define the sample residual (error) term as

$$e_i = y_i - a - bx_i \tag{13.7}$$

where e_i is used to measure the distance from the point (x_i, y_i) to the line, as indicated in Figure 13.3.

What we need now is a mathematical procedure for determining the sample regression line that best fits the data. The most reasonable approach is to find the values of a and b such that the estimated values of dependent variable $\hat{y}$ (in the equation $\hat{y} = a + bx$) are as close as possible to the observed values y.

The Method of Least Squares

A method of minimizing *the sum of squared deviations* is used as a criterion for finding values of a and b. The smaller e_i is, the closer $\hat{y}$ is to the actual y_i value. Put another way, the smaller e_i is, the better the fit of the regression line is.

Because a small value for e is desirable, we wish to find values for a and b that will make e as small as possible. In other words, we find the line of best fit in regression analysis by determining the values of a and b that minimize *the sum of the squared residuals*. This procedure is known as the **method of least squares.** It is accomplished as follows:

$$\text{Minimize} \sum_{i=1}^{n} e_i^2 = \sum_{i=1}^{n} (y_i - \hat{y}_i)^2 \tag{13.8}$$

The sample regression line determined by minimizing $\sum_{i=1}^{n} e_i^2$ is called the *least-squares regression line.* Because $\hat{y}_i = a + bx_i$, minimizing

$$\sum_{i=1}^{n} e_i^2 = \sum_{i=1}^{n} (y_i - \hat{y}_i)^2$$

is equivalent to minimizing

$$\sum_{i=1}^{n} e_i^2 = \sum_{i=1}^{n} (y_i - a - bx_i)^2 \qquad (13.9)$$

That is, we find a and b such that the sum of squared deviations $\sum_{i=1}^{n} e_i^2$, taken over the sample values, is at a minimum. We estimate a and b by the two **normal equations.** (The derivation of the normal equations is shown in Appendix 13A.)

$$\sum_{i=1}^{n} y_i = na + b \sum_{i=1}^{n} x_i \qquad (13.10a)$$

$$\sum_{i=1}^{n} x_i y_i = a \sum_{i=1}^{n} x_i + b \sum_{i=1}^{n} x_i^2 \qquad (13.10b)$$

Equations 13.10a and 13.10b can be regarded as a two-equation simultaneous equation system. The two unknowns are the estimates a and b (not y and x), because we must choose a and b from among an infinite possible set of values, given the sample of values of y_i and x_i.

Estimation of Intercept and Slope

To estimate the intercept, we divide Equation 13.10a by n and rearrange terms.

$$a = \bar{y} - b\bar{x} = \frac{\sum_{i=1}^{n} y_i - b\left(\sum_{i=1}^{n} x_i\right)}{n} \qquad (13.11)$$

Equation 13.11 implies that the intercept of a simple regression is the mean of y ($\bar{y}$) minus the slope (b) times the mean of x ($\bar{x}$). Here b is yet to be estimated.

To estimate the slope b, we substitute Equation 13.11 into 13.10b, and, letting $\sum_{i=1}^{n} x_i = n\bar{x}$, we obtain

$$b = \frac{\sum_{i=1}^{n} x_i y_i - n\bar{x}\bar{y}}{\sum_{i=1}^{n} x_i^2 - n\bar{x}^2} = \frac{n\sum_{i=1}^{n} x_i y_i - \left(\sum_{i=1}^{n} x_i\right)\left(\sum_{i=1}^{n} y_i\right)}{n\left(\sum_{i=1}^{n} x_i^2\right) - \left(\sum_{i=1}^{n} x_i\right)^2} \qquad (13.12)$$

Although the formulas given in Equations 13.11 and 13.12 are useful, in a practical sense it is just as easy to use Equations 13.10a and 13.10b directly. These two equations require only the solution of two equations (linear) in two unknowns.

Alternatively, we replace x_i by its deviation from $(x_i - \bar{x})$ in Equation 13.10b and obtain

$$\sum_{i=1}^{n} (x_i - \bar{x}) y_i = a \sum_{i=1}^{n} (x_i - \bar{x}) + b \sum_{i=1}^{n} (x_i - \bar{x})^2$$

Because the first term on the right-hand side of this equation is zero, the equation immediately implies that[4]

$$b = \frac{\displaystyle\sum_{i=1}^{n} (x_i - \bar{x})(y_i - \bar{y})}{\displaystyle\sum_{i=1}^{n} (x_i - \bar{x})^2} = \frac{\displaystyle\sum_{i=1}^{n} (x_i - \bar{x}) y_i}{\displaystyle\sum_{i=1}^{n} (x_i - \bar{x})^2} = \frac{s_{xy}}{s_x^2}$$

$$= \sum_{i=1}^{n} (x_i - \bar{x})(y_i - \bar{y})/n \bigg/ \sum_{i=1}^{n} (x_i - \bar{x})^2/n$$

(13.13)

where s_{xy} represents the sample covariance between x and y (as discussed in Section 6.9 of Chapter 6) and s_x^2 represents the sample variance of x.

EXAMPLE 13.1   *Relationship Between Height and Weight*

Using the sample data of Table 13.3, we will illustrate how Equations 13.11 and 13.13 can be used to estimate the least-squares regression line and its parameters.[5] Columns (1) and (2) of Table 13.4 give the hypothetical data of heights and weights for 6 children.

The sums in columns (5) and (6) of Table 13.4 give us the information we need to calculate b via Equation 13.13.

$$b = \frac{\displaystyle\sum_{i=1}^{n} (x_i - \bar{x})(y_i - \bar{y})}{\displaystyle\sum_{i=1}^{n} (x_i - \bar{x})^2}$$

$$= \frac{39.5}{17.50} = 2.2571$$

This implies that each 1-in. increase in height spells a 2.2571-lb increase in weight.

[4]The second equality of Equation 13.13 holds because

$$\sum_{i=1}^{n} (x_i - \bar{x})(y_i - \bar{y}) = \sum_{i=1}^{n} (x_i - \bar{x}) y_i + \bar{y} \sum_{i=1}^{n} (x_i - \bar{x})$$

$$= \sum_{i=1}^{n} (x_i - \bar{x}) y_i$$

[5]In general, a sample of 6 would not be sufficient. We use a small sample here for computational ease only.

Table 13.4 Procedure for Calculating a and b

(1) x_i (inches)	(2) y_i (pounds)	(3) $(x_i - \bar{x})$	(4) $(y_i - \bar{y})$	(5) $(x_i - \bar{x})(y_i - \bar{y})$	(6) $(x_i - \bar{x})^2$
55	92	−2.5	−6.1667	15.4168	6.25
56	95	−1.5	−3.1667	4.7501	2.25
57	99	−.5	.8333	−.4167	.25
58	97	.5	−1.1667	−.5834	.25
59	102	1.5	3.8333	5.7499	2.25
60	104	2.5	5.8333	14.5833	6.25
Sum 345	589	0	0	39.5	17.50
Mean 57.5	98.1667				

Using this value of b and the means of x and y shown in Table 13.4, we obtain the following value of a:

$$a = \bar{y} - b\bar{x} = 98.1667 - (2.2571)(57.5) = -31.6166$$

Hence the least-squares regression line for this example is

$$\hat{y}_i = -31.6166 + 2.2571x_i \qquad (13.14)$$

where $\hat{y}_i$ represents the estimated regression line, as indicated in Figure 13.4. Substituting $x_i = 0, 1, 2, 3, 4,$ and 5 into Equation 13.14, we obtain 6 estimated values of $\hat{y}_i(y_i's)$.

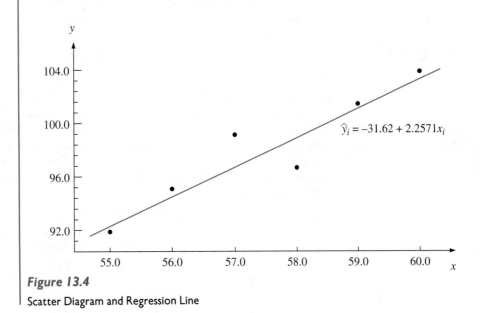

Figure 13.4

Scatter Diagram and Regression Line

$$\hat{y}_1 = -31.6166 + (2.2571)(55) = 92.5239$$

$$\hat{y}_2 = -31.6166 + (2.2571)(56) = 94.7810$$

$$\hat{y}_3 = -31.6166 + (2.2571)(57) = 97.0381$$

$$\hat{y}_4 = -31.6166 + (2.2571)(58) = 99.2952$$

$$\hat{y}_5 = -31.6166 + (2.2571)(59) = 101.5523$$

$$\hat{y}_6 = -31.6166 + (2.2571)(60) = 103.8094$$

The regression line of Figure 13.4 was determined by the method of least squares, so there is no other line that could be drawn such that the sum of squared residuals between the points and the line (measured in a vertical direction) would be smaller than this line. The residuals and the estimated values of y_i for the 6 sample points are summarized in Table 13.5.

Table 13.5 Observations of y_i, $\hat{y}_i$, and e_i

y_i	$\hat{y}_i$	$e_i = y_i - \hat{y}_i$
92	92.5239	.5239
95	94.7810	.2190
97	99.2952	−2.2952
99	97.0381	1.9619
102	101.5523	.4477
104	103.8094	.1906

EXAMPLE 13.2   *The Relationship Between the Price of Gasoline and the Price of Crude Oil*

The Organization of Petroleum Exporting Countries (OPEC) has tried to control the price of crude oil since 1973. The price of crude oil rose dramatically from the mid 1970s to the mid 1980s. As a result, motorists were confronted with a similar upward spiral of gasoline prices. The following table presents a gallon of regular leaded gasoline and a barrel of crude oil in terms of the average value at the point of production during 1975–1988 (data from the *Statistical Abstract of the United States,* 1990, pp. 483, 485).

Price of Gasoline and Crude Oil

Year, i	Gasoline y (cents/gallon)	Crude Oil x ($/barrel)
1975	57	7.67
1976	59	8.19
1977	62	8.57
1978	63	9.00
1979	86	12.64

Price of Gasoline and Crude Oil

Year, i	Gasoline y (cents/gallon)	Crude Oil x (\$/barrel)
1980	119	21.59
1981	133	31.77
1982	122	28.52
1983	116	26.19
1984	113	25.88
1985	112	24.09
1986	86	12.51
1987	90	15.40
1988	90	12.57

To investigate the relationship between the price of a gallon of gasoline and the price of a barrel of crude oil, we estimate the following regressive line

$$y_i = a + bx_i \tag{12.5}$$

where y_i and x_i represent a gallon of gasoline and a barrel of crude oil in i^{th} year respectively.

Based on the data listed in the table, we first obtain

$$\sum_{i=1}^{14} y_i = 1{,}308, \sum_{i=1}^{14} x_i = 244.59, \sum_{i=1}^{14} y_i^2 = 131{,}038,$$

$$\sum_{i=1}^{14} x_i^2 = 5{,}216.6545, \text{ and } \sum_{i=1}^{14} x_i y_i = 25{,}633.56.$$

Substituting this information into Equations 13.12 and 13.11, we obtain slope and intercept estimates as

$$b = \frac{14(25{,}633.56) - (244.59)(1{,}308)}{14(5{,}216.6545) - (244.59)^2}$$

$$= 2.9485$$

$$a = \frac{1{,}308}{14} - \frac{(2.9485)(244.59)}{14}$$

$$= 41.9161$$

Substituting estimated a and b into Equation 13.5, we obtain the estimated regression line as

$$\hat{y}_i = 41.9161 + 2.9485x_i$$

13.4   STANDARD ASSUMPTIONS FOR LINEAR REGRESSION

To obtain some desirable properties for the estimators of a regression relationship, we often make five standard assumptions for the standard population regression $Y_i = \alpha + \beta x_i + \epsilon_i$. We shall discuss first the assumptions and then their implications.

Assumption A. Either the x_i's are fixed numbers (set, for example, by the experimenter) or they are random variables that are statistically independent of the random variable ϵ_i whose values have been observed (random).

Assumption B. The random variable ϵ_i is assumed to be normally distributed.

Assumption C. The random variable ϵ_i is assumed to have a mean of zero; that is, $E(\epsilon_i) = 0$ for $i = 1, 2, \ldots, n$. This assumption implies that the mean value of Y given X, $E(Y|X)$, is $E(Y|X = x_i) = \alpha + \beta x_i$. This assumption implies that there are no omitted variables associated with the population regression specification. (The issue of specification error associated with regression analysis will be discussed in Chapter 16.)

Assumption D. The random variables ϵ_i are assumed to be statistically independent of one another, so that $E(\epsilon_i \epsilon_j) = 0$ for $i \neq j$. This assumption implies that no correlation exists among residuals. (If the residuals are correlated over time for time-series data, then we call these kinds of residuals **autocorrelated residuals.** This issue will be discussed in Chapter 16.)

Assumption E. The random variables ϵ_i all have constant variance, say σ_ϵ^2, so $E(\epsilon_i^2) = \sigma_\epsilon^2$ for $i = 1, 2, \ldots, n$. In other words, the population residual variance is constant over all values of x_i.

Now let's consider the implications of these five assumptions. Assumption A holds if x consists of a fixed number, because the covariance of a random variable and a constant is always zero. In addition, it should be noted that the constant has no variation from its fixed value. When x_i is a random variable, this assumption may be violated. If x_i cannot be measured precisely, then there exists an error-in-variable problem; x_i and ϵ_i are not independent of one another.[6]

Assumptions B through E concern the error term (ϵ_i) in the regression equation. Assumption B assumes that the difference between the y_i's and their conditional expectations ($\alpha + \beta x_i$) is normally distributed. This assumption is needed only when statistical tests of significance are conducted. Assumption C means that for a given x_i, the difference between y_i and its conditional mean ($\alpha + \beta x_i$) is sometimes positive and sometimes negative but on average is zero.

Assumption D means that the error of one point in the population cannot be related systematically to the error of any other point in the population. In other

---

[6]For instance, if in economic or business research, current instead of permanent income is used as the independent variable in estimating consumption function, then there are proxy errors associated with income measurements, as discussed in Appendix 14A. If the regression equation is part of interdependent equations, then the x_i's and ϵ_i's also are not independent of each other. However, we will take assumption A as given.

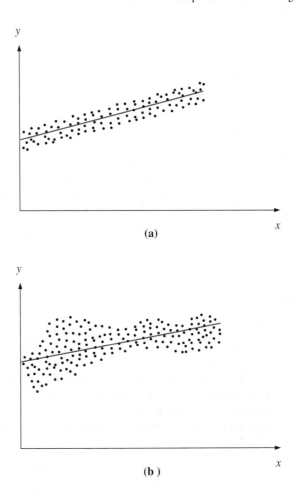

Figure 13.5

(a) Constant and
(b) Nonconstant
Residual Variance

words, knowledge about the magnitude and sign of one or more errors does not help us predict the magnitude and sign of any other error. This assumption is frequently violated in time-series analysis, which is discussed in detail in Chapter 18.

Finally, Assumption E means that the random errors all have the same variance. Figure 13.5 shows what the error terms should look like with a constant variance and with one that varies.

In summary, assumptions B through E imply that the random variable ϵ_i is normally, identically, and independently distributed with mean zero and variance σ_ϵ^2. If all the assumptions are true, then the estimators of α and β as determined by the least-squares method are **best linear unbiased estimators (BLUE)**. Essentially, BLUE means that the estimates of the parameters are best because the error variances of least-squares estimators are smaller than those of any other unbiased estimators. *Linear* means that the estimators are a linear function of the observed values of the dependent variable Y. The estimators are said to be unbiased because the expected value of each sample coefficient is equal to the population parameter.

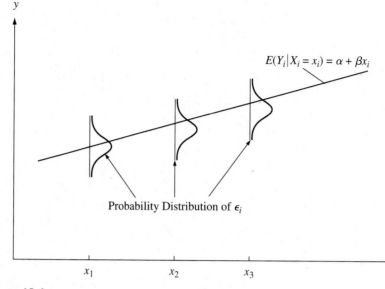

Figure 13.6

The Regression Line and the Probability Distribution of e_i

The implications of assumptions B, C, and E are apparent in Figure 13.6, which shows distributions of errors for 3 particular values of x: x_1, x_2, and x_3. Note that the relative frequency distributions of errors are normally distributed with mean zero and constant variance σ^2. The straight line shown in Figure 13.6 plots $E(Y_i \mid X_i = x_i)$ as

$$E(Y_i \mid X_i = x_i) = \alpha + \beta x_i$$

13.5   THE STANDARD ERROR OF ESTIMATE AND THE COEFFICIENT OF DETERMINATION

Two alternative measures can be used to measure the goodness of fit for a regression. The **standard error of residuals** is a measure of the absolute fit of the sample points of the sample regression line. The **coefficient of determination** is an index of the relative goodness of fit of a sample regression line. To discuss these two goodness of fit measures, we first need to present some of the components for measuring the variability of y_i in regression analysis.

Variance Decomposition

In regression analysis, two means are associated with the dependent variable y_i:

Overall mean ($\bar{y}$)
Conditional mean ($\hat{y}_i = a + bx_i$)

Based on these two different means we can break down the total deviation ($y_i - \bar{y}$) into unexplained deviation ($y_i - \hat{y}_i$) and explained deviation $\hat{y}_i - \bar{y}$ as

$$y_i - \bar{y} = (y_i - \hat{y}_i) + (\hat{y}_i - \bar{y}) \; . \tag{13.15}$$

Total   Unexplained   Explained
Deviation   Deviation   Deviation

Equation 13.15 implies that the deviation of y_i from its overall mean ($\bar{y}$) can be dissected into two components, ($y_i - \hat{y}_i$) and ($\hat{y}_1 - \bar{y}$). The deviation ($y_i - \hat{y}_i$) cannot be explained (or accounted for) by the regression line because when x_i changes, both y_i and $\hat{y}_i$ change; hence it is called the unexplained deviation. However, the deviation ($\hat{y}_1 - \bar{y}$) can be explained by the regression line because when x_i changes, $\bar{y}$ remains constant; thus it is called the explained deviation. The relationship is illustrated in Figure 13.7. By squaring each deviation and summing over all observations of Equation 13.15, it can be shown (see Appendix 13B) that

$$\sum_{i=1}^{n} (y_i - \bar{y})^2 = \sum_{i=1}^{n} (y_i - \hat{y}_i)^2 + \sum_{i=1}^{n} (\hat{y}_i - \bar{y})^2 \tag{13.16}$$

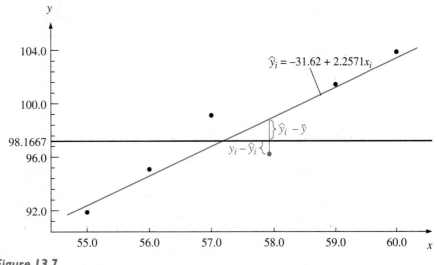

Figure 13.7

Estimated Regression Line

This equation implies that the **total variation** of the dependent variable y_i can be dissected into **unexplained variation** and **explained variation.** Alternative terms used to describe these components follow.

$$\sum_{i=1}^{n} (y_i - \bar{y})^2 = \text{total variation, sum of squares total (SST)}$$

$$\sum_{i=1}^{n} (y_i - \hat{y}_i)^2 = \text{unexplained variation, sum of squares error (SSE)}$$

$$\sum_{i=1}^{n} (\hat{y}_i - \bar{y})^2 = \text{explained variation, sum of squares due to regression (SSR)}$$

In summary, we have

$$
\begin{array}{ccccc}
\text{SST} & = & \text{SSE} & + & \text{SSR} \\
\text{Total} & & \text{Unexplained} & & \text{Explained} \\
\text{Variation} & & \text{Variation} & & \text{Variation}
\end{array}
\qquad (13.17)
$$

On the basis of Equations 13.16 and 13.17, we can define and discuss both the standard error of residuals and the coefficient of determination. We now use our height–weight example to calculate SST and SSE, as shown in Table 13.6. (Note that Table 13.6 is an ANOVA table.)

Table 13.6 Analysis of Variance

(1) Actual y_i	(2) Estimated $\hat{y}_i$	(3) $(y_i - \hat{y}_i)^2 = e_i^2$	(4) $(y_i - \bar{y})^2$
92	92.5239	.27447	38.0282
95	94.7810	.04796	10.0280
99	97.0381	3.84905	.6944
97	99.2952	5.26794	1.3612
102	101.5523	.20043	14.6942
104	103.8094	.03633	34.0274
Total		SSE = 9.67618	SST = 98.8334

Sources of Variation	Sum of Squares	Degrees of Freedom	Mean Square
Due to regression	$SSR = \sum_{i=1}^{n} (\hat{y}_i - \bar{y})^2$	k	SSR/k
Residuals	$SSE = \sum_{i=1}^{n} (y_i - \hat{y}_i)^2$	$n - k - 1$	SSE/$(n - k - 1)$
Total	$SST = \sum_{i=1}^{n} (y_i - \bar{y})^2$	$n - 1$	SST/$(n - 1)$

Source of Variation	Sum of Squares	Degrees of Freedom	Mean Square
Due to regression	89.1572	1	89.1572
Residuals	9.6762	4	2.4191
Total	98.8334	5	19.7667

If we divide both sides of Equation 13.16 by $(n - 1)$, we have

$$\frac{\sum_{i=1}^{n} (y_i - \bar{y})^2}{n - 1} = \frac{\sum_{i=1}^{n} (y_i - \hat{y})^2}{n - 1} + \frac{\sum_{i=1}^{n} (\hat{y}_i - \bar{y})^2}{n - 1} \qquad (13.18)$$

It can be shown that Equation 13.18 can be rewritten as[7]

$$\frac{\sum_{i=1}^{n} (y_i - \bar{y})^2}{n - 1} = \frac{\sum_{i=1}^{n} (y_i - \hat{y}_i)^2}{n - 1} + \frac{b^2 \sum_{i=1}^{n} (x_i - \bar{x})^2}{n - 1} \qquad (13.19)$$

$$\begin{array}{ccc} \text{Total} & = & \text{residual} & + & (\text{slope})^2 \ (\text{variance of} \\ \text{variance} & & \text{variance} & & \text{independent} \\ & & & & \text{variable}) \end{array}$$

The residual variance defined in Equation 13.19 is not an unbiased estimate of the population residual variance, which is discussed in the next section.

In Table 13.6, k represents the number of independent variables. In simple regression analysis, k is equal to 1. In the upper portion of Table 13.6, columns (1) and (2) represent the actual and estimated weights listed in Table 13.5. Column (3) represents the squared residuals, and column (4) represents the square of actual observations deviated from the overall mean $\bar{y}$. The lower portion of Table 13.6 represents the results of dissecting the variation, a technique used to calculate the standard error of residuals (estimates) and the coefficient of determination.

Standard Error of Residuals (Estimate)

The first measure of goodness of fit in regression analysis is called the **sample standard deviation of error term** (s_e).

$$s_e = \sqrt{\frac{\text{SSE}}{n - 2}} \qquad (13.20)$$

[7]Because

$$\sum_{i=1}^{n} (\hat{y}_i - \bar{y})^2 = \sum_{i=1}^{n} [a + bx_i - (a + b\bar{x})]^2 = b^2 \sum_{i=1}^{n} (x_i - \bar{x})^2$$

where SSE $= \Sigma_{i=1}^{n} (y_i - \hat{y}_i)^2$. Here s_e is a sample statistic about the goodness of fit of the sample regression line, and s_e^2 represents an unbiased estimate of the variance of the error terms (σ_e^2) about the population regression line. From Chapter 9 we know that an unbiased sample variance is calculated by dividing the sum of squared deviations by the degrees of freedom, $n - 2$.

Note that the number of elements that can be chosen freely is called the **degrees of freedom.** In this case there are two sample statistics (a and b) that we must calculate before we can compute the value of $\hat{y}$ (because $\hat{y} = a + bx$). Therefore, only ($n - 2$) observations are "free" to vary if a and b are held constant.

From Table 13.6, s_e for our familiar example involving student height and weight is calculated as follows:

$$s_e = \sqrt{\frac{\text{SSE}}{n - 2}} = \sqrt{\frac{9.6762}{4}} = 1.5553$$

The value of s_e can be used to describe the distribution of $\hat{y}_i$ in a manner similar to that which we used in the standard deviation of y (that is, s_y) to describe the distribution of y. In addition, s_e can be used to describe the distributions of a and b. All these concepts and their applications will be discussed in the next chapter.

The Coefficient of Determination

Alternatively, we can use either Equation 13.16 or Equation 13.17 to calculate a relative measure of goodness of fit. If we divide both sides of Equation 13.17 by SST, we obtain

$$\frac{\text{SST}}{\text{SST}} = 1.0 = \frac{\text{SSE}}{\text{SST}} + \frac{\text{SSR}}{\text{SST}}$$

Using this equation, we can derive the coefficient of determination (R^2) as

$$R^2 = \frac{\text{SSR}}{\text{SST}} = 1 - \frac{\text{SSE}}{\text{SST}} \tag{13.21}$$

Because SSE is the unexplained variation in y, the ratio SSE/SST is the proportion of the total variation of the dependent variable y_i that *cannot* be explained by the regression relation. Similarly, the ratio SSR/SST is the proportion of the total variation that *can* be explained by the regression line. Equation 13.21 is used to explain the relationship between SSR/SST and SSE/SST. In summary, R^2 is used to measure the explanatory power of the independent variable x. In our height and weight example,

$$R^2 = \frac{\text{SSR}}{\text{SST}} = \frac{89.1572}{98.8334} = .9021$$

The R^2 indicated in Equation 13.21 does not adjust for the degrees of freedom. We have already seen (Table 13.6) that in order to obtain the unbiased s_y^2 and s_e^2, we must divide SST and SSR by the degrees of freedom ($n - 1$) and ($n - k - 1$), respectively. Using these concepts, we can define the *adjusted* coefficient of determination $\overline{R}^2$ as

$$\overline{R}^2 = 1 - \frac{\text{SSE}/(n - k - 1)}{\text{SST}/(n - 1)} \tag{13.22}$$

For our example, $n = 6$, $k = 1$, SST = 98.8334, and SSE = 89.1573. The adjusted coefficient of determination is

$$\overline{R}^2 = 1 - \frac{9.6792/4}{98.8334/5} = 1 - \frac{2.4191}{19.7667}$$

$$= .8776$$

The magnitude of $\overline{R}^2$ is always less than the magnitude of R^2, because $\overline{R}^2$ has been adjusted for the degrees of freedom.

A MINITAB solution using the height and weight data given in Table 13.4 is shown in Figure 13.8. This output contains nearly all the calculations performed so far. In particular,

$b = 2.2571$ $\text{SSE}/(n - k - 1) = 2.419$

$a = -31.62$ $s_e = 1.555$

$\text{SSR} = 89.157$ $R^2 = .902$

$\text{SSE} = 9.676$ $\overline{R}^2 = .878$

$\text{SSR}/k = 89.157$

Figure 13.8

MINITAB Output
of Table 13.4

```
MTB > READ C1 C2
DATA>    92 55
DATA>    95 56
DATA>    99 57
DATA>    97 58
DATA>    102 59
DATA>    104 60
DATA>    END
      6 ROWS READ
MTB > REGRESS C1 1 C2;
SUBC> DW.

The regression equation is
C1 = - 31.6 + 2.26 C2

Predictor      Coef      Stdev    t-ratio        p
Constant     -31.62      21.39      -1.48    0.213
C2            2.2571     0.3718       6.07    0.004

s = 1.555      R-sq = 90.2%      R-sq(adj) = 87.8%

Analysis of Variance

SOURCE        DF         SS         MS         F        p
Regression     1     89.157     89.157     36.86    0.004
Error          4      9.676      2.419
Total          5     98.833


Durbin-Watson statistic = 3.03

MTB > PAPER
```

13.6 THE BIVARIATE NORMAL DISTRIBUTION AND CORRELATION ANALYSIS

In **correlation analysis,** we assume a population where both X and Y vary jointly. Correlation analysis doesn't imply causality as regression analysis does.[8] If both X and Y are normally distributed, then we shall call this joint distribution a **bivariate normal distribution.**[9] In Chapter 6 we discussed the relationship between two variables in terms of covariance—for example, $\text{Cov}(X, Y)$. Now we will explore correlation analysis.

Both the covariance and the correlation coefficient are designed to measure the degree of a linear relationship between a pair of variables. The covariance is an absolute measure and the correlation coefficient a relative measure in determining the relationship between two variables. The population relationship between two variables can be defined as

$$\sigma_{XY} = \text{Cov}(X, Y) = E[(X - \mu_X)(Y - \mu_Y)] \text{ and} \qquad (13.23a)$$

$$\rho = \sigma_{XY}/\sigma_X \sigma_Y \qquad (13.23b)$$

where μ_X and μ_Y are population means of X and Y, respectively, and σ_X and σ_Y are population standard deviations of X and Y, respectively. $\text{Cov}(X, Y)$ was discussed in Chapter 6. Equation 13.23b represents the population correlation coefficient ρ, which is standardized by dividing $\text{Cov}(X, Y)$ by the product of the population standard deviation of X (that is, σ_X) and the population standard deviation of Y (that is, σ_Y).

Three values of the correlation coefficient ρ that can serve as benchmarks for interpreting a correlation coefficient are $\rho = 1$, $\rho = -1$, and $\rho = 0$. $\rho = 1$ means that two variables X and Y exist in a perfect positive linear relationship; $\rho = -1$ means that two variables X and Y exist in a perfect negative linear relationship; and $\rho = 0$ means that X and Y are not linearly related—that is, they are independent random variables. The association between two variables increases as the magnitude of the correlation coefficient approaches 1. If the absolute value of ρ is less than 1, then the larger (in absolute value) the correlation, the stronger the linear association between two random variables.

The Sample Correlation Coefficient

Sample data of random variables X and Y are used to estimate the population correlation coefficient ρ. The sample statistic associated with ρ is the sample correlation coefficient; it is denoted by the letter r.

$$r = \frac{S_{xy}}{S_x S_y} \qquad (13.24)$$

[8]Strictly speaking, regression implies causality only under some "prediction" cases.

[9]The bivariate normal density function will be discussed in Appendix 13C.

where

$$s_{xy} = \frac{1}{n-1} \sum_{i=1}^{n} (x_i - \bar{x})(y_i - \bar{y}) = \text{sample covariance}$$

$$s_x = \left[\frac{1}{n-1} \sum_{i=1}^{n} (x_i - \bar{x})^2 \right]^{1/2} = \text{sample standard deviation of } x$$

$$s_y = \left[\frac{1}{n-1} \sum_{i=1}^{n} (y_i - \bar{y})^2 \right]^{1/2} = \text{sample standard deviation of } y$$

To illustrate the procedure for calculating the sample correlation coefficient, we consider once again the data for our standard height and weight example. Following Tables 13.4 and 13.6, we obtain

$$s_{xy} = (39.5/5) = 7.9$$

$$s_x = \left[\frac{(17.5)}{5} \right]^{1/2} = 1.8708$$

$$s_y = \left[\frac{(98.8334)}{5} \right]^{1/2} = 4.4460$$

Substituting these numbers into Equation 13.24 yields

$$r = \frac{7.9}{(1.8708)(4.4460)} = .9498$$

The Relationship Between r and b

We can explore the relationship between the value of r and the value of the slope b by comparing Equations 13.13 and 13.24, which are reproduced here.

$$b = s_{xy}/s_x^2 \qquad (13.13)$$

$$r = s_{xy}/s_x s_y \qquad (13.24)$$

Equation 13.24 can be rewritten as

$$r = [(s_{xy})/s_x^2][(s_x)/s_y] = b(s_x/s_y)$$

Because both s_x and s_y are always positive, the sign of r is identical to the sign of b. In other words, a positive correlation must correspond to a regression line with positive slope, and a negative r must correspond to a negative slope.

The magnitude of r is determined by the magnitudes of both b and s_x/s_y. In other words, $b = 1$ does not necessarily imply that $r = 1$, unless $s_x/s_y = 1$. Similarly, $r = 1$ does not necessarily imply that $b = 1$, unless $s_y/s_x = 1$.

The Relationship Between r and R²

The correlation coefficient (r) is used to measure the relationship between x and y, and the coefficient of determination is used to measure the percentage of the variation of y that is attributable to the variation of x. Hence it is useful to investigate the relationship between the correlation coefficient r and the coefficient of determination R^2. Squaring both sides of Equation 13.24, we have

$$\text{Coefficient of determination } R^2 = \frac{\text{SSR}}{\text{SST}} \tag{13.21}$$

where

$$\text{SST} = \sum_{n=1}^{n} (y_i - \bar{y})^2$$

$$\text{SSR} = \sum_{i=1}^{n} (\hat{y}_i - \bar{y})^2$$

It has been shown (footnote 7) that

$$\text{SSR} = b^2 \sum_{i=1}^{n} (x_i - \bar{x})^2 \tag{13.25}$$

Substituting Equation 13.25 and the definition of SST into Equation 13.21, we obtain

$$R^2 = b^2 \sum_{i=1}^{n} (x_i - \bar{x})^2 \bigg/ \sum_{i=1}^{n} (y_i - \bar{y})^2$$

$$= b^2 s_x^2 / s_y^2 \tag{13.26}$$

$$= r^2$$

From Equation 13.26 we can conclude that $R^2 = r^2$. In our example, $R^2 = .9021$ and $r^2 = (.9498)^2 = .9021$.

From Equation 13.25 and the definitions of SST, SSE, and s_e, we can redefine Equation 13.20 as

$$s_e = \sqrt{\text{SSE}/(n - 2)} \tag{13.27}$$

where

$$\text{SSE} = \sum_{i=1}^{n} (y_i - \bar{y})^2 - b^2 \sum_{i=1}^{n} (x_i - \bar{x})^2$$

$$= \sum_{i=1}^{n} y_i^2 - n\left(\sum_{i=1}^{n} y_i/n\right)^2 - b^2\left[\sum_{i=1}^{n} x_i^2 - n\left(\sum_{i=1}^{n} x_i/n\right)^2\right]$$

EXAMPLE 13.3   *The Effect of R&D Spending on a Company's Value*

Wallin and Gilman (1986) use a simple linear regression analysis to investigate the effect of research and development (R&D) spending on a company's value.[10] Data for the 20 largest R&D spenders in terms of the 1981–1982 averages are presented

[10]C. C. Wallin and J. J. Gilman (1986). "Determining the Optimum Level for R&D Spending," *Research Management,* Vol. 14, No. 5, Sept./Oct., 19–24.

in Table 13.7. In this table y and x represent the price/earnings (P/E) ratio and R&D expenditures/sales (R/S) ratio, respectively. Figure 13.9 illustrates the MINITAB simple linear regression output in terms of the data in Table 13.7.

Figure 13.9 can be divided into three parts. First, in the data input part, C_1 and C_2 represent y and x respectively. Second, the output part includes (1) the correlation coefficient between C_1 and $C_2 = .726$, (2) a scatter diagram of plotting C_1 and C_2, and (3) regressing C_1 against C_2.

Table 13.7 P/E Ratio and R/S Ratio for Top 20 R&D Spenders (based on the 1981–1982 average)

Company	P/E Ratio, y	R/S Ratio, x
1	5.6	.003
2	7.2	.004
3	8.1	.009
4	9.9	.021
5	6.0	.023
6	8.2	.030
7	6.3	.035
8	10.0	.037
9	8.5	.044
10	13.2	.051
11	8.4	.058
12	11.1	.058
13	11.1	.067
14	13.2	.080
15	13.4	.080
16	11.5	.083
17	9.8	.091
18	16.1	.092
19	7.0	.064
20	5.9	.028

Source: C. C. Wallin and J. J. Gilman (1986). "Determining the Optimum Level for R&D Spending," *Research Management,* Vol. 14, No. 5, Sept./Oct. 1986, 19–24 (adapted from Figure 1, p. 20).

Figure 13.9
MINITAB Output of Regression $y(C_1)$ on $x(C_2)$

```
MTB > READ C1 C2
DATA> 5.6 .003
DATA> 7.2 .004
DATA> 8.1 .009
DATA> 9.9 .021
DATA> 6.0 .023
DATA> 8.2 .030
DATA> 6.3 .035
DATA> 10.0 .037
DATA> 8.5 .044
```

(*continued*)

Figure 13.9

(Continued)

```
DATA> 13.2 .051
DATA> 8.4 .058
DATA> 11.1 .058
DATA> 11.1 .067
DATA> 13.2 .080
DATA> 13.4 .080
DATA> 11.5 .083
DATA> 9.8 .091
DATA> 16.1 .092
DATA> 7.0 .064
DATA> 5.9 .028
DATA> END
     20 ROWS READ
MTB > CORRELATION C1 C2

Correlation of C1 and C2 = 0.726

MTB > PLOT C1 C2


C1      -                                              *
        -
        -
        -
  14.0+
        -                              *             2
        -
        -
        -                         *    *      *
  10.5+
        -           *       *                    *
        -
        -      *        *      *      *
        -
   7.0+  *                          *
        -           *      *
        -      *        *
        -
        +---------+---------+---------+---------+---------+------C2
      0.000     0.020     0.040     0.060     0.080     0.100


MTB > BRIEF 1
MTB > REGRESS C1 1 C2;

SUBC> DW.

The regression equation is
C1 = 5.98 + 74.1 C2

Predictor        Coef      Stdev     t-ratio         p
Constant       5.9772     0.9174        6.52     0.000
C2             74.07      16.52         4.48     0.000

s = 2.074      R-sq = 52.7%     R-sq(adj) = 50.1%

Analysis of Variance

SOURCE         DF         SS          MS          F         p
Regression      1      86.404      86.404      20.09     0.000
Error          18      77.414       4.301
Total          19     163.818


Durbin-Watson statistic = 2.58

MTB > PAPER
```

From the estimate that $r = .726$ and the pattern of the scatter diagram, we can conclude that the P/E ratio is correlated highly with the R/S ratio. The estimated regression line can be defined as

$$\bar{y} = 5.98 + 74.1x$$

In sum, the MINITAB output for sample statistics that have been discussed in this chapter are listed here.

$b = 74.07$ $a = 5.9772$ SSR $= 86.404$
SSE $= 77.414$ MSE $= 86.404$ MSR $= 4.301$
$s_e = 2.074$ $R^2 = .527$ $\bar{R}^2 = .501$.

Other sample statistic outputs in Figure 13.9 will be discussed in the next chapter.

EXAMPLE 13.4   *The Regression Relationship Between Number of Cars and Size of Household*

Say we have random samples of 10 households showing the numbers of cars per household listed in Table 13.8. From Table 13.8, we can obtain the following statistics:

$$\bar{y} = 2.4 \qquad \bar{x} = 3.6 \qquad \Sigma xy = 99 \qquad \Sigma x^2 = 150$$

$$\Sigma y^2 = 68 \qquad \sum_{i=1}^{10} (x_i - \bar{x})(y_i - \bar{y}) = 12.6$$

$$\sum_{i=1}^{10} (x_i - \bar{x})^2 = 20.4 \qquad \sum_{i=1}^{10} (y - \bar{y})^2 = 10.4$$

Table 13.8   Numbers of Cars per Household

Household	Cars, y	People, x
1	4	6
2	1	2
3	3	4
4	2	3
5	2	4
6	3	4
7	4	6
8	1	3
9	2	2
10	2	2
Total	24	36

Following Equation 13.12, we can estimate the regression slope as

$$b = \frac{\sum\limits_{i=1}^{10} (x_i - \bar{x})(y_i - \bar{y})}{\sum\limits_{i=1}^{10} (x_i - \bar{x})^2} = \frac{12.6}{20.4} = .6176 = .62$$

This implies that, on the average, the number of cars for each household increases by approximately .62 when the number of people in the household increases by 1. Equation 13.11 yields an estimate of the intercept.

$$a = \frac{24}{10} - .62\left(\frac{36}{10}\right) = .168$$

The estimated regression line is

$$\hat{y} = .168 + .62x$$

From Equations 13.26 and 13.27, we can estimate R^2 and s_e as

$$R^2 = (.6176)^2\left(\frac{20.4}{10.4}\right) = .7482$$

$$s_e = \sqrt{[10.4 - (.6176)^2(20.4)]/8}$$
$$= .5721$$

Other related statistical analysis will be done in Example 14.1.

The MINITAB output of Example 13.4 is presented in Figure 13.10, which displays most of the results we have calculated in this example. Some of the estimates of this output will be investigated in the next chapter.

Figure 13.10

MINITAB Output
for Example 13.4

```
MTB > READ C1 C2
DATA> 4 6
DATA> 1 2
DATA> 3 4
DATA> 2 3
DATA> 2 4
DATA> 3 4
DATA> 4 6
DATA> 1 3
DATA> 2 2
DATA> 2 2
DATA> END
     10 ROWS READ
MTB > REGRESS C1 1 C2;
SUBC> DW.

The regression equation is
C1 = 0.176 + 0.618 C2

Predictor      Coef       Stdev     t-ratio        p
Constant     0.1765      0.4905        0.36    0.728
C2           0.6176      0.1266        4.88    0.000
```

Figure 13.10
(Continued)

```
s = 0.5720      R-sq = 74.8%      R-sq(adj) = 71.7%

Analysis of Variance

SOURCE        DF         SS          MS        F        p
Regression     1      7.7824      7.7824     23.78    0.000
Error          8      2.6176      0.3272
Total          9     10.4000


Durbin-Watson statistic = 2.44

MTB > PAPER
```

Summary

In this chapter, we discussed the basic concepts of simple linear regression and the correlation coefficient. Both population and sample regression lines were defined. The least-squares method of estimating the intercept and slope of a regression line were also discussed. Coefficient of determination of a regression analysis was defined. In the next chapter, the ideas and analyses introduced in this chapter will be used for further analysis. And applications of simple regression in business and economic decisions will be explored.

Appendix 13A Derivation of Normal Equations

In this appendix, we derive the normal equations that are used to obtain the least-squares estimates of population regression parameters. For convenience, we denote the function to be minimized as

$$F = \sum_{i=1}^{n} e_i^2 = \sum_{i=1}^{n} (y_i - a - bx_i)^2 \tag{13A.1}$$

Because this function is to be minimized with respect to a and b, it is necessary to take the partial derivatives of F with respect to these two variables. The partial derivatives are

$$\frac{\partial F}{\partial a} = \sum_{i=1}^{n} 2(y_i - a - bx_i)(-1)$$

$$\frac{\partial F}{\partial b} = \sum_{i=1}^{n} 2(y_i - a - bx_i)(-x_i)$$

Setting these partial derivatives equal to zero yields the following two normal equations.

$$\sum_{i=1}^{n} y_i = na + b \sum_{i=1}^{n} x_i$$

$$\sum_{i=1}^{n} x_i y_i = a \sum_{i=1}^{n} x_i + b \sum_{i=1}^{n} x_i^2$$

(13A.2)

Note that setting the first partial equal to zero is identical to requiring that the sum of the residuals be zero, because the term in parentheses is the residual $e_i = (y_i - a - bx_i)$. Equations (13A.2) can be used to solve for a and b as discussed in the text.

Appendix 13B The Derivation of Equation 13.16

The left-hand side of Equation 13.16 can be written as

$$\sum_{i=1}^{n} (y_i - \bar{y})^2 = \sum_{i=1}^{n} (y_i - \hat{y}_i)^2 + 2 \sum_{i=1}^{n} (y_i - \hat{y}_i)(\hat{y}_i - \bar{y}) + \sum_{i=1}^{n} (\hat{y}_i - \bar{y})^2 \quad (13B.1)$$

In addition, we know that

$$\sum_{i=1}^{n} (y_i - \hat{y}_i)(\hat{y}_i - \bar{y}) = \sum_{i=1}^{n} [a + bx_i - (a + b\bar{x})][y_i - \hat{y}_i]$$

$$= b \sum_{i=1}^{n} (x_i - \bar{x})(y_i - \hat{y}_i)$$

$$= -b\bar{x} \sum_{i=1}^{n} (y_i - \hat{y}) + b \sum_{i=1}^{n} (y_i - \hat{y}_i)(x_i)$$

Because assumptions C and A discussed in Section 13.4 imply that

$$\sum_{i=1}^{n} (y_i - \hat{y}) = \sum_{i=1}^{n} e_i = 0 \quad \text{and} \quad \sum_{i=1}^{n} (y_i - \hat{y}_i)(x_i) = \sum_{i=1}^{n} e_i x_i = 0$$

Hence

$$\sum_{i=1}^{n} (y_i - \hat{y}_i)(\hat{y}_i - \bar{y}) = 0 + 0 = 0 \quad (13B.2)$$

Substituting Equation 13B.2 into Equation 13B.1, we obtain Equation 13.16.

Appendix 13C The Bivariate Normal Density Function

In correlation analysis, we assume a population where both X and Y vary jointly. It is called a joint distribution of two variables. If both X and Y are normally distributed, then we call this known distribution a **bivariate normal distribution.**

Following Appendix 7A, we can define the probability density function (p.d.f.) of the normally distributed random variables X and Y as

$$f(X) = \frac{1}{\sigma_X \sqrt{2\pi}} \exp\left[\frac{-(X - \mu_X)^2}{2\sigma_X^2}\right], \quad -\infty < X < \infty \qquad (13C.1)$$

$$f(Y) = \frac{1}{\sigma_Y \sqrt{2\pi}} \exp\left[\frac{-(Y - \mu_Y)}{2\sigma_Y^2}\right], \quad -\infty < Y < \infty \qquad (13C.2)$$

where μ_X and μ_Y are population means for X and Y, respectively; σ_X and σ_Y are population standard deviations of X and Y, respectively; $\pi = 3.1416$; and exp represents the exponential function.

If ρ represents the population correlation between X and Y, then the p.d.f. of the bivariate normal distribution can be defined as

$$f(X,Y) = \frac{1}{2\pi\sigma_X\sigma_Y\sqrt{1-\rho^2}} \exp(-q/2),$$
$$-\infty < X < \infty, \; -\infty < Y < \infty \qquad (13C.3)$$

where $\sigma_X > 0$, $\sigma_Y > 0$, and $-1 < \rho < 1$,

$$q = \frac{1}{1-\rho^2}\left[\left(\left(\frac{X-\mu_X}{\sigma_X}\right)^2 - 2\rho\left(\frac{X-\mu_X}{\sigma_X}\right)\left(\frac{Y-\mu_Y}{\sigma_Y}\right) + \left(\frac{Y-\mu_Y}{\sigma_Y}\right)^2\right)\right]$$

It can be shown that the conditional mean of Y, given X, is linear in x and given by

$$E(Y|X) = \mu_Y + \rho\left\{\frac{\sigma_Y}{\sigma_X}\right\}(X - \mu_X) \qquad (13C.4)$$

It is also clear that given X, we can define the conditional variance of Y as

$$\sigma^2(Y|X) = \sigma_Y^2(1 - \rho^2) \qquad (13C.5)$$

Equation 13C.4 can be regarded as describing the population linear regression line. For example, if we have a bivariate normal distribution of heights of brothers and sisters, we can see that they vary together and there is no cause-and-effect relationship. Accordingly, a linear regression in terms of the bivariate normal distribution variable is treated as though there were a two-way relationship instead of an existing causal relationship. It should be noted that regression implies a causal relationship only under a "prediction" case.

Equation (13C.3) represents a joint p.d.f. for X and Y. If $\rho = 0$, then Equation 13C.3 becomes

$$f(X,Y) = f(X)f(Y) \qquad (13C.6)$$

This implies that the joint p.d.f. of X and Y is equal to the p.d.f. of X times the p.d.f. of Y. We also know that both X and Y are normally distributed. Therefore, X is independent of Y.

EXAMPLE 13C.1   *Using a Mathematics Aptitude Test to Predict Grade in Statistics*

Let X and Y represent scores in a mathematics aptitude test and numerical grade in elementary statistics, respectively. In addition, we assume that the parameters in Equation 13C.3 are

$$\mu_X = 550 \qquad \sigma_X = 40 \qquad \mu = 80 \qquad \sigma_Y = 4 \qquad \rho = .7$$

Substituting this information into Equations 13B.4 and 13B.5, respectively, we obtain

$$E(Y|X) = 80 + .7(40/4)(X - 550)$$
$$= 41.5 + .07X \tag{13C.7}$$
$$\sigma^2(Y|X) = (16)(1 - .49) = 8.16 \tag{13C.8}$$

If we know nothing about the aptitude test score of a particular student (say, John), we have to use the distribution of Y to predict his elementary statistics grade.

$$95\% \text{ interval} = 80 \pm (1.96)(4) = 80 \pm 7.84$$

That is, we predict with 95 percent probability that John's grade will fall between 87.84 and 72.16.

Alternatively, suppose we know that John's mathematics aptitude score is 650. In this case, we can use Equations 13C.7 and 13C.8 to predict John's grade in elementary statistics.

$$E(Y|X = 700) = 41.5 + (.07)(650) = 87$$

and

$$\sigma^2(Y|X = 700) = 8.16$$

We can now base our interval on a normal probability distribution with a mean of 87 and a standard deviation of 2.86.

$$95\% \text{ interval} = 87 \pm (1.96)(2.86) = 87 \pm 5.61$$

That is, we predict with 95 percent probability that John's grade will fall between 92.61 and 81.39.

Two things have happened to this interval. First, the center has shifted upward to take into account the fact that John's mathematics aptitude score is above average. Second, the width of the interval has been narrowed from $87.84 - 72.16 = 15.68$ grade points to $92.61 - 81.39 = 11.22$ grade points. In this sense, the information about John's mathematics aptitude score has made us less uncertain about his grade in statistics. This issue is discussed in further detail in Section 14.4 in the next chapter.

Questions and Problems

1. Discuss the standard assumptions for linear regression analysis.

2. A study by the New York/New Jersey Port Authority on the effects of train ticket prices on the number of passengers produced the following results.

Ticket Price	Passengers/Hour
$6.00	500
$6.50	490
$7.00	475
$7.50	450
$8.00	400
$8.50	350

  a. Which variable should be the independent variable and which the dependent variable?
  b. Plot the data.
  c. Use the method of least squares to estimate the slope and intercept.

3. A Department of Agriculture research team has investigated the relationship between the wheat harvest and the amount of fertilizer used.

Pounds of fertilizer per acre	20	30	40	50	60
Bushels of wheat per acre	100	111	120	135	145

  a. Plot the data.
  b. Use the method of least squares to estimate the slope and intercept.
  c. Predict the number of bushels of wheat that will be grown if 35 lb of fertilizer are used.

4. As vice president in charge of marketing, Bob Seller is interested in the relationship between dollars spent on advertising and the number of widgets his company sells. He has collected the following data on advertising dollars and numbers of widgets sold:

Advertising dollars (thousands)	10	15	25	70	100
Widgets sold (thousands)	100	120	145	250	400

  a. Which variable should be the dependent variable and which the independent variable?
  b. Plot the data.
  c. Use the method of least squares to estimate the slope and intercept.
  d. Predict the sale of widgets if $175,000 is spent on advertising.

5. Financial economists are often interested in measuring the relationship between the return on an individual stock and the return on the S&P 500. This model is usually referred to as the market model. Use the MINITAB program and the following rates of return for Ford stock and the S&P 500 in the table to:
  a. Plot the data. (Hint: Follow the procedure presented in Figure 13.7.)
  b. Use the method of least squares to estimate the slope and intercept. (Hint: Follow the procedures presented in Figure 13.8.)
  c. Calculate the standard error of the estimates.
  d. Calculate the coefficient of determination.

Year	Ford	S&P 500
70	.4260	.0010
71	.2933	.1080
72	.1717	.1557
73	−.4512	−.1737
74	−.0968	−.2964
75	.3960	.3149
76	.4614	.1918
77	−.2067	−.1153
78	−.0026	.0105
79	−.1479	.1228
80	−.2938	.2586
81	−.1025	−.0994
82	1.3212	.1549
83	.1286	.1706
84	.0980	.0115
85	.3237	.2633
86	.0081	.1462
87	.3961	.0203
88	−.2995	.1240
89	−.0767	.2725
90	−.3209	−.0656

6. Explain whether you would expect a positive relationship, a negative relationship, or no relationship to exist for the following pairs of data. If you think there is a relationship, identify the dependent variable.
   a. The height of a mother and that of her son
   b. The income and age of female accountants
   c. The height and weight of a gorilla
   d. The cost of a car and the cost of insuring that car
   e. The time it takes a woman to run a marathon and the number of hours she spends training

7. Mary Jones, a professor of statistics, has collected the following sample of hours spent studying for her course and grades received on the midterm exam.

Sampled student	1	2	3	4	5	6	7	8	9
Hours of study	22	18	30	22	29	35	18	21	40
Exam grade	63	59	85	70	90	93	72	75	98

   a. Plot the data.
   b. Use the method of least squares to estimate α and β.
   c. Use the regression equation to predict the grade of a student who spent 25 hours studying.

8. An English professor has estimated the following relationship between English SAT scores (x) and score in the freshman English course (y).

   $$y = 30 + .12x \qquad R^2 = .35$$
   $$(.79) \ (.05)$$

   where standard deviations are shown in parentheses. The average SAT score for these students was 550.
   a. What is the students' average score in this course?
   b. Use the regression equation to predict the English course score for a student with an English SAT score of 400. of 500. of 600. of 700. of 800.
   c. If there is a 50-point difference in the SAT scores of two students at this school, what is the predicted difference in their course scores?

9. Elmore Truesdale, vice president in charge of strategic pricing, is trying to find the relationship between the price of widgets and the quantity of widgets sold. Mr. Truesdale has collected the following data.

Price	$12.50	12.00	11.50
Widgets sold (thousands)	125	135	140
Price	$11.00	10.50	10.00
Widgets sold (thousands)	148	170	185

   a. Plot the data.
   b. Use the method of least squares to estimate α and β.
   c. Use the regression model to predict how many widgets would be sold if the price of widgets were $9.80.

10. The following table gives data on personal consumption C and disposable income Y^d in the United States.

Year	C	Y^d
1976	1803.9	2001.0
1977	1883.8	2066.6
1978	1961.0	2167.4
1979	2004.4	2212.6
1980	2000.4	2214.3
1981	2024.2	2248.6
1982	2050.7	2261.5
1983	2145.9	2334.6
1984	2239.9	2468.4

    a. Plot the data, using C as the dependent variable.
    b. Use the method of least squares to calculate α and β.
    c. Interpret α and β.
    d. Calculate the standard error of the estimates.

11. The following table shows the annual rates of return for several assets and the rate of inflation.

Year	Common Stocks	Corporate Bonds	Treasury Bonds	Rate of Inflation
1967	24.0%	− 5.0%	− 9.2%	3.0%
1968	11.1	2.6	− .3	4.7
1969	− 8.5	− 8.1	− 5.1	6.1
1970	4.0	18.4	12.1	5.5
1971	14.3	11.0	13.2	3.4
1972	19.0	7.3	5.7	3.4
1973	− 14.7	1.1	− 1.1	8.8
1974	− 26.5	− 3.1	4.4	12.2
1975	37.2	14.6	9.2	7.0
1976	23.8	18.6	16.8	4.8
1977	− 7.2	1.7	− .7	6.8
1978	6.6	− .1	− 1.2	9.0
1979	18.4	− 4.2	− 1.2	13.3
1980	32.4	− 2.6	− 4.0	12.4
1981	− 4.9	− 1.0	1.8	8.9
1982	21.4	43.8	40.3	3.9
1983	22.5	4.7	.7	3.8
1984	6.3	16.4	15.4	4.0
1985	32.2	30.9	31.0	3.8
1986	18.6	18.5	23.4	1.1

a. Use the method of least squares to estimate the relationship between the rate of return on common stocks and the rate of inflation by using the MINITAB program.

b. Do common stocks serve as a hedge against inflation?

c. Repeat parts (a) and (b), using corporate bond returns.

d. Repeat parts (a) and (b), using Treasury bond returns.

12. Briefly explain the difference between a dependent variable and an independent variable in regression analysis.

13. What is causality? What is correlation? What is the relationship among casualty, correlation, and regression analysis?

14. Explain the difference between a population and a subpopulation.

15. What is a scatter diagram? Briefly explain the concept of a regression line in the context of a scatter diagram.

16. The "market model" equation in finance is

$$R_{j,t} = \alpha_j + \beta_j R_{m,t} + e_{j,t}$$

where

$R_{j,t}$ = return on stock j in month t
$R_{m,t}$ = return on the S&P 500 in month t
$e_{j,t}$ = error term

a. What is the independent variable?

b. What is the dependent variable?

c. What are the regression coefficients?

17. Suppose you collect data on household consumption and income in the United States and estimate the regression equation $C = 500 + .8Y$, where C = consumption and Y = income.

a. Plot the relationship between consumption and income that this equation reflects.

b. Explain the relationship between consumption and income.

18. Suppose you estimate the regression equation $y = 5 + .6x$.

a. What is the dependent variable?

b. What is the independent variable?

c. What is the intercept?

d. What is the slope?

19. What is a sample? What is a population? Briefly explain how a sample can be used to estimate population parameters.

20. Briefly explain what we mean by the method of least squares.

21. You are given the following information on the heights and weights of 5 people.

Weight (pounds)	Height (inches)
180	72
165	66
130	62
220	78
110	60

a. If you are interested in finding the relationship between height and weight, which variable should be the dependent variable?

b. Use the method of least squares to estimate the slope and intercept.

22. Use the data given in question 21 and the results you got there to plot the regression line. Calculate the estimated values of y and the error from the regression.

23. The consumption function can be estimated by regressing private consumption on GNP. Use the data given in Table 2.2 to estimate the consumption function.

24. What do we mean when we say that estimates of α and β determined by the least-squares method are BLUE?

25. Briefly explain what we mean by a direct and by an inverse relationship.

26. The accompanying scatter diagram shows the numbers of hours 5 students studied and their midterm scores.

Midterm Scores

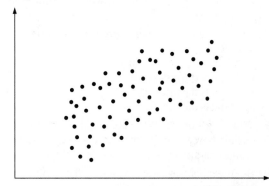

Hours Studied

a. Which is the dependent variable and which the independent variable?
b. Is there a direct or an inverse relationship between hours studied and midterm score?
c. Explain how we use regression analysis to estimate the relationship between hours studied and midterm score.

27. Suppose you are a safety consultant for the Department of Transportation. You are interested in the relationship between the number of miles a trucker drives per year and the number of accidents he or she has per year. You collect the following information from 6 truckers.

Trucker	Miles Driven	Accidents
1	90,000	3
2	119,000	4
3	87,000	2
4	135,000	6
5	150,000	5
6	92,000	3

a. Draw a scatter diagram showing the relationship between miles driven and number of accidents.
b. Is there a direct or an inverse relationship between miles driven and number of accidents?

28. Use the method of least squares to estimate the intercept and slope of the regression line for the data given in question 27.

29. Use the data from question 27 and your results from question 28 to estimate the number of accidents for each trucker. Also calculate the errors from the regression line.

30. Suppose a labor economist at the Department of Labor estimates the following relationship between years of experience and earnings of accountants.

Earnings $= 22,000 + 3,200$(years of experience)

Estimate the earnings for the following 5 accountants.

Accountant	Years of Experience
Ramon	10.2
Mary	6.5
Nguyen	3.4
Ted	5.3
Johanna	12.7

31. Now suppose the actual earnings of the 5 accountants in question 30 are as follows. Calculate the error from the regression.

Accountant	Actual Earnings
Ramon	$63,000
Mary	37,000
Nguyen	32,000
Ted	41,000
Johanna	71,000

32. In order to determine whether a company should encourage its employees to live close to work, an efficiency expert collects data on the number of latenesses per month and the number of miles an employee lives from work. The relationship is given in the accompanying scatter diagram. From this scatter diagram, describe the relationship between miles from work and number of latenesses. Will the use of a regression line be helpful in estimating this relationship?

Days Late

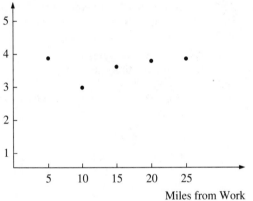

33. The manager of the Tow Time Auto Club would like to know the relationship between the age of a car and the number of service calls per year. He collects the following information.

x Car's Age (years)	y Service Calls per Year
.5	0
1	2
2.5	1
3.5	5
4.2	8
5.6	7

a. Draw a scatter diagram for these data.
b. Use the method of least squares to estimate the parameters α and β.

34. Use the results from question 33 to predict the number of service calls for a car that is 3 years old and for one that is 6 years old.

35. Explain what we mean by the goodness of fit for a regression. Give two measures we can use to assess the goodness of fit.

36. Suppose you have the following information about number of dollars spent on advertising, X, and amount of car sales, Y: $Cov(X,Y) = 500$, $Var(X) = 250$, $Var(Y) = 1,000$, mean of $X = 1,000$, and mean of $Y = 2,500$. Use this information to estimate the parameters α and β.

37. Use the information given in question 36 to find the correlation between advertising dollars and sales.

38. Briefly define total deviation, unexplained deviation, and explained deviation.

39. Use the data and results from question 21 to calculate SSE, SSR, SST, and the coefficient of determination. Interpret what the coefficient tells us.

40. Use the data and results from question 21 to calculate the standard error of the residuals.

41. Use the data and results of questions 30 and 31 to calculate SSE, SSR, SST, and the coefficient of determination.

42. Calculate the standard error of the residual using the results from question 41.

43. Look at the scatter diagram given in question 32. Would you expect a regression on that data to produce a high or a low coefficient of determination?

44. Explain the difference between the coefficient of determination and the adjusted coefficient of determination. Which do you believe provides a better measure of the goodness of fit of a regression?

45. Briefly explain what the slope of a regression line tells us.

46. Look at the following graph and identify the explained error, the unexplained error, and the total error.

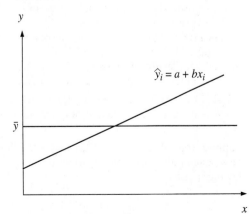

47. Use the data given in Table 2B.2 in Appendix B of Chapter 2 to estimate the relationship between GM stock's rate of return and the rate of return for the S&P 500. Calculate α, β, and the coefficient of determination.

48. Redo question 47, but this time estimate the relationship between the rates of return for Ford and the S&P 500.

49. Use the data given in Table 2B.1 to estimate the regression for DPS regressed on EPS for General Motors. Calculate the coefficient of determination.

50. Use the data given in Table 2B.1 to estimate the regression for PPS regressed on EPS for Ford. Calculate the coefficient of determination.

51. Use the data given in Table 2B.1 to calculate the correlation coefficient between Ford's and GM's EPS.

52. Suppose you are interested in finding the relationship between bond prices and interest rates. You run a regression of bond prices against the prime lending rate and find that the slope of the regression line is negative. What does this tell you about the relationship between bond prices and interest rates? Is this relationship consistent with standard financial theory?

53. What type of correlation (positive, negative, or zero) would you expect from the following pairs of variables?
   a. A company's earnings per share and its dividends per share
   b. A company's earnings per share and its price per share
   c. GM's EPS and the auto industry's average EPS
   d. Education and salary of an employee
   e. Advertising dollars spent and volume of sales
   f. Bond prices and interest rates
   g. The price charged for bread and the quantity of bread sold
   h. The hem length of dresses in France and the value of the Dow Jones Industrial Average

54. What is the relationship between the correlation coefficient r and the coefficient of determination R^2?

55. Suppose you collect data for IBM's sales and dollars spent on advertising and then compute the following statistics.

Cov(sales, advertising $) = 22
Var(sales) = 10
Var(advertising $) = 64
Mean sales = 100
Mean advertising $ = 20

   a. Compute the correlation coefficient between advertising dollars and sales.
   b. Calculate the coefficient of determination that would result from a regression of sales on advertising dollars.
   c. Calculate the regression parameters α and β.

56. Suppose you estimate a regression and compute SSE = 17.57 and SSR = 102.76. Calculate SST, R^2, and r by using this information.

57. The Department of Accounting at a university is interested in the relationship between SAT score and graduating grade point average (GPA). The accompanying table presents a summary of the data it has collected.

SAT, x	GPA y	xy
600	3.2	1920
420	2.5	1050
750	3.9	2925
650	3.6	2340
550	3.4	1870
680	3.7	2516
$\Sigma x = 3650$	$\Sigma y = 20.3$	$\Sigma xy = 12621$

x^2	y^2
360000	10.24
176400	6.25
562500	15.21
422500	12.96
302500	11.56
462400	13.69
$\Sigma x^2 = 2286300$	$\Sigma y^2 = 69.91$

   a. Draw a scatter diagram for these data.
   b. Estimate the regression parameters α and β.

58. Use the data given in question 57 to calculate r and R^2. Also use your regression results to estimate the graduating GPA for someone who scores 620 on the SAT.

59. The Department of Education at a university is interested in the relationship between the number of years of education and a person's salary. It collects the following information.

Person	Education (years)	Salary
1	8	$21,000
2	12	24,000
3	13	19,500
4	14	40,000
5	16	72,000

a. Draw a scatter diagram for these data.
b. Calculate the regression parameters α and β.

60. Use the data and your results from question 59 to estimate the earnings of someone with 15 years of education. Also compute r and R^2.

61. A market researcher is interested in who buys Fun Time Cereal. In order to analyze this problem, she collects data on the age of the consumer and how many boxes that person consumes each month.

Age	Boxes per Month
8	6
10	8
16	5
22	4
35	2
45	0

a. Compute the correlation between age and number of boxes of cereal consumed (and presumably purchased).
b. Use the method of least squares to estimate α and β.

62. Use the information given in question 61 to calculate the standard error of the residual. Briefly explain how we can use the standard error of the residual as a measure of the goodness of fit.

63. Use the data given in question 23 of Chapter 2 to compute the correlation between the dollar/pound exchange rate and the dollar/yen exchange rate.

64. Use the data given in question 24 of Chapter 2 to compute the correlation coefficient between J&J's current ratio and the industry's.

65. Use the data given in question 22 of Chapter 2 to estimate the regression coefficients for a regression of J&J's return on total asset (ROA) on the pharmaceutical industry's current ratio.

66. Use your results from question 65 to compute the standard error of the estimate.

67. Repeat questions 64–66, using J&J's inventory turnover.

68. Repeat questions 64–66, using J&J's ROA.

69. Repeat questions 64–66, using J&J's price earnings ratio.

70. You are given the following information on unemployment in the United States and in New Jersey (Data from New Jersey Economic Indicators, March 1990).

Number Unemployed (in thousands)

Year	United States	New Jersey
1970	4,093	138
1971	5,016	172
1972	4,882	182
1973	4,365	180
1974	5,156	204
1975	7,929	334
1976	7,406	346
1977	6,991	317
1978	6,202	248
1979	6,137	247
1980	7,637	260
1981	8,273	263
1982	10,678	326
1983	10,717	288
1984	8,539	236
1985	8,312	217
1986	8,237	197
1987	7,425	160
1988	6,701	151
1989	6,528	163

If you are interested in the relationship between unemployment in the United States and in New Jersey, which unemployment figure should be your independent variable? Use the MINITAB program to estimate a model showing the relationship between unemployment in the United States and in New Jersey.

71. Use the information and results from question 70 to compute the standard error of the estimate and the coefficient of determination.

72. Use the data given in question 70 to find the correlation coefficient between unemployment in the United States and in New Jersey for 1980–1989.

73. Consider the following table. Fill in the values missing from the table, using the least-squares method.

x	y	xy	x^2	y^2	$\hat{y}$	e	e^2	$(y - \bar{y})^2$
5	50							
7	35							
9	25							
11	20							
13	15							
15	10							

74. Use your results in question 73 to find the coefficient of determination and the standard error of the estimate.

75. Suppose you estimate a regression and compute SST = 217.47 and SSR = 121.73. Use this information to calculate SSE, R^2, and r.

76. Suppose you estimate a regression and compute SST = 1017.17 and SSE = 302.33. Use this information to calculate SSR, R^2, and r.

CHAPTER 14

Simple Linear Regression and Correlation: Analyses and Applications

CHAPTER OUTLINE

Key Terms

unbiased estimate
standard deviation of error terms
standard errors of estimate
prediction
forecast
conditional prediction
forecast error
confidence belt

confidence interval
prediction interval
market model
mean response
individual response
measurement errors
proxy errors

14.1   INTRODUCTION

This chapter clarifies and expands on the material presented in Chapter 13 by providing calculations, analyses, and applications to business and economics.

First, statistics used to test the significance of the intercept (a), slope (b), and simple correlation coefficient (r) are derived, and the use of these statistics is demonstrated. Second, confidence intervals for alternative prediction methods in terms of simple regression are investigated. Third, applications of simple regression in business are explored in some detail. Finally, an example is offered to show how the statistical computer programs MINITAB and SAS can be used to do simple regression analyses.

14.2   TESTS OF THE SIGNIFICANCE OF α AND β

Chapter 13 detailed basic concepts and estimation procedures for the linear regression line, regression parameters, and correlation coefficients. Now we will discuss statistical tests involving regression parameters, intercept (a), and slope (b). (The sample regression coefficients a and b are estimates of population regression coefficients α and β, just as $\bar{x}$ is the estimate of μ.) In addition, these coefficients have sampling distributions (just as $\bar{x}$ has a sampling distribution). With the usual regression assumptions, the sampling distributions of a and b have the following properties.

1. a and b are **unbiased estimates** of α and β; that is,[1] $E(a) = \alpha$ and $E(b) = \beta$. This means that the expected value of a is equal to the population parameter α and that the expected value of b is β. On average, then, the value obtained from the estimators is equal to the population parameters. Some estimates will be too low and some will be too high, but there is no systematic bias.

2. The sampling distributions of a and b are normally distributed with means α and β and variances s_a^2 and s_b^2.

$$s_a^2 = s_e^2 \frac{\sum_{i=1}^{n} x_i^2}{n \sum_{i=1}^{n} (x_i - \bar{x})^2} \tag{14.1}$$

$$s_b^2 = \frac{s_e^2}{\sum_{i=1}^{n} (x_i - \bar{x})^2} \tag{14.2}$$

[1]If there are measurement errors or proxy errors associated with the independent variable, then b is no longer an unbiased estimate for β. See Appendix 14A.

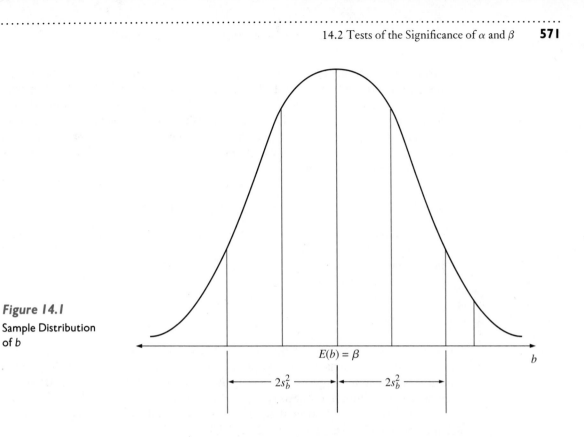

Figure 14.1

Sample Distribution of b

where s_e is the estimate of the **standard deviation of error terms** for a regression, as defined in Equation 13.20. s_a and s_b are **standard errors of estimate** for a and b. The sample distribution of b is presented in Figure 14.1.

Hypothesis Testing and Confidence Interval for β and α

The slope β is the parameter of most interest to business, economic, and other statisticians, because it can be used to measure the relationship between dependent and independent variables. The slope β measures the change in y that results from a 1-unit change in x. If β is equal to $\frac{1}{2}$, then $E(Y|X)$ changes by plus $\frac{1}{2}$ unit when x is increased by 1 unit.

For example, let the population regression function be

$$E(Y|X = x) = \mu_{Yx} = \alpha + \beta x$$

where μ_{Yx} is the population mean of Y, given x. (Once again, assume x is height and Y is weight.) Then β shows the increase in weight when there is a unit increase (a 1-inch increase) in height. As another example, let Y be consumption and x be income. Then β shows the increase in consumption when there is a unit increase in income. If β is equal to .75, then consumption is expected to go up 75 cents when income increases by 1 dollar.

In business and economic research, we need a guideline to help us determine whether the independent variable X is useful in predicting the value of Y. Suppose the population relationship is such that $\beta = 0$. This means the population regression line must be horizontal; that is, $\hat{Y} = \overline{Y}$. When $\beta = 0$, the value of X is of no help in predicting Y: no matter how much X changes, there is no change in Y (on the average). Thus determining whether $\beta = 0$ often proves beneficial, but how is such a determination made?

Two-Tailed z Test Versus Two-Tailed t Test for β

When researchers want to test whether the slope is different from zero, the alternative hypothesis (H_1) is that the slope is different from zero; it doesn't matter whether the slope is positive or negative. The null hypothesis (H_0) in such cases is that the slope is zero.

$$H_0 : \beta = 0$$

$$H_1 : \beta \neq 0$$

If b is normally distributed with $E(b) = \beta$ and variance s_b^2 (these were described earlier in this section as the two properties of b), we can graph the sampling distribution as shown in Figure 14.1. We may use either the z statistic or the t statistic to test whether the null hypothesis (H_0: $\beta = 0$) is true.

To perform this test, we will again use the sample data of heights and weights from Chapter 13, restated now in Table 14.1.

The worksheet for calculating a and b for the data of Table 14.1 is given in Table 14.2. From Table 14.2, we can obtain

$$s_x^2 = \frac{17.5}{5} = 3.5$$

$$s_y^2 = \frac{98.8334}{5} = 19.7667$$

$$s_{xy} = \frac{39.5}{5} = 7.9$$

$$r_{x,y} = \frac{7.9}{\sqrt{(3.5)(19.7667)}} = .9498$$

$$b = \frac{s_{xy}}{s_x^2} = \frac{7.9}{3.5} = 2.2571$$

Table 14.1 Weight and Height Data

Height, x_i (inches)	Weight, y_i (pounds)
55	92
56	95
57	99
58	97
59	102
60	104

Table 14.2 Worksheet for Calculation of the Coefficients a and b

y_i	x_i	$(y_i - \bar{y})$	$(x_i - \bar{x})$	x_i^2	$(y_i - \bar{y})^2$	$(x_i - \bar{x})^2$	$(x_i - \bar{x})(y_i - \bar{y})$
92	55	−6.1667	−2.5	3025	38.0282	6.25	15.4168
95	56	−3.1667	−1.5	3136	10.0280	2.25	4.7501
99	57	.8333	− .5	3249	.6944	.25	−.4167
97	58	−1.1667	.5	3364	1.3612	.25	−.5834
102	59	3.8333	1.5	3481	14.6942	2.25	5.7499
104	60	5.8333	2.5	3600	34.0274	6.25	14.5833
Sum 589	345	−.0002	0	19,855	98.8334	17.50	39.50
Mean 98.1667	57.5	—	—	—	—	—	—

The sample slope estimate (b) equals 2.2571 lb/in. To test $E(b) = \beta = 0$, we follow the hypothesis-testing technique discussed in Chapter 11. If the sample size is large, we can find the z statistic.

$$z = [b - E(b)]/S_b = (b - 0)/S_b = b/S_b \qquad (14.3)$$

From Table 13.6 and Table 14.2, we obtain $s_e^2 = 9.67618/(6 - 2) = 2.4191$ and $\sum_{i=1}^{n}(x_i - \bar{x})^2 = 17.5$. Substituting these estimations into Equation 14.2 yields

$$s_b = \sqrt{\frac{2.4191}{17.5}} = .3718$$

Then the z statistic becomes $2.2571/.3718 = 6.0707$, which shows that $b = 2.2571$ lb is 6.0707 standard deviations away from $E(b) = \beta = 0$. Under the significance level $\alpha = .05$, from Table 3 of Appendix A we have $Z_{.05} = 1.96$. Because $6.0707 > 1.96$, we conclude that it is highly improbable that $b = 2.2571$ lb/in. came from a population with $\beta = 0$, and we reject H_0. That is, we accept H_1, which is $\beta \neq 0$, and conclude that the independent variable *is* useful in predicting the dependent variable.

It should be noted that when the sample size is as small as in this example, we should use a t statistic instead of a z statistic because we are using the sample estimate of σ_e—that is, s_e—to calculate s_b. The value of s_b is a measure of the amount of sampling error in the regression coefficient b, just as s_x was a measure of the sampling error of $\bar{x}$. By using the t test, we can redefine Equation 14.3 as

$$t_{n-2} = \frac{b - 0}{s_b} \tag{14.4}$$

The t statistic, t_{n-2}, which follows a t distribution with $(n - 2)$ degrees of freedom, was discussed in Chapter 9. If the sample size is large, then the difference between the t statistic and the z statistic indicated in Equation 14.3 is small enough for us to use the z statistic in testing the null hypothesis.

The t statistic associated with β in terms of Equation 14.4 is

$$t_{n-2} = \frac{2.2571}{.3718} = 6.0707$$

Using this information, we can perform a two-tailed t test as follows:

$$H_0: \beta = 0 \quad \text{versus} \quad H_1: \beta \neq 0$$

Before we perform the null hypothesis test, however, we must specify the significance level α. We choose .01. Because a two-tailed test is used, the regression region on the right tail and left tail has an area of .01/2, or .005. When the degrees of freedom is $\upsilon = (n - 2) = 4$, then $t_{.01/2,4} = t_{.005,4} = 4.604$. The estimated t_{n-2} is 6.0707, which is larger than the absolute value of both $-t_{.005,4}$ and $t_{.005,4}$, as indicated in Figure 14.2. Hence we can conclude that the estimated slope is significantly different from zero when α is equal to 1 percent under the two-tailed test. In other words, the regression line does improve our ability to estimate the dependent variable, weight.

Figure 14.2

Two-Tailed Test
of Estimated
Slope ($\hat{b}$)

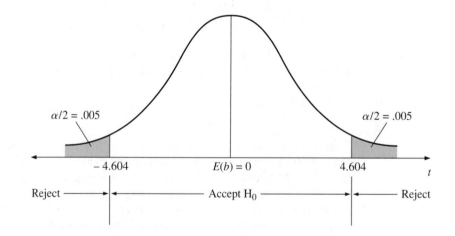

Two-Tailed t Test for α

Similarly to the null hypothesis test for β, we can define the hypotheses in a t test for α as

$$H_0: \alpha = 0 \quad \text{versus} \quad H_1: \alpha \neq 0$$

A t statistic to test whether the population intercept, α, is significantly different from zero is

$$t_{n-2} = \frac{a - 0}{s_a}, \tag{14.5}$$

where s_a is the standard deviation of the sample intercept a. Using the data of Table 14.2, we can estimate the intercept a and its standard error s_a as follows:

$$a = \bar{y} - b\bar{x} = 98.1667 - (2.2571)(57.5) = -31.6166$$

$$s_a = \sqrt{\frac{(2.4191)(19{,}855)}{(6)(17.5)}}$$
$$= 21.3879$$

To test whether the intercept, α, is equal to zero, we divide $s_a = 21.3879$ into $a = -31.6166$, obtaining

$$t = \frac{-31.6166}{21.3879} = -1.4782$$

Because -1.4782 is longer than $-2.776 = t_{.05/2,4} = t_{.025,4}$ (from Table A4 in Appendix A at the end of this book), we conclude that the estimated intercept, a, is not significantly different from zero under a two-tailed t test with $\alpha = .05$.

One-Tailed t Test for β

A researcher sometimes uses a one-tailed hypothesis test where the alternative test is that the slope is greater than zero or less than zero.

$$H_0: \beta = 0 \quad \text{versus} \quad H_1: \beta > 0 \text{ or } \beta < 0$$

For $(n - 2) = 6 - 2 = 4$ degrees of freedom, the probability that t is larger than 6.0707 falls below .005 (see Table 4 in Appendix A). Thus it is highly unlikely that $b = 2.2571$ will occur by chance when $\beta = 0$, and we can reject H_0 and accept H_1.

For the one-tailed test, the critical values are $t_{.005,4} = 4.604$ and $t_{.01,4} = 3.747$. Again, $t_4 = 6.0707$ is larger than both 3.747 and 4.604. Hence we can also conclude that the estimated slope is significantly different from zero when $\alpha = .5$ percent or $\alpha = 1$ percent under a one-tailed test. Incidentally, using the non-negative one-tailed test makes more sense, because it is not reasonable to expect an inverse relationship between height and weight.

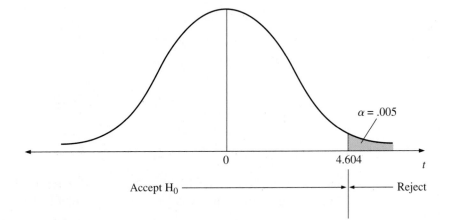

Figure 14.3

One-Tailed Test for
Regression Slope

Figure 14.3 shows the critical *t*-value for $\alpha = .05$ percent with 4 degrees of free-dom. Our estimated *t*-value is equal to 6.0707 and it is larger than 4.604, so it falls within the rejection region when $\alpha = .5$ percent.

Confidence Interval for β

On the basis of the confidence interval concepts discussed in Chapter 10, we obtain the confidence interval for b as

$$b - t(\alpha/2, n - 2)s_b \le \beta \le b + t(\alpha/2, n - 2)s_b \qquad (14.6)$$

A 95 percent confidence interval of a two-tailed test, given that $n = 6$, $t(.025, 4) = 2.776$, and $s_b = .3718$, is

$$2.2571 - (2.776)(.3718) \le \beta \le 2.2571 + (2.776)(.3718)$$
$$1.2250 \le \beta \le 3.2892$$

Thus an increase in weight of between 1.2250 and 3.2892 lb for each 1-in. increase in height can be expected. It should be noted that this result is based on only 6 obser-vations and that, all other things being equal, precision would be greater if *n* were larger.

Similarly, the 99 percent confidence interval of a two-tailed test, given $n = 6$, $t(._{005,4}) = 4.604$, and $s_b = .3718$, is

$$2.2571 - (4.604)(.3718) \le \beta \le 2.2571 + (4.604)(.3718)$$
$$.5453 \le \beta \le 3.9689$$

Figure 14.4 shows only the 99 percent confidence intervals for the population regression slope, calculated from the height and weight data in Table 14.1. Note that the 99 percent confidence interval is wider than the 95 percent confidence interval.

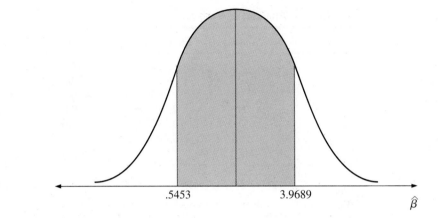

Figure 14.4

Confidence Interval for Regression Slope

.5453   3.9689

$\hat{\beta}$

The F Test Versus the t Test

Besides using the t statistic to test whether regression slope is significantly different from zero, we can also use the F statistic to test whether the regression slope is significantly different from zero. In this section we will discuss how the F statistic is used to perform the test and how the F test is related to the t test.

Procedure for Using the F Test

From Equation 13.17 and Table 13.6 in the last chapter, we have

$$\underset{\substack{\text{Total} \\ \text{Variation}}}{\text{SST}} = \underset{\substack{\text{Unexplained} \\ \text{Variation}}}{\text{SSE}} + \underset{\substack{\text{Explained} \\ \text{Variation}}}{\text{SSR}} \qquad (14.7)$$

Recall that the degrees of freedom associated with SST and SSE are $(n - 1)$ and $(n - 2)$, respectively. For convenience, we repeat Table 13.6 here as Table 14.3. From Table 14.3, we know that the degrees of freedom for SST must equal the sum of SSE and SSR; therefore, the number of degrees of freedom for SSR equals 1 (the number of independent variables). As we noted in Chapters 12 and 13, a sum of squares divided by its degrees of freedom is called a mean square. There are three different mean squares for a regression analysis, as indicated in Table 14.3.

Table 14.3 shows that an ANOVA table can be constructed for both regression analysis and analysis of variance. If the estimated slope b is not significantly different from zero, then

$$\hat{y}_i = a + bx_i$$
$$= \bar{y} + b(x_i - \bar{x}) \qquad (14.8)$$
$$= \bar{y}$$

Table 14.3 Analyses of Variance

	Actual y_i	Estimated $\hat{y}_i$	$(\hat{y}_i - y_i)^2$	$(y_i - \bar{y})^2$
	92	92.5239	.27447	38.0282
	95	94.7810	.04796	10.0280
	99	97.0381	3.84905	.6944
	97	99.2952	5.26794	1.3612
	102	101.5523	.20043	14.6942
	104	103.8094	.03633	34.0274
	Total		9.67618	98.8334

Sources of Variation	Sum of Squares	Degrees of Freedom	Mean Square
Regression	$SSR = \displaystyle\sum_{i=1}^{n} (\hat{y}_i - \bar{y})^2$	k	SSR/k
Residual	$SSE = \displaystyle\sum_{i=1}^{n} (y_i - \hat{y}_i)^2$	$n - k - 1$	$SSE/(n - k - 1)$
Total	$SST = \displaystyle\sum_{i=1}^{n} (y_i - \bar{y})^2$	$n - 1$	$SST/(n - 1)$
Regression	89.1572	1	89.1572
Residual	9.6762	4	2.4191
Total	98.8334	5	19.7667

This implies that $SSR = \sum_{i=1}^{n}(\hat{y}_i - \bar{y})^2$ is small and that SST approaches SSE. Therefore, we can use the ratio F of Equation 14.9 to test whether the estimated slope b is significantly different from zero.

$$F_{(1,n-2)} = \frac{SSR/1}{SSE/(n-2)} = \frac{MSR}{MSE} \tag{14.9}$$

where

MSR = mean square regression (due to regression)
MSE = s_e^2 = mean square error (residuals)

From Chapters 9, 12, and 13, we know that the statistic $F_{(1,n-2)}$ is an F distribution with 1 and $(n - 2)$ degrees of freedom. Using the empirical results of Table 14.3, we calculate the F-value as

$$F = \frac{89.1572/1}{9.6762/4} = 36.8563$$

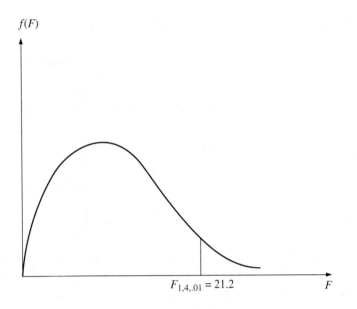

Figure 14.5

Critical Value for F Test

$f(F)$

F

$F_{1,4,.01} = 21.2$

To use this estimated F-value to test whether $b = 2.2571$ is significantly different from zero, we first choose a significance level of α and then use the F table, Table A6 of Appendix A, to determine the critical value. If we choose a significance level of $\alpha = .01$, the critical value is $F_{1,4} = 21.2$.

As shown in Figure 14.5, the decision rule is to reject H_0—that is, to accept the hypothesis that the regression line does contribute to an explanation of the variation of y—if the calculated value of F based on our height and weight example exceeds 21.2 (see Figure 14.5). Because the sample value of F, 36.8563, is larger than the critical value, 21.2, we reject the null hypothesis and conclude that the height does indeed help explain the variation of weight.

The Relationship Between the F Test and the t Test

The F test on the variation ratio between mean square regression (MSR) and mean square error (MSE), as defined in Equation 14.9, is comparable to the t test on the significance of the slope: both test whether the slope is significantly different from zero. Actually, the t test and the F test on the variance ratio between MSR and MSE are equivalent tests for the significance of the linear relationship between two variables x and y. This will be explained further in Appendix 14B.

The advantage of using the F test is that it can be generalized to test a set of regression coefficients associated with multiple regression, which is discussed in the next chapter. In addition, the F test can also be used to investigate other important topics in regression analysis (Chapter 16). However, the t test, rather than the F test, is used to test whether β differs from a specific value other than zero.

14.3   TEST OF THE SIGNIFICANCE OF ρ

So far in this chapter we have used the z statistic and the t statistic to test H_0: $\beta = 0$. There is also a t test and a z test to test H_0: $\rho = 0$. In other words, these tests are used to determine whether ρ, the population correlation coefficient between two variables, is statistically significantly different from zero.

t Test for Testing $\rho > 0$ or $\rho = 0$

If x and y are bivariate normally distributed, then we can use the t-distributed random variable t_{n-2} to test whether ρ is statistically significantly different from zero.

$$t_{n-2} = \frac{r\sqrt{n-2}}{\sqrt{1-r^2}} \tag{14.10}$$

where r is the sample correlation coefficient between x and y, and n represents the number of observations. Using our example of children's height and weight, $r = .9498$, and $n = 6$, we find that

$$t_4 = \frac{.9498\sqrt{6-2}}{\sqrt{1-(.9498)^2}} = \frac{1.8996}{.3129} = 6.0709$$

Table A4 of Appendix A gives the critical values for 4 degrees of freedom, at $\alpha = .05$ and $\alpha = .025$, as

$$t_{4,.05} = 2.132 \quad \text{and} \quad t_{4,.025} = 2.776$$

Therefore, the null hypothesis of no relationship between x and y can be rejected against the alternative that the true correlation is positive at both 5 percent and 2.5 percent significance levels. The height and weight data, then, contain fairly strong evidence supporting the hypothesis of a positive (linear) association between students' heights and weights.

The t-value for testing H_0: $\rho = 0$ is equal to that used for testing H_0: $\beta = 0$ unless there are rounding errors. That is, $t_{n-2} = b/s_b$.[2]

---

[2]From Equation 13.26, we have

$$r = b(s_x/s_y) \text{ and } r^2 = b^2(r_x^2/s_y^2)$$

Substituting these two equations into Equation 14.10 and rearranging the terms, we obtain

$$t_{n-2} \frac{b}{\sqrt{\dfrac{s_y^2 - b^2 s_x^2}{(n-2)(s_x^2)}}} = \frac{b}{\sqrt{s_e^2 / \displaystyle\sum_{i=1}^{n}(x_i - \bar{x})^2}} = \frac{b}{s_b}$$

z Test for Testing $\rho = 0$ or $\rho =$ Constant

The t-value of r indicated in Equation 14.10 cannot be used for making confidence statements about sample correlations. In addition, this approach is not suitable for testing a null hypothesis other than $\rho = 0$, such as $\rho = .20$, for example, or $\rho_1 - \rho_2 = 0$. A convenient and sufficiently accurate solution of these problems was provided by Fisher (1921), who derived a transformation from r to a quantity, h, distributed almost normally with variance[3]

$$s_h^2 = \frac{1}{n - 3} \qquad (14.11)$$

where

$$h = \frac{1}{2}\left[\log_e \left(\frac{1 + r}{1 - r} \right) \right] \qquad (14.12)$$

Note that the variance of h is independent of the value of the correlation in the population from which the sample is drawn (in contrast to the variance of r, which is dependent on the population). If $r = 0$, then $h = 0$; therefore, the null hypothesis of testing that $h = 0$ is identical to testing that $r = 0$. In our example,

$$r = .9498, h = \frac{1}{2}\log_e \frac{(1 + .9498)}{(1 - .9498)} = 1.8297$$

Using Equation 14.11, we have

$$s_h = \sqrt{\frac{1}{6 - 3}} = .5773$$

Dividing .5773 into 1.8297, we obtain

$$z = 1.8297/.5773 = 3.1694$$

We are testing the hypothesis that $\rho = 0$ against the alternative that $\rho \neq 0$. From the t-distribution table, Table A4 in Appendix A, we find the critical value is $Z_{.005} = 2.576$, which corresponds to degrees of freedom (df) $= \infty$. Approximately this value can also be taken from the normal distribution table, Table A3 of Appendix A. These results imply that the correlation coefficient .9498 is significantly different from zero at the $\alpha = 1$ percent significance level. Finally, the method employed in Equation 14.6 can be used to obtain confidence intervals for h.

A 99 percent confidence interval for h, given that $h = 1.8297$, $z = .005 = 2.576$, and $s_h = .5773$, is

$$1.8297 - (2.576)(.5773) \leq h \leq 1.8297 + (2.576)(.5773)$$

or

$$.3426 \leq h \leq 3.3168.$$

---

[3]R. A. Fisher (1921). "On the Probable Error of a Correlation Coefficient Deduced from a Small Sample," *Mentor,* Vol. 1, part 4, 3–32.

Because this interval does not include $h = 0$, it also does not include $r = 0$. Again, this implies that the correlation coefficient .9498 is significantly different from zero at $\alpha = 1$ percent.

14.4   CONFIDENCE INTERVAL FOR THE MEAN RESPONSE AND PREDICTION INTERVAL FOR THE INDIVIDUAL RESPONSE

In this section, we discuss both point estimates and confidence intervals for the mean response. We also consider point estimates and prediction intervals for the individual response.

Point Estimates of the Mean Response and the Individual Response

One of the important uses of a sample regression line is to obtain **predictions** (or **forecasts**) for the dependent variable, conditional on an assumed value of the independent variable. This kind of prediction is called a **conditional prediction** (or conditional forecast). Suppose the independent variable is equal to some specified value x_{n+1} and that the linear relationship between y_i and x_i continues to hold.[4] Then the corresponding value of the dependent variable Y_{n+1} is

$$Y_{n+1,i} = \alpha + \beta x_{n+1,i} + \epsilon_{n+1,i} \tag{14.13}$$

which, given x_{n+1}, has the conditional expectation

$$E(Y_{n+1} | x_{n+1}) = \alpha + \beta x_{n+1} \tag{14.14}$$

Equation 14.14 can be used to estimate the conditional expectation $E(Y_{n+1}/x_{n+1})$ when the independent variable is fixed at x_{n+1}; Equation 14.13 can be used to estimate the actual value for a given independent variable x_{n+1}. In other words, Equation 14.14 is used to estimate the **mean response,** and Equation 14.13 to estimate the **individual response.** For both problems we can obtain both the point estimate and the interval estimate.

To obtain the best point estimate, we should first estimate the sample regression line

$$y_i = a + b x_i + e_i \tag{14.15}$$

Then we can substitute the given value x_{n+1} into the estimated Equation 14.15 and obtain

$$\hat{y}_{n+1} = a + b x_{n+1} \tag{14.16}$$

---

[4] x_{n+1} can be a given value or forecasted value. If a regression is used to describe a time-series relationship, then x_{n+1} is a forecasted value. This issue will be discussed in detail later in this chapter.

This is the best point estimate for forecasts of both conditional expectation (mean response) and actual value (individual response). The forecast of conditional expectation value is equal to the forecast of actual expectation value. However, they are interpreted differently. This different interpretation will become important when we investigate the process of making interval estimates.

Interval Estimates of Forecasts

To construct a confidence interval for forecasts, it is necessary to know the distribution, the mean, and the variance of $\hat{y}_{n+1}$. The distribution of $\hat{y}_{n+1}$ is a t distribution with $(n-2)$ degrees of freedom. The variance associated with $\hat{y}_{n+1}$ can be classified into three cases. Let's examine them individually.

Case 14.1 Conditional Expectation (Mean Response) with $x_{n+1} = \bar{x}$

From the definitions of the intercept of a regression and the sample regression line, we have

$$\hat{y}_{n+1} = \bar{y} - b\bar{x} + bx = \bar{y} + b(x - \bar{x})$$

If $x = \bar{x}$, then we have $\hat{y}_{n+1} = \bar{y}$. Following Appendix 14C, we obtain the estimate of the variance for y_{n+1} as

$$s^2(\hat{y}_{n+1}) = s^2(\bar{y}) = s_e^2/n \tag{14.17}$$

Case 14.2 Conditional Expectation (Mean Response) with $x_{n+1} \neq \bar{x}$

In this case, the forecast value can be defined as

$$\hat{y}_{n+1} = \bar{y} + b(x_{n+1} - \bar{x})$$

Following Appendix 14C, we obtain the estimate of the variance for y_{n+1} as

$$s^2(\hat{y}_{n+1}) = s^2[\bar{y} + b(x_{n+1} - \bar{x})] \tag{14.18}$$

$$= s_e^2 \left[\frac{1}{n} + \frac{(x_{n+1} - \bar{x})^2}{\sum_{i=1}^{n}(x_i - \bar{x})^2} \right]$$

This is the variance for the estimated dependent variable ($\hat{y}_{n+1}$).

Case 14.3 Actual Value of $y_{n+1,i}$ (Individual Response)

In this case, we want to predict the actual value of y_{n+1}. The procedure for finding the variance is to find the variance of the difference $\hat{y}_{n+1} - y_{n+1,i}$—in other words, the **forecast error.** The sample variance of residual $(\hat{y}_{n+1} - y_{n+1,i})$ can be defined as

$$s^2(\hat{y}_{n+1,i}) = s^2(\hat{y}_{n+1} - y_{n+1,i}) = s_e^2 \left(\frac{1}{n} + \frac{(x_{n+1} - \bar{x})^2}{\sum\limits_{i=1}^{n}(x_i - \bar{x})^2} \right) + s_e^2$$

$$= s_e^2 \left(1 + \frac{1}{n} + \frac{(x_{n+1} - \bar{x})^2}{\sum\limits_{i=1}^{n}(x_i - \bar{x})^2} \right) \tag{14.19}$$

Using Equations 14.17, 14.18, and 14.19, we can obtain a confidence interval for the mean response and for the individual response as follows. For case 1 of the mean response, the confidence interval is

$$\hat{y}_{n+1} \pm t_{n-2,\alpha/2} \frac{s_e}{\sqrt{n}} \tag{14.20}$$

For case 2 of the mean response, the confidence interval is

$$\hat{y}_{n+1} \pm t_{n-2,\alpha/2}s_e \sqrt{\frac{1}{n} + \frac{(x_{n+1} - \bar{x})^2}{\sum\limits_{i=1}^{n}(x_i - \bar{x})^2}} \tag{14.21}$$

For case 3, the individual response of actual value $y_{n+1,i}$, the prediction interval is

$$\hat{y}_{n+1,i} \pm t_{n-2,\alpha/2}s_e \sqrt{1 + \frac{1}{n} + \frac{(x_{n+1} - \bar{x})^2}{\sum\limits_{i=1}^{n}(x_i - \bar{x})^2}} \tag{14.22}$$

To show how Equations 14.20, 14.21, and 14.22 can be applied, we will now use the height and weight example to estimate the variances and the prediction intervals.

Calculating Standard Errors

Table 14.4 is the worksheet for 3 alternative forecasts that we generate by using the data of Table 14.1. From Table 14.2,

$$\bar{x} = 57.5, \qquad \sum_{i=1}^{n}(x_i - \bar{x})^2 = 17.5$$

Table 14.4 Worksheet for Calculating 3 Alternative Standard Errors

(1) x	(2) y	(3) $\hat{y}_{n+1}$	(4) s_e^2	(5) $s(\bar{y})$	(6) $s(y_{n+1})$	(7) $s(\hat{y}_{n+1,i})$
55	92	92.5239	2.4191	.6350	1.1257	1.9200
56	95	94.7810	2.4191	.6350	.8452	1.7702
57	99	97.0381	2.4191	.6350	.6617	1.6903
58	97	99.2952	2.4191	.6350	.6617	1.6903
59	102	101.5523	2.4191	.6350	.8452	1.7702
60	104	103.8094	2.4191	.6350	1.1257	1.9200

Let $x_{n+1} = 55, 56, 57, 58, 59,$ and 60. Substituting these data into Equation 14.16, we obtain $\hat{y}_{n+1}$ as indicated in column (3) of Table 14.4. Substituting $s_e^2 = 2.4191$ and $n = 6$ into Equation 14.17, we obtain $s(\bar{y})$ as indicated in column (5) of Table 14.4.

To calculate $s^2(\hat{y}_{n+1}|x_{n+1} \neq \bar{x})$ and $s^2(\hat{y}_{n+1,i})$, we substitute related information into Equations 14.21 and 14.22 as follows:

1. $x_{n+1} = 55$

$$s^2(\hat{y}_{n+1}) = s_e^2 \left(\frac{1}{n} + \frac{(x_{n+1} - \bar{x})^2}{\sum_{i=1}^{n}(x_i - \bar{x})^2} \right)$$

$$= (2.4191)\left(\frac{1}{6} + \frac{(55 - 57.5)^2}{17.5} \right) = 1.2671$$

$$s^2(\hat{y}_{n+1,i}) = s_e^2 \left(1 + \frac{1}{n} + \frac{(x_{n+1} - \bar{x})^2}{\sum_{i=1}^{n}(x_i - \bar{x})^2} \right)$$

$$= 1.2671 + 2.4191 = 3.6862$$

2. $x_{n+1} = 56$

$$s^2(\hat{y}_{n+1}) = (2.4191)\left(\frac{1}{6} + \frac{(56 - 57.5)^2}{17.5} \right) = .7144$$

$$s^2(\hat{y}_{n+1,i}) = 2.4191 + .7144 = 3.1335$$

3. $x_{n+1} = 57$

$$s^2(\hat{y}_{n+1}) = 2.4191\left(\frac{1}{6} + \frac{(57 - 57.5)^2}{17.5} \right) = .4379$$

$$s^2(\hat{y}_{n+1,i}) = 2.4191 + .4379 = 2.8570$$

4. $x_{n+1} = 58$

$$s^2(\hat{y}_{n+1}) = (2.4191)\left(\frac{1}{6} + \frac{(58 - 57.5)^2}{17.5}\right) = .4379$$

$$s^2(\hat{y}_{n+1,i}) = 2.4191 + .4379 = 2.8570$$

5. $x_{n+1} = 59$

$$s^2(\hat{y}_{n+1}) = (2.4191)\left(\frac{1}{6} + \frac{(59 - 57.5)^2}{17.5}\right) = .7144$$

$$s^2(\hat{y}_{n+1,i}) = 2.4191 + .7144 = 3.1335$$

6. $x_{n+1} = 60$

$$s^2(\hat{y}_{n+1}) = (2.4191)\left(\frac{1}{6} + \frac{(60 - 57.5)^2}{17.5}\right) = 1.2671$$

$$s^2(\hat{y}_{n+1,i}) = 2.4191 + 1.2671 = 3.6862$$

Alternative estimates of $s(\hat{y}_{n+1})$ and $s(\hat{y}_{n+1,i})$ are listed in columns (6) and (7) of Table 14.4, respectively.

Confidence Interval for the Mean Response and Prediction Interval for the Individual Response

Let $\alpha = .05$; then $t_{.025,4} = 2.776$. Substituting all related information into Equations 14.20, 14.21, and 14.22, we can obtain 95 percent confidence interval estimates for the mean response and individual response as shown in the cases that follow.

Case 14.4 The Mean Response with $x_{n+1} = \bar{x}$

Because $x_{n+1} = \bar{x} = 57.5$ and

$$\hat{y}_{n+1} = -31.6166 + (2.2571)(57.5) = 98.1667$$

we can use $s_e/\sqrt{n} = .6350$ as indicated in column (5) of Table 14.4 to define the 95 percent confidence interval in terms of Equation 14.20 as

$$98.1667 - (2.776)(.6350) < E(Y_{n+1}|\bar{x}) < 98.1667 + (2.776)(.6350)$$

$$96.4038 < E(Y_{n+1}|\bar{x}) < 99.9295$$

How do we interpret this interval? We say that if 100 random samples of size 6 is selected, and the confidence intervals of Equation 14.20 are constructed, we should expect 95 percent of those intervals to contain $E(Y_{n+1}|\bar{x} = 57.5)$. The confidence interval calculated here is one of the 100 such intervals. This confidence interval is indicated in Figure 14.6, where A and B represent the upper bound and lower bound, respectively.

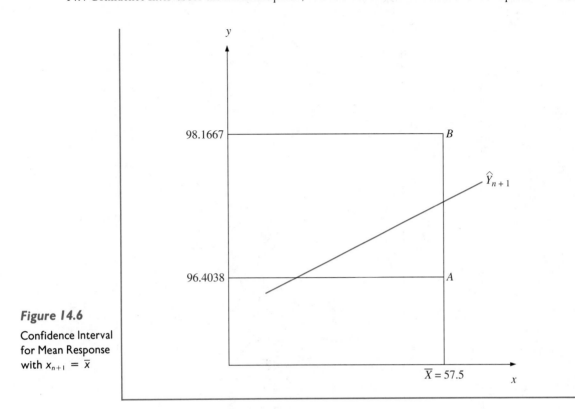

Figure 14.6

Confidence Interval for Mean Response with $x_{n+1} = \bar{x}$

Case 14.5 The Mean Response with $x_{n+1} \neq \bar{x}$

In this case, the confidence intervals depend on the values of x and $\hat{y}_{n+1}$. Using alternative standard errors as indicated in column (6) of Table 14.4, we find that the 95 percent confidence intervals for this case are

$x = 55$:
$$92.5239 - (2.776)(1.1257) < E(Y_{n+1}|x_{n+1}) < 92.5239 + (2.776)(1.1257)$$
$$89.3990 \ < E(Y_{n+1}|x_{n+1}) < 95.6488$$

$x = 56$: $\qquad 92.4347 \ < E(Y_{n+1}|x_{n+1}) < 97.1273$

$x = 57$: $\qquad 95.2012 \ < E(Y_{n+1}|x_{n+1}) < 98.8750$

$x = 58$: $\qquad 97.4583 \ < E(Y_{n+1}|x_{n+1}) < 101.1321$

$x = 59$: $\qquad 99.2060 \ < E(Y_{n+1}|x_{n+1}) < 103.8986$

$x = 60$: $\qquad 100.6845 \ < E(Y_{n+1}|x_{n+1}) < 106.9343$

How are these intervals interpreted? If 100 samples of size 6 are selected and 6 confidence intervals of Equation 14.21 corresponding to $x = 55, 56, \ldots, 60$ are constructed, we should expect 95 of them to contain $E(Y_{n+1}|x_{n+1})$ for a given x. Each

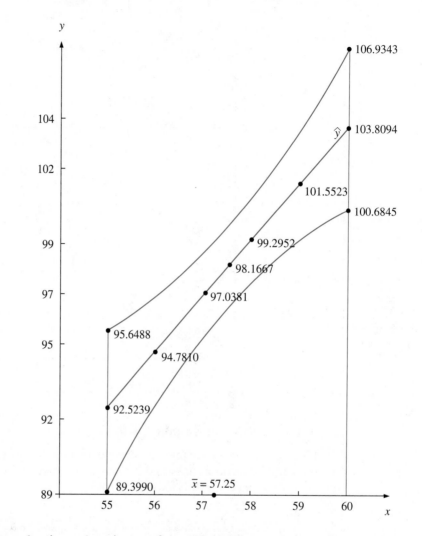

Figure 14.7

Confidence Interval for Mean Response with $x_{n+1} \neq \bar{x}$

interval estimate here is one of the 100 such intervals for a given x. These confidence intervals are graphically presented in Figure 14.7.

When we connect the points, we get a **confidence belt** that is symmetric in width around the value $\bar{x} \neq x$. Note that this confidence belt was constructed in a single sample. Each time a new sample is selected, there is a new confidence belt.

Case 14.6 The Individual Response

The standard error here is larger than that of Case 14.5 by an amount s_e^2. Using the alternative standard errors indicated in column (7) of Table 14.4, we find that the 95 percent prediction intervals for this case are

$x = 55$:
$$92.5239 - (2.776)(1.9200) < E(Y_{n+1}|x_{n+1}) < 92.5239 + (2.776)(1.9200)$$
$$89.1940 \ < E(Y_{n+1}|x_{n+1}) < 97.8538$$

$x = 56$: $\qquad\qquad\qquad 89.8669 \ < E(Y_{n+1}|x_{n+1}) < 99.6951$

$x = 57$: $\qquad\qquad\qquad 92.3458 \ < E(Y_{n+1}|x_{n+1}) < 101.7304$

$x = 58$: $\qquad\qquad\qquad 94.6029 \ < E(Y_{n+1}|x_{n+1}) < 103.9875$

$x = 59$: $\qquad\qquad\qquad 96.6382 \ < E(Y_{n+1}|x_{n+1}) < 106.4664$

$x = 60$: $\qquad\qquad\qquad 98.4795 \ < E(Y_{n+1}|x_{n+1}) < 109.1393$

The interpretations of these prediction intervals are similar to those of the confidence intervals of Case 14.5, but these intervals are wider. They are shown in Figure 14.8.

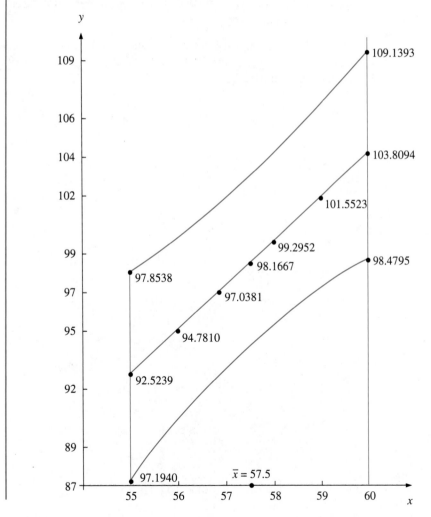

Figure 14.8

Prediction Interval for Individual Response

As in Figure 14.7, we can construct a confidence belt by connecting the points. The confidence belt of Case 14.6 is for the forecasts of the actual values of the students' weights. The confidence belt of Case 14.5 is for the forecasts of the conditional expectation of the students' weights. The confidence interval for the actual-value forecast is larger than that of the conditional expectation by 1 unit of standard error of estimate (s_e).

If x_{n+1} is equal to the previous sample mean $\bar{x}$, and if n is large, then the standard deviation of actual value, $\hat{y}_{n+1,i}$, as indicated in Equation 14.19, approaches the standard error of the estimate s_e. This result should not be too surprising, for we know that the larger the sample, and the less a given value x_{n+1} deviates from $\bar{x}$, the more faith we have in the sampling results and in the subsequent forecast.

Using MINITAB to Calculate Confidence Interval and Prediction Interval

MINITAB output of Table 14.4 is presented in Figure 14.9. In Figure 14.9, $C_1 = x$, $C_2 = y$, fit $= \hat{y}_{n+1}$, standard deviation fit $= s(\hat{y}_{n+1})$, and MS of error $= s_e^2$. As discussed previously,

$$s(\hat{y}_{n+1,i}) = \sqrt{s_e^2 + s^2(\hat{y}_{n+1})}$$

```
MTB > READ C1 C2
DATA> 55 92
DATA> 56 95
DATA> 57 99
DATA> 58 97
DATA> 59 102
DATA> 60 104
DATA> END
      6 ROWS READ
MTB > REGRESS C2 1 C1;
SUBC> DW;
SUBC> PREDICT 61.

The regression equation is
C2 = - 31.6 + 2.26 C1

Predictor      Coef      Stdev    t-ratio        p
Constant     -31.62      21.39      -1.48    0.213
C1           2.2571     0.3718       6.07    0.004

s = 1.555      R-sq = 90.2%      R-sq(adj) = 87.8%

Analysis of Variance

SOURCE        DF          SS         MS        F        p
Regression     1      89.157     89.157    36.86    0.004
Error          4       9.676      2.419
Total          5      98.833

Obs.     C1         C2        Fit  Stdev.Fit  Residual   St.Resid
  1    55.0     92.000     92.524      1.126    -0.524      -0.49
  2    56.0     95.000     94.781      0.845     0.219       0.17
  3    57.0     99.000     97.038      0.662     1.962       1.39
  4    58.0     97.000     99.295      0.662    -2.295      -1.63
  5    59.0    102.000    101.552      0.845     0.448       0.34
  6    60.0    104.000    103.810      1.126     0.190       0.18
```

Figure 14.9

MINITAB Output of Table 14.4

Figure 14.9
(Continued)

```
Durbin-Watson statistic = 3.03

    Fit  Stdev.Fit        95% C.I.           95% P.I.
106.067    1.448   (102.045,110.088)   (100.165,111.968)
```

Finally, for $x = 61$, the fit, standard deviation fit, 95 percent confidence interval (C.I.), and 95 percent prediction interval (P.I.) are presented in the last row of Figure 14.9.

EXAMPLE 14.1 *Forecasting the Average Number of Cars in a Household of 3 People*

We return to the data on cars and people per household used in Example 13.2 of Chapter 13 to forecast the average number of cars in a household of 3 people. These data are given in Table 14.5.

Table 14.5 Numbers of Cars per Household

Household	Cars	People
1	4	6
2	1	2
3	3	4
4	2	3
5	2	4
6	3	4
7	4	6
8	1	3
9	2	2
10	2	4
Total	24	38

From Table 14.5, we can obtain the following statistics:

$$\bar{y} = 2.4 \quad \bar{x} = 3.8 \quad \Sigma xy = 103 \quad \Sigma x^2 = 162 \quad \Sigma y^2 = 68$$
$$\Sigma(x - \bar{x})(y - \bar{y}) = 11.8 \quad \Sigma(x - \bar{x})^2 = 17.6 \quad \Sigma(y - \bar{y})^2 = 10.4$$

From Example 13.2, we estimate the intercept and slope as

$$b = \frac{11.8}{17.6} = .67$$
$$a = 2.4 - (.67)(3.8) = -.146$$
$$\hat{y} = -.146 + .67x$$

The standard error of slope can be calculated as follows:

$$s_e = \sqrt{\frac{\Sigma(y - \hat{y})^2}{n - 2}} = \sqrt{\frac{2.489}{8}} = .5578$$

$$\text{Standard error of } b = \frac{s_e}{\sqrt{\Sigma(x - \bar{x})^2}} = \frac{.5578}{\sqrt{17.6}} = .1329$$

Dividing .67 by .1329, we obtain the t statistic for b.

$$t = \frac{.67}{.1329} = 5.041$$

From Table A4 in Appendix A, we obtain $t_{8,.025} = 2.306$. Because 5.041 is larger than 2.036, we conclude that the estimated b is significantly different from zero at the 95 percent level of significance. A family of 3 people will have an average of 1.864 cars $[-.146 + .67(3) = 1.864]$. The 95 percent confidence interval for the average number of cars in a family of 3 people is constructed in accordance with Equation 14.21 as

$$\hat{y}_{n+1} \pm t_{n-1,\alpha/2} s(\hat{y}_{n+1})$$

where

$\hat{y}_{n+1}$ = the mean value of y at the $(n + 1)$th level of x

$$s(y_{n+1}) = s_e \sqrt{\frac{1}{n} + \frac{(x_{n+1} - \bar{x})^2}{\Sigma(x - \bar{x})^2}}$$

$$= (.5578) \sqrt{\frac{1}{10} + \frac{(3 - 3.8)^2}{17.6}}$$

$$= (.5578)(.3693) = .2060$$

The 95 percent confidence interval is

$$1.864 \pm (2.306)(.2060) = 1.864 \pm .4750 = 1.389 \text{ to } 2.339$$

On the basis of sample data, we are 95 percent confident that the average number of cars in a household of 3 people will be in the interval of 1.389 cars to 2.339 cars.

14.5   BUSINESS APPLICATIONS

In this section, we employ both data on stock rates of return and auditing data to show how simple linear regression and correlation analyses can be used in various real-world business applications.

■ APPLICATION 14.1   The Relationship Between Layoff Rate and the Unemployment Compensation Subsidy Rate

To test whether the unemployment compensation subsidy causes firms to lay off more people than they would if they knew that they would not receive an outside subsidy for layoffs, Tropel (1983) used the data of unemployment compensation

Table 14.6 Subsidy Rate and Layoff Rate for 11 Industries

Industry	Subsidy Rate (%), x	Layoff Rate, y
Apparel	57	12.54
Chemicals	32	1.78
Construction	31	7.10
Electrical machinery	29	8.38
Fabricated metals	27	11.72
Food	36	5.10
Machinery	32	4.44
Misc. manufacturing	61	9.82
Primary metals	23	7.34
Retail	27	1.98
Wholesale trade	33	1.86

Source: R. H. Tropel (1983), "On Layoffs and Unemployment Insurance." *American Economic Review,* Vol. 83, 541–559. Reprinted by permission of the publisher.

subsidy rate x (as a percentage of total revenues) and the layoff rate y (number of workers per 1,000) to do regression analysis for 11 industries, as indicated in Table 14.6.

The model of regressing y against x can be defined as

$$y_i = a + bx_i + e_i$$

where y_i and x_i are subsidy rate and layoff rate for i^{it} industries, respectively. The MINITAB output of this regression is presented in Figure 14.10. From Figure 14.10, we find that the estimated slope $b = .14468$ and the t value associated with

Figure 14.10

MINITAB Regression Output for Application 14.1

```
MTB > READ C1 C2
DATA> 57 12.54
DATA> 32 1.78
DATA> 31 7.10
DATA> 29 8.38
DATA> 27 11.72
DATA> 36 5.10
DATA> 32 4.44
DATA> 61 9.82
DATA> 23 7.34
DATA> 27 1.98
DATA> 33 1.86
DATA> END
      11 ROWS READ
MTB > REGRESS C2 1 C1;
SUBC> DW.

The regression equation is
C2 = 1.45 + 0.145 C1

Predictor       Coef       Stdev      t-ratio        p
Constant        1.448      3.471        0.42      0.686
C1              0.14468    0.09339      1.55      0.156

s = 3.625     R-sq = 21.1%     R-sq(adj) = 12.3%
```

(continued)

Figure 14.10

(Continued)

```
Analysis of Variance

SOURCE        DF          SS          MS         F       p
Regression    1        31.53       31.53      2.40    0.156
Error         9       118.24       13.14
Total        10       149.77

Durbin-Watson statistic = 1.74

MTB > PAPER
```

this slope is 1.55. From Table A4 in Appendix A, we find $t_{10,.025} = 2.228$, which is larger than 1.55. Hence we conclude that the subsidy rate does not contribute information for the prediction of the layoff rate at $\alpha = .05$.

■ | **APPLICATION 14.2** Market Model Estimation and Analysis

In Chapters 2 through 4, annual rates of return for General Motors and annual market rates of return during 1970–1990 were analyzed in detail. Now let's see how the market rates of return are used to explain the variations in rates of return for GM. The regression relationship can be defined as

$$R_{g,t} = a + bR_{m,t} + e_{g,t} \qquad (14.23)$$

where

$R_{g,t}$ = annual rate of return for GM in period t
$R_{m,t}$ = market rate of return in period t

Equation 14.23 is called the **market model** in financial analysis. It is often used to investigate the relationship between rates of return for individual securities and market rates of return. Further implications of the market model will be discussed in Chapter 21.

First we use data listed in columns (3) and (4) of Table 14.7 to estimate the market model for GM. MINITAB output of the market model for GM is presented in Figure 14.11. From Figure 14.11, we can define the estimated sample regression line as

$$\hat{R}_{g,t} = .0083 + .9800R_{m,t} \qquad (14.24)$$

The positive value for b implies that rates of return for GM increase when market rates of return increase. The $\hat{b} = .9800$ means that a 1 percent increase in market rates of return, $R_{m,t}$, in 1 year is associated with an increase in the annual rate of return for GM during the next year of about .98 percent.

To obtain the goodness-of-fit measures and other statistics, we need to calculate the components of total variation for $R_{g,t}$. The analysis of variance in Figure 14.11 reveals that the total variation in $R_{g,t}$ is

$$SST = \sum_{t=1}^{21} (R_{g,t} - \overline{R}_g)^2 = 2.21126$$

Table 14.7 Rates of Return for GM and Market Rates of Return (1970–1990)

```
MTB > PRINT C2-C4

ROW    C2        C3        C4

  1    70     0.2137    0.0010
  2    71     0.0422    0.1080
  3    72     0.0631    0.1557
  4    73    -0.3667   -0.1737
  5    74    -0.2597   -0.2964
  6    75     0.9522    0.3149
  7    76     0.4584    0.1918
  8    77    -0.1124   -0.1153
  9    78    -0.0498    0.0105
 10    79     0.0288    0.1228
 11    80    -0.0410    0.2586
 12    81    -0.0911   -0.0994
 13    82     0.6826    0.1549
 14    83     0.2373    0.1706
 15    84     0.1176    0.0115
 16    85    -0.0383    0.2633
 17    86     0.0088    0.1462
 18    87     0.0058    0.0203
 19    88     0.4418    0.1240
 20    89    -0.4581    0.2725
 21    90    -0.1154   -0.0656

MTB > DESCRIBE C3

              N      MEAN    MEDIAN    TRMEAN    STDEV    SEMEAN
C3           21    0.0819    0.0088    0.0645   0.3325    0.0726

              MIN      MAX        Q1        Q3
C3        -0.4581   0.9522   -0.1018    0.2255

MTB > DESCRIBE C4

              N      MEAN    MEDIAN    TRMEAN    STDEV    SEMEAN
C4           21    0.0751    0.1228    0.0820   0.1605    0.0350

              MIN      MAX        Q1        Q3
C4        -0.2964   0.3149   -0.0323    0.1812
```

This SST can be decomposed as

$$\text{Explained variation (SSR)} = \sum_{t=1}^{21} (\hat{R}_{g,t} - \overline{R}_{g,t})^2$$

$$= .49469$$

and

$$\text{Unexplained variation (SSE)} = \sum_{t=1}^{21} (R_{g,t} - \hat{R}_{g,t})^2 = \sum_{t=1}^{21} \hat{e}_{g,t}^2$$

$$= 1.71657$$

Drawing on our information about SST, SSR, SSE, and the related degrees of freedom, we construct the ANOVA table presented in Table 14.8.

Figure 14.11

MINITAB Output
of Market Model
for GM

```
MTB > BRIEF 3
MTB > REGRESS C3 1 C4;
SUBC> DW;
SUBC> PREDICT 0.12.

The regression equation is
GM = 0.0083 + 0.980 MARKET

Predictor      Coef      Stdev     t-ratio      p
Constant    0.00834    0.07273      0.11     0.910
MARKET      0.9800     0.4188       2.34     0.030

s = 0.3006      R-sq = 22.4%     R-sq(adj) = 18.3%

Analysis of Variance

SOURCE        DF        SS          MS        F        p
Regression     1     0.49469     0.49469    5.48    0.030
Error         19     1.71657     0.09035
Total         20     2.21126

Obs.   MARKET        GM       Fit  Stdev.Fit  Residual   St.Resid
  1     0.001     0.2137    0.0093    0.0726    0.2044     0.70
  2     0.108     0.0422    0.1142    0.0670   -0.0720    -0.25
  3     0.156     0.0631    0.1609    0.0738   -0.0978    -0.34
  4    -0.174    -0.3667   -0.1619    0.1231   -0.2048    -0.75
  5    -0.296    -0.2597   -0.2821    0.1688    0.0224     0.09 X
  6     0.315     0.9522    0.3169    0.1200    0.6353     2.31R
  7     0.192     0.4584    0.1963    0.0818    0.2621     0.91
  8    -0.115    -0.1124   -0.1047    0.1032   -0.0077    -0.03
  9     0.010    -0.0498    0.0186    0.0709   -0.0684    -0.23
 10     0.123     0.0288    0.1287    0.0686   -0.0999    -0.34
 11     0.259    -0.0410    0.2618    0.1010   -0.3028    -1.07
 12    -0.099    -0.0911   -0.0891    0.0982   -0.0020    -0.01
 13     0.155     0.6826    0.1601    0.0736    0.5225     1.79
 14     0.171     0.2373    0.1755    0.0768    0.0618     0.21
 15     0.012     0.1176    0.0196    0.0708    0.0980     0.34
 16     0.263    -0.0383    0.2664    0.1026   -0.3047    -1.08
 17     0.146     0.0088    0.1516    0.0720   -0.1428    -0.49
 18     0.020     0.0058    0.0282    0.0695   -0.0224    -0.08
 19     0.124     0.4418    0.1299    0.0687    0.3119     1.07
 20     0.273    -0.4581    0.2754    0.1055   -0.7335    -2.61R
 21    -0.066    -0.1154   -0.0559    0.0882   -0.0595    -0.21

R denotes an obs. with a large st. resid.
X denotes an obs. whose X value gives it large influence.

Durbin-Watson statistic = 1.87


   Fit  Stdev.Fit       95% C.I.          95% P.I.
0.1259    0.0682   (-0.0169, 0.2688)  (-0.5193, 0.7712)
```

Table 14.8 ANOVA for Equation 14.23

Sources of Variation	Sum of Squares	Degrees of Freedom	Mean Square
Regression	.49469	1	.49469
Error	1.71657	19	.09035
Total	2.21126	20	.11056

Substituting information indicated in Tables 14.7 and 14.8 into Equations 13.20, 13.21, 14.2, 14.4, and 14.9, we obtain

$$S_e = \sqrt{\sum_{t=1}^{n} \hat{e}_{g,t}^2/n - 2}$$

$$= \sqrt{1.71657/19} = .3006$$

$$R^2 = \frac{.49469}{2.21126} = .224$$

$$S_b = S_e \Big/ \sqrt{\sum_{t=1}^{21} (R_{m,t} - \overline{R}_m)^2}$$

$$= .3006/.7178 = .4188$$

$$t_{19} = .9800/.4188 = 2.34$$

$$F_{1,19} = \frac{SSR}{SSE/19} = \frac{.49469}{.09035}$$

$$= 5.48$$

The standard error of residuals, $s_e = .3006$, can be used to measure the absolute goodness of fit for the estimated market model, $\hat{R}_{g,t}$. And the coefficient of determination $R^2 = .224$ can be used to measure the relative goodness of fit for $\hat{R}_{g,t}$. The estimated R^2 implies that 22.4 percent of the variation in rates of return for GM has been explained by the variation of market rates of return. $t_{19} = 2.34$ can be used to test the following null hypothesis:

$$H_0: \beta = 0 \quad \text{versus} \quad H_1: \beta \neq 0$$

Using a significance level of $\alpha = 1$ percent, we have $t_{.025,19} = 2.0930$. Because $2.34 > 2.0930$, we can reject the null hypothesis and conclude that there is a linear relationship between $r_{g,t}$ and $R_{m,t}$ at $\alpha = 5$ percent.

Equation 14.23 can be used to forecast the rate of return for GM in 1991. If we assume that $R_{m,n+1} = .1200$ for 1991, then the forecasted rate of return for GM in 1991 is

$$\hat{R}_{g,1991} = .00834 + (.9800)(.1200)$$

$$= .1259$$

We use Equation 14.19 to calculate the standard error prediction (s_p).

$$s_p^2 = s_e^2 \left(1 + \frac{1}{21} + \frac{(r_{m,n+1} - \overline{R}_m)^2}{\sum_{t=1}^{21} (R_{m,t} - \overline{R})^2}\right)$$

$$= (.0904)\left(1 + \frac{1}{21} + \frac{(.1200 - .0751)^2}{.5152}\right)$$

$$= (.0904)(1.0515) = .0951$$

where s_p represents the standard error of the forecasted rate of return for GM in 1991. Using this information, we can estimate a 95 percent forecast interval for the rate of return for GM in 1991. From Table A4 in Appendix A, the value of $t_{.025,19}$ is 2.0930. Using related information, we get the following 95 percent interval estimate:

$$.1259 \pm (2.0930)\sqrt{(.0951)} = .1259 \pm .6454$$
$$= (-.5195, .7713)$$

Using this method, we can expect that the interval determined will include the true value of the rate of return in 1991 for GM 95 percent of the time. This prediction interval is almost identical to that obtained via MINITAB (Figure 14.11).

■ | **APPLICATION 14.3** Relationship Between Audited and Book Inventory Value

Accountants often use the audit sampling approach to do statistical auditing. Auditors use a simple regression model like that indicated in Equation 14.23 to estimate the relationship between client-reported account values and audited account values:

$$y_i = a + bx_i + e_i \qquad (14.24)$$

where

y_i = ith audited account value
x_i = ith reported account value

Using as an example the inventory valuation demonstration data indicated in Table 14.9, we will show how regression analysis can be used to estimate the mean per-unit audited account value of the client population ($\overline{Y}$) as defined in Equation 14.24.

$$\overline{Y} = a + b\mu_x = \bar{y} + b(\mu_x - \bar{x}) = \bar{y} - b\bar{x} + b\mu_x \qquad (14.25)$$

where

$\bar{y}$ = mean of y_i
$\bar{x}$ = mean of x_i
μ_x = mean per-unit reported account value (a population parameter)

The use of Equation 14.25 is similar to the case of mean response with $x_{n+1} \neq \bar{x}$ in the last section, constructing interval estimates of forecasts.

The 30 sample inventory accounts are randomly drawn from a population with the following population information:

Number of accounts = N = 2,000
Reported aggregate account value = X = \$400,000
Mean per-unit reported account value = $\bar{x}$ = \$200

From the data listed in Table 14.9, we obtain

$$\hat{b} = .9122 \quad \text{and} \quad s_e = \$8.8180$$

Table 14.9 Inventory Valuation Demonstration Data

(1) Sample Item Number, n_i	(2) Account Number	(3) Reported Account Value, x_i	(4) Audited Account Value, y_i
1	2545	$ 161.21	$ 168.69
2	3988	183.68	174.53
3	3825	246.80	255.70
4	2613	207.28	208.46
5	3071	169.52	180.12
6	2848	180.26	189.76
7	3207	221.28	227.55
8	2109	185.58	174.61
9	2299	236.34	243.62
10	3052	202.44	209.35
11	2486	184.76	198.66
12	2822	191.21	198.51
13	3818	198.86	219.76
14	3674	192.65	208.46
15	2304	210.83	214.12
16	3206	208.59	219.41
17	3659	205.98	215.83
18	3544	148.35	172.39
19	3666	197.77	192.84
20	3937	238.25	249.08
21	3187	244.85	231.89
22	2622	192.28	191.72
23	2530	179.93	172.80
24	2320	180.81	187.11
25	2943	194.53	192.32
26	3670	216.40	221.92
27	3506	201.34	219.25
28	2416	212.00	204.39
29	2135	190.21	201.51
30	3181	188.29	205.40
		$5,972.28	$6,149.76
		$\bar{x} = \$199.076$	$\bar{y} = \$204.992$

Source: Andrew D. Bailey, Jr. (1981), *Statistical Auditing: Review, Concepts and Problems.* (San Diego: Harcourt), pp. 124–125. Copyright © 1981 by Harcourt Brace Jovanovich, Inc., reprinted by permission of the publisher.

Substituting $\hat{b} = .9122$, $\bar{y} = \$204.992$, $\bar{x} = \$199.076$, and $\mu_x = \$200.00$ into Equation 14.25, we have

$$\bar{Y} = \$204.992 + (.9122)(\$200.00 - \$199.076)$$
$$= \$205.8349$$

Following Equation 14.21, we can construct the confidence interval associated with μ_Y as

$$\overline{Y} - t_{(\alpha/2,n-2)}S_e \left[\frac{1}{n} + \frac{(\mu x - \overline{x})^2}{\sum\limits_{i=1}^{n} (x_i - \overline{x})^2} \right]^{1/2} < \mu_Y$$

$$< \overline{Y} + t_{(\alpha/2,n-2)}S_e \left[\frac{1}{n} + \frac{(\mu x - \overline{x})^2}{\sum\limits_{i=1}^{n} (x_i - \overline{x})^2} \right]^{1/2} \qquad (14.26)$$

Let $\alpha = .10$; then $t_{.05,28} = 1.701$. Substituting $\overline{Y}$, s_e, n, μ_x, $\overline{x}$, $\Sigma(x_i - \overline{x})^2$, and $t_{.05,28}$ into Equation 14.26, we have

$$\$205.8349 - (1.701)(1.6113) < \mu_Y < \$205.8349 + (1.701)(1.6113)$$

$$\$203.0941 < \mu_Y < \$208.5757$$

We can expect that the interval determined will include the true value of the per-unit audited inventory account value 90 percent of the time.

■ | **APPLICATION 14.4** Hamburger Sales: Predicting Profits[5]

Healthy Hamburgers has a chain of 12 stores in Northern Illinois. Sales figures and profits for the stores are given in Table 14.10. Our task is to obtain a regression line for the data and, assuming sales of $10 million, to predict profit for one store.

A worksheet for calculating regression coefficients is presented in Table 14.11. Substituting data from Table 14.11 into Equations 13.12 and 13.11, we obtain slope and intercept estimates.

$$b = \frac{n(\Sigma x_i y_i) - (\Sigma x_i)(\Sigma y_i)}{n(\Sigma x_i^2) - (\Sigma x_i)^2} = \frac{12(35.29) - 132(2.7)}{12(1796) - (132)^2} = .01593$$

$$a = \frac{\Sigma y - b(\Sigma x)}{n} = \frac{271 - .01593(132)}{12} = .0506$$

Thus the regression line can be defined as

$$\hat{y} = .0506 + .01593x$$

For sales of $x = 10$ (that is, $10 million), estimated profit is

$$\hat{y}_{n+1} = .0506 + .01593(10) = .2099, \text{ or } \$209,900$$

Here we estimate standard error of the estimate in accordance with Equation 13.27.

$$\sqrt{\frac{\Sigma y_i^2 - (\Sigma y_i)^2/n - b^2 [\Sigma x_i^2 - (\Sigma x_i)^2/n]}{n - 2}}$$

[5]This application is drawn from William J. Stevenson (1986), *Production/Operations Management,* 2nd ed. (Homewood, IL: Irwin) pp. 137–141. Reprinted by permission of Richard D. Irwin.

Table 14.10 Sales and Profit for 12 Healthy Hamburgers Stores

Sales, x_i (millions)	Profits, y_i (millions)
$7	$.15
2	.10
6	.13
4	.15
14	.25
15	.27
16	.24
12	.20
14	.27
20	.44
15	.34
7	.17

Table 14.11 Worksheet for Calculating Regression Coefficients for Healthy Hamburgers

x	y	xy	x^2	y^2
7	.15	1.05	49	.0225
2	.10	.20	4	.0100
6	.13	.78	36	.0169
4	.15	.60	16	.0225
14	.25	3.50	196	.0625
15	.27	4.05	225	.0729
16	.24	3.84	256	.0576
12	.20	2.40	144	.0400
14	.27	3.78	196	.0729
20	.44	8.80	400	.1936
15	.34	5.10	225	.1156
7	.17	1.19	49	.0289
132	2.71	35.29	1796	.7159

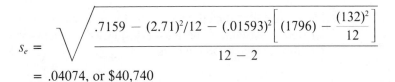

$$s_e = \sqrt{\dfrac{.7159 - (2.71)^2/12 - (.01593)^2\left[(1796) - \dfrac{(132)^2}{12}\right]}{12 - 2}}$$

$$= .04074, \text{ or } \$40{,}740$$

The prediction interval of $\hat{y}_{n+1,i}$ for the given value of x (that is, $x_{n+1,i}$), can be obtained from Equation 14.22.

If we substitute $s_e = .04074$, $n = 12$, $x_{n+1} = 10$, $\Sigma x^2 = 1796$, and $\Sigma x = 132$ into the standard error of regression (s_{reg}) portion of Equation 14.22, we obtain

$$s_{reg} = (.04074) \sqrt{1 + \frac{1}{12} + \frac{(10 - 11)^2}{1796 - (132)^2/12}} = .04245$$

Because $t_{12-2,.05/2} = 2.23$ (from Table A4 in Appendix A), a 95 percent confidence interval for predicted y, y_{n+1}, is

$$\hat{y}_{n+1,i} \pm t_{n-2,\alpha/2}(s_{reg}) = .2099 \pm 2.23(.04245)$$

$$= .2099 \pm .0947$$

or

$$.1152 \text{ to } .3046$$

That is, estimated profit on sales of $10 million is $209,900, and on the basis of sample data, we are 95 percent confident that actual profit will be in the range of $115,200 to $304,600.

Now we will test whether the estimated slope $b = .01593$ is significantly different from zero. We can estimate the standard deviation of b in accordance with Equation 14.2 as

$$s_b = s_e/\sqrt{\Sigma x^2 - (\Sigma x)^2/n}$$

$$= (.04074) \sqrt{\frac{1}{1796 - (132)^2/12}} = .0022$$

By dividing .0022 into .01593, we obtain the t-value for b.

$$t_{12-2} = \frac{b}{s_b} = \frac{.01593}{.0022} = 7.24$$

Table A4 in Appendix A reveals that $t_{10,.025} = 2.23$. Because 7.24 is larger than 2.23, we can conclude that there is a strong relationship between the two variables (profit and sales) for 12 Healthy Hamburgers stores.

■ **APPLICATION 14.5** The Impact of Cable TV Penetration on Network Share TV Revenues

To investigate the effect of cable TV penetration on network share of advertising revenue, Krugman and Rust (1987) used the data listed in Table 14.12 to do the following regression analysis

$$y_t = a + bx_t + \epsilon_t$$

where

y_t = network share of TV advertising revenue in period t
x_t = the percentage of U.S. TV households subscribing to cable TV in period t

Table 14.12 Percentage of U.S. TV Households (*x*) and Network Share of TV Revenues (*y*) during 1980–1985

	U.S. Cable TV Households (%), *x*	Network Share of Television Revenues, *y*
1980	21.1	98.9
1981	23.7	97.9
1982	25.8	96.5
1983	30.0	95.2
1984	35.7	94.0
1985	41.1	91.9

Source: D. M. Krugman and R. T. Rust, "The Impact of Cable Penetration on Network Viewing," *Journal of Marketing Research,* Vol. 27, No. 9, Oct./Nov. 1987, 9–12.

The MINITAB regression output is presented in Figure 14.12. From this output, the estimated simple linear regression can be written as

$$y_t = 106 - .335x_t$$
$$(147.22)\,(-14.18)$$

t-values are presented in the parentheses.

Figure 14.12

MINITAB Output for Application 14.5

```
MTB > READ C1 C2
DATA> 21.1 98.9
DATA> 23.7 97.9
DATA> 25.8 96.5
DATA> 30.0 95.2
DATA> 35.7 94.0
DATA> 41.1 91.9
DATA> END
        6 ROWS READ
MTB > BRIEF 1
MTB > REGRESS C2 1 C1;
SUBC> DW;
SUBC> PREDICT 54.

The regression equation is
C2 = 106 - 0.335 C1

Predictor      Coef      Stdev     t-ratio        p
Constant     105.634     0.718     147.22     0.000
C1          -0.33486    0.02362    -14.18     0.000

s = 0.4030     R-sq = 98.0%     R-sq(adj) = 97.6%

Analysis of Variance

SOURCE       DF        SS         MS         F         p
Regression    1      32.644     32.644    200.98    0.000
Error         4       0.650      0.162
Total         5      33.293
```

(*continued*)

Figure 14.12
(Continued)

```
Durbin-Watson statistic = 1.69

   Fit  Stdev.Fit        95% C.I.          95% P.I.
87.551     0.600   ( 85.885, 89.218)  ( 85.544, 89.559) XX

X  denotes a row with X values away from the center
XX denotes a row with very extreme X values
```

From Table A4 in Appendix A, we find that $t_{5,.005} = 4.604$. This critical value of t is smaller than both 147.22 and 14.18; hence we can conclude that both the estimated intercept and the slope are significantly different from 0 at $\alpha = .01$.

Finally, we find that the confidence interval for prediction is (85.885, 89.218) and the prediction interval for $x = 54$ is (85.544, 89.559).

14.6 USING COMPUTER PROGRAMS TO DO SIMPLE REGRESSION ANALYSIS

In general, regression analysis is done on an electronic computer. Without the help of a computer, the arithmetic involved would be very time-consuming.

Most modern computing facilities have available prewritten computer program packages such as the MINITAB, the SAS, the SPSS, and the BMDP for carrying out regression analysis. These packages are available for use with personal computers. The user need only input the data and specify the model that is to be fitted for empirical analyses. All these packages can produce all the information we discussed in this chapter and in Chapter 13. In addition, these packages enable the user to select options that produce much more numerical and graphical output. In Chapters 13 and 14, we have applied the MINITAB to run simple linear regression. Now we show how the SAS program can be used to run simple linear regression.

Drawing on a set of sample market sales data called "Territory Data for Click Ball Point Pens" (see Table 14.13), we will show how the SAS computer program can be used to do the simple regression analysis discussed in this and the last chapter. This set of data represents the annual territory sales of Click, a national manufacturer of ball point pens, and other related variables. The company intends to use this set of data to investigate the effectiveness of the firm's marketing efforts. The company uses regional wholesalers to distribute Click pens, and it supplements their efforts with company sales representatives and spot TV advertising. The data to be analyzed are sales (y), advertising (x_1), number of sales representatives (x_2), and wholesaler efficiency index (x_3), where 4 = outstanding, 3 = good, 2 = average, and 1 = poor.

First we input all data listed in Table 14.11 into the SAS regression program. Then we specify the models to be analyzed.

$$y_i = a_0 + a_1 x_{1i} + e_i, \quad i = 1, 2, \ldots, 40 \tag{14.27}$$

$$y_i = b_0 + b_1 x_{2i} + e_i, \quad i = 1, 2, \ldots, 40 \tag{14.28}$$

$$y_i = c_0 + c_1 x_{3i} + e_i, \quad i = 1, 2, \ldots, 40 \tag{14.29}$$

Table 14.13 Territory Data for Click Ball Point Pens

Territory	Sales, Y (thousands)	Advertising, X_1 (TV spots per month)	Number of Sales Representatives, X_2	Wholesaler Efficiency Index, X_3
005	260.3	5	3	4
019	286.1	7	5	2
033	279.4	6	3	3
039	410.8	9	4	4
061	438.2	12	6	1
082	315.3	8	3	4
091	565.1	11	7	3
101	570.0	16	8	2
115	426.1	13	4	3
118	315.0	7	3	4
133	403.6	10	6	1
149	220.5	4	4	1
162	343.6	9	4	3
164	644.6	17	8	4
178	520.4	19	7	2
187	329.5	9	3	2
189	426.0	11	6	4
205	343.2	8	3	3
222	450.4	13	5	4
237	421.8	14	5	2
242	245.6	7	4	4
251	503.3	16	6	3
260	375.7	9	5	3
266	265.5	5	3	3
279	620.6	18	6	4
298	450.5	18	5	3
306	270.1	5	3	2
332	368.0	7	6	2
347	556.1	12	7	1
358	570.0	13	6	4
362	318.5	8	4	3
370	260.2	6	3	2
391	667.0	16	8	2
408	618.3	19	8	2
412	525.3	17	7	4
430	332.2	10	4	3
442	393.2	12	5	3
467	283.5	8	3	3
471	376.2	10	5	4
488	481.8	12	5	2

Source: C. A. Gilbert, in G. A. Churchill, Jr., *Marketing Research: Methodological Foundations,* 3rd ed., 1983, p. 563. Copyright © 1983 by the Dryden Press, reprinted by permission of the publisher.

After we specify these three simple regression models on the SAS regression programs, we can have the SAS program do 3 simple regression analyses. Their outputs are presented in Figures 14.13 and 14.14. Figure 14.13 presents the scatter diagrams; Figures 14.14a and 14.14b present the regression outputs.

Figures 14.13a, 14.13b and 14.13c are scatter diagrams showing how (a) y and x_1, (b) y and x_2, and (c) y and x_3 are related. Figures 14.4a and 14.14b show the estimated regression coefficients a_1, b_1, and c_1 with t statistics 11.43, 11.524 and .012, respectively. From Table A3 in Appendix A, using interpolation, we find that $t_{38,.025} = 2.025$; we can conclude that both a_1 and b_1 are significantly different from zero at $\alpha = .05$. However, c_1 is not significantly different from zero at $\alpha = .05$.

By using the output listed in Figures 14.13 and 14.14, we can do related analyses in terms of the concepts and methodologies we learned in Chapters 13 and 14.

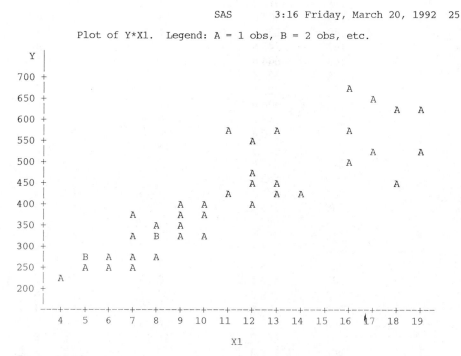

Figure 14.13a

Scatter Diagram Showing the Relationship Between y and x_1

Figure 14.13b

Scatter Diagram
Showing the Relation-
ship Between y and x_2

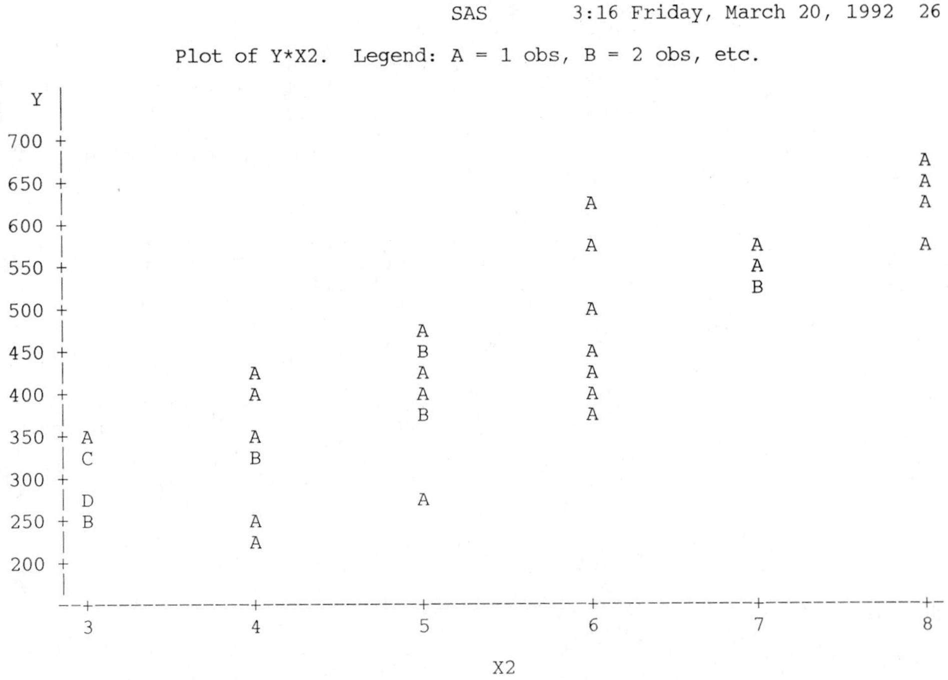

```
                                        SAS        3:16 Friday, March 20, 1992  26

                        Plot of Y*X2.   Legend: A = 1 obs, B = 2 obs, etc.
   Y |
     |
 700 +                                                                          A
     |                                                                          A
 650 +                                                                          A
     |                                          A
 600 +                                          A            A                  A
     |                                          A            A
 550 +                                                       A
     |                                                       B
 500 +                                          A
     |                             A
 450 +                             B            A
     |              A              A            A
 400 +              A              A            A
     |                             B            A
 350 + A                           A
     | C                           B
 300 +
     | D                A
 250 + B                A
     |                  A
 200 +
     |
     --+---------------+---------------+---------------+---------------+---------------+--
       3               4               5               6               7               8

                                        X2
```

Figure 14.13c

Scatter Diagram
Showing the Relation-
ship Between y and x_3

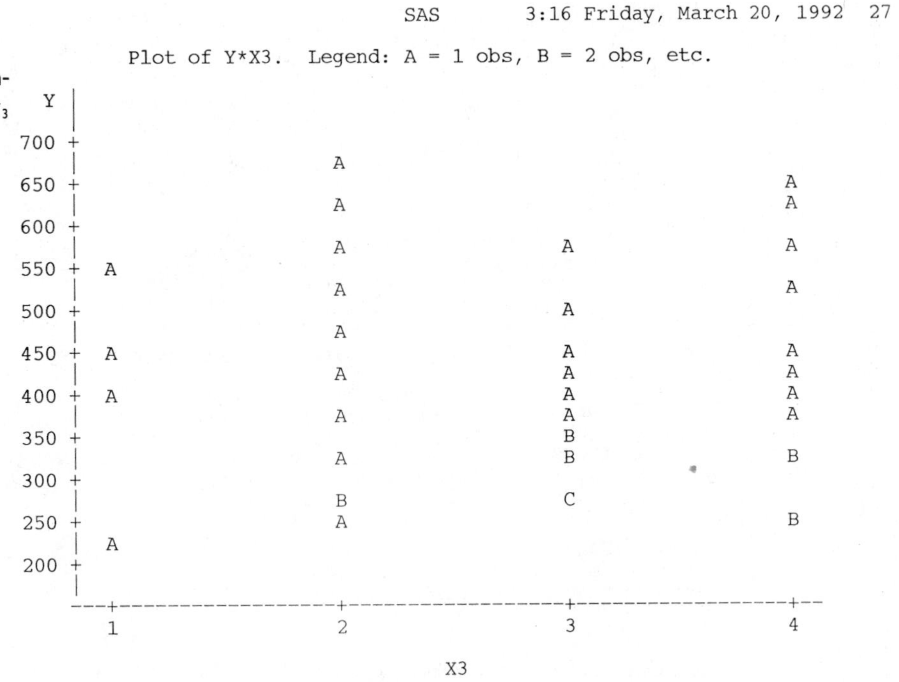

```
                                        SAS        3:16 Friday, March 20, 1992  27

                        Plot of Y*X3.   Legend: A = 1 obs, B = 2 obs, etc.
   Y |
     |
 700 +                    A
     |
 650 +                    A                                            A
     |                    A                                            A
 600 +                    A                        A                   A
     | A
 550 +                    A                                            A
     |                    A
 500 +                                             A
     |                    A
 450 + A                                           A                   A
     |                    A                         A                   A
 400 + A                                            A                   A
     |                    A                         A                   A
 350 +                                              B
     | A                                            B                   B
 300 +                    A                         B                   B
     |                    B                         C
 250 +                    A                                             B
     | A
 200 +
     |
     --+----------------+----------------+----------------+--
       1                2                3                4

                                        X3
```

Dependent Variable: Y

Analysis of Variance

Source	DF	Sum of Squares	Mean Square	F Value	Prob>F
Model	1	463451.00888	463451.00888	130.644	0.0001
Error	38	134802.01487	3547.42144		
C Total	39	598253.02375			

Root MSE	59.56023	R-square	0.7747
Dep Mean	411.28750	Adj R-sq	0.7687
C.V.	14.48141		

Parameter Estimates

Variable	DF	Parameter Estimate	Standard Error	T for H0: Parameter=0	Prob > |T|
INTERCEP	1	135.433596	25.90650568	5.228	0.0001
X1	1	25.307698	2.21415038	11.430	0.0001

Durbin-Watson D	1.721
(For Number of Obs.)	40
1st Order Autocorrelation	0.133

Figure 14.14a

SAS Output of $y_i = a + b_1 x_{1i} + e_i$

Dependent Variable: Y

Analysis of Variance

Source	DF	Sum of Squares	Mean Square	F Value	Prob>F
Model	1	465161.12840	465161.12840	132.811	0.0001
Error	38	133091.89535	3502.41830		
C Total	39	598253.02375			

Root MSE	59.18123	R-square	0.7775
Dep Mean	411.28750	Adj R-sq	0.7717
C.V.	14.38926		

Parameter Estimates

Variable	DF	Parameter Estimate	Standard Error	T for H0: Parameter=0	Prob > |T|
INTERCEP	1	80.065802	30.22585829	2.649	0.0117
X2	1	66.244340	5.74818946	11.524	0.0001

Durbin-Watson D	2.041
(For Number of Obs.)	40
1st Order Autocorrelation	-0.040

Figure 14.14b

SAS Output of $y_i = a + b_2 x_{2i} + e_i$ and $y_i = a + b_3 x_{3i} + e_i$

Figure 14.14b
(Continued)

Dependent Variable: Y

Analysis of Variance

Source	DF	Sum of Squares	Mean Square	F Value	Prob>F
Model	1	2.19822	2.19822	0.000	0.9906
Error	38	598250.82553	15743.44278		
C Total	39	598253.02375			

Root MSE	125.47288	R-square	0.0000	
Dep Mean	411.28750	Adj R-sq	-0.0263	
C.V.	30.50734			

Parameter Estimates

Variable	DF	Parameter Estimate	Standard Error	T for H0: Parameter=0	Prob > |T|
INTERCEP	1	410.606023	60.98903066	6.732	0.0001
X3	1	0.241231	20.41491700	0.012	0.9906

Durbin-Watson D	1.724
(For Number of Obs.)	40
1st Order Autocorrelation	0.115

Summary

In this chapter and Chapter 13, we used simple regression analysis and correlation analysis to determine the relationship between two variables. In Chapter 13, we discussed estimation of the intercept and slope. To determine whether the regression does a good job of explaining the dependent variable, we investigated in detail two goodness-of-fit measures: the standard error of residuals and the coefficient of determination.

Another measure of association, the sample correlation coefficient, was discussed in Chapter 13, along with the relationship between the correlation coefficient r and the slope estimate b and that between r and the coefficient of determination R^2.

Chapter 14 showed how researchers can use the standard error and the parameter value to construct a t test to determine whether the parameter values are equal to 0. We also discussed the F test and confidence intervals and point estimates for forecasting. The MINITAB Program is used to do this kind of analysis. Applications of regression analysis drawn from finance, accounting, and marketing rounded out the picture. Finally, we saw how SAS statistical computer programs can be used to do simple regression analysis.

Appendix 14A   Impact of Measurement Error and Proxy Error on Slope Estimates

The data collected for business and economics research are sometimes subject to errors in measurement. Recall from Chapter 2 that there are two classifications of data. Primary data are collected by the researcher specifically for a study. Secondary data are applicable to the study in question but were collected for some other reason. Survey data collected by a researcher to determine voting preference in an election are primary data. Stock prices appearing in the *Wall Street Journal* are secondary data; they were not collected for a particular study. Both primary and secondary data are subject to **measurement error,** such as computer programming errors, errors resulting from inaccurate measuring equipment, and deviations from sample statistics and population parameters. In addition to measurement error, **proxy error** can occur when a researcher uses data that do not match their theoretical definition. In other words, proxy error is the error caused by using one measurement in place of (as a proxy for) another measurement. For example, accounting income from the income statement is frequently used as a proxy for economic income to determine company value. However, accounting income is subject to changing accounting methods, and this characteristic can affect the measurement of the trends in a firm's earning power. In economics, current income (GNP) is often used as a proxy for permanent income in investigating the consumption function. Permanent income, which equals current income adjusted for transitory income, should be used instead.

If the independent variable of the regression, x_i, is subject to either measurement error or proxy error, then the observed x_i can be defined as

$$x_i^* = x_i + \eta_i, \tag{14A.1}$$

where x_i is the true value of the independent variable and x_i^* is the observed value of x_i measured with errors. It is assumed that the measurement error, η_i, is independent of x_i. That is, $\text{Cov}(\eta_i, x_i) = 0$ in this case, and the observed linear regression becomes

$$y_i = a + b(x_i + \eta_i) + (e_i - b\eta_i) \tag{14A.2}$$

If we let $e_i - b\eta_i = e_i^*$, then Equation 14A.2 can be rewritten as

$$y_i = a + bx_i^* + e_i^* \tag{14A.3}$$

In Equation 14A.3, the independent variable, x_i^*, is no longer uncorrelated with the residual e_i^*.[6]

---

[6] $\text{Cov}(x_i^*, e_i^*) = \text{Cov}[(x_i + \eta_i), (e_i - b\eta_i)]$

$\qquad\qquad = \text{Cov}(x_i, e_i) + \text{Cov}(x_i, -b\eta_i) + \text{Cov}(\eta_i, e_i) + \text{Cov}(\eta_i, -b\eta_i)$

$\qquad\qquad = -b\,\text{Var}(\eta_i) \neq 0$

We can illustrate Equation 14A.1 or Equation 14A.2 by using a simple version of Friedman's (1957) theory of consumption function. In this kind of consumption function, the consumer's income is assumed to consist of a permanent component and a transitory component. The transitory component is that part of income that the consumer considers accidental. The consumption decision is determined by the permanent component. In Equation 14A.2, x_i^* represents current income, x_i represents permanent income, and $(x_i^* - x_i)$ represents transitory income. Hence Equation 14A.1 represents a current income of the consumption function.

Equation 14A.3 violates one of the assumptions of standard linear regression analysis: the error term is not independent of the observed independent variable, x_i^*. Therefore the estimated slope, b, is no longer an unbiased estimator for β. That is,

$$E(\hat{b}) = \frac{\beta}{1 + \dfrac{(n-1)\sigma_\eta^2}{n\sigma_x^2}} \tag{14A.4}$$

Equation 14A.4 implies that if the independent variable, x_i, is measured with errors, then the ordinary least-squares estimate of β, which is b, will be a downward-biased estimate of β. When σ_η^2 approaches zero, $E(b)$ approaches β. If σ_η^2/σ_x^2 is known, then an unbiased estimate of β is

$$b' = \left(1 + \frac{n-1}{n}\frac{\sigma_\eta^2}{\sigma_x^2}\right)b \tag{14A.5}$$

However, the ratio σ_η^2/σ_x^2 is seldom known. For example, in business and economics, we often use accounting income as a proxy for economic income in regression analysis. Therefore, the problem of proxy or measurement error looms large. For demonstration purposes, if $n = 10$ and $\sigma_\eta^2/\sigma_x^2 = 1/3$, then

$$E(b) = \frac{\beta}{1 + (9/10)(1/3)} = .77\beta$$

This example shows how proxy error can make the slope estimate downward-biased.

Appendix 14B The Relationship Between the F Test and the t Test

As we saw in Section 9.3, the t distribution with v degrees of freedom (t_v) can be defined as

$$t_v = \frac{Z}{\sqrt{U/v}} \tag{14B.1}$$

where Z is a standard normal variable, U is a chi-square random variable with v degrees of freedom, and Z and U are independent.

From Equation (14B.1), we can obtain

$$t_v^2 = \frac{Z^2}{U/v}$$

From Section 9.4, we know that Z^2 is a chi-square distribution with 1 degree of freedom. Hence t_v^2 represents a ratio between two independent chi-square distributions. From Section 9.5, we know that t_v^2 represents an F distribution with 1 and v degrees of freedom. From this result, we can conclude that the calculated F_c value should always equal the square of the calculated t_c value. Here $t_c = 6.0707$, so $t_c^2 = (6.0707)^2 = 36.8534$, which differs from $F_c = 36.8555$ only because of rounding errors.

Appendix 14C Derivation of Variance for Alternative Forecasts

Derivation of Equation 14.17

$$\text{Var}(\hat{y}_{n+1}) = \text{Var}(\bar{y})$$

$$= \text{Var}\left(\sum_{i=1}^{n} y_i/n\right)$$

$$= \frac{1}{n^2} \text{Var}(y_1 + \cdots + y_n)$$

$$= \frac{1}{n^2} [\sigma_e^2 + \cdots + \sigma_e]^2$$

$$= \frac{\sigma_\epsilon^2}{n}$$

If we use the sample estimate s_e^2 for σ_ϵ^2, the estimate of $\text{Var}(y_{n+1})$ becomes

$$s^2(\hat{y}_{n+1}) = \frac{s_e^2}{n} \tag{14C.1}$$

Derivation of Equation 14.18

$$\text{Var}(\hat{y}_{n+1}) = \text{Var}[\bar{y} + b(x_i - \bar{x})]$$

$$= \text{Var}(\bar{y}) + \text{Var}[b(x_i - \bar{x})]$$

$$= \frac{\sigma_\epsilon^2}{n} + (x_i - \bar{x})^2 \text{Var}(b)$$

$$= \frac{\sigma_\epsilon^2}{n} + (x_i - \bar{x})^2 \frac{\sigma_\epsilon^2}{\sum\limits_{i=1}^{n} (x_i - \bar{x})^2}$$

If we use the sample estimate s_e for σ_ϵ^2, the estimate of $Var(y_{n+1})$ becomes

$$s^2(\hat{y}_{n+1}) = s_e^2\left(\frac{1}{n} + \frac{(x_{n+1} - \bar{x})}{\sum\limits_{i=1}^{n}(x_i - \bar{x})}\right) \tag{14C.2}$$

Derivation of Equation 14.19

$$Var(\hat{y}_{n+1} - y_{n+1,i}) = Var(\hat{y}_{n+1}) + Var(y_{n+1,i}) \tag{14C.3}$$

The sample estimate of $Var(y_{n+1,i})$ can be defined as

$$\hat{\sigma}^2(y_{n+1,i}) = s_e^2 \tag{14C.4}$$

Using Equations 14C.2 and 14C.4, we obtain the sample estimate of $Var(\hat{y}_{n+1} - y_{n+1,i})$ as

$$s^2(\hat{y}_{n+1} - y_{n+1,i}) = s_e^2\left(1 + \frac{1}{n} + \frac{(x_{n+1} - \bar{x})^2}{\sum\limits_{i=1}^{n}(x_i - \bar{x})^2}\right)$$

Questions and Problems

1. Here x is the number of units of a product produced during a certain period, and y represents total variable costs incurred during the period.

x	0	1	2	3	4	5	6
y	1	2	3	5	8	11	12

   a. Find the estimated equation for the regression of y on x.
   b. Find the predicted value of y given $x = 8$.
   c. Find the standard error of estimate.
   d. Find the coefficient of determination r^2.

2. Use again the data given for question 1.
   a. Find the standard deviation of the regression line's slope s_b.
   b. Find an interval that you can be 95 percent confident will contain b, the slope of the population regression line.

3. In a regression problem, $n = 30$, $\Sigma x_i = 15$, $\Sigma y_i = 30$, $\Sigma x_i y_i = 30$, $\Sigma x_i^2 = 10$, and $\Sigma y_i^2 = 160$.
   a. Find the regression line $\hat{y} = a + bx$.
   b. Estimate the variance σ_{yx}^2.
   c. Test H_0: $b = 0$ against H_1: $b \neq 0$. Let $\alpha = .05$.

4. In a regression analysis, $n = 25$, $\Sigma X_i = 75$, $\Sigma Y_i = 50$, $\Sigma X_i^2 = 625$, $\Sigma X_i Y_i = 30$, and $\Sigma Y_i^2 = 228$.
   a. Find the regression equation.

   b. Find s_{yx}^2, s_a^2, and s_b^2.
   c. Test whether $b = 0$. Let $\alpha = .01$. Use H_1: $b \neq 0$.
   d. Find a 95 percent confidence interval for a.

5. For each of the following sets of quantities, find the sample correlation coefficient. Test the hypothesis that $\rho = 0$. Let $\alpha = .05$. Use H_1: $\rho \neq 0$.
   a. $n = 11$, $\Sigma y^2 = 400$, $\Sigma(x-\bar{x})(y-\bar{y}) = 400$, $\Sigma x^2 = 625$.
   b. $n = 18$, $\Sigma y^2 = 100$, $\Sigma(x-\bar{x})(y-\bar{y}) = 36$, $\Sigma x^2 = 36$.

6. Find the sample correlation coefficient and test H_0: $\rho = 0$ for the data of question 4. Use H_1: $\rho \neq 0$ and $\alpha = .05$.

7. Records were kept on the scores 14 job applicants got on a manual-dexterity test and on their production output after a week on the job.
   a. Use the following data to estimate the correlation between test score and production output.
   b. Is there a significant correlation? Let $\alpha = .05$.

Score	112	72	61	50	48	117	13
Output	153	83	36	93	86	121	20
Score	19	13	43	84	31	124	66
Output	26	16	62	103	30	120	84

8. The managers of a weight-loss clinic wish to confirm their belief that there is a relationship between the weight of a person entering the program and the number of pounds lost. The table presents data for 8 of the clinic's clients.

x (beginning weight)	142	306	261	177
y (weight loss)	15	146	73	50

x (beginning weight)	205	165	289	154
y (weight loss)	25	15	36	12

Use the following regression information for the relationship between beginning weight and pounds lost to complete parts (a) and (b).

$\hat{y} = -67.78 + .54x$

$r^2 = .58$ $s_e = 31.60$ $s_b = .186$

$\Sigma x^2 = 28,911.87$ $\Sigma y^2 = 14,362.00$

$\Sigma xy = 15,557.48$

a. Test the hypothesis that there is no relationship between beginning weight and pounds lost, using the t test on the slope of the regression line. Let $\alpha = .05$.

b. Repeat the test in part (a), but use the t test on r, the correlation coefficient.

9. A certain firm provides expense accounts for its executives. Using the following data, the firm's human resources department ran a regression analysis to determine whether a relationship exists between annual salary and amount claimed in expenses each year.

x (salary in $1,000s)	40	38	22	25	30	35	40
y (expenses in $1,000s)	.8	1.6	.6	.9	1.2	2.2	1.1

The regression results were

$\hat{y} = .227 + .030x$ $r = .398$

$s_{y|x} = .547$ $s_b = .0306$

$\Sigma y^2 = 1.78$ $\Sigma x^2 = 320.86$ $\Sigma xy = 9.63$

a. Test the hypothesis that there is no relationship between annual salary and expenses claimed, using the t test on the regression coefficient b. Use $\alpha = .05$.

b. Repeat the test of part (a), but use the t test on r.

10. Briefly explain what we mean when we say that α and β are unbiased. Why is unbiasedness an important property in an estimator?

11. Explain the purpose of constructing confidence intervals for the parameters α and β.

12. What assumptions do we make about the distributions of a and b? Why is this important in constructing confidence intervals?

13. When we are estimating the regression $y = \alpha + \beta x$, which parameter is of greater interest, α or β? Why?

14. In testing the significance of the parameters, why do we sometimes use a two-tailed test and sometimes use a one-tailed test?

15. In testing the significance of the parameters, why do we sometimes use a z test and sometimes use a t test?

16. Compare the width of a 90 percent confidence interval with that of a 99 percent confidence interval. Which is wider? Why?

17. Compare the standard error of the estimate discussed in Chapter 13 with the standard error of the regression coefficient discussed in this chapter.

18. What null hypothesis do we generally use in testing the significance of the slope coefficient b?

19. Ralph Farmer of the Department of Agriculture is interested in the relationship between the amount of fertilizer used and the number of bushels of wheat harvested. He collects the following information on 6 farmers.

x (pounds of fertilizer)	y (bushels of wheat)
100	1000
150	1250
180	1710
200	2100
222	2500

Use MINITAB to do the following.

a. Draw a scatter diagram for the data.

b. Estimate the regression parameters for α and β.

c. Calculate the standard error of the estimate and the standard error of the coefficient b.

d. Calculate the t-value for the coefficient of b.

e. If 210 lb of fertilizer is used, what amount of wheat can be expected to be harvested? What is the 95 percent confidence interval? (Hint: Follow the procedure used in Figure 14.10.)

20. Use the data from question 19 to construct a 95 percent confidence interval for b. Also construct a 90 percent confidence interval for b. Which interval is larger?

21. Using the data given in question 19, compute SST, SSE, SSR, and R^2. Use an F test to test the significance of the regression.

22. A recent study of Departments of Labor in all 50 states indicates that the amount (in thousands of dollars) spent on job placement for the unemployed and the number of people employed has a slope of 1.7 and a standard error of the regression coefficient of .43. Test the significance of the slope coefficient at the 95 percent confidence level.

Use the following data to answer questions 23–30. The table gives monthly rates of return for 3-month Treasury bills; the value-weighted New York Stock Exchange Index; and Chrysler, Ford, and GM stock.

Month	T-Bill, R_f	NYSE, R_m	Chrysler, R_1
87.01	.004414	.12823	.29054
87.02	.004543	.04100	−.01309
87.03	.004543	.02469	.18037
87.04	.004583	−.01483	.03846
87.05	.004599	.00644	−.11111
87.06	.004607	.04797	.01103
87.07	.004623	.04682	.19414
87.08	.004904	.03688	.09816
87.09	.005193	−.02085	−.06983
87.10	.004977	−.21643	−.35649
87.11	.004623	−.07547	−.23944
87.12	.004688	.06851	.10494

Month	Ford, R_2	GM, R_3
87.01	.33378	.14015
87.02	.02689	.00831
87.03	.10475	.04690
87.04	.08741	.15200
87.05	.00137	−.03889
87.06	.08941	−.03079
87.07	.03409	.07564
87.08	.06273	.04923
87.09	.09259	−.09783
87.10	.21939	−.29518
87.11	.05795	−.01496
87.12	.05975	.08869

23. In finance, we are often interested in how the return of one stock is related to some market index such as the NYSE. The model we usually estimate to understand this relationship is known as the market model and is given by the equation

$$R_{j,t} = \alpha_j + \beta_j R_{m,t} + e_{j,t}$$

where

$R_{j,t}$ = return on stock j in month t
$R_{m,t}$ = return on some market index in month t
α_j = intercept of the regression line
β_j = slope of the regression line
$e_{j,t}$ = a random error term

Use MINITAB to do the following.

a. Draw a scatter diagram for Ford and the NYSE index.
b. Estimate the parameters α and β.
c. Compute SSE, SSR, SST, R^2, and the standard error of the estimate for this regression.
d. Compute the standard error for b, and use a t test to test the significance of the slope of the regression.

24. In finance, we sometimes choose to estimate the capital asset pricing (CAPM) version of the market model, which is given by the equation

$$R_{j,t} - R_{f,t} = \alpha_j + \beta_j[R_{m,t} - R_{f,t}] + e_{j,t}$$

where $R_{f,t}$ is the return on a risk-free asset (such as T-bills) in month t. Repeat parts (a)–(d) of question 23 for the CAPM version of the market model using the MINITAB program.

25. Using R^2 as the measure of goodness of fit, compare the market model estimated in question 23 with the CAPM version estimated in question 24.

26. Find a 95 percent confidence interval for the slope coefficients you calculated in questions 23 and 24. Which estimate of β has the wider confidence interval?

27. Suppose we are interested in testing whether β is equal to 1. Then we would test H_0: $\beta = 1$ against H_1: $\beta \neq 1$. Using the model given in question 23, test this hypothesis.

28. Repeat questions 23–27, using GM stock's rates of return.

29. Repeat questions 23–27, using Chrysler stock's rates of return.

30. Suppose we are interested in the relationship between the return on the risk-free asset (T-bills) and the return on the NYSE index.
    a. Estimate the intercept and the slope for a regression of $R_{m,t}$ on $R_{f,t}$.
    b. Compute the standard error of the regression and use a t test to test the significance of b.
    c. Calculate a 99 percent confidence interval for b.

31. When we are interested in the relationship between a dependent variable and time, we sometimes use a time-trend regression. That is, we use a dependent variable that consists only of the day, month, or year of our observations. A time-trend regression is given by the equation

    $$y_t = \alpha + \beta t + e_t$$

    where t represents time, $t = 1, 2, 3, \ldots, T$. Suppose we are interested in how Johnson & Johnson's inventory turnover has changed over time. We collect data on J&J's inventory turnover for a 20-year period from 1969 to 1988.

    a. Draw a scatter diagram for these data.
    b. Estimate the regression coefficients α and β.
    c. Calculate SSR, SSE, SST, and R^2, and the standard error of the estimate.
    d. Compute the standard error of b, and use a t test to test the significance of b.

t	J&J's Inventory Turnover	t	J&J's Inventory Turnover
1	3.19	11	2.71
2	3.02	12	2.70
3	2.96	13	2.78
4	3.10	14	2.38
5	2.92	15	2.28
6	2.28	16	2.37
7	2.77	17	2.45
8	2.76	18	2.33
9	2.84	19	2.27
10	2.76	20	2.32

32. Use the information and your calculations from question 31 to construct a 90 percent and a 99 percent confidence interval for b.

33. When estimating the relationship between the price of a good and the quantity of the good sold (the demand curve), economists sometimes choose to transform the price and quantity data by taking the natural logarithm of both. When this is done, the slope coefficient β can be interpreted as the price elasticity of demand (the sensitivity of quantity to changes in price). Consider the following information on the price and quantity of So-Good Candy Bars.

Price	Quantity
$1.50	100
1.25	135
1.00	175
.75	225
.50	300
.25	500

    a. Estimate the elasticity of demand for these data.
    b. Use a t test to test the significance of b.
    c. Construct a 95 percent confidence interval for the price elasticity.

34. The batting instructor of the Minnesota Twins is interested in the relationship between number of hours of batting practice and batting average. He collects the following data on 8 players.

Hours of Batting Practice per Week	Batting Average
5	.265
8	.277
9	.254
10	.320
11	.301
9	.260
7	.230
6	.272

a. Draw a scatter diagram for these data.
b. Compute the regression parameters α and β.
c. Compute the standard error of b, and use a t test to test the significance of the slope of the regression.
d. Construct a 95 percent confidence interval for b.

35. Suppose you estimate a regression of y against x and find that $b = 1.3$ and $\sigma_b = .4$ (standard error of the regression, which is known)
a. Construct a 90 percent and a 99 percent confidence interval for b.
b. Now suppose σ_b is not known. How would this change the way you construct your confidence interval for b? Assume that 15 observations were used to estimate the regression.

36. Use the data from question 21 of Chapter 13 to compute the standard error of b. Use a t test to test the significance of b. Construct a 99 percent confidence interval for b.

37. Use the data from question 33 of Chapter 13 to compute the standard error of b. Use a t test to test whether $b = 1$. Construct a 90 percent confidence interval for b.

38. Use the data from question 27 of Chapter 13 to compute the standard errors of a and b. Construct a 95 percent confidence interval for both a and b.

39. Use the information given in question 57 of Chapter 13, and perform a t test to test the significance of a and b.

40. Investment advisors sometimes recommend holding gold as part of an investor's portfolio, because the value of gold appears to be negatively related to that of the stock market. Thus when the stock market goes down in value, the value of gold goes up in value, and some of the investor's losses in the market are offset by gains in the value of her or his gold. The accompanying table shows data on annual rates of return for a gold mutual fund and for the S&P 500.

Year	Gold Mutual Fund	S&P 500
1979	151.30	18.16
1980	70.70	31.48
1981	−18.90	−4.85
1982	47.30	20.37
1983	8.20	22.30
1984	−25.30	5.97
1985	−11.00	31.05
1986	30.10	18.75
1987	51.50	5.24
1988	11.30	16.58

Use MINITAB to answer the following questions.
a. Estimate the slope of the regression of the rates of return of the gold mutual fund against those of the S&P 500.
b. Use a t test to test the hypothesis that $b < 0$.
c. If you expect the rate of return to be 20 percent next year (1989), what is the rate of return of the gold mutual fund you expect next year? What is the 95 percent confidence interval for your expectation?

41. Use the data from question 40 to construct a 95 percent confidence interval for b. Is it possible for the true b to be negative?

42. Briefly explain how we can use regression analysis to forecast values of y.

43. Explain why we often construct interval estimates of forecasts.

44. How is the size of the forecast interval affected when we use values of x that are much greater or much less than the mean value of x for forecasting?

45. Use the data given in question 31 to forecast Johnson & Johnson's inventory turnover for 1989 and 1990. Construct a 95 percent confidence interval for both of these forecasts.

46. Use the regression estimated in question 40 to forecast the return for the gold mutual fund in 1989 and 1990. Assume that the best forecast for the return of the S&P 500 in 1989 and 1990 is the mean of the S&P 500's returns for the previous 5 years. Construct a 99 percent confidence interval for both of these forecasts.

47. Use the data and the regression given in question 19 to forecast the number of bushels of wheat that will be harvested if 250 lb of fertilizer are used. Construct a 90 percent and a 99 percent confidence belt for the regression line. For which interval is the confidence belt wider? Explain.

48. Use the regression results from questions 23 and 24 to forecast the return for Ford in January 1988, using both the standard market model and the CAPM version of the market model. Assume that the return for the NYSE is 12 percent in January 1988.

49. Construct a 95 percent confidence interval for the forecasts produced in question 48. Which model has the larger interval?

50. What is proxy error? Give some examples of proxy error in economics, accounting, and finance.

51. Briefly explain how proxy error of x affects the results from a standard linear regression.

52. Use your results from the regression given in question 65 of Chapter 13 to test the significance of b via a t test. Also construct a 90 percent confidence interval for b. (Hint: Calculations from question 66 in Chapter 13 also may help.)

53. Again using your results from question 65 of Chapter 13, forecast the value of J&J's current ratio. Assume that the best forecast of the industry's current ratio is the mean of that ratio. Construct a 99 percent confidence interval for this forecast. (Hint: Your results from question 52 may be helpful.)

54. Redo questions 52 and 53, using J&J's inventory turnover.

55. Redo questions 52 and 53, using J&J's return on assets.

56. Redo questions 52 and 53, using J&J's price/earnings ratio.

57. Suppose you estimate the following simple regression:

$$\hat{y} = 50 + 2.23x$$
$$\text{SSE} = 22,300$$
$$n = 23$$
$$\Sigma(x - \bar{x})^2 = 2,700$$

a. On the basis of the information provided, test the significance of the slope at the 99 percent confidence level.

b. Construct a 90 percent confidence interval for the slope coefficient.

58. Suppose a researcher is interested in the relationship between the dollar volume of sales and the number of miles customers live from the store. She collects data on dollar volume of sales per customer (y) and the miles a customer lives from the store (x) for 28 customers. The following relationship is then estimated:

$$\hat{y} = 75 - .85x$$
$$s_b = .32$$
$$n = 28$$

a. Interpret the meaning of the slope coefficient.

b. Test whether the slope coefficient is significant at the 95 percent level of confidence.

59. Using the information from question 58, construct a 90 percent confidence interval for b.

60. In finance we are sometimes interested in hedging the risk associated with future price changes by using futures contracts. A futures contract allows the buyer of the contract to buy the commodity at a later date at a price that is agreed upon now. By purchasing the correct number of contracts, an investor can reduce or even eliminate his or her risk. The correct number of contracts to purchase is known as the hedge ratio, and it can be estimated by regression analysis. The regression to be estimated is

$$\Delta S = \alpha + \beta \Delta f + e$$

where

ΔS = change in the spot price of the commodity
Δf = change in the futures price of the commodity
β = hedge ratio (number of futures contracts used for hedging)

Suppose you collect 30 daily spot and futures prices over a 1-year period and estimate β to be 3.32 with a standard error of 1.12. Construct a 95 percent confidence interval around the hedge ratio β.

61. Estimate the mathematical model, using the least-squares method and the information given.

x	y	xy x^2 y^2 $\hat{y}$ e e^2 $(y - \bar{y})^2$
10	12.8	
20	18.9	
30	21	
40	38	
50	40	

62. Estimate the mathematical model, using the least-squares method and the information given.

x	y	xy x^2 y^2 $\hat{y}$ e e^2 $(y - \bar{y})^2$
10	52	
20	48	
30	31	
40	28	
50	10	

63. Evaluate the goodness of fit for the following graph.

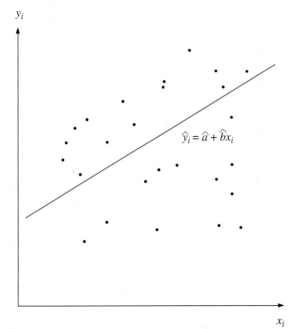

64. The following table summarizes the sales X and advertising expenditures Y (both in millions) for Rivera Company.

X	10	20	30	40	50	60
Y	70	210	230	340	360	530

a. Estimate the regression, using X as the explanatory variable.

b. What will be the expected sales for next year if the company allocates $70 million to advertising?

c. Perform a test to see whether advertising expenditures have a positive impact on sales.

65. The table on page 620 lists the administrative and enrollment breakdowns for the schools of each municipality in Middlesex County, New Jersey. Using total enrollment as the dependent variable and number of administrators as the dependent variable, run a simple regression by using information from both 1982–83 and 1990–91.

66. The table below shows the undergraduate GPA and quantitative scores on the GRE of 10 students. Explain the MINITAB output on page 621.

Student	GPA	Quantitative Scores on GRE
1	4.00	630
2	2.62	590
3	3.30	580
4	3.15	490
5	3.54	720
6	3.21	690
7	3.57	700
8	3.61	690
9	2.90	520
10	3.05	540

Administrative and Enrollment Breakdown

District	1982–83 Total Enrollment	Number of Administrators	1990–91 Total Enrollment	Number of Administrators
Carteret	2,962	29	2,525	28
Cranbury Twp	266	1	312	2
Dunellen Boro	901	5	826	6
East Brunswick Twp	7,652	36	6,657	43
Edison Twp	10,349	53	10,966	52
Highland Park Boro	1,625	10	1,441	13
Jamesburg Boro	497	3	410	2
Metuchen Boro	1,879	17	1,590	22
Middlesex Bor.	2,151	13	1,720	14
Middlesex Co-Ed Ser Comm	56	3	213	8
Middlesex City Vocational	4,181	19	3,314	27
Milltown Boro	708	4	628	5
Monroe Twp	2,545	19	2,485	22
New Brunswick City	4,286	33	4,086	32
North Brunswick Twp	3,319	25	3,996	24
Old Bridge Twp	9,120	48	8,037	50
Perth Amboy City	5,774	32	6,274	42
Piscataway Twp	6,155	38	5,637	39
Sayreville Boro	4,391	26	4,245	24
South Amboy City	903	8	948	7
South Brunswick Twp	3,125	20	3,871	37
South Plainfield Boro	3,381	24	3,001	20
South River Boro	1,650	10	1,502	10
Spotswood Boro	1,689	17	1,385	13
Woodbridge Twp	11,726	83	10,724	82
Middlesex County	91,291	576	86,793	624
Franklin	4,330	38	4,155	37

Source: *Home News,* December 15, 1991. Reprinted with permission of the publisher.

```
MTB > READ C1 C2
DATA> 4.00 630
DATA> 2.62 590
DATA> 3.30 580
DATA> 3.15 490
DATA> 3.54 720
DATA> 3.21 690
DATA> 3.57 700
DATA> 3.61 690
DATA> 2.90 520
DATA> 3.05 540
DATA> END
       10 ROWS READ
MTB > BRIEF 3
MTB > REGRESS C2 1 C1;
SUBC> DW;
SUBC> PREDICT C1.

The regression equation is
C2 = 229 + 117 C1

Predictor      Coef      Stdev     t-ratio       p
Constant      229.2      201.4       1.14      0.288
C1           117.09      60.71       1.93      0.090

s = 72.65      R-sq = 31.7%     R-sq(adj) = 23.2%

Analysis of Variance

SOURCE        DF         SS          MS         F        p
Regression     1       19630       19630      3.72     0.090
Error          8       42220        5278
Total          9       61850

Obs.     C1        C2       Fit  Stdev.Fit  Residual   St.Resid
  1     4.00     630.0     697.5     48.6      -67.5      -1.25
  2     2.62     590.0     536.0     47.0       54.0       0.98
  3     3.30     580.0     615.6     23.0      -35.6      -0.52
  4     3.15     490.0     598.0     24.6     -108.0      -1.58
  5     3.54     720.0     643.7     27.4       76.3       1.13
  6     3.21     690.0     605.0     23.5       85.0       1.24
  7     3.57     700.0     647.2     28.4       52.8       0.79
  8     3.61     690.0     651.9     29.9       38.1       0.58
  9     2.90     520.0     568.8     33.2      -48.8      -0.75
 10     3.05     540.0     586.3     27.4      -46.3      -0.69


Durbin-Watson statistic = 1.68

   Fit    Stdev.Fit        95% C.I.            95% P.I.
  697.5      48.6    ( 585.5,  809.6)   ( 496.0,  899.1)
  536.0      47.0    ( 427.6,  644.3)   ( 336.4,  735.5)
  615.6      23.0    ( 562.6,  668.6)   ( 439.8,  791.3)
  598.0      24.6    ( 541.3,  654.8)   ( 421.1,  774.9)
  643.7      27.4    ( 580.6,  706.8)   ( 464.6,  822.8)
  605.0      23.5    ( 550.7,  659.4)   ( 428.9,  781.2)
  647.2      28.4    ( 581.7,  712.7)   ( 467.3,  827.1)
  651.9      29.9    ( 582.9,  720.8)   ( 470.7,  833.1)
  568.8      33.2    ( 492.1,  645.4)   ( 384.5,  753.0)
  586.3      27.4    ( 523.2,  649.4)   ( 407.2,  765.4)


MTB > PAPER
```

MINITAB Output for Question 66

CHAPTER 15

Multiple Linear Regression

Key Terms

multiple linear regression
multiple regression analysis
conditional mean
partial regression coefficient
three-dimensional regression graph
regression plane
autocorrelation
serial correlation
perfect collinearity

multicollinearity
residual standard error
coefficient of determination
mean response
actual value
individual response
conditional prediction
cross-section regression
stepwise regression

15.1 INTRODUCTION

Chapters 13 and 14 examined in detail the simple regression model with one independent variable (such as amount of fertilizer) and one dependent variable (such as yield of corn). In many cases, however, more than one factor can affect the outcome under study. In addition to fertilizer, rainfall and temperature certainly influence the yield of corn. In business, not only rates of return for the stock market at large affect the return on General Motors or Ford stock. Other variables, such as leverage ratio, payout ratio, and dividend yield also contribute. Therefore, regression analysis with more than one independent variable is an important analytical tool.

The model that extends a simple regression to use with two or more independent variables is called a **multiple linear regression.** Simple linear regression analysis (see Chapters 13 and 14) helps us determine the relationship between two variables or predict the value of one variable from our knowledge of another. **Multiple regression analysis,** in contrast, is a technique for determining the relationship between a dependent variable and more than one independent variable. In addition, it can be used to employ several independent variables to predict the value of a dependent variable.

In this chapter, we first discuss the assumptions of the multiple regression model. Then we consider the method of least-squares estimation for a multiple regression model, the standard error of the residual estimate, and the coefficient of determination. Tests on sets and individual regression coefficients and forecasts in terms of a multiple regression are also investigated. Finally, we consider applications of the multiple regression model in business and economics.

15.2 THE MODEL AND ITS ASSUMPTIONS

In this section we first review the simple regression model and extend it to a multiple regression model. Then we define and analyze the regression plane for two independent variables. Finally, the important assumptions we must make to use the multiple regression model are explored in some detail.

The Multiple Regression Model

In multiple regression, simple regression is extended by introducing more than one independent variable. Recall from Chapter 13 that a simple linear regression model can be defined as $Y_i = \alpha + \beta x_i + \epsilon_i$ and its estimate as $y_i = a + bx_i + e_i$. The sample intercept a and the sample slope b are estimates for α and β, respectively.

The normal equations used to estimate unknown parameters α and β are

$$na + b \sum_{i=1}^{n} x_i = \sum_{i=1}^{n} y_i$$

and

$$a \sum_{i=1}^{n} x_i + b \sum_{i=1}^{n} x_i^2 = \sum_{i=1}^{n} x_i y_i$$

The foregoing equations from simple linear regression are the starting point for our exploration of multiple regression in this chapter.

Suppose an individual's annual salary (Y) depends on the number of years of education (X_1) and the number of years of work experience (X_2) the individual has had. The population regression model is

$$Y_i = \alpha + \beta_1 X_{1i} + \beta_2 X_{2i} + \epsilon_i \tag{15.1}$$

and its estimate is

$$y_i = a + b_1 x_{1i} + b_2 x_{2i} + e_i \tag{15.2}$$

where Equations 15.1 and 15.2 represent the multiple population regression line and the multiple sample regression line, respectively. In Equation 15.1, α is the intercept of the regression; β_1 is the slope that represents the conditional relationship between Y and X_1, assuming X_2 is fixed; and β_2 is the slope that represents the conditional relationship between Y and X_2, assuming X_1 is fixed. If the model defined in Equation 15.1 is linear, then the relationship between Y and each of the independent variables can be described by a straight line. In other words, the **conditional mean** of the dependent variable is given by the following population regression equation:

$$E(Y_i | X_1 = x_1, X_2 = x_2) = \alpha + \beta_1 x_1 + \beta_2 x_2$$

The coefficients β_1 and β_2 are called **partial regression coefficients.** They indicate only the partial influence of each independent variable when the influence of all other independent variables is held constant. Just as in simple regression, the multiple sample regression line of Equation 15.2 can be used to estimate the multiple population regression line of Equation 15.1.

The Regression Plane for Two Explanatory Variables

Let us say that the stock price per share (y) can be modeled as a function of both dividend per share (x_1) and retained earnings (x_2) per share.[1]

$$y_i = a + b_1 x_{1i} + b_2 x_{2i} + e_i$$

where $y_i(P_i)$ = stock price per share for the ith firm, $x_{1i}(D_i)$ = dividend per share for the ith firm, and $x_{2i}(RE_i)$ = retained earnings per share for the ith firm. (Retained earnings per share equals earnings per share minus dividend per share.) The first goal of the analysis is to obtain the estimated multiple regression model

$$\hat{y}_i = a + b_1 x_{1i} + b_2 x_{2i} \tag{15.2a}$$

[1]Practical examples based on Equation 15.2 will be explored in the applications section of this chapter.

The value of b_1 indicates that after the influence of the retained earnings per share is taken into account, a \$1 increase in the dividend per share (D_i) will increase the mean value of the price per share (P_i) by b_1, other things being equal. Similarly, a \$1 increase in retained earnings per share will increase the mean price per share by b_2. If there is only one explanatory variable, the estimated regression equation generates a straight line, as we saw in Chapter 13. There are two explanatory variables in Equation 15.2a, so it represents a **regression plane (three-dimensional regression graph).** On this three-variable regression plane, a combination of three observations (one for the value of y, one for x_1, and one for x_2) represents a single point. These points can be depicted on a three-dimensional scatter diagram. In Figure 15.1, the best-fitted regression plane would pass near the actual sample observation points indicated by the symbol $\times$, some falling above the plane and some below in such a way as to minimize L in

$$L = \sum_{i=1}^{n} (y_i - \hat{y})^2 \tag{15.3}$$

where y_i and $\hat{y}_i$ are as defined in Equations 15.2 and 15.2a, respectively.[2]

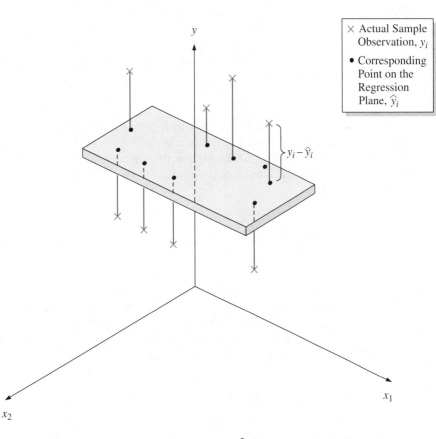

Figure 15.1

Regression Plane with $y_i(P_i)$ as Dependent Variable and with $x_{1i}(D_i)$ and $x_{2i}(RE_i)$ as Independent Variables

[2]Using Equation 15.3 to estimate regression parameters will be discussed in Section 15.3.

If there are k independent variables, then Equation 15.1 can be generalized to

$$Y_i = \alpha + \beta_1 X_{1i} + \beta_2 X_{2i} + \cdots + \beta_k X_{ki} + \epsilon_i \qquad (15.4)$$

The following section explains how regression parameters are estimated via the least-squares estimation method discussed in Chapter 13.

Assumptions for the Multiple Regression Model

As in simple regression analysis, we need five assumptions to perform a regression analysis of the model defined in Equation 15.4.

1. The error term ϵ_i is distributed with conditional mean zero and variance σ_ϵ^2 for $i = 1, 2, \ldots, n$.

2. The error term ϵ_i is independent of each of the k independent variables X_1, X_2, $\ldots$, X_k. In other words, there are no measurement errors associated with any independent variables (see Appendix 14A).

3. Any two errors ϵ_i and ϵ_j are not correlated with one another; that is, their covariance is zero: $\text{Cov}(\epsilon_i, \epsilon_j) = 0$ for $i \neq j$. This assumption means that there is no **autocorrelation (serial correlation)** among residual terms. This issue is discussed further in Chapter 16.

4. The independent variables are not *perfectly* related to each other in a linear function. In other words, it is not possible to find a set of numbers $d_0, d_1, d_2, \ldots, d_k$ such that

$$d_0 + d_1 X_{1i} + d_2 X_{2i} + \cdots + d_k X_{ki} = 0, \qquad i = 1, 2, \ldots, n$$

In practice, the linear relationship among independent variables is usually not perfect. When a perfect linear relationship occurs, a condition known as **perfect collinearity** exists. **Multicollinearity** is the condition in which two variables are highly correlated. This issue is discussed in greater detail in Chapter 16.

15.3 ESTIMATING MULTIPLE REGRESSION PARAMETERS

To estimate the best-fitted regression plane, we use the least-squares method to estimate the regression parameters. The principle of using the least-squares method for estimating the parameters of one population regression model is demonstrated in Equation 15.3 and Figure 15.1. Taking Equation 15.2 as an example, we estimate the coefficients a, b_1, and b_2 by minimizing

$$L = \sum_{i=1}^{n} e_i^2 = \sum_{i=1}^{n} (y_i - a - b_1 x_{1i} - b_2 x_{2i})^2$$

Using the same principle and technique (Appendix 13A), we can obtain the normal equations for estimating a, b_1, and b_2.[3]

$$na + b_1 \sum_{i=1}^{n} x_{1i} + b_2 \sum_{i=1}^{n} x_{2i} = \sum_{i=1}^{n} y_i$$

$$a \sum_{i=1}^{n} x_{1i} + b_1 \sum_{i=1}^{n} x_{1i}^2 + b_2 \sum_{i=1}^{n} x_{1i}x_{2i} = \sum_{i=1}^{n} x_{1i}y_i \qquad (15.5)$$

$$a \sum_{i=1}^{n} x_{2i} + b_i \sum_{i=1}^{n} x_{1i}x_{2i} + b_2 \sum_{i=1}^{n} x_{2i}^2 = \sum_{i=1}^{n} x_{2i}y_i$$

If we substitute $(x_{1i} - \bar{x}_1)$, $(x_{2i} - \bar{x}_2)$ and $(y_i - \bar{y})$ for x_{1i}, x_{2i}, and y_i, then the normal equations reduce to[4]

$$b_1 \sum_{i=1}^{n} x_{1i}'^2 + b_2 \sum_{i=1}^{n} x_{1i}'x_{2i}' = \sum_{i=1}^{n} x_{1i}'y_i'$$

$$\qquad (15.6)$$

$$b_i \sum_{i=1}^{n} x_{1i}'x_{2i}' + b_2 \sum_{i=1}^{n} x_{2i}'^2 = \sum_{i=1}^{n} x_{2i}'y_i'$$

There are two equations and two unknowns, b_1 and b_2, associated with this equation system. Hence we can solve b_1 and b_2 uniquely by substitution.

$$b_1 = \frac{\left(\sum_{i=1}^{n} x_{1i}'y_i'\right)\left(\sum_{i=1}^{n} x_{2i}'^2\right) - \left(\sum_{i=1}^{n} x_{2i}'y_i'\right)\left(\sum_{i=1}^{n} x_{1i}'x_{2i}'\right)}{\left(\sum_{i=1}^{n} x_{1i}'^2\right)\left(\sum_{i=1}^{n} x_{2i}'^2\right) - \left(\sum_{i=1}^{n} x_{1i}'x_{2i}'\right)^2} \qquad (15.7)$$

$$b_2 = \frac{\left(\sum_{i=1}^{n} x_{1i}'^2\right)\left(\sum_{i=1}^{n} x_{2i}'y_i'\right) - \left(\sum_{i=1}^{n} x_{1i}'x_{2i}'\right)\left(\sum_{i=1}^{n} x_{1i}'y_i'\right)}{\left(\sum_{i=1}^{n} x_{1i}'^2\right)\left(\sum_{i=1}^{n} x_{2i}'^2\right) - \left(\sum_{i=1}^{n} x_{1i}'x_{2i}'\right)^2} \qquad (15.8)$$

From the estimated b_1 and b_2, we obtain the estimated regression line

$$\hat{y}_i' = b_1 x_{1i}' + b_2 x_{2i}' \qquad (15.9)$$

---

[3]Equation 15.5 is a three-equation simultaneous equation system with three unknowns. The values of these three unknowns can be obtained by solving this system of simultaneous equations, by using the formula derived in this section, or by using an appropriate computer package (see Section 15.8).

[4]In this new coordinate system, $\Sigma_{i=1}^{n}x_{1i}$, $\Sigma_{i=1}^{n}x_{2i}$, and $\Sigma_{i=1}^{n}y_i$ become $\Sigma_{i=1}^{n}(x_{1i} - \bar{x}_1) = 0$, $\Sigma_{i=1}^{n}(x_{2i} - \bar{x}_2) = 0$, and $\Sigma_{i=1}^{n}(y_i - \bar{y}) = 0$. If we set $x_{1i}' = x_{1i} - \bar{x}_1$, $x_{2i}' = x_{2i} - \bar{x}_2$, and $y_i' = y_i - \bar{y}$, then Equations 15.5 reduce to Equations 15.6.

It can be shown that the intercept of Equation 15.2 is estimated as[5]

$$a = \bar{y} - b_1\bar{x}_1 - b_2\bar{x}_2 \tag{15.10}$$

EXAMPLE 15.1   *Annual Salary, Years of Education, and Years of Work Experience*

Let us use the hypothetical data given in Table 15.1 to demonstrate the procedure for estimating a multiple regression. In Table 15.1, y represents an individual's annual salary (in thousands of dollars), x_1 represents that individual's years of education, and x_2 represents her or his years of work experience.

Table 15.1 Data for Example 15.1

	x_{1i}	x_{2i}	y_i
	5	7	15
	10	5	17
	9	14	26
	13	8	24
	15	6	27
Total	52	40	109.0
Mean	10.4	8	21.8

From the data of Table 15.1, we estimate the regression line

$$\hat{y}_i = a + b_1x_{1i} + b_2x_{2i} \tag{15.11}$$

The worksheet for estimating this regression line is given in Table 15.2. (This table is included to show how computers calculate mean, variance, and covariance. You do not need to remember the procedure.)

Substituting information from Table 15.2 into Equations 15.7, 15.8, and 15.9, we obtain

$$\hat{b}_1 = \frac{(62.4)(50) - (36)(-11)}{(59.2)(50) - (-11)^2} = \frac{3516}{2839} = 1.2385$$

$$\hat{b}_2 = \frac{(59.2)(36) - (-11)(62.4)}{(59.2)(50) - (-11)^2} = \frac{2817.6}{2839} = .99246$$

$$\hat{a} = 21.8 - (1.2385)(10.4) - (.99246)(8) = .980$$

---

[5]Using the definitions of $\hat{y}_i'$, x_{1i}', and x_{2i}', we can rewrite Equation 15.9 as

$$(\hat{y}_i - \bar{y}) = b_1(x_{1i} - \bar{x}_1) + b_2(x_{2i} - \bar{x}_2)$$

which becomes

$$\hat{y}_i = (\bar{y} - b_1\bar{x}_1 - b_2\bar{x}_2) + b_1x_{1i} + b_2x_{2i} \tag{15.9'}$$

Table 15.2 Worksheet for Estimating a Regression Line (Example 15.1)

	x_{1i}	x_{2i}	y	a	b	c	aa	bb	cc
	5	7	15	−5.4	−1	−6.8	29.16	1	46.24
	10	5	17	−.4	−3	−4.8	.16	9	23.04
	9	14	26	−1.4	6	4.2	1.96	36	17.64
	13	8	24	2.6	0	2.2	6.76	0	4.84
	15	6	27	4.6	−2	5.2	21.16	4	27.04
Mean	10.4	8	21.8						
Total	52	40	109	0	0	0	59.2	50	118.8

	$(x_{1i} - \bar{x}_1)(y_i - \bar{y})$ ac	$(x_{2i} - \bar{x}_2)(y_i - \bar{y})$ bc	$(x_{1i} - \bar{x}_1)(x_{2i} - \bar{x}_2)$ ab
	36.72	6.8	5.4
	1.92	14.4	1.2
	−5.88	25.2	−8.4
	5.72	0	0
	23.92	−10.4	−9.2
Total	62.4	36	−11

> Hence the regression line of Equation 15.11 becomes
>
> $$\hat{y}_i = .980 + 1.2385x_{1i} + .9925x_{2i} \tag{15.12}$$

The next section shows how to compute standard errors of estimates and the coefficients of determination.

15.4 THE RESIDUAL STANDARD ERROR AND THE COEFFICIENT OF DETERMINATION

As in the case of simple regression, the standard error of estimate can be used as an absolute measure, and the coefficient of determination as a relative measure, of how well the multiple regression equation fits the observed data. The interpretations of these two goodness-of-fit measures are analogous to those discussed in Chapter 13.

The Residual Standard Error

Just like simple regression, multiple regression can be used to break down the total variation of a dependent variable y_i into unexplained variation and explained variation.

$$\sum_{i=1}^{n} (y_i - \bar{y})^2 = \sum_{i=1}^{n} (y_i - \hat{y}_i)^2 + \sum_{i=1}^{n} (\hat{y}_i - \bar{y})^2$$

Sum of Squares Total (SST)	Sum of Squares Error (SSE)	Sum of Squares due to Regresssion (SSR)	(15.13)

Equation 15.13 is identical to Equation 13.16 except that the estimated dependent variable ($\hat{y}_i$) of multiple regression is determined by two or more independent variables. SSR and SSE are the explained and unexplained sums of squares, respectively.

Using the definition of sum of squares error, we can define the estimate of the standard deviation of error terms, sometimes called the **residual standard error,** as

$$s_e = \sqrt{\frac{\sum_{i=1}^{n} (y_i - \hat{y}_i)^2}{n - 3}} \qquad (15.14)$$

Because there are three parameters—a, b_1, and b_2—for Equation 15.2 that we must estimate before calculating the residual, the number of degress of freedom is ($n - 3$). In other words, ($n - 3$) sample values are "free" to vary. More generally, the number of degrees of freedom for estimating the residual standard error for Equation 15.4 is [$n - (k + 1)$].

EXAMPLE 15.2   *Computing y_i, e_i, and e_i^2*

Using the data presented in Example 15.1, we can estimate y_i, e_i and e_i^2, as shown in Table 15.3.

Here $\hat{y}_i$ is obtained by substituting x_{1i} and x_{2i} into Equation 15.12. For example, $14.1198 = .980 + 1.2385(5) + .9925(7)$; $\hat{e}_i = y_i - \hat{y}_i$.

$$\sum_{i=1}^{5} (y_i - \hat{y}_i)^2 = (15 - 14.1198)^2 + (17 - 18.3272)^2$$
$$+ (26 - 26.0209)^2 + (24 - 25.0200)^2$$
$$+ (27 - 25.5120)^2 = 5.7912$$

Hence,

$$s_e = \sqrt{\frac{5.7912}{5 - 3}} = 1.7016$$

s_e is one of the important components in determining the distribution of estimated a, b_1, and b_2 and fitted dependent variable ($\hat{y}$).

Table 15.3 Actual Values, Predicted Values, and Residuals for Annual Salary Regression

	Actual Value, y_i	Predicted Value, $\hat{y}_i$	Residuals e_i	e_i^2
	15	14.1198	.8802	.7748
	17	18.3272	−1.3272	1.7615
	26	26.0209	−.0209	.0004
	24	25.0200	−1.0200	1.0404
	27	25.5120	1.4880	2.2141
Total	109	—	—	5.7912

The Coefficient of Determination

We can use Equation 15.13 to calculate a relative measure of the goodness of fit for a multiple regression.

$$R^2 = \frac{\sum_{i=1}^{n} (\hat{y}_i - \bar{y})^2}{\sum_{i=1}^{n} (y_i - \bar{y})^2} = \frac{\text{explained variation of } y \text{ (SSR)}}{\text{total variation of } y \text{ (SST)}}$$

$$= 1 - \frac{\text{SSE}}{\text{SST}}$$

(15.15)

The **coefficient of determination** R^2 is the proportion of total variation in y (SST) that is explained by the intercept and the independent variable x_1 and x_2. Note that both R^2 and s_e can be used to measure the goodness of fit for a regression. However, R^2 is a relative measure and s_e an absolute measure. Now we use the ANOVA table given in Table 15.4 to calculate the relationship between R^2 and s_e for the general multiple regression model in Equation 15.4.

Table 15.4 Notation of Analysis of Variance Table

(1) Source of Variation	(2) Sum of Squares	(3) Degrees of Freedom	(4) Mean Square
Due to regression	$\text{SSR} = \sum_{i=1}^{n} (\hat{y}_i - \bar{y})^2$	k	SSR/k
Residual	$\text{SSE} = \sum_{i=1}^{n} (y_i - \hat{y}_i)^2$	$n - k - 1$	SSE/$(n - k - 1)$
Total	$\text{SST} = \sum_{i=1}^{n} (y_i - \bar{y})^2$	$n - 1$	SST/$(n - 1)$

There are four columns in Table 15.4. Column (1) represents the sources of variation, column (2) alternative sums of squares that are identical to those discussed in Equation 15.13, column (3) degrees of freedom associated with each source of variation, and column (4) the mean squares. Note that alternative mean squares represent alternative variance estimates. Mean square due to the regression are also called explained variance; mean square due to the residuals are also called unexplained variance; and mean square due to the total variation can also be called variance of the dependent variable. Using those estimates, we can obtain an adjusted (or corrected) coefficient of determination $\overline{R}^2$.

$$\overline{R}^2 = 1 - \frac{SSE/(n-k-1)}{SST/(n-1)} = 1 - (1-R^2)\frac{n-1}{n-k-1} \qquad (15.16)$$

The difference between R^2 and $\overline{R}^2$ is that $\overline{R}^2$ is adjusted for degrees of freedom for both SSE and SST. $\overline{R}^2$ is always smaller than R^2. If the sample size becomes large, however, $\overline{R}^2$ approaches R^2. $\overline{R}^2$ can generally help us avoid overestimating the goodness of fit for a regression relationship by adding more independent variables (relevant or not) to a regression equation. Note that the standard error of estimate (Equation 15.14) also has been adjusted for the degrees of freedom $(n-k-1)$.

If we divide components in Equation 15.13 by their related degrees of freedom, then it can be shown that

$$\underbrace{\frac{\sum_{i=1}^{n}(y_i-\overline{y})^2}{n-1}}_{\substack{\text{Total}\\\text{Variance}}} \neq \underbrace{\frac{\sum_{i=1}^{n}(y_i-\hat{y}_i)^2}{k}}_{\substack{\text{Unexplained}\\\text{Variance}}} + \underbrace{\frac{\sum_{i=1}^{n}(\hat{y}_i-\overline{y})^2}{n-k}}_{\substack{\text{Explained}\\\text{Variance}}} \qquad (15.17)$$

so the unadjusted coefficient of determination can be redefined as

$$\overline{R}^2 = 1 - \frac{\text{explained variance}}{\text{total variance}} \qquad (15.16')$$

Using the example of the last section, we can calculate the analysis of variance of Table 15.4 as shown in Table 15.5.

From Table 15.5, we can calculate R^2 and $\overline{R}^2$.

$$R^2 = 113.0088/118.8 = .95125$$

$$\overline{R}^2 = 1 - \frac{2.8956}{29.7} = 1 - .09749 = .90251$$

Table 15.5 Analysis of Variance Results

Source of Variation	Sum of Squares	Degrees of Freedom	Mean Square
Due to regression	113.0088	$k = 2$	56.5044
Residual	5.7912	$5 - 2 - 1 = 2$	2.8956
Total	118.8	$5 - 1 = 4$	29.7

Both R^2 and $\overline{R}^2$ imply that more than 90 percent of the variation of annual salary can be explained by years of education and years of work experience. However, $\overline{R}^2$ is 4.874 percent smaller than that of R^2.

15.5   TESTS ON SETS AND INDIVIDUAL REGRESSION COEFFICIENTS

After having estimated the regression model, we would like to know whether the dependent variable is related to the independent variables. To find out, we can test whether an individual regression coefficient or a set of regression coefficients is significantly different from zero. As we saw in Chapter 14, the t statistic is to test an individual coefficient, the F statistic to test linear restrictions on the parameters or regression coefficients. For this purpose, we need to assume that ϵ_i is normally distributed.

Logically, we perform the joint test first. If the joint test is not significant, then there is no need for the individual tests, and we normally abandon or modify the model. If the joint test is rejected, we must find out which regression coefficients are significant, so we perform individual tests.

Test on Sets of Regression Coefficients

Until now, our discussion has been limited to point estimation of multiple regression coefficients, the coefficient of determination, and the standard error of estimate. Now we will discuss how to use the F statistic to test whether all true population regression (slope) coefficients equal zero. The F test rather than the t test is used. The null hypothesis for our case is

$$H_0: \beta_1 = \beta_2 = \cdots = \beta_k = 0 \qquad (15.18)$$
$$H_1: \text{At least one } \beta \text{ is not zero.}$$

If the null hypothesis is not true, then each $\hat{y}_i$ will differ from $\overline{y}$ substantially, and the explained variation $\sum_{i=1}^{n}(\hat{y}_i - \overline{y})^2$ will be large relative to the unexplained residual variation $\sum_{i=1}^{n}(y_i - \hat{y}_i)^2$. In other words, the R^2 indicated in Equation 15.15 is relatively large. Thus we can construct the F ratio as indicated in Equation 15.19 to test whether the null hypothesis can be rejected.

$$F_{k,n-k-1} = \frac{\sum\limits_{i=1}^{n}(\hat{y}_i - \overline{y})^2/k}{\sum\limits_{i=1}^{n}(y_i - \hat{y}_i)^2/(n - k - 1)} \qquad (15.19)$$

The F ratio we have constructed is the ratio of two mean square errors, as we noted in the last section, and they are two unbiased estimates of variances. Following the definition of the F distribution established in Chapters 9 and 14, we know that the F ratio has an F distribution with k and $(n - k - 1)$ 9 degrees of freedom. This F ratio enables us to test whether at least one of the regression coefficients is significantly different from zero.

Consider the case $k = 2$. If there is no regression relationship (that is, if $\beta_1 = \beta_2 = 0$) and because

$$\hat{y}_i = a + b_1 x_1 + b_2 x_{2i}$$
$$= \bar{y} + b_1(x_{1i} - \bar{x}_1) + b_2(x_{2i} - \bar{x}_2)$$

the $\hat{y}_i$ will be close or equal to $\bar{y}$, so the F-value will be smaller or close to zero. Thus we cannot reject the null hypothesis that all regression coefficients are insignificantly different from zero.

Substituting related data from Table 15.5 into Equation 15.19, we obtain

$$F = \frac{113.0088/2}{5.7912/2} = \frac{56.5044}{2.8956}$$
$$= 19.514$$

From Table A6 of Appendix A, we find that the critical value for a significance level of $\alpha = .05$ is $F_{.05,2,2} = 19.0$, which is smaller than 19.514. Therefore, we can conclude that at least one of the regression coefficients is significantly different from zero. Thus there is a regression relationship in the population, and the improvement of explanatory power achieved by fitting a regression plane is not due to chance. In other words, the null hypothesis that years of education and years of work experience contribute nothing to an individual's annual salary is rejected at a 5 percent level of significance.

Finally, the relationship between the R^2 indicated in Equation 15.15 and the F statistic in Equation 15.19 can be shown to be[6]

$$F_{k,n-k-1} = \frac{n - k - 1}{k} \cdot \frac{R^2}{1 - R^2}$$

Hypothesis Tests for Individual Regression Coefficients

In the last section, we used the F statistic to do a joint test about a regression relationship. Now we want to use the t statistic to test whether multiple regression coefficients are significantly different from zero.

Hypothesis Testing Specification

We follow the procedure of the last chapter to define the null hypothesis and alternative hypothesis for testing individual multiple regression coefficients.

---

[6]Because $R^2 = 1 - \text{SSE/SST} = \text{SSR/SST}$,

$$\frac{R^2}{1 - R^2} = \frac{\text{SSR/SST}}{\text{SSE/SST}} = \frac{\text{SSR}}{\text{SSE}}$$

1. Two-tailed test

$$H_0: \beta_j = 0 \qquad (j = 1, 2, \ldots, k) \tag{15.20}$$
$$H_1: \beta_j \neq 0$$

2. One-tailed test

$$H_0: \beta_j = 0 \qquad\qquad (j = 1, 2, \ldots, k) \tag{15.21}$$
$$H_1: \beta_j > 0 \text{ or } \beta_j < 0$$

Let's look at Equation 15.12 as an example. For convenience, the estimated regression line is repeated here.

$$\hat{y}_i = .980 + 1.2385 x_1 + .9925 x_2$$

In this equation, besides the estimated intercept (α) and slopes (β_1 and β_2), we have estimated the standard error of estimate for $\hat{y}_i$ as $s_e = 1.7016$. To perform the null hypothesis test, we need to know the sample distribution of b_j and the t statistic as defined in the equation

$$t_{n-k-1} = (b_j - 0)/s_{bj} \tag{15.22}$$

where t_{n-k-1} represents a t statistic with $(n - k - 1)$ degrees of freedom, $k =$ the number of independent variables, and s_{bj} represents the standard error associated with b_j. The concepts and procedure used to calculate s_{bj} are similar to those used for simple regression. However, s_{bj} is quite tedious to calculate by hand; fortunately, its value is readily available in the computer output of any standard regression analysis program. Thus in practice we find t simply by finding the ratio of the coefficient to its estimated standard error. When the calculated value of t exceeds the critical value $t_{\alpha,n-k-1}$ indicated in the t-distribution table, the null hypothesis of no significance can be rejected. We conclude that the jth independent variable x_j does have an important influence on the dependent variable y_i after the influence of all other independent variables in the model is taken into account.

Performing the t Test for Multiple Regression Slopes

To perform the t test for multiple regression coefficients b_1 and b_2, we estimate the sample variance of the coefficients b_1 and b_2 in accordance with Equations 15.23 and 15.24[7]

$$\mathrm{Var}(b_1) = s_{b_1}^2 = \frac{s_e^2}{(1 - r^2) \displaystyle\sum_{i=1}^{n} (x_{1i} - \bar{x}_1)^2} \tag{15.23}$$

[7]Derivations of Equations 15.23 and 15.24 can be found in Appendix 15A. Note that these two equations are generally estimated by computer packages (see Section 15.8). Manual approaches are presented here to show how sample variances of multiple regression slopes are actually calculated.

$$= \frac{s_e^2 \left(\sum_{i=1}^{n} x_{2i}'^2 \right)}{\left(\sum_{i=1}^{n} x_{1i}'^2 \right) \left(\sum_{i=1}^{n} x_{2i}'^2 \right) - \left(\sum_{i=1}^{n} x_{1i}' x_{2i}' \right)^2}$$

$$\mathrm{Var}(b_2) = s_{b_2}^2 = \frac{s_e^2}{(1 - r^2) \sum_{i=1}^{n} (x_{2i} - \bar{x}_2)^2} \tag{15.24}$$

$$= \frac{s_e^2 \left(\sum_{i=1}^{n} x_{1i}'^2 \right)}{\left(\sum_{i=1}^{n} x_{1i}'^2 \right) \left(\sum_{i=1}^{n} x_{2i}'^2 \right) - \left(\sum_{i=1}^{n} x_{1i}' x_{2i}' \right)^2}$$

where r represents the correlation coefficient between x_{1i} and x_{2i}. If the magnitude of r is great, a collinearity problem might exist. This issue will be discussed in detail in the next chapter.

Substituting the required numerical values obtained from Tables 15.2 and 15.3, we calculate sample variances of b_1 and b_2 for Equation 15.16.

$$S_{b_1}^2 = \frac{(2.8956)(50)}{(59.2)(50) - (-11)^2}$$

$$= \frac{(2.8956)(50)}{2839}$$

$$= .05100$$

and

$$s_{b_2}^2 = \frac{(2.8956)(59.2)}{2839}$$

$$= .06038$$

Then $s_{b_1} = .2258$ and $s_{b_2} = .2457$. Dividing b_1 and b_2 by s_{b_1} and s_{b_2}, we obtain t-values for b_1 and b_2.

$$t_{b_1} = \frac{1.2385}{.2258} = 5.4849$$

$$t_{b_2} = \frac{.9925}{.2457} = 4.0395$$

Because $n = 5$ and $k = 2$, from Table A4 in Appendix A the critical value for a one-tailed test on either coefficient (at a significance level of $\alpha = .05$) is

$$t_{\alpha, n-k-1} = t_{.05, 2} = 2.920$$

We choose a one-tailed test because *a priori* theoretical propositions were that both x_1 and x_2 were positively related to y. Comparing 5.4849 and 4.0395 with 2.920, we conclude that both years of education and years of work experience are significantly related to an individual's annual salary.

Figure 15.2 presents all the estimates and hypothesis-testing information we have discussed in the last three sections. This example certainly proves that multiple regression analysis can be more efficiently performed by using the MINITAB computer program.

Figure 15.2

MINITAB Output of Multiple Regression in Terms of Data Given in Table 15.1

```
MTB > READ C1-C3
DATA> 5 7 15
DATA> 10 5 17
DATA> 9 14 26
DATA> 13 8 24
DATA> 15 6 27
DATA> END
        5 ROWS READ
MTB > BRIEF 3
MTB > REGRESS C3 2 C1 C2;
SUBC> DW.

The regression equation is
GM = 0.98 + 1.24 C1 + 0.992 C2

Predictor       Coef       Stdev     t-ratio          p
Constant       0.980       3.439        0.29      0.802
C1            1.2385       0.2258        5.48      0.032
C2            0.9925       0.2457        4.04      0.056

s = 1.702      R-sq = 95.1%      R-sq(adj) = 90.3%

Analysis of Variance

SOURCE         DF          SS          MS         F        p
Regression      2     113.009      56.504     19.51    0.049
Error           2       5.791       2.896
Total           4     118.800

SOURCE         DF      SEQ SS
C1              1      65.773
C2              1      47.236

Obs.      C1         GM      Fit Stdev.Fit   Residual   St.Resid
  1      5.0     15.000   14.120     1.499      0.880       1.09
  2     10.0     17.000   18.327     1.076     -1.327      -1.01
  3      9.0     26.000   26.021     1.632     -0.021      -0.04
  4     13.0     24.000   25.020     0.961     -1.020      -0.73
  5     15.0     27.000   25.512     1.301      1.488       1.36

Durbin-Watson statistic = 2.39
```

15.6 CONFIDENCE INTERVAL FOR THE MEAN RESPONSE AND PREDICTION INTERVAL FOR THE INDIVIDUAL RESPONSE

Point Estimates of the Mean and the Individual Responses

One of the important uses of the multiple regression line is to obtain predictions and forecasts for the dependent variable, given an assumed set of values of the independent variables. This kind of prediction is called the **conditional prediction** (forecast), just as in simple regression (see Section 14.5, of which this model is an extension). Suppose the independent variables are equal to some specified values $x_{1,n+1}$ and $x_{2,n+1}$ and that the linear relationship among y_n, $x_{1,n}$, and $x_{2,n}$ continues to hold.[8] Then the corresponding value of the dependent variable y_{n+1} is

$$Y_{n+1} = \alpha + \beta_1 X_{1,n+1} + \beta_2 X_{2,n+1} + \epsilon_{n+1} \tag{15.25}$$

which, given $x_{1,n+1}$ and $x_{2,n+1}$, has expectation

$$E(Y_{n+1} | x_{1,n+1}, x_{2,n+1}) = \alpha + \beta_1 x_{1,n+1} + \beta_2 x_{2,n+1} \tag{15.26}$$

Equation 15.26 yields the **mean response** $E(Y_{n+1} | x_{1,n+1}, x_{2,n+1})$ that we want to estimate when the independent variables are fixed at $x_{1,n+1}$ and $x_{2,n+1}$. Equation 15.25 yields the **actual value** (or **individual response**) that we want to predict.

To obtain the best point estimate, we first estimate the sample regression line as defined in Equation 15.2. Then we substitute the given values $x_{1,n+1}$ and $x_{2,n+1}$ into the estimated Equation 15.12, obtaining

$$\hat{y}_{n+1} = a + b_1 x_{1,n+1} + b_2 x_{2,n+1} \tag{15.27}$$

This is the best point estimate for both conditional-expectation and actual-value forecasts. In other words, the forecast of conditional expectation value is equal to the forecast of actual value. However, the forecasts are interpreted differently. The importance of these different interpretations will emerge when we investigate the process of making interval estimates.

Interval Estimates of Forecasts

To construct a confidence interval for forecasts, it is necessary to know the distribution, mean, and variance of $\hat{y}_{n+1}$. The distribution of $\hat{y}_{n+1}$ is a t distribution with $(n-3)$ degrees of freedom. The variance associated with $\hat{y}_{n+1}$ may be classified into three cases. First, we deal with a case in which the conditional mean $(\hat{y}_{n+1})$ is equal to the unconditional mean $(\bar{y})$. In the second and third cases, we deal with the conditional mean. However, case 2 involves the mean response and case 3 the individual response.

---

[8] $x_{1,n+1}$ and $x_{2,n+1}$ can be either given values or forecasted values. When a regression is used to describe a time-series relationship, they are forecasted values.

Case 15.1 Conditional Expectation (Mean Response) with $x_{1,n+1} = \bar{x}_1$ and $\bar{x}_{2,n+1} = \bar{x}_2$

From the definitions of the intercept of a regression and the sample regression line, we have

$$\hat{y}_{n+1} = (\bar{y} - b_1\bar{x}_1 - b_2\bar{x}_2) + b_1 x_{1,n+1} + b_2 x_{2,n+1}$$
$$= \bar{y} + b_1(x_{1,n+1} - \bar{x}_1) + b_2(x_{2,n+1} - \bar{x}_2)$$

If $x_{1,n} = \bar{x}_1$ and $x_{2,n} = \bar{x}_2$, then $\hat{y}_{n+1} = \bar{y}$. Following Appendix 14C, we obtain the estimate of the variance for y_{n+1} as

$$s^2(\hat{y}_{n+1}) = s^2(\bar{y}) = s_e^2/n \tag{15.28}$$

Case 15.2 Conditional Expectation (Mean Response) with $x_{1,n+1} \neq \bar{x}_1$ and $x_{2,n+1} \neq \bar{x}_2$

In this case, the forecast value can be defined as

$$\hat{y}_{n+1} = \bar{y} + b_1(x_{1,n+1} - \bar{x}_1) + b_2(x_{2,n+1} - \bar{x}_2) \tag{15.29}$$

Following Appendix 15B, we obtain the estimate of the variance for $\hat{y}_{n+1}$ in terms of sample standard variance of estimates s_e^2 as

$$s_1^2 = s^2(\hat{y}_{n+1}) = s_e^2 \left[\frac{1}{n} + \frac{(x_{1,n+1} - \bar{x}_1)^2}{(1 - r^2)C_1^2} + \frac{(x_{2,n+1} - \bar{x}_2)^2}{(1 - r^2)C_2^2} \right.$$
$$\left. - \frac{2(x_{1,n+1} - \bar{x}_1)(x_{2,n+1} - \bar{x}_2)r}{(1 - r^2)C_1 C_2} \right] \tag{15.30}$$

where $C_1 = \sqrt{\sum_{i=1}^{n}(x_{1,i} - \bar{x}_1)^2}$, $C_2 = \sqrt{\sum_{i=1}^{n}(x_{2,i} - \bar{x}_2)^2}$ and r = correlation coefficient between $x_{1,i}$ and $x_{2,i}$.

Case 15.3 Actual Value (Individual Response) of y_{n+1}

After we have derived the sample variance for $\hat{y}_{n+1}$, we derive the sample variance for individual response (observation), $y_{n+1,i}$ (which deviates from $\hat{y}_{n+1}$ by a random error e).

$$y_{n+1,i} = \hat{y}_{n+1} + e_i$$

The variance of an individual observation, $y_{n+1,i}$, includes the variance of the observation about the regression line (s_e^2) as well as $s^2(\hat{y}_{n+1,i})$. Because $\hat{y}_{n+1}$ and e_i are independent, $s^2(y_{n+1,i}) = s^2(\hat{y}_{n+1,i}) + s_e^2$. More explicitly,

$$s^2(\hat{y}_{n+1,i}) = s_1^2 + s_e^2 = s_2^2 \tag{15.31}$$

where s_1^2 is defined in Equation 15.30.

Using Equations 15.28, 15.30, and 15.31, we can obtain a confidence interval for prediction as follows:

1. For prediction of the conditional expectation with $x_{1,n+1} = \bar{x}_1$ and $x_{2,n+1} = \bar{x}_2$, the confidence interval is

$$\hat{y}_{n+1} \pm t_{n-3,\alpha/2} \frac{s_e}{\sqrt{n}} \tag{15.32}$$

2. For prediction of the conditional expectation with $x_{1,n+1} \neq \bar{x}_1$ and $x_{2,n+1} \neq \bar{x}_2$, the confidence interval is

$$\hat{y}_{n+1} \pm (t_{n-3,\alpha/2})s_1 \tag{15.33}$$

where s_1 is defined in Equation 15.30.

3. For prediction of the actual value y_{n+1}, the prediction interval is

$$\hat{y}_{n+1,i} \pm (t_{n-3,\alpha/2})s_2 \tag{15.34}$$

where s_2 is defined in Equation 15.31.

To show how Equation 15.34 is applied in constructing the confidence interval for forecasting the actual value of y_{n+1}, let's use the annual salary example (Table 15.2) to find the 95 percent prediction interval for annual salary, y_{n+1}, when a person has 6 years of education and 5 years of work experience. The predicted annual salary can be computed from Equation 15.12.

$$\hat{y}_{n+1,i} = .980 + (1.2385)(6) + (.9925)(5)$$

$$= 13.3735 \text{ (in thousands of dollars)}$$

From Table 15.2, we have

$$\sum_{i=1}^{n} (x_{1i} - \bar{x}_1)^2 = 59.2, \sum_{i=1}^{n} (x_{2i} - \bar{x}_2)^2 = 50, n = 5$$

$$\sum_{i=1}^{n} (x_{1i} - \bar{x}_1)(x_{2i} - \bar{x}_2) = -11, \bar{x}_1 = 10.4, \bar{x}_2 = 8$$

Using this information, we calculate

$$C_1 = \sqrt{59.2} = 7.6942, \qquad C_2 = \sqrt{50} = 7.0711$$

$$r = \frac{\sum_{i=1}^{n} (x_{1i} - \bar{x}_1)(x_{2i} - \bar{x}_2)}{\sqrt{\sum_{i=1}^{n} (x_{1i} - \bar{x}_1)^2 \sum_{i=1}^{n} (x_{2i} - \bar{x}_2)^2}} = \frac{-11}{(7.6942)(7.0711)}$$

$$= -.2022, r^2 = .0409$$

$$(x_{1,n+1} - \bar{x}_1)^2 = (6 - 10.4)^2 = 19.36$$

$$(x_{2,n+1} - \bar{x}_2)^2 = (5 - 8)^2 = 9$$

From Table 15.3, we have $s_e^2 = 5.7912/(5 - 3) = 2.8956$. Substituting this information into Equation 15.31 yields

$$s_2^2 = (2.8956) \left[1 + \frac{1}{5} + \frac{19.36}{(1 - .0409)(59.2)} + \frac{9}{(1 - .0409)(50)} \right.$$
$$\left. - \frac{2(-.2022)(-4.4)(-3)}{(1 - .0409)(7.6942)(7.0711)} \right]$$
$$= (2.8956)(1.83) = 5.2989$$

We will use $n = 5$, $s_2 = \sqrt{5.2989} = 2.3019$, and $\hat{y}_{n+1,i} = 13.3735$. From Table A4, in Appendix A, we have $t_{2,.025} = 4.303$. Substituting this information into Equation 15.34, we find that the annual salary is predicted with 95 percent confidence by the interval

$$13.3735 \pm (4.303)(2.3019) = 13.3735 \pm 9.9051$$
$$3.4684 \leq y_{n+1,i} \leq 23.2786$$

When n is large, we can modify this expression by replacing t with the appropriate normal deviate z.

MINITAB output showing prediction results of $x_{1,n+1,i} = 6$ and $x_{2,n+1,i} = 5$ is presented in Figure 15.3. The prediction interval shown in the last row of Figure 15.3 is (3.466, 23.280), which is similar to what we calculated before.

Figure 15.3

MINITAB Output of $y_{n+1,i}$

```
MTB > READ C1-C3
DATA> 5 7 15
DATA> 10 5 17
DATA> 9 14 26
DATA> 13 8 24
DATA> 15 6 27
DATA> END
      5 ROWS READ
MTB > REGRESS C3 2 C1 C2;
SUBC> DW;
SUBC> PREDICT 6 5.

The regression equation is
GM = 0.98 + 1.24 C1 + 0.992 C2

Predictor      Coef      Stdev    t-ratio        p
Constant      0.980      3.439       0.29    0.802
C1           1.2385     0.2258       5.48    0.032
C2           0.9925     0.2457       4.04    0.056

s = 1.702      R-sq = 95.1%     R-sq(adj) = 90.3%

Analysis of Variance

SOURCE        DF          SS         MS        F        p
Regression     2     113.009     56.504    19.51    0.049
Error          2       5.791      2.896
Total          4     118.800
```

(continued)

Figure 15.3

(Continued)

```
SOURCE        DF      SEQ SS
C1            1       65.773
C2            1       47.236

Obs.     C1         GM       Fit  Stdev.Fit   Residual   St.Resid
  1     5.0     15.000    14.120    1.499      0.880      1.09
  2    10.0     17.000    18.327    1.076     -1.327     -1.01
  3     9.0     26.000    26.021    1.632     -0.021     -0.04
  4    13.0     24.000    25.020    0.961     -1.020     -0.73
  5    15.0     27.000    25.512    1.301      1.488      1.36


Durbin-Watson statistic = 2.39

    Fit   Stdev.Fit        95% C.I.           95% P.I.
  13.373    1.551     ( 6.699, 20.047)   ( 3.466, 23.280)
```

In the next two sections, we will explore applications of multiple regression in business and economics. Section 15.8 explicitly treats the use of SAS and MINITAB computer programs to do multiple regression analyses.

15.7   BUSINESS AND ECONOMIC APPLICATIONS

Multiple regression analysis has been widely used in decision making in business and economics. Five examples are discussed in this section.

■ | **APPLICATION 15.1**   Overall Job-Worth of Performance for Certain Army Jobs

Bobko and Donnelly (1988) employed multiple regression to estimate overall job-worth to the army of certain army jobs from attributes of those jobs.[9] Their final regression prediction model is

$$y_i = b_0 + b_1 x_{1i} + b_2 x_{2i} + b_3 x_{3i} + b_4 x_{4i} + b_5 x_{5i} + b_6 x_{6i} + b_7 x_{7i} + e_i$$

where

y_i = job-value judgments of overall worth for the ith individual
x_{1i} = performance level for the ith job
x_{2i} = combat probability for the ith job
x_{3i} = enlistment bonus for the ith job
x_{4i} = reenlistment bonus for the ith job
x_{5i} = aptitude required for entry into the ith job
x_{6i} = cost of error for the ith job
x_{7i} = job variety for the ith job

[9]P. Bobko and L. Donnelly (1988), "Identifying Correlations of Job-Level, Overall Worth Estimates: Application in a Public Sector Organization," *Human Performance* 3, 187–204.

Table 15.6 Best Subset Regression of Overall Worth on Job-Level Predictors

Source	df	Sum of Squares	F	p
Regression	7	16.007	274.12	.0001
Error (residual)	87	.726		

Variable	Regression Weight	F	p
Performance level	.013	1666.22	.0001
Combat probability	.039	21.19	.0001
Enlistment bonus	.034	18.52	.0001
Reenlistment bonus	.016	15.73	.0001
Aptitude	.013	26.01	.0001
Cost of error	.029	5.61	.0201
Task variety	.016	2.51	.1166

Source: Bobko and Donnelly (1988), *Human Performance.*
Note: Adjusted R^2 = .953; n = 95 mean estimates of overall worth.

Bobko and Donnelly estimated this multiple regression model using data obtained from interviews. Their regression results are presented in Table 15.6. As would be expected, performance level was the single best predictor of 95 estimates of judgments of overall worth. The other job-level correlates were combat probability, enlistment bonus, reenlistment bonus, aptitude, cost of error, and task variety. The first six of these predictors had statistically significant regression weights (coefficients), p-value $< .05$, indicating their unique contribution to the prediction of overall worth estimates. However, task variety was not statistically significant.

APPLICATION 15.2 The Relationship Between Individual Stock Rates of Return, Payout Ratio, and Market Rates of Return

To demonstrate multiple regression analysis, a time-series regression for first quarter 1981 to first quarter 1991 is run, the dependent variable being the rate of return for the IBM stock ($R_{j,t}$) and the independent variables being the payout ratio (dividend per share/earnings per share) for IBM ($P_{j,t}$) and the rates of return on the S&P 500 Index, $R_{m,t}$. The results are as follows:

$$R_{j,t} = \alpha_j + \gamma_j P_{j,t} + \beta_j R_{m,t} + \epsilon_{j,t}$$

Fortunately, the results do not have to be calculated by hand but can be obtained by using MINITAB. The MINITAB results are presented in Table 15.7. The parameter value for the market rates of return is .9193, which is called the beta coefficient. A 1 percent increase in the market rate of return will lead to a .9193 percent change in the rate of return of the IBM stock, given the payout ratio—that is, the rate of return of IBM stock is less volatile than that of the market. The payout ratio has a coefficient of $-.14341$. This result implies that a 1 percent increase in

Table 15.7 $R_{j,t} = \alpha_j + \gamma_j P_{j,t} + \beta_j R_{m,t} + \epsilon_{j,t}$

Variable	Coefficient	Standard Error	t-Ratio	p-Value
Constant	.0847	.0391	2.17	.036
Market rate of return	.9193	.1631	5.64	.000
Payout ratio	−.1434	.0659	−2.18	.036

$R^2 = .472$
$\overline{R}^2 = .444$
F-value $= 26.64$
Observations   41

the payout ratio will lead to a .1434 percent decrease in the mean rate of return on IBM stock, given the market rate of return.

The independent variables are statistically significant at the 5 percent level. The t-value for the market is 5.64, and the associated p-value is .000. This strongly suggests that the population coefficient for the market is not equal to zero. The t statistic for the payout ratio, which is calculated by dividing the parameter value (−.1434) by the standard error (.0659), is −2.18. Its p-value is .036, which means that the lowest level of significance at which the null hypothesis can be rejected is 3.6 percent.

R^2 for the regression is .472. In other words, the independent variables explain about 47 percent of the variation in the rate of return on IBM stock. The adjusted R-square, $\overline{R}^2$, which takes into account overfitting in the sample, is equal to .444.

The F-value, which tests the hypothesis that the population coefficients of the independent variables are both zero against the alternative that they are not, is equal to 26.64. The degrees of freedom associated with this F-value are $\nu_1 = 2$ and $\nu_2 = 38$. From Table A6 in Appendix A, we find that the critical value for the F test is $F_{.01,2,30} = 5.39$ and $F_{.01,2,40} = 5.18$. Because the F-value for the regression is greater than the critical value 5.18, the null hypothesis should be rejected.

■ **APPLICATION 15.3** Analyzing the Determination of Price per Share

To further demonstrate multiple regression techniques, let us say that a **cross-section regression** is run. In a cross-section regression, all data come from a single period. The dependent variable in this regression is the price per share (P_j) of the 30 firms used to compile the Dow Jones Industrial Average for the year 1983. The independent variables are the dividend per share (D_j) and the retained earnings per share (RE$_j$) for the 30 firms. (Retained earnings per share is defined as earnings per share minus dividend per share. Price per share is the high–low average of 1983; dividend per share and retained earnings per share are based on 1983 annual balance sheet and income statement.) The sample regression relationship is

$$P_j = a + b_1 D_j + b_2 \text{RE}_j + e_j \qquad (j = 1, 2, \ldots, 30)$$

Table 15.8 $P_j = a + b_1 D_j + b_2 RE_j + e_j$

Variable	Coefficient	Standard Error	*T* Ratio	*p*-value
Constant	22.77	5.49	4.15	.000
DPS	10.73	2.51	4.28	.000
RE	1.43	.623	2.29	.015

$R^2 = .544$
$\overline{R}^2 = .510$
F-value $= 16.11$
Observations   30

Empirical results are presented in Table 15.8. The constant term is highly significant with a *t*-value of 4.15. This result means that the intercept term is statistically different from zero. The dividend-per-share variable is highly significant with a *t*-value of 4.28 and a *p*-value of .000. Thus we can reject the null hypothesis that the population coefficient is equal to zero and accept the alternative hypothesis that it differs from zero and makes a contribution to price per share. The coefficient for this variable is 10.73; mean price per share increases $10.73 when the dividend per share increases by $1.00, given retained earnings.

The coefficient for the retained-earnings variable has a *t*-value of 2.29 and a *p*-value of .015. This is the lowest level of significance at which the null hypothesis can be rejected; thus the null hypothesis is rejected at both a 10 and a 5 percent level. The coefficient for retained earnings is 1.43. When the retained earnings increases by $1.00, the price per share tends to rise by $1.43.

The value of R^2 is .54, which means that the model explains 54 percent of the observed fluctuations in the price per share. The adjusted *R*-square, $\overline{R}^2$, is .51. The *F*-value for the regression is 16.10. The number of degrees of freedom for the regression and residual are 2 and 27, respectively. The critical value for *F* at a 1 percent level of significance is 5.49. Because the regression *F*-value is greater than the critical value, the null hypothesis that the coefficients are equal to zero is rejected. It should be noted that R^2 tends to be *smaller* with cross-sectional data than with time-series data; this is because greater amounts of "noise" (fluctuation in observations) exist in different firms.

■ | *APPLICATION 15.4* Multiple Regression Approach to Evaluating Real Estate Property

To show how the multiple regression technique can be used by real estate appraisers, Andrews and Ferguson (1986) used the data in Table 15.9 to do the multiple regression analysis

$$y_i = b_0 + b_1 x_{1i} + b_2 x_{2i} + e_i$$

Table 15.9 Sale Price, House Size, and Condition Rating

Sale Price, y (thousands of dollars)	Home Size, x_1 (hundreds of sq. ft.)	Condition Rating, x_2 (1 to 10)
60.0	23	5
32.7	11	2
57.7	20	9
45.5	17	3
47.0	15	8
55.3	21	4
64.5	24	7
42.6	13	6
54.5	19	7
57.5	25	2

Source: R. L. Andrews and J. T. Ferguson, "Integrating Judgment with a Regression Appraisal." *The Real Estate Appraiser and Analyst,* Vol. 52, No. 2, Spring 1986 (Table 1).

where

y_i = sale price for i^{th} house

x_{1i} = home size for i^{th} house

x_{2i} = condition rating for i^{th} house

MINITAB regression outputs in terms of Table 15.9 are presented in Figure 15.4. From this output, the estimated regression is

$$\hat{y}_i = 9.782 + 1.87094x_{1i} + 1.2781x_{2i}$$
$$(6.00) \qquad (24.56) \qquad\quad (8.85)$$

t-values are in parenthesis.

Figure 15.4

MINITAB Output for Table 15.9

```
MTB > READ C1-C3
DATA> 60.0 23 5
DATA> 32.7 11 2
DATA> 57.7 20 9
DATA> 45.5 17 3
DATA> 47.0 15 8
DATA> 55.3 21 4
DATA> 64.5 24 7
DATA> 42.6 13 6
DATA> 54.5 19 7
DATA> 57.5 25 2
DATA> END
      10 ROWS READ
MTB > BRIEF 1
MTB > REGRESS C1 2 C2 C3;
SUBC> DW.

The regression equation is
C1 = 9.78 + 1.87 C2 + 1.28 GM
```

Figure 15.4

(Continued)

```
Predictor      Coef       Stdev     t-ratio       p
Constant      9.782       1.630        6.00    0.000
C2          1.87094     0.07617       24.56    0.000
GM           1.2781      0.1444        8.85    0.000

s = 1.081      R-sq = 99.0%     R-sq(adj) = 98.7%

Analysis of Variance

SOURCE        DF         SS          MS         F        p
Regression     2       819.33      409.66    350.87   0.000
Error          7         8.17        1.17
Total          9       827.50


Durbin-Watson statistic = 1.56


MTB > CORRELATION C1-C3

              C1         C2
C2          0.938
GM          0.373      0.043

MTB > PAPER
```

From Table A4 in Appendix A, we find that $t_{.025,7} = 2.365$. Because t-values for 3 regression parameters are larger than 2.365, all estimated parameters are significantly different from 0 at $\alpha = .05$. This estimated regression can be used to estimate the sale price for a house. For example, if $x_1 = 18$ and $x_2 = 5$, the predicted sale price is

$$\hat{y}_i = 9.781 + (1.87094)(18) + (1.2781)(5)$$

$$= 49.8484$$

This implies that the estimated sale price is \$49,848.4 if the home size is 18,000 square feet and the condition rating is 5.

■ **APPLICATION 15.5** Multiple Regression Approach to Doing Cost Analysis

To show how the multiple regression technique can be used to do cost analysis by accountants, we look at Benston's research. Benston (1966) used a set of sample data (as shown in Table 15.10) from a firm's accounting and production records to provide cost information about the firm's shipping department to do the multiple regression analysis

$$y_t = b_0 + b_1 x_{1t} + b_2 x_{2t} + b_3 x_{3t} + e$$

where

y_t = hours of labor in t^{th} week

x_{1t} = thousands of pounds shipped in t^{th} week

x_{2t} = percentage of units shipped by truck in t^{th} week

x_{3t} = average number of pounds per shipment in t^{th} week

Table 15.10 Hours of Labor and Related Factors Cause Costs to Be Incurred

Week	Hours of Labor, y	Thousands of Pounds Shipped, x_1	Percentage of Units Shipped by Truck, x_2	Average Number of Pounds per Shipment, x_3
1	100	5.1	90	20
2	85	3.8	99	22
3	108	5.3	58	19
4	116	7.5	16	15
5	92	4.5	54	20
6	63	3.3	42	26
7	79	5.3	12	25
8	101	5.9	32	21
9	88	4.0	56	24
10	71	4.2	64	29
11	122	6.8	78	10
12	85	3.9	90	30
13	50	3.8	74	28
14	114	7.5	89	14
15	104	4.5	90	21
16	111	6.0	40	20
17	110	8.1	55	16
18	100	2.9	64	19
19	82	4.0	35	23
20	85	4.8	58	25

Source: G. J. Benston (1966), "Multiple Regression Analysis of Cost Behavior," *Accounting Review,* Vol. 41, No. 4, 657–672. Reprinted by permission of the publisher.

MINITAB regression output is presented in Figure 15.5. From p-values indicated in Figure 15.5, we find that b_0 and b_3 are significantly different from 0 at $\alpha = .01$. Hence we can conclude that the only important variable in determining the hours of labor required in the shipping department is the average number of pounds per shipment.

Figure 15.5

MINITAB
Output for
Application 15.5

```
MTB > READ C1-C4
DATA> 100 5.1 90 20
DATA> 85 3.8 99 22
DATA> 108 5.3 58 19
DATA> 116 7.5 16 15
DATA> 92 4.5 54 20
DATA> 63 3.3 42 26
DATA> 79 5.3 12 25
DATA> 101 5.9 32 21
DATA> 88 4.0 56 24
DATA> 71 4.2 64 29
DATA> 122 6.8 78 10
DATA> 85 3.9 90 30
DATA> 50 3.8 74 28
DATA> 114 7.5 89 14
DATA> 104 4.5 90 21
DATA> 111 6.0 40 20
DATA> 110 8.1 55 16
DATA> 100 2.9 64 19
DATA> 82 4.0 35 23
DATA> 85 4.8 58 25
DATA> END
     20 ROWS READ
MTB > REGRESS C1 3 C2 C3 C4;
SUBC> DW.

The regression equation is
C1 = 132 + 2.73 C2 + 0.0472 C3 - 2.59 C4

Predictor       Coef      Stdev    t-ratio         p
Constant      131.92      25.69       5.13     0.000
C2             2.726      2.275       1.20     0.248
C3           0.04722    0.09335       0.51     0.620
C4           -2.5874     0.6428      -4.03     0.001

s = 9.810      R-sq = 77.0%     R-sq(adj) = 72.7%

Analysis of Variance

SOURCE          DF          SS          MS         F         p
Regression       3      5158.3      1719.4     17.87     0.000
Error           16      1539.9        96.2
Total           19      6698.2

SOURCE          DF      SEQ SS
C2               1      3400.6
C3               1       198.4
C4               1      1559.3

Unusual Observations
Obs.       C2          C1      Fit Stdev.Fit  Residual   St.Resid
  13     3.80       50.00    73.33      3.92    -23.33      -2.59R

R denotes an obs. with a large st. resid.

Durbin-Watson statistic = 2.43

MTB > PAPER
```

15.8 USING COMPUTER PROGRAMS TO DO MULTIPLE REGRESSION ANALYSES

SAS Program for Multiple Regression Analysis

In an example taken from Churchill's *Marketing Research,* data for the sales of Click ballpoint pens (y), advertising (x_1, measured in TV spots per month), number of sales representatives (x_2), and a wholesaler efficiency index (x_3) were presented in Table 14.10 of the last chapter.

In Section 14.6, we investigated only the relationship between two variables (y and x_1, y and x_2, and y and x_3). Now we will expand that analysis by using the following three regression models:[10]

$$y_i = a + b_1 x_{1i} + e_i \tag{a}$$

$$y_i = a + b_1 x_{1i} + b_2 x_{2i} + e_i \tag{b}$$

$$y_i = a + b_1 x_{1i} + b_2 x_{2i} + b_3 x_{3i} + e_i \tag{c}$$

Equation a can be used to investigate the relationship between y and x_1, which was discussed in Section 14.6.

Figure 15.6a

SAS Output for Regression Results of $y_i = a + b_1 x_{1i} + e_i$

```
Model: MODEL1
Dependent Variable: Y

                        Analysis of Variance

                        Sum of          Mean
Source        DF        Squares         Square      F Value    Prob>F

Model          1    463451.00888   463451.00888    130.644     0.0001
Error         38    134802.01487     3547.42144
C Total       39    598253.02375


      Root MSE        59.56023     R-square     0.7747
      Dep Mean       411.28750     Adj R-sq     0.7687
      C.V.            14.48141


                        Parameter Estimates

                    Parameter     Standard     T for H0:
    Variable   DF    Estimate       Error     Parameter=0    Prob > |T|

    INTERCEP    1   135.433596   25.90650568      5.228        0.0001
    X1          1    25.307698    2.21415038     11.430        0.0001



Durbin-Watson D              1.721
(For Number of Obs.)            40
1st Order Autocorrelation    0.133
```

[10]In these regressions we hold the price of a ballpoint pen and the income of a consumer constant, because this is a set of cross-sectional data.

Equation b can be used to analyze whether the second explanatory variable, x_2, improves the equation's power to explain the variation of sales. Equation c can be used to analyze whether the third explanatory variable, x_3, further improves that explanatory power. Part of the output of the SAS program for Equations a, b, and c is presented in Figures 15.6a, 15.6b, and 15.6c. Figure 15.6a shows the regression results of Equation a, Figure 15.6b the regression results of Equation b, and Figure 15.6c the regression results of Equation c. Using these results, we will review and summarize simple regression and multiple regression results that have been discussed in Chapters 13, 14, and 15.

Figure 15.6b

SAS Output for Regression Results of
$y_i = a + b_1 x_{1i} + b_2 x_{2i} + e_i$

Dependent Variable: Y

Analysis of Variance

Source	DF	Sum of Squares	Mean Square	F Value	Prob>F
Model	2	522778.45899	261389.22949	128.141	0.0001
Error	37	75474.56476	2039.85310		
C Total	39	598253.02375			

Root MSE	45.16473	R-square	0.8738
Dep Mean	411.28750	Adj R-sq	0.8670
C.V.	10.98130		

Parameter Estimates

Variable	DF	Parameter Estimate	Standard Error	T for H0: Parameter=0	Prob > |T|
INTERCEP	1	69.328469	23.15546229	2.994	0.0049
X1	1	14.156185	2.66360071	5.315	0.0001
X2	1	37.531322	6.95929855	5.393	0.0001

Durbin-Watson D	2.125
(For Number of Obs.)	40
1st Order Autocorrelation	-0.083

Figure 15.6c

SAS Output for Regression Results of
$y_i = a + b_1 x_{1i} + b_2 x_{2i} + b_3 x_{3i} + e_i$

Dependent Variable: Y

Analysis of Variance

Source	DF	Sum of Squares	Mean Square	F Value	Prob>F
Model	3	527209.08074	175736.36025	89.051	0.0001
Error	36	71043.94301	1973.44286		
C Total	39	598253.02375			

Root MSE	44.42345	R-square	0.8812
Dep Mean	411.28750	Adj R-sq	0.8714
C.V.	10.80107		

(continued)

Figure 15.6c
(Continued)

Parameter Estimates

Variable	DF	Parameter Estimate	Standard Error	T for H0: Parameter=0	Prob > |T|
INTERCEP	1	31.150390	34.17504533	0.911	0.3681
X1	1	12.968162	2.73723213	4.738	0.0001
X2	1	41.245624	7.28010741	5.666	0.0001
X3	1	11.524255	7.69117684	1.498	0.1428

```
Durbin-Watson D               2.104
(For Number of Obs.)            40
1st Order Autocorrelation    -0.083
```

Computer outputs of Figures 15.6a, 15.6b and 15.6c present the following results of simple and multiple regression.

1. Estimated intercept and slopes
2. F-values for the whole regression
3. t-values for individual regression coefficients
4. ANOVA of regression
5. R^2 and $\overline{R}^2$
6. p-values
7. Durbin–Watson D and first-order autocorrelation (these two statistics are discussed in Section 16.4 of Chapter 16)
8. Standard error of residual estimate (mean square of error)
9. Root MSE $= \sqrt{\text{MSE}}$ error. For example, for Equation b, Root MSE $= \sqrt{2039.85310} = 45.16473$. The root MSE estimate can be used to measure the performance of prediction. We will explore this in further detail in Chapter 18 when we discuss time-series analysis.

These SAS regression outputs give us almost all the sample statistics we have examined so far. Now let's consider the practical implications of Equations a, b, and c. In Section 14.6 of the last chapter, we discussed the estimated regression of Equation a.

Equation b specifies a regression model in which sales is the dependent variable and the independent variables are number of TV spots x_1 and number of sales representatives x_2. The fitted regression equation is

$$\hat{y} = 69.3 + 14.2x_1 + 37.5x_2 \qquad F = 128.141$$
$$(2.994) \quad (5.315) \quad (5.393)$$

Here t-values are indicated in parentheses.

This regression indicates that when the number of TV spots increases by 1 unit, sales increase by $14,200 on average. And when the number of sales representatives increases by 1 person, sales increase by $37,500 on average.

The F-value for the regression of Equation b is 128.141. There are 40 observations and 2 independent variables, so the number of degrees of freedom in the model is $40 - 2 - 1 = 37$. By interpolation, it can be shown that the critical value of $F_{.05,2,37}$ is 3.25 (Table A6 in Appendix A). Because the F-value for the regression is greater than the critical value, the hypothesis that the coefficients are equal to zero is rejected. From the t-values associated with estimated regression coefficients, we find that the estimated intercept and slopes are significant at $\alpha = .01$.

Because the t-values of b_2 are significantly different from zero, we conclude that adding the number of sales representatives improves the equation's power to explain sales. This conclusion can also be drawn from the fact that $\overline{R}^2$ has increased from .7687 to .8670.

The fitted regression of Equation c is

$$\hat{y} = 31.1504 + 12.9682x_1 + 41.2456x_2 + 11.5243x_3 \qquad F = 89.051$$
$$\phantom{\hat{y} = 31.1504 +} (.911) \qquad (4.738) \qquad (5.666) \qquad (1.498)$$

Again, t-values are indicated in parentheses.

Following Section 15.5, we first test the whole set of regression coefficients in terms of the F statistic. From Table A6 in Appendix A, by interpolation, we find that $F_{.01,3,36} = 2.88$. $F = 89.051$ is much larger than 2.88. This implies that we reject the following null hypothesis of our joint test:

$$H_0: \beta_1 = \beta_2 = \beta_3 = 0$$

Now we can use t statistics to test which individual coefficient is significantly different from zero. From Table A4 in Appendix A, by interpolation, we find that the critical value of t statistic is $t_{.005,36} = 2.72$. By comparing this critical value with 4.738, 5.666, and 1.498, we conclude that b_1 and b_2 are significantly different from zero and that b_3 is not significantly different from zero at $\alpha = .01$. In other words, the wholesaler efficiency index does not increase the explanatory power of Equation c.

MINITAB Program for Multiple Regression Prediction

MINITAB is used to run the regression defined in Figure 15.6c and presented in Figure 15.7. Besides regression parameters, we also predict y by assuming $x_1 = 13$, $x_2 = 9$, and $x_3 = 5$. The results are listed in the last row of Figure 15.7. They are

1. $\hat{y}_{n+1,i} = 628.57$
2. $s(\hat{y}_{n+1}) = 34.92$
3. $s(\hat{y}_{n+1,i}) = \sqrt{s^2(\hat{y}_{n+1} + s_e^2)}$
$$= \sqrt{(34.92)^2 + 1973}$$
$$= 56.50$$
4. 95 percent confidence interval: (557.73, 699.40)
5. 95 percent prediction interval: (513.94, 743.19)

Figure 15.7

MINITAB Output of $y_i = a + b_1 x_{1i} + b_2 x_{2i} + b_3 x_{3i} + e_i$

```
MTB > PRINT C1-C4

ROW     C1      C2   C3   C4

  1    260.3     5    3    4
  2    286.1     7    5    2
  3    279.4     6    3    3
  4    410.8     9    4    4
  5    438.2    12    6    1
  6    315.3     8    3    4
  7    565.1    11    7    3
  8    570.0    16    8    2
  9    426.1    13    4    3
 10    315.0     7    3    4
 11    403.6    10    6    1
 12    220.5     4    4    1
 13    343.6     9    4    3
 14    644.6    17    8    4
 15    520.4    19    7    2
 16    329.5     9    3    2
 17    426.0    11    6    4
 18    343.2     8    3    3
 19    450.4    13    5    4
 20    421.8    14    5    2
 21    245.6     7    4    4
 22    503.3    16    6    3
 23    375.7     9    5    3
 24    265.5     5    3    3
 25    620.6    18    6    4
 26    450.5    18    5    3
 27    270.1     5    3    2
 28    368.0     7    6    2
 29    556.1    12    7    1
 30    570.0    13    6    4
 31    318.5     8    4    3
 32    260.2     6    3    2
 33    667.0    16    8    2
 34    618.3    19    8    2
 35    525.3    17    7    4
 36    332.2    10    4    3
 37    393.2    12    5    3
 38    283.5     8    3    3
 39    376.2    10    5    4
 40    481.8    12    5    2
```

Figure 15.7
(Continued)

```
MTB > NAME C1'Y' C2'X1' C3'X2' C4'X3'
MTB > BRIEF 3
MTB > REGRESS C1 3 C2 C3 C4;
SUBC> DW;
SUBC> PREDICT 13 9 5.

The regression equation is
Y = 31.2 + 13.0 X1 + 41.2 X2 + 11.5 X3

Predictor        Coef       Stdev     t-ratio          p
Constant        31.15       34.18        0.91      0.368
X1             12.968       2.737        4.74      0.000
X2             41.246       7.280        5.67      0.000
X3             11.524       7.691        1.50      0.143

s = 44.42       R-sq = 88.1%     R-sq(adj) = 87.1%

Analysis of Variance

SOURCE          DF          SS          MS          F          p
Regression       3      527209      175736      89.05      0.000
Error           36       71044        1973
Total           39      598253

SOURCE          DF      SEQ SS
X1               1      463451
X2               1       59327
X3               1        4431

Obs.      X1          Y        Fit Stdev.Fit   Residual    St.Resid
  1      5.0      260.30     265.83     14.96      -5.53       -0.13
  2      7.0      286.10     351.20     12.82     -65.10       -1.53
  3      6.0      279.40     267.27     11.35      12.13        0.28
  4      9.0      410.80     358.94     11.54      51.86        1.21
  5     12.0      438.20     445.77     15.11      -7.57       -0.18
  6      8.0      315.30     304.73     13.18      10.57        0.25
  7     11.0      565.10     497.09     16.43      68.01        1.65
  8     16.0      570.00     591.65     15.24     -21.65       -0.52
  9     13.0      426.10     399.29     13.88      26.81        0.64
 10      7.0      315.00     291.76     13.24      23.24        0.55
 11     10.0      403.60     419.83     15.63     -16.23       -0.39
 12      4.0      220.50     259.53     18.78     -39.03       -0.97
 13      9.0      343.60     347.42      8.26      -3.82       -0.09
 14     17.0      644.60     627.67     18.77      16.93        0.42
 15     19.0      520.40     589.31     17.23     -68.91       -1.68
 16      9.0      329.50     294.65     15.87      34.85        0.84
 17     11.0      426.00     467.37     14.98     -41.37       -0.99
 18      8.0      343.20     293.21     11.61      49.99        1.17
 19     13.0      450.40     452.06     11.57      -1.66       -0.04
 20     14.0      421.80     441.98     13.88     -20.18       -0.48
 21      7.0      245.60     333.01     13.61     -87.41       -2.07R
 22     16.0      503.30     520.69     11.52     -17.39       -0.41
 23      9.0      375.70     388.66      9.07     -12.96       -0.30
 24      5.0      265.50     254.30     12.18      11.20        0.26
 25     18.0      620.60     558.15     16.71      62.45        1.52
```

(continued)

Figure 15.7
(Continued)

```
26   18.0   450.50   505.38   20.34   -54.88   -1.39
27    5.0   270.10   242.78   13.82    27.32    0.65
28    7.0   368.00   392.45   17.60   -24.45   -0.60
29   12.0   556.10   487.01   16.82    69.09    1.68
30   13.0   570.00   493.31   12.85    76.69    1.80
31    8.0   318.50   334.45    8.63   -15.95   -0.37
32    6.0   260.20   255.74   13.55     4.46    0.11
33   16.0   667.00   591.65   15.24    75.35    1.81
34   19.0   618.30   630.56   16.53   -12.26   -0.30
35   17.0   525.30   586.43   15.37   -61.13   -1.47
36   10.0   332.20   360.39    8.77   -28.19   -0.65
37   12.0   393.20   427.57    7.61   -34.37   -0.79
38    8.0   283.50   293.21   11.61    -9.71   -0.23
39   10.0   376.20   413.16   12.25   -36.96   -0.87
40   12.0   481.80   416.04   10.48    65.76    1.52
```
R denotes an obs. with a large st. resid.

Durbin-Watson statistic = 2.10

```
    Fit   Stdev.Fit       95% C.I.           95% P.I.
 628.57       34.92   ( 557.73, 699.40)  ( 513.94, 743.19) XX
```

X denotes a row with X values away from the center
XX denotes a row with very extreme X values

MTB > PAPER

Stepwise Regression Analysis

In this example, we want to use **stepwise regression** to establish a statistical model to predict the sales of Click ballpoint pens (y). We are considering three possible explanatory variables: advertising (x_1) measured in TV spots per month, the number of sales representatives (x_2), and a wholesaler efficiency index (x_3). The question is what variables should be included in the statistical model to explain the sales. The stepwise regression method suggests the following steps.

Step 1:

Run simple regression on each explanatory variable, and choose the model that explains the highest amount of variation in y. The regression results obtained are presented in Figures 14.14a and 14.14b. The R^2-value in each computer report is used to determine which variable enters the model first. Upon comparing R^2 values for the three models, we conclude that x_2, which has the highest R^2-value (.7775), should enter the model first.

Independent Variable	R^2	F-value
x_1	.7747	130.644
x_2	.7775	132.811
x_3	.0000	.000

Step 2:

The second variable to enter should be the variable that, in conjunction with the first variable, explains the greatest amount of variation in y.

Independent Variables	R^2	F-value
x_2 x_1	.8738	128.141
x_2 x_3	.807	77.46

The R^2-values and F-values in the foregoing table are obtained from Figure 15.6b and Figure 15.8. The table shows the results when x_1 and x_3 are combined with x_2 to explain the variation in y. The combination of x_1 and x_2 clearly yields a higher R^2 (.8738). This suggests that x_1 should be the second variable to enter.

Figure 15.8

MINITAB Output of $y_i = a + b_1 x_2 + b_3 x_3 + e_i$

```
MTB > BRIEF 2
MTB > REGRESS C1 2 C3 C4;
SUBC> DW.

The regression equation is
C1 = 5.2 + 68.7 C3 + 22.1 C4

Predictor      Coef      Stdev    t-ratio        p
Constant       5.19      42.40       0.12    0.903
C3           68.744       5.523      12.45    0.000
C4           22.079       9.252       2.39    0.022

s = 55.83      R-sq = 80.7%    R-sq(adj) = 79.7%

Analysis of Variance

SOURCE        DF          SS        MS        F        p
Regression     2      482914    241457    77.46    0.000
Error         37      115339      3117
Total         39      598253

SOURCE        DF      SEQ SS
C3             1      465161
C4             1       17753

Unusual Observations
Obs.      C3         C1      Fit Stdev.Fit  Residual   St.Resid
 21     4.00     245.60   368.49    14.28   -122.89      -2.28R
 25     6.00     620.60   505.97    15.79    114.63       2.14R

R denotes an obs. with a large st. resid.

Durbin-Watson statistic = 2.22

MTB > PAPER
```

Step 3:

In this step, we want to decide whether another variable should enter the model to explain y. Note that every time an additional variable is included in a model, R^2 increases. The question is whether the increase in R^2 justifies inclusion of the variable. We apply an F test to answer this question.

$$F = \frac{(R_f^2 - R_R^2)/(k_f - k_r)}{(1 - R_R^2)/(N - k_f - 1)}$$

where

$R_f^2 = R^2$ of the model with the new variable
$R_R^2 = R^2$ of the model without the new variable
$k_f =$ number of the variables in the model with the new variable
$k_R =$ number of the variables in the model without the new variable

To determine whether x_3 should be included in the model, we need to compare the R^2 of the model with x_3 and R^2 of the model without x_3.

Independent Variables	R^2
x_2 x_1	.8738
x_2 x_1 x_3	.8812

Using the foregoing formula, we compute

$$F = \frac{(.8812 - .8738)/(3 - 2)}{(1 - .8812)/(40 - 3 - 1)} = 2.24 < F_{.05,1,36} = 4.11$$

Because including x_3 does not increase R^2 significantly, the null hypothesis that x_3 should not be included is not rejected in this case. Our conclusion from the stepwise regression analysis is that the best model should include only x_1 and x_2 as explanatory variables.

Some computer packages are programmed to perform the whole complicated stepwise regression in response to one simple command. Figure 15.9 shows the output of a stepwise regression analysis using MINITAB.

Figure 15.9

Stepwise Regression Analysis

```
MTB > STEPWISE REGRESSION C1 3 C2 C3 C4

 STEPWISE REGRESSION OF     C1    ON  3 PREDICTORS, WITH N =   40

        STEP         1        2
     CONSTANT      80.07    69.33

     C3            66.2     37.5
     T-RATIO       11.52    5.39

     C2                     14.2
     T-RATIO                5.31
```

Figure 15.9
(Continued)

```
S                59.2      45.2
R-SQ            77.75     87.38
 MORE? (YES, NO, SUBCOMMAND, OR HELP)
SUBC> YES

 NO VARIABLES ENTERED OR REMOVED

 MORE? (YES, NO, SUBCOMMAND, OR HELP)
SUBC> NO
MTB > PAPER
```

Summary

In this chapter we examined multiple regression analysis, which describes the relationship between a dependent variable and two or more independent variables. Methods of estimating multiple regression (slope) coefficients and their standard errors were discussed in depth. The residual standard error and the coefficient of determination were also explored in some detail.

Both t tests and F tests for testing regression relationships were discussed in this chapter. We investigated the confidence interval for the mean response and the prediction interval for the individual response. And finally, we saw how multiple regression analyses can be used in business and economics decision making.

Appendix 15A Derivation of the Sampling Variance of the Least-Squares Slope Estimations

Using the definition of the simple correlation coefficient given in Equation 13.24 in Chapter 13, we can obtain the correlation coefficient between x_1 an x_2 as

$$r = \frac{\sum_{i=1}^{n} (x_{1i} - \bar{x}_1)(x_{2i} - \bar{x}_2)}{C_1 C_2} \tag{15A.1}$$

where

$$C_1 = \sqrt{\sum_{i=1}^{n} (x_{1i} - \bar{x}_1)^2} \tag{15A.2a}$$

$$C_2 = \sqrt{\sum_{i=1}^{n} (x_{2i} - \bar{x}_2)^2} \tag{15A.2b}$$

Substituting 15A.1, 15A.2a, and 15A.2b into Equation 15.7 yields

$$
\begin{aligned}
b_1 &= \frac{C_2^2 \left[\sum_{i=1}^{n} (x_{1i} - \bar{x}_1)(y_i') \right] - rC_1C_2 \left[\sum_{i=1}^{n} (x_{2i} - \bar{x}_2)(y_i') \right]}{C_1^2 C_2^2 - r^2 C_1^2 C_2^2} \\
&= \sum_{i=1}^{n} \left[\frac{(x_{1i} - \bar{x}_1)(y_i') - (rC_1/C_2)(x_{2i} - \bar{x}_2)(y_i')}{(1 - r^2)C_1^2} \right] \\
&= \sum_{i=1}^{n} \left[\frac{(x_{1i} - \bar{x}_1) - (rC_1/C_2)(x_{2i} - \bar{x}_2)}{(1 - r^2)C_1^2} \right] y_i'
\end{aligned}
\qquad (15A.3)
$$

Substituting

$$
\sum_{i=1}^{n} (x_{1i} - \bar{x}_1)y_i' = \sum_{i=1}^{n} (x_{1i} - \bar{x}_1) y_i
$$

and

$$
\sum_{i=1}^{n} (x_{2i} - \bar{x}_2)y_i' = \sum_{i=1}^{n} (x_{2i} - \bar{x}_2)y_i
$$

into Equation 15A.3, and letting the coefficient of y_i equal B_{1i}, we obtain

$$
b_1 = \sum_{i=1}^{n} B_{1i}y_i
\qquad (15A.4)
$$

Similarly,

$$
\begin{aligned}
b_2 &= \sum_{i=1}^{n} \left(\frac{(x_{2i} - \bar{x}_2) - \dfrac{rC_2}{C_1}(x_{1i} - \bar{x}_i)}{(1 - r^2)C_2^2} \right) y_i \\
&= \sum_{i=1}^{n} B_{2i}y_i
\end{aligned}
\qquad (15A.5)
$$

Substituting Equation 15.2 into Equations 15A.4 and 15A.5, we get

$$
b_1 = a \sum_{i=1}^{n} B_{1i} + b_1 \sum_{i=1}^{n} B_{1i}x_{1i} + b_2 \sum_{i=1}^{n} B_{1i}x_{2i} + \sum_{i=1}^{n} B_{1i}e_i
$$

and

$$
b_2 = a \sum_{i=1}^{n} B_{2i} + b_1 \sum_{i=1}^{n} B_{2i}x_{1i} + b_2 \sum_{i=1}^{n} B_{2i}x_{2i} + \sum_{i=1}^{n} B_{2i}e_i
$$

It can easily be shown that $\Sigma_{i=1}^n B_{1i} = 0$, $\Sigma_{i=1}^n B_{2i} = 0$, $\Sigma_{i=1}^n B_{1i}x_{1i} = 1$, $\Sigma_{i=1}^n B_{2i}x_{2i} = 1$, $\Sigma_{i=1}^n B_{2i}x_{1i} = 0$, and $\Sigma_{i=1}^n B_{1i}x_{2i} = 0$. Therefore, these two equations imply that

$$b_1 - E(b_1) = b_1 - \beta_1 = \sum_{i=1}^n B_{1i}e_i \qquad (15A.6)$$

and

$$b_2 - E(b_2) = b_2 - \beta_2 = \sum_{i=1}^n B_{2i}e_i \qquad (15A.7)$$

From Equation 15A.6 we obtain

$$\text{Var}(b_1) = E\left[\left(\sum_{i=1}^n B_{1i}e_i\right)^2\right] - \left[E\left(\sum_{i=1}^n B_{1i}e_i\right)\right]^2$$

$$= \sum_{i=1}^n \sum_{j=1}^n B_{1i}B_{1j}E(e_ie_j) - \left(\sum_{i=1}^n B_{1i}\right)^2 [E(e_i)]^2 = s_e^2 \sum_{i=1}^n B_{1i}^2 \qquad (15A.8)$$

In Equation 15.8, the last equality holds because $E(e_i) = 0$ and $E(e_ie_j) = 0$ when $i \neq j$.

From the definition of B_{1i} in Equation 15A.4, we have

$$B_{1i}^2 = \frac{(x_{1i} - \bar{x}_1)^2 + r^2C_1^2/C_2^2(x_{2i} - \bar{x}_2)^2 - 2r(C_1/C_2)(x_{1i} - \bar{x}_1)(x_{2i} - \bar{x}_2)}{(1 - r^2)^2 C_1^4}$$

$$(15A.9)$$

And from Equations 15A.9, 15A.1, 15A.2a, and 15A.2b,

$$\sum_{i=1}^n B_{1i}^2 = \frac{C_1^2 + r^2(C_1^2/C_2^2)(C_2^2) - 2r(C^1/C_2)(r_1C_1C_2)}{(1 - r^2)^2 C_1^4}$$

$$= \frac{C_1^2[1 - 2r + r^2]}{(1 - r^2)^2/C_1^4} \qquad (15A.10)$$

$$= \frac{1}{(1 - r^2)C_1^2}$$

Substituting Equation 15A.10 into Equation 15A.8 yields

$$\text{Var}(b_1) = \frac{s_e^2}{(1 - r^2)\sum_{i=1}^n (x_{1i} - \bar{x}_1)^2}. \qquad (15A.11)$$

Similarly, it can be proved that

$$\text{Var}(b_2) = \frac{s_e^2}{(1 - r^2)\sum_{i=1}^n (x_{2i} - \bar{x}_2)^2} \qquad (15A.12)$$

Equations 15A.11 and 15A.12 are Equations 15.23 and 15.24, respectively.

If the correlation coefficient between x_1 and x_2—that is, r—is equal to zero, then $\text{Var}(b_1)$ and $\text{Var}(b_2)$ reduce to

$$\text{Var}(b_1) = \frac{s_e^2}{\displaystyle\sum_{i=1}^{n} (x_{1i} - \bar{x}_1)^2} \quad \text{and} \quad \text{Var}(b_2) = \frac{s_e^2}{\displaystyle\sum_{i=1}^{n} (x_{2i} - \bar{x}_2)^2}$$

This implies that the sample variance of multiple regression slopes reduces to a simple regression case, as indicated in Equation 14.2.

Appendix 15B Derivation of Equation 15.30

From Equation 15.27, we have

$$
\begin{aligned}
\hat{y}_{n+1} &= a + b_1 x_{1,n+1} + b_2 x_{2,n+1} \\
&= \bar{y} + b_1(x_{1,n+1} - \bar{x}_1) + b_2(x_{2,n+1} - \bar{x}_2)
\end{aligned}
\tag{15B.1}
$$

Hence, we obtain

$$
\begin{aligned}
\text{Var}(\hat{y}_{n+1}) &= \text{Var}[\bar{y} + b_1(x_{1,n+1} - \bar{x}_1) + b_2(x_{2,n+1} - \bar{x}_2)] \\
&= \text{Var}(\bar{y}) + \text{Var}[b_1(x_{1,n+1} - \bar{x}_1) + \text{Var}[b_2(x_{2,n+1} - \bar{x}_2)] \\
&\quad + 2\,\text{Cov}[b_1(x_{1,n+1} - \bar{x}_1), b_2(x_{1,n+1} - \bar{x}_2)] \\
&= \frac{\sigma_e^2}{n} + (x_{1,n+1} - \bar{x}_1)^2 \text{Var}(b_1) + (x_{2,n+1} - \bar{x}_2)^2 \text{Var}(b_2) \\
&\quad + 2(x_{1,n+1} - \bar{x}_1)(x_{2,n+1} - \bar{x}_2)\text{Cov}(b_1, b_2)
\end{aligned}
\tag{15B.2}
$$

From Equations 15A.6 and 15A.7 in Appendix 15A, we can show that

$$
\begin{aligned}
\text{Cov}(b_1, b_2) &= E(b_1 - \beta_1)(b_2 - \beta_2) \\
&= \sum_{i=1}^{n} \sum_{j=1}^{n} B_{1i} B_{2j} E(e_i e_j)
\end{aligned}
\tag{15B.3}
$$

Because $E(e_i e_j) = 0$ when $i \neq j$, Equation 15B.3 reduces to

$$
\text{Cov}(b_1, b_2) = s_e^2 \sum_{i=1}^{n} B_{1i} B_{2i}
\tag{15B.4}
$$

where B_{1i} and B_{2i} are defined in Equations 15A.4 and 15A.5. It can be shown that

$$
\sum_{i=1}^{n} B_{1i} B_{2i} = \frac{-r}{(1 - r^2)C_1 C_2}
\tag{15B.5}
$$

and substituting Equations 15B.5 into Equation 15B.4 yields

$$Cov(b_1, b_2) = \frac{-rs_e^2}{(1 - r^2)\sqrt{\sum_{i=1}^{n}(x_{1,i} - \bar{x}_1)^2 \sum_{i=1}^{n}(x_{2,i} - \bar{x}_2)^2}} \quad (15B.6)$$

Substituting Equations 15A.9, 15A.10, and 15B.5 into Equation 15B.2, we have

$$Var(y_{n+1}) = \left(\frac{1}{n} + \frac{(x_{1,n+1} - \bar{x}_1)^2}{(1 - r^2)\sum_{i=1}^{n}(x_{1,i} - \bar{x}_1)^2} + \frac{(x_{2,n+1} - \bar{x}_2)^2}{(1 - r^2)\sum_{i=1}^{n}(x_{2,i} - \bar{x}_2)^2} \right.$$
$$\left. - \frac{2(x_{1,n+1} - \bar{x}_1)(x_{2,n+1} - \bar{x}_2)r}{(1 - r^2)\sqrt{\sum_{i=1}^{n}(x_{1,i} - \bar{x}_1)^2 \sum_{i=1}^{n}(x_{2,i} - \bar{x}_2)^2}} \right) s_e^2 \quad (15B.7)$$

This is Equation 15.30.

Questions and Problems

1. Compare simple regression to multiple regression. When would you use simple regression? When would you use multiple regression?

2. In simple regression, the geometric interpretation is to fit a line that best describes the relationship between x and y. What is the geometric interpretation of multiple regression when there are two independent variables?

3. Discuss the differences between the assumptions of the simple and the multiple linear regression models.

4. We can test the significance of a simple regression either by using a t test to test the slope coefficient or by using an F test to test the significance of the model. How does our approach differ when we are testing the significance of a multiple regression?

5. Explain how the number of degrees of freedom available for estimating σ^2 of the error term is related to the number of variables in the regression.

6. Briefly compare the concepts of simple correlation, partial correlation, and multiple correlation.

7. Compare the ways the regression coefficients are interpreted in simple regression and in multiple regression.

8. What is a partial regression coefficient? How do we measure it?

9. What is multicollinearity? Why is it a problem in multiple regression?

10. Suppose an NFL scout is interested in what physical attributes make for a good quarterback. He collects data on the height and weight of 8 quarterbacks and their performance ratings for the year. The data are summarized in the following table.

Performance Rating, y	Height (inches), x_1	Weight (pounds), x_2
94.3	73	210
83.3	69	185
92.3	77	225
72.4	75	215
69.5	71	190
65.8	70	180
101.2	76	212
77.4	73	195

Use the MINITAB program to answer the following questions.
  a. Estimate the regression coefficients α, β_1, and β_2 and interpret the results.
  b. Compute the t-values for the coefficients and test the significance of β_1 and β_2.

11. Using the information given in question 10, compute SSR, SSE, SST, and R^2. Also use an F test to test the significance of the model.

12. Using the results from question 10, forecast the performance rating for a quarterback who is 6 feet 1 inch tall and weighs 200 pounds. Construct a 95 percent confidence interval around this forecast.

13. The chairperson of the finance department at Rutgers University would like to find the relationship between undergraduate grade point average (UGPA) and GMAT scores on graduate grade point average (GGPA). She collects the following data on 6 students.

UGPA, x_1	GMAT, x_2	GGPA, y
3.45	485	3.62
3.10	500	3.75
3.00	525	3.81
2.95	560	3.88
3.11	575	3.85
2.87	625	3.95

  a. Calculate the regression parameters for α, β_1, and β_2.
  b. Compute the standard errors of the regression coefficients. Use a t test to test the significance of b_1 and b_2.

14. Suppose we were interested in testing the joint significance of b_1 and b_2 in terms of data from question 13. That is, the null hypothesis is H_0: $\beta_1 = \beta_2 = 0$.
  a. Explain how we would conduct such a test.
  b. Test the joint significance of β_1 and β_2.

15. Use the data and results from question 13 to construct 90 percent confidence intervals for b_1 and b_2.

16. Suppose a student has a 3.85 undergraduate GPA and a GMAT score of 575.
  a. Forecast this student's graduate GPA.
  b. Construct a 90 percent confidence interval for this forecast.

17. Suppose a labor economist is interested in the effect of experience and education on income. He obtains the following regression.

$$INCOME = 24,000 + 1,000(EXPER) + 500(EDUC)$$

where

INCOME	= income measured in dollars
EXPER	= years of experience
EDUC	= years of education

Interpret the regression coefficients for EXPER and EDUC.

18. Suppose you calculate $s_{b_1} = 325$ and $s_{b_2} = 285$, and you know that 50 observations were used to estimate the model. Test the significance of the regression coefficients in question 17.

19. An agent for Decade 100 Real Estate Company is interested in developing a model that explains the value of a piece of real estate. She collects data on the following variables:

Number of bedrooms	Sales price
Number of bathrooms	Age of house
Miles from main highway	Size of lot

  a. Which variables should be the independent variables?
  b. Write down a multiple regression equation that might be of interest to this realtor, and explain to her what signs to expect for the regression coefficients and how to interpret the regression coefficients.
  c. Explain the usefulness of this model.
  d. Will employing confidence intervals for forecasted values be useful in this analysis? Explain.

20. Suppose you estimate a regression using 20 observations and 16 independent variables. You compute R^2 to be .98. Explain why R^2 may not be an appropriate measure of the goodness of fit. Can you think of a better one?

21. Suppose a travel consultant is interested in the relationship between people's incomes and the amount of money they spend for vacations. He chooses to estimate the regression

$$E(VAC) = \alpha + \beta_1(WSAL) + \beta_2(MSAL)$$

where

VAC = dollars spent on vacation
WSAL = weekly salary
MSAL = monthly salary

Do you think he will encounter any difficulties in estimating this model?

22. Thomas Chen, an education professor, is interested in the relationship among final exam scores, midterm exam scores, and hours studied for the final. He collects the following data.

Final Exam Score, y	Midterm Exam Score, x_1	Hours Studied, x_2
75	74	5
83	89	8
72	65	9
88	92	4
95	90	10

a. Estimate the regression coefficients for α, β_1, and β_2.
b. Compute the standard error of the estimated R^2 and of the adjusted R^2.

23. Using the data and your results from question 22, test the joint significance of β_1 and β_2. Also construct a 99 percent confidence interval for β_1 and β_2.

24. Using the data and your results from question 22, forecast the final exam score for a student who scored 97 on the midterm and studied $6\frac{1}{2}$ hours for the final. Construct a 90 percent confidence interval for this forecast.

25. In multiple regression, we can test the significance of the individual regression coefficients by using a t test, or we can test the joint significance of the coefficients by using an F test. Is it possible for the t tests to be significant while the F test is insignificant? Explain.

26. An economist at the National Academy of Movie Theater Owners wants to estimate the demand for movie tickets. He chooses to estimate the equation.

$$QT_t = \alpha + \beta_1 PT_t + \beta_2(GNP_t) + \epsilon_t$$

where

QT$_t$ = quantity of movie tickets purchased in year t
PT$_t$ = average price of movie tickets in year t
GNP$_t$ = gross national product in year t (in billions of dollars)

What signs do you expect for the coefficients on price and GNP to have?

27. Suppose the economist of question 26 collects the following data.

Year	QT	PT	GNP
1986	1,000	$7.00	1,000
1987	1,100	7.25	1,250
1988	1,200	6.75	1,175
1989	1,300	6.50	1,800
1990	1,400	6.50	2,000
1991	1,500	6.25	2,250

Use MINITAB to answer the following questions.
a. Estimate the demand for movie tickets.
b. Do the coefficients carry the corrrect signs?
c. If you were going to use a t test to test the significance of b_1 and b_2, should you use a one-tailed or a two-tailed test?
d. Use a t test to test the significance of b_1 and b_2.

28. Use your results from question 27 to compute R^2 and $\overline{R}^2$. Also use an F test to test the joint significance of the regression.

29. Construct 95 percent confidence intervals for the coefficients b_1 and b_2 from the regression in question 27.

30. Suppose you have obtained the following 1992 and 1993 forecasts of GNP and ticket prices.

Year	GNP (in billions of dollars)	Prices
1992	2,572	7.25
1993	3,000	8.00

a. Forecast the quantity of tickets sold for 1992 and 1993.
b. Construct 90 percent confidence intervals for these forecasts. What information do these confidence intervals provide?

31. Suppose the economist in question 26 is interested in estimating the price and income elasticity of demand for movie tickets. He can do this by taking the natural logarithms of QT, PT, and GNP and reestimating the multiple regression. Using the data from question 27, estimate the price and income elasticity for movie tickets and interpret your results.

32. Use a t test to test the significance of the estimated elasticities.

33. Use an F test to test the joint significance of the price and income elasticity.

34. An investment analyst is interested in developing an equation to forecast the earnings per share of a company. He collects the following data for 5 companies.

Company	EPS	Sales in $	Advertising Expense in $	Cost in $
1	1.00	100	80	50
2	2.00	175	120	28
3	1.50	89	72	30
4	3.00	225	175	20
5	3.25	300	240	25

Use MINITAB to answer the following questions. (Hint: Follow the procedure used in Figure 15.7.)
a. Formulate a suitable regression model to explain EPS.
b. Are there any variables you may want to omit from the regression? If so, why?

35. Suppose the analyst of question 34 decides on the following regression:

$$EPS = \alpha + \beta_1(SALES) + \beta_2(COST) + \epsilon$$

a. Estimate the intercept and slope coefficients.
b. Use an F test to test the joint significance of the slope coefficients.
c. Compute the standard error for a, b_1, and b_2, and use a t test to test their significance.

36. Construct 90 percent confidence intervals for α, β_1, and β_2, using the results from question 35.

37. Forecast the EPS for a company with $400 in sales and a cost of $65. Construct a 99 percent confidence interval for this forecast.

38. You estimate a regression using a computer package that generates the following output.

Coefficient	Estimate	Standard Error
Intercept	12.53	6.54
X_1	−9.37	5.25
X_2	14.75	4.36
X_3	.27	.09

a. Compute the t-values for the coefficients.
b. Say the sample used to estimate the regression consisted of 27 observations. Are the coefficients significant?

39. Use the foregoing information to construct 95 percent confidence intervals for the parameters.

40. Buford Lightfoot, a stock market analyst, is interested in finding a model to describe the returns for different stocks. He estimates the following regression:

$$R_{it} = \alpha + \beta_1 R_{m,t} + \beta_2 I_{i,t} + \epsilon_t$$

where

$R_{m,t}$ = return on the S&P 500 in month t
$R_{i,t}$ = return on stock i in month t
$I_{i,t}$ = index for stock i's industry in month t

The results of this regression are

Coefficient	Estimate	Standard Error
Intercept	−3.45	2.32
R_m	1.32	.65
I_i	−.32	.10

Interpret the results of the regression and compute the t-values for the coefficients.

41. Construct a 90 percent confidence interval for the parameter estimates from question 40. Assume $n = 30$.

42. Say you know that the return on the S&P 500 will be 3 percent next month and that the industry index next month will be 2. Forecast stock i's return.

43. Suppose you fit the model

$$y = \alpha + \beta_1 x_1 + \beta_2 x_2 + \beta_3 x_1 x_2 + \beta_4 x_1^2 + \beta_5 x_2^2 + \epsilon$$

using 35 data points and obtain SSE = .56 and R^2 = .85. Test the null hypothesis that all β's are equal to zero against the alternative hypothesis that at least one of the β's is nonzero. Conduct this test at the 95 percent confidence level.

44. Again consider question 43. Examine SSE and R^2, and explain whether the model provides a good fit.

45. Suppose you estimate the model

$$z = \alpha + \beta_1 x_1 + \beta_2 x_2 + \epsilon$$

using 25 observations and obtain

$$\Sigma(z_i - \hat{z}_i)^2 = 2.45 \quad \text{and} \quad \Sigma(z_i - \bar{z})^2 = 3.65$$

Compute R^2. Does the model provide a good fit?

46. Explain why, given the same independent variables, the confidence interval for the mean value of y is always narrower than the corresponding confidence interval for any other value of y.

47. You are given the following information for x and y.

Cov(x,y) = 6.3   Mean of y = 75
Var(x) = 4.2 $\Sigma(x_i - \bar{x})^2 = 42$
Var(y) = 2,772.7 $s_e = 50$
Mean of x = 100 $n = 12$

a. Compute the least-squares estimates for the slope and intercept.
b. Compute the t-value for b and construct a 99 percent confidence interval for β.

48. Use the data from question 47 to compute the sample correlation coefficient r between x and y. Use a t test to test the significance of r.

49. Suppose x and y are bivariately normally distributed. Use a t test to test the significance of the sample correlation coefficient r if $r = .79$ and $n = 12$.

50. The admissions officer at Poindexter U. would like to determine the effect of high school GPA and SAT scores on undergraduate GPA. He collects the following data on 6 students.

HSGPA, x_1	SAT, x_2	UGPA, y
2.6	585	2.02
2.9	525	2.75
3.0	475	3.10
2.8	620	2.95
3.1	525	3.25
3.87	650	3.95

a. Calculate the regression parameters for α, β_1, and β_2.
b. Compute the standard errors of the regression coefficients. Use a t test to test the significance of b_1 and b_2.

51. Suppose we are interested in testing the joint significance of b_1 and b_2. That is, the null hypothesis is H_0: $\beta_1 = \beta_2 = 0$. Test the joint significance of β_1 and β_2.

52. Use the data and results from question 50 to construct 90 percent confidence intervals for β_1 and β_2.

53. Suppose a student with a 3.85 high school GPA and an SAT score of 555 applies for admission to Poindexter U.
a. Forecast this student's undergraduate GPA.
b. Construct a 90 percent confidence interval for this forecast.

54. You have been hired as an economist for the Federal Reserve Bank of New York. Your job is to forecast future interest rates. Summarize the theory of the interest rate, and formulate a mathematical model that can be used to forecast the interest rate.

55. You have been hired as a consultant for AT&T. Formulate a mathematical model that could be used to estimate the demand for telephone service.

56. You have been hired as an economist for the Public Utility Commission of Wisconsin. The agency needs an estimate of the demand for electricity in order to determine what rates the electric companies can charge. Formulate a mathematical model that can be used to estimate the demand for electricity.

57. Researchers interested in determining the relationship between a firm's annual sales and its expenditures on research and development (x_1), television advertising (x_2), and all other advertising (x_3) run a regression analysis of 23 firms in the same industry. The results are

$$\hat{y} = -2.3 + 5.8x_1 + 4.2x_2 + 7.4x_3$$
$$\phantom{\hat{y} = -2.3 +} (1.20) (1.31) (1.56)$$

The quantities in parentheses are the standard errors of the net regression coefficients. The standard error of estimate $S_{y.123}$ is 124. The standard deviation of the dependent variable S_y is 325.

a. Interpret the net regression coefficient b_1.
b. Test, at the 1 percent level of significance, whether each of the net regression coefficients is significantly different from zero.
c. What is the expected effect when highly correlated independent variables are included in a multiple regression equation?
d. Calculate the coefficient of multiple correlation and the coefficient of multiple determination.
e. Estimate the average annual sales for a firm that has research and development expenditures of $6 million, television advertising expenditures of $10 million, and all other advertising expenditures of $7 million.

58. A statistician is interested in using product price and the amount of advertising to predict the sales of a product. Several combinations of price and advertising are tried, with the following results. (Sales are given in ten-thousands of dollars, advertising in thousands of dollars.)

Sales, y	12 8 9 14 6 11 10 8
(ten-thousands of dollars)	
Price, x_1	4 4 5 5 6 6 7 7
Advertising, x_2	3 0 5 7 3 8 6 8
(thousands of dollars)	

Determine the estimated multiple regression line and the value of r^2.

59. Use the following equation to answer parts (a) through (d).

$$\hat{y} = -1.67 + 2.46x_1 - 5.48x_2$$

a. What is the meaning of the numbers -1.67, 2.46, and -5.48?
b. Graph the relationship between $\hat{y}$ and x_1 for $x_2 = 10$.
c. Graph the relationship between $\hat{y}$ and x_1 for $x_2 = 20$.
d. What is the difference between the lines you graphed in parts (b) and (c)?

60. An economist states that wages should be inversely related to the rate of unemployment and should be positively related to prices. Test these claims at the .10 Type I error level, using the following data in a multiple regression specification. Report a p-value (the significance) for each coefficient.

Average dollar wage	266 255 235 220 207 189 175
Unemployment rate	9.7 7.6 7.1 5.8 6.1 7.1 7.7
Price index	289 272 246 217 195 181 170

Average dollar wage	163 154 145 136 127 120
Unemployment rate	8.5 5.6 4.9 5.6 5.9 4.9
Price index	161 147 133 125 121 116

61. How are simple and multiple regression similar? How are they different?

62. The closing prices of six stocks on the last day of last month seem to be quite highly correlated with their latest reported earnings-per-share figures and with the percentage of earnings growth they experienced in the past year. The figures are given in the accompanying table. Use Equation 13.6 or a computer to show that the regression equation is $\hat{y} = .72 + 5.94x_1 + 1.08x_2$.

Closing Price per Share, y	Latest Earnings per Share, x_1	Percentage Earnings Growth, x_2
$10	$1.50	3
18	2.00	10
22	2.00	8
30	2.50	6
30	3.00	10
40	7.00	−1

63. A regression equation was found to be $\hat{y} = 1.5 + .2x_1 + 3.1x_2$. R^2 for this equation was .95. The values of x_1 ranged from -10 to 15, and those of x_2 from -20 to -50. Which of the following statements are true?
a. Variable x_2 is more strongly correlated with y than is x_1 because its regression coefficient is larger.
b. When $x_1 = 0$ and $x_2 = 0$, the predicted value of y is 1.5, and this value has legitimate physical meaning because the data values for x_1 and x_2 span zero.
c. A very high proportion of the squared error of prediction incurred by using $\bar{y}$ as the predictor can be eliminated by using the regression in making predictions.

64. A regression equation was found to be $\hat{y} = 10 + 14x_1 - 7x_2$. Which of the following statements are true?

   a. A 1-unit increase in x_1 causes y to increase by 14 units.

   b. Variable y is more highly correlated with x_1 than with x_2 because the coefficient of x_1 is positive.

   c. If the value of x_2 is large enough, negative predictions of y will be obtained.

65. A regression analysis has two independent variables (x_1 and x_2).

   a. What does it mean if x_1 and x_2 are independent of each other? In that case, what is the correlation between them?

   b. Is saying that x_1 and x_2 are independent variables the same as saying that they are independent of each other? Explain.

CHAPTER 16

Other Topics in Applied Regression Analysis

CHAPTER OUTLINE

Key Terms

multicollinearity
coefficient of determination
heteroscedasticity
variance inflationary factor
autocorrelation
Durbin–Watson statistic
first-order autocorrelation
specification error

proxies
law of diminishing returns
log-log linear model
log-linear model
the Durbin H
dummy variable
interaction

16.1   INTRODUCTION

In Chapters 13, 14, and 15 we discussed in some detail the technique of regression analysis and its applications. The main objectives in fitting a regression equation are (1) to estimate the regression coefficients and related parameters and (2) to predict the value of the dependent variable in terms of that of the independent variable (or variables). Several alternative specifications are possible in this kind of applied regression analysis, and a number of problems may occur.

In this chapter, we examine some of the problems associated with applying the multiple regression model. We also explore such related topics as lagged dependent variables and nonlinear regressions. Problems with the error term—specifically, violations of the assumptions of the regression model that were cited in Chapters 14 and 15—can arise when we are running a regression. In this chapter, we discuss the detection of these problems, which include errors that are correlated and errors whose means and variance are not constant. Another problem we may encounter is a high correlation between independent variables. This problem can increase the value of standard errors and reduce the t statistics of the parameters, leading to incorrect inferences in hypothesis testing.

Other topics we address in this chapter include specification bias and model building. We also show how a nonlinear functional form can be transformed into a linear regression analysis. In some cases, for example, both independent and dependent variables can be transformed by using logarithms, and the nonlinear relationship then becomes a linear relationship. Furthermore, a regression can have a lagged dependent variable as one of the independent variables when there is a relationship between previous observations in a time series and the value in the present period. In addition, regression with dummy and interaction variables is discussed in detail. Finally, the effect of alternative business strategies is investigated.

16.2   MULTICOLLINEARITY

Definition and Effect

The term **multicollinearity** refers to the effect, on the precision of regression parameter estimates, of two or more of the independent variables being highly correlated. For example, multicollinearity would be a problem if we were studying the cross-sectional relationship by regressing price per share against dividend per share and retained earnings per share, as discussed in Application 15.2 in Chapter 15. Because dividend per share and retained earnings per share are highly correlated, the precision of the least-squares estimated regression coefficient might be affected.

If a set of independent variables is perfectly correlated, the least-squares approach cannot be used to estimate the regression coefficients: the normal equations are not solvable. If independent variables move together, it is impossible to distinguish the separate effects of these variables on y. Perfect multicollinearity would occur, for example, if the following independent variables were specified to model the expenditures on food for a cross section of individuals.

$$x_1 = \text{average income in dollars}$$

$$x_2 = \text{average income in cents}$$

The variables x_1 and x_2 are perfectly correlated because $x_2 = 100$ times x_1 for each of the individuals in the data set. If both of these variables were included in a regression model, least-squares results would not be obtainable because the two variables measure the same thing. Remember that the regression coefficient of x_2 is a slope term that measures the change in the dependent variable that is associated with a 1-unit change in x_2, other variables being held constant. But here it is impossible to keep the rest of the variables constant, because x_1 changes in the same direction and with the same magnitude as x_2. The solution? Simply delete one of the variables and run the regression again.

Unfortunately, most of the problems researchers face are not so easy to detect. Observations are more likely to be *highly* correlated than perfectly correlated. In such cases, least-squares estimates can be obtained but are difficult to interpret.

For example, suppose national income in period t, y_t is modeled with independent variables x_1 = output of manufactured goods in period t and x_{2t} = output of durable goods in period t as defined in Equation 16.1.

$$y_t = a + b_1 x_{1t} + b_2 x_{2t} + e_t \tag{16.1}$$

If the simple correlation r between x_{1t} and x_{2t} is .90, it can be concluded that the explanatory values of the two variables overlap considerably, probably because they are highly correlated and tend to measure the same thing.

In Equation 16.1, the first coefficient of two highly correlated variables is the slope term b_1, which measures the change in the national income that is due to a 1-unit change in the output of manufactured goods, the output of durable goods being held constant. When one of the correlated variables changes, the other is likely to change in the same direction and with approximately the same magnitude. However, the standard error of the coefficient will tend to be great, leading to lower t-values for the coefficient.[1] This increase in the standard error results from the fact that estimates are sensitive to any changes in observations or model specification.

From Equations 15.23 and 15.24, the sample variances of b_1 and b_2 ($s_{b_1}^2$ and $s_{b_2}^2$) of Equation 16.1 can be defined as

$$s_{b_1}^2 = \frac{s_e^2}{(1 - r^2) \sum\limits_{i=1}^{n} (x_{1t} - \bar{x}_1)^2} \tag{16.2}$$

$$s_{b_2}^2 = \frac{s_e^2}{(1 - r^2) \sum\limits_{i=1}^{n} (x_{2t} - \bar{x}_2)^2} \tag{16.3}$$

---

[1] If x_1 and x_2 are highly correlated, then the regression can give weight to either x_1 or x_2 and it won't matter. Sampling idiosyncracies determine the choice. Hence the large sampling error occurs.

where s_e^2 = sample variance of e_i; $\bar{x}_1$ and $\bar{x}_2$ are means of x_{1i} and x_{2i}, respectively; and r is the correlation coefficient between x_{1i} and x_{2i}. In Equations 16.2 and 16.3, the factor $(1 - r^2)$ can be used to measure the impact of collinearity on $s_{b_1}^2$ and $s_{b_2}^2$. If x_{1i} and x_{2i} are uncorrelated, then $(1 - r^2) = 1$. If x_{1i} and x_{2i} are perfectly correlated $(r^2 = 1)$, then $(1 - r^2) = 0$. In this case the denominators of both $s_{b_1}^2$ and $s_{b_2}^2$ vanish, and both $s_{b_1}^2$ and $s_{b_2}^2$ equal infinity. In our national-income example, $r^2 = 81$ and $(1 - r^2) = .19$. Therefore, the precision of estimated $s_{b_1}^2$ and $s_{b_2}^2$ is greatly affected by the collinearity between x_{1i} and x_{2i}.

Rules of Thumb in Determining the Degree of Collinearity

Several rules of thumb are helpful when we are testing for multicollinearity. These rules involve inspection of the correlation between the independent variables. First, multicollinearity is a problem if the correlation coefficient between any two independent variables is greater than .80 or .90. If there are more than two independent variables in a regression, as indicated in Equation 16.4, then the simple correlation coefficient between any two independent variables is not sufficient to detect the existence of multicollinearity of a regression.

$$y_i = \alpha + \beta_1 x_{1i} + \beta_2 x_{2i} + \beta_3 x_{3i} + \epsilon_i \tag{16.4}$$

We must also consider that simple correlation coefficients generally fail to take into account the possible correlation between any one independent variable and all others taken as a group. Therefore, it is customary to regress each of the independent variables against all others and to note whether any of the resulting R^2-values are near 1. Using Equation 16.4 as an example, we can define three multiple regressions in terms of x_{1i}, x_{2i}, and x_{3i}.

$$x_{1i} = a_0 + a_1 x_{2i} + a_2 x_{3i} \tag{16.5a}$$

$$x_{2i} = b_0 + b_1 x_{1i} + b_2 x_{3i} \tag{16.5b}$$

$$x_{3i} = c_0 + c_1 x_{1i} + c_2 x_{2i} \tag{16.5c}$$

R_i^2 $(i = 1, 2, 3)$ of these three regressions represents the **coefficient of determination** for the ith independent variable. It can be used to determine whether multicollinearity plagues Equation 16.4.

In sum, to check for multicollinearity, we first calculate the three simple correlation coefficients between x_1 and x_2 (r_{12}), between x_1 and x_3 (r_{13}), and between x_2 and x_3 (r_{23}). Then we find R^2 associated with Equations 16.5a, 16.5b, and 16.5c. In a sense these two methods are similar, but the first is easier to understand. This estimated information can be used to determine the existence of multicollinearity.

One way to measure collinearity is to use the measurements of $(1 - R^2)$ indicated in Equation 16.2 or 16.3 to construct a **variance inflationary factor** (VIF) for each explanatory variable in Equation 16.4.

$$\text{VIF}_i = \frac{1}{1 - R_i^2} \tag{16.6}$$

where R_i^2 is as defined in Equation 16.5. If there are only two independent variables,

R_i is merely the correlation coefficient. If a set of independent variables is uncorrelated, then VIF_i is equal to 1. If R_i^2 approaches 1, both VIF and the standard deviations of the slopes ($s_{b_1}^2$ and $s_{b_2}^2$) approach infinity. Researchers have used $VIF_i = 10$ as a critical-value rule of thumb to determine whether too much correlation exists between the ith independent variable and other independent variables.[2] The corresponding R_i values of VIF_i at least 10 is now illustrated. Thus, $1/(1 - R_i^2) \geq 10$ implies $1 - R_i^2 \leq .1$. This implies that $.95 \leq R_i \leq 1$ or $-1 \leq R_i \leq -.95$.

EXAMPLE 16.1 *Analyzing the Determination of Price per Share*

Data on dividend per share (DPS), price per share (PPS), and retained earnings (RE) per share from 1983 to 1985 for the 30 firms employed to compile the Dow Jones Industrial Average are used to estimate Equation 16.7. Regression results are shown in Table 16.1.

$$PPS_{i,t} = a + b(DPS_{i,t}) + c(RE_{i,t}) + e_{i,t} \tag{16.7}$$

where $PPS_{i,t}$, $DPS_{i,t}$, and $RE_{i,t}$ repreesnt price per share, dividends per share, and retained earnings per share for the ith firm in the tth year, respectively. This crosssectional model states that the price per share is a function of dividends per share and retained earnings per share. The results of the regressions seem to be satisfactory with all of the independent variables and appear significant at the 5 percent level ($F_{.05,2.27} = 3.35$, from Table A6). However, we must examine the correlations between $DPS_{i,t}$ and $RE_{i,t}$ to determine whether multicollinearity may be a problem.

The 1985 regression results indicated in column (4) of Table 16.1 are used to determine the degree of multicollinearity. To do this analysis, we assemble the cor-

Table 16.1 Regression Results of Equation 16.7

(1)	(2) 1983	(3) 1984	(4) 1985
Constant	22.773	11.336	21.529
	(4.15)	(2.33)	(2.81)
DPS	10.733	12.434	12.553
	(4.28)	(4.41)	(3.16)
RE	1.431	3.088	3.014
	(2.30)	(2.39)	(1.81)
R^2	.54	.70	.54
F-statistic	16.11	31.44	15.97

[2]See D. W. Marquardt, (1980), "You Should Standardize the Prediction Variables in Your Regression Models," discussion of "A Critique of Some Ridge Regression Methods," by G. Smith and F. Campbell, *Journal of the American Statistical Association* 75, 87–91.

Table 16.2 A Simple Correlation Matrix

Variable	PPS, y	DPS, x_1	RE, x_2
PPS, y	$r_{y,y} = 1.000$	$r_{y,1} = .6972$	$r_{y,2} = .6105$
DPS, x_1	—	$r_{1,1} = 1.000$	$r_{1,2} = .6064$
RE, x_2	—	—	$r_{2,2} = 1.000$

relation coefficients among PPS, DPS, and RE in Table 16.2. We note from the correlation matrix that each variable is perfectly correlated with itself; hence we find three entries equal to 1.000 along the diagonal of the table. It is the boxed entry that is of crucial importance. Substituting $r_{1,2} = .6064$ into Equation 16.6, we obtain $\text{VIF}_j = 1/[1 - (.6064)^2] = 1.5815$. Because 1.5815 is much smaller than 10, we conclude that the degree of collinearity for this regression is relatively unimportant.

16.3 HETEROSCEDASTICITY

Definition and Concept

Heteroscedasticity arises when the variances of the error terms of a regression model are not constant over different sample observations. For example, heteroscedasticity would be a problem in a study of sales for a cross section of firms in an industry, because the error terms for large firms would be likely to have larger variances than those for small firms. In other words, the high volatility in sales for larger firms might pose problems for the researcher. The probable error terms are shown in Figure 16.1. This figure indicates that the magnitude of error terms is a function of firm size. For example, it may be the case that $\sigma_i^2 < \sigma_j^2 < \sigma_k^2$.

Another commonly cited example of heteroscedasticity is the relationship between family expenditures and income. High-income families are likely to exhibit a higher variance in spending than lower-income families. Figure 16.2 shows a reasonable plot of error versus level of expenditures. This figure indicates that expenditures for consumers with lower income have smaller error terms.

Heteroscedasticity poses a problem when we are estimating parameters in the regression model, because the least-squares estimation procedure places more weight on observations that have large errors and variances. Thus the regression line is adjusted to give a good fit for the large-variance portion of the observations but largely ignores the small-variance part of the data. The result is that the variances of the estimates do not have a minimum variance.

Sales for Firms

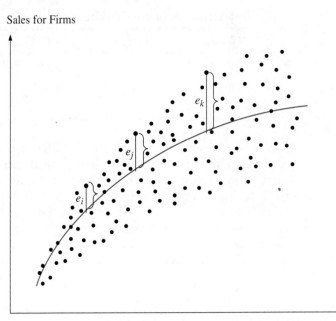

Figure 16.1

A Possible Relationship Between Sales and Firm Size

Size of Firms

Expenditure for
Individual Consumers

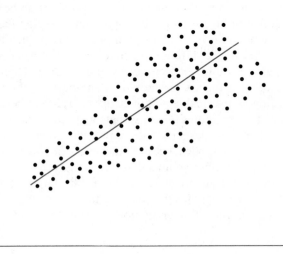

Figure 16.2

Cross-sectional Relationship Between Expenditure and Income

Income for
Individual Consumers

Evaluating the Existence of Heteroscedasticity

The easiest way to check for heteroscedasticity is to look at a plot of the residuals against the independent variables or the expected values. To estimate the error term, we first compute the predicted dependent value by the regression model

$$\hat{y}_i = a + b_1 x_{1i} + b_2 x_{2i} + \cdots + b_k x_{ki} \tag{16.8}$$

Then we calculate the error term by taking the actual value for y_i and subtracting the predicted value: $e_i = y_i - \hat{y}_i$. In practice, we generally plot the residuals range e_i against independent variables on a series of graphs or examine the predicted values $\hat{y}_i$. If the residuals appear to be random and the width of the scatter diagram seems constant throughout the data—that is, if no pattern is apparent—then no heteroscedasticity is present.

A somewhat more involved evaluation for the existence of heteroscedasticity consists of the following steps.

1. Run a standard regression.
2. Calculate the residuals, $e_i = \hat{y}_i - y_i$.
3. Run a regression using the square of the residuals as the dependent variable and the estimated dependent variable $\hat{y}_i$ as the independent variable.
4. Estimate nR^2, where n is the sample size and R^2 is the coefficient of determination.
5. Use the χ^2 statistic with 1 degree of freedom to test whether nR^2 is significantly different from zero.

The use of this method to analyze the existence of heteroscedasticity is illustrated in the following example.

EXAMPLE 16.2   *Residual Heteroscedasticity Analysis for Price per Share*

The regression results for Equation 16.7 were run for the 30 Dow Jones industrials for 1983, 1984, and 1985. These results appear in Table 16.1.

To check for heteroscedasticity, we calculate the residuals of the regression and plot them against the predicted values $\hat{y}_{i,t}$. Residuals (e) and predicted values ($\hat{y}$) for all three years are listed in Table 16.3. Figure 16.3 shows the plots of residuals from the least-squares regression against $\hat{y}_{i,t}$ for 1983. Similar plots for 1984 and 1985 are given in Figures 16.4 and 16.5. As can be seen in all three plots, there are patterns in the residuals. The pattern for 1985 is stronger than those for 1983 and 1984. We might conclude that the residuals plotted against predicted values are not random and do violate the standard assumptions of the regression model.

In addition to making this visual inspection of residuals, we must run a regression with the squared error terms as the dependent variable and $\hat{y}_i$ as the independent variable. Let's look at the results for 1985.

$$e_{i,t}^2 = c + f\hat{y}_{i,t} + \text{residuals} \tag{16.9}$$

Table 16.3 Residuals (e) and Predicted Values ($\hat{y}$) for 30 Dow Jones Industrials (1983, 1984, and 1985)

```
MTB > PRINT C1-C6

ROW    83e    83y^    84e    84y^    85e    85y^

  1   11.2    0.3    0.9    7.2    4.4   11.2
  2   15.6   14.8    9.5    8.0   -4.3   12.8
  3   -6.2   20.8    2.3   10.1   -0.3   24.0
  4    5.8   22.7   -6.9   26.4  -11.0   24.3
  5  -16.2   33.8   -8.1   26.6   17.9   28.8
  6  -12.9   33.8  -14.4   26.6    6.2   32.3
  7   12.7   34.3   -3.7   29.8  -12.1   37.1
  8   -7.4   34.8   -5.4   31.6   -9.4   39.4
  9    7.9   37.0    5.7   31.9  -13.1   39.7
 10   -0.6   37.8   19.7   32.0    2.5   41.3
 11   -9.8   40.2    4.8   32.2    1.4   43.1
 12   -4.1   40.3   11.1   35.9    8.2   44.8
 13    4.8   40.4  -11.3   37.3   35.9   45.0
 14   -8.3   40.9    1.9   38.4  -17.1   48.4
 15   -5.5   41.4   -3.1   39.3   19.8   48.7
 16   -4.6   41.9   14.3   39.6  -10.0   49.0
 17   -5.2   42.4   -8.4   40.2    1.5   49.1
 18   12.0   46.6   -1.3   41.5   -0.2   50.9
 19    2.7   48.0   -8.4   42.6   -1.1   53.9
 20    5.7   51.2   -8.7   43.2  -12.2   56.4
 21    7.3   51.7    2.4   45.5   12.0   57.8
 22  -17.1   51.8   10.6   46.0  -19.9   58.0
 23   -0.8   52.8  -16.3   47.6   14.5   58.3
 24   -8.2   55.1    6.7   50.3    3.8   64.0
 25  -21.7   57.5   -3.1   53.6   -4.4   64.4
 26  -22.5   59.9   -7.2   56.7  -35.5   65.5
 27   20.9   61.6   16.2   62.5   17.4   72.3
 28    8.6   65.8  -18.5   63.5  -18.8   73.9
 29   51.8   70.2   40.2   82.9   59.8   95.7
 30  -15.9   77.4  -21.3   99.6  -35.9  106.2

MTB > PAPER
```

Figure 16.3

Plots of Residuals Against the Predicted Values, y (1983)

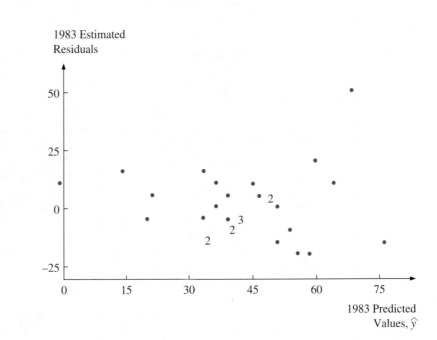

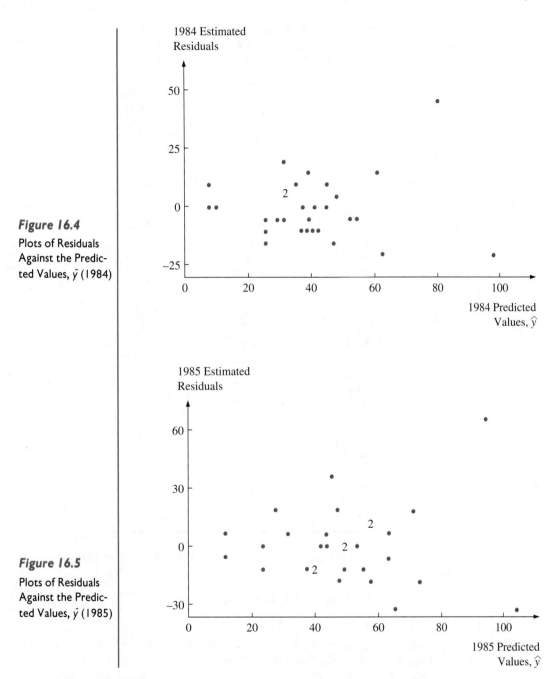

Figure 16.4

Plots of Residuals Against the Predicted Values, $\hat{y}$ (1984)

Figure 16.5

Plots of Residuals Against the Predicted Values, $\hat{y}$ (1985)

Table 16.4 $\hat{e}_{i,t}^2$ and $\hat{y}_{i,t}$ for a Test of Heterosce-dasticity for Equation 16.7 in Terms of 1985 Data

$\hat{y}_{i,t}$	$\hat{e}_{i,t}$	$\hat{e}_{i,t}^2$
24.2806	−11.0306	121.6741
56.3864	−12.1984	148.8009
50.9328	−.1828	.033415
72.3072	17.4428	304.2512
64.0412	3.8338	14.69802
23.9555	−.3305	.109230
48.6921	19.8079	392.3529
57.7841	11.9659	143.1827
58.0467	−19.9217	396.8741
73.9102	−18.7852	352.8837
65.5484	−35.5484	1263.688
39.7382	−13.1132	171.9560
48.3659	−17.1159	292.9540
53.8913	−1.1413	1.302565
11.2178	4.4072	19.42341
32.282	6.218	38.66352
64.4438	−4.4438	19.74735
58.2638	14.4862	209.8499
43.1089	1.3911	1.935159
28.8115	17.9385	321.7897
95.6623	59.9377	3580.550
106.2	−35.8637	1286.204
12.7873	−4.2873	18.38094
41.2735	2.4765	6.133052
49.1375	1.4875	2.212656
37.105	−12.105	146.5310
48.9577	−9.9577	99.15578
39.3571	−9.3571	87.55532
44.9646	35.9104	1289.556
44.8209	8.1791	66.89767

Using $\hat{e}_{i,t}^2$ and $\hat{y}_{i,t}$ for 1985 as presented in Table 16.4, we estimate Equation 16.9 and obtain

$$\hat{e}_{i,t}^2 = 635.379 + 19.95672\hat{y}_{i,t} \qquad R^2 = .3450$$

$$(t = 3.8399)$$

Because there are $n = 30$ sets of observations, the test is based on $nR^2 = (30)(.3450) = 10.35$. From Table A5 in Appendix A, we find that for a test at the 5 percent level, $\chi_{1,.05}^2 = 3.84$. Therefore, we can conclude that the residuals in the regression of price per share on earnings per share and retained earnings per share have the same variance and should be rejected. One way to deal with this problem is to use a two-stage procedure to estimate the parameters of regression models. In the first stage, we estimate the parameters of Equation 16.7 and the predicted value $\hat{y}_{i,t}$, of the dependent variable. Predicted values of $y_{i,t}$ for 1985 are listed in Table 16.4. In the second stage, we estimate a transformed Equation 16.7:

$$\frac{y_{i,t}}{\hat{y}_{i,t}} = a\,\frac{1}{\hat{y}_{i,t}} + b\,\frac{x_{1i,t}}{\hat{y}_{i,t}} + c\,\frac{x_{2i,t}}{y_{i,t}} + e'_{i,t} \qquad (16.7')$$

where $e'_{i,t}$ is an error term that approximates constant variance. Using 1985 data, we estimate Equation 16.7' and its results.

$$\frac{y_{i,t}}{\hat{y}_{i,t}} = 16.446\left(\frac{1}{\hat{y}_{i,t}}\right) + 16.062\,\frac{x_{1i,t}}{\hat{y}_{i,t}} + 1.715\,\frac{x_{2i,t}}{\hat{y}_{i,t}}$$

$$(4.013) \qquad\qquad (5.7813) \qquad (1.959)$$

From the t-values indicated in parentheses, we find that the t-values for second-stage estimates are more efficient than those for the one-stage estimates indicated in Table 16.1.

16.4   AUTOCORRELATION

Basic Concept

One of the assumptions of the regression model is that the errors are uncorrelated; in other words, the correlation between error terms is equal to zero. We are quite likely to encounter only uncorrelated errors when dealing with cross-sectional data. However, **autocorrelation**—the correlation of an error term and a lagged version of itself—is likely to occur with time-series data because errors made in a particular time period are readily carried over to future time periods. For example, an underestimate of the GNP in one year can generate more underestimates in future time periods.

Figure 16.6 shows examples of autocorrelation. The error in period t is graphed on the y axis, the error in period $t - 1$ on the x axis. If no autocorrelation exists, the plot is random, as shown in Figure 16.6a. The upward slope of the diagram in

Figure 16.6

(a) No Autocorrelation, (b) Positive Autocorrelation, and (c) Negative Autocorrelation

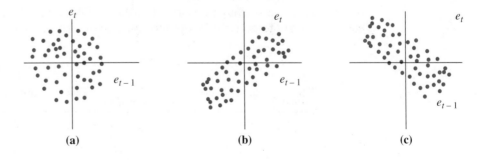

(a)   (b)   (c)

Figure 16.6b signals positive autocorrelation. This implies that a large error in period $t - 1$ will be associated with a large error in period t. Negative autocorrelation is apparent in Figure 16.6c, where high errors in the previous period tend to result in low errors in the next period. Again, if plotting reveals a pattern, the regression assumption that the residuals are random and not correlated through time has been violated.

The Durbin–Watson Statistic

Detecting autocorrelation by inspecting errors is difficult; thus the **Durbin–Watson statistic** (DW) is generally used to detect first-order autocorrelation. **First-order autocorrelation** occurs when correlation between errors is separated by one period. Here we will discuss both first-order autocorrelation and the Durbin–Watson (DW) statistic in detail.

If e_t and e_{t-1} are residual terms in periods t and $t - 1$, respectively, then first-order correlation r_1 for these error terms is defined as follows:

$$r_1 = \frac{\sum_{t=2}^{n} (e_t - \bar{e}_t)(e_{t-1} - \bar{e}_{t-1})}{\sqrt{\sum_{t=2}^{n} (e_t - \bar{e}_t)^2 \sum_{t=2}^{n} (e_{t-1} - \bar{e}_{t-1})^2}} \tag{16.10}$$

where $\bar{e}_t$ and $\bar{e}_{t-1}$ are means of e_t and e_{t-1}, respectively.

The DW statistic can be used to test the null hypothesis that no first-order autocorrelation exists among the residuals of a regression. The statistic, calculated from the residuals, is

$$DW = \frac{\sum_{t=2}^{n} (e_t - e_{t-1})^2}{\sum_{t=1}^{n} e_t^2} \tag{16.11}$$

where e_t and e_{t-1} are error terms in periods t and $t - 1$, respectively. This statistic is calculated by summing the difference in the error terms separated by one period and dividing by the squared error term. Figure 16.7 shows that the DW statistic falls between zero and 4. If no autocorrelation exists, DW equals 2, because the difference in the error terms is directly proportional to the error term in period t. Positive

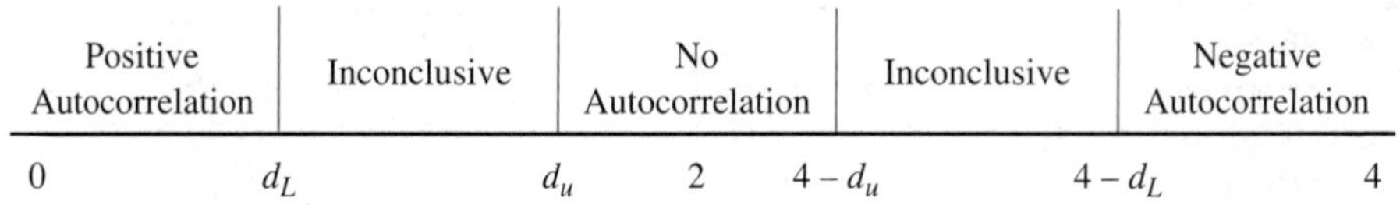

Positive Autocorrelation	Inconclusive	No Autocorrelation	Inconclusive	Negative Autocorrelation
0 $\qquad d_L$		$d_u \quad 2 \quad 4 - d_u$		$4 - d_L \qquad 4$

Figure 16.7

Critical Values of the Durbin–Watson Statistic

autocorrelation exists if DW is low, because the difference between the error term in period t and in period $t - 1$ tends to be very small. Negatively correlated errors are closer to 4 because the difference in the error terms tends to be large.

We can use the DW statistics listed in the Table A9 of Appendix A to determine whether the null hypothesis is accepted or rejected. These tables give two values, d_L and d_U, for different sample sizes n and number of independent variables k. If the DW statistic falls between d_U and $4 - d_U$, the null hypothesis that the correlation between lagged errors is equal to zero is accepted. If the statistic is less than d_L, that null hypothesis is rejected in favor of positive autocorrelation. Negative autocorrelation exists if the DW is greater than $4 - d_L$. Did you notice that there is an indeterminate zone in which no judgment can be made? This zone is between d_L and d_U, or $4 - d_U$ and $4 - d_L$. (The probability level of d_U, d_L depends on whether we are performing a one- or a two-tailed test.)

EXAMPLE 16.3 *How to Detect First-Order Autocorrelation*

Annual rates of return for both GM and Ford and market rates of return during 1970 to 1990, which can be found in Table 4.15 in Chapter 4, are used to estimate the market model of Equation 16.12 for GM and Ford.

$$R_{i,t} = a_i + b_i R_{m,t} + e_{i,t} \tag{16.12}$$

where $R_{i,t}$ is the rate of return for the ith firm in period t, $R_{m,t}$ is the market rate of return, a_i and b_i are regression parameters, and $e_{i,t}$ is the error term.

MINITAB outputs of the estimated market models for GM and Ford are presented in Figures 16.8 and 16.9. These two outputs indicate that the beta coefficients b_i for GM and Ford are .9800 and .7767, respectively. In addition, we see that the DW statistics for GM and Ford are 1.87 and 1.65. Remember, this test determines whether there is evidence of autocorrelation among the residuals. In this example, there are one independent variable and 21 observations. Looking these values up in Table A9 of Appendix A for a level of significance of 5 percent

Figure 16.8

MINITAB Output of Market Model for GM

```
MTB > BRIEF 3
MTB > REGRESS C3 1 C4;
SUBC> DW.

The regression equation is
C3 = 0.0083 + 0.980 C4

Predictor       Coef      Stdev     t-ratio        p
Constant      0.00834    0.07273      0.11      0.910
C4            0.9800     0.4188       2.34      0.030

s = 0.3006      R-sq = 22.4%     R-sq(adj) = 18.3%

Analysis of Variance

SOURCE        DF         SS          MS         F        p
Regression     1      0.49469     0.49469     5.48     0.030
Error         19      1.71657     0.09035
Total         20      2.21126

Obs.     C4        C3      Fit  Stdev.Fit  Residual   St.Resid
  1    0.001    0.2137   0.0093    0.0726    0.2044      0.70
  2    0.108    0.0422   0.1142    0.0670   -0.0720     -0.25
```

(continued)

Figure 16.8

(Continued)

```
 3    0.156    0.0631    0.1609    0.0738   -0.0978   -0.34
 4   -0.174   -0.3667   -0.1619    0.1231   -0.2048   -0.75
 5   -0.296   -0.2597   -0.2821    0.1688    0.0224    0.09 X
 6    0.315    0.9522    0.3169    0.1200    0.6353    2.31R
 7    0.192    0.4584    0.1963    0.0818    0.2621    0.91
 8   -0.115   -0.1124   -0.1047    0.1032   -0.0077   -0.03
 9    0.010   -0.0498    0.0186    0.0709   -0.0684   -0.23
10    0.123    0.0288    0.1287    0.0686   -0.0999   -0.34
11    0.259   -0.0410    0.2618    0.1010   -0.3028   -1.07
12   -0.099   -0.0911   -0.0891    0.0982   -0.0020   -0.01
13    0.155    0.6826    0.1601    0.0736    0.5225    1.79
14    0.171    0.2373    0.1755    0.0768    0.0618    0.21
15    0.012    0.1176    0.0196    0.0708    0.0980    0.34
16    0.263   -0.0383    0.2664    0.1026   -0.3047   -1.08
17    0.146    0.0088    0.1516    0.0720   -0.1428   -0.49
18    0.020    0.0058    0.0282    0.0695   -0.0224   -0.08
19    0.124    0.4418    0.1299    0.0687    0.3119    1.07
20    0.273   -0.4581    0.2754    0.1055   -0.7335   -2.61R
21   -0.066   -0.1154   -0.0559    0.0882   -0.0595   -0.21
```

```
R denotes an obs. with a large st. resid.
X denotes an obs. whose X value gives it large influence.
```

```
Durbin-Watson statistic = 1.87
```

Figure 16.9

MINITAB Output of Market Model for Ford

```
MTB > REGRESS C3 1 C4;
SUBC> DW.

The regression equation is
Ford = 0.0284 + 0.777 Market

Predictor       Coef      Stdev    t-ratio       p
Constant     0.02838    0.07822       0.36    0.721
Market        0.7767     0.4504       1.72    0.101

s = 0.3232      R-sq = 13.5%     R-sq(adj) = 9.0%

Analysis of Variance

SOURCE        DF         SS         MS       F        p
Regression     1     0.3107     0.3107    2.97    0.101
Error         19     1.9852     0.1045
Total         20     2.2959

Obs.   Market       Ford       Fit  Stdev.Fit   Residual   St.Resid
  1    0.001     0.4260     0.0292     0.0780     0.3968       1.27
  2    0.108     0.2953     0.1123     0.0721     0.1810       0.57
  3    0.156     0.1717     0.1493     0.0793     0.0224       0.07
  4   -0.174    -0.4512    -0.1065     0.1324    -0.3447      -1.17
  5   -0.296    -0.0968    -0.2018     0.1816     0.1050       0.39 X
  6    0.315     0.3960     0.2730     0.1290     0.1230       0.42
  7    0.192     0.4614     0.1773     0.0880     0.2841       0.91
  8   -0.115    -0.2067    -0.0612     0.1110    -0.1455      -0.48
  9    0.010    -0.0055     0.0365     0.0763    -0.0420      -0.13
 10    0.123    -0.1452     0.1238     0.0737    -0.2690      -0.85
 11    0.259    -0.2896     0.2292     0.1087    -0.5188      -1.70
 12   -0.099     0.0407    -0.0488     0.1056     0.0895       0.29
 13    0.155     0.9686     0.1487     0.0792     0.8199       2.62R
 14    0.171     0.1286     0.1609     0.0826    -0.0323      -0.10
 15    0.012     0.0980     0.0373     0.0761     0.0607       0.19
 16    0.263     0.3237     0.2329     0.1103     0.0908       0.30
 17    0.146     0.0081     0.1419     0.0775    -0.1338      -0.43
 18    0.020     0.3961     0.0441     0.0747     0.3520       1.12
 19    0.124    -0.2995     0.1247     0.0739    -0.4242      -1.35
 20    0.273    -0.0767     0.2400     0.1135    -0.3167      -1.05
 21   -0.066    -0.3209    -0.0226     0.0948    -0.2983      -0.97
```

```
R denotes an obs. with a large st. resid.
X denotes an obs. whose X value gives it large influence.
```

```
Durbin-Watson statistic = 1.65
```

(under a two-tailed test, α = 10 percent), we find that d_L is 1.22 and d_U is 1.42. Because both DW values are greater than d_U and less than $4 - d_U$, we conclude that there is no evidence of autocorrelation in either regression.

Dividend per share (DPS_t) and earnings per share (EPS_t) for both GM and Ford during 1970–1990 (presented in Table 2B.1 in Chapter 2) are used to estimate the regression specified in Equation 16.13.

$$DPS_{i,t} = a_i + b_i EPS_{i,t} + e_{i,t} \qquad (16.13)$$

where $DPS_{i,t}$ and $EPS_{i,t}$ are dividend per share and earnings per share for the ith firm in period t.

MINITAB output for estimated Equation 16.13 is presented in Figures 16.10 and 16.11. From Figures 16.10 and 16.11, we know that DPS is highly correlated with EPS for both GM and Ford. And the DW statistics are 1.20 and .59, respectively.

Figure 16.10

MINITAB Output of Equation 16.13 for GM

```
MTB > PRINT C1-C6

ROW    C1    C2    EPS     DPS   LAGEPS  LAGDPS

 1     70    70    2.09    3.40    5.95    4.30
 2     71    71    6.72    3.40    2.09    3.40
 3     72    72    7.51    4.45    6.72    3.40
 4     73    73    8.34    5.25    7.51    4.45
 5     74    74    3.27    3.40    8.34    5.25
 6     75    75    4.32    2.40    3.27    3.40
 7     76    76   10.08    5.55    4.32    2.40
 8     77    77   11.62    6.80   10.08    5.55
 9     78    78   12.24    6.00   11.62    6.80
10     79    79   10.04    5.30   12.24    6.00
11     80    80   -2.65    2.95   10.04    5.30
12     81    81    1.07    2.40   -2.65    2.95
13     82    82    3.09    2.40    1.07    2.40
14     83    83   11.84    2.80    3.09    2.40
15     84    84   14.22    4.75   11.84    2.80
16     85    85   12.28    5.00   14.22    4.75
17     86    86    8.22    5.00   12.28    5.00
18     87    87   10.06    5.00    8.22    5.00
19     88    88   13.64    5.00   10.06    5.00
20     89    89    6.33    3.00   13.64    5.00
21     90    90   -4.09    3.00    6.33    3.00


MTB > BRIEF 2
MTB > REGRESS C4 1 C3;
SUBC> DW.

The regression equation is
DPS = 2.87 + 0.179 EPS

Predictor       Coef       Stdev     t-ratio        p
Constant      2.8740      0.3649        7.88    0.000
EPS           0.17902     0.04162       4.30    0.000

s = 0.9668      R-sq = 49.3%     R-sq(adj) = 46.7%

Analysis of Variance

SOURCE        DF          SS          MS         F        p
Regression     1      17.299      17.299     18.51    0.000
Error         19      17.761       0.935
Total         20      35.060
```

(continued)

Figure 16.10

(Continued)

```
Unusual Observations
Obs.      EPS       DPS      Fit Stdev.Fit  Residual  St.Resid
 14      11.8     2.800    4.994    0.287    -2.194    -2.38R

R denotes an obs. with a large st. resid.



Durbin-Watson statistic = 1.20
```

Figure 16.11

MINITAB Output
of Equation 16.13
for Ford

```
MTB > PRINT C1-C6

ROW    C1    C2     EPS    DPS   LAGEPS  LAGDPS

  1    70    70    4.77   2.40    5.03    2.40
  2    71    71    6.18   2.50    4.77    2.40
  3    72    72    8.52   2.68    6.18    2.50
  4    73    73    9.13   3.20    8.52    2.68
  5    74    74    3.86   3.20    9.13    3.20
  6    75    75    3.46   2.60    3.86    3.20
  7    76    76   10.45   2.80    3.46    2.60
  8    77    77   14.16   3.04   10.45    2.80
  9    78    78   13.35   3.50   14.16    3.04
 10    79    79    9.75   3.90   13.35    3.50
 11    80    80  -12.83   2.60    9.75    3.90
 12    81    81   -8.81   1.20  -12.83    2.60
 13    82    82   -5.46   0.00   -8.81    1.20
 14    83    83   10.29   0.50   -5.46    0.00
 15    84    84   15.79   2.00   10.29    0.50
 16    85    85   13.63   2.40   15.79    2.00
 17    86    86   12.32   2.22   13.63    2.40
 18    87    87   18.10   3.15   12.32    2.22
 19    88    88   10.96   2.30   18.10    3.15
 20    89    89    8.22   3.00   10.96    2.30
 21    90    90    1.86   3.00    8.22    3.00

MTB > REGRESS C4 1 C3;
SUBC> DW.

The regression equation is
DPS = 2.18 + 0.0437 EPS

Predictor      Coef       Stdev     t-ratio       p
Constant     2.1777      0.2630       8.28     0.000
EPS         0.04373     0.02507       1.74     0.097

s = 0.8941     R-sq = 13.8%     R-sq(adj) = 9.3%

Analysis of Variance

SOURCE        DF        SS·          MS        F       p
Regression     1     2.4315      2.4315     3.04    0.097
Error         19    15.1888      0.7994
Total         20    17.6203

Unusual Observations
Obs.     EPS      DPS      Fit Stdev.Fit  Residual  St.Resid
 11    -12.8    2.600    1.617    0.535     0.983     1.37 X
 13     -5.5    0.000    1.939    0.369    -1.939    -2.38R
 14     10.3    0.500    2.628    0.212    -2.128    -2.45R

R denotes an obs. with a large st. resid.
X denotes an obs. whose X value gives it large influence.

Durbin-Watson statistic = 0.59
```

In a two-tailed test, we look up in the Durbin–Watson table critical values for a 5 percent level of significance, the number of observations 21, and the number of independent variables 1. The critical values are 1.22 and 1.42. Remember that if the DW falls between the two values, the test is inconclusive. If it is less than 1.22, positive autocorrelation is a problem. The DW of GM is equal to 1.20, and the DW of Ford is below that value, so we conclude that positive autocorrelation exists among the residuals in the regression model of Equation 16.13 for Ford.

When results of Equation 16.13 for GM and Ford, as indicated in Figures 16.10 and 16.11, imply that the residuals of regression might be autocorrelated, least-squares estimates and inferences based on them can be very unreliable. Under these circumstances, a modified model of Equation 16.13 can be used to adjust for the autocorrelation.

$$\text{DPS}_{i,t} - \hat{\rho}\text{DPS}_{i,t-1} = a_i(1 - \hat{\rho}) + b_i(\text{EPS}_{i,t} - \text{EPS}_{i,t-1}) + e'_{i,t} \quad (16.13a)$$

where $e'_{i,t} = e_{i,t} - \hat{\rho}e_{i,t-1}$, $\hat{\rho} = 1 - d/2 = $ estimated first-order autocorrelation. MINITAB output in terms of Equation 16.13a for Ford is presented in Figure 16.12. From this figure, we find that

$$\text{DPS}_{i,t} - \hat{\rho}\text{DPS}_{i,t-1} = .6694 + .0426(\text{EPS}_{i,t} - \hat{\rho}\text{EPS}_{i,t-1})$$

$$(t = 1.96) \qquad \text{DW} = 1.23$$

This result implies that the DW statistic has improved substantially.

Figure 16.12

MINITAB Output of Equation 16.13a for Ford

```
MTB > LET K1=1-0.59/2
MTB > LET C7=C5*K1
MTB > LET C8=C6*K1
MTB > LET C9=C3-C7
MTB > LET C10=C4-C8
MTB > REGRESS C10 1 C9;
SUBC> DW.

The regression equation is
C10 = 0.669 + 0.0426 C9

Predictor      Coef       Stdev     t-ratio        p
Constant     0.6694      0.1459        4.59    0.000
C9          0.04263     0.02170        1.96    0.064

s = 0.6391     R-sq = 16.9%     R-sq(adj) = 12.5%

Analysis of Variance

SOURCE         DF          SS          MS        F        p
Regression      1      1.5751      1.5751     3.86    0.064
Error          19      7.7596      0.4084
Total          20      9.3348

Unusual Observations
Obs.     C9       C10      Fit Stdev.Fit  Residual   St.Resid
 11    -19.7    -0.150   -0.170     0.491     0.021      0.05 X
 12      0.2    -0.633    0.679     0.144    -1.312     -2.11R
 13      0.8    -0.846    0.701     0.142    -1.547     -2.48R
R denotes an obs. with a large st. resid.
X denotes an obs. whose X value gives it large influence.

Durbin-Watson statistic = 1.23

MTB > PAPER
```

16.5 MODEL SPECIFICATION AND SPECIFICATION BIAS (Optional)

A **specification error** is the error associated with either omitting a relevant variable from a regression model or including an irrelevant variable in it. When specifying a regression model (that is, when determining which variables should be included in the model), we must make two decisions: *which* variables to include and *what* functional form—a log form, a squared term, or a lagged term—to use.

There is often a theoretical basis for selecting the independent variables for a regression model. For example, economic theory states that the demand for a product is a function of price, cost of substitute goods, income, and consumer tastes. In practice, of course, it may be impossible for the researcher to obtain information on all of these items, so **proxies** for the variables are used instead. Because it may be difficult to obtain data on the price of related goods, for example, some type of price index can be used as a proxy.

Model building is more art than science. The researcher tries to include all the variables that affect the outcome of the dependent variable, but no specification can perfectly determine the movements and attributes of the variable in question. The best the researcher can do is search for variables that seem consistent with underlying theory, practice, and common sense. Model specification is of great importance: if significant explanatory variables are left out, the model's worth is compromised even though least-squares estimates of the parameters are obtained. Here again, good judgment and reliance on theory must guide the researcher.

EXAMPLE 16.4 *Impact of the Omission of Variables on Estimated Regression Coefficients*

Suppose we omit retained earnings (RE) from Equation 16.7 for the year 1985. The equation becomes

$$\text{PPS}_i = a' + b'\text{DPS}_i \tag{16.7'}$$

where

$$b' = \frac{\text{Cov}(\text{DPS}_i, \text{DPS}_i)}{\text{Var}(\text{DPS}_i)} = \frac{22.5700}{1.3336} = 16.9241$$

$$a' = \overline{\text{PPS}} - \hat{b}'\overline{\text{DPS}} = 49.8770 - (16.9241)(1.9580) = 16.740$$

By regressing RE_i on DPS_i, we obtain the auxiliary regression

$$\text{RE}_i = b_0 + b_1\text{DPS}_i \tag{16.14}$$

where

$$b_1 = \frac{\text{Cov}(\text{RE}_i, \text{DPS}_i)}{\text{Var}(\text{DPS}_i)} = \frac{1.9337}{1.3336} = 1.4500$$

$$b_0 = \overline{\text{RE}}_i - \hat{b}_1\overline{\text{DPS}}_i = 1.2496 - (1.4500)(1.9580) = -1.589$$

From the specification analysis of Theil (1971),[3] the relationship among b, b', c, and b_1 can be defined as

$$b' = b + b_1 c \qquad (16.15)$$

where b' and b_1 are estimated in accordance with Equations 16.7' and 16.14, respectively, and b and c are estimated by using Equation 16.7. Substituting into the foregoing equation the estimated b and c (from Table 16.1), $b_1 = 1.4500$, and $b' = 16.9241$, we obtain

$$16.9241 \cong 12.5530 + (1.4500)(3.0144) = 16.9239$$

This implies that the misspecification error when RE_i is omitted from Equation 16.7 is 4.3709 (16.9239 − 12.5530) for b.

16.6   NONLINEAR MODELS (Optional)

Thus far we have assumed that there is a linear relationship between the dependent variable and a set of independent variables. This assumption yields a convenient approximation of the phenomena being modeled. However, there are times when other functional forms of the independent variable provide a better depiction of reality. In this section we discuss nonlinear models, including quadratic and log-linear models. We will continue to use the same regression concepts, such as hypothesis testing and confidence intervals, in our analysis.

The Quadratic Model

A quadratic model takes the form

$$y_i = a_0 + a_1 x_i + a_2 x_i^2 + e_i \qquad (16.16)$$

The only difference between this model and the models previously specified is the squared term in this model. The square of the variable is calculated and used as another independent variable, and a regression on the data is run. The quadratic term traces a parabola, as shown in Figure 16.13. If the parameter for the squared term has a positive value, the parabola opens upward. A negative parameter implies a downward-opening parabola. The linear model uses only the part of the data that is available to fit the curve. An example is shown in Figure 16.14. Here, only the upward-sloping part is used by the linear model to fit the data.

Again, we must exercise judgment and common sense to determine whether a quadratic term is needed in the model. We may have some idea how the dependent variable will react to changes in the independent variable, and graphs of the data may give us more information to help us specify the model. A quadratic model

[3] H. Theil (1971). *Principles of Econometrics,* (New York: Wiley).

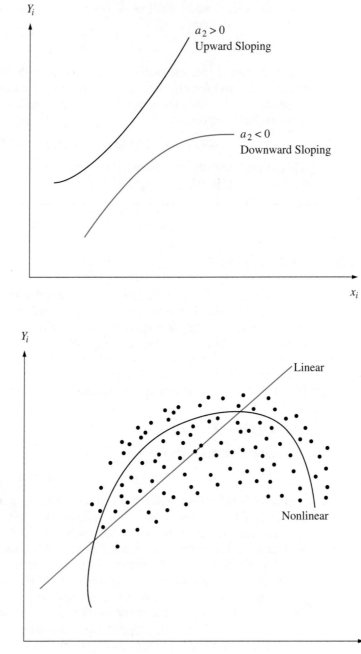

Figure 16.13

Two Different Types of Non-linear Curve

Figure 16.14

Linear Curve Versus Non-linear Curve

might be of interest in a production function where output of a product is the dependent variable and an input is the independent variable. The **law of diminishing returns** states that after a certain point, the marginal product of the variable inputs declines when additional units of a variable input are added to fixed inputs. In agriculture for example, doubling the fertilizer doubles the output of corn at low levels of fertilizer use. However, further increases in fertilizer increase the output only marginally. If a regression were to be run, the sign of the quadratic term would be negative, reflecting the fact that the output increases at a decreasing rate. It should be noted that having x and x^2 in the regression introduces a certain degree of collinearity.

EXAMPLE 16.5 *A Nonlinear Market Model*

Suppose the following regression model is run.

$$R_{i,t} = a + b_1 R_{m,t} + b_2 R_{m,t}^2 + e_{i,t} \qquad (16.17)$$

where $R_{i,t}$ is the rate of return on Allied Signal stock for first quarter 1980 to first quarter 1991, and $R_{m,t}$ is the rate of return on the market index (the S&P 500). The estimated results we get when we add a quadratic term for $R_{m,t}$ are

$$R_{i,t} = .0105 + 1.1467 R_{m,t} - 2.728 R_{m,t}^2, \qquad R^2 = .596$$
$$\quad (.61) \qquad (7.20) \qquad (-2.29)$$

where t-values are in parentheses. From Table A4 in Appendix A, by interpolation, we find that $t_{.025,38} = 2.025$. Because both 7.20 and 2.29 are larger than 2.025, we conclude that both estimated b_1 and b_2 are significantly different from 0 at $\alpha = .05$. These results imply that $R_{m,t}^2$ should be included in the regression because the t-value associated with this quadratic term is statistically significant at $\alpha = .05$.

The Log-Linear and the Log-Log Linear Model

A common transformed linear model involves the e-based logarithmic transformation of variables such as the one shown in Equation 16.18.[4]

$$\log_e Y_i = \alpha + \beta_1(\log_e x_{1i}) + \beta_2(\log_e x_{2i}) + \cdots + \beta_n(\log_e x_{ni}) + \epsilon_i \quad (16.18)$$

The **log-log linear model** is a linear model with logarithmic transformation made on both dependent and independent variables. If only the dependent variable is being lognormally transformed, then we call this linear model a **log-linear model.** As in the quadratic case, a visual inspection of the data may help determine whether a model should be specified in a log form.

[4]Equation 16.18 is obtained by taking the logarithmic transformation of a model of the equation

$$Y_i = \alpha_0 x_{1i}^{\beta_1} x_{2i}^{\beta_2} \cdots x_{ni}^{\beta_n}$$

and letting $\alpha = \log \alpha_0$.

The coefficients of a log-log linear model are elasticity coefficients, which give the percentage change in the dependent variable that is due to a 1 percent change in the independent variable. For example, suppose the demand relationship

$$Q = aP_1^b X^c P_2^d$$

has been specified, where Q is quantity purchased, P_1 is price, X is income, P_2 is the price of a competing good, and a, b, c, and d are parameters. Then b, c, and d are elasticities of P_1, X, and P_2, respectively.[5]

EXAMPLE 16.6   *The Relationship Between Cylinder Volume and Miles per Gallon*

To study the relationship between cylinder volume and miles per gallon, we use the 1986 EPA mileage guide which gives the engine size and estimated city miles per gallon ratings for 11 gasoline-fueled subcompact and compact cars. That data, as given in Table 16.5, was used to estimate the following regression relationships:

$$y_i = a + bx_i + e_i \tag{16.19a}$$

$$\log_e y_i = a' + b' \log_e x_i + e_i' \tag{16.19b}$$

Table 16.5 Cylinder Volume and Miles per Gallon for 11 Different Kinds of Cars

Car	Cylinder Volume, x	Miles per Gallon, y
VW Golf	97	37
Chevy Cavalier	173	19
Plymouth Horizon	97	31
Pontiac Firebird	151	23
Corvette	350	17
Honda Accord	119	27
Dodge Omni	97	31
Renault Alliance	85	35
Olds Firenza	173	19
Nissan Sentra	97	31
Ford Escort	114	32

Source: 1986 Gas Mileage Guide, EPA Fuel Economy Estimates, U.S. Dept. of Energy. *Wards Automotive Yearbook,* 1986.

[5]For example, the elasticity coefficient, e_p, is defined as $(dQ/dP_1)(P_1/Q)$. The first derivative is

$$dQ/dP_1 = abP_1^{b-1}X^c P_2^d$$

Substituting Q and this equation into the definition of e_p, we obtain

$$e_p = abP_1^{b-1}X^c P_2^d \left(\frac{P_1}{aP_1^b X^c P_2^d} \right)$$

$$= b$$

where

y_i = miles per gallon for i^{th} kind of car

x_i = cylinder volume for i^{th} kind of car

Equation 16.19a is a linear model and Equation 16.19b is a log-log linear model that is similar to Equation 16.18. MINITAB regression outputs for Equation 16.19a and 16.19b are presented in Figures 16.15 and 16.16 respectively. From these outputs, the estimates regression lines are

$$\hat{y}_i = 37.677 - .07241x_i \qquad R^2 = .634$$
$$(12.95) \quad (-3.95)$$

$$\log \hat{y}_i = 6.2020 - .60133 \log x_i \qquad R^2 = .841$$
$$(14.61) \quad (-6.90)$$

t-values are in parentheses.

Figure 16.15

MINITAB Output
of Equation 16.19a

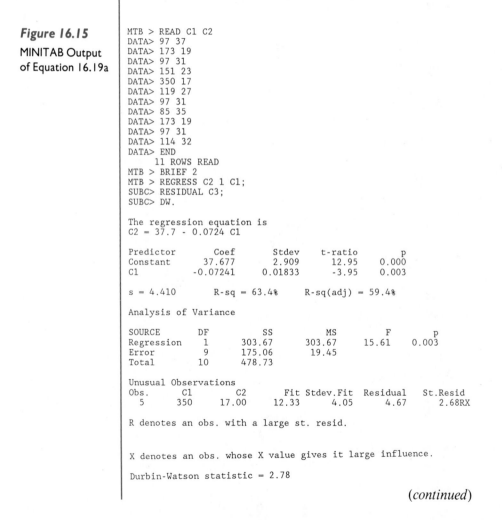

```
MTB > READ C1 C2
DATA> 97 37
DATA> 173 19
DATA> 97 31
DATA> 151 23
DATA> 350 17
DATA> 119 27
DATA> 97 31
DATA> 85 35
DATA> 173 19
DATA> 97 31
DATA> 114 32
DATA> END
      11 ROWS READ
MTB > BRIEF 2
MTB > REGRESS C2 1 C1;
SUBC> RESIDUAL C3;
SUBC> DW.

The regression equation is
C2 = 37.7 - 0.0724 C1

Predictor      Coef      Stdev    t-ratio        p
Constant     37.677      2.909      12.95    0.000
C1          -0.07241    0.01833      -3.95    0.003

s = 4.410      R-sq = 63.4%     R-sq(adj) = 59.4%

Analysis of Variance

SOURCE        DF         SS         MS        F        p
Regression     1     303.67     303.67    15.61    0.003
Error          9     175.06      19.45
Total         10     478.73

Unusual Observations
Obs.      C1        C2     Fit Stdev.Fit  Residual   St.Resid
  5      350     17.00   12.33      4.05      4.67     2.68RX

R denotes an obs. with a large st. resid.


X denotes an obs. whose X value gives it large influence.

Durbin-Watson statistic = 2.78
```

(continued)

Figure 16.15

(Continued)

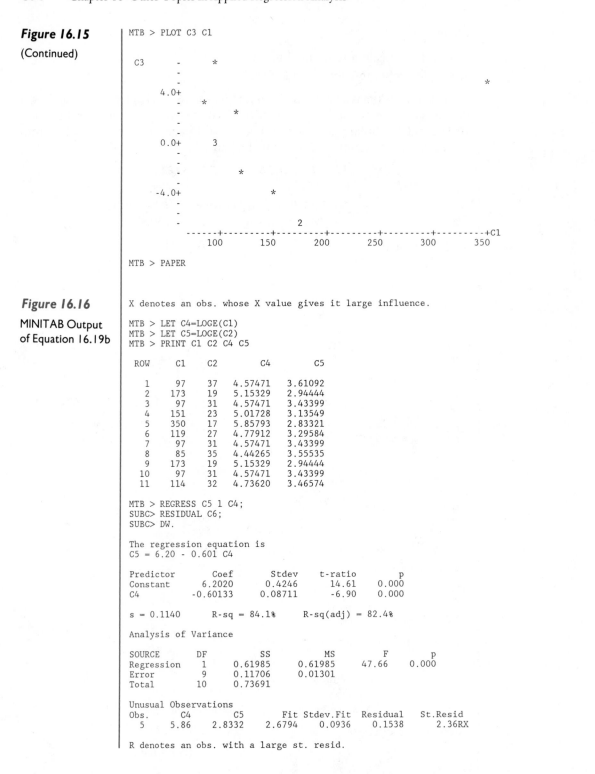

```
MTB > PLOT C3 C1

 C3        -        *
           -
           -
      4.0+                                                              *
           -    *
           -         *
           -
           -
      0.0+      3
           -
           -
           -    *
           -
     -4.0+              *
           -
           -
           -                2
           ------+---------+---------+---------+---------+---------+C1
               100       150       200       250       300       350

MTB > PAPER
```

Figure 16.16

MINITAB Output
of Equation 16.19b

```
X denotes an obs. whose X value gives it large influence.

MTB > LET C4=LOGE(C1)
MTB > LET C5=LOGE(C2)
MTB > PRINT C1 C2 C4 C5

  ROW    C1    C2       C4        C5

    1    97    37    4.57471    3.61092
    2   173    19    5.15329    2.94444
    3    97    31    4.57471    3.43399
    4   151    23    5.01728    3.13549
    5   350    17    5.85793    2.83321
    6   119    27    4.77912    3.29584
    7    97    31    4.57471    3.43399
    8    85    35    4.44265    3.55535
    9   173    19    5.15329    2.94444
   10    97    31    4.57471    3.43399
   11   114    32    4.73620    3.46574

MTB > REGRESS C5 1 C4;
SUBC> RESIDUAL C6;
SUBC> DW.

The regression equation is
C5 = 6.20 - 0.601 C4

Predictor       Coef      Stdev    t-ratio        p
Constant      6.2020     0.4246      14.61    0.000
C4           -0.60133    0.08711      -6.90    0.000

s = 0.1140     R-sq = 84.1%     R-sq(adj) = 82.4%

Analysis of Variance

SOURCE        DF        SS         MS        F        p
Regression     1    0.61985    0.61985    47.66    0.000
Error          9    0.11706    0.01301
Total         10    0.73691

Unusual Observations
Obs.     C4      C5    Fit Stdev.Fit  Residual    St.Resid
  5    5.86  2.8332  2.6794    0.0936    0.1538      2.36RX

R denotes an obs. with a large st. resid.
```

Figure 16.16

(Continued)

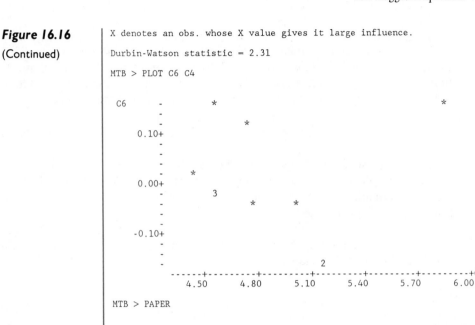

```
X denotes an obs. whose X value gives it large influence.

Durbin-Watson statistic = 2.31

MTB > PLOT C6 C4


  C6      -              *                              *
          -
          -                  *
    0.10+
          -
          -
          -
          -        *
    0.00+
          -          3
          -              *       *
          -
          -
   -0.10+
          -
          -
          -                        2
        ------+---------+---------+---------+---------+---------+C4
          4.50      4.80      5.10      5.40      5.70      6.00

MTB > PAPER
```

From Table A4 in Appendix A, we find $t_{.005,9} = 3.250$. Because all estimated t-values are larger than 3.250, all estimated parameters are significantly different from 0 at $\alpha = .01$. The bottom portion of Figure 16.15 presents the scatter diagram of residuals against the independent variable x, which shows that there are some patterns and, therefore, heteroscedasticity in the residuals. The scatter diagram presented in the bottom portion of Figure 16.16, however, shows that there is no heteroscedasticity in the residuals of log-log linear regression. These results indicate that the logarithmic transformation will make the residuals become homoscedastic. Also note that the R^2 of the log-log linear model is .841, which is larger than that of the linear model (.634). Finally, $\hat{b}' = -.60133$ implies that with a 1 percent increase in cylinder volume, miles per gallon will decrease by .60133 percent.

16.7 *LAGGED DEPENDENT VARIABLES (Optional)*

In all the models we have discussed, the dependent variable was a function of independent variables in period t. However, for time-series data, we often want to lag the dependent variable by one period to estimate the effect on the variable from a previous period. The model is

$$Y_t = \alpha + \beta_1 X_{1t} + \beta_2 X_{2t} + \cdots + \beta_k X_{kt} + \gamma Y_{t-1} + \epsilon_t \qquad (16.20)$$

Here the dependent variable is a function of the X's and of the dependent variable lagged one period.

A regression can be run on the data to estimate the coefficients of Equation 16.20. However, our interpretation of these estimated coefficients must be modified to take into account the long-run effect. The short-run (current) effect is that a 1-unit increase in X_k leads to a β_k-unit increase in Y. This is the usual interpretation of regression coefficients. The long-run effect of regression coefficients is

$$\beta_i^L = \frac{\beta_i}{1 - \gamma}, \qquad i = 1, 2, \ldots, k \qquad (16.21)$$

where β_i^L represents the long-run coefficient that takes the lagged effect into account, and γ is the coefficient associated with the lagged dependent variable as defined in Equation 16.20. In a moment, we will offer two examples to show how Equation 16.21 is calculated.

When a lagged dependent variable is used in the regression, the Durbin–Watson statistic is not a reliable indicator of autocorrelation. Another statistic—the **Durbin H**—is used instead. This statistic is

$$\mathrm{DH} = (1 - d/2) \sqrt{\frac{n}{1 - nV(\gamma)}} \qquad (16.22)$$

where d is the Durbin–Watson statistic defined in Equation 16.11, and $V(\gamma)$ is the least-squares estimate of the variance of the coefficient of the lagged variable. Under the null hypothesis, H is normally distributed with mean zero and variance 1. Therefore, the Z statistic of normal distribution can be used to do the test.

EXAMPLE 16.7   *The Relationship Between Dividend per Share and Earnings per Share*

MINITAB outputs of two regressions of Equation 16.23 for GM and Ford are presented in Figures 16.17 and 16.18, respectively.

$$\mathrm{DPS}_{i,t} = \alpha + \beta \mathrm{EPS}_{i,t} + \gamma \mathrm{DPS}_{i,t-1} + \epsilon_{i,t} \qquad (16.23)$$

Equation 16.23 is obtained by adding a variable for lagged dividend per share (DPS_{t-1}) to the right-hand side of Equation 16.13.

The results shown in Figures 16.17 and 16.18 indicate that t statistics of the γ coefficient for GM and Ford are 2.69 and 8.09, respectively. Therefore, lagged dividend per share is important in explaining dividend per share in period t for both GM and Ford.

Long-run coefficients (accumulated effect over all future periods) associated with EPS in terms of Equation 16.21 are calculated as follows:

$$\frac{.1591}{1 - .3974} = .2640 \qquad \text{long-run coefficient for GM}$$

$$\frac{.0719}{1 - .8629} = .5244 \qquad \text{long-run coefficient for IBM}$$

Figure 16.17

MINITAB Output of Equation 16.23 for GM

```
MTB > REGRESS C4 2 C3 C6;
SUBC> DW.

The regression equation is
DPS = 1.34 + 0.159 EPS + 0.397 LAGDPS

Predictor       Coef      Stdev     t-ratio       p
Constant       1.3407     0.6512       2.06     0.054
EPS           0.15911    0.03684       4.32     0.000
LAGDPS         0.3974     0.1475       2.69     0.015

s = 0.8386     R-sq = 63.9%     R-sq(adj) = 59.9%

Analysis of Variance

SOURCE        DF         SS         MS        F         p
Regression     2       22.402     11.201    15.93     0.000
Error         18       12.658      0.703
Total         20       35.060

SOURCE        DF       SEQ SS
EPS            1       17.299
LAGDPS         1        5.103

Obs.    EPS       DPS       Fit   Stdev.Fit   Residual   St.Resid
  1     2.1      3.400     3.382    0.263       0.018       0.02
  2     6.7      3.400     3.761    0.218      -0.361      -0.45
  3     7.5      4.450     3.887    0.221       0.563       0.70
  4     8.3      5.250     4.436    0.190       0.814       1.00
  5     3.3      3.400     3.947    0.293      -0.547      -0.70
  6     4.3      2.400     3.379    0.232      -0.979      -1.22
  7    10.1      5.550     3.898    0.359       1.652       2.18R
  8    11.6      6.800     5.395    0.294       1.405       1.79
  9    12.2      6.000     5.991    0.430       0.009       0.01
 10    10.0      5.300     5.323    0.321      -0.023      -0.03
 11    -2.7      2.950     3.025    0.461      -0.075      -0.11
 12     1.1      2.400     2.683    0.319      -0.283      -0.37
 13     3.1      2.400     2.786    0.334      -0.386      -0.50
 14    11.8      2.800     4.178    0.392      -1.378      -1.86
 15    14.2      4.750     4.716    0.408       0.034       0.05
 16    12.3      5.000     5.182    0.263      -0.182      -0.23
 17     8.2      5.000     4.636    0.216       0.364       0.45
 18    10.1      5.000     4.928    0.231       0.072       0.09
 19    13.6      5.000     5.498    0.305      -0.498      -0.64
 20     6.3      3.000     4.335    0.222      -1.335      -1.65
 21    -4.1      3.000     1.882    0.456       1.118       1.59

R denotes an obs. with a large st. resid.

Durbin-Watson statistic = 1.78
```

Figure 16.18

MINITAB Output of Equation 16.23 for Ford

```
MTB > REGRESS C4 2 C3 C6;
SUBC> DW.

The regression equation is
DPS = - 0.140 + 0.0718 EPS + 0.863 LAGDPS

Predictor       Coef      Stdev     t-ratio       p
Constant      -0.1399     0.3129      -0.45     0.660
EPS           0.07185    0.01246       5.76     0.000
LAGDPS         0.8629     0.1067       8.09     0.000

s = 0.4268     R-sq = 81.4%     R-sq(adj) = 79.3%

Analysis of Variance

SOURCE        DF         SS         MS        F         p
Regression     2      14.3414     7.1707    39.36     0.000
Error         18       3.2790     0.1822
Total         20      17.6203
```

(continued)

Figure 16.18

(Continued)

```
SOURCE       DF      SEQ SS
EPS          1       2.4315
LAGDPS       1      11.9098

Obs.    EPS       DPS      Fit  Stdev.Fit  Residual  St.Resid
  1     4.8    2.4000   2.2737    0.0980    0.1263      0.30
  2     6.2    2.5000   2.3750    0.0941    0.1250      0.30
  3     8.5    2.6800   2.6294    0.0953    0.0506      0.12
  4     9.1    3.2000   2.8286    0.1014    0.3714      0.90
  5     3.9    3.2000   2.8987    0.1216    0.3013      0.74
  6     3.5    2.6000   2.8699    0.1224   -0.2699     -0.66
  7    10.4    2.8000   2.8544    0.1053   -0.0544     -0.13
  8    14.2    3.0400   3.2935    0.1404   -0.2535     -0.63
  9    13.4    3.5000   3.4424    0.1466    0.0576      0.14
 10     9.8    3.9000   3.5807    0.1560    0.3193      0.80
 11   -12.8    2.6000   2.3036    0.2691    0.2964      0.89
 12    -8.8    1.2000   1.4706    0.2150   -0.2706     -0.73
 13    -5.5    0.0000   0.5033    0.2501   -0.5033     -1.46
 14    10.3    0.5000   0.5994    0.2704   -0.0994     -0.30
 15    15.8    2.0000   1.4260    0.2269    0.5740      1.59
 16    13.6    2.4000   2.5651    0.1248   -0.1651     -0.40
 17    12.3    2.2200   2.8162    0.1133   -0.5962     -1.45
 18    18.1    3.1500   3.0761    0.1625    0.0739      0.19
 19    11.0    2.3000   3.3656    0.1363   -1.0656     -2.63R
 20     8.2    3.0000   2.4353    0.0951    0.5647      1.36
 21     1.9    3.0000   2.5824    0.1188    0.4176      1.02

R denotes an obs. with a large st. resid.

Durbin-Watson statistic = 2.09
```

These long-run coefficients imply that a $1.00 increase in EPS will spell a total increase of $.2640 and $.5244 in DPS for GM and IBM, respectively. Total dividend increases are much higher than short-run (current) increases of $.195 and $.052 for GM and IBM.

EXAMPLE 16.8   *Consumption Function Analysis*

In this example, a consumption function is specified with a lagged dependent variable. A consumption function measures the change in consumption that is attributable to a 1-unit change in income. If the slope term in the regression for income is .75, for example, individuals tend to spend 75 cents out of every additional dollar earned. (The slope term is called the *marginal propensity to consume,* or MPC.) A regression of Equation 16.24 is run with personal consumption (C_t) in the United States from 1948 to 1990 as the dependent variable, and with disposable income (DI_t) and personal consumption lagged one period as the independent variables.

$$C_t = \alpha + \beta_1 DI_t + \beta_2 C_{t-1} + \epsilon_t \tag{16.24}$$

The data used to run this regression are listed in Table 16.6, and the results are presented in Table 16.7. The critical *t*-value used to do the test is $t_{.005,40} = 2.704$.

With a *t*-value of $-.435$, the constant is not statistically different from zero. Disposable income has a coefficient of .460. This implies that individuals will consume about 46 cents out of every additional dollar in income. At 5.675, the *t* statistic is significant at every level of significance. The lagged consumption variable is also

Table 16.6 Personal Consumption and Disposable Income (in billions of dollars)

Year	C_t	DI_t	C_{t-1}
1948	4650	5000	4625
1949	4661	4915	4650
1950	4834	5220	4661
1951	4853	5308	4834
1952	4915	5379	4853
1953	5029	5515	4915
1954	5066	5505	5029
1955	5287	5714	5066
1956	5349	5881	5287
1957	5370	5909	5349
1958	5357	5908	5370
1959	5531	6027	5357
1960	5561	6036	5531
1961	5579	6113	5561
1962	5729	6271	5579
1963	5855	6378	5729
1964	6099	6727	5855
1965	6362	7027	6099
1966	6607	7280	6362
1967	6730	7513	6607
1968	7003	7728	6730
1969	7185	7891	7003
1970	7275	8134	7185
1971	7409	8322	7275
1972	7726	8562	7409
1973	7972	9042	7726
1974	7826	8867	7972
1975	7926	8944	7826
1976	8272	9175	7926
1977	8551	9381	8272
1978	8808	9735	8551
1979	8904	9829	8808
1980	8783	9722	8904
1981	8794	9769	8783
1982	8818	9725	8794
1983	9139	9930	8818
1984	9489	10419	9139
1985	9840	10625	9489
1986	10123	10905	9840
1987	10311	10946	10123
1988	10580	11368	10311
1989	10678	11531	10580
1990	10668	11508	10678

Table 16.7 Results of Regression of Equation 16.24

Variable	Coefficient	Standard Error	t-value	p-value
Constant	-24.971	57.451	$-.435$	.6662
DI_t	.460	.081	5.675	.0001
C_{t-1}	.509	.091	5.667	.0001

$R^2 = .9977$
$\overline{R}^2 = .9976$
Observations   43
First-order autocorrelation $(\hat{\rho}) = .529$
DW $= .937$

highly significant; it has a coefficient of .509. If disposable income increases by 1 unit in the current period, the expected increase in consumption is .460 in the current period, is $.460 \times .509 = .234$ the next year, is $(.509)^2 \times .460 = .119$ two years later, and so on. The total increase on all future consumption in terms of Equation 16.22 is $.460/(1 - .509) = .937$. Note that the long-run coefficient, .937, is much larger than the short-run coefficient, .460.

Substituting $n = 43$, $d = .937$ and $V(\hat{\beta}_2) = (.090)^2 = .0081$ into Equation 16.22, we obtain

$$ DH = \left(1 - \frac{.937}{2}\right) \sqrt{\frac{43}{1 - 43(.0081)}} = 4.3174 $$

Using Table A3 in Appendix A, we find that DWH $= 4.3174$ is larger than $z = 3$ $(\alpha = .0014)$. Hence there is autocorrelation associated with this consumption function.

To adjust for the impact of autocorrelation, we can use a modified regression model to estimate the consumption function. It is

$$ C_t - \hat{\rho}C_{t-1} = \alpha(1 - \hat{\rho}) + \beta_1(DI_t - \rho DI_{t-1}) + \beta_2(C_{t-1} - \hat{\rho}C_{t-2}) + \epsilon_t' \quad (16.24a) $$

where $\epsilon_t' = \epsilon_t - \hat{\rho}\epsilon_{t-1}$.

Plugging the data listed in Table 16.6 and $\hat{\rho} = .529$ into Equation 16.24a yields the results presented in Table 16.8. These results are more appropriate for null hypothesis testing than are those indicated in Table 16.7.

Table 16.8 Results of Regression of Equation 16.24a

Variable	Coefficient	Standard Error	t-value	p-value
Constant	-25.961	47.586	$-.546$	.5885
DI_t	.660	.082	8.022	.0001
C_{t-1}	.289	.090	3.215	.0026

16.8 DUMMY VARIABLES

So far we have used data that could take on any number of values. In this section, we will examine an independent variable that can take on either of just two values: 1 and 0. This binary variable is called a **dummy variable,** and it enables us to include information that is not quantitative. For example, a regression that models individuals' income might include a dummy variable for sex of the worker. The independent dummy variable for sex could take on the value 1 for a male worker and the value 0 for a female worker. (The assignment of dummy variables is arbitrary; we could—and in this day and age probably *should*—have reversed the assignment: 1 for female, 0 for male.) The regression is

$$Y_i = \alpha + \beta_1 X_{1i} + \beta_2 X_{2i} + \cdots + \beta_k X_{ki} + \gamma D_{1i} + \epsilon_i \qquad (16.25)$$

where the betas ($\beta_1, \beta_2, \ldots, \beta_k$) are the coefficients for the quantitative variables, and γ is the parameter for the dummy variable:

$$D_1 = 1 \quad \text{if the worker is a male}$$

$$D_1 = 0 \quad \text{if the worker is a female}$$

Dummy variables indicate whether a shift in the intercept term is attributable to the characteristic of the dummy. The dummy variable having a statistically significant coefficient of 1,214, for example, would indicate that the intercept for males is $1,214 higher than the intercept for females. Figure 16.19 plots salary and years of experience. The female intercept is given by α. If the intercept for the dummy were negative, the male intercept would be lower than the female. The dummy variable in Equation 16.25 deals with the intercept term, not the slope term, so the dummy indicates only a shift in the intercept term. In other words, a positive and statistically significant coefficient for the dummy variable indicates that even

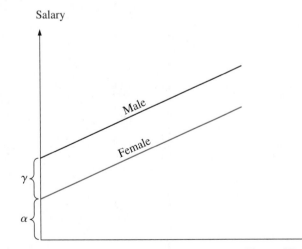

Figure 16.19

The Relationship Between Salary and Years of Experience

though the salaries for males and females are affected by the same factors in the same way—that is, they have the same slope coefficients—males begin with a higher level of earnings and maintain that difference across all values for years of experience.

EXAMPLE 16.9 *Analysis of the Money Supply*

In October 1979, the Federal Reserve's Board of Governors switched from targeting interest rates to targeting the money supply. Before this period, the Fed adhered to a policy of increasing the money supply by an amount that would keep interest rates stable; hence it "targeted" interest rates. After October 1979, the Fed focused on increasing the money supply at a fixed rate and let interest rates seek their own equilibrium level.

$$M_{3t} = \alpha + \beta_1 GNP_t + \beta_2 PRIME_t + \gamma DUM_t + \epsilon_t \qquad (16.26)$$

In this model, we investigate whether the Fed's policy change had an effect on the money supply. M_3 is the money supply, GNP is the gross national product, PRIME is the prime interest rate, and DUM is a dummy variable in which 1 equals the years 1979–1985 and 0 the years 1959–1978. A significant positive sign would indicate that the money supply was greater after the change. A negative sign would indicate that the money supply was less.

The regression of Equation 16.26 is run using the annual data for 1959–1990 presented in Table 16.9. The regression results appear in Table 16.10.

Table 16.9 GNP, Prime Rate, and M_3 (1959–1990)

```
MTB > PRINT C1-C6
```

ROW	YEAR	GNP	PRIMERt	DUMMY	GNPPRIME	M3
1	59	1629.1	4.48	1	7298.4	140.0
2	60	1665.3	4.82	1	8026.7	140.7
3	61	1708.7	4.50	1	7689.1	145.2
4	62	1799.4	4.50	1	8097.3	147.9
5	63	1873.3	4.50	1	8429.9	153.4
6	64	1973.3	4.50	1	8879.9	160.4
7	65	2087.6	4.54	1	9477.7	167.9
8	66	2208.3	5.63	1	12432.7	172.1
9	67	2271.4	5.61	1	12742.6	183.3
10	68	2365.6	6.30	1	14903.3	197.5
11	69	2423.3	7.96	1	19289.5	204.0
12	70	2416.2	7.91	1	19112.1	214.5
13	71	2484.8	5.72	1	14213.1	228.4
14	72	2608.5	5.25	1	13694.6	249.3
15	73	2744.1	8.03	1	22035.1	262.9
16	74	2729.3	10.81	1	29503.7	274.4
17	75	2695.0	7.86	1	21182.7	287.6
18	76	2826.7	6.84	1	19334.6	306.4
19	77	2958.6	6.83	1	20207.2	331.3
20	78	3115.2	9.06	1	28223.7	358.5
21	79	3192.4	12.67	0	40447.7	382.9
22	80	3187.1	15.27	0	48667.0	408.9
23	81	3248.8	18.87	0	61304.9	436.5
24	82	3166.0	14.86	0	47046.8	474.5
25	83	3279.1	10.79	0	35381.5	521.2
26	84	3501.4	12.04	0	42156.9	552.1

Table 16.9

(Continued)

27	85	3618.7	9.93	0	35933.7	620.1
28	86	3717.9	8.33	0	30970.1	724.7
29	87	3845.3	8.22	0	31608.4	750.4
30	88	4016.9	9.32	0	37437.5	787.5
31	89	4117.7	10.87	0	44759.4	794.8
32	90	4155.8	10.01	0	41599.6	825.5

Table 16.10 Results of Regression of Equation 16.26

	Coefficient	t-value	p-value
Constant	-212.74	-4.84	.000
GNP	.2356	13.57	.000
PRIME	-19.066	-6.23	.000
DUM	198.41	6.60	.000
$R^2 = .924$	$F = 176.1$	DW $= .17$	

The relationship between GNP and the money supply is extremely strong. There is a negative relationship between the money supply and the prime interest rate. In addition, the dummy variable has a significant t-value at $\alpha = 1$ percent, indicating that the money supply did increase after the Federal Reserve Board changed its policy.

16.9   *REGRESSION WITH INTERACTION VARIABLES*

The regression models specified thus far assume that there is no interaction between the independent variables. This assumption is not always realistic. In many situations, the relationship between one of the independent variables and the dependent variable is dependent on the value of another independent variable. This situation reflects **interaction.**

For example, suppose the following multiple regression model is specified.

$$\text{CROP}_t = \alpha + \beta_1 \text{RAIN}_t + \beta_2 \text{FERT}_t + \epsilon_t \qquad (16.27)$$

In this model, the amount of corn a farmer produces (CROP_t) is a function of the amount of rain received in a growing season (RAIN_t) and the amount of fertilizer used (FERT_t). Note that there is no interaction in this model; the fertilizer affects the output of corn, but this effect doesn't depend on how much rain fell (see Figure 16.20). In Figure 16.20, fertilizer is graphed on the x axis, crop production on the y axis. The rate of increase in crop production is constant for any change in the amount of rain.

However, interaction results if more rain makes the fertilizer more productive and increases corn production. We can model this interaction between the two variables by adding an interaction term.

$$\text{CORN}_t = \alpha + \beta_1 \text{RAIN}_t + \beta_2 \text{FERT}_t + \beta_3 (\text{FERT}_t \times \text{RAIN}_t) + \epsilon_t \quad (16.28)$$

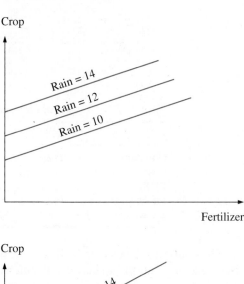

Crop

Fertilizer

Figure 16.20

Impact of Fertilizer
on Output With-
out Interaction
Effect

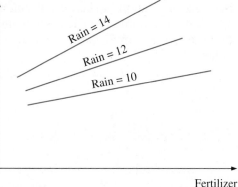

Crop

Fertilizer

Figure 16.21

Impact of Fertilizer
on Output with
Interaction Effect

To create an interaction term, we multiply the two observations whose interaction
we wish to investigate. This term measures whether additional rain makes fertilizer
more productive. In the model shown in Equation 16.27, the change in the corn
production that results from a change in the amount of fertilizer is given by the
slope term β_2. A 1-unit change in the amount of fertilizer leads to a β_2-unit change
in crop production. In the interaction model of Equation 16.28, the change in the
corn production that is associated with a 1-unit change in the amount of fertilizer
is equal to $(\beta_2 + \beta_3 \text{RAIN}_i)$. If the interaction term has a positive sign, then the rain
makes the fertilizer more effective. This effectiveness is shown in Figure 16.21,
where the dependent variable is graphed on the y axis and fertilizer on the x axis.
As the amount of rain increases, the slope of the line increases, indicating that fertilizer has a greater impact when more rain is present. In general, the interaction
term tests whether the slope parameter for one variable changes as a function of the
other variable. The t statistic is used to determine the statistical significance of the
coefficient associated with the interaction term.

EXAMPLE 16.10   *Analysis of the Money Supply, with Interaction*

Equation 16.26 with interaction can be written as

$$M_{3t} = \alpha + \beta_1 GNP_t + \beta_2 PRIME_t + \beta_3(GNP_t)(PRIME_t) + \gamma DUM_t + \epsilon_t \ (16.29)$$

MINITAB results for this equation are presented in Figure 16.22. This output indicates that the t statistic associated with the interaction term is 3.08 and that the p-value associated with the interaction term is .005. Hence the coefficient associated with the interaction term is significantly different from 0 at $\alpha = 1$ percent.

Figure 16.22

MINITAB Output
for Equation 16.29

```
MTB > BRIEF 3
MTB > REGRESS C6 4 C2 C3 C4 C5;
SUBC> DW.
* NOTE * GNPPRIME is highly correlated with other  predictor variables

The regression equation is
M3 = 198 + 0.132 GNP - 71.1 PRIMERt - 127 DUMMY + 0.0186 GNPPRIME

Predictor        Coef       Stdev     t-ratio        p
Constant       197.93       92.18        2.15    0.041
GNP            0.13247     0.03680        3.60    0.001
PRIMERt         -71.14       17.12       -4.15    0.000
DUMMY         -126.55       35.20       -3.60    0.001
GNPPRIME      0.018589    0.006038        3.08    0.005

s = 35.31       R-sq = 97.8%     R-sq(adj) = 97.5%

Analysis of Variance

SOURCE         DF          SS          MS         F        p
Regression      4     1493207      373302    299.46    0.000
Error          27       33658        1247
Total          31     1526864

SOURCE         DF      SEQ SS
GNP             1     1391385
PRIMERt         1       19320
DUMMY           1       70686
GNPPRIME        1       11815

Obs.     GNP        M3      Fit Stdev.Fit   Residual   St.Resid
  1     1629    140.00   104.17     15.00      35.83       1.12
  2     1665    140.70    98.32     13.56      42.38       1.30
  3     1709    145.20   120.55     13.94      24.65       0.76
  4     1799    147.90   140.15     12.87       7.75       0.24
  5     1873    153.40   156.13     12.06      -2.73      -0.08
  6     1973    160.40   177.74     11.09     -17.34      -0.52
  7     2088    167.90   201.15     10.10     -33.25      -0.98
  8     2208    172.10   194.53      8.46     -22.43      -0.65
  9     2271    183.30   210.07      8.32     -26.77      -0.78
 10     2366    197.50   213.63      8.67     -16.13      -0.47
 11     2423    204.00   184.72     11.55      19.28       0.58
 12     2416    214.50   184.04     11.48      30.46       0.91
 13     2485    228.40   257.85      8.63     -29.45      -0.86
 14     2608    249.30   298.03      9.58     -48.73      -1.43
 15     2744    262.90   273.28     11.15     -10.38      -0.31
 16     2729    274.40   212.39     15.73      62.01       1.96
 17     2695    287.60   263.02     10.77      24.58       0.73
 18     2827    306.40   318.67     11.02     -12.27      -0.37
 19     2959    331.30   353.08     12.39     -21.78      -0.66
 20     3115    358.50   364.21     16.42      -5.71      -0.18
 21     3192    382.90   471.41     14.63     -88.51      -2.75R
 22     3187    408.90   438.54     15.66     -29.64      -0.94
 23     3249    436.50   425.55     25.21      10.95       0.44 X
```

(continued)

Figure 16.22

(Continued)

```
24    3166    474.50    434.80    15.60     39.70     1.25
25    3279    521.20    522.45    15.03     -1.25    -0.04
26    3501    552.10    588.93    10.62    -36.83    -1.09
27    3619    620.10    638.88    11.81    -18.78    -0.56
28    3718    724.70    673.57    16.93     51.13     1.65
29    3845    750.40    710.14    16.79     40.26     1.30
30    4017    787.50    762.98    14.34     24.52     0.76
31    4118    794.80    802.18    21.93     -7.38    -0.27
32    4156    825.50    809.66    18.51     15.84     0.53

R denotes an obs. with a large st. resid.
X denotes an obs. whose X value gives it large influence.

Durbin-Watson statistic = 1.17

MTB > PAPER
```

16.10   REGRESSION APPROACH TO INVESTIGATING THE EFFECT OF ALTERNATIVE BUSINESS STRATEGIES[6]

Johnson *et al.* (1989) used multiple regression with dummy variables to investigate the relationship between business strategy and wages within the context of a significant environmental change, deregulation of the airline industry (1978–1984). Their regression results were

$$
\log_e \text{Wages} = \underset{(62.19)}{4.5848^{***}} + \underset{(1.18)}{.003 \text{ PROFITS}} - \underset{(-.19)}{.000 \text{ DEBT}}
$$

$$
+ \underset{(2.38)}{.1348 \text{ PERCENT UNION}^*} - \underset{(-.78)}{.1250 \text{ LOAD FACTOR}} + \underset{(1.24)}{.0000 \text{ SALES}}
$$

$$
- \underset{(-.75)}{.1650 \text{ FUEL COST}} - \underset{(-1.88)}{.0940 \text{ COST}^*} - \underset{(-1.84)}{.1271 \text{ FOCUS}^*}
$$

$$
- \underset{(-2.28)}{.0952 \text{ STUCK}^*} \tag{16.30}
$$

Equation 16.30 is a log-linear model discussed in Section 16.6. In this estimated multiple regression, t-values appear in parentheses beneath the coefficients $R^2 = .18$, $n = 92$. *** means $p < .001$; ** means $p < .01$; * means $p < .05$, one-sided test.

In this equation, cost, focus, and stuck are business strategic variables as defined in Table 16.11. Table 16.11 presents four alternative business strategies. They are (1) the cost leadership strategy (cost), to maintain the lowest position in the industry; (2) the product differentiation strategy (Diff.), to create a unique product or

---

[6]This section is essentially based on N. B. Johnson *et al.* (1989), "Deregulation, Business Strategy and Wages in the Airline Industry," *Industrial Relations* 28, 3, 419–430.

Table 16.11 Strategic Classification and Mean Wages by Regulating Period

Regulation			Deregulation		
Airline	Deflated Average Wage	Strategy	Airline	Deflated Average Wage	Strategy
U.S. Air	14,099	Focus	National	12,345	Cost
Delta	12,133	Cost	U.S. Air	11,911	Diff.
Ozark	12,094	Focus	American	11,797	Diff.
Frontier	11,959	Focus	Delta	11,400	Cost[a]
Texas Air	11,827	Focus	TWA	11,397	Diff.
American	11,711	Diff.	United	11,240	Diff.
National	11,643	Cost	Western	11,199	Stuck
Eastern	11,374	Stuck	Northwest	11,085	Cost
Piedmont	11,332	Focus	Braniff	11,019	Stuck
TWA	11,201	Diff.	Pan Am	11,017	Stuck
Western	11,104	Stuck	Ozark	11,014	Focus
United	11,010	Diff.	Pacific SW	10,842	Focus
Continental	10,911	Focus	Frontier	10,808	Focus
Pan Am	10,831	Focus	Eastern	10,785	Stuck
Northwest	10,722	Cost	Texas Air	10,423	Cost
Braniff	10,696	Focus	Republic	10,098	Focus
			Continental	9,799	Stuck/Cost[b]
			Piedmont	9,706	Focus
			Southwest	8,902	Focus
			People	4,105	Cost

Source: Based on "Deregulation, Business Strategy and Wages in the Airline Industry," *Industrial Relations,* Vol. 28, No. 3, pp. 419–430, reprinted with permission from Basil Blackwell Ltd.
[a]Delta was coded as a Differentiator in alternative regressions.
[b]Continental was coded as Stuck in 1978–1982 and as Cost in 1983–1984.

industrywide service through brand image (Coca-Cola is a good example), customer service, technology (Polaroid cameras) or other distinguishing features; (3) the focus business strategy (focus), to cater to a narrow strategic target with the aim of being more effective or efficient than those that are competing on a national basis; and (4) the stuck-in-the-middle (stuck) strategy, where no clearly defined strategic position exists.

Results of Equation 16.30 indicate that estimated coefficients associated with COST, FOCUS, and STUCK are all significant at $p = .05$; therefore, we can conclude that different business strategies did affect wages in the airline industry during the years 1978–1984.

Summary

In this chapter, we extended the concepts and issues of simple and multiple regression that were discussed in Chapters 13 through 15. Specifically, we investigated other topics in regression analysis, such as multicollinearity, heteroscedasticity, autocorrelation, and misspecification. We also examined nonlinear regression, regression with lagged dependent variables, dummy variables, and interaction variables. Related economics and business examples were used to demonstrate how the new models and techniques presented in this chapter can be used to analyze data.

Questions and Problems

1. Consider the hypothesis that poverty is a function of race and sex. Sample data on the subject are collected and coded using the three dummy variables P, R, and S. The P dummy represents poverty ($P = 1$ for poverty), the R dummy represents race ($R = 1$ for black) and the S dummy represents sex ($S = 1$ for female). Suppose the following multiple regression equation is esimated

$$\hat{P} = .05 + .23R + .45S$$

   a. Interpret the coefficients of R and S.
   b. Is there any problem associated with this interpretation of the equation? In other words, is there a violation of the assumptions of linear probability models?

2. The relationship between drug abuse and crime has been described by the regression

$$y = a_0 + a_1 x_1 + a_2 x_2 + a_3 x_3$$

   where

   x_1 = per-gram retaii price of heroin
   x_2 = average temperature
   x_3 = time trend
   y = crime

   Say the data for murder are $a_0 = 51.66$, $a_1 = 1.45$ (2.89), $a_2 = .04 (.22)$, and $a_3 = .05 (.07)$. The values in parentheses are the t-values. $R^2 = .523$.
   a. Interpret the multiple regression equation.
   b. What problems may be associated with the interpretation of this equation? Explain

3. Protski, Inc., an audit firm, wants to develop a multiple regression model that can explain the value of a house Y, measured in thousands of dollars, by the age of the house X_1, its square footage X_2, the number of bathrooms X_3, the absence (0) or presence (1) of an attached garage D_1, and the absence (0) or presence (1) of a view D_2. A random sample of 20 houses is used to gather observations.

Here are the results (standard errors are in parentheses):

$$Y = 63.53 - .5827X_1 + .00956X_2 + .81X_3$$
$$(38.12) \quad (.4907) \quad\quad (.01967) (11 \ .75)$$
$$- 4.98D_1 + 13.07D_2$$
$$(19.01) \quad\quad (17.69)$$

The error sum of squares is 7,892; the total sum of squares is 9,665.
   a. Comment on the significance of the regression coefficients.
   b. Comment on the overall significance of this regression.

4. a. Define autocorrelation. State which assumptions of the regression model are violated when autocorrelation exists.
   b. What is the difference between positive and negative autocorrelation?
   c. Describe a technique used to detect autocorrelation.

5. A firm with a nationwide system of bus facilities wants to develop a regression model that can explain its profit Y, measured in thousands of dollars per year, by its annual sales of bus repair and maintenance services (X_1), its annual sales of bus equipment (X_2), and its annual sales of bus advertising panels (X_3). A random sample of 12 of its facilities yields these results (standard errors are in parentheses):

$$Y = -2.29 - .0279X_1 + .0885X_2 + 3.753X_3$$
$$(13.65) \quad (.1439) \quad\quad (.0161) \quad\quad (2.402)$$

The error sum of squares is 879.5; the total sum of squares is 5,981.3.
   a. Comment on the significance of the regression coefficients.
   b. Comment on the overall significance of this regression.
   c. Do you see evidence of a possible violation of crucial assumptions?

6. Dividends per share (DPS), price per share (PPS), and retained earnings (RE) for the 30 Dow Jones industrials for 1984 give us the following multiple regression model.

$$PPS_i = 11.336 + 12.434DPS_i + 3.0875RE_i$$
$$(2.33) \quad (4.41) \quad \quad (2.39)$$
$$R^2 = .70 \quad F = 31.44$$

Correlation Matrix

Variables	PPS_i, Y	DPS_i, X_1	RE_i, X_2
PPS_i, Y	1	.79761	.69761
DPS_i, X_1	—	1	.62539
RE_i, X_2	—	—	1

a. Interpret the multiple regression equation.
b. Interpret the correlation matrix.

7. From Table 16.1, we can define the empirical relationship among PPS_i, DPS_i, and RE_i as

$$PPS_i = 11.336 + 12.434DPS_i + 3.0875RE_i$$

We also have

Cov(PPS,DPS) = 20.174
Cov(RE,DPS) = 1.6529
Var(DPS) = 1.2120
$\overline{PPS} = 40.958 \quad \overline{DPS} = 1.8862 \quad \overline{RE} = 1.9930$

a. Calculate the coefficients of the new equation $PPS_i = \alpha_1 + \beta_1 DPS_i$.
b. Regress RE_i on DPS_i and obtain the equation $RE_i = b_0 + b_1 DPS_i$. Calculate b_0 and b_1.
c. Relate β_1' to β_1, b_1, and β_2 and estimate the specification bias associated with β_1'.

8. What is multicollinearity? What problems does it cause? How can we detect multicollinearity? When we detect multicollinearity, what should we do?

9. What is autocorrelation? What problems does autocorrelation cause? How can we detect autocorrelation?

10. What is heteroscedasticity? What problems does it cause? How can we detect heteroscedasticity?

11. What is specification bias? What problems does specification bias lead to? How can we avoid specification bias?

12. What is a nonlinear regression model? Why do we sometimes choose to estimate a nonlinear model?

13. What is a lagged dependent variable? Why do we use lagged dependent variables in a regression?

14. What is a dummy variable? What does the coefficient on the dummy variable measure? Give some examples drawn from economics, finance, and accounting of times when we would want to use a dummy variable in a regression.

15. Suppose we are interested in measuring the differences in earnings among whites, blacks, Hispanics, and Asians. How many dummy variables should we use in our regression?

16. What are interaction variables? When would we choose to use interaction variables? What does the coefficient of the interaction variable tell us?

17. Suppose you have a sample of 40 observations and 3 explanatory variables and you want to test for autocorrelation. What can you say about autocorrelation if you have the following Durbin–Watson statistics?

a. $d = 1.30$ b. $d = 1.00$ c. $d = 2.25$
d. $d = 1.95$ e. $d = 3.55$

18. When we use a lagged dependent variable in our regression, R^2 is generally much higher than when such a variable is not included. Can you think of any reasons why?

19. Suppose you are interested in how stock returns differ in different months of the year. You decide to use dummy variables to examine this difference. If you choose to use 12 dummy variables, what problem will you encounter? What is the solution to this problem?

20. Look at the following scatter diagrams and explain whether heteroscedasticity appears to be a problem in either of them.

a. *y*

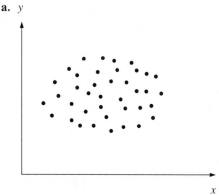

b. y

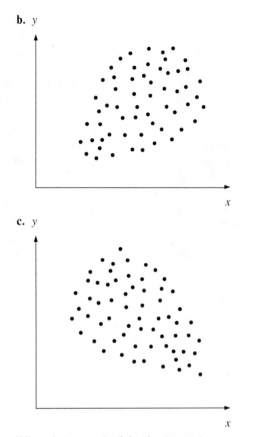

c. y

21. When heteroscedasticity is detected, we sometimes use a weighted regression in which the dependent and independent variables are weighted by the variances of their error terms. Thus the estimated regression becomes $y_i/s_e = \beta_1 x_1/s_e + \beta_2 x_2/s_e + e_i/s_e$, where s_e is the standard error of residuals. Explain intuitively why this may produce better regression results.

22. What assumptions concerning the slope coefficient β must we make when we use dummy variables in a regression?

23. You are interested in the relationship between y and three possible explanatory variables x_1, x_2, and x_3. You are given the following correlation matrix.

	y	x_1	x_2	x_3
y	1.00	.85	.86	.99
x_1		1.00	.32	.85
x_2			1.00	.50
x_3				1.00

Given this information, do you think multicollinearity will be a problem? If so, between which variables?

24. Suppose you have been hired by a lawyer who is interested in showing that a company discriminates against women in the wages it pays. You estimate the regression

$$WAGE_i = 20,000 + 5,000EXPER_i$$
$$+ 200EDUC_i - 3,000SEX_i$$

where

$WAGE_i$ = wage for person i
$EXPER_i$ = years of experience for person i
$EDUC_i$ = years of education for person i
SEX_i = dummy variable (1 for female, 0 for male)

a. Interpret the coefficients for experience and education.

b. Interpret the coefficient for sex. Does discrimination exist?

25. Suppose you also calculated the standard errors for the coefficients for experience, education, and sex as 2,000, 85, and 2,500, respectively. How would your answer to part (b) of question 24 change?

26. In order to forecast the value of a variable, we sometimes use a nonlinear trend regression such as

$$y_t = \alpha + \beta_1 t + \beta_2 t^2 + e_t \qquad (A)$$

where t = time. Briefly explain why this model may be better than a model such as

$$y_t = \alpha + \beta_1 t + e_t \qquad (B)$$

27. Suppose you are given the following data for Abbott Laboratories sales.

Year	Sales	Year	Sales
1968	351.0	1978	1467.6
1969	403.9	1978	1683.2
1970	457.5	1980	2038.2
1971	458.1	1981	2342.5
1972	521.8	1982	2602.4
1973	620.4	1983	2927.9
1974	765.4	1984	3104.0
1975	940.6	1985	3360.3
1976	1084.8	1986	3870.7
1977	1244.9	1987	4387.9

Use MINITAB to do the following.
a. Use model A from question 26 to estimate the relationship between sales and time (t).
b. Use model B from question 26 to estimate the relationship among sales, time (t), and the square of time (t^2).

28. Use MINITAB and the data given in question 27 and the equations given in question 26.
a. Draw a graph showing the actual amount of sales and the estimate of sales based on equation A.
b. Draw a graph showing the actual amount of sales and the estimate of sales based on equation B.
c. Referring to the graphs you drew in parts (a) and (b), compare the two models used for forecasting. Which model does a better job?

29. Redo question 28, this time using only data from 1978 to 1987. Does your answer to part (c) change? If so, account for this result.

30. The following are error terms from a regression.

Year	e	Year	e
1970	1.2	1982	$-.50$
1971	$-.3$	1983	$-.20$
1972	2.4	1984	1.10
1973	-1.0	1985	2.10
1974	.4	1986	-1.50
1975	$-.5$	1987	2.20
1976	$-.4$	1988	.50
1977	2.3	1989	-3.10
1978	-2.7	1990	4.20
1979	.1	1991	-1.10
1980	-3.00	1992	1.80
1981	2.40		

Compute the Durbin–Watson d statistic. Does autocorrelation appear to be a problem?

31. A biologist is interested in the effect of temperature and humidity on cell growth. She collects the following information from 8 samples.

Sample	Temperature	Humidity	Number of Cells
1	50°	20%	100
2	55	30	125
3	60	40	175
4	60	50	200
5	70	45	218

Sample	Temperature	Humidity	Number of Cells
6	75	70	235
7	80	65	220
8	85	80	250

Use the MINITAB program to estimate the relationship among number of cells, temperature, and humidity. Use an interaction variable to estimate the interaction effect of temperature and humidity on cell growth. Interpret your results.

32. Use a t test to test the significance of the coefficient on the interaction variable in question 31. Construct a 90 percent confidence interval for this coefficient.

33. A popular belief in some financial circles is that most of the movement of the stock market takes place in January. Suppose you are interested in testing this "January effect" on General Motors stock. Explain how you could do this by using a dummy variable.

34. A college admissions officer is interested in knowing whether there is a difference between males' and females' math SAT scores. She collects the following information on the math SAT score, high school grade point average, and sex of 6 students.

Student	Y Math SAT Score	X High School GPA	Sex
1	620	3.10	M
2	525	3.85	F
3	650	3.25	M
4	550	3.89	F
5	700	3.60	M
6	675	4.00	F

Use a dummy variable to test whether there is a difference between the math SAT scores of males and females. Be sure to interpret the results.

35. The batting instructor of the Toronto Blue Jays would like to see whether playing ball in the winter (winter ball) has any effect on a player's season batting average. He collects the following information on 6 players.

Player	Y Season Batting Average	X Hours of Batting Practice	Played Winter Ball
1	.300	12	yes
2	.275	11	no
3	.250	8	no
4	.325	20	yes
5	.265	8	yes
6	.350	25	yes

Use a dummy variable to test whether playing winter ball improves a player's regular-season batting average.

36. Suppose you estimate a multiple regression and find the t statistics on the coefficients to be insignificant, whereas the F statistic indicates that the coefficients are jointly significant. What problem have you probably encountered?

37. Suppose a labor economist is interested in seeing whether there is a difference in earnings and education between people in the northeast and people in other parts of the country. He estimates the following regression.

$$EARN_i = 18,500 + 2,325EDUC_i + 1,725DUM_i$$

where DUM_i is a dummy variable equal to 1 if the person is from the northeast and equal to 0 zero if the person is from anywhere else.

Interpret the foregoing regression. Do people from the northeast earn more than people from other parts of the country?

38. A financial analyst is interested in the relationship between dividend per share (DPS) and earnings per share (EPS). He collects information on these two variables and estimates the regression

$$DPS_i = \alpha + \beta EPS_i + e_i$$

From this regression he computes the error from the regression for each company.

Company	EPS	e
1	1.00	.05
2	2.10	.35
3	1.20	.10
4	3.00	.72
5	1.25	.25
6	5.20	1.21
7	8.00	2.12

Does heteroscedasticity appear to be a problem in this regression? (Hint: First use MINITAB to plot e_i against EPS.)

39. You are interested in examining the relationship between consumption and income (the consumption function) during two different periods: the period before the Vietnam War and the period during and after the Vietnam War. Explain how you would do this.

40. The following scatter diagram shows the relationship between average total cost and quantity of output. Suggest a multiple regression model to describe this relationship.

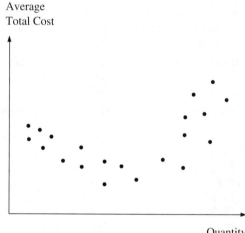

Average
Total Cost

Quantity
of Output

41. You are given the following error terms, which are the result of a regression of sales versus advertising expenditures for the Huessy Corporation over a 20-year period.

Year	e_t	Year	e_t
1	.075	11	.065
2	.080	12	.180
3	−.100	13	−.120
4	−.070	14	−.070
5	.500	15	.050
6	−.230	16	−.230
7	−.007	17	−.107
8	.088	18	.288
9	−.101	19	−.131
10	−.007	20	−.007

a. Compute the Durbin–Watson statistic.
b. Does autocorrelation exist?

Debt/Equity Ratio

Year	Chrys.	Ford	GM	Indus.
69	1.23	.76	.45	.74
70	1.23	.81	.44	.79
71	1.20	.89	.69	.91
72	1.21	.95	.56	.89
73	1.24	1.02	.62	.92
74	1.53	1.27	.63	1.09
75	1.60	1.21	.66	1.10
76	1.51	1.22	.70	1.08
77	1.62	1.28	.69	1.12
78	1.39	1.28	.74	1.14
79	2.65	1.26	.68	1.23
80	13.41	1.84	.94	2.42
81	7.04	2.13	1.20	2.07
82	5.32	2.61	1.26	7.03
83	3.96	2.16	1.20	2.14
84	1.74	1.79	1.15	1.66
85	1.99	1.58	1.16	2.57
86	1.71	1.55	1.37	2.07
87	2.07	1.43	1.63	1.90
88	5.41	5.66	3.60	4.61

Return on Assets

Year	Chrys.	Ford	GM	Indus.
69	.02	.06	.11	.08
70	.00	.05	.04	.04
71	.02	.06	.11	.07
72	.04	.07	.12	.09
73	.04	.07	.12	.09
74	−.01	.03	.05	.03
75	−.03	.02	.06	.03
76	.05	.06	.12	.08
77	.02	.09	.12	.09
78	−.03	.07	.11	.07
79	−.17	.05	.09	.05
80	−.26	−.06	−.02	−.05
81	−.08	−.05	.01	−.02
82	−.02	−.03	.02	−.01
83	.04	.08	.08	.07
84	.16	.11	.09	.10
85	.13	.08	.06	.07
86	.10	.09	.04	.06
87	.06	.10	.04	.06
88	.02	.04	.03	.03

Use MINITAB and the following information to answer questions 42–54. To find out whether there is a relationship between the amount of financial leverage a firm uses and the return on the firm's assets, you collect information on the debt/equity ratio and return on assets for the "big three" auto makers and the average for the auto industry.

42. Estimate the regression of Chrysler's return on assets against its debt/equity ratio. Compute the Durbin–Watson statistic. Does autocorrelation exist?

43. Redo question 42 using the data for Ford.

44. Redo question 42 using the data for GM.

45. Redo question 42 using the data for the auto industry.

46. Using Chrysler's data, compute r_1, the correlation coefficient between e_t and e_{t-1}, for the error terms computed in equation 35.

47. Redo question 46 using the data for Ford.

48. Redo question 46 using the data for GM.

49. Redo question 46 using the data for the auto industry.

50. Compare your computations from questions 46–49 with the Durbin–Watson statistics you calculated in questions 42–45.

51. Suppose you believe that in addition to there being a relationship between the return on assets and the debt/equity ratio, the return on assets in the previous period may also play an important part in determining the return on assets in the current period. You estimate the equation

$$ROA_t = a + bDE_t + ROA_{t-1} + e_t$$

  a. Use the data for Chrysler to estimate the foregoing equation.
  b. Compare your results to the simple regressions you computed in question 42.

52. Redo question 51 using the data for Ford.

53. Redo question 51 using the data for GM.

54. Redo question 51 using the data for the auto industry.

55. Suppose you have a sample of 50 observations and 4 explanatory variables, and you want to test for autocorrelation. What can you say about autocorrelation if you have the following Durbin–Watson statistics?

a. $d = 1.90$ b. $d = .90$ c. $d = 2.55$
d. $d = 1.75$ e. $d = 3.45$

56. A financial analyst is interested in the relationship between earnings per share (EPS) and sales. She collects information on these two variables. Then she estimates the regression

$$EPS_i = \alpha + \beta SALES_i + e_i$$

and, from it, computes the error from the regression for each company.

Company	Sales	e
1	1,200	400.05
2	2,210	500.35
3	3,201	−50.10
4	3,400	−200.72
5	4,525	300.25
6	5,320	−100.21
7	6,001	−87.25

Does heteroscedasticity appear to be a problem in this regression?

57. Suppose we have the following two versions of the market model, which shows the relationship between the rate of return on a stock and the rate of return on some market index.

Standard market model: $R_{i,t} = \alpha + \beta R_{m,t} + e_{i,t}$

Quadratic market

model: $R_{i,t} = \alpha + \beta R_{m,t} + \gamma R_{m,t}^2 + e_{i,t}$

a. Suppose the quadratic market model is the correct form to estimate but we estimate the standard market model instead. What is the effect on our parameter estimates?
b. Suppose the standard market model is the correct form to estimate but we estimate the quadratic market model instead. What is the effect on our parameter estimates?

58. An economist estimates the equation

$$C_t = 1,000 + .75Y_t + .10C_{t-1}$$

where

C_t = consumption in time period t
Y_t = income in time period t
C_{t-1} = consumption in time period $t - 1$

Interpret the coefficients of the regression model.

59. A farmer is interested in measuring the relationship among the number of bushels of corn grown, the amount of rainfall, and the amount of fertilizer used. Give two different equations that he could use to find this relationship.

60. Suppose the farmer in question 59 collects the following data.

Bushels of Corn	Rainfall (inches)	Fertilizer (pounds)
1,000	4	10
1,211	5	9
1,600	7	12
900	2	4
2,000	9	15

a. Estimate the models you suggested in question 59.
b. Compare the results of the two models. Which model do you believe is better? Explain.

61. A marketing manager believes that advertising expenditures are effective in increasing sales only to a certain extent. He discovers that when the advertising expenditures exceed a certain level, sales respond accordingly. Propose two mathematical models that can be used to describe this type of data.

62. An economist wants to study the factors that determine the hourly wage rate. He comes up with the regression

$$y = a + bx_1 + cx_2 + dx_3$$

where

y = hourly wage rate
x_1 = age of the employee
x_2 = years of experience
x_3 = years of schooling

What problem might the economist encounter in estimating this model?

63. An economist conducts research on the relationship between household spending and income. She collects data on 60 households' spending and income during 1986. After using regression analysis, she obtains the following results:

$$CONSUMPTION = 4.906 + .756 \times INCOME$$
$$R^2 = .58 \qquad DW = .32$$

The researcher claims that there is a serial correlation problem because of the low DW statistic. Do you believe this conclusion is correct?

64. Which of the following models is nonlinear in parameters where y and x are variables and a, b, and c are parameters?

a. $y = a x_1^b x_2^c$
c. $y = a + bx + cx^2$
b. $y = a + bx^c$
d. $y = a + b(1/x)$

Use the following information to answer questions 65–70. What determines the voting behavior in a presidential election? An economist believes that people "vote their pockets." He argues that economic condition is the greatest concern of voters. Therefore, the percentage of votes the incumbent gets depends on macroeconomic variables such as the inflation rate (IR), the unemployment rate (UR), and the growth rate of disposable income (DI). The following table contains these data.

y	DI	UR	IR
45	1.2	8.3	3.5
48	1.8	7.4	3.8
49.5	2.0	7.1	3.9
48.8	1.9	6.5	4.2
50.4	2.2	6.2	4.6
51.3	2.4	5.9	4.9
52.4	2.7	5.7	5.0
47.6	1.9	7.0	3.4
54.1	3.0	5.1	5.1
50.0	2.3	6.1	3.4

In the regression, we are interested in what percentage of votes the incumbent receives. We decide to use the following model.

$$y = a + b_1 DI + b_2 UR + b_3 IR$$

where

y = percentage of votes the incumbent receives

DI = rate of increase in disposable income
UR = unemployment rate
IR = inflation rate

65. What are the hypothesized signs of b_1, b_2, and b_3?

66. Estimate the model using regression analysis. Conduct a test to see whether the coefficients of UR and IR are significant at the 5 percent level.

67. Use an F test to test the hypothesis that $b_2 = b_3 = 0$ at the 5 percent level.

68. When Jimmy Carter was a candidate for the presidency, he coined a new term, the Misery Index, where Misery Index = UR + IR. Run a regression with only DI and the Misery Index. Conduct a test at the 5 percent level to see whether the coefficient on the misery index is significant in explaining y.

69. Are you convinced that DI is the only significant variable that affects y in our model? Before you run the regression, do you expect to obtain a low or a high R^2?

70. A political scientist wants to add an important variable to the equation associated with question 64 to catch the effect of war on voting behavior. He argues that as a result of patriotism, the incumbent receives a higher percentage of votes during a war than during peacetime. Suggest a way to specify the new model in order to catch the patriotism effect.

71. A financial analyst is interested in the relationship between earnings per share (EPS) and sales for Addison Company. He collects information on these two variables for a 10-year period. He estimates the regression

$$EPS_i = \alpha + \beta SALES_i + e_i$$

and from it, computes the error from the regression for each company.

Year	e_t	Year	e_t
1	130.05	6	−10.11
2	−540.35	7	83.35
3	150.10	8	−90.30
4	−240.32	9	34.04
5	100.24	10	−127.20

Does autocorrelation appear to be a problem in this regression?

PART V

Selected Topics in Statistical Analysis for Business and Economics

In the previous 16 chapters of this book, we have studied descriptive statistics, probability and important distributions, statistical inference based on samples, and regression and correlation analyses. In this last part of the book, we discuss the application of selected statistical methods in business and economics. These methods include nonparametric statistics, time-series analysis and forecasting, index numbers and stock market indexes, sampling surveys, and statistical decision theory.

CHAPTER 17

Nonparametric Statistics

CHAPTER OUTLINE

Key Terms

parametric test

classical test

nonparametric test

distribution-free test

nonparametric statistics

matched-pairs sign test

Wilcoxon matched-pairs sign-rank test

ranks

Wilcoxon's W statistic

Mann–Whitney U test

Wilcoxon rank-sum test

U statistic

Kruskal–Wallis test

Spearman rank correlation test

runs test

run

number of runs

mean absolute relative prediction error

17.1 INTRODUCTION

In previous chapters, we discussed alternative tests of hypotheses. These tests were generally concerned with statistical measures such as the mean, variance, or proportion of a population. A mean, variance, or proportion is referred to as a parameter in statistics. To test these parameters, we generally assume that the sample observations were drawn from a normally distributed population. The assumption of normality is especially critical when the sample size is small. Tests such as the Z, t, and F tests discussed in Chapter 11 depend on assumptions about the parameters of the population, so all these tests are **parametric tests** or **classical tests.** A parametric test is generally a test based on a parametric model.

Recently, a number of useful hypothesis-testing techniques that do not make restrictive distribution assumptions about the parameters of the population have been developed. Such testing procedures are referred to as **nonparametric tests** or **distribution-free tests.** Distribution-free tests are valid over a wide range of distributions of populations. (However, these nonparametric tests do require certain assumptions, such as independent sample observations.) For example, a sample is taken to test the effectiveness of a new toothpaste in reducing plaque. Samples of 10 people using a new brand and 10 people using the leading brand are compared. If the distribution of plaque reduction is skewed, a test based on the assumption of normality is no longer appropriate, and a nonparametric approach is necessary. Furthermore, the nonparametric test can be used to reduce the effect of outliers.

The main advantage nonparametric tests offer is that they do not require us to assume that the sample observations were drawn from a normal distribution. In addition, nonparametric tests are easier than parametric tests to conduct and to understand. The main disadvantages of nonparametric tests are that they ignore a certain amount of information and that they are not so efficient as parametric tests.

The main purpose of this chapter is to introduce some additional **nonparametric statistics** and explore their applications in testing hypotheses. Actually, in Chapter 12 we discussed two nonparametric methods for hypothesis testing: the chi-square test for goodness of fit and the chi-square test for independence.[1] This chapter focuses on the development and use of six more nonparametric tests:

1. The matched-pairs sign test
2. The Wilcoxon matched-pairs signed-rank test
3. The Mann–Whitney U test (rank-sum test)
4. The Kruskal–Wallis test
5. The Spearman rank correlation test
6. The number-of-runs test

After discussing these nonparametric methods, we will examine five examples of the use of nonparametric statistical methods in business decision making.

---

[1]The first test is concerned with how well a set of data fits a hypothesized probability distribution. The second seeks to determine whether a relationship exists between two variables. These two tests are generally large-sample tests.

17.2 THE MATCHED-PAIRS SIGN TEST

We begin our discussion of nonparametric tests with one of the easiest to employ—the **matched-pairs sign test.** The sign test is used to test the central tendency of a population distribution and is most frequently employed in analyzing matched-pairs data. A sign test uses the sign of the difference between two numbers rather than the actual quantitative measurements.

We will illustrate the matched-pairs sign test in terms of data obtained from a sample survey. Table 17.1 shows some of the data derived from a survey that sought to determine whether economists believed a Democratic president or a Republican president would have a more positive effect on the economy. Prior to a presidential election, 55 economists were surveyed and asked to rank, on a scale from 1 to 10, the likelihood that either a Democratic or a Republican president would have a positive impact on the economy.

Columns (2) and (3) in the table show the economists' rankings of the potential for a chief executive from each of the political parties having a positive impact on the economy; 10 represents the greatest positive impact. The last column indicates only the sign of the difference, either + or −. If there is no difference between the rankings, a 0 is displayed. A plus sign means a higher numerical score was assigned to the Democratic presidential candidate than to the Republican candidate, a minus sign means the reverse, and a zero denotes a tie score.

The null hypothesis of our test is that there is no tendency to prefer one political party over the other in assessing the president's potential impact on the economy. To implement this hypothesis test, we compare only the numbers for economists who have a preference for one political party. Hence we do not include, in our test, data for economists who predicted that both political parties would do equally well.

Table 17.1 Assessing Political Party Preferences

(1) Economist	(2) Score for Democrat	(3) Score for Republican	(4) Sign of Difference
1	8	6	+
2	9	4	+
3	9	6	+
4	4	5	−
5	7	8	−
6	9	3	+
7	8	5	+
8	9	6	+
9	7	7	0
10	9	8	+
.	.	.	.
.	.	.	.

Among the 55 economists surveyed, 33 stated that a president from the Democratic Party would have the greater positive impact on the economy, 17 stated that a Republican president would have the greater impact, and 5 stated that the two parties would do equally well. Because tied cases are excluded in a sign test, our analysis will include 33 plus signs and 17 minus signs.

We want to test the null hypothesis of no difference in the impact on the economy by the political parties in question; that is, we want to test the hypothesis that plus and minus signs are equally likely to occur. We would expect an equal number of plus and minus signs if the null hypothesis were true. On the other hand, either too many pluses or too many minuses will be grounds for rejection of the null hypothesis. If we use p^* to denote the probability of obtaining a plus sign, we can state the hypotheses as

H_0: There are no differences in the parties' impact on the economy. ($p^* = .50$)

H_1: There are differences in the parties' impact on the economy. ($p^* \neq .50$)

We use the large-sample method of the normal approximation to the binomial distribution (see Chapter 7). If the observed proportion of plus signs is $\hat{p}$, then the mean and standard deviation of the sampling distribution of $\hat{p}$ are

$$\mu_{\hat{p}} = p^* \qquad (17.1)$$

$$\sigma_{\hat{p}} = \sqrt{\frac{p^*q^*}{n}}$$

Our sample consists of 33 plus signs and 17 minus signs, so we substitute $n = 50$ into Equation 17.1. This yields

$$p^* = .5$$

and

$$\sqrt{p^*q^*/n} = \sqrt{(.5)(.5)/50} = .071$$

Because our sample consists of 50 observations and the observed proportion of plus signs is $\hat{p} = 33/50 = .66$, our test statistic, Z, can be approximated by a standard normal distribution.

$$Z = \frac{\hat{p} - p^*}{\sqrt{p^*q^*/n}} = \frac{.66 - .50}{.071} = 2.254$$

Hence, assuming that we test the hypothesis at the 5 percent level of significance, we would reject the null hypothesis if $Z < -1.96$ or $Z > 1.96$. Accordingly, our results dictate that we reject the null hypothesis that plus and minus signs are equally likely to occur. That is, because the number of plus signs is greater than the number of minus signs, we conclude that the economists in the sample believe that a president from the Democratic Party is more likely than a Republican president to have a positive influence on the economy.

As we noted in Chapter 11, the critical interval estimate for $p*$ (the true proportion of positive signs), rather than the Z-value, can be used to do the null hypothesis test. The critical interval estimates for $p*$ are

$$\hat{p} + 1.96\sqrt{p*q*/n} = .66 + (1.96)(.071) = .521$$
$$\hat{p} - 1.96\sqrt{p*q*/n} = .66 - (1.96)(.071) = .799$$

That is, $.521 < p < .799$. This interval does not contain $p* = .5$, so we reject the null hypothesis.

17.3   THE WILCOXON MATCHED-PAIRS SIGNED-RANK TEST

The **Wilcoxon matched-pairs signed-rank test** is preferable when the differences between the matched pairs can be quantitatively determined, rather than merely assigned signs. In other words, the Wilcoxon test provides a method of incorporating information about the relative size of the differences between the matched pairs in terms of **ranks.**

To illustrate how to employ the Wilcoxon matched-pairs signed rank test, we will use a sample of net income figures for Lawrence Inc., which we assume to be emerging from a major corporate reorganization (see Table 17.2). Data are given for the 10 market regions in which Lawrence sells its product. Columns (2) and (3) of Table 17.2 show the net income figures (in millions of dollars) for each region for the year before corporate restructuring and the year after, respectively.

Table 17.2 Net Income Figures for Lawrence Inc.

(1) Market Region	(2) Net Income Before Reorganization	(3) Net Income After Reorganization	(4) Difference, $d = (3) - (2)$	(5) Rank of $|d|$	(6) Signed +	(7) Signed −
1	41	62	21	7	7	
2	34	49	15	6	6	
3	43	39	−4	4		4
4	29	28	−1	1		1
5	55	55	0			
6	63	66	3	2.5	2.5	
7	35	47	12	5	5	
8	42	72	30	9	9	
9	57	84	27	8	8	
10	45	42	−3	2.5		2.5
					37.5	7.5

As in the sign test we conducted in Section 17.2, we calculate the difference in net income for each region before and after the reorganization. This difference is entered in column (4). Next we determine the absolute values of the differences for each region and rank the regions accordingly from 1 to n, where n is the number of regions in our example. The smallest absolute difference is assigned the rank 1. When the difference within a particular region is 0, no ranking is assigned. Hence, in our example, the data for region 5 are no longer included in our test. When absolute differences are tied, the mean rank value is assigned to those differences. In our example, because regions 6 and 10 are tied for the rank of 3, both are assigned the rank of 2.5, which is the average of 2 and 3. These ranks are entered in column (6) or column (7), depending on their sign (+ or −) in column (4). Positive-signed ranks are listed in column (6), negative-signed ranks in column (7). Again, because the difference from region 5 is 0, only 9 samples need be included in our test.

Our table also displays the sum of the ranks in columns (6) and (7). These sums are critical to our null hypothesis, which is[2]

$$H_0\text{: Sum of plus ranks} = \text{sum of minus ranks} \tag{17.2}$$
$$\Sigma \text{ rank}(+) = \Sigma \text{ rank}(-)$$

In other words, the null hypothesis implies that the population of positive and negative differences is distributed around the mean of zero. The test statistic, referred to as **Wilcoxon's W statistic,** is the smaller of the sum of the plus ranks (W^+) and the sum of the negative ranks (W^-).

$$W^+ = \sum_{i=1}^{n} R_i^+ \tag{17.3}$$

$$W^- = \sum_{i=1}^{n} R_i^- \tag{17.3}$$

For samples of $n \leq 20$, we use Table A11 in Appendix A to obtain the critical values of the test statistic W. Note that Table A11 represents the maximum value that W can have and still be considered significant at various levels of significance.

In our example, because one observation of the difference is 0, the effective sample size $n = 10 - 1 = 9$. From Table 17.2, we obtain $W^+ = 37.5$ and $W^- = 7.5$. The two-tailed value in Table A11 corresponding to $n = 9$ and $\alpha = .05$ is 6. Consequently, we are to accept H_0 if $W^- \geq 6$. Because the value of W^- is larger than 6, we cannot reject H_0 and must conclude that the net income levels before and after corporate reorganization do not differ significantly.

Kruskal and Wallis have shown that when n is large (at least 25), W is approximately normally distributed with mean μ_W and standard deviation σ_W defined as follows[3]

---

[2]Technically, this is not a null hypothesis, because it is stated in sample—not population—terms.

[3]See W. H. Kruskal and W. A. Wallis (1952), "Use of Ranks in One-Criterion Variance Analysis," *Journal of the American Statistical Association* 47, 152, 583–621.

$$\mu_W = n(n + 1)/4$$
$$\sigma_W = \sqrt{n(n + 1)(2n + 1)/24}$$

(17.4)

This implies that we can compute $Z = (W - \mu_W)/\sigma_W$ and perform the standard Z test, which we examined in detail in Chapter 11. Note that the power of the signed-rank test discussed in this section is higher than that of the sign test discussed in the last section. The efficiency of the matched-pairs sign test compares to that of the Wilcoxon matched-pairs signed-rank test as $3/\pi$ to $2/\pi$. Finally, note that the t test rather than the Z test is appropriate when the sample size is smaller than 25.

The MINITAB output of Table 17.2 is shown in Figure 17.1. To discuss the four tests presented in Figure 17.1, we will rewrite the null hypothesis of Equation 17.2 as

H_0: The population differences are centered at d_0. (17.2′)

```
MTB > READ INTO C1 C2
DATA> 41 62
DATA> 34 49
DATA> 43 39
DATA> 29 28
DATA> 55 55
DATA> 63 66
DATA> 35 47
DATA> 42 72
DATA> 57 74
DATA> 45 42
DATA> END
    10 ROWS READ
MTB > SUBTRACT C2 FROM C1, PUT INTO C3
MTB > WTEST C3

TEST OF MEDIAN = 0.000000 VERSUS MEDIAN N.E. 0.000000

                  N FOR   WILCOXON            ESTIMATED
             N    TEST    STATISTIC  P-VALUE    MEDIAN
EPS         10      9        7.5      0.086     -8.500
MTB > WTEST OF CENTER=1 USING C3;
SUBC> ALTERNATIVE=1.

TEST OF MEDIAN = 1.000 VERSUS MEDIAN G.T. 1.000

                  N FOR   WILCOXON            ESTIMATED
             N    TEST    STATISTIC  P-VALUE    MEDIAN
EPS         10      9        5.0      0.984     -8.500
MTB > WTEST OF CENTER=1 USING C3;
SUBC> ALTERNATIVE=-1.

TEST OF MEDIAN = 1.000 VERSUS MEDIAN L.T. 1.000

                  N FOR   WILCOXON            ESTIMATED
             N    TEST    STATISTIC  P-VALUE    MEDIAN
EPS         10      9        5.0      0.022     -8.500


MTB > WTEST OF CENTER=1 USING C3.

TEST OF MEDIAN = 1.000 VERSUS MEDIAN N.E. 1.000

                  N FOR   WILCOXON            ESTIMATED
             N    TEST    STATISTIC  P-VALUE    MEDIAN
EPS         10      9        5.0      0.044     -8.500
MTB > PAPER
```

Figure 17.1

MINITAB Output
of Table 17.2

The first test is to test $d_0 = 0$, and its p-value is .086. Thus it is not significant at $\alpha = .05$, which is identical to what we found before. Both the second and third tests are one-tailed tests. Their alternative tests are $d_0 > 1$ and $d_0 < 1$, respectively. From $p = .984$ and $p = .022$, we conclude that only the third test is significant at $\alpha = .05$. The fourth test is to test $d_0 = 1$. It is significant at $\alpha = .05$. From the third and fourth tests, we conclude that the net income increases by at least 1 million after the company is reorganized.

17.4 MANN–WHITNEY U TEST (WILCOXON RANK-SUM TEST)

We will now consider another nonparametric technique that involves comparing data from two samples. The **Mann–Whitney U test,** also referred to as the **Wilcoxon rank-sum test,** tests whether two independent samples have been drawn from two populations that have the same relative frequency distribution. Unlike a sign test, the Mann–Whitney U test directly considers the rankings of the observations in each sample.

To illustrate the procedure for the Mann–Whitney U test we will refer to Table 17.3, which shows the research and development expenditures of 15 companies in each of two major industries, A and B.

The first step in performing the Mann–Whitney U test is to combine the two samples. In Table 17.4 we rank the firms according to the dollar value of the expenditures, 1 representing least R&D expenditure and 30 representing greatest R&D

Table 17.3 R&D Expenditures of Two Major Industries (in millions of dollars)

Industry A	Rank for Industry A	Industry B	Rank for Industry B
40	1	54	8
41	2	52	6
43	3	69	15
46	4	70	16
47	5	71	17
53	7	72	18
55	9	73	19
56	10	76	20
61	11	77	21
63	12	78	22
64	13	82	25
68	14	83	26
79	23	84	27
80	24	88	29
85	28	89	30
	166		299

Table 17.4 Ranking by Dollar Value of Expenditure

Rank	R&D Expenditure	Industry
1	40	A
2	41	A
3	43	A
4	46	A
5	47	A
6	52	A
7	53	B
8	54	A
9	55	A
10	56	A
11	61	B
12	63	A
13	64	A
14	68	A
15	69	B
16	70	B
17	71	B
18	72	B
19	73	B
20	76	B
21	77	B
22	78	B
23	79	A
24	80	A
25	82	B
26	83	B
27	84	B
28	85	A
29	88	B
30	89	B

expenditure. Note that firms continue to be designated by industry in Table 17.4. The next step is to sum the ranks of the sample observations listed in Table 17.4.

Referring to the data for industry A as sample 1, we can calculate the sum of ranks of items in sample 1, designated $R_1 = 166$, as shown in the second column of Table 17.3. We designate the sum of ranks of items in sample 2 (the data for industry B) as R_2. Accordingly, $R_2 = 299$, as indicated in the last column of Table 17.3. In general, if several variables are tied, then we assign each the average of the ranks.

If the null hypothesis is true—in other words, if the samples from the two industries were drawn from the same population—then we would expect that the totals of these two ranks (R_1 and R_2) would be approximately equal. In order to test this hypothesis, we calculate a **U statistic.** The U statistic is a test statistic that depends on the number of observations in the samples as well as on the total of the ranks for one of the samples—in this case, R_1.

$$U_1 = n_1 n_2 + \frac{n_1(n_1 + 1)}{2} - R_1 \qquad (17.5)$$

where n_1 is the number of observations in sample 1 and n_2 is the number of observations in sample 2. This test statistic could also be stated in terms of R_2:

$$U_2 = n_1 n_2 + \frac{n_2(n_2 + 1)}{2} - R_2 \qquad (17.6)$$

The U statistic measures the difference between the ranked observations of the two samples and provides evidence on the difference between their population distributions. Either very small or very large U values provide evidence of the separation of the matched observations of the two samples.

It can be shown that the sample distribution of U has the following mean and standard deviation.

$$\mu_U = \frac{n_1 n_2}{2} \qquad (17.7)$$

$$\sigma_U = \sqrt{\frac{n_1 n_2(n_1 + n_2 + 1)}{12}} \qquad (17.8)$$

Moreover, when the numbers of observations in both samples n_1 and n_2 are in excess of approximately 10 observations, the sampling distributions of ranking dollar value of expenditure approach a normal distribution.

For our example, the U statistic, the mean, and the standard deviation are calculated as follows:

$$U = (15)(15) + \frac{15(15 + 1)}{2} - 166 = 179$$

$$\mu_U = \frac{(15)(15)}{2} = 112.5$$

$$\sigma_U = \sqrt{\frac{(15)(15)(15 + 15 + 1)}{12}} = 24.11$$

Using this information, we can calculate the standardized normal variate.

$$Z = \frac{U - \mu_U}{\sigma_U} = \frac{179 - 112.5}{24.11} = 2.76$$

Again assuming a two-tailed test at the 5 percent level of significance, we find from Table A3 in Appendix A that the critical value for Z is 1.96. Because our calculated Z-value exceeds the critical value, we can reject the null hypothesis that the sample observations were drawn from the same population.

```
MTB > READ C1 C2
DATA> 40 54
DATA> 41 52
DATA> 43 69
DATA> 46 70
DATA> 47 71
DATA> 53 72
DATA> 55 73
DATA> 56 76
DATA> 61 77
DATA> 63 78
DATA> 64 82
DATA> 68 83
DATA> 79 84
DATA> 80 88
DATA> 85 89
DATA> END
     15 ROWS READ
MTB > MANN-WHITNEY C1 C2

Mann-Whitney Confidence Interval and Test

C1          N =  15     Median =        56.00
C2          N =  15     Median =        76.00
Point estimate for ETA1-ETA2 is        -17.00
95.4 pct c.i. for ETA1-ETA2 is (-28.00,-6.00)
W = 166.0
Test of ETA1 = ETA2  vs.  ETA1 n.e. ETA2 is significant at 0.0062

MTB > PAPER
```

Figure 17.2
MINITAB Output
of Table 17.3

A MINITAB solution for this example is shown in Figure 17.2. The Mann–Whitney statistic is denoted as W, which is the same as the first sample ranks. R_2 is obtained by using the identity

$$R_1 + R_2 = \frac{n(n + 1)}{2}$$

where $n = n_1 + n_2$. The values of U_1 and U_2 are then easily calculated. The p-value of .0062 in Figure 17.2 indicates that we should reject the null hypothesis that the two samples are equal at $\alpha = .05$.

When samples n_1 and n_2 are both ≤ 10, Table A12 in Appendix A may be used to obtain the critical values of test statistic R_1 for both one- and two-tailed tests at various levels of significance. The producer commodity price indexes for January 1985 and January 1986 for 6 product categories listed in the table are used to show how the Wilcoxon rank-sum test can be performed (data from *Standard & Poor's Statistical Service, Current Statistics,* Jan. 1987, pp. 12–13). Combined ranks are shown in parentheses.

Product Category	January 1985	January 1986
Processed poultry	198.8 (4)	192.4 (3)
Concrete ingredients	331.0 (8)	339.0 (9)
Lumber	343.0 (10)	329.6 (7)

Product Category	January 1985	January 1986
Gas fuels	1,073.0 (12)	1,034.3 (11)
Drugs and pharmaceuticals	247.4 (5)	265.9 (6)
Synthetic fibers	157.6 (2)	151.1 (1)

From the combined ranks information in the table, we find

$$R_1 = 4 + 8 + 10 + 12 + 5 + 2 = 41$$

$$R_2 = 3 + 9 + 7 + 11 + 6 + 1 = 37$$

As a check on the ranking procedure, by substituting $R_1 = 41$, $R_2 = 37$, and $n = 12$ into the identity discussed previously, we obtain

$$41 + 37 = \frac{12(12 + 1)}{2} = 78$$

Because both n_1 and $n_2 \leq 10$, Table A12 is used to obtain the critical value R_1 statistic. With $n_1 = 6$ and $n_2 = 6$, we observe (Table A12) that at the .05 level of significance, the lower and upper critical values for the two-tailed test are, respectively, 26 and 52. Because the observed value of the test statistic $R_1 = 41$ falls between the critical values, the null hypothesis cannot be rejected. In other words, the probability distribution of economic indexes did not change during the period of January 1985 and January 1986.

17.5 KRUSKAL–WALLIS TEST FOR m INDEPENDENT SAMPLES

The **Kruskal–Wallis test** is a one-factor analysis of variance by ranks. It is a non-parametric test that represents a generalization of the two-sample Mann–Whitney U rank-sum test to situations where more than two populations are involved. Unlike one-factor analyses of variance (see Chapter 12), the Kruskal–Wallis test makes no assumptions about the population distribution.

This test is based on a test statistic calculated from ranks established by pooling the observations from c independent simple random samples (where $c > 2$). The null hypothesis is that the populations are identically distributed or, alternatively, that the samples were drawn from c identical populations. Let's follow the procedure through an example.

Assume that simple samples of executive vice presidents in a certain industry were drawn from firms classified into three size categories (large, medium, and small). After being assured of the confidentiality of their replies, the 20 executives were asked to rate the overall quality of their Board of Directors' performance in setting general corporate policy during the past 3-year period on a scale from 0 to 100, with 0 denoting the lowest rating and 100 the highest. The scores, classified by size of firm, and the rankings of the pooled sample scores are shown in Table 17.5.

Table 17.5 Calculations for the Kruskal–Wallis Test Scores and Ranks Classified by Size of Firm

Large		Medium		Small	
Score	*Rank*	*Score*	*Rank*	*Score*	*Rank*
79	12	69	6	83	14
96	20	78	11	66	5
86	16	85	15	51	1
88	17	62	3	94	19
76	10	63	4	71	7
91	18	73	8	61	2
81	13			74	9
$n_1 = 7$		$n_2 = 6$		$n_3 = 7$	
$R_1 = 106$		$R_2 = 47$		$R_3 = 57$	

The result was the following pooled ranking, with the lowest score that was actually given represented by 1 and the highest by 20.

Score	51	61	62	63	66	69	71	73	74	76	78	79
Rank	1	2	3	4	5	6	7	8	9	10	11	12

Score	81	83	85	86	88	91	94	96
Rank	13	14	15	16	17	18	19	20

The Kruskal–Wallis test statistic, K, compares the variations of the ranks of the sample groups.

$$K = \frac{12}{n(n+1)} \sum_{i=1}^{c} \frac{R_i^2}{n_i} - 3(n+1) \qquad (17.9)$$

where

$$n_i = \text{number of observations in the } i\text{th sample}$$
$$n = n_1 + n_2 + \cdots + n_c = \text{total number of observations in the } c \text{ samples}$$
$$R_i = \text{sum of the ranks for the } i\text{th sample}$$

Table 17.5 gives the sample sizes and rank sums for each sample group. Substituting into the foregoing formula, we compute the K statistic in the present example.

$$K = \frac{12}{20(20+1)} \left(\frac{106^2}{7} + \frac{47^2}{6} + \frac{57^2}{7} \right) - 3(20+1) = 6.64$$

As a check on calculations at this point, make sure the ranks sum to $n(n+1)/2 = (20)(21)/2 = 210$; here $106 + 47 + 57 = 210$.

It can be shown that the sampling distribution of K is approximately the same as the chi-square distribution with $v = c - 1$ degrees of freedom (where c is the number of sample groups). In this example, where there are 3 sample groups, the number of degrees of freedom is $v = c - 1 = 3 - 1 = 2$. Testing the null hypothesis at the 5 percent level of significance ($\alpha = .05$) and using Table A5 of Appendix

A, we find the critical value of χ^2 to be $\chi^2_{2,.05} = 5.991$. Hence our rule for the one-tailed test is

If $K > 5.991$, reject the null hypothesis.

If $K \leq 5.991$, do not reject the null hypothesis.

Because $K = 6.64$ is greater than the critical value of 5.991, we reject the null hypothesis of identically distributed populations. Therefore, we conclude that there are significant differences by size of firm in the scores assigned by these 3 samples of executive vice presidents.

MINITAB output of the Kruskal–Wallis statistics of Table 17.5 is presented in Figure 17.3. Note that the Board of Directors' performance scores are stored in $C1$, whereas column $C2$ contains the sample number of each observation (1, 2, or 3). The value of the Kruskal–Wallis statistic is called H and agrees with the previous result. The *p*-value associated with H is .037.

```
MTB > READ C1-C2
DATA> 79 1
DATA> 96 1
DATA> 86 1
DATA> 88 1
DATA> 76 1
DATA> 91 1
DATA> 81 1
DATA> 69 2
DATA> 78 2
DATA> 85 2
DATA> 62 2
DATA> 63 2
DATA> 73 2
DATA> 83 3
DATA> 66 3
DATA> 51 3
DATA> 94 3
DATA> 71 3
DATA> 61 3
DATA> 74 3
DATA> END
      20 ROWS READ
MTB > KRUSKAL-WALLIS C1 C2

LEVEL    NOBS    MEDIAN   AVE. RANK    Z VALUE
   1       7     86.00      15.1        2.58
   2       6     71.00       7.8       -1.32
   3       7     71.00       8.1       -1.31
OVERALL   20                10.5

H = 6.64   d.f. = 2   p = 0.037

MTB > PAPER
```

Figure 17.3
MINITAB Output
of Table 17.5

17.6   SPEARMAN RANK CORRELATION TEST

The **Spearman rank correlation test** is a nonparametric method of correlation designed to measure the strength of association between two sets of ranked data. As we have learned, nonparametric procedures can be useful in correlation analysis where the basic data are not available in the form of numerical magnitudes but where rankings can be assigned. If two variables of interest can be ranked in separate ordered series, a rank correlation coefficient can be computed. We will consider two different cases, the first representing perfect direct correlation between two series; the second, perfect inverse correlation.

Table 17.6 displays data on the rankings of a simple random sample of 10 students according to learning abilities in mathematics and physics. Clearly, this represents a case in which it would be almost impossible to obtain precise quantitative measures of these abilities but in which rankings may be feasible. In rank correlation analysis, the rankings may be assigned in order from high to low, with 1 representing the highest rating, 2 the next highest, and so on, or from low to high, with 1 representing the lowest rank, 2 the next lowest, and so on. The computed rank correlation coefficient is the same regardless of the rank ordering used.

The rank correlation coefficient (γ_s) is computed by the formula

$$r_s = 1 - \frac{6 \Sigma \, d^2}{n(n^2 - 1)} \qquad (17.10)$$

where

d = difference between the ranks for the paired observations
n = number of paired observations

The calculations of the rank correlation coefficients for the two extreme cases are shown in Tables 17.6 and 17.7. In the first table there is a perfect direct correlation in the rankings; that is, the student who ranks highest in mathematics ability is also best in physics. In the second table, there is perfect inverse correlation; that is, the student who ranks highest in mathematics is worst in physics.

In the case of perfect correlation between the ranks, $r_s = 1$; in perfect inverse correlation, $r_s = -1$. An r_s-value of zero indicates no correlation between rankings. Tied ranks are handled in the calculations by averaging. Substituting $\Sigma d^2 = 0$ and $\Sigma d^2 = 33$ into Equation 17.10, we obtain

$$r_s = 1 - \frac{6 \Sigma \, d^2}{n(n^2 - 1)} = 1 - \frac{6(0)}{10(10^2 - 1)} = 1$$

$$r_s = 1 - \frac{6 \Sigma \, d^2}{n(n^2 - 1)} = 1 - \frac{6(330)}{10(10^2 - 1)} = -1$$

The significance of rank correlation is tested in the same way as for the sample correlation coefficient r. We compute the statistic

$$t = \frac{r_s}{\sqrt{(1 - r_s^2)/(n - 2)}} \qquad (17.11)$$

Table 17.6 Rank Correlation of Mathematics Learning Ability with Physics Learning Ability (Perfect Direct Correlation)

Student	Rank in Mathematics Ability, X	Rank in Physics Ability, Y	Difference in Ranks, $d = X - Y$	$d^2 = (X - Y)^2$
A	1	1	0	0
B	2	2	0	0
C	3	3	0	0
D	4	4	0	0
E	5	5	0	0
F	6	6	0	0
G	7	7	0	0
H	8	8	0	0
I	9	9	0	0
J	10	10	0	0

Table 17.7 Rank Correlation of Mathematics Learning Ability with Physics Learning Ability (Perfect Inverse Correlation)

Student	Rank in Mathematics Ability, X	Rank in Physics Ability, Y	Difference in Ranks $d = X - Y$	$d^2 = (X - Y)^2$
A	1	10	-9	81
B	2	9	-7	49
C	3	8	-5	25
D	4	7	-3	9
E	5	6	-1	1
F	6	5	1	1
G	7	4	3	9
H	8	3	5	25
I	9	2	7	49
J	10	1	9	81
				330

which has a t distribution with $(n - 2)$ degrees of freedom. If $r_s = .90$, then

$$t = \frac{.90}{\sqrt{(1 - .81)/(10 - 2)}} = 5.84$$

Assuming we are using a two-tailed test of the null hypothesis of zero correlation in the ranked data of the population, the critical t at a 5 percent level of significance with 8 degrees of freedom is equal to 2.306, as indicated in Figure 17.4. We reject the hypothesis of no rank correlation and conclude that a positive linear relationship exists between rank in mathematics learning ability and in physics learning ability.

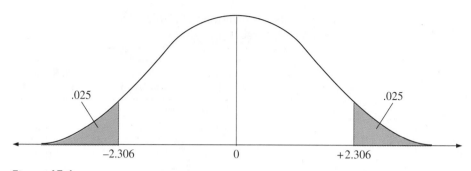

Figure 17.4

Sampling Distribution of Rank Correlation with $\alpha = .05$ and 8 Degrees of Freedom

17.7   *THE NUMBER-OF-RUNS TEST*

In economics and finance, we are often interested in examining the randomness of series of data. For example, if the movements of stock prices are random over time, it is impossible to forecast future stock prices accurately. Hence it is not possible to earn abnormal profits by using data on past stock prices. This hypothesis has come to be known in finance as the *efficient market hypothesis.* In this section, we discuss a nonparametric procedure known as a **runs test** that can be used to examine the randomness of a series.

To test whether the collected sample data are random, we can use a nonparametric test statistic known as the **number of runs** to perform statistical inference. A **run** is a sequence of identical occurrences preceded and followed by different occurrences or by none at all. For example, suppose you toss a coin 15 times, recording each appearance of heads (H) or tails (T), and get

$$\overline{\text{HHHH}}\,\underline{\text{TTTT}}\,\overline{\text{HHHH}}\,\underline{\text{TTT}}$$

In this sequence, there are four runs: two runs of heads and two runs of tails.

To illustrate how a number-of-runs test works, let's consider a series of observations generated by the tossing of a coin. The probability of tossing a head or a tail is $\frac{1}{2}$, or 50 percent, so if we tossed the coin enough times, we would expect about half of the tosses to be heads and about half to be tails. However, even though this series is random, the sequence in which the head and tails appear could be HTHTHTHTHTHTHTHTHTHTHTHTHTHTHTHTHTHTHTHTH or HHHTHTTTHTTTHHTTTHTHH or any of many others. The first series consists of 41 observations and 41 runs, the second of 20 observations but only 11 runs. The runs test looks at the number of runs and answers this question: Are there too many or too few runs for the series to be considered random? The hypotheses we are interested in, then, are

H_0: The series of observations are random.

H_1: The series of observations are not random.

The methods we use to do this hypothesis testing depends on whether the number of observations is smaller than 40 or, on the other hand, equal to or larger than 40. If the sample is smaller than 40, we use the Wald–Wolfowitz two-sample runs test presented in Table A10 in Appendix A.

If the number of observations is large enough (40 or more), the distribution is approximately normal, and we can use the normally distributed random variable Z defined in Equation 17.12.

$$Z = \frac{R - \mu_R}{\sigma_R} \qquad (17.12)$$

where

R = number of runs in our series
μ_R = mean value of $R = 2n_1 n_2 / n + 1$

σ_R = standard deviation of R; $\sigma_R = \sqrt{\dfrac{2n_1 n_2 (2n_1 n_2 - n)}{n^2(n - 1)}}$

n_1 = number of times we observe the first value
n_2 = number of times we observe the second value
$n = n_1 + n_2$

To illustrate how the number-of-runs test works, let's return to our previous example. Here, we generated two series of observations, by tossing a coin 41 times and 20 times, respectively.

Series 1

H T H T H T H T H T H T H T H T H T H T H T
H T H T H T H T H T H T H T H T H T H T H

This series consists of 44 observations and 41 runs.

Series 2

HHH T H TT H TTT HH TTT H T HH

This series consists of 20 observations and 11 runs.

We can use Equation 17.12 to test the randomness of the first series because $n > 40$. $R = 41$, $n_1 = 21$, $n_2 = 20$,

$$\mu_R = 2(21)(20)/(21 + 20) + 1$$
$$= 21.49$$

and

$$\sigma_R = \sqrt{\frac{2(21)(20)[2(21)(20) - 41]}{(41)^2(41 - 1)}}$$

$$= 3.16$$

so Z, our test statistic, is $Z = 41 - 21.49/3.16 = 6.17$. From Table A3 in Appendix A, we find $Z_{.025} = 1.96$. Because our Z-value is 6.17, we are able to reject the null hypothesis of randomness of the series at the $\alpha = .05$ level.

Because $n < 40$ for the second series, we should use Table A10 in Appendix A to perform the test. Parts 1 and 2 of Table A10 present the critical values of the runs test at the .05 level of significance for the second series $n_1 = 10$, $n_2 = 10$, and $R = 11$. From Table A10 in Appendix A, we find that we would reject the null hypothesis at the .05 level if $R \geq 16$ or if $R \leq 6$ for the two-tailed test. Because the observed number of runs is 11, we cannot reject the null hypothesis that the series is random.

Finally, we compare the efficiency of some nonparametric tests discussed in this chapter with parametric tests as shown in the following table.

Application	Parametric Test	Nonparametric Test	Efficiency of Nonparametric Test with Normal Population
Two dependent samples	t test or z test	Sign test	.63
		Wilcoxon signed-ranks	.95
Two independent samples	t test or z test	Wilcoxon rank-sum	.95
Several independent samples	Analysis of variance (F test)	Kruskal–Wallis test	.95
Correlation	Linear correlation	Rank correlation	.91
Randomness	No parametric test	Runs test	No basis for comparison

From this table we know that the efficiency of nonparametric tests is always lower than that of parametric tests if the population is distributed normally. The range of efficiency is from .63 to .95.

17.8 BUSINESS APPLICATIONS

In this section we present five applications that show how nonparametric statistics can be used in business decision making.

APPLICATION 17.1 Testing Randomness of Stock Rates of Return

The number-of-runs test can be applied to a series of stock rates of return to see whether the stock rates of return are random or exhibit a pattern that could be exploited for earning abnormal profits.

The annual data on stock rates of return for GM, Ford, and the market for the period 1970–1990 are listed in Table 17.8. The numbers of runs are presented to the right of each variable. Table 17.8 indicates that the numbers of runs for rates of

Table 17.8 Rates of Return for GM and Ford and Market Rates of Return (1970–1990)

Observations	GM $R_{i,t}$	GM Run	Ford $R_{i,t}$	Ford Run	Market $R_{m,t}$	Market Run
1	.2137		.4260		.0010	
2	.0422	1	.2933	1	.1080	1
3	.0631		.1717		.1557	
4	−.3667		−.4512		−.1737	
		2		2		2
5	−.2597		−.0968		−.2964	
6	.9522		.3960		.3149	
		3		3		3
7	.4584		.4614		.1918	
8	−.1124		−.2067		−.1153	4
		4				
9	−.0498		−.0026		.0105	
10	.0288	5	−.1479	4	.1228	5
11	−.0410		−.2938		.2586	
		6				
12	−.0911		−.1025		−.0994	6
13	.6826		1.3212		.1549	
14	.2373	7	.1286		.1706	
15	.1176		.0980		.0115	
				5		
16	−.0383	8	.3237		.2633	
						7
17	.0088		.0081		.1462	
18	.0058	9	.3961		.0203	
19	.4418		−.2995		.1240	
20	−.4581		−.0767	6	.2725	
		10				
21	−.1154		−.3209		−.0656	8

return for GM, rates of return for Ford, and market rates of return are 10, 6, and 8, respectively. If we assume that n_1 and n_2 represent "number of minus signs" and "number of plus signs," respectively, then n_1 and n_2 for all three variables are

	$R_{i,t}$ (GM)	$R_{i,t}$ (Ford)	$R_{m,t}$ (market)
n_1	9	10	5
n_2	12	11	16

To do the test, we need to find the critical values from Table A10 in Appendix A of this book. At a 5 percent level of significance for a two-tailed test, the critical values for $n_1 = 9$ and $n_2 = 12$ is either $R \geq 16$ or $R \leq 6$; the critical values for $n_1 = 10$ and $n_2 = 11$ is either $R \geq 17$ or $R \leq 6$; the critical values for $n_1 = 5$ and n_2

= 16 is $R \leq 4$. The calculated numbers of runs for $R_{i,t}$ (GM), $R_{i,t}$ (Ford), and the market rates of return are 10, 6, and 8, respectively, so we cannot reject the null hypothesis that rate of return for GM and the market rates of return are random. However, we can reject the hypothesis that rate of return for Ford is random.

MINITAB output in terms of data in Table 17.8 is shown in Figure 17.5, which indicates that all three series of data are random at $\alpha = .05$. In this MINITAB output, K represents the mean of each series. For example, the mean of the rate of return for GM is .0819. Based on the mean, we find that there are 7 observations above K and 14 observations below K. From this information, we find that there are 8 runs associated with the rate of return of GM. Similarly, we can calculate related information for the rate of return of Ford and the market rate of return.

Figure 17.5

MINITAB Output of Table 17.8

```
MTB > PRINT C1-C5

ROW    C1    C2      GM      FORD    MARKET

  1    70    70    0.2137   0.4260   0.0010
  2    71    71    0.0422   0.2933   0.1080
  3    72    72    0.0631   0.1717   0.1557
  4    73    73   -0.3667  -0.4512  -0.1737
  5    74    74   -0.2597  -0.0968  -0.2964
  6    75    75    0.9522   0.3960   0.3149
  7    76    76    0.4584   0.4614   0.1918
  8    77    77   -0.1124  -0.2067  -0.1153
  9    78    78   -0.0498  -0.0026   0.0105
 10    79    79    0.0288  -0.1479   0.1228
 11    80    80   -0.0410  -0.2938   0.2586
 12    81    81   -0.0911  -0.1025  -0.0994
 13    82    82    0.6826   1.3212   0.1549
 14    83    83    0.2373   0.1286   0.1706
 15    84    84    0.1176   0.0980   0.0115
 16    85    85   -0.0383   0.3237   0.2633
 17    86    86    0.0088   0.0081   0.1462
 18    87    87    0.0058   0.3961   0.0203
 19    88    88    0.4418  -0.2995   0.1240
 20    89    89   -0.4581  -0.0767   0.2725
 21    90    90   -0.1154  -0.3209  -0.0656

MTB > RUNS C3

    GM

    K =      0.0819


    THE OBSERVED NO. OF RUNS =    8
    THE EXPECTED NO. OF RUNS =   10.3333
     7 OBSERVATIONS ABOVE K    14 BELOW
               THE TEST IS SIGNIFICANT AT  0.2370
               CANNOT REJECT AT ALPHA = 0.05


MTB > RUNS C4

    FORD

    K =      0.0965


    THE OBSERVED NO. OF RUNS =    8
    THE EXPECTED NO. OF RUNS =   11.4762
    10 OBSERVATIONS ABOVE K    11 BELOW
               THE TEST IS SIGNIFICANT AT  0.1190
               CANNOT REJECT AT ALPHA = 0.05
```

Figure 17.5
(Continued)

```
MTB > RUNS C5

    MARKET

    K =      0.0751


    THE OBSERVED NO. OF RUNS =   13
    THE EXPECTED NO. OF RUNS =   11.2857
    12 OBSERVATIONS ABOVE K     9 BELOW
             THE TEST IS SIGNIFICANT AT   0.4330
             CANNOT REJECT AT ALPHA = 0.05

MTB > PAPER
```

If we assume that n_1 and n_2 represent the "number of observations above the mean" and the "number of observations below the mean," respectively, the n_1 and n_2 for all three variables are

	$R_{i,t}$ (GM)	$R_{i,t}$ (Ford)	$R_{m,t}$ (market)
n_1	7	10	12
n_2	14	11	9

At a 5 percent level of significance for a two-tailed test, the critical values for all three variables from Table A10 are

	Lower tail ($\alpha = .025$)	Upper tail ($\alpha = .025$)
$R_{i,t}$ (GM)	5	15
$R_{i,t}$ (Ford)	6	17
$R_{m,t}$ (market)	6	16

Figure 17.5 indicates that the observed number of runs for $R_{i,t}$ (GM), $R_{i,t}$ (Ford), and $R_{m,t}$ (market) are 8, 8, and 13 respectively. Because all these runs are larger than lower tail and smaller than upper tail, we cannot reject the hypothesis that all three kinds of returns are random. This conclusion is consistent with that of the MINITAB output.

■ | **APPLICATION 17.2** Comparing Errors in Earnings Forecasts by Firm Size: Management Forecasts Versus Analysts' Forecasts

Here we investigate the accuracy of management forecasts relative to analysts' forecasts. Jaggi (1980)[4] used sample data from 1971–1974 in terms of either five industries (Table 17.9) or six different firm sizes (Table 17.10). Using the Wilcoxon matched-pairs signed-rank test we discussed in Section 17.3, Jaggi tested the statistical differences between management forecasts and analysts' forecasts by using either Z-values or Wilcoxon t-values as indicated in both Tables 17.9 and 17.10,

[4]This example is adapted from results given by B. Jaggi (1980), "Further Evidence on the Accuracy of Management Forecasts Vis-à-Vis Analysts' Forecasts," *Accounting Review* 55, 96–101.

Table 17.9 Comparison of Forecast Errors by Industry

Industry	Number of Forecasts	Mean Absolute Relative Prediction Error (percentage)		Wilcoxon Matched-Pairs Signed-Rank Test		Level of Significance (two-tailed)
		Management	*Analyst*	*z-value*	*t-value*	
Banking and finance	12	21.8	23.0		38	.15
Utilities	22	24.8	33.3		68	.09
Manufacturing	82	30.4	29.8	−.021		.45
Chemicals	22	25.0	35.8		32	.01
Services (transportation and recreation)	18	21.8	23.0		20	.01

Source: Adapted from B. Jaggi (1980), "Further Evidence on the Accuracy of Management Forecasts Vis-à-Vis Analysts' Forecasts," *Accounting Review* 55, 96–101.

Table 17.10 Comparison of Forecast Errors by Firm Size

Firm Size (revenue in millions of dollars)	Number of Firms	Mean Absolute Relative Prediction Error (percentage)		Wilcoxon Matched-Pairs Signed-Rank Test		Level of Significance (two-tailed)
		Management	*Analyst*	*z-value*	*t-value*	
0–99	8	43.2	37.9		9	.12
100–299	39	32.2	37.0	−2.02		.02
300–499	22	14.7	20.6		51	.02
500–999	37	27.4	28.0	1.45		.07
1000–1999	28	23.1	23.3	−1.98		.02
2000–above	22	28.2	29.8		117	.14

Source: Adapted from B. Jaggi (1980), "Further Evidence on the Accuracy of Management Forecasts Vis-à-Vis Analysts' Forecasts," *Accounting Review* 55, 96–101.

The formula for calculating these two testing statistics is presented in Equation 17.4. Jaggi used Z-values or t-values when the sample size is larger (or smaller) than 30. Note that t-values are Wilcoxon's W statistics defined in Equation 17.13. In these two tables, the **mean absolute relative prediction error** (MARPE)[5] is defined as

$$\text{MARPE} = \frac{|\hat{E}_t - E_t|}{E_t} \qquad (17.13)$$

where $\hat{E}_t$ represents the forecast for time period t and E_t represents actual reported earnings for time period t.

[5]Other methods of comparing the predicted and observed values will be discussed in the next chapter.

Using a 5 percent level of significance, Jaggi found that only chemicals and services industries and subgroups 2, 3, and 5 are statistically significant. In other words, only in two industries and in firms of these three sizes are management forecasts statistically different from analysts' forecasts.

■ **APPLICATION 17.3** Studying the Relationship Between Respondent and Nonrespondent in Mail Surveys

Finn *et al.* applied the Spearman Rank Correlation test to the relationship between respondent and nonrespondent in mail surveys about consumer willingness to buy products made in ten different countries (see Table 17.11).[6] Respondent rank order is obtained by averaging 387 respondent random sampling results in mail surveys. Nonrespondent rank order is obtained by a telephone interview to 10 randomly selected nonrespondents which was by mail.

Substituting information from Table 17.11 into Equation 17.10, we obtain the correlation coefficient.

$$r_s = 1 - \frac{6\Sigma\, d^2}{n(n^2 - 1)} = 1 - \frac{6(2)}{10(10^2 - 1)} = .988$$

Table 17.11 Ranking of Consumer Willingness to Buy Products Made in Indicated Countries

Country	Order[a] for Respondent	Rank Order for Nonrespondent	Difference in Ranks, $d = x - y$	$d^2 = (x - y)^2$
United Kingdom	1	1	0	0
Japan	2	3	-1	1
France	3	2	1	1
Taiwan	4	4	0	0
Brazil	5	5	0	0
India	6	6	0	0
Iran	7	7	0	0
Angola	8	8	0	0
U.S.S.R.	9	9	0	0
Cuba	10	10	0	0
				2

Source: D. W. Finn *et al.* (1983), "An Examination of the Effects of Sample Composition Bias in a Mail Survey," *Journal of Market Research* 25 (October), 331–338. Reprinted by permission of the American Marketing Association.

[a]Data collected in Spring 1977.

[6]D. W. Finn, C. K. Wang, and C. W. Lamb (1983), "An Examination of the Effects of Sample Composition Bias in a Mail Survey," *Journal of Market Research* 25 (October), 331–338.

Then, substituting $r_s = .988$ into Equation 17.11, we obtain the t-value.

$$t = \frac{r_s}{\sqrt{(1 - r_s^2)/(n - 2)}} = \frac{.988}{\sqrt{[1 - (.988)^2)]/(10 - 2)}} = 18.09$$

From Table A4 in Appendix A we find $t_{.005,8} = 3.3554$. Because 18.09 is much larger than 3.3554, we conclude that rank order for respondent and rank order for nonrespondent are highly correlated. In other words, the opinions elicited from the two groups are almost identical.

MINITAB output of Table 17.11 is presented in Figure 17.6. The correlation coefficient it shows is .988, which is identical to previous results.

Figure 17.6

MINITAB Output of Table 17.11

```
MTB > READ INTO C1-C2
DATA> 1 1
DATA> 2 3
DATA> 3 2
DATA> 4 4
DATA> 5 5
DATA> 6 6
DATA> 7 7
DATA> 8 8
DATA> 9 9
DATA> 10 10
DATA> END
     10 ROWS READ
MTB > RANK C1, PUT INTO C3
MTB > RANK C2, PUT INTO C4
MTB > CORRELATION C3 C4

Correlation of GM and FORD = 0.988

MTB > PAPER
```

■ **APPLICATION 17.4** Testing the Randomness of the Pattern Exhibited by Quality Control Data Over Time

If the process is in control, the distribution of sample values should be randomly distributed above and below the center line of a control chart, as we noted in Section 10.9 of Chapter 10. To test whether the pattern of, say, 10 sample observations over time appears to be random, we can use a two-tailed hypothesis test.

H_0: The sequence of sample values is random.

H_1: The sequence of sample values is not random.

Figure 17.7 shows a run of 3 consecutive points down and another run of 7 consecutive points up. Hence there are two runs with $n_1 = 3$ and $n_2 = 7$. The total number of observations, n, is smaller than 40, so this is a small-sample case and we can use the Wald–Wolfowitz two-sample runs-test table to do the test. Table A10 indicates that the critical value of R for $n_1 = 3$ and $n_2 = 7$ is equal to 2. The number of runs for the data presented in Figure 17.7 is also 2. Therefore, we can reject the null hypothesis of randomness at $\alpha = 5$ percent. In other words, we conclude that the production process is not random and is out of control.

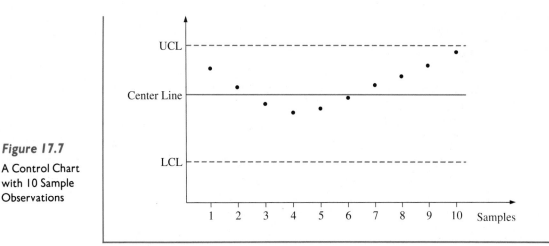

Figure 17.7

A Control Chart
with 10 Sample
Observations

■ | ***APPLICATION 17.5*** Comparing Cash Compensation for 3 Different Groups of Corporate Executives

In Application 12.1 in Chapter 12 we used the analysis of variance approach to test whether there is a difference among 3 different groups of corporate executives' cash compensation. This set of data is repeated in Table 17.12.

Now we use the Kruskal–Wallis test instead of the analysis of variance test to determine whether there is any difference in the 1986 total cash compensation for the 3 groups of corporate executives.

Table 17.12 1986 Total Cash Compensation for 3 Groups of Executives

Banks and Bank Holding Companies	Utilities	Office Equipment and Computers
$ 755	$520	$438
712	295	828
845	553	622
985	950	453
1,300	930	562
1,143	428	348
733	510	405
1,189	864	938

Source: "Executive Compensation Scoreboard," *Business Week,* May 4, 1987, 59–94, by special permission, copyright © 87 by McGraw-Hill, Inc.

Figure 17.8

MINITAB Output
of Application 17.5

```
MTB > READ C1 C2
DATA> 755 1
DATA> 520 2
DATA> 438 3
DATA> 712 1
DATA> 295 2
DATA> 828 3
DATA> 845 1
DATA> 553 2
DATA> 622 3
DATA> 985 1
DATA> 950 2
DATA> 453 3
DATA> 1300 1
DATA> 930 2
DATA> 562 3
DATA> 1143 1
DATA> 428 2
DATA> 348 3
DATA> 733 1
DATA> 510 2
DATA> 405 3
DATA> 1189 1
DATA> 864 2
DATA> 938 3
DATA> END
     24 ROWS READ
MTB > KRUSKAL-WALLIS C1 C2

LEVEL     NOBS     MEDIAN    AVE. RANK    Z VALUE
   1         8      915.0       18.1        2.76
   2         8      536.5       10.5       -0.98
   3         8      507.5        8.9       -1.78
OVERALL     24                  12.5

H = 7.80   d.f. = 2   p = 0.020
```

The MINITAB output of the Kruskal–Wallis statistic is shown in Figure 17.8. The K statistic (H) as defined in Equation 17.9 equals 7.80, which is significant at $\alpha = .020$.

Summary

In this chapter, we discussed six non-parametric statistical tests that do not require the assumption of normality in the distribution of the population. We also gave examples of the use, in business and economic decision making, of the matched-pairs sign test, the Wilcoxon matched-pairs signed-rank test, the Mann–Whitney U test, the Kruskal–Wallis test, the Spearman rank correlation test, and the number-of-runs test.

Questions and Problems

1. Mr. John is a central New Jersey real estate sales-man. He claims that the median selling price of houses in the area is about $100,000. To check this claim, you randomly select 10 houses that recently sold in this area and record the following prices (in thousands of dollars).

   120 115 100 113 103 97 90 111 95 88

   Using the sign test, determine whether the sales-man's claim is reasonable. (Test at the .05 level of significance.)

2. Use the sign test to test, at the 5 percent level, whether the following data come from a population with median 10. Use H_1: median $\neq$ 10.

   13.7 8.1 15.9 12.3 3.4 17.2 13.1 17.2
   12.0 25.9 17.5 12.9 13.6 11.5 7.7 16.1
   9.4 11.2 12.7 10.8

3. Use the number-of-runs test to determine at the 5 percent level of significance, whether it can be con-cluded that these binary sequences are not random.
   a. 0 0 1 1 1 0 1 0 1 1 1 0 0 0 1 0 0 1 1 0 1 1 1
   b. 0 0 1 1 0 0 1 1 0 0 1 1 1 1 0 0 0 0 1 0 1 0 1 0
   c. 1 0 1 1 1 0 1 0 1 1 1 0 0 0 0 1 0 1 0 0 1 1 1 1 0 1 1 0 1 1 0 0

4. The following sequence indicates whether a man-ufacturer's daily production of video cassette recorders (VCRs) was above (X) or below (Y) the long-term median number of defective VCRs.

   X Y Y X X X X Y Y X Y Y X X X X Y Y X Y X X
   Y Y X

   a. Does this series suggest a departure from ran-domness? (Test at α = .05.)
   b. Why would the production manager care whether this series suggested randomness?

5. The following sequence indicates whether an acci-dent occurred at a given intersection during the rush "hour" 7:00 A.M. to 9:30 A.M. (Y indicates an accident, N no accident.)

   N Y N N Y Y N N N Y N Y Y N N N Y N Y Y
   N N

   a. Is there an indication of departure from ran-domness at the .05 Type I error level?
   b. Why would a transportation official be con-cerned about the nonrandom occurrence of accidents?

6. At the 5 percent level of significance, can we con-clude that this sequence of symbols is a random series?

   $+ - - + + + - + + - + - + - - - +$
   $+ - - + + + - - +$

7. Find the Spearman rank correlation for the scores for friendliness and response time given to brand Y computers by 7 users. Test for significance, assuming that the critical value for r_s when n = 7 is $\pm$ .893.

User	Friendliness	Response Time
1	66	79
2	75	69
3	71	84
4	61	78
5	48	65
6	90	82
7	80	90

8. In a rank correlation problem with n = 50 obser-vations, the sum of the squared differences between the rank observations is d_i^2 = 12,500. Calculate r.

9. The ratings assigned by a personnel manager and his assistant to several job applicants are given in the accompanying table. Use Spearman's rank correlation measure r to show whether the person-nel manager and his assistant disagree on how they rank the applicants. Let α = .05.

	Applicant					
	1	2	3	4	5	6
Manager	9	11	10	5	3	8
Assistant manager	7	6	5	8	9	11

	Applicant				
	7	8	9	10	11
Manager	2	1	4	3	2
Assistant manager	10	4	6	1	7

10. A company wishes to compare typing accuracy on two kinds of computer keyboards. Fifteen experienced typists type the same 600 words. Keyboard X_1 is used 7 times and keyboard X_2 8 times, with the following results:

Number of Errors

Board X_1	13	9	16	15	10	11	12	
Board X_2	15	9	18	12	14	17	20	19

Compute the Mann–Whitney U statistic, and state the assumption for a U test. What does a large calculated U-value indicate?

11. The owner of a convenience store has noticed that the register shortages are higher at the eastside store than at the westside store. The following dollar shortages were reported for 10 randomly selected days.

Eastside Store

8.40	16.00	4.50	20.35	12.45
10.35	7.55	11.30	3.50	7.20

Westside Store

2.75	9/00	7.00	15.55	13.00
4.75	10.80	12.00	1.90	30.00

Compute the Mann–Whitney U statistic, and state the assumption for a U test. Use a 5 percent significance level.

12. The manufacturer of a new shaving cream tests 3 new advertising campaigns in a total of 21 markets. Sales in the third week after introduction are given in the accompanying table.

Shaving Cream Sales
(cases per thousand of
population)

A	B	C
38	26	40
42	30	36
27	18	32
60	42	37
36	24	42
54	30	46
40	26	38

Use the Kruskal–Wallis test to determine whether the median sales levels for the 3 campaigns are different at the 5 percent level of significance.

13. Compare parametric tests to nonparametric tests. What are the assumptions of each type of test? Give some examples of each type of test.

14. The "weak form" of the efficient market hypothesis says that historical stock prices cannot be used to earn abnormal profits—that is, stock prices move randomly. Go to the library, collect 30 days of indexes from the Dow Jones Industrial Average, and use a number-of-runs test to test whether these indexes move randomly at a 10 percent level of significance.

15. Toss a coin 50 times and use a number-of-runs test to see whether the outcome of tossing a coin is random at a 5 percent level of significance.

Use the following information to answer questions 16–19. A psychologist conducting research on the differences in aptitude between males and females found 10 pairs of twins wherein one of the twins was male and the other female. Results of the tests on their math and verbal skills are shown in the table.

	Male		**Female**	
Twin	**Math Score**	**Verbal Score**	**Math Score**	**Verbal Score**
1	93	87	83	91
2	80	75	92	87
3	75	75	82	65
4	65	55	62	78
5	87	67	79	95
6	98	87	90	90
7	85	86	72	83
8	90	95	99	96
9	85	83	78	82
10	95	98	87	90

16. Use a Wilcoxon matched-pairs signed-rank test to test the hypothesis that there is a difference between the scores for males and those for females on the math portion of the test. Do a 5 percent test.

17. Use a Wilcoxon matched-pairs signed-rank test to test the hypothesis that there is a difference between the scores of males and those of females on the verbal portion of the test. Use the MINI-

TAB program. (Hint: Follow the procedures presented in Figure 17.1.) Do a 5 percent test.

18. Use a Wilcoxon matched-pairs signed-rank test to test the hypothesis that there is a difference between the math and verbal scores for the male twins. Do a 5 percent test.

19. Use a Wilcoxon matched-pairs signed-rank test to test the hypothesis that there is a difference between the math and verbal scores for the female twins. Use the MINITAB program. (Hint: Refer to Figure 17.1.) Do a 5 percent test

20. A statistics professor is interested in whether there is any difference between the scores in the first-period statistics class and those in her third-period statistics class. She collects the information shown in the accompanying table. Use a rank-sum test to test at a 10 percent level whether the scores in the first-period class and those in the third-period class are different. Use the MINITAB program. (Hint: Refer to Figure 17.2.)

Period 1	Period 3	Period 1	Period 3
85	84	88	95
72	93	90	88
93	87	95	64
65	80	86	63
88	55	92	68
90	95	98	70
55	75	62	71
82	72	70	67
75	76	71	85
89	88	65	90
62	80	73	72
60	75	55	77
80	62	46	86
54	60	69	88
63	90	70	90

21. The prices for ABC Company's stock over a 12-day period follow. Test, at the $\alpha = .05$ level, whether changes in the stock price are random.

83.20	79.21	89.82
81.15	78.30	90.10
79.32	79.65	89.75
80.10	80.27	92.25

22. The following array shows price changes for silver. A + denotes an increase in price from the previous day; a − denotes a decrease in price from the previous day.

$$+ + + + + - - - + - + + + - - - -$$
$$+ - - + - + - + - -$$

Test, at the 10 percent level of significance, whether changes in the price of silver are random.

23. Suppose we want to test whether the number of times black or red comes up on a roulette wheel is random. In 100 spins of the wheel, we find that black comes up 48 times and red comes up 52 times. The number of runs in this sample is 48. Test, at the 1 percent level of significance, whether the appearance of red or black is random.

24. Suppose you toss a coin 50 times and receive 22 heads and 28 tails with 25 runs. Is the tossing of a head or tail random? Test this hypothesis at the 5 percent level of significance.

25. You are given the following 4 samples of price changes for a stock. Count the number of runs in each sample.
   a. $+ + + - - +$ c. $+ + + + + -$
   b. $+ - + + - -$ d. $- - + + - -$

26. Suppose you collect the following information on salaries for two groups of college professors: professors in the sciences and professors in the humanities.

Science Professors	Humanities Professors
$52,500	$29,200
$68,270	$42,700
$55,000	$51,000
$48,900	$37,000
$75,000	$41,000

   a. Compute the ranks for each group.
   b. Do we lose information by converting numerical information into ranks?
   c. Are means and variances of ranks meaningful?

27. In a TV newsroom, 25 economists were asked one by one whether each was optimistic about the future of the economy. The answers (in order) were

$$+ + + + - - - - - - + + - - + - + + + -$$
$$- - + + + + - - - +$$

The alert TV audience suspects that when the economists answered the question, each was influenced by the answer of the economist who responded right before him or her. Do a test to see whether this suspicion can be verified. Use a 5 percent test.

28. A personnel manager is considering keeping 2 out of 10 summer interns on, in a regular job, after the summer is over. Before this personnel manager accepts the subjective evaluation of the candidates, she wants to see some consensus between two evaluators. The rankings of the candidates by two different senior managers who work with them are summarized here. Do you think there is some kind of consensus between the two managers? Do a 5 percent test.

Candidate	A B C D E F G H I J
Ranking by first manager	1 3 5 7 9 2 4 6 8 10
Ranking by second manager	2 4 6 8 10 1 3 5 7 9

29. Two senior managers were asked to score the performance of 10 job candidates in question 28. The scores assigned are believed to reflect the subjective judgments of the two senior managers, and they are believed not to follow a normal distribution. The scores obtained by the 10 candidates were

Candidate	A B C D E F G H I J
Score by first manager	72 76 77 78 79 89 90 87 88 82
Score by second manager	71 74 83 84 85 73 92 86 73 95

Would you say the second manager is tougher in scoring? Do a 5 percent test.

30. A consumer organization wants to know whether you get what you pay for when you buy a stereo system. Its crew ranked 10 stereo systems and listed their ranks and prices in the accompanying table. Do the data support the alternative hypothesis that you get what you pay for? Do a 5 percent test.

Rank	Price	Rank	Price
1	200	6	250
2	220	7	280
3	230	8	240
4	190	9	300
5	170	10	350

31. Two kinds of emission controls were installed and tested in 30 cars of the same make. The following table shows the test results, in the unit of emission. Can you argue that the two kinds of emission controls have a different effect? Do a 10 percent test. Assume that the data do not come from a normal distribution.

Emission Control A			Emission Control B		
16	17	15	18	17	16
17	15	12	19	20	11
22	21	11	18	22	9
23	22	11	19	25	27
10	10	26	11	14	12

32. The following data are the win–loss record of a professional baseball team during the last 34 games. Do you think the team is "streaky"? Do a 5 percent test. The + sign represents a win, the − a loss.

+ + − − + + − − + − + + + + − − − − +
+ + + − + − + + − + + + − + +

33. A market analyst stopped people in the local shopping mall and asked them to rate two kinds of shampoo on a scale of 1 to 4, 1 being the lowest. The ratings of the two different brands of shampoo are listed here. Do a test to determine whether brand B is better than brand A. Use 5 percent.

Brand A	Brand B	Brand A	Brand B	Brand A	Brand B	Brand A	Brand B
1	4	2	4	2	3	3	4
1	2	2	4	2	1	2	4
2	2	1	2	2	2	3	4
2	1	2	3	4	4	3	4
4	2	2	3	3	4	2	3
3	2	2	4	1	3	1	2
3	2	2	4	4	2	3	4
2	2	1	4				

Use the following information to answer questions 34–37. It is believed that American League pitchers should have higher earned-run averages because of the designated hitter rule. A baseball analyst collected 10 pitchers' earned-run averages from each of the 4 divisions in the major leagues and ranked them in the following table. (A smaller rank means a smaller earned run average.)

| National League | | American League | |
East	West	East	West
1	3	4	5
2	40	39	38
6	7	9	18
16	13	21	22
24	20	27	28
33	34	37	35
8	10	11	12
15	14	17	19
23	25	26	29
30	31	32	36

34. Do a test to determine whether the American League pitchers have higher earned-run averages. Use 10 percent.

35. Do a test to determine whether the 4 different leagues have the same earned-run average. Use 10 percent.

36. Do a test to determine whether the two divisions in the National League are equal in earned-run average. Use 10 percent.

37. Do a test to determine whether the American League West has a higher earned-run average than the National League West. Use 10 percent.

38. "Pitching wins the Pennant" is one of the important theories in baseball. A statistician wants to know whether this is true. He used last year's results, which are given in the accompanying table, as the sample. From the performance of these 12 teams, can he argue that winning percentage and earned-run average are negatively correlated? Do a test at a 10 percent level of acceptance, assuming that the data do not follow a normal distribution.

Winning Percentage	Earned-Run Average
.634	3.24
.611	3.35
.598	3.36
.573	3.21
.531	3.87
.521	3.69
.479	3.98
.469	4.23
.427	4.59
.402	4.21
.389	3.72
.366	3.98

Use MINITAB and the following information to answer questions 39–42. A new production manager who believes that music can improve productivity arranges for music to be played on two out of three assembly lines. The first assembly line has no music. The second assembly line hears classical music. The third assembly line hears "easy-listening" music. The productivity in the 10 working days after the experiment began is summarized in the following table. The data are believed not to follow a normal distribution.

| Assembly Lines | | |
1	2	3
110	114	130
157	159	139
121	120	96
103	160	140
149	142	116
123	130	142
142	112	99
134	119	133
124	127	111
118	116	105

39. Can the production manager say that playing music (of whatever type) is better than not playing music? Do a 5 percent test. (Hint: Refer to Figure 17.2.)

40. Is playing classical music better than playing easy-listening music? Do a 5 percent test. (Hint: Refer to Figure 17.2.)

41. Some people say it makes no difference whether music is played. Can you refute this argument? Do a 5 percent test. (Hint: Refer to Figure 17.2.)

42. Is playing classical music better than not playing any music? Do a 5 percent test. (Hint: Refer to Figure 17.2.)

43. Assume that the rank (quality) and price of 5 stereos are as shown in in the following table. Show that the Spearman rank correlation coefficients are 1 and -1, respectively.

Price	Quality	Price	Quality
1	1	1	5
2	2	2	4
3	3	3	3
4	4	4	2
5	5	5	1

44. Is seniority related to hourly wage? A personnel manager collects data on the 10 employees' hourly wages and their seniority and ranks them in the following table. The rankings are created by assigning large ranks to higher numbers.

Hourly Wage	Seniority	Hourly Wage	Seniority
1	2	6	5
2	3	7	6
3	1	8	9
4	4	9	10
5	7	10	8

Do the data support the alternative hypothesis that hourly wage and seniority are related? Do a 10 percent test.

45. Can money buy a championship? A baseball writer is interested in knowing whether high salaries can produce good performance. He collected the average salaries of 12 teams and their winning percentages. The data are ranked in the accompanying table.

Ranking of Winning Percentage	Ranking of Average Salaries
1	3
2	2
3	4
4	1
5	5
6	6

Ranking of Winning Percentage	Ranking of Average Salaries
7	9
8	10
9	11
10	12
11	8
12	7

Do the data support the argument that winning takes money? Do a 5 percent test.

46. The debt/equity ratio is computed by dividing total debt by total assets. It is used to measure how much leverage a firm uses. A financial analyst feels that the debt/equity ratio in industry A is higher than that in industry B. He randomly selected 20 firms from industries A and B, obtaining the following numbers.

A	B	A	B
.76	.23	.78	.67
.92	.78	.34	.23
.54	.76	.73	.24
.74	.32	.54	.34
.75	.13	.43	.22

Do a 5 percent test to decide whether industry A's debt/equity ratio is higher. Assume the data do not follow a normal distribution.

47. The New Land Food Corporation is considering retiring 10 machines. For replacement, it can order some machines of the same model or it can switch to the new model. The company has decided to try out one of the new machines before it makes a large investment in many of them. The daily productivity figures for the new machine and an old machine for the last 12 working days are recorded in the following table. Do a test to determine whether the new machine is better. Use 5 percent.

Day	1	2	3	4	5	6	7	8	9	10	11	12
New	23	21	34	24	34	33	22	32	21	15	34	33
Old	21	17	32	23	24	29	34	33	23	23	34	23

48. The dean of a business school believes that a good knowledge of economics helps business students do well in other business courses. He randomly pulled out 15 students' grades in economics and related business courses and ranked them in the accompanying table.

Economics	Business	Economics	Business
7	1	8	10
3	6	13	11
2	3	9	8
4	5	1	2
15	14	11	15
5	4	12	12
10	9	6	7
14	13		

Do a test to determine whether economics grade and grades in other business courses are related. Use 5 percent.

Use the following information to answer questions 49–53. The personnel manager wants to know whether employees strictly follow the rule for their lunch break. He suspects that those who are more senior in the company tend to take a longer lunch. He classifies the employees into three categories according to their seniority in the company. The amounts of lunch time (in minutes) are summarized in the following table. Assume these data do not follow a normal distribution.

Time with the Company

Group 1, Less Than 1 Month	Group 2, Between 1 and 6 Months	Group 3, More Than 6 Months
43	50	63
39	55	66
42	54	59
49	49	55
50	51	57
39	60	62
47	57	53
48	55	59
39	42	56
47	38	78

49. Do the data support the hypothesis that the three different groups of employees do not spend the same amount of time at lunch? Do a 10 percent test.

50. Do the data support the hypothesis that group 1 spends less time at lunch than group 2. Do a 5 percent test.

51. Do the data support the hypothesis that group 2 spends less time at lunch than group 3. Do a 5 percent test.

52. Now assume that the data for the three different groups are actually data for the same group of individuals being observed at different seniority levels in the company. That is, group 1 represents a sample of employees who have worked for less than 1 month; group 2 represents these same employees when they have worked for the company between 1 and 6 months. Under these circumstances, will you change the way you did questions 50 and 51?

53. Using the same assumption about the data, do a test to determine whether a person who uses more time for lunch during his first month of work will spend more time at lunch during the next 5 months at work. Use 5 percent.

54. Two experts in the insurance field rank 10 insurance companies in terms of their financial soundness. The rankings of the two experts are recorded here.

Company	A	B	C	D	E	F	G	H	I	J
Analyst I	2	3	5	7	10	1	6	4	8	9
Analyst II	3	4	7	6	8	2	10	1	5	9

Do the data support the hypothesis that there is some kind of consensus between A and B? Do a 5 percent test.

Use the following information to answer questions 55–58. A movie theatre opens 3 ticket windows operated by 3 ticket sellers. If the sellers are equally effective, they should sell about the same numbers of tickets. The table gives the ticket sales of the ticket sellers for the last 10 shows. Assume the data do not follow a normal distribution.

	Seller		
Show	A	B	C
1	340	330	350
2	310	320	301
3	300	279	295
4	234	245	235
5	257	256	273
6	297	296	313
7	316	317	354
8	277	232	243
9	241	250	253
10	281	271	248

55. Consider sellers A and B. Compute the rank correlation between their sales.

56. Do a test to determine whether sellers A, B, and C are equally effective in selling tickets. Use 5 percent.

57. Do a test to determine whether sellers A and B are equally effective in selling tickets. Use 5 percent.

58. Do a test to determine whether sellers B and C are equally effective in selling tickets. Use 10 percent.

59. The plant manager in a manufacturing company wants to know whether the morning productivity is higher than the afternoon productivity. He collected the productivity numbers for 10 workers on a certain day and summarized them in the following table. Assume that the productivity numbers do not follow a normal distribution. Do a test to determine whether morning productivity is higher. Use 5 percent.

Morning Productivity	Afternoon Productivity
34	35
32	33
29	31
30	36
42	41
45	33
44	42
43	39
32	31
45	40

a. Do the test assuming that each of the 10 pairs of numbers belongs to a certain worker.

b. Do the test assuming that we do not know to whom the 10 numbers in the morning and the 10 numbers in the afternoon belong.

60. A basketball player shot 32 times in a game, and her coach recorded the results of the 32 shots:

$$+ - + + - + + + + - - - - + - - -$$
$$- + + + + + + - - - + - + + -$$

where $+$ means "score" and $-$ indicates "miss." The coach says that the player is "streaky." Do the data support what the coach says? Do a 5 percent test.

61. A college professor believes that those students who do well on the midterm exam tend to do well on the final. He ranked the midterm and final exams of 30 students in his class.

Midterm	Final	Midterm	Final
1	1	16	18
2	4	17	17
3	5	18	16
4	3	19	19
5	2	20	20
6	6	21	22
7	14	22	21
8	7	23	23
9	13	24	27
10	8	25	24
11	12	26	25
12	9	27	26
13	11	28	28
14	10	29	29
15	15	30	30

Do a test to see whether you can verify the professor's theory. Use 5 percent.

62. Movie critics are generally thought to be very subjective. At the end of the year, two critics rank 20 films as shown in the table.

Critic A	Critic B	Critic A	Critic B
1	5	11	11
2	3	12	10
3	1	13	12
4	6	14	14
5	2	15	18
6	7	16	19
7	8	17	15
8	4	18	20
9	9	19	17
10	13	20	16

Can you say that the two movie critics have different views? Do a 5 percent test.

63. Do question 70 in Chapter 12, assuming that the data do not follow a normal distribution. Use 5 percent.

Use the following information to answer questions 64–66. As a result of the rising costs in health insurance, an insurance company decides to help its clients improve their health and cut down their bills. Thirty large companies that bought the group health insurance were picked to institute a "quit smoking" program and/or an exercise program. The amounts by which insurance claims were reduced are ranked in the following table.

"Quit Smoking" Program	Exercise Program	Both Programs
1	3	5
4	2	6
12	10	7
16	14	8
17	15	9
18	19	11
24	21	13
27	25	20
28	26	22
29	30	23

64. Can you argue that the three different groups are equally effective in reducing insurance claims? Do a 5 percent test.

65. Do a 5 percent test to determine whether the group that implemented both programs is more effective than the group that used only the "stop smoking" program.

66. Do a 5 percent test to determine whether the group using both programs is more effective than the group using only the exercise program.

67. An advertising agency tries to show the effect of a successful advertising campaign. It talked to a food company that is producing a new product, spaghetti sauce. The arrangement is to market the product in three different ways: no frills, a low-budget campaign, and a big-budget campaign. The three spaghetti sauces have the same ingredients and are sold side by side in different packages. The sales in 10 supermarkets during the first month are recorded in the following table.

No Frills	Low Budget	High Budget
301	402	423
326	355	364
337	348	359
362	351	340
383	372	389
321	356	354
362	375	378

No Frills	Low Budget	High Budget
357	358	356
334	335	338
310	312	332

a. Do a 5 percent test to determine whether the three approaches result in different sales. Assume the sales do not follow a normal distribution.

b. Do a 5 percent test to determine whether the high-budget campaign is better than the low-budget campaign.

68. A statistician wants to compare the prices of beef in New York and Los Angeles. He collected beef prices from 10 supermarkets in each city and recorded the prices.

Beef Price (in dollars per pound)

New York	Los Angeles	New York	Los Angeles
3.05	2.95	4.21	4.09
4.35	3.33	3.72	3.85
3.37	3.45	3.29	3.65
3.42	3.50	3.95	3.89
4.05	4.32	4.95	3.76

Can you argue that beef prices are different in the two cities? Do a 5 percent test.

69. The Better Business Bureau suspects that High-Cost Gas Station is consistently charging higher prices for repairs than other garages. The BBB sent 10 cars to High-Cost for an estimate of repair costs. Then it sent the same 10 cars to Expressway Gas Station. The results are reported here.

Car	High-Cost	Expressway
1	$678	$579
2	784	321
3	653	654
4	673	642
5	732	738
6	758	721
7	632	621
8	654	311
9	521	411
10	432	421

Can you argue that High-Cost Gas Station is charging higher prices than Expressway? Do a 10 percent test.

70. A statistics instructor is curious about whether a relationship exists between the scores of students who purchased the student workbook and those of students who did not. Because he did not require students to buy the workbook, some bought it and others didn't. The grades are recorded here and are believed not to follow a normal distribution.

Bought Workbook	Did Not Buy Workbook
75	73
74	65
80	64
81	63
82	62
77	78
76	79
72	71
70	69
67	68
83	87
84	
85	
86	

Do a test to see whether buying the workbook improved the grades. Use 5 percent. Assume that the data do not follow a normal distribution.

Use the following information to answer questions 71 and 72. A national survey was conducted to study the increase in the cost of personal health insurance. Forty people from 3 regions were asked about the increase in the last year. The results are compiled in the following table; the data do not follow a normal distribution.

Northeast	West Coast	Southeast
$320	$232	$254
279	257	231
283	264	242
281	232	220
273	243	253
274	221	223
267	215	210
279	275	263
292	262	275
284	211	227
273	267	262
275	268	231
258	291	
252	250	

71. Do a test at the 5 percent level of significance to determine whether the average increases in the 3 regions are the same.

72. Do a test at the 5 percent level of significance to determine whether the average increase in the Northeast is higher than that on the West Coast.

73. The Dow Jones Industrial Averages from 1961 to 1986 are recorded in the following table. (Data from *Dow Jones Investor's Handbook* (1986), *Wall Street Journal,* January 2, 1987). Do a 5 percent test to determine whether the Dow Jones Industrial Average is a random series of data.

Year	DJIA	Year	DJIA
1961	731	1974	759
1962	652	1975	802
1963	763	1976	975
1964	874	1977	835
1965	969	1978	805
1966	786	1979	839
1967	905	1980	964
1968	944	1981	899
1969	800	1982	1047
1970	839	1983	1259
1971	885	1984	1212
1972	951	1985	1547
1973	924	1986	1896

Use the following information to answer questions 74–76. Ron Moy has taught corporate finance for years, but this is the first year he has included the use of the computer to his course. Actually, one class did not use the computer. A second class had to use the computer. The third class was introduced to the computer but had the option not to use it. The test results that follow are grades on standardized tests and are believed not to follow a normal distribution.

Class 1	Class 2	Class 3
76	78	79
72	75	73
69	63	68
71	81	88
33	87	62
67	65	86
62	77	66
60	80	70
59	65	78
58	69	65

74. Do a test to determine whether the classes have different averages. Use 5 percent.

75. Perhaps not surprisingly, a colleague from the computer science department strongly argues for using the computer. Can you prove for him that class 2 is better than class 1? Do a 5 percent test.

76. A psychology professor argues that we should not impose any restrictions on the students. Not all students benefit from using the computer. He suggests that the best method is giving the students freedom of choice. Do a 5 percent test comparing class 3 with class 2. Do the data support the psychology professor's viewpoint?

77. A supervisor in the local factory feels that productivity during overtime is lower than productivity during normal working time. In order to verify his belief, he collected data on productivity per hour during both normal time and overtime. The productivity rates, in units per hour, are recorded in the table.

Overtime			Normal Time		
249	253	257	250	251	255
252	254	256	260	259	258
261	262	265	264	266	267
263	268	269	273	274	275
270	271	272	278	277	276

Do a 5 percent test to determine whether normal-time productivity is higher than overtime productivity.

78. A psychologist believes that a person who watches the evening news one day is more likely than others to watch it the next day. To show that this is true, he kept track of his wife's TV viewing pattern without letting her know about the experiment. After 35 days of observation, he recorded the following TV viewing pattern:

W W O O O W W W W W O O O W W W W O
W W W W O O O O O W W W W W W W W

where W means "watch" and O means "miss." Do the data support the psychologist's theory? Do a 5 percent test.

79. A company produces 3 kinds of fruit baskets that are sold at the same price. The fruit baskets are displayed in 9 supermarkets in the local area. The sales are recorded in the following table.

	Baskets		
Supermarket	A	B	C
1	40	43	45
2	44	41	47
3	42	48	61
4	74	60	63
5	73	59	64
6	72	75	58
7	69	70	57
8	68	54	67
9	49	52	53

Do a 5 percent test to determine whether the 3 different baskets are equally popular. Assume the data do not follow a normal distribution.

80. An old proverb in baseball is "pitching wins pennants." Fourteen teams' pitching, in terms of earned-run average (ERA), and their standing in the American League West are given in the following table (data from *Associated Press,* June 17, 1990).

Team	Standing	ERA
Oakland	1	2.94
Chicago	2	3.01
Minnesota	3	4.16
California	4	3.49
Seattle	5	3.90
Texas	6	4.32
Kansas City	7	4.01

Do a 5 percent test to determine whether there is a negative relationship between ERA and team standing.

81. Use the Spearman rank correlation test to investigate the relationship between market rates of return, in terms of the S&P 500, and the annual rate of return on 3-month Treasury bills during 1970–1990, which are indicated in the following table.

	Rate of Return	
Year	3-Month T-Bill	S&P 500
1970	6.46	.0010
1971	4.35	.1080
1972	4.07	.1557
1973	7.04	−.1737
1974	7.89	−.2964

| | Rate of Return | | | | Rate of Return | |
Year	3-Month T-Bill	S&P 500		Year	3-Month T-Bill	S&P 500
1975	5.84	.3149		1983	8.63	.1706
1976	4.99	.1918		1984	9.58	.0115
1977	5.27	−.1153		1985	7.48	.2633
1978	7.22	.0105		1986	5.98	.1462
1979	10.04	.1228		1987	5.82	.0203
1980	11.51	.2586		1988	6.69	.1240
1981	14.03	−.0994		1989	8.12	.2725
1982	10.69	.1549		1990	7.51	−.0656

CHAPTER 18

Time-Series: Analysis, Model, and Forecasting

CHAPTER OUTLINE

Key Terms

time-series data
cross-section data
trend component
seasonal component
cyclical component
irregular component
percentage of moving average
seasonal index
seasonal index method
leading indicators
coincident indicators

lagging indicators
exponential smoothing
exponential smoothing constant
exponential smoothing
mean squared error
Holt–Winters forecasting model
autoregressive forecasting model
X-11 model
trend–cycle component
trading-day component

18.1   INTRODUCTION

In the first 17 chapters of this book, we used both time-series and cross-section data to show how statistical analysis techniques can be used in economic and business decision making.

Time-series data are any set of data from a quantifiable (or qualitative) event that are recorded *over time.* For example, we read newspapers everyday and can obtain the Dow Jones Industrial Average (DJIA) index over time. The series of DJIA index values, ordered through time, constitutes time-series data. Other types of time-series data are based on the rate of inflation, the consumer price index, the balance of trade, and the annual profit of a firm.

Cross-section data are observations made on individuals, groups of individuals, objects, or geographic areas *at a particular time.* For example, price per share for N firms in 1991 is a set of cross-section data. On the other hand, price per share for General Motors over time, $P_t\,(t = 1, 2, \ldots, T)$, is a set of time-series data.

The purpose of this chapter is to describe components of time-series analyses and to discuss alternative methods of economic and business forecasting in terms of time-series data. First, a classical description of three time-series components is offered. Then the moving average and seasonally adjusted time series are explored. Time trend regression, exponential smoothing and forecasting, and the Holt–Winters forecasting model for nonseasonal series are investigated in detail. Finally, the autoregressive forecasting model is discussed in some detail. The X-11 model for decomposing time-series components is discussed in Appendix 18A. Appendix 18B addresses the Holt–Winters forecasting model for seasonal series.

18.2   THE CLASSICAL TIME-SERIES COMPONENT MODEL

Several factors result in the interdependence of time-series data over time; these factors are trend, seasonal, and business cycle factors. For example, the current earnings of a growing company tend to be greater than its earnings in the period just ended, and, of course, the expected earnings in the next period will be greater than the current earnings. Therefore, the correlation between any adjacent earnings is positive, and this is due to the trend factor. Seasonal factors also contribute to the interdependence of time-series data. Retail sales in the fourth quarter account for a major portion of total annual sales of department stores. This seasonal factor ensures that the sales volume in the fourth quarter of each year is highly correlated with the fourth-quarter sales volume of any other year. The business cycle is another cause of inderdependency in a time-series model. In short, it is traditionally assumed that the total variation in a time series is composed of four basic components: a **trend component,** a **seasonal component,** a **cyclical component,** and an **irregular component.** We will now discuss these four components in some detail.

Table 18.1 Earnings per Share of Philip Morris

Year	EPS
1977	1.399
1978	1.698
1979	2.043
1980	2.315
1981	2.635
1982	3.115
1983	3.590
1984	3.620
1985	5.235
1986	6.200
1987	7.840

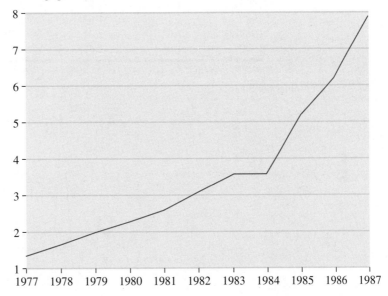

Earnings per Share

Figure 18.1

Earnings per Share of Philip Morris

The Trend Component

A trend is a pattern that exhibits a tendency either to grow or to decrease fairly steadily over time. For example, the earnings per share (EPS) of Philip Morris exhibits two separate trends (or a quadratic trend) over time (see Table 18.1 and Figure 18.1). One of the trends is from 1977 to 1984, the other from 1984 to 1987.

The Seasonal Component

The phenomenon of seasonality is common in the business world. Retailers can rely on greater sales volume in December than in any other month; stock returns are typically higher in January than in most other months—the "January effect."

Table 18.2 Quarterly Earnings per Share of IBM Corporation

	Quarter			
Year	1	2	3	4
1980	1.17	1.31	1.51	2.11
1981	1.25	1.37	1.18	1.83
1982	1.33	1.81	1.75	2.5
1983	1.62	2.22	2.14	3.06
1984	1.97	2.65	2.6	3.55
1985	1.61	2.3	2.4	4.36
1986	1.65	2.12	1.76	2.28
1987	1.3	1.95	2.0	3.47
1988	2.1	1.63	2.1	3.97
1989	1.61	2.31	1.51	1.04
1990	1.81	2.45		

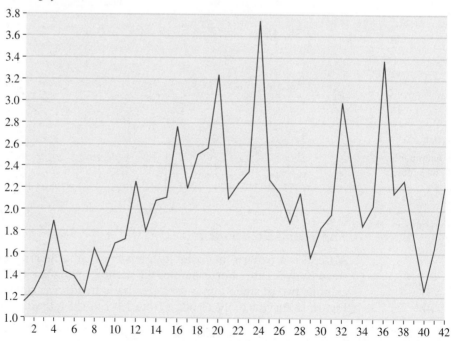

Earnings per Share

Figure 18.2

Quarterly Earnings per Share of IBM

Time (First Quarter 1980 to Second Quarter 1990)

Table 18.2 and Figure 18.2 show earnings per share of IBM Corporation over a period of 42 quarters (first quarter 1980 to second quarter 1990). The table offers evidence of seasonal behavior for the first 36 quarters. The fourth-quarter figures tend to be relatively high, whereas those in the first quarter are relatively low. This seasonal behavior is quite clear in Figure 18.2, where an obvious pattern almost repeats itself each year.

The Cyclical Component and Business Cycles

Cyclical patterns are long-term oscillatory patterns that are unrelated to seasonal behavior. They are not necessarily regular but instead follow rather smooth patterns of upswings and downswings, each swing lasting more than 2 or 3 years. Figure 18.3 demonstrates the cyclical pattern of the S&P 500 Composite Index during the period of 1976–1988, which will be discussed in detail in the next chapter. Figure 18.4 shows the cyclical patterns of 3-month interest rates of return on Eurodollar deposits, U.S. certificates of deposit (CDs) and Treasury bills during the period of first quarter 1985 to fourth quarter 1988.[1]

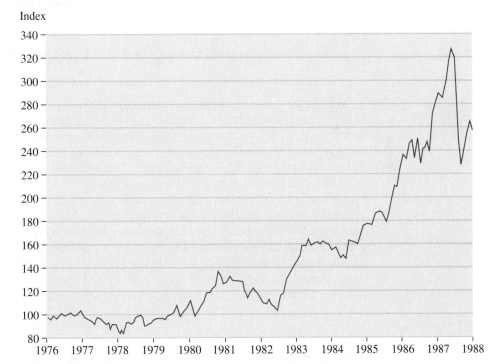

Figure 18.3
S&P 500
Composite Index,
76/1–88/3

[1]A Eurodollar is any dollar on deposit outside the United States. In the bottom portion of Figure 18.4, "spreads" are the differences between two different kinds of interest rates. For example, the Eurodollar rate is .18 percent higher than the U.S. CD rate in the first quarter of 1986.

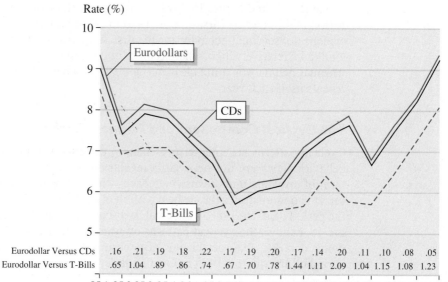

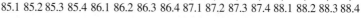

	85.1	85.2	85.3	85.4	86.1	86.2	86.3	86.4	87.1	87.2	87.3	87.4	88.1	88.2	88.3	88.4
Eurodollar Versus CDs	.16	.21	.19	.18	.22	.17	.19	.20	.17	.14	.20	.11	.10	.08	.05	
Eurodollar Versus T-Bills	.65	1.04	.89	.86	.74	.67	.70	.78	1.44	1.11	2.09	1.04	1.15	1.08	1.23	

Year.Quarter

Figure 18.4

Three-Month Rates on Eurodollar Deposits, U.S. CDs, and U.S. T-Bills, 1985–1988 (Quarterly Data)

The National Bureau of Economic Research (NBER) and the U.S. Department of Commerce have specified a number of time series as statistical business indicators of cyclical revivals and recessions. These time series have been classified into three groups.[2] The first group is the so-called **leading indicators,** such as the S&P index of the prices of 500 common stocks. These series have usually reached their cyclical turning points prior to the analogous turns in economic activity. The second group is the **coincident indicators,** such as unemployment rate, the index of industrial production, and GNP in current dollars. The third group is the **lagging indicators,** such as index of labor cost per unit of output in manufacturing, business expenditures, and new plant and equipment. A particular indicator series is considered a leading, a coinciding, or a lagging indicator of overall economic activity, depending on whether the cyclical component of the series exhibits a tendency to precede, match, or follow the cyclical behavior of the economy at large.

The Irregular Component

The last component of the variation in a time series is the irregular element introduced by the unexpected event. For example, the announcement of a takeover bid may cause the price of the target company's stock to jump up 20 percent or more

[2]Index numbers are essential elements for these business indicators. Therefore, we discuss these business indicators after we discuss index numbers and stock market indexes in the next chapter.

in a single day. Fears of an outbreak of war in the Middle East and concerns about trade deficits and antitakeover legislation contributed to a spectacular decline in the stock market on October 19, 1987. And Iraq's invasion of Kuwait on August 3, 1990, caused worldwide stock markets to drop more than 10 percent within a week. These irregular elements arise suddenly and have a temporary impact on time-series behavior.

EXAMPLE 18.1   *Graphical Presentation of Time-Series Components*

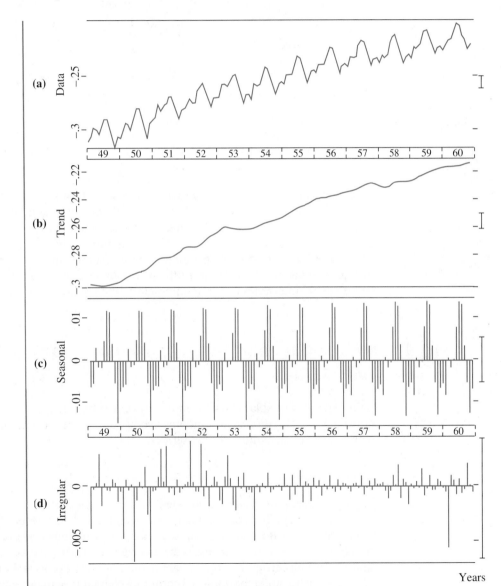

Figure 18.5
Time-Series
Decomposition

Source: H. Levenback and J. P. Cleary (1984) *The Modern Forecaster* (New York: Lifetime Learning Publications), p. 50.

In Figure 18.5, Levenbach and Cleary show how a set of time-series data can be broken down into three components. Figure 18.5a is a plot of the original series of data. Figure 18.5b presents the trend component (long-term trend plus cyclical effects) of the series. The data obviously exhibit an upward trend. Figure 18.5c presents the seasonal component of the data, and the irregular component appears in Figure 18.5d.

Overall, a set of time-series data, x_t, can be described by using the additive model of Equation 18.1 or the multiplicative model of Equation 18.2.

$$x_t = T_t + C_t + S_t + I_t \qquad (18.1)$$

$$x_t = T_t S_t C_t I_t \qquad (18.2)$$

where

T_t = trend component
C_t = cyclical component
S_t = seasonal component
I_t = irregular component

For long-term planning and decision making in terms of time-series components, business executives concentrate primarily on forecasting the trend movement. For intermediate-term planning—say, from about 2 to about 5 years—fluctuations in the business cycle are of critical importance too. For short-term planning, and for purposes of operational decisions and control, seasonal variations must also be taken into account. In the next four sections of this chapter, we will analyze these time-series factors (components) and see how they can be forecasted.

18.3   MOVING AVERAGE AND SEASONALLY ADJUSTED TIME SERIES

In this section, we explain how the moving-average method is used to smooth time-series data. We also discuss how moving average and related techniques can be used to obtain seasonally adjusted time-series data.

Moving Averages

Moving averages are usually associated with data smoothing. Smoothing a time series reduces the effects of seasonality and irregularity. As a result, the smoothed data reveal more information about seasonal trends and business cycles. The most common moving-average method is the unweighted moving average, in which each value of the data carries the same weight in the smoothing process. For a time series $x_1, \ldots, x_n$ the formula for doing a 3-term unweighted moving average is

$$z_t = \left(\frac{1}{3}\right) \sum_{i=0}^{2} x_{t-i} \qquad (t = 3, \ldots, n) \qquad (18.3)$$

Similarly, the k-term unweighted moving average is written

$$z_t = (1/k) \sum_{i=0}^{k-1} x_{t-i} \qquad (t = k, \ldots, n) \qquad (18.4)$$

Alternatively, the weighted moving average can be used to replace the unweighted moving average. A k-term weighted moving average can be defined as

$$z_t = \sum_{i=0}^{k-1} w_{t-i} x_{t-i} \qquad (t = k, \ldots, n) \qquad (18.5)$$

where $\sum_{i=0}^{k-1} w_{t-i} = 1$

The w_{t-i}'s are known as weights and they sum to unity. If the w_{t-i}'s do not sum to unity, they can be normalized with a new set of weights (w_{t-i}^*) that sum to unity. The unweighted moving average is a special case of the weighted moving average with $w_i = 1/k$ for all i. An example of a weighted-average calculation appears in Table 18.3. Here columns (1) and (2) represent observation value (x_{t-i}) and weight (w_i), respectively. Column (3) represents $x_{t-i} w_{t-i}$. From Table 18.3, we obtain

$$z_t = \sum_{i=0}^{3} x_{t-i} w_{t-i} = .0501$$

One of the important applications of moving averages is to deseasonalize seasonal time series data which will be discussed in the next section.

Table 18.3 Weighted Average

(1) Observation Value, x_{t-i}	(2) Weight, w_{t-i}	(3) $x_{t-i} w_{t-i}$
.035	.10	.0035
.002	.30	.0006
.100	.25	.0250
.060	.35	.0210
	1.00	.0501 weighted average

Seasonal Index and Seasonally Adjusted Time Series

In Section 18.2, we noted that many business and economic time series contain a strong seasonal component. This component generally needs to be removed for either monthly or quarterly data. This section demonstrates how the moving-average procedure is used to remove the seasonal component and to do related analysis.

Suppose we have a quarterly time series, x_t, with a seasonal component. Then we can apply Equation 18.6, which is obtained by letting $k = 4$ in Equation 18.4, to remove the seasonal component.

$$z_t = \left(\frac{1}{4}\right) \sum_{i=0}^{3} x_{t-i} \qquad (t = 4, \ldots, n) \tag{18.6}$$

EXAMPLE 18.2 *Seasonally Adjusted Quarterly Earnings per Share of Johnson & Johnson*

For the data on quarterly earnings per share of J&J Corporation given in Table 18.4, the first number in the series of the 4-quarter moving average is

$$\frac{.3 + .2717 + .2967 + .215}{4} = .27085$$

and the second number is

$$\frac{.2717 + .2967 + .215 + .325}{4} = .2771$$

The complete series appears in column (3) of Table 18.4.

Table 18.4 Actual (x_t) and Centered 4-Point Moving Average (z_t^*) Earnings per Share of Johnson & Johnson

(1) t	(2) Earnings per Share, x_t	(3) 4-Point Moving Average, z_t	(4) Centered 4-Point, Moving Average, z_t^*
1	.3		
2	.2717		
		.27085	
3	.2967		.273975
		.2771	
4	.215		.286262
		.295425	
5	.325		.301462
		.3075	
6	.345		.310625
		.31375	
7	.345		.325
		.33625	
8	.24		.340625
		.345	
9	.415		.33375
		.3225	
10	.38		.31875
		.315	
11	.255		.306875
		.29875	

Table 18.4
(Continued)

(1) t	(2) Earnings per Share, x_t	(3) 4-Point Moving Average, z_t	(4) Centered 4-Point, Moving Average, z_t^*
12	.21		.2975
		.29625	
13	.35		.31375
		.33125	
14	.37		.32625
		.32125	
15	.395		.32625
		.33125	
16	.17		.324375
		.3175	
17	.39		.315
		.3125	
18	.315		.328125
		.34375	
19	.375		.35375
		.36375	
20	.295		.3775
		.39125	
21	.47		.39875
		.40625	
22	.425		.413125
		.42	
23	.435		.314375
		.20875	
24	.35		.220625
		.2325	
25	$-.375$[a]		.24125
		.25	
26	.52		.240625
		.23125	
27	.505		.363125
		.495	
28	.275		.51125
		.5275	
29	. 68		.540625
		.55375	
30	.65		.57875
		.60375	
31	.61		.62
		.63625	
32	.475		.653125
		.67	
33	.81		.6825
		.695	
34	.785		.705
		.715	

Table 18.4
(Continued)

(1) t	(2) Earnings per Share, x_t	(3) 4-Point Moving Average, z_t	(4) Centered 4-Point, Moving Average, z_t^*
35	.71		.7325
		.75	
36	.555		.763125
		.77625	
37	.95		.7875
		.79875	
38	.89		.805625
		.8125	
39	.8		.785
		.7575	
40	.61		.77875
		.8	
41	.73		
42	1.06		

[a]J&J experienced negative earnings this quarter as a result of the Tylenol poisoning tragedy.

This 4-quarter moving-averages time series is free from seasonality because it is always based on values such that each "season" is represented in each single observation of the new series (see Figure 18.6). However, the location in time of the members of the series of moving averages does not correspond precisely with that of the members of the original series. Actually, the first 4-quarter moving average would be centered midway between the second-quarter and third-quarter dates. Hence the 4-quarter moving-averages series indicated in Equation 18.6 should be rewritten either as

$$z_{t-.5} = \left(\frac{1}{4}\right) \sum_{i=2}^{-1} x_{t-i} \quad (t = 3, 4, \ldots, n-2) \quad (18.7)$$

or

$$z_{t+.5} = \left(\frac{1}{4}\right) \sum_{i=-1}^{2} x_{t+i} \quad (t = 2, 3, \ldots, n-2) \quad (18.7a)$$

Then the location-adjusted (centered) moving-averages series can be written as

$$z_t^* = \frac{z_{t-.5} + z_{t+.5}}{2} \quad (t = 3, 4, \ldots, n-2) \quad (18.8)$$

When

$$t = 3, z_3^* = z_{2.5} + z_{3.5} = \frac{x_1 + 2x_2 + 2x_3 + 2x_4 + x_5}{8}$$

EPS Versus Adjusted EPS

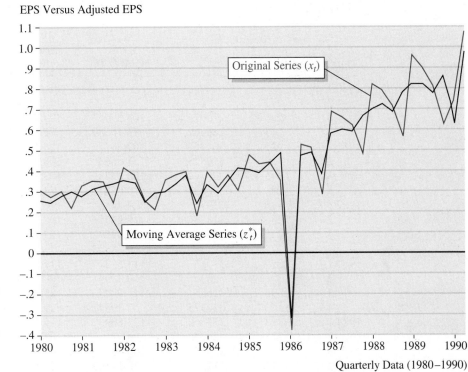

Figure 18.6

Earnings per Share Versus Moving-Average EPS for Johnson & Johnson

Quarterly Data (1980–1990)

The location-adjusted moving averages, z_t^*, are given in column (4) of Table 18.4. Both x_t and z_t^* are presented in Figure 18.6.

We can use the location-adjusted moving-averages data obtained from Equation 18.8 to calculate seasonally adjusted series if we assume that the seasonal pattern through time is very stable. To do this, we need first to divide original data (x_t) by the location-adjusted moving averages (z_t^*) to obtain the **percentage of moving average.** That is,

$$\text{Percentage of moving average (PMA)} = 100 \left(\frac{x_t}{z_t^*} \right) \qquad (18.9)$$

The PMA of earnings per share for Johnson & Johnson is presented in column (4) of Table 18.5.

In our case, the first observation of PMA is

$$100 \left(\frac{x_3}{z_3^*} \right) = 100 \left(\frac{.2967}{.273975} \right) = 108.2945$$

We assume that for any given quarter, in each year, the effect of seasonality is to raise or lower the observation by a constant proportionate amount **(seasonal index)** compared with what it would have been in the absence of seasonal influences. Then we use the so-called **seasonal index method** to remove the seasonal component.

Table 18.5 Seasonal Adjustment of Earnings per Share of Johnson & Johnson by the Seasonal Index Method

(1) Date	(2) EPS, x_t	(3) z_t^*	(4) $100(x_t/z_t^*)$	(5) Seasonal Index	(6) Adjusted EPS [Col. (2) ÷ Col. (5)] × 100
1980.1	.3			116.1149	.258364
2	.2717			108.9414	.249400
3	.2967	.273975	108.2945	103.7882	.285870
4	.215	.286262	75.10589	71.15542	.302155
1981.1	.325	.301462	107.8077	116.1149	.279895
2	.345	.310625	111.0663	108.9414	.316684
3	.345	.325	106.1538	103.7882	.332407
4	.24	.340625	70.45871	71.15542	.337289
1982.1	.415	.33375	124.3445	116.1149	.357404
2	.38	.31875	119.2156	108.9414	.348811
3	.255	.306875	83.09572	103.7882	.245692
4	.21	.2975	70.58823	71.15542	.295128
1983.1	.35	.31375	111.5537	116.1149	.301425
2	.37	.32625	113.4099	108.9414	.339632
3	.395	.32625	121.0727	103.7882	.380582
4	.17	.324375	52.40847	71.15542	.238913
1984.1	.39	.315	123.8095	116.1149	.335874
2	.315	.328125	96	108.9414	.289146
3	.375	.35375	106.0070	103.7882	.361312
4	.295	.3775	78.14569	71.15542	.414585
1985.1	.47	.39875	117.8683	116.1149	.404771
2	.425	.413125	102.8744	108.9414	.390177
3	.435	.314375	138.3697	103.7882	.419122
4	.35	.220625	158.6402	71.15542	.491881
1986.1	−.375	.24125	−155.440	116.1149	−.32295
2	.52	.240625	216.1038	108.9414	.477320
3	.505	.363125	139.0705	103.7882	.486567
4	.275	.51125	53.78973	71.15542	.386477
1987.1	.68	.540625	125.7803	116.1149	.585626
2	.65	.57875	112.3110	108.9414	.596651
3	.61	.62	98.38709	103.7882	.587735
4	.475	.653125	72.72727	71.15542	.667552
1988.1	.81	.6825	118.6813	116.1149	.697584
2	.785	.705	111.3475	108.9414	.720570
3	.71	.7325	96.92832	103.7882	.684085
4	.555	.763125	72.72727	71.15542	.779982
1989.1	.95	.7875	120.6349	116.1149	.818155
2	.89	.805625	110.4732	108.9414	.816952
3	.8	.785	101.9108	103.7882	.770800
4	.61	.77875	78.33065	71.15542	.857278
1990.1	.73			116.1149	.628687
2	1.06			108.9414	.973000

Ratio of (x_t/z_t^*) 100

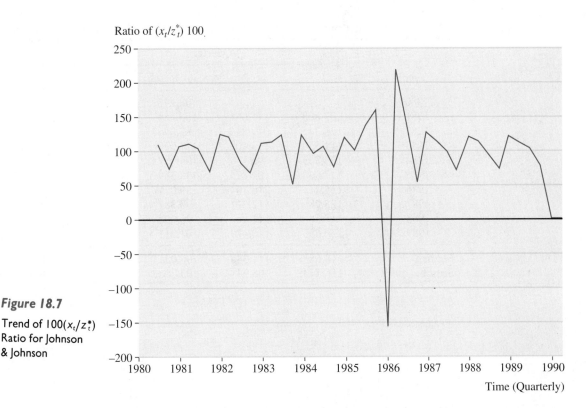

Figure 18.7

Trend of $100(x_t/z_t^*)$ Ratio for Johnson & Johnson

Time (Quarterly)

Let's explore the logic of and procedure for calculating the seasonal index listed in column (5) of Table 18.5. By dividing z_t^* into x_t, we can explicitly write the percentage of moving average as

$$100 \left(\frac{x_t}{z_t^*} \right) = \frac{100 T_t C_t S_t I_t}{T_t C_t} = 100 S_t I_t \qquad (18.10)$$

The $100 S_t I_t$ series for earnings per share of Johnson & Johnson is presented in Figure 18.7. This series contains both seasonal and irregular components. The next step is to remove the effect of irregular movements from $100(x_t/z_t^*)$. We do this by taking the median of the percentage of moving-average figures for the same quarter as indicated in Table 18.6. The medians for the first through the fourth quarters are 118.68, 111.348, 106.081, and 72.7273, respectively. The total of these medians is 408.8363. It is desirable that the total of the 4 indexes be 400, in order that they average 100 percent, so we multiply each of them by an adjustment factor (400/ 408.8363) to make the sum of the 4-quarter seasonal indexes equal 1. The seasonal index is presented in column (5) of Table 18.5.[3] Dividing the seasonal index into the original quarterly data and multiplying the result by 100, we obtain the adjusted series presented in column (6) of Table 18.5 and in Figure 18.8.

---

[3]The mean instead of the median can also be used to calculate the seasonal index.

Table 18.6 Calculation of Seasonal Indexes of EPS for Johnson & Johnson Corporation

	Quarter				
Year	*1*	*2*	*3*	*4*	**Sums**
1980			108.295	75.1059	
1981	107.808	111.066	106.154	70.4587	
1982	124.345	119.216	83.0957	70.5882	
1983	111.554	113.409	121.073	52.408	
1984	123.809	96	106.007	78.1457	
1985	117.868	102.874	138.369	158.64	
1986	−155.44	216.104	139.071	53.7897	
1987	125.78	112.311	98.387	72.7273	
1988	118.68	111.348	96.928	72.7273	
1989	120.635	110.473	101.911	78.3307	
Median	118.68	111.348	106.081	72.7273	408.8363
Seasonal index	116.1149	108.9414	103.7882	71.15542	400

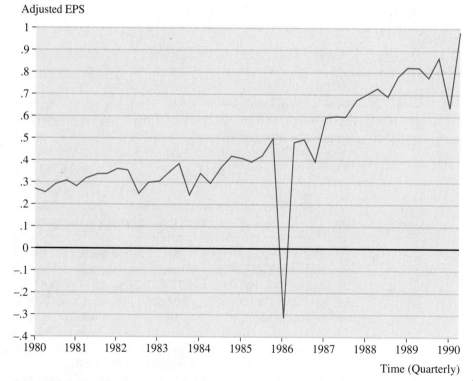

Adjusted EPS

Figure 18.8

Adjusted Earnings per Share (EPS) of Johnson & Johnson

This seasonal index method of seasonal adjustment shows us one possible and simple way to solve the problem of eliminating the seasonal component. In practice, however, it generally can be solved by computer. Important government monthly and quarterly economic data such as consumer price indexes and employment and unemployment rates have strong seasonal components, and government agencies generally publish these data in both unadjusted and adjusted forms. The seasonal adjustment procedure used in official United States government publications is the Census X-11 method, which is based upon the moving-averages method. The X-11 model for decomposing time-series components is discussed in Appendix 18A of this chapter. In the next section, we will look at time trend regression.

18.4 LINEAR AND LOG-LINEAR TIME TREND REGRESSIONS

If a time series is expected to change linearly over time, the simple linear regression model defined in Equation 18.11 can be used to relate the time series, x_t, to time t, and the least-squares line is used to forecast future values of x_t.

$$x_t = \alpha + \beta t + \epsilon_t \tag{18.11}$$

If the relationship between x_t and t is multiplicative instead of additive, then transforming x_t by taking the natural logarithm enables us to make the relationship linear. For example, let x_0 and x_t be the sales of a firm in the base year and in year t, respectively. Then the underlying relationship is

$$x_t = x_0 e^{gt}$$

where x_0 is the base-year sales figure, g is the growth rate, and t is the length of time in terms of number of periods. Then, via the natural logarithm transformation, we obtain

$$\begin{aligned} \log_e x_t &= \log_e(x_0 e^{gt}) \\ &= \log_e x_0 + gt \end{aligned} \tag{18.12}$$

where $\log_e$ is the natural logarithm operator. Equation 18.12 can be defined as a log-linear regression model,[4]

$$\log_e x_t = \alpha' + \beta' t + \epsilon_t' \tag{18.13}$$

where

$\alpha' = \log_e x_0$

$\beta' = g =$ growth rate of a firm's sales

---

[4]In this regression, we implicitly assume that x_t is lognormally distributed and that $\log_e x_t$ is normally distributed. The relationship between the normal and lognormal distributions was discussed in Section 7.4 of Chapter 7.

Table 18.7 Ford's Annual Sales

Year	Sales, x_t (in millions)	t
1968	$14,075.10	1
1969	$14,755.60	2
1970	$14,979.90	3
1971	$16,433.00	4
1972	$20,194.40	5
1973	$23,015.10	6
1974	$23,620.60	7
1975	$24,009.11	8
1976	$28,839.61	9
1977	$37,841.51	10
1978	$42,784.11	11
1979	$43,513.71	12
1980	$37,085.51	13
1981	$38,247.11	14
1982	$37,067.21	15
1983	$44,454.61	16
1984	$52,366.41	17
1985	$52,774.41	18
1986	$62,868.30	19
1987	$72,797.20	20
1988	$82,193.00	21
1989	$82,879.40	22
1990	$81,844.00	23

Ford's annual sales data (1968–1990), presented in Table 18.7, are used to show how Equation 18.12 can be employed to forecast Ford's future sales, and Equation 18.13 to estimate the growth rate of Ford's historical sales.

EXAMPLE 18.3 *Forecasting Sales and Estimating Growth Rate*

Suppose Ford Motor Company is interested in forecasting its sales revenues for each of the next 6 years. The sales manager of the company would also like to estimate the historical growth rate of sales revenue.

To make forecasts and assess their reliability, we must construct a time-series model for the sales revenue data listed in Table 18.7. A plot of the data (Figure 18.9) reveals a linearly increasing trend. Therefore, the linear time trend regression defined in Equation 18.12 can be used to do forecasting. By the method of least squares (see Section 13.3), we obtain the least-squares model in terms of sales (x_t) and time intervals (t) as

$$\hat{x}_t = \hat{\alpha} + \hat{\beta}t = 3239.9485 + 3167.1018t$$

with $R^2 = .9117$.

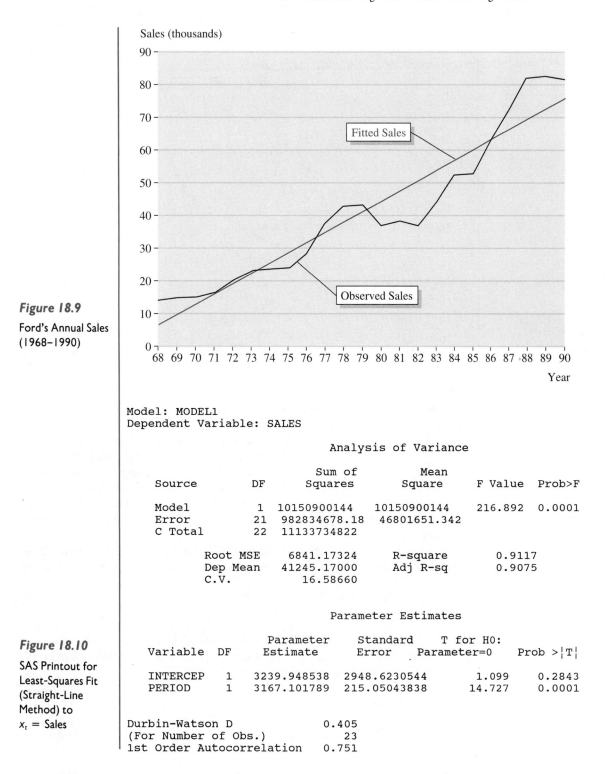

Figure 18.9

Ford's Annual Sales (1968–1990)

```
Model: MODEL1
Dependent Variable: SALES

                        Analysis of Variance

                        Sum of          Mean
    Source      DF      Squares         Square      F Value    Prob>F

    Model        1    10150900144    10150900144    216.892    0.0001
    Error       21    982834678.18   46801651.342
    C Total     22    11133734822

         Root MSE      6841.17324     R-square      0.9117
         Dep Mean     41245.17000     Adj R-sq      0.9075
         C.V.            16.58660


                        Parameter Estimates

                    Parameter     Standard     T for H0:
    Variable   DF    Estimate       Error    Parameter=0    Prob >|T|

    INTERCEP    1   3239.948538   2948.6230544      1.099      0.2843
    PERIOD      1   3167.101789    215.05043838     14.727      0.0001


Durbin-Watson D                   0.405
(For Number of Obs.)                 23
1st Order Autocorrelation         0.751
```

Figure 18.10

SAS Printout for Least-Squares Fit (Straight-Line Method) to x_t = Sales

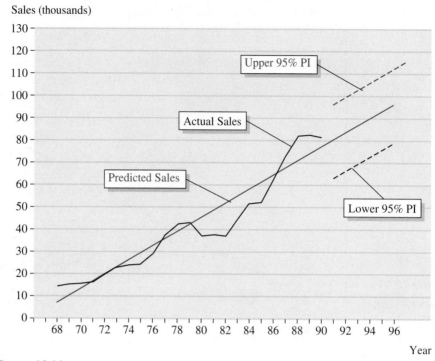

Figure 18.11

Observed (Years 1–23) and Forecast (Years 24–30) Sales Using the Straight-Line Model

This least-squares line is shown in Figure 18.9, and the SAS printout is given in Figure 18.10. We can now forecast sales for years 24–29. The forecasts of sales and the corresponding 95 percent prediction intervals are shown in Figure 18.11. Although it is not easily perceptible in the figure, the prediction interval widens as we attempt to forecast further into the future (see the printout in Figure 18.10). This agrees with the intuitive notion that short-term forecasts should be more reliable than long-term forecasts.

To estimate the growth rate for Ford's sales during the period 1968–1990, we use data listed in Table 18.7 to fit the log-linear regression of Equation 18.13 and obtain

$$\log_e \hat{x}_t = \hat{\alpha}' + \hat{\beta}'t = 9.4834 + .0827t$$

$$(.0516) \quad (.0038) \quad R^2 = .958$$

Figures in parentheses are standard errors. This result implies that the estimated growth rate $\hat{g} = \hat{\beta}' = 8.27$ percent. In other words, the annual growth rate of Ford's sales was 8.27 percent during the period 1968–1990.

EXPONENTIAL SMOOTHING AND FORECASTING

Simple Exponential Smoothing and Forecasting

Smoothing techniques are often used to forecast future values of a time series. One problem that arises in using a moving average to forecast time series is that values at the ends of the series are lost, as shown in Section 18.3. Therefore, we must subjectively extend the graph of the moving average into the future. No exact calculation of a forecast is available, because generating the moving average at a future time period t requires that we know one or more future values of the series. A technique that leads to forecasts that can be explicitly calculated is called **exponential smoothing.** To use the exponential smoothing technique in forecasting, we need only past and current values of the time series.

To obtain an exponentially smoothed series, we first need to choose a weight α between 0 and 1, called the **exponential smoothing constant.** The exponentially smoothed series, denoted s_t, is then calculated as follows:

$$
\begin{aligned}
s_1 &= x_1 \\
s_2 &= \alpha x_2 + (1 - \alpha)s_1 \\
s_3 &= \alpha x_3 + (1 - \alpha)s_2 \\
&\vdots \qquad \vdots \\
s_t &= \alpha x_t + (1 - \alpha)s_{t-1}
\end{aligned}
\tag{18.14}
$$

We can see that the exponentially smoothed value at time t is simply a weighted average of the current time-series value x_t and the exponentially smoothed value at the previous time period, s_{t-1}. Then we can use s_t to do forecasting as follows:

$$
\hat{x}_{t+1} = s_t = \alpha x_t + (1 - \alpha)s_{t-1}
\tag{18.15}
$$

where $\hat{x}_{t+1}$ is the next period's forecast value. In other words, $\hat{x}_{t+1}$ is expressed in terms of the smoothing constant times x_t plus $(1 - \alpha)$ times s_{t-1}.

If the manager of a company in 1990 ($t = 1$) knows only that current sales of his or her company equal $x_1 = 5,000$ units and that current sales have been forecasted as $s_0 = 5,100$ units, then he or she can use Equation 18.15 to forecast 1991 sales. If we choose $\alpha = .30$ as a smoothing constant, then the sales for 1991 are forecasted in terms of Equation 18.15 as

$$
\hat{x}_2 = s_1 = (.30)(5,000) + (1 - .30)(5,100) = 5,070 \text{ units}
$$

Rewriting Equation 18.15 as

$$
\hat{x}_{t+1} = s_t = s_{t-1} + \alpha(x_t - s_{t-1})
\tag{18.16}
$$

implies that simple exponential smoothing is the weighted average of s_{t-1} and the forecast error $(x_t - s_{t-1})$ with weights of 1 and α, respectively. The term **exponential smoothing** refers to the fact that s_t can be expressed as a weighted average with exponentially decreasing weights, as we now illustrate.

We substitute the expressions for s_{t-1} and s_{t-2} into the expression for s_t as defined in Equation 18.15 and obtain

$$s_{t-1} = \alpha x_{t-1} + (1 - \alpha)s_{t-2}$$
$$s_{t-2} = \alpha x_{t-2} + (1 - \alpha)s_{t-3}$$

Repeatedly substituting s_{t-2} and s_{t-3} into Equation 18.15 reveals that

$$
\begin{aligned}
s_t &= \alpha x_t + (1 - \alpha)s_{t-1} \\
&= \alpha x_t + \alpha(1 - \alpha)x_{t-1} + (1 - \alpha)^2 s_{t-2} \\
&= \alpha x_t + \alpha(1 - \alpha)x_{t-1} + \alpha(1 - \alpha)^2 x_{t-2} + (1 - \alpha)^3 s_{t-3}
\end{aligned}
$$

Continuous substitution for s_{t-k}, where $k = 2, 3, \ldots, t$, yields

$$s_t = \left[\alpha \sum_{k=0}^{t-1} (1 - \alpha)^k x_{t-k} \right] + (1 - \alpha)^t s_0 \qquad (0 < \alpha < 1) \qquad (18.17)$$

where s_0 is an initial estimate of the smoothed value.

The sum of weights approaches unity as t approaches infinity; hence we use the term *average*.[5] The weights decrease geometrically with increasing k, so the most recent values of x_t are assigned the greatest weight. All the previous values of x_t are included in the expression for s_t. Because α is less than unity, the most remote values of x_t are associated with the smallest weights. The selection of α depends on the sensitivity of the response required by the model. For example, a small α is used to represent the small sensitivity of the response, and it implies that a single change won't affect the moving average much. The smaller the value of α, the slower the response. Note that the method discussed in this section is good only for short-term forecasting.

In the next example, we draw on annual earnings per share (EPS) data for both Johnson & Johnson (J&J) and International Business Machines (IBM) to show how the simple exponential smoothing method defined in Equation 18.15 can be used to do data analysis.

EXAMPLE 18.4   *Simple Exponential Smoothing of EPS for Both J&J and IBM*

Consider the EPS for both J&J and IBM from 1980 to 1989 as shown in the second column of Table 18.8. Using $\alpha = .3$, we calculate the exponentially smoothed series presented in the third column of Table 18.8 as follows:

---

[5]Let $0 < \alpha \le 1$, as $t \ge \infty$, $(1 - \alpha)^t \ge 0$. Let

$$y = \alpha + \alpha(1 - \alpha) + \alpha(1 - \alpha)^2 + \cdots + \alpha(1 - \alpha)^{t-1} \qquad (A)$$
$$(1 - \alpha)y = \alpha(1 - \alpha) + \alpha(1 - \alpha)^2 + \cdots + \alpha(1 - \alpha)^t + \cdots \qquad (B)$$

Subtracting Equation B from Equation A yields $y = 1 - (1 - \alpha)^t$. Because $\alpha < 1$, y approaches 1 if t approaches infinity. This implies that $\alpha + \alpha(1 - \alpha) + \alpha(1 - \alpha)^2 + \cdots = 1$.

Table 18.8 Simple Exponential Smoothing ($\alpha = .3$) of EPS for J&J and IBM

t	x_t	s_t
J&J		
1980	1.0834	1.0834
1981	1.255	1.13488
1982	1.26	1.172416
1983	1.285	1.206191
1984	1.375	1.256833
1985	1.68	1.383783
1986	.925	1.246148
1987	2.415	1.596804
1988	2.86	1.975762
1989	3.25	2.358033
IBM		
1980	6.1	6.1
1981	5.63	5.959
1982	7.39	6.3883
1983	9.04	7.18381
1984	10.77	8.259667
1985	10.67	8.982766
1986	7.81	8.630936
1987	8.72	8.657655
1988	9.8	9.000359
1989	6.47	8.241251

J&J

$s_1 = x_1 = 1.0834$

$s_2 = .3(1.255) + .7(1.0834) = 1.13488$

$\vdots$

$s_{10} = .3(3.25) + .7(1.358033) = 2.358033$

IBM

$s_1 = x_1 = 6.1$

$s_2 = .3(1.255) + .7(6.1) = 5.959$

$\vdots$

$s_{10} = .3(6.47) + .7(9.000359)$
$\qquad = 8.241251$

We see from the table that the most recent estimates of smoothed EPS for J&J and IBM are

$$s_n = s_{10} = 2.358033 \quad \text{(J\&J)}$$

$$s_n = s_{10} = 8.241251 \quad \text{(IBM)}$$

These values are then used as the forecast of EPS for both J&J and IBM for future years. The observed series and these forecasts for J&J and IBM are graphed in Figures 18.12 and 18.13, respectively.

Finally, note that the choice of the smoothing constant (α) affects the precision of the forecast. In practice, we can try several different values to see which would

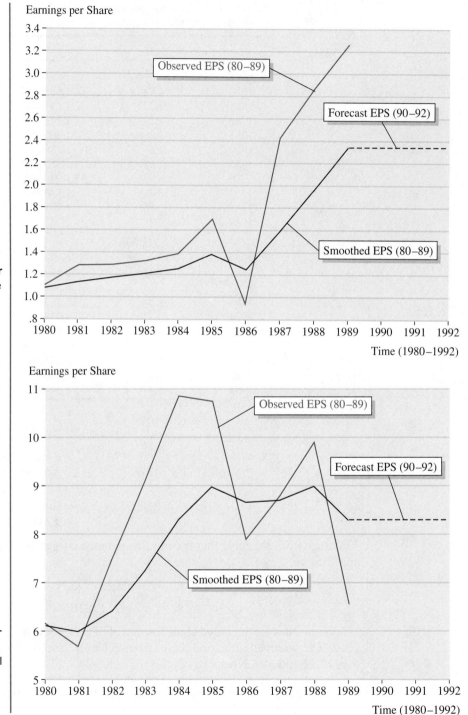

Figure 18.12

Annual Earnings per Share of J&J (Simple Exponential Smoothing)

Figure 18.13

Annual Earnings per Share of IBM (Simple Exponential Smoothing)

have been most successful in predicting historical movement in the time series. For example, we might compute the smoothed series for values of α of .3, .4, .5, and .7 and calculate the forecast **mean squared error** (MSE) for these four different α-values.

$$\text{MSE} = \frac{\sum_{t=1}^{n}(x_t - \hat{x}_t)^2}{n} \qquad (18.18)$$

where x_t and $\hat{x}_t$ are actual value and forecast value, respectively. The value of α for which this MSE is smallest is then used in the prediction of future values.

The Holt–Winters Forecasting Model for Nonseasonal Series[6]

The simple exponential smoothing technique discussed in the previous section does not recognize the trend in the time series. In this section we will generalize the simple exponential smoothing model defined in Equation 18.15 by explicitly recognizing the trend in a time series. The **Holt–Winters forecasting model** consists of both an exponentially smoothed component (s_t) and a trend component (T_t). The trend component is used in calculating the exponentially smoothed value. Here s_t and T_t can be written as

$$s_t = \alpha x_t + (1 - \alpha)(s_{t-1} + T_{t-1}) \qquad (18.19a)$$

$$T_t = \beta(s_t - s_{t-1}) + (1 - \beta)T_{t-1} \qquad (18.19b)$$

where α and β are two smoothing constants, each of which is between 0 and 1. We estimate the trend component of the series by using a weighted average of the most recent change in the smoothed component [represented by $(s_t - s_{t-1})$] and the time trend estimate (represented by T_{t-1}) from the previous period. The procedure for calculating the Holt–Winters components is as follows:

1. Choose an exponential smoothing constant α between 0 and 1. Small values of α give less weight to the current values of the time series and more weight to the past. Large values of α give more weight to the current values of the series.

2. Choose a trend smoothing constant β between 0 and 1. Small values of β give less weight to the current changes in the level of the series and more weight to the past trend. Larger choices assign more weight to the most recent trend of the series.

3. Estimate the first observation of trend T_1 by one of the following two alternative methods.

   Method 1:

   Let $T_1 = 0$. If there are a large number of observations in the time series, this method provides an adequate initial estimate for the trend.

---

[6]The Holt–Winters forecasting model for seasonal series will be discussed in Appendix 18B.

Method 2:

Use the first 5 (or so) observations to estimate the initial trend by following the linear time trend regression line

$$x_t = a + bt + e_t$$

Then use the estimated slope $\hat{b}$ as the first trend observation; that is, $T_1 = \hat{b}$.

4. Calculate the components s_t and T_t from the time series as follows:

$$s_1 = x_1$$
$$T_1 = 0 \text{ or } \hat{b}$$
$$s_2 = \alpha x_2 + (1 - \alpha)(s_1 + T_1)$$
$$T_2 = \beta(s_2 - s_1) + (1 - \beta)T_1$$
$$\vdots$$
$$s_t = \alpha x_t + (1 - \alpha)(s_{t-1} + T_{t-1})$$
$$T_t = \beta(s_t - s_{t-1}) + (1 - \beta)T_{t-1}$$

The data on earnings per share of J&J and IBM listed in Table 18.8 show how the forecasting model defined in Equations 18.19a and 18.19b can be used to do data analysis.

EXAMPLE 18.5 *Using the Holt–Winters Model to Estimate the EPS of J&J and IBM*

Now let's use the Holt–Winters model to do the exponential smoothing for the EPS data for both J&J and IBM listed in Table 18.9. We begin by using the first 5 observations to estimate the first term of the trend component. The estimated slopes for the EPS of J&J and IBM are 0 and 1.275, respectively. Let $\alpha = .3$ and $\beta = .2$. Following the formula for the Holt–Winters components listed in step 4, we calculate

J&J	IBM
$s_1 = x_1 = 1.0834$	$s_1 = x_1 = 6.1$
$T_1 = 0$	$T_1 = 1.275$
$s_2 = .3(1.255) + .7(1.0834 + 0)$	$s_2 = .3(5.63) + .7(6.1 + 1.275)$
$\quad = 1.13488$	$\quad = 6.8515$
$T_2 = .2(1.13488 - 1.0834) + .8(0)$	$T_2 = .2(6.8515 - 6.1) + .8(1.275)$
$\quad = .010$	$\quad = 1.170$
$\vdots$	$\vdots$

The remaining calculations are carried out in precisely the same way. All s_t- and T_t-values for both J&J and IBM are given in Table 18.9.

How are these estimates of EPS level and trend used to forecast future observations? Given a series $x_1, x_2, \ldots, x_n$, the most recent EPS level and trend estimates

Table 18.9 EPS for J&J and IBM and Their Smoothed Series in Terms of the Holt–Winters Forecasting Model

t	x_t	s_t	T_t
		J&J	
1980	1.0834		
1981	1.255	1.0834	0
1982	1.26	1.13488	.0107
1983	1.285	1.179623	.017
1983	1.285	1.223266	.022
1984	1.375	1.223266	.022
1985	1.68	1.284520	.030
1986	.925	1.424326	.052
1987	2.415	1.311031	.019
1988	2.86	1.655563	.084
1989	3.25	2.075801	.151
		2.534020	.213
		IBM	
1980	6.1	6.1	1.275
1981	5.63	6.8515	1.170
1982	7.39	7.83226	1.132
1983	9.04	8.987256	1.137
1984	10.77	10.31791	1.176
1985	10.67	11.24650	1.126
1986	7.81	11.00392	.852
1987	8.72	10.91548	.664
1988	9.8	11.04584	.558
1989	6.47	10.06335	.250

are s_n and T_n, respectively. To do forecasting, we assume that the latest trend will continue from the most recent level. In general, standing at time n and looking m time periods into the future, we define the prediction for the m period ahead as

$$\hat{x}_{t+m} = s_t + mT_t \qquad (18.20)$$

If $T_t = 0$, then this prediction reduces to the simple exponential smoothing prediction discussed in Example 18.3. On the basis of this formula and the information given in Table 18.9, we calculate the future predictions for both J&J and IBM as

J&J	**IBM**
$s_{1990} = 2.534020 + .213 = 2.74702$	$s_{1990} = 10.06335 + .250 = 10.31335$
$s_{1991} = 2.534020 + (2)(.213)$	$s_{1991} = 10.06335 + (2)(.250)$
$\quad = 2.96002$	$\quad = 10.56335$
$s_{1992} = 2.534020 + (3)(.213)$	$s_{1992} = 10.06335 + (3)(.250)$
$\quad = 3.17302$	$\quad = 10.81335$

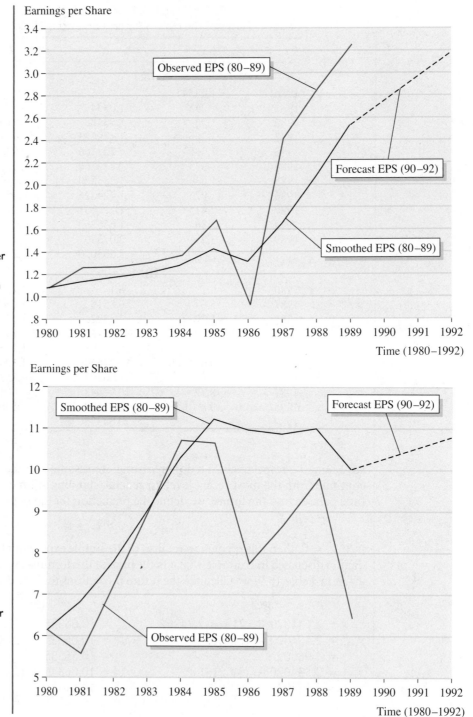

Figure 18.14

Annual Earnings per Share of J&J with Forecasts Based on the Holt–Winters Model

Figure 18.15

Annual Earnings per Share of IBM with Forecasts Based on the Holt–Winters Model

Figures 18.14 and 18.15 show the data series and three forecasts for J&J and IBM, respectively.

Finally, note that the choice of smoothing constants (α and β) affects the precision of a forecast. In practice, we can try several different values of α and β to see which would have been most successful in predicting historical movement in the time series. Again, the forecast mean squared error as defined in Equation 18.18 can be used as a benchmark in deciding what values of α and β are appropriate for forecasting future observations.

18.6 AUTOREGRESSIVE FORECASTING MODEL

A time-series analysis always reveals some degree of correlation between elements. For example, a certain firm's current sales may be correlated with sales in the previous period and even with sales in several prior periods. Under these circumstances, we can regress the time series x_t on some combination of its past values to derive a forecasting equation.

Suppose we attempt to predict the value of x_t by using previous observation. The prediction equation is

$$\hat{x}_t = a_0 + a_1 x_{t-1} \tag{18.21}$$

where a_0 and a_1 are the least-squares regression estimates. This is called a first-order **autoregressive forecasting model,** AR(1). If the current value of a time series depends on the two most recent observations, we can use the model

$$\hat{x}_t = a_0 + a_1 x_{t-1} + a_2 x_{t-2} \tag{18.22}$$

where a_0, a_1, and a_2 are least-squares regression estimates. This is called a second-order autoregressive model, AR(2). Generally, the autoregressive model of order p, AR(P), can be expressed as

$$\hat{x}_t = a_0 + a_1 x_{t-1} + a_2 x_{t-2} + \cdots + a_p x_{t-p} \tag{18.23}$$

where a_0, a_1, a_2, $\ldots$, a_p are least-squares regression estimates.

In the next example, quarterly data on Johnson & Johnson's sales are employed to show how the autoregressive model can be used in forecasting.

EXAMPLE 18.6 *Sales Forecast for Johnson & Johnson*

Quarterly sales data for Johnson & Johnson from first quarter 1980 through second quarter 1990 are presented in Table 18.10 and Figure 18.16.

Table 18.10 Quarterly Sales Data for Johnson & Johnson (first quarter 1980 to second quarter 1990)

Quarter	S_t	S_{t-1}	S_{t-2}	S_{t-3}
1	1188.845			
2	1206.616	1188.845		
3	1195.811	1206.616	1188.845	
4	1246.126	1195.811	1206.616	1188.845
5	1336	1246.126	1195.811	1206.616
6	1134.2	1336	1246.126	1195.811
7	1338	1334.2	1336	1246.126
8	1390.8	1338	1334.2	1336
9	1451.9	1390.8	1338	1334.2
10	1450.4	1451.9	1390.8	1338
11	1476.8	1450.4	1451.9	1390.8
12	1381.8	1476.8	1450.4	1451.9
13	1512.598	1381.8	1476.8	1450.4
14	1512.2	1512.598	1381.8	1476.8
15	1475.7	1512.2	1512.598	1381.8
16	1472.398	1475.7	1512.2	1512.598
17	1522.5	1472.398	1475.7	1512.2
18	1547.298	1522.5	1472.398	1475.7
19	1502.398	1547.298	1522.5	1472.398
20	1552.298	1502.398	1547.298	1522.5
21	1588	1552.298	1502.398	1547.298
22	1572.899	1588	1552.298	1502.398
23	1602.799	1572.899	1588	1552.298
24	1657.599	1602.799	1572.899	1588
25	1742.5	1657.599	1602.799	1572.899
26	1732.299	1742.5	1657.599	1602.799
27	1782.299	1732.299	1742.5	1657.599
28	1745.799	1782.299	1732.299	1742.5
29	1981	1745.799	1782.299	1732.299
30	2010	1981	1745.799	1782.299
31	1993	2010	1981	1745.799
32	2028	1993	2010	1981
33	2311	2028	1993	2010
34	2290	2311	2028	1993
35	2199	2290	2311	2028
36	2200	2199	2290	2311
37	2445	2200	2199	2290
38	2390	2445	2200	2199
39	2451	2390	2445	2200
40	2471	2451	2390	2445
41	2838	2471	2451	2390
42	2825	2838	2471	2451

Quarterly Sales (thousands)

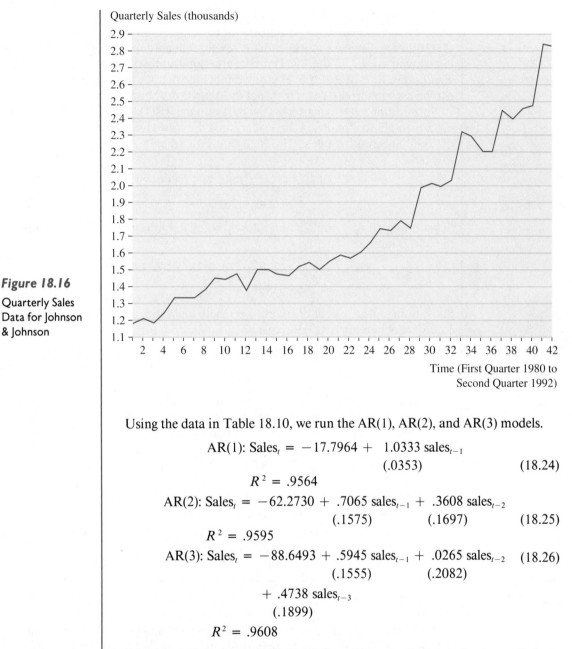

Figure 18.16

Quarterly Sales Data for Johnson & Johnson

Time (First Quarter 1980 to Second Quarter 1992)

Using the data in Table 18.10, we run the AR(1), AR(2), and AR(3) models.

$$\text{AR(1): Sales}_t = -17.7964 + 1.0333 \text{ sales}_{t-1}$$
$$(.0353) \tag{18.24}$$
$$R^2 = .9564$$

$$\text{AR(2): Sales}_t = -62.2730 + .7065 \text{ sales}_{t-1} + .3608 \text{ sales}_{t-2}$$
$$(.1575) \qquad (.1697) \tag{18.25}$$
$$R^2 = .9595$$

$$\text{AR(3): Sales}_t = -88.6493 + .5945 \text{ sales}_{t-1} + .0265 \text{ sales}_{t-2} \tag{18.26}$$
$$(.1555) \qquad (.2082)$$
$$+ .4738 \text{ sales}_{t-3}$$
$$(.1899)$$
$$R^2 = .9608$$

In Equations 18.24, 18.25, and 18.26, figures in parentheses under the coefficients are standard errors.

Table 18.10 makes it clear that the observations used to run AR(1), AR(2), and AR(3) are 41, 40, and 39, respectively. Therefore, by the central limit theorem, the parameter estimators divided by their standard errors approximate standard normal distributions.

From the standard error indicated in the parentheses and the parameter estimator, we can calculate the Z statistic for each regression slope. Looking up these Z statistics in Table A3 of Appendix A reveals that all slopes except the slope associated with $sales_{t-2}$ in the AR(3) model are significantly different from zero at the significance level of $\alpha = .05$. Hence we conclude that the autoregressive processes can be used to forecast quarterly sales of Johnson & Johnson.

Substituting related quarterly sales data into the AR(1), AR(2), and AR(3) models, we obtain the following three alternative forecasted sales for the third quarter of 1990. Substituting $sales_{t-1} = 2825$ into Equation 18.24, we obtain the AR(1) forecast.

$$Sales_{43} = -17.7964 + 1.0333(2825)$$
$$= 2901.2761$$

Substituting $sales_{t-1} = 2825$ and $sales_{t-2} = 2838$ into Equation 18.25, we obtain the AR(2) forecast.

$$Sales_{43} = -62.2730 + .7065(2825) + .3608(2838)$$
$$= 2957.5399$$

Substituting $sales_{t-1} = 2825$, $sales_{t-2} = 2838$, and $sales_{t-3} = 2471$ into Equation 18.26, we obtain the AR(3) forecast.

$$Sales_{43} = -88.6493 + .5945(2825) + .0265(2838) + .4738(2471)$$
$$= 2836.78$$

To determine which model we should choose, we can use the mean absolute relative prediction error (MARPE) defined in Equation 17.13 in Chapter 17 to see which one gives us the smallest error. Equation 17.13 is

$$MARPE = \frac{|\hat{S}_t - S_t|}{S_t} \tag{17.13}$$

where $\hat{S}_t$ represents the sales forecast for time period t and S_t represents actual reported sales for time period t.

Summary

In this chapter, we examined time-series component analysis and several methods of forecasting. The major components of a time series are the trend, cyclical, seasonal, and irregular components. To analyze these time-series components, we used the moving-average method to obtain seasonally adjusted time series. After investigating the analysis of time-series components, we discussed several forecasting models in detail. These forecasting models are linear time trend regression, sim ple exponential smoothing, the Holt–Winters forecasting model without seasonality, the Holt–Winters forecasting model with seasonality, and autoregressive forecasting.

Many factors determine the power of any forecasting model. They include the time horizon of the forecast, the stability of variance of data, and the presence of a trend, seasonal, or cyclical component.

Appendix 18A The X-11 Model for Decomposing Time-Series Components[7]

The classical method of analyzing the relative contribution of each trend–cycle, seasonal, and irregular component to changes in an original set of time-series data can be improved by using the **X-11 model.**

The X-11 model has a long history of application by government and business forecasters. The U.S. Bureau of the Census designed it to analyze historical time series and to determine seasonal adjustments and growth trends.[8] It first decomposes the time-series data into trend–cycle (C), seasonal (S), trading-day (TD), and irregular (I) components and then uses the recategorized data to construct a seasonally adjusted series (Section 18.3 in the text). The X-11 program is based on the premise that seasonal fluctuations can be measured in an original series of economic data and separated from trend, cyclical, trading-day, and irregular fluctuations. The seasonal component (S) reflects an intra-year pattern of variation—one that is repeated from year to year or one that evolves. The **trend–cycle component** (C) includes the long-term trend and the business cycle. The **trading-day component** (TD) consists of variations that are attributed to the composition of the calendar. The **irregular component** (I) is composed of residual variations that reflect the effect of random or unexplained events in the time series. Shiskin (1967) has shown that the relationship among these six variables can be defined as follows:

$$\overline{O}_t^2 = \overline{I}_t^2 + \overline{C}_t^2 + \overline{S}_t^2 + \overline{P}_t^2 + \overline{TD}_t^2 \qquad (18A.1)$$

where the bar over each variable represents the mean of the absolute changes. For example, $\overline{O}_t$ represents the average of $|O_2 - O_1|, |O_3 - O_2|, \dots, |O_t - O_{t-1}|$. O_t = original series; I_t = final irregular series; C_t = final trend–cycle; S_t = final seasonal factors; P_t = prior monthly adjustment factors (not applicable to the quarterly model); and TD_t = final trading-day adjustment factors (not applicable to the quarterly model). In general, the sum of squares of the means of the absolute

---

[7]This appendix is essentially drawn from J. A. Gentry and C. F. Lee, "Measuring and Interpreting Time, Firm and Ledger Effect," in Cheng F. Lee (1983), *Financial Analysis and Planning: Theory and Application, A Book of Readings.*

[8]See J. Shiskin *et al.* (1967), "The X-11 Variant of the Census Method II Seasonal Adjustment Program." Technical paper No. 15, U.S. Department of Commerce.

changes does not exactly equal $\overline{O}_t^2$, and $(\overline{O}_t')^2$ is substituted, where $(\overline{O}_t')^2 = \overline{I}_t^2 + \overline{C}_t^2 + \overline{S}_t^2$. In addition, the relative contribution of the changes in each component for each time span is given by the ratio $\overline{I}_t^2/(\overline{O}_t')^2$, $\overline{C}_t^2/(\overline{O}_t')^2$, or $\overline{S}_t^2/(\overline{O}_t')^2$.

EXAMPLE 18A.1 Using the X-11 Model to Analyze Caterpillar's Quarterly Sales Data

Let's look more closely at the statistical computation of the relative contributions of the C, S, and I components to the percentage change in the original time series. The quarterly sales of Caterpillar Tractor Company from first quarter 1969 to the fourth quarter 1980 are the data used in this example. These original sales data are given in Table 18A.1 and are graphically presented in the upper portion of Figure 18A.1.

The first step in calculating the relative contribution of each component for a 1-quarter time span is to determine the absolute change in the original sales series (O_t) that took place during that quarter—that is, $|O_1 - O_2|$. Sales in the first and second quarters of 1969 (O_1 and O_2) were $500.4 million and $558.9 million, respectively, so the absolute change in sales between the first and second quarters was $58.4 million. The absolute difference in sales between the third and fourth quarters, $|O_3 - O_4|$, was $23.1 million, $|\$432.7 - \$459.6|$. Thus for the original sales series (O_t), the X-11 routine is used to calculate the absolute change in sales for each of the 36 pairs of successive quarters: $|O_1 - O_2|, |O_2 - O_3|, |O_3 - O_4|, \ldots, |O_{34} - O_{35}|, |O_{35} - O_{36}|$. The mean of the changes in the original sales series $\overline{O}_t$ was $109.12 million; this is shown in Table 18A.2.

The X-11 routine also involves calculating the absolute change in the original sales for a time span of 2, 3, and 4 quarters. Because the procedure is similar for

Table 18A.1 Original Quarterly Sales Data for Caterpillar Tractor, First Quarter 1969 to Fourth Quarter 1980 (in millions of dollars)

Original Series Year	Quarterly Sales Data				
	1st Quarter	*2nd Quarter*	*3rd Quarter*	*4th Quarter*	**Total**
1969	500.4	558.9	482.7	459.6	2001.6
1970	524.6	537.0	579.1	487.1	2127.8
1971	564.4	585.1	522.3	503.4	2175.2
1972	620.8	653.6	678.5	649.3	2602.2
1973	751.8	800.2	823.4	807.0	3182.4
1974	822.4	956.8	1081.7	1221.2	4082.1
1975	1125.8	1328.7	1293.0	1216.2	4963.7
1976	1199.8	1266.6	1312.9	1263.0	5042.3
1977	1363.5	1454.6	1513.2	1517.6	5848.9
1978	1630.1	1843.7	1816.8	1928.6	7219.2
1979	1923.7	2136.7	2232.2	1320.6	7613.2
1980	2100.4	2316.3	2085.7	2095.4	8597.8

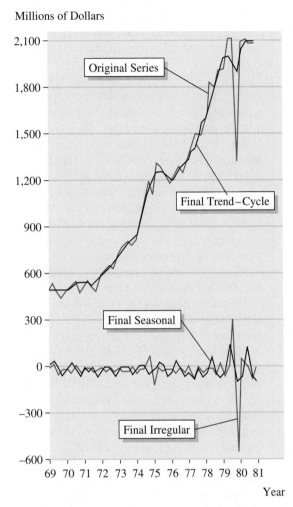

Millions of Dollars

Figure 18A.1

Original Sales and
the X-11 Final
Component Series
of Caterpillar,
1969–1980

Source: J. A. Gentry and C. F. Lee, "Measuring and Interpreting Time, Firm and Ledger Effect," in
Cheng F. Lee (1983), *Financial Analysis and Planning: Theory and Application, A Book of Readings.*

Table 18A.2 Mean of the Absolute Changes in Sales Related to Trend–Cycle,
Seasonal, and Irregular Components for 1-, 2-, 3-, and 4-Quarter Time Spans Without
Regard to Sign

	Mean Values (in millions of dollars)			
Span in Quarters	*Original*	*Trend–Cycle*	*Seasonal*	*Irregular*
1	109.12	42.05	53.83	72.36
2	141.93	114.51	64.72	60.22
3	153.38	146.92	52.22	61.98
4	190.84	191.05	6.49	69.67

each time span, we will use the 4-quarter time span to illustrate it. The absolute change in sales every 4 quarters is calculated with this model. All possible 4-quarter combinations of changes in sales are computed: $|O_1 - O_5|$, $|O_5 - O_9|$, $|O_9 - O_{13}|$, ..., $|O_{29} - O_{33}|$, $|O_2 - O_6|$, $|O_6 - O_{10}|$, ..., $|O_{30} - O_{34}|$, $|O_3 - O_7|$, ..., $|O_{31} - O_{35}|$, $|O_4 - O_8|$, ..., $|O_{32} - O_{36}|$. The same procedure is utilized to calculate a 2- and a 3-quarter time span. The means of the changes in the original series (O_t) for a 2-, a 3-, and a 4-quarter time span were $141.93 million, $153.38 million, and $190.84 million. These values are also presented in Table 18A.2.

The next step in the process is to calculate the mean absolute change in the final adjusted time series for the C, I, and S components. Using the X-11 method, we compute a final adjusted table for each component. A brief reivew of the process used to calculate the final estimated C, I, and S components follows.

The moving average used to estimate the C component is selected on the basis of the amplitude of the irregular variations in the data relative to the amplitude of long-term systematic variations. The routine selects a moving average that provides a suitable compromise between the need to smooth the irregular with a long-term, inflexible moving average and the need to reproduce accurately the systematic element with a short-term, flexible moving average.[9]

Selection of the appropriate moving average for estimating the trend–cycle (C) component is made on the basis of a preliminary estimate of the $\overline{I}/\overline{C}$ rate (the ratio of the mean absolute quarter-to-quarter change in the irregular component to that in the trend–cycle component). A 13-term Henderson average (given in the form of Equation 18.5) of the preliminary seasonally adjusted series is used as the preliminary estimate of C, and the ratio of the preliminary seasonally adjusted series to the 13-term average is used as the estimate of the I component (Shiskin, p. 3). The extreme values of the series are replaced through a smoothing routine. Finally, a 5-term Henderson curve given in the form of Equation 18.5 is used to modify the seasonally adjusted series to obtain the final trend–cycle (C) and irregular (I) series (Shiskin, pp. 3–4). The final trend–cycle, 5-term Henderson curve is presented graphically in Figure 18A.1. Figure 18A.1 shows that the C component tracks the original time series reasonably closely.

The S/I ratios for each quarter are smoothed by a 3×5-term moving average (a 3-term average of a 5-term average) to estimate final seasonal factors. Because the statistical calculations of the final C, S, and I components are lengthy and complex, the numerous tables generated by the model are not presented. The final S and I series appear in Figure 18A.1. The irregular component is substantially more volatile than the seasonal component for Caterpillar sales. A strike in fourth quarter 1979 caused a substantial deviation from the original series and had a profound effect on the I component. A summary of the mean absolute changes in sales in the C, S, and I series for 1-, 2-, 3-, and 4-quarter time series is presented in Table 18A.2. The calculation of these mean absolute changes follows the same procedure used

---

[9]See Julius Shiskin, Allan H. Young, and John C. Musgrave (1967), "The X-11 Variant of the Census Method II Seasonal Adjustment Program." Technical paper No. 15, U.S. Department of Commerce, February.

in computing the change in the original sales series. These mean values in Table 18A.2 provide the base for computing the relative contribution of each component to changes in the original series. A revision to Equation 18A.1 is shown in Equation 18A.2, which specifies the relationships we use in calculating the relative contributions of the C, S, and I components.

$$(\overline{O'})^2 = \overline{I}^2 + \overline{C}^2 + \overline{S}^2 \qquad (18A.2)$$

The calculations utilize the data in Table 18A.2.

The following example illustrates computation of the relative contribution of each component to changes in the original Caterpillar sales series for a 1-quarter time span. Substituting the values of $\overline{I}$, $\overline{C}$, and $\overline{S}$ from Table 18A.2 into Equation 18A.2 produces

$$(72.358)^2 + (42.051)^2 + (53.831)^2 = 9901.842064 = (99.508)^2$$

For a 1-quarter time span, the relative contributions of the three components are

$$I \text{ component} = \frac{\overline{I}^2}{(\overline{O'})^2} = \frac{(72.358)^2}{(99.508)^2} = 52.88\%$$

$$C \text{ component} = \frac{\overline{C}^2}{(\overline{O'})^2} = \frac{(42.051)^2}{(99.508)^2} = 17.86\%$$

$$S \text{ component} = \frac{\overline{S}^2}{(\overline{O'})^2} = \frac{(53.831)^2}{(99.508)^2} = \frac{29.27\%}{100.00\%}$$

These data indicate that the irregular component contributed 52.88 percent of the change in the original sales series for Caterpillar Tractor Company. Another 17.86 percent of the change in the original sales series was related to the trend–cycle component, and the seasonal component was responsible for 29.27 percent of the change. In summary, permanent information signals contributed 47 percent of the change in past quarterly sales of Caterpillar, and random events accounted for 53 percent.

The relative contribution of each component to changes in Caterpillar's original sales series for 1-, 2-, 3-, and 4-quarter time spans are shown in Table 18A.3. For the 2-, 3-, and 4-quarter time spans, the irregular component composes approxi-

Table 18A.3 Relative Contributions of Components to Changes in Caterpillar Sales for 1-, 2-, 3-, and 4-Quarter Time Spans

	Relative Contribution (in percent)			
Span in Quarters	*Trend–Cycle*	*Seasonal*	*Irregular*	**Total**
1	17.86	29.27	52.88	100.00
2	46.94	28.44	24.62	100.00
3	68.50	13.08	18.42	100.00
4	82.58	.15	17.27	100.00

mately 25 percent, 18 percent, and 17 percent, respectively, of the change in sales. The trend–cycle component increased as the length of the time span increased. The seasonal component declined as the time span increased. It ended at almost zero for a 4-quarter time span. This change over time in the relative contribution of each S, C, and I component is referred to as the time effect. For further information related to the X-11 model, see Makridakis *et al.* Levenback and Cleary also cite reasons for using the X-11 model.[10]

Appendix 18B The Holt–Winters Forecasting Model for Seasonal Series

In this appendix, we will generalize the Holt–Winters forecasting model discussed in Section 18.5 to take into account the existence of seasonality. As in the nonseasonal case, we will use x_t, s_t, and T_t to denote, respectively, the observed value and the level and trend estimates at time t. F_t is used to denote the seasonal factor, so if the time series contains L periods per year, the seasonal factor for the corresponding period in the previous period will be F_{t-L}. The Holt–Winters method for seasonal series can be expressed by the following three equations:

$$s_t = \alpha \left(\frac{x_t}{F_{t-L}} \right) + (1 - \alpha)(s_{t-1} + T_{t-1}) \qquad (18B.1)$$

$$T_t = \beta(s_t - s_{t-1}) + (1 - \beta)T_{t-1} \qquad (18B.2)$$

$$F_t = \gamma \left(\frac{x_t}{s_t} \right) + (1 - \gamma)F_{t-L} \qquad (18B.3)$$

where α, β, and γ are smoothing constants whose values are set between 0 and 1.

In Equation 18B.1, the term $s_{t-1} + T_{t-1}$ represents an estimate of the level at time t, formed 1 time period earlier. This estimate is updated when the new observation x_t becomes available. However, here it is necessary to remove the influence of seasonality from that observation by deflating it by the latest available estimate, F_{t-L}, of the seasonal factor for that period. The updating equation for trend, Equation 18B.2, is identical to that used previously, Equation 18.19b in the text.

Finally, the seasonal factor is estimated by Equation 18B.3. The most recent estimate of the factor, available from the previous year, is F_{t-L}. However, dividing the new observation x_t by the level estimate s_t suggests a seasonal factor x_t/s_t. The new estimate of the seasonal factor is then a weighted average of these two quantities.

The procedure for forecasting via the Holt–Winters forecasting model for sea-

---

[10]S. Makridakis, S. C. Wheelwright, and V. E. McGee (1983), *Forecasting Methods and Applications* (New York: Wiley). Also see Chapter 9, "Seasonal Adjustment and Cycle Forecasting," of Hans Levenbach and James P. Cleary (1984), *The Modern Forecaster: The Forecasting Process Through Data Analysis* (New York: Lifetime Learning Publications).

sonal series is similar to that for nonseasonal series. Here the forecast for a particular month includes the effect of all three smoothing equations. The forecast for m periods ahead is

$$\hat{x}_{t+m} = (s_t + mT_t)(F_{t+m-L}) \tag{18B.4}$$

If no seasonality exists—that is, if $F_{t+m-L} = 1$, then this equation reduces to Equation 18.20 in the text.

We will use quarterly data listed in Table 18.4 in the text for Johnson & Johnson (J&J) during the period first quarter 1980 through second quarter 1990 to demonstrate how Equations 18B.1, 18B.2, 18B.3, and 18B.4 are used to do exponential smoothing and forecasting.

EXAMPLE 18B.1   *The Holt–Winters Forecasting Model for J&J's Quarterly EPS*

Table 18.4 and Figure 18.6 in the text make it clear that Johnson & Johnson's quarterly EPS in the period 1980–1990 exhibited significant seasonality. The fourth-quarter EPS especially appeared to be considerably lower than those for the other three quarters.

The Holt–Winters forecasting model with seasonality is used to determine the smoothed value, s_t, and the predicted value, $\hat{x}_t$, for each time period. The smoothing constants are $\alpha = .2$, $\beta = .3$, and $\gamma = .3$.

First, we use the first three years of data to determine the seasonal indexes. Working with Equation 18.9, we present the percentage of moving average (PMA) in terms of the first three years' data in column (4) of Table 18B.1. Table 18B.2 shows

Table 18B.1 Seasonal Index and Seasonally Adjusted EPS for J&J in Terms of the First 36 Quarters' Data

(1) Date	(2) x_t	(3) z_t^*	(4) x_t/z^*	(5) Seasonal Index	(6) Seasonally Adjusted EPS, d_t
1980.1	.3			1.130064	.265471
2	.2717	.273975	1.082946	1.128077	.240852
3	.2967	.2862625	.751059	1.037570	.285956
4	.215	.3014625	1.078078	.704288	.305272
1981.1	.325	.310625	1.110664	1.130064	.287594
2	.345	.325	1.061538	1.128077	.305830
3	.345	.340625	.704587	1.037570	.332507
4	.24	.33	1.257576	.704288	.340769
1982.1	.415	.31125	1.220884	1.130064	.367235
2	.38			1.128077	.336856
3	.225			1.037570	.216852
4	.21			.704288	.298173

Table 18B.2 Calculation of Seasonal Indexes of EPS for J&J

| Year | Quarter | | | | Sums |
	1	*2*	*3*	*4*	**Sums**
1980			1.082945	.751058	
1981	1.078077	1.110663	1.061538	.704587	
1982	1.257575	1.220883			
Median	1.167826	1.165773	1.072241	.727823	4.133665
Seasonal Index	1.130064	1.128077	1.037570	.704288	4

the procedure for calculating the seasonal index in terms of the first three years' data. These indexes are

$$\text{Quarter } 1 = 1.30064 \qquad \text{Quarter } 2 = 1.128077$$
$$\text{Quarter } 3 = 1.037570 \qquad \text{Quarter } 4 = .704288$$

and these are the four values of F_t in 1979.

The data from the first three years were seasonally adjusted to obtain d_t; see column (6) of Table 18B.1. Drawing a least-squares line through these 12 values by means of simple time trend linear regression produces

$$\hat{d}_t = .275977 + .003482t$$

The value $\hat{b} = .003482$ becomes the initial trend estimate of T_0. Finally, the initial smoothed value for fourth quarter 1979 is

$$s_0 = [a + b(0)](\text{initial seasonal index for fourth quarter})$$
$$= (.275977)(.704288) = .194367$$

This estimate of s_0 becomes the forecast value for each of the quarters in 1980, as indicated in column (6) of Table 18B.3.

The calculation of Table 18B.3 in terms of $t = 10$ is shown as follows:

1. $x_{10} = .38$
2. Substituting related information into Equation 18B.1 yields

$$s_{10} = .2\left(\frac{x_{10}}{F_{10-4}}\right) + .8(s_9 + T_9)$$

$$= .2\left(\frac{x_{10}}{F_6}\right) + .8(s_9 + T_9)$$

$$= .2\left(\frac{.38}{1.161223}\right) + .8(.341176 + .015552)$$

$$= .350830$$

Table 18B.3 Solution Using Holt–Winters Model with Seasonality ($\alpha = .2, \beta = .3,$ $\gamma = .3$)

t	x_t	T_t	F_t	s_t	$\hat{x}_t$	$x_t - \hat{x}_t$
			1.130064			
			1.128077			
			1.03757			
		.003482	.704288	.194367		
1	.3	.007539	1.216830	.211373	.194367	.105632
2	.2717	.008855	1.154676	.223301	.194367	.077332
3	.2967	.012083	1.092720	.242916	.194367	.102332
4	.215	.015100	.736347	.265054	.194367	.020632
5	.325	.014315	1.203080	.277541	.340901	−.01590
6	.345	.014731	1.161223	.293242	.337000	.007999
7	.345	.015196	1.099288	.309524	.336530	.008469
8	.24	.015269	.737006	.324963	.239107	.000892
9	.415	.015552	1.207070	.341176	.409327	.005672
10	.38	.013783	1.137799	.350830	.414241	−.03424
11	.225	.004186	.972431	.332626	.400815	−.17581
12	.21	.001074	.708896	.326438	.248233	−.03823
13	.35	−.00117	1.173072	.320001	.395330	−.04533
14	.37	−.00079	1.143230	.320095	.362756	.007243
15	.395	.004416	1.032669	.336678	.310496	.084503
16	.17	−.00166	.655186	.320838	.241801	−.07180
17	.39	−.00086	1.184691	.321834	.374419	.015580
18	.315	−.00358	1.103258	.311883	.366944	−.05194
19	.375	−.00029	1.075243	.319262	.318366	.056633
20	.295	.007578	.714987	.345221	.208980	.086019
21	.47	.010214	1.219233	.361585	.417959	.052040
22	.425	.011019	1.112749	.374484	.410191	.014808
23	.435	.012162	1.087874	.389314	.414510	.020489
24	.35	.107445	.751036	.419085	.287051	.062948
25	−.375	−.02720	.462445	.287710	.532233	−.90723
26	.52	−.01479	1.295702	.301870	.289882	.230117
27	.505	−.00416	1.231274	.322503	.312304	.192695
28	.275	−.00129	.777324	.327903	.239083	.035916
29	.68	.067334	.691031	.555374	.151038	.528961
30	.65	.060071	1.232807	.598498	.806845	−.15684
31	.61	.050282	1.154252	.625940	.810880	−.20088
32	.475	.046373	.758996	.663192	.525644	−.05064
33	.81	.074129	.786682	.802084	.490332	.319667
34	.785	.059761	1.147274	.828322	1.080202	−.29520
35	.71	.043383	1.063528	.833490	1.025073	−.31507
36	.555	.034644	.727700	.847745	.665544	−.11054
37	.95	.054157	.851490	.947432	.694161	.255838
38	.89	.040607	1.082257	.956422	1.149099	−.25909
39	.8	.025918	.997616	.948066	1.060369	−.26036
40	.61	.017774	.702665	.946839	.708769	−.09876
41	.73	.011336	.828242	.943154	.821359	−.09135
42	1.06	.012833	1.089009	.959480	1.033006	.026993

3. Substituting related information into Equation 18B.2 yields

$$T_{10} = .3(s_{10} - s_9) + .7T_9$$
$$= .3(.350830 - .341176) + .7(.015552)$$
$$= .013783$$

4. Substituting related information into Equation 18B.3 yields

$$F_{10} = .3\left(\frac{x_{10}}{s_{10}}\right) + .7F_6$$
$$= .3\left(\frac{.38}{.350830}\right) + .7(1.161223)$$
$$= 1.137799$$

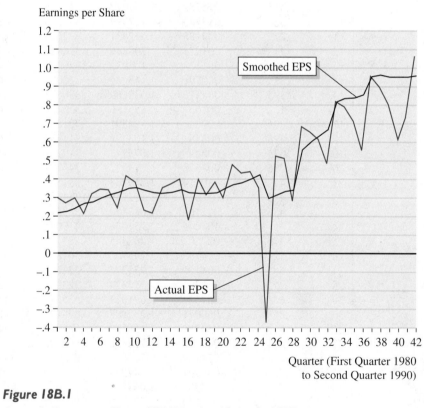

Figure 18B.1

Quarterly Earnings per Share of J&J (Actual and Smoothed EPS)

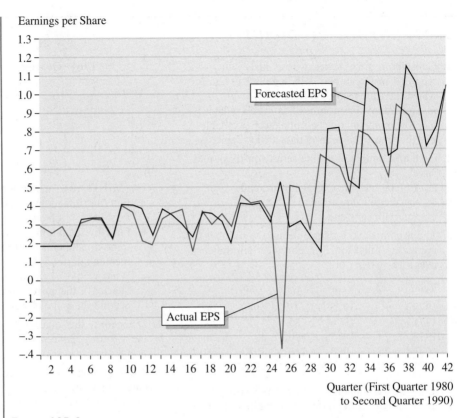

Earnings per Share

Figure 18B.2

Quarterly Earnings per Share of J&J (Actual and Forecasted EPS)

Similarly, we can calculate all other values of s_t, T_t, and F_t, which are listed in columns (5), (3), and (4), respectively. Figure 18B.1 presents actual data and smoothed data s_t.

Using Equation 18B.4, we estimate $\hat{x}_{t+1}$ ($t = 5, 6, \ldots, 42$); it is shown in column (6) of Table 18B.3. For example,

$$\hat{x}_{11} = (s_{10} + T_{10})(F_7) = (.350830 + .013783)(1.099288)$$

$$= .400815$$

Figure 18B.2 presents actual data and forecasted data $(\hat{x}_t)$. If we let $m \geqslant 1$, then we can forecast future observations. For example, to forecast the EPS of J&J in the third quarter of 1990, we let $m = 1$. Finally, in the last column of Table 18B.3, we present the residual in period t, $(x_t - \hat{x}_t)$.

Questions and Problems

1. Consider a time series whose first value was recorded in December 1945. The last period for which there are records is June 1984.
   a. How many full months of data are available?
   b. How many full quarters of data are available?
   c. How many full years of data are available?

2. Give an example of a time series you think may have
   a. A moderately increasing linear trend
   b. A decreasing linear trend
   c. A curvilinear trend

3. The accompanying data indicate the number of mergers (x_t) that took place in a certain industry over a 15-year period.

Year	x_t	Year	x_t	Year	x_t
1970	15	1975	41	1980	148
1971	17	1976	85	1981	203
1972	24	1977	90	1982	249
1973	26	1978	110	1983	280
1974	30	1979	125	1984	307

   a. Plot these data on a frequency polygon.
   b. What type of trend (linear or nonlinear) might best be fitted to this time series?
   c. Is there evidence of seasonal variation in this series?

4. When a 5-month moving average is found for a time series, how many months do not have averages associated with them (a) at the beginning of the time series and (b) at the end of the time series?

5. Find the 3-year moving-average values for the merger time series described in question 3.

6. Find a 4-year moving-average series for the merger data given in question 3. Center the average on the years.

7. Fit a least-squares trend line to the merger data given in question 3. Let $t = 1$ for 1970.

8. The following quarterly data show the number of cameras (in hundreds) returned to a particular manufacturer for warranty service over the past 5 years.

Year	Quarter			
	I	II	III	IV
5	.6	.4	.3	.6
4	.9	.6	.5	.8
3	1.6	1.8	1.8	1.6
2	1.3	1.1	1.0	1.3
1	1.5	1.3	1.1	1.5

Use MINITAB to answer the following questions.
   a. Plot this time series with time on the horizontal axis. Let $t = 1$ be the first quarter 5 years ago.
   b. Find the equation of the least-squares linear trend line that fits this time series. Let $t = 1$ be the first quarter 5 years ago.
   c. What would be the trend line for the second quarter of the current year—that is, 2 periods beyond the end of the actual date?

9. Determine the quarterly seasonal indexes for the warranty service time series described in question 8.

10. A cab company has supplied the accompanying data, which show the number of accidents involving its cabs over the past 5 years.

Year	Winter	Spring	Summer	Fall
5 years ago	7	5	4	6
4 years ago	7	7	5	7
3 years ago	11	10	6	9
2 years ago	22	11	7	10
Last year	16	12	9	12

Find the four seasonal indexes for accidents.

11. Actual billings for the Weygant Corporation were $135,478 in March, and the March seasonal index for this corporation's billings is 104. What is the seasonally adjusted March billing figure? What would be the expected annual billings based on the March figure?

12. The accompanying time series represents the number of patients received in a clinic emergency room. The seasonal indexes for each quarter are

also given. Find the seasonally adjusted figures for the time series. Do these seasonal indexes tell the emergency room manager how many staff members to have on hand and what supplies to order for each quarter?

	Quarter			
	I	**II**	**III**	**IV**
Patient visits	8,220	6,150	5,316	6,834
Seasonal index	115	73	85	110

13. What are time-series data? Why would we ever be interested in looking at time-series data? Give some examples of time-series data.

14. What is a seasonal factor? Why is seasonality sometimes a problem in modeling time-series data? Give some examples of seasonal effects.

15. Why do we sometimes need special techniques to analyze time-series data?

16. What is a business cycle? Why must businesses be able to forecast business cycles?

17. Define the four components of a time series.

18. Explain why it is easier to forecast when the time series contains seasonal effects rather than a cyclical effect?

19. Which of the components would you expect to exist in each of the following time series?
    a. The quarterly earnings of Ford for the years 1981 through 1990
    b. The monthly sales of Sears for 1990
    c. The U.S. unemployment rate for each year from 1981 through 1990
    d. The U.S. unemployment rate for each month in 1990

20. What are the advantages and disadvantages of using a simple moving-average technique for forecasting?

21. What are the advantages and disadvantages of using a linear trend for forecasting?

22. What are the advantages and disadvantages of using a nonlinear trend for forecasting?

23. What is exponential smoothing? What are the advantages and disadvantages of using exponential smoothing for forecasting?

24. What is an autoregressive process? What are the advantages and disadvantages of using an autoregressive process for forecasting?

25. What is the X-11 model? What is it used for? Briefly explain how the X-11 model is used in forecasting.

26. If you were asked to forecast the population of your town over the next 5 years, how would you do it? What information would you ask for?

27. Three time-series graphs follow. Try to identify the components of each time series.

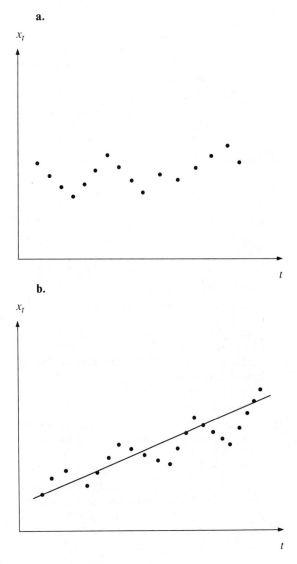

a.

x_t

t

b.

x_t

t

c.

x_t

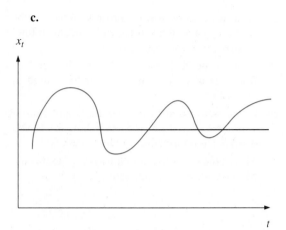

t

28. Look at Figure 18.1. What are the components of this time series? If you were asked to forecast this time series, what method would you use?

29. Look at Figure 18.2. What are the components of this time series? If you were asked to forecast this time series, what method would you use?

30. Look at Figure 18.3. What are the components of this time series? If you were asked to forecast the S&P 500 index, what method would you use?

31. If you were asked to forecast the size of the entering class at a college, how would you do this? What information would be useful in conducting your forecast?

32. You are told that the number of sales per month for a store follows an AR(1) process of the form

$$x_t = 125 + .6x_{t-1} + e_t$$

where e_t is normally distributed with zero mean and constant variance. Say $x_{30} = 1,000$, $x_{31} = 1,125$, and $x_{32} = 1,227$. Forecast sales for the following time periods.
a. x_{33}, x_{34}, and x_{35} at $t = 31$
b. x_{34} and x_{35} at $t = 30$

33. You are told that the number of sales per month for a store follows an AR(2) process of the form

$$x_t = 15 + .6x_{t-1} - .2x_{t-2} + e_t$$

where e_t has a mean of zero and $E[e_t e_{t'}] = 0$ when $t \neq t'$. The values for the last 3 periods are $x_{101} = 823$, $x_{102} = 927$, and $x_{103} = 992$. Forecast sales for the next three periods $t = 104$, 105, and 106.

Use the following information to answer questions 34–42. You are given the following return information for 3-month T-bills, the NYSE Index, Chrysler, Ford, and GM for the 3-year period from January 1985 through December 1987.

Month	T-Bill R_f	NYSE Index R_m	Chrysler R_1
85.01	.006280	.07950	.03906
85.02	.006687	.01661	.00376
85.03	.006886	−.00037	.05243
85.04	.006432	−.00277	−.00358
85.05	.006057	.05872	.02518
85.06	.005634	.01719	.03158
85.07	.005737	−.00351	−.00685
85.08	.005785	−.00463	.02414
85.09	.005753	−.03667	−.03030
85.10	.005801	.04462	.11189
85.11	.005865	.06884	.07233
85.12	.005753	.04554	.09971
86.01	.005729	.00737	−.01072
86.02	.005721	.07375	.23035
86.03	.005321	.05560	.19273
86.04	.004920	−.01322	−.19220
86.05	.004993	.05147	.02759
86.06	.005041	.01509	.03020
86.07	.004736	−.05480	−.06229
86.08	.004495	.07312	.08392
86.09	.004237	−.07957	−.06193
86.10	.004213	.05402	.06944
86.11	.004350	.01857	.02273
86.12	.004495	−.02677	−.05143
87.01	.004414	.12823	.29054
87.02	.004543	.04100	−.01309
87.03	.004543	.02469	.18037
87.04	.004583	−.01483	.03846
87.05	.004599	.00644	−.11111
87.06	.004607	.04797	.01103
87.07	.004623	.04682	.19414
87.08	.004904	.03688	.09816
87.09	.005193	−.02085	−.06983
87.10	.004977	−.21643	−.35649
87.11	.004623	−.07547	−.23944
87.12	.004688	.06851	.10494

Month	Ford R_2	GM R_3
85.01	.07945	.06061
85.02	−.08461	−.02857
85.03	−.05042	−.08176
85.04	−.02124	−.07363
85.05	.06422	.07763
85.06	.03736	.00524
85.07	.00222	−.01736
85.08	−.01401	−.03003
85.09	.00568	−.00557
85.10	.06667	−.00373
85.11	.16129	.06929
85.12	.07407	.03084
86.01	.09181	.05151
86.02	.14571	.06757
86.03	.14111	.10932
86.04	−.06626	−.07246
86.05	.06446	.01250
86.06	.02717	−.02665
86.07	−.01950	−.12238
86.08	.11682	.07523
86.09	−.11297	−.05903
86.10	.09481	.04981
86.11	.01961	.04218
86.12	−.03846	−.09434
87.01	.33378	.14015
87.02	.02689	.00831
87.03	.10475	.04690
87.04	.08741	.15200
87.05	−.00137	−.03889
87.06	.08941	−.03079
87.07	.03409	.07564
87.08	.06273	.04923
87.09	−.09259	−.09783
87.10	−.21939	−.29518
87.11	−.05795	−.01496
87.12	.05975	.08869

34. Use MINITAB to slot the return data for T-bills against time. (Let $t = 1$ be the first month.) Can you identify any of the components of the time series?

35. Compute a simple 3-period moving average for the return on T-bills. Forecast the value for January 1988, using this method.

36. With the MINITAB program, use an AR(1) model to describe the time-series behavior of T-bills. Forecast the value for January 1988, using the AR(1) procedure.

37. Using only data from January 1985 through November 1987, forecast the value for December 1987, using both the 3-period moving average and the AR(1) model. Compare your results. Which model forecasts better?

38. Repeat question 37, using the data for the NYSE index.

39. Repeat question 37, using the data for Chrysler.

40. Repeat question 37, using the data for Ford.

41. Repeat question 37, using the data for GM.

42. Compare the two methods you used for forecasting in questions 34–41. Is one method superior to the other in all cases?

43. Suppose you are an investment analyst and are interested in estimating the future dividend for Hamby Corp. You know that Hamby's dividends grow at an exponential rate—that is,

$$D_t = D_0(1 + g)^t$$

where D_t is the dividend in year t, D_0 is the dividend this year, and g is the growth rate of dividends (assumed to be constant). Is there any way to transform this model into a linear regression?

44. Suppose you are given the following dividend information for Hamby Corp. Forecast the dividend for years 6, 7, 8, 9, and 10, using the method you proposed in question 43.

Year	D
0	1.25
1	1.32
2	1.37
3	1.45
4	1.53
5	1.60

45. Again use the data given in question 44, but this time apply a linear time trend. Plot the estimates from this regression and from your results in question 44.

46. Suppose you have the following information about a company's EPS. What would be the best method for modeling this company's EPS? Forecast the EPS for years 6, 7, 8, 9, and 10.

Year	EPS
0	$3.25
1	3.65
2	4.03
3	4.45
4	4.87
5	5.09

47. Explain why we use *t* as an explanatory variable in a linear time trend model when it is not time that causes the dependent variable to change.

48. Suppose you are given the following sales information for Julian Corp. Estimate the growth rate of sales for Julian Corp. Use this information to forecast the company's sales for year 10.

Year	Sales
0	1,250,625
1	1,321,001
2	1,372,435
3	1,458,020
4	1,531,035
5	1,600,995

49. Evaluate the following statement: "Because sales have increased at a steady rate over the last 10 years, the best way to forecast future sales is to use a linear time trend."

50. Go to the library and obtain the earnings per share for General Motors for the years 1979 through 1988. Use the data for earnings in 1979 through 1988 to obtain a forecasting equation.

51. Indicate which component of a time series will be affected by each of the following events.
   a. A hurricane that results in the postponement of consumer purchases
   b. A downturn in business activity
   c. The annual Columbus Day sale at a department store
   d. A flood at a wholesale warehouse that results in a delay in the shipment of clothing to a local department store
   e. A general increase in the demand for video cameras

52. You are given the following sales information (in millions of dollars) on Acme Widget Company:

Year	Sales ($)	Year	Sales ($)
1985	3.2	1989	4.8
1986	4.5	1990	5.1
1987	3.9	1991	5.6
1988	4.2		

   a. Use a line chart to graph sales.
   b. Estimate the relationship between sales and time, using a time trend regression.

Use the following information on total nonfarm payrolls in New Jersey from 1965 to 1989, which is taken from *New Jersey Economic Indicators,* March 1990, to answer questions 53–57.

Year	Total Nonfarm Payrolls	Year	Total Nonfarm Payrolls
1965	2,257.8	1978	2,961.9
1966	2,359.1	1979	3,027.2
1967	2,421.5	1980	3,060.4
1968	2,485.2	1980	3,060.4
1969	2,569.6	1981	3,089.9
1970	2,606.2	1982	3,092.7
1971	2,607.6	1983	3,165.1
1972	2,674.4	1984	3,329.3
1973	2,760.8	1985	3,414.1
1974	2,783.4	1986	3,489.9
1975	2,699.9	1987	3,581.6
1976	2,753.7	1988	3,659.5
1977	2,836.9	1989	3,709.8

53. Use the MINITAB program to plot the data for nonfarm income, and identify the components of the time series.

54. Compute the 3-year moving average for nonfarm income. Use this information to forecast nonfarm income in 1990 and in 1991.

55. Use the MINITAB program to do a time trend regression to forecast nonfarm income in 1990 and in 1991.

56. Use a first-order autoregressive process to forecast nonfarm income in 1990 and in 1991.

57. Compare the different forecasts of nonfarm income that you made in questions 54–56.

Use the following employment data (in thousands) for the United States and for New Jersey to answer questions 58–65.

Employment

Year	United States	New Jersey
1970	78,678	2,859
1971	79,367	2,840
1972	82,153	2,935
1973	85,064	3,011
1974	86,794	3,023
1975	85,846	2,929
1976	88,752	2,973
1977	92,017	3,065
1978	96,048	3,209
1979	98,824	3,323
1980	99,303	3,334
1981	100,397	3,330
1982	99,526	3,306
1983	100,834	3,385
1984	105,005	3,589
1985	107,150	3,621
1986	109,597	3,712
1987	112,440	3,806
1988	114,968	3,824
1989	117,342	3,826

58. Graph the employment for the United States, and try to identify the components of the time series.

59. Compute the 4-year moving average for employment in the United States. Use this information to forecast employment in the United States in 1990.

60. Use a time trend regression to forecast employment in the United States in 1990, 1991, and 1992.

61. Use a first-order autoregressive model to forecast employment in the United States in 1990, 1991, and 1992.

62. Do you think the first-order AR(1) is a good model to use to explain the data?

63. Compare the different forecasts generated for 1990 by the methods you used in questions 59–62. Which method do you think is best? Why?

64. Plot the New Jersey employment data. Do you think the linear trend model provides a good approximation of the data? Use the data to forecast the employment in 1990.

65. Compare your forecasts for New Jersey with your forecasts for the United States. Which set of data is harder to forecast? Why?

Use the following data on the labor force in thousands of people in the United States and in New Jersey to answer questions 66–70.

Labor Force

Year	United States	New Jersey
1970	82,771	2,996
1971	84,382	3,012
1972	87,034	3,117
1973	89,429	3,190
1974	91,949	3,226
1975	93,775	3,264
1976	96,158	3,318
1977	99,009	3,383
1978	102,251	3,457
1979	104,962	3,570
1980	106,940	3,594
1981	108,670	3,593
1982	110,204	3,632
1983	111,550	3,673
1984	113,544	3,825
1985	115,461	3,839
1986	117,834	3,908
1987	119,865	3,966
1988	121,669	3,975
1989	123,869	3,989

66. Plot the labor force in the United States and in New Jersey, and try to identify the components of the time series. Which labor force data appear to be more stable?

67. Compute the 5-year moving averages for the labor force in the United States and in New Jersey.

68. Use a linear time trend regression to estimate the labor force in the United States and in New Jersey in 1990, 1991, and 1992.

69. Use an exponential trend model to forecast the labor force in the United States and in New Jersey for 1990–1993.

70. What are the growth rates of the United States and New Jersey labor forces? Does the linear model or the exponential trend model give a faster growth estimate?

71. Suppose you generate the following data by tossing a coin 50 times. Let the initial value be $50. If you toss a head, increase the value by $.50. If you toss a tail, decrease the value by $.50. Graph the data. Does this series of data exhibit any time-series pattern? What time-series pattern would you expect it to exhibit?

72. Can you use any regression or time-series method to forecast the values in periods 50, 51, and 52 in question 71?

73. What is the best forecast for the value at period 51?

74. Suppose you adjusted the data generated in question 71 by adding $.25 to every fourth coin toss. Graph these data. Does this new series exhibit any time-series pattern? What time-series pattern would you expect it to exhibit?

75. What is the best forecast for the time series generated in question 74?

76. Johnson & Johnson's quarterly sales, in millions of dollars, from first quarter 1990 to first quarter 1991 are

First quarter 1990	2809
Second quarter 1990	2825
Third quarter 1990	2775
Fourth quarter 1990	2794
First quarter 1991	3149

Use this set of data and the data in Table 18.10 to run an autoregression model with 1, 2, and 3 lags from first quarter 1980 to fourth quarter 1990. Use MINITAB. Then use actual sales data for first quarter 1991 to calculate the prediction error as defined in Equation 17.13.

CHAPTER 19

Index Numbers and Stock Market Indexes

CHAPTER OUTLINE

Key Terms

index number
simple aggregative price index
price relative
consumer price index
base year
simple relative price index
weighted relative price index
Laspeyres price index
weighted aggregative price index
Paasche price index
GNP deflator
Fisher's ideal price index

quantity index
Paasche quantity index
Fisher's ideal quantity index
FRB index of industrial production
value index
stock market index
market-value-weighted index
S&P 500 index
price-weighted index
Dow Jones Industrial Average
equally weighted index
indicators of economic activity

19.1 INTRODUCTION

Business executives and government officials often make judgments that involve summarizing how business, economic, and financial variables change with time or place. Examples of variation over *time* include variation in gross national production, variation in the price of consumer goods, and variation in stock market prices. As an example of variation with changes in *place,* consider a company that wishes to transfer an executive from Chicago to San Francisco. What should be the executive's minimum salary increase to compensate for the higher cost of living in San Francisco?

In all these cases, we need to have a single composite figure to summarize the average difference between two time periods or between the two cities. Index numbers can be used to answer questions of this type. An **index number** is a summary measure that compares related items over time or place. In other words, index numbers enable us to express the level of an activity or phenomenon in relation to its level at another time or place.

In this chapter we will investigate how alternative index numbers are compiled and used in business, economics, and finance analyses. First, a discussion of price indexes, quantity indexes, and value indexes lays the foundation for an understanding of economic and financial indexes. Then we develop several types of stock market indexes and examine the major indexes provided in the daily financial news. Finally, applications of index numbers in business and economics are discussed.

19.2   PRICE INDEXES

In this section we first develop a **simple aggregative price index** based on a single good and then expand the concept to a combination of several goods. We also address some of the problems associated with price indexes and explore techniques that have been developed to deal with these problems.

Simple Aggregative Price Index

In its simplest form, an index number is nothing more than a percentage figure that expresses the relationship between two numbers, one of the numbers being used as the base. For example, in a time series of prices of a particular commodity, we can express the prices as percentages by dividing each figure by the price in the base period. These percentages are referred to as **price relatives.**

An understanding of the simple aggregative price index can perhaps best be facilitated through the use of an example of price relatives. Assume that the price of eggs has risen over 3 consecutive years as follows:

$$1989: \quad P_0 = \$1.00$$
$$1990: \quad P_1 = \$1.20$$
$$1991: \quad P_2 = \$1.50$$

To illustrate the change in prices, we calculate the ratio of prices with respect to a **base year,** the year from which the future price changes are measured. The appropriate base year depends on relevant economic factors. The base years of U.S. government indexes are shifted forward approximately every decade to reflect changes in economic conditions over time. In our example, the base year is 1989.

$$1989: \quad P_0/P_0 = 1.00/1.00 = 1.00 = I_0$$
$$1990: \quad P_1/P_0 = 1.20/1.00 = 1.20 = I_1$$
$$1991: \quad P_2/P_0 = 1.50/1.00 = 1.50 = I_2$$

$I_1 = 1.20$ means that the price of eggs increased 20 percent between 1989 and 1990. Likewise, $I_2 = 1.50$ means that between 1989 and 1991 there was a 50 percent increase in the price of eggs. Because 1989 is the base year, I_0 must equal 1.00.

Government price indexes are usually expressed on a basis of 100, and to be consistent with this practice, we will multiply each of the foregoing indexes by 100. Hence the price indexes for eggs from 1989 to 1991 are

$$I_0: \quad 1.00 \times 100 = 100$$
$$I_1: \quad 1.20 \times 100 = 120$$
$$I_2: \quad 1.50 \times 100 = 150$$

Each price of eggs is a price relative—the ratio of the price in a given year to the price in the base year.

The previous example of a price index was expressed in terms of only one commodity, eggs. A more realistic approach would be to include a group of commodities, as does the **consumer price index** (CPI). The CPI measures the cost of a market basket of some 2,000 consumer goods and services purchased by a "typical" urban family. The composition of this basket is food, clothing, housing, fuels, transportation, and medical care. For the sake of simplicity and ease of understanding, our market basket will consist of just four commodities:

One dozen eggs   One pound of butter
One gallon of milk   One loaf of bread

We will also assume that the same amount of each good was purchased in each of our 3 consecutive years. Table 19.1 illustrates the prices, for the years 1989 to 1991, of the individual goods that make up our market basket.

The table indicates that the same market basket of commodities cost $4.00 in 1989, $5.00 in 1990, and $6.00 in 1991. However, because we are interested in determining the price index of the market basket for the various years, we simply calculate the price relatives (ratios of prices between two different periods) of the group of commodities, using 1989 as the base year. The price indexes for 1989 through 1991 are

$$1989: \quad 4.00/4.00 \times 100 = 100$$
$$1990: \quad 5.00/4.00 \times 100 = 125$$
$$1991: \quad 6.00/4.00 \times 100 = 150$$

Table 19.1 Price and Quantity for Four Commodities, 1989–1991

Commodity	Quantity	1989 P_0	1990 P_1	1991 P_2
Eggs	One dozen	1.00	1.20	1.50
Milk	One gallon	1.50	1.75	2.00
Butter	One pound	1.10	1.35	1.60
Bread	One loaf	.40	.70	.90
		4.00	5.00	6.00

The indexes indicate that the cost of this *specific* list of goods increased 25 percent between 1989 and 1990 and 50 percent between 1989 and 1991.

Formally, we can write the simple price index as follows:

$$I_t = \frac{\sum\limits_{i=1}^{4} P_{ti}}{\sum\limits_{i=1}^{4} P_{0i}} \times 100 \tag{19.1}$$

where

I_t = price index for the year t

P_{it} = price of the ith commodity in the year t

P_{i0} = price of the ith commodity in the base year 0

Note that the quantity for the ith good can be regarded as $Q_{0i} = 1$ for all i.

Simple Average of Price Relatives

Two disadvantages of the simple aggregate price index give rise to a need for the simple average of relatives. These two disadvantages are

1. The units used to state the prices of the commodities affect the price index. For example, if the price of eggs were stated in half-dozens rather than in dozens, then the price indexes would be

    1989: $3.50/3.50 \times 100 = 100$

    1990: $4.40/3.50 \times 100 = 125.7$

    1991: $5.25/3.50 \times 100 = 150$

    These indexes are not identical to those calculated from the price of eggs per dozen.

2. The index does not consider the relative importance of the commodities, and it is unduly influenced by the price variation of high-priced commodities. For example, if our market basket were enlarged to include a shirt that cost $16 in 1989, $20 in 1990, and $10 in 1991, calculating our price indexes in terms of Equation 19.1 would yield

    $I_{89} = (4.00 + 16.00)/(4.00 + 16.00) \times 100 = 20/20 \times 100 = 100$

    $I_{90} = (5.00 + 20.00)/(4.00 + 16.00) \times 100 = 25/20 \times 100 = 125$

    $I_{91} = (6.00 + 10.00)/(4.00 + 16.00) \times 100 = 16/20 \times 100 = 80$

    The shirt makes up a majority of the index. The decrease in the price of a shirt makes the index decrease by 20 percent, even though that shirt is not the most important item for consumers.

Table 19.2 Calculation of the Simple Average of Price Relatives

Commodity	1989 P_0/P_0	1990 P_1/P_0	1991 P_2/P_0
Eggs	100	$1.20/1.00 \times 100 = 120$	$1.50/1.00 \times 100 = 150$
Milk	100	$1.75/1.5 \times 100 = 117$	$2.00/1.50 \times 100 = 133$
Butter	100	$1.35/1.10 \times 100 = 123$	$1.60/1.10 \times 100 = 145$
Bread	100	$.70/.40 \times 100 = 175$	$.90/.40 \times 100 = 225$
Shirt	100	$20.00/16.00 \times 100 = 125$	$10.00/16.00 \times 100 = 63$
	500%	660%	716%

Because of these limitations, we need an index that removes the bias due to the difference in measurement and takes the relative importance of the commodity into account. We can improve on our index by taking an average of the price relatives.

Using the data from Table 19.2, we can calculate the **simple relative price index** in period t as follows:

$$I_t = \frac{\sum_{i=1}^{n} (P_{ti}/P_{0i} \times 100)}{n} \qquad (19.2)$$

Hence the averages of the price relatives for 1990 and 1991 are 660/5, or 132, and 716/5, or 143.2, respectively.

Each item in the index is weighted by $1/P_0$, which makes all items equally important. Thus we have removed the influence of different units of measurement for the various commodities. However, this kind of index still doesn't take the relative importance of the commodity into account.

Weighted Relative Price Index

One major disadvantage of the simple relative price index is that it treats all commodities as equal. An index should reflect the value of some commodities in relation to the value of others. We need a **weighted relative price index**—one that takes into consideration the weights, or worth, of various commodities. We base the value of commodity weights on the quantity purchased. In other words, the total dollars spent on the commodities determine their weight in the index.

Suppose the following are the amounts spent on our market basket, V_{0i}, and the related prices and quantities of those commodities, in 1989.

Commodity	V_{0i}	P_{0i}	Q_{0i}
Eggs	$150	$1.00	150
Milk	450	1.50	300
Butter	220	1.10	200
Bread	440	.40	1100
Shirts	160	16.00	10
	$1420		

Table 19.3 Calculation of Individual Weighted Price Relatives for 1989–1991

Commodity	Weight	$(P_{0i}/P_{0i})V_{0i}$	$(P_{1i}/P_{0i})V_{0i}$	$(P_{2i}/P_{0i})V_{0i}$
Eggs	150	150	(1.2)150 = 180	(1.5)150 = 225
Milk	450	450	(1.17)450 = 527	(1.33)450 = 599
Butter	220	220	(1.23)220 = 271	(1.45)220 = 319
Bread	440	440	(1.75)440 = 770	(2.25)440 = 990
Shirts	160	160	(1.25)160 = 200	(.63)160 = 101
	1420	1420	1948	2234

That is, the value of the ith commodity purchased in 1989 is

$$V_{0i} = P_{0i} \times Q_{0i} \tag{19.3}$$

Hence Q_{01} is the quantity of eggs purchased in the base year (1989).

Accordingly, the purpose of the weighted relative price index is to show how much we need to spend in subsequent years to buy the same amount of commodities as we bought in the base year.

Formally, we derive the weighted relative for the ith commodity in period t as follows:

$$(P_{ti}/P_{0i})V_{0i} = (P_{ti}/P_{0i})P_{0i} \times Q_{0i} = P_{ti} \times Q_{0i} \tag{19.4}$$

Summing this for the five commodities in our market basket gives us the price index in period t, I_t, as

$$I_t = \frac{\sum_{i=1}^{5}(P_{ti}/P_{0i})V_{0i}}{\sum_{i=1}^{5}V_{0i}} \times 100 \tag{19.5}$$

Table 19.3 shows the individual weighted price relatives for 1989 to 1991. Thus the weighted relative price indexes are

1989: (1420/1420)(100) = 100

1990: (1948/1420)(100) = 137

1991: (2234/1420)(100) = 157

This weighted relative price index is higher than the simple relative price index, because bread has a high importance rating (that is now taken into account) and because bread underwent a substantial price increase.

Weighted Aggregative Price Index

In this section we discuss 3 weighted aggregative price indexes: the Laspeyres price index, the Paasche price index, and Fisher's ideal price index.

The Laspeyres Price Index

We can rewrite the weighted price index given in Equation 19.5 by using Equation 19.4.

$$
\begin{aligned}
I_t &= \frac{\sum_{i=1}^{5} \left(\frac{P_{ti}}{P_{0i}} \right) V_{0i}}{\sum_{i=1}^{5} V_{0i}} \times 100 \\[2ex]
&= \frac{(P_{t1}/P_{01})(P_{01}Q_{01}) + \cdots + (P_{t5}/P_{05})(P_{05}Q_{05})}{P_{01} \times Q_{01} + \cdots + P_{05} \times Q_{05}} \times 100 \\[2ex]
&= \frac{P_{t1} \times Q_{01} + \cdots + P_{t5} \times Q_{05}}{P_{01} \times Q_{01} + \cdots + P_{05} \times Q_{05}} \times 100 \\[2ex]
&= \frac{\sum_{i=1}^{5} P_{ti}Q_{0i}}{\sum_{i=1}^{5} P_{0i}Q_{0i}} \times 100
\end{aligned}
\tag{19.6}
$$

Through this derivation, we see that this index is the same index as in Equation 19.5. That is, this price index is weighted on the basis of the base-year quantities. In other words, the numerator is the value of the expenditures in year t that are necessary to buy the same quantity of the commodities as was purchased in the base year. This greatly simplifies updating the index, particularly in that most aggregate business indexes contain a large number of items. The denominator is the value of the expenditures required to buy a given amount in the base year. This formula is referred to as the **Laspeyres price index.** The CPI is a Laspeyres index. The disadvantage of this index is that it tends to give more weight to those items that show a dramatic price increase. A sharp increase in a particular commodity's price is typically accompanied by a decrease in the demand (measured by Q) for this item (consumers may be substituting another item). The Laspeyres index fails to adjust for this situation, but even so, its advantages outweigh its disadvantages.

From the data listed in Tables 19.1, 19.2, and 19.3, and from assumptions, we have the price and quantity information listed in Table 19.4.

Substituting data from Table 19.4 into Equation 19.6, we obtain the price indexes for 1990 and 1991.

$$
I_{90} = \frac{180 + 525 + 270 + 770 + 200}{150 + 450 + 220 + 440 + 160} = \frac{1945}{1420} = 137
$$

$$
I_{91} = \frac{225 + 600 + 320 + 990 + 100}{1420} = \frac{2235}{1420} = 157
$$

These figures imply that the **weighted aggregative price indexes** estimated from 5 commodities for 1990 and 1991 are 37 percent and 57 percent higher than those of 1989, respectively.

Table 19.4 Price and Quantity for Five Commodities, 1989–1991

	1989		1990		1991	
Commodity	*Price*	*Quantity*	*Price*	*Quantity*	*Price*	*Quantity*
Eggs	1.00	150	1.20	160	1.50	180
Milk	1.50	300	1.75	250	2.00	300
Butter	1.10	200	1.35	180	1.60	250
Bread	.40	1100	.70	1000	.90	1050
Shirts	16.00	10	20.00	15	10.00	20

The Paasche Price Index

The only difference between a **Paasche price index** and a Laspeyres index is that a Paasche index employs the current-year quantities (Q_t) rather than the base-year quantities (Q_0). Formally,

$$I_t = \frac{\sum_{i=1}^{n} P_{ti}Q_{ti}}{\sum_{i=1}^{n} P_{0i}Q_{ti}} \tag{19.7}$$

In other words, the numerator determines the amount of money necessary to purchase a given amount of commodities in the current year at current-year prices. Accordingly, the denominator determines the amount of money required to buy the current-year quantities at base-year prices. The gross national product (GNP) deflator is a Paasche index. This **GNP deflator** is broader than the CPI because it includes not only consumer goods and services but also investment goods, goods and services purchased by government, and goods and services that enter into world trade.

The complexity of updating the reference-year quantities for a Paasche index makes it difficult (and often impossible) to apply. Furthermore, because it reflects changes in both price and quantity, we cannot use it to reflect price changes between two periods. In addition, it tends to understate price increases and to overstate price decreases because it simultaneously reflects the quantity changes in the demand (Q). Its obvious advantage is that it uses current-year quantities, which provide a realistic and up-to-date estimate of total expense.

Substituting related data from Table 19.4 into Equation 19.7, we obtain the Paasche indexes for 1990 and 1991.

$$I_{90} = \frac{192 + 437.5 + 243 + 700 + 300}{160 + 375 + 198 + 400 + 240} = \frac{1872.5}{1373} = 136$$

$$I_{91} = \frac{270 + 600 + 400 + 945 + 200}{180 + 450 + 275 + 420 + 320} = \frac{2415}{1645} = 147$$

Fisher's Ideal Price Index

Fisher's ideal price index offers a compromise between the Laspeyres price index and the Paasche price index. This index is found by multiplying the square root of the Laspeyres index by the Paasche price index. Using Equations 19.6 and 19.7, we obtain Fisher's ideal price index (FI).

$$
\text{FI}_t = \sqrt{\left(\frac{\sum_{i=1}^{n} P_{ti} Q_{ti}}{\sum_{i=1}^{n} P_{0i} Q_{ti}} \right) \left(\frac{\sum_{i=1}^{n} P_{ti} Q_{0i}}{\sum_{i=1}^{n} P_{0i} Q_{0i}} \right)} \tag{19.8}
$$

Hence Fisher's ideal price index lies between the Laspeyres price index and the Paasche price index.

By substituting related information from this section, we find that

$$
\text{FI}_{90} = \sqrt{(136)(137)} = 136
$$
$$
\text{FI}_{91} = \sqrt{(147)(157)} = 152
$$

As we expected, the Fisher index is larger than the Paasche index and smaller than the Laspeyres index.

19.3 QUANTITY INDEXES

A **quantity index** measures a change in quantity from a base year to a particular year; such quantities include the volume of industrial production, the physical volume of imports and exports, quantities of goods and services consumed, and the volume of stock transactions. In this section we will discuss two major kinds of quantity indexes—weighted aggregative quantity indexes and weighted relative quantity indexes.

Laspeyres Quantity Index

The Laspeyres quantity index is derived by simply interchanging the P's and Q's in the Laspeyres price index.

$$
I_t = \frac{\sum_{i=1}^{n} Q_{ti} P_{0i}}{\sum_{i=1}^{n} Q_{0i} P_{0i}} \times 100 \tag{19.9}
$$

This index represents the total cost of the quantities in the year in question at base-year prices as a percentage of the total cost of the base-year quantities. Because prices are kept constant, any change in the index is due to the change in quantities between the base year and the year in question.

Table 19.5 Worksheet for Calculating the Laspeyres Quantity Index

| Commodity | 1989 | | 1990 | 1991 | | | |
	Q_{0i}	P_{0i}	Q_{1i}	Q_{2i}	$Q_{0i}P_{0i}$	$Q_{1i}P_{0i}$	$Q_{2i}P_{0i}$
Automobiles	40	1,000	50	60	40,000	50,000	60,000
Computers	30	500	40	50	15,000	20,000	25,000
Televisions	10	200	20	30	2,000	4,000	6,000
					57,000	74,000	91,000

From the data in Table 19.5 we can compute the Laspeyres quantity indexes for 1989 to 1991.

$$I_{89} = 57,000/57,000 \times 100 = 100$$

$$I_{90} = 74,000/57,000 \times 100 = 130$$

$$I_{91} = 91,000/57,000 \times 100 = 160$$

Hence the indexes for 1990 and 1991 indicate that the cost of the three commodities increased 30 percent and 60 percent, respectively. The price has been held constant, so the change in the index is due to changes in the quantities of the commodities for the period in question. In other words, the 1990 and 1991 indexes show that the quantities of the goods increased 30 percent and 60 percent, respectively, from the 1989 base year.

Paasche Quantity Index

The same relationship that exists between the Laspeyres price and quantity indexes also exists between the Paasche price and quantity indexes: interchanging the P's and Q's in the Paasche price index creates the Paasche quantity index.

$$I_t = \frac{\sum_{i=1}^{n} Q_{ti}P_{ti}}{\sum_{i=1}^{n} Q_{0i}P_{ti}} \times 100 \tag{19.10}$$

In other words, the Paasche quantity index represents the total cost of the purchased quantities in a given year as a percentage of what the base-year quantities would have cost had they been purchased during that year.

From the data in Table 19.6 we can calculate the Paasche quantity indexes for 1989 to 1991.

$$I_{89} = 100,500/100,500 \times 100 = 100$$

$$I_{90} = 135,000/105,000 \times 100 = 129$$

$$I_{91} = 191,000/124,000 \times 100 = 154$$

Table 19.6 Worksheet for Calculating the Paasche Quantity Index

Commodity	1989		1990		1991		$Q_{0i}P_{0i}$	$Q_{0i}P_{1i}$	$Q_{1i}P_{1i}$	$Q_{0i}P_{2i}$	$Q_{2i}P_{2i}$
	Q_{0i}	P_{0i}	Q_{1i}	P_{1i}	Q_{2i}	P_{2i}					
Automobiles	40	2000	50	2100	60	2500	80,000	84,000	105,000	100,000	150,000
Computers	30	600	40	600	50	700	18,000	18,000	24,000	21,000	35,000
Televisions	10	250	20	300	20	300	2,500	3,000	6,000	3,000	6,000
							100,500	105,000	135,000	124,000	191,000

Prices are held fixed in the equation, so a change between numerator and denominator reflects a change in the quantities between the two years. In other words, 1990 and 1991 saw a 23 percent and a 54 percent increase in quantity, respectively, over the 1989 quantity.

Fisher's Ideal Quantity Index

As you may have guessed, there is a compromise between the Laspeyres quantity index and the Paasche quantity index. This compromise is called **Fisher's ideal quantity index** (FIQ). It is computed as follows:

$$\text{FIQ} = \sqrt{(\text{Laspeyres quantity index})(\text{Paasche quantity index})} \quad (19.11)$$

Like Fisher's price index, Fisher's quantity index lies between the corresponding Laspeyres and Paasche indexes.

Substituting related information into Equation 19.11 yields

$$\text{FIQ}_{90} = \sqrt{(130)(129)} = 129$$
$$\text{FIQ}_{91} = \sqrt{(160)(154)} = 157$$

FRB Index of Industrial Production

Probably the most widely used and best-known quantity index in the United States is the Federal Reserve Board (FRB) index of industrial production. This index measures changes in the physical volume of output of manufacturing, mining, and utilities. The **FRB index of industrial production** is closely watched by business executives, economists, and financial analysts as a major indicator of the physical output of the economy. It is one of the roughly coincident indicators used by the National Bureau of Economic Research as a cyclical indicator.[1]

The FRB index of industrial production is the weighted arithmetic mean of the quantity relatives as defined as

$$I_t = \frac{\sum_{i=1}^{n} \left(\frac{Q_{ti}}{Q_{0i}} \times 100 \right) Q_{0i}P_{0i}}{\sum_{i=1}^{n} Q_{0i}P_{0i}}$$

[1] See Application 19.2 in Section 19.6.

where Q_{ti}/Q_{0i} is the quantity relative and $Q_{0i}P_{0i}$ is the weight. From the proof of Equation 19.6, we know this equation is identical to Equation 19.9.

Numerous problems plague the use of both quantity relative (Q_{ti}/Q_{0i}) and value weights ($Q_{0i}P_{0i}$). Because many industries cannot easily provide physical output data for the quantity relatives, such related data as shipments and employee-hours worked that tend to move parallel to output are sometimes used instead. Value-added data instead of final product data are also used as weights to avoid the problem of double counting. For example, if the value of the final product were used for a tire company that sells its tires to an automobile company, and the value of the final product of the automobile company were also used, the tires that went into making the automobile would be counted twice. A firm's value added is conceptually equivalent to the total of its factor-of-production payments: rent, interest, wages, and profits.

19.4 VALUE INDEX

A **value index** measures the total cost of the purchased quantities in a given year at the prices prevailing during that year compared to the cost of the purchased quantities in the base period at base-year prices. The value index is

$$I_t = \frac{\sum_{i=1}^{n} Q_{ti}P_{ti}}{\sum_{i=1}^{n} Q_{0i}P_{0i}}\,(100) \qquad (19.12)$$

Thus the value index reflects simultaneous changes in quantities and prices for the period in question.

From the data in Table 19.7 we can compute year-to-year changes in the value index.

For 1989–1990,
$$I = (129{,}000/57{,}000)(100) = 196$$
For 1990–1991,
$$I = (191{,}000/129{,}000)(100) = 148$$

Table 19.7 Worksheet for Calculating the Value Index

Commodity	1989 $P_{0i}Q_{0i}$	1990 $P_{1i}Q_{1i}$	1991 $P_{2i}Q_{2i}$
Automobiles	40,000	100,000	150,000
Computers	15,000	24,000	35,000
Televisions	2,000	5,000	6,000
	57,000	129,000	191,000

Again, the value index reflects changes from year to year that are due to changes both in prices and in quantities.

19.5 STOCK MARKET INDEXES

A **stock market index** is a statistical measure that shows how the prices of a group of stocks change over time. A stock market index encompasses either all or only a portion of stocks in its market. Stock market indexes employ different weighting schemes, so we can use this basis to categorize the indexes by type. The three most common types of stock market indexes are market-value-weighted indexes, price-weighted indexes, and equally weighted indexes. Price per share in current period (P_0), price per share in next period (P_1), number of shares outstanding in current period (Q_0), and number of shares outstanding in next period (Q_1) are listed in Table 19.8. These data are used to illustrate the various weighting schemes and to provide information about the weights applied to the major stock market indexes.

Table 19.8 Price per Share and Outstanding Shares for Stocks, A, B, and C

Stock	P_0	P_1	Q_0	Q_1
A	100	100	60	55
B	50	60	90	100
C	20	40	250	300

Market-Value-Weighted Index

The **market-value-weighted index** is similar to the value index given in Equation 19.12 and discussed in Section 19.4. We compute the index by taking the ratio of the market value of the outstanding shares at time t to their market value at the initial period. From Table 19.8, we calculate the market-value-weighted index as follows:

$$I = \frac{\sum_{i=1}^{3} Q_{ti}P_{ti}}{\sum_{i=1}^{3} Q_{0i}P_{0i}} = \frac{55(100) + 100(60) + 300(40)}{60(100) + 90(50) + 250(20)}(100)$$

$$= \frac{23,500}{15,500}(100) = 152$$

This figure implies that the market-value-weighted index increased by 52 percent from the base period to the current period.

Standard and Poor's 500 Composite Index is an example of a market-value-weighted index. The **S&P 500 index** comprises industrial firms, utilities, transportation firms, and financial firms. Changes in the index are based on changes in the firms' total market value with respect to a base year. Currently, the base period (1941–1943 = 10) for the S&P 500 index is stated formally as follows:

$$\text{S\&P 500 index} = \frac{\sum_{i=1}^{500} P_{ti} Q_{ti}}{\sum_{i=1}^{500} P_{0i} Q_{0i}} (10) \qquad (19.13)$$

where

P_{0i} = per-share stock price at base year 0
P_{ti} = per-share stock price at index data t
Q_{0i} = number of shares for firm i at base year 0
Q_{ti} = number of shares for firm i at index year t

S&P 500 Index

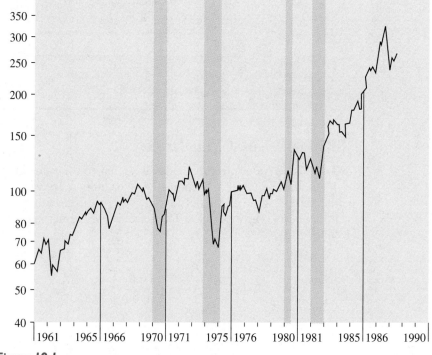

Figure 19.1

The S&P 500, 1961–1988 (monthly averages: 1941–1943 = 10; shaded areas represent periods of business recessions)
Source: Analyst's Outlook, Standard & Poor's, 1988, p. 186.

The index is multiplied by an index set equal to 10. The specification of this index is identical to that of the value index indicated in Equation 19.12. The fluctuation of the S&P 500 index during 1961–1988 is presented in Figure 19.1.

The New York Stock Exchange (NYSE) also publishes a market index, which differs in only two respects from the S&P 500 index. First, the NYSE index includes the stocks of *all* firms listed on the NYSE, whereas the S&P 500 index includes only a portion of the firms on the exchange. In addition, the NYSE index uses a base index of 50 (as opposed to 10), which was chosen to represent an approximate price of an average share in December 1985.

Price-Weighted Index

The **price-weighted index** shows the change in the average price of the stocks that are included in the index. Using the data from Table 19.8, we can compute the price-weighted index as follows:

$$I = \frac{(100 + 60 + 40)/3}{(100 + 50 + 20)/3}(100)$$

$$= \frac{200/3}{170/3}(100)$$

$$= 117.65$$

The closest thing to a true price-weighted stock market index is the **Dow Jones Industrial Average** (DJIA). Simply stated, the DJIA is an arithmetic average of the stock prices that make up the index. The DJIA originally assumed a single share of each stock in the index, and the total of the stock prices was divided by the number of stocks that made up the index:[2]

$$DJIA_t = \frac{\dfrac{\sum_{i=1}^{30} P_{ti}}{30}}{\dfrac{\sum_{i=1}^{30} P_{0i}}{30}} \tag{19.14}$$

Today the index is adjusted for stock splits and the issuance of stock dividends:

$$DJIA_t = \frac{\sum_{i=1}^{30} \dfrac{P_{ti}}{AD_t}}{\sum_{i=1}^{30} P_{0i}} \tag{19.14a}$$

[2]There are 30 blue-chip firms included in this index. Their names appear in Table 9.1.

Table 19.9 Adjustment of DJIA Divisor to Allow for a Stock Split

	Before Split	After 2-for-1 Stock Split by Stock A
Stock	*Price*	*Price*
A	60	30
B	30	30
C	20	20
D	10	10
	120	90

Average before split $= \dfrac{120}{4} = 30$

Adjusted divisor $= \dfrac{\text{sum of prices after the split}}{\text{average before split}} = \dfrac{90}{30} = 3$

Average after split $= \dfrac{90}{3} = 30$

Before-split divisor $= 4$ After-split divisor $= 3$

where P_{it} = the closing price of stock i on day t, and AD_t = the adjusted divisor on day t. This index is similar to the simple price index given in Equation 19.1 except for the stock splits adjusted over time. The adjustment process is illustrated in Table 19.9.

Alternatively, the average after split can be calculated as

$$\text{Average} = \frac{30 \times 2 + 30 + 20 + 10}{4} = 30$$

This average is identical to that obtained by using the adjusted-divisor approach.

As Table 19.9 shows, the adjustment process is designed to keep the index value the same as it would have been if the split had not occurred. Similar adjustments have been made when it has been found necessary to replace one of the component stocks with the stock of another company, thus preserving the consistency and comparability of index values at different points in time.

Nevertheless, the adjustment process used for the DJIA has its share of critics. Because price weighting itself causes high-priced stocks to dominate the series, the same effect can cause a shift in this balance when rapidly growing firms split their stock. For example, a 20 percent increase in the price of stock A in Table 19.9 would in itself have caused a 10 percent increase in the value of the sample index before the split, whereas a 20 percent increase in the price of stock B would have caused only a 5 percent increase in the index value. After the 2-for-1 split of stock A, a 20 percent increase in either stock A or stock B would have the same effect on the index value (a 6.7 percent increase); a downward shift in the importance of stock A relative to that of the other stocks in the sample has occurred. This effect could relegate the stock of the fastest-growing companies to a position of *least* importance in determining index values.

Equally Weighted Index

The **equally weighted index** is based on the supposition that an equal amount is invested in each of the stocks included in the index. Hence, in computing the index, we will assume that $1,000 is invested in each of the 3 stocks. Using the price information listed in Table 19.8, we find the following numbers of shares purchased in the initial period.

Stock	Number of Shares at Initial Period
A	10 (1,000/100)
B	20 (1,000/50)
C	50 (1,000/20)

Using this information and the price information listed in Table 19.8, we can calculate the equally weighted index.

$$I = \frac{10(100) + 20(60) + 50(40)}{10(100) + 20(50) + 50(20)} (100)$$

$$= \frac{4,200}{3,000} (100) = 140$$

The equally weighted index is based on the changes in the price of the individual stocks, given that an equal amount of money is initially invested in each stock. In other words, the index keeps the number of shares constant, while providing for changes in the per-share price. This is similar to the Laspeyres price index (Equation 19.6), except that the initial quantity should be determined by the equal-amount-net-of-investment assumption. One of the two Wilshire 5000 equity indexes is an equally weighted index. The market-value-weighted Wilshire 5000 equity index is discussed in the next section.

Wilshire 5000 Equity Index

The Wilshire 5000 equity index, which includes 5,000 stocks, is compiled by both market-value-weighted and equally weighted approaches. The market-value-weighted approach is identical to the value index given in Equation 19.12. The equally weighted approach is identical to that discussed in the last section. This index is being used increasingly, because it contains most equity securities available for investment, including all NYSE and AMEX issues and the most active stocks traded on the over-the-counter (OTC) market.

The following formula is used to compute the market-value-weighted Wilshire 5000 equity index.

$$I_t = I_{t-1} \left[\sum_{j=1}^{N} (S_{jt})P_{jt} \bigg/ \sum_{j=1}^{N} (S_{jt-1})P_{jt-1} \right] \tag{19.15}$$

where

I_t = index value for the tth period

N = number of stocks in the index

P_{jt} = price of the jth security for the tth period

S_{jt} = shares outstanding of the jth security for the tth period

P_{jt-1} = price of the jth security for the $(t-1)$th period

S_{jt-1} = shares outstanding of the jth security for the $(t-1)$th period

In the event that P_{jt} is not available for a given security, that security is dropped from the summations. If P_{jt-1} is not available but P_{jt} is—that is, if a security has just resumed trading—the last available price is substituted for P_{jt-1}.

As an example, we present in Table 19.10 monthly equity values for the Wilshire 5000 equity index from January 1989 to January 1992. In this table, only the value-weighted index appears. Figure 19.2 is a graph of the Wilshire 5000 equity index from January 1970 to December 1991. The monthly Wilshire equity index (Table 19.10) can be used to calculate the market rates of return by the method discussed in Appendix 2B in Chapter 2. Monthly rates of return calculated from the data of Table 19.10 in terms of the value-weighted index are presented in Table 19.11. In Table 19.11, price appreciation represents the percentage change of index, and the total return is equal to price appreciation plus the dividend yield. Note that the Wilshire index can be used in place of the S&P 500 as an NBER leading economic indicator, a topic discussed in Section 19.6 of this chapter.

Table 19.10 Value-Weighted Wilshire 5000 Equity Index

Date	Wilshire Index[a]	Date	Wilshire Index[a]
1/31/89	2,917.261	8/31/90	3,053.601
2/28/89	2,857.863	9/28/90	2,879.335
3/31/89	2,915.072	10/31/90	2,833.986
4/28/89	3,053.132	11/30/90	3,015.022
5/31/89	3,162.609	12/31/90	3,101.355
6/30/89	3,137.008	1/31/91	3,245.346
7/31/89	3,377.403	2/28/91	3,484.851
8/31/89	3,440.843	3/28/91	3,583.671
9/29/89	3,426.656	4/30/91	3,587.924
10/31/89	3,320.354	5/31/91	3,719.297
11/30/89	3,367.637	6/28/91	3,545.470
12/29/89	3,419.879	7/31/91	3,705.893
1/31/90	3,163.301	8/30/91	3,795.043
2/28/90	3,201.205	9/30/91	3,743.976
3/30/90	3,273.458	10/31/91	3,807.081
4/30/90	3,172.327	11/29/91	3,649.992
5/31/90	3,448.484	12/31/91	4,041.102
6/29/90	3,424.366	01/31/92	4,027.770
7/31/90	3,384.365		

[a]12/31/90 base = 1,404.596

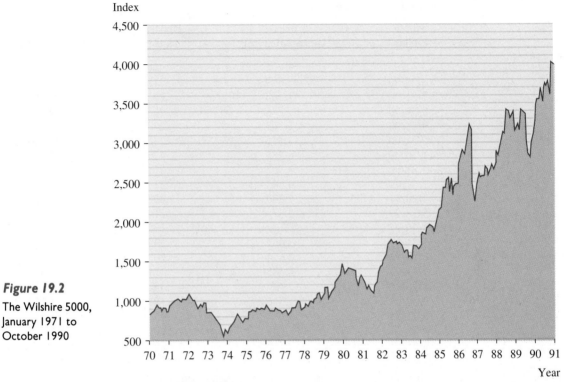

Figure 19.2

The Wilshire 5000, January 1971 to October 1990

Source: Wilshire 5000 Equity Index, Wilshire 5000® a registered Service Mark of Wilshire Associates Incorporated, Santa Monica, California.

Table 19.11 Monthly Returns for Value-Weighted Wilshire 5000 Equity Index

Month Ending	Price Appreciation (%)[a]	Dividends Yield (%)	Total Return (%)
1/31/89	6.531	.281	6.812
2/28/89	−2.036	.368	−1.668
3/31/89	2.002	.272	2.274
4/28/89	4.736	.180	4.916
5/31/89	3.586	.475	4.061
6/30/89	−.810	.233	−.577
7/31/89	7.663	.213	7.876
8/31/89	1.878	.394	2.272
9/29/89	−.412	.239	−.173
10/31/89	−3.102	.182	−2.920
11/30/89	1.424	.342	1.766
12/29/89	1.551	.265	1.816
1/31/90	−7.503	.163	−7.340
2/28/90	1.198	.390	1.588
3/30/90	2.257	.242	2.499
4/30/90	−3.090	.210	−2.880

(continued)

Table 19.11 (Continued)

Month Ending	Price Appreciation (%)[a]	Dividends Yield (%)	Total Return (%)
5/31/90	8.705	.426	9.131
6/29/90	−.699	.216	−.483
7/31/90	−1.168	.202	−.966
8/31/90	−9.773	.363	−9.410
9/28/90	−5.707	.213	−5.494
10/31/90	−1.575	.235	−1.340
11/30/90	6.388	.429	6.817
12/31/90	2.863	.311	3.174
1/31/91	4.643	.215	4.858
2/28/91	7.380	.400	7.780
3/28/91	2.836	.210	3.046
4/30/91	.119	.198	.317
5/31/91	3.662	.348	4.010
6/28/91	−4.674	.209	−4.465
7/31/91	4.525	.171	4.696
8/30/91	2.406	.356	2.762
9/30/91	−1.346	.198	−1.148
10/31/91	1.686	.151	1.837
11/29/91	−4.126	.307	−3.819
12/31/91	10.715	.265	10.980
01/31/92	−.330	.132	−.198

[a]Represents monthly percentage change in Wilshire 5000 equity index.

19.6   BUSINESS AND ECONOMIC APPLICATIONS

■   **APPLICATION 19.1**   Deflation of Value Series by Price Indexes

Corporate executives and economists are often interested in dividing time-series data expressed in monetary terms into two components—a "real," or quantity, component and a price component. Together, these two components express a value series in terms of price × quantity. Hence if we are interested only in the quantity component of a time series, we can find it simply by dividing the value series by price. Often we use a price index as a deflator. With the available information, it is important to use an index that is relevant to the time series in question. For example, one might use the consumer price index (CPI) to calculate real weekly wage and then use the GNP deflator to calculate real GNP.

If we want to calculate the real weekly wage for a large city for 1987–1989, we can use the consumer price index to deflate the nominal weekly wage and obtain the real wage rate as indicated in Table 19.12.

Table 19.13 shows nominal GNP and the implicit price deflator for GNP and real GNP for the period 1982–1989. Figure 19.3 plots both nominal and real GNP data.

Table 19.12 Calculation of Real Weekly Wages for Factory Workers in a Large City, 1987–1989

(1) Year	(2) Average Weekly Wage	(3) Consumer Price Index (1982–1984 = 100)	(4) Real Weekly Wage
1987	$320	113.6	$281.69
1988	$360	118.3	304.31
1989	$400	124.0	322.58

Source: 1990 *Economic Report of the President*, p. 262.

Table 19.13 Nominal GNP, GNP Deflators, and Real GNP (1980–1990)

Year	Nominal GNP	GNP Deflators	Real GNP
1980	2,731.3	85.7	3,187.1
1981	3,053.9	94.0	3,248.8
1982	3,166.0	100.0	3,166.0
1983	3,407.0	103.9	3,279.1
1984	3,771.0	107.7	3,501.4
1985	4,013.1	110.9	3,618.7
1986	4,231.0	113.8	3,717.9
1987	4,514.4	117.4	3,845.3
1988	4,872.5	121.3	4,016.9
1989	5,200.7	126.3	4,117.7
1990	5,464.9	131.5	4,155.8

Calculating real GNP gives insight into the "real" growth in the economy. We do not want to be misled by the trend of GNP when prices generally increase at a rate greater than zero. A price component (price deflator) provides information about changes in the nominal component of output. Formally, we compute real GNP as follows:

$$\text{Real GNP} = \frac{\text{nominal GNP}}{\text{GNP deflator}/100} \tag{19.16}$$

Equation 19.16 is used to calculate real GNP for the period 1980–1990; the results are presented in the last column of Table 19.13.

To calculate the growth rate for nominal GNP (X_t), GNP deflator (Y_t), and real GNP (Z_t) from the data listed in Table 19.13, we run the following regressions.

$$\log_e X_t = a_0 + a_1 t + e_{1t} \tag{19.17a}$$

$$\log_e Y_t = b_0 + b_1 t + e_{2t} \tag{19.17b}$$

$$\log_e Z_t = c_0 + c_1 t + e_{3t} \tag{19.17c}$$

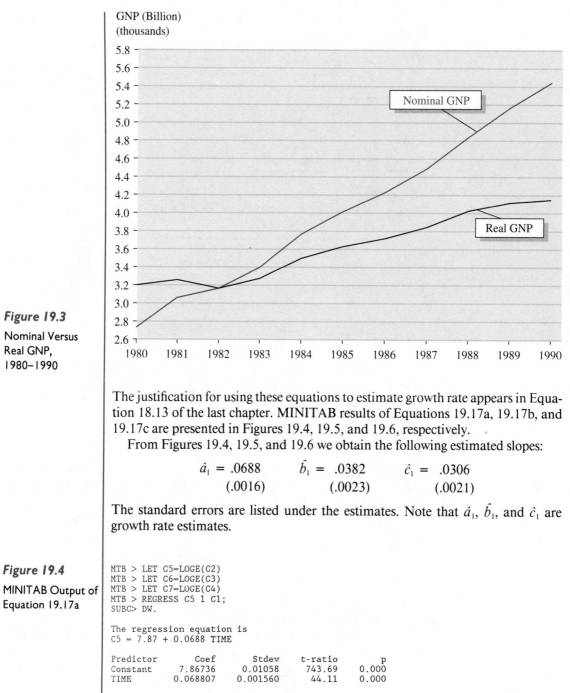

Figure 19.3

Nominal Versus
Real GNP,
1980–1990

The justification for using these equations to estimate growth rate appears in Equation 18.13 of the last chapter. MINITAB results of Equations 19.17a, 19.17b, and 19.17c are presented in Figures 19.4, 19.5, and 19.6, respectively.

From Figures 19.4, 19.5, and 19.6 we obtain the following estimated slopes:

$$\hat{a}_1 = .0688 \qquad \hat{b}_1 = .0382 \qquad \hat{c}_1 = .0306$$
$$(.0016) \qquad\quad (.0023) \qquad\quad (.0021)$$

The standard errors are listed under the estimates. Note that $\hat{a}_1$, $\hat{b}_1$, and $\hat{c}_1$ are growth rate estimates.

Figure 19.4

MINITAB Output of
Equation 19.17a

```
MTB > LET C5=LOGE(C2)
MTB > LET C6=LOGE(C3)
MTB > LET C7=LOGE(C4)
MTB > REGRESS C5 1 C1;
SUBC> DW.

The regression equation is
C5 = 7.87 + 0.0688 TIME

Predictor      Coef      Stdev     t-ratio      p
Constant    7.86736    0.01058     743.69    0.000
TIME        0.068807   0.001560     44.11    0.000

s = 0.01636    R-sq = 99.5%    R-sq(adj) = 99.5%

Analysis of Variance
```

Figure 19.4

(Continued)

```
SOURCE        DF          SS          MS         F        p
Regression     1      0.52078     0.52078    1946.01    0.000
Error          9      0.00241     0.00027
Total         10      0.52319


Obs.    TIME         C5      Fit  Stdev.Fit   Residual   St.Resid
  1     1.0     7.91253   7.93616    0.00923   -0.02363     -1.75
  2     2.0     8.02417   8.00497    0.00795    0.01920      1.34
  3     3.0     8.06022   8.07378    0.00680   -0.01355     -0.91
  4     4.0     8.13359   8.14258    0.00584   -0.00900     -0.59
  5     5.0     8.23510   8.21139    0.00517    0.02370      1.53
  6     6.0     8.29732   8.28020    0.00493    0.01712      1.10
  7     7.0     8.35019   8.34900    0.00517    0.00119      0.08
  8     8.0     8.41503   8.41781    0.00584   -0.00278     -0.18
  9     9.0     8.49136   8.48662    0.00680    0.00475      0.32
 10    10.0     8.55655   8.55542    0.00795    0.00112      0.08
 11    11.0     8.60610   8.62423    0.00923   -0.01813     -1.34


Durbin-Watson statistic = 1.97
```

Figure 19.5

MINITAB Output
of Equation 19.17b

```
MTB > REGRESS C6 1 C1;
SUBC> DW.

The regression equation is
C6 = 4.47 + 0.0382 TIME

Predictor       Coef      Stdev     t-ratio       p
Constant     4.46601    0.01531     291.66     0.000
TIME        0.038190   0.002258      16.92     0.000

s = 0.02368     R-sq = 97.0%     R-sq(adj) = 96.6%

Analysis of Variance

SOURCE        DF          SS          MS         F        p
Regression     1      0.16043     0.16043     286.15    0.000
Error          9      0.00505     0.00056
Total         10      0.16548


Obs.    TIME         C6      Fit  Stdev.Fit   Residual   St.Resid
  1     1.0     4.45085   4.50420    0.01336   -0.05335    -2.73R
  2     2.0     4.54329   4.54239    0.01151    0.00090      0.04
  3     3.0     4.60517   4.58058    0.00984    0.02459      1.14
  4     4.0     4.64343   4.61877    0.00845    0.02466      1.11
  5     5.0     4.67935   4.65696    0.00749    0.02239      1.00
  6     6.0     4.70863   4.69515    0.00714    0.01348      0.60
  7     7.0     4.73444   4.73334    0.00749    0.00110      0.05
  8     8.0     4.76559   4.77153    0.00845   -0.00595     -0.27
  9     9.0     4.79827   4.80972    0.00984   -0.01146     -0.53
 10    10.0     4.83866   4.84791    0.01151   -0.00925     -0.45
 11    11.0     4.87901   4.88611    0.01336   -0.00710     -0.36

R denotes an obs. with a large st. resid.

Durbin-Watson statistic = 0.76

MTB > PAPER
```

Figure 19.6

MINITAB Output
of Equation 19.17c

```
MTB > REGRESS C7 1 C1;
SUBC> DW.

The regression equation is
C7 = 8.01 + 0.0306 TIME

Predictor       Coef      Stdev     t-ratio       p
Constant     8.00652    0.01417     564.91     0.000
TIME        0.030615   0.002090      14.65     0.000

s = 0.02192     R-sq = 96.0%     R-sq(adj) = 95.5%
```

(continued)

Figure 19.6

(Continued)

```
Analysis of Variance

SOURCE        DF        SS         MS         F        p
Regression     1     0.10310    0.10310    214.64    0.000
Error          9     0.00432    0.00048
Total         10     0.10743

Obs.    TIME        C7       Fit  Stdev.Fit  Residual   St.Resid
  1      1.0     8.06687   8.03714   0.01236    0.02973     1.64
  2      2.0     8.08604   8.06775   0.01066    0.01829     0.95
  3      3.0     8.06022   8.09837   0.00911   -0.03814    -1.91
  4      4.0     8.09532   8.12898   0.00782   -0.03366    -1.64
  5      5.0     8.16092   8.15960   0.00693    0.00132     0.06
  6      6.0     8.19387   8.19021   0.00661    0.00366     0.18
  7      7.0     8.22091   8.22083   0.00693    0.00009     0.00
  8      8.0     8.25461   8.25144   0.00782    0.00316     0.15
  9      9.0     8.29827   8.28206   0.00911    0.01621     0.81
 10     10.0     8.32305   8.31267   0.01066    0.01038     0.54
 11     11.0     8.33226   8.34329   0.01236   -0.01103    -0.61


Durbin-Watson statistic = 1.21

MTB > PAPER
```

From these estimates, we know that the growth rate for nominal GNP is approximately equal to the growth rate of the GNP deflator plus the growth rate of real GNP.

■ **APPLICATION 19.2** Using Business and Economic Index Numbers to Predict Business Cycles

The National Bureau of Economic Analysis publishes a series of 46 categories of economic data reflecting movements in the business cycle (see Table 19.14). These are called **indicators** of economic activity. Of the 26 indicators, 12 make up an index of leading indicators. A leading indicator's "lead" must be greater than 3 months, and a lagging indicator follows economic activity by more than 3 months. Indicators of economic activity that lead or lag economic activity by 3 months or less are called coincident indicators.

The 12 components of the index of leading indicators are

1. Average weekly hours of manufacturing workers
2. Average weekly initial claims for unemployment insurance
3. The real value of manufacturers' new orders for consumer goods and materials
4. The index of new business formations
5. The index of 500 common stock prices
6. Contracts and orders for plant and equipment in 1972 dollars
7. The index of new private-housing starts authorized by local building permits
8. The ratio of price to unit labor cost, manufacturing
9. The net change in inventories on hand and on order in 1972 dollars

Table 19.14 Cyclical Indicators: Short List of the National Bureau of Economic Research

Leading Indicators
  Average hourly workweek, production workers, manufacturing
  Average weekly initial claims, state unemployment insurance
  Index of net business formation
  New orders, durable-goods industries
  Contracts and orders, plant and equipment
  Index of new building permits, private housing units
  Change in book value, manufacturing and trade inventories
  Index of industrial materials prices
  Index of stock prices, 500 common stocks
  Corporate profits after taxes (quarterly)
  Index: ratio of price to unit labor cost, manufacturing
  Change in consumer installment debt

Roughly Coincident Indicators
  GNP in current dollars
  GNP in 1958 dollars
  Index of industrial production
  Personal income
  Manufacturing and trade sales
  Sales of retail stores
  Employees on nonagricultural payrolls
  Unemployment rate, total

Lagging Indicators
  Unemployment rate, persons unemployed 15 weeks or over
  Business expenditures, new plant and equipment
  Book value, manufacturing and trade inventories
  Index of labor cost per unit of output in manufacturing
  Commerical and industrial loans outstanding in large commercial banks
  Banks rates on short-term business loans

Source: U.S. Department of Commerce.

10. The change in sensitive materials prices

11. The change in total consumer and business credit outstanding

12. Corporate profits after taxes (quarterly)

Table 19.15 presents information on the dates and durations of business cycles as determined by the National Bureau of Economic Research (NBER). It is particularly interesting to compare the dates and durations of business cycles to movements in the stock market. Comparing the graphs of the S&P 500 (Figure 19.1) and the Wilshire 5000 stock indexes (Figure 19.2) to the business cycle information reported in Table 19.15 reveals the relationship between the business cycle and stock market movements. For example, from February 1961 to December 1969, the United States sustained 106 months of economic growth (largely due to gov-

Table 19.15 NBER Business Cycle Reference Dates and Durations

Trough	Peak	Contractions	Expansions	Trough to Trough	Peak to Peak
December 1854	June 1857	NA	30	NA	NA
December 1858	October 1860	18	22	48	40
June 1861	April 1865	8	46	30	54
December 1867	June 1869	32	18	78	50
December 1870	October 1873	18	34	36	52
March 1879	March 1882	65	36	99	101
May 1885	March 1887	38	22	74	60
April 1888	July 1890	13	27	35	40
May 1891	January 1893	10	20	37	30
June 1894	December 1895	17	18	37	35
June 1897	June 1899	18	24	36	42
December 1900	September 1902	18	21	42	39
August 1904	May 1907	23	33	44	56
June 1908	January 1910	13	19	46	32
January 1912	January 1913	24	12	43	36
December 1914	August 1918	23	44	35	67
March 1919	January 1920	7	10	51	17
July 1921	May 1923	18	22	28	40
July 1924	October 1926	14	27	36	41
November 1927	August 1929	13	21	40	34
March 1933	May 1937	43	50	64	93
June 1938	February 1945	13	80	63	93
October 1945	November 1948	8	37	88	45
October 1949	July 1953	11	45	48	56
May 1954	August 1957	10	39	55	49
April 1958	April 1960	8	24	47	32
February 1961	December 1969	10	106	34	116
November 1970	November 1973	11	36	117	47
March 1975	January 1980	16	58	52	74
July 1980	July 1981	6	12	64	18
November 1982	?	16	80[a]	28	96[a]
?		NA	NA	89[a]	NA

[a]The 80-month duration of the last expansion, the 96-month duration of the last peak-to-peak cycle, and the 89-month duration of the last trough-to-trough cycle are conservative estimates. They assume a peak in July 1989 and, for the last of these, a trough 9 months later. Wartime expansions and cycles containing wartime expansions are underlined.

ernment expenditures on the Vietnam War). Figure 19.1 shows that over the same period, the S&P 500 index moved higher, indicating a direct link between economic growth and stock market movements. That is, when the economy is growing, stock prices tend to increase. When the economy is in recession, stock prices decline. In general, stock price movements *lead* changes in economic condition. Therefore, the stock market is one of the leading indicators of the business cycle.

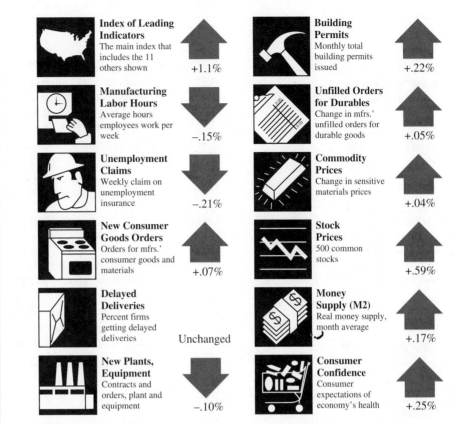

Index of Leading Indicators
The main index that includes the 11 others shown   +1.1%

Building Permits
Monthly total building permits issued   +.22%

Manufacturing Labor Hours
Average hours employees work per week   −.15%

Unfilled Orders for Durables
Change in mfrs.' unfilled orders for durable goods   +.05%

Unemployment Claims
Weekly claim on unemployment insurance   −.21%

Commodity Prices
Change in sensitive materials prices   +.04%

New Consumer Goods Orders
Orders for mfrs.' consumer goods and materials   +.07%

Stock Prices
500 common stocks   +.59%

Delayed Deliveries
Percent firms getting delayed deliveries   Unchanged

Money Supply (M2)
Real money supply, month average   +.17%

New Plants, Equipment
Contracts and orders, plant and equipment   −.10%

Consumer Confidence
Consumer expectations of economy's health   +.25%

Source: Bureau of Economic Analysis. *Knight-Ridder Tribune News.*

Figure 19.7

Statistics That Forecast the Economy
Source: *Home News,* March 30, 1991. Reprinted by permission of Knight-Ridder Tribune News.

In the business section of the *Home News* on June 1, 1991, a headline read, "Leading Indicators Up Again." In support of this headline, the newspaper identified the 12 statistics listed in Figure 19.7 as "statistics that forecast the economy." Of these 12 statistics, 8 increased, 1 experienced no change, and 3 decreased. This example essentially shows that the cyclical indicators in Table 19.14 can be used to forecast economic activities.

Summary

In this chapter, we examined different index numbers. Price, quantity, and value indexes were explored. Then stock market indexes and applications of index numbers to calculate real wage income and real GNP were discussed in detail. Finally, we demonstrated how the NBER economic indicators are used to forecast business cycles.

Questions and Problems

1. Find the following sales indexes for the accompanying retail sales volume (in thousands of dollars) with the base year indicated.

Year	Sales ($)	Year	Sales ($)
1984	85,390	1987	92,289
1985	86,745	1988	97,725
1986	88,452		

   a. 1984
   b. 1988

2. An alloy is made up of 27 percent metal A, 34 percent metal B, and 39 percent metal C. During the past year, prices of the metals have changed as shown in the table. Find the simple aggregative index for the cost of the alloy.

	Price per Pound	
Metal	**1988**	**1989**
A	2.08	3.48
B	7.61	7.78
C	4.49	3.80

3. Find the simple aggregative index number for the following data.

	1972		**1989**	
Item	**Unit Price**	**Unit Sold**	**Unit Price**	**Unit Sold**
A	$ 2.50	400	$ 4.00	650
B	16.00	150	19.50	175
C	9.50	250	16.00	350

   a. Find the Laspeyres index number.
   b. Find the Paasche index number.
   c. Find the ideal index number.
   d. Find a physical-volume index for 1989, weighting quantities with 1972 base-year prices.

4. Use the data in question 2.
   a. Find the weighted aggregative index for the cost of the alloy.
   b. How can you explain the values of these index numbers in view of the fact that only 2 out of 3 constituent metal prices rose?

5. Say you earned $10,725 in 1975 when the CPI in your city was 150 and earned $29,500 in 1989 when your city's CPI hit 358. Express your 1989 purchasing power as a percentage of your 1975 purchasing power.

6. Assume the CPI for your city had the following values.

   Year 1982 1983 1984 1985 1986
   Value 210 238 265 300 335

   a. Find the purchasing power of the dollar in each of these years as a proportion of the 1986 dollar.
   b. Explain the meaning of these figures. (*Hint:* What percentage of 1986 goods could you buy in these years?)

7. Suppose you have a stock market indicator made up of 5 common stocks selling at the prices per share indicated in the following table.

Stock	Price per Share
A	$106.00
B	87.75
C	49.50
D	32.75
E	23.50

   a. Find the market average.
   b. Suppose stock A split 2 for 1 and stock D split 3 for 1. Find the new denominator for your average.
   c. After the splits, the prices settled to the values shown in the following table. Find the indicator's new value.

Stock	Price per Share
A	$54.00
B	90.75
C	53.25
D	11.75
E	23.25

8. For the accompanying data for the retail price of selected appliances, find the Laspeyres retail price index for each year, using 1967 as the base.

Appliance	Average Unit Price 1967	1978	1991	Thousands of Units Sold, 1967
A	255	268	310	5930
B	310	327	323	1950
C	223	250	265	2010
D	37	39	42	890

9. Which of the following index numbers could be found by using the data in question 8?
   a. Simple aggregative index
   b. Laspeyres index
   c. Fisher's ideal index
   d. Weighted aggregative index
   e. Paasche index

10. What is an index number? Give some examples of index numbers. Why are they useful?

11. What is the difference between an index constructed with simple averages and an index constructed with weighted averages? If we construct an index of stock prices in which all the companies are small and approximately the same size, is there any difference between a weighted-average index and a simple-average index? What if we construct an index using companies of many different sizes?

12. What is the difference between the ways the Paasche and Laspeyres price indexes are computed?

13. Under what conditions would you expect Paasche and Laspeyres indexes to be significantly different?

14. What is Fisher's ideal price index? Why might it be better to use than a Paasche or Laspeyres index?

15. One commonly reported index in business is the index of leading economic indicators. What is the purpose of this index? If you were asked to construct your own index of leading economic indicators, what information would you use? How would you construct it?

16. Explain how a price-weighted stock index must be adjusted to reflect the stock split of a company.

17. Some people argue that price indexes do not reflect the improvements in quality of the products we buy. Would this limitation cause estimates of inflation to be too high or too low?

18. What is a base year? What is the value of the index in the base year?

19. What is the consumer price index? What does it measure? How is it constructed?

20. Use the data in Tables 19.5 and 19.6 to compute the Laspeyres index and compare the result to the Paasche index. Also compute Fisher's ideal quantity index.

21. What is a real component? What is a price component? How are the two related?

22. What is the Wilshire 5000 equity index? What securities are included in this index? How is it computed?

23. You are given the following information on prices and quantities for widgets.

Year	Price	Quantity
1985	$1.00	1,000
1986	1.25	1,800
1987	1.31	2,000
1988	1.44	2,222
1989	1.51	2,325
1990	1.85	3,100

Using 1987 as the base year, compute the Laspeyres price index for widgets.

24. Drawing on the data given in question 23, compute the Paasche price index, again using 1987 as the base year.

25. Using your calculations in questions 23 and 24, compute Fisher's ideal price index.

26. On the Island of Crusoe, there are only two goods: coconuts and fish. Suppose you have the following information on the prices and quantities of fish and coconuts.

Year	Fish Q	P	Coconuts Q	P
1987	100	$3.00	75	$1.00
1988	110	2.90	80	1.02
1989	99	3.12	79	1.05
1990	121	3.45	88	1.15

Using 1988 as the base year, compute the Paasche price index.

27. Using the data from question 26, compute the Laspeyres price index.

28. Using the data from question 26, compute Fisher's ideal price index.

29. Using the Laspeyres, Paasche, and Fisher's indexes you computed in questions 26–28, compute the percentage change in price for each year (inflation rate), and compare the results for the three indexes. Which one gives the highest rate? Which one the lowest?

30. Three commonly reported measures of price level are the consumer price index, the producer price index, and the GNP deflator. Explain why these measures may yield different inflation rates.

31. Here are some price indexes for four different types of collectibles.

Collectible	1980	1985	1990
Baseball cards	100	210	275
Paintings	100	195	325
Jewelry	100	250	245
Gold coins	100	199	202

a. What is the base year?
b. Which of the collectibles increased the most in price from 1980 to 1990?
c. Which of the collectibles increased the most in price from 1985 to 1990?
d. Which of the collectibles increased the most in price from 1980 to 1985?

32. Consider the following price and market-value information for five stocks in 1990.

Stock	Price	Market Value
A	$100	$1,000,000
B	50	3,000,000
C	72	500,000
D	35	1,250,000
E	27	300,000

Compute an equally weighted price average and a value-weighted price average. Explain why the two indexes differ.

33. Now suppose you have the following price and market-value information for the same stocks of question 32 in 1991.

Stock	Price	Market Value
A	$105	$1,500,000
B	30	2,000,000
C	72	800,000
D	25	1,850,000
E	57	900,000

Compute an equally weighted and a value-weighted relative price index.

34. Suppose you are given the following information about wages and prices for five years.

Year	Average Hourly Wage	Consumer Price Index
1986	$ 8.22	100
1987	9.37	110
1988	10.01	108
1989	11.27	135
1990	15.43	200

a. Compute the change in real wages between 1986 and 1987.
b. Are workers any better off in 1987 than they were in 1986?

35. Use the data given in question 34 to compute the change in real wages between 1986 and 1990. Are workers any better off?

36. You are given the following cost indexes for three categories of consumer expenditures.

Year	Housing	Food	Transportation
1980	127.2	129.3	151.2
1985	145.6	141.2	170.6
1990	166.7	171.2	200.3

Compute the percentage change in housing, food, and transportation costs between 1980 and 1985. Which expenditure underwent the greatest price increase?

37. Repeat question 36 for the period 1985 through 1990 and the period 1980 through 1990.

38. Agricultural economists have data on the price per bushel of wheat, corn, barley, and hops for the last 20 years. They want to obtain a measure for the aggregative price movements of grain over this period. In deciding how to produce this price index, what factors should they take into consideration?

39. The following table shows the price of hamburger over the last 10 years.

Year	Price per Pound
1	$1.82
2	1.95
3	1.86
4	2.10

Year	Price per Pound
5	2.45
6	3.10
7	2.60
8	2.45
9	2.75
10	2.58

a. Form a price index, using year 1 as the base.
b. Form a price index, using year 6 as the base.

40. The following table shows the price of gold over the last 10 weeks.

Week	Price per Ounce
1	$410.82
2	401.95
3	391.56
4	382.10
5	392.45
6	403.10
7	412.60
8	392.45
9	399.75
10	402.58

a. Form a price index, using week 1 as the base.
b. Form a price index, using week 5 as the base.

41. The following table shows the average price, taken monthly, of Widget Company stock over the last year.

Month	Price	Month	Price
1	$41\frac{3}{8}$	7	$50\frac{1}{4}$
2	$42\frac{1}{4}$	8	$49\frac{3}{4}$
3	$43\frac{1}{8}$	9	$52\frac{7}{8}$
4	$45\frac{3}{4}$	10	$53\frac{5}{8}$
5	$47\frac{1}{2}$	11	$54\frac{1}{8}$
6	$49\frac{3}{8}$	12	$51\frac{5}{8}$

a. Form a price index, using week 1 as the base.
b. Form a price index, using week 7 as the base.

42. The following table shows the price and volume of shares (in thousands) traded for ABC Company and XYZ Company in the first 10 weeks of 1991.

	ABC Company		XYZ Company	
Week	Price	Volume	Price	Volume
1	$12\frac{2}{8}$	3.8	$25\frac{1}{8}$	6.4
2	$11\frac{3}{8}$	2.9	$24\frac{1}{4}$	7.1
3	$13\frac{5}{8}$	3.2	$24\frac{7}{8}$	6.9
4	$12\frac{7}{8}$	3.1	$26\frac{5}{8}$	7.4
5	$13\frac{1}{8}$	3.5	$26\frac{5}{8}$	7.7
6	$14\frac{2}{8}$	4.1	$25\frac{3}{8}$	6.9
7	$13\frac{7}{8}$	3.9	$27\frac{1}{8}$	7.4
8	$14\frac{1}{8}$	4.4	$27\frac{3}{8}$	6.8
9	$14\frac{7}{8}$	4.1	$28\frac{7}{8}$	7.2
10	$13\frac{7}{8}$	4.3	$29\frac{1}{8}$	7.4

Compute the market-value weighted index using week 1 as the base.

43. Use the data given in question 42, and use week 1 as the base, to compute
a. The Laspeyres aggregative quantity index
b. The Paasche aggregative quantity index
c. Fisher's ideal quantity index

44. Redo question 43, this time using week 5 as the base. Compare the results to your results in question 43.

45. The following table shows the price per pound and the volume (in thousands of pounds) for chicken and beef at Eat More Grocery Store.

	Chicken		Beef	
Week	Price	Volume	Price	Volume
1	$1.25	15	$1.86	11
2	1.35	14	1.99	12
3	1.19	17	2.10	11
4	1.45	18	2.05	11
5	1.29	19	1.79	15
6	1.39	22	2.09	18
7	1.45	15	2.29	15
8	1.09	25	2.39	14
9	1.39	19	2.45	13
10	1.49	15	2.89	9

Use week 1 as the base to:

a. Compute the Laspeyres aggregative price index.
b. Compute the Paasche aggregative price index.
c. Compute Fisher's ideal price index.

46. Use the data in question 45, and use week 1 as the base, to compute
   a. The Laspeyres aggregative quantity index
   b. The Paasche aggregative quantity index
   c. Fisher's ideal quantity index

47. Use the week 10 data in question 46 to show that the ratio of the Laspeyres price index to the Laspeyres quantity index is equal to the ratio of the Paasche price index to the Paasche quantity index.

48. Show that the ratio of the Laspeyres price index to the Laspeyres quantity index is equal to the ratio of the Paasche price index to the Paasche quantity index.

49. Briefly explain the differences between aggregative price indexes and aggregative quantity indexes.

50. The following table shows the price and volume (in thousands) for shirts and pants at Snappy Dresser Department Store.

| | *Shirts* | | *Pants* | |
Month	Price	Volume	Price	Volume
1	$27.00	50	$62.00	35
2	23.27	61	54.25	37
3	28.95	49	57.00	38
4	32.45	42	48.27	54
5	22.45	50	49.75	60
6	19.95	75	52.20	55
7	21.23	62	47.50	58
8	27.22	55	45.10	62
9	22.95	57	40.00	75
10	19.90	70	35.00	77
11	24.95	62	44.21	63
12	22.95	66	48.95	64

Compute the value indexes using week 12 as the base.

51. Use the data in question 50, and use week 1 as the base, to compute the value indexes.

52. Redo question 51, this time using month 6 as the base.

53. Using the business section of any newspaper, find the current value of the S&P 500 index. What stocks are included in the S&P 500? How is the index computed? Is the S&P 500 index a better or a worse measure of stock market activity than the Wilshire 5000?

54. Using the business section of any newspaper, find the most recent value of the Dow Jones Industrial Average (DJIA). What stocks are included in the DJIA? Compare the DJIA to the Wilshire 5000 and the S&P 500 as a measure of stock market activity.

55. Explain the problems that would arise from comparing price indexes for computers over the last three decades.

56. The consumer price index (CPI) and the producer price index (PPI) are often reported as measures of the inflation rate. What problems appear using these indexes to measure the inflation rate? Do these two measures really indicate the "true" cost of living?

57. You are given the following information on prices and quantities for Knick-Knacks.

Year	Price	Quantity
1985	$21.00	12,300
1986	18.25	13,000
1987	19.31	21,300
1988	22.44	20,212
1989	23.51	24,345
1990	21.85	32,300

Using 1985 as the base year, compute the value price index for Knick-Knacks.

58. Using the data from question 57, compute the Paasche price index, again using 1985 as the base year.

59. Using your calculations from 57 and 58, compute the Laspeyres index. Why are you getting the same answer as in question 58?

60. Are the Laspeyres quantity index and the Paasche quantity index equal? Why?

61. Suppose you are given the following information about wages and prices for five years.

Year	Annual Salary	Consumer Price Index
1987	$38,202	95
1988	39,837	100
1989	41,001	108
1990	41,327	125
1991	55,943	200

a. Compute the change in real salaries between 1988 and 1989.

b. Are workers any better off in 1989 than they were in 1988?

62. Use the data given in question 61 to compute the change in real wages between 1987 and 1991. Are workers any better off?

63. Use the data presented in the table (right) to calculate the growth rate for nominal GNP, GNP deflator, and real GNP using MINITAB. (Hint: Refer to Equation 19.17.)

Year	Nominal GNP	GNP Deflators	Real GNP
1976	1,676.2	59.3	2,826.7
1977	1,991.1	67.3	2,958.6
1978	2,249.2	72.2	3,115.2
1979	2,509.2	78.6	3,192.4
1980	2,731.3	85.7	3,187.1
1981	3,053.9	94.0	3,248.8
1982	3,166.0	100.0	3,166.0
1983	3,407.0	103.9	3,279.1
1984	3,771.0	107.7	3,501.4
1985	4,013.1	110.9	3,618.7
1986	4,231.0	113.8	3,717.9
1987	4,514.4	117.4	3,845.3
1988	4,872.5	121.3	4,016.9
1989	5,200.7	126.3	4,117.7
1990	5,464.9	131.5	4,155.8

CHAPTER 20

Sampling Surveys: Methods and Applications

CHAPTER OUTLINE

Key Terms

sampling error
nonsampling errors
sample selection bias
response bias
measurement error
nonresponses
self-selection bias

simple random sampling
random number table
finite sample adjustment factor
stratified random sampling
optimal allocation of sample
two-stage cluster sampling
jackknife method

20.1 INTRODUCTION

In statistics we are interested in information about a population. For example, we might be interested in how the residents of a community feel about the construction of a new high school. There are two ways to obtain information about how the residents feel about this issue. We could take a census, and simply ask each and every resident about his or her attitude toward such a project. Or we could take a smaller sample of the residents and try to draw inferences about the community's feelings from the feelings that members of this sample express.

In Part III of this book, we investigated sampling in terms of only simple random sampling, in which each potential sample of N members has an equal chance of being chosen. However, survey sampling is more likely to require elaborate sampling designs for the selection of sample members; these designs are discussed in Sections 20.3 and 20.5. We also focus in this chapter on the problem confronting researchers who want to discover something about a population that is not very large. (In previous chapters, we generally assumed that the number of population members was very large compared with the number of sample members.) And we discuss the advantages and disadvantages of sampling and show how sampling is applied to decision making in business and economics.

As we noted earlier, there are four reasons why sampling may be preferred to taking a census. First, sampling is more economical. Second, it is preferable when information needs to be gathered quickly. A third reason for using a sample instead of a census is that the population of interest may be very large. A fourth reason is quality control, which we discussed in detail in Sections 10.8 and 10.9 of Chapter 10.

This chapter discusses techniques for designing a sampling experiment, and it adds examples and applications to our earlier (Chapter 1) discussion of sampling. First, the basic sampling methods of simple random sampling and stratified sampling are explained. Then we address determining the sample size and discuss sampling and nonsampling errors. Cluster sampling and two-phase sampling are also investigated, and applications of sampling methods are demonstrated. Finally, we compare ratio estimation and regression estimation. Appendix 20A shows how the jackknife method is used to remove the bias of a sample estimator.

20.2 SAMPLING AND NONSAMPLING ERRORS

There are several advantages of sampling over taking a census. However, working with a sample taken from a population does not enable us to determine the precise value of the population parameter, such as population mean or variance. This kind of error is due to sampling error. **Sampling error** is the difference between the sample estimate and its population parameter that is due entirely to the fact that sample instead of census data are used to estimate the parameter. For instance, say we were interested in determining the mean income of lawyers in a particular law firm. Had we used a simple random sample consisting of 50 lawyers, the difference between the mean income of this sample and population mean income would be the sampling error.[1] If by chance our sample consisted entirely of partners in the law firm, our resulting error would be quite large. Now although it is unlikely that a random sample of 50 lawyers from a law firm consisting of 475 associates and 50 partners would result in the selection of only partners, it is theoretically possible. And if it happened, our inferences about the mean income of this law firm would be based

---

[1]We will return to this example in Example 20.3 in the next section of this chapter.

on the income of the partners alone—and hence would be greatly overstated. Parts of the errors in estimating the population parameters that result from "the luck of the draw" are not sampling errors any more.

Errors that are unconnected with the pure random sampling procedure used fall into the category of **nonsampling errors.** Selection of the wrong population is one example of nonsampling error. For example, a researcher who tries to draw inferences about the views of Americans on gun control would fall victim to nonsampling error if he chose to sample only members of the National Rifle Association. This kind of nonsampling error is called **sample selection bias.**

Another source of nonsampling error is **response bias.** Poorly worded questionnaires and improper interviewing techniques may distort the responses of individuals so much that they do not accurately reflect the respondents' true opinions. Furthermore, respondents may have an incentive to distort the truth—say, to exaggerate their incomes if they think their friends will have access the survey or to understate their incomes if they think the IRS will find out. This kind of error is called **measurement error.**

A third possible source of nonsampling error is **nonresponses.** Individuals who choose not to respond to a survey may have very different views from those who do respond. For example, automobile owners who are dissatisfied with their cars may be more likely than satisfied customers to respond to a questionnaire on customer satisfaction. This kind of bias is called **self-selection bias.**

Because nonsampling errors can have a great impact on the results of a survey, the researcher must design the study carefully to minimize these errors. In a study on the obtaining of periodic market information on small business for four large cities, Keon and Assael (1982) found that most of the time, nonsampling errors count for more than 90 percent of the total errors![2]

20.3 *SIMPLE AND STRATIFIED RANDOM SAMPLING*

Designing the Sampling Study

The first step in sampling is to design a study that will yield the information the researcher needs. Designing the study involves determining what questions to address and identifying the population that will make possible achievement of the study's goals. Poor planning may add to the costs of the study or even invalidate the results.

There are six main steps in survey sampling.

1. Determine what information is required for the study.

2. Construct a population list to be used for the sampling survey.

3. Decide what method to use in selecting the sample.

---

[2]J. Keon and H. Assael (1982), "Nonsampling vs. Sampling Errors in Survey Research," *Journal of Marketing,* Spring 1982, 114–123.

4. Determine the sample size.

5. Decide by what method to infer population parameters from sample data.

6. Draw appropriate conclusions from the sample information.

We will discuss the last four of these steps.

Statistical Inferences in Terms of Simple Random Sampling

Once the researcher has determined the questions to be addressed and the population to be studied, the data collection process begins. The issue now is how the sample members should be selected from the population.

The easiest sampling technique is **simple random sampling,** in which each member of a population has an equal and independent chance of being chosen. For example, suppose a population consists of 100 balls, numbered from 1 to 100, that represent households to be surveyed for their annual income. If we were interested in a random sample of 10 balls from this population, we could simply place all 100 balls in a bag, mix the balls thoroughly, and draw 10 of them.

Alternatively, as noted in Section 8.2, we could use a table of random numbers to achieve the same objective more efficiently.

Random Number Tables

Random numbers were discussed in detail in Chapter 8. We now consider **random number tables.** There are several random number tables, but we chose to reproduce, as Table A8 in Appendix A, part of the Rand Corporation table, which has 1 million digits.

Let us illustrate the use of random number tables with an example. Suppose there are 800 students at the Rutgers University School of Business, and we wish to select a random sample of 15 students to estimate their average grade. A list of students is compiled, and each is assigned a serial number from 001 to 800.

Because 800 is a 3-digit number, we should list only 3 digits of the random numbers. The procedure is started at some arbitrary point on Table A8—say, the top of the third column—and the last 3 digits of the random numbers are read off. Thus the first 20 are

769	463	779	850
630	179	596	562
240	238	742	384
610	061	976	951
127	201	033	221

Hence we select the following 15 numbers for our sample.

769	463	779
630	179	596
240	238	742
610	061	033
127	201	562

As you have noticed, only numbers less than 800 were selected. If we needed to choose 50 students, we would continuously choose numbers smaller than 800 until 50 students had been selected. During the selection procedure, we didn't replace any number that had already been chosen. This is known as sampling without replacement. Sampling with replacement, which allows the possibility of an individual being included in the sample more than once, will not be discussed here.

After selecting the random numbers from the table, we could make them *all* usable by subtracting 800 from those greater than 800. Hence, for the random number 976, the $976 - 800 = 176$th student is selected.

Confidence Interval for Population Mean

We will use mean and standard deviation statistics and the central limit theorem (introduced in Chapter 8) to draw inferences about the total population on the basis of information gathered from a random sample. To use the central limit theorem, we must assume that the sample is sufficiently large. However, when the population size N is finite, we use an adjustment factor to obtain the unbiased estimator.

Let $x_1, x_2, \ldots, x_n$ denote the values observed from a simple random sample of size n taken from a population of N numbers with mean μ. Then estimating the population mean involves the following steps:

1. Calculate the sample mean $\bar{x}$.

$$\bar{x} = \frac{1}{n} \sum_{i=1}^{n} x_i \qquad (20.1)$$

$\bar{x}$ is an unbiased estimator of the population mean μ if $E(x_i) = \mu$ for all i.

2. Calculate the variance of $\bar{x}$.

$$s^2 = \frac{1}{n-1} \sum_{i=1}^{n} (x_i - \bar{x})^2 \qquad (20.2)$$

The sample variance s^2 is a biased estimator of the population variance σ^2 when the sample size is finite. Hence the unbiased estimated variance for the sample mean is[3]

$$\hat{\sigma}_{\bar{x}}^2 = \frac{s^2}{n} \cdot \frac{(N-n)}{N-1} \qquad (20.3)$$

where N and n are population size and sample size, respectively. $\frac{N-n}{N-1}$ is called the **finite sample adjustment factor.** $\sigma_{\bar{x}}$ is the standard deviation of $\bar{x}$. If N is large, then the adjustment factor approximately equals 1.

3. Because we have taken a sample of n from the population consisting of N members, we cannot be certain of the true population mean. However, if the sample

---

[3]This result is obtained under the assumption that $\sigma^2 = \sum_{i=1}^{N} (x_i - \mu)^2 / N$. If $\sigma^2 = \sum_{i=1}^{N} (x_i - \mu)^2 / N - 1$, then the finite sample adjustment factor will be $(N - n)/N$. This kind of finite sampling adjustment factor will be used in both stratified random sampling and two-staged cluster sampling.

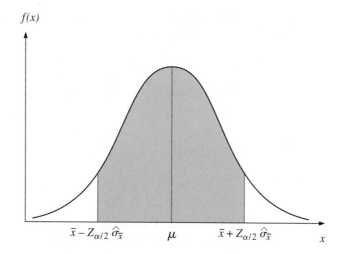

Figure 20.1

Confidence Interval Estimate for Population Mean

is large enough to permit use of the central limit theorem, we can construct a $100(1 - \alpha)$ percent confidence interval for the population mean. The confidence interval will be

$$\bar{x} - Z_{\alpha/2}\hat{\sigma}_{\bar{x}} < \mu < \bar{x} + Z_{\alpha/2}\hat{\sigma}_{\bar{x}} \qquad (20.4)$$

where $Z_{\alpha/2}$ is the number for which,

$$P[Z > Z_{\alpha/2}] = \alpha/2$$

and the random variable Z follows a standard normal distribution.

In other words, Equation 20.4 enables us to construct an interval estimate for the true mean of the population, as illustrated in Figure 20.1.

EXAMPLE 20.1   *Simple Random Sampling to Determine Household Income*

Suppose an investment adviser is trying to decide whether a small retirement community consisting of 1,000 residents represents a promising source of potential clients. To determine the potential business, the investment adviser decides to analyze the size of the residents' investment portfolios. A random sample of 75 residents, who were able to respond anonymously, produces a sample mean of $375,000 with a sample standard deviation of $120,000.

We can use this information to construct a 95 percent confidence interval for the mean value of the investment portfolio.

$$
\begin{aligned}
N &= 1{,}000 \\
n &= 75 \\
\bar{x} &= \$375{,}000 \\
s &= \$120{,}000
\end{aligned}
$$

First, we need to produce an unbiased estimate of the standard deviation of the sample mean from Equation 20.3.

$$\hat{\sigma}_{\bar{x}}^2 = \frac{s^2}{n} \cdot \frac{N - n}{N - 1}$$

$$= \frac{(120{,}000)^2}{75} \cdot \frac{1000 - 75}{1000}$$

$$= 177{,}772{,}800$$

$$\hat{\sigma}_{\bar{x}} = \sqrt{13{,}333.15} = 13{,}327$$

A 95 percent confidence interval can be constructed as follows:

$$\bar{x} - Z_{.025}\hat{\sigma}_{\bar{x}} < \mu < \bar{x} + Z_{.025}\hat{\sigma}_{\bar{x}}$$

The value for $Z_{.025}$ can be found in Table A3 of Appendix A. It is $Z_{.025} = 1.96$, so the 95 percent confidence interval for the mean value of the investment portfolio for this population is

$$375{,}000 - (1.96)(13{,}333.15) < \mu < 375{,}000 + (1.96)(\$13{,}333.15)$$

or

$$\$348{,}867.03 < \mu < \$401{,}132.97$$

Given the information from the sample, we may expect, with 95 percent confidence, that the true population mean μ falls between $348,867.03 and $401,132.97.

Confidence Interval for Population Proportion

The same approach we used to calculate a confidence interval for a random sample with mean x can be applied to sample proportions. Again, we follow these three steps:

1. Compute the sample proportion $\hat{p}$, which is an unbiased estimator of the population proportion p.
2. Compute an unbiased estimator for the variance of the estimator.

$$\hat{\sigma}_{\hat{p}}^2 = \frac{\hat{p}(1 - \hat{p})}{n} \cdot \frac{N - n}{N - 1} \tag{20.5}$$

3. If the sample is large enough, use the central limit theorem to construct a $100(1 - \alpha)$ percent confidence interval.

$$\hat{p} - Z_{\alpha/2}\hat{\sigma}_{\hat{p}} < p < \hat{p} + Z_{\alpha/2}\hat{\sigma}_{\hat{p}} \tag{20.6}$$

Equation 20.6 can be interpreted to mean we may expect, with $100(1 - \alpha)$ percent confidence, that the population proportion p falls within this interval.

EXAMPLE 20.2 *Simple Random Sampling to Determine the Proportion of College-Bound High School Seniors*

Suppose we want to determine the proportion of college-bound high school seniors in a class of 500. A survey of 30 randomly selected students reveals that 19 will be attending college. Given this information, we want to estimate the population proportion p.

$$N = 500$$
$$n = 30$$
$$\hat{p} = \tfrac{19}{30} = .6333$$

To estimate our confidence interval, we need to calculate an unbiased estimate of the variance of the population proportion.

$$
\begin{aligned}
\hat{\sigma}_{\hat{p}}^2 &= \frac{\hat{p}(1 - \hat{p})}{n} \cdot \frac{N - n}{N} \\
&= \frac{.6333(1 - .6333)}{30} \cdot \frac{500 - 30}{500} \\
&= .00729 \\
\hat{\sigma}_{\hat{p}} &= \sqrt{.00729} = .0854
\end{aligned}
$$

The $100(1 - \alpha)$ percent confidence interval in terms of Equation 20.6 is

$$\hat{p} - Z_{\alpha/2}\hat{\sigma}_{\hat{p}} < p < \hat{p} + Z_{\alpha/2}\hat{\sigma}_{\hat{p}}$$

To construct a 90 percent confidence interval, we use the $Z_{.05}$-value given in Table A3 of Appendix A. It is $Z_{.05} = 1.645$, so the 90 percent confidence interval can be given as

$$.6333 - (1.645)(.0854) < p < .6333 + (1.645)(.0854)$$
$$.4928 < p < .7738$$

This implies that we may expect, with 90 percent confidence, that the true proportion of high school seniors who will be attending college falls between 49.28 percent and 77.38 percent.

Stratified Random Sampling

There are times when simple random sampling is not the best sampling method. In some cases, it may be more appropriate to divide the population into groups, or *strata*. **Stratified random sampling** is the selection of independent simple random samples from each stratum of the population.

A stratified random sample may be preferable to a simple random sample when there is reason to believe that different groups within the population have markedly different views. For example, suppose a researcher at Johnson & Johnson is interested in employees' opinions on a child care program. He believes that the views of female employees are important and that their views may be quite different from those of male employees. If the company has a high percentage of male employees, a simple random sample may not guarantee that the sample percentage of female employees is the same as the population percentage. In this instance a stratified random sampling is called for, wherein the population is first divided into male and female subgroups, or strata, and a simple random sample is taken from each stratum.

To conduct a stratified random sampling survey, we begin by dividing the total population of N members into H mutually exclusive and collectively exhaustive groups. Each of these H strata contains its own population consisting of N_1, $N_2, \ldots, N_H$ members. Because the strata are mutually exclusive and collectively exhaustive, we know that

$$N_1 + N_2 + \cdots + N_H = N$$

Our approach in stratified random sampling is to treat each stratum as a separate population, and our sampling survey consists of sampling each stratum separately. Using this technique, we will take a sample of $n_1, n_2, \ldots, n_H$ from each stratum. The samples taken from the strata do not have to be the same size. The total sample taken is the sum of all these samples; that is,

$$n = n_1 + n_2 + \cdots + n_H$$

The techniques for stratified random sampling can be used to

1. Produce an unbiased estimator for the population mean μ.
2. Produce an unbiased estimator for the variance of the sample mean.
3. Construct a $100(1 - \alpha)$ percent confidence interval for the population mean.

To produce an unbiased estimator for the population mean, we take a weighted average of the sample means in the individual strata.

$$\bar{x}_{st} = \sum_{j=1}^{H} W_j \bar{x}_j \tag{20.7}$$

where

$\bar{x}_{st}$ = sample mean for the overall population from stratified sampling

$W_j = \dfrac{N_j}{N}$ = proportion of the jth stratum

$\bar{x}_j$ = sample mean for the jth stratum

Next we need to find an unbiased estimator of the variance of the sample mean. An unbiased estimator of the variance of the sample mean for the jth stratum is

$$\hat{\sigma}_{\bar{x}_j}^2 = \frac{s_j^2}{n_j} \cdot \frac{N_j - n_j}{N_j}$$

where s_j^2 is the sample variance for the jth stratum. Note that this formula is identical to the estimator we used in simple random sampling. Because each stratum is treated as a separate population, the variance for each is calculated in the same way as in simple random sampling.

To find an unbiased estimator of the variance of the estimator μ, we again take a weighted average of the variances of the individual strata.[4]

$$\hat{\sigma}_{\bar{x}_{st}}^2 = \sum_{j=1}^{H} W_j^2 \hat{\sigma}_{\bar{x}_j}^2 \tag{20.8}$$

To construct confidence intervals for the population mean, we again need a sample size large enough for us to assume normality. The confidence interval is

$$\bar{x} - Z_{\alpha/2}\hat{\sigma}_{\bar{x}_{st}} < \mu < \bar{x} + Z_{\alpha/2}\hat{\sigma}_{\bar{x}_{st}}$$

EXAMPLE 20.3 *Stratified Random Sampling to Determine the Mean Income of Lawyers*

Suppose a researcher is interested in determining the mean income of lawyers at a large New York City law firm. The firm consists of 525 lawyers, 475 associates, and 50 partners. Because there are relatively few partners in the population, the researcher believes a simple random sample might understate the earnings of the partners. He decides to undertake a stratified random sample of 75 lawyers: 50 associates and 25 partners. The sample means and standard deviations are

Associates	Partners
$\bar{x}_1 = \$62,750$	$\bar{x}_2 = \$271,860$
$S_1 = \$11,620$	$S_2 = \$\ 80,210$
$n_1 = \ 50$	$n_2 = \ 25$
$N_1 = \ 475$	$N_2 = \ 50$

An unbiased estimator for the population mean income can be calculated as

$$\bar{x} = \sum_{j=1}^{2} W_j \bar{x}_j$$

where

$$W_j = \frac{N_j}{N}$$

$$\bar{x} = \frac{475}{525}(62,750) + \frac{50}{525}(271,860)$$

$$= \$82,665$$

---

[4] $\bar{x}_{st} = \sum_{j=1}^{H} W_j \bar{x}_j$, where $W_j = N_j/N$. Because the samples N_j are selected by random sampling and are independent of each other, Equation 20.8 holds.

To produce an unbiased estimator of the variance of the estimator for μ, we first need to produce unbiased variance estimators for all the strata.

$$\sigma_{\bar{x}_1}^2 = \frac{(11{,}620)^2}{50} \cdot \frac{475 - 50}{475}$$

$$= 2{,}416{,}226$$

$$\sigma_{\bar{x}_2}^2 = \frac{(80{,}210)^2}{25} \cdot \frac{50 - 25}{50}$$

$$= 128{,}672{,}882$$

$$\hat{\sigma}_{\bar{x}}^2 = \sum_{j=1}^{2} W_j^2 \hat{\sigma}_{\bar{x}_j}^2$$

$$= \left(\frac{475}{525}\right)^2 (2{,}416{,}226) + \left(\frac{50}{525}\right)^2 (128{,}672{,}882)$$

$$= 3{,}145{,}009$$

$$\sigma_{\bar{x}} = \sqrt{2{,}145{,}009} = 1{,}773$$

For a 95 percent confidence interval, we use the $Z_{.025}$-value from Table A3 in Appendix A. It is $Z_{.025} = 1.96$, so the 95 percent confidence interval for the income of lawyers at this law firm is

$$82{,}665 - (1.96)(1{,}773) < \mu < 82{,}665 + (1.96)(1{,}773)$$
$$\$79{,}190 < \mu < \$86{,}140$$

We can say with 95 percent confidence that the population mean μ falls between $79,190 and $86,140.

20.4   DETERMINING THE SAMPLE SIZE

We have discussed the advantages of sampling overtaking a census, and we have looked at two important sampling methods. One fundamental question remains unanswered: How large should the sample be?

Sample Size for Simple Random Sampling

For simple random sampling, the sample size n can be found via Equation 20.3, the formula for finding the variance of the estimator $\bar{x}$.

$$\sigma_{\bar{x}}^2 = \frac{\sigma^2}{n} \cdot \frac{N - n}{(N - 1)}$$

By solving for n, we determine the sample size. We then multiply both sides of the equation by $(N - 1)n$.

$$(N - 1)n\sigma_{\bar{x}}^2 = N\sigma^2 - n\sigma^2$$

Next we add $n\sigma^2$ to both sides of the equation and rearrange.

$$[(N - 1)\sigma_{\bar{x}}^2 + \sigma^2]n = N\sigma^2$$

Dividing both sides by the bracketed term yields

$$n = \frac{N\sigma^2}{(N - 1)\sigma_{\bar{x}}^2 + \sigma^2} \qquad (20.9)$$

If the population variance σ^2 is known, Equation 20.9 can be used to determine the sample size necessary to achieve any specified value for the level of precision $\bar{x}$, $\sigma_{\bar{x}}^2$. Equation 20.9 makes apparent the inverse relationship between $\sigma_{\bar{x}}^2$ and n; that is, the smaller the variance of the estimator that we desire, the larger our sample needs to be.

EXAMPLE 20.4 *Sample Size for Accounts Receivable, Case 1*

Crow Company's accountant decides that the best way to determine the company's mean accounts receivable is to take a simple random sample of the 1,025 accounts. Assume that the population variance σ^2 is \$2,425. What size sample should the accountant take if she would like to have a level of precision, as measured by $\sigma_{\bar{x}}^2$, of \$75?

$$n = \frac{N\sigma^2}{(N - 1)\sigma_{\bar{x}}^2 + \sigma^2}$$

$$= \frac{1025(2,425)}{(1025 - 1)(75) + 2,425} = 31.37$$

That is, a simple random sample of 32 accounts receivable will produce the desired result. Note that we rounded the sample size to the nearest whole number.

EXAMPLE 20.5 *Sample Size for Accounts Receivable, Case 2*

Use the information from Example 20.4, but assume the accountant would like $\sigma_{\bar{x}}^2 = \$50$.

$$\frac{1,025(2,425)}{(1,025 - 1)(50) + 2,425} = 46.35$$

The accountant must increase the sample size to 47 if she wishes to reduce $\sigma_{\bar{x}}^2$ to \$50 or—what is the same thing—to improve her precision from \$75 to \$50.

EXAMPLE 20.6   *Sample Size for Accounts Receivable, Case 3*

Suppose our accountant is not sure what value for $\sigma_{\bar{x}}^2$ would be appropriate. However, she would like to produce a sample in which the 95 percent confidence interval extends $10 on each side of the sample mean. She can do this by noting that $Z_{\alpha/2} \cdot \sigma_{\bar{x}}$ = length of the confidence interval on each side of the sample mean. Thus, for a 95 percent confidence interval extending $10 on each side of the mean, $1.96\sigma_{\bar{x}} = \$10$. Solving for $\sigma_{\bar{x}}$, we get

$$\sigma_{\bar{x}} = \frac{10}{1.96} = 5.10$$

$$\sigma_{\bar{x}}^2 = (5.10)^2 = 26.01$$

so

$$n = \frac{1025(2,425)}{(1025 - 1)(26.01) + 2,425}$$

$$= 85.54$$

In this instance, the accountant randomly samples 86 accounts receivable.

If we define the absolute value of the difference between sample mean $\bar{x}$ and population mean μ ($d = |\bar{x} - \mu|$) as a precision measure, then it can be shown that the relationship among precision, the level of reliability (the Z-value of a normal distribution), and standard error $\sigma_{\bar{x}}$ is $d = z\sigma_{\bar{x}}$. From this relationship, it can be shown that the sample size can be defined as

$$n = \frac{(z\sigma)^2}{d^2} \tag{20.10a}$$

$$n = \frac{N(z\sigma)^2}{Nd^2 + (z\sigma)^2} \tag{20.10b}$$

where Equations 20.10a and 20.10b are for sampling with replacement and without replacement, respectively.

Using Equation 20.10b, we can calculate the sample size for Example 20.6 as

$$n = \frac{(1025)(1.96)^2(2425)}{(1025)(10)^2 + (1.96)^2(2425)} = 85.40$$

The sample size calculated from Equation 20.10b is almost identical to that of Example 20.6.

Now let's consider simple random sampling for estimating the population proportion p. Let $\hat{p}$ be the random variable that represents the sample proportion. Then, from Equation 20.5, we can solve this equation for sample size. We obtain

$$n = \frac{N\hat{p}(1 - \hat{p})}{(N - 1)\sigma_p^2 + \hat{p}(1 - \hat{p})} \tag{20.11}$$

The sample size obtained from Equation 20.11 does not connect with the desired degree of precision. To directly relate the desired precision with the estimate of sample size, we first let d represent a "margin of error" in estimating sample proportion $\hat{P}$. Then we also let α represent the risk that actual error is larger than d. In other words,

$$P(|\hat{p} - p| \geq d) = \alpha$$

The formula for a sample size that uses the information of d is derived as follows. First, we let

$$d^2 = z^2 \frac{\hat{p}(1 - \hat{p})}{n} \cdot \frac{N - n}{N - 1}$$

where z is the abscissa of the normal curve that cuts off an area α at the tails. Solving for n yields

$$n = \frac{Nz^2\hat{p}(1 - \hat{p})}{(N - 1)d^2 + z^2\hat{p}(1 - \hat{p})} \qquad (20.12)$$

Both Equations 20.11 and 20.12 involve the unknown population proportion P whose estimation is the objective of the study. A conservative estimate of n is obtained by choosing for P the value nearest to $\frac{1}{2}$ in the range in which P is thought likely to lie.

EXAMPLE 20.7   *Nielson Survey About the Evening News on NBC*

Suppose the Nielson organization is planning to make a simple random sampling to estimate what percentage of American TV viewers watch the 6:30 evening news on NBC. What is the sample size needed for $d = .04$?

The potential number of TV watchers, N, is very large, so Equation 20.11 can be approximated by

$$n' = \frac{[z^2\hat{p}(1 - \hat{p})]/d^2}{1 + 1/N\left(\dfrac{z^2\hat{p}(1 - \hat{p})}{d^2} - 1\right)} \doteq \frac{z^2\hat{p}(1 - \hat{p})}{d^2} \qquad (20.12')$$

Substituting $d = .04$, $\hat{p} = .5$, and $z = 1.96$ into Equation 20.12', we obtain

$$n' = \frac{(1.96)(.5)(.5)}{(.04)^2} = 306.25$$

A simple random sample of 307 will suffice.

Sample Size for Stratified Random Sampling

We can also derive a formula for the sample size needed in stratified random sampling. As in the case of simple random sampling, our required sample size depends on the variance of the population and the desired level of precision. However, the size of the sample in stratified random sampling also depends on one other factor: how we allocate the total sample among the strata. There are two possible approaches.

1. *Proportional allocation.* In cases where the sampling will be distributed proportionally, the proportions are determined by the relative sizes of the strata. For example, if the total population is 1,000 and the total population of the first stratum is 400, 40 percent of the total sample is assigned to the first stratum.

$$n_j = \frac{N_j}{N} n$$

For a proportional allocation from a stratified sample, the sample size n can be determined by the formula[5]

$$n = \frac{\sum\limits_{j=1}^{H} N_j s_j^2}{N \sigma_{\bar{x}_{st}}^2 + \frac{1}{N} \sum\limits_{j=1}^{H} N_j s_j^2} \tag{20.13}$$

where $\sigma_{\bar{x}_{st}}^2$ is the desired variance and s_j^2 is the sample variance in the *j*th stratum. From Equation 20.13, we can see that there is an inverse relationship between the sample size n and the degree of precision we desire, as measured by $\sigma_{\bar{x}}^2$.

2. *Optimal allocation with similar variable cost in each stratum.* Sometimes, the sample size for each stratum is dictated not by the relative size of the strata but by the allocation that yields the most precise estimates in the sense that standard errors of point estimates are minimal. In other words, in a sampling survey we generally have allocated a fixed budget. This fixed budget includes fixed costs and variable costs that are similar for each stratum. Given a fixed budget, C, we want to select a sample of size n among different strata in such a way as to minimize the variance of the sample estimate. This kind of allocation of sample size is called **optimal allocation of sample.**

The optimal proportion of the sample that should be given to the *j*th stratum is

$$n_j = \frac{N_j s_j}{\sum\limits_{j=1}^{H} N_i s_j} \cdot n \tag{20.14}$$

where s_j is the sample standard deviation for the *j*th stratum. For an optimal allocation of a stratified sample, the total sample n is given by the formula

---

[5]The derivation of the sample size for proportional allocation defined in Equation 20.13 and that of the sample size for optimal allocation with similar variable cost in each stratum, defined in Equation 20.15, can be found in T. Yamane (1967), *Elementary Sampling Theory,* (Englewood Cliffs, N.J.: Prentice Hall), Chapter 6.

$$n = \frac{1/N \left(\sum_{j=1}^{H} N_j s_j \right)^2}{N \sigma_{\bar{x}_{st}}^2 + 1/N \sum_{j=1}^{H} N_j s_j^2} \tag{20.15}$$

where $\sigma_{\bar{x}_{st}}^2$ is the desired variance for the sample mean.

EXAMPLE 20.8 *Sample Size for Stratified Random Sampling*

Let's return to our hypothetical New York City law firm and use the information given in Example 20.3 to determine the sample size when the desired sample standard deviation, $\sigma_{\bar{x}_{st}}$, is $1,900.

Substituting $N_1 = 475$, $N_2 = 50$, $N = 525$, and $\sigma_{\bar{x}_{st}} = 1,900$ into Equation 20.13, we can determine the sample size for a proportional allocation as

$$n = \frac{475(11,620)^2 + 50(80,210)^2}{525(1,900)^2 + \dfrac{1}{525}[475(11,620)^2 + 50(80,210)^2]}$$

$$= 146.69$$

For the degree of precision we desire, we should sample 129 of the lawyers in the firm if we plan to use a proportional allocation in our stratified sampling.

For an optimal allocation, the sample size is

$$n = \frac{\dfrac{1}{525}[475(11,620) + 50(80,210)]^2}{525(1,900)^2 + \dfrac{1}{525}[475(11,620)^2 + 50(80,210)^2]}$$

$$= 65.77$$

Obviously, using an optimal allocation in our stratified random sampling greatly reduces the size of our sample—from 147 to 66.

20.5 TWO-STAGE CLUSTER SAMPLING

When a researcher is interested in surveying a population that is dispersed throughout a large geographic region, neither simple nor stratified random sampling may be the best method for constructing the survey. Although a simple or stratified random sample still produces good estimates for determining population parameters, the expense of sampling across a large geographic region often dictates the need for alternative sampling techniques. Under these conditions, and also the cost when there are no reliable elements in the population to construct a sampling list, cluster sampling is preferred. This is because **two-stage cluster sampling** treats each cluster as a sampling unit.

In cluster sampling, we divide the population into clusters—geographically compact units such as congressional districts at the state level or political wards within a city. After dividing our population into clusters, we take a simple random sample of clusters and conduct a census in each of the sampled clusters. In other words, every individual in each of the sampled clusters is contacted. The advantage of cluster sampling over simple or stratified random sampling should be obvious. In conducting a census on a random sample of clusters, we can greatly reduce our costs by sampling in geographically compact areas. However, it should be noted that cluster sampling increases the sampling variance.

The technique used for cluster sampling parallels the approaches used in other sampling methods. To conduct a survey using the two-stage cluster sampling approach, we take the following steps.[6]

1. Divide the population into M clusters. For example, New York City might be divided into M voting districts.

2. Take a simple random sample of m sample clusters. Then take a simple random sample from each cluster. In the first stage, a random sample of m voting districts is selected. In other words, instead of selecting families one at a time, we have selected m groups of families, and, in our present case, each group of families lives in the same voting district. Then, in the second stage, random samples of $n_1, n_2, \ldots, n_m$ families are selected from each of the m districts' mth population observations, $N_1, N_2, \ldots, N_m$. Thus our sample size $n = n_1 + n_2 + \cdots + n_m$.

3. Compute an unbiased estimator of the population total.

$$\hat{X} = \frac{M}{m} \sum_{i=1}^{m} \frac{N_i}{n_i} \sum_{j=1}^{n_i} x_{ij}$$

$$= \frac{M}{m} \sum_{i=1}^{m} \hat{X}_i$$

(20.16)

where

x_{ij} = the observation in the jth sample with sample size n_i and the ith cluster with sample size m

m = number of sampled clusters

N_i = number of population members in cluster i

M_i = number of population clusters

4. By dividing N into Equation 20.16, obtain the estimated population mean $\hat{\mu}$ as

$$\hat{\mu} = \frac{\sum_{i=1}^{m} N_i \bar{x}_i}{(\bar{N})(m)}$$

(20.17)

[6]See T. Yamane (1967), *Elementary Sampling Theory*, Chapter 8.

where

$$\bar{x}_i = \frac{\sum_{j=1}^{n_i} x_{ij}}{n_i}$$

$$\bar{N} = \frac{N}{M}$$

It can be shown that the variance associated with $\hat{\mu}$ is

$$\sigma_{\hat{\mu}}^2 = \text{Var}(\hat{\mu}) = \frac{(M-m)}{Mm\bar{N}^2} \frac{\sum_{i=1}^{m} n_i^2(\bar{x}_i - \bar{x})^2}{m-1}$$

$$+ \frac{1}{Mm\bar{N}^2} \sum_{i=1}^{m} N_i^2 \frac{(N_i - n_i)}{N_i} \frac{\sum_{j=1}^{n_i}(x_{ij} - \bar{x}_i)^2}{(n_i - 1)} \qquad (20.18)$$

where

$$\bar{x} = \frac{\sum_{i=1}^{m} \bar{x}_i}{m}$$

If $n_i = N_i$, then Equation 20.18 reduces to

$$\sigma_{\hat{\mu}}^2 = \text{Var}(\hat{\mu}) = \frac{(M-m)}{Mm\bar{N}^2} \frac{\sum_{i=1}^{m} N_i^2(\bar{x}_i - \bar{x})^2}{(m-1)} \qquad (20.19)$$

If $\bar{N}$ is not available, we can use $\bar{n} = \sum_{i=1}^{m} n_i/m$ to substitute for $\bar{N}$ in both Equations 20.17 and 20.19.

5. Again, if the sample size is large, construct a $100(1 - \alpha)$ percent confidence interval.

$$\hat{\mu} - Z_{\alpha/2}\sigma_{\hat{\mu}} < \mu < \hat{\mu} + Z_{\alpha/2}\sigma_{\hat{\mu}} \qquad (20.20)$$

EXAMPLE 20.9 *Population Mean Estimate for Average Family Income*

A simple random sample of 25 voting districts is taken from a city with a total of 300 voting districts. Each family in the sample voting districts is surveyed to obtain information about family income. The sample data are listed in Table 20.1.

Table 20.1 Sample Family Income Data

Sample Voting District, i	Mean Income (thousands of dollars), $\bar{x}_i$	Number of Families, N_i	(A) $N_i\bar{x}_i$	(B) $N_i^2(\bar{x}_i - \bar{x})^2$
1	30.62	30	918.6	89956.80518
2	28.96	25	724	84937.2736
3	21.56	28	603.68	284742.6203
4	25.18	32	805.76	244039.1616
5	33.56	27	906.12	36311.28424
6	26.89	39	1048.71	286627.8896
7	24.56	40	982.4	412554.4284
8	29.67	38	1127.46	173063.3216
9	40.12	31	1243.72	237.9491353
10	53.16	33	1754.28	171312.5477
11	42.56	35	1489.6	4621.824256
12	56.37	37	2085.69	339701.0667
13	29.45	29	854.05	104885.5586
14	50.66	40	2026.4	161359.6764
15	48.29	45	2173.05	119203.0865
16	42.39	43	1822.77	5808.451854
17	56.17	37	2078.29	331129.8125
18	45.89	29	1330.81	23378.28768
19	53.84	27	1453.68	127452.4272
20	49.26	28	1379.28	58557.80496
21	39.45	32	1262.4	1396.008714
22	42.57	45	1915.65	7719.028164
23	51.38	39	2003.82	176176.2949
24	55.66	29	1614.14	190296.2639
25	31.59	42	1326.78	143761.6989
Sum	1009.8	860	34931.14	3579230.573

$$\bar{x} = 40.617604$$

From the data of Table 20.1, we have $m = 25$ and $M = 300$. The total number of families in the sample is

$$\sum_{i=1}^{m} N_i = 30 + 25 + \cdots + 42 = 860$$

To obtain estimated average family income, we also need

$$\sum_{i=1}^{m} N_i\bar{x}_i = 918.6 + 724.0 + \cdots + 1{,}326.87 = 34{,}931.14$$

Substituting these figures into Equation 20.17, we obtain

$$\hat{\mu} = \frac{\sum\limits_{i=1}^{m} N_i \bar{x}_i}{Nm} = \frac{34{,}931.14}{860} = 40.6176$$

On the basis of the sample data, we estimate that annual family income is $40,617.6.

In order to obtain an interval estimate, we need

$$\bar{N} = \frac{\sum\limits_{i=1}^{m} N_i}{m} = \frac{860}{25} = 34.4$$

Also,

$$\frac{\sum\limits_{i=1}^{m} N_i^2 (\bar{x}_i - \bar{x})^2}{m - 1} = \frac{(30)^2(30.62 - 40.62)^2 + \cdots + (31.59)^2(31.59 - 40.62)^2}{24}$$

$$= 149{,}134.607$$

so

$$\mathrm{Var}(\hat{\mu}) = \frac{(275)(149{,}134.607)}{(300)(25)(34.4)^2}$$

$$= 4.6210$$

Taking square roots, we obtain $\sigma_{\hat{\mu}} = 2.1497$. For a 95 percent confidence interval, $z_{\alpha/2} = z_{.025} = 1.96$, so a 95 percent confidence interval for the population mean is

$$40.6176 - (1.96)(2.1497) < \mu < 40.6176 + (1.96)(2.1497)$$

or

$$36.4042 < \mu < 44.8310$$

20.6 RATIO ESTIMATES VERSUS REGRESSION ESTIMATES

So far in this chapter, we have been interested in making inferences about a population that are based on samples from that population. In this section we consider two methods—the ratio method and the regression method—that can be used in conjunction with sampling to improve the parameter estimates based on sample data. (We discuss only the simple random sampling case, but both methods can also be used with stratified random sampling.) Note that the ratio method, which is easier than the regression method, is often used in sampling surveys.

Ratio Method

In the ratio method an auxiliary variate y_i, which is correlated with x_i, is obtained for each unit in the sample. The population total Y of the y_i must be known. The value of y_i can be highly correlated with x_i (as are, say, sales and earnings), or y_i can be the value of x_i at some previous time when a complete census was taken.

To use the ratio method to estimate X, the population total of the x_i, we need to know Y, the population total of the y_i. The estimate of X, $\hat{X}_r$, is

$$\hat{X}_r = \frac{x}{y} Y = \frac{\bar{x}}{\bar{y}} Y \qquad (20.21)$$

where

y = sample total of y_i
x = sample total of x_i
$\bar{y}$ = sample mean of y_i = y/n
$\bar{x}$ = sample mean of x_i = x/n
Y = population total of y_i

If y_i is the value of x_i at some previous period, in the ratio method we use the sample to estimate the relative change during the time interval.

EXAMPLE 20.10   *Ratio Estimate of Sales Prediction*

Suppose a sales manager at Bono Corporation is interested in estimating the 1991 total sales at the company's 100 stores. To do this, he collects information for 1990 and 1991 sales from a simple random sample of 30 stores. The results of this sampling are presented in Table 20.2.

An estimate of 1991 sales can be produced in two ways. To use the ratio method, the sales manager needs to know overall Y (total sales in 1990). If 1990 total sales were \$14.3 million, then the ratio method yields the following prediction of 1991 sales.

$$\hat{X}_r = (\bar{x}/\bar{y})Y$$

$$= \frac{159.28}{149.70} (\$14,300,000)$$

$$= \$15,215,124$$

The second approach to estimating X is to use the sample mean per store to get the population total.

$$\hat{X}_r = N\bar{x}$$

$$= 100(159.28) = \$15,928,000.$$

This method doesn't utilize the 1990 sample information; hence it is not as precise as the estimate that uses the 1990 sample information. This method, however, can be useful when 1990 sample information is not available.

Table 20.2 Sales for 1990 and 1991 (thousands of dollars)

Store	1990 Sales	1991 Sales
1	100.2	107.4
2	74.3	82.5
3	88.6	75.6
4	210.7	223.4
5	109.5	125.6
6	110.6	111.5
7	62.4	53.5
8	88.3	89.6
9	237.6	245.3
10	196.4	188.7
11	147.6	168.9
12	185.6	200.7
13	95.7	95.8
14	100.3	98.6
15	127.6	135.4
16	130.2	170.6
17	210.3	221.4
18	213.6	262.1
19	220.5	275.3
20	250.6	248.5
21	275.8	300.3
22	125.6	121.2
23	130.5	131.7
24	180.7	191.8
25	89.3	94.2
26	75.6	78.3
27	185.5	191.2
28	184.3	190.2
29	150.6	165.3
30	132.4	133.7
Mean	149.6966	159.2766
Ratio	1.063996	

Regression Method

A second approach to increasing the precision of estimates of population parameters based on sampling is the regression method. As we saw in Chapters 13 and 14, simple linear regression enables us to relate two variables that are correlated with one another.

Suppose we are interested in $\overline{X}$, the population mean of X. To produce an estimate of $\overline{X}$, we use our knowledge of the fact that an auxiliary variable y_i is correlated with x_i. Again, we can employ simple random samples of y_i and x_i to produce

$\bar{y}$ and $\bar{x}$, the sample means of y and x. In the regression method, the estimate of the population mean $\overline{X}$ is

$$\bar{x}_{lr} = \bar{x} + b(\overline{Y} - \bar{y}) \tag{20.22}$$

where

$\bar{y}$ = sample mean of Y
$\overline{Y}$ = population mean of Y (known)

EXAMPLE 20.11 *Regression Estimate of Sales Prediction*

To illustrate the regression method, we will use the data given in Table 20.2. A simple linear regression of y_i (sales in 1991) and x_i (sales in 1990) is run. The results are

$$x_i = -7.45 + 1.11y_i$$

To use the regression method, we simply substitute the slope estimate $b = 1.11$ and $\overline{Y} = 14,300,000/100 = 143,000$ into Equation 20.22.

$$\bar{x}_{lr} = \bar{x} + b(\overline{Y} - \bar{y})$$
$$= 159.28 + 1.11(143 - 149.70)$$
$$= 151.84$$

Again, we could produce an estimate of the population total as

$$\hat{X} = N\bar{x}_{lr}$$
$$= 100(151.84)$$
$$= \$15,184,000$$

Comparison of the Ratio and Regression Methods

You may have noticed some similarities between the regression method and the ratio method. Both methods use an auxiliary variate y_i, which is correlated with x_i. In fact, the ratio method is a special case of the regression method. These two methods are identical when the regression line passes through the origin.

EXAMPLE 20.12 *Labor Force Sampling*

One important economic application of sampling is in determining structural changes in the labor force. The federal government conducts a census at the start of each decade. However, the Commerce Department's Bureau of the Census uses sampling to update census figures continually and to provide economists and other policy makers with labor force statistics such as the unemployment rate, employee wages, and the age, sex, and race of the labor force. The issue is how best to produce the labor force estimates.

The most widely used survey on the structure of the labor force is the current population survey (CPS). The CPS is a monthly survey that deals primarily with labor force data for the noninstitutional civilian population. Questions related to labor force participation are asked of each member in every sample household. In addition, supplementary questions regarding monetary income and work experience for the previous year are asked every March.

The present CPS sample was selected from the 1980 census files and consists of 60,000 occupied households. All 50 states and the District of Columbia are represented in the current CPS sample's 729 areas, which include 1,973 counties, independent cities, and minor civil divisions.

The estimates that the samples yield of the total noninstitutional civilian population of the United States by age, sex, and race are used to update census information. Through stratified sampling or two-stage cluster sampling and a technique such as the ratio method or the regression method, the Bureau of the Census is able to provide monthly estimates about the labor force.

20.7 BUSINESS AND ECONOMIC APPLICATIONS

APPLICATION 20.1 Sampling in an IRS Audit

We have discussed several ways of taking samples from a population. In this section we apply these techniques to the accounting problem of auditing accounts receivable.

Suppose an Internal Revenue Service auditor is interested in determining whether the number of accounts receivable reported on Leclair Company's tax return is correct. Because Leclair has a total of 1,675 accounts receivable, the auditor has decided to sample the accounts receivable rather than to conduct a census.

Using a simple random sampling of 100 accounts receivable, the auditor finds a sample mean of $127.84 and a sample standard deviation of $42.62. Leclair has reported that the mean value of its accounts receivable is $94.25. Should the auditor suspect that the company is underreporting its accounts receivable?

Because the auditor has information only on the sample mean, he decides to construct a 95 percent confidence interval for the mean value of accounts receivable. If the mean value of accounts receivable reported by Leclair Company falls outside this interval, the auditor will consider this reason to investigate the company's tax return further.

To construct a 95 percent confidence interval for the mean μ, the auditor uses $N = 1,675$, $n = 100$, $\bar{x} = \$127.84$, and $s = \$42.62$. He needs unbiased estimators for the population mean and the variance of $\bar{x}$. The sample mean $\bar{x}$ can be used as an unbiased estimator of the mean μ, and $\hat{\sigma}_{\bar{x}}^2$ can be calculated as follows:

$$
\begin{aligned}
\hat{\sigma}_{\bar{x}}^2 &= \frac{s^2}{n} \cdot \frac{N - n}{N - 1} \\
&= \frac{(42.62)^2}{100} \cdot \frac{1{,}675 - 100}{1{,}674} \\
&= 17.09 \\
\hat{\sigma}_{\bar{x}} &= \sqrt{17.09} = 4.13
\end{aligned}
$$

The 95 percent confidence interval is

$$
\bar{x} - Z_{\alpha/2}\hat{\sigma}_{\bar{x}} < \mu < \bar{x} + Z_{\alpha/2}\hat{\sigma}_{\bar{x}}
$$

The value for $Z_{\alpha/2}$ can be found in Table A3 in Appendix A. It is $Z_{\alpha/2} = Z_{.025} = 1.96$, so

$$
127.84 - (1.96)(4.13) < \mu < 127.84 + (1.96)(4.13)
$$
$$
119.74 < \mu < 135.93
$$

Because the mean value of accounts receivable falls outside the 95 percent confidence interval, the auditor's suspicions are aroused, prompting an investigation of Leclair Company's tax returns.

■ | **APPLICATION 20.2** Sampling Survey for 1977 Generic Drug Substitution Law in Wisconsin[7]

In 1977 the state of Wisconsin passed a law that permitted the substitution of generic drugs for brand-name drugs when prescriptions were being filled. Consumers had simply to request the substitution, and the pharmacist was legally bound to make it. The legislation was designed to save customers money on their prescriptions. Thus it proved disconcerting to the Department of Health and Social Services when, some two and a half years after its passage, few customers were taking advantage of the law by asking for generic drugs.

Several hypotheses were advanced to explain why this was happening, including suggestions that the law had not been well publicized, that consumers were not aware of its existence, that consumers had unfavorable attitudes toward generic drugs, and that a number of personal and situational factors (such as age, income, household size, and education) were affecting customers' use of the law. It was decided that the best way to collect the needed information was through self-administered questionnaires. Because of the difficulty of obtaining an accurate mailing list, the questionnaires were to be delivered by hand but returned by mail. Further,

---

[7]This application is drawn from G. A. Churchill, Jr. (1983), *Marketing Research: Methodological Foundations*, 3d ed., (Chicago: Dryden), pp. 441–442. Copyright © 1983 by The Dryden Press, reprinted by permission of the publisher.

investigators decided to determine the feasibility of the data collection and sampling plans by initially confining the study to Madison, the state capital, and to the main campus of the University of Wisconsin. They realized, of course, that restricting the original investigation in this manner would probably introduce some bias into the results because of a number of demographic differences between Madison and the remainder of the state.

The 1,000 households to be surveyed were selected in the following manner. First, detailed maps were used to divide the city into aldermanic districts. To ensure even geographic representation, samples were drawn from each aldermanic district. This was done by randomly choosing city blocks within each aldermanic district and then randomly selecting 10 households in each of the selected blocks to receive questionnaires.

Consider aldermanic district 11, for example, which was located on the city's west side. The study was to be limited to residents of Madison who were over 18. The total adult population in district 11 was 5,115, and the total adult population in the city was 122,016. The proportion of the total sample that was to come from aldermanic district 11 was thus 5,115/122,016 = .0419. This meant that 42 households [1,000(.0419) = 42] were to be included in the district 11 sample.

The 42 households were selected from 5 blocks within the district as follows: All blocks within the district were numbered. Then 5 blocks were randomly selected from this larger set of blocks. On each of the first 4 blocks that were selected, 10 households were interviewed, and 2 households were interviewed on the fifth block. The households were selected by first going around the block to count the number of dwelling units. Each field worker was to begin the count at the southwest corner, following the detailed instructions that were provided. Suppose, for example, that there were 50 dwelling units in a selected block. The field worker was then instructed to generate a random start between 1 and 5, using the table of random numbers each carried. If the number was, say, 2, the field worker was to drop off questionnaires at the second, seventh, and twelfth households, and so on, in the initial numbering scheme.

Summary

In this chapter, we explained why sampling surveys are needed for analyzing business and economic data. Then we looked at three sampling methods: simple random sampling, stratified random sampling, and two-stage cluster sampling. Next we investigated how ratio and regression methods are used to estimate the total value and the mean value of a population on the basis of sample information. Finally, we explored some applications of sampling surveys.

Appendix 20A The Jackknife Method for Removing Bias from a Sample Estimate

In this appendix we discuss the jackknife method, which can be used in conjunction with sampling to remove the bias of an estimator and to produce confidence intervals.

The jackknife is a general technique that can be applied to any linear estimator. It works by using the original sample to create a new set of "pseudovalues." The jackknife procedure involves the following steps.

1. The n sample values are divided into m subsets, and m is set equal to n in many applications. For example, removing one piece of data at a time leaves $m = n$ subsets of data with $(n - 1)$ observations in each set.

2. An estimate based on all the data is calculated. Call this value x_{All}.

3. An estimate based on all the data except the data from the first of the m subsets is calculated; call it x_{-1}. Estimates of $x_{-2}, x_{-3}, \ldots, x_{-m}$ are also calculated.

4. The "pseudovalue" x_1 is calculated as

$$x_1 = x_{All} + (m - 1)(x_{All} - x_{-1}) \qquad (20A.1)$$

Likewise, we can calculate $x_2, x_3, \ldots, x_m$. These pseudovalues will constitute a "pseudosample" that acts like a random sample. Alternatively, Equation 20A.1 can be rewritten as

$$x_1 = mx_{All} - (m - 1)x_{-1} \qquad (20A.2)$$

5. The mean $\bar{x}$ and the standard deviation s of the pseudosample can now be calculated and used to produce confidence intervals. For example, a 90 percent confidence interval for the population mean μ can be defined as

$$\mu = \bar{x} \pm t_{.05} \frac{s}{\sqrt{m}} \qquad (20A.3)$$

where $t_{.05}$ is the t statistic with the significance level $\alpha = .05$.

It may not be clear from an introduction of the jackknife technique why this procedure is preferable to a simple or a stratified random sample. It has been shown that when the original estimate is biased but is asymptotically unbiased—that is, unbiased in large samples—jackknifing often eliminates the bias. Also, the jackknife procedure makes it possible to compute confidence intervals for the population parameters when the samples taken are small and the population standard deviation is unknown.

EXAMPLE 20A.1 *Removing the Bias of Accounts Receivable Estimates*

Suppose an auditor is interested in determining the mean growth rate of uncollectible accounts receivable. This figure will help the auditor find out whether a store has an abnormally high increase in uncollectibles.

To obtain these estimates, the auditor randomly samples the uncollectibles of 6 department stores in 1989 and 1990. The results of this sample are given in Table 20A.1. The auditor is interested in constructing a 95 percent confidence interval for the ratio in uncollectibles. In Table 20A.2, we present the ratio in uncollectibles, u, where

$$u = \frac{1990 \text{ uncollectibles}}{1989 \text{ uncollectibles}}$$

The issue now before us is how we compute a confidence interval for u by using the jackknife method.

The first step in the jackknife procedure is to calculate x_{All}, the observation based on all the data. From our example, this is the ratio of total uncollectibles in 1990 to total uncollectibles in 1989.

$$x_{All} = \frac{914,500}{825,000} = 1.108$$

Table 20A.1 Random Sample of Uncollectibles for 6 Stores

Store	1989 Uncollectibles	1990 Uncollectibles
AAA	$200,000	$225,000
BBB	$ 84,000	$ 92,000
CCC	$127,000	$152,000
DDD	$ 12,000	$ 13,500
EEE	$375,000	$390,000
FFF	$ 27,000	$ 42,000
Total	$825,000	$914,500

Table 20A.2 U Ratios for 6 Different Stores

Store	U
AAA	1.125
BBB	1.095
CCC	1.197
DDD	1.125
EEE	1.040
FFF	1.556

Next we compute x_{-1}, using all the data except the data of AAA Company.

$$x_{-1} = \frac{914{,}500 - 225{,}000}{825{,}000 - 200{,}000} = \frac{689{,}500}{625{,}000} = 1.103$$

Similarly, we compute x_{-2} by deleting the data of BBB Company, x_{-3} by deleting the data of CCC Company, and so on.

$$x_{-2} = \frac{914{,}500 - 92{,}000}{825{,}000 - 84{,}000} = 1.110$$

$$x_{-3} = \frac{914{,}500 - 152{,}000}{825{,}000 - 127{,}000} = 1.092$$

$$x_{-4} = \frac{914{,}500 - 13{,}500}{825{,}000 - 12{,}000} = 1.108$$

$$x_{-5} = \frac{914{,}500 - 390{,}000}{825{,}000 - 375{,}000} = 1.166$$

$$x_{-6} = \frac{914{,}500 - 42{,}000}{825{,}000 - 27{,}000} = 1.093$$

Using this information, we calculate our pseudovalues $x_1, x_2, \ldots, x_6$ in accordance with Equation 20A.1.

$$\begin{aligned}
x_1 &= x_{\text{All}} + (m - 1)(x_{\text{All}} - x_{-1}) \\
&= 1.108 + 5(1.108 - 1.103) \\
&= 1.133 \\
x_2 &= 1.108 + 5(1.108 - 1.110) \\
&= 1.098 \\
x_3 &= 1.108 + 5(1.108 - 1.092) \\
&= 1.188 \\
x_4 &= 1.108 + 5(1.108 - 1.108) \\
&= 1.108 \\
x_5 &= 1.108 + 5(1.108 - 1.166) \\
&= .818 \\
x_6 &= 1.108 + 5(1.108 - 1.093) \\
&= 1.183
\end{aligned}$$

We can now use these pseudovalues, $x_1, \ldots, x_6$, as our pseudosample to construct our confidence interval. First we compute the sample mean $\bar{x}$ of the pseudosample. Then we compute the sample standard deviation s of the pseudosample. Finally, we construct a confidence interval, using Student's t distribution (it is given in Table A4 of Appendix A).

$$\bar{x} = \frac{x_1 + x_2 + x_3 + x_4 + x_5 + x_6}{6}$$

$$= \frac{6.528}{6}$$

$$= 1.088$$

$$s = \left\{ \frac{1}{n-1} (x_1 - \bar{x})^2 + (x_2 - \bar{x})^2 + \cdots + (x_6 - \bar{x})^2 \right\}^{1/2}$$

$$= \left\{ \frac{1}{5} [(1.133 - 1.088)^2 + (1.098 - 1.088)^2 + (1.188 - 1.088)^2 \right.$$

$$\left. + (1.108 - 1.088)^2 + (.818 - 1.088)^2 + (1.183 - 1.088)^2] \right\}^{1/2}$$

$$= .137$$

For a 95 percent confidence interval, we use $t_{.025}$ with 5 degrees of freedom.

$$t_{.025,5} = 2.57$$

Then the 95 percent confidence interval in terms of Equation 20A.3 can be computed as

$$\bar{x} - t_{.025} \frac{s}{\sqrt{n}} < \mu < \bar{x} + t_{.025} \frac{s}{\sqrt{n}}$$

$$1.088 - 2.57(.137/\sqrt{6}) < \mu < 1.088 + 2.57(.137/\sqrt{6})$$

$$.944 < \mu < 1.232$$

The 95 percent confidence interval for the increase in uncollectibles is the interval between .930 and 1.245.

Questions and Problems

1. Under what conditions might stratified random sampling or cluster sampling be preferable to simple random sampling? Explain.

2. Suppose we are interested in the proportion of college seniors in a class of 900 who will be attending graduate school. A survey of 50 seniors reveals that 15 will be attending graduate school. We want to estimate the population proportion p, given this information. $N = 900$, $n = 50$, $\alpha = 10$ percent.

3. Suppose you are a financial consultant trying to determine whether a group of 1,500 country club members represents a good source of potential clients for a real estate firm in New Jersey. To determine the potential business, you decide to analyze the size of the club members' purchases of homes. A random sample of 90 members produces a sample mean of $280,000 with a sample standard deviation of $75,000. Using this information, construct a 95 percent confidence interval for the mean purchases $N = 1,500$, $n = 90$, $\bar{X} = \$280,000$, $s = \$75,000$.

4. If X is a normal variable with known variance equal to 650, how large a sample must we take to be 90 percent confident that the sample mean will not differ from the true mean by more than ± 5 units?

5. If a normal population is known to have σ equal to 10, how large a sample should we take in order to be 90 percent confident that the sample mean will not differ from the population mean by more than $\pm .75$ units?

6. A sales manager wants to know what proportion of her accounts are inactive. How many accounts should she examine if she wants her confidence interval to be no more than $w = .08$ (w being the desired maximum width of the confidence interval) with 95 percent confidence.

7. A company manager wishes to estimate the mean length of time μ it takes company crews to do certain jobs. She wants to estimate μ within ± 7 minutes with 95 percent confidence. Because the value of σ, the population standard duration, is unknown, she took a preliminary sample of $n = 20$ jobs and found that the 20 job completion times had a standard deviation of $s = 18$ minutes. How much larger should she make her sample to obtain the desired confidence interval?

8. The claims manager for an insurance company would like to know the mean amount of automobile insurance repair claims paid by his company. He took a sample of $n = 25$ claims and found $\bar{x} = \$950$ and $s = \$280$. How much larger should his sample be if he wants to estimate the mean payment to within ± 50 with 90 percent confidence?

9. A marketing manager is interested in the number of trips per month that people take to a nearby shopping center. Denote the number of trips per month by X. The manager feels that the variance of X is 16 for women and 9 for men. He decides to stratify by sex, and there are 100 males and 200 females in the population of interest. He takes a random sample of 10 males and another random sample of 15 females and gets the following results.

Males	4	2	1	1	4	7	10	5	7	3
Females	6	9	12	4	2	10	9	7	5	8
	16	14	10	8						

a. Estimate the mean of each of the two strata, and determine the variance of each of the sample means.

b. Estimate the mean of the entire population, and determine the variance of your estimate.

c. Pool the two samples and find the sample mean for all 25 observations. Is this the same as the estimate you found in part (a)? Explain.

10. Is the plan in question 9 stratified sampling with proportional allocation? When might proportional allocation *not* be the best form of allocation in a stratified sampling plan?

11. The number of cars licensed in a particular state last year was 4.8 million; the number of cars licensed in a neighboring state was 3.9 million. For the current year, the officials of the neighboring state estimate that there will be 4.65 million cars registered. Can you use this information to estimate the number of cars that will be registered in the first state in the current year?

12. In questions 9, suppose the marketing manager is interested in the *total* number of trips to the shopping center in a given month, not in the average number of trips per person for that month.

a. Estimate the population totals for the two strata, and determine the sample variance of estimation in each case.

b. Estimate the total for the entire population of males and females, and determine the variance of estimation.

13. Outline the basic steps used in designing a sampling study. What problems may you encounter if your study is poorly planned?

14. What is simple random sampling? What are the advantages and disadvantages of this technique compared to other sampling techniques?

15. What is sampling error? What is a nonsampling error? Give some examples of each.

16. What is two-stage cluster sampling? What are the benefits and disadvantages of this technique compared to other sampling techniques?

17. Suppose you are working for a political consulting firm that is trying to forecast the outcome of a presidential election. Because of the time and cost involved in conducting a simple random sample across all 50 states, you've been asked to devise a sampling strategy to predict the outcome of the

election. Given the time and money constraints, what sampling technique will you propose?

18. Use the information given in Example 20.1 to determine the size of the sample if you want a level of precision of $\hat{\sigma}_{\bar{x}} = \$12,000$.

19. Again using the data from Example 20.2 and question 18, construct a 95 percent confidence interval.

20. Use the data in Example 20.2 to determine the size of the sample you need if you want a level of precision of .05.

21. Suppose you have decided to conduct a sampling survey on the salaries of baseball players. There are essentially two types of players, those not eligible for free agency and those eligible for free agency. If you believe that the salaries of players who are eligible for free agency will differ from those of players who are not, what type of sampling method should you use?

22. Explain how we can use the ratio method to improve the parameter estimates based on sample data.

23. Explain how we can use the regression method to improve the parameter estimates based on sample data.

24. What is the jackknife method? What advantages does it offer? Briefly explain how we use the jackknife method.

25. Suppose you have decided to conduct a study to determine whether accounting majors should be required to take statistics. Briefly explain how you would set up this study. What problems might you encounter? How would you deal with these problems?

26. Using the business section of the *Wall Street Journal* or the *New York Times,* obtain a list of all stocks traded on the American Stock Exchange. Use a random sample of 15 stocks to compute the mean percentage increase in the prices of these stocks over the last month. Compare your result to the actual change in the AMEX index over that period.

27. The Citizens for Fair Taxes are interested in the average property tax paid by the 2,000 residents of their city. A random sample of 50 of these households had a mean property tax of $1,472 with a standard deviation of $311.

a. Find an estimate of the variance of the sample mean.

b. Find a 95 percent confidence interval for the population.

28. The Mom and Pop Grocery Store has 115 employees. In a random sample of 30 of these employees, the mean number of days that an employee was late each year was 14 days, and the sample standard deviation was 3.4 days. Find a 99 percent confidence interval for the mean number of days late each year.

29. The academic advisor to the football team of Rah Rah University is interested in the mean number of hours that players spend studying during the football season. Of the 100 members of the football team, 35 were randomly sampled and found to study an average of 22.5 hours per week with a standard deviation of 8.1 hours. Find a 90 percent confidence interval for the mean number of hours that the football players study.

30. Suppose a quality control expert is interested in drawing a random sample of 100 light bulbs from a case of 10,000. Explain how a table of random numbers could be used to do this.

31. Suppose the quality control expert of question 30 knows from past experience that the number of defective light bulbs in a case of 10,000 has a population standard deviation of 221. She would like to compute a 99 percent confidence interval for the population mean with a standard deviation of 50. How many light bulbs should she sample?

32. A city is required to report to the state the mean property tax of its citizens. From previous years, officials know that the population standard deviation is likely to be $975. A 99 percent confidence interval is desired with a standard deviation of $500. How many of the city's 1,800 households should be sampled?

33. An advertising executive is interested in how viewers looked upon a television ad by McDonald's in New York City (either favorably or unfavorably). Briefly explain how the executive could analyze this question by using sampling. Is one method of sampling preferable to another?

34. In order to correctly assess property taxes in the state, New Jersey has decided to require that all municipalities report the average home price in their districts. From past years, one municipality

estimates that the population standard deviation for its 2,500 homes is $51,721. If the town would like to produce a 95 percent confidence interval with a level of precision of $10,000, how many homes should be sampled?

35. The Students for an Affordable Education have asked you to estimate the average amount of money spent per semester on textbooks. To produce this estimate, you have decided to randomly sample 25 members of your Introduction to Economics course. Are there any problems associated with this sample? What other sampling techniques might you use?

36. An auditor would like to estimate the total value of a corporation's accounts receivable. From previous years the auditor has found the population standard deviation to be $125 for the 1,000 accounts receivable. If the auditor would like to have a level of precision of $100, with a 95 percent confidence interval, how large a sample should he select?

37. Suppose the auditor in question 36 decides to divide the accounts receivable into strata. He would like a level of precision of $10. Determine the total number of sample observations under (a) proportional allocation, (b) optimal allocation.

Stratum	Population Size	Estimated Standard Deviation
1	250	$85
2	325	125
3	225	50
4	200	100

38. A first-year chemistry class consists of 150 students. A random sample of 50 of these students reveals that 31 are majoring in engineering. Find a 95 percent confidence interval for the proportion of students in this class who are majoring in engineering.

39. Explain whether the confidence interval gets wider or narrower when
   a. The confidence interval is 99 percent instead of 95 percent.
   b. The number of observations in the sample decreases from 100 to 50.
   c. The population standard deviation is smaller.

40. The student government of your school has asked you to survey students in order to determine how

many hours the library should be open. You have decided to conduct a stratified sample by class year: first-year, sophomore, etc. What factors must you account for in determining the number of sample observations in each stratum?

41. A movie studio executive wants to poll a sample of movie goers to determine how viewers will react to a new movie. Briefly explain how the sample should be designed.

42. A quality control engineer at the National Bullet Company wants to test for the number of defective bullets (duds) in a case of 1,000. From past experience, he knows that population standard deviation per box is 20 bullets. If he would like to estimate the mean number of duds with a level of precision of 5 bullets, with a 95 percent confidence interval, how large a sample should he take?

43. Suppose you want to estimate a population mean μ. From your sample you find that the sample mean is 200, the sample standard deviation is 30, the total population is 500, and the sample drawn is 30. Find a 90 percent confidence interval for the population mean.

44. Suppose you want to estimate a population proportion p. The total sample consists of 1,000, your sample consists of 100, and the sample proportion is .42. Find a 95 percent confidence interval for the population proportion.

45. A survey based on a stratified sample produced the following information.

	Stratum			
	1	**2**	**3**	**4**
N	1,000	2,000	2,200	3,200
n	40	50	42	75
$\bar{x}$	27.6	30.4	18.7	32.5
s^2	4.2	7.1	6.4	2.8
$\hat{p}$	.5	.6	.3	.6

where $\bar{x}$, s^2, and $\hat{p}$ are the sample mean, sample variance, and sample proportion, respectively.
   a. Find a 90 percent confidence interval for the population mean.
   b. Find a 95 percent confidence interval for the population proportion.

46. The results of a sample survey based on cluster sampling follow.

	Cluster				
	1	**2**	**3**	**4**	**5**
n	3	7	8	2	9
x_i	6.2	8.3	5.4	7.3	9.1

$M = 2,500$ and $N = 100$

Find a 90 percent confidence interval for the population mean.

47. A survey based on a stratified sample produced the following information.

	Stratum				
	1	**2**	**3**	**4**	**5**
N	1,900	3,000	3,500	4,200	4,100
n	30	51	72	65	83
$\bar{x}$	22.6	40.5	28.7	22.5	25.4
s^2	5.2	4.1	7.4	3.8	5.1
$\hat{p}$	.3	.1	.2	.3	.4

where $\bar{x}$, s^2, and $\hat{p}$ are the sample mean, sample variance, and sample proportion, respectively.

a. Find a 90 percent confidence interval for the population mean.

b. Find a 95 percent confidence interval for the population proportion.

46. The results of a sample survey based on cluster sampling follow.

	Cluster				
	1	**2**	**3**	**4**	**5**
n	2	5	7	3	8
x_i	4.2	2.3	8.4	5.3	6.1

$M = 5,500$ and $N = 300$

Find a 95 percent confidence interval for the population mean.

49. A supermarket manager wants to know what type of people purchase health foods and wishes to determine why segments of the population resist buying these products. Discuss how you would go about setting up a study to provide this information. What difficulties might you encounter?

50. A real estate developer wants to determine which features of a house have been most influential in determining its selling price. Describe how you would set up a study to provide this information. What difficulties might you encounter?

51. Suppose the *American Economic Review* is interested in knowing whether student subscribers and regular subscribers differ in their viewpoints on the articles offered in the journal. Explain how you would set up such a study.

52. A city consists of a total of 2 million residents and is divided into 3 boroughs that have 750,000, 900,000 and 350,000 residents, respectively. The city council is considering building a new baseball stadium. If the project is undertaken, it will be financed by an increase in taxes. In order to determine how the city feels about the new stadium, independent random samples of 500 adults from each borough were taken. The numbers in favor of the stadium were found to be 325, 201, and 400, respectively.

a. Using an unbiased estimation procedure, find an estimate of all adults in the city who favor the stadium.

b. Find a 90 percent confidence interval for this population proportion.

53. Suppose a large Wall Street law firm has 500 lawyers of whom 95 are partners and 405 are associates. A random sample of 15 partners finds that 11 own their own homes, and a random sample of 25 associates finds that 15 own their own homes.

a. Find an estimate of the proportion of all lawyers in this firm who own their own homes, using an unbiased estimation procedure.

b. Find a 95 percent confidence interval for all lawyers in this firm who own their own homes.

54. Refer to question 53. Suppose a random sample of the 12 partners reveals that 6 of them graduated from Ivy League schools, and a random sample of 20 associates finds that 14 graduated from Ivy League schools.

a. Find an estimate of the proportion of all lawyers in this firm who graduated from Ivy League schools.

b. Construct a 99 percent confidence interval for all lawyers in this firm who graduated from Ivy League schools.

55. The president of a local union is interested in the mean value of bonuses awarded to a company's employees. The company has 35 divisions, and a simple random sample of 5 of these is taken. The following table gives the results of this sample.

Division Sampled	Number of Employees	Mean Bonus
1	55	$ 70
2	97	101
3	60	40
4	35	89
5	72	56

a. Find a point estimate of the population mean bonus per employee.

b. Find a 90 percent confidence interval for the population mean.

56. A personnel manager is interested in the average age of the company's 872 employees. Suppose he takes a simple random sample of 35 of these employees and finds the sample standard deviation of their ages to be 12.3 years. The personnel manager wants to obtain a 95 percent confidence interval for the population mean age with a level of precision of 2.4 years on each side of the sample mean. How many sample observations must he take?

57. A company has a fleet of 322 automobiles. A random sample of 35 of the cars finds that the sample standard deviation of annual repair costs is $272. Company planners want to construct, for the overall mean of annual repair costs, a 90 percent confidence interval that extends $100 on either side of the sample mean. How many additional sample observations must they take?

58. Mention some situations in which two-stage cluster sampling should be used.

59. A clothing store has an inventory of 920 different items. In order to estimate the total dollar value of inventories, an auditor takes a simple random sample of the items. On the basis of last year's data, the population standard deviation is estimated to be $97. The auditor would like to produce, for the population total, a 95 percent confidence interval that extends $15,000 on each side of the sample estimate. How large a sample size is necessary to meet this requirement?

60. Suppose a corporation is interested in the proportion of employees who favor a new child care program. The corporation has 750 employees from which it wants to take a simple random sample. The planners would like to make the sample large enough so that they can produce a 90 percent confidence interval that extends no more than 7 percent on each side of the sample proportion in favor of the new program. How large a sample should they take? Assume that the sample standard deviation is .24.

61. A movie theater chain has 35 theaters in California, 50 in New York, and 45 in Pennsylvania. Management is considering adding a new snack item to its concession stands. In order to determine whether this new snack will be a success, management tested the product in 10 theaters in California, 12 in New York, and 9 in Pennsylvania for 1 month. The sample means and standard deviations for the numbers of purchases are shown here.

	California	New York	Pennsylvania
Mean	100.4	98.2	77.6
Standard deviation	40.3	21.7	45.1

a. Use an unbiased estimation procedure to find an estimate of the mean number of purchases per movie theater in a month for all 130 movie theaters.

b. Find an estimate of the variance of the estimator in part (a), using an unbiased estimation procedure.

c. Find a 90 percent confidence interval for the population mean number of purchases per theater.

62. Suppose that in question 61 we are interested in knowing how large a sample to take for:

a. A proportional allocation.

b. An optimal allocation, assuming the stratum population standard deviations are the same as the corresponding sample values.

63. A delivery company has a fleet of 502 trucks. A random sample of 35 of these trucks finds that the sample standard deviation of annual repair costs is $753. If you would like to construct, for the overall mean of annual repair costs, a 95 percent confidence interval that extends $250 on either side of the sample mean, how many additional sample observations must you take?

64. A fast-food chain has 25 restaurants in Alabama (Ala.), 20 in Louisiana (La.), 25 in Texas (Tex.), and 32 in Arkansas (Ark.). Management is considering adding a hamburger item to its menu. In order to determine whether this burger will be a success, management tested the product in 15 restaurants in Alabama, 11 in Louisiana, 18 in Texas, and 5 in Arkansas for 1 week. The sample means and standard deviations for the numbers of purchases are shown here.

	Ala.	La.	Tex.	Ark.
$\bar{x}$	127.5	221.3	99.7	127.6
s	83.3	43.8	27.6	70.2

where $\bar{x}$ and s are the mean and standard deviation, respectively.

a. Use an unbiased estimation procedure to find an estimate of the mean number of purchases of the burgers for all 102 restaurants

b. Find an estimate of the variance of the estimator in part (a), using an unbiased estimation procedure.

c. Find a 99 percent confidence interval for the population mean number of burgers sold per restaurant.

65. Suppose that in question 64, a sample of 35 restaurants is to be taken. Determine how the sample should be allocated among the three states for

a. A proportional allocation.

b. An optimal allocation, assuming the stratum population standard deviations are the same as the corresponding sample values.

66. A company that operates three different types of factories is interested in the number of defective products produced. The following table gives the results of a sampling study done on the number of defective parts.

Number of Defective Parts in Factories of

	Type 1	Type 2	Type 3
N_i	75	90	100
n_i	10	15	20
$\bar{x}_i$	12.3	11.5	16.4
s_i	4.7	7.5	3.4

a. Find an estimate of the total number of defective parts, using an unbiased estimation procedure.

b. Find a 95 percent confidence interval for this total.

67. Use the information given in question 66 to find an estimate of the mean number of defective parts, using an unbiased estimation procedure. Also find a 99 percent confidence interval for this value.

68. Refer to question 66. Suppose a sample of 30 factories is to be taken. Determine how many factories of each type the company should select when it is using

a. A proportional allocation.

b. An optimal allocation, assuming the stratum population standard deviations are identical to the corresponding sample values.

69. An auditor is interested in estimating the population mean value for Aloha Company's accounts receivable. The population has been divided into three strata. The accompanying table gives information on the strata and the estimated standard deviations.

Stratum	Population Size	Estimated Standard Deviation
A	600	$175
B	1,005	220
C	700	195

$\sigma_{\bar{x}} = 12.75$. Assume you would like a 95 percent confidence interval to extend $25 on each side of the estimate.

a. Determine the total sample size for a proportional allocation.

b. Determine the sample size for an optimal allocation.

70. A company is interested in estimating the value of accounts receivable for its 75 stores. To do this, the company collects information on 1990 and 1991 accounts receivable from a simple random sample of 10 stores. The results of this sampling study are given in the table.

Store	1990 Accounts Receivable	1991 Accounts Receivable
1	$15,600	$16,200
2	9,510	8,900
3	27,000	29,000
4	18,000	19,200
5	32,000	32,500
6	22,200	19,000
7	25,200	28,500
8	15,000	17,000
9	19,600	24,000
10	9,900	11,100

Suppose 1990 accounts receivable were $1.4 million. Use the ratio method to estimate the accounts receivable for 1991.

71. Refer to question 70. Use the MINITAB program in terms of the regression approach to forecast accounts receivable in 1991. Compare your results to those you got with the ratio method in question 70.

72. A company is interested in estimating sales for its 200 stores. To do this, the company collects information on 1990 and 1991 sales from a simple random sample of 12 stores. The results of this sampling study are given in the table.

Store	1990 Sales	1991 Sales
1	$155,600	$160,200
2	89,540	98,900
3	275,400	282,000
4	180,900	190,200
5	325,000	320,500
6	222,200	199,900
7	250,400	278,400
8	151,000	178,000
9	190,600	240,000
10	99,900	115,100
11	311,000	354,000
12	272,500	295,000

Suppose 1990 sales were $27 million. Use the ratio method to estimate the sales for 1991.

73. Refer to question 72. Use the MINITAB program in terms of the regression approach to forecast sales in 1991. Compare your results to those you got with the ratio method in question 70.

74. An auditor is interested in estimating a company's bad accounts. He collects information on bad accounts for the company's 200 stores, using a random sample of 25 stores. The results are given in the table.

	1990 Bad Accounts	1991 Bad Accounts
Mean	$127,000	$135,000

The total amount of bad accounts in 1990 was $24 million. Use the ratio method to forecast the bad accounts in 1991.

75. Refer to question 70. Suppose the company would like to construct a 90 percent confidence interval for the increase in accounts receivable. Use the jackknife method to construct this confidence interval.

76. Refer to queston 72. Use the jackknife method to construct a 95 percent confidence interval for the increase in sales.

77. Suppose a corporation is interested in the proportion of employees who favor a new "flex time" work schedule. The corporation has 420 employees from which the planners wish to take a simple random sample. The planners would like to make the sample large enough so that they can produce a 95 percent confidence interval that extends no more than 5 percent on each side of the sample proportion in favor of the new program. How large a sample should they take?

78. A company operating three factories that use different types of production is interested in the number of defective products produced. The following table gives the results of a sampling study done on the number of defective parts.

	Number of Defective Parts in Production of		
	Type 1	**Type 2**	**Type 3**
N_i	55	40	150
n_i	12	9	40
$\bar{x}_i$	22.3	35.4	26.3
s_i	10.7	9.5	13.4

a. Find an estimate of the total number of defective parts. Using an unbiased estimation procedure.

b. Find a 90 percent confidence interval for this total.

79. Use the information given in question 78 to find an estimate of the mean number of defective parts, using an unbiased estimation procedure. Also find a 99 percent confidence interval for this value.

80. Refer to question 78. Suppose a sample of 50 factories is to be taken. Determine how many factories of each type the company should select when it is using

   a. A proportional allocation.

   b. An optimal allocation, assuming the stratum population standard deviations are identical to the corresponding sample values.

81. A quality control engineer at the Brite Lite Light Bulb Company wants to test for the number of defective light bulbs in a case of 1,200. From past experience, he knows that the population standard deviation per box is 25 bulbs. He would like to estimate the mean number of defective bulbs with a level of precision of 8 bulbs. How large a sample should he take?

82. A market research group takes a random sample of 5 of a city's 41 voting districts. Each household in each sampled district is questioned on the number of hours its members watch television per day. The results of the sample follow.

District	Number of Households	Mean Number of Hours of TV per Day
1	52	3
2	31	5
3	17	6
4	28	4
5	41	7

Find a 90 percent confidence interval for the population mean number of hours of television watched.

83. Suppose you want to estimate the proportion of a population of voters. A random sample reveals that the sample proportion is .42; $N = 2,500$ and $n = 500$. Find a 99 percent confidence interval for the population proportion.

84. The results of a sample survey based on cluster sampling are

	Cluster				
	1	**2**	**3**	**4**	**5**
n	3	2	8	9	6
x_i	50.2	42.3	38.4	51.3	36.1

$M = 4,000$ and $N = 250$

Find a 90 percent confidence interval for the population mean.

85. You have been hired to design a sampling procedure for testing a new anti-acne drug. One hundred people will be used in the study; half will receive the new drug, and half will receive the standard acne drug. Explain how you would decide which people get which drug. Be specific.

CHAPTER 21

Statistical Decision Theory: Methods and Applications

CHAPTER OUTLINE

Key Terms

statistical decision theory
Bayesian decision statistics
decision theory
actions
alternatives
states of nature
event
outcomes
payoff
probability
prior probability
posterior probability
maximin criterion
minimax regret criterion

worst outcome
expected monetary value
utility analysis
total utility
marginal utility
utility function
risk-averse
risk-neutral
risk lover
expected utility
decision node
event node
systematic risk
market risk

capital market line

Sharpe investment performance
measure

capital asset pricing model

Treynor investment performance
measure

statistical distribution method

market portfolio

lending portfolio

borrowing portfolio

market risk premium

present value

net present value

21.1   INTRODUCTION

In business, decision making is at the heart of management. Using statistics as a guide, this chapter introduces and examines decision making in business and economics in terms of statistical decision theory. The branch of statistics called *statistical decision theory* is sometimes termed *Bayesian decision statistics,* in honor of research presented over 200 years ago by the English philosopher the Reverend Thomas Bayes (1702–1761). Nevertheless, statistical decision theory is a new branch of statistics. Propelled by research by Howard Raiffa, John Pratt, and Leonard Savage (among others), it developed rapidly in the 1950s, and it now occupies an important place in statistical literature. In contrast to classical statistics, where the focus is on estimation, constructing intervals, and hypothesis testing, statistical decision theory focuses on the process of making a decision. In other words, it is concerned with the situation in which an individual, group, or corporation has several feasible alternative courses of action in an uncertain environment.

This chapter discusses methods for selecting the best management alternatives by using statistical decision theory. Here are a few examples of statistical decision problems associated with business decision making:

1. Manufacturers must decide what products to produce.

2. Portfolio managers must decide what investments to purchase while maintaining a portfolio consistent with investors' risk–return preference.

3. Oil company managers must determine, with the help of geologists, where to drill.

4. Corporate managers must choose from among alternative investment projects under conditions of uncertainty.

In this final chapter of the book, we will discuss a variety of statistical methods, collectively referred to as **decision theory,** for dealing with such decision-making problems. First, we present the four key elements of making a decision on the basis of extreme values, expected monetary value, and utility analysis. Then we explore Bayes' strategies and decision trees of expected monetary values in terms of statistical concepts and methodology. We also propose a mean and variance trade-off analysis to replace the expected utility analysis in business decision making. Finally, the mean and variance method is applied in the context of a capital budgeting decision. Appendix 21A discusses how the spreadsheet can be used to do decision-tree analysis. Appendix 21B presents the graphical derivation of the capital market line; Appendix 21C discusses present value and the net present value (NPV) decision rule; and the derivation of standard deviation for NPV is presented in Appendix 21D.

21.2   FOUR KEY ELEMENTS OF A DECISION

Four elements are needed to analyze a decision-making problem: actions, states of nature, outcomes, and probabilities.

1. The choices available to the decision maker are called **actions** (or sometimes **alternatives**). For instance, in a person's decision whether to carry an umbrella, the possible actions are to carry the umbrella and not to carry the umbrella. Although our approach assumes that the decision maker can specify a finite number of mutually exclusive and exhaustive actions, it is also possible to analyze problems with an infinite number of outcomes.

2. The uncertain elements in a problem are referred to as **states of nature.** The states of nature are simply **events.** Like actions, states of nature can be either finite or infinite. The states of nature (events) in our umbrella decision are rain and no rain.

3. An **outcome** is a consequence for each combination of an action and an event (state of nature). The possible outcomes in our umbrella decision are stay dry, be burdened unnecessarily, get wet, and be dry and free. The reward or penalty attached to each outcome is termed the **payoff,** which in business decisions is usually expressed in monetary terms. The relationship among the actions, events, and outcomes involved in a decision process can be presented in a decision tree (see Figure 21.1).[1]

4. **Probability,** which we discussed in Chapter 5, is the chance that an event will occur. The probability of each event may be set by referring to historical data, expert opinion, or any other factor (including personal judgment) the decision

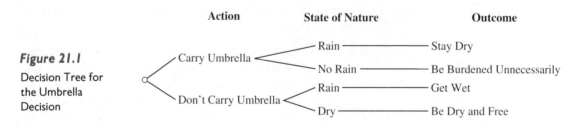

Figure 21.1

Decision Tree for the Umbrella Decision

[1]Decision trees were introduced in Section 5.3 of Chapter 5. They will be discussed in terms of expected monetary values in Section 21.6.

maker wishes to use. These probabilities are initial probabilities, so they are called **prior probabilities.** (Revised probabilities based on the prior probabilities are called **posterior probabilities.** They will be discussed in detail in Section 21.4.) In our umbrella example, the probability is the chance of rain as assigned in accordance with the weather forecast.

Armed with all the foregoing information, the decision maker can decide whether to carry an umbrella.

In the next section, we first discuss alternative business decision rules without probabilities and then introduce the probability variable into the decision process.

21.3 DECISIONS BASED ON EXTREME VALUES

Most of the decision rules discussed in this chapter require the specification of probabilities for the states of nature. However, two strategies do not. The maximin criterion maximizes the minimum payoff; the minimax regret criterion minimizes the difference between the optimal payoff and the actual outcome.

Maximin Criterion

Using this criterion, we first consider the worst possible outcome for each action and then choose the highest of the minimum payoffs (hence "maximin"). The **worst outcome** is simply the smallest payoff that could conceivably result whatever state of nature happens. Thus the **maximin criterion** is a decision rule for born pessimists. Pessimists tend to assume nature is against them no matter what activity they choose. (Some even go so far as to be convinced that they can guarantee sunny weather by carrying an umbrella!)

EXAMPLE 21.1 *Applying the Maximin Criterion to Different Possible Economic Conditions*

Suppose a firm can produce any one of four products, 1, 2, 3, and 4. The four products, then, are the possible actions. The state of nature (which in this case is the state of the economy) has three possibilities: recession, flat, and boom. The payoff, or the outcome, is whatever profits result. This information is presented in Table 21.1.

In this example, the minimum payoff is −$10 for product 1, −$15 for product 2, −$7 for product 3, and −$3 for product 4. Product 4 has the lowest negative payoff in terms of the maximin criterion. Therefore, if it subscribes to the maximin criterion, this pessimistic firm will choose to produce product 4.

Table 21.1 Data for Example 21.1

Product	State of Nature			Minimum Payoff
	Recession	*Flat*	*Boom*	
1	−$10	−$5	$20	−$10
2	−$15	0	$15	−$15
3	−$7	−$1	$25	−$7
4	−$3	$2	$17	−$3

EXAMPLE 21.2 *Applying the Maximin Criterion in Rock Music Promotion*

Suppose a rock music promoter has the option of booking a major group at the local indoor stadium, which has a seating capacity of 17,000, or at a 100,000-seat out-door stadium. Obviously, rain is a major concern for the promoter. The possible actions are holding the concert indoors and holding it outdoors; the states of nature are rain and no rain. The payoffs are the profits. These factors are summarized in Table 21.2. By the maximin criterion, the concert should be held indoors because the minimum profits for that strategy are $100,000, whereas they are 0 for holding it outdoors.

Table 21.2 Data for Example 21.2

	State of Nature		Minimum Payoff
	Rain	*No Rain*	
Indoors	$100,000	$100,000	$100,000
Outdoors	0	$100,000,000	0

The main problem with the maximin criterion is that it does not take into account the probabilities of the states of nature. In the concert example, suppose the concert is to be held in Arizona, where the likelihood of rainfall is quite low. Because the probability of rain is small, it may be better to hold the concert outside.

Another problem with this method is that it does not take other alternatives into consideration. For example, assume that an investor must choose between two stocks and that the states of nature are a recession, a flat economy, and a boom (see Table 21.3). Rates of return are the payoffs. The maximin criterion would dictate choosing stock B, because its minimum payoff is 2 percent. However, although stock A has lower rates of return (−5 percent) for the recession, its rates of return are much higher when the economy is flat or booming. Of course, the chances of recession, flat economy, and boom are not necessarily equal. Hence the decision

Table 21.3 Choice of Two Stocks in Terms of Rates of Return by the Maximin Criterion

| Stock | State of Nature | | | Minimum Payoff |
	Recession	*Flat*	*Boom*	
A (growth type)	−5%	8%	15%	−5%
B (income type)	2%	4%	7%	2%

method described in Table 21.3 is relatively restrictive. Later in this chapter, we will explicitly take into account the probability of occurrence of different states of nature.

Minimax Regret Criterion

In the **minimax regret criterion,** the best action is the one that minimizes the maximum regrets for each decision. When the decision maker aims to maximize the benefit, the regret equals the difference between the optimal payoff and the actual payoff. In Example 21.1, the best outcome in the recession is a loss of $3 for product 4, so the regret for product 1 if a recession occurs is −$3 − (−$10) = $7, and for product 2 the regret is −$3 − (−$15) = $12. In the flat state of nature, the regret for product 1 is $2 − (−$5) = $7, and in the boom state of nature, the regret for product 1 is $25 − $20 = $5. The regrets for all four products under the three economic conditions are presented in Table 21.4.

The best product under this criterion is the one that minimizes the maximum regret (hence "minimax"). Thus product 3 is the best choice because its regret, 4, is the smallest.

Table 21.4 Regret Table

| Product | State of Nature | | | Maximum Regret |
	Recession	*Flat*	*Boom*	
1	7	7	5	7
2	12	2	10	12
3	4	3	0	4
4	0	0	8	8

EXAMPLE 21.3 *Applying the Minimax Regret Criterion*

The following table gives the regrets for the concert example.

| | State of Nature | | Maximum Regret |
	Rain	**No Rain**	
Indoor	0	900,000	900,000
Outdoor	100,000	0	100,000

> By this criterion the best choice is the outdoor concert, because it has the minimum regret ($100,000). Different methods, it seems, can lead to different answers.

The methods presented in this chapter are only aids in decision making. The decision maker must rely partly on personal judgment when making decisions. In our concert example, for instance, the promoter must also take into consideration such factors as the probability of rain, the expected attendance, and ticket prices.

In the following sections, we will discuss adding probability information to statistical decision theory.

21.4 EXPECTED MONETARY VALUE AND UTILITY ANALYSIS

In this section, we discuss both the expected monetary value criterion and utility analysis through probability information. We also examine the application of these techniques in decision making.

The Expected Monetary Value Criterion

The monetary value at a particular state of nature is calculated by multiplying the probability of the action by the payoff of that particular action. If a decision maker has H possible actions, $A_1, A_2, \ldots, A_H$, and is faced with M states of nature, then the **expected monetary value** associated with the ith action, $EMV(a_i)$, can be obtained by summing the monetary value over all states of nature.

$$EMV(A_i) = \sum_{j=1}^{M} P_j M_{ij} \, (i = 1, 2, \ldots, H) \tag{21.1}$$

where

P_j = probability associated with state of nature j with $\Sigma_{j=1}^{M} P_j = 1$

M_{ij} = payoff corresponding to the ith action and the jth state of nature

For example, suppose economists estimate the probability of a recession to be .2, that of a flat economy to be .5, and that of a boom to be .3, as shown in Table 21.5. The payoff (profit) and EMV related to each product listed in the last column of Table 21.5 can be calculated as

$$EMV_1 = (.2)(-20) + (.5)(0) + (.3)(15) = \$.5$$
$$EMV_2 = (.2)(-5) + (.5)(-2) + (.3)(5) = -\$.5$$
$$EMV_3 = (.2)(-10) + (.5)(1) + (.3)(25) = \$6.0$$
$$EMV_4 = (.2)(-1) + (.5)(5) + (.3)(0) = \$2.3$$

Table 21.5 State of Economy and Payoff

	State of Nature			
	Recession	*Flat*	*Boom*	**EMV**
Probability of occurring	.2	.5	.3	
Product 1	−$20	$0	$15	$.5
Product 2	−$5	−$2	$5	−$.5
Product 3	−$10	$1	$25	$6.0
Product 4	−$1	$5	$0	$2.3

The best alternative is the one that maximizes the EMV. In this example, the product with the highest EMV is product 3, with an EMV of $6.0.

EXAMPLE 21.4 *Applying the EMV Criterion to Pricing a New Product*

A marketing manager who is responsible for pricing a new product must decide which of the following three alternative pricing strategies to use.

A_1 (skim-pricing strategy): $15.50/unit

A_2 (intermediate price): $12.00/unit

A_3 (penetration strategy): $ 8.50/unit

The payoff results given in Table 21.6 are total net income associated with each state of nature. In addition, the probability that each state of nature will occur is given at the bottom of the table.

The payoff (net income) is calculated as follows:

$$\text{EMV}(A_1) = (.7)(100) + (.2)(60) + (.1)(-60) = \$76$$
$$\text{EMV}(A_2) = (.7)(60) + (.2)(110) + (.1)(-30) = \$61$$
$$\text{EMV}(A_3) = (.7)(-50) + (.2)(0) + (.1)(90) = -\$26$$

Alternative A_1 offers the highest EMV. It is the best choice if the decision maker's goal is to maximize expected return. However, if the decision maker wants to min-

Table 21.6 State of Nature and Payoff (thousands of dollars)

	State of Nature		
Alternative	*Light Demand,* S_1	*Moderate Demand,* S_2	*Heavy Demand,* S_3
A_1	100	60	−60
A_2	60	110	−30
A_3	−50	0	90
Probability of occurring	.70	.20	.10

imize potential loss, then Alternative A_2, with a maximum loss of $30, is the best choice. In Section 21.5, after we discuss Bayes' strategies, this example will be extended to allow sample information.

EXAMPLE 21.5 *Applying the EMV Criterion to Selecting a Stock*

A portfolio manager predicts the following probabilities (P_i) for the rates of return on four different stocks associated with three different economic conditions. The rates of return for the *j*th stock ($j = 1, 2, 3, 4$) are the payoffs.

Using Equation 21.1, we can calculate the EMV for each stock; all are listed in the last column of Table 21.7. By the EMV criterion, the best choice is stock 4, with a rate of return of

$$(.3)(.20) + (.5)(.03) + (.2)(-.25) = 2.5\%$$

Table 21.7 States of Market and Payoffs

	State of Nature			
	Boom	*Flat*	*Recession*	**EMV**
P_i	.3	.5	.2	
R_1	10%	0%	−5%	2%
R_2	15%	−2%	−10%	1.5%
R_3	7%	2%	−5%	2.1%
R_4	20%	3%	−25%	2.5%

The problem with the EMV criterion is that it does not take the element of risk into consideration. The following example illustrates this. Assume that a coin is flipped. In game 1, a head pays $2 and a tail pays $0. In game 2, a head pays $1 million and a tail pays −$750,000.

	State of Nature		
Game	Heads	Tails	EMV
1	$2	0	$1
2	$1 million	−$750,000	$125,000

In the first game, the EMV is ($2)(.5) + (0)(.5) = $1; in the second game, it is ($1,000,000)(.5) + (−$750,000)(.5) = $125,000. By the EMV criterion, the best choice is game 2. However, a very high degree of risk is associated with game 2: $750,000 will be lost if a tail results. In contrast, there is no possibility of loss in game 1. Anyone who chooses game 2 must be prepared to accept a negative expected payoff as the price for the chance of earning a large payoff. In doing so, he is expressing a preference for risk. By contrast, anyone who chooses game 1 accepts a lower expected payoff in order to eliminate the chance of experiencing a large loss—and thus expresses an aversion to risk. The next section shows how we can employ utility analysis to take individual attitude toward risk into account. In Section 21.7, we will use mean and variance trade-off analysis to take this kind of investigation further.

Utility Analysis

In all the decisions we have looked at, the decision criterion of choice was the maximization of expected monetary value. That is, an individual or corporation believes that the action offering the highest expected monetary value is the preferred course. However, this kind of decision rule does not allow for risk. For example, investors who, in spreading their investments over a portfolio of stocks, accept a lower expected return in order to reduce the chance of a large loss are expressing an aversion to risk. Hence the investors' or the managers' attitudes toward risk are important in their decision making.

Utility analysis gives us information on the decision makers' attitude toward risk. Utility measures the satisfaction a consumer or decision maker derives from consumption or the income associated with investment. For example, Table 21.8 gives the utility function for a consumer who consumes ice cream. In this case, the utility function relates the scoops of ice cream consumed to the utility generated from this consumption.

It does not matter what units are used to measure utility. The total utility for the first scoop of ice cream for this consumer could be 100, 140 for the second, and so on. The **total utility** increases as the scoops increase; however, this total utility increases at a decreasing rate. This implies that **marginal utility** decreases as the number of scoops of ice cream increases. That is, as a person eats more and more ice cream, the extra satisfaction received from each additional scoop decreases. In Table 21.8 the marginal utility of any row is equal to the difference of the utility of that row and the utility of the previous row, for example, $4 = 14 - 10$.

Utility functions can also be used to do utility analysis in business decisions. In this case the **utility function,** a curve relating utility to payoff, can be used to determine whether an investor is risk-averse, risk-neutral, or a risk lover. Various types

Table 21.8 Utility Function

Scoops	Utility	Marginal Utility
1	10	10
2	14	4
3	17	3
4	19	2
5	20	1
6	20.5	.5

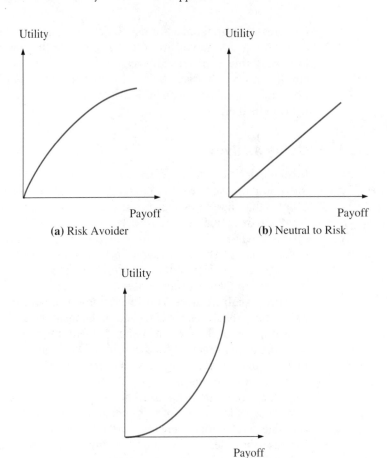

(a) Risk Avoider

(b) Neutral to Risk

(c) Risk Lover

Figure 21.2

Various Types of
Utility Functions

of utility functions are described in Figure 21.2. Here the payoff presented on the horizontal axis can originate from either positive or negative value. A **risk-averse** investor has a utility function wherein utility increases at a decreasing rate as payoff increases. In other words, a risk avoider prefers a small but certain monetary gain to a gamble that has a higher expected monetary value but may involve a large but unlikely gain or a large but unlikely loss. A **risk-neutral** investor has a utility function wherein utility increases at a constant rate. For an individual neutral to risk, every increase of, say, $100 has an associated constant increase in utility. This type of individual uses the criterion of maximizing expected monetary value in decision making, because doing so maximizes expected utility. A **risk lover**'s utility function has utility increasing at an increasing rate. This type of person willingly accepts gambles having a smaller expected monetary value than an alternative payoff that is a "sure thing."

We will use the following example to analyze how the utility function operates within the decision-making process and how it affects the decision.

Table 21.9 Alternative Investment Payoffs

Investment Opportunities	Payoffs	Probability	Utility
Risky	−$100	.5	0
	$200	.5	1
Risk-free	$50	1.0	?

Suppose an investor faces investment opportunities with the payoffs −$100, $200, and $50, as indicated in Table 21.9. We are interested in the investor's utility level for each situation. The different utility levels the investor can reach lead to different decisions. For simplicity, we attach a utility of 0 to the payoff of −$100 and a utility of 1 to the $200 payoff, leaving us with the utility for the $50 payoff. In order to link the utility for the $50 payoff with the information on the decision maker's preference for risk, we then ask, "Would the investor prefer to receive $50 with certainty or to gamble, possibly gaining $200 but just as likely to lose $100?"

Drawing the information given in Table 21.9, we can calculate the expected utility of the payoff as

$$.5U(-\$100) + .5U(\$200) = .5(0) + .5(1) = .5$$

Case I: Risk-averse
$$U(\$50) > .5$$

Case II: Risk-neutral
$$U(\$50) = .5$$

Case III: Risk lover
$$U(\$50) < .5$$

Thus we know that if the utility for the $50 payoff is greater than .5, the investor is risk-averse. By similar reasoning, if the utility for the $50 payoff is less than (equal to) .5, the investor is a risk lover (is risk-neutral). We graph these three alternative utility curves in Figure 21.3, which is a more detailed version of Figure 21.2.

The expected monetary value (EMV) for risky investment opportunities with two possible uncertain investment payoffs as indicated in Table 21.9 is

$$(-\$100)(.5) + (\$200)(.5) = \$50$$

If we use the EMV approach to analyze this problem, we implicitly assume that the investor is risk-neutral. If we employ utility analysis, we consider not only the risk-neutral case but also the cases of both the risk avoider and the risk lover. In short, the EMV approach uses expected *objective* dollar utility as the decision criterion, and utility analysis uses expected *subjective* utility.

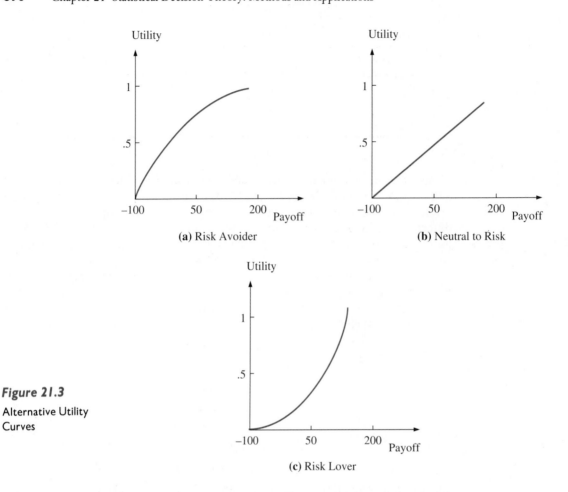

Figure 21.3

Alternative Utility
Curves

(a) Risk Avoider

(b) Neutral to Risk

(c) Risk Lover

The process we went through in this example tells us several things:

1. The utility function affects the decision-making process.

2. If the utility of the expected payoff is greater than, equal to, or less than the expected utility of the payoff, the decision maker will be risk-averse, risk-neutral, or risk-loving, respectively.[2]

3. If the utility function is strictly concave, linear, or convex, the decision maker will be risk-averse, risk-neutral, or risk-loving, respectively.

4. If marginal utility is decreasing, constant, or increasing, the decision maker will be risk-averse, risk-neutral, or risk-loving, respectively.

Let's look at another example of the use of utility analysis in decision making.

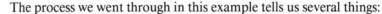

[2]In our case, the expected utility of the payoff is 0.5. The utility of the expected payoff for a risk-averse investor is larger than .5; the utility of the expected payoff for a risk-neutral investor is .5; the utility of the expected payoff for a risk lover is smaller than .5.

EXAMPLE 21.6 *Different Attitudes Toward Risk*

Assume that an investor's utility function can be defined by $U(x) = \sqrt{x}$, where x = return (payoff). Now, he faces the choice of $x = 14.5$ with certainty, or $x = 25$ with probability .5 and $x = 4$ with probability .5. We want to determine whether this is a risk-averse person.

From the foregoing example, we know that if the investor's utility of the expected payoff is greater than the expected utility of the payoff, if the utility function is strictly concave, or if the marginal utility is diminishing, then the investor is said to be risk-averse. In fact, the last two criteria are the same. To determine whether this individual is risk-averse, we simply take the ordinary derivatives of $U(x)$ with respect to x two times; then we can mathematically show that the individual is risk-averse if x is larger than zero.[3] Such an investor will take the certain payoff of $x = 14.5$.

Another approach to determining an individual's risk preference is to calculate the expected utility of the payoff and the utility of the expected payoff as indicated in Table 21.10. In the table, $U(x)$, the utility given x, is calculated by substituting the values for x of 4 and 25, respectively, into the utility function $U(x) = \sqrt{x}$. $E[U(x)]$, the expected utility given x, is equal to $pU(x)$. Finally, $E(x)$, the expected value of x, and $U[E(x)]$, the utility from the expected value of x, are calculated as follows:

$$E(x) = (.5)(4) + (.5)(25) = 14.5$$
$$U[(E(x)] = \sqrt{14.5} = 3.808$$

From this example, we can see that the utility of the expected payoff, 3.808, is greater than the expected utility of the payoff, 3.5 $(1 + 2.5)$. Again, we can see that the investor is risk-averse; the utility received from the expected value of x is greater than the expected value of the utility of x. What this means is that the investor will get greater utility from receiving the expected value of x with certainty than he'd get if he took the gamble.

An investor is risk-averse if he prefers a safe investment over a risky investment with a higher expected value. For example, if an investor prefers to invest in risk-free Treasury bills rather than investing in a stock with a higher expected value, he is considered risk-averse.

---

[3] $\dfrac{dU(x)}{dx} = \dfrac{d(x)^{1/2}}{dx} = \tfrac{1}{2}x^{-1/2}$

$\dfrac{d^2U(x)}{dx^2} = \dfrac{d(\frac{1}{2}x^{-1/2})}{dx} = -\tfrac{1}{4}x^{-3/2}$

If $x > 0$, then we know the second derivative of the utility function is negative. This condition represents the curve as concave, as shown in Figure 21.3a.

Table 21.10 Expected Utility of the Payoff and Utility of the Expected Payoff

x	$U(x) = \sqrt{x}$	P	$E[U(x)]$	$E(x)$	$U[E(x)]$
4	2	.5	1	2	
25	5	.5	2.5	12.5	
—	—	1.0	3.5	14.5	3.808

Having determined the appropriate utilities, we need only solve the decision-making problem by finding that course of action with the highest **expected utility.** Employing utility analysis concepts, we can modify the expected monetary value criterion defined in Equation 21.1 to be the expected utility criterion.

$$E[U(A_i)] = \sum_{j=1}^{m} P_j U_{ij} \ (i = 1, 2, \ldots, n)$$ (21.2)

where

$E[U(A_i)]$ = expected utility of action i
$\quad P_j$ = probability associated with state of nature j
$\quad U_{ij}$ = utility corresponding to the ith action and the jth state of nature

In Equation 21.2, we also assume that $\Sigma_{j=1}^{m} P_j = 1$. If the decision maker is risk-neutral (indifferent to risk), the expected utility criterion and the expected monetary value criterion are equivalent.

21.5   *BAYES STRATEGIES*

We studied Bayes' theorem in Chapter 5. This theorem enables us to work out the probability for one event that is conditional on another event. Bayesian analysis can be used in the decision-making process. The difference between Bayes analysis, the maximin criterion, and the minimax regret criterion is that in Bayes analysis, probabilities of the states of nature must be specified.

Recall Bayes' theorem:

$$P(E_2|E_1) = \frac{P(E_1|E_2)P(E_2)}{P(E_1)}$$ (5.19)

where $P(E_2|E_1)$ and $P(E_1|E_2)$ are conditional probabilities of event 2 (E_2) given event 1 (E_1) and of E_1 and given E_2. $P(E_1)$ and $P(E_2)$ are unconditional probabilities of E_1 and E_2, respectively. In terms of Bayesian statistic, $P(E_2)$ is the initial or prior probability of E_2 and is modified to the posterior probability, $P(E_2|E_1)$, given the sample information that event E_1 has occurred. We can incorporate different states of nature into Equation 5.19 to obtain the generalized Bayes model defined in Equation 21.3.

$$P(S_i|I) = \frac{P(I \cap S_i)}{P(I)} = \frac{P(I|S_i)P(S_i)}{\sum\limits_{j=1}^{m} P(I|S_j) P(S_j)} \qquad (21.3)$$

where $P(S_i|I)$ is the probability of state of nature S_i given sample information I. $P(S_i)$ is the probability of state of nature S_i *not* incorporating sample information I, and it is called a prior probability of S_i. We also assume that there exist m states of nature.

EXAMPLE 21.7 *Bayesian Approach in Forecasting Interest Rates*

Suppose that macroeconomists are hired to predict interest rates. Past results for economic prognosticators are presented in the following table.

| | **Interest Rate Outcome** | |
Belief	**Up**	**Down**
Strong credit market	.60	.30
Weak credit market	.40	.70

When economists believed that credit markets would be strong, interest rates went up and went down with 60 percent and 30 percent chances, respectively. Thus

$$P \text{ (strong market/up)} \quad = .60$$
$$P \text{ (strong market/down)} = .30$$

Now suppose the economists believe that the probability that rates will rise is .7 and that the probability of lower rates is .3.

$$P(\text{up}) = .7 \qquad P(\text{down}) = .3$$

Following Equation 21.3, we find that in the case of two states of nature, the probability that interest rates will rise, given a strong credit market assessment by economists, is

$$P(\text{up}|\text{strong market}) =$$
$$\frac{P(\text{strong market}|\text{up})P(\text{up})}{P(\text{strong market}|\text{up})[P(\text{up})] + P(\text{strong market}|\text{down})[P(\text{down})]}$$
$$= \frac{.60(.70)}{.6(.7) + .3(.3)} = \frac{.42}{.51} = .82$$

The probability that interest rates will rise, given a weak credit market assessment, is

$$P(\text{up/weak market}) = \frac{.4(.7)}{(.4)(.7) + (.7)(.3)} = .57$$

Corporate executives can use these probabilities to assess future interest rates.

21.6 DECISION TREES AND EXPECTED MONETARY VALUES

In this section we use the expected monetary value (EMV) criterion in decision tree form to select the best alternative in business decision making. As a general approach to structuring complex decisions, a decision tree helps direct the user to a solution. It is a graphical tool that describes the types of actions available to the decision maker and the resulting events.

The decision tree approach to capital budget decision making is used to analyze investment opportunities involving a sequence of investment decisions over time. To best illustrate the use of the decision tree, we will develop a problem involving numerous decisions.

First, we must enumerate some of the basic rules for implementing this method. The decision maker should try to include only important decisions or events. The decision tree model requires the decision maker to make subjective estimates when assessing probabilities. And it is important to develop the tree in chronological order to ensure the proper sequence of events and decisions.

A decision point, or decision node, is represented by a box. The available alternatives are represented by branches out of this node. A circle represents an **event node,** and branches from this type of node represent possible events.

The expected monetary value (EMV) is calculated for each event node by multiplying probabilities by conditional profits and summing them. The EMV is then placed in the event node and represents the expected value of all branches arising from that node.

A decision tree is shown in Figure 21.4. The states of nature are high, medium, and low levels of GNP, and their probabilities are .2, .5, and .3, respectively. For product 1, the expected value is $(.2 \times 100) + (.5 \times 5) + (.3 \times -30) = 13.5$. The highest EMV (14) is that of product 3.

Each square on the decision tree denotes a decision that must be made. The circles indicate the states of nature. The square represents the decision to produce product 1, 2, or 3. As we have said, the states of nature that can occur are high, medium, and low levels of economic performance. Now let's look at two examples of how decision trees using objective (dollar payoff) utility instead of subjective utility are employed in business decision making.

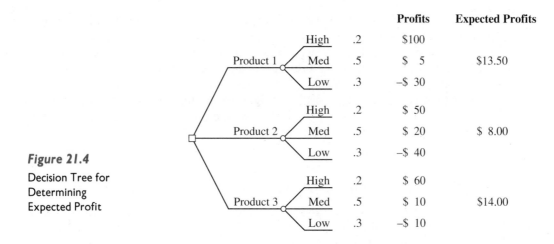

		Profits	Expected Profits
High	.2	$100	
Med	.5	$ 5	$13.50
Low	.3	–$ 30	

Figure 21.4

Decision Tree for Determining Expected Profit

EXAMPLE 21.8   *A Decision Tree for Testing a Drilling Site*

An oil company is trying to decide whether to test for the presence of oil or to drill for oil (see Figure 21.5). If oil is struck, the revenues are $1 million, with a cost of $100,000 to drill. The firm has to decide whether to test first for the presence of oil. Without testing, the probability of striking oil is .1. Thus the firm's expected value without testing is 0 ($1 million times .1, less drilling fees of $100,000). The cost of

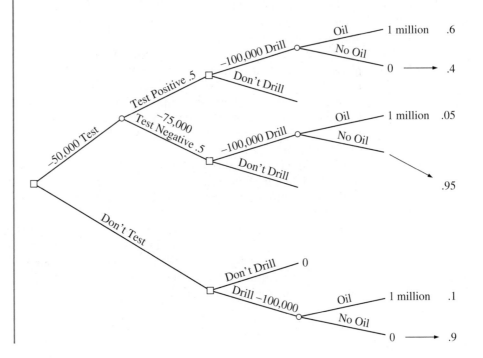

Figure 21.5

Decision Tree for Example 21.8

the oil test is $50,000. If the test is positive, there is a .6 probability that oil will be struck; if the test is negative, the probability of striking oil is .05.

The expected gain for a negative test result is .05(1,000,000) − 100,000 − 50,000 = −$100,000. If the test is positive, the expected profit is (1,000,000)(.6) − 100,000 − 50,000 = $450,000. If the firm's test is negative, the firm should not drill; however, if the test is positive, it should drill. This analysis makes it clear that the test should be done. In Appendix 21A we show how the spreadsheet can be used for decision tree analysis for drilling oil.

EXAMPLE 21.9   *A Decision Tree for Capital Budgeting*

A firm currently sells paper and paperboard packaging materials. Company planners predict that, with the advent of plastic shrink-film packaging, their line of products may be obsolete within a decade. They must quickly decide on a short-term plan of action from among four alternatives: (1) do nothing, (2) establish a tie-in with a machinery company that manufactures plastic packaging, (3) acquire such a company, or (4) take on the research and development of plastic packaging. These four alternatives are the first four branches arising out of the event node in Figure 21.6. If the company planners do nothing, the firm's short-term profits will be about the same as in the previous year. If they decide to establish a tie-in with another firm, they foresee one of two events occurring: there is a 90 percent chance of successful introduction of their new plastics line and a 10 percent possibility of failure. If they decide on acquisition, they foresee a 10 percent chance of problems with antitrust laws, a 30 percent possibility of an unsuccessful introduction of the plastics line, and a 60 percent chance of success. If they decide to manufacture a whole plastics line on their own, they foresee many more problems. They anticipate a 10 percent chance of having trouble developing their own machines, a 10 percent chance of having problems with suppliers in developing a total packaging system for their customers, a 30 percent chance that customers will not purchase their systems, and a 50 percent chance of success in the development and introduction of the plastics line.

Conditional profit is the amount of profit the firm can expect to make by adopting each of the preceding sets of alternatives and consequent events.

In Figure 21.6 the expected monetary values are shown in the event nodes. The firm's financial planner can use EMV to decide which action to take, selecting the decision node with the highest EMV (in this case, establishing a tie-in, which has an EMV of 76.6). The slash marks indicate elimination of nonoptimal decision branches from consideration. If the probabilities associated with events change, then the EMVs associated with the alternatives may change—and with them the selection of the optimal alternative.

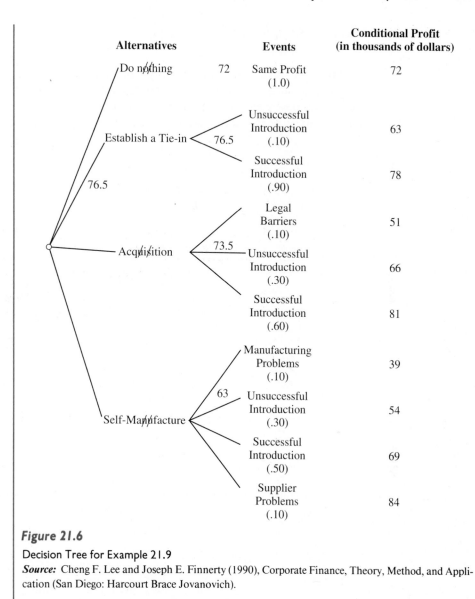

Figure 21.6

Decision Tree for Example 21.9
Source: Cheng F. Lee and Joseph E. Finnerty (1990), Corporate Finance, Theory, Method, and Application (San Diego: Harcourt Brace Jovanovich).

For Example 21.9, we greatly simplified the number of possible alternatives and events. In fact, decision trees are more useful for more complex problems—that is, for problems containing more possibilities or problems in which management must make a sequence of decisions rather than a single decision.

EXAMPLE 21.10   *Utilizing Sample Information to Improve the Determination of Pricing Policy[4]*

In Example 21.4 we determined the price of a new product without using sample information. Such sample information as test market results, however, can be very helpful. Suppose past product performances can give some indication about the relationship between test market results and product performance nationally. Let

Z_1 = disappointing or only slightly successful test market

Z_2 = moderately successful market

Z_3 = highly successful market

Table 21.11 Conditional Probabilities of Each Test Market Result, Given Each State of Nature

Test Market Result	State of Nature		
	Light Demand S_1	*Moderate Demand* S_2	*Heavy Demand* S_3
Z_1	.5	.2	.2
Z_2	.3	.7	.6
Z_3	.2	.1	.2
	1.0	1.0	1.0

Table 21.12 Revision of Prior Probabilities in Light of Possible Test Market Result

(1) j	(2) State of Nature S_j	(3) Prior Probability $P(S_j)$	(4) Conditional Probability $P(Z_k \mid S_j)$	(5) = (3) × (4) Joint Probability $P(S_j)P(Z_k \mid S_j)$	(6) = (5) ÷ Sum of (5) Posterior Probability $P(S_j \mid Z_k)$
Z_1	S_1	.7	.5	.35	.854
	S_2	.2	.2	.04	.097
	S_3	.1	.2	.02	.049
				.41	1.000
Z_2	S_1	.7	.3	.21	.512
	S_2	.2	.7	.14	.342
	S_3	.1	.6	.06	.146
				.41	1.000
Z_3	S_1	.7	.2	.14	.778
	S_2	.2	.1	.02	.111
	S_3	.1	.2	.02	.111
				.18	1.000

[4]This example is similar to an example given in Gilbert A. Churchill, Jr. (1983), *Market Research: Methodological Foundations,* 3d ed. (Chicago: Dryden) pp. 37–42.

Table 21.13 Expected Value of Each Alternative, Given Each Research Outcome

Z_1, disappointing or only slightly successful test market
$\quad$ EV$(A_1) = (100)(.854) + (60)(.097) + (-50)(.049) = 88.77$
$\quad$ EV$(A_2) = (60)(.854) + (110)(.097) + (0)(.049) = 61.91$
$\quad$ EV$(A_3) = (-60)(.854) + (-30)(.097) + (90)(.049) = -49.74$
Z_2, moderately successful test market
$\quad$ EV$(A_1) = (100)(.512) + (60)(.342) + (-50)(.146) = 64.42$
$\quad$ EV$(A_2) = (60)(.512) + (110)(.342) + (0)(.146) = 68.34$
$\quad$ EV$(A_3) = (-60)(.512) + (-30)(.342) + (90)(.146) = -27.84$
Z_3, highly successful test market
$\quad$ EV$(A_1) = (100)(.778) + (60)(.111) + (-50)(.111) = 78.91$
$\quad$ EV$(A_2) = (60)(.778) + (110)(.111) + (0)(.111) = 58.89$
$\quad$ EV$(A_3) = (-60)(.778) + (-30)(.111) + (90)(.111) = -40.02$

$\qquad$ By using Bayes' theorem (Equation 21.3) and supposing that past experiences provided the estimate of conditional probabilities given in Table 21.11, we find the revised prior probabilities $P(S_1) = .7$, $P(S_2) = .2$, and $P(S_3) = .1$, as presented in Table 21.12. Conditional probabilities from Table 21.11 are presented in column (4) of Table 21.12. Using Equation 21.3, we calculate the posterior probabilities $P(S_j | Z_k)$ presented in column (6) of Table 21.12. Using the information on states of nature and payoffs listed in Table 21.6 and the posterior probability information listed in Table 21.12, we find the expected value of each alternative, given each research outcome (see Table 21.13).

$\qquad$ The probability of obtaining each test market result—that is, the probability of each Z_k—is given as

$$P(Z_k) = \sum_{j=1}^{n} P(S_j)P(Z_k | S_j)$$

and for $k = 1$, for example, the probability is

$$P(Z_1) = P(S_1)P(Z_1 | S_1) + P(S_2)P(Z_1 | S_2) + P(S_3)P(Z_1 | S_3)$$
$$= (.7)(.5) + (.2)(.2) + (.1)(.2)$$
$$= .41$$

$\qquad$ The probability is given as the sum of the elements in column (5) of Table 21.12. Table 21.12 thus indicates that the probabilities asociated with these test markets are $P(Z_1) = .41$, $P(Z_2) = .41$, and $P(Z_3) = .18$. (The probabilities sum to 1, as they should, because one of the three test market outcomes must result.) The expected value of the test-marketing procedure is found by weighting each expected value of the optimal action, given each research result, by the probability of receiving that expected value. The expected value of the proposed research is thus found to be

$$\text{EV(research)} = (88.75)(.41) + (68.34)(.29) + (78.91)(.18)$$
$$= 78.61$$

This value is $2.61 over the expected value of the optimal action without research, which, as indicated in Example 21.4, is $76. Hence the market research should be undertaken if the research cost is less than $2.61.

21.7 MEAN AND VARIANCE TRADE-OFF ANALYSIS

The Mean–Variance Rule and the Dominance Principle

The expected utility rule we discussed in Section 21.6 is theoretically the best criterion available, but sometimes it is very hard to implement. We frequently do not know the investor's utility function, and furthermore, the decision maker, as in the case of a manager, must act on behalf of many stockholders with different utility functions. Hence the expected utility rule is often replaced by a more practical mean–variance decision criterion that assumes that the decision maker has a risk-average utility function.

According to the mean–variance rule, the expected return (mean) measures an investment's profitability, whereas the variance (or standard deviation) of returns measures its risk. Consider the following four alternative projects, with the means $\bar{x}$ and standard deviations σ_x specified.

Investment Project	$\bar{x}$	σ_x
A	$ 9	$ 90
B	8	90
C	8	100
D	10	120

To discuss the implications of the trade-off between risk and return and of the dominance principle, we plot this set of data in Figure 21.7. A pairwise comparison of the investment projects shows that project A dominates projects B and C; it has the highest return, and its risk is equal to that of project B and lower than that of project C. However, there is no clear-cut decision between projects A and D, projects B and D, or projects C and D. Here the investor needs to consider the trade-off between profit and risk in terms of his or her attitude toward risk.

In the analysis of stock investments, average rates of return and their variance (or standard deviation) are used to represent investments' profitability and risk, respectively. The variance of rates of return can be decomposed into two components by the market model[5] defined in Equation 21.4.

$$R_{i,t} = \alpha_i + \beta_i R_{m,t} + e_{i,t} \tag{21.4}$$

where $R_{i,t}$ and $R_{m,t}$ are rates of return for the ith security (portfolio) and market rates of return, respectively.

---

[5]We discussed the market model in Chapter 14 and elsewhere.

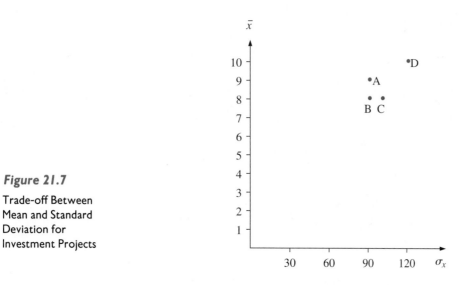

Figure 21.7

Trade-off Between
Mean and Standard
Deviation for
Investment Projects

Following Equation 13.19 of Chapter 13

$$\sigma_i^2 = \beta_i^2 \sigma_m^2 + \sigma_{ei}^2 \tag{21.5}$$

where

σ_i^2 = variance of $R_{i,t}$
σ_m^2 = variance of market rates of return
σ_{ei}^2 = residual variance of rates of return for the ith security

In investment analysis, we define σ_i^2, $\beta_i^2 \sigma_m^2$, and σ_{ei}^2 as total risk, systematic risk, and unsystematic risk, respectively.

Systematic risk is the part of total risk that results from the basic variability of stock prices. It accounts for the tendency of stock prices to move together with the general market. The other portion of total risk is unsystematic risk, which is the result of variations peculiar to the firm or industry—for example, a labor strike or resource shortage.

Systematic risk, also referred to as **market risk,** reflects the fluctuations and changes in general market conditions. Some stocks and portfolios are very sensitive to movements in the market; others exhibit more independence and stability. A measure of a stock's or a portfolio's relative sensitivity to the market, assigned on the basis of its past record, is designated by the lower-case Greek letter beta (β).

EXAMPLE 21.11 *Market Model and Risk Decomposition for GM*

The annual rate of return for GM and the market rate of return for 1970 to 1990 are used to estimate the market model in accordance with Equation 21.4 and in terms of MINITAB. The results are shown in Figure 21.8. From Figure 21.8, we

```
MTB > PRINT C2-C4

ROW    C2       C3        C4

 1     70     0.2137    0.0010
 2     71     0.0422    0.1080
 3     72     0.0631    0.1557
 4     73    -0.3667   -0.1737
 5     74    -0.2597   -0.2964
 6     75     0.9522    0.3149
 7     76     0.4584    0.1918
 8     77    -0.1124   -0.1153
 9     78    -0.0498    0.0105
10     79     0.0288    0.1228
11     80    -0.0410    0.2586
12     81    -0.0911   -0.0994
13     82     0.6826    0.1549
14     83     0.2373    0.1706
15     84     0.1176    0.0115
16     85    -0.0383    0.2633
17     86     0.0088    0.1462
18     87     0.0058    0.0203
19     88     0.4418    0.1240
20     89    -0.4581    0.2725
21     90    -0.1154   -0.0656

MTB > REGRESS C3 1 C4;
SUBC> DW.

The regression equation is
C3 = 0.0083 + 0.980 C4

Predictor        Coef        Stdev      t-ratio         p
Constant       0.00834      0.07273       0.11       0.910
C4             0.9800       0.4188        2.34       0.030

s = 0.3006      R-sq = 22.4%      R-sq(adj) = 18.3%

Analysis of Variance

SOURCE         DF        SS          MS          F         p
Regression      1      0.49469     0.49469      5.48     0.030
Error          19      1.71657     0.09035
Total          20      2.21126

Durbin-Watson statistic = 1.87
```

Figure 21.8

MINITAB Output
of the Market
Model for GM

find that the beta coefficient for the market model is .9800. From the analysis of variance data in Figure 21.8, we obtain the total risk (σ_i^2), systematic risk ($\beta_i^2 \sigma_m^2$) and unsystematic risk (σ_{ei}^2) as follows:

$$\sigma_i^2 = \frac{2.21126}{20} = .11056$$

$$\beta_i^2 \sigma_m^2 = \frac{.4969}{20} = .02485$$

$$\sigma_{ei}^2 = \frac{1.71657}{19} = .09035$$

$$\beta_i^2 \sigma_m^2 + \sigma_{ei}^2 = .02485 + .09035 = .1152 \doteq .11056$$

The Capital Market Line

From the rates of return and alternative measurments of risk, we can derive either the trade-off between expected return and total risk or the trade-off between expected return and systematic risk. In these cases, the utility function used to measure a decision maker's attitude toward risk is the Von Neumann and Morgenstein (VNM) type. The use of the term *utility* in the VNM type of utility function differs from its use by traditional economists. The VNM type of utility function is applied in situations where money payoffs are inappropriate as a measuring device. In traditional economics, utility reflects the inherent satisfaction delivered by a commodity and is measured in terms of psychic gains and losses. Von Neumann and Morgenstein, on the other hand, conceived of utility as a measure of value that provides a basis for making choices in the assessment of situations involving risk. This approach integrates the EMV and the utility analyses discussed in Section 21.4.[6]

The **capital market line** used to describe the trade-off between expected return and total risk is[7]

$$E(R_i) = R_f + [E(R_m) - R_f]\frac{\sigma_i}{\sigma_m}$$ (21.6)

where

R_f = risk-free rate
$E(R_m)$ = expected return on the market portfolio
$E(R_p)$ = expected return on the *i*th portfolio
σ_i, σ_m = standard deviations of the portfolio and the market, respectively

The capital market line (CML) defined in Equation 21.6 implies that the expected rates of return for portfolio *i* equal the risk-free rate plus total market risk,

$$\frac{[(E(R_m) - R_f]\sigma_i}{\sigma_m}.$$

The total portfolio risk premium is equal to price per market risk,

$$\frac{E(R_m) - R_f}{\sigma_m},$$

times total risk associated with portfolio *i*—that is, σ_i. By using the concept of CML, we can define the **Sharpe investment performance measure** for the *i*th portfolio as

$$SP_i = \frac{\bar{R}_i - R_f}{\sigma_i}$$ (21.7)

[6]In other words, the utility function can be defined as

$$U[E(R),\sigma]$$

where $E(R)$ and σ represent expected rates of return and the standard deviation of rates of return, respectively. By assuming that the investors are risk avoiders, we have $[\partial U/\partial E(R)] > 0$ and $(\partial U/\partial \sigma) < 0$. In other words, investors prefer return and dislike risk.

[7]The graphical derivation of this model appears in Appendix 21B.

EXAMPLE 21.12   *Using the Sharpe Investment Performance Measure to Determine Investment Performance*

An investor is considering investing in either mutual fund A or mutual fund B. For past performance, he calculates for both funds the average returns and variances listed in Table 21.14. It is assumed that the T-bill rate is 8 percent, which the firm uses as the risk-free rate.

The Sharpe performance measure, then, gives

$$SP_A = \frac{.20 - .08}{.08} = 1.5$$

$$SP_B = \frac{.15 - .08}{.05} = 1.4$$

These calculations reveal that mutual fund A will give a slightly better performance and thus is the better alternative of the two investments.

Table 21.14 Return and Standard Deviation for Mutual Funds

	Mutual Fund A	**Mutual Fund B**
Average return, $\overline{R}_i$	20%	15%
Standard deviation, σ_i	8%	5%

The Capital Asset Pricing Model

The capital market line (Equation 21.6) is used to describe the trade-off between expected rate of return and total risk. The trade-off between expected rate of return and systematic risk defined in Equation 21.8 is called the **capital asset pricing model** (CAPM).

$$E(R_i) = R_f + \beta_i(E(R_m) - R_f) \tag{21.8}$$

where

$E(R_i)$ = expected rate of return for asset i
R_f = risk-free rate
β_i = measure of systematic risk (beta) for asset i
$E(R_m)$ = expected return on the market portfolio

Equation 21.8 implies that β_i is the systematic risk for determining the price of the individual asset and the portfolio. Figure 21.9 illustrates graphically the relationship between $E(R_i)$ and β_i that is defined in Equation 21.8. (Professor William Sharpe won the Nobel prize in economics mainly because he derived this model.)

The reason why the CAPM can be regarded as part of decision theory is that it is based on a utility function in terms of expected rates of return and the standard

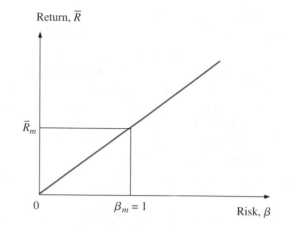

Figure 21.9

The Capital Asset Pricing Model (CAPM)

deviation of rates of return. Expected rates of return and the standard deviation of rates of return are essentially based on the monetary value of the investment. In Section 21.4, we used expected monetary values and utility analysis to make decisions. Here we treat risk (the standard deviation of rates of return) as an explicit factor, whereas in Section 21.4 we treated it as an implicit factor in determining the value of an investment.

With the capital asset pricing model, we must assume that all utility-maximizing investors will attempt to position themselves somewhere along the CML and will attempt to put some portion of their wealth into the market portfolio of risky assets.

The CAPM implies that the market portfolio is the only relevant portfolio of risky assets. Hence the relevant risk measurement of any individual security is its covariance with the market portfolio—that is, the systematic risk of the security.

The relationship between the capital market line (CML) and the CAPM can be shown by starting with the definition of the beta coefficient:

$$\beta_i = \frac{\text{Cov}(R_i, R_m)}{\text{Var}(R_m)} = \frac{\rho_{i,m}\sigma_i\sigma_m}{\sigma_m^2} = \frac{\rho_{i,m}\sigma_i}{\sigma_m} \qquad (21.9)$$

where

σ_i = standard deviation of the ith security's rate of return
σ_m = standard deviation of the market rate of return
$\rho_{i,m}$ = correlation coefficient of R_i and R_m

If $\rho_{i,m} = 1$, then Equation 21.8 reduces to

$$E(R_i) = R_f + \frac{\sigma_i}{\sigma_m}[(E(R_m) - R_f] \qquad (21.6)$$

If $\rho_{i,m} = 1$, then this implies that the portfolio in question is the efficient portfolio, or, for an individual security, it implies that the returns and risks associated with the asset are similar to those associated with the market as a whole. The implications of this comparison, in turn, are that

1. Equation 21.8 is a generalized case of Equation 21.6, because Equation 21.8 includes the correlation coefficient, whereas Equation 21.6 assumes that the correlation coefficient is equal to 1.

2. The capital asset pricing model (CAPM) instead of the capital market line (CML) should be used to price an individual security or an inefficient portfolio. To use the CML to price an inefficient portfolio would be to price unsystematic risk.

3. The CML prices the risk premium in terms of total risk, and the security market line (SML) prices the risk premium in terms of systematic risk.

In order to apply the CAPM, we need to estimate the beta coefficient, the risk-free rate, and the market risk premium. Estimates of these quantities can be obtained from time-series data as shown in Example 21.11 (see also Chapter 14).

The capital asset pricing model (CAPM) defined in Equation 21.8 implies that rates of return for the ith security (or portfolio) equal the risk-free rate plus the security's (or portfolio's) risk premium $[E(R_m) - R_f]\beta_i$. This risk premium is equal to systematic risk β_i times the expected market risk premium, $E(R_m - R_f)$. By using the concept of CAPM, we can define the **Treynor investment performance measure**[8] for the ith security (or portfolio) as follows:

$$TP_i = \frac{\overline{R}_i - R_f}{\beta_i} \tag{21.10}$$

If in Equation 21.10 we also know that the beta coefficients for mutual funds A and B are $\beta_A = 1.8$ and $\beta_B = 1.2$, respectively, then the Treynor investment measures for these two mutual funds are

$$TM_A = \frac{.20 - .08}{1.8} = .0667$$

$$TM_B = \frac{.15 - .08}{1.2} = .0583$$

Like the Sharpe performance measure, the Treynor measure indicates that mutual fund A is the better alternative of the two investments.

---

[8]The derivation and justification of this investment performance measure can be found in J. Treynor (1965), "How to Rate Management of Investment Fund," *Harvard Business Review* 43, 63–75.

21.8 THE MEAN AND VARIANCE METHOD FOR CAPITAL BUDGETING DECISIONS

The capital budgeting decision is the manager's decision to undertake a certain project instead of other projects.

Capital budgeting frequently incorporates the concept of probability theory. Consider two projects (project X and project Y) and three states of the economy for any given time (prosperity, normal, and recession). For each of these states, a probability of occurrence can be calculated and an estimate made of its return; see Table 21.15. We can calculate the expected returns $\overline{R}$ for projects X and Y as follows:

$$\overline{R} = \sum_{i=1}^{m} R_i P_i \qquad (21.11)$$

where R_i is the return for the ith state of nature and P_i is the probability associated with the ith state of nature. Substituting the information given in Table 21.1 into Equation 21.11, we obtain

$$\overline{R}_X = 6.25\% + 7.50\% + 1.25\% = 15.00\%$$
$$\overline{R}_Y = 10\% + 7.50\% - 2.50\% = 15.00\%$$

The standard deviation for these returns can be found by using

$$\sigma = \sqrt{\sum_{i=1}^{n} (R_i - \overline{R})^2 p_i} \qquad (21.12)$$

Table 21.15 Means and Standard Deviation

State of Economy	Probability, P_i	Return, R_i	$R_i P_i$
Project X			
Prosperity	.25	25%	6.25%
Normal	.50	15%	7.50%
Recession	.25	5%	1.25%
	1.00		15.00%
Standard deviation = σ_X = 7.07%			
Project Y			
Prosperity	.25	40%	10%
Normal	.50	15%	7.5%
Recession	.25	−10%	−2.5%
	1.00		15.00%
Standard deviation = σ_Y = 17.68%			

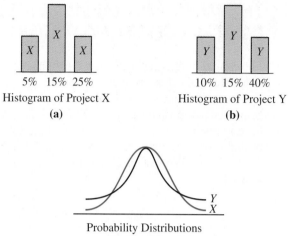

Histogram of Project X   Histogram of Project Y
        (a)   (b)

Figure 21.10

Histograms and
Probability Distri-
butions of Projects
X and Y

Probability Distributions
of Projects X and Y
(c)

Substituting into Equation 21.12 the information from Table 21.15, $\overline{R}_X = .15$ and $\overline{R}_Y = 0.15$, we obtain

$$\sigma_X = [(.25 - .15)^2(.25) + (.15 - .15)^2(.50) + (.05 - .15)^2(.25)]^{1/2}$$

$$= 7.07\%$$

$$\sigma_Y = [(.40 - .15)^2(.25) + (.15 - .15)^2(.50) + (-.10 - .15)^2(.25)]^{1/2}$$

$$= 17.68\%$$

The data given in Table 21.1 can be used to draw histograms of both projects (see Figure 21.9). If we assume that rates of return k are continuously and normally distributed, then Figure 21.10a can be drawn approximately as Figure 21.10b.

The concept of statistical probability distribution can be combined with capital budgeting to derive the **statistical distribution method** for selecting risky investment projects. The expected return for both projects is 15 percent, but because project Y has a flatter distribution with a wider range of values, it is the riskier project. Project X has a normal distribution with a larger collection of values closer to the 15 percent expected rate of return and is therefore more stable.

Statistical Distribution of Cash Flow

Accounting concepts make it possible to define the net cash flows as

$$\text{Net } C_t = \text{CF}_t - \tau_c(\text{CF}_t - d_t - I_t)$$

where

$\text{CF}_t = Q_t(P_t - V_t) = d_t$
$\text{CF}_t = $ cash flow
$\quad Q = $ quantity produced and sold
$\quad P = $ price per unit
$\quad V = $ variable costs
$\quad d_t = $ depreciation
$\quad \tau_c = $ tax rate
$\quad I_t = $ interest expense

For this equation, net cash flow is a random number because Q, P, and V are not known with certainty. We can assume that Net C_t has a normal distribution with mean $\overline{C}_t$ and variance σ_t^2, which was defined in Equation 7.17.

If two projects have the same expected cash flow, or return, as determined by the expected value defined in Equation 21.10, we might be indifferent between the projects if we were to make our choice on the basis of return alone. If, however, we also take risk (variance) into account, we will get a more accurate picture of what type of cash flow or return distribution to expect.

With the introduction of risk, a firm is not necessarily indifferent between two investment proposals that are equal in net present value (NPV).[9] We should estimate both NPV and its standard deviation σ_{NPV} when we perform capital budgeting analysis under uncertainty. Net present value under uncertainty can be defined as

$$\text{NPV} = \sum_{t=1}^{N} \frac{\tilde{C}_t}{(1 + R_f)^t} + \frac{S_t}{(1 + R_f)^N} - I_0 \tag{21.13}$$

where

$\tilde{C}_t = $ uncertain net cash flow in period t
$R_f = $ risk-free discount rate
$S_t = $ salvage value of facilities
$I_0 = $ initial outlay (investment)

The mean of the NPV distribution and its standard deviation can be defined as follows for mutually independent cash flows.

$$\text{NPV} = \sum_{t=1}^{N} \frac{\overline{C}_t}{(1 + R_f)^t} + \frac{S_t}{(1 + R_f)^N} - I_0 \tag{21.14}$$

$$\sigma_{\text{NPV}} = \left(\sum_{t=1}^{n} \frac{\sigma_t^2}{(1 + R_f)^{2t}} \right)^{1/2} \tag{21.15}$$

The generalized case for Equations 21.14 and 21.15 is explored in Appendix 21C.

[9]See Appendix 21C for the definition of NPV.

EXAMPLE 21.13   *The Mean and Variance Approach for Capital Budgeting Decisions*

A firm is considering the introduction of two new product lines, A and B, that have the same life and have the cash flows, standard deviations of cash flows, and salvage values shown in Table 21.16. Assume a discount rate of 10 percent. Both projects have the same expected NPV.

$$\overline{NPV}_A = \overline{NPV}_B = \sum_{t=1}^{5} \frac{\overline{C}_t}{(1 + R_f)^t}$$

$$= 20(PVIF_{10\%,1}) + 20(PVIF_{10\%,2}) + 20(PVIF_{10\%,3})$$
$$+ 20(PVIF_{10\%,4}) + 20(PVIF_{10\%,5}) - 60 + 5(.6209)$$

$$= 20(.9091) + 20(.08264) + 20(.7513) + 20(.6830)$$
$$+ 20(.6209) - 60 + 5(.6209)$$

$$= 18.90$$

where PVIF = present value interest factor (see Appendix 21.C for the calculation).

However, because the standard deviation of A's cash flows is greater than that of B's, project A is riskier than project B. This difference can be explicitly evaluated only by using the statistical distribution method. To compare the riskiness of the two projects, we calculate the standard deviation of their NPVs. We will assume

Table 21.16 Data for Example 21.10 (in thousands)

	Project A	Project B
Initial investment	$60	$60
Cash Flows		
Year 1	$20	$20
Standard Deviation	$ 4	$ 2
Year 2	$20	$20
Standard Deviation	$ 4	$ 2
Year 3	$20	$20
Standard Deviation	$ 4	$ 2
Year 4	$20	$20
Standard Deviation	$ 4	$ 2
Year 5	$20	$20
Standard Deviation	$ 4	$ 2
Salvage Value	$ 5	$ 5

that cash flows between different periods are perfectly positively correlated. The σ_{NPV} can then be defined (see Appendix 21D) as

$$\sigma_{NPV} = \sum_{t=1}^{N} \frac{\sigma_t}{(1 + R_f)^t} \tag{21.16}$$

$$\sigma_{NPV_A} = (\$4)(PVIF_{10\%,1}) + (\$4)(PVIF_{10\%,2}) + \cdots + (\$4)(PVIF_{10\%,5})$$

$$= (4)(.9091) + (4)(.8264) + (.7513)4 + 4(.6830) + 4(.6209)$$

$$= 15.16, \text{ or } \$15,160$$

$$\sigma_{NPV_B} = (\$2)(PVIF_{10\%,1}) + (\$2)(PVIF_{10\%,2}) + \cdots + (\$2)(PVIF_{10\%,5})$$

$$= (2)(.9091) + (2)(.8264) + (2)(.7513) + (2)(.6830) + (2)(.6209)$$

$$= 7.58, \text{ or } \$7,580$$

With the same NPV, project B's cash flows would fluctuate $7,580 per year, and project A's $15,160. Therefore, B is to be preferred given the same returns, because it is less risky.

Summary

In this chapter, we examined the concepts and applications of statistical decision theory and saw that it is different from the classical statistics we have worked with in the last 20 chapters. In the context of statistical decision theory, we discussed elements of decision making under uncertainty. Decisions based on extreme values, expected monetary values and utility measurement, Bayes strategies, and decision trees were explored. In addition, we developed the Von Neumann and Morgenstein utility and risk aversion concepts in order to discuss trade-offs between risk and return. The capital asset pricing model and the statistical distribution method for project selection were also investigated.

Appendix 21A Using the Spreadsheet in Decision-Tree Analysis

J. M. Jones (1986, *European Journal of Operation Research,* pp. 385–400) showed how the Lotus 1-2-3 spreadsheet package can be used to construct an entire decision tree. Using the information presented in Figure 21.5, we use Lotus 1-2-3 to construct the decision tree in Figure 21A.1. In this figure, D represents the decision node and C represents the event (chance) node which correspond to □ and ○ in Figure 21.5, respectively. Figure 21A.1 illustrates all the information we have discussed so far in a more systematic fashion.

There are 3 steps in applying a spreadsheet to decision-tree analysis. Data from Example 21.8 is used to show how these 3 steps can be executed.

Figure 21A.1

Decision Tree with Drilling Cost of $100,000

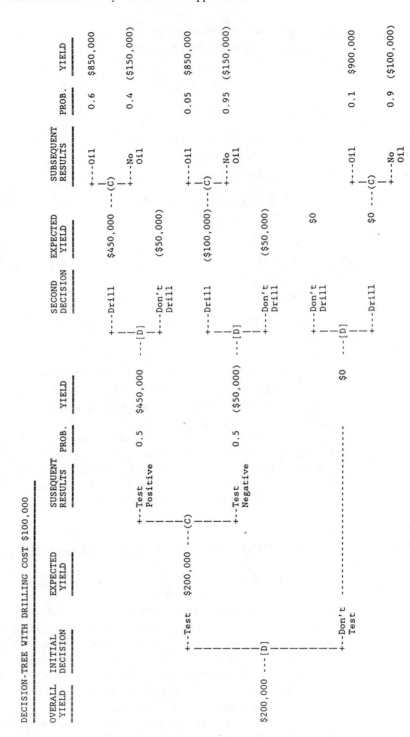

1. Building the Decision Tree on the Spreadsheet (Lotus 1-2-3)

In the Lotus 1-2-3 spreadsheet we know that the cells contain either numbers or labels; we can build the tree on spreadsheet by adopting the following conventions.

a. Denote decision nodes by $\boxed{\text{D}}$.

b. Denote chance nodes by C.

c. Denote the decision emanating from decision nodes and chance outcomes emanating from chance nodes by appropriate labels.

d. Provide a connective structure for the tree using vertical and horizontal dashed line segments.

2. Solving the Tree

The two main tasks involved in solving the decision tree are averaging out and folding back. Because the Lotus 1-2-3 spreadsheet has excellent computational abilities, these two tasks can be done easily. The process is as follows:

a. Create a "master table" within the spreadsheet that incorporates all of the values used as input in the process of developing the tree (see Table 21A.1).

b. Calculate all yields in the tips of the tree from the master table values. (These yields are the net of all costs involved.)

c. Calculate all probabilities needed in the tree from the corresponding values in the master table and put them in the appropriate places of the tree.

d. Use the built-in calculating capabilities of the spreadsheet program to perform the averaging out and folding back process.

3. Sensitive Analysis

The input values in the master table may be subject to change. We can change the values in the master table and then get results under different situations. For example, if we change the drilling cost from $100,000 to $50,000, the overall yield changes from $200,000 to $225,000. The new decision tree is shown in Figure 21A.2.

Table 21A.1 Numbers for the Oil Drilling Problems

Test cost			$50,000
Drilling cost			$100,000
Payoff for successful drilling			$1,000,000
Probability for Test Results			
Positive	.5		
Negative	.5		
Probability for Drilling Results			
Test positive		No test	
Success	.6	Success	.1
Fail		Fail	.9
Test negative			
Success	.05		
Fail	.95		

Figure 21A.2

Decision Tree with Drilling Cost of $50,000

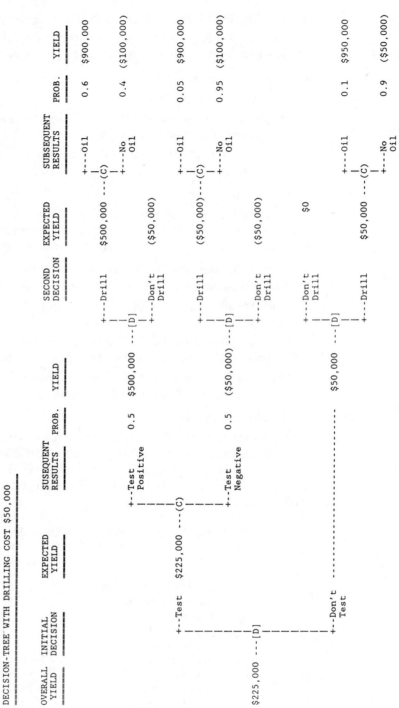

Appendix 21B   Graphical Derivation of the Capital Market Line

The term *risk-free assets* in general refers to government securities such as treasury bills (T-bills). These assets are backed by the federal government and are default-free. In other words, T-bills are riskless; the cash flow from them is certain. An investor's portfolio can be composed of different sets of portfolio opportunities, which may include risk-free assets with a return of R_f, shown on the vertical axis of the risk and return space in Figure 21B.1.

The opportunity to invest in risk-free assets that yield a return of R_f frees the investor to create portfolio combinations that include some risky assets. The investor is able to achieve any combination of risk and return that lies along the line connecting R_f and a point tangent to M_p, the **market portfolio.** All the portfolios along the line $R_f M_p C$ are preferred to the risky portfolio opportunities on the curve $AM_p B$, because they all have higher expected returns and some risk. Therefore, the points of the line $R_f M_p C$ represent the best attainable combinations of risk and return.

At point R_f, an investor has all available funds invested in the riskless asset and expects to receive the return of R_f. The portfolios along the line $R_f M_p$ contain combinations of investments in the risk-free asset and investments in a portfolio of risky assets, M_p. In a sense, the investors who hold these portfolios lend the government money at the risk-free rate R_f—hence the name **lending portfolio.**

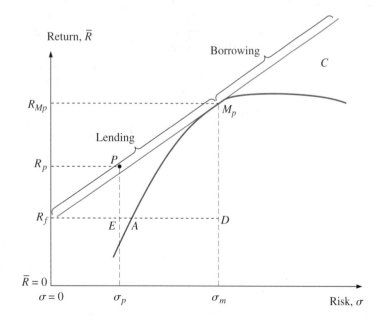

Figure 21B.1

The Capital Market Line

At point M_p, the investor holds only risky assets, having put all her wealth into the risky asset, or market, portfolio. At M_p, investors receive a rate of return R_m and undertake risk σ_m.

If it is assumed that investors can borrow money at the risk-free rate R_f and invest this money in the risky portfolio M_p, they will be able to construct portfolios with higher rates of return but higher risks along the line extending beyond M_p. The extent of movement along the line M_pC is regulated by the margin requirements imposed by the various branches of government. The margin requirements stipulate the minimum amount of money investors must pay to buy stock. The higher the margin requirement, the shorter the line M_pC. The amount of money that investors can borrow may also be limited by the credit-worthiness of the borrowers. The portfolios along segment M_pC are called **borrowing portfolios,** because the investor must borrow funds in order to achieve these combinations of risk and return. The new efficient frontier becomes R_fM_pC and is referred to as the capital market line (CML). The capital market line describes the relationship between expected return and total risk.

Equation 21.6 can be derived geometrically. An investor has three choices in terms of investments. She may invest in R_f, the riskless asset; in the market portfolio M_p; or in any other efficient portfolio along the efficient frontier, such as portfolio P in Figure 21A.1.

If the investor puts her money into the riskless asset, she can receive a return of R_f. If she invests in the market portfolio, she can expect an average return of R_m and risk of σ_m. If she invests in portfolio P, she can expect an average return of R_p with risk of σ_p. The difference between R_m and $R_f (R_m - R_f)$, is called the **market risk premium.**

The investor in portfolio P takes on a risk of σ_p; her risk premium is $(R_p - R_f)$, which is less than the risk of an investor who holds portfolio M_p.

By geometric theory, triangles R_fPE and R_fM_pD are similar—that is, they are directly proportional. Therefore,[10]

$$E(R_p) - R_f = [E(R_m) - R_f]\frac{\sigma_p}{\sigma_m} \tag{21B.1}$$

[10]Because $\Delta PR_fE \approx \Delta M_pR_fD$, it follows that

$$\frac{E(R_p) - R_f}{E(R_m) - R_f} = \frac{\sigma_p}{\sigma_m}$$

From this equation, we obtain Equation 21A.1.

Appendix 21C Present Value and Net Present Value

In this appendix, we review the concepts of present value and net present value.

Present Value

Because many investment projects will generate returns for several years into the future, it is important to assess the present (current) value of future payments. Suppose a payment is to be received in t years' time and the risk-free annual interest rate for a period of t years is r_t. We know that the future value at the end of t years is $(1 + r_t)^t$ per dollar. Conversely, it follows that the **present value** of a dollar received at the end of t years is

$$\text{Present value per dollar} = \frac{1}{(1 + r_t)^t} \qquad (21C.1)$$

For example, say $1,000 is to be received in 4 years' time. At an annual interest rate of 8 percent, the present value of this future receipt is

$$\frac{1,000}{(1.08)^4} = \$735.03$$

Net Present Value

More generally, we can consider a stream of annual receipts, which may be positive or negative. Suppose that, in dollars, we are to receive C_0 now, C_1 in 1 year's time, C_2 in 2 years' time, and so on, until finally we receive C_N in year N. Again, let r_t denote the annual rate of interest for a period of t years. Then, to find the net present value of this stream of receipts, we simply add the individual present values, obtaining

$$\text{NPV} = C_0 + \frac{C_1}{(1 + r_1)^1} + \frac{C_2}{(1 + r_2)^2} + \cdots + \frac{C_N}{(1 + r_N)^N} \qquad (21C.2)$$
$$= \sum_{t=0}^{N} \frac{C_t}{(1 + r_t)^t}$$

Typically, the rate of interest r_t depends on the period t. When a constant rate, r, is assumed for each period, the **net present value** formula, Equation 21B.2, simplifies to

$$\text{NPV} = \sum_{t=0}^{N} \frac{C_t}{(1 + r)^t} \qquad (21C.3)$$

EXAMPLE 21C.1   *NPV Criteria for Capital Budgeting Decisions*

A corporation must choose between two projects. Each project requires an immediate investment, and additional costs will be incurred in the next year. The returns from these projects will be spread over a period of 4 years. The following table shows the dollar amounts involved.

		Year 0	Year 1	Year 2	Year 3	Year 4
Project A	Costs	80,000	20,000	0	0	0
	Returns	0	20,000	30,000	50,000	50,000
Project B	Costs	50,000	50,000	0	0	0
	Returns	0	40,000	60,000	30,000	10,000

At first glance, these data might suggest that for project A total returns exceed total costs by $50,000, whereas the amount of this excess for project B is only $40,000, signaling a preference for project A. However, such an argument neglects the timing of the returns. See what happens when we calculate the present values of the net receipts for each project, assuming an annual interest rate of 8 percent over the period.

		Year 0	Year 1	Year 2	Year 3	Year 4
Project A	Net Returns	−80,000	0	30,000	50,000	50,000
	Present Values	−80,000	0	25,720	39,692	36,751
Project B	Net Returns	−50,000	−10,000	60,000	30,000	10,000
	Present Values	−50,000	−9,259	51,440	23,815	7,350

We must compare the sums of these present values when evaluating the projects. For project A, substituting $r = .08$ into Equation 21C.3 yields

$$\text{NPV} = -80,000 + \frac{0}{(1.08)^1} + \frac{30,000}{(1.08)^2} + \frac{50,000}{(1.08)^3} + \frac{50,000}{(1.08)^4}$$

$$= -80,000 + 0 + 25,720 + 39,692 + 36,751$$

$$= \$22,163$$

Similarly, for project B,

$$\text{NPV} = -50,000 - \frac{10,000}{(1.08)^1} + \frac{60,000}{(1.08)^2} + \frac{30,000}{(1.08)^3} + \frac{10,000}{(1.08)^4}$$

$$= -50,000 - 9,259 + 51,440 + 23,815 + 7,350$$

$$= \$23,346$$

It emerges, then, that if future returns are discounted at an annual rate of 8 percent, the net present value is higher for project B than for project A. Project B is preferable because it provides the firm with larger cash flows in the early years, giving the firm a greater opportunity to reinvest the funds.

Appendix 21D Derivation of Standard Deviation for NPV

In Section 21.8 we discussed calculation of the standard deviation of NPV where cash flows are perfectly positively correlated or where they are independent of each other. Now we develop a general formula for the standard deviation of NPV for use in all cash flow relationships.

The general equation for the standard deviation of NPV, σ_{NPV}, with a mean of

$$\overline{\mathrm{NPV}} = \sum_{t=1}^{N} \frac{\overline{C}_t}{(1 + R_f)^t} + \frac{\overline{S}_t}{(1 + R_f)^N} - I_0$$

is

$$\sigma_{\mathrm{NPV}} = \left(\sum_{t=1}^{N} \frac{\sigma_t^2}{(1 + R_f)^{2t}} + \sum_{t=1}^{N} \sum_{\tau=1}^{N} W_t W_\tau \mathrm{Cov}(C_\tau, C_t) \right)^{1/2}, \quad \tau \neq t \quad (21\mathrm{D}.1)$$

where

σ_t^2 = variance of cash flows in the tth period

W_t, W_τ = discount factor for the tth and the τth period, respectively; that is,

$$W_t = \frac{1}{(1 + R_f)^t} \quad \text{and} \quad W_\tau = \frac{1}{(1 + R_f)^\tau}$$

$\mathrm{Cov}(C_\tau, C_t)$ = covariability between cash flows C_τ and C_t

Cash flows between periods t and τ are generally related. Therefore, $\mathrm{Cov}(C_\tau, C_t)$ is an important factor in the estimation of σ_{NPV}. The magnitude, sign, and degree of the relationships of these cash flows depend on the economic operating conditions and on the nature of the product or service being produced. If there are only three periods, then all terms within the parentheses in Equation 21D.1 can be presented as in Table 21D.1. The summation of the diagonal elements $[W_1^2 \sigma_1^2, W_2^2 \sigma_2^2, W_3^2 \sigma_3^2]$ of Table 21D.1 results in the first part of Equation 21D.1, or

$$\sum_{t=1}^{N} \frac{\sigma_t^2}{(1 + R_f)^{2t}}$$

Table 21D.1 Variance–Covariance Matrix

$W_1^2 \sigma_1^2$	$W_1 W_2 \mathrm{Cov}(C_1, C_2)$	$W_1 W_3 \mathrm{Cov}(C_1, C_3)$
$W_2 W_1 \mathrm{Cov}(C_2, C_1)$	$W_2^2 \sigma_2^2$	$W_2 W_3 \mathrm{Cov}(C_2, C_3)$
$W_3 W_1 \mathrm{Cov}(C_3, C_1)$	$W_2 W_3 \mathrm{Cov}(C_2, C_3)$	$W_3^2 \sigma_3^2$

The summation of all other elements in Table 21D.1 gives the second portion of Equation 21D.1, or

$$\sum_{t=1}^{N} \sum_{\tau=1}^{N} W_t W_\tau \text{Cov}(C_\tau, C_t), \quad t \neq \tau$$

Equation 21D.1 is the general equation for σ_{NPV}. Both Equation 21.16 for σ_{NPV} under perfectly positively correlated cash flows and Equation 21.15 for independent cash flows are special cases derived from the general Equation 21D.1. If $\rho_{12} = \rho_{13} = \rho_{23} = 1$, then $\text{Cov}(C_1, C_2) = \sigma_1 \sigma_2$, $\text{Cov}(C_1, C_3) = \sigma_1 \sigma_3$, and $\text{Cov}(C_2, C_3) = \sigma_2 \sigma_3$. Therefore, Equation 21D.1 reduces to

$$\sigma_{\text{NPV}} = \left(\frac{\sigma_1^2}{(1 + R_f)^2} + \frac{\sigma_2^2}{(1 + R_f)^4} + \frac{\sigma_3^2}{(1 + R_f)^6} + \frac{2\sigma_1 \sigma_2}{(1 + R_f)^3} + \frac{2\sigma_1 \sigma_3}{(1 + R_f)^4} \right.$$

$$\left. + \frac{2\sigma_2 \sigma_3}{(1 + R_f)^5} \right)^{1/2}$$

$$= \sum_{t=1}^{3} \frac{\sigma_t}{(1 + R_f)^t}$$

which is Equation 21.15.

Questions and Problems

1. What are the basic elements of decision making? Define those elements separately. If we don't know the probabilities for the states of nature, can we still make a decision?

2. John faces the following decision problem.

Study Hours per Day	High Confidence	Average Confidence	Low Confidence
0	60	40	30
5	80	60	50
10	90	80	60

With 5 hours of study per day, John estimates that there are three different numbers of points he can get on the midterm: 80 with high confidence, 60 with average confidence, and 50 with low confidence. He also has two other possible actions: studying 0 hours per day and studying 10 hours per day. Estimate the points he can get on the midterm in each of the three states: high, average, and low. Try to use the maximin criterion to choose the best action and specify the most points he can get on the midterm.

3. In question 2, rebuild the table by using the minimax criterion and specify the best action and the best points. Is the best action the same as that of question 2? If yes, is this by chance or is it always true?

4. Reconsider the table in question 2 in the following way:

Study Hours per Day	High	Average	Low
0	.2	.5	.3
5	.2	.6	.2
10	.4	.5	.1

If John studies 5 hours per day, then he estimates the probabilities in 3 confidence levels as .2 for high, .6 for average, and .2 for low. The same interpretations apply to the other two levels.

a. What are the probabilities for the high level, given 0, 5, and 10 studying hours per day?

b. Suppose the probability for each action is $\frac{1}{3}$. What are the probabilities for 0, 5, and 10 hours of study per day, given the high level?

5. a. Using the expected monetary value (EMV) criterion, try to construct a table to determine which action is best.

b. Does applying the EMV criterion yield the same best action as applying the maximin criterion or applying minimax regret criterion?

6. Assume the following utilities for the different midterm points in question 2. The point scores 30, 40, 50, 60, 80, and 90 have utilities of 2, 5, 7.5, 9.5, 11, and 12 units, respectively. Is this a risk-averse, risk-neutral, or risk-loving type of utility function? (The implications of this assessment are worth pondering!)

7. Define the terms *risk-averse, risk-neutral,* and *risk-taking.*

8. a. Reconstruct an expected utility table by using the table in question 4 and the assumption for utility units in question 6. If John makes his decision in accordance with the criterion of largest expected utility, which action will he choose?

b. Redefine *risk averse, risk-neutral,* and *risk-taking* in terms of the expected utility concept.

9. Given the utility function $U(W) = 10W^{1/2}$, wherein W = payoff.

a. Graph the function.

b. Does the function exhibit risk aversion? What is your criterion?

c. How will changing the constant term 10 to an arbitrary number a affect the answer to parts (a) and (b)? (That is, by assuming that a is greater or less than 0, what different result will we get?)

10. Mr. Clark has $100 and would like to try his luck in an Atlantic City casino. Suppose he is faced with a 60/40 chance of losing $20 or winning $15. Further suppose that for a fee of $10, he can buy insurance that completely removes the risk.

a. If Mr. Clark's utility function is logarithmic for $U(W) = \ln W$, is he a risk-averse person? How do you know?

b. Will Mr. Clark buy the insurance or take the gamble?

c. Say the risk increases to a 70/30 chance of losing $20 or winning $15. How much of a premium will Mr. Clark now pay for insurance to remove the risk completely (assuming he remains risk-averse)?

d. Say Mr. Clark's initial wealth increases to $150. What change does this bring about in the risk premium he pays to remove the risk completely (assuming he remains risk-averse)?

11. Lottery A offers a 70 percent chance of winning $45 and a 30 percent chance of losing $100. Lottery B offers a 60 percent chance of winning $55 and a 40 percent chance of losing $85. Lottery C offers an 80 percent chance of winning $30 and a 20 percent chance of losing $110. Without knowing any additional information, such as the utility function, which lottery will you choose? By what criterion?

12. Use the R_i and R_m information given in the table to calculate systematic risk, which was defined in Equation 21.4 in the text.

Quarterly Rates of Return for IBM (R_i) and Market Rates (R_m), Second Quarter 1981 to Second Quarter 1991

		Market Return	IBM Return
1981			
	2	−.03522	−.05835
	3	−.11454	−.04993
	4	.054828	.066697
1982	1	−.08641	.065670
	2	−.02098	.029037
	3	.098622	.224494
	4	.167912	.323475
1983	1	.087599	.066077
	2	.099045	.191154
	3	−.01213	.062993
	4	−.00686	−.03093
1984	1	−.03486	−.05778
	2	−.03769	−.06403
	3	.084345	.185342
	4	.006863	−.00020
1985	1	.080243	.040406
	2	.061939	−.01692
	3	−.05092	.009898
	4	.160369	.264177
1986	1	.130726	−.01864
	2	.049979	−.02574
	3	−.07781	−.07440
	4	.046904	−.09962

(continued)

(Continued)

		Market Return	IBM Return
1987	1	.204525	.260208
	2	.042166	.089758
	3	.058651	−.06553
	4	−.23232	−.22653
1988	1	.047883	−.05865
	2	.056433	.193728
	3	−.00581	−.08557
	4	.021367	.065872
1989	1	.061752	−.09558
	2	.078373	.036288
	3	.098025	−.01264
	4	.012172	−.12736
1990	1	−.03808	.140345
	2	.053185	.118586
	3	−.14515	−.08438
	4	.078974	.073654
1991	1	.136272	.018451
	2	−.01082	−.13646

13. a. Given $R_f = 5$ percent and $E(R_m) = 10$ percent, plot the security market line (SML) (write the equation).

  b. If $\beta_i = 2$, and $E(R_i) = 12$ percent, will a wise investor *purchase* stock i? Why or why not?

14. Discuss some of the different methods of decision making, and explain when you would use each one.

15. Describe why knowing which outcome you prefer is not adequate for making a choice under uncertainty.

16. Describe why good decisions sometimes result in bad outcomes.

17. What is the maximin criterion? When is it best to use the maximin criterion?

18. What is the minimax regret criterion? When is using this criterion best?

19. Using Example 21.10 in the text as an example, explain how Bayesian analysis can be applied in decision making.

20. What is the expected monetary value (EMV) criterion? Briefly explain how this criterion is applied.

21. Suppose you are interested in evaluating a stock's price. You have analyzed the probability that the stock will go up on any given day as 1/3, the probability that the stock will go down on any given day as 1/3, and the probability that the stock's price will not change on any given day as 1/3. Use a decision tree to show the possible stock price movements for 3 days.

22. Briefly explain what the dominance principle is and how it can be used in risk-and-return analysis.

23. Draw the capital market line. Write down the equation for the capital market line. Explain what the capital market line tells us.

24. What is the capital asset pricing model (CAPM)? What are the assumptions of this model? What does it tell us?

25. What is the CML? What is the SML? How are they similar? How are they different?

26. An investor wants to choose among three investment alternatives: a passbook savings account, a government bond fund, and a growth stock fund. The payoffs for a $20,000 investment are given in the following table.

	State of Nature		
Investment	Low Growth	Normal Growth	High Growth
Savings account	$1,000	$1,000	$1,000
Bond fund	$1,500	$1,000	$ 800
Stock fund	$ 500	$1,250	$1,500

  a. Which investment does applying the minimax regret criterion instruct us to choose?

  b. Which investment does applying the maximin criterion instruct us to choose?

27. Now suppose the investor in question 26 assigns probabilities of .3 to low growth, .4 to normal growth, and .3 to high growth. Use the expected monetary value criterion to determine which investment should be chosen.

28. You are given the following information on the market and on XYZ Company's stock.

$$\text{Return on market} = 10\%$$
$$\text{Risk-free interest rate} = 6\%$$
$$\text{Beta for XYZ stock} = 1.5$$

Compute the expected return on XYZ's stock.

29. You are trying to decide whether you should study a lot or a little for your statistics midterm. You construct the following grade–payoff table.

State of Nature

Action	Easy Test	Hard test
Study a lot	98	95
Don't study	90	55

   a. Use the minimax regret criterion to determine how much to study.

   b. Use the maximin criterion to determine how much to study.

30. A studio that has just produced a new movie must decide when to release it. The possible actions are

A_1: release the movie in the spring
A_2: release the movie in the fall
A_3: release the movie at Christmas time
A_4: release the movie in the summer

The possible states of nature are

S_1: low movie attendance
S_2: average movie attendance
S_3: high movie attendance

The payoff table is

State of Nature

Action	S_1	S_2	S_3
A_1	−20	10	20
A_2	10	10	10
A_3	25	35	45
A_4	20	19	40

   a. Which of these actions will the studio choose if it uses the minimax criterion?

   b. Which of these actions will the studio choose if it uses the maximin criterion?

31. Suppose that in question 30, you have assigned the probabilities of .2 to low movie attendance, .5 to average movie attendance, and .3 to high movie attendance. Use the expected monetary value criterion to determine when the studio should release the movie.

32. What is a decision tree? Briefly explain how a decision tree can be used in decision theory.

33. Consider the following payoff table, where the cell entries are in dollars.

Alternative

Outcome	A	B	C	D
1	5	7	5	4
2	3	2	2	7
3	9	8	7	5
4	7	10	6	4

Can any alternatives be eliminated by using dominance?

34. Use an example to show the similarities and differences between the Sharpe and the Treynor investment performance measures.

35. A local deli prepares fresh potato salad for its customers every day. The unsold salad has to be thrown away. The demand for potato salad can be classified as low (100 pounds), medium (200 pounds), or high (300 pounds). The production runs being considered are 100, 200, and 300 pounds. The payoffs for all combinations of production and demand are shown here.

Demand

Production	100	200	300
100	400	300	− 100
200	− 400	600	700
300	− 800	− 300	1500

   a. What is the maximin solution of this problem?

   b. What is the expected monetary value of each action if the probabilities of demand being low, medium, and high are .3, .3 and .4, respectively?

Use the following information to answer questions 34–40. A manufacturer is planning its production for the next 6 months. It has to decide how much of an important ingredient to keep in inventory. The demand for the ingredient may be low, medium, or high. The manufacturer is considering holding either a low or a high amount of inventory. The possible payoffs for all the combinations of inventory holding and demand are shown here.

	Demand		
Inventory Holding	Low S_1	Medium S_2	High S_3
Low A_1	200	300	300
High A_2	100	200	500
Probability	.3	.2	.5

36. What is the minimax regret solution to this problem?

37. What is the expected payoff of each action?

38. An economics consultant predicts that the demand for the ingredient will be low in the next 6 months. In the past few years, this economist has provided forecasting about the demand for the ingredient. The track record of the consultant is summarized by the following conditional probability distribution:

$$p(H \mid S_1) = .6 \quad p(H \mid S_2) = .4 \quad p(H \mid S_3) = .1$$

$$p(M \mid S_1) = .2 \quad p(M \mid S_2) = .2 \quad p(M \mid S_3) = .1$$

$$p(L \mid S_1) = .2 \quad p(L \mid S_2) = .5 \quad p(L \mid S_3) = .8$$

Assume that this time, the economist predicts a low demand for the future. Find the posterior distribution of S_1, of S_2, and of S_3.

39. Use the foregoing information to evaluate the action of accumulating a high inventory. (Obtain the expected monetary value by using posterior probability.)

40. Write out the decision tree for this question.

Use the following information to answer questions 41–45. A company is considering what size copying machine it should lease. The copier comes in three different sizes: small, medium, and large. A larger machine can handle more work, but it also costs more. The demand for the machine in the next year is uncertain. The cost of leasing a smaller machine is lower than the cost of leasing a larger one. However, at those times when the small machine could not handle the high demand, the company would have to lease a second machine and pay a significantly higher cost than if it had leased a larger machine in the first place. The possible costs of leasing the three different copiers under conditions of low and high demand are presented in the following table.

	Future Demand	
Size of Copier	Low (S_1)	High (S_2)
Small	400	800
Medium	500	900
Large	600	600
Pr(S)	.4	.6

41. Find the maximin solution.

42. Find the minimax regret solution.

43. Would you lease a medium-sized machine under any circumstances? Why or why not?

44. What are the expected costs of leasing a large machine?

45. An economic consultant predicts that the demand for the machine will be high the next year and suggests that the company should therefore lease a large machine. Assume that the consultant has the following track record of predicting demand.

$$\Pr(I_1 \mid S_1) = .8 \qquad \Pr(I_2 \mid S_1) = .2$$

$$\Pr(I_1 \mid S_2) = .2 \qquad \Pr(I_2 \mid S_2) = .8$$

where I_1 indicates that the consultant predicts S_1, and I_2 indicates that the consultant predicts S_2. Do you agree with the consultant's advice?

46. A limousine chauffeur is going to take a guest from the hotel to the airport. To catch the flight, the chauffeur has to arrive at the airport in 30 minutes. There are two routes to the airport: a local route and the highway. The chauffeur has found that it always takes him 25 minutes to get to the airport when he takes the local route. When he takes the highway, the time consumed depends on the traffic. When the highway is jammed, it takes him 36 minutes to get to the airport. When the highway is clear, the trip takes only 10 minutes. There is a .10 probablity that the highway will be jammed.
  a. Which route should the chauffeur take on the way to the airport?
  b. Which way should the chauffeur take when he is coming back to the hotel?

47. Does the risk-averse decision maker ever take any risk?

Use the following information to answer questions 48–50. Assume an investor has the utility function $W^{1/3}$, where W is wealth. The state government issues a lottery ticket that pays the winner $300. The lottery ticket costs $1. The chance of winning the lottery is 1/200. The investor has $270 in original wealth.

48. What is the expected value of this lottery? What is the investor's expected utility if he buys the lottery?

49. Use the lottery case to show that the investor is risk-averse.

50. Will the investor buy the lottery ticket if his utility function is W^3?

51. The owner of a personal computer company is considering whether to install a large or a small new assembly line. The possible payoffs (in thousands of dollars) depend on the state of the economy and are presented in the following table.

Size of Assembly Line	State	
	Boom	**Recession**
Large	20	5
Small	10	8
Probability	.3	.7

a. Find the minimax regret solution.
b. Find the expected payoff of installing a large assemebly line when the probability of boom and the probability of recession are .3 and .7, respectively.
c. If the beginning wealth of the company is 20 and the utility function of the owner is $W^{1/2}$, what is the expected utility of the two options?

52. Mr. Montero is deciding how to invest the money for his son's tuition, which is due 2 months from today. He can put the money in a 2-month certificate of deposit (CD) earning a 10 percent annual rate of interest, or he can put the money in a 1-month CD now and earn 9 percent. If he puts the money in the 1-month CD, then a month later he will have to invest it in another 1-month CD. The rate 1 month from today is uncertain. Both CDs are protected by FDIC insurance. Assume that Mr. Montero knows the 1-month rate in the next month follows a normal distribution with a mean of 11 percent and a standard deviation of 2 percent. What will be his choice if he is a risk averter? What will be his choice if he is risk-neutral?

Use the following information to answer questions 53–55. Ms. Jones is thinking of investing in a new project that will cost $1,000 to start. There are two ways to raise this $1,000. She can take out $1,000 from her own pocket or take out $500 and invite a friend to share the investment. The investment will generate the following revenues, depending on the outcome of the investment.

	State 1	State 2	State 3
Revenue	$800	$1,000	$1,500

The revenue will be equally split between Ms. Jones and her friend if they share the project. It is estimated that the probabilities of states 1, 2, and 3 are 1/3, 1/3, and 1/3, respectively.

53. a. What is the expected net gain of the project if Ms. Jones undertakes the investment alone?
b. What is the expected net gain of the project for Ms. Jones if the investment is shared?

54. Ms. Jones hired Dr. Lee, an economics consultant, to evaluate the probabilities of states 1, 2, and 3. Suppose this consultant has the following track record.

	Actuality, P(I_i | S_i)		
	S_1	S_2	S_3
I_1	.8	.2	.2
I_2	.1	.6	.2
I_3	.1	.2	.6

a. Obtain the posterior distribution when Dr. Lee predicts I_1.
b. Evaluate the expected payoff for Ms. Jones of sharing the project.

55. Ms. Jones has the utility function $U = f(W) = W^{1/2}$, and her initial wealth is $1,000. Should she invest in the project? If so, should she invest alone? Use the original probability function to answer this question.

56. The owner of the New Land Food Corporation is considering a new project that has the following possible payoffs (in thousands of dollars).

Profits	Probability	Profits	Probability
200	.1	0	.15
1,000	.25	−5,000	.2
5,000	.2	−1,000	.1

The owner's current assets are worth about $5,000. His utility function is $U = W^{1/2}$.

a. What is the expected value of this project?

b. What is the expected utility of this project?

57. The owner of North America Toy company is considering enlarging its production capacity to meet increasing future demand. The company can either expand its old plant or establish a new plant. The possible payoffs of these two actions are related to the increase in future demand and are shown in the following table.

Demand (in thousands of dollars)

Action	Low	Medium	High
Expansion	150	250	250
New plant	0	250	500
Probability	1/3	1/3	1/3

a. Write out the decision tree, and determine which action the owner should take if he uses the minimax regret approach.

b. Assume the net worth of the firm at this stage is $500 thousands. The utility function of the owner is $U(W) = 400 - 4000/W^{1/2}$. Which action will generate the higher expected utility?

58. Ace Corporation is sending a shipment of crystal balls from North Carolina to California. There is a chance that the shipment may be damaged, so Ace Corporation is thinking of buying insurance to cover the shipment. The possible costs that buying and not buying insurance may entail are presented in the following table.

	Shipment Wrecked	Shipment Safe
Buying	$100	$100
Not buying	$5,000	0
Probability	.01	.99

What is the expected value of the insurance policy? If the insurance policy is not worth the cost of insurance ($100), why do people buy it?

59. An ice cream stand at the beach wants to order some ice cream for the coming weekend. The demand for ice cream depends on the weather. The ice cream has to be ordered in 50-gallon units.

The profits that selling ice cream yields under different combinations of state and ice cream order are presented here.

	Bad Weather S_1	Good Weather S_2
50 gallons	$500	$500
100 gallons	$300	$900

Historically, the probabilities of S_1 and S_2 during this time of the year are about .4 and .6, respectively.

a. Find the minimax regret solution.

b. Find the highest-expected-value solution.

c. Compare the utility of ordering 50 gallons versus 100 gallons if the ice cream stand owner's utility function is

$$U = f(E,S) = 50E - 25S$$

where E is the expected wealth and S is the standard deviation of wealth.

Use the following information to answer questions 60–63. A new company is formed to invest in a new project. This company is going to raise the needed capital, $100,000, by issuing $50,000 bonds and $50,000 stock. The bondholder is guaranteed a 10 percent interest rate regardless of the performance of the company. The stockholder will receive whatever is left after bondholders are paid. An investor is thinking of investing $40,000 in the company for 1 year. A year later, she will pull out of the investment. She can put the money in any combination of bonds and stock. The possible payoffs of the project (in thousands of dollars) are recorded here.

	Recession S_1	Stable Economy S_2	Boom S_3
Earnings before interest	5	10	20
Interest	5	5	5
Earnings after interest	0	5	15
Value of all stocks	50	55	65
Probability	.4	.3	.3

60. The investor can choose to put all of her $40,000 in either bonds or stock. What is the expected value for each of these two options at the end of the year? What is the standard deviation of these two options?

61. Assume that the investor puts her $40,000 in a portfolio consisting x percent of bonds and $(1 - x)$ percent of stock. What is the expected value of this portfolio?

62. Assume that the investor's initial wealth is $40,000 and that her utility function is $U = W^{1/2}$. What is the expected utility of the portfolio described in question 61?

63. A stock analyst has just released a report saying that the economy will be good in the coming year. His track record is

$$Pr(good | S_1) = .2 \qquad Pr(bad | S_1) = .8$$
$$Pr(good | S_2) = .5 \qquad Pr(bad | S_2) = .5$$
$$Pr(good | S_3) = .8 \qquad Pr(bad | S_3) = .2$$

Evaluate the portfolio made up entirely of stock and that made up entirely of bonds.

64. President Reagan was interested in establishing a "defensive wall" to blunt the threat of a nuclear missile attack against this country. The plan was commonly known as "Star Wars." Assume a defense contractor comes up with two defense systems. System A will destroy 60 percent of the missiles launched against this country, but 40 percent of the missiles will get through. System B has a 60 percent probability of destroying all the missiles launched but a 40 percent chance of letting all the missiles get through. Should you use the expected value approach to choose which system to support? Make your choice using common sense.

65. Peter Campbell plans to invest in real estate income property. He plans to hold that property for 7 years. He is considering 2 income properties, A and B. Their initial investment, cash flows, standard deviations of cash flows, and resale values are as follows.

Property A (in thousands of dollars)

Year	Cash Flow	Standard Deviation
0	−500	0
1	18	1
2	18	1.1
3	18	1.3
4	18	1.4
5	18	1.5
6	18	1.7
7	18	1.9
7	400 (resale value)	0

Property B (in thousands of dollars)

Year	Cash Flow	Standard Deviation
0	−500	0
1	18	.9
2	18	1.1
3	18	1.4
4	18	1.5
5	18	1.6
6	18	1.8
7	18	1.9
7	400 (resale value)	0

Use Lotus 1-2-3 to calculate the NPV and σ_{NPV} of each property. (Assume a discount rate of 12 percent.) According to the results, which property should Peter choose?

66. The MINITAB output on page 928 is a market model for Ford stock return. Please explain the result in accordance with Equations 21.4 and 21.5. (Hint: Refer to Example 21.11.)

67. Assuming that 2 conditions in Table 21A.1 change as follows: (1) test cost changes to $25,000 and (2) under the condition that the test is positive, the probability of successful drilling is .7 (failure is .3), use the methods introduced in Appendix 21A to analyze the oil drilling problem again.

MINITAB Output of Market Model for Ford (for Question 66)

```
MTB > PRINT C2-C4

 ROW   Year     Ford    Market

   1     70    0.4260   0.0010
   2     71    0.2933   0.1080
   3     72    0.1717   0.1557
   4     73   -0.4512  -0.1737
   5     74   -0.0968  -0.2964
   6     75    0.3960   0.3149
   7     76    0.4614   0.1918
   8     77   -0.2067  -0.1153
   9     78   -0.0026   0.0105
  10     79   -0.1479   0.1228
  11     80   -0.2938   0.2586
  12     81   -0.1025  -0.0994
  13     82    1.3212   0.1549
  14     83    0.1286   0.1706
  15     84    0.0980   0.0115
  16     85    0.3237   0.2633
  17     86    0.0081   0.1462
  18     87    0.3961   0.0203
  19     88   -0.2995   0.1240
  20     89   -0.0767   0.2725
  21     90   -0.3209  -0.0656

MTB > BRIEF 2
MTB > REGRESS C3 1 C4;
SUBC> DW.

The regression equation is
Ford = 0.0306 + 0.878 Market

Predictor      Coef      Stdev    t-ratio        p
Constant     0.03057    0.09090      0.34    0.740
Market        0.8777     0.5234      1.68    0.110

s = 0.3756      R-sq = 12.9%     R-sq(adj) = 8.3%

Analysis of Variance

SOURCE       DF          SS          MS        F        p
Regression    1      0.3968      0.3968     2.81    0.110
Error        19      2.6809      0.1411
Total        20      3.0777

Unusual Observations
Obs.   Market     Ford       Fit Stdev.Fit  Residual   St.Resid
   5   -0.296  -0.0968   -0.2296    0.2110    0.1328    0.43 X
  13    0.155   1.3212    0.1665    0.0920    1.1547    3.17R

R denotes an obs. with a large st. resid.
X denotes an obs. whose X value gives it large influence.

Durbin-Watson statistic = 1.80

MTB > PAPER
```

Appendix A Statistical Tables

Table AI Probability Function of the Binomial Distribution

The table shows the probability of x successes in n independent trials, each with probability of success p. For example, the probability of 4 successes in 8 independent trials, each with probability of success .35, is .1875.

n	x	p									
		.05	.10	.15	.20	.25	.30	.35	.40	.45	.50
1	0	.9500	.9000	.8500	.8000	.7500	.7000	.6500	.6000	.5500	.5000
	1	.0500	.1000	.1500	.2000	.2500	.3000	.3500	.4000	.4500	.5000
2	0	.9025	.8100	.7225	.6400	.5625	.4900	.4225	.3600	.3025	.2500
	1	.0950	.1800	.2550	.3200	.3750	.4200	.4550	.4800	.4950	.5000
	2	.0025	.0100	.0225	.0400	.0625	.0900	.1225	.1600	.2025	.2500
3	0	.8574	.7290	.6141	.5120	.4219	.3430	.2746	.2160	.1664	.1250
	1	.1354	.2430	.3251	.3840	.4219	.4410	.4436	.4320	.4084	.3750
	2	.0071	.0270	.0574	.0960	.1406	.1890	.2389	.2880	.3341	.3750
	3	.0001	.0010	.0034	.0080	.0156	.0270	.0429	.0640	.0911	.1250
4	0	.8145	.6561	.5220	.4096	.3164	.2401	.1785	.1296	.0915	.0625
	1	.1715	.2916	.3685	.4096	.4219	.4116	.3845	.3456	.2995	.2500
	2	.0135	.0486	.0975	.1536	.2109	.2646	.3105	.3456	.3675	.3750
	3	.0005	.0036	.0115	.0256	.0469	.0756	.1115	.1536	.2005	.2500
	4	.0000	.0001	.0005	.0016	.0039	.0081	.0150	.0256	.0410	.0625
5	0	.7738	.5905	.4437	.3277	.2373	.1681	.1160	.0778	.0503	.0312
	1	.2036	.3280	.3915	.4096	.3955	.3602	.3124	.2592	.2059	.1562
	2	.0214	.0729	.1382	.2048	.2637	.3087	.3364	.3456	.3369	.3125
	3	.0011	.0081	.0244	.0512	.0879	.1323	.1811	.2304	.2757	.3125
	4	.0000	.0004	.0022	.0064	.0146	.0284	.0488	.0768	.1128	.1562
	5	.0000	.0000	.0001	.0003	.0010	.0024	.0053	.0102	.0185	.0312
6	0	.7351	.5314	.3771	.2621	.1780	.1176	.0754	.0467	.0277	.0156
	1	.2321	.3543	.3993	.3932	.3560	.3025	.2437	.1866	.1359	.0938
	2	.0305	.0984	.1762	.2458	.2966	.3241	.3280	.3110	.2780	.2344
	3	.0021	.0146	.0415	.0819	.1318	.1852	.2355	.2765	.3032	.3125
	4	.0001	.0012	.0055	.0154	.0330	.0595	.0951	.1382	.1861	.2344
	5	.0000	.0001	.0004	.0015	.0044	.0102	.0205	.0369	.0609	.0938
	6	.0000	.0000	.0000	.0001	.0002	.0007	.0018	.0041	.0083	.0156
7	0	.6983	.4783	.3206	.2097	.1335	.0824	.0490	.0280	.0152	.0078
	1	.2573	.3720	.3960	.3670	.3115	.2471	.1848	.1306	.0872	.0547
	2	.0406	.1240	.2097	.2753	.3115	.3177	.2985	.2613	.2140	.1641

Table AI (Continued)

n	x	.05	.10	.15	.20	.25	.30	.35	.40	.45	.50
							p				
	3	.0036	.0230	.0617	.1147	.1730	.2269	.2679	.2903	.2918	.2734
	4	.0002	.0026	.0109	.0287	.0577	.0972	.1442	.1935	.2388	.2734
	5	.0000	.0002	.0012	.0043	.0115	.0250	.0466	.0774	.1172	.1641
	6	.0000	.0000	.0001	.0004	.0013	.0036	.0084	.0172	.0320	.0547
	7	.0000	.0000	.0000	.0000	.0001	.0002	.0006	.0016	.0037	.0078
8	0	.6634	.4305	.2725	.1678	.1001	.0576	.0319	.0168	.0084	.0039
	1	.2793	.3826	.3847	.3355	.2670	.1977	.1373	.0896	.0548	.0312
	2	.0515	.1488	.2376	.2936	.3115	.2965	.2587	.2090	.1569	.1094
	3	.0054	.0331	.0839	.1468	.2076	.2541	.2786	.2787	.2568	.2188
	4	.0004	.0046	.0815	.0459	.0865	.1361	.1875	.2322	.2627	.2734
	5	.0000	.0004	.0026	.0092	.0231	.0467	.0808	.1239	.1719	.2188
	6	.0000	.0000	.0002	.0011	.0038	.0100	.0217	.0413	.0703	.1094
	7	.0000	.0000	.0000	.0001	.0004	.0012	.0033	.0079	.0164	.0312
	8	.0000	.0000	.0000	.0000	.0000	.0001	.0002	.0007	.0017	.0039
9	0	.6302	.3874	.2316	.1342	.0751	.0404	.0207	.0101	.0046	.0020
	1	.2985	.3874	.3679	.3020	.2253	.1556	.1004	.0605	.0339	.0176
	2	.0629	.1722	.2597	.3020	.3003	.2668	.2162	.1612	.1110	.0703
	3	.0077	.0446	.1069	.1762	.2336	.2668	.2716	.2508	.2119	.1641
	4	.0006	.0074	.0283	.0661	.1168	.1715	.2194	.2508	.2600	.2461
	5	.0000	.0008	.0050	.0165	.0389	.0735	.1181	.1672	.2128	.2461
	6	.0000	.0001	.0006	.0028	.0087	.0210	.0424	.0743	.1160	.1641
	7	.0000	.0000	.0000	.0003	.0012	.0039	.0098	.0212	.0407	.0703
	8	.0000	.0000	.0000	.0000	.0001	.0004	.0013	.0035	.0083	.0176
	9	.0000	.0000	.0000	.0000	.0000	.0000	.0001	.0003	.0008	.0020
10	0	.5987	.3487	.1969	.1074	.0563	.0282	.0135	.0060	.0025	.0010
	1	.3151	.3874	.3474	.2684	.1877	.1211	.0725	.0403	.0207	.0098
	2	.0746	.1937	.2759	.3020	.2816	.2335	.1757	.1209	.0763	.0439
	3	.0105	.0574	.1298	.2013	.2503	.2668	.2522	.2150	.1665	.1172
	4	.0010	.0112	.0401	.0881	.1460	.2001	.2377	.2508	.2384	.2051
	5	.0001	.0015	.0085	.0264	.0584	.1029	.1536	.2007	.2340	.2461
	6	.0000	.0001	.0012	.0055	.0162	.0368	.0689	.1115	.1596	.2051
	7	.0000	.0000	.0001	.0008	.0031	.0090	.0212	.0425	.0746	.1172
	8	.0000	.0000	.0000	.0001	.0004	.0014	.0043	.0106	.0229	.0439
	9	.0000	.0000	.0000	.0000	.0000	.0001	.0005	.0016	.0042	.0098
	10	.0000	.0000	.0000	.0000	.0000	.0000	.0000	.0001	.0003	.0010
11	0	.5688	.3138	.1673	.0859	.0422	.0198	.0088	.0036	.0014	.0005
	1	.3293	.3835	.3248	.2362	.1549	.0932	.0518	.0266	.0125	.0054
	2	.0867	.2131	.2866	.2953	.2581	.1998	.1395	.0887	.0513	.0269
	3	.0137	.0710	.1517	.2215	.2581	.2568	.2254	.1774	.1259	.0806
	4	.0014	.0158	.0536	.1107	.1721	.2201	.2428	.2365	.2060	.1611
	5	.0001	.0025	.0132	.0388	.0803	.1321	.1830	.2207	.2360	.2256
	6	.0000	.0003	.0023	.0097	.0268	.0566	.0985	.1471	.1931	.2256
	7	.0000	.0000	.0003	.0017	.0064	.0173	.0379	.0701	.1128	.1611

Table AI (Continued)

n	x	.05	.10	.15	.20	.25	.30	.35	.40	.45	.50
						p					
	8	.0000	.0000	.0000	.0002	.0011	.0037	.0102	.0234	.0462	.0806
	9	.0000	.0000	.0000	.0000	.0001	.0005	.0018	.0052	.0126	.0269
	10	.0000	.0000	.0000	.0000	.0000	.0000	.0002	.0007	.0021	.0054
	11	.0000	.0000	.0000	.0000	.0000	.0000	.0000	.0000	.0002	.0005
12	0	.5404	.2824	.1422	.0687	.0317	.0138	.0057	.0022	.0008	.0002
	1	.3413	.3766	.3012	.2062	.1267	.0712	.0368	.0174	.0075	.0029
	2	.0988	.2301	.2924	.2835	.2323	.1678	.1088	.0639	.0339	.0161
	3	.0173	.0852	.1720	.2362	.2581	.2397	.1954	.1419	.0923	.0537
	4	.0021	.0213	.0683	.1329	.1936	.2311	.2367	.2128	.1700	.1208
	5	.0002	.0038	.0193	.0532	.1032	.1585	.2039	.2270	.2225	.1934
	6	.0000	.0005	.0040	.0155	.0401	.0792	.1281	.1766	.2124	.2256
	7	.0000	.0000	.0006	.0033	.0115	.0291	.0591	.1009	.1489	.1934
	8	.0000	.0000	.0001	.0005	.0024	.0078	.0199	.0420	.0762	.1208
	9	.0000	.0000	.0000	.0001	.0004	.0015	.0048	.0125	.0277	.0537
	10	.0000	.0000	.0000	.0000	.0000	.0002	.0008	.0025	.0068	.0161
	11	.0000	.0000	.0000	.0000	.0000	.0000	.0001	.0003	.0010	.0029
	12	.0000	.0000	.0000	.0000	.0000	.0000	.0000	.0000	.0001	.0002
13	0	.5133	.2542	.1209	.0550	.0238	.0097	.0037	.0013	.0004	.0001
	1	.3512	.3672	.2774	.1787	.1029	.0540	.0259	.0113	.0045	.0016
	2	.1109	.2448	.2937	.2680	.2059	.1388	.0836	.0453	.0220	.0095
	3	.0214	.0997	.1900	.2457	.2517	.2181	.1651	.1107	.0660	.0349
	4	.0028	.0277	.0838	.1535	.2097	.2337	.2222	.1845	.1350	.0873
	5	.0003	.0055	.0266	.0691	.1258	.1803	.2154	.2214	.1989	.1571
	6	.0000	.0008	.0063	.0230	.0559	.1030	.1546	.1968	.2169	.2095
	7	.0000	.0001	.0011	.0058	.0186	.0442	.0833	.1312	.1775	.2095
	8	.0000	.0000	.0001	.0011	.0047	.0142	.0336	.0656	.1089	.1571
	9	.0000	.0000	.0000	.0001	.0009	.0034	.0101	.0243	.0495	.0873
	10	.0000	.0000	.0000	.0000	.0001	.0006	.0022	.0065	.0162	.0349
	11	.0000	.0000	.0000	.0000	.0000	.0001	.0003	.0012	.0036	.0095
	12	.0000	.0000	.0000	.0000	.0000	.0000	.0000	.0001	.0005	.0016
	13	.0000	.0000	.0000	.0000	.0000	.0000	.0000	.0000	.0000	.0001
14	0	.4877	.2288	.1028	.0440	.0178	.0068	.0024	.0008	.0002	.0001
	1	.3593	.3559	.2539	.1539	.0832	.0407	.0181	.0073	.0027	.0009
	2	.1229	.2570	.2912	.2501	.1802	.1134	.0634	.0317	.0141	.0056
	3	.0259	.1142	.2056	.2501	.2402	.1943	.1366	.0845	.0462	.0222
	4	.0037	.0348	.0998	.1720	.2202	.2290	.2022	.1549	.1040	.0611
	5	.0004	.0078	.0352	.0860	.1468	.1963	.2178	.2066	.1701	.1222
	6	.0000	.0013	.0093	.0322	.0734	.1262	.1759	.2066	.2088	.1833
	7	.0000	.0002	.0019	.0092	.0280	.0618	.1082	.1574	.1952	.2095
	8	.0000	.0000	.0003	.0020	.0082	.0232	.0510	.0918	.1398	.1833
	9	.0000	.0000	.0000	.0003	.0018	.0066	.0183	.0408	.0762	.1222
	10	.0000	.0000	.0000	.0000	.0003	.0014	.0049	.0136	.0312	.0611
	11	.0000	.0000	.0000	.0000	.0000	.0002	.0010	.0033	.0093	.0222

Table AI (Continued)

n	x	.05	.10	.15	.20	.25	.30	.35	.40	.45	.50
							p				
	12	.0000	.0000	.0000	.0000	.0000	.0000	.0001	.0005	.0019	.0056
	13	.0000	.0000	.0000	.0000	.0000	.0000	.0000	.0001	.0002	.0009
	14	.0000	.0000	.0000	.0000	.0000	.0000	.0000	.0000	.0000	.0001
15	0	.4633	.2059	.0874	.0352	.0134	.0047	.0016	.0005	.0001	.0000
	1	.3658	.3432	.2312	.1319	.0668	.0305	.0126	.0047	.0016	.0005
	2	.1348	.2669	.2856	.2309	.1559	.0916	.0476	.0219	.0090	.0032
	3	.0307	.1285	.2184	.2501	.2252	.1700	.1110	.0634	.0318	.0139
	4	.0049	.0428	.1156	.1876	.2252	.2186	.1792	.1268	.0780	.0417
	5	.0006	.0105	.0449	.1032	.1651	.2061	.2123	.1859	.1404	.0916
	6	.0000	.0019	.0132	.0430	.0917	.1472	.1906	.2066	.1914	.1527
	7	.0000	.0003	.0030	.0138	.0393	.0811	.1319	.1771	.2013	.1964
	8	.0000	.0000	.0005	.0035	.0131	.0348	.0710	.1181	.1647	.1964
	9	.0000	.0000	.0001	.0007	.0034	.0116	.0298	.0612	.1048	.1527
	10	.0000	.0000	.0000	.0001	.0007	.0030	.0096	.0245	.0515	.0916
	11	.0000	.0000	.0000	.0000	.0001	.0006	.0024	.0074	.0191	.0417
	12	.0000	.0000	.0000	.0000	.0000	.0001	.0004	.0016	.0052	.0139
	13	.0000	.0000	.0000	.0000	.0000	.0000	.0001	.0003	.0010	.0032
	14	.0000	.0000	.0000	.0000	.0000	.0000	.0000	.0000	.0001	.0005
	15	.0000	.0000	.0000	.0000	.0000	.0000	.0000	.0000	.0000	.0000
16	0	.4401	.1853	.0743	.0281	.0100	.0033	.0010	.0003	.0001	.0000
	1	.3706	.3294	.2097	.1126	.0535	.0228	.0087	.0030	.0009	.0002
	2	.1463	.2745	.2775	.2111	.1336	.0732	.0353	.0150	.0056	.0018
	3	.0359	.1423	.2285	.2463	.2079	.1465	.0888	.0468	.0215	.0085
	4	.0061	.0514	.1311	.2001	.2252	.2040	.1553	.1014	.0572	.0278
	5	.0008	.0137	.0555	.1201	.1802	.2099	.2008	.1623	.1123	.0667
	6	.0001	.0028	.0180	.0550	.1101	.1649	.1982	.1983	.1684	.1222
	7	.0000	.0004	.0045	.0197	.0524	.1010	.1524	.1889	.1969	.1746
	8	.0000	.0001	.0009	.0055	.0197	.0487	.0923	.1417	.1812	.1964
	9	.0000	.0000	.0001	.0012	.0058	.0185	.0442	.0840	.1318	.1746
	10	.0000	.0000	.0000	.0002	.0014	.0056	.0167	.0392	.0755	.1222
	11	.0000	.0000	.0000	.0000	.0002	.0013	.0049	.0142	.0337	.0667
	12	.0000	.0000	.0000	.0000	.0000	.0002	.0011	.0040	.0115	.0278
	13	.0000	.0000	.0000	.0000	.0000	.0000	.0002	.0008	.0029	.0085
	14	.0000	.0000	.0000	.0000	.0000	.0000	.0000	.0001	.0005	.0018
	15	.0000	.0000	.0000	.0000	.0000	.0000	.0000	.0000	.0001	.0002
	16	.0000	.0000	.0000	.0000	.0000	.0000	.0000	.0000	.0000	.0000
17	0	.4181	.1668	.0631	.0225	.0075	.0023	.0007	.0002	.0000	.0000
	1	.3741	.3150	.1893	.0957	.0426	.0169	.0060	.0019	.0005	.0001
	2	.1575	.2800	.2673	.1914	.1136	.0581	.0260	.0102	.0035	.0010
	3	.0415	.1556	.2359	.2393	.1893	.1245	.0701	.0341	.0144	.0052
	4	.0076	.0605	.1457	.2093	.2209	.1868	.1320	.0796	.0411	.0182
	5	.0010	.0175	.0668	.1361	.1914	.2081	.1849	.1379	.0875	.0472

Table A1 (Continued)

n	x	.05	.10	.15	.20	.25	.30	.35	.40	.45	.50
							p				
	6	.0001	.0039	.0236	.0680	.1276	.1784	.1991	.1839	.1432	.0944
	7	.0000	.0007	.0065	.0267	.0668	.1201	.1685	.1927	.1841	.1484
	8	.0000	.0001	.0014	.0084	.0279	.0644	.1134	.1606	.1883	.1855
	9	.0000	.0000	.0003	.0021	.0093	.0276	.0611	.1070	.1540	.1855
	10	.0000	.0000	.0000	.0004	.0025	.0095	.0263	.0571	.1008	.1484
	11	.0000	.0000	.0000	.0001	.0005	.0026	.0090	.0242	.0525	.0944
	12	.0000	.0000	.0000	.0000	.0001	.0006	.0024	.0081	.0215	.0472
	13	.0000	.0000	.0000	.0000	.0000	.0001	.0005	.0021	.0068	.0182
	14	.0000	.0000	.0000	.0000	.0000	.0000	.0001	.0004	.0016	.0052
	15	.0000	.0000	.0000	.0000	.0000	.0000	.0000	.0001	.0003	.0010
	16	.0000	.0000	.0000	.0000	.0000	.0000	.0000	.0000	.0000	.0001
	17	.0000	.0000	.0000	.0000	.0000	.0000	.0000	.0000	.0000	.0000
18	0	.3972	.1501	.0536	.0180	.0056	.0016	.0004	.0001	.0000	.0000
	1	.3763	.3002	.1704	.0811	.0338	.0126	.0042	.0012	.0003	.0001
	2	.1683	.2835	.2556	.1723	.0958	.0458	.0190	.0069	.0022	.0006
	3	.0473	.1680	.2406	.2297	.1704	.1046	.0547	.0246	.0095	.0031
	4	.0093	.0700	.1592	.2153	.2130	.1681	.1104	.0614	.0291	.0117
	5	.0014	.0218	.0787	.1507	.1988	.2017	.1664	.1146	.0666	.0327
	6	.0002	.0052	.0301	.0816	.1436	.1873	.1941	.1655	.1181	.0708
	7	.0000	.0010	.0091	.0350	.0820	.1376	.1792	.1892	.1657	.1214
	8	.0000	.0002	.0022	.0120	.0376	.0811	.1327	.1734	.1864	.1669
	9	.0000	.0000	.0004	.0033	.0139	.0386	.0794	.1284	.1694	.1855
	10	.0000	.0000	.0001	.0008	.0042	.0149	.0385	.0771	.1248	.1669
	11	.0000	.0000	.0000	.0001	.0010	.0046	.0151	.0374	.0742	.1214
	12	.0000	.0000	.0000	.0000	.0002	.0012	.0047	.0145	.0354	.0708
	13	.0000	.0000	.0000	.0000	.0000	.0002	.0012	.0044	.0134	.0327
	14	.0000	.0000	.0000	.0000	.0000	.0000	.0002	.0011	.0039	.0117
	15	.0000	.0000	.0000	.0000	.0000	.0000	.0000	.0002	.0009	.0031
	16	.0000	.0000	.0000	.0000	.0000	.0000	.0000	.0000	.0001	.0006
	17	.0000	.0000	.0000	.0000	.0000	.0000	.0000	.0000	.0000	.0001
	18	.0000	.0000	.0000	.0000	.0000	.0000	.0000	.0000	.0000	.0000
19	0	.3774	.1351	.0456	.0144	.0042	.0011	.0003	.0001	.0000	.0000
	1	.3774	.2852	.1529	.0685	.0268	.0093	.0029	.0008	.0002	.0000
	2	.1787	.2852	.2428	.1540	.0803	.0358	.0138	.0046	.0013	.0003
	3	.0533	.1796	.2428	.2182	.1517	.0869	.0422	.0175	.0062	.0018
	4	.0112	.0798	.1714	.2182	.2023	.1491	.0909	.0467	.0203	.0074
	5	.0018	.0266	.0907	.1636	.2023	.1916	.1468	.0933	.0497	.0222
	6	.0002	.0069	.0374	.0955	.1574	.1916	.1844	.1451	.0949	.0518
	7	.0000	.0014	.0122	.0443	.0974	.1525	.1844	.1797	.1443	.0961
	8	.0000	.0002	.0032	.0166	.0487	.0981	.1489	.1797	.1771	.1442
	9	.0000	.0000	.0007	.0051	.0198	.0514	.0980	.1464	.1771	.1762
	10	.0000	.0000	.0001	.0013	.0066	.0220	.0528	.0976	.1449	.1762

Table A1 (Continued)

n	x	.05	.10	.15	.20	.25	.30	.35	.40	.45	.50
							p				
	11	.0000	.0000	.0000	.0003	.0018	.0077	.0233	.0532	.0970	.1442
	12	.0000	.0000	.0000	.0000	.0004	.0022	.0083	.0237	.0529	.0961
	13	.0000	.0000	.0000	.0000	.0001	.0005	.0024	.0085	.0233	.0518
	14	.0000	.0000	.0000	.0000	.0000	.0001	.0006	.0024	.0082	.0222
	15	.0000	.0000	.0000	.0000	.0000	.0000	.0001	.0005	.0022	.0074
	16	.0000	.0000	.0000	.0000	.0000	.0000	.0000	.0001	.0005	.0018
	17	.0000	.0000	.0000	.0000	.0000	.0000	.0000	.0000	.0001	.0003
	18	.0000	.0000	.0000	.0000	.0000	.0000	.0000	.0000	.0000	.0000
	19	.0000	.0000	.0000	.0000	.0000	.0000	.0000	.0000	.0000	.0000
20	0	.3585	.1216	.0388	.0115	.0032	.0008	.0002	.0000	.0000	.0000
	1	.3774	.2702	.1368	.0576	.0211	.0068	.0020	.0005	.0001	.0000
	2	.1887	.2852	.2293	.1369	.0669	.0278	.0100	.0031	.0008	.0002
	3	.0596	.1901	.2428	.2054	.1339	.0716	.0323	.0123	.0040	.0011
	4	.0133	.0898	.1821	.2182	.1897	.1304	.0738	.0350	.0139	.0046
	5	.0022	.0319	.1028	.1746	.2023	.1789	.1272	.0746	.0365	.0148
	6	.0003	.0089	.0454	.1091	.1686	.1916	.1712	.1244	.0746	.0370
	7	.0000	.0020	.0160	.0545	.1124	.1643	.1844	.1659	.1221	.0739
	8	.0000	.0004	.0046	.0222	.0609	.1144	.1614	.1797	.1623	.1201
	9	.0000	.0001	.0011	.0074	.0271	.0654	.1158	.1597	.1771	.1602
	10	.0000	.0000	.0002	.0020	.0099	.0308	.0686	.1171	.1593	.1762
	11	.0000	.0000	.0000	.0005	.0030	.0120	.0336	.0710	.1185	.1602
	12	.0000	.0000	.0000	.0001	.0008	.0039	.0136	.0355	.0727	.1201
	13	.0000	.0000	.0000	.0000	.0002	.0010	.0045	.0146	.0366	.0739
	14	.0000	.0000	.0000	.0000	.0000	.0002	.0012	.0049	.0150	.0370
	15	.0000	.0000	.0000	.0000	.0000	.0000	.0003	.0013	.0049	.0148
	16	.0000	.0000	.0000	.0000	.0000	.0000	.0000	.0003	.0013	.0046
	17	.0000	.0000	.0000	.0000	.0000	.0000	.0000	.0000	.0002	.0011
	18	.0000	.0000	.0000	.0000	.0000	.0000	.0000	.0000	.0000	.0002
	19	.0000	.0000	.0000	.0000	.0000	.0000	.0000	.0000	.0000	.0000
	20	.0000	.0000	.0000	.0000	.0000	.0000	.0000	.0000	.0000	.0000

Source: Reprinted from *Tables of the Binomial Probability Distribution* (1950), courtesy of the National Institute of Standards and Technology, Technology Administration, U.S. Department of Commerce.

Table A2 Poisson Probabilities

For a given value of λ, entry indicates the probability of obtaining a specified value of X

X	.1	.2	.3	.4	.5	.6	.7	.8	.9	1.0
0	.9048	.8187	.7408	.6703	.6065	.5488	.4966	.4493	.4066	.3679
1	.0905	.1637	.2222	.2681	.3033	.3293	.3476	.3595	.3659	.3679
2	.0045	.0164	.0333	.0536	.0758	.0988	.1217	.1438	.1647	.1839
3	.0002	.0011	.0033	.0072	.0126	.0198	.0284	.0383	.0494	.0613
4	.0000	.0001	.0003	.0007	.0016	.0030	.0050	.0077	.0111	.0153
5	.0000	.0000	.0000	.0001	.0002	.0004	.0007	.0012	.0020	.0031
6	.0000	.0000	.0000	.0000	.0000	.0000	.0001	.0002	.0003	.0005
7	.0000	.0000	.0000	.0000	.0000	.0000	.0000	.0000	.0000	.0001

X	1.1	1.2	1.3	1.4	1.5	1.6	1.7	1.8	1.9	2.0
0	.3329	.3012	.2725	.2466	.2231	.2019	.1827	.1653	.1496	.1353
1	.3662	.3614	.3543	.3452	.3347	.3230	.3106	.2975	.2842	.2707
2	.2014	.2169	.2303	.2417	.2510	.2584	.2640	.2678	.2700	.2707
3	.0738	.0867	.0998	.1128	.1255	.1378	.1496	.1607	.1710	.1804
4	.0203	.0260	.0324	.0395	.0471	.0551	.0636	.0723	.0812	.0902
5	.0045	.0062	.0084	.0111	.0141	.0176	.0216	.0260	.0309	.0361
6	.0008	.0012	.0018	.0026	.0035	.0047	.0061	.0078	.0098	.0120
7	.0001	.0002	.0003	.0005	.0008	.0011	.0015	.0020	.0027	.0034
8	.0000	.0000	.0001	.0001	.0001	.0002	.0003	.0005	.0006	.0009
9	.0000	.0000	.0000	.0000	.0000	.0000	.0001	.0001	.0001	.0002

X	2.1	2.2	2.3	2.4	2.5	2.6	2.7	2.8	2.9	3.0
0	.1225	.1108	.1003	.0907	.0821	.0743	.0672	.0608	.0550	.0498
1	.2572	.2438	.2306	.2177	.2052	.1931	.1815	.1703	.1596	.1494
2	.2700	.2681	.2652	.2613	.2565	.2510	.2450	.2384	.2314	.2240
3	.1890	.1966	.2033	.2090	.2138	.2176	.2205	.2225	.2237	.2240
4	.0992	.1082	.1169	.1254	.1336	.1414	.1488	.1557	.1622	.1680
5	.0417	.0476	.0538	.0602	.0668	.0735	.0804	.0872	.0940	.1008
6	.0146	.0174	.0206	.0241	.0278	.0319	.0362	.0407	.0455	.0504
7	.0044	.0055	.0068	.0083	.0099	.0118	.0139	.0163	.0188	.0216
8	.0011	.0015	.0019	.0025	.0031	.0038	.0047	.0057	.0068	.0081
9	.0003	.0004	.0005	.0007	.0009	.0011	.0014	.0018	.0022	.0027
10	.0001	.0001	.0001	.0002	.0002	.0003	.0004	.0005	.0006	.0008
11	.0000	.0000	.0000	.0000	.0000	.0001	.0001	.0001	.0002	.0002
12	.0000	.0000	.0000	.0000	.0000	.0000	.0000	.0000	.0000	.0001

X	3.1	3.2	3.3	3.4	3.5	3.6	3.7	3.8	3.9	4.0
0	.0450	.0408	.0369	.0334	.0302	.0273	.0247	.0224	.0202	.0183
1	.1397	.1304	.1217	.1135	.1057	.0984	.0915	.0850	.0789	.0733
2	.2165	.2087	.2008	.1929	.1850	.1771	.1692	.1615	.1539	.1465

Table A2 (Continued)

					λ					
X	3.1	3.2	3.3	3.4	3.5	3.6	3.7	3.8	3.9	4.0
3	.2237	.2226	.2209	.2186	.2158	.2125	.2087	.2046	.2001	.1954
4	.1734	.1781	.1823	.1858	.1888	.1912	.1931	.1944	.1951	.1954
5	.1075	.1140	.1203	.1264	.1322	.1377	.1429	.1477	.1522	.1563
6	.0555	.0608	.0662	.0716	.0771	.0826	.0881	.0936	.0989	.1042
7	.0246	.2078	.0312	.0348	.0385	.0425	.0466	.0508	.0551	.0595
8	.0095	.0111	.0129	.0148	.0169	.0191	.0215	.0241	.0269	.0298
9	.0033	.0040	.0047	.0056	.0066	.0076	.0089	.0102	.0116	.0132
10	.0010	.0013	.0016	.0019	.0023	.0028	.0033	.0039	.0045	.0053
11	.0003	.0004	.0005	.0006	.0007	.0009	.0011	.0013	.0016	.0019
12	.0001	.0001	.0001	.0002	.0002	.0003	.0003	.0004	.0005	.0006
13	.0000	.0000	.0000	.0000	.0001	.0001	.0001	.0001	.0002	.0002
14	.0000	.0000	.0000	.0000	.0000	.0000	.0000	.0000	.0000	.0001

					λ					
X	4.1	4.2	4.3	4.4	4.5	4.6	4.7	4.8	4.9	5.0
0	.0166	.0150	.0136	.0123	.0111	.0101	.0091	.0082	.0074	.0067
1	.0679	.0630	.0583	.0540	.0500	.0462	.0427	.0395	.0365	.0337
2	.1393	.1323	.1254	.1188	.1125	.1063	.1005	.0948	.0894	.0842
3	.1904	.1852	.1798	.1743	.1687	.1631	.1574	.1517	.1460	.1404
4	.1951	.1944	.1933	.1917	.1898	.1875	.1849	.1820	.1789	.1755
5	.1600	.1633	.1662	.1687	.1708	.1725	.1738	.1747	.1753	.1755
6	.1093	.1143	.1191	.1237	.1281	.1323	.1362	.1398	.1432	.1462
7	.0640	.0686	.0732	.0778	.0824	.0869	.0914	.0959	.1002	.1044
8	.0328	.0360	.0393	.0428	.0463	.0500	.0537	.0575	.0614	.0653
9	.0150	.0168	.0188	.0209	.0232	.0255	.0280	.0307	.0334	.0363
10	.0061	.0071	.0081	.0092	.0104	.0118	.0132	.0147	.0164	.0181
11	.0023	.0027	.0032	.0037	.0043	.0049	.0056	.0064	.0073	.0082
12	.0008	.0009	.0011	.0014	.0016	.0019	.0022	.0026	.0030	.0034
13	.0002	.0003	.0004	.0005	.0006	.0007	.0008	.0009	.0011	.0013
14	.0001	.0001	.0001	.0001	.0002	.0002	.0003	.0003	.0004	.0005
15	.0000	.0000	.0000	.0000	.0001	.0001	.0001	.0001	.0001	.0002

					λ					
X	5.1	5.2	5.3	5.4	5.5	5.6	5.7	5.8	5.9	6.0
0	.0061	.0055	.0050	.0045	.0041	.0037	.0033	.0030	.0027	.0025
1	.0311	.0287	.0265	.0244	.0225	.0207	.0191	.0176	.0162	.0149
2	.0793	.0746	.0701	.0659	.0618	.0580	.0544	.0509	.0477	.0446
3	.1348	.1293	.1239	.1185	.1133	.1082	.1033	.0985	.0938	.0892
4	.1719	.1681	.1641	.1600	.1558	.1515	.1472	.1428	.1383	.1339
5	.1753	.1748	.1740	.1728	.1714	.1697	.1678	.1656	.1632	.1606
6	.1490	.1515	.1537	.1555	.1571	.1584	.1594	.1601	.1605	.1606

Table A2 (Continued)

					λ					
X	5.1	5.2	5.3	5.4	5.5	5.6	5.7	5.8	5.9	6.0
7	.1086	.1125	.1163	.1200	.1234	.1267	.1298	.1326	.1353	.1377
8	.0692	.0731	.0771	.0810	.0849	.0887	.0925	.0962	.0998	.1033
9	.0392	.0423	.0454	.0486	.0519	.0552	.0586	.0620	.0654	.0688
10	.0200	.0220	.0241	.0262	.0285	.0309	.0334	.0359	.0386	.0413
11	.0093	.0104	.0116	.0129	.0143	.0157	.0173	.0190	.0207	.0225
12	.0039	.0045	.0051	.0058	.0065	.0073	.0082	.0092	.0102	.0113
13	.0015	.0018	.0021	.0024	.0028	.0032	.0036	.0041	.0046	.0052
14	.0006	.0007	.0008	.0009	.0011	.0013	.0015	.0017	.0019	.0022
15	.0002	.0002	.0003	.0003	.0004	.0005	.0006	.0007	.0008	.0009
16	.0001	.0001	.0001	.0001	.0001	.0002	.0002	.0002	.0003	.0003
17	.0000	.0000	.0000	.0000	.0000	.0000	.0001	.0001	.0001	.0001

					λ					
X	6.1	6.2	6.3	6.4	6.5	6.6	6.7	6.8	6.9	7.0
0	.0022	.0020	.0018	.0017	.0015	.0014	.0012	.0011	.0010	.0009
1	.0137	.0126	.0116	.0106	.0098	.0090	.0082	.0076	.0070	.0064
2	.0417	.0390	.0364	.0340	.0318	.0296	.0276	.0258	.0240	.0223
3	.0848	.0806	.0765	.0726	.0688	.0652	.0617	.0584	.0552	.0521
4	.1294	.1249	.1205	.1162	.1118	.1076	.1034	.0992	.0952	.0912
5	.1579	.1549	.1519	.1487	.1454	.1420	.1385	.1349	.1314	.1277
6	.1605	.1601	.1595	.1586	.1575	.1562	.1546	.1529	.1511	.1490
7	.1399	.1418	.1435	.1450	.1462	.1472	.1480	.1486	.1489	.1490
8	.1066	.1099	.1130	.1160	.1188	.1215	.1240	.1263	.1284	.1304
9	.0723	.0757	.0791	.0825	.0858	.0891	.0923	.0954	.0985	.1014
10	.0441	.0469	.0498	.0528	.0558	.0588	.0618	.0649	.0679	.0710
11	.0245	.0265	.0285	.0307	.0330	.0353	.0377	.0401	.0426	.0452
12	.0124	.0137	.0150	.0164	.0179	.0194	.0210	.0227	.0245	.0264
13	.0058	.0065	.0073	.0081	.0089	.0098	.0108	.0119	.0130	.0142
14	.0025	.0029	.0033	.0037	.0041	.0046	.0052	.0058	.0064	.0071
15	.0010	.0012	.0014	.0016	.0018	.0020	.0023	.0026	.0029	.0033
16	.0004	.0005	.0005	.0006	.0007	.0008	.0010	.0011	.0013	.0014
17	.0001	.0002	.0002	.0002	.0003	.0003	.0004	.0004	.0005	.0006
18	.0000	.0001	.0001	.0001	.0001	.0001	.0001	.0002	.0002	.0002
19	.0000	.0000	.0000	.0000	.0000	.0000	.0000	.0001	.0001	.0001

					λ					
X	7.1	7.2	7.3	7.4	7.5	7.6	7.7	7.8	7.9	8.0
0	.0008	.0007	.0007	.0006	.0006	.0005	.0005	.0004	.0004	.0003
1	.0059	.0054	.0049	.0045	.0041	.0038	.0035	.0032	.0029	.0027
2	.0208	.0194	.0180	.0167	.0156	.0145	.0134	.0125	.0116	.0107
3	.0492	.0464	.0438	.0413	.0389	.0366	.0345	.0324	.0305	.0286
4	.0874	.0836	.0799	.0764	.0729	.0696	.0663	.0632	.0602	.0573

Table A2 (Continued)

					λ					
X	7.1	7.2	7.3	7.4	7.5	7.6	7.7	7.8	7.9	8.0
5	.1241	.1204	.1167	.1130	.1094	.1057	.1021	.0986	.0951	.0916
6	.1468	.1445	.1420	.1394	.1367	.1339	.1311	.1282	.1252	.1221
7	.1489	.1486	.1481	.1474	.1465	.1454	.1442	.1428	.1413	.1396
8	.1321	.1337	.1351	.1363	.1373	.1382	.1388	.1392	.1395	.1396
9	.1042	.1070	.1096	.1121	.1144	.1167	.1187	.1207	.1224	.1241
10	.0740	.0770	.0800	.0829	.0858	.0887	.0914	.0941	.0967	.0993
11	.0478	.0504	.0531	.0558	.0585	.0613	.0640	.0667	.0695	.0722
12	0283	.0303	.0323	.0344	.0366	.0388	.0411	.0434	.0457	.0481
13	.0154	.0168	.0181	.0196	.0211	.0227	.0243	.0260	.0278	.0296
14	.0078	.0086	.0095	.0104	.0113	.0123	.0134	.0145	.0157	.0169
15	.0037	.0041	.0046	.0051	.0057	.0062	.0069	.0075	.0083	.0090
16	.0016	.0019	.0021	.0024	.0026	.0030	.0033	.0037	.0041	.0045
17	.0007	.0008	.0009	.0010	.0012	.0013	.0015	.0017	.0019	.0021
18	.0003	.0003	.0004	.0004	.0005	.0006	.0006	.0007	.0008	.0009
19	.0001	.0001	.0001	.0002	.0002	.0002	.0003	.0003	.0003	.0004
20	.0000	.0000	.0001	.0001	.0001	.0001	.0001	.0001	.0001	.0002
21	.0000	.0000	.0000	.0000	.0000	.0000	.0000	.0000	.0001	.0001

					λ					
X	8.1	8.2	8.3	8.4	8.5	8.6	8.7	8.8	8.9	9.0
0	.0003	.0003	.0002	.0002	.0002	.0002	.0002	.0002	.0001	.0001
1	.0025	.0023	.0021	.0019	.0017	.0016	.0014	.0013	.0012	.0011
2	.0100	.0092	.0086	.0079	.0074	.0068	.0063	.0058	.0054	.0050
3	.0269	.0252	.0237	.0222	.0208	.0195	.0183	.0171	.0160	.0150
4	.0544	.0517	.0491	.0466	.0443	.0420	.0398	.0377	.0357	.0337
5	.0882	.0849	.0816	.0784	.0752	.0722	.0692	.0663	.0635	.0607
6	.1191	.1160	.1128	.1097	.1066	.1034	.1003	.0972	.0941	.0911
7	.1378	.1358	.1338	.1317	.1294	.1271	.1247	.1222	.1197	.1171
8	.1395	.1392	.1388	.1382	.1375	.1366	.1356	.1344	.1332	.1318
9	.1256	.1269	.1280	.1290	.1299	.1306	.1311	.1315	.1317	.1318
10	.1017	.1040	.1063	.1084	.1104	.1123	.1140	.1157	.1172	.1186
11	.0749	.0776	.0802	.0828	.0853	.0878	.0902	.0925	.0948	.0970
12	.0505	.0530	.0555	.0579	.0604	.0629	.0654	.0679	.0703	.0728
13	.0315	.0334	.0354	.0374	.0395	.0416	.0438	.0459	.0481	.0504
14	.0182	.0196	.0210	.0225	.0240	.0256	.0272	.0289	.0306	.0324
15	.0098	.0107	.0116	.0126	.0136	.0147	.0158	.0169	.0182	.0194
16	.0050	.0055	.0060	.0066	.0072	.0079	.0086	.0093	.0101	.0109
17	.0024	.0026	.0029	.0033	.0036	.0040	.0044	.0048	.0053	.0058
18	.0011	.0012	.0014	.0015	.0017	.0019	.0021	.0024	.0026	.0029
19	.0005	.0005	.0006	.0007	.0008	.0009	.0010	.0011	.0012	.0014
20	.0002	.0002	.0002	.0003	.0003	.0004	.0004	.0005	.0005	.0006
21	.0001	.0001	.0001	.0001	.0001	.0002	.0002	.0002	.0002	.0003
22	.0000	.0000	.0000	.0000	.0001	.0001	.0001	.0001	.0001	.0001

Table A2 (Continued)

X	9.1	9.2	9.3	9.4	λ 9.5	9.6	9.7	9.8	9.9	10
0	.0001	.0001	.0001	.0001	.0001	.0001	.0001	.0001	.0001	.0000
1	.0010	.0009	.0009	.0008	.0007	.0007	.0006	.0005	.0005	.0005
2	.0046	.0043	.0040	.0037	.0034	.0031	.0029	.0027	.0025	.0023
3	.0140	.0131	.0123	.0115	.0107	.0100	.0093	.0087	.0081	.0076
4	.0319	.0302	.0285	.0269	.0254	.0240	.0226	.0213	.0201	.0189
5	.0581	.0555	.0530	.0506	.0483	.0460	.0439	.0418	.0398	.0378
6	.0881	.0851	.0822	.0793	.0764	.0736	.0709	.0682	.0656	.0631
7	.1145	.1118	.1091	.1064	.1037	.1010	.0982	.0955	.0928	.0901
8	.1302	.1286	.1269	.1251	.1232	.1212	.1191	.1170	.1148	.1126
9	.1317	.1315	.1311	.1306	.1300	.1293	.1284	.1274	.1263	.1251
10	.1198	.1210	.1219	.1228	.1235	.1241	.1245	.1249	.1250	.1251
11	.0991	.1012	.1031	.1049	.1067	.1083	.1098	.1112	.1125	.1137
12	.0752	.0776	.0799	.0822	.0844	.0866	.0888	.0908	.0928	.0948
13	.0526	.0549	.0572	.0594	.0617	.0640	.0662	.0685	.0707	.0729
14	.0342	.0361	.0380	.0399	.0419	.0439	.0459	.0479	.0500	.0521
15	.0208	.0221	.0235	.0250	.0265	.0281	.0297	.0313	.0330	.0347
16	.0118	.0127	.0137	.0147	.0157	.0168	.0180	.0192	.0204	.0217
17	.0063	.0069	.0075	.0081	.0088	.0095	.0103	.0111	.0119	.0128
18	.0032	.0035	.0039	.0042	.0046	.0051	.0055	.0060	.0065	.0071
19	.0015	.0017	.0019	.0021	.0023	.0026	.0028	.0031	.0034	.0037
20	.0007	.0008	.0009	.0010	.0011	.0012	.0014	.0015	.0017	.0019
21	.0003	.0003	.0004	.0004	.0005	.0006	.0006	.0007	.0008	.0009
22	.0001	.0001	.0002	.0002	.0002	.0002	.0003	.0003	.0004	.0004
23	.0000	.0001	.0001	.0001	.0001	.0001	.0001	.0001	.0002	.0002
24	.0000	.0000	.0000	.0000	.0000	.0000	.0000	.0001	.0001	.0001

X	11	12	13	14	λ 15	16	17	18	19	20
0	.0000	.0000	.0000	.0000	.0000	.0000	.0000	.0000	.0000	.0000
1	.0002	.0001	.0000	.0000	.0000	.0000	.0000	.0000	.0000	.0000
2	.0010	.0004	.0002	.0001	.0000	.0000	.0000	.0000	.0000	.0000
3	.0037	.0018	.0008	.0004	.0002	.0001	.0000	.0000	.0000	.0000
4	.0102	.0053	.0027	.0013	.0006	.0003	.0001	.0001	.0000	.0000
5	.0224	.0127	.0070	.0037	.0019	.0010	.0005	.0002	.0001	.0001
6	.0411	.0255	.0152	.0087	.0048	.0026	.0014	.0007	.0004	.0002
7	.0646	.0437	.0281	.0174	.0104	.0060	.0034	.0018	.0010	.0005
8	.0888	.0655	.0457	.0304	.0194	.0120	.0072	.0042	.0024	.0013
9	.1085	.0874	.0661	.0473	.0324	.0213	.0135	.0083	.0050	.0029
10	.1194	.1048	.0859	.0663	.0486	.0341	.0230	.0150	.0095	.0058
11	.1194	.1144	.1015	.0844	.0663	.0496	.0355	.0245	.0164	.0106
12	.1094	.1144	.1099	.0984	.0829	.0661	.0504	.0368	.0259	.0176
13	.0926	.1056	.1099	.1060	.0956	.0814	.0658	.0509	.0378	.0271
14	.0728	.0905	.1021	.1060	.1024	.0930	.0800	.0655	.0514	.0387

Table A2 (Continued)

X	11	12	13	14	λ 15	16	17	18	19	20
15	.0534	.0724	.0885	.0989	.1024	.0992	.0906	.0786	.0650	.0516
16	.0367	.0543	.0719	.0866	.0960	.0992	.0963	.0884	.0772	.0646
17	.0237	.0383	.0550	.0713	.0847	.0934	.0963	.0936	.0863	.0760
18	.0145	.0256	.0397	.0554	.0706	.0830	.0909	.0936	.0911	.0844
19	.0084	.0161	.0272	.0409	.0557	.0699	.0814	.0887	.0911	.0888
20	.0046	.0097	.0177	.0286	.0418	.0559	.0692	.0798	.0866	.0888
21	.0024	.0055	.0109	.0191	.0299	.0426	.0560	.0684	.0783	.0846
22	.0012	.0030	.0065	.0121	.0204	.0310	.0433	.0560	.0676	.0769
23	.0006	.0016	.0037	.0074	.0133	.0216	.0320	.0438	.0559	.0669
24	.0003	.0008	.0020	.0043	.0083	.0144	.0226	.0328	.0442	.0557
25	.0001	.0004	.0010	.0024	.0050	.0092	.0154	.0237	.0336	.0446
26	.0000	.0002	.0005	.0013	.0029	.0057	.0101	.0164	.0246	.0343
27	.0000	.0001	.0002	.0007	.0016	.0034	.0063	.0109	.0173	.0254
28	.0000	.0000	.0001	.0003	.0009	.0019	.0038	.0070	.0117	.0181
29	.0000	.0000	.0001	.0002	.0004	.0011	.0023	.0044	.0077	.0125
30	.0000	.0000	.0000	.0001	.0002	.0006	.0013	.0026	.0049	.0083
31	.0000	.0000	.0000	.0000	.0001	.0003	.0007	.0015	.0030	.0054
32	.0000	.0000	.0000	.0000	.0001	.0001	.0004	.0009	.0018	.0034
33	.0000	.0000	.0000	.0000	.0000	.0001	.0002	.0005	.0010	.0020
34	.0000	.0000	.0000	.0000	.0000	.0000	.0001	.0002	.0006	.0012
35	.0000	.0000	.0000	.0000	.0000	.0000	.0000	.0001	.0003	.0007
36	.0000	.0000	.0000	.0000	.0000	.0000	.0000	.0001	.0002	.0004
37	.0000	.0000	.0000	.0000	.0000	.0000	.0000	.0000	.0001	.0002
38	.0000	.0000	.0000	.0000	.0000	.0000	.0000	.0000	.0000	.0001
39	.0000	.0000	.0000	.0000	.0000	.0000	.0000	.0000	.0000	.0001

Source: Extracted from William H. Beyer, ed., *CRC Basic Statistical Tables* (Cleveland, Ohio: The Chemical Rubber Co., 1971).

Table A3 The Standardized Normal Distribution

The entries in this table are the probabilities that a standard normal random variable is between 0 and z (the shaded area).

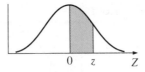

Second Decimal Place in z

z	.00	.01	.02	.03	.04	.05	.06	.07	.08	.09
.0	.0000	.0040	.0080	.0120	.0160	.0199	.0239	.0279	.0319	.0359
.1	.0398	.0438	.0478	.0517	.0557	.0596	.0636	.0675	.0714	.0753
.2	.0793	.0832	.0871	.0910	.0948	.0987	.1026	.1064	.1103	.1141
.3	.1179	.1217	.1255	.1293	.1331	.1368	.1406	.1443	.1480	.1517
.4	.1554	.1591	.1628	.1664	.1700	.1736	.1772	.1808	.1844	.1879
.5	.1915	.1950	.1985	.2019	.2054	.2088	.2123	.2157	.2190	.2224
.6	.2257	.2291	.2324	.2357	.2389	.2422	.2454	.2486	.2517	.2549
.7	.2580	.2611	.2642	.2673	.2704	.2734	.2764	.2794	.2823	.2852
.8	.2881	.2910	.2939	.2967	.2995	.3023	.3051	.3078	.3106	.3133
.9	.3159	.3186	.3212	.3238	.3264	.3289	.3315	.3340	.3365	.3389
1.0	.3413	.3438	.3461	.3485	.3508	.3531	.3554	.3577	.3599	.3621
1.1	.3643	.3665	.3686	.3708	.3729	.3749	.3770	.3790	.3810	.3830
1.2	.3849	.3869	.3888	.3907	.3925	.3944	.3962	.3980	.3997	.4015
1.3	.4032	.4049	.4066	.4082	.4099	.4115	.4131	.4147	.4162	.4177
1.4	.4192	.4207	.4222	.4236	.4251	.4265	.4279	.4292	.4306	.4319
1.5	.4332	.4345	.4357	.4370	.4382	.4394	.4406	.4418	.4429	.4441
1.6	.4452	.4463	.4474	.4484	.4495	.4505	.4515	.4525	.4535	.4545
1.7	.4554	.4564	.4573	.4582	.4591	.4599	.4608	.4616	.4625	.4633
1.8	.4641	.4649	.4656	.4664	.4671	.4678	.4686	.4693	.4699	.4706
1.9	.4713	.4719	.4726	.4732	.4738	.4744	.4750	.4756	.4761	.4767
2.0	.4772	.4778	.4783	.4788	.4793	.4798	.4803	.4808	.4812	.4817
2.1	.4821	.4826	.4830	.4834	.4838	.4842	.4846	.4850	.4854	.4857
2.2	.4861	.4864	.4868	.4871	.4875	.4878	.4881	.4884	.4887	.4890
2.3	.4893	.4896	.4898	.4901	.4904	.4906	.4909	.4911	.4913	.4916
2.4	.4918	.4920	.4922	.4925	.4927	.4929	.4931	.4932	.4934	.4936
2.5	.4938	.4940	.4941	.4943	.4945	.4946	.4948	.4949	.4951	.4952
2.6	.4953	.4955	.4956	.4957	.4959	.4960	.4961	.4962	.4963	.4974
2.7	.4965	.4966	.4967	.4968	.4969	.4970	.4971	.4972	.4973	.4974
2.8	.4974	.4975	.4976	.4977	.4977	.4978	.4979	.4979	.4980	.4981
2.9	.4981	.4982	.4982	.4983	.4984	.4984	.4985	.4985	.4986	.4986
3.0	.4987	.4987	.4987	.4988	.4988	.4989	.4989	.4989	.4990	.4990
3.1	.4990	.4991	.4991	.4991	.4992	.4992	.4992	.4992	.4993	.4993
3.2	.4993	.4993	.4994	.4994	.4994	.4994	.4994	.4995	.4995	.4995
3.3	.4995	.4995	.4995	.4996	.4996	.4996	.4996	.4996	.4996	.4997
3.4	.4997	.4997	.4997	.4997	.4997	.4997	.4997	.4997	.4997	.4998
3.5	.4998									
4.0	.49997									
4.5	.499997									
5.0	.4999997									

Source: Reprinted from *Standard Mathematical Tables,* 15th ed., © CRC Press, Inc., Boca Raton, FL.

Table A4 Critical Values of t

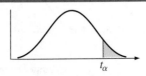

Degrees of Freedom ν	$t_{.100}$	$t_{.050}$	$t_{.025}$	$t_{.010}$	$t_{.005}$
1	3.078	6.314	12.706	31.821	63.657
2	1.886	2.920	4.303	6.965	9.925
3	1.638	2.353	3.182	4.541	5.841
4	1.533	2.132	2.776	3.747	4.604
5	1.476	2.015	2.571	3.365	4.032
6	1.440	1.943	2.447	3.143	3.707
7	1.415	1.895	2.365	2.998	3.499
8	1.397	1.860	2.306	2.896	3.355
9	1.383	1.833	2.262	2.821	3.250
10	1.372	1.812	2.228	2.764	3.169
11	1.363	1.796	2.201	2.718	3.106
12	1.356	1.782	2.179	2.681	3.055
13	1.350	1.771	2.160	2.650	3.012
14	1.345	1.761	2.145	2.624	2.977
15	1.341	1.753	2.131	2.602	2.947
16	1.337	1.746	2.120	2.583	2.921
17	1.333	1.740	2.110	2.567	2.898
18	1.330	1.734	2.101	2.552	2.878
19	1.328	1.729	2.093	2.539	2.861
20	1.325	1.725	2.086	2.528	2.845
21	1.323	1.721	2.080	2.518	2.831
22	1.321	1.717	2.074	2.508	2.819
23	1.319	1.714	2.069	2.500	2.808
24	1.318	1.711	2.064	2.492	2.797
25	1.316	1.708	2.060	2.485	2.787
26	1.315	1.706	2.056	2.479	2.779
27	1.314	1.703	2.052	2.473	2.771
28	1.313	1.701	2.048	2.467	2.763
29	1.311	1.699	2.045	2.462	2.756
30	1.310	1.697	2.042	2.457	2.750
40	1.303	1.684	2.021	2.423	2.704
60	1.296	1.671	2.000	2.390	2.660
120	1.289	1.658	1.980	2.358	2.617
∞	1.282	1.645	1.960	2.326	2.576

Source: From M. Merrington, "Table of Percentage Points of the *t*-Distribution," *Biometrika,* 1941, 32,300. Reproduced by permission of the *Biometrika* trustees.

Table A5 Critical Values of χ^2

Degrees of Freedom ν	$\chi^2_{.995}$	$\chi^2_{.990}$	$\chi^2_{.975}$	$\chi^2_{.950}$	$\chi^2_{.900}$
1	.0000393	.0001571	.0009821	.0039321	.0157908
2	.0100251	.0201007	.0506356	.102587	.210720
3	.0717212	.114832	.215795	.351846	.584375
4	.206990	.297110	.484419	.710721	1.063623
5	.411740	.554300	.831211	1.145476	1.61031
6	.675727	.872085	1.237347	1.63539	2.20413
7	.989265	1.239043	1.68987	2.16735	2.83311
8	1.344419	1.646482	2.17973	2.73264	3.48954
9	1.734926	2.087912	2.70039	3.32511	4.16816
10	2.15585	2.55821	3.24697	3.94030	4.86518
11	2.60321	3.05347	3.81575	4.57481	5.57779
12	3.07382	3.57056	4.40379	5.22603	6.30380
13	3.56503	4.10691	5.00874	5.89186	7.04150
14	4.07468	4.66043	5.62872	6.57063	7.78953
15	4.60094	5.22935	6.26214	7.26094	8.54675
16	5.14224	5.81221	6.90766	7.96164	9.31223
17	5.69724	6.40776	7.56418	8.67176	10.0852
18	6.26481	7.01491	8.23075	9.39046	10.8649
19	6.84398	7.63273	8.90655	10.1170	11.6509
20	7.43386	8.26040	9.59083	10.8508	12.4426
21	8.03366	8.89720	10.28293	11.5913	13.2396
22	8.64272	9.54249	10.9823	12.3380	14.0415
23	9.26042	10.19567	11.6885	13.0905	14.8479
24	9.88623	10.8564	12.4011	13.8484	15.6587
25	10.5197	11.5240	13.1197	14.6114	16.4734
26	11.1603	12.1981	13.8439	15.3791	17.2919
27	11.8076	12.8786	14.5733	16.1513	18.1138
28	12.4613	13.5648	15.3079	16.9279	18.9392
29	13.1211	14.2565	16.0471	17.7083	19.7677
30	13.7867	14.9535	16.7908	18.4926	20.5992
40	20.7065	22.1643	24.4331	26.5093	29.0505
50	27.9907	29.7067	32.3574	34.7642	37.6886
60	35.5346	37.4848	40.4817	43.1879	46.4589
70	43.2752	45.4418	48.7576	51.7393	55.3290
80	51.1720	53.5400	57.1532	60.3915	64.2778
90	59.1963	61.7541	65.6466	69.1260	73.2912
100	67.3276	70.0648	74.2219	77.9295	82.3581

Table A5 (Continued)

Degrees of Freedom ν	$\chi^2_{.100}$	$\chi^2_{.050}$	$\chi^2_{.025}$	$\chi^2_{.010}$	$\chi^2_{.005}$
1	2.70554	3.84146	5.02389	6.63490	7.87944
2	4.60517	5.99147	7.37776	9.21034	10.5966
3	6.25139	7.81473	9.34840	11.3449	12.8381
4	7.77944	9.48773	11.1433	13.2767	14.8602
5	9.23635	11.0705	12.8325	15.0863	16.7496
6	10.6446	12.5916	14.4494	16.8119	18.5476
7	12.0170	14.0671	16.0128	18.4753	20.2777
8	13.3616	15.5073	17.5346	20.0902	21.9550
9	14.6837	16.9190	19.0228	21.6660	23.5893
10	15.9871	18.3070	20.4831	23.2093	25.1882
11	17.2750	19.6751	21.9200	24.7250	26.7569
12	18.5494	21.0261	23.3367	26.2170	28.2995
13	19.8119	22.3621	24.7356	27.6883	29.8194
14	21.0642	23.6848	26.1190	29.1413	31.3193
15	22.3072	24.9958	27.4884	30.5779	32.8013
16	23.5418	26.2962	28.8454	31.9999	34.2672
17	24.7690	27.5871	30.1910	33.4087	35.7185
18	25.9894	28.8693	31.5264	34.8053	37.1564
19	27.2036	30.1435	32.8523	36.1908	38.5822
20	28.4120	31.4104	34.1696	37.5662	39.9968
21	29.6151	32.6705	35.4789	38.9321	41.4010
22	30.8133	33.9244	36.7807	40.2894	42.7956
23	32.0069	35.1725	38.0757	41.6384	44.1813
24	33.1963	36.4151	39.3641	42.9798	45.5585
25	34.3816	37.6525	40.6465	44.3141	46.9278
26	35.5631	38.8852	41.9232	45.6417	48.2899
27	36.7412	40.1133	43.1944	46.9630	49.6449
28	37.9159	41.3372	44.4607	48.2782	50.9933
29	39.0875	42.5569	45.7222	49.5879	52.3356
30	40.2560	43.7729	46.9792	50.8922	53.6720
40	51.8050	55.7585	59.3417	63.6907	66.7659
50	63.1671	67.5048	71.4202	76.1539	79.4900
60	74.3970	79.0819	83.2976	88.3794	91.9517
70	85.5271	90.5312	95.0231	100.425	104.215
80	96.5782	101.879	106.629	112.329	116.321
90	107.565	113.145	118.136	124.116	128.229
100	118.498	124.342	129.561	135.807	140.169

Source: From C. M. Thompson, "Tables of the Percentage Points of the χ^2-Distribution," *Biometrika*, 1941, *32*, 188–189. Reproduced by permission of the *Biometrika* trustees.

Table A6 Critical Values of F

For a particular combination of numerator and denominator degrees of freedom, entry represents the critical values of F corresponding to a specified upper tail area (α).

Denominator

v_2	\multicolumn{19}{c}{Numerator v_1}
	1	2	3	4	5	6	7	8	9	10	12	15	20	24	30	40	60	120	∞
---	---	---	---	---	---	---	---	---	---	---	---	---	---	---	---	---	---	---	---
1	161.4	199.5	215.7	224.6	230.2	234.0	236.8	238.9	240.5	241.9	243.9	245.9	248.0	249.1	250.1	251.1	252.2	253.3	254.3
2	18.51	19.00	19.16	19.25	19.30	19.33	19.35	19.37	19.38	19.40	19.41	19.43	19.45	19.45	19.46	19.47	19.48	19.49	19.50
3	10.13	9.55	9.28	9.12	9.01	8.94	8.89	8.85	8.81	8.79	8.74	8.70	8.66	8.64	8.62	8.59	8.57	8.55	8.53
4	7.71	6.94	6.59	6.39	6.26	6.16	6.09	6.04	6.00	5.96	5.91	5.86	5.80	5.77	5.75	5.72	5.69	5.66	5.63
5	6.61	5.79	5.41	5.19	5.05	4.95	4.88	4.82	4.77	4.74	4.68	4.62	4.56	4.53	4.50	4.46	4.43	4.40	4.36
6	5.99	5.14	4.76	4.53	4.39	4.28	4.21	4.15	4.10	4.06	4.00	3.94	3.87	3.84	3.81	3.77	3.74	3.70	3.67
7	5.59	4.74	4.35	4.12	3.97	3.87	3.79	3.73	3.68	3.64	3.57	3.51	3.44	3.41	3.38	3.34	3.30	3.27	3.23
8	5.32	4.46	4.07	3.84	3.69	3.58	3.50	3.44	3.39	3.35	3.28	3.22	3.15	3.12	3.08	3.04	3.01	2.97	2.93
9	5.12	4.26	3.86	3.63	3.48	3.37	3.29	3.23	3.18	3.14	3.07	3.01	2.94	2.90	2.86	2.83	2.79	2.75	2.71
10	4.96	4.10	3.71	3.48	3.33	3.22	3.14	3.07	3.02	2.98	2.91	2.85	2.77	2.74	2.70	2.66	2.62	2.58	2.54
11	4.84	3.98	3.59	3.36	3.20	3.09	3.01	2.95	2.90	2.85	2.79	2.72	2.65	2.61	2.57	2.53	2.49	2.45	2.40
12	4.75	3.89	3.49	3.26	3.11	3.00	2.91	2.85	2.80	2.75	2.69	2.62	2.54	2.51	2.47	2.43	2.38	2.34	2.30
13	4.67	3.81	3.41	3.18	3.03	2.92	2.83	2.77	2.71	2.67	2.60	2.53	2.46	2.42	2.38	2.34	2.30	2.25	2.21
14	4.60	3.74	3.34	3.11	2.96	2.85	2.76	2.70	2.65	2.60	2.53	2.46	2.39	2.35	2.31	2.27	2.22	2.18	2.13
15	4.54	3.68	3.29	3.06	2.90	2.79	2.71	2.64	2.59	2.54	2.48	2.40	2.33	2.29	2.25	2.20	2.16	2.11	2.07
16	4.49	3.63	3.24	3.01	2.85	2.74	2.66	2.59	2.54	2.49	2.42	2.35	2.28	2.24	2.19	2.15	2.11	2.06	2.01
17	4.45	3.59	3.20	2.96	2.81	2.70	2.61	2.55	2.49	2.45	2.38	2.31	2.23	2.19	2.15	2.10	2.06	2.01	1.96
18	4.41	3.55	3.16	2.93	2.77	2.66	2.58	2.51	2.46	2.41	2.34	2.27	2.19	2.15	2.11	2.06	2.02	1.97	1.92
19	4.38	3.52	3.13	2.90	2.74	2.63	2.54	2.48	2.42	2.38	2.31	2.23	2.16	2.11	2.07	2.03	1.98	1.93	1.88
20	4.35	3.49	3.10	2.87	2.71	2.60	2.51	2.45	2.39	2.35	2.28	2.20	2.12	2.08	2.04	1.99	1.95	1.90	1.84
21	4.32	3.47	3.07	2.84	2.68	2.57	2.49	2.42	2.37	2.32	2.25	2.18	2.10	2.05	2.01	1.96	1.92	1.87	1.81
22	4.30	3.44	3.05	2.82	2.66	2.55	2.46	2.40	2.34	2.30	2.23	2.15	2.07	2.03	1.98	1.94	1.89	1.84	1.78
23	4.28	3.42	3.03	2.80	2.64	2.53	2.44	2.37	2.32	2.27	2.20	2.13	2.05	2.01	1.96	1.91	1.86	1.81	1.76
24	4.26	3.40	3.01	2.78	2.62	2.51	2.42	2.36	2.30	2.25	2.18	2.11	2.03	1.98	1.94	1.89	1.84	1.79	1.73
25	4.24	3.39	2.99	2.76	2.60	2.49	2.40	2.34	2.28	2.24	2.16	2.09	2.01	1.96	1.92	1.87	1.82	1.77	1.71
26	4.23	3.37	2.98	2.74	2.59	2.47	2.39	2.32	2.27	2.22	2.15	2.07	1.99	1.95	1.90	1.85	1.80	1.75	1.69
27	4.21	3.35	2.96	2.73	2.57	2.46	2.37	2.31	2.25	2.20	2.13	2.06	1.97	1.93	1.88	1.84	1.79	1.73	1.67
28	4.20	3.34	2.95	2.71	2.56	2.45	2.36	2.29	2.24	2.19	2.12	2.04	1.96	1.91	1.87	1.82	1.77	1.71	1.65
29	4.18	3.33	2.93	2.70	2.55	2.43	2.35	2.28	2.22	2.18	2.10	2.03	1.94	1.90	1.85	1.81	1.75	1.70	1.64
30	4.17	3.32	2.92	2.69	2.53	2.42	2.33	2.27	2.21	2.16	2.09	2.01	1.93	1.89	1.84	1.79	1.74	1.68	1.62
40	4.08	3.23	2.84	2.61	2.45	2.34	2.25	2.18	2.12	2.08	2.00	1.92	1.84	1.79	1.74	1.69	1.64	1.58	1.51
60	4.00	3.15	2.76	2.53	2.37	2.25	2.17	2.10	2.04	1.99	1.92	1.84	1.75	1.70	1.65	1.59	1.53	1.47	1.39
120	3.92	3.07	2.68	2.45	2.29	2.17	2.09	2.02	1.96	1.91	1.83	1.75	1.66	1.61	1.55	1.50	1.43	1.35	1.25
∞	3.84	3.00	2.60	2.37	2.21	2.10	2.01	1.94	1.88	1.83	1.75	1.67	1.57	1.52	1.46	1.39	1.32	1.22	1.00

Table A6 (Continued)

F_{α}, ν_1, ν_2

$\alpha = .025$

Numerator ν_1

Denominator ν_2	1	2	3	4	5	6	7	8	9	10	12	15	20	24	30	40	60	120	∞
1	647.8	799.5	864.2	899.6	921.8	937.1	948.2	956.7	963.3	968.6	976.7	984.9	993.1	997.2	1001	1006	1010	1014	1018
2	38.51	39.00	39.17	39.25	39.30	39.33	39.36	39.37	39.39	39.40	39.41	39.43	39.45	39.46	39.46	39.47	39.48	39.49	39.50
3	17.44	16.04	15.44	15.10	14.88	14.73	14.62	14.54	14.47	14.42	14.34	14.25	14.17	14.12	14.08	14.04	13.99	13.95	13.90
4	12.22	10.65	9.98	9.60	9.36	9.20	9.07	8.98	8.90	8.84	8.75	8.66	8.56	8.51	8.46	8.41	8.36	8.31	8.26
5	10.01	8.43	7.76	7.39	7.15	6.98	6.85	6.76	6.68	6.62	6.52	6.43	6.33	6.28	6.23	6.18	6.12	6.07	6.02
6	8.81	7.26	6.60	6.23	5.99	5.82	5.70	5.60	5.52	5.46	5.37	5.27	5.17	5.12	5.07	5.01	4.96	4.90	4.85
7	8.07	6.54	5.89	5.52	5.29	5.12	4.99	4.90	4.82	4.76	4.67	4.57	4.47	4.42	4.36	4.31	4.25	4.20	4.14
8	7.57	6.06	5.42	5.05	4.82	4.65	4.53	4.43	4.36	4.30	4.20	4.10	4.00	3.95	3.89	3.84	3.78	3.73	3.67
9	7.21	5.71	5.08	4.72	4.48	4.32	4.20	4.10	4.03	3.96	3.87	3.77	3.67	3.61	3.56	3.51	3.45	3.39	3.33
10	6.94	5.46	4.83	4.47	4.24	4.07	3.95	3.85	3.78	3.72	3.62	3.52	3.42	3.37	3.31	3.26	3.20	3.14	3.08
11	6.72	5.26	4.63	4.28	4.04	3.88	3.76	3.66	3.59	3.53	3.43	3.33	3.23	3.17	3.12	3.06	3.00	2.94	2.88
12	6.55	5.10	4.47	4.12	3.89	3.73	3.61	3.51	3.44	3.37	3.28	3.18	3.07	3.02	2.96	2.91	2.85	2.79	2.72
13	6.41	4.97	4.35	4.00	3.77	3.60	3.48	3.39	3.31	3.25	3.15	3.05	2.95	2.89	2.84	2.78	2.72	2.66	2.60
14	6.30	4.86	4.24	3.89	3.66	3.50	3.38	3.29	3.21	3.15	3.05	2.95	2.84	2.79	2.73	2.67	2.61	2.55	2.49
15	6.20	4.77	4.15	3.80	3.58	3.41	3.29	3.20	3.12	3.06	2.96	2.86	2.76	2.70	2.64	2.59	2.52	2.46	2.40
16	6.12	4.69	4.08	3.73	3.50	3.34	3.22	3.12	3.05	2.99	2.89	2.79	2.68	2.63	2.57	2.51	2.45	2.38	2.32
17	6.04	4.62	4.01	3.66	3.44	3.28	3.16	3.06	2.98	2.92	2.82	2.72	2.62	2.56	2.50	2.44	2.38	2.32	2.25
18	5.98	4.56	3.95	3.61	3.38	3.22	3.10	3.01	2.93	2.87	2.77	2.67	2.56	2.50	2.44	2.38	2.32	2.26	2.19
19	5.92	4.51	3.90	3.56	3.33	3.17	3.05	2.96	2.88	2.82	2.72	2.62	2.51	2.45	2.39	2.33	2.27	2.20	2.13
20	5.87	4.46	3.86	3.51	3.29	3.13	3.01	2.91	2.84	2.77	2.68	2.57	2.46	2.41	2.35	2.29	2.22	2.16	2.09
21	5.83	4.42	3.82	3.48	3.25	3.09	2.97	2.87	2.80	2.73	2.64	2.53	2.42	2.37	2.31	2.25	2.18	2.11	2.04
22	5.79	4.38	3.78	3.44	3.22	3.05	2.93	2.84	2.76	2.70	2.60	2.50	2.39	2.33	2.27	2.21	2.14	2.08	2.00
23	5.75	4.35	3.75	3.41	3.18	3.02	2.90	2.81	2.73	2.67	2.57	2.47	2.36	2.30	2.24	2.18	2.11	2.04	1.97
24	5.72	4.32	3.72	3.38	3.15	2.99	2.87	2.78	2.70	2.64	2.54	2.44	2.33	2.27	2.21	2.15	2.08	2.01	1.94
25	5.69	4.29	3.69	3.35	3.13	2.97	2.85	2.75	2.68	2.61	2.51	2.41	2.30	2.24	2.18	2.12	2.05	1.98	1.91
26	5.66	4.27	3.67	3.33	3.10	2.94	2.82	2.73	2.65	2.59	2.49	2.39	2.28	2.22	2.16	2.09	2.03	1.95	1.88
27	5.63	4.24	3.65	3.31	3.08	2.92	2.80	2.71	2.63	2.57	2.47	2.36	2.25	2.19	2.13	2.07	2.00	1.93	1.85
28	5.61	4.22	3.63	3.29	3.06	2.90	2.78	2.69	2.61	2.55	2.45	2.34	2.23	2.17	2.11	2.05	1.98	1.91	1.83
29	5.59	4.20	3.61	3.27	3.04	2.88	2.76	2.67	2.59	2.53	2.43	2.32	2.21	2.15	2.09	2.03	1.96	1.89	1.81
30	5.57	4.18	3.59	3.25	3.03	2.87	2.75	2.65	2.57	2.51	2.41	2.31	2.20	2.14	2.07	2.01	1.94	1.87	1.79
40	5.42	4.05	3.46	3.13	2.90	2.74	2.62	2.53	2.45	2.39	2.29	2.18	2.07	2.01	1.94	1.88	1.80	1.72	1.64
60	5.29	3.93	3.34	3.01	2.79	2.63	2.51	2.41	2.33	2.27	2.17	2.06	1.94	1.88	1.82	1.74	1.67	1.58	1.48
120	5.15	3.80	3.23	2.89	2.67	2.52	2.39	2.30	2.22	2.16	2.05	1.94	1.82	1.76	1.69	1.61	1.53	1.43	1.31
∞	5.02	3.69	3.12	2.79	2.57	2.41	2.29	2.19	2.11	2.05	1.94	1.83	1.71	1.64	1.57	1.48	1.39	1.27	1.00

$\alpha = .01$

$F_{\alpha}, v_1, v_2 \quad F$

Denominator

v_2	\multicolumn{19}{c}{Numerator v_1}																		
	1	**2**	**3**	**4**	**5**	**6**	**7**	**8**	**9**	**10**	**12**	**15**	**20**	**24**	**30**	**40**	**60**	**120**	**∞**
1	4052	4999.5	5403	5625	5764	5859	5928	5982	6022	6056	6106	6157	6209	6235	6261	6287	6313	6339	6366
2	98.50	99.00	99.17	99.25	99.30	99.33	99.36	99.37	99.39	99.40	99.42	99.43	99.45	99.46	99.47	99.47	99.48	99.49	99.50
3	34.12	30.82	29.46	28.71	28.24	27.91	27.67	27.49	27.35	27.23	27.05	26.87	26.69	26.60	26.50	26.41	26.32	26.22	26.13
4	21.20	18.00	16.69	15.98	15.52	15.21	14.98	14.80	14.66	14.55	14.37	14.20	14.02	13.93	13.84	13.75	13.65	13.56	13.46
5	16.26	13.27	12.06	11.39	10.97	10.67	10.46	10.29	10.16	10.05	9.89	9.72	9.55	9.47	9.38	9.29	9.20	9.11	9.02
6	13.75	10.92	9.78	9.15	8.75	8.47	8.26	8.10	7.98	7.87	7.72	7.56	7.40	7.31	7.23	7.14	7.06	6.97	6.88
7	12.25	9.55	8.45	7.85	7.46	7.19	6.99	6.84	6.72	6.62	6.47	6.31	6.16	6.07	5.99	5.91	5.82	5.74	5.65
8	11.26	8.65	7.59	7.01	6.63	6.37	6.18	6.03	5.91	5.81	5.67	5.52	5.36	5.28	5.20	5.12	5.03	4.95	4.86
9	10.56	8.02	6.99	6.42	6.06	5.80	5.61	5.47	5.35	5.26	5.11	4.96	4.81	4.73	4.65	4.57	4.48	4.40	4.31
10	10.04	7.56	6.55	5.99	5.64	5.39	5.20	5.06	4.94	4.85	4.71	4.56	4.41	4.33	4.25	4.17	4.08	4.00	3.91
11	9.65	7.21	6.22	5.67	5.32	5.07	4.89	4.74	4.63	4.54	4.40	4.25	4.10	4.02	3.94	3.86	3.78	3.69	3.60
12	9.33	6.93	5.95	5.41	5.06	4.82	4.64	4.50	4.39	4.30	4.16	4.01	3.86	3.78	3.70	3.62	3.54	3.45	3.36
13	9.07	6.70	5.74	5.21	4.86	4.62	4.44	4.30	4.19	4.10	3.96	3.82	3.66	3.59	3.51	3.43	3.34	3.25	3.17
14	8.86	6.51	5.56	5.04	4.69	4.46	4.28	4.14	4.03	3.94	3.80	3.66	3.51	3.43	3.35	3.27	3.18	3.09	3.00
15	8.68	6.36	5.42	4.89	4.56	4.32	4.14	4.00	3.89	3.80	3.67	3.52	3.37	3.29	3.21	3.13	3.05	2.96	2.87
16	8.53	6.23	5.29	4.77	4.44	4.20	4.03	3.89	3.78	3.69	3.55	3.41	3.26	3.18	3.10	3.02	2.93	2.84	2.75
17	8.40	6.11	5.18	4.67	4.34	4.10	3.93	3.79	3.68	3.59	3.46	3.31	3.16	3.08	3.00	2.92	2.83	2.75	2.65
18	8.29	6.01	5.09	4.58	4.25	4.01	3.84	3.71	3.60	3.51	3.37	3.23	3.08	3.00	2.92	2.84	2.75	2.66	2.57
19	8.18	5.93	5.01	4.50	4.17	3.94	3.77	3.63	3.52	3.43	3.30	3.15	3.00	2.92	2.84	2.76	2.67	2.58	2.49
20	8.10	5.85	4.94	4.43	4.10	3.87	3.70	3.56	3.46	3.37	3.23	3.09	2.94	2.86	2.78	2.69	2.61	2.52	2.42
21	8.02	5.78	4.87	4.37	4.04	3.81	3.64	3.51	3.40	3.31	3.17	3.03	2.88	2.80	2.72	2.64	2.55	2.46	2.36
22	7.95	5.72	4.82	4.31	3.99	3.76	3.59	3.45	3.35	3.26	3.12	2.98	2.83	2.75	2.67	2.58	2.50	2.40	2.31
23	7.88	5.66	4.76	4.26	3.94	3.71	3.54	3.41	3.30	3.21	3.07	2.93	2.78	2.70	2.62	2.54	2.45	2.35	2.26
24	7.82	5.61	4.72	4.22	3.90	3.67	3.50	3.36	3.26	3.17	3.03	2.89	2.74	2.66	2.58	2.49	2.40	2.31	2.21
25	7.77	5.57	4.68	4.18	3.85	3.63	3.46	3.32	3.22	3.13	2.99	2.85	2.70	2.62	2.54	2.45	2.36	2.27	2.17
26	7.72	5.53	4.64	4.14	3.82	3.59	3.42	3.29	3.18	3.09	2.96	2.81	2.66	2.58	2.50	2.42	2.33	2.23	2.13
27	7.68	5.49	4.60	4.11	3.78	3.56	3.39	3.26	3.15	3.06	2.93	2.78	2.63	2.55	2.47	2.38	2.29	2.20	2.10
28	7.64	5.45	4.57	4.07	3.75	3.53	3.36	3.23	3.12	3.03	2.90	2.75	2.60	2.52	2.44	2.35	2.26	2.17	2.06
29	7.60	5.42	4.54	4.04	3.73	3.50	3.33	3.20	3.09	3.00	2.87	2.73	2.57	2.49	2.41	2.33	2.23	2.14	2.03
30	7.56	5.39	4.51	4.02	3.70	3.47	3.30	3.17	3.07	2.98	2.84	2.70	2.55	2.47	2.39	2.30	2.21	2.11	2.01
40	7.31	5.18	4.31	3.83	3.51	3.29	3.12	2.99	2.89	2.80	2.66	2.52	2.37	2.29	2.20	2.11	2.02	1.92	1.80
60	7.08	4.98	4.13	3.65	3.34	3.12	2.95	2.82	2.72	2.63	2.50	2.35	2.20	2.12	2.03	1.94	1.84	1.73	1.60
120	6.85	4.79	3.95	3.48	3.17	2.96	2.79	2.66	2.56	2.47	2.34	2.19	2.03	1.95	1.86	1.76	1.66	1.53	1.38
∞	6.63	4.61	3.78	3.32	3.02	2.80	2.64	2.51	2.41	2.32	2.18	2.04	1.88	1.79	1.70	1.59	1.47	1.32	1.00

Table A6 (Continued)

$\alpha = .005$

F_α, ν_1, ν_2 F

Numerator ν_1

Denominator ν_2	1	2	3	4	5	6	7	8	9	10	12	15	20	24	30	40	60	120	∞
1	16211	20000	21615	22500	23056	23437	23715	23925	24091	24224	24426	24630	24836	24940	25044	25148	25253	25359	25465
2	198.5	199.0	199.2	199.2	199.3	199.3	199.4	199.4	199.4	199.4	199.4	199.4	199.4	199.5	199.5	199.5	199.5	199.5	199.5
3	55.55	49.80	47.47	46.19	45.39	44.84	44.43	44.13	43.88	43.69	43.39	43.08	42.78	42.62	42.47	42.31	42.15	41.99	41.83
4	31.33	26.28	24.26	23.15	22.46	21.97	21.62	21.35	21.14	20.97	20.70	20.44	20.17	20.03	19.89	19.75	19.61	19.47	19.32
5	22.78	18.31	16.53	15.56	14.94	14.51	14.20	13.96	13.77	13.62	13.38	13.15	12.90	12.78	12.66	12.53	12.40	12.27	12.14
6	18.63	14.54	12.92	12.03	11.46	11.07	10.79	10.57	10.39	10.25	10.03	9.81	9.59	9.47	9.36	9.24	9.12	9.00	8.88
7	16.24	12.40	10.88	10.05	9.52	9.16	8.89	8.68	8.51	8.38	8.18	7.97	7.75	7.65	7.53	7.42	7.31	7.19	7.08
8	14.69	11.04	9.60	8.81	8.30	7.95	7.69	7.50	7.34	7.21	7.01	6.81	6.61	6.50	6.40	6.29	6.18	6.06	5.95
9	13.61	10.11	8.72	7.96	7.47	7.13	6.88	6.69	6.54	6.42	6.23	6.03	5.83	5.73	5.62	5.52	5.41	5.30	5.19
10	12.83	9.43	8.08	7.34	6.87	6.54	6.30	6.12	5.97	5.85	5.66	5.47	5.27	5.17	5.07	4.97	4.86	4.75	4.64
11	12.23	8.91	7.60	6.88	6.42	6.10	5.86	5.68	5.54	5.42	5.24	5.05	4.86	4.76	4.65	4.55	4.44	4.34	4.23
12	11.75	8.51	7.23	6.52	6.07	5.76	5.52	5.35	5.20	5.09	4.91	4.72	4.53	4.43	4.33	4.23	4.12	4.01	3.90
13	11.37	8.19	6.93	6.23	5.79	5.48	5.25	5.08	4.94	4.82	4.64	4.46	4.27	4.17	4.07	3.97	3.87	3.76	3.65
14	11.06	7.92	6.68	6.00	5.56	5.26	5.03	4.86	4.72	4.60	4.43	4.25	4.06	3.96	3.86	3.76	3.66	3.55	3.44
15	10.80	7.70	6.48	5.80	5.37	5.07	4.85	4.67	4.54	4.42	4.25	4.07	3.88	3.79	3.69	3.58	3.48	3.37	3.26
16	10.58	7.51	6.30	5.64	5.21	4.91	4.69	4.52	4.38	4.27	4.10	3.92	3.73	3.64	3.54	3.44	3.33	3.22	3.11
17	10.38	7.35	6.16	5.50	5.07	4.78	4.56	4.39	4.25	4.14	3.97	3.79	3.61	3.51	3.41	3.31	3.21	3.10	2.98
18	10.22	7.21	6.03	5.37	4.96	4.66	4.44	4.28	4.14	4.03	3.86	3.68	3.50	3.40	3.30	3.20	3.10	2.99	2.87
19	10.07	7.09	5.92	5.27	4.85	4.56	4.34	4.18	4.04	3.93	3.76	3.59	3.40	3.31	3.21	3.11	3.00	2.89	2.78
20	9.94	6.99	5.82	5.17	4.76	4.47	4.26	4.09	3.96	3.85	3.68	3.50	3.32	3.22	3.12	3.02	2.92	2.81	2.69
21	9.83	6.89	5.73	5.09	4.68	4.39	4.18	4.01	3.88	3.77	3.60	3.43	3.24	3.15	3.05	2.95	2.84	2.73	2.61
22	9.73	6.81	5.65	5.02	4.61	4.32	4.11	3.94	3.81	3.70	3.54	3.36	3.18	3.08	2.98	2.88	2.77	2.66	2.55
23	9.63	6.73	5.58	4.95	4.54	4.26	4.05	3.88	3.75	3.64	3.47	3.30	3.12	3.02	2.92	2.82	2.71	2.60	2.48
24	9.55	6.66	5.52	4.89	4.49	4.20	3.99	3.83	3.69	3.59	3.42	3.25	3.06	2.97	2.87	2.77	2.66	2.55	2.43
25	9.48	6.60	5.46	4.84	4.43	4.15	3.94	3.78	3.64	3.54	3.37	3.20	3.01	2.92	2.82	2.72	2.61	2.50	2.38
26	9.41	6.54	5.41	4.79	4.38	4.10	3.89	3.73	3.60	3.49	3.33	3.15	2.97	2.87	2.77	2.67	2.56	2.45	2.33
27	9.34	6.49	5.36	4.74	4.34	4.06	3.85	3.69	3.56	3.45	3.28	3.11	2.93	2.83	2.73	2.63	2.52	2.41	2.29
28	9.28	6.44	5.32	4.70	4.30	4.02	3.81	3.65	3.52	3.41	3.25	3.07	2.89	2.79	2.69	2.59	2.48	2.37	2.25
29	9.23	6.40	5.28	4.66	4.26	3.98	3.77	3.61	3.48	3.38	3.21	3.04	2.86	2.76	2.66	2.56	2.45	2.33	2.21
30	9.18	6.35	5.24	4.62	4.23	3.95	3.74	3.58	3.45	3.34	3.18	3.01	2.82	2.73	2.63	2.52	2.42	2.30	2.18
40	8.83	6.07	4.98	4.37	3.99	3.71	3.51	3.35	3.22	3.12	2.95	2.78	2.60	2.50	2.40	2.30	2.18	2.06	1.93
60	8.49	5.79	4.73	4.14	3.76	3.49	3.29	3.13	3.01	2.90	2.74	2.57	2.39	2.29	2.19	2.08	1.96	1.83	1.69
120	8.18	5.54	4.50	3.92	3.55	3.28	3.09	2.93	2.81	2.71	2.54	2.37	2.19	2.09	1.98	1.87	1.75	1.61	1.43
∞	7.88	5.30	4.28	3.72	3.35	3.09	2.90	2.74	2.62	2.52	2.36	2.19	2.00	1.90	1.79	1.67	1.53	1.36	1.00

Source: Reprinted from E. S. Pearson and H. O. Hartley, eds., *Biometrika Tables for Statisticians,* 3rd ed., 1966. Reprinted by permission of the *Biometrika* trustees.

Table A7 Exponential Function

c	e^{-c}	c	e^{-c}	c	e^{-c}
.00	1.000000	2.35	.095369	4.70	.009095
.05	.951229	2.40	.090718	4.75	.008652
.10	.904837	2.45	.086294	4.80	.008230
.15	.860708	2.50	.082085	4.85	.007828
.20	.818731	2.55	.078082	4.90	.007447
.25	.778801	2.60	.074274	4.95	.007083
.30	.740818	2.65	.070651	5.00	.006738
.35	.704688	2.70	.067206	5.05	.006409
.40	.670320	2.75	.063928	5.10	.006097
.45	.637628	2.80	.060810	5.15	.005799
.50	.606531	2.85	.057844	5.20	.005517
.55	.576950	2.90	.055023	5.25	.005248
.60	.548812	2.95	.052340	5.30	.004992
.65	.522046	3.00	.049787	5.35	.004748
.70	.496585	3.05	.047359	5.40	.004517
.75	.472367	3.10	.045049	5.45	.004296
.80	.449329	3.15	.042852	5.50	.004087
.85	.427415	3.20	.040762	5.55	.003887
.90	.406570	3.25	.038774	5.60	.003698
.95	.386741	3.30	.036883	5.65	.003518
1.00	.367879	3.35	.035084	5.70	.003346
1.05	.349938	3.40	.033373	5.75	.003183
1.10	.332871	3.45	.031746	5.80	.003028
1.15	.316637	3.50	.030197	5.85	.002880
1.20	.301194	3.55	.028725	5.90	.002739
1.25	.286505	3.60	.027324	5.95	.002606
1.30	.272532	3.65	.025991	6.00	.002479
1.35	.259240	3.70	.024724	6.05	.002358
1.40	.246597	3.75	.023518	6.10	.002243
1.45	.234570	3.80	.022371	6.15	.002133
1.50	.223130	3.85	.021280	6.20	.002029
1.55	.212248	3.90	.020242	6.25	.001930
1.60	.201897	3.95	.019255	6.30	.001836
1.65	.192050	4.00	.018316	6.35	.001747
1.70	.182684	4.05	.017422	6.40	.001661
1.75	.173774	4.10	.016573	6.45	.001581
1.80	.165299	4.15	.015764	6.50	.001503
1.85	.157237	4.20	.014996	6.55	.001430
1.90	.149569	4.25	.014264	6.60	.001360
1.95	.142274	4.30	.013569	6.65	.001294
2.00	.135335	4.35	.012907	6.70	.001231
2.05	.128735	4.40	.012277	6.75	.001171
2.10	.122456	4.45	.011679	6.80	.001114
2.15	.116484	4.50	.011109	6.85	.001059
2.20	.110803	4.55	.010567	6.90	.001008
2.25	.105399	4.60	.010052	6.95	.000959
2.30	.100259	4.65	.009562	7.00	.000912

Table A7 (Continued)

c	e^{-c}	c	e^{-c}	c	e^{-c}
7.05	.000867	8.05	.000319	9.05	.000117
7.10	.000825	8.10	.000304	9.10	.000112
7.15	.000785	8.15	.000289	9.15	.000106
7.20	.000747	8.20	.000275	9.20	.000101
7.25	.000710	8.25	.000261	9.25	.000096
7.30	.000676	8.30	.000249	9.30	.000091
7.35	.000643	8.35	.000236	9.35	.000087
7.40	.000611	8.40	.000225	9.40	.000083
7.45	.000581	8.45	.000214	9.45	.000079
7.50	.000553	8.50	.000204	9.50	.000075
7.55	.000526	8.55	.000194	9.55	.000071
7.60	.000501	8.60	.000184	9.60	.000068
7.65	.000476	8.65	.000175	9.65	.000064
7.70	.000453	8.70	.000167	9.70	.000061
7.75	.000431	8.75	.000158	9.75	.000058
7.80	.000410	8.80	.000151	9.80	.000056
7.85	.000390	8.85	.000143	9.85	.000053
7.90	.000371	8.90	.000136	9.90	.000050
7.95	.000353	8.95	.000130	9.95	.000048
8.00	.000336	9.00	.000123	10.00	.000045

Table A8 Random Numbers

12651	61646	11769	75109	86996	97669	25757	32535	07122	76763
81769	74436	02630	72310	45049	18029	07469	42341	98173	79260
36737	98863	77240	76251	00654	64688	09343	70278	67331	98729
82861	54371	76610	94934	72748	44124	05610	53750	95938	01485
21325	15732	24127	37431	09723	63529	73977	95218	96074	42138
74146	47887	62463	23045	41490	07954	22597	60012	98866	90959
90759	64410	54179	66075	61051	75385	51378	08360	95946	95547
55683	98078	02238	91540	21219	17720	87817	41705	95785	12563
79686	17969	76061	83748	55920	83612	41540	86492	06447	60568
70333	00201	86201	69716	78185	62154	77930	67663	29529	75116
14042	53536	07779	04157	41172	36473	42123	43929	50533	33437
59911	08256	06596	48416	69770	68797	56080	14223	59199	30162
62368	62623	62742	14891	39247	52242	98832	69533	91174	57979
57529	97751	54976	48957	74599	08759	78494	52785	68526	64618
15469	90574	78033	66885	13936	42117	71831	22961	94225	31816
18625	23674	53850	32827	81647	80820	00420	63555	74489	80141
74626	68394	88562	70745	23701	45630	65891	58220	35442	60414
11119	16519	27384	90199	79210	76965	99546	30323	31664	22845
41101	17336	48951	53674	17880	45260	08575	49321	36191	17095
32123	91576	84221	78902	82010	30847	62329	63898	23268	74283
26091	68409	69704	82267	14751	13151	93115	01437	56945	89661
67680	79790	48462	59278	44185	29616	76531	19589	83139	28454
15184	19260	14073	07026	25264	08388	27182	22557	61501	67481
58010	45039	57181	10238	36874	28546	37444	80824	63981	39942
56425	53996	86245	32623	78858	08143	60377	42925	42815	11159
82630	84066	13592	60642	17904	99718	63432	88642	37858	25431
14927	40909	23900	48761	44860	92467	31742	87142	03607	32059
23740	22505	07489	85986	74420	21744	97711	36648	35620	97949
32990	97446	03711	63824	07953	85965	87089	11687	92414	67257
05310	24058	91946	78437	34365	82469	12430	84754	19354	72745
21839	39937	27534	88913	49055	19218	47712	67677	51889	70926
08833	42549	93981	94051	28382	83725	72643	64233	97252	17133
58336	11139	47479	00931	91560	95372	97642	33856	54825	55680
62032	91144	75478	47431	52726	30289	42411	91886	51818	78292
45171	30557	53116	04118	58301	24375	65609	85810	18620	49198

Table A8 (Continued)

91611	62656	60128	35609	63698	78356	50682	22505	01692	36291
55472	63819	86314	49174	93582	73604	78614	78849	23096	72825
18573	09729	74091	53994	10970	86557	65661	41854	26037	53296
60866	02955	90288	82136	83644	94455	06560	78029	98768	71296
45043	55608	82767	60890	74646	79485	13619	98868	40857	19415
17831	09737	79473	75945	28394	79334	70577	38048	03607	06932
40137	03981	07585	18128	11178	32601	27994	05641	22600	86064
77776	31343	14576	97706	16039	47517	43300	59080	80392	63189
69605	44104	40103	95635	05635	81673	68657	09559	23510	95875
19916	52934	26499	09821	97331	80993	61299	36979	73599	35055
02606	58552	07678	56619	65325	30705	99582	53390	46357	13244
65183	73160	87131	35530	47946	09854	18080	02321	05809	04893
10740	98914	44916	11322	89717	88189	30143	52687	19420	60061
98642	89822	71691	51573	83666	61642	46683	33761	47542	23551
60139	25601	93663	25547	02654	94829	48672	28736	84994	13071

Source: From *A Million Random Digits with 100,000 Normal Deviates*, RAND (New York: The Fress Press) Copyright 1955 and 1983 by RAND. Used by permission.

Table A9 Cutoff Points for the Distribution of the Durbin–Watson Test Statistic

	$k = 1$		$k = 2$		$k = 3$		$k = 4$		$k = 5$	
n	d_L	d_U	d_L	d_U	d_L	d_U	d_L	d_U	d_L	d_U
					$\alpha = .05$					
15	1.08	1.36	.95	1.54	.82	1.75	.69	1.97	.56	2.21
16	1.10	1.37	.98	1.54	.86	1.73	.74	1.93	.62	2.15
17	1.13	1.38	1.02	1.54	.90	1.71	.78	1.90	.67	2.10
18	1.16	1.39	1.05	1.53	.93	1.69	.82	1.87	.71	2.06
19	1.18	1.40	1.08	1.53	.97	1.68	.86	1.85	.75	2.02
20	1.20	1.41	1.10	1.54	1.00	1.68	.90	1.83	.79	1.99
21	1.22	1.42	1.13	1.54	1.03	1.67	.93	1.81	.83	1.96
22	1.24	1.43	1.15	1.54	1.05	1.66	.96	1.80	.86	1.94
23	1.26	1.44	1.17	1.54	1.08	1.66	.99	1.79	.90	1.92
24	1.27	1.45	1.19	1.55	1.10	1.66	1.01	1.78	.93	1.90
25	1.29	1.45	1.21	1.55	1.12	1.66	1.04	1.77	.95	1.89
26	1.30	1.46	1.22	1.55	1.14	1.65	1.06	1.76	.98	1.88
27	1.32	1.47	1.24	1.56	1.16	1.65	1.08	1.76	1.01	1.86
28	1.33	1.48	1.26	1.56	1.18	1.65	1.10	1.75	1.03	1.85
29	1.34	1.48	1.27	1.56	1.20	1.65	1.12	1.74	1.05	1.84
30	1.35	1.49	1.28	1.57	1.21	1.65	1.14	1.74	1.07	1.83
31	1.36	1.50	1.30	1.57	1.23	1.65	1.16	1.74	1.09	1.83
32	1.37	1.50	1.31	1.57	1.24	1.65	1.18	1.73	1.11	1.82
33	1.38	1.51	1.32	1.58	1.26	1.65	1.19	1.73	1.13	1.81
34	1.39	1.51	1.33	1.58	1.27	1.65	1.21	1.73	1.15	1.81
35	1.40	1.52	1.34	1.58	1.28	1.65	1.22	1.73	1.16	1.80
36	1.41	1.52	1.35	1.59	1.29	1.65	1.24	1.73	1.18	1.80
37	1.42	1.53	1.36	1.59	1.31	1.66	1.25	1.72	1.19	1.80
38	1.43	1.54	1.37	1.59	1.32	1.66	1.26	1.72	1.21	1.79
39	1.43	1.54	1.38	1.60	1.33	1.66	1.27	1.72	1.22	1.79
40	1.44	1.54	1.39	1.60	1.34	1.66	1.29	1.72	1.23	1.79
45	1.48	1.57	1.43	1.62	1.38	1.67	1.34	1.72	1.29	1.78
50	1.50	1.59	1.46	1.63	1.42	1.67	1.38	1.72	1.34	1.77
55	1.53	1.60	1.49	1.64	1.45	1.68	1.41	1.72	1.38	1.77
60	1.55	1.62	1.51	1.65	1.48	1.69	1.44	1.73	1.41	1.77
65	1.57	1.63	1.54	1.66	1.50	1.70	1.47	1.73	1.44	1.77
70	1.58	1.64	1.55	1.67	1.52	1.70	1.49	1.74	1.46	1.77
75	1.60	1.65	1.57	1.68	1.54	1.71	1.51	1.74	1.49	1.77
80	1.61	1.66	1.59	1.69	1.56	1.72	1.53	1.74	1.51	1.77
85	1.62	1.67	1.60	1.70	1.57	1.72	1.55	1.75	1.52	1.77
90	1.63	1.68	1.61	1.70	1.59	1.73	1.57	1.75	1.54	1.78
95	1.64	1.69	1.62	1.71	1.60	1.73	1.58	1.75	1.56	1.78
100	1.65	1.69	1.63	1.72	1.61	1.74	1.59	1.76	1.57	1.78

Table A9 (Continued)

n	k = 1 d_L	k = 1 d_U	k = 2 d_L	k = 2 d_U	k = 3 d_L	k = 3 d_U	k = 4 d_L	k = 4 d_U	k = 5 d_L	k = 5 d_U
					$\alpha = .01$					
15	.81	1.07	.70	1.25	.59	1.46	.49	1.70	.39	1.96
16	.84	1.09	.74	1.25	.63	1.44	.53	1.66	.44	1.90
17	.87	1.10	.77	1.25	.67	1.43	.57	1.63	.48	1.85
18	.90	1.12	.80	1.26	.71	1.42	.61	1.60	.52	1.80
19	.93	1.13	.83	1.26	.74	1.41	.65	1.58	.56	1.77
20	.95	1.15	.86	1.27	.77	1.41	.68	1.57	.60	1.74
21	.97	1.16	.89	1.27	.80	1.41	.72	1.55	.63	1.71
22	1.00	1.17	.91	1.28	.83	1.40	.75	1.54	.66	1.69
23	1.02	1.19	.94	1.29	.86	1.40	.77	1.53	.70	1.67
24	1.04	1.20	.96	1.30	.88	1.41	.80	1.53	.72	1.66
25	1.05	1.21	.98	1.30	.90	1.41	.83	1.52	.75	1.65
26	1.07	1.22	1.00	1.31	.93	1.41	.85	1.52	.78	1.64
27	1.09	1.23	1.02	1.32	.95	.141	.88	1.51	.81	1.63
28	1.10	1.24	1.04	1.32	.97	1.41	.90	1.51	.83	1.62
29	1.12	1.25	1.05	1.33	.99	1.42	.92	1.51	.85	1.61
30	1.13	1.26	1.07	1.34	1.01	1.42	.94	1.51	.88	1.61
31	1.15	1.27	1.08	1.34	1.02	1.42	.96	1.51	.90	1.60
32	1.16	1.28	1.10	1.35	1.04	1.43	.98	1.51	.92	1.60
33	1.17	1.29	1.11	1.36	1.05	1.43	1.00	1.51	.94	1.59
34	1.18	1.30	1.13	1.36	1.07	1.43	1.01	1.51	.95	1.59
35	1.19	1.31	1.14	1.37	1.08	1.44	1.03	1.51	.97	1.59
36	1.21	1.32	1.15	1.38	1.10	1.44	1.04	1.51	.99	1.59
37	1.22	1.32	1.16	1.38	1.11	1.45	1.06	1.51	1.00	1.59
38	1.23	1.33	1.18	1.39	1.12	1.45	1.07	1.52	1.02	1.58
39	1.24	1.34	1.19	1.39	1.14	1.45	1.09	1.52	1.03	1.58
40	1.25	1.34	1.20	1.40	1.15	1.46	1.10	1.52	1.05	1.58
45	1.29	1.38	1.24	1.42	1.20	1.48	1.16	1.53	1.11	1.58
50	1.32	1.40	1.28	1.45	1.24	1.49	1.20	1.54	1.16	1.59
55	1.36	1.43	1.32	1.47	1.28	1.51	1.25	1.55	1.21	1.59
60	1.38	1.45	1.35	1.48	1.32	1.52	1.28	1.56	1.25	1.60
65	1.41	1.47	1.38	1.50	1.35	1.53	1.31	1.57	1.28	1.61
70	1.43	1.49	1.40	1.52	1.37	1.55	1.34	1.58	1.31	1.61
75	1.45	1.50	1.42	1.53	1.39	1.56	1.37	1.59	1.34	1.62
80	1.47	1.52	1.44	1.54	1.42	1.57	1.39	1.60	1.36	1.62
85	1.48	1.53	1.46	1.55	1.43	1.58	1.41	1.60	1.39	1.63
90	1.50	1.54	1.47	1.56	1.45	1.59	1.43	1.61	1.41	1.64
95	1.51	1.55	1.49	1.57	1.47	1.60	1.45	1.62	1.42	1.64
100	1.52	1.56	1.50	1.58	1.48	1.60	1.46	1.63	1.44	1.65

Source: From J. Durbin and G. S. Watson, "Testing for Serial Correlation in Least Squares Regression, II," *Biometrika,* 1951, *30,* 159–178. Reproduced by permission of the *Biometrika* trustees.

Table A10 Lower and Upper Critical Values R for the Runs Test for Randomness

n_1 \ n_2	2	3	4	5	6	7	8	9	10	11	12	13	14	15	16	17	18	19	20
Lower Tail (α = .025)																			
2											2	2	2	2	2	2	2	2	2
3				2	2	2	2	2	2	2	2	2	2	3	3	3	3	3	3
4			2	2	2	2	3	3	3	3	3	3	3	3	4	4	4	4	4
5		2	2	2	3	3	3	3	3	4	4	4	4	4	4	4	5	5	5
6		2	2	3	3	3	3	4	4	4	4	5	5	5	5	5	5	6	6
7		2	2	3	3	3	4	4	5	5	5	5	5	6	6	6	6	6	6
8		2	3	3	3	4	4	5	5	5	6	6	6	6	6	7	7	7	7
9		2	3	3	4	4	5	5	5	6	6	6	7	7	7	7	8	8	8
10		2	3	3	4	5	5	5	6	6	7	7	7	7	8	8	8	8	9
11		2	3	4	4	5	5	6	6	7	7	7	8	8	8	9	9	9	9
12	2	2	3	4	4	5	6	6	7	7	7	8	8	8	9	9	9	10	10
13	2	2	3	4	5	5	6	6	7	7	8	8	9	9	9	10	10	10	10
14	2	2	3	4	5	5	6	7	7	8	8	9	9	9	10	10	10	11	11
15	2	3	3	4	5	6	6	7	7	8	8	9	9	10	10	11	11	11	12
16	2	3	4	4	5	6	6	7	8	8	9	9	10	10	11	11	11	12	12
17	2	3	4	4	5	6	7	7	8	9	9	10	10	11	11	11	12	12	13
18	2	3	4	5	5	6	7	8	8	9	9	10	10	11	11	12	12	13	13
19	2	3	4	5	6	6	7	8	8	9	10	10	11	11	12	12	13	13	13
20	2	3	4	5	6	6	7	8	9	9	10	10	11	12	12	13	13	13	14
Upper Tail (α = .025)																			
2																			
3																			
4				9	9														
5			9	10	10	11	11												
6			9	10	11	12	12	13	13	13	13								
7				11	12	13	13	14	14	14	14	15	15	15					
8				11	12	13	14	14	15	15	16	16	16	16	17	17	17	17	17
9					13	14	14	15	16	16	16	17	17	18	18	18	18	18	18
10					13	14	15	16	16	17	17	18	18	18	19	19	19	20	20
11					13	14	15	16	17	17	18	19	19	19	20	20	20	21	21
12					13	14	16	16	17	18	19	19	20	20	21	21	21	22	22
13						15	16	17	18	19	19	20	20	21	21	22	22	23	23
14						15	16	17	18	19	20	20	21	22	22	23	23	23	24
15						15	16	18	18	19	20	21	22	22	23	23	24	24	25
16							17	18	19	20	21	21	22	23	23	24	25	25	25
17							17	18	19	20	21	22	23	23	24	25	25	26	26
18							17	18	19	20	21	22	23	24	25	25	26	26	27
19							17	18	20	21	22	23	23	24	25	26	26	27	27
20							17	18	20	21	22	23	24	25	25	26	27	27	28

Source: Adapted from F. S. Swed and C. Eisenhart, *Ann. Math. Statist.,* 14, 1943, 83–86.

Table A11 Critical Values of W in the Wilcoxon Matched-Pairs Signed-Rank Test

For sample size n, the table shows, for selected probabilities, α, the numbers W_α such that $P(W \leq W_\alpha) = \alpha$, where the distribution of the random variable W is that of the Wilcoxon test statistic under the null hypothesis.

	α				
n	*.005*	*.010*	*.025*	*.050*	*.100*
4	0	0	0	0	1
5	0	0	0	1	3
6	0	0	1	3	4
7	0	1	3	4	6
8	1	2	4	6	9
9	2	4	6	9	11
10	4	6	9	11	15
11	6	8	11	14	18
12	8	10	14	18	22
13	10	13	18	22	27
14	13	16	22	26	32
15	16	20	26	31	37
16	20	24	30	36	43
17	24	28	35	42	49
18	28	33	41	48	56
19	33	38	47	54	63
20	38	44	53	61	70

Table A12 Lower and Upper Critical Values R_{n_1} and R_{n_2} of the Wilcoxon Rank-Sum Test

	α		n_1						
n_2	One-Tailed	Two-Tailed	4	5	6	7	8	9	10
4	.05	.10	11,25						
	.025	.05	10,26						
	.01	.02	—,—						
	.005	.01	—,—						
5	.05	.10	12,28	19,36					
	.025	.05	11,29	17,38					
	.01	.02	10,30	16,39					
	.005	.01	—,—	15,40					
6	.05	.10	13,31	20,40	28,50				
	.025	.05	12,32	18,42	26,52				
	.01	.02	11,33	17,43	24,54				
	.005	.01	10,34	16,44	23,55				
7	.05	.10	14,34	21,44	29,55	39,66			
	.025	.05	13,35	20,45	27,57	36,69			
	.01	.02	11,37	18,47	25,59	34,71			
	.005	.01	10,38	16,49	24,60	32,73			
8	.05	.10	15,37	23,47	31,59	41,71	51,85		
	.025	.05	14,38	21,49	29,61	38,74	49,87		
	.01	.02	12,40	19,51	27,63	35,77	45,91		
	.005	.01	11,41	17,53	25,65	34,78	43,93		
9	.05	.10	16,40	24,51	33,63	43,76	54,90	66,105	
	.025	.05	14,42	22,53	31,65	40,79	51,93	62,109	
	.01	.02	13,43	20,55	28,68	37,82	47,97	59,112	
	.005	.01	11,45	18,57	26,70	35,84	45,99	56,115	
10	.05	.10	17,43	26,54	35,67	45,81	56,96	69,111	82,128
	.025	.05	15,45	23,57	32,70	42,84	53,99	65,115	78,132
	.01	.02	13,47	21,59	29,73	39,87	49,103	61,119	74,136
	.005	.01	12,48	19,61	27,75	37,89	47,105	58,122	71,139

Source: Adapted from Table 1 of F. Wilcoxon and R. A. Wilcox, *Some Rapid Approximate Statistical Procedures.* Copyright © 1949, 1964 Lederle Laboratories, Division of American Cyanamid Company. All rights reserved. Reprinted with permission.

Table A13 Factors for Control Charts

	$\overline{X}$-charts			S-Charts					R-charts					
n	A	A_2	A_3	c_4	B_3	B_4	B_5	B_6	d_2	d_3	D_1	D_2	D_3	D_4
2	2.121	1.880	2.659	.7979	0	3.267	0	2.606	1.128	.853	0	3.686	0	3.267
3	1.732	1.023	1.954	.8862	0	2.568	0	2.276	1.693	.888	0	4.358	0	2.574
4	1.500	.729	1.628	.9213	0	2.266	0	2.088	2.059	.880	0	4.698	0	2.282
5	1.342	.577	1.427	.9400	0	2.089	0	1.964	2.326	.864	0	4.918	0	2.114
6	1.225	.483	1.287	.9515	.030	1.970	.029	1.874	2.534	.848	0	5.078	0	2.004
7	1.134	.419	1.182	.9594	.118	1.882	.113	1.806	2.704	.833	.204	5.204	.076	1.924
8	1.061	.373	1.099	.9650	.185	1.815	.179	1.751	2.847	.820	.388	5.306	.136	1.864
9	1.000	.337	1.032	.9693	.239	1.761	.232	1.707	2.970	.808	.547	5.393	.184	1.816
10	.949	.308	.975	.9727	.284	1.716	.276	1.669	3.078	.797	.687	5.469	.223	1.777
11	.905	.285	.927	.9754	.321	1.679	.313	1.637	3.173	.787	.811	5.535	.256	1.744
12	.866	.266	.886	.9776	.354	1.646	.346	1.610	3.258	.778	.922	5.594	.283	1.717
13	.832	.249	.850	.9794	.382	1.618	.374	1.585	3.336	.770	1.025	5.647	.307	1.693
14	.802	.235	.817	.9810	.406	1.594	.399	1.563	3.407	.763	1.118	5.696	.328	1.672
15	.775	.223	.789	.9823	.428	1.572	.421	1.544	3.472	.756	1.203	5.741	.347	1.653
16	.750	.212	.763	.9835	.448	1.552	.440	1.526	3.532	.750	1.282	5.782	.363	1.637
17	.728	.203	.739	.9845	.466	1.534	.458	1.511	3.588	.744	1.356	5.820	.378	1.622
18	.707	.194	.718	.9854	.482	1.518	.475	1.496	3.640	.739	1.424	5.856	.391	1.608
19	.688	.187	.698	.9862	.497	1.503	.490	1.483	3.689	.734	1.487	5.891	.403	1.597
20	.671	.180	.680	.9869	.510	1.490	.504	1.470	3.735	.729	1.549	5.921	.415	1.585
21	.655	.173	.663	.9876	.523	1.477	.516	1.459	3.778	.724	1.605	5.951	.425	1.575
22	.640	.167	.647	.9882	.534	1.466	.528	1.448	3.819	.720	1.659	5.979	.434	1.566
23	.626	.162	.633	.9887	.545	1.455	.539	1.438	3.858	.716	1.710	6.006	.443	1.557
24	.612	.157	.619	.9892	.555	1.445	.549	1.429	3.895	.712	1.759	6.031	.451	1.548
25	.600	.153	.606	.9896	.565	1.435	.559	1.420	3.931	.708	1.806	6.056	.459	4.541

Source: Copyright American Society for Testing and Materials. Reprinted with permission.

Table A14 Present Value of $1 $P = S_n(1 + r)^{-n}$

Years Hence	1%	2%	4%	6%	8%	10%	12%	14%	15%	16%	18%	20%	22%	24%	25%	26%	28%	30%	35%	40%	45%	50%
1	.990	.980	.962	.943	.926	.909	.893	.877	.870	.862	.847	.833	.820	.806	.800	.794	.781	.769	.741	.714	.690	.667
2	.980	.961	.925	.890	.857	.826	.797	.769	.756	.743	.718	.694	.672	.650	.640	.630	.610	.592	.549	.510	.476	.444
3	.971	.942	.889	.840	.794	.751	.712	.675	.658	.641	.609	.579	.551	.524	.512	.500	.477	.455	.406	.364	.328	.296
4	.961	.924	.855	.792	.735	.683	.636	.592	.572	.552	.516	.482	.451	.423	.410	.397	.373	.350	.301	.260	.226	.198
5	.951	.906	.822	.747	.681	.621	.567	.519	.497	.476	.437	.402	.370	.341	.328	.315	.291	.269	.223	.186	.156	.132
6	.942	.888	.790	.705	.630	.564	.507	.456	.432	.410	.370	.335	.303	.275	.262	.250	.227	.207	.165	.133	.108	.088
7	.933	.871	.760	.665	.583	.513	.452	.400	.376	.354	.314	.279	.249	.222	.210	.198	.178	.159	.122	.095	.074	.059
8	.923	.853	.731	.627	.540	.467	.404	.351	.327	.305	.266	.233	.204	.179	.168	.157	.139	.123	.091	.068	.051	.039
9	.914	.837	.703	.592	.500	.424	.361	.308	.284	.263	.225	.194	.167	.144	.134	.125	.108	.094	.067	.048	.035	.026
10	.905	.820	.676	.558	.463	.386	.322	.270	.247	.227	.191	.162	.137	.116	.107	.099	.085	.073	.050	.035	.024	.017
11	.896	.804	.650	.527	.429	.350	.287	.237	.215	.195	.162	.135	.112	.094	.086	.079	.066	.056	.037	.025	.017	.012
12	.887	.788	.625	.497	.397	.319	.257	.208	.187	.168	.137	.112	.092	.076	.069	.062	.052	.043	.027	.018	.012	.008
13	.879	.773	.601	.469	.368	.290	.229	.182	.163	.145	.116	.093	.075	.061	.055	.050	.040	.033	.020	.013	.008	.005
14	.870	.758	.577	.442	.340	.263	.205	.160	.141	.125	.099	.078	.062	.049	.044	.039	.032	.025	.015	.009	.006	.003
15	.861	.743	.555	.417	.315	.239	.183	.140	.123	.108	.084	.065	.051	.040	.035	.031	.025	.020	.011	.006	.004	.002
16	.853	.728	.534	.394	.292	.218	.163	.123	.107	.093	.071	.054	.042	.032	.028	.025	.019	.015	.008	.005	.003	.002
17	.844	.714	.513	.371	.270	.198	.146	.108	.093	.080	.060	.045	.034	.026	.023	.020	.015	.012	.006	.003	.002	.001
18	.836	.700	.494	.350	.250	.180	.130	.095	.081	.069	.051	.038	.028	.021	.018	.016	.012	.009	.005	.002	.001	.001
19	.828	.686	.475	.331	.232	.164	.116	.083	.070	.060	.043	.031	.023	.017	.014	.012	.009	.007	.003	.002	.001	
20	.820	.673	.456	.312	.215	.149	.104	.073	.061	.051	.037	.026	.019	.014	.012	.010	.007	.005	.002	.001	.001	
21	.811	.660	.439	.294	.199	.135	.093	.064	.053	.044	.031	.022	.015	.011	.009	.008	.006	.004	.002	.001		
22	.803	.647	.422	.278	.184	.123	.083	.056	.046	.038	.026	.018	.013	.009	.007	.006	.004	.003	.001	.001		
23	.795	.634	.406	.262	.170	.112	.074	.049	.040	.033	.022	.015	.010	.007	.006	.005	.003	.002	.001			
24	.788	.622	.390	.247	.158	.102	.066	.043	.035	.028	.019	.013	.008	.006	.005	.004	.003	.002	.001			
25	.780	.610	.375	.233	.146	.092	.059	.038	.030	.024	.016	.010	.007	.005	.004	.003	.002	.001	.001			
26	.772	.598	.361	.220	.135	.084	.053	.033	.026	.021	.014	.009	.006	.004	.003	.002	.002	.001				
27	.764	.586	.347	.207	.125	.076	.047	.029	.023	.018	.011	.007	.005	.003	.002	.002	.001	.001				
28	.757	.574	.333	.196	.116	.069	.042	.026	.020	.016	.010	.006	.004	.002	.002	.002	.001	.001				
29	.749	.563	.321	.185	.107	.063	.037	.022	.017	.014	.008	.005	.003	.002	.002	.001	.001	.001				
30	.742	.552	.308	.174	.099	.057	.033	.020	.015	.012	.007	.004	.003	.002	.001	.001	.001	.001				
40	.672	.453	.208	.097	.046	.022	.011	.005	.004	.003	.001	.001										
50	.608	.372	.141	.054	.021	.009	.003	.001	.001	.001												

Source: From Jerome Bracken and Charles J. Christenson, Tables for Use in Analyzing Business Decisions, 1965, reprinted by permission of Richard D. Irwin, Inc.

A31

Appendix B Description of Data Sets

The following 8 data sets used in the text are available on an IBM-compatible floppy disk, for instructors who request it. In addition, the data sets will be updated on disk each year.

1. Annual Macroeconomic Data (1950–1990)

The macroeconomic data included in this data set are GNP (Gross National Product), CPI (Consumer Price Index), yield of 3-month T-bills, prime rate, private consumption, private investment, net exports, and government expenditures. The data set is also given in Table 2.2 in the text.

2. Financial Ratios for Three Auto Companies (1966–1990)

The three auto companies are Chrysler, Ford, and General Motors. The financial ratios included are the current ratio, inventory turnover, debt ratio (total debt/total assets), profit margin (net income/sales), return on assets (net income/total assets), P/E ratio, payout ratio [dividend per share (DPS)/earnings per share (EPS)]. These data are also used in Appendixes 2C, 3A, and 4C in the text.

3. EPS, DPS, PPS, and Rates of Return for GM and Ford (1969–1990)

The data included in the first part of this data set are earnings per share (EPS), dividend per share (DPS), and price per share (PPS). At the far right of the data set is the S&P 500 Index. In the second part of the data set are the annual rates of return for GM, Ford, and the market. This information can also be found in Appendix 2B, Tables 2B.1 and 2B.2.

4. Annual Ford Sales Data (1968–1990)

This set gives annual sales data for Ford from 1968 to 1990. The data are also presented in Table 18.7 in the text.

5. Quarterly EPS and Sales Data for Johnson & Johnson and IBM (1980–1990)

Included are quarterly EPS and sales data for Johnson & Johnson and IBM from the first quarter of 1980 to the second quarter of 1990. The EPS and sales data for Johnson & Johnson can also be found in Tables 18.5 and 8.10, respectively, in the text.

6. Monthly Rates of Return for Dow Jones 30 Companies (December 1987 to June 1990)

This data set includes the monthly rates of return for the Dow Jones 30 and the S&P 500. The information is also given in Section 9.8 of the text.

The names of these 30 companies are

Allied Signal	International Paper
Aluminum Co. of America	McDonald's
American Express	Merck
American Telephone & Telegraph	Minnesota Mining & Manufacturing
Bethlehem Steel	Navistar International
Boeing	Philip Morris
Chevron	Primerica
Coca-Cola	Procter & Gamble
Du Pont (E. I.) De Nemours	Sears, Roebuck
Eastman Kodak	Texaco
Exxon	USX
General Electric	Union Carbide
General Motors	United Technologies
Goodyear Tire & Rubber	Westinghouse Electric
IBM	Woolworth (F. W.)

7. Monthly Wilshire 5000 Equity Index (January 1989 to January 1991)

The information given here is the value-weighted monthly Wilshire 5000 Equity Index. It can also be found in Table 19.10 in the text.

8. Monthly Rates of Return for the Value-Weighted Wilshire 5000 Equity Index (January 1989 to January 1991)

Included are the percentage change in price appreciation, dividend yield, and total rates of return of the value-weighted Wilshire 5000 Equity Index. The data are also given in Table 19.11 in the text.

Appendix C Introduction to MINITAB: Microcomputer Version

MINITAB is a user-friendly statistics package. Students who are beginners in statistics will find that the application of MINITAB in their statistics course will assist them in grasping statistics without greatly increasing their study load.

This appendix provides a brief description of the basic functions of MINITAB. With this basic knowledge, students can start to work with MINITAB.

Appendix.C.1 General Description

MINITAB is a command-driven statistics package with more than 200 commands available. Data are stored and processed in a worksheet, a table with rows and columns.

MINITAB will accept words typed in upper- or lowercase letters or a combination of the two.

Appendix.C.2 Data Input

For MINITAB to perform statistical computations, data must first be input. Data for each variable in your data set are stored in columns. There are two ways to enter data: READ and SET commands.

READ Commands

The form of the command is

    READ C1

where C1 represents column 1. After typing this line, press RETURN or ENTER. Then type the data, one number per line. The computer will prompt you after each entered line with

    DATA⟩

When all the data have been input, type

    END

For example, suppose you have the following four observations: 25, 33, 41, and 58. These data can be entered as follows:

    MTB) READ C1
    DATA) 25
    DATA) 33
    DATA) 41
    DATA) 58
    DATA) END

If two or more groups of data (or variables) are to be read, you could type

    READ C1 C2

or

    READ C1 C2 C3

or

    READ C1-C3

For example, suppose you have four observations for three variables:

    variable 1: 13 17 21 11
    variable 2: 23 27 32 20
    variable 3: 35 31 42 37

The data can be entered as follows:

    MTB) READ C1-C3
    DATA) 13 23 35
    DATA) 17 27 31
    DATA) 21 32 42
    DATA) 11 20 37
    DATA) END

SET COMMAND

This command allows you to enter numbers consecutively on one or more lines for each variable. For instance, in the first example of the READ command, the data can be entered as follows:

    MTB) SET C1
    DATA) 25 33 41 58
    DATA) END

The second example of READ command can be entered as follows:

    MTB) SET C1
    DATA) 13 17 21 11
    DATA) END

```
MTB) SET C2
DATA) 23 27 32 20
DATA) END

MTB) SET C3
DATA) 35 31 42 37
DATA)END
```

Appendix.C.3 Data Corrections

Suppose that after entering the data you find an error. The following three instructions allow you to correct errors:

1. The LET command enables you to replace an erroneous entry. For example,

LET C2(3) = 7

changes the third value in column 2 to 7.

2. The DELETE command simply erases the data you specify. For example,

DELETE 3:6 C3

deletes rows 3 through 6 from column 3.

3. The INSERT command lets you add new material. For example,

```
INSERT BETWEEN ROWS 4 AND 5 OF COLUMN C2 C3
DATA) 13 17
DATA) END
```

inserts a new row of data between rows 4 and 5. The new data are 13 and 17.

Appendix.C.4 Output

To check to see if entered data have been input correctly on the screen, type

PRINT C1

or

PRINT C1 C2 C3

or

PRINT C1-C3

To print out the results of your statistical operations on paper, type PAPER before you enter the print commands. To stop printing, type NOPAPER after you get your printout.

Appendix.C.5 Saving Data

To save a data set, type SAVE 'a:filename'. Once the data set has been saved, you can retrieve it at any time with the following command:

RETRIEVE 'a:filename'

Appendix.C.6 Other Commands

To create new variables from existing ones, use the LET command. For example,

LET C4 = C2 + C3

creates a variable that is the sum of the values stored in the second and third columns, and that variable is stored in column 4.

The MINITAB symbols for common arithmetic operations are

+ add
− subtract
* multiply
/ divide
** exponent (raise to a power)

To erase entire columns, type

ERASE C1 C2 C3

or

ERASE C1-C3

When you have completed your work in MINITAB, type

STOP

Appendix D Introduction to SAS: Microcomputer Version

SAS is a powerful statistics package for manipulating data and performing varied statistical analyses. It is designed for larger data bases.

This appendix will provide a brief description of the basic functions of SAS. With this basic knowledge, students can easily understand SAS programs written by others and start to write SAS programs themselves.

Appendix.D.1 General Description

SAS is normally run in batch mode, rather than interactively as MINITAB is. Therefore, we first put all the command statements into a file (or program), and then submit the entire file to SAS for processing. SAS will perform the requested analyses and return two files: a log of the program (the SASLOG), with notes and error messages, and a list of the results from the analyses. Note that *every* statement in the SAS program must end with a semicolon. SAS will accept words typed in upper- or lowercase letters or a combination of the two.

Appendix.D.2 Data Input

For SAS to perform statistical computations, data must first be input. The first command in any SAS program is the DATA command. For example, the command

    DATA SALES;

will create a data set called "SALES". To read in the data, we then use the INPUT command, which tells SAS how the data values are arranged on the data lines and what the variable names are. An example of an INPUT command is

    INPUT SALES REGION $;

This command informs SAS that you are going to read in two variables, called SALES and REGION. The listing of variable names in the INPUT statement tells SAS that the data are arranged on the data lines in the order listed, with at least one space between values. The dollar sign after REGION tells SAS that the variable region contains alphabetic characters.

After the INPUT command comes the CARDS statement, which tells SAS where to start reading, followed by the data to be read. For example, we may have sales and cost data for several regions listed as follows:

REGION	COST	SALES
EAST	4325	5647
WEST	5941	7103
SOUTH	2387	3492
NORTH	3762	4481

Our entire data input will be as follows:

```
DATA SALES;
INPUT SALES COST REGION $;
CARDS;
5647 4325 EAST
7103 5941 WEST
3492 2387 SOUTH
4481 3762 NORTH
;
```

Appendix.D.3 Data Modifying

To create new variables from existing ones, simply specify the appropriate formula, using the following symbols for the standard arithmetic operations:

+ add
− subtract
* multiply
/ divide
** raise to a power

Here are some examples:

```
PROFIT = SALES − COST;
MARGIN = (SALES − COST)/COST;
```

Appendix.D.4 Analyzing the Data

SAS procedures (nicknamed PROCs) are used to process data in SAS data sets. There are procedures for all kinds of analyses from printing the input data to simple statistics to more complicated statistical analyses. The SAS procedures are written after the data lines. The following will introduce three SAS procedures: PROC PRINT, PROC ANOVA, and PROC REG.

I. PROC PRINT;

The PROC PRINT statement asks SAS to print out the data values in the data set just created. The word PROC signals the beginning of a PROC step, a series of statements that describes the analysis to be performed. The word *print* names the SAS procedure we want to use.

2. PROC ANOVA;

This procedure statement must be followed by a statement identifying the treatment variable and the model. Two examples are as follows:

Completely Randomized Design

```
DATA EXAMPLE1;
INPUT X T;
CARDS;
25 1
27 1
31 1
32 2
35 2
30 2
27 3
32 3
48 3
18 4
23 4
29 4
;
PROC ANOVA;
CLASS T;
MODEL X = T;
RUN;
```

Randomized Block Design

```
DATA EXAMPLE2;
INPUT X T B;
CARDS;
25 1 1
27 1 2
31 1 3
32 2 1
35 2 2
30 2 3
27 3 1
32 3 2
48 3 3
18 4 1
23 4 2
29 4 3
;
PROC ANOVA;
```

CLASS T B;
MODEL X = T B;
RUN;

3. PROC REG

The REG procedure is used to perform both simple and multiple regression. The procedure statement is followed by the MODEL statement, which specifies the dependent and independent variables. For example, the commands and data input for the multiple regression for Table 14.13 are as follows:

```
DATA TAB1413;
INPUT Y X1 X2 X3;
CARDS;
260.3   5    3   4
286.1   7    5   2
279.4   6    3   3
410.8   9    4   4
438.2   12   6   1
315.3   8    3   4
565.1   11   7   3
570.0   16   8   2
426.1   13   4   3
315.0   7    3   4
403.6   10   6   1
220.5   4    4   1
343.6   9    4   3
644.6   17   8   4
520.4   19   7   2
329.5   9    3   2
426.0   11   6   4
343.2   8    3   3
450.4   13   5   4
421.8   14   5   2
245.6   7    4   4
503.3   16   6   3
375.7   9    5   3
265.5   5    3   3
620.6   18   6   4
450.5   18   5   3
270.1   5    3   2
368.0   7    6   2
556.1   12   7   1
570.0   13   6   4
318.5   8    4   3
260.2   6    3   2
667.0   16   8   2
```

```
618.3   19   8   2
525.3   17   7   4
332.2   10   4   3
393.2   12   5   3
283.5    8   3   3
376.2   10   5   4
481.8   12   5   2
;
PROC REG;
   MODEL Y = X1 X2 X3/DW;
RUN;
```

The regression results are as follows:

```
Model: MODEL1
Dependent Variable: Y
```

Analysis of Variance

Source	DF	Sum of Squares	Mean Square	F Value	Prob>F
Model	3	527209.08074	175736.36025	89.051	0.0001
Error	36	71043.94301	1973.44286		
C Total	39	598253.02375			

Root MSE	44.42345	R-square	0.8812	
Dep Mean	411.28750	Adj R-sq	0.8714	
C.V.	10.80107			

Parameter Estimates

Variable	DF	Parameter Estimate	Standard Error	T for H0: Parameter=0	Prob > |T|
INTERCEP	1	31.150390	34.17504533	0.911	0.3681
X1	1	12.968162	2.73723213	4.738	0.0001
X2	1	41.245624	7.28010741	5.666	0.0001
X3	1	11.524255	7.69117684	1.498	0.1428

```
Durbin-Watson D             2.104
(For Number of Obs.)         40
1st Order Autocorrelation  -0.083
```

Answers to Selected Odd-Numbered Questions and Problems

Chapter 1

1. Statistical inference allows us to infer the value of various unknown items. For example, statistical inference can be used to determine the proportion of people who will buy a new product or the future revenues from new product sales.
3. Deductive analysis uses general information to draw conclusions about specific examples. Inductive analysis uses specific cases to draw conclusions about a general population.
5. Because the population represents all items of interest, it should always be as large or larger than the sample size.
7. When testing the market for a new product, the researcher is interested in knowing the opinion of all potential buyers by looking at the responses of a few people. This represents statistical inference.
9. She could compute some measure of the class average and some measure of how spread out the scores are around the class average.
11. Inferential statistics.
13. Election polls generally rely on inferential statistics because surveying individual voters is too costly and time consuming.
15. The average return of AT&T stock over the last five years gives us a benchmark for comparison. If the return is expected to be above its average, it may be a good time to buy AT&T stock because we expect it to exceed its average return.
19. Because the DJIA consists of 30 major industrial stocks, it represents a sample of all stocks.
21. Sample.
23. Inferential. Descriptive.

25. **a.** Descriptive.   **b.** Inferential.
27. Sample.

Chapter 2

1. A primary source of data is data that is collected specifically for the particular analysis, for example, an election poll or the Nielson television ratings.
3. A sample is when we select a few observations from a population of interest. A census occurs when we analyze all observations from a population of interest. A sample is useful because it usually requires less cost and time than a census. A census is appropriate when the population is fairly small.
5. **a.** Line charts are constructed by graphing data points and drawing lines to connect the points. Line charts are good for presenting data that is observed over time (time series data).
   b. A component-parts line chart is a line chart that shows the individual parts or components of the data.
   c. A bar chart is useful for presenting small amounts of data. We use the term *bar chart* because the data are presented as bars.
11. Figure 2.3 indicates that the DJIA returns were lower than the returns for Tri-Continental and the S&P 500 for nearly all the periods examined with the exception of the 5-year period. Tri-Continental produced the best returns for nearly all periods. The return on the S&P 500 had the second best returns for most of the periods.
13. The inventory turnovers for Ford and GM have

followed each other closely over the 25-year period. The exceptions are the years 1987 and 1988 when Ford's inventory turnover was much higher than GM's.

15. A pie chart is best for showing how a total amount (the pie) is divided up (pieces of the pie).

19. A pie chart would best show the distribution of sales.

37. a. The team with the greatest number of yards, San Francisco, has the best pass offense.
b. The team with the least number of yards, Philadelphia, has the best pass defense.
c. Washington has the best rush offense.
d. San Francisco has the best rush defense.

39. a. The greatest reason teen-agers drink is that they are upset (41%).
b. 41% + 25% = 66%.

43. The graph shows net bond fund purchases in 1991. From the graph we can see that a large amount of money has flowed into the bond funds in July, August, and September, after a dip in bond fund purchases in June.

45. a. The two pie charts show the percentage of homeowners for different levels of income.
b. Most first-time homeowners were in the $20,000 to $49,999 income class.

49. The high percentage of people who consider transportation to be the main cause of pollution in New Jersey and the high percentage of people in question 48 who consider traffic congestion to be a problem indicate that transportation in the state is a major concern to residents.

51. a. More women.
b. 19.3% − 10.8% = 8.5% increase.
c. 16% − 8.6% = 7.4% increase.

Chapter 3

1. See Section 3.2 in the text. For grouped data, scores are merged into classes, each with an associated frequency.

3. See Section 3.3 in the text. Cumulative frequency measures total observations up through a certain class.

5. See Section 3.2 in the text.

7. Refer to Table 3.10.

9. a.

Class (mid point) (%)	Frequency	Relative Frequency	Cumulative Frequency
0	1	.059	1
1	3	.176	4
2	3	.176	7
3	1	.059	8
4	4	.235	12
5	2	.118	14
6	2	.118	16
7	0	0	16
8	0	0	16
9	0	0	16
10	0	0	16
11	1	.059	17

15. MTB > STEM-AND-LEAF C1

Stem-and-leaf of TB/RATE   N = 21
Leaf Unit = .10

```
   3      4 039
   7      5 2889
   9      6 46
  (5)     7 02458
   7      8 16
   5      9 5
   4     10 06
   2     11 5
   1     12
   1     13
   1     14 0
```

MTB > STEM-AND-LEAF C2

Stem-and-leaf of PRIMRATE   N = 21
Leaf Unit = 1.0

```
   2      0 55
   6      0 6677
  (6)     0 888999
   9      1 0000
   5      1 22
   3      1 45
   1      1
   1      1 8
```

17.

Class	Cumulative Frequency
$351,000 ~ 400,000	3
$401,000 ~ 450,000	4
$451,000 ~ 500,000	5

Class	Cumulative Frequency
$501,000 ~ 550,000	7
$551,000 ~ 600,000	7
$601,000 ~ 650,000	7
$651,000 ~ 700,000	8
$701,000 and over	11

21. See text. The Lorenz curve follows a cumulative frequency distribution of the data.

23.

Class	Tally	Frequency
51 ~ 60	/	1
61 ~ 70	/////	5
71 ~ 80	///	3
81 ~ 90	///	3
91 ~ 100	///	3

27. It means that the income distribution is characterized by absolute inequality.

29.

Class	Cumulative Frequency
1–5	2
6–10	9
11–15	17
16–20	24
21–25	26
26–30	28
31–35	30

35. Modestia has a more equal distribution of income.

39.

Class	Relative Frequency	Cumulative Relative Frequency
401–450	.182	.182
451–500	.227	.409
501–550	.341	.750
550–600	.136	.886
601–650	.091	.977
651–700	.023	1.000

43.

Company	Price Change
1	−6
2	−120
3	−50
4	−30
5	−20
6	5
7	−20
8	30
9	−30
10	0
11	−15
12	−20
13	−5
14	2
15	−10
16	−13
17	0
18	−30
19	−10

45.

Class	Cumulative Frequency
−120 ~ −91	1
−90 ~ −61	1
−60 ~ −31	2
−30 ~ −1	14
0 ~ 30	19

Chapter 4

1. $\overline{X} = 83.78$ (83.7). The median is the middle score, which is 90. The mode is the most frequently occurring value, which is 100.

3. Descriptive statistics summarize data and allow comparisons to be made. The mean, median, mode, and variance are all descriptive statistics.

5. Original cost = $340. New cost = $255. % Price decrease = 25%.

7. a. Mean 1960–1979: $\overline{X}_{DJIA} = 4.1605$; $\overline{X}_{S\&P500} = 4.225$. Mean 1970–1989: $\overline{X}_{DJIA} = 9.0705$; $\overline{X}_{S\&P500} = 8.209$. Mean 1981–1990: $\overline{X}_{DJIA} = 12.675$; $\overline{X}_{S\&P500} = 9.983$. Standard deviation 1960–1979: std dev$_{DJIA}$ = 1639; std dev$_{S\&P500}$ = 16.02. Standard deviation 1970–1989: std dev$_{DJIA}$ = 17.36; std dev$_{S\&P500}$ = 16.13. Standard deviation 1981–1990: std dev$_{DJIA}$ = 16.12; std dev$_{S\&P500}$ = 12.85.

b. Geometric mean 1960–1979: DJIA = .0288 or 2.88%; S&P 500 = .0298 or 2.98%. Geometric mean 1970–1989: DJIA = .0771 or 7.71%; S&P 500 = .0696 or 6.96%. Geometric mean

1981–1990: DJIA = .1163 or 11.63%; S&P 500 = .0929 or 9.29%.

9. **a.** $\overline{X}_A$ = 11%; $\overline{X}_B$ = 11.32%; std dev A = .774; std dev B = 1.243; CV_A = .0704; CV_B = .1098.

b. Because stock B has a higher mean and a higher standard deviation, it has greater risk (as measured by the standard deviation) and greater expected return (as measured by the mean) than stock A. To compare, we could look at the coefficient of variation to see which stock has the greatest risk per unit of reward. Because stock B has a greater coefficient of variation, it has more risk per unit of expected return and may, therefore, be considered less desirable.

13. **a.** Both the arithmetic and geometric means are methods for computing the data's central tendency. The arithmetic mean is found by summing up the data and dividing by the number of observations. The geometric mean is found by multiplying the data together and taking the nth root. The geometric mean is especially useful in finance to compute average rates of return because it incorporates the concept of compound interest.

b. Arith mean = .075; geometric mean = .0627.

c. As discussed in the text, the geometric mean return is lower than the arithmetic return.

15. The mean would be higher because the salaries of a few superstars would raise the overall average.

17. Because we square the deviations from the mean when we compute the variance, the variance and the mean are in different units. After taking the square root of the variance, the standard deviation will be in the same units as the mean.

19. A population consists of all observations of interest, while a sample is a smaller subset of the population. We use a different divisor to compute the sample standard deviation because the adjustment gives us an unbiased estimator of the standard deviation.

23. **a.** $\overline{X}_A$ = 8%; S_A^2 = .50; S_A = .7071; $\overline{X}_B$ = 10%; S_B^2 = 50; S_B = 7.071; CV_A = .7071/8 = .088; CV_B = 7.071/10 = .7071.

b. The stock is difficult to compare because A has both a lower mean and a lower standard deviation (less risk). By looking at the coefficient of variation, we can see the amount of risk you must take per unit of return. Here, A has a

much lower price paid (risk) per unit of return and is probably a better buy.

25. **a.** $\overline{X}$ = 2.96; S^2 = 2.36.

b. The mode is the score most frequently repeated, which is four days in this case. The median is the middle score, which is three days (the same number of people were absent more than three days as were absent less than three days).

c. Just by looking at the table, we can see that more people were absent four or five days than were absent 0 days or 1 day, so distribution is not symmetric.

27. $\overline{X}$ = .0755; S^2 = .0262; S = .1618; μ_3 = −.0021; CS = −.507. The returns are negatively skewed.

29. **a.** Variance measures the average squared deviation from the mean. Standard deviation is the square root of the variance and therefore measures the same thing. By taking the square root, however, the standard deviation will be in the same units as the mean. Mean absolute deviation is the average absolute deviation from the mean. Because the deviations are not squared, values that are farther from the mean are not given additional weight as they are in the computation of the variance. The range is just the difference between the highest and lowest observations.

b. The range would be the easiest to compute, while the other three measures are about equally difficult.

31. **a.** Symmetric.

b. Negatively skewed.

c. Positively skewed.

33. XYZ stock has a lower mean return than ABC stock but also a lower level of risk as measured by the standard deviation. One way to compare is to look at the risk taken per unit of return, that is, to use the coefficient of variation. CV_{XYZ} = .25; CV_{ABC} = .33. So, the price you pay for each unit of return is lower for XYZ stock and thus makes XYZ better when this criterion is considered.

35. Because standard deviation measures how spread out the values are around the mean, it can be important in quality control. For example, suppose a case of light bulbs averages one defective bulb. There is a big difference between a company

that produces 100 cases of bulbs with 50 cases having no defective bulbs and 50 cases having two defective bulbs versus a company with 99 cases of bulbs with no defects and one case with 100 defective bulbs. Even though the means are the same, the first company clearly has better quality control.

37. Just because you have the same chance of falling below the median as above the median doesn't mean you should ignore variability. If the values are all close together, then there will be a low variability; that is, there will be little chance that you will do exceptionally well or exceptionally poorly. Likewise, a high variability implies an all or nothing return on your stock, which is probably not desired by most people.

39. $\overline{X} = 1.62$; $S^2 = .1508$; $S = .388$; $\mu_3 = -.0046$; CS $= -.0787$.

41. $CV_A = .333$; $CV_B = .357$. Choose A because the risk per unit of reward is lower.

43. $\mu_{A3} = .6$; $CS_A = .096$; $\mu_{B3} = -138$; $CS_B = -.196$.

45. $\overline{X} = 1.30$; $S^2 = .0698$; $S = .2642$; $CV = .2032$.

49. 1969–1978: $\overline{X} = 6.00$; $S^2 = .7087$; $S = .842$; $CV = .140$. 1979–1990: $\overline{X} = 9.68$; $S^2 = 6.911$; $S = 2.629$. Both the mean and the standard deviation for inventory turnover rose in the 1979–1990 period. Once again, the increase is probably the result of increased foreign competition.

51. Current ratio 1969–1990: $\overline{X} = 1.65$; $S^2 = .1324$; $S = .3639$; $CV = .2205$. 1969–1978: $\overline{X} = 1.97$; $S^2 = .0246$; $S = .1568$; $CV = .0796$. 1979–1990: $\overline{X} = 1.38$; $S^2 = .0571$; $S = .2390$; $CV = .1732$. Inventory turnover 1969–1990: $\overline{X} = 9.21$; $S^2 = 8.758$; $S = 2.959$; $CV = .321$. 1969–1978: $\overline{X} = 6.69$; $S^2 = 1.403$; $S = 1.184$; $CV = .177$. 1979–1990: $\overline{X} = 11.31$; $S^2 = 5.021$; $S = 2.241$; $CV = .198$. Once again, we see a deterioration in the mean current ratio and mean inventory turnover.

55. In comparing dispersion, the coefficient of variation would be preferred because it is unit free. In this case, PHS has the greater dispersion.

57. All three populations will have identical numerical values.

59. a. Median $= 205$.
    b. $\overline{X} = 208.7$; $S^2 = 4292.2$; $S = 65.5$.

61. a. Median $= 2.6$; $\overline{X} = 2.59$.
    b. $S^2 = 7.38$; S $= 2.72$.

c. MAD $= 1.58$. For first observation, $Z = -2.75$. The rest of the Z-values are

-4.9	-2.75367
1.5	$-.40073$
2.1	$-.18014$
2.3	$-.10661$
2.5	$-.03308$
2.6	.003676
2.6	.003676
3.2	.224264
3.7	.408088
4.1	.555147
5.3	.996323
6.1	1.290441

Chapter 5

1. Because there are four queens in a deck of 52 cards, the chance of drawing a queen is 4/52. When the card is replaced, the probability is $(4/52)(4/52) = .005917$. When the card is not replaced, the probability is $(4/52)(3/51) = .004525$.

3. The three events are independent. Therefore the answer can be obtained as $1/6 \times 1/6 \times 1/6 = 1/216$.

5. a. $A \cup B = \{a, b, c, d, e\}$.
   $A \cap B = \{e\}$
   $A^c = \{b, d\}$. $B^c = \{a, c\}$.
   $A^c \cap B = \{b, d\}$
   $(A \cup B)^c = \{ \ \}$

7. $1 - (_{365}P_{20}/365^{20}) = .411$

9. a. The largest mean is 12. If every roll yields a pair of sixes, the largest mean occurs.
   b. The smallest mean is 2. If every roll yields a pair of ones, the smallest mean occurs.
   c. The smallest possible standard deviation is 0; that means that every time we roll the pair of dice, we obtain the same outcome.

11. 1/38.

13. Pr(sales will result in a sales commission) $= 60\%$.

15. There are 1,450 people who score less than 600, therefore, Prob(score less than 600) $= 1,450/1,900$.

19. Yes!, if we may assume that every time he goes to the plate the chance of getting a hit does not change (always .3).

21. Define TV = owning a TV and COMP = owning a computer.
    a. Pr(TV ∩ COMP) = 1/4.
    b. Pr(ONLY COMP) = 15%.
    c. Pr(TV | COMP) = 62.5%.
    d. Pr(COMP) = 40%.

25. 13/204.

27. **a.** 30%.
    b. 70%.

29. Pr(older than 20 ∩ holding a job) = 4%.

	Older Than 20	Teen-ager
Holding a job	16%	36%
In school	24%	24%
Total	40%	60%

31. **a.** $n(S)$ = the number of students taking statistics; $n(MI)$ = the number of students taking microeconomics; $n(MA)$ = the number of students taking macroeconomics. $n(S ∩ MI ∩ MA)$ = 100.
    b. Pr(S | MI) = 200/450.

35. Because he talks to the customers twice, his chance for selling the insurance to each customer is 80%. For 3 customers, the chance of selling none is 20% × 20% × 20% = .8%.

37. Pr(conservative | recession) = 3/5.

39. Pr(strongly motivated | benefitted) = 2/3.

41. $_5C_2$ = 10.

43. **a.** Pr(old | blue collar) = 3/8.
    b. Pr(white collar | teenager) = 3/5.

45. An important job in statistics is to make inferences on unknown parameters using a limited amount of data. This kind of exercise is subject to errors by chance. Probability theory helps us understand how much uncertainty is involved in a certain inferential procedure.

47. A composite event is the union of a simple event. For example, rolling a die and obtaining 5 is a simple event. Rolling a die and obtaining an odd number is a composite event.

49. When two events do not occur at the same time they are mutually exclusive.

51. A joint probability is the probability of two events occurring at the same time. A marginal probability is the probability of one event occurring regardless of the other events.

53. Pr(H) = 1/2; Pr(5) = 1/6; Pr(H ∩ 5) = 1/12.

55. (1/13)(1/13)(1/13)(1/13) = 1/28,561.

57. **a.** $A ∩ B$ = { }; $A ∪ B$ = {1, 2, 3, 4, 5, 6}.
    b. $A ∩ B$ = {1, 3}; $A ∪ B$ = {1, 3, 5}.
    c. $A ∩ B$ = {2}; $A ∪ B$ = {1, 2, 3, 4, 5}.
    d. $A ∩ B$ = {3, 4}; $A ∪ B$ = {1, 2, 3, 4, 5, 6}.

59. There are only 4 cases in which you have a royal flush. There are $_{52}C_5$ ways of getting any 5 cards. Pr(Royal Flush) = $4/_{52}C_5$ ≈ .0000015. To get a royal flush of spades, there is only 1 chance. Therefore the probability is ≈ .0000004.

61. Pr(H, H, H) = 1/8. Pr(T, T, T) = 1/8. Pr(H, H, H) + Pr(T, T, T) = 1/4.

63. 12.

65. **a.** .133.
    b. .533
    c. Pr(growth | increased) ≈ .714.

67. Pr(T ∩ getting a one) = 1/12.

69. $(70\%)^{12}$ = .0138. The assumption is that the probability of rolling a strike remains the same for all 12 rolls.

71. 3 × 5 = 15.

73. 12 × 5 = 60.

75. $\binom{6}{4} × \binom{5}{3}$ = 150.

77. Pr(4 $250 weeks) = 6.25%; Pr(3 $250 weeks) = 25%; Pr(3 or 4 $250 weeks) = 31.25%.

79. Yes. The question is not whether a person will die eventually. The question is what is the probability that an insured will die during the period of policy effectiveness.

81. **a.** Not drawing a spade.
    b. Inflation of at least 5% per year.
    c. GNP growth of no more than 4% per year.

Chapter 6

1. $\frac{21}{230} \frac{43}{230} \frac{23}{230} \frac{48}{230} \frac{31}{230} \frac{29}{230} \frac{35}{230}$. $F(1)$ = 21/230; $F(2)$ = 64/230; $F(3)$ = 87/230; $F(4)$ = 135/230; $F(5)$ = 166/230; $F(6)$ = 195/230; $F(7)$ = 1.

3. Mean = 3.16; Var = 2.1944; std dev = 1.4814.

5. Pr(X = 2) ≈ .276; $E(X)$ = 1.5; Var(X) = 1.275.

7. Pr(X = 4) = .054; PR(X = 5) = .013; Pr(X = 6) = .00217; PR($X ≥ 4$) = .0698.

9. $\frac{\binom{5}{1}\binom{7}{3}}{\binom{12}{4}}$ = .354.

11. $\Pr(X = 0) = .0111$; $\Pr(X = 1) = .0500$; $\Pr(X = 2) = .1124$. Therefore $\Pr(X \leq 2) = .1735$.

13. Use binomial distribution. $\Pr(X \geq 4) = .7102$.

15. Use binomial distribution. $\Pr(X = 5) = .00243$.

17. Use Poisson distribution. $\Pr(X = 4) = .0016$.

19. Use Poisson distribution. $\Pr(X > 1) = .0902$.

21. Obviously a binomial distribution problem. $\Pr(X < 20) = \Pr(X = 0) + \Pr(X = 1) + \cdots + (\Pr(X = 19)$ where, for example, $\Pr(X = 0) = \binom{400}{0}(.1)^0(1 - .1)^{400-0}$.

23. Use binomial distribution. $\Pr(X > 2) = .1271$.

25. **a.** $\Pr$(winning the 2nd bet) $= 1/38$.
    b. Use binomial distribution. $\Pr(X = 2) = .0020$.

27. $\Pr(X < Z) = .5578$.

29. $\Pr(X = 5) = .2007$.

31. $\lambda = (.01)(30) = .3$, since time = assumed 30 days. $\Pr(X = 3) = .0033$.

33. $\lambda = 15$, since time = 5 days. $\Pr(X = 0) = .0000003$.

35. $\lambda = 2$; $\Pr(X = 0) = .1353$.

37. $\Pr(X > 2) = 58.86\%$.

39. Hypergeometric distribution

$$\frac{\binom{10}{3}\binom{30}{2}}{\binom{40}{5}} = .1388.$$

41. See definition in the book.

43. Binomial distribution. $\Pr(X = 7) = .2270$.

45. See textbook.

47. Expected value = 1.95. The variance = 1.6475.

49. $\Pr(X = 5) = .0016$; $\Pr(X \leq 5) = .0020$.

51. $\Pr(X = 2) = .2605$; $\Pr(X \leq 2) = .8651$.

53. $\binom{10}{3}(30\%)^3(1 - 30\%)^{10-3} = .2668$; $\Pr(X \leq 3) = .6496$.

55. $\lambda = 8$; $\Pr(X = 10) = .0993$; $\Pr(X = 3) = .0286$.

57. $C_5 = \max[0, Su^3d^2 - E] = \max[0, 100(1.10)^3(.95)^2 - 101] = 19.12$.

59. $C_5 = \max[0, 100(1.10)^3(.85)^2 - 101] = 0$. So, a lower d reduces the value of the call in period 5.

61. $C_5 = \max[0, 110(1.10)^3(.95)^2 - 101] = 31.14$. So, higher s raises C_5.

63. $\dfrac{\binom{25}{4}\binom{32}{6}}{\binom{57}{10}} = .2655$

65. In each draw the probability of drawing a red card is 50%. Use binomial distribution. $\Pr(X \geq 3) = \Pr(X = 3) + \Pr(X = 4) + \Pr(X = 5) = .5$.

67. $\Pr(X = 4) = .0039$.

69. $\Pr(X = 2) = .2109$.

71. $\Pr(x = 8) = .02216$. The probability is very small; it implies that these people know something.

73. **a.** Correlation coefficient = .3924.
    b. From Figure 6.12 we know that Merck's mean monthly return is higher than IBM's, and the standard deviation of the former is smaller than the latter. From a normal investor's point of view, Merck is a better choice because it has higher return but lower risk.

Chapter 7

1. $\Pr(-1.87 < Z < .8) = .7574$.

3. **a.** .0495. **b.** .9913. **c.** .0016.
   d. .0853. **e.** .1445. **f.** .9498.

5. Manual calculation.
   a. .8133. **b.** .992. **c.** .1093.
   d. .0668. **e.** .6129. **f.** .0303.

7. 3.15.

9. $f(y) = \dfrac{1}{y\sigma\sqrt{2\pi}} \exp\left[-\dfrac{1}{2\sigma^2}(\ln y - \mu)^2 \right] y > 0.$

11. .0032.

13. $20,000 \times .00036 = 7.2$, about 7.

15. $8.08 million.

17. .0062.

19. $\mu = 12.06$.

21. $\Pr(x > 40) = 84.13\%$.

23. **a.** $\Pr(X > 16) = .9525$.
    b. $\Pr(X < 16) = .0475$.

25. $\Pr(X > 20) = .0322$.

27. $\Pr(X < 20) = .9015$.

29. Log-normal can describe the data with skewness.

31. **a.** $\Pr(Z < -1) = 15.87\%$.
    b. $\Pr(Z < -1.5) = .0668$.
    c. $\Pr(Z < 2) = 97.72\%$.
    d. $\Pr(Z < 3) = .9987$.
    e. $\Pr(Z < .5) = .6915$.
    f. $\Pr(Z < 2.5) = .9938$.

33. **a.** $\Pr(Z > Z_0) = .1$, $Z_0 = 1.28$.
    b. $\Pr(Z > Z_0) = .75$, $Z_0 = -.67$.
    c. $\Pr(-Z_0 < Z < Z_0) = .95$, $Z_0 = 1.96$.
    d. $\Pr(Z < Z_0) = .95$, $Z_0 = 1.64$.
    e. $\Pr(-Z_0 < Z < Z_0) = .9$, $Z_0 = 1.64$.
    f. $\Pr(-Z_0 < Z < Z_0) = 1.0$, $Z_0 = \infty$.

35. When the sample size becomes large, sometimes it is difficult to use binomial or Poisson distribution to obtain the probability. At this time, normal approximation to these distributions offers an easy way to measure the probability. If x is a binomial,

then when n is large, $x \sim N(np, \sqrt{np(1-p)^2})$. If x is a Poisson distribution, then when n is large, $x \sim N(\lambda, \sqrt{\lambda^2})$.

37. a. .9927.   **b.** .0073.   **c.** .9854.

39. a. 0.   **b.** 84.13%.
   c. .9544.   **d.** 84.13%.

41. a. 50%.   **b.** .3707.

43. $C = 6.949$.

45. $C = \$6.77$. So, a shorter time to expiration reduces the value of the call.

47. $C = \$8.47$. So, a higher risk-free rate raises the value of the call.

51. $\Pr(x \geq 18) = .978463$, using binomial; $\Pr(x \geq 18) = .9706$, using binomial; normal is faster.

53. a. 2.7182.   **b.** 1.6487.   **c.** .6065.
   d. .0821.   **e.** 22.198   **f.** .3679.
   g. 1.0513.   **h.** 1.3771.   **i.** 445.8578.
   j. .0045.

55. The cumulative distribution function $F(a)$ is defined as $\Pr(x \leq a)$. It will be useful for computing probability.

57. a. .2514.   **b.** 0(.0002).

59. $C = \$17.86$.

61. $np = 300$, $np(1-p) = 120$.

63. a. .5028.   **b.** .0207.

65. a. .0062.   **b.** .1587.

67. a.

Rank	Percent
1	100/1700 = 5.88%
2	300/1700 = 17.65%
3	900/1700 = 52.94%
4	300/1700 = 17.65%
5	100/1700 = 5.88%

   b. Mean = 3. Variance = .8234. Std dev = .9074.

69. If the distribution is more concentrated, the cumulative density will approach 1 more quickly. Therefore, these curves represent the following: A: $(1, .5)$; B: $(1, 1)$; C: $(1, 2)$.

Chapter 8

1. a. 15.87%.   **b.** .0228.
   c. The average of 4 pieces of data ($\overline{X}_4$) has a much smaller standard deviation than X. Thus there is a smaller chance of seeing $\overline{X}$ take an extreme value as high as 495.

3. .975.

5.

$\overline{X}_2$	$\Pr(\overline{X}_2)$
0	.04
1	.02
2	.09
3	.24
4	.16
5	.2
6	.25

9. .9236.

11. $\Pr(\overline{X} > ?) = .95$. We may solve ? as 2.9344.

13. A census can give us a much better view of the population than a sample.

15. 1,100 registered voters; the voters who are over 65.

17. Nonsampling error is the error resulting from a wrongly designed statistic or sampling method. In this case, the error cannot be removed just by collecting more sample.

19. See the textbook.

21. a. 950.   **b.** 64.95.

23. a. 24.   **b.** .35.

25. .9319.

27. .0418.

29. .67.

31. a. 7.98%.   **b.** $(.00202)^2$.   **c.** .1379.

33. 50%.

35. a. 6.   **b.** 1/9.

37. .9793.

39. a. 25%.   **b.** .0009375.
   c. .031.   **d.** .4463.

40. .0207.

41. a. 1.   **b.** 1122.   **c.** 1544.49.

43. Mean = 25.2; variance = 6.96.

45. The mean is 3.5. The probability of each pair occurring is $p(i, j) = 1/36$ if $i = j$; $p(i, j) = 1/18$ if $i \neq j$.

47. The sample mean is 1.5. $P(i\ j) = 1/9$ if $i = j$; $P(i\ j) = 2/9$ if $i \neq j$.

49. The sample mean is the mathematical mean of a group of samples. Because the sample mean is used to make inferences on the population mean, we are interested in its distribution.

51. .0207.

53. .0125.

55. 120.

57. Put the 5 numbers in a jar and randomly select 4 numbers at a time. Replace the number after each selection. Then compute the mean and standard deviation for each selection.

59. One.

61. 50%.

63. .0384.

65. $E(\overline{X}) = 3.5$. $\text{Std}(\overline{X}) = 1.08$.

67. $\Pr(\frac{x}{n} > \frac{1}{2}) = 1$. $n \cdot p = 140$.

69. 0.

71. 15.87 cars.

73. Because the mean is 20%, there is a 50% probability that at least 20% of the students tested will fail the test.

75. $\sigma = 63.25$; mean $= 20$.

77. No, the probability is too small.

79.

Points	$P(X)$	Points	$P(X)$
2	1/36	8	5/36
3	2/36	9	4/36
4	3/36	10	3/36
5	4/36	11	2/36
6	5/36	12	1/36
7	6/36		

81. The distribution is a binomial distribution.

$\overline{X}$	$\Pr(\overline{X})$
8	.32768
7.4	.4096
6.8	.2048
6.2	.0512
5.6	.0064
5	.0003

83. $E(\text{points}) = 36/8$; $\text{Var(points)} = .75$; std dev $= .87$.

Chapter 9

1. CDF for the uniform distribution is linear between a and b. For a value of a, the CDF equals 0; for a value of b, the CDF equals 1.

3. $E(s^2) = \sigma^2$; $s^2 = \sigma^2/n$; $\sigma^2 = 20(20) = 400$; $\text{Var}(s^2) = (2\sigma^4)/(n-1) = [2(400)^2]/(20-1) = 16{,}842.1$.

5. To test the hypothesis of equality of population variances, we take the ratio of the 2 sample variances: $F = s_x^2/s_y^2$.

9. 3/4.

13. a.
$$f(y) = \begin{cases} \dfrac{1}{32-3} & \text{if } 3 \le y \le 32 \\ 0 & \text{otherwise} \end{cases}$$

b.
$$F(y) = \begin{cases} 0 & \text{if } y \le 3 \\ \dfrac{x-3}{32-3} & \text{if } 3 \le y \le 32 \\ 1 & \text{if } y > 32 \end{cases}$$

c. $\mu = 17.5$; $\sigma^2 = 70.08$.

15. a. $\Pr(x \le 35) = .444$.
b. $\Pr(x > 30) = .667$.
c. $\Pr(30 \le x \le 45) = .333$.

17. a. $\Pr(x > 4) = .0000$.
b. $\Pr(x > .7) = .030$.
c. $\Pr(x > .50) = .082$.

19. a. $P(3 \le x \le 5) = .179$.
b. $P(5 \le x \le 10) = .153$.
c. $P(2 \le x \le 7) = .417$.

21. $E(y) = 1/3$.

23. a. $t_{.05,10} = 1.81$.
b. $t_{.025,4} = 2.78$.
c. $t_{.10,7} = 1.41$.

25. a. $t_0 = -1.32$.
b. $t_0 = 2.14$.
c. $t_0 = -2.57$.

27. a. .05.   **b.** .05.   **c.** .95.

29. a. 46.98.   **b.** 100.42.
c. 15.99.   **d.** 37.57.

31. a. .95.   **b.** .10.   **c.** .005.   **d.** .90.

33. a. 2.98.   **b.** 26.87.   **c.** 2.96.   **d.** 5.27.

35. a. .05.   **b.** .05.   **c.** .975.   **d.** .01.

37. a. .998.   **b.** .0005.   **c.** .0005.

39. a. .4353.   **b.** .3325.   **c.** .2764.

41. As the degrees of freedom increase, the $P(\chi^2 \le 9)$ gets larger.

43. .6429.

45. $\bar{x} = 2.5$, $\sigma^2 = .75$.

47. .4724.

49. $\Pr(X > 5) = .8825$, $\Pr(3 < X < 10) = .1489$.

51. .1738.

53. .1889.

55. .6765.

57. .3012.

59. .5276.

61. a. 230.   **b.** 31.75.   **c.** .1818.

63. .2835.

65. The exponential distribution looks at the amount of time until the first occurrence; the Poisson distribution looks at the number of occurrences of an event over time.

67. a. .3033.   **b.** .0677.   **c.** 2.

Chapter 10

1. a. $\bar{x} = 4$; $s_x = 6$; $s_x^2 = 36$; $s_{\bar{x}} = s_x/\sqrt{n} = 2$.

   b. $\bar{x} = 4$; $s_x = \sqrt{12}$; $s_x^2 = 12$; $s_{\bar{x}} = s_x/\sqrt{n} = \sqrt{3}/2$.

   c. $\bar{x} = 20$; $s_x = 10$; $s_x^2 = 100$; $s_{\bar{x}} = s_x/\sqrt{n} = 2$.

3. a. 2.57.   **b.** 2.17.   **c.** 3.08.

7. a. 2.57.

   b. 1.96 (check the normal distribution table).

   c. Yes.

9. $n = 164$.

11. $z(\sigma/\sqrt{n}) = .9\%$; $20.4\% \pm .9\% = (19.5\%\ 21.3\%)$.

13. See the textbook.

15. See the textbook.

17. If he repeats the experiment by 100 times, for 95 times, the confidence interval will include the average earning.

19. (162094, 189150).

21. (167901, 183423).

23. (.494, .606).

25. (1134, 1406).

27. a. (745, 967).   **b.** (763, 949).

29. a. (822.6, 926.4).   **b.** (582.2, 1166.8).

31. (.384, .566).

33. (1033.4, 1166.6).

35. (1041.2, 1058.8).

37. $\bar{x} = 16.9$; $S_x = .45$; $s_{\bar{x}} = .05$; $16.9 \pm 1.96 \cdot (.05) = (16.8, 17.0)$.

39. $\bar{x} = 72.8$; $S_x = 1.2$; $S_{\bar{x}} = 1.2/\sqrt{40}$; $(72.3, 73.3)$.

41. (71.1, 72.5).

43. (19.05, 20.95).

45. (21609.26, 23390.74).

47. (23.74, 24.26).

49. (.468, .632).

51. (5.68, 6.32).

53. (34344, 35656).

57. (.38, .50).

59. (5.25, 6.63).

61. (.36, .56).

63. (.13, .27).

65. (117, 123).

67. a. 129.   **b.** (123, 135).

69. (24710, 25290).

71. (.499, .561).

73. (.42, .50).

75. a. (.428, .465).   **b.** (.45, .49).

77. $n = 27$.

79. $n = 9604$.

81. a. 169.5.   **b.** (153.27, 185.73).

83. (2268, 2516).

85. (.67, .73).

87. (12.046, 12.054).

89. (30.75, 37.25).

91. $22.92 < \sigma < 27.59$.

93. (83.82, 86.4).

95. $58.86 < \sigma < 61.19$.

97. a. Convenience lot: A quantity of material that can be conveniently handled as a segment of product, for example, a case, a day's product, or a car load.

   b. Single versus double sampling plans: Single-sampling plans draw only one random sample from each lot, and every item in the sample is examined and classified as either good or defective; double sampling plans provide for the taking of a second sample when the results of a first sample are marginal.

   c. Upper and lower control limits are two limits drawn in a control chart to describe whether sample values will fall between these two limits.

   d. Acceptance sampling may be defined as sampling procedures in which a decision is made to accept an entire lot of product or service based on only a sampling of the lot.

99. $\hat{p} = .023$; $s_{\hat{p}} = .01499$; $\text{UCL}_p = .068$; $\text{LCL}_p = -.022$.

101. $\hat{p} = .0208$; $s_{\hat{p}} = .0285$.

Chapter 11

1. a. Rej H_0.   **b.** Rej H_0.   **c.** Rej H_0.

3. Accept H_0.

5. Accept H_0.

7. Rej H_0: $\mu \leq 12983$.

9. a. Rej H_0: $\mu = 180$.

   b. P value $.091 < .1$, reject H_0; P value $.046 < .05$, reject H_0.

11. See text for the definition of hypothesis testing. We are interested in hypothesis testing mainly for the purpose of decision making. It is not likely that we may prove a hypothesis true; we may only come up with evidence to support a hypothesis.

13. a. H_0: Innocent.   **b.** H_0: Dangerous.
   c. H_0: Unsafe.   **d.** H_0: Unsafe.

15. When we are interested in knowing whether the parameter of interest is greater than or less than a certain value, we have a one-tail test. When we are interested in knowing whether the parameter of interest is different from a certain value, we have a two-tail test.

17. Simple hypothesis: When the parameter of interest in both the null and the alternative hypotheses takes only one value, e.g., H_0: $\theta = 5$, H_1: $\theta = 11$. Composite hypothesis: When the parameter of interest in the alternative hypothesis can take more than one value, e.g., H_0: $\theta = 5$, H_1: $\theta \neq 5$.

19. Rejection region is determined by the null hypothesis. If we assume that the null hypothesis is true, then we obtain a region of, say, 5% for rejecting the null hypothesis. By doing this, the type one error is controlled under 5%.

21. In a two-tail test for $u = \mu_Q$, we divide the rejection region in two. Therefore, it is more difficult to reject a certain null hypothesis with half the size of the rejection region defined by a one-sided test.

23. Rej H_0: $\mu = 0$.

25. Rej H_0: $\mu = 100$.

29. Rej H_0: $\mu = 500$.

31. Accept H_0: $\mu = 500$.

33. Rej H_0: $\mu = 10$.

35. Rej H_0: $\mu = 1,000$.

37. Yes. Now that we have a small sample, we need to do a t test when we do not know the true value of the variance.

39. Rej H_0: $\mu = 200$.

41. Accept H_0: $\mu = 35$.

43. Accept H_0: $\mu \leq 30$.

45. Accept H_0: $P \leq .75$.

47. Accept H_0: $P = .95$.

49. Rej H_0: $p \geq .9$.

51. Accept H_0: $p \leq .52$.

53. Accept H_0: $\mu = 4$.

55. Rej H_0: $\mu_A - \mu_B = 0$.

57. Accept H_0: $\mu_A - \mu_B = 0$.

59. Accept H_0: $\sigma^2 \leq .5$.

61. Accept H_0: $P_1 - P_2 = 0$.

63. Rej H_0: $P_1 - P_2 \leq 0$.

65. Accept H_0: $P_1 - P_2 \geq 0$.

67. Rej H_0: $\sigma^2 \leq 225$.

71. H_0: presumed the cage is dangerous.

73. When the null hypothesis is true, the probability of obtaining one correct answer is 1/2. To have 6 of 10 correct answers, we may use the binomial distribution formula.

$$\binom{10}{6}\left(\frac{1}{2}\right)^6\left(1 - \frac{1}{2}\right)^{10-6} = .21$$

75. $Pr(x \geq 9 \mid P = 1/2) = .01$.

77.

P	β	$1 - \beta$
.5	.945	.055
.6	.833	.167
.7	.617	.383
.8	.322	.678
.9	.070	.930
1.0	0	1.

79. Rej H_0: $P \leq .5$.

81. Accept H_0: $P = .001$.

83. Accept H_0: $\mu \geq 70$.

85. Accept H_0: $P = .7$.

87. Accept H_0: $P \geq .01$.

89. Rej H_0: $P \geq .1$.

91. Accept H_0: $P = .5$.

93. Rej H_0: $P_1 - P_2 \leq 0$.

95. Accept H_0: $P_1 - P_2 \leq 0$.

97. Reject H_0.

Chapter 12

1. $F = 40/9.5 = 4.211 > F_{3,8}(5\%) = 4.07$, reject H_0.

3. $F = 8.025/3.074 = 2.611 < 3.88 = F_{2,12}(.05)$, accept H_0.

5. First: Testing H_0: $\sigma_1^2 = \sigma_2^2$; $F = S_2^2/S_1^2 = 2.04 < 2.4 = F_{15,15}(5\%)$, accept H_0. Second: Testing H_0: $\mu_1 = \mu_2$; $t = .33 < 2.101 = t_{18}$, reject H_0.

7. One-way ANOVA studies whether the treatment means are equal. Two-way ANOVA studies whether the treatment means are equal after controlling the variation resulting from a second factor.

9. No, the two-factor ANOVA gives a better result if there is a second factor that causes additional variation.

11. $F = 6.842/25.478 = .269 < 3.59 = F_{2,17}(5\%)$, accept H_0.

13. $F = 1.667/13.467 = .124 < 3.88 = F_{2,12}(5\%)$, accept H_0.

15. $F = 93.6/18.03 = 5.19 > 3.88 = F_{2,12}(5\%)$, reject H_0.

17. $F = 24.27/25.2 = .963 < 3.88 = F_{2,12}(5\%)$.

19. $F = 127.16/70.81 = 1.796 < 3.68 = F_{2,15}(5\%)$, accept H_0.

21. $F = 44.59/5.46 = 8.17 > 3.40 = F_{2,24}(5\%)$, reject H_0.

23. $F = 16.59/4.22 = 3.93 > 3.55 = F_{2,18}(5\%)$, reject H_0.

25. $\Sigma[(f_o - f_e)^2/f_e] = 4.76 < 7.82 = \chi_3^2(5\%)$, accept H_0: equally popular.

27. **a.** $\Sigma[(f_o - f_e)^2/f_e] = 24.26 > 7.82 = \chi_2^2(5\%)$, reject H_0: same frequency.
   b. $\Sigma[(f_o - f_e)^2/f_e] = .38 < 5.99 = \chi_2^2(5\%)$, accept H_0.

29. Collect the numbers that were drawn in the previous lottery games. If the game is fair, we would see that all numbers have the same frequencies of being drawn. Use the chi-squares test to examine whether they have different tendencies of being drawn.

31. $\Sigma[(f_o - f_e)^2/f_e] = 3.30 < 7.82 = \chi_3^2(5\%)$, accept H_0.

33. $\Sigma[(f_o - f_e)^2/f_e] = 531 > 9.48 = \chi_4^2(5\%)$, reject H_0.

35. $\Sigma(f_o - f_e)^2/f_e = 57.86 > 12.59 = \chi_6^2(5\%)$, reject H_0.

37. $\Sigma[(f_o - f_e)^2/f_e] = 5.94 < 12.59 = \chi_6^2(5\%)$, accept H_0.

39. $\Sigma(f_o - f_e)^2/f_e = .81 < 9.48 = \chi_4^2(5\%)$, accept H_0.

41. $\Sigma(f_o - f_e)^2/f_e = .9 < 9.48 = \chi_4^2(5\%)$, accept H_0.

43. $\Sigma(f_o - f_e)^2/f_e = .37 < 5.99 = \chi_2^2(5\%)$, accept H_0.

45. $\Sigma(f_o - f_e)^2/f_e = 13.19 > 9.48 = \chi_4^2(5\%)$, reject H_0.

47. $\Sigma(f_o - f_e)^2/f_e = 12.02 < 12.59 = \chi_6^2(5\%)$, accept H_0.

49. $\Sigma(f_o - f_e)^2/f_e = 3.75 < 7.82 = \chi_3^2(5\%)$, accept H_0.

51. $F = .123 < 3.88 < F_{2,12}(5\%)$, accept H_0.

53. $\Sigma(f_o - f_e)^2/f_e = 0 < 7.82 = \chi_3^2(5\%)$, accept H_0.

55. $\Sigma(f_o - f_e)^2/f_e = 80.1 > 12.59 = \chi_6^2(5\%)$, reject H_0.

57. $F = 4.389/1.744 = 2.516 < 3.68 = F_{2,15}(5\%)$, accept H_0.

59. $\Sigma[(f_o - f_e)^2/f_e] = 12.67 < 14.067$, accept H_0.

61. $\Sigma[(f_o - f_e)^2/f_e] = 6.51 > 5.99$, reject H_0.

63. $\Sigma[(f_o - f_e)^2/f_e] = 33.8 > 12.59$, reject H_0.

65. $F = 54.056/15.389 = 3.513 < 3.68$, accept H_0.

67. $\Sigma(f_o - f_e)^2/f_e = 180 > 5.99$, reject H_0.

69. $F = 32.45/3.16 = 10.28 > 3.4$, reject H_0.

71. $F = 8.73/0.81 = 10.78 > 3.55 = F_{2,18}(5\%)$, reject H_0.

73. $\Sigma[(f_o - f_e)^2/f_e] = 51.05 > 9.48 = \chi_4^2(5\%)$, reject H_0.

75. $\Sigma(f_o - f_e)^2/f_e = 19.46 > 12.59 = \chi_6^2(5\%)$, reject H_0.

77. $\Sigma[(f_o - f_e)^2/f_e] = 4.15 < 16.92 = \chi_9^2(5\%)$, accept H_0.

79. $F = 80.153/45.38 = 1.766 < 3.49 = F_{3,12}(5\%)$, accept H_0.

81. $\Sigma(f_o - f_e)^2/f_e = 85.71 > 7.82 = \chi_6^2(5\%)$, reject H_0.

83. $\Sigma(f_o - f_e)^2/f_e = 573.79 > 9.48 = \chi_4^2(5\%)$, reject H_0.

85. $\Sigma[(f_o - f_e)^2/f_e] = 13.5 > 9.48 = \chi_4^2(5\%)$, reject H_0.

87. $\Sigma(f_o - f_e)^2/f_e = 26.34 > 9.48 = \chi_4^2(5\%)$, reject H_0.

89. $\Sigma[(f_o - f_e)^2/f_e] = 4 < 5.99 = \chi_2^2(5\%)$, accept H_0.

91. $\Sigma[(f_o - f_e)^2/f_e] = 1.17 < 5.99 = \chi_2^2(5\%)$, accept H_0.

93. $\Sigma[(f_o - f_e)^2/f_e] = .625 < 5.99 = \chi_2^2(5\%)$, accept H_0.

95. $\Sigma(f_o - f_e)^2/f_e = 23.69 > 9.48 = \chi_4^2(5\%)$, reject H_0.

97. $\Sigma[(f_o - f_e)^2/f_e] = 6.27 < 9.48 = \chi_4^2(5\%)$, accept H_0.

Chapter 13

1. Assumption A: Either the x_i are fixed numbers or they are random numbers that are statistically independent of the error term ϵ_i. Assumption B: ϵ_i are normally distributed. Assumption C: $E(\epsilon_i) = 0$. Assumption D: $E(\epsilon_i\epsilon_j) = 0, \forall i \neq j$. Assumption E: $E(\epsilon_i^2) = \sigma_\epsilon^2$.

3. **b.** $b = 1.14, a = 76.6$.
   c. For $X = 35$ lb, $Y = 76.6 + 1.14(35) = 116.5$.

5. **b.** $b = .865, a = .031, \hat{Y}_i = .031 + .865(X_i)$.
   c. SSE $= 2.68$, SE $= .38$.
   d. SST $= 3.08, R^2 = .130$.

7. **b.** $b = 1.61, a = 36.26$.

9. **b.** $b = -23.6, a = 416$; for 9.80, $Y = 416 - 23.6(9.80) = 184.72$.

11. **a.** $b = -1.630, a = 21.76$.
    b. Because the slope coefficient is less than 1, com-

mon stocks are not a hedge against inflation in this time period.

13. For causality to exist, X must cause Y. For example, the number of calories a person consumes is one factor that affects a person's weight. Correlation is simply a statistical relationship between two things, where causality may or may not be present. In regression analysis, we usually try to have a causal relationship.

15. A scatter diagram is a plot of the X and Y variables. A regression line is the line fit through the scatter diagram that best explains the relationship between X and Y.

19. A sample represents a subset of the population of interest. A population consists of the entire group of interest. By taking a sample of the population and applying various statistical rules we can make educated guesses of population parameters such as the mean and variance.

21. **a.** Because weight depends on a person's height, it should be the dependent variable.
    b. $b = 5.71, a = -224.996$.

23. The consumption function is $C = -166.17 + .688(Y)$.

25. A direct relationship means that as X increases in value, we expect Y to increase in value. An inverse relationship means X and Y tend to move in opposite directions.

29. $\hat{Y}_i = -1.78 + .0005(X_i)$, $e_i = Y_i - \hat{Y}_i$

$\hat{Y}$	e
2.72	.28
2.57	−.57
4.17	−.17
4.97	1.03
5.72	−.72
2.82	.18

31. Error = actual − estimated; Bob = 8,360; Mary = −5,800; Sue = −880; Ted = 2,040; Anne = 8,360.

33. **b.** $b = 1.51, a = -.519$.

35. Goodness of fit for a regression measures how well the regression line explains the relationship between X and Y. R^2 and the standard error of the regression are two measures of goodness of fit.

37. $r_{XY} = [\text{Cov}(X, Y)]/\sigma_X\sigma_Y$, (missing a variance for Y, Var = 1,000) = $500/\sqrt{250}\sqrt{1000}$ = 1.

39. SSE = 269.04; SST = 7,420; SSR = 7,150.96.

41. SST = 1,176,800,000; SSE = 178,355,200; SSR = 998,444,800; R^2 = .85.

43. It should be a scatter plot of question 32. You would expect a low coefficient of determination because there is a very weak relationship between miles worked and days late.

45. The slope of the regression lines measures how much Y will change when there is a one-unit change in X.

47. $b = .98, a = .0084$, SST = 2.211, SSE = 1.716, R^2 = .224.

49. $b = .178, a = 2.90$, SST = 35.08, SSE = 17.885, R^2 = .49.

51. $r_{XY} = 33.07$, $\text{Std}_X = 7.792$, $r_{XY} = .836$.

53. **a.** Positive **b.** Positive **c.** Positive
    d. Positive **e.** Positive **f.** Negative
    g. Negative
    h. Zero, although some researchers have found a negative correlation between the two for some strange reason.

55. **a.** $r_{XY} = .87$.
    b. $R^2 = .757$.
    c. $\beta = .344, \alpha = 93.12$.

57. $\beta = .0041, \alpha = .886$.

59. **b.** $\beta = 5,613.64; \alpha = -35,431.82$.

61. **a.** $r_{XY} = -.959$. **b.** $\beta = -.187, \alpha = 8.405$.

63. $r_{xy} = .922$.

65. See the table in question 64. $b = 2.442, a = -2.830$.

67. $r_{xy} = -.388, b = -1.0389, a = 4.8412, S_e = .19055$.

69. $r_{xy} = .214, b = .1905, a = 13.561, S_e = 6.7675$.

71. $S_e = 53.65$; SST = 81,212.94; SSE = 51,817.7; $R^2 = .3620$.

73. $\beta = -3.78, \alpha = 63.633$.

75. SSE = 95.74, $R^2 = .560, r = .748$.

Chapter 14

1. **a.** $b = 2, a = 0$.
   b. For $X = 8, \hat{Y} = 0 + 2(8) = 16$.
   c. $S_e = \sqrt{4/(7 - 2)} = .894$.
   d. $R^2 = .966$.

3. **a.** $b = 6, a = -2$.
   b. $S_e^2 = 4.375$.
   c. $t_{.025,28} = 2.0484$, accept H_0.

5. a. $S_{XY} = (\Sigma X_i Y_i)/(n-1) = 400/(11-1) = 40$;
$r_{XY} = .80$; $t_{n-1} = (r\sqrt{n-2}/\sqrt{1-r^2} = (.80\sqrt{11-2})/\sqrt{1-.80^2} = 4.0$; $t(.05/2, 10) = 2.23$, reject H_0: $\rho = 0$.

b. $r_{XY} = .600$; $t_{n-1} = (.60\sqrt{18-2})/\sqrt{1-.60^2} = 3.0$; $t(.05/2, 17) = 2.11$, so, reject H_0.

7. a. $r_{XY} = 1,456.75/\sqrt{1405.15}\sqrt{1861.56} = .901$.

b. $t_{n-1} = (.901\sqrt{14-2})/\sqrt{1-.901^2} = 7.19$, $t(.025, 13) = 2.16$, reject H_0: $\rho = 0$.

9. a. $t = .980$, $t(.025, 7-2) = 2.571$, accept H_0.

b. $t_{n-1} = 1.057$, $t(.025, 7-1) = 2.447$, accept H_0.

11. Because the estimates of α and β are not known with certainty, we construct confidence intervals so that we will have a better idea of the range of values the true parameter value can take.

13. Usually β is of greater interest because it measures the relationship between X and Y. That is, when X changes, how much do we expect Y to change?

15. When the standard deviation is known or when the sample is large enough, the standardized variable will be distributed as a standard normal or z distribution. When the standard deviation is unknown and when the sample is small, the standardized variable will be distributed as student-t.

17. The standard error of the estimate is a measure of the goodness of fit for the entire regression, whereas the standard error of the regression coefficient is a measure of the goodness of fit of that coefficient. In simple regression the two are closely related and in principle measure similar things.

19. b. $b = 12.40$, $a = -400.96$.

c. $S_b = 1.99$.

d. $t = 6.23$.

21. $F(1, n-2) = 38.97$, $F(1, 3, .01) = 34.1$, reject H_0.

23. a. $\beta = 1.324$, $\alpha = .0276$.

c. $R^2 = .748$.

d. $t = 5.347$, reject H_0: $\beta = 0$.

25. Based on R^2, both models perform equally well.

27. $t = 1.33$, cannot reject the null hypothesis.

29. b. $\beta = 1.904$, $\alpha = -.000936$.

c. SST $= .3773$, SSE $= .079973$, SSR $= .2973$, $R^2 = .788$.

d. $S_b = .3122$, $t = 1.903/.3122 = 6.09$, reject H_0.

31. b. $\alpha = 3.1042$, $\beta = -.04235$.

c. SST $= 1.7167$, SSE $= .5238$, SSR $= 1.1929$, $R^2 = .6948$, $S_e = .1706$.

d. $S_b = .0066$, $t = -.04235/.0066 = -6.42$, reject H_0.

33. a. $b = -.861$ (elasticity), $a = 5.08$.

b. $t = -.861 - 0/.0633 = -13.60$, $t(.01/2, 4) = 4.604$, reject H_0: $b = 0$.

c. $-1.037 \le \beta \le -.685$.

35. a. 90% interval, 99% interval, $.642 \le \beta \le 1.958$, $.268 \le \beta \le 2.332$.

b. If σ_b is not known and you have only 15 observations, you would use the student-t distribution instead of the Z distribution.

37. $b = 1.51$, $a = -.52$, $t = 1.51 - 0/.396 = 3.81$, $t(.05/2, 6-2) = 2.776$, so reject H_0: $b = 0$ at the 5% level.

39. $b = .0041$, $a = .889$, $.0023 \le \beta \le .0059$, $-.202 \le \alpha \le 1.98$.

41. $-.183 \le \beta \le 2.777$, yes.

43. Because forecasts are simply educated guesses, we construct intervals around the forecasts so that we know the possible values the forecast can undertake.

45. $1.99 \le y_{n+2} \le 2.35$.

47. $b = 12.40$, $a = -400.96$, $S_e = 188.54$; 90% interval is 2699.04 ± 166.43; 99% interval is 2699.04 ± 413.13. For us to be more certain, the 99% interval must be wider.

49. Standard market model: $-.102 \le y_{n+1} \le .474$; CAPM market model: $-.106 \le y_{n+1} \le .470$. The confidence intervals are about the same for both models.

51. When we have proxy error on the X variable, one of the assumptions of linear regression is violated. This leads to biased parameter estimates.

53. $2.540 \le y_{n+1} \le 3.014$.

55. $.092 \le y_{n+1} \le .138$.

57. a. $t = (2.23 - 0)/.627 = 3.56$, $t(.01/2, 23-2) = 2.831$, reject H_0: $b = 0$.

b. $.455 \le \beta \le 4.00$.

59. $-1.40 \le \beta \le -.30$.

61. $b = .735$, $a = 4.09$.

63. Although there appears to be a positive relationship between X and Y, the relationship is relatively weak; thus, measures of the goodness of fit such as R^2 will tend to be small.

65. $b = .0057$, $a = 3.56$.

Chapter 15

1. In simple regression, we are relating the dependent variable to one explanatory variable. In multiple regression, we assume that the independent variable can be explained by more than one variable. In most real-life situations we usually would use multiple regression rather than simple regression.

3. The assumptions of the multiple regression model are the same as for simple regression, with the added assumption that the independent variables are not perfectly related to one another in a linear fashion.

5. The number of degrees of freedom is equal to the number of coefficients estimated, including the intercept term.

7. In simple regression, the slope coefficient measures the relationship between x and y. In multiple regression, each slope coefficient measures the relationship of that explanatory variable with y, assuming that all other explanatory variables are held constant.

9. Multicollinearity occurs when a combination of independent variables are linearly related. When there is perfect multicollinearity, it will be impossible to estimate the parameters of the model. When there is less than perfect multicollinearity, the least squares method will not be able to sort out the partial influences of the independent variable with the dependent variable.

11. **a.** SSE $= 691.79$; SST $= 1,159.72$; SSR $= 467.93$; $R^2 = .403$; $F = 1.69$; $F(.05, 2, 5) = 5.79$; accept H_0: $b_1 = b_2 = 0$.

13. **a.** $b_1 = -.315$, $b_2 = .0011$, $a = 4.18$.
    b. $S_{b_1}^2 = .0051$, $S_{b_2}^2 = 7.81 \times 10^{-8}$; for b_1, $t = -4.39$, $t(.01/2, 6 - 2) = 4.604$, accept H_0: $b_1 = 0$; for b_2, $t = 3.93$, accept H_0: $b_2 = 0$.

15. $-.468 \le \beta_1 \le -.162$, $.0005 \le \beta_2 \le .0017$.

17. The coefficient on experience indicates that for every one year of experience, we expect income to increase by $1,000. The coefficient on education indicates that for every one year of education, we expect income to increase by $500.

19. **a.** Bedrooms, miles, bathrooms, age, and size should all be independent variables.
    b. Sales price$_i$ $= \alpha + \beta_1$bedrooms$_i$ + $\beta_2$bathrooms$_i$ + $\beta_3$miles$_i$ + $\beta_4$age$_i$ + $\beta_5$size$_i$ + ϵ_i. Expect coefficients for bedrooms, bath-

rooms, and size to be positive. Expect coefficients on miles and age to be negative.
    c. A model like this would allow Ms. Shady to make a good guess regarding the selling price of a house.
    d. Confidence intervals will allow Ms. Shady to place an upper and lower bound on the house's selling price.

21. Because monthly salary is four times a person's weekly salary, there will be a problem of multicollinearity.

23. $t_{b_1} = .733 - 0/.205 = 3.58$, $t_{b_2} = 1.039 - 0/.938 = 1.11$, $t(.10/2, 5 - 2 - 1) = 2.92$, reject H_0: $b_1 = 0$ at 10% level for two-tailed test, accept H_0: $b_2 = 0$, $-1.30 \le \beta_1 \le 2.77$, $-8.27 \le \beta_2 \le 10.35$.

25. It is possible for the t-tests to be significant while the F-test is insignificant.

27. **a.** $b_1 = -152.34$, $b_2 = .26$, $a = 1861.34$.
    b. Yes, $b < 0$, $b_2 > 0$.
    c. Because we have beliefs about the sign of the coefficient, we should use a one-tailed test.
    d. $t_{b_1} = (-152.337 - 0)/130.50 = -1.17$, $t_{b_2} = (.26 - 0)/.095 = 2.74$, $t(.05, 6 - 2 - 1) = 2.353$, accept H_0: $\beta_1 = 0$, reject H_0: $\beta_2 = 0$ at .05 level for a one-tailed test.

29. $-.042 \le \beta_2 \le .562$.

31. $b_1 = -.910$, $b_2 = .313$, $b_1 = $ price elasticity $= \dfrac{\%\Delta Q}{\%\Delta P}$, it measures the responsiveness of quantity to a change in price. A value of $-.910$ indicates that a 1% price increase leads to a .91% decrease in quantity. $b_2 = $ income elasticity $= \dfrac{\%\Delta Q}{\%\Delta \text{Inc.}}$, it measures how sensitive quantity is to a change in income. A value of .313 indicates that a 1% increase in income leads to a .31% increase in quantity.

33. $F(.05, 2, 3) = 9.55$, reject H_0: $\beta_1 = \beta_2 = 0$ at 5% level.

35. **a.** $b_1 = .0077$, $b_2 = -.0310$, $a = 1.72$.
    b. $F(.05, 2, 2) = 19.0$, reject H_0: $\beta_1 = \beta_2 = 0$ at 5% level.
    c. $S_{b_2} = .011$, $t_{b_1} = (.0077 - 0)/.0014 = 5.5$, $t_{b_2} = (-.031 - 0)/.011 = -2.82$, $t(.05/2, 2) = 4.303$, reject H_0: $\beta_1 = 0$ at 5% level for a two-tailed test. $t(.10, 2) = 1.886$, reject H_0: $\beta_2 = 0$ at 10% level for a one-tailed test.

37. $\hat{\text{EPS}} = 1.72 + .0077(400) - .031(65) = \$2.785$, $2.785 \pm 9.925(.123)$, 1.564 to 4.006.

39. $-1.00 \le \alpha \le 26.06$, $-20.23 \le \beta_1 \le 1.49$, 5.73 $\le \beta_2 \le 23.77$, $.084 \le \beta_3 \le .456$.

41. $.213 \le \beta_1 \le 2.43$, $-.490 \le \beta_2 \le -.150$.

43. The test can be conducted using an F-test. $F_{K,n-K-1} = 32.83$, $F(.05, 5, 29) = 2.53$, reject H_0: $\beta_1 = \beta_2 = \beta_3 = \beta_4 = \beta_5 = 0$.

45. By most standards, an $R^2 = .329$ does not represent a particularly good fit, with the model explaining only about 33% of the total variation in Z. However, in many real-world cases, being able to explain 33% of the total variation is exceptional, as is the case in explaining stock returns.

47. a. $b = 1.5$, $a = -75$.
  b. $t_b = (1.5 - 0)/7.72 = .194$, $-22.96 \le \beta_1 \le 25.96$.

49. $t = 4.07$, $t = (.01/2, 12 - 2) = 3.169$, reject H_0: $r = 0$ at 1% level for a two-tailed test.

51. $F_{K,n-K-1} = 9.29$, $F(.05, 2, (6 - 2 - 1)) = 9.55$, accept H_0: $\beta_1 = \beta_2 = 0$.

53. a. $\hat{y} = 4.14$.
  b. $4.14 \pm 2.353(.522)$, 2.91 to 5.37.

55. Demand is usually assumed to depend on price and income. If we consider GNP to be the income of the entire country, we might specify the following model: $D_t = \alpha + \beta_1 P_t + \beta_2 GNP_t + \epsilon_t$, where D_t = demand in period t, P_t = price in period t, GNP_t = GNP in period t.

57. a. A \$1 increase in R&D is expected to increase annual sales by \$5.80. A \$1 increase in TV advertising is expected to increase sales by \$4.20. Finally, a \$1 increase in all other advertising is expected to increase sales by \$7.40.
  b. $t_1 = (5.8 - 0)/1.20 = 4.83$, $t_2 = (4.2 - 0)/1.31 = 3.21$, $t_3 = (7.4 - 0)/1.56 = 4.74$, $t(.01/2, 23 - 3 - 1) = 2.861$. All coefficients are significant.
  c. This is a case of multicollinearity.
  d. $R^2 = .874$, $R = .935$.
  e. $\hat{y} = 126.3$.

61. Both simple regression and multiple regression attempt to show a relationship between a dependent variable and an explanatory variable. Because multiple regression allows for more than one explanatory variable, it is more general and therefore more useful than simple regression.

63. a. False. The size of the coefficient is determined by the relative size of x_1 and y and x_2 and y; it has nothing to do with the degree of correlation.

b. True.
  c. True. The high R^2 indicates that the regression line provides a good measure of the relationship between y and x_1 and x_2.

65. a. If n_1 and n_2 are independent, then $Cov(\chi_1, \chi_2) = 0$. If χ_1 and χ_2 are independent, their correlation will be zero.
  b. Independence between χ_1 and χ_2 implies $r = 0$; however, if $r = 0$, it does not imply independence.

Chapter 16

1. a. The coefficient of R indicates that a black person has a 23% greater chance of being in poverty. Likewise, a female has a 45% greater chance of being in poverty.
  b. Not discussed.
  c. Not discussed.
  d. One problem with the linear probability model is that predicted values of the dependent variable may be greater than 1 or less than 0, which is outside the bounds for P_i.

3. a. To test the significance we need to compute the t-values: $t_{x1} = 1.19$, $t_{x2} = .49$, $t_{x3} = .07$, $t_{x4} = .26$, $t_{x5} = .74$. None of the coefficients on the explanatory variables is significant.
  b. Can use an F-test to test overall significance. $SSR = SST - SSE = 9,665 - 7,892 = 1,773$; $F_{K,n-K-1} = .629$, not significant.

5. a. $t_{x1} = .19$, $t_{x2} = 5.5$, $t_{x3} = 1.56$. Only the coefficient on x_2 is significant.
  b. $F_{k,n-k-1} = 15.47$. $F(.01, 3, 8) = 7.59$, so regression is significant at the 1% level.
  c. The low t-values for x_1 and x_3 combined with the overall significance of the regression may indicate the presence of multicollinearity.

7. a. $\beta_1 = 16.64$, $\alpha_1 = 9.572$, $PPS_i = 9.572 + 16.64 \, DPS_i$.
  b. $b_1 = 1.364$, $b_o = -.580$, $RE_i = -.580 + 1.364 \, DPS_i$.
  c. $\beta_1' = 16.645$, bias $= 16.645 - 12.434 = 4.211$.

9. Autocorrelation occurs when current and lagged error terms are correlated. Autocorrelation reduces the efficiency of the least squares estimates. It can be detected by using the Durbin–Watson test or Durbin's–H statistic.

11. Specification bias occurs when either the model is incorrectly formulated or we have errors in the estimation of the explanatory variables. This leads to biased parameter estimates. Specification bias can be avoided by using business and economic theory to specify an appropriate mathematical relationship. Also, care should be taken in the measurement of the explanatory variables.

13. A lagged dependent variable is used in a regression when we believe the current value of y is related to a past value of y—in this case, a lagged value of y as an explanatory variable.

15. To examine the difference in earnings among 4 groups, we would use only 3 dummy variables. The effect on the fourth group would be measured by the intercept term. If we used 4 dummy variables, perfect multicollinearity would exist because the 4 dummies would represent a linear combination of the intercept term, which is simply a column of ones.

17. Reject the null hypothesis of no autocorrelation if $DW < d_L$ and $DW > 4 - d_L$ (for positive autocorrelation). Null hypothesis accepted if $DW > d_U$ and $4 - d_U > DW$ (for negative autocorrelation). Test is inconclusive if DW is between $4 - d_U$ and $4 - d_L$, and DW between d_L and d_U.
   a. At the 5% level for $n = 40$, $k = 3$, $d_L = 1.34$, $d_U = 1.66$, and $DW = 1.30$: $DW < d_L$ reject H_0 in favor of the alternative hypothesis of positive autocorrelation.
   b. $DW < d_L \rightarrow$ positive autocorrelation.
   c. $DW > d_U \rightarrow$ accept H_0: no autocorrelation.
   d. $DW > d_U \rightarrow$ accept H_0.
   e. $DW > 4 - d_L \rightarrow$ reject H_0: negative autocorrelation.

19. If you use 12 dummy variables to analyze the different stock returns in different months, you will have a multicollinearity problem. You can avoid this by using only 11 dummy variables and allowing the intercept term to account for the extra month.

21. One problem with heteroscedasticity is that the least squares method tends to place more weight on the observations with the larger variance. By weighting each observation by $1/\sigma_i$, we attempt to give a more even weight to all observations regardless of the size of their variance.

23. A correlation between x_1 and x_3 of .85 may indicate that multicollinearity will be a problem if both variables are used as explanatory variables in the same regression.

25. $t_{b1} = 2.5$, $t_{b2} = 2.35$, $t_{b3} = 1.2$. Because the coefficient on sex is not statistically significant, we cannot conclude that discrimination exists.

27. a. $b = 9.93$, $a = 306.70$.
   b. $b = 207.84$, $a = -450.66$. Solve this question by using the MINITAB program.

31. To create an interaction variable, we multiply temperature and humidity together. The coefficient on this new variable will show us how temperature combined with humidity affect cell growth. The estimated regression is

$$\text{Cell}_i = -224.59 + 5.28 \text{ Temp}_i + 6.47 \text{ Hum}_i$$
$$(1.87) \qquad\qquad (1.74)$$
$$-.074(\text{Temp}_i \times \text{Hum}_i)_i$$
$$(.024)$$

The results indicate that a 1° increase in temperature will lead to an expected change of $5.28 - .074$ Hum_i in cell growth. Likewise, a 1% change in humidity will lead to an expected change of $6.47 - .074 \text{ Temp}_i$ in cell growth.

33. Because we are interested only in the month of January relative to the other 11 months, we could create a dummy variable equal to 1 in January and 0 for the other 11 months. The coefficient on this dummy variable would tell us if there are any differences between the returns in January relative to the other 11 months. The regression could be $R_{it} = \alpha + \beta_1 R_{Mt} + \beta_2 DUM_t + \epsilon_t$.

35. To test the effect of winter ball on batting average, we use a dummy variable equal to 1 if the player played winter ball and 0 if he did not. The coefficient for winter ball is .0156 and the t-value is $t = (.0156 - 0)/.0083 = 1.88$, $t(.10, 6 - 2 - 1) = 1.638$, so the coefficient is significant at the 10% level for a one-tailed test. Therefore, it appears that playing winter ball helps a player's batting average.

37. The coefficient on education predicts a $2,325 increase in earnings for each additional year of education. The coefficient on the dummy variable predicts that a person from the northeast will earn $1,725 more than a similarly educated person from other parts of the country.

39. The model can be estimated by using a dummy variable equal to 1 after the war and 0 before the war. The coefficient on the dummy variable will indicate the difference between consumption after the war relative to consumption before the war.

41. DW = 3.03/1.06 = 2.86
   a. For $k = 1$, $n = 20$, $d_L = 1.20$, and $d_U = 1.41$; $4 - d_L = 2.59$; $4 - d_L = 2.80$; $4 - d_L < DW < 4$, so reject H_0; negative autocorrelation.
   b. Reject H_0 in favor of negative autocorrelation.

49. $r_1 = .0105/\sqrt{(.0228)(.02275)} = .4610$.

55. $d_L = 1.38$, $d_U = 1.72$.
   a. DW = 1.90, $d_U < DW < 2$, accept H_0: no autocorrelation.
   b. DW = .90, DW $< d_L$, reject H_0 in favor of H_1: positive autocorrelation.
   c. DW = 2.55, $4 - d_L < DW < 4$, reject H_0 in favor of H_1: negative autocorrelation.
   d. DW = 1.75, $d_U < DW < 2$, accept H_0: no autocorrelation.
   e. $d = 3.45$, $4 - d_L < DW < 4$, reject H_0 in favor of H_1: negative autocorrelation.

57. a. In this case we have omitted an explanatory variable; this will cause our parameter estimates to be biased.
   b. In this case we are including an irrelevant explanatory variable that will reduce the efficiency of our estimates but not lead to any bias.

59. $CORN_t = \alpha + \beta_1 RAIN_t + \beta_2 FERT_t + \epsilon_t$ or with an interaction variable $CORN_t = \alpha + \beta_1 RAIN_t + \beta_2 FERT_t + \beta_3 (RAIN_t \times FERT_t) + \epsilon_t$.

61. $SALES_t = \alpha + \beta_1 ADVER_t + \beta_2 ADVER_t^2 + \epsilon_t$: This model will capture the nonlinear relationship between sales and advertising. $SALES_t = \alpha + \beta_1 ADVER_t + \beta_2 (DUM_t \times ADVER_t) + \epsilon_t$: Where $DUM_t = 1$ if advertising is above the specified level and 0 otherwise. This model uses a dummy variable times advertising dollars to see if the slope of the regression line changes after a certain amount of advertising dollars are spent.

63. Because serial or autocorrelation is a problem associated with time series data, it is not appropriate because this is cross-sectional data.

65. We would expect a faster growth rate of disposable income to lead to more votes, so b_1 should be positive. Because inflation and unemployment are undesirable, higher numbers for these variables should reduce votes, so b_2 and b_3 should be negative.

67. $F_{k, n-k-1} = \dfrac{R^2}{1 - R^2} \dfrac{n - k - 1}{k}$

$= \dfrac{.987}{1 - .987} \left(\dfrac{(10 - 3 - 1)}{3} \right)$

$= 151.85$

$F(.01, 3, 6) = 9.78$, so regression is significant at the 1% level.

69. Because the coefficient is only marginally insignificant, it appears that the missey index may be important. Before the regression was run you would expect a lower R^2 than we actually received because inflation and unemployment tend to be inversely related. Because of this relationship we would expect that the missey index might not contain much information about the votes received.

71. Because we have only 10 years of data, the use of the Durbin–Watson test is not appropriate. If we scan the data, however, we can see a pattern of alternating negative and positive error terms indicating negative autocorrelation.

Chapter 17

1. If the median housing price is $100 thousand, we should have 50% of the population valued over $100 thousand. Our sample shows 5 of 9 houses exceeds 50%

$$\bar{P} = \dfrac{5}{9} \quad \dfrac{\bar{P} - P}{\sqrt{\dfrac{P(1 - P)}{n}}} = .33 < 1.96$$

Accept H_0: $P = 1/2$.
3. a. $(R - \mu_R)/\sigma_R = -.13 > -1.96$, accept H_0.
   b. $(R - \mu_R)/\sigma_R = -.34 > -1.96$, accept H_0.
   c. $(R - \mu_R)/\sigma_R = .46 < 1.96$, accept H_0.
5. a. $Z = .62 < 1.96$, accept H_0: random walk.
   b. If there is a momentum in the occurring of accident, it means some precautionary measures should be taken whenever one occurs.
7. Use the assumption and rank the data: $r_s = .64$, because $-.89 < .64 < .89$, accept H_0.
9. $r_s = .027$, $t = .08 < 2.26$, accept H_0.
11. H_0: Eastside is not higher than westside. H_1: Eastside is higher than westside. $(u_1 - \mu_u)/\sigma_u = -.15 > -1.64$, accept H_0.

13. Parametric tests assume the data used comes from a certain distribution. The nonparametric test does not use this assumption. See text for example.

15. Toss coins and record the outcomes, then apply the run test.

17. $(18 - 27.5)/9.81 = -.97 > -1.96$, accept H_0: no difference.

19. $(14.5 - 22.5)/8.44 = -.95$, accept H_0: no difference.

21. Define $+$ as price increase and $-$ as price decrease. Then the data can be transformed to $- - + - - + + + + - +$. $(R - \mu_R)/\sigma_R = (6 - 6.45)/1.56 = -.29 > -1.96$. Accept H_0: random. Note: For $n < 20$, it is better that we use the Wald–Wolfowitz test, which is not introduced in this book. We still use the sign test for this question.

23. $(R - \mu)/\sigma_R = -.59 > -2.57$, accept H_0: random.

25. a. $R = 3$. **b.** $R = 4$. **c.** $R = 2$. **d.** $R = 3$.

27. $R = 11, n_1 = 15, n_2 = 10, n = 25, Z = -1.86 > -1.96$, accept H_0.

29. Use the Wilcoxon sign-rank test. $w^+ = 25.5, (w^+ - \mu_w)/\sigma_w = -.20$; accept H_0: The second manager is not tougher. Note: We suggest $n \geq 25$ for using the Wilcoxon sign-rank test.

31. Pool the data together and rank the data. Add ranks for group B and obtain $R_B = 243, U_A = 102$: $(U_A - \mu)/\sigma = -.44$; accept H_0: same effect.

33. $\dfrac{\bar{P} - P}{\sqrt{\dfrac{P(1 - P)}{n}}} = 2.75 > 1.64.$

Reject H_0: B is not better.

35. $\dfrac{12}{n(n + 1)}\left[\dfrac{R_1^2}{n_1} + \dfrac{R_2^2}{n_2} + \dfrac{R_3^2}{n_3} + \dfrac{R_4^2}{n_4}\right] - 3(n + 1)$
$= 2.9$

Accept H_0: Four leagues have about the same earned run average.

37. Rerank the data for the two west divisions and obtain the rank sum for the National League West; obtain $R_1 = 94$. $[U_1 - E(U_1)]/\sigma_u = .84$, accept H_0.

39. $(90 - 100)/22.73 = -.44 < -1.645$, accept H_0: Playing music is not any better.

41. No! From the test results in question 39, we know that the test statistic is .44, which is insignificantly different from zero.

43. $1 - (6 \cdot 0)/[5(5^2 - 1)] = 1; 1 - (6 \cdot 40)/[5(5^2 - 1)] = -1.$

45. $\dfrac{r_s - 0}{\sqrt{\dfrac{1 - r_s^2}{n - 2}}} = 3.89 > 1.81$

Reject H_0: Money does not buy winning.

47. $(u_1 - \mu)/\sigma = -.23 > -1.64$, accept H_0: The new machine is not better.

49. $\dfrac{12}{n(n + 1)}\left[\dfrac{R_1^2}{n_1} + \dfrac{R_2^2}{n_2} + \dfrac{R_3^2}{n_3}\right] - 3(n + 1)$
$= 18.26 > \chi^2_{.10,2} = 4.605$

Reject H_0: The three groups spend the same time on lunch.

51. $(86.5 - 50)/13.228 = 2.76 > 1.64$, reject H_0: Group 2 did not spend less time than group 1.

53. $r_s = -.19, t = -.55 < 2.306$, accept H_0: no positive correlation.

55. $r_s = .9, t = -0.55 < 2.306.$

57. $(u_1 - \mu)/\sigma = -.38 > -1.96$, accept H_0: equally effective.

59. a. Use the sign test.

$\dfrac{\bar{P} - P}{\sqrt{\dfrac{(1 - P)P}{n}}} = .63$

Accept H_0: Morning productivity is not higher.
b. $[(U_2 - \mu)/\sigma = .57$, accept H_0.

61. $R_s = .97, t = 21.1 > 1.701$, reject H_0: no positive correlation.

63. Since $P = .176 > .05$, accept H_0: The three locations have the same amount of sales.

65. $(U_1 - \mu)/\sigma = -1.29 > -1.64$, accept H_0: Implementing both programs is not more effective than the stop smoking program only.

67. a. $\dfrac{12}{30(30 + 1)}\left[\dfrac{116^2}{10} + \dfrac{163.5^2}{10} + \dfrac{185.5^2}{10}\right]$
$- 3(30 + 1) = 3.26 < 5.99$

Accept H_0: equal sales.
b. $R_1 = 112.5, R_2 = 97.5, (42.5 - 50)/13.23 = -.57$, accept H_0: The high budget is not better.

69. $(26.5 - 50)/13.23 = -1.77 > -1.28$, reject H_0: The high cost gas station is not charging higher.

71. $\dfrac{12}{n(n + 1)}\left[\dfrac{R_1^2}{n_1} + \dfrac{R_2^2}{n_2} + \dfrac{R_3^2}{n_3}\right] - 3(n + 1)$
$= 16.74 > 5.99$

Reject H_0: The increases in the three regions are the same.

73. The signs of increase $(+)$ and decrease $(-)$ are summarized as follows: $- + + + - + + + - +$ $+ + - + + + - - + + - + + - + +$. Runs $= 14: (R - \mu_R)/\sigma_R = .91$.
Accept H_0: no momentum.

75. $(80.5 - 50)/13.23 = 2.31 > 1.64$, reject H_0: Class two is not better.

77. $(143 - 112.5)/24.11 = 1.27 < 1.64$, accept H_0: Normal time productivity is not higher.

79. $\dfrac{12}{27(27 + 1)} \left[\dfrac{136^2}{9} + \dfrac{117^2}{9} + \dfrac{125^2}{9} \right] - 3(27 + 1) = .32 < 5.99$

Accept H_0: equally popular.

81. $r_s = 1 - \dfrac{6 \cdot \Sigma d^2}{n(n^2 - 1)} = .047$

$\dfrac{r_s - 0}{\sqrt{\dfrac{1 - r_s^2}{n - 2}}} = \dfrac{.047 - 0}{\sqrt{\dfrac{1 - .047^2}{21 - 2}}} = .21$

Accept H_0: no relation between interest rates and S&P 500 return.

Chapter 18

1. **a.** 463 months.   **b.** 154 quarters.
   c. 38 years.

3. **b.** Curve.   **c.** No.

5.

Year	Mergers: 3-Year MA	
1970	15	
1971	17	18.66666
1972	24	22.33333
1973	26	26.66666
1974	30	32.33333
1975	41	52
1976	85	72
1977	90	95
1978	110	108.3333
1979	125	127.6666
1980	148	158.6666
1981	203	200
1982	249	244
1983	280	278.6666
1984	307	

7. Regression output $y = -54.82 + 21.44x$; $R^2 = .91$; where $x = $ time and $y = $ mergers.

11. $\$135,478/1.04 = 130,267$; $\$130,267 \times 12 = 1,563,207$.

13. See text.

15. Special techniques are needed for time-series analysis mainly because the time-series data are not independent of each other. Most statistical methods discussed in this textbook assume that the data are random samples. In the case of time-series data, a piece of sample is often affected by the value that precedes it.

17. See text.

19. **a.** Trend, cycle, seasonal, and irregular.
    b. Seasonal and irregular.
    c. Cycle and irregular.
    d. Seasonal and irregular.

21. Advantage: simple, sometimes (especially in the short run) the trend is a linear one. Disadvantage: linear trend may not be appropriate for the data.

23. See text.

25. See text.

27. **a.** Some kind of cyclical factor.
    b. A trend and cycle factors.
    c. Cyclical factors.

29. Trend, seasonal, and irregular—use Holt–Winters exponential smoothing method.

31. If the college admits mostly local students, I would obtain numbers of the potential future students, namely, today's high school students in the local areas. I would build a regression model relating freshman class size to the number of high school sophomores or seniors.

33. $\hat{x}_{104} = 15 + .6x_{103} - .2x_{102} = 15 + .6(992) - .2(927) = 424.8$. $\hat{x}_{105} = 15 + .6\hat{x}_{104} - .2x_{103} = 15 + .6(424.8) - .2(992) = 71.48$. $\hat{x}_{106} = 15 + .6\hat{x}_{105} - .2\hat{x}_{104} = 15 + .6(71.48) - .2(992) = -140.51$. The return on T-bills declines during 1985 and 1986, then stabilizes during 1987. It is hard to detect any obvious systematic component.

37. AR(1): .004615. 3-month MA: .004931. Actual: .004688. The AR(1) model gives a better forecast.

39. $R_{1,t} = .015207 + .171238 \cdot R_{1,t-1}$. The AR(1) forecast is $-.02579$. The moving average forecast is $-.2219$.

41. $R_{3,t} = -.00094 + .051385 R_{3,t-1}$. The AR(1) forecast is $-.00171$. The moving average forecast is $-.1359$.

43. Take log on both sides of the function: log D_t = log D_o + t log $(1 + g)$.

47. The time factor is not intended to "explain" the causality; it is intended to catch a linear trend movement along time.

49. When sales increases at a steady rate, the best model for predicting the sales is an exponential model.

51. a. Irregular.
  b. Trend or cyclical.
  c. Seasonal.
  d. Irregular.
  e. Trend.

55. y = 2209.539 + 55.874t; for 1990, y = 2209.539 + 55.874 (26) = 3662.263; 1991, y = 2209.539 + 55.874 (27) = 3718.137.

57. The forecast in question 54 is too low compared with the other forecasts. This is because the moving average method does not deal with the increasing trend.

61. We take the first difference on the original data. Then, we run a regression of AR(1) to obtain y_t = 1699 + .204 y_{t-1}, where y_t is the first-order difference of the employment data. So $\hat{y}_{1990}$ = 1699 + .204(2374) = 1699 + 484 = 2183; $\hat{y}_{1991}$ = 1699 + .204(2183) = 2144; $\hat{y}_{1992}$ = 1699 + .204(2144) = 2136.

Time	Forecast
1990	117342 + 2183 = 119525
1991	119525 + 2144 = 121669
1992	121699 + 2136 = 123805

71. The data generated will follow a random walk process. That means that the data exhibit no meaningful pattern.

73. The best forecast is the value at period 51.

75. The best forecast for the time series is the value generated in the last period. If the value to be forecasted is a 4th coin toss, then add .25 to the value generated one period before.

Chapter 19

1.

	1984	1985	1986	1987	1988
a.	1	1.016	1.036	1.081	1.144
b.	.874	.888	.905	.944	1

3. a. Laspeyres Index = 147.6.
  b. Paasche Index = 149.8.
  c. Ideal Index = 148.7.
  d. Laspeyres Quantity Index = 134.2.

5. The purchasing power in 1975 is 10725/150 = 71.5. The purchasing power in 1989 is 29500/358 = 82.4. 82.4/71.5 = 115.2%.

7. a. Market average = 59.9.
  b. New denominator (adjusted divisor): sum of prices after the split/old index value prior to split = 224.7/59.9 = 3.75.
  c. 62.1.

9. (a), (b), (d).

11. In simple average, each component of the index is weighted equally. In weighted averages, each component is weighted using the values of the component. If all companies are small and approximately the same size, there will be no difference between a simple and a weighted average index.

13. When there is a drastic change in the quantity.

15. The purpose of the index is to predict whether the general economic condition of the country is getting better or worse. If you are to construct your own index, then you should use the information that is most related to your own business.

17. Inflation rates will be overstated.

19. See text.

21. If a statistic is denominated in dollars, then it can be considered as a multiplication of a real component and a price component.

23.

Year	Laspeyres Price Index
1985	76.34
1986	95.42
1987	100
1988	109.92
1989	115.27
1990	141.22

25.

Year	Fisher
1985	76.34
1986	95.42
1987	100
1988	109.92
1989	115.27
1990	141.22

27.

Year	Laspeyres
1987	102.35

1988	100
1989	106.64
1990	117.7

29. Laspeyres indexes gave a higher rate in this case. The inflation rate can be obtained (using 1988 as a base year) as:

	Paasche	Las-peyres	Fisher
1987	−2.32	−2.35	−2.33
1988	0	0	0
1989	6.57	6.64	6.60
1990	17.7	17.7	17.7 (from 1988 to 1990)

31. a. 1980.   **b.** Painting.
   c. Painting.   **d.** Jewelry.
33. Equal weighted = 57.8, value weighted = 52.86.
35. 7.72 − 8.22 = −.5. No.

37.

	Housing	Food	Transportation
Percentage changes between 85 and 90	14.49%	21.25%	17.4%

The biggest increase (30) is in food.

	Housing	Food	Transportation
Percentage changes between 80 and 90	31.05%	32.05%	32.47%

39.

Year	Index (base = 1)	Index (base = 6)
1	100	58.71
2	107.14	62.90
3	102.2	60.00
4	115.38	67.74
5	134.62	79.03
6	170.33	100

7	142.86	83.87
8	134.62	79.03
9	151.1	88.71
10	141.76	83.23

41. Base = 1, base = 7.

41.375	100	82.3383
42.25	102.1148	84.0796
43.125	104.2296	85.8208
45.75	110.5740	91.0447
47.5	114.8036	94.5273
49.375	119.3353	98.2587
50.25	121.4501	100
49.75	120.2416	99.0049
52.875	127.7945	105.2238
53.625	129.6072	106.7164
54.125	130.8157	107.7114
51.625	124.7734	102.7363

43.

Week	Indexes
1	100
2	95.32
3	103.01
4	105.69
5	106.35
6	106.02
7	109.7
8	111.04
9	117.06
10	115.05

45.

Week	LI	PI	FI
1	100	100	100
2	107.47	107.43	107.45
3	104.44	103.88	104.16
4	112.98	113.24	113.11
5	99.57	99.44	99.5
6	111.81	111.84	111.82
7	119.71	120.26	119.99
8	108.75	105.97	107.35
9	121.91	121.55	121.73
10	138.08	136.26	137.17

47. 138.08/90.51 = 1.5255, 136.26/89.32 = 1.5255.
49. A price index uses quantity as weights and measures the changes in price. A quantity index uses price as weights and measures the changes in quantity.

51.

Month	Index
1	100
2	97.35
3	101.83
4	112.77
5	116.69
6	124.07
7	115.66
8	121.97
9	122.39
10	116.14
11	123.07
12	132.03

53. See text and newspaper.

55. It is very difficult to compare the computer prices using any index because computer quality has improved dramatically during the last three decades. It would be like comparing apples and oranges.

57.

Year	Price	Quantity	Index
85	21	12,300	100
86	18.25	13,000	91.85
87	19.31	21,300	159.23
88	22.44	20,212	175.59
89	23.51	24,345	221.58
90	21.85	32,300	273.23

59.

Year	Index
85	100
86	86.90
87	91.95
88	106.86
89	111.95
90	104.05

There is only one commodity in the construction of the index.

61.

Year	Annual Salary	CPI	Real Wage
1987	38,202	95	40,213
1988	39,837	100	39,837
1989	41,001	108	37,964
1990	41,327	125	33,062
1991	55,943	200	27,972

a. Change in real salaries between 1988 and 1989: $37,964 - 39,837 = -1873$.

b. No.

63. a. The growth rate for nominal GNP is .0796.

b. The growth rate for GNP deflator is .0529. The growth rate for real GNP is .0267.

Chapter 20

1. In stratified sampling, we divide the entire population into different groups or strata. After this division, a simple random sample of each strata is taken. Stratified sampling is preferred to simple random sampling when we believe that different subsets of the population may have different beliefs. In cluster sampling, we divide the entire population into different groups or clusters and take a census on a random sample of clusters. Cluster sampling is good when the population covers a large geographic region, thus making simple random sampling prohibitively expensive.

3. $\$264,977 \leq x \leq \$295,023$.

5. $n = 481.07 = 482$ (round up).

7. $n = 28.96$ or approximately 29.

9. a. $\overline{X}_{male} = 4.40$, $\overline{X}_{female} = 8.87$, $\overline{X}_{pop} = 7.08$.

b. Var(Pop) = 16.743.

c. $\hat{\sigma}_{\overline{x}}^2 = .488$.

11. If the ratio of licensed cars in the two states remains relatively constant over time, then information about one state can be used to predict the number of licensed cars in the other state. Ratio = 1.23, so $\hat{x} = 5.72$ million.

13. (i) Determine the relevant information required. (ii) Construct a population list for the relevant population. (iii) Decide on the appropriate sampling method to use. (iv) Determine the appropriate sample size. (v) Determine the method used to infer population parameters. (vi) Draw conclusions from the sample information. Poorly designed sample studies can add to the costs of the study or invalidate the results.

15. Sampling error represents the difference between the true population value and the value inferred by the sample; that is, it is the error that results based on the sample chosen. Nonsampling error represents errors not related to the sample chosen, such as sampling the wrong population, inaccurate responses from subjects, or a measurement error.

17. One possible approach is to use cluster sampling and to take a census of only a handful of states. This would reduce the cost of traveling to all 50 states.

19. $351,480 \leq \mu \leq 398,520$.

21. If you believe that the salaries for the two groups differ, you would want to use stratified sampling.

23. Regression can also be used to predict the value of x by assuming a linear relationship between x and y.

25. One approach would be to survey accounting departments at other colleges. Some of the problems associated with such a study include the costs of surveying other schools, the possibility of non-responses, and whether a census or sample should be taken.

27. **a.** $\hat{\sigma}_x^2 = 1,886.06; \hat{\sigma}_{\bar{x}} = 43.43$.
    b. $\$1,386.88 \leq \mu \leq \$1,557.12$.

29. $\hat{\sigma}_{\bar{x}}^2 = 1.218, \hat{\sigma}_{\bar{x}} = 1.104, 20.66 \leq \mu \leq 24.34$.

31. $n = 127.99$.

33. Because New York City is so large, it would not be practical to conduct a census on all people who saw the ad. There are several sampling methods that might be appropriate in this case. If the executive believes that the sex and age of the viewers would lead them to have different opinions of the ad, he might wish to use stratified sampling. Each stratum might consist of different age groups of people of the same sex. If the executive believes that the views will be based more on socio-economic background, cluster sampling of various neighborhoods might be appropriate.

35. The problem with this approach is that you may have incorrectly identified the population of interest. Students taking an Introduction to Economics course may not be a representative sample. To obtain a more representative sample, you will probably wish to survey students who are studying courses other than economics and who will provide a representative sample from different class years.

37. **a.** Proportional allocation: $n = 86.31$.
    b. Optimal allocation: $n = 79.23$.

39. **a.** The confidence interval will get wider because in order for us to be more confident, we must have a larger range of values for the true population parameter to take.

b. The smaller the number of observations in our sample, the less certain we will be and hence the wider the interval.

c. The smaller the population standard deviation, the narrower the interval because the smaller population standard deviation the greater the chances that our sample is reflective of the values of the population.

41. Because the opinion a person has about a movie is likely to depend on the age and sex of the person, stratified sampling may be the best approach.

43. $191.22 \leq \mu \leq 208.73$.

45. **a.** $27.54 \leq \mu \leq 28.06$.
    b. $.443 \leq P \leq .577$.

47. **a.** $27.43 \leq \mu \leq 28.09$.
    b. $.2036 \leq P \leq .3324$.

49. To determine what types of people purchase health foods, the manager would begin by dividing the population into different groups that he believes may differ in their views on health foods. For example, he may believe that the age, sex, income, and education level may determine a person's views. He could then divide these groups into strata and use a stratified sampling approach. The stratified approach would have two advantages. First, it would ensure that all groups were fairly represented. Second, the information from each stratum would provide information, and tests could be conducted between the different strata to see if views differed.

51. Stratified sampling would probably be appropriate here. To conduct this study, the researcher simply needs to obtain a list of student and regular subscribers and randomly sample each group.

53. **a.** $\hat{P} = .625$.   **b.** $.466 \leq P \leq .784$.

55. $S = 2.7202$.

57. $68.24 \leq \mu \leq 77.18$.

59. $n = 118.5$.

61. **a.** $\bar{X} = 91.66$.   **b.** $\hat{\sigma}_{\bar{x}} = 5.87$.
    c. $82.00 \leq \mu \leq 101.32$.

63. $n = 32.65$.

65. **a.** Proportional allocation. $n_1 = 8.58, n_2 = 6.86, n_3 = 8.58, n_4 = 10.98$.
    b. Optimal allocation. $n_1 = 12.36, n_2 = 5.20, n_3 = 4.10, n_4 = 13.34$.

67. $11.62 \leq \mu \leq 15.54$.

69. **a.** Proportional allocation: $n = 225.45$.
    b. Optimal allocation: $n = 223.5$.

71. $\hat{x} = 1,310,360.30$.

73. $\hat{x} = 200(150,001.97) = 30,000,394$.

75. $.996 \leq \mu \leq 1.126$.

77. $n = 196.9$.

79. $21.50 \leq \mu \leq 30.37$.

81. $n = 9.7$.

83. $.368 \leq p \leq .472$.

85. One way to conduct the study is to form two groups that have similar people in each. For example, if there are 50 females in the study, you might like to have 25 use the new drug and 25 not use the new drug.

Chapter 21

1. See text.

3.

	Good	Average	Bad
0	30	40	30
5	10	20	10
10	0	0	0

5. Tables 21.5 and 21.6 are both good examples.

7. See text.

9. a.

W	U(W)
0	0
1	10
2	14.14
3	17.32
4	20

b. Yes. As W increases, the increase in utility is decreasing.

c. If a is positive, it will not change the conclusion in (b). If a is negative, the utility function is risk taking.

11. We use the expected monetary value criteria: lottery A, $45 \times .7 - 100 \times .3 = 1.5$; lottery B, $55 \times .6 - 85 \times .4 = -1$; lottery C, $30 \times .8 - 110 \times .2 = 2$; pick C.

13. a. $E(R_i) = 5\% + \beta_i(10\% - 5\%)$.

b. If $\beta_i = 2$, $E(R_i) = 10\%$. Yes, because the actual return is higher than the expected return.

15. At uncertainty, in addition to knowing which outcome is preferred, we usually need to know the probability of each outcome occurring.

17. It is a conservative strategy. In step one, we pick all worst possible outcomes. In step two, we pick the best outcome of all worst outcomes.

19. Carefully go over Example 21.10 and identify the ingredients for Bayes theorem formula.

23. See text.

25. See text.

27. $E(\text{saving}) = 1,000$; $E(\text{bond}) = 1,090$; $E(\text{stock}) = 1,100$.

29. a. Regret matrix

0	0
8	40

The maximum regret of each action is circled. Pick "study a lot" because it gives the minimum regret among all.

b. Maximin: choose "study a lot."

98	95
90	55

33. Yes. C should never be picked because B dominates C. (Namely, B gives you better results or as good a result as C can.)

35. a. Maximin: product 100 pounds

400	300	-100
-400	600	700
-800	-300	1,500

b. $E(a_1) = 170$, $E(a_2) = 340$, $E(a_3) = 270$. Choose to produce 200 pounds.

37. Expected payoff: $E(\text{low}) = 270$, $E(\text{high}) = 320$, choose high inventory.

39. EMV $= 115$.

41. Minimax: pick large machine.

400	800
500	900
600	600

43. No, the small machine option dominates the medium size machine option.

45. EMV(small) $= 5,200/7$; EMV(medium) $= 5,900/7$; EMV(large) $= 4,200/7$. Choose the large machine; expected cost is lower.

47. No. A risk averter will take risk if he is awarded with enough compensation for taking the risk.

49. An easy way to find whether an investor is a risk averter is to see if he will take a fair game. If the lottery costs 1.5 (the EMV of the lottery), we may compute the expected utility of buying and not buying lottery. Buy lottery: $EU = (300 + 270 - 1.5)^{1/3}(1/200) + (270 - 1.5)^{1/3}(199/200) = 6.46$. Don't buy lottery: $EU = 270^{1/3} = 6.463$. Because not buying lottery gives a higher expected utility, the investor will not take the fair bet and is a risk averter.

51. a. Regret:

$$\begin{matrix} 0 & 3 \\ 10 & 0 \end{matrix}$$ Minimax: buy large.

b. 9.5.

c. EU(large) = 5.397, EU(small) = 5.347.

53. a. Expected revenue = 3,300/3; expected net gain = 100.

b. 50.

55. Expected utility of the joint project: 32.33. Expected utility of invest alone: 32.88. Investing alone is better.

57. b. $U(w) = 400 - 4,000/2^{1/2}$; expansion: $E[U(w)] = 250.3$; new plant: $E[U(w)] = 249.5$ expansion.

59. a. Regret matrix:

$$\begin{matrix} & 0 & 400 \\ \text{Minimax} & 200 & 0 \end{matrix}$$

b. 500 ← 50 gallons, 660 ← 100 gallons.

c. S of 50 gallons is 0; S of 100 gallons is 294. 50 gallons: 25,000; 100 gallons: 25,650.

61. $44.8 \cdot (1 - x) + 44 \cdot x$.

63. Pr(good) = .47, Pr(S_1 | good) = .17, Pr(S_2 | good) = .32, Pr(S_3 | good) = .51; portfolio of all stocks: EMV = 59.25; portfolio of all debts = 44.

65. Peter should choose project A because its risk is lower.

Index

Examples and Applications

Chapter 8 Applications

A Case for a Large Sample 298 / A Case for a Small Sample 298 / Sampling Distributions of Radial Tires' Life 309 / Audit Sampling 317 / Patient Waiting Time 319

Chapter 9 Examples

An Application of the Uniform Distribution in Quality Control 336 / Using the t Distribution to Analyze Audit Sampling Information 339 / "No More Than 8 Items in This Line, *Please!*" 349

Chapter 10 Examples

Sample Mean and Sample Variance: Point Estimate 370 / Population Mean: Point Estimate 371 / Confidence Intervals in Terms of 20 Samples 378 / Sandbags We Can Have *Real* Confidence In: 95 Percent and 99 Percent Confidence Intervals 381 / 95 Percent Confidence Interval for the Sandbag Sample with a Smaller Sample Size 382 / 95 Percent Confidence Interval for the Mean External Audit Fees for 32 Diverse Companies 383 / 95 Percent Confidence Interval for the Average Weight of Football Players 386 / 90 Percent Confidence Interval for the Average Weight of Football Players 386 / Estimate for Waiting Time at a Bank 387 / 95 Percent Confidence Interval for the True Mean Incremental Profit of "Successful" Trade Promotion 387 / 95 Percent Confidence Interval for Voting Proportion 389 / 95 Percent Confidence Interval for Commodity Preference Proportion 389 / 95 Percent Confidence Interval for the Proportion of Working Adults Who Use Computer Equipment 390 / Confidence Intervals for σ^2 392

Chapter 10 Applications

$\overline{X}$-Chart, $\overline{R}$-Chart, and S-Chart for Consolidated Auto Supply Company 400 / P-Chart for Quality Control at the Newton Branch Post Office 405 / Using Interval Estimates to Evaluate Donors and Donations Models 406 / Shoppers' Attitudes Toward Shoplifting and Shoplifting Prevention Devices 409

Chapter 11 Examples

Testing the Average Weight of Cat Food per Bag 434 / Comparing Unleaded Gasoline Prices at Texaco and Shell Stations 437 / Average Mileage of a Moving Van 444 / Competitive Versus Coordinative Bargaining Strategies 446 / The Effect of a Moderator on the Number of Ideas Generated 447 / The Promotability of Company Employees 449 / Defects in Canned Food 450 / Variability in Customer Waiting Time 451

Chapter 11 Applications

Rates of Return for GM Versus Those for Ford 453 / Analysis of the Bank Risk Premium 455 / Analysis of Rates of Return for Retail Firms 456 / Hypothesis Testing Approach to Interpret the Quality Control Chart 457 / Comparison of Organizational Values at Two Different Companies 457

Chapter 12 Examples

Uniform Distribution: Market Shares of Different Types of Cars 498 / Binomial Distribution: Correct Picks in a Football Pool 499 / Poisson Distribution: Number of Patient Arrivals 500